International Conference on Research in Mechanical Engineering Sciences (RiMES 2017)

MATEC Web of Conferences Volume 144 (2018)

Manipal, India
21 - 23 November 2017

Editors:

P.B. Raghuvir
T.M. Mathew

ISBN: 978-1-5108-5581-6

Printed from e-media with permission by:

Curran Associates, Inc.
57 Morehouse Lane
Red Hook, NY 12571

CURRAN ASSOCIATES INC. proceedings.com

Some format issues inherent in the e-media version may also appear in this print version.

Printed by Curran Associates, Inc. (2018)

For additional information, please contact EDP Sciences – Web of Conferences
at the address below.

EDP Sciences – Web of Conferences
17, Avenue du Hoggar
Parc d'Activité de Courtabœuf
BP 112
F-91944 Les Ulis Cedex A
France

Phone: +33 (0) 1 69 18 75 75
Fax: +33 (0) 1 69 28 84 91

contact-edps@webofconferences.org

Additional copies of this publication are available from:

Curran Associates, Inc.
57 Morehouse Lane
Red Hook, NY 12571 USA
Phone: 845-758-0400
Fax: 845-758-2633
Email: curran@proceedings.com
Web: www.proceedings.com

TABLE OF CONTENTS

PREFACE

The International Conference on Research in Mechanical Engineering Sciences (RiMES 2017) was organized as one of the events of the Diamond Jubilee Celebrations of Manipal Institute of Technology, Manipal Academy of Higher Education, Manipal, India. The aim of the conference was to serve as a platform for academicians and experts from industries for sharing knowledge and results in the field of Engineering and Technology along with managerial aspects, to bring together the mechanical engineering and allied community to explore, disseminate and strengthen initiatives in new directions.

We received about 150 articles, which were screened for plagiarism and were reviewed by experts nationally and internationally. Among the papers received only 94 were accepted for oral presentation and publication in this proceedings. Articles are grouped within 5 chapters, which are the broad areas of the conference. On behalf of the organizing committee of RiMES 2017, we would like to express our gratitude to all the participants, for their contributions to the conference. We also thank the authors for submitting papers and for being patient enough to revise it multiple times based on recommendations received from the reviewers.

Thanks to the members of the technical committee, who accepted to get the papers reviewed by experts on a voluntary basis and also for providing valuable suggestions in the course of preparation of this conference.

We will be failing in our duty if we do not thank the Organizing Committee. We have received excellent feedback, from all the participants, which will help us in future to organize such conference in a better way.

We place on record our sincere gratitude to the chairperson of each and every conference session, for smoothly conducting the oral presentation of the papers and coordinating the discussions among the enthusiastic participants. The discussions and deliberations during the sessions are sure to be cherished by the participants.

The keynote lecture as well as the session lectures were well attended and appreciated by the conference participants. Our heartfelt acknowledgment to them as well.

We are deeply indebted to EDP Sciences for excellently bringing out this conference proceedings through the MATEC Web of Conferences.

Dr. Vijay G. S. (Convenor – RiMES 2017)

Dr. Gowrishanker M. C. (Co-Convenor – RiMES 2017)

Prof. Suhas Y. Nayak. (Co-Convenor – RiMES 2017)

Statement of Peer review

In submitting conference proceedings to MATEC Web of Conferences, I certify to the Publisher that I adhere to the **Policy on Publishing Integrity** of the journal in order to safeguard good scientific practice in publishing.

1. All articles have been subjected to peer review administered by the proceedings editors.
2. Reviews have been conducted by expert referees, who have been requested to provide unbiased and constructive comments aimed, whenever possible, at improving the work.
3. Proceedings editors have taken all reasonable steps to ensure the quality of the materials they publish and their decision to accept or reject a paper for publication has been based only on the merits of the work and the relevance to the journal.

Title, date and place of the conference

Title: International Conference on Research in
Mechanical Engineering Sciences (RiMES 2017)

Dates: 21, 22 and 23 December, 2017

Place: Manipal Institute of Technology, Manipal, Karnataka, India

Proceedings editor(s):

(1) Dr. Raghuvir Pai B.
Professor, Department of Mechanical and Manufacturing Engineering,
Manipal Institute of Technology, Manipal.

(2) Dr. Mathew T. Mathew
University of Illinois College of Medicine at Rockford.

Date and editor's signature

01-12-2017

MATEC Web of Conferences **144**, 01001 (2018) https://doi.org/10.1051/matecconf/201814401001
RiMES 2017

Computational study on the effect of a conical spring on handling of buses at low speed

Aravind Rajagopalan[1*] and *Hariharan Sankara Subramanian*[1]

[1]*Automobile Engineering, School of Mechanical Engineering, SASTRA, Thanjvaur, India*

Abstract. Handling and ride characteristic are dependent to a large extent on the characteristic of a vehicle's suspension system. This work explores the effect of the use of conical spring in place of conventional cylindrical profiled helical spring design in the handling of a bus at low speeds through full vehicle multi-body simulations. The bus was modelled using standard template available in ADAMS™ software package. The vehicle inertial properties were verified against properties in literature. The conventional spring characteristic (L) from ADAMS™ database was taken as reference and compared it with a non-linear characteristic (NL) based on literature data. The planned maneuover was to execute a right turn based on standard road dimension inputs from IRC 86:1983 at a constant speed of 30 km/hour with acceleration controlled by software module. Chassis displacements, displacements of spring were tracked to understand handling and ride quality. The variation of chassis displacements showed a significant improvement in ride characteristic of vehicle with most vibrations being damped in NL at time lower than the L characteristic suspension. All through the study, lateral acceleration was well within the rollover threshold and tire interaction forces did not exhibit any significant changes.

Keywords: Non-Linear Spring, Progressive Spring, Conical Spring, Bus, multi-body dynamic simulation

1 Introduction

The handling of a heavy vehicle has remained the focus of studies since the early 1950s and more documented research is available in public domain from tests conducted in USA from 1960s. Vehicles were tested in a standardised test track and the significant parameter influencing the dynamic stability, handling and ride were identified. The data collected as part of these studies formed the basis of development of an integrated software tools using multi body technique to analyse vehicle handing. Of the parameters identified, the suspension systems are known to influence both the ride characteristic and the handling of the vehicles [1].

The handling of the vehicle is dependent on the suspension subsystem, as it connects the vehicles sprung mass with un-sprung mass which comprised of the tyre, thus resulting in loading of tyres. Modern development have identified the use of conical spring in the place

* Corresponding author: aravi1995.05@gmail.com

of conventional helical spring as the conical spring having potential of more favourable characteristics than helical spring like desirable buckling characteristic, with a compromise on space occupied in packaging of vehicle. The Automotive Research Association of India (ARAI) is the nodal body framing rules for vehicle standardization in India and roads in India are constructed based on standards developed by the Indian Road Congress (IRC). The rules framed by these two nodal bodies form major input to vehicle design limitations for vehicles in India. The objective of this work was to examine the variations of the behaviour of a reference vehicle when the conventional spring is replaced with conical spring behaviour. In order to compare the results, the variations in the geometry of the springs is limited by modelling two springs closer to one another and a standardized road manoeuvre was chosen.

2 Nomenclature

IRC	-	Indian Road Congress
L	-	Linear Suspension behaviour
NL	-	Linear Suspension behaviour
FN	-	Normal Force on tyre at the tyre-road interaction point
FL	-	Lateral Force on tyres at the tyre-road interaction point
FL	-	Front Left tyre to driver
FR	-	Front Right tyre to driver
RL	-	Rear Left tyre to driver
RR	-	Rear Right tyre to driver

3 Methodology

A multi-body simulation of a bus model using ADAMS™ was planned for this study. A standard template of a bus readily available in ADAMS™ was modified with the properties from literature [2]. The bus model was assumed to be un-laden.

3.1 Suspension Model

The vehicle has an independent suspension, and the vehicle inertial properties were altered to values as per literature [2] for conventional vehicles. Due to this change a theoretical springs were needed to simulate the behaviour of the linear leaf spring system. The pre-load in the system is 36kN in the front axle and 17671N in the rear axle.

3.1.1 Mathematical model for Linear Behaviour:

This type of spring predominately obeys the equation, $P = k*y$, where y is the deflection of the spring, P is the axial load on the spring and k is the stiffness of the spring, as a result Force vs. Displacement plot for this spring is a straight line as show in Fig. **1**. The spring was designed using the formulae given in[3] by using leaf spring material as AISI 5160 alloy steel[4].

For the shown plot it was noted that the stiffness of the spring was much closer to that of the required spring rate. For any force exceeding the force of 82kN it noted that the deformation is large, in order to consider this behaviour, the following plot was programmed into the system.

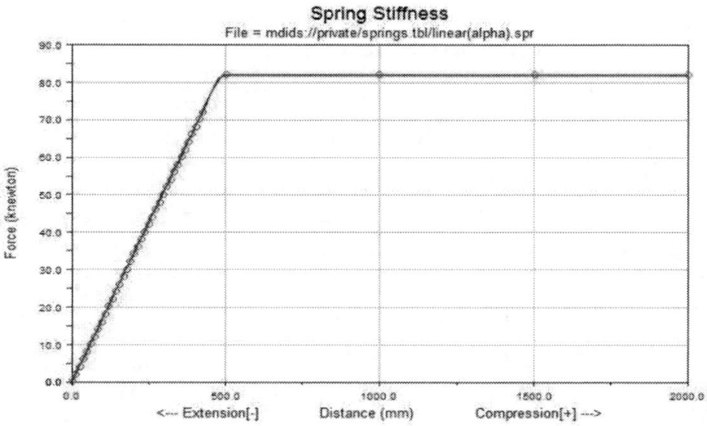

Fig. 1 Linear spring characteristic

3.1.2 Mathematical model for Non-Linear behaviour

The Non-Linear spring characteristic, Force vs. Displacement plot is a curved plot as show in the Figure 2. The spring was designed using the formulae given in [5] by using the material in reference [4]. Similarly like in the case of Linear system the model was capped at 82kN. To keep the variation to a minimal limit both the L and NL spring design were made with the same material and the diameters of the spring was close by.

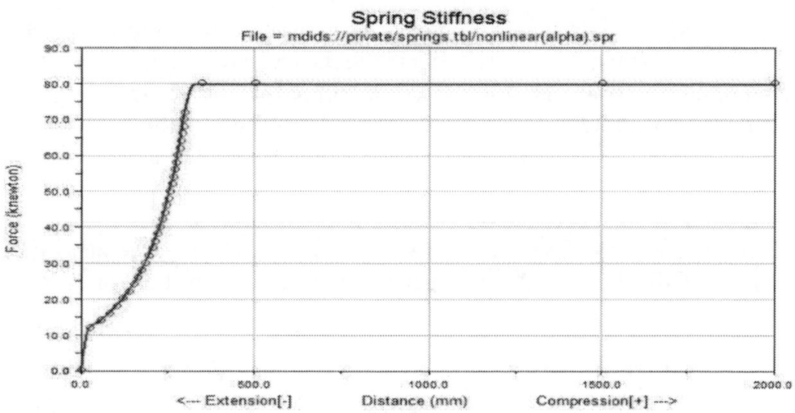

Fig. 2 Non-Linear Spring Characteristic

3.2 Vehicle Description

A full vehicle model was input in ADAMSTM. The vehicle had an engine with gear box attached, brakes in front and rear wheels. An overview of the vehicle model used is summarized in Table 1.

Table 1. Vehicle Specification

Engine	
Type	Rear Mounted Engine
Idle RPM	1000
Max. RPM	2200
Brakes	
Max. Front Brake Torque	5880Nm
Max. Rear Brake Torque	3670Nm
Tyre Specifications	
Front tyre radius	0.32m
Rear tyre radius	0.32m
Tyre file	msc truck pac 2002:tir
Power Train	
Gear ratio	6 speed configuration

3.3 Simulation Conditions

The track of the test is designed in such a way so as to allow the settling time for bus to settle down to the applied linear acceleration and to stabilise its steering characteristics, after a steady state is reached, the bus is turned to a radius specified as per IRC guidelines [6]. The analysis was carried on for 30 seconds, which was in three phases as shown in Figure 3.

Phase I: consists of a straight line driving of the vehicle for 10 seconds. The objective of this segment is to ensure that the entire vehicle reaches a steady state linear motion, where in there are no unnecessary fluctuations in the output parameters.

Transition Phase: It is in this phase that an external steering input is given to the steering wheel which causes the bus to turn. In this phase all the fluctuations are bound to happen and the reading for peaks and falls are present.

Phase II: In this phase the vehicle has turned into the turn and has stabilised to the new condition of motion, thus causing a nearly steady state values in all the measured output parameters. The IRC-86:1983, states the speed of the vehicle for the dimension and purpose of road.

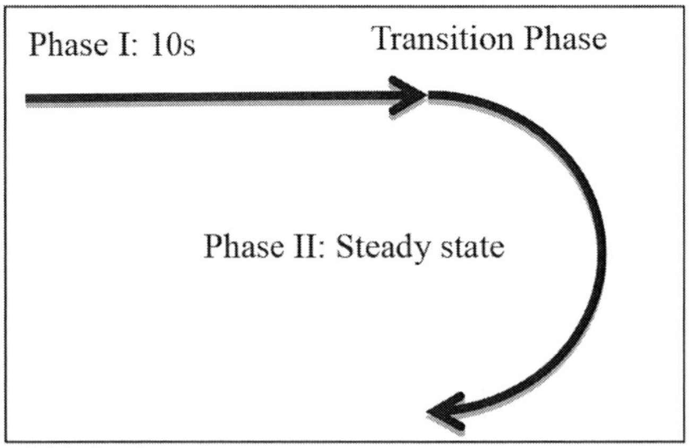

Fig. 3. Description of right hand turn manoeuvre

4 Results and Discussion

In order to understand the variation caused by changing the spring, all important suspension related parameter were studied in steering, chassis, springs and tires as per reference.

4.1 Chassis and Spring Displacement:

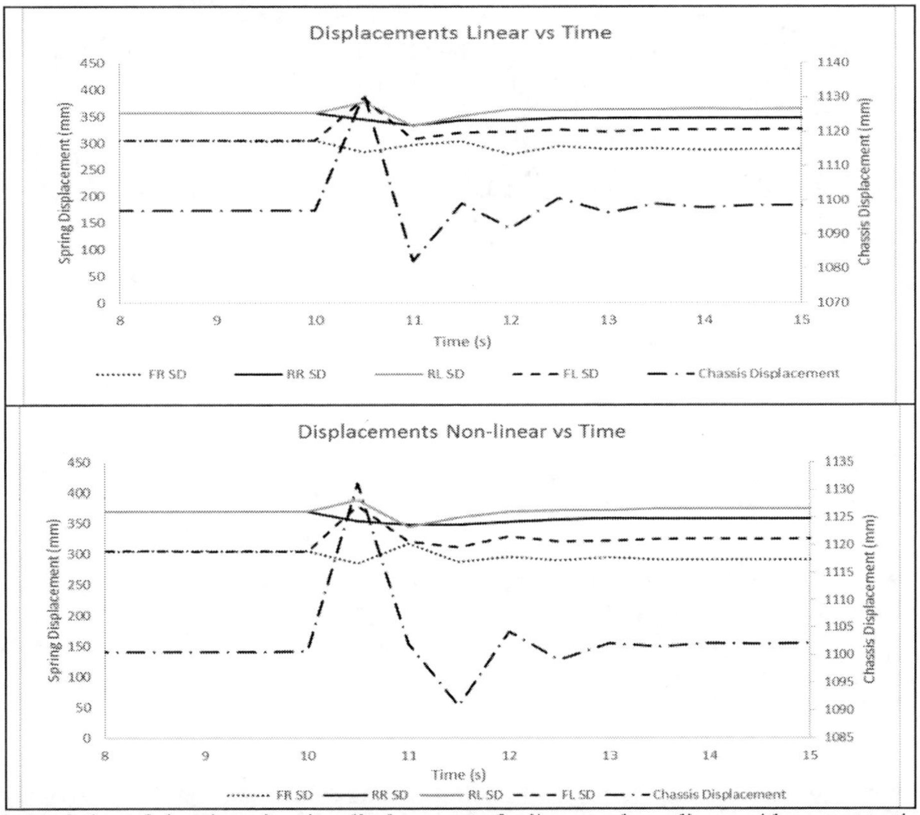

Fig. 4. Variation of chassis and spring displacements for linear and non-linear with respect to time.

The results summarized in Figure 4 indicate that, after the commencement of the turn, the FR and the RR spring displacements are lower that the FL and the RL spring displacements. This phenomenon implies weight transfer to the right, which is an expected behaviour of bus taking a turn. From the Chassis displacement plot it is observed that the Non-Linear model settles at 13.5s for an excitation at 10s whereas the Linear model settles at 14s for the same, moreover form the plot is noted that after the excitation there are 7 peaks whereas in Non-Linear model there are only 5 peaks of excitation before settling. The amplitude of initial excitation however has not changed between the models but the successive excitation amplitudes have reduced considerably.

From the above two graphs it is clear that there is no significant difference in the amplitude of the start of the vibration, but in the case of NL model, the damping was faster. Post turn fluctuation in the individual spring displacement is lower in NL model which could result in the better damping of the overall chassis system. Due to the graded stiffness nature of the NL spring the overall ground clearance of the vehicle is dropped down by a margin of 2mm.

4.2 Chassis Lateral Acceleration:

The maximum lateral acceleration observed was 0.478m/s^2. From literature, the limit for a static rollover condition was 0.4g (3.924m/s^2) acceleration [7]. It was be noted that the variation of difference in the variation of L to NL model is too low to say that their behaviour is different form one another. From the Figure 5, the lateral acceleration in all three phases was within threshold.

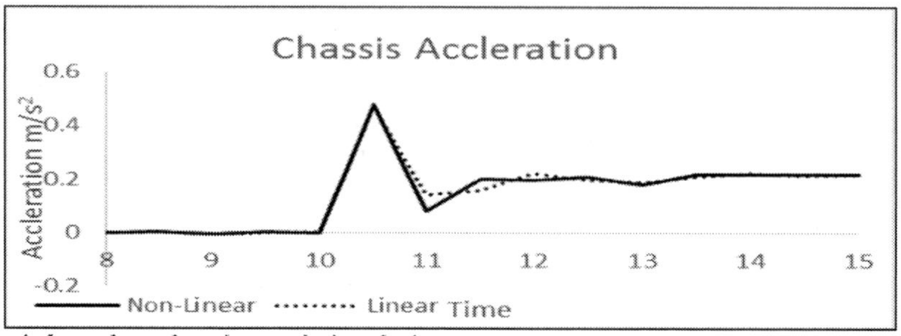

Fig. 5. Chassis lateral acceleration variation during manoeuvre.

4.3 Normal Force and Lateral Force:

Lateral force and Normal forces for the model was studied at the tyre contact points of the vehicle. Knowing the interaction of forces at each wheels helps in determining the motion of each wheel thus the motion of the overall vehicle.

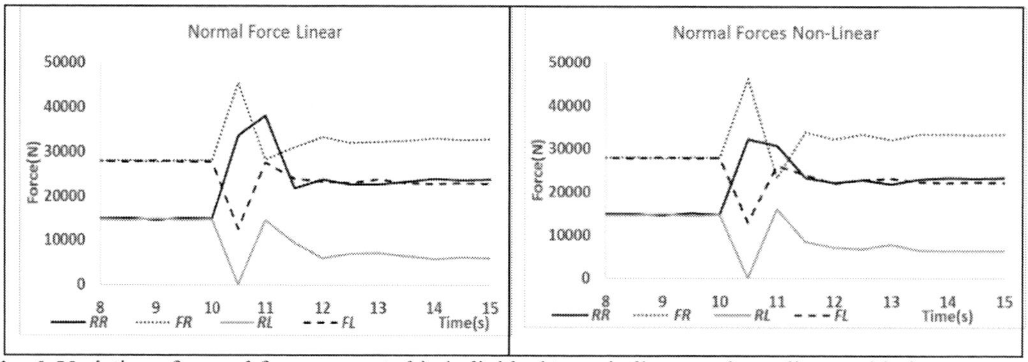

Fig. 6. Variation of normal force generated in individual tyres in linear and non-linear with time

From Figure 6, it was noted form the FR and RR wheels in both the models have a higher value than their counterparts, this is a proof that the vehicle model is in accordance to the physical models. It noted form the above graphs that the RL and FR wheels have a higher Normal force and there is a marginal drop in the RR and the FL wheels in the non-linear model compared to the linear mode. It was observed that normal wheel forces did not change direction (i.e. no lift-off or Rollover).

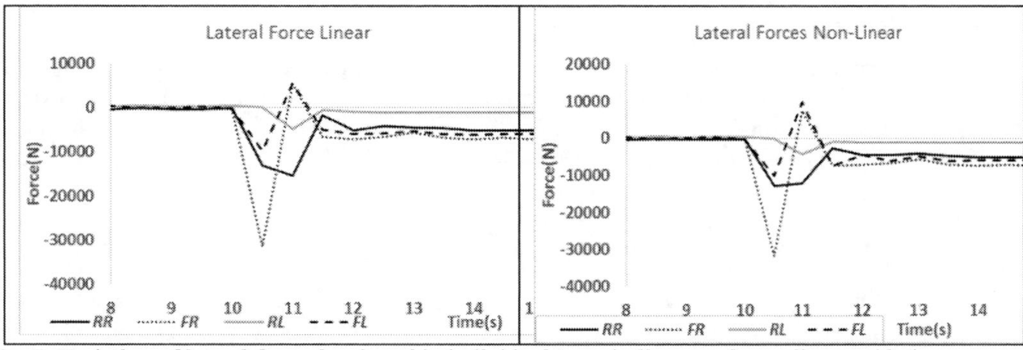

Fig. 7. Variation of lateral force developed in individual tyres in linear and non-linear with time.

Comparing the lateral force graphs (Figure 7) of both the models it is observed that there is no significant changes in the lateral force patterns. The lateral forces in the FR and FL reach a positive value before returning to the negative domain, this is an indication for that fact that the vehicles steering is adjusted to ensure the vehicle is in the proper path.

Overall the lateral force generation at the tyre in both the models is relatively the same. To study tyre loads differently, the ratio between the Lateral Force and Normal Force is studied for determining the overall handling characteristics of the tyre as per reference[8], summarized in Figure 8.

Even for significant changes in normal forces the lateral force to normal force ratio relatively remains constant throughout the study. Form the graphs it is noted that the FL and RR diagonal has a higher lateral force compared to RL and FR diagonal in case of the Non-linear model.

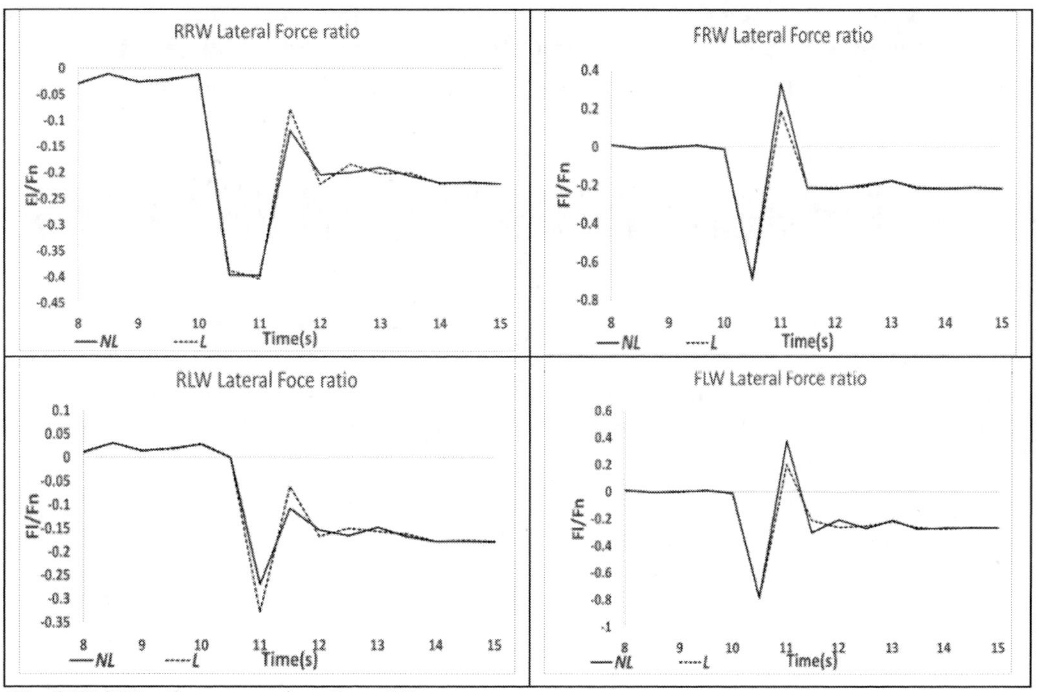

Fig. 8 FL/FN ratios comparison L vs. NL.

4.4 Limitations:

The models have been evaluated for low speed turn with lateral acceleration well below roll-over threshold to examine a typical driving under low speed of vehicle. No extreme manoeuvres were considered. The spring properties have been assumed for a generic material and it remains valid for a comparison study.

5 Conclusions

It was concluded that changing of cylindrical to conical profile spring provided a smoother ride without compromising on the levels of handling and safety as provided in cylindrical profile spring.

References

1. P. S. Fancher, R. D. Ervin, C. B. Winkler, and T. D. Gillespie, "A factbook of the mechanical properties of the components for single-unit and articulated heavy trucks, UMTRI-86-12," 1986.
2. G. J. Heydinger, R. A. Bixel, W. R. Garrott, M. Pyne, J. G. Howe, and D. A. Guenther, "Measured Vehicle Inertial Parameters- NHTSA ' s Data Through November 1998," SAE -1999-01-1336, no. November 1998, 1999.
3. FacultyofMechanicalEngineering, P.S.G. Design Data book. Coimbatore: Kalaikathir acchagam, coimbatore, India, 2008.
4. AZoM, "AISI 5160 Alloy Steel (UNS G51600)." [Online]. Available: http://www.azom.com/article.aspx?articleID=6743. [Accessed: 27-Jul-2017].
5. I. L. J. A. Den Boer, "Nonlinear dynamic behavior of a conical spring with top mass," 2009.
6. The Indian Road Congress, Geometric Design Standards for Urban Roads in Plains, IRC: 86-1983. Delhi: The Indian Roads congress , New Delhi 110011, 1983.
7. C. B. Winkler, Rollover of Heavy Commercial Vehicles. 1999.
8. R. D. Ervin, C. B. Winkler, J. E. Bernard, and R. K. Gupta, "Effects of tire properties on truck and bus handling," Washington D.C., U.S.A., 1976.

Theoretical analysis of steady state operating forces in control valves

Hubballi Basavaraj[1*] and *Sondur Vilas*[2]

[1]*Visvesavraya Technological University, Jain College of Engineering Belagavi, India*
[2]*Visvesvaraya Technological University, Belagavi, India*

Abstract: The controlling components, such as valves are used to regulate controlled fluid power. It is not always possible to calculate valve forces accurately, and with some types of valves even the existence of certain types of forces cannot be predicted with certainty. In many cases, however, the analysis can be made fairly completely and accurately. The assumption of steady state conditions is valid for the valve alone, but transient effects in the rest of the system may be large. These effects are particularly important with regard to the instability of valves, where the system may react on the valve in such a way as to make it squeal or oscillate, sometimes with large amplitude. The origin of the steady state flow force understood from a brief qualitative explanation. The following paper will summarize much of what is known about valve forces in the spool type controlling element.

1 Introduction

In order to determine the transient or dynamic charcteristic of a hydraulic control circuit, the complete charcteristics of each component must be known. The basic information is required for designing of hydraulic control components but also for the proper selection of the right components and control circuits which will result in optimum overall system performance. Control valves of the types with which are consist of arrangements of orifices, variable and fixed, which are so connected that they can vary the restrictions which they offer to the flow of liquid from sources to load, in the desired fashion, as the moving member of the valve is positioned by some eternal agent.

The hydraulic components must closely match the elements preceding them in the chain of command and then must perform the assigned output function precisely during both steady-state and transient operation. In order to make the proper selection, there must be complete familiarity with the various types of hydraulic control components as well as their steady-state and dynamic characteristics. Every component type has its own particular features and limitations, such as hysteresis, dead band, threshold signal level and nonlinearities.

* Corresponding author: bvhubliabhi@gmail.com

2 Literature Review

J F.blackburn, et al. [1] shown the effect of different spool designs on the forces with some experimental work. The operating member of a control valve–spool is subjected to a number of different forces while it is in operation. These forces have various origins and various characteristics, and it is necessary to know their magnitudes and natures in order to design for successful operation of the valve. In many cases, these forces are large enough to cause erratic operation, or even to prevent operation completely, if suitable precautions are not taken.

R. Amirante, et al. [2] analyze the behaviour of the flow force peak values in the initial spool opening of an open center directional control valve and to establish a connection to the behaviour of a closed center valve. In order to evaluate the different behaviours of the valve, this analysis will be conducted at different flow rates. R. Amirante, et al.[3] analysis made to evaluate the valve fluid dynamic performance, exploiting computational fluid dynamics (CFD) techniques, in order to give the reliable indications needed to define the valve design criteria and avoid expensive experimental tests. M. Batoli, [4] studied the position of flow force peak value on a closed center directional control valve. M. Borghi at al. [5] provided the fluid dynamic similarity based procedure to estimate analytically the flow force peak value at different pressure drops and flow rates. In many literatures analyzed the effects of different spool edge shapes produce on the flow force profile during the axial spool movement [6-9].

3 Methodology

The main objective of this paper is about the discussion of forces acts on the directional control valve spool. It is analyzed that, maximum force is greatly decreased by the use of modified construction (Half round baffle). Further substantial reductions can be made by minor adjustments of the dimensional parameters and by improved workmanship.

The operating member of a control valve-the spool is subjected to a number of different forces while it is in operation. Alomg with the force supplied by the valve-operating device, but in addition other forces may either aid or oppose the operating force. These forces have their magnitudes and natures in order to design for successful operation of the valve. In many cases, these forces are large enough to cause erratic operation, or even prevent operation completely, if suitable precautions are not taken.

The forces which act upon a valve spool may be classified according to their direction of action as peripheral, lateral and axial. The first class is usually negligibly small. Lateral forces are not directly important since they act normal to the direction of motion of the spool, but in practice they may be very large so that the friction of the spool against the sleeve becomes excessively great. This friction may be so large in serious cases that the spool becomes immovable, hence the name “hydraulic lock”. Axial forces are naturally important since they directly affect the operation of the valve. They are often considerably greater than the inertial and frictional forces, and hence, they directly determine the design of the valve-operating device. In addition they often cause the valve to oscillate or “sing” and this tendency alone may make an otherwise good valve useless valve for a particular application.

3.1 Lateral Forces—Friction and Hydraulic Lock

Valve sticking or excessive friction, although usually greatly aggravated by the presence of dirt in the oil, frequently occurs in spite of all precautions to exclude dirt. It has long been known that the condition is worst with plane pistons and that it can be greatly alleviated by adding peripheral grooves around the piston lands. The standard explanation for the effect of the grooves is that they act to equalize the pressure distribution around the land. It is of interest to develop a quantitative theory of the pressure distribution and the resulting land force for various land configurations.

The following assumptions will be made that fully developed laminar flow is present throughout the length of the flow path.

1) Cylinder bore is perfectly straight but the spool land is not.
2) Inertial and gravitational effects will also be neglected.
3) All the flow is parallel to the axis and that the peripheral component of flow is negligibly small.

First consider the case of a truly cylindrical spool in a true bore, with the axes of the spool and bore parallel but non-coincident. Consider the flows dq_1 and dq_2 through two elements of the peripheral clearance space at the top and bottom of the spool, the spool being displaced upward as in Fig. 1., each element has a peripheral width, $dz = a\, d\theta$, normal to the plane of the figure. Since for each elementary conduit the cross-sectional area, $y_1\, dz$ or $y_2\, dz$, is constant over the length, l, of the spool land, the pressure gradient $dp/dx = (P_1 - P_2)/l$ = constant for both conduits.

Thus the curve of pressure versus x is a staright line between the points P_1 and P_2 at the ends of the spool for each of the two conduits.

The downward force on the spool is

$$df_1 = \int_0^l p\, dz\, dx = \frac{P_1+P_2}{2}\, l\, dz \qquad (1)$$

The upward force, df_2, is exactly the same; the df's balance by pairs all around the spool and the net lateral force is zero for any case in which the surface of the spool is parallel with the adjacent wall of the cylinder.

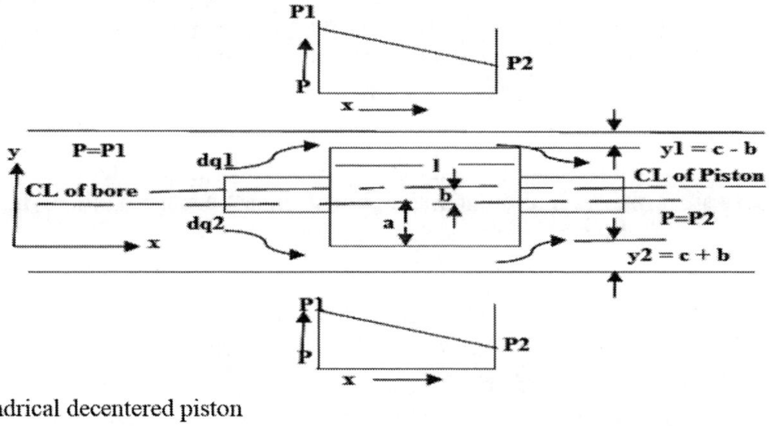

Fig. 1. Cylindrical decentered piston

The net force is also zero for any configuration possessing axial symmetry, but this is not necessarily true for decentered spools, where the axial symmetry is destroyed. The simplest case of this type is that of a decentered conical shown in Fig. 2. Here the cross-sectional area and, therefore the pressure gradient are no longer constant over the length of the spool, and the force per unit peripheral width, dz, must be found by a double integration.

The resulting formulas are:

For an elementary length, dx, of the conduit of width dz and height y,

$$\frac{dp}{dx} = -\frac{12\mu\, dq}{y^3\, dz} \qquad (2)$$

where μ is the absolute viscosity and dq/dz is the flow through a conduit of width dz. With integration and insertion of the boundary conditions,
When x=0, $y = C_1$, and $p = P_1$
And when x = l, $y = C_1 + t$, and $p = P_2 = P_1 - \Delta P$

$$\text{We get } p = P_1 - \frac{\Delta p(C_1 + t)^2}{t(2C_1 + t)}\left(1 - \frac{C_1^2}{y^2}\right) \tag{3}$$

where t is the radial taper of the piston. Thus, the variations of pressure with (y and hence with) x is parabolic.

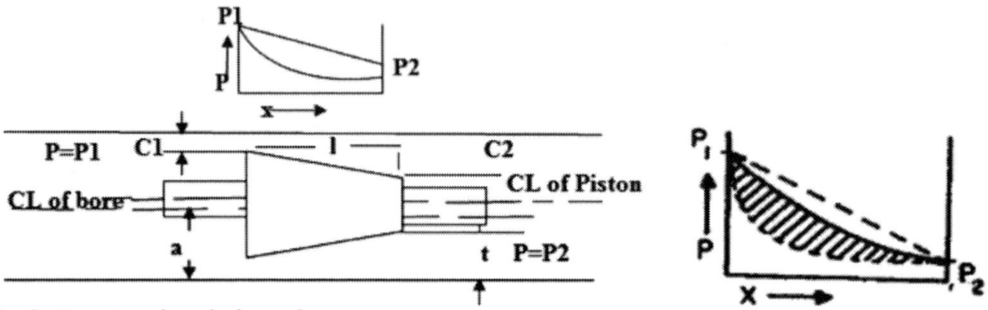

Fig. 2. Decentered conical spool

The actual deviation of this parabolic pressure curve from the linear curve of a uniform conduit depends essentially upon the taper and upon the closeness of approach of the large end to the cylinder wall. For the upper duct of fig, the large end approaches closely, the pressure gradient near the large end is high, and the p-x curve drops rapidly from the P_1 point. For the lower duct, the ratio of the y's at the upstream and downstream ends is much nearer unity, and the departure from linearity is much less.
The radial force on an elementary area, dx dz, of the piston will be p dx dz . Integration gives the total radial force on the elementary strip of width dz as

$$df = P_1 - \left(\frac{(C_1+t)\Delta P}{(2C_1+t)}\right) l\, dz \tag{4}$$

The total lateral force F on the spool is obtained by integrating df around the periphery, when account is taken of the fact that only the component of df parallel to the displacement b is of-interest since the normal components will cancel because of the symmetry of the configuration. Thus, if df is the component of df parallel to b,

$$dF = -la\left(P_1 - \Delta P \frac{C+t+b\cos\theta}{2C+t+2b\cos\theta}\right)\cos\theta \tag{5}$$

where C is the radial clearance at the large end with the spool centered. Integration with respect to θ from 0 to 2π gives the total side-thrust on the piston:

$$F = \frac{\pi a t\, \Delta p}{2b}\left(1 - \frac{2C+t}{\sqrt{(2C+t)^2 - 4b^2}}\right) \tag{6}$$

Since the second term in the parentheses is greater than 1, the force will be negative, that is, away from the larger opening, and is therefore a decentring force. Thus the equilibrium of a centered tapered spool with the higher pressure applied to the larger end is unstable, and the spool will be forced into contact with the wall. Conversely, if the higher pressure is applied to the smaller end, the spool will center itself in the bore.

Calculations have been made for one land of a very small pilot spool which stuck persistently. The resulting curves for F versus b are plotted in Fig. 3.

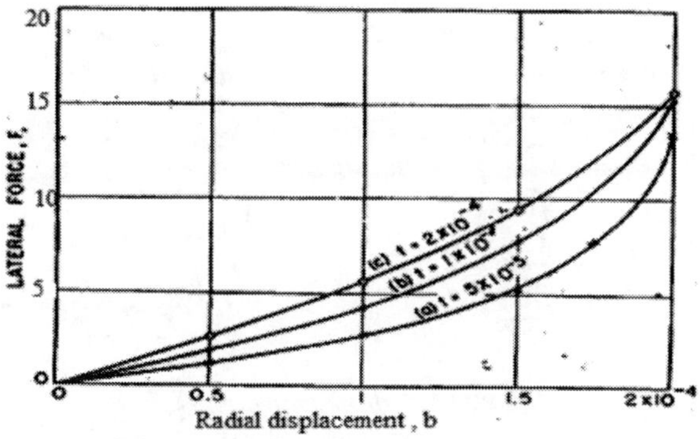

Fig. 3. Lateral force versus displacement for a small spool land

The values of the various quantities involved are: Axial length, l = 2.4mm, radius, a = 1.6mm, radial clearance at larger end, C=0.0051 mm, Pressure drop, Δp = 1000 psi, Radial taper, t = 0.002 mm. for curve (a), = 0.0025 mm, for curve (b),= 0.005 mm, for curve (c)

From Fig. 3, it can be seen that the peripheral effects of varying the amount of tapers are: (1) the rate of increase of decentering force for small displacements from center decreases rapidly with decreasing taper, and (2) the force holding the spool against the wall (when b = 0.0051mm, in the present case) changes comparatively little.

3.2 Axial Forces

The axial forces that operate on a valve spool include the inertial force, which is readily calculated and is usually small, the frictional forces, which are not accurately calculable but which should also be small in a properly designed and properly applied valve, and various forces which owe their existence to the fact that fluid is flowing through the valve and exerting force on the spool. Knowledge of these flows–produced forces is still incomplete and every new basic valve design presents new problems, but for certain common types of valves, the principal forces are known, and their magnitudes can be calculated with good accuracy [10].

 https://doi.org/10.1051/matecconf/201814401002

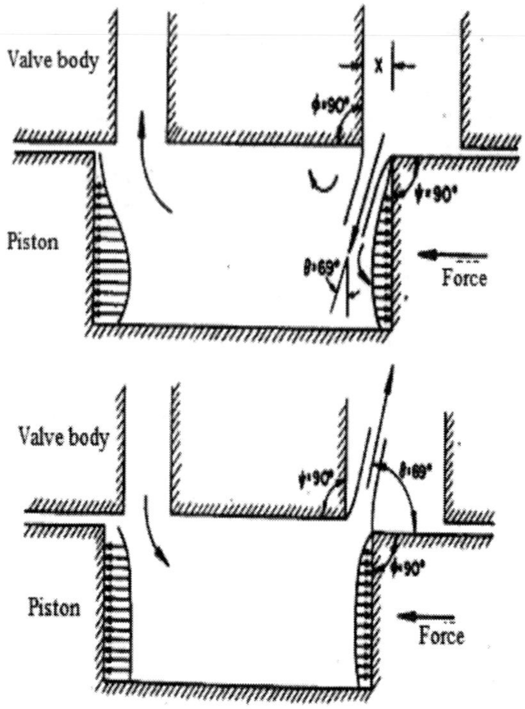

(a) Jet Entering Piston Chamber (b) Jet leaving piston chamber

Fig. 4. Origin of the steady state flow force

In Fig. 4, which represents one metering orifice of a spool valve, with the upstream and downstream chambers and the connecting ducts, the cross-sectional area of the stream near the orifice is much smaller than it is everywhere else, and therefore, the velocity of flow is small except near the orifice edges.

A quantitative solution of this case can be found as follows. Assuming:

1) The fluid is nonviscous and incompressible.
2) The peripheral width of the orifice is large compared with its axial length so that the flow can be considered to be two-dimensional.
3) The flow is assumed to be quasi-irrotational in a region immediately upstream from the orifice.

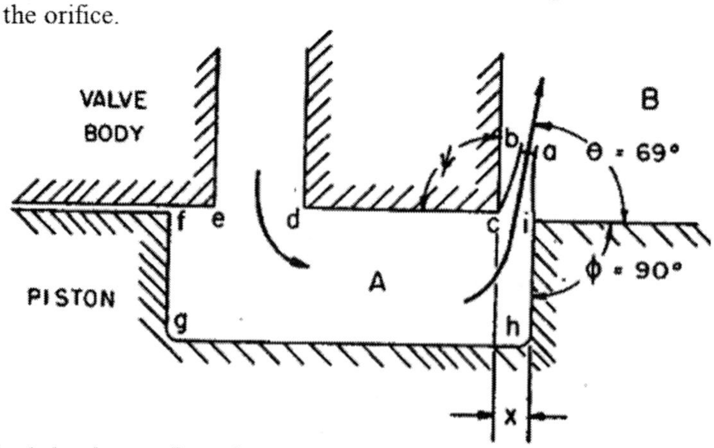

Fig. 5. Square-land chamber configuration

With the assumptions that the flow is two-dimensional, irrotational, nonviscous, and incompressible, the solution of the flow pattern in the region upstream from the orifice becomes a solution of Laplace's equation for the configuration shown in Fig. 5. this solution has been found by vonmises, and it is found that, for a square-land valve when the valve opening, x, is small compared with the other dimensions of the upstream chamber, the angle θ, which the axis of the stream makes with the spool axis is 69°, this value has been experimentally verified both on actual valves and on a two-dimensional valve model with glass sides [1]. For lands with angles other than 90, the value of θ can also be found either experimentally or theoretically. When θ is known, the axial force on the piston can be derived in the following manner.

The axial force on the spool equals the axial component of the net rate of efflux of momentum through the boundary a-b-c-d-e-f-g-h-i-a in Fig. 5, where a-b is the vena contracta of the jet. In an actual valve, the area of a-b is much smaller than that of d-e where the fluid enters the upstream chamber, A. Since the velocities are inversely proportional to the areas, the influx of momentum through d-e is negligible small compared with the efflux at a-b, which is equal to QUρ, where Q = total rate of flow, U = velocity of jet at vena contracta

ρ=density of fluid.

By Bernoulli's equation, $u=\sqrt{(2\Delta p)}/\rho$

And the net axial force is:

$$F_{AB} = F_{fg} - F_{hi} = QU\rho \cos\theta \tag{7}$$

This equation can be transformed into the more useful form as:

$$F_{AB} = C_q\, wx\, \Delta p \cos\theta \tag{8}$$

where C_q = coefficient of discharge = $\frac{Q}{wx\sqrt{(2\Delta p/\rho)}}$

w = peripheral width of orifice, x = axial length of orifice, Δp = pressure difference between chamber A and chamber B.

Since θ is always less than 90° for valves of the type under discussion, F_{AB} of Eq.8. is always positive and tends to close the valve. If the direction of flow is reversed so that the jet flows into chamber A instead of out of it, the equation is still valid, and if the jet angle θ is the same for both cases, the force will be the same. It should be noted that for Fig. 5, the jet angle depends upon the angle ϕ but is independent of ψ; for the reversed flow, the opposite is true. In either case, it is the configuration of the upstream chamber that determines the force. Eq. 7. and 8 apply only to an ideal valve with perfectly sharp corners and zero radial clearance. Since such an ideal valve cannot be built, it is important to investigate the effects of these imperfections.

For an orifice with radial clearance, Cr, as shown in Fig. 6. The Eq.8., becomes:

$$F_{AB} = 2C_q w\, \Delta p\, \sqrt{x^2 + C_r^2} \cos\theta \tag{9}$$

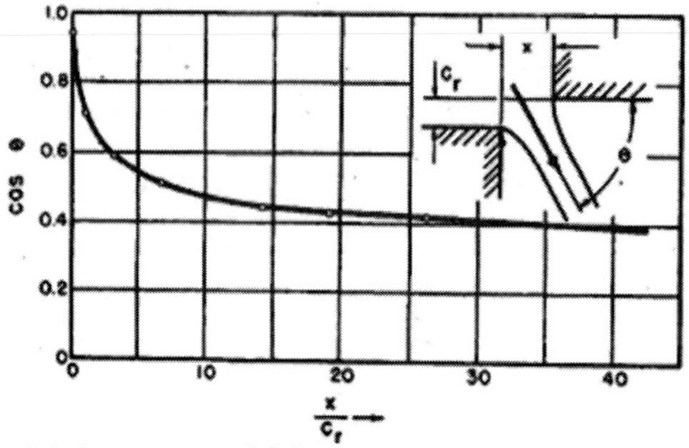

Fig. 6. Effect of radial clearance on cosθ [1]

Here, cosθ is no longer a constant but varies with the quantity x/C_r, as shown by the curve. If the net force on the spool is plotted against the displacement for a constant value of pressure drop. The curve for an ideal valve is shown dashed in Fig. 7. It can be seen that that the effect of radial clearance is to make the curve rise abnormally rapidly for small openings, after which it sags downward and approaches the ideal curve asymptotically.

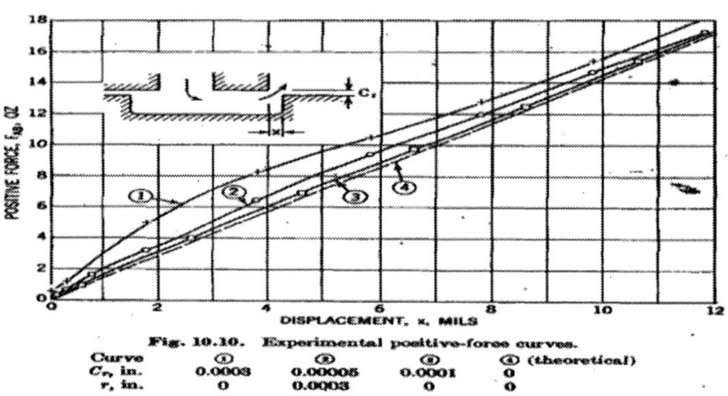

Fig. 7. Experimental positive-force curves [1]

The effect of finite radius of the land edges is difficult to calculate theoretically but is very similar to that of radius clearance. Qualitatively, the variation of θ can be visualized with reference to Fig. 8.

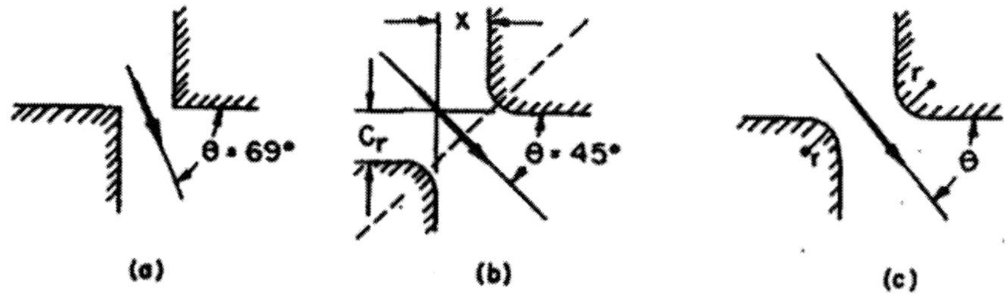

Fig. 8. Land Configurations

In Fig. 8(a), the ideal case of zero radial clearance and sharp land edges is represented. The figure is skew-symmetrical about the center of the opening; it is the skewness of the symmetry that causes the axis of the jet to deviate from the vertical and it is this deviation and the resulting x-component of momentum that gives rise to the axial force on the spool.

Fig. 8(b) represents a valve with the axial displacement, x, set equal to the radial clearance, C_r. The figure is symmetrical about the two 45° lines, and the jet angle, θ, is 45° because symmetry. This is true also for any value of radius r of the land edges is long as the r's are the same for both edges. If the valve opening, x is varied, however, the symmetry will be destroyed, and θ will approach 69°; for small values of x, the jet will approach parallelism with the axis, and the force will be greater than that for an ideal valve.

In a valve with zero radial clearance but finite curvature of the land edges, as in Fig. 8(c), the symmetry is skewed as for the ideal valve but not to such a great extent; the angle θ is less than 69°; the piston force is greater than for an ideal valve; and θ varies with x . For large values of x, θ approaches 69°; for negative values of x, θ approaches 0 as x approaches –r. In an actual valve in which neither C_r nor r is zero, θ approaches zero at small or negative values of x, and the slope of the curve of force versus x increases at most by a factor of 1/0.36 =2.78. This is not particularly serious, of course, since in most cases the curve drops off and approaches that for an ideal valve before the actual value of the force is very large; that is, the phenomenon occurs essentially only for small values of x.

In an actual four-way vale, there are two identical orifices in serious and, therefore, there is twice the force on the piston, equation 9 then becomes

$$F_T = 2F_{AB} = 2Q\sqrt{\rho P_v}\cos\theta \quad (10)$$

The radial component of the force on the spool can be easily cancelled by the simple scheme of dividing the port peripherally into two or more equal parts equally spaced around the sleeve.The above analysis of the steady state flow force on a spool valve is assumed to be accurate and adequate for practical application, since it has been verified by a large amount of experimental investigations and operating experience [1]. In so far as the configurations of the metering orifices are similar, this method can be applied to valves of other constructions.

3.3 Flow-Force Compensation

Steady state flow forces of the type discussed above have been recognized for years, and various designs have been suggested for reducing or eliminating them. One which works excellently for the valve shown in Fig. 9. In the valve construction to be described, the axial component of efflux of momentum from the piston chamber is made greater than the influx so that a negative (opening) force is developed which h is proportional to x for a given pressure drop, and this negative force is balanced by the positive force generated at a conventional square-land orifice. It should perhaps be noted that this method can be used to give a zero-force port, but since the special port construction is somewhat more difficult to manufacture than a conventional square port and since it may lead to the transient instability, the balancing scheme is preferable.

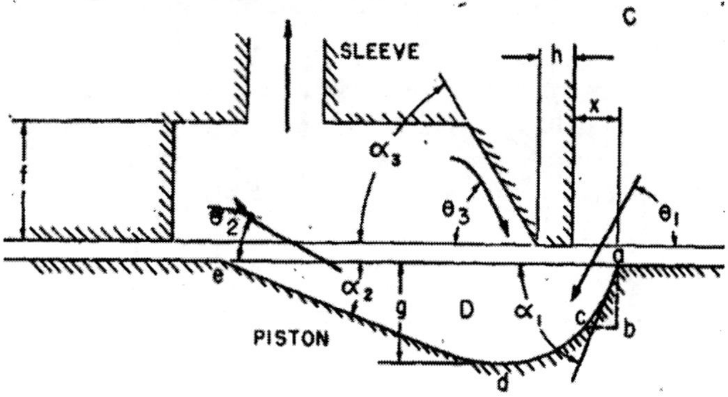

Fig. 9. Negative Force port

The construction of the negative–force port is shown schematically in Fig.9. It differes from a conventional port in two ways; the chamber in the spool is shaped some what like a turbine bucket, and the sleeve is cut to form an extension of the downstream chamber. Here, the jet through the orifice enters the spool chamber at an angle of θ_1, is reflected from the chamber wall near d, and leaves at an angle of θ_2. If the cross section of the flow were constant and the frictional force negligible, the resulting force on the spool would be

$$F_{CD} = QU\rho(\cos\theta_1 \cos\theta_2) \quad (11)$$

Thus, the negative force of Eq.11 may easily become a negative force, since $\cos\theta_2$ is ordinarily considerably larger than $\cos\theta_1$. Furthermore, an eddy is produced which moves clockwise, and an additional flow enters the piston chamber at an angle of θ_3, which increases the negative force still more. The total resultant force on the piston is thus rather complicated. It depends upon numerous variables such as α_1, α_2, α_3, ρ, μ, ω and d, where ρ is the density and μ the viscosity of the fluid, ω the peripheral width of the port, and d the piston diameter. A theoretical derivation of the force would be practically impossible.

The port width, w, occupies only a portion of the circumference of the spool in many valves. Although it is usually large compared with x so that the flow can be treated as two-dimensional at the orifice, it is small compared with the dimensions of the chambers. When the incomplete jet of fluid hits the spool near the point d and is reflected upward, it tends to spread out normal to the plane of the figure, and the condition of constancy of cross section assumed in Eq.11 no longer holds. The actual exit angle, θ_2, thus tends to be greater than α_2, and the jet tends to separate from the piston surface. If $\alpha_1 = \theta_1$, the angle of incidence is zero, and the negative force is a maximum; α_1 should normally be made equal to 69°, the influx angle for a square land. It is found, however, that a square step such as a-b-c in Fig.9 has little effect on the force if it is not too deep, and it is sometimes desirable for manufacturing reasons that such a step be provided. In the case for which w = 360°, there would be no lateral spreading.

3.4 Modified Spool Design (Half round baffle type) for Flow-Force compensation

One possible measure of the effect of the modified construction would be the ratio of the maximum forces on the spool of a conventional and a modified valve. It is possible that, this maximum force is greatly decreased by the use of modified construction. Further substantial reductions can be made by minor adjustments of the dimensional parameters and by improved workmanship. A sketch of a four-way valve of this construction is given in Fig. 10.

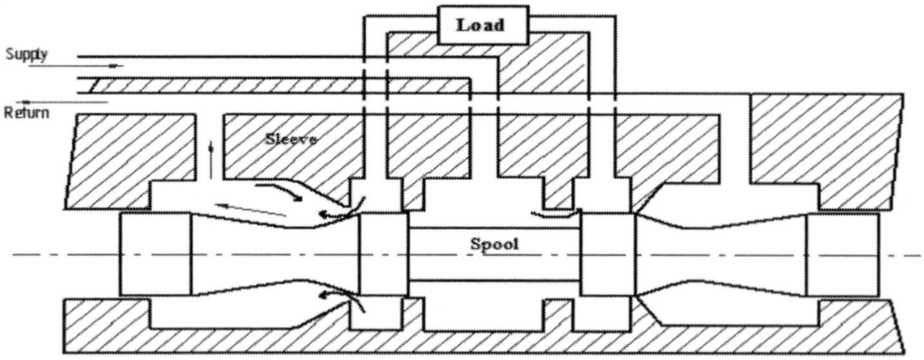

Fig.10. Four-way valve with force compensation (Half round baffle)

In practice, the force compensation is excellent over most of the range valve openings, and the nonlinearity near the origin. In addition, the above design is much easier to manufacture because the dimensions are not so critical, and the compensation appears to be relatively immune to wear of the metering edges and therefore will change little as the valve ages. Finally, the most attractive feature of this design is that it allows the degree of compensation to be adjusted; if the valve is somewhat overcompensated; the half round baffles can be rotated so as to permit part of the incoming jet to slip by and hit the opposite side of the valve, thereby reducing the compensation. It is easy to find an angle at which the compensation ever deteriorates because of wear or other causes, the baffles can easily be readjusted. This method has been thoroughly tested by many authors with the different geometries in the field.

The method of flow–force compensation that has been outlined is good in theory, and should be excellent in practice for long-stroke valves. Even if the compensation is initially excellent, the inevitable erosion of the land corners by dirt in the oil acts to make it less effective after the valve has been in service for a while, and this tends to restrict its application to short–life or expendable devices. Furthermore, even if wear and life are not important factors, the radical nonlinearity of the relationship between the valve force and displacement at small openings makes it difficult to use the valve in many high-performance systems.

The principal disadvantages of this design are that it is usually applicable only to a very restricted range of operating conditions, that it is very difficult to design and mechanize, and particularly, it is very desirable to have as strong an actuator as possible in order to have some margin of force to take care of excessive friction or other to have some margin of force to take care of excessive friction and other sponging effects. Also, even if the force available from the actuator cannot be increased, it is still very desirable to have the actuator as stiff as possible on the interests of valve stability.

4 Conclusions

Most of the designs that have been proposed for the compensation of flow forces fall into one or the other of two categories: either the incoming jet is deflected in such a way that it carries out of the piston chamber the momentum that it brought in (or more), or else this momentum is caught and transferred to the sleeve by some sort of vane or other member which projects into the spool chamber. In concluding this, it is pertinent to point out that compensation of flow forces is not always necessary, or even desirable. If plenty of force is available to actuate the valve, compensation is unnecessary and merely adds complication and expense.

It is not always possible to calculate valve forces accurately, and with some types of valve even the existence of certain types of forces cannot be predicted with certainty. In many cases, however, the analysis can be made fairly completely and accurately. The most of derivations will apply primarily to spool valves, but the formulas will apply to other construction as well, although they may have to be modified appropriately. A three dimensional analysis of a spool type control valve is essential to determine correctly the flow force and the flow rate values, especially for the designs and avoids expensive tests.

References

1. J F.blackburn, et al, “Fluid Power Control”, John Wiley and Sons Inc, Newyork London,(1960)

2. R. Amirante, G. Del Vescovo and A. Lippolis "Flow forces analysis of an open center hydraulic directional control valve sliding spool", Energy Conversion and Management Elsevier ,47 114–131, (2006)
3. R. Amirante, G. Del Vescovo and A. Lippolis "Evaluation of the Flow forces on an open center directional control valve by means of a computational fluid dynamic analysis", Energy Conversion and Management Elsevier, 47, 1748–1760 (2006)
4. M. Batoli, "Theoretical and experimental analysis of flow forces on a hydraulic directional control valve", Pneumatica: 3, 6-10 (1996)
5. M.Borghi, M.Milani, R.Paoluzzi, "Stationary axial flow force analysis on compensated spool valves", International Journal of Fluid Power; 1, 17–25, (2000)
6. A.Macor, S.Dato, "Flow force reduction in a hydraulic directional control valve by means of a restricted section shaping". In: Proceedings of the 55 Congresso nazionale ATI, Bari- (2000)
7. A.Macor, "Experimental analysis on a directional valve with a flat notch metering section", In: Proceedings of the 57 Congresso Nazionale ATI, Pisa (2002).
8. G.Del Vescovo, A. Lippolis, "Flow forces analysis on an electro piloted hydraulic directional control valve" In: Proceedings of the 3rd International Congress MiniHydro, (2001)
9. G.Del Vescovo, A. Lippolis, "CFD analysis of flow forces on spool valves", In: Proceedings of the First International Conference on Computational Methods in Fluid Power Technology, Melbourne, 26–28, (2003)
10. B V. Hubballi and Dr V B. Sondur, "Directional Control Spool Valve Performance Criteria and Analysis of Flow-Reaction Forces", International Journal of Emerging Technology and Advanced Engineering, Volume 5, 86-93, (2015)

MATEC Web of Conferences **144**, 01003 (2018) https://doi.org/10.1051/matecconf/201814401003
RiMES 2017

Evaluation of brake parameters in copper discs of various thicknesses and speeds using Neodymium – Iron – Boron Magnets

G. L. Anantha Krishna[1] and *K. M. Sathish Kumar*[1*]

[1]*Department of Mechanical Engineering, BMS Institute of Technology & Management, Bengaluru, 560064, India*

Abstract. Neodymium – Iron – Boron (NdFeB) permanent magnets of 12.5 mm thickness and 50 mm diameter are chosen for analyses because of their higher remanence and coercivity. Experimental analyses were carried out with Copper discs of thickness 4 mm, 6 mm and 8 mm at 2000 rpm, 3000 rpm, 4000 rpm and 5000 rpm. Experiments were conducted with three different positions of magnets such as 2 coaxial magnets, single magnet and single magnet with sudden application conditions. The brake parameters recorded are % speed reduction, deceleration and time taken. In 2 coaxial magnets condition, brake parameters are better in 6 mm thick disc. In single magnet condition, the brake parameters in 6 mm thick disc are found to be more consistent than 4 mm and 8 mm thick discs. In single magnet with sudden application condition, in 4 mm thick disc, the brake parameters are found better. During analysis, very high repulsion was experienced by magnet with 8 mm thick Copper disc at all the above mentioned speeds in single magnet with sudden application condition. For high speed train applications, single magnet condition with 6mm thick disc may be suitable. For high speed automotive applications, single magnet with sudden application condition with 4 mm thick disc may be suitable.

1 Introduction

Conventional friction brakes have numerous problems such as brake squealing noise [1 – 10], thermal cracking and thermo mechanical failures [11 – 15], fade and recovery behavior of friction material [16 – 17], effects due to formulation of ingredients of materials of brake lining [18 – 23], brake judder [24 – 25], brake grabbing, disc distortion etc.

The eddy current principle states that a non – stationary magnetic field induces swirl like electric currents in a conductive body subjected to its influence. These eddy currents occur either when a moving conductive body crosses a magnetic field or by imposing a varying magnetic field upon a stationary conductor.

The eddy currents generate a magnetic field whose flux is opposite to the applied one, in an attempt to reduce the net magnetic flux to zero. This magnetic interaction generates the force necessary to cause a deceleration of the conductor body. This idea can be

* Corresponding author: sathishmech@bmsit.in

implemented in high speed trains and high speed automotive application. During braking, the metal wheels are exposed to a magnetic field from an electromagnet or permanent magnets, generating eddy currents in the wheels. The magnetic interaction between the applied field and the eddy currents acts to slow the wheels down. The faster the wheels are spinning, the stronger the effect, meaning that as the train slows the braking force is reduced, producing a smooth stopping motion. The advantage associated with this lies in the absence of moving parts, being contactless system there is no wear of brake pads. The drag force is given by $F_d = \alpha\sigma\delta B^2 AV$ [26] where F_d is drag force, α is a constant, σ is conductivity of conductor, B is flux density, A is area created by permanent magnet, V is velocity of conducting plate. If a disc made of conducting material associated with the shaft is considered, it is a moving conductor. The necessary magnetic field can be obtained by means of an electromagnet or permanent magnet. Braking torque is directly proportional to the speed of the moving conductor.

A permanent magnet is an object made from a material that is magnetized and creates its own persistent magnetic field. An electromagnet is made from a coil wire that acts as a magnet when an electric current passes through it but stops being a magnet when the current stops.

Rare earth magnets offer high values of flux density at very high values of magnetizing force. Consequently, very short magnet lengths are needed and hence volume of magnet will be small. Various rare earth magnets in use are Neodymium – Iron – Boron (NdFeB) magnets [27], Samarium–cobalt magnet, Alnico magnets. A neodymium magnet is the most widely used type of rare earth magnet. It is a permanent magnet made from an alloy of neodymium, iron and boron (NdFeB). NdFeB magnets have higher remanence, much higher coercivity and energy product.

Flux density is a measure of how concentrated the magnetic field is. Permeability is a measure of how easily a magnetic field can be set up in a material. When the induced moment is opposite to the applied magnetic field, such materials are called diamagnetic materials, e.g. copper. In diamagnetic substances, the flux grows slightly more slowly with the field than it does in the vacuum.

The disadvantage of electromagnetic brakes is its compatibility with signaling and train control systems. Electromagnetic forces are generated when the eddy current brake is activated. These forces could result in interference or potentially even irreparable damage to the signaling systems [28]. Also, electromagnetic braking system is very expensive.

As the drag force is dependent on thickness of conductor, conductivity of conductor, velocity of conductor NdFeB magnets having higher energy product, remanence and coercivity, author tried to analyses experimentally the effect of thickness, speed on brake parameters such as % speed reduction, deceleration and time taken using Copper as conducting material and NdFeB permanent magnets.

2 Methodology

To carry out the experiments, a special purpose test rig was developed and fabricated. The test rig is as shown in Fig. 1. The base is made of Aluminum on which a DC motor of 10,000 rpm is mounted. Stainless steel shaft is coupled to motor on which discs of thickness 4 mm, 6 mm and 8 mm made of Copper are mounted. The diameter of the discs chosen is 200 mm. The speeds at which experiments were conducted are 2000 rpm, 3000 rpm, 4000 rpm and 5000 rpm to simulate speeds of 75 km/hr, 113 km/hr, 150.8 km/hr and 188.5 km/hr respectively. The speed of discs is controlled using dimmerstat. The magnets are mounted on threaded shafts made of gunmetal. Permanent magnets are moved towards and away from rotating disc using stepper motors. The magnets move towards rotating disc at a velocity of 4 mm/second. The values of Magnetic flux density of magnets and speed of discs are acquired through Hall Effect sensor and optical speed sensor respectively which are mounted close to the disc. The Arduino Uno microcontroller board based on the ATmega328 is used for data acquisition and processing. Ultrasonic ranging module HC - SR04 provides 20 mm – 4000 mm non-contact measurement function to measure gap between disc and magnet is used.

The NdFeB magnets of 12.5 mm thickness and 50 mm diameter are chosen for investigations. Various parameters such as disc speed, magnetic flux density, gap between disc and magnets, time taken for speed reduction and deceleration are acquired through data acquisition system with a special software developed using C#.

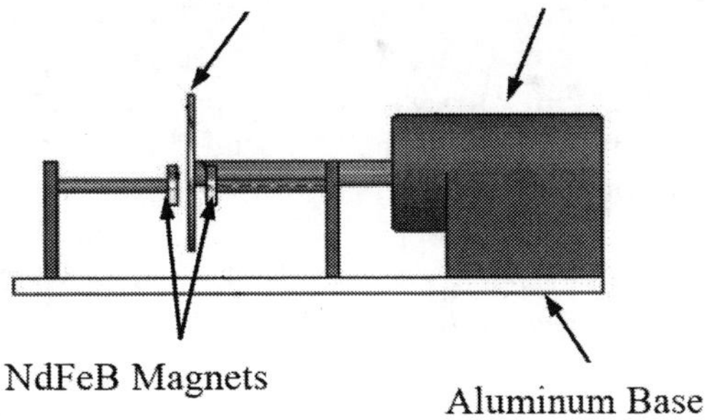

Fig. 1. Schematic of test rig

Experiments were conducted with different positions of magnets and movement of magnets towards rotating disc. The experiments were conducted with gradual movement of magnets towards disc and sudden movement towards disc. The gradual movement of magnets is to simulate braking sequence in high speed train applications whereas sudden application is to simulate braking sequence in high speed automotive application. The different positions of magnets with which experiments were carried out are as shown in Fig. 2.

- 2 coaxial magnets on either side of rotating disc and move at a velocity of 4 mm/second towards rotating disc.
- Single magnet on one side of rotating disc and moves with a velocity of 4 mm/second towards rotating disc.
- Single magnet on one side of rotating disc and moves towards rotating disc suddenly to simulate the actual brake application condition.

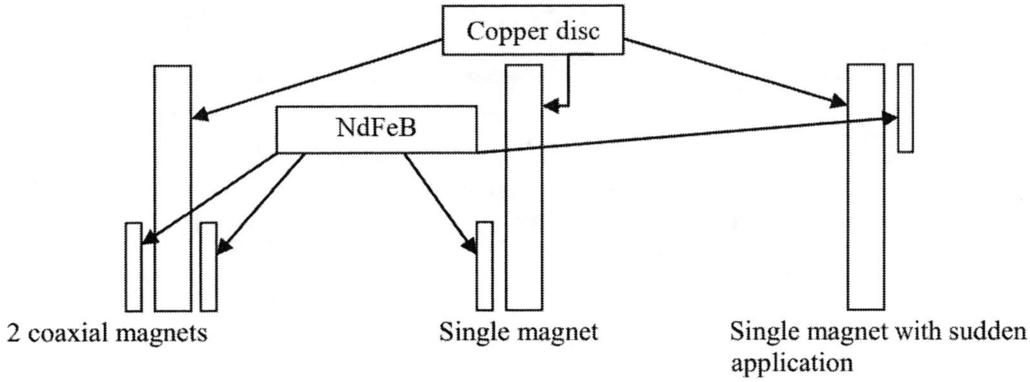

Fig. 2. Schematic of different positions of magnets

The electrical conductivity, permeability and susceptibility properties of Copper are 5.96 × 10^7 Siemens/m, 1.256629 × 10^{-6} Henry/m and -9.2 × 10^{-6} respectively. As Copper is diamagnetic material, induced magnetization is slightly slow.

3 Results and discussions

In the present work the experiments were carried out with permanent magnets made of Neodymium – Iron – Boron (NdFeB) magnets of 12.5 mm thickness and 50 mm diameter. All the discs are of 200 mm diameter. The speeds at which experiments were conducted are 2000 rpm, 3000 rpm, 4000 rpm and 5000 rpm to simulate speeds of 75 km/hr, 113 km/hr, 150.8 km/hr and 188.5 km/hr respectively. Repulsion was experienced by magnets when brought closer than 5 mm from the rotating disc. Hence, the gap between disc and magnet was maintained at 5 mm for all the conditions of positions of magnets. From experimental investigations it was observed that the induced magnetization begins when distance between magnets and rotating disc is 35 mm. Induced magnetization increases as the distance between magnets and disc decreases.

Experiments were carried out to study the influence of various parameters such as conductivity, permeability, susceptibility of Copper, speed of Copper disc and thickness of Copper discs on braking parameters such as % speed reduction, deceleration and time taken for speed reduction. The time taken is estimated when the distance between magnets and rotating disc is 5 mm. Deceleration considered is the maximum value recorded during experimentation. Investigations were carried out with Copper discs of 4 mm, 6 mm and 8 mm thickness at 2000 rpm, 3000 rpm, 4000 rpm and 5000 rpm.

3.1 Two coaxial magnets condition

Experiments under this condition was carried out to understand the influence of two magnets on induced magnetization when positioned on the same axis but on either side of rotating disc. Fig. 3, Fig. 4 and Fig. 5 indicates comparisons of % speed reduction, time taken for speed reduction and deceleration values for various thicknesses of Copper discs at 2000 rpm, 3000 rpm, 4000 rpm and 5000 rpm respectively in 2 coaxial magnets condition.

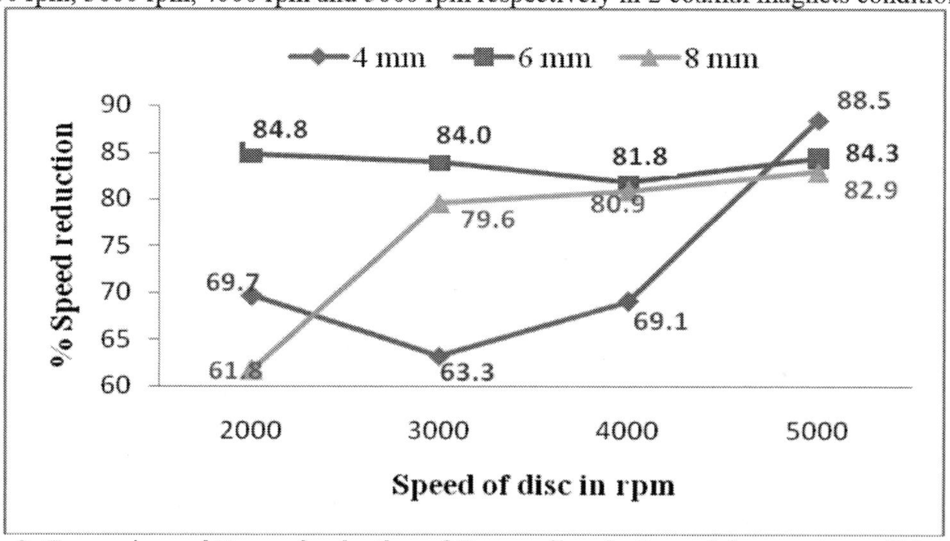

Fig. 3. Comparison of % speed reduction of Copper discs in 2 coaxial magnets condition

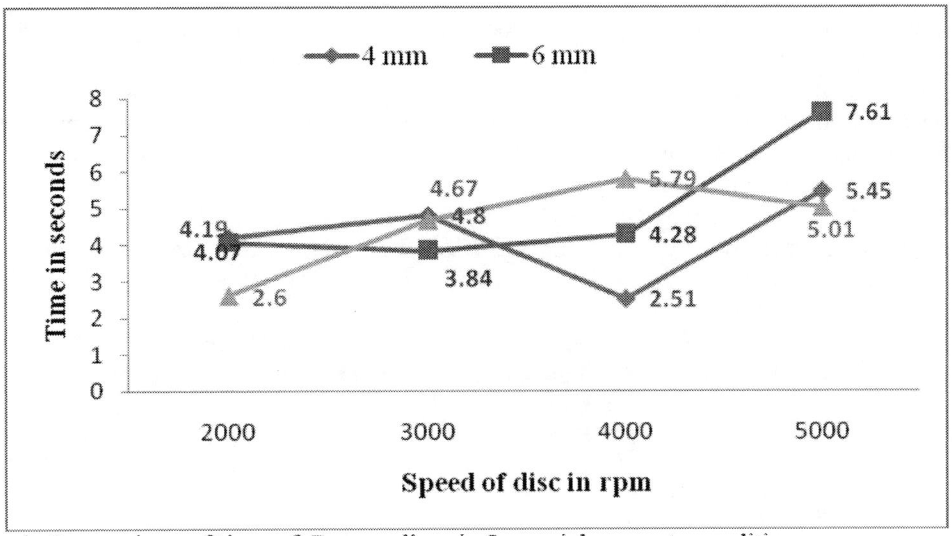

Fig. 4. Comparison of time of Copper discs in 2 coaxial magnets condition

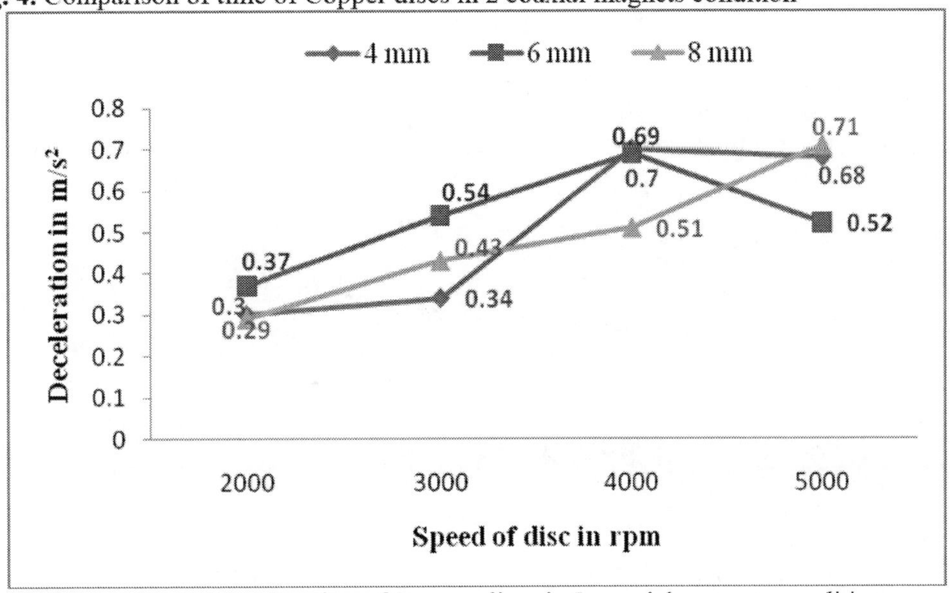

Fig. 5. Comparison of deceleration of Copper discs in 2 coaxial magnets condition

Referring to Fig. 3, it can be observed that, in 6 mm thick disc, % speed reductions are 84.8%, 84%, 81.8% and 84.3% at 2000 rpm, 3000 rpm, 4000 rpm and 5000 rpm respectively. These % speed reductions are higher compared to that of 4 mm and 8 mm thick discs. Also, the % speed reductions are uniform at all the above said speeds. It can be observed that % speed reduction is more in 6 mm thick disc than that in 4 mm thick disc. This is due to the fact that the drag force is directly proportional to the thickness of conductor. However, in 8 mm thick disc there is a decrease in % speed reduction. This may due to skin effect wherein in very fast changing fields, the magnetic field does not penetrate completely into the interior of the material.

Time taken was estimated when the distance between magnets and disc was 5 mm as it was observed that induced magnetization increased as the distance between magnets and disc decreased. Referring to Fig. 4, it can be observed that in 6 mm thick disc, the time taken for speed reductions are 4.07 s, 3.84 s, 4.28 s and 7.61 s at 2000 rpm, 3000 rpm, 4000 rpm and 5000 rpm respectively. The time taken is less at 2000 rpm, 3000 rpm, 4000 rpm compared to the time taken in 4 mm and 8 mm thick discs. However, at 5000 rpm, the time taken is high at 7.61 s. This may be due to slow growth of flux in the disc.

Deceleration considered is the maximum value recorded during the experimentation. Referring to Fig. 5, it can be observed that the deceleration in 6 mm thick disc is higher compared to the decelerations of 4 mm and 8 mm thick discs. The decelerations recorded for 6 mm thick disc are 0.37 ms^{-2}, 0.54 ms^{-2}, 0.69 ms^{-2} and 0.52 ms^{-2} at 2000 rpm, 3000 rpm, 4000 rpm and 5000 rpm respectively. It can be observed that at 5000 rpm, the time taken is more compared to 4 mm and 8 mm thick discs. This is due to slow flux growth and more time taken and hence lesser deceleration.

It may be observed that the brake parameters such as % speed reduction, time taken to achieve speed reduction and deceleration of 6 mm thick disc were found to be better compared with those of 4 mm and 8 mm thick discs. It can be observed that the deceleration is tolerable and speed reduction is gradual.

3.2 Single magnet condition

In this condition of experimentation, only one magnet is moved towards rotating disc at 4 mm/sec velocity. This condition of position of magnet was chosen to understand the behavior on induced magnetization. Fig. 6, Fig. 7 and Fig. 8 indicates comparisons of % speed reduction, time taken for speed reduction and deceleration values for various thicknesses of Copper discs at 2000 rpm, 3000 rpm, 4000 rpm and 5000 rpm respectively in single magnet condition.

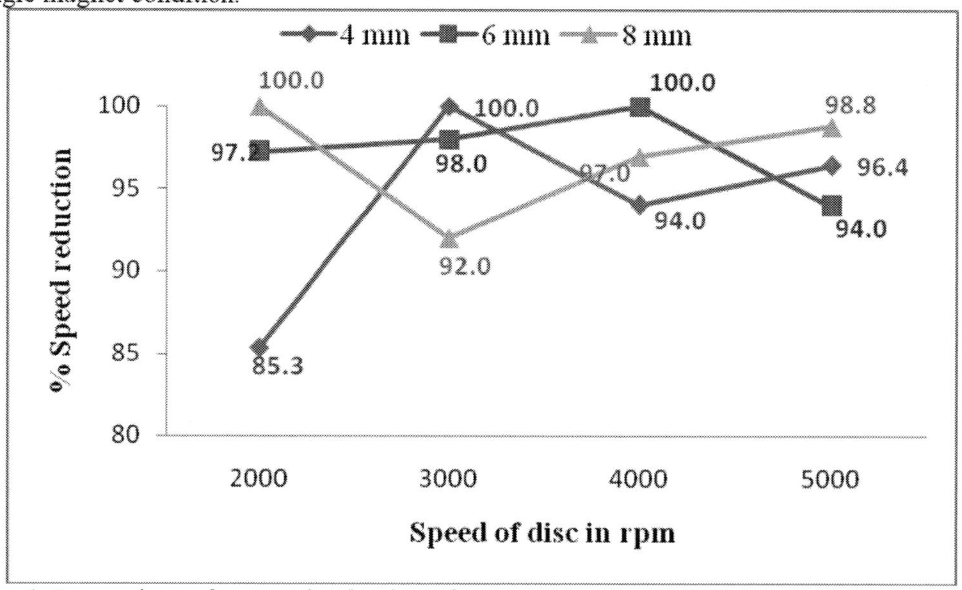

Fig. 6. Comparison of % speed reduction of Copper discs in single magnet condition

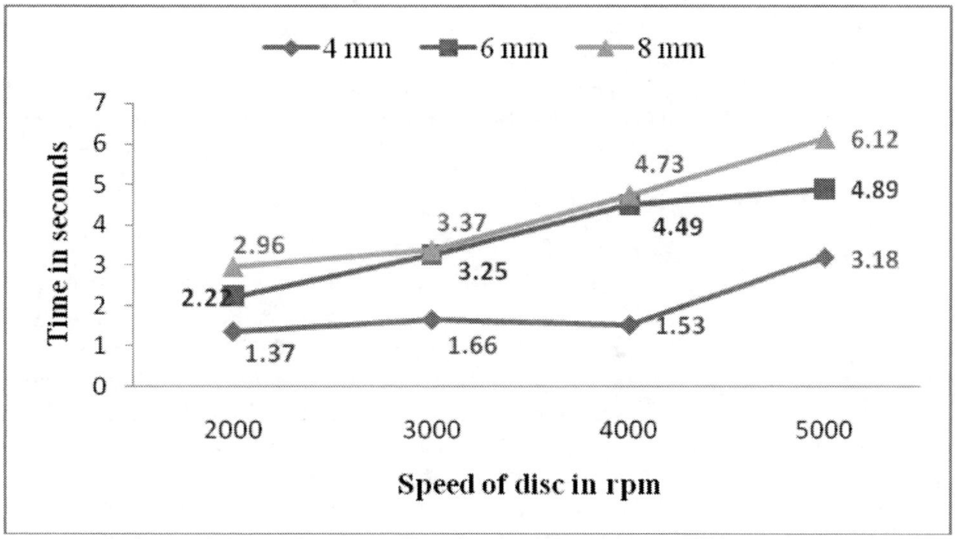

Fig. 7. Comparison of time of Copper discs in single magnet condition

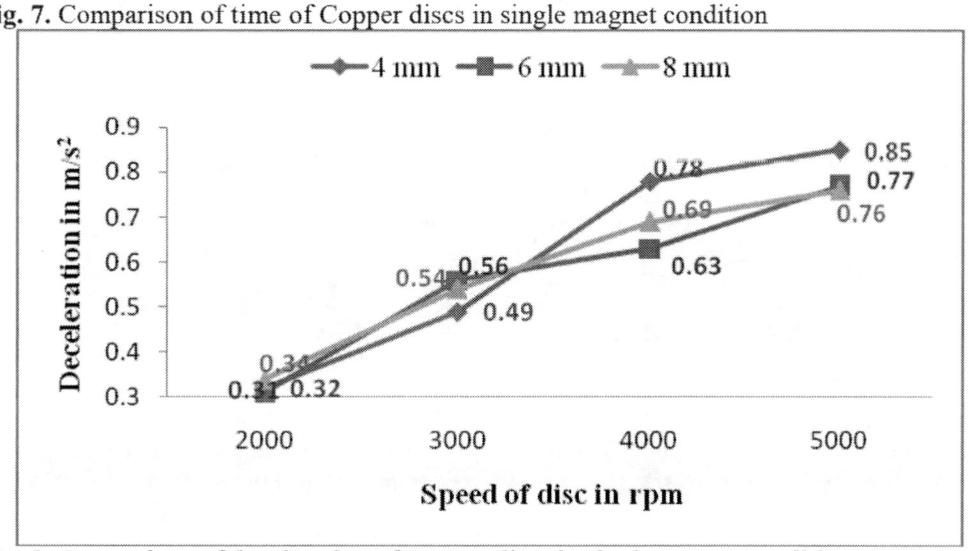

Fig. 8. Comparison of deceleration of Copper discs in single magnet condition

Referring to Fig. 6, it can be observed that in 6 mm thick disc, % speed reductions are 97.2%, 98%, 100% and 94.0% at 2000 rpm, 3000 rpm, 4000 rpm and 5000 rpm respectively. Even though the % speed reduction in 4 mm thick disc and 8 mm thick disc are comparable to that of 6 mm thick disc, the % speed reduction is higher and more uniform in 6 mm thick disc. This may be due to skin effect in 4 mm thick disc and magnetization in 8 mm thick disc. Magnetization $M = \frac{m}{v}$, as volume increases, magnetization decreases.

Referring to Fig. 7, it can be observed that in 4 mm thick disc, the time taken for speed reductions are 1.37 s, 1.66 s, 1.53 s and 3.184 s at 2000 rpm, 3000 rpm, 4000 rpm and 5000 rpm respectively. The time taken is less compared to the time taken in 6 mm and 8 mm thick discs. This is due to the reason that speed reduction is faster in 4 mm thick disc than that in 6 mm and 8 mm thick discs at 2000 rpm, 3000 rpm, 4000 rpm and 5000 rpm.

Referring to Fig. 8, it can be observed that the deceleration in 4 mm thick disc is found to be comparable with the decelerations of 6 mm and 8 mm thick discs. The decelerations recorded for 4 mm thick disc are 0.32 ms^{-2}, 0.49 ms^{-2}, 0.78 ms^{-2} and 0.85 ms^{-2} at 2000 rpm, 3000 rpm, 4000 rpm and 5000 rpm respectively.

From the point of view of higher speed reduction achieved, less time taken and higher deceleration, 4 mm thick Copper disc can be a better option in single magnet condition.

3.3 Comparison between two coaxial magnets condition and single magnet condition

In 2 coaxial magnets % speed reduction is less and time for speed reduction is more. This may be due to nullifying effect of magnetic flux when two magnets come closer to disc. When like poles are placed near each other, their magnetic fields cancel each other, and there is a neutral point where the combined field strength is zero. Also, there is repulsion between magnets which may be causing the reduction in induced magnetization. However, speed reduction is more gradual compared to single magnet condition.

In single magnet condition, as the Copper is diamagnetic material, flux growth is slow. Even though the flux growth is slow which is evident from higher time taken for speed reduction, % of speed reduction is more in single magnet condition compared to 2 coaxial magnets condition. It may be observed that brake parameters such as % speed reduction, time taken for speed reduction and deceleration are better in single magnet condition.

From the point of view of % speed reduction, comparable time taken and moderate deceleration, both 4 mm thick and 6 mm thick Copper discs in single magnet condition can be used for high speed train applications. Since the mechanism requirements in single magnet condition are less, single magnet condition will be more economical. Also, because 4 mm thick disc is lesser in weight, more economical and less space consuming compared with 6 mm and 8 mm discs, 4 mm thick disc is economical and advantageous.

3.4 Single magnet with sudden application condition

In this condition of experimentation, only one magnet is moved towards rotating disc suddenly. This is to simulate actual braking condition in automobiles. This condition of position of magnet was chosen to understand the behavior on induced magnetization when magnet is brought closer to the rotating disc at a rapid pace. Fig. 9, Fig. 10 and Fig. 11 indicates comparisons of % speed reduction, time taken for speed reduction and deceleration values for various thicknesses of Copper discs at 2000 rpm, 3000 rpm, 4000 rpm and 5000 rpm respectively in single magnet with sudden application condition.

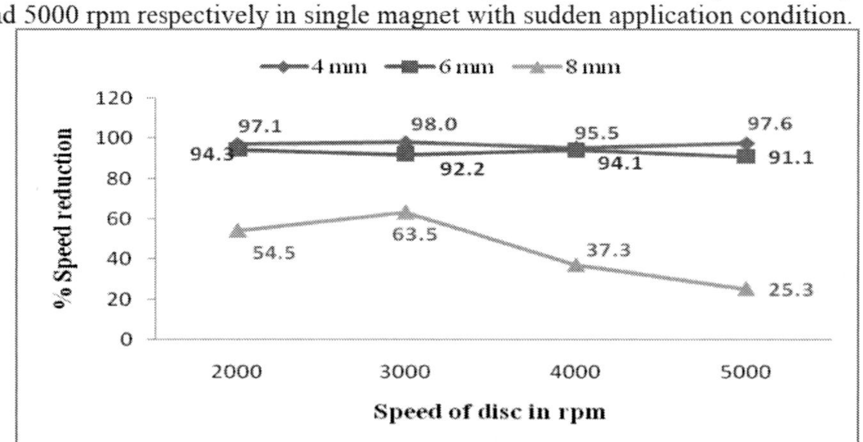

Fig. 9. Comparison of % speed reduction of Copper discs in single magnet with sudden application condition

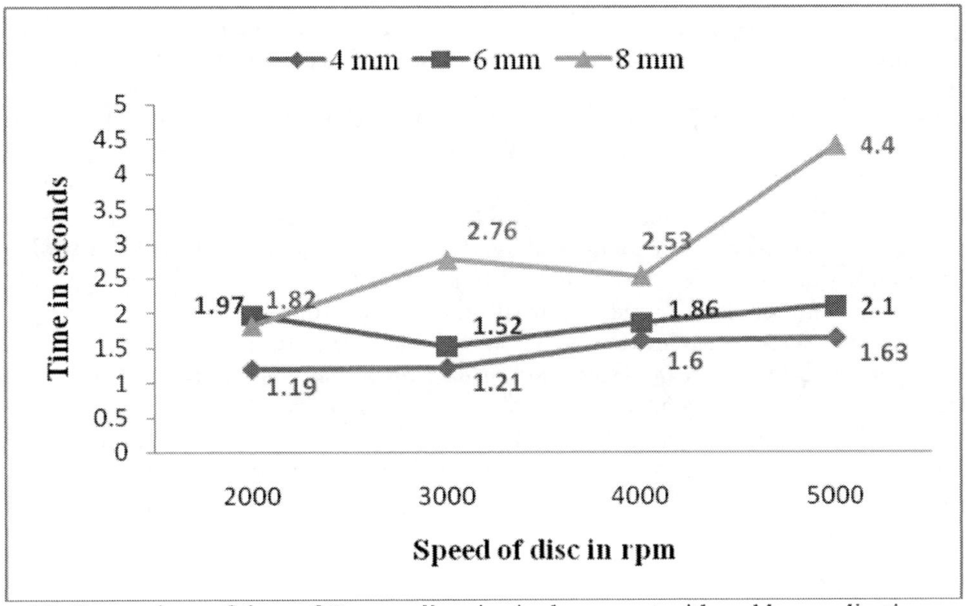

Fig. 10. Comparison of time of Copper discs in single magnet with sudden application condition

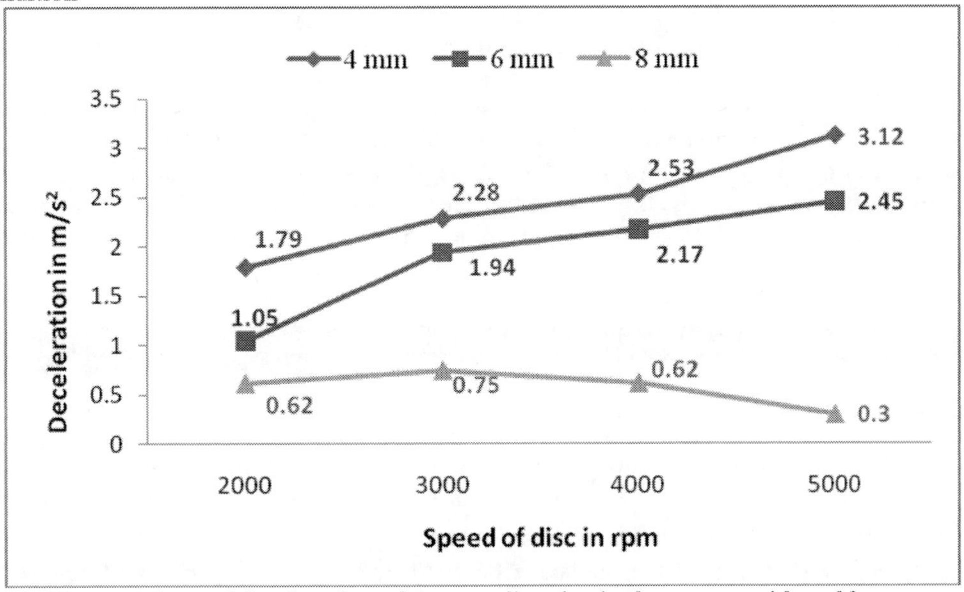

Fig. 11. Comparison of deceleration of Copper discs in single magnet with sudden application condition

Referring to Fig. 9, it can be observed that in 4 mm thick disc, % speed reductions are 97.1%, 98%, 95.5% and 97.6% at 2000 rpm, 3000 rpm, 4000 rpm and 5000 rpm respectively. These % speed reductions are higher and more uniform for all speeds compared to that of 6 mm and 8 mm thick discs. It was observed that a very high repulsion was experienced with 8 mm thick disc at all the aforesaid speeds. This may be due to high magnetic induction which repelled the original magnetic field. Also, may be due to Copper being diamagnetic in nature.

Referring to Fig. 10, it can be observed that in 4 mm thick disc, the time taken for speed reductions are 1.19 s, 1.21 s, 1.6 s and 1.63 s at 2000 rpm, 3000 rpm, 4000 rpm and 5000 rpm respectively. The time taken is less compared to the time taken in 6 mm and 8 mm thick discs. This may be due to better interaction between the properties of Copper and induced magnetization.

Referring to Fig. 11, it can be observed that the decelerations in 4 mm thick disc are 1.79 ms^{-2}, 2.28 ms^{-2}, 2.53 ms^{-2} and 3.12 ms^{-2} at 2000 rpm, 3000 rpm, 4000 rpm and 5000 rpm respectively. The deceleration values of 4 mm thick disc are higher compared to the decelerations of 6 mm and 8 mm thick discs. From experimental investigations it was found that very high repulsion was experienced by 8 mm thick discs at 4000 rpm and 5000 rpm. The magnet could not travel close to the rotating disc because of high repulsion.

From the experimental investigations it was observed that the brake parameters were better in 4 mm thick disc compared to 6 mm and 8 mm thick discs. As very high repulsion was experienced by 8 mm thick disc, 8 mm thick disc may not be suitable for both high speed trains and high speed automotive applications. Because of higher % of speed reduction, higher deceleration and less time taken for speed reduction, 4 mm thick disc may be more suitable for high speed automotive applications in single magnet with sudden application condition.

4 Conclusions

From the experimental investigations it is found that in single magnet condition with 4 mm thick disc, higher speed reduction, time taken is less and higher deceleration are achieved compared to that in 2 coaxial magnets condition. Hence, 4 mm thick disc in single magnet condition may be more suitable for high speed train applications.

From the experimental investigations in single magnet with sudden application condition, it is found that deceleration in 4 mm thick disc is higher and time taken is lesser compared to that in 6 mm and 8 mm thick discs. Because of higher % of speed reduction, higher decelerations and less time taken for speed reduction, 4 mm thick disc may be more suitable for high speed automotive applications. Single magnet with sudden application condition may also be used as braking system for high speed cars as there is a necessity of speed reduction only and not stopping. Since it is a contactless braking, the problems of temperature rise and loss of friction coefficient may be greatly eliminated.

References

1. Kung SW, Dunlop KB, Ballinger RS. "Complex Eigenvalue analysis for reducing low frequency brake squeal". *Technical report* 2000 – 01 – 0444, Warrendale (PA); *SAE (2000)*
2. Dunlop KB, Richle MA, Longhouse RE. "An investigative overview of automotive disc brake noise", *SAE paper* (1999).
3. Mario Triches Junior, Samir N.Y.Gerges, Roberto Jordan, "Analysis of brake squeal noise using the finite element method: A parametric study". *Applied Acoustics* Vol.69, pp.147 -162(2008)
4. Andreas Wagner, Gottfried Spelsberg – Korspeter, Peter Hagedorn, "Structural optimisation of an asymmetric automotive brake disc with cooling channels to avoid squeal", *Journal of Sound and Vibration*, pp.1 – 11,(2013)
5. R. Sedaghati, "Benchmark case studies in structural design optimisation using the force method", *International journal of solids and structures* Issue 42, pp.5848 – 5871,(2005)

6. Y. D. Kwon, S. B. Kwon, S. B. Jin, J. Y. Kim. "Convergence enhanced genetic algorithm with successive zooming method for solving continuous optimisation problems". *Computers and structures* Issue 81,pp.1715 – 1725,(2003)
7. N. Ali, K. Behdinan, Z. Fawaz. "Applicability and viability of a GA based finite element analysis architecture for structural design optimisation". *Computers and structures* Issue 81, pp.2259 – 2271,(2003)
8. Sherif H. "Investigation on effect of surface topography of pad / disc assembly on squeal generation". *Wear,* Issue 257 (7 – 8), (2004), pp.687 – 695
9. H. Hetzler, K.Willner, "On the influence of contact tribology on brake squeal", *Tribology International*, Vol.11, pp.237 – 246, (2011)
10. A. Akay, O. Giannini, F. Massi, A. Sestieri, "Disc brake squeal characterization through simplified test rigs", *Mechanical systems and signal processing*, Vol.16, pp.2590 – 2607, (2009)
11. AliBelhocine, MostefaBouchetara, "Thermomechanicalmodelling of dry contacts in automotive disc brake", *International journal of Thermal Sciences* Vol.60, pp.161 – 170, (2012)
12. Thomas J. Mackin et al, "Thermal cracking in disc brakes", *Engineering Failure Analysis*, Issue 9, pp.63 – 76, (2002)
13. H. S. Qi, A. J. Day, "Investigation of disc / pad interface temperatures in friction braking", *Wear* Issue 262, pp.505 – 513, (2007)
14. H. S. Qi, A. J. Day, K. H. Kuan, G. Rosala. A contribution towards understanding brake interface temperatures". In: *proceedings of the international conference 'braking 2004 – vehicle braking and chassis control*; Leeds, UK, 7 – 9 July, (2004)
15. T. Valvano, K. Lee. "An analytical method to predict thermal distortion of a brake rotor". *Society of Automotive Engineers*, Inc, 2000 – 01 – 0445,(2000)
16. P. J. Blau, J. C. Mclaughlin, *Tribol. Int* 36,pp.709 – 715,(2003)
17. J. Bijwe, Nidhi, N. Majumdar, B. K. Satapathy, "Influence of modified phenolic resins on the fade and recovery of friction materials", *Wear*, Issue 259, pp.1068 – 1078, (2005)
18. H. Smales, "Friction materials – black art or Science?"*J.Automob.Eng.*Issue 209, pp.151 – 157,(1995)
19. Mukesh Kumar, Jayashree Bijwe, "Studies on reduced scale tribometer to investigate the effects of metal additives on friction coefficient – temperature sensitivity in brake materials", *Wear*, Vol.269, pp.838 – 846, (2010),
20. "Brake lining quality control procedure", SAE J661a, in: *SAE Handbook*, SAE, Detroit, MI, pp.31 – 55, (1978)
21. "Replacement brake lining assemblies", E.C.E Regulation no. – 90 *INTEREUROPE Regulation ltd*, UN, (1993)
22. U. S. Hong, S. L. Jung, K. H. Cho, M. H. Cho, S. J. Kim, H. Jang. "Wear mechanism of multipurpose friction materials with different phenolic resin matrices", *Wear*, Issue 266 (7 – 8), pp.739 – 774, (2009)
23. RukiyeErtan, NurettinYavuz, "An experimental study on the effects of manufacturing parameters on the tribological properties of brake lining materials", *Wear*, Issue 268, pp.1524 – 1532, (2010)
24. Osman TahaSen, Jason T. Dreyer, Rajendra Singh, "Low frequency dynamics of a translating friction element in the presence of frictional guides, as motivated by a brake

vibration problem", *Journal of Sound and Vibration,* Issue 332, pp.5766 – 5788. (2013)

25. S. Lignon, J-J. Sinou, L. Jezequel, "Stability analysis and μ - synthesis control of brake systems", *Journal of Sound and Vibration*, Issue 298, pp.1073 – 1087, (2006)
26. Der – Ming Ma, Jaw – KuenShiau: *The design of Eddy current magnet brakes*. Transactions of the Canadian Society for Mechanical Engineering", Vol.35, No.1, (2011).
27. Marc T. Thompson: *Practical Issues in the use of NdFeB Permanent Magnets in Maglev, Motors, Bearings and Eddy Current Brakes*, Proceedings of IEEE, Vol.97, Issue 11, pp.1758 – 1767, (2009)
28. Schykowski, Jennifer: *Eddy – current Braking: A long road to success*. Railway Gazette, DVV media UK Ltd.(2013)

Conceptual design of planetary gearbox system for constant generator speed in hydro power plant

Bhargav[1], *M. A. Parameshwaran*[2], *Sivaraj S.*[2] and *Nithin Venkataram*[1*]

[1]*Dept. of Mechanical and Manufacturing Engineering, Ramaiah University of Applied Sciences Bangalore, India*
[2]*MAGTORQ Pvt. Ltd., Hosur, Tamil Nadu, India*

Abstract. Micro Hydro Power Plant (MHPP) is emerging as one of the most clean, renewable and reliable energy technology for harnessing power. In MHPP hydro governors are avoided, that results in turbine speed fluctuation. MHPP requires either speed or torque amplification of generator for constant power generation. To achieve this, planetary gear transmission system is explored for MHPP due to its higher efficiency and compact size. A conceptual planetary gearbox system is developed for MHPP to maintain constant generator speed. The conceptual gearbox is designed, modelled and analysed using ADAMS software. Simulation results are found to be in close agreement with analytical results. Hence, conceptual design of planetary gearbox can be used to govern constant generator speed. In this paper, a MHPP which generate constant power of 5 kW at constant generator speed of 1490 rpm is analysed and validated

1 Introduction

Electric energy is an essential requirement and its availability plays an important role in day to day life. The gap between the demand and supply of electricity is increasing rapidly and it is not possible to meet the demand, because of which most of the rural areas is still starved by the grid supply. Hence there is a great need to produce power for remote rural areas. Micro Hydro Power Plant (MHPP) is one of the emerging solutions for rural electrification, since it is built near valley, small rivers which are located far away from the national grid [1]. The power generating capacity of MHPP ranges between 5 kW to 100 kW [2]. In MHPP, the powerhouse is directly placed on the runway stream with minimal arrangement for directing water towards powerhouse and hence hydro governor can be avoided. Thus MHPP needs no reservoir for power generation. The water from the runway stream is used for power generation and goes back to runway stream through tailrace. Thus MHPP has a minimal impact on local ecosystem when compared to large hydro power plant [3]. Power house of MHPP expects minimum maintenance, since it is located far from the national grid. In addition, the power house needs a good control system for maintaining constant power generation. A good control system will have a uniform transmission, which

* Corresponding author: nithin31@gmail.com

results in constant turbine and generator speed. In the absence of hydro governor, the uneven flow of water will lead to variable turbine speed, which results in variable generator speed. Thus implementation of MHPP is a challenging task due to lack of technical development for maintaining constant generator speed. Hence a new design of transmission system is required to govern constant generator speed for variable input turbine speed [4].

The speed control in power production is possible through a transmission system. In MHPP, two types of transmission can be used, direct transmission and indirect transmission. In direct transmission turbine shaft is directly coupled to generator shaft, which yields in very high efficiency. However, both turbine and generator needs to be designed as per the requirement which is uneconomic. Hence it is not suitable for MHPP. To match both turbine speed and generator speed, indirect transmission system is used. In MHPP, the turbine speed ranges between 100-500 rpm and standard generator runs at much higher rotation per minute, so a speed multiplier is used. From literature, it is observed that there are various indirect transmission systems such as chain, belt and gear drive available for MHPP. Chain and belt drive has low efficiency and load sharing capacity, hence it is used in low very transmission ratio application for power generating capacity under 3 kW [5].

In addition to this, the gear transmissions are used in power transmission system. Table 1 shows the comparison between gear and planetary transmission. The planetary gear drive has high power to weight ratio with high load carrying capacity [6]. Also, more than one speed ratio is achieved when compare to fixed axis gearbox. Planetary gearbox has good efficiency, reliability and compactness hence it is preferred for MHPP [7].

Table 1. Comparison between gear drive planetary gear drive

Type of transmission system	Efficiency	Application	Features
Gear transmission	75-85%.	Low transmission	Higher transmission ratio, no of stages increases, increasing overall dimensions
Planetary gear transmission	95%	High transmission	Compact and high load sharing capacity

In MHPP, generator operates at higher speed than the turbine. Fig. 1 shows conventional planetary drive transmission. Conventional planetary gear drive either amplifies or reduces speed depending upon the input conditions. When ring gear is fixed, input is given to turbine coupled to carrier and output is extracted from sun gear coupled to generator, planetary gear transmission acts as speed multiplier. Similarly, if the input motion is given to sun gear and output is extracted from carrier, planetary gear transmission act as speed reducer. If the input speed varies, so does the output speed by a factor of transmission ratio. There is no control over variable turbine speed for maintaining constant output speed i.e. if turbine speed varies the generator speed will also vary. Hence there is a need to develop a novel concept to maintain constant generator speed.

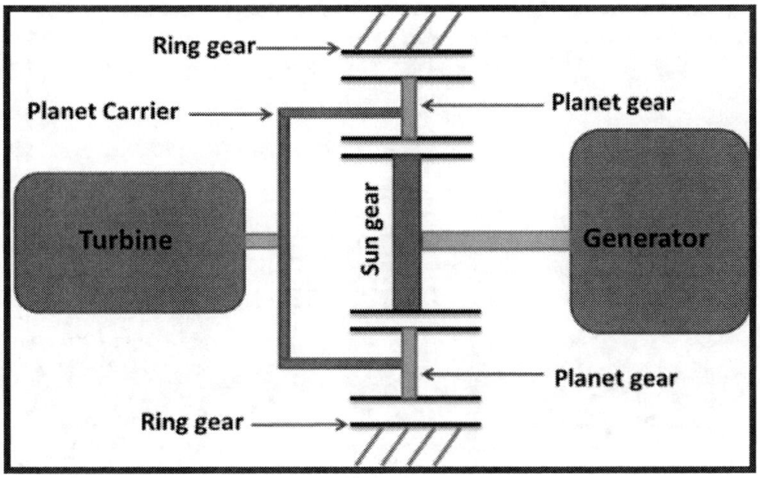

Fig. 1. Existing concept used in planetary gear drive for MHPP

2 Conceptual design

A conceptual design is developed to maintain constant generator speed. The design consists of planetary gearbox in which ring gear is fixed, input is given to turbine coupled carrier and output is extracted from sun gear coupled to generator, the input speed amplifies depending on transmission ratio. On the other hand, if the ring gear is allowed to rotate in the direction of the planet gear, the output speed equals to sum of amplified speed due to carrier and amplified speed due to ring gear. The generalized relation between sun, carrier and ring gear speed is provided in eqn.1.

$$N_{sun} = N_{sun\text{-}carrier} + N_{sun\text{-}ring\ gear} \quad (1)$$

When the ring gear rotates in the direction opposite to the planet gear, the ring gear motion opposes the planet gear motion resulting in reduced output. In other words output speed equals difference in speed due to carrier and speed due to ring gear. The generalized relation is provided in eqn. 2.

$$N_{sun} = N_{sun\text{-}carrier} - N_{sun\text{-}ring\ gear} \quad (2)$$

From the eqn. 1 and 2, it is observed that the constant speed can be achieved by controlling the speed of ring gear. Based on this understanding, a concept is designed for maintaining constant generator speed. The conceptual design consists of single stage planetary gearbox, with auxiliary motor as shown in Fig. 2. The planetary gearbox consists of sun gear, planet carrier and ring gear. The ring gear is connected to auxiliary unit. The auxiliary unit consists of a reduction helical and spur gearbox, which is coupled to an auxiliary motor.

Planetary gearbox system is coupled between turbine and generator. Turbine shaft is connected to the planet carrier, which acts as input to the planetary gearbox. Output of planetary gear is extracted from sun gear which is connected to generator.

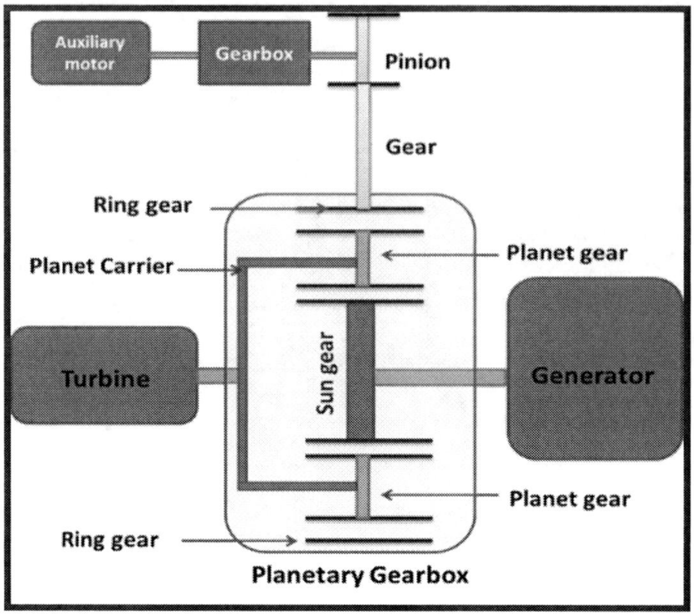

Fig. 2. Line diagram of planetary gearbox system to govern constant generator speed

3 Analytical calculations

3.1 Design requirement

The developed conceptual design of planetary gearbox system is an indirection transmission system. The planetary gearbox is designed for transmitting turbine speed to generator. It is assumed that turbine rotates at a speed of 150 rpm and standard generator for generating 5 kW, 50 Hz rotates at 1500 rpm. Hence single stage planetary gearbox with 3 planet gear is design for 1:9.9 speed ratio.

3.2 Design of planetary gearbox

Table 2. Tabulation method for arriving at no. of teeth's in gear (Maitra, 2001)

Planet Carrier	Sun Gear	Planet Gear	Ring Gear
0	+1	(Zs / Zp)	- (Zs / Zi)
0	x	- (Zs / Zp)* x	- (Zs / Zi)* x
y	x+y	y - (Zs / Zp)* x	y - (Zs/Zi)* x

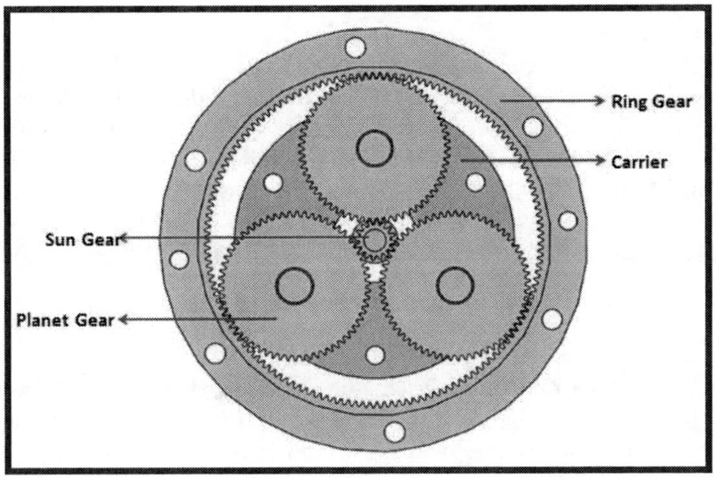

Fig. 3. Schematic of planetary gear drive

Using table 2, number of teeth on sun, planet and ring gear is calculated. From 1st row, it is observed that when a sun gear rotates in clockwise direction, the planet and ring gear rotates in an anti-clock wise direction. Similarly, in third row, if the planet carrier rotates at y rpm then sun gear rotates at x+y rpm. The planetary gear has three planets revolving and rotating around sun gear. The sun gear and the planet gear are enclosed by ring gear. Three planet gears are coupled to output shaft through a carrier. The planetary gear assembly is shown in Fig. 3.

When ring gear is fixed, the eqn. 3, is equated to zero

$$y - (Z_s/Z_i)x = 0 \tag{3}$$

$$Z_i = Z_s + Z_p \tag{4}$$

By considering, minimum number of teeth on sun gear as 16 teeth, and by using eqn. 4, the number of teeth on ring gear is calculated to be 143, and number of teeth on planet is found to be 63. The module of the planetary gearbox is found by using eqn. 5.

$$m = (2*M_t/4*\sigma*\pi^2*z*y)^{1/3} \tag{5}$$

By substituting the above data in eqn. 5, module is found to be 1.25mm. Torque at sun gear is, found by using eqn. 6

$$P = 2\pi NT_s/60 \tag{6}$$

The power provided at sun gear is considered to be 5 kW. By substituting the value of power in eqn. 6, we get torque at sun gear as 32,030 N-mm.

From energy eqn. 7, 8 and 9, torque at ring and carrier is calculated

$$T_SN_S + T_CN_C + T_RN_R = 0 \tag{7}$$

$$T_SN_S + T_CN_C = 0 \tag{8}$$

$$T_S + T_C + T_R = 0 \quad (9)$$

Since the ring gear is fixed, there will be no motion and hence the speed at ring gear is equal to zero. By substituting these values in eqn. 7, 8 and 9, the torque on carrier are calculated to be -318,300 N-mm, the negative indicates that it is a resisting torque developed by the carrier. On similar calculation, the torque on the ring gear is found to be 286,270 N-mm.

3.3 Effect of auxiliary motor on generator speed

In this section a generalized equation is proposed to study the effect of auxiliary motor on the generator speed is carried out. When the generator speed varies from the required speed, the difference in the speed is compensated by using an auxiliary motor. The auxiliary motor in turns drives the gearbox, which is in mesh with ring gear of planetary gearbox. Therefore, to find out the effect of auxiliary motor on the generator speed, the carrier of the planetary gearbox is fixed. And input speed is given to auxiliary motor.

Ring gear speed= Auxiliary motor speed / Auxiliary helical gear ratio spur gear ratio* (10)

*Output generator speed= Ring gear speed * gear ratio* $_{Ring-Sun}$ (11)

The generalized equation for speed compensation by the auxiliary motor to maintain the constant speed are given by eqn. 10 and 11.

4 Modeling and Multi Body Dynamic Simulations

The arrived dimension of the planetary gearbox and auxiliary gearbox is modeled in modeling software. The modeled system is imported into a multibody dynamics ADAMS software to verify the analytical calculation through kinematic analysis. In addition to this the dynamic analysis is carried out to evaluate the torque and speed required by the auxiliary motor to maintain constant generator speed.

4.1 Multi body dynamic simulations of planetary gearbox system

The model is applied with an input speed of 150 rpm (15.70 rad/s) at carrier of the planetary gearbox as shown in Fig. 4 and simulated for 1 second with 5000 steps. This simulation is carried out for validating the transmission ratio obtained through analytical calculation. Therefore the ring gear is fixed which indicates that the auxiliary motor is in off condition.

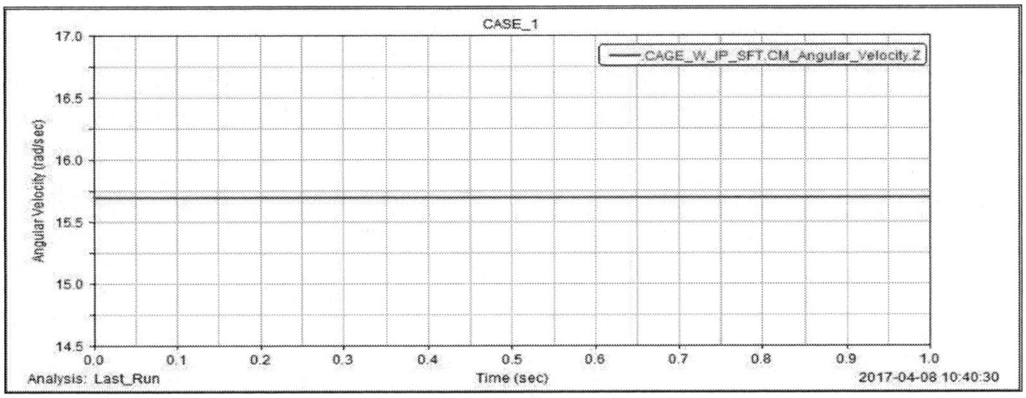

Fig. 4. Speed characteristic of the carrier of planetary gearbox system

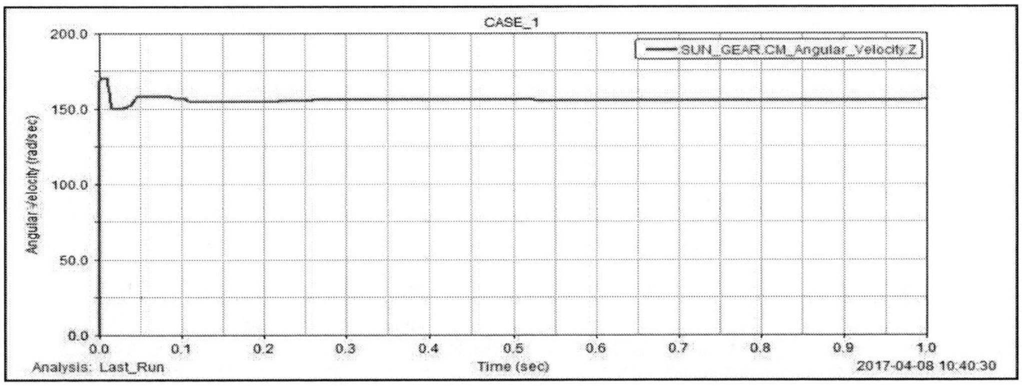

Fig. 5. Speed characteristic of the sun gear of planetary gearbox system

The output is extracted at sun gear and is presented in Fig. 5. From the output plot it can be observed that the sun gear is rotating at 1490 rpm (156.30 rad/s). When input and output speed ratio is calculated from simulation, it is found to be approximately 1:9.9.

The modeled planetary gearbox is simulated for dynamic condition for analyzing torque requirement. In the dynamic simulation, the input torque of 32,030 N-mm is applied at sun gear as per analytical calculation and simulated for 1 second with 5000 steps as shown in Fig. 6. The torque on carrier is measured and is found to be 318,000 N-mm and torque variation is shown in Fig. 7. Similarly, the torque variation of ring gear is shown in Fig. 8 and is found to be 285,000 N-mm.

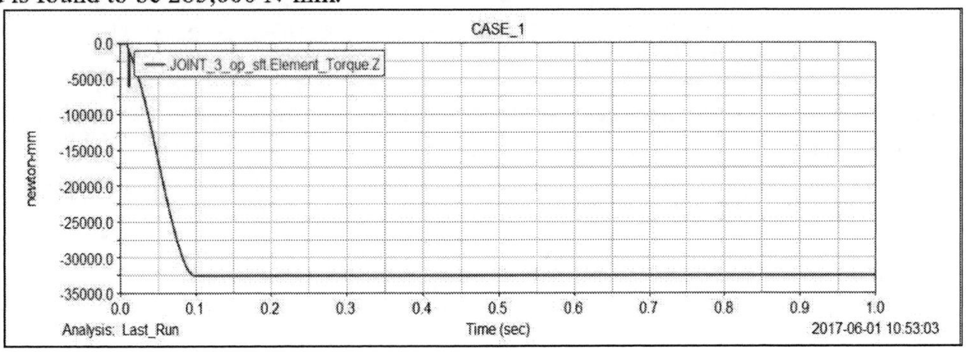

Fig. 6. Torque characteristic of the sun gear of planetary gearbox system

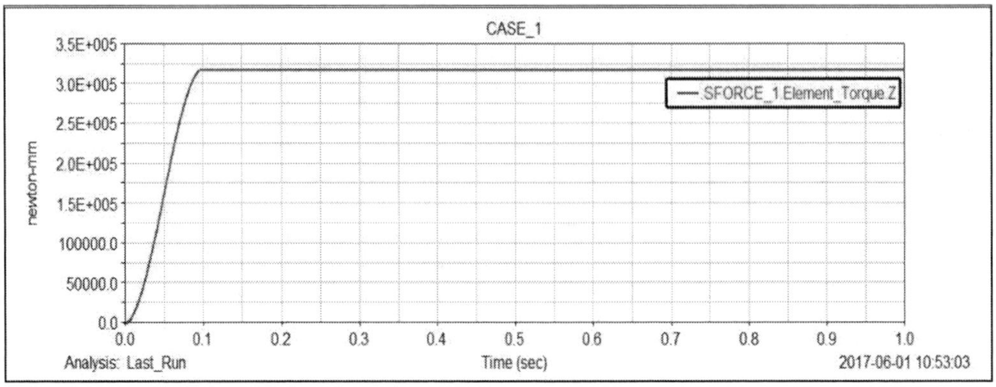

Fig. 7. Torque characteristic of the carrier

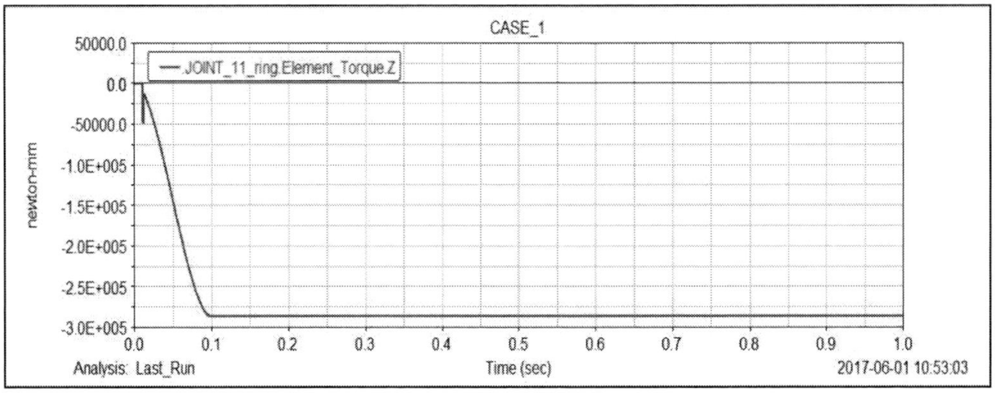

Fig. 8. Torque characteristic of the ring gear of planetary gearbox system

4.2 Simulation of conceptual design to determine the effect of auxiliary motor speed on generator speed

The aim of auxiliary motor and gearbox is to maintain constant speed of the generator. To find the effect of auxiliary motor on the sun gear coupled with generator, the simulation is carried out by fixing the carrier of the planetary gearbox. The input speed of 150rpm (15.70 rad/s) is provided to auxiliary motor and corresponding output speed is measured at sun gear which is coupled with generator. Fig. 9 shows the input auxiliary motor speed of 15.70 rad/s (150 rpm). From simulation results, the speed of ring gear is measured and found to be 0.866 rad/s (8.26 rpm) which is shown in Fig. 10. The variation in the speed is due to start up condition and can be neglected.

When the auxiliary motor is rotating at 15.70 rad/s (150 rpm), the speed at sun gear coupled with generator is measured to be 7.73 rad/s (73.89 rpm) and is shown in Fig. 11. From the figure it can be observed that the speed of sun gear is not uniform, hence average speed is considered. Therefore the speed reduction ratio from auxiliary motor to sun gear coupled with generator is 2.03 time.

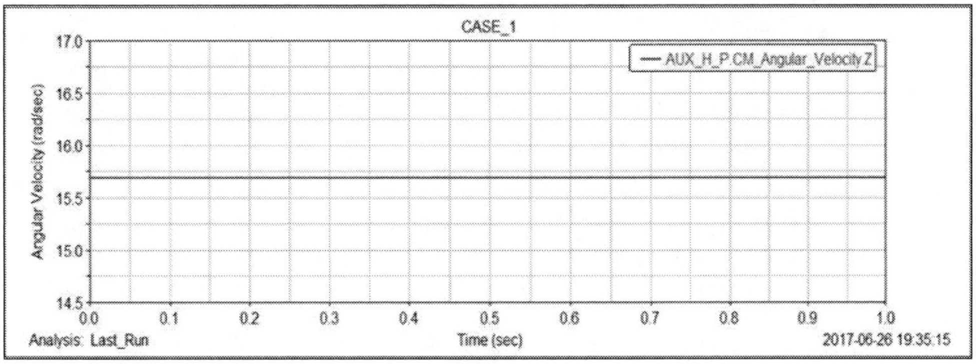

Fig. 9. Auxiliary motor speed

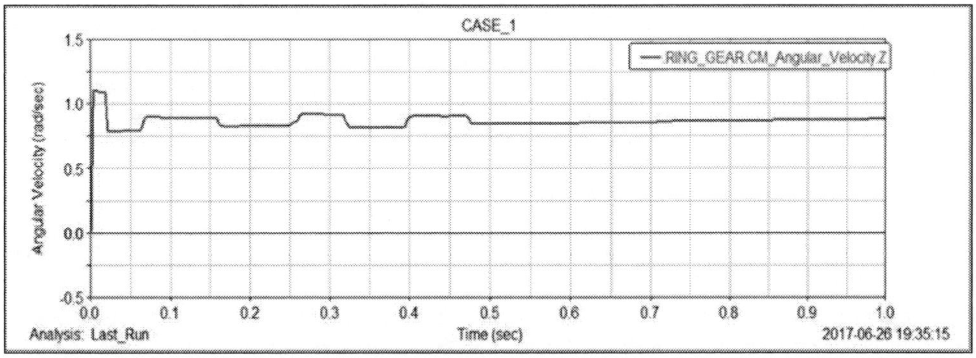

Fig. 10. Speed characteristic of the ring gear of planetary gearbox system

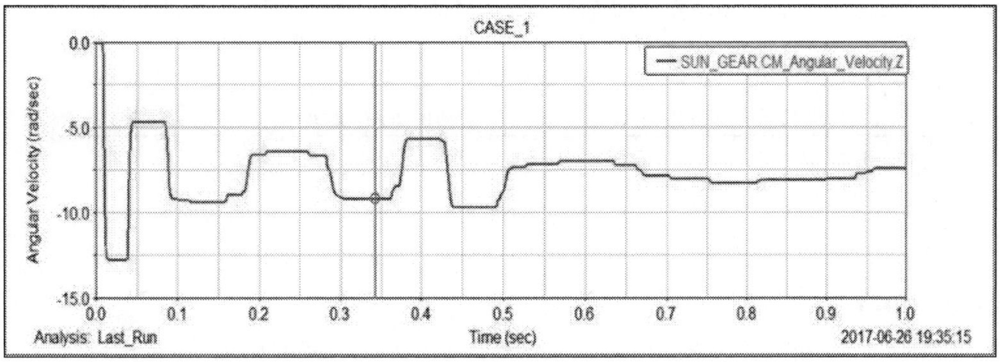

Fig. 11. Speed characteristic of sun gear of planetary gearbox system

5 Results and discussion

The table 3, shows the comparison between simulation and analytical result. When auxiliary motor is at off condition, the ring gear rpm is zero indicating that it is fixed and generator speed is 1490 rpm. Hence it can be observed that the planetary gearbox is designed for a gear ratio of 1: 9.9. The error between the simulation and analytical result is found to be well within 1%.

Table 3. Comparison of kinematic analysis result of designed planetary gear system

Particulars	Turbine motor (rpm)	Auxiliary motor (rpm)	Carrier (rpm)	Generator Sun gear (rpm)	Ring gear (rpm)
Simulation Result	150	-	150	1489.72	0
Analytical Result	150	-	150	1490.62	0

The table 4, shows results of torque comparison of planetary gearbox when the auxiliary gearbox is under off condition. The error between the simulation and analytical result is found to be well within 1%.

Table 4. Comparison of torque characteristics

Particulars	Turbine motor (rpm)	Auxiliary motor (rpm)	Carrier (N-mm)	Sun gear (N-mm)	Ring gear (N-mm)
Simulation Result	150	-	310670	32030	278650
Analytical Result	150	-	318300	32030	286270

The table 5, shows the comparison of simulation and analytical results when the auxiliary motor is running at 150 rpm. Analytical results are calculated using the eqn. 10 and 11. The error between the simulation and analytical result is found to be well within 1%.

Table 5. Kinematic analysis results of Effect of Auxiliary motor speed on generator speed

Particulars	Turbine motor	Auxiliary motor	Carrier	Sun gear	Ring gear
	Speed in rpm				
Simulation Result	-	150	0	73.89	8.26
Analytical Result	-	150	0	74.09	8.29

6 Conclusions

Conceptual design of planetary gearbox system is developed for maintaining constant generator speed for variable input turbine speed. The planetary gearbox is designed for 1:9.9 speed ratio. Speed ratio between the auxiliary motor and sun gear coupled with generator is found to be 2.03:1. The analytical and simulation result are within the acceptable limits. Hence the conceptual design can be used to in micro hydro power plant for power generation.

References

1. Doolla and T.S. Bhatti, J. Electr. Pow. Syst. Res. 76, 889-896, (2006)
2. A. M. A. Haidar, Md. F. M. Senan, A. Noman and T. Radman, J. Renew. Sust. Energ. Rev., 16, 518-524, (2012)

3. B. A. Nasir, Energ. Proc., 50, 19-29, (2014)
4. J. A. Laghari, H. Mokhlis, A. H. A. Bakar and H. Mohammad, J. Renew. Sust. Energ. Rev., 20, 279-293, (2013)
5. C. Jaliu, D. Diaconescu and R.Saulescu, , Renew. Energ. Pow. Qual. J, 1, 650-653, (2008)
6. R. Saulescu, C. Jaliu, D.V. Diaconescu and O. Climescu, Proc. of ASME 2011 Int. Des. Eng. Tech. Conf. and Comp. Info. Eng. Conf., 1-10, (2011)
7. G. M. Maitra, Handbook of Gear Design, Tata McGraw-Hill Pub. Comp. Ltd., (2001)

Numerical simulation of time-invariant error and its effect on planetary gearbox dynamics

Nithin Venkataram[1*], *Rahul M. Cadambi*[1], *and Arun R. Rao*[1]

[1]*Dept. of Mechanical and Manufacturing Engineering, Ramaiah University of Applied Sciences, Bangalore, India*

Abstract. Planetary gearbox is used in high precision applications such as robotic arm, control system of antenna, positioning and radar tracking systems. Planetary gearbox have high torque-to-weight ratio, high degree of control over the speed range and better efficiency. Most of the literatures assume that the gearbox are free from errors. Errors significantly affect the dynamic characteristics of the gearbox. The major challenge is to model these errors and study its behaviour under dynamic condition. The simulation results of time domain signal when converted to frequency domain signal, it shows the presence of error in the gearbox. Also, simulation result indicates a non-uniform motion of planetary gearbox in the presence of errors.

1 Introduction

In applications such as position control system in robotics, radars, antenna and tracking system, the most important criteria is speed reduction and torque amplification. Generally in these applications, there are numerous problem related to accuracy in controlling and positioning. Although servo motors are in use, their applications are limited due to no torque amplification. These drawbacks can be overcome by the use of mechanical gear drives. These mechanical gear drives has a major advantage of speed reduction and torque amplification.

In recent years, planetary gearbox has received great attention when compared to fixed axis gearbox. In fixed axis gearbox, gears revolve about an axis and only one speed reduction ratio is achieved. In addition to this, in fixed axis gearbox, the power to weight ratio is comparatively low. To overcome these shortcomings, the planetary gearbox is used. In addition to this, just by increasing the number of planet gears in planetary gearbox, the load carrying capacity drastically increases. Also, the torsional rigidity is higher than other fixed axis gearbox. In addition, it is possible to obtain variable speed reduction by fixing either sun, planet or ring gear.

The challenge lies in addressing issues related to improving the accuracy of the position control system and its dynamics. In dynamic condition, when gears in the planetary gear meshes, it transfers uniform angular motion and torque, which is possible only in an ideal

[*] Corresponding author: nithin31@gmail.com

case. The main reason for the deviation from ideal condition is due to errors present in the system.

In dynamic condition, contact loss is one of the important parameter to be analysed [1] for tuned and mistuned planetary gearbox. The contact loss between the two spur gear pair is studied through non-linear dynamic analysis [2] through experimental work. A model is developed to study the influence of damping co-efficient and vibrational parameters of a gearbox system [3]. The gearbox, which is coupled to speed control drive unit introduces speed change and so does the change in frequency and vibration characteristics. These vibration characteristics are analysed by condition monitoring of planetary gearbox to identify errors under dynamic conditions [4]. In addition to these, the dynamic response of the planetary gear significantly changes with stiffness of the pins [5] and eccentricity of gears [6].

Most of the literature assumes that the gearboxes are free from any defect or errors. But in actual condition the planetary gearbox are not free from defects. The challenge lies in identifying these errors in planetary gearbox under dynamic conditions. This paper focuses on modelling of errors associated with gear manufacturing and its effect on planetary gearbox dynamics.

2 Planetary gearbox design

The majority of the high precision application in which planetary gearbox are used have a speed reduction and torque amplification. To analyse the effect of errors in a planetary gearbox, a single stage planetary gearbox with 5:1 reduction ratio is designed from the first principle. The line diagram of the designed planetary gearbox is shown in Fig. 1.

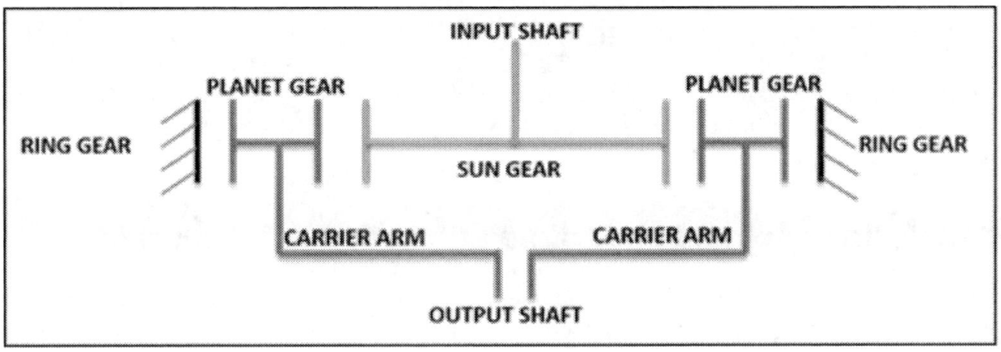

Fig. 1. Line diagram of a single stage planetary gearbox

The designed planetary gearbox consists of a sun gear which is coupled to input shaft. Around sun gear, three planet gears revolve and simultaneously meshes with internal ring gear. These three planet gears are connected to carrier arm which is coupled to output shaft of the planetary gearbox. The planetary gearbox parameters are tabulated in table 1.

Table 1. Planetary gearbox parameter

Parameter	**Sun gear**	**Planet gear**	**Ring Gear**
No. of teeth	24	36	96
Module	1	1	1
Pressure angle (degree)	20	20	20
Pitch circle diameter (mm)	24	36	98.5
Root diameter (mm)	21.5	36.75	94

3 DIN gear standards

Errors are classified into three categories: time-invariant assembly error, time-invariant assembly dependent error and time-variant assembly dependent error [7].

The planetary gearbox and its components are manufactured as per DIN standards. This section provides details about the various manufacturing tolerance level that can be achieved. As per DIN 3961, the tolerance level is considered from functionality and manufacturing point of view. The functionality is sub-classified based on uniformity of transmission of motion, quiet running and static load carrying capacity. For application such as position control system, antenna and radar tracking system the main functionality is uniformity of transmission of motion. As per DIN 3961 [8], the critical errors that are responsible for non-uniformity in transmission of motion are single flank, two flank and total pitch errors. Among these errors, the effect of single flank error is identified as it is a time-invariant error. Therefore, the present paper focus is on analysing single flank error on planetary gearbox under dynamic condition. Considering the gear parameters from table 1, single flank error is calculated as,

$$\text{Profile form deviation} = 1.5 + 0.25*(m_n+9*\sqrt{m_n}) \quad (1)$$

$$\text{Profile angle deviation} = 2.5+0.25*(m_n +3*\sqrt{m_n}) \quad (2)$$

$$\text{Total pitch deviation} = 7.25*(d^{(1/3)} / z^{(1/7)}) \quad (3)$$

$$\text{Total profile deviation} = \sqrt{(\text{profile angle deviation})^2 + (\text{profile form deviation})^2} \quad (4)$$

$$\text{Single flank error} = 0.8*(\text{Toltal pitch deviation} + \text{Total profile deviation}) \quad (5)$$

Substituting the value of reference diameter (d) for planet gear as 36, number of teeth (z) as 36 and module (m) as 1 mm in eqn. 1 to 5 [8], the single flank error is found to be 43 μm for class 8 gear quality.

4 Modelling and rigid body simulation of single flank error

In previous section, value of single flank error is calculated through numerical calculation. In this section, single flank error is modelled on a single planet gear and its effect under dynamic condition is simulated. The modelled single flank error on a planet is shown in Fig. 2. The error is indicated by measuring at pitch circle diameter i.e. (1.437-0.043 = 1.395 mm).

Planetary gearbox components are modelled and assembled in Solidworks 2012. Model is imported into MSc ADAMS 2013 environment for the simulation and is shown in Fig. 3. Contact properties between the sun, planet and ring gear are modelled in ADAMS [9]. In ADAMS the ring gear is fixed, revolutary joint is provided for sun, ring and carrier of the planetary gear. The input motion of 104.7 rad/s is provided for the sun. Simulation is carried out for 1 second and 2000 steps.

From the simulation, the time domain signal of contact force vs. time is plotted and is shown in Fig. 4. The plot shows the variation of contact force between the sun and planet gears. The variation of force is due to engaging and disengaging of gear tooth.

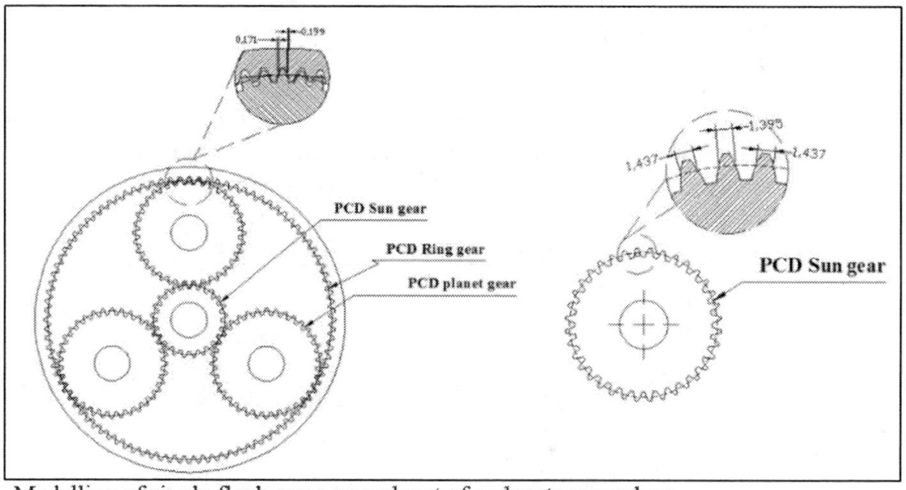

Fig. 2. Modelling of single flank error on a planet of a planetary gearbox

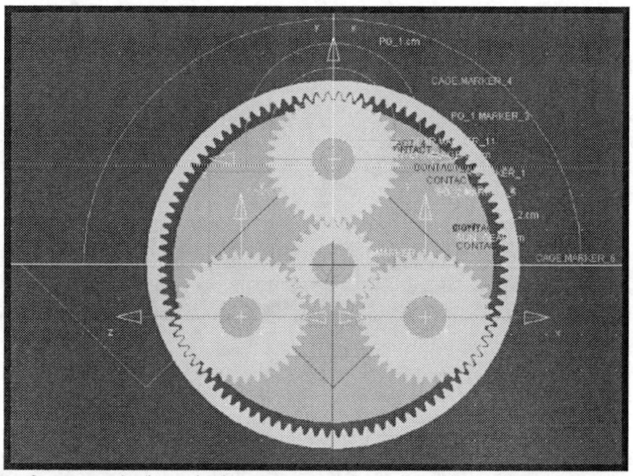

Fig. 3. Constraints for the error induced planetary gearbox in ADAMS

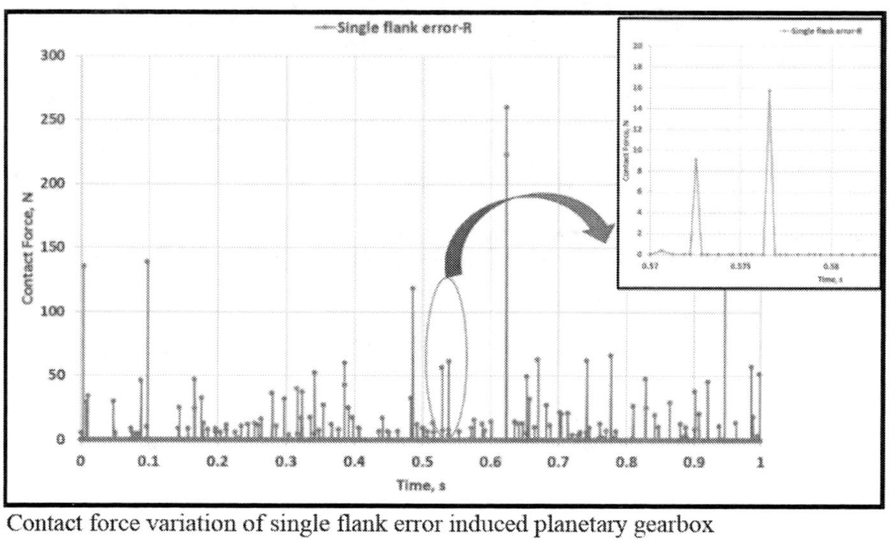

Fig. 4. Contact force variation of single flank error induced planetary gearbox

This time domain signal of contact force vs. time is converted to frequency domain signal for mesh frequencies and its harmonics through an FFT analysis. Fig. 5, shows the mesh frequency and its harmonics of the error induced planetary gearbox. The presence of these errors does not cause any change in mesh frequencies. Frequencies are plotted up to 1000 Hz and gear meshing frequencies are indicated by three distinct peak at 320 Hz, 640 Hz & 960 Hz. The presence of error is indicated by sidebands around meshing frequencies which are presented in Fig. 6.

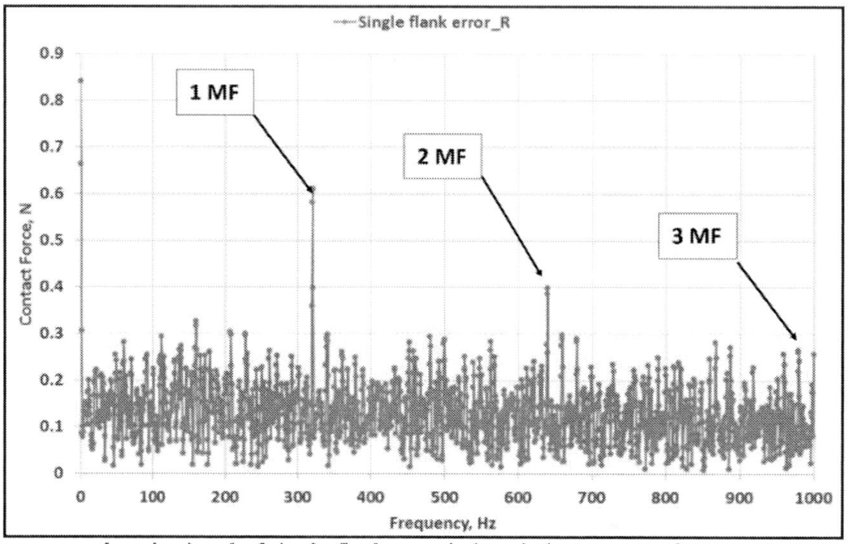

Fig. 5: Frequency domain signal of single flank error induced planetary gearbox

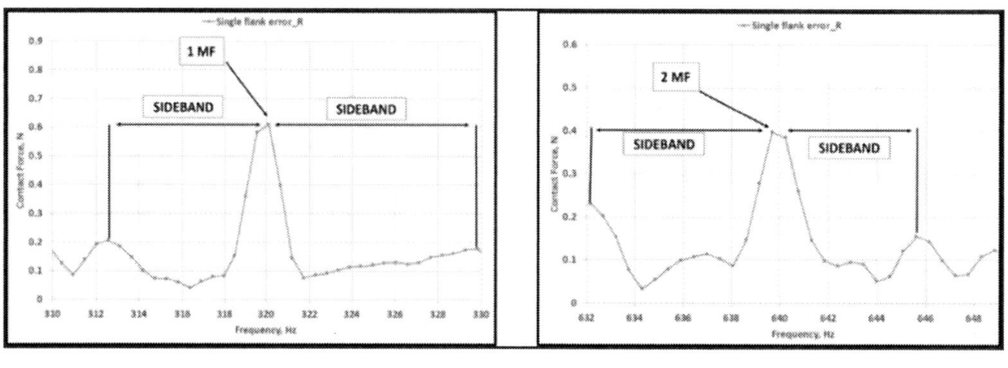

Fig. 6. Sidebands in the vicinity of mesh frequencies

These simulation result addresses the presence of single flank error and its effect on dynamic condition. Analysis of these results are discussed in section 7.

5 Error modelling and rigid body simulation for transmission error

When the planetary gearbox transfers motion from input to output shaft, it should be of uniform motion without any losses. This is possible only in an ideal condition which is unrealistic in nature. Various type of errors related to gear manufacturing have significant influence on these losses. These losses are generally known as transmission error of a gearbox. From literature, it is inferred that there are challenges in addressing, how these errors affect the transmission error of a planetary gearbox.

Transmission error is the difference between theoretical and measured angular velocity of any gear mesh. The term transmission error is represented in form of a mathematical expression as shown in eqn. 6,

$$\text{Transmission Error} = \theta_s - (i) * \theta_c \tag{6}$$

Where:

θ_s is angular rotation of input gear (sun gear), rad
θ_c is angular rotation of output gear (Carrier gear), rad
i is the velocity ration of the planetary gear

In this section, transmission error is studied without considering the effect of tooth defection. To simulate the transmission error, the sun gear (input angle) is rotated by 3.14 rad/s and simulated for 1 second with 5000 steps. It infers that sun gear rotates by 3.14 rad in a second with 5000 equal interval steps. According to gear ratio, angle rotated by carrier should be 1/5th of the input angle. The difference between the input and output angle gives the transmission error. The complete simulation is carried out under no-load condition. The variation of transmission error in planetary gearbox due to single flank error is shown in Fig. 7.

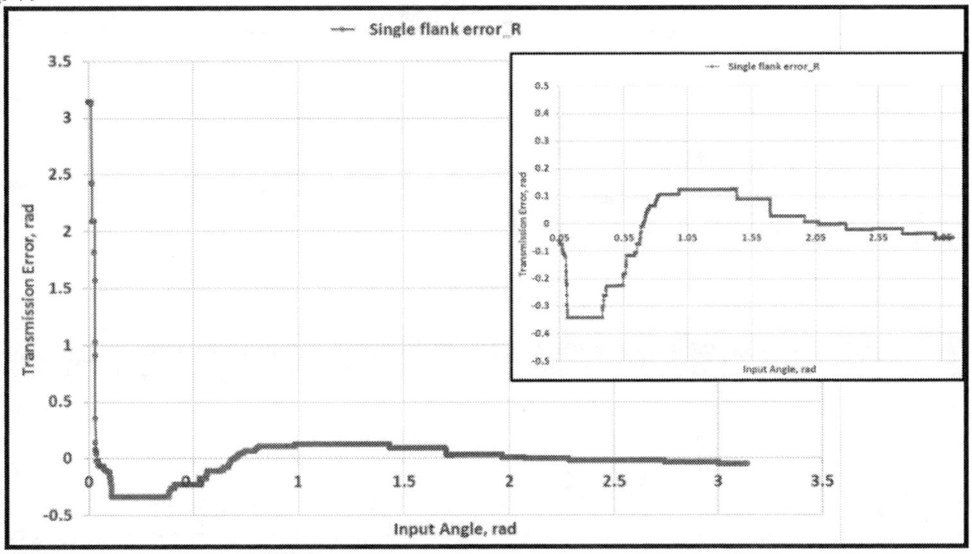

Fig. 7. Transmission error variation

6 Flexible body simulation of single flank error

In previous section, the effect of single flank error on transmission error is studies through rigid body analysis, i.e, without considering any tooth deflection. In this section, the designed planetary gearbox is analysed for transmission error by considering gear tooth deformation. The planetary gearbox is modelled as plane stress condition as there are no stresses developed in out plane section as shown in Fig. 8. In addition to this, shafts are assumed to be rigid.

The element size arrived from the convergence study is 0.4 mm. CPS4R [10] element is selected, which has 4-node bilinear, reduced integration with hourglass control. This element is a 3 DOF with two translational and one rotational degree of freedom. The material property assigned is as provided in table 2.

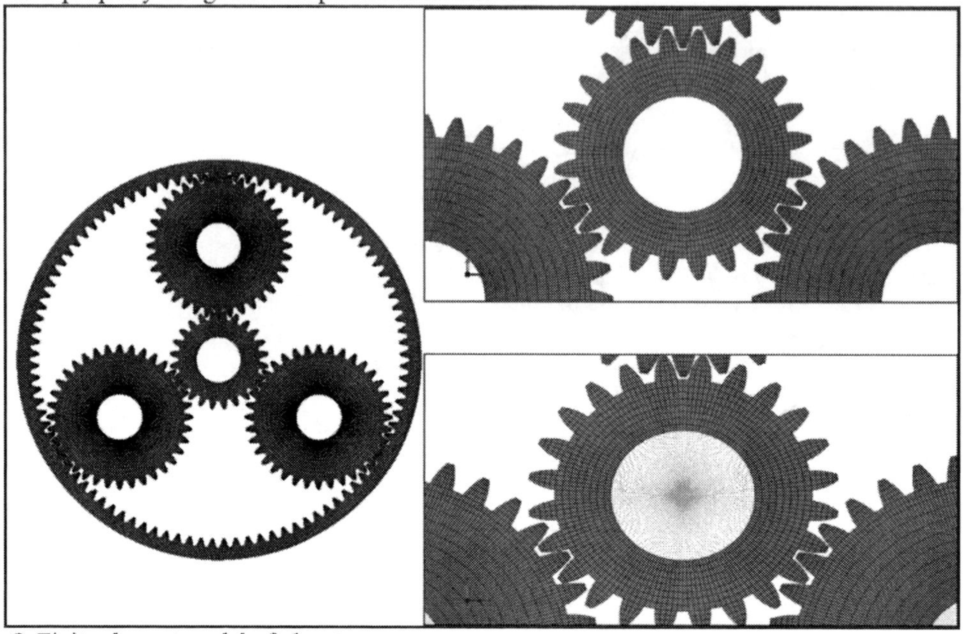

Fig. 8. Finite element model of planetary gear

Table 2. Steel material properties

Material Properties	*Values*
Young`s Modulus	210000 N/mm^2
Density	7.58e-09 tonne/mm^3
Poisson`s ratio	0.3

Two types of forces arise during gear teeth contact: normal and tangential. For tangential component, a co-efficient of friction 0.03 is applied through penalty method. In the case of normal contact, a hard contact is applied with no separation of mating surface.

As multiple tooth contact are involved, a non-linear quasi-static analysis is carried out. Reference points representing the rigid shafts are used for applying loads and fixing boundary condition in simulation. The ring gear is fixed, input torque and resisting torque are applied on input and carrier arm. Planet gears are allowed to rotate about the axis of sun gear. One pitch rotation angle is provided as input for the sun gear. The complete analysis is divided into two step process. Each step is further divided into 1000 sub-step with 1e-5 increment value.

Input and output angles are measured during post processing of results. Difference between them gives the transmission error due to fault induced in planetary gearbox and is shown in Fig. 9.

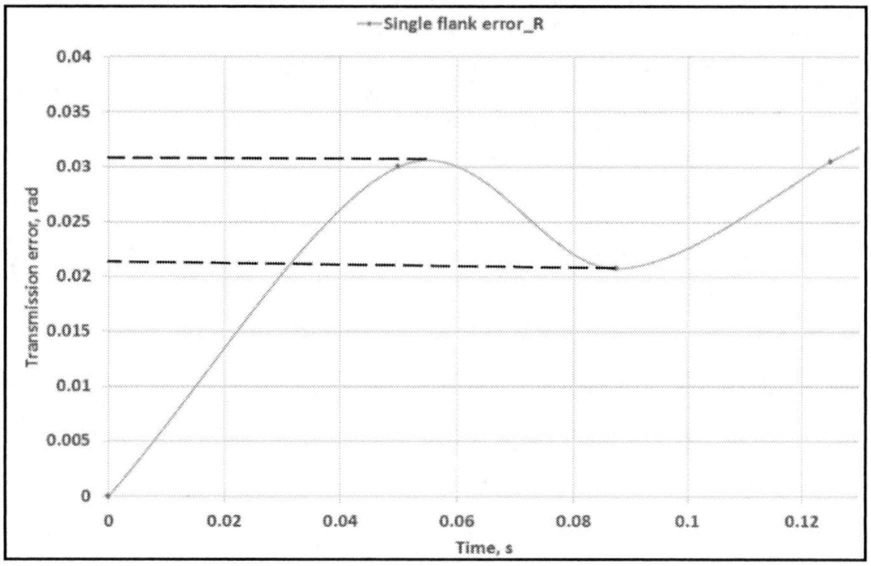

Fig. 9. Transmission error plot of error induced planetary gearbox

7 Results and discussion

In this section, the effect of single flank error on dynamic response of planetary gearbox is analyzed using time and frequency domain signal. The variation in tooth thickness due to single flank error is shown in Fig. 2. This variation is due to the single flank error of 43μm modelled on single gear tooth. The single flank error causes the variation of contact forces which is presented in Fig. 4. This is due to the fact that there is a delay in modelled gear tooth to come in contact with other meshing tooth.

When a time domain signal is converted to a frequency domain signal, the single flank error does not cause any change in mesh frequency and its harmonics. But when observed closely around the mesh frequencies, small peaks are present and these peaks are known as sidebands. These sidebands are shown in Fig. 6 and frequencies are known as pass frequencies. These pass frequencies are exactly 7 Hz on either side of mesh frequencies. These frequencies are calculated based on a literature [11].

A close observation of Fig. 2 shows the clearance between the gears are not uniform, this causes non-uniformity in motion. This non-uniformity in motion is shown in Fig. 7. In an ideal condition, the transmission error should be zero. On the other hand, the presence of single flank error varies the transmission error from -0.35 to 0.12 radians for an input of 3.1415 radians without considering any tooth deflection. This shift is due to the variation between the teeth as shown in Fig. 2.

The transmission error in a planetary gearbox is analysed using flexible body analysis with considering tooth deflection. In this case, the transmission error varies from -0.021 radian to +0.031 radians for an input of one pitch rotation.

8 Conclusion

The effect of single flank error on uniform transfer of motion is analyzed through rigid and flexible body analysis. The single flank error of DIN class 8 is modelled and simulated for frequency domain signal through rigid body analysis. In the presence of 43 μm single flank error is indicated by equally spaced sidebands around the mesh frequencies and its harmonics. Amplitude of the sideband depends on intensity of error. Simulation results of mesh frequencies and sidebands are in good agreement with analytical calculation of existing literature [10]. And the transmission error value varies from -0.3 to 0.12 radians. Similarly, in flexible body analysis the transmission error value varies from 0.021 to 0.031 radians. Finally, it is concluded that the presence of time invariant single flank error has significant influence on planetary gearbox dynamics and transmission error. And transmission error in a planetary gearbox is significantly increases with tooth deflection and manufacturing error.

References

1. J. Lin, R. G. Parker., J. Vib. and Acoust., **3**, 121 (1999)
2. R. G. Parker, S. M. Vijayakar, & T. Imajo., J. Sound Vib., **3**, 237 (2000)
3. W. Bartelmus, J. Mech. Sys. and Sig. Proc., **5**, 5 (2001)
4. W. Bartelmus, F. Chaari, R. Zimroz & M. Haddar., E. J. Mech. A Solids., **4**, 29 (2010)
5. C. Zhu., X. Y. Xu., T. C. Lim., X. S. Du & M. Y. Liu, J. Mech. Engg. Sci., **1,** 227 (2012)
6. F. Charri, T. Fakhfakh & R. Hbaieb. J. Adv. Manu. Tech., **27**, (2006)
7. A. Bodas, A. Kahraman, JSME Series C, **47(3)**, (2004)
8. DIN 3961, Tolerance for cylindrical teeth, Germany (2002)
9. D. D. A. Browski, J. Adamczyk, & H. P. Mora, *Diagnostyka- Applied Structural Health, Usuage and Condition Monitoring*. **2**, 62 (2012)
10. ABAQUS/CAE 6.11 User's Manual
11. Y. Lei, J. Lin, M.J. Zuo & Z. He., J. Int. Mea. Confed., **1**, 48 (2014)

Analysis of various parameters affecting weir design and fabrication of Weirs: An Overview

B. Rao *Vanishree*[1] and R. *Manjula*[1*]

[1]*Sahyadri college of Engineering And Management, Adyar, Mangaluru-575009, Karnataka, India*

Abstract. Several geometric parameters influence weir discharge and it is explained in our present study. The PK weir is more efficient than rectangular labyrinth weir due to reduction in entrance losses of PK weir inlet key geometry. PK weir is a type of labyrinth weir with variation in angle. Therefore, in this paper we also discuss about labyrinth weir. PKW are compatible for constructing on existing or new spillway with lesser space requirement and structural stability. PKW are installed in order to increase the discharge capacity for a given footprint area. PKW is used to increase the overflow capacity of existing dams such as, Hazelmare dam (South Africa), Dartmouth dam (Australia). Trapezoidal piano key weirs have higher discharge efficiency than rectangular weirs and it is due to interference wedge. C_d of TPKW is influenced by L/W but W_i/W_o has least effect.

1 Introduction

Weirs are ordinarily used to avoid flooding, measure water discharge and help render waterways more safe by watercraft. A few dams have little spillway limits contrasted with configuration surges. Weir structures are utilized as a part of channel, flow, and store applications for flow estimation, flow redirection, and discharge control, and they are regularly intended for nothing flow conditions i.e., the surface of tail water is underneath the weir crest. Weir is utilized as flow control.

1.1 Submergence

At the point when tail water surpasses crest height the weir is submerged [1]. Submergence will probably be of concern when low-head control structures are introduced on mellow sloping waterways or channels with tightening influences and so on. When all is said in done strong solid weirs are utilized, yet these days, they are made of free stones, for example, gabion weir, since the last can better meet normal and natural prerequisites. It limits sedimentation and eutrophication is an impoundment. Weir submergence is found in abundance in-flow vegetation downstream. At low submergence levels, raised tail water

* Corresponding author: manjurguku@gmail.com

does not influence free flow add up to upstream head (H_o), known as particular submergence. At the point when this farthest point is surpassed, upstream aggregate head (H*) builds in respect to H_o for a given weir discharge (Q), expanding potential upstream flooding. In customary submergence, tail water is more than weir crest which is caused by downstream control. In neighborhood submergence, lifted tail water happens when bay cycle discharge surpasses free flow limit of outlet cycles.

1.2 Labyrinth weir

Weir crest is a prominent characteristic of labyrinth weir. Labyrinth weir is not having linear plan form. Weir crest length reduces upstream head for a given discharge. A labyrinth weir has advantage when width of a channel is limited. Compact disc for labyrinth weir is lesser than direct weir. Because of increment in weir length, discharge builds 3-4 times [2] and discharge limit (Q) is relative to weir length (L) [3]. For labyrinth weirs, as the crest length expands, discharge increments for a given upstream head. They are utilized as control structures. It can be utilized where scope of discharges that have constrained variety in upstream water level. It is using pressurized water more effective spillway structure can oblige expanded base-flow repository stockpiling volumes without trading off surge wellbeing, which is helpful for water system, metropolitan and hydropower applications. Plan direction for applications (repository), in which moving toward flow field is non-uniform and isn't opposite to the weir hub[4] ,varying approach flow edges, flow joining, projection initiated flow constrictions, and the quantity of cycles can altogether impact weir discharge proficiency. Thought of labyrinth side weir was first exhibited by Emiroglu. It was accounted for that discharge co-proficient of labyrinth side weir is 1.5 to 4.5 times higher than typical rectangular side weir in a rectangular channel. Trapezoidal weir of labyrinth format is ideal contrasted with triangular weir [5]. Group of bends for outlining labyrinth weirs was built up [6]. Plan method for labyrinth spillways, where discharge coefficient esteems for labyrinth points between 6° and 35° are proposed [2] and their conditions are substantial for scope of widths amongst T_s and $2T_s$, T_s=wall thickness and $H/P < 0.9$. Nappe is a fly like flow ignoring the weir which shows sticking, circulated air through, mostly circulated air through or suffocated (non-circulated air through) conditions which influence the discharge. Most Labyrinth weir plan techniques accessible are lab tried on flumes, in which the moving toward flow field is generally uniform and is opposite to the weir [2, 5, 7, and 8]. The labyrinth weir plan technique exhibited by Crookston [7] uses general weir condition and introduces coefficient of discharge for quarter-round and half-round labyrinth weirs for $6° < \alpha < 35°$ is, $Q = 2/3 * C_d * L * \sqrt{2g} * (H_t)3/2$, where upstream head ,$H_t = V^2/(2g+h)$. Labyrinth weir can be introduced in repository spillways in numerous arrangements viz., somewhat or completely anticipating, and adjusted channel, flush and arced. Houston [9] detailed that for a 2-cycle somewhat anticipating labyrinth weir introduction expanded Q by 10.4% in respect to a flush introduction with adjusted projections. The cycle arrangement of labyrinth weir however might be arced to better adjust the cycle introductions to the focalizing approach flow vectors and to additionally expand the weir crest length [10] and discharge efficiency [11]. Water driven productivity of Prado spillway could be expanded had the cycle design been bended to enhance the arrangement to the moving toward flow, introduction [12]. Be that as it may, for most in-channel applications, arced labyrinth weirs are not reasonable. The air entrainment rate is influenced by parameters like weir included edge θ and weir ledge slope Φ. A specialist [13] showed 3 instruments of oxygen exchange for free falling water fly diving into an accepting pool.

1. In the fly itself amid the fall, measure of oxygen exchange is little.

2. By means of free surface of getting water, relies upon power of surface tumult in accepting pool.

3. In air-water biphasic flow that exists inside accepting pool due to air diversion administered via air excitement rate by bubble contact time in water. The last instrument was most huge supporter of oxygenation process.

Tsang's [14] proposed 4 air stimulation instruments which are smooth, rough, oscillating and dis-integrated. Disintegrated planes have favorable position because of more prominent surface zone; however entrainment rate and air pocket infiltration are reduced as a result of vitality misfortune to encompassing environment amid fall. Numerous analysts [15-21] have completed research facility examinations concerning weir air circulation. Researchers [20-21] considered oxygen exchange proficiency of triangular and rectangular labyrinth weir sub-current rates fluctuating from 1 to 4lps, demonstrated that weirs enhance air circulation qualities. Nappe might be insecure and create flow surging in labyrinth weir. Nappe obstruction signifies diminishment in labyrinth weir discharge proficiency caused by impacting or interfacing nappe's close upstream summit. Nearby submergence and non-opposite flow lines are in charge of labyrinth weir discharge coefficients being less then direct weirs (sideways weirs have bring down C_d esteems contrasted with straight weirs). For smaller H_t, nappe communication happens downstream and no change is head-discharge connections. As H_t increments, impacting nappe's make confined semi angled water powered hop condition and constrained discharge conveying limit close upstream end, which brings about nearby submergence. Neighborhood submergence starts close upstream zeniths of outlet and increments with H_t until the point when it is submerged autonomous of downstream tail water. Pressure driven results of impacting labyrinth weir nappe's incorporate standing waves and air building in downstream cycles. Size of district impacted by nearby submergence and nappe crash differs with α, A, crest shape, P, H_t and nappe air circulation. Labyrinth weirs are ordinarily intended for, yet not restricted to $H_t/P < 1$, where H_t is add up to upstream head and P is weir stature. Geometric parameters influencing water powered plan are sidewall edge (α), number of cycles (N), crest shape, upstream profundity, weir stature, weir length [12]. For considerable approach speed, estimation of discharge coefficient fluctuates with upstream head. Submergence connections in view of flow reduction factor Q_s/Q_1 and for sharp crested weirs as a component of submergence ratio H_d/H^* [22-26]. For labyrinth weir submergence, submergence flow magnification factor was bigger than free-flow magnification factor and presumed that submergence impacts are less noteworthy for labyrinth weirs than direct weirs. Head magnification factor can be changed over to flow lessening factor and the other way around. Disadvantages of customary labyrinth weirs is high constructional costs, can't be effortlessly put on crest of basic solid gravity dams, reduction of discharge limit with expanding the nappe stature when the weir included edge is picked too little and as number of cycle's increments due to nappe crash and neighborhood submergence. To expand the dependability of the structure, base of the keys are sloped. Outline methods proposed for labyrinth weirs are not pertinent for PK weirs. PK weir is a productive arrangement.

1.3 Piano Key Weir

PKW is a type of labyrinth weir [27] with productive hydraulic performance, particularly at low heads. It has upstream and downstream overhangs of variable width to lessen footprint area [28] to enable easy weir location on a top of a dam crest of existing dams and new gravity dams unlike conventional weirs, where submergence isn't a factor. The first scale display examines demonstrated that this present PKW's discharge capacity [29] and efficiency is up to four times than a conventional ogee-crested weir at rise to head, crest length and same width [30] and 10% higher than labyrinth weir of same crest impression [31].The PKW has inlet and outlet base inclines, 2 transversal dividers with channel and outlets. PKW has sloped up-and downstream key floors. PK weir geometry impacts hydraulic efficiency and performance [32]. Its principle geometric parameters are weir height P, the quantity of PKW units N_u, PKW unit width W_u, lateral or side crest length B, inlet and outlet key widths W_i and W_o, upstream and downstream overhang lengths B_o and B_i, wall thickness T_s [33]. Expanding the inlet or outlet width ratio W_i/W_o, the upstream overhang length B_o and the weir stature P, shape of upstream overhang, the developed length to width ratio (L/W) were considered to improve the discharge limit and performance [34-35]. Investigation of execution of PKW in perspective of air circulation limit and floating debris section was proposed [30].The floating debris will effectively disregard as the water level expands [30]. The first model size PKW was worked by Electricité de France (EDF) amid 2006 to build the discharge limit of the Golours dam spillway [36]. They are high performance and practical answer for new dam ventures, in which space and accessible repository portion are restricted, contrasted with design discharge to be released. PK weir would likely deliver higher aggregate weir discharge attributable to its smaller compact geometry (large weir length for a given base area). The surge discharge estimations and security issues of countless dams require spillway recovery since a few dams have little spillway limits. PKW is a novel type of weir created by Blanc and Lempérière (2011)[28, 37]. EDF favored the utilization of PKW in dam recovery in France. Later on, two PKW's were based on St. Marc and Gloriettes dams in 2008 and 2009[38, 39] and at Etroit dam to overhaul spillway limit [40]. Sri Lanka's first spillway was developed at Giritale in 2013. PK weirs limit upstream immersion because of its improved low head surge discharge productivity. 1-D show was produced at college of Liège to enhance its outline [41]. Most of the dams claimed by EDF have short hydrological records. A few elements considered are: cost, expulsion of huge amounts of cement or adding new loads to existing structures, energy dissemination structures, lessening energy generation amid development works. Advantages of PKW are the expansion of the aggregate successful crest length for a given width [28]. It is utilized to build discharge limit with respect to a given head or abatement the set out toward a given discharge, because of generally vast anticipated spillway length. Henceforth, it altogether decreases the afflux and limits danger of upstream flooding. It permits high crest level which can expand the capacity limit in the store. Moreover, past prudent thought, PKW's are free flow spillways and have a high level of safety and reliability beyond cost consideration. Favorable position of PKW as contrasted and a labyrinth weir are reduced footprint base area and because of inside slopes in the alveoli, decreases the force following up on sidelong dividers and consequently the basic structural cost. In 2011, EDF began to extend information on some optional parameters, for example, the overhangs, the crest shape, and the dam stature. The program incorporates testing of trapezoidal shape with a side divider point of the parallel dividers to contrast water powered execution with rectangle shape. The majority of the analyses were completed under free-flow conditions. It was discovered that, discharge limit of PK weir could be enhanced by including plan changes (adjusted noses on upstream summits, parapet divider, enhanced crest shapes) [32].

Extra weir length can be created with a PKW, by cantilevering both the up-and downstream keys past the basic impression. Classifications of PKW based on geometry are: Type A, in which up-and downstream finished overhangs are indistinguishable, Type B in which there is only upstream overhang, Type C in which there is only downstream overhang, Type D in which there is sloped floor and has no overhang. Type-A PK weir are using pressurized water more efficient than rectangular labyrinth weir with same crest length [42]. Type B geometry is more proficient than Type A for PK weirs [43]. A few specialists [44, 45] have proposed for ideal outline W_i/W_o=1.2 where W_i and W_o are widths of downstream and upstream summits. Hein [45] found that PKW geometry is for the most part more powerful than trapezoidal labyrinth weir for nappe aeration on account of overhang geometry. Trapezoidal labyrinth weirs regularly give more discharge limit per unit weir length than PK weirs [32]. An absence of understanding endures in regards to prevailing flow conditions, because of expansive number of geometric parameters and absence of thorough examination of the discharge coefficient [31, 46]. Some institutionalized plans are accessible for the PKW [47]. A scientist [46] empowered to portray the flow on the structure as far as velocity, pressure, free surface levels and flow patterns. Submergence (S) is the ratio of aggregate downstream head (H_d) to add up to submerged flow upstream head (H^*), S= H_d/H^*. As H_d value rises, it approaches H^*, where weir never again goes about as control structure. Use of parapet dividers demonstrated increment in discharge limit up to 15% on Etroit Dam[38], yet the effectiveness of these structures has just been contemplated on single PKW setup with or without parapet dividers changing aggregate weir stature ratio $P/(B-B_o)$ from 0.58 to 0.65. In a similar scope of slopes, if the PKW height is adjusted by changing slope of channel and outlet base inclines, instigates comparable varieties in discharge limit [48]. It isn't clear to the point that, PKW with parapet dividers is in reality more effective that standard PKW having same aggregate height.

2 Literature survey

Construction of PKW can be made using reinforced concrete, steel or other mixed materials. PKW prototype can be fabricated using PVC, acrylic sheet. Water is allowed to flow to stabilize the weir and allow thermal contraction of acrylic sheet. Care should be taken by reducing reinforcement quality, improve concrete durability through concrete cover increase, increase PKW stiffness, structural response to dynamic loads due to flow vibrations while using reinforced concrete walls. Laboratory scale labyrinth weir can be fabricated using high density polyethylene (HDPE).

2.1 LABYRINTH WEIR

As crest effectiveness diminishes, C_{dl} turns out to be less imperative with expanding head which is seen in trapezoidal labyrinth weirs with impact of angle α between lateral crest and principle stream direction [2]. Compared to α = 12° (downstream cycle sidewall angle) the reduction in efficiency was more rapid for α=6°, where upstream cycle flow area was significantly larger [11]. For labyrinth weirs, diminishing of Cd with head is essential for small values of α and it causes increment of L/W. Weir efficiency is distinguished by C_{dl} gain with reduction of α. To circumvent lateral crest submergence; increment the inlet channel width till outlet width is sufficient ([30], Michael's), the inlet slope or inlet height utilizing crest appendage to build C_d [30, 34]. An expansion in inlet channel cross-section expands head needed to create control section. As number of cycles (N) builds, impact of abutment-induced flow on discharge efficiency diminishes. L/B, θ affects discharge capacity of labyrinth side weir. Q for labyrinth side weir increases with reducing θ. As θ increases, cycle orientation is improved and spillway efficiency increases.

2.1.1. Effect of orientation and configuration

Discharge capacity of arced labyrinth weirs are significantly higher than horse shoe weir (~690 and 380% for α=6°, α=12° respectively), because of significantly longer crest length. Flush discharge efficiency is reduced by as much as 11% for α=12^0, in-channel labyrinth weir configuration. To enhance discharge efficiency of rounded inlet, abutment wall radius should be increased. Rounded inlet and projecting geometries are more productive than in-channel labyrinth weir at lesser Ht/P and are less proficient (~ 5-7%) at higher H_t/P values. For $H_t/P > 0.15$, α=120^0, projecting labyrinth weirs are ~ 2 to 5% less effective with respect to in-channel and for α=6°, it is 10% less proficient for Ht/P >0.25. Projecting configuration shows increased discharge efficiency as compared to in-channel labyrinth weir which is limited to $H_t/P<0.5$. Arced configuration of a labyrinth weir improves discharge efficiency of 5-11%. Discharge limit of arced labyrinth weirs are essentially higher than horse shoe weir (~690 and 380% for α=6°, α=12° separately), on account of significantly longer crest length. Arced design configuration requires larger base area than projecting configuration setup. Flush orientation has minimum efficiency while arced orientation has most noteworthy efficiency.

2.1.2 Effect of nappe

For H/P=0.1, low head; H/P=0.2, transition head; 0.35<H/P <0.5, high head. At lower value of heads, transition from partially clinging to leaping to springing nappe, depends on rising head, crest thickness and crest shape. For H/P= 0.05, leaping nappe remains in contact with the lateral crest. For $0.09 \leq H/P \leq 0.1$, springing free nappe is observed and it continues till it reaches high head. At low heads, inlet slopes has no effect on nappe shape and transition from leaping to springing nappe occurs at $0.11 \leq H/P \leq 0.12$. Overpressure head increases with the upstream head. Flow behavior of upstream head is different from downstream and lateral head. It is observed that, $0.16 \leq H/P \leq 0.17$, nappe is fully aerated. Downstream nappe slope has significant effect on nappe shape. For moderately low heads ($H/P \leq 0.2$) streamlines are homogenously distributed. For H/P > 0.2, streamline distribution is less optimal. C_d value increases at H_t/P ~0.25 due to nappe shifting from non-aerated to aerated. H*/Ho v/s H_d/H_o is a function of H_o/P, and as it builds up, weirs become less susceptible to submergence. H^*/H_o v/s H_d/H_o is accurate compared to Ville Monte relationship in predicting total upstream head on submerged labyrinth weir, for different side wall angles. As sidewall angle is increased submerged labyrinth weir approaches linear weir. Tullis observed that as C_d diminishes, nappe height increases. For L/b = 0.5, variation of C_d is insignificant against F_1, but it is vice versa for L/b=1. Length of side weir and Froude number F_1, crest height of side weir affect the secondary flow. Increase of secondary flow maybe created due to lateral flow and it is due to growth of deviation angle and kinetic energy towards side of weir.

2.1.3. Effect of velocity profiles

For H/P=0.1 and 0.2, velocity profiles are relatively uniform along lateral and the inlet center. Velocities increases, for H/P=0.35 and 0.5 along sidewall and inlet centers.

2.1.4. Effect of recirculation zone, turbulence

Non-rectangular front part appearance of upstream overhangs diminishes re-circulation zone size and thus discharge limit increments. As approach flow angle drifts from right angle to labyrinth weir axis it produces wake induced turbulence close to upstream apex.

2.1.5. Effect of submergence

I. Compared to linear weir, labyrinth weir is less influenced by submergence which is consistent with Taylor's observations (1968). At the same submergence, flow reduction of labyrinth weir is less than linear weir. As per our review, when H* approximately reaches H_o, for S<~0.48 (Modular Submergence), tail water submergence effects don't influence free flow head discharge relationship till total head exceeds 48% of upstream total head. Local submergence is contributes to nappe interference.

II. Local submergence of labyrinth weir is observed at $H_t/P = 0.6$, and C_d is significantly less than linear weir, but it has high discharge capacity of 270% and 180% for $\alpha=6^0, 12^0$ respectively.

2.1.6. Effect of jet overcrossing

As the head increases there is loss of effective crest length, also efficiency decreases due to lateral jet over crossing.

2.2 PK WEIR

Weir equations:

Because of unpredictability of PKW geometry, weir equation head v/s weir length as,

$$Q= C_{dw} * W * \sqrt{2gH^3}, \text{ where } W=1.5\,(W_i+W_o) + 3*T_s \quad (1)$$

Co-efficient of discharge is also affected by ratio of $(h_1\text{-}p)/p$, weir included angles.

$$C_{pkw}=f\,(L/W,\ a_1/b_1,\ P/a_1,\ P/P^1,\ T_s/P,\ T_s/R,\ H/a_1,\ S_{in},\ S_{out},\ F_1,\ R,\ w) \quad (2)$$

Total discharge,

$$Q=f(\rho,g,v,\sigma,H,L,P,P^1,\ W,\ a_1,\ b_1,\ B_1,\ S_{in}\ ,\ S_{out},\ T_s,\ R,\alpha) \quad (3)$$

Discharge coefficient of crest given by Swamee (1988) is,

$$C_{dl}= \frac{3}{2} *C_{dw} * W/L \quad (4)$$

$$\frac{L}{W} = \frac{a1+b1+2B}{a1+b1} \quad (5)$$

In General,

$$H^*/H_o = A^*(H_d/H_o)^B + C^* e^{D(H_d/H_o)} \quad (6)$$

Crest length amplification proportion is signification. For same proportion H/a, most elevated estimations of 'r' discharge enhancement proportion, are identified with higher estimations of L/W. As H (H/a) builds, efficiency diminishes. Smaller estimations of P, reduce efficiency. Jet over-crossing is because of H, b, P and outlet suffocating because of powerlessness to clear stream-wise stream. For given H/P, L/W isn't imperative. 'r' is figured for a given estimation of width (W) and highest value of height of weir. T_s/P and T_s/R are important for low heads. For Gloriettes model, L/W=5.6, has lowest specific discharge [49].

Surface tension effects may become significant for H/P<0.06[50]. The smaller the weir dimension more important is H/T_s at transition from leaping to springing nappe [51].To eliminate surface tension effect and to ensure that free surface is undular, the ratio of H/P should be optimal.

2.2.1. Effect of velocity profiles

Lateral discharge is a function of water height over lateral crest and velocity in inlet direction, and it can be reduced since inertia effect is related to velocity in direction of inlet key [52].

By increasing slope of inlet and outlet, PKW efficiency decreases. For H/P < 0.15 there is reduction in discharge coefficient at low heads. TPKW's design constraint was to maximize side wall angle α, keeping L/W constant, which is a prominent parameter for PKW performance. Upstream and downstream length (B) of TPKW influences the building cost. Better hydraulic performance is observed by half – round and quarter-round shapes compared to flat-top shape, which is validated by Anderson. It is suggested that half and quarter round shapes should be built on lateral crest, flat top shapes should be built on upstream and downstream crest. No matter the position of rounded face, quarter-round shape on lateral crest has same discharge efficiency. Sharp and broad crested weir is less efficient than Ogee crest weir and PKW is more efficient than ogee crest by 30% -60% for type A, 10-70% for type B and 5-45% for type C. It was found that [53], for weir included angle 45^0 and 60^0, C_d value is greater than for angles, $\theta=90^0$, 120^0, 150^0.

PKW is identified with lessening inlet flow constriction, approach velocity and head loss entering the inlet owing to increment in wetted perimeter related to apex overhang. In outlet keys, increment in flow area and wetted edge enables water to leave more efficiently. To build outlet cycle discharge limit by diminishing submergence, incorporates expanding downstream apron slope, bringing down downstream apron elevation or enhancing downstream cycle footprint.

2.2.2. Effect of submergence

For PK weir, as submergence level rises, it stops behaving as head discharge control structure, which is at lower H_d/H_o values, as H_o/P increases. For larger values of S, submerged PKW turns out to be less effective than labyrinth weir. With respect to labyrinth weir submergence behavior, scope of H_d/H_o for which PKW is less sensitive to submergence effects expanded with increasing H_o/P. Little increment in discharge efficiency happens as PKW changes between free flow and modular submergence. (Oamane 2011 found similar observation).

2.2.3. Effect of suction

PKW behaves like sharp crested weir for low heads since suction effect is starting, and hence discharge increases.

2.2.4. Effect of parapet walls

Flow submergence is retarded due to presence of raised side walls at low heads and it increases discharge capacity upto 15%, especially at downstream crest. The efficiency of parapet wall depends on weir height and W_i/W_o [52]. Hence there is no direct effect on discharge capacity of the key slopes. Weir height without parapet wall is important to reduce longitudinal velocity. On increasing the height by parapet wall, velocity and lateral discharge are not changed but is induced by water height over lateral crest.

2.2.5. Effect of dimensionless parameters

If Froude's number $F_1<1$, the flow is said to be subcritical. C_d decreases with increase in $(h_1\text{-}p)/p$, increase in Froude number; decreases weir included angle. Reynolds number (R) and Weber number (w_1) are maintained sufficiently large in order to avoid liquid viscosity and surface tension effects.

3 Conclusions

When compared with other solutions, the PKW has a number of advantages, no change of normal and maximum operating levels, minor impact on dam structure, no impact on existing gate and other mechanical components, reliability of free flow spillway, requiring little maintenance and low cost solution involving the construction of simple concrete units [49]. Experimental investigations were performed on models of various types of PKWs. The results indicate that an increase of the downstream key width results in an increase of the discharge capacity. Consequently, decrease in the upstream key width results in an increase of the local submergence of the upstream key and a decrease of its discharge efficiency. The upstream weir overhangs significantly increase the discharge coefficient. The upstream weir overhang increases the inlet flow area and the wetted perimeter, resulting in reduced inlet velocities, flow contraction and energy loss [54]. The model studies clearly shows that the number of keys and the crest shape of the weirs were two parameters that had a significant bearing on model performance. The PK spillway shows great promise in significantly improving the existing flood discharge capacities of spillways at dams such as Nachchaduwa, which experienced critical conditions in 2011 and 2014 [55]. The main influence of a parapet wall on a PKW crest is an increase of the total PKW height. The use of a parapet wall, increasing the total weir height to approach its optimal value, has been shown to increase the discharge capacity of the PKW. The main effect of the parapet walls is to increase the inlet height and, hence to reduce the longitudinal velocity, resulting in an increase of the lateral discharge. However, the height of the parapet wall has to be limited to maintain the interest of overhang use, limiting the head losses at the inlet key entrance. Increasing the parapet wall height so decreases the PKW discharge capacity until the one of a labyrinth weir with the same horizontal shape. PKW is more convenient and cost-effective to use standard PKW without parapet walls. However, parapet walls provide a good opportunity for future PKW rehabilitations, enabling, in some cases, an increase in the discharge of the initial PKW by up to 20% by a limited increase of the maximal reservoir level [52]. The discharge capacity of TPKW is greater than the rectangular plan form. Among the geometrical ratios, W_i/W_0 and L/W were found to have the lowest and highest influence on increasing the C_d value respectively. For a given head (H), the weir discharge increases for higher P values remarkably. Regardless of the L/W value, the peak value for C_d corresponds to H/P ratio. Operation of a weir under nappe or transition flow regime with high efficiency is preferable [56].

Notations

A= inside apex width

B = lateral crest length or weir sidewall length

B_i = downstream overhang length

B_o =upstream overhang length

L = developed weir crest length

N_u =number of PKW units

P = weir wall height

S_{in} =inlet key bottom floor slope

S_{out} =outlet key bottom floor slope

T_s= vertical wall thickness of the weir

W = weir channel width

W_i = inlet key width

W_o= outlet key width

C_d = discharge coefficient

g = acceleration due to gravity

h = piezometric head

H_d/H_o = ratio of submerged total downstream head and the total free-flow head relative to the weir crest

H^*/H_o = ratio of total submerged head and the total free-flow head measured at a common flow rate relative to the weir crest elevation.

Q = discharge over weir

S = submergence factor

V = flow velocity

w = cycle width for a weir

α = sidewall angle

H_t/P = H/P = headwater ratio

a1, a_1 = upstream key width

b1, b_1 = downstream key width

B_1 = length of side weir

P` = approach flow depth

ν = kinematic viscosity

ρ = density of water

σ = surface tension

θ = weir included angle

α` = upstream weir side wall angle = $\alpha+\theta/2\theta = \Theta/N$

Θ = arced labyrinth weir angle

N = weir cycle number

C_{dl} = discharge coefficient of crest

C_{dw} = discharge coefficient of weir

F_1 = Froude's number

w_1 = Weber number

R = Reynolds' number

Φ = weir ledge slope

References

1. Kirkpatrick, K.W. "Discharge coefficients for spillways at TVA dams." Trans. Am. Soc. Civ. Eng., 122,190-210, (1957).
2. Tullis, B.P., Amanian, N., and Waldron, D. "Design of labyrinth weir spillways." J. Hydraul. Eng. 121(3), 247-255, (1995).
3. Henderson, F.M. "Open channel flow." Channel controls, MacMillan, New York, 174-176. (1966).
4. Copel and, R., and Fletcher, B. "Model Study of Prado Spillway, California, Hydraulic model investigation Rep. ERDC /CHL TR-00-17, U.S. Army corps of Engineers Research and Development Centre Vicksburg, MS (2000).
5. Hay, N., and Taylor, G. "Performance and design of labyrinth weirs." J. Hydrol Eng., 96(11), 2337-2357, (1970).
6. Darvas, L.A. "Discussion of performance and design of labyrinth weirs by Hay and Taylor." J. Hydro. Div., 97 (8), 1246-1251 (1971).
7. Crookston, B.M. "Labyrinth weirs." Ph.D. dissertation, Utah State Univ., Logan, UT (2010).
8. Magalhães, A., and Lorena, M. "Hydraulic design of labyrinth weirs." Rep. No. 736 National laboratory of civil Engineering, Lisbon, Portugal (1989).
9. Houston, K. "Hydraulic model study of Hyrum Dam auxiliary labyrinth spillways." Rep. No. GR-82-13, U.S Bureau of Reclamation Denver, Co. (1983).
10. Cassidy, J. "Labyrinth crests spillway planning design and construction." Int. conf. on hydraulic aspects of Floods and Flood control, city Univ., London (1983).
11. Page, D., Garcia, V., and Ninot, C. "Aliviaderos en laberinto. presa de Maria Cristina." Ingenieria Civil, 146, 5-20 (in Spanish) (2007).
12. Falvey, H.T. "Hydraulic design of labyrinth weirs." ASCE Reston VA (2003).
13. Gameson, A.L.H. "Weirs and aeration of rivers." J. Inst. Water Eng., 11 (1957).
14. Tsang, C.C. "Hydraulic and aeration performance of labyrinth weirs." Ph.D., thesis, Univ. of London, (1987).

15. Van-der kroon, G.T.N., and Schram A.H. "Weir aeration. I: H2O." 22, 528-537 (1969a). Van-der Kroon, G.T.N., and Schram, A.H. "weir aeration. II: H2O." 22, 538-545 (1969b).
16. Apted, R.W., and Novak, P. "Some studies of oxygen uptake at weirs." Proc. 15th congress, IAHR Paper B23.V (1973).
17. Avery, S., and Novak, P. "Oxygen transfer at hydraulic structures." J. Hydraul. Div., Am. Soc. Civ. Eng., 104 (11) 1521-1540 (1978).
18. Thene, J.R. "Gas transfer at weirs using the hydrocarbon gas tracer method with headspace analysis." MSc thesis, Univ. of Minnesota, (1988).
19. Labocha, M., Corsi, R.L., and Zytner, R.G. "Parameters influencing oxygen uptake at clarifier weirs." Water Environ. Res., 68(6), 988-994 (1996).
20. Wormleaton, P.R., and Soufiani, E. "Aeration performance of triangular plan form labyrinth weirs." J. Environ. Eng., 124(8), 709-719 (1998).
21. Wormleaton, P.R., and Tsang, C.C. "Aeration performance of rectangular plan form labyrinth weir." J. Environ. Eng., 126(5), 456-465 (2000).
22. Bazin, H. "Experiences nouvelles sur l' écoulement en déversoir." Vol.VII, Annales des Points et Chausses, Memories et documents (1894).
23. Cox, G.N. "The submerged weir as a measuring device." Bulletin engineering experiment station series no. 67, Univ. of Winconsin, Madison, wis. (1928).
24. Francis, J.B. "Experiments on the flow of water over submerged weirs." Trans. Am. Soc. Civ. Eng., 13, 303-312, (1884).
25. Fteley, A., and Stearns, F.P. "Description of some experiments on flow of water made during the construction of works for conveying water of Sudbury River to Boston." Trans. Am. Soc. Civ. Eng., 12, 101-108, (1883).
26. Ville Monte, J.R. "submerged weir discharge studies." Eng. New-Rec., 139 (26), 54-56, (1947).
27. Tullis, B.P., Young, J.C and Chandler, M.A. "Head-discharge relationships for submerged labyrinth weirs." J. of Hydraul. Eng., 133 (3), 248-254, (2007).
28. Lempérière, F. and Ouamane, A. The piano keys weir: a new cost-effective solution for spillways. Hydro power and Dams Issue 5, 144-149, (2003).
29. Lempérière, F., Ouamane, A. "The piano keys weir: a new cost-effective solution for spillways." Int. J. Hydropower and Dams 10 (5), 144-149 (2003).
30. Ouamane, A. and Lempérière F. "Design of a new economic shape of weir'. Proceedings of the international symposium on Dams in the societies of the 21st century, 463-470. Barcelona, Spain (2006).
31. Anderson, R.M. and Tullis, B.P. "Influence of PKW geometry on discharge." In: Liege, S.E., Laugier, F., Boillat, J.Ḹ., Pirotton, M., Reverchon, B. and Schleiss, A.J., Eds., Proc. Int. Work. Labyrinth and PK weirs PKW 2011, CRC press, London, 75-80, (2011).
32. Anderson, R.M. "Piano key weir head discharge relationships." M.Sc. Thesis, Utah state Univ. Logan, UT (2011).
33. Pralong, J., Vermeulen, J., Laugier, F., Erpicum, S., Boillat, J.L. "A naming convention for the piano key weir geometrical parameters." Proc. Int. Workshop labyrinth and piano key weirs Liège B, 256-263, (2011).
34. Ouamane, A., Lempérière, F. "Nouvelle conception de déversoir pour l'accroissement de la capacitè des retenues des barrages [New weir conception to increase dam

reservoir capacities].” Proc. Coll. Int. Protection et Préservation des Resources en Eau, Blida, Algérie, 59-67 [in French], (2006b).

35. Le Doucen, O., Leite Ribeiro, M., Boillat, J.-L., Schleiss, A Laugier, F. “Etude Paramétrique de la capacité des PK-weirs [Parametric study of PK-weir capacity].” Proc. Modèles physiques hydrauliques : Outils indispensables du XXIe siècle, Lyon, 155-160. SHF, Paris (in French), (2009).
36. Laugie, F. “Design and construction of the first piano key weir spillway at the Goulours Dam.” Hydropower and Dams 14(5), 94-101, (2007).
37. Blanc, P., Lempérière, F. “Labyrinth spillways have a promising future.” Hydropower and Dams 8(4), 129-131, (2001).
38. Bieri, M., Leite Ribeiro, M., Boillat, J.-L., Schleiss, A., Laugier, F., Delorme, F., Villard, J.-F. “Rehabilitation de la capacitè d’evacuation des crues: Intégration de <<PK-Weirs>> sur des barrages existants (Rehabilitation of floods discharge capacity: Integration of <<PK –Weirs>> on existing dams).” Proc. Colloque CFBR-SHF: Dimensionnement et fonctionnement des évacuateurs de crues. SHF, Paris (in French), (2009).
39. Laugier, F., Lochu, A., Gille, C., Leite Ribeiro, M., Boillat, J.-L. “Design and construction of a labyrinth PKW spillway at Saint–Marc Dam, France.” Hydropower and Dams, 16(5), 100-107, (2009).
40. Vermeulen, J., Laugier, F., Faramond, L. and Gilles, C. “Lessons learnt from design and construction of EDF first Piano Key weirs.” Pro. International workshop on labyrinths and piano key weirs PKW 2011, CPC press, 215-224, (2011).
41. Erpicum, S., Machiels, o., Archambeau, P., Dewals, B.J., Pirotton, M. “ID numerical approach to model the flow over a Piano key weir.” Proc. Sim Hydro 2010 Nice. SHF, Paris. (2010).
42. Anderson, R.M., and Tullis, B.P. “Comparison of Piano key and rectangular labyrinth weir hydraulics.” J. Hydraul. Eng. 138 (4), 358-361. (2012).
43. Lempérière, F. “New labyrinth weirs triple the spillways discharge.” (http://www.hydrocoop.org). (2010).
44. Lempérière, F., Jun, G. “Low cost increase of dam storage and flood mitigation: the Piano keys weir.” Proc. 19th congress Int.comm. Irrig. Drain. Beijng, 1-19, (2005).
45. Hien, T.C., Son, H.T., Khanh, M.H.T. Results of some piano keys weir hydraulic model test in Vietnam. Proc. 22nd I COLD Congress Barcelona Q87 (R 39), 581-596 (2006).
46. Machiels, O., Erpicum, S., Dewals, B., Archambeau, P. and Pirotton, M. “Experimental observation of flow characteristics over a PKW.” J. Hydraulic res, 49, 359-366, (2011).
47. Machiels, O., Erpicum, S., Archambeau, P., Denals, B.J., Pirotton, M. “Piano key weir preliminary design method: Application to a new dam project.” Proc. Int. conf. Labyrinth and Piano key weirs Liege B, 199-206 (2011).
48. Machiels, O., Erpicum, S., Archambeau, P., Dewals, B.J., and Pirotton, M. “Influence of Piano key weir height on its discharge capacity.” Labyrinth 2nd piano key weirs, CRC Press, London, 59-66. (2011).
49. Marcelo LEITE RIBEIRO, Martin BIERI, Jean-Louis BOILLAT, Anton J. SCHLEISS, F. DELORME, F. LAUGIER. “Hydraulic capacity improvement of existing spillways-Design of piano key weirs.” VINGT TROISIEME CONGRES DES GRANDS BARRAGES Brasilia, (2009).

50. Novak, P., Moffat, A.I.B., Nalluri, C. and Narayanan R. Hydraulic structures Unwin Hyman, London, UK [Google Scholar] (1990).
51. Johnson, M.C. "Discharge coefficient analysis for flat-topped and sharp-crested layers." Irrigation Science, 19(3), 133-137, (2000).
52. O. Machiels; P. Archambeau; B. Dewals, and M. Pirotton. "Parapet wall effect on piano key weir efficiency." J. of Irrigation and drainage Eng., 506, (2013).
53. Emiroglu ME, Kaya N, Agaccioglu H. "Discharge capacity of labyrinth side weir located on a straight Channel." Journal of irrigation and drainage engineering 2010; 136 (1): 37-46.
54. Abdorreza, Kabiri, Samani, and Amir Javaheri. "Discharge coefficients for free and submerged flow over piano key weirs." J. Hydraul. Res., 50:1, 114-120 (2012).
55. R.A.R.V.Krishantha, S.S. Wickramasuriya. "Climate change adaptation: A Hydraulic model study to improve the spillway discharge of genital reservoir." G78-1-4799-1740-2/15/$31.00©2015 IEEE. (2015).
56. A.Mehboudi, J. Attari, S.A. Hosseini. "Experimental study of discharge coefficient for trapezoidal piano key weirs. http://dx.doi.org/10.1016/j.flowmeasinst.2016.06.005 (2016).

Experimental investigation of torsional vibration isolation using Magneto Rheological Elastomer

Praveen Shenoy K[1*], *Abhishek Kumar Singh*[1], *Sai Aditya Raman K*[1], and *K. V. Gangadharan*[1]
[1]*Centre for System Design, National Institute of Technology Karnataka, Surathkal, Mangalore-575025, Karnataka, India*

Abstract: Rotating systems suffer from lateral and torsional vibrations which have detrimental effect on the roto-dynamic performance. Many available technologies such as vibration isolators and vibration absorbers deal with the torsional vibrations to a certain extent, however passive isolators and absorbers find less application when the input conditions are dynamic. The present work discusses use of a smart material called as Magneto Rheological Elastomer (MRE), whose properties can be changed based on magnetic field input, as a potential isolator for torsional vibrations under dynamic loading conditions. Carbonyl Iron Particles (CIP) of average size 5 μm were mixed with RTV Silicone rubber to form the MRE. The effect of magnetic field on the system parameters was comprehended under impulse loading conditions using a custom built in-house system. Series arrangement of accelerometers were used to differentiate between the torsional and the bending modes of vibration of the system. Impact hammer tests were carried out on the torsional system to study its response, in the presence and absence of magnetic field. The tests revealed a shift in torsional frequency in the presence of magnetic field which elucidates the ability of MRE to work as a potential vibration isolator for torsional systems.

Keywords

Torsional Vibrations, Isolation, Impact hammer tests, Magneto Rheological elastomer

1 Introduction

Torsional vibrations have been a subject of study due to its role in causing catastrophic failure in many of the rotating members. All systems subjected to rotation involve torsional oscillations during start up or shut down of prime mover. During this, the system may pass through the torsional natural frequencies of the system. Since the time for which the system dwells in the frequency is less, the possibility for a failure is minimum. However prolonged modes of operation may lead to an imminent failure of the system. Hence an in-depth study of reduction of torsional vibrations in the rotating systems has been one of the prominent aspects in the design of rotating systems 1 2.

[*] Corresponding Author: praveen.me14f12@nitk.edu.in

Reducing vibration in systems has been carried out in various ways. Vibration absorbers and vibration isolation concepts have been used effectively in arresting translatory as well as rotational vibrations of systems. Absorbers employ the fact that adding a secondary mass to the primary system of interest reduces the amplitude of vibration on the primary system. Isolators on the other hand rely heavily on the system's inherent properties (mainly stiffness and damping) to reduce the amplitudes of critical oscillations. The present paper focusses on the latter.

Vibration isolation has been efficiently carried out in systems in various ways. It involves changing the stiffness or the damping of the system between the source of disturbance (Base) and the source of interest (Mass/Inertia). Based on the controllability, isolators are divided into "Passive, Active and Semi-active" systems. Passive isolation systems have been successfully implemented in many applications for it is economical and very easy to be built into. However it fails to work for systems with varying properties of stiffness and damping. Active isolators resolve this issue as they have a feedback supplied to the actuation unit. Due to their bulky nature and sophistication, they have been popular only within a limited branch of dynamics. Semi-active isolators bridge the gap between the two. They incorporate the simple yet effective characteristic of the passive systems and the ability to change the properties of the systems as that of the active systems.

One aspect that sets the semi active and the active vibration systems apart is the ability of the material to change its inherent property. Such materials are called Smart Materials. Many smart materials such as piezo materials, shape memory alloys, magnetostrictive materials and newer forms of composites have been used in the effective mitigation of vibration amplitudes. One such material is called Magneto Rheological Elastomer also known as MRE. MRE is the solid form in the family of smart materials called as Magneto Rheological materials. Its fluid analogue, the Magneto Rheological Fluids have been used as fluid dampers in suspension systems for their excellent damping capabilities. MRE on the other hand have been used for vibration isolation under dynamic loading conditions. It consists of a viscoelastic matrix material in which the iron particles are imbibed either randomly (Isotropic) or along the direction of induced magnetic field during its curing process (Anisotropic). The elastomer upon curing forms into a solid form and on application of magnetic field, the iron particles arrange themselves as chains in the direction of the applied field. Due to this, the material undergoes changes in its properties. This has been used as isolators in systems with varying ranges of natural frequency making it to be efficiently used in semi-active and active vibration isolation systems.

MRE have been extensively studied for its isolation properties. Since the initial work carried out by Ginder et al. 3, numerous papers deal with MRE's isolation capabilities. However the isolation capabilities of MRE have been studied in the translatory modes of vibration. Liao et al. 4 worked on real time tunable vibration isolators with multiple layers of MRE. Jeong et al. 5 carried out designing a MRE elastomer differential mount which was used in between the engine and the car body. MRE has also been studied in seismic vibration isolation. Jung et al. 6 carried out base vibration tests on MRE with various forms of input being given to the structure. Behrooz et al. 7 8 studied the variable stiffness and damping in base isolation. An increase of 57 % in the shear performance was observed. It has to be noted that the frequency of operation for which the isolation studies have found to be effective is in the lower range (i.e., in the range of 1 to 100 Hz). Hegde et al. 9 examined the effect of particle size on the dynamic damping of the MRE. Transmissibility ratio for two particle sizes were compared along with the loss factor. It was found that the smaller particle sized MRE showed better improvement on loss factor compared to pure silicone due to agglomeration of the particles.

Research has been carried out in using MRE as a potential semi-active vibration absorber in powertrain systems. Hoang et al. 11 carried out numerical simulations on a vibration absorber. An adaptive tunable vibration absorber (ATVA) based on MRE was designed and developed. The effectiveness of ATVA was evaluated in terms of shift in the resonant frequencies was numerically investigated on powertrain systems.

Experimental modal analysis provides an excellent way to determine many of the modal parameters. The modal parameters are determined using the Frequency Response Functions (FRFs) which is the ratio of the output response to the input force which is given to the system. The input force may be given in-terms of modal shakers or using an impact hammer which gives impulse input. The present paper deals with understanding the torsional frequency variation under the influence of magnetic field using FRF response.

2 Methodology

2.1 MRE Preparation

An RTV based silicone material is used as the matrix with Carbonyl Iron Particles (CIP) of an average diameter of 5 μm mixed into it. The volume fraction of the matrix to the CIP is 73:27. The details of the constituent materials are as shown in Table 1:

Table 1: MRE composition

Sl No	Constituents	Make	Density ($^{kg}/_{m^3}$)
1	RTV Based Silicone rubber	Dow Corning	884.57
2	Carbonyl Iron Particles	BASF, Type CN	7850

The procedure involves mixing the RTV silicone with the CIP in the mentioned concentration. The mixture is stirred continuously till the constituent materials get mixed thoroughly. This mixture is then poured into the mould and is degassed until the air bubbles are removed. This is subjected to curing for a period of 20 hours under constant pressure at room temperature.

2.2 Experimental Setup

The schematic of the experimental setup is as shown in Fig. 1. It consists of two Aluminum shafts connected via the MRE coupler and one Mild Steel disk mounted on each of the shafts at the free end. To measure the torque, two HBM T22, 10Nm Torque transducers (TT) was used on either side of the isolator. The Torque transducers and the bearings were firmly supported on the rigid base. Further an YMC Piezotronics Inc. single axis accelerometer was used to measure the frequency response of the base. In order to extract the bending and the torsional modes of the system separately, four triaxial accelerometers (Measurement Specialities Inc), two on either disks were used. A PCB Piezotronics impact hammer was used to provide the impulse force. To induce magnetic field, permanent magnets (type N52, Neodymium magnets) were used. A Lake Shore, (Model 410) Gaussmeter was used to measure the generated magnetic field. The data obtained from the sensors was read through NI 9234, a National Instrument Data Acquisition unit into a NI PXIe-1082 which was further read through LabVIEW 2014 (Fig. 2)

Prior to finding out the bending and the torsional modes of vibrations, a modal analysis was carried out on the rigid base member (Fig. 3) using Roving hammer test. This was done to analyse the predominant frequencies of the base which could have a likely effect on the system's natural frequency. This is proceeded with finding the bending and the torsional modes of vibrations using the Serial Accelerometer Arrangement (SAA) and the Parallel

Accelerometer Arrangement (PAA) of the two tri-axial accelerometers 12 as shown in Fig. 4. The frequencies corresponding to the peak amplitudes from the response graphs determine the torsional and the bending modes of vibrations. Once the torsional modes was found, the effect of field on the torsional natural frequencies and the amplitude was studied. To impart torsional oscillations on the system, an extension is attached on the inertial element (disc), and an impact is given on end of the extension as shown in Fig. 2. Similar approach has been carried out to impart torsional oscillations on systems 13 14. The frequency resolution is 0.1 Hz with the number of averages considered is 10.

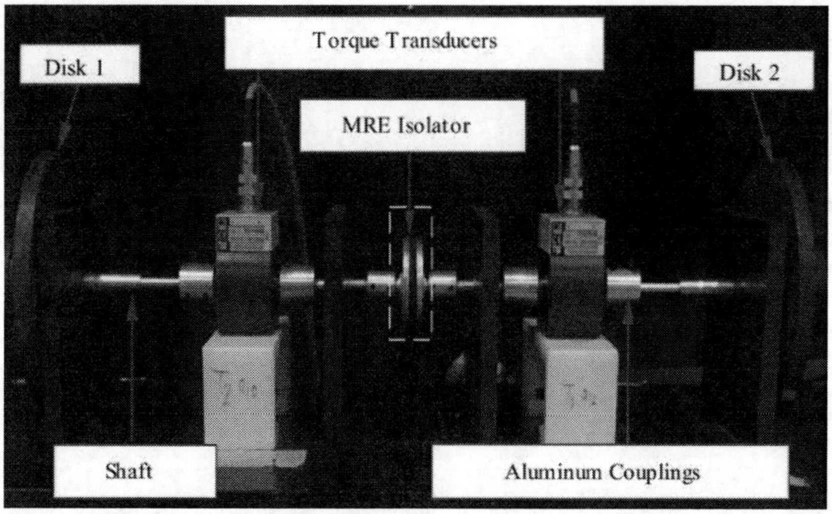

Fig. 1. Torsional 2 DoF system

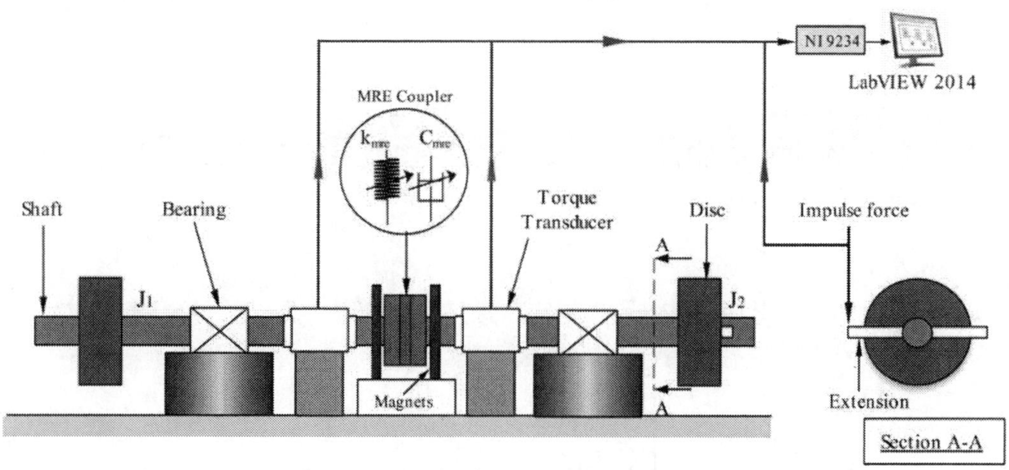

Fig. 2. Schematic representation of experimental setup

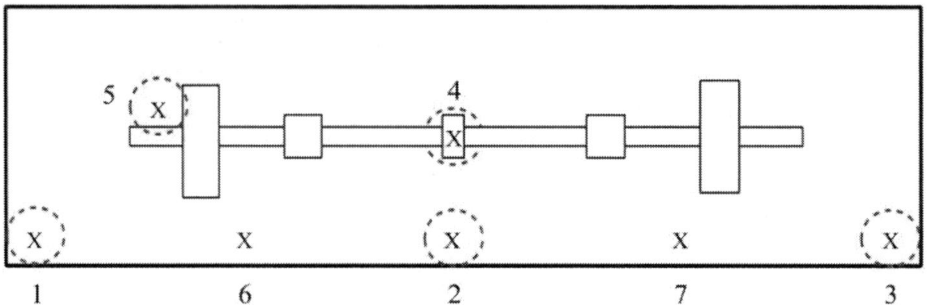

Fig. 3. Accelerometer position and impact Locations

Kuang et al. 12 developed a two accelerometer system to evaluate the bending and torsional natural frequencies of a shaft-rotor system. Two accelerometers in parallel and serial combinations have been used and the corresponding natural frequencies in bending and torsion were calculated. Similar approach is applied in the present set-up to obtain the torsional natural frequencies of the system.

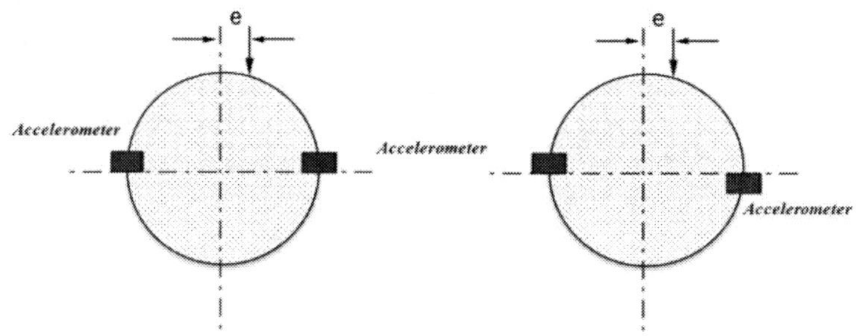

Fig. 4. Parallel and Serial arrangements of Accelerometers

For the serial and parallel arrangement of the accelerometers, the equations 12 of the harmonic response is given as,

$$P_{ij}(\omega) = 2\left\{\sum_{r=1}^{n} \frac{\phi_{ri}\phi_{rj}}{\omega_r^{\,2}\left[1-\left(\frac{\omega}{\omega_r}\right)^2\right]}\right\} F_j(\omega) \tag{1}$$

$$S_{ij}(\omega) = 2r\left\{\sum_{r=1}^{m} \frac{\theta_{ri}\theta_{rj}}{\Omega_r^{\,2}\left[1-\left(\frac{\omega}{\Omega_r}\right)^2\right]}\right\} T_j(\omega) \tag{2}$$

wherein:

P_{ij} *and* S_{ij} are the harmonic response in PAA and SAA measurement

$\phi_{ri}\phi_{rj}$ *and* $\theta_{ri}\theta_{rj}$ Are the linear displacement and the angular displacement terms

ω_r *and* Ω_r are the bending mode and torsional mode natural frequencies

F_j *and* T_j are the harmonic force and harmonic torque applied on the disks

For a single degree of freedom system the harmonic signal in torsional mode is given as,

$$S(\omega) = 2r\left\{\sum_{r=1}^{m} \frac{\theta^2}{\Omega_r^{\,2}\left[1-\left(\frac{\omega}{\Omega_r}\right)^2\right]}\right\} T_j(\omega) \tag{3}$$

From the above equation, the FRF function for the torsional modes of frequencies can be obtained as,

$$FRF = \frac{S(\omega)}{T_j(\omega)} = 2r\left\{\sum_{r=1}^{m} \frac{\theta^2}{\Omega_r^{\,2}\left[1-\left(\frac{\omega}{\Omega_r}\right)^2\right]}\right\} \tag{4}$$

3 Results and discussion

3.1 Impact test results on the Torsional vibration system base

As discussed in the previous section, the primary aim was to identify if any of the natural frequencies of the base structure are close to torsional natural frequencies of interest. Table 2 indicates the lowest frequencies at which peaks were observed in inertance FRFs obtained at various positions (1-5) when impact was given at locations (i) to (vii) shown in Fig. 3.

Table 2. Lowest identified critical frequencies based on FRF curves for seven input and five response positions. (All frequencies are in Hz)

	Accelerometer positions				
Impact position	**1**	**2**	**3**	**4**	**5**
(i)	145	600	600	535	820
(ii)	270	550	820	550	315
(iii)	260	300	580	555	820
(iv)	250	290	600	600	820
(v)	115	110	110	350	120
(vi)	280	100	280	100	500
(vii)	270	320	475	560	480

From the table it is seen that the least frequency component obtained is 100 Hz and 110 Hz with the mountings on the base. This implies that the base frequency components do not affect the torsional natural frequencies which are expected to be within 50 Hz corresponding to the working torsional frequency range of many of the rotating systems. It is seen that the addition of the mountings increases the natural frequency of the base to 115 Hz. This is due to the fact that the mountings result in increase of the stiffness of the base structure.

3.2 Determination of torsional and bending modes using accelerometer pair measurements

Fig. 5 and Fig. 6 show the torsional and bending modes of the disk-shaft system with the serial and parallel arrangement of the accelerometers. As seen, the serial arrangement of the accelerometers reveal the torsional mode of vibration which occurs at approximately 13.5 Hz (780 rpm) which is the speed of interest for many of the torsional systems.

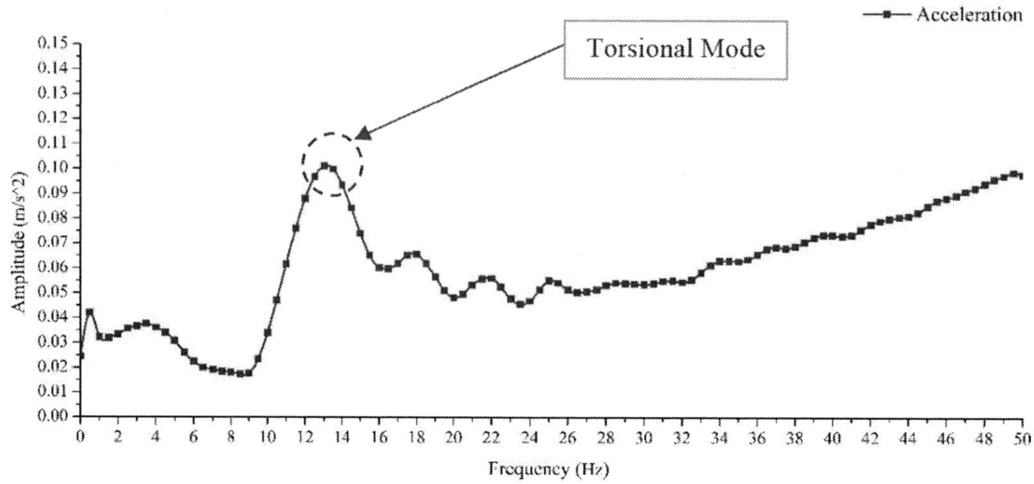

Fig. 5. Accelerometers reading for serial arrangement

In the parallel arrangement of accelerometers, bending mode of system is enhanced while torsional mode is cancelled due to phase lag in accelerometers. It is seen that the bending mode occurs at approximately 42 Hz which is found to be away from the torsional mode of vibration. From this it is seen that under the frequency range of interest (i.e., 1 Hz to 50 Hz), there is only the torsional natural frequency seen and with this, the effect of magnetic field on the stiffness is investigated.

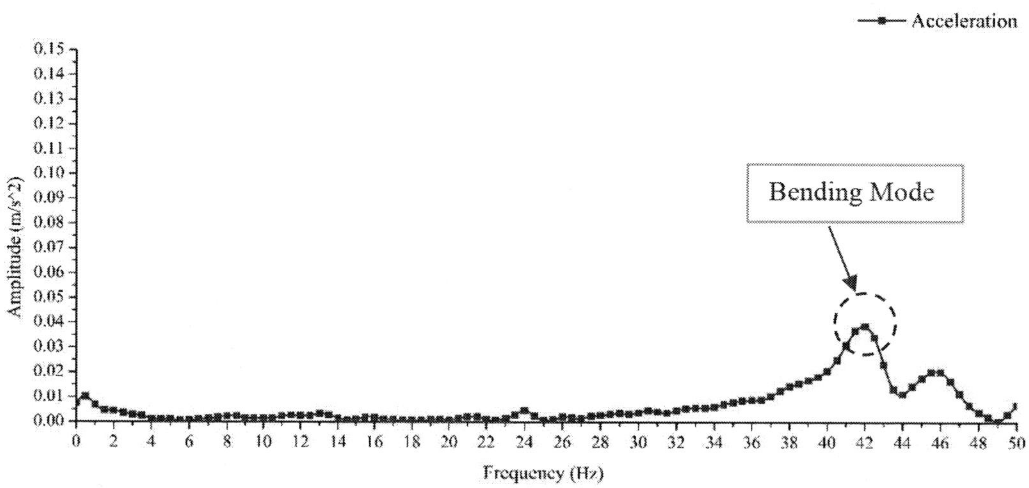

Fig. 6. Accelerometers reading for parallel arrangement

3.3 Variation in torsional natural frequency under the influence of magnetic field

The system is subjected to impact hammer tests and the response is analysed. Two torque transducers, one measuring the input torque and the other measuring the output (both with respect to the input impact force), is used to obtain the FRF. Fig. 7 and Fig. 8 show the response of the system in the absence and presence of magnetic field for the input (Torque transducer-1) and the output (Torque transducer-2) transducers.

The FRF graph for TT-1 is as shown in Fig. 7. The initial DC component is neglected. From the graph it is seen that under 0T (no field condition), the FRF peaks up at approximately 13.5 Hz, which represents the torsional natural frequency of the system. However under the influence of magnetic field (i.e. 0.15T), it is seen that the FRF shows a frequency shift to 14.5 Hz along with a reduction in amplitude, H(ω). Similar case is seen in Fig. 8 with the torque transducer-2.

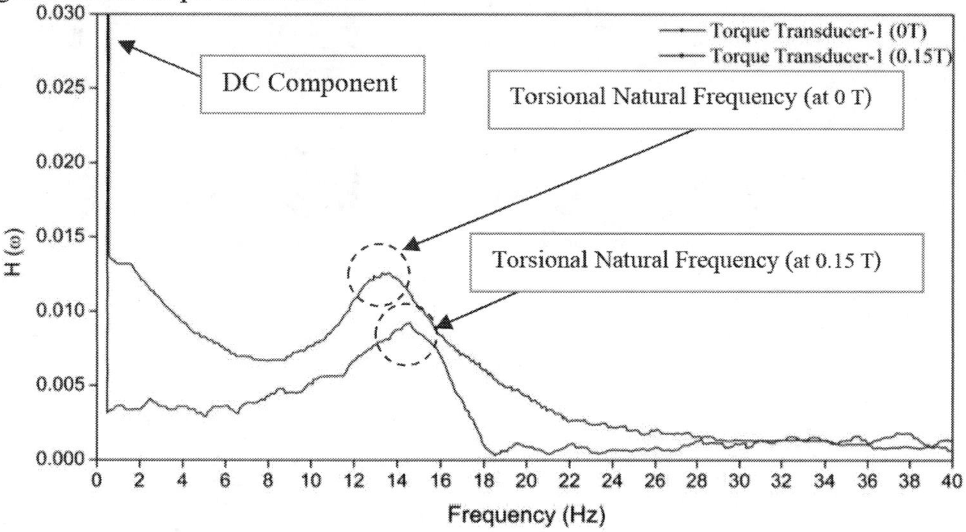

Fig. 7. FRF for Torque transducer-1

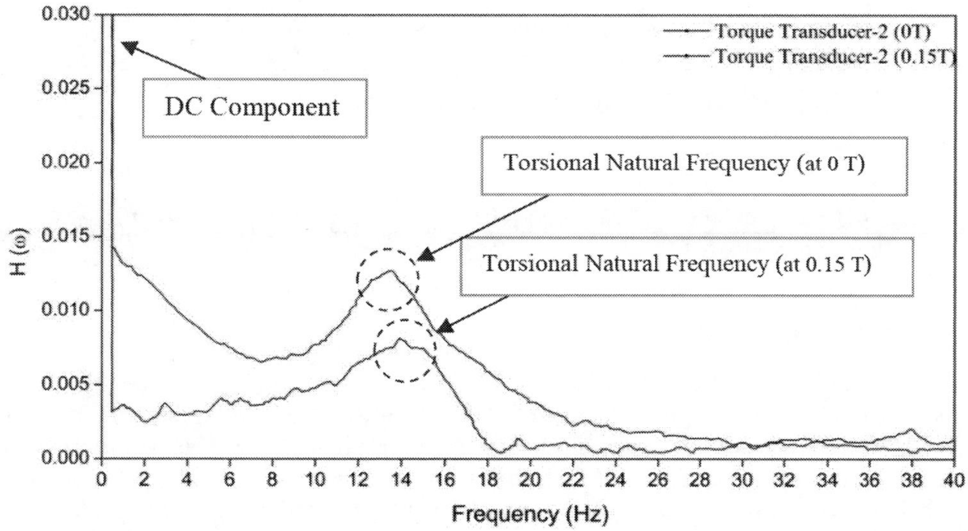

Fig. 8. FRF for Torque transducer-2

From the graphs it is seen that the absolute shift in the frequency is approximately 7 % when the field is increased to 0.15 T. The shift in the torsional natural frequency is attributed to the influence of magnetic field on the properties of MRE. Under the influence of the magnetic field, there is an enhancement of dynamic stiffness of the elastomer which results in increase of the natural frequency of the MRE.

This enhancement in stiffness can be attributed to the compressive forces generated on the matrix material due to CIP (Fig. 9). For simplicity a unit cell of the CIP-Matrix is considered 15 16 17. Due to the magnetic field lines, the CIPs get aligned along the magnetic field and this in-turn forms a compressive force on the intermediate matrix which in turn increase the stiffness of the unit cell of the CIP-Matrix. When the bulk of the elastomer is considered, this results in the overall stiffness increase of the material.

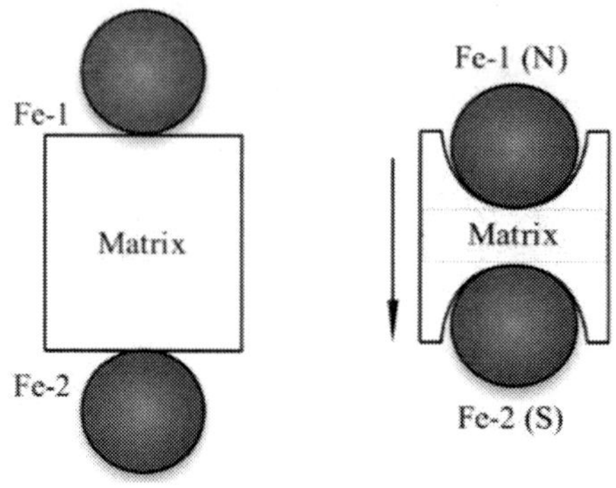

Fig. 9. Unit cell representation of the MRE under non-magnetized and magnetized condition

It is also seen from the FRF plots that there is a reduction in amplitudes at the resonant frequency in the 0 T and 0.15 T curves. Table 3 shows the comparison between the FRF amplitudes between the two conditions.

$$Absolute\ Reduction = \frac{Amplitude\ (0\ T) - Amplitude\ (0.15\ T)}{Amplitude\ (0.15\ T)} \times 100 \quad (5)$$

Table 3. FRF amplitude comparison under 0 T and 0.15 T

Torque Transducers	Amplitude (at 0 T frequency)	Absolute Reduction (%)
TT 1 (Input Side – 0 T)	0.0126	35.6
TT 1 (Input Side – 0.15 T)	0.00812	
TT 2 (Input Side – 0 T)	0.0127	40.6
TT 2 (Input Side – 0.15 T)	0.00754	

4 Conclusions

The isolation capabilities of Magneto Rheological Elastomer has been investigated experimentally using the impact hammer technique. It was seen that in the presence of field, the natural frequency of the system shifted from 13.5 Hz to 14.5 Hz. This was ascribed to the increase in the stiffness of the MRE in the presence of field. The well-established di-pole mechanism was used to explain the effect. Also there was a reduction in the amplitude of 0T field, in the presence of field. This proved that the MRE, under torsional shear, worked as an effective isolator.

The authors acknowledge the support from SOLVE: The Virtual Lab @ NITK (Grant number: No.F.16-35/2009-DL, Ministry of Human Resources Development) and experimental facility provided by Centre for System Design (CSD): A Centre of excellence (http://csd.nitk.ac.in/) at National Institute of Technology Karnataka, India

References

1. Szenasi, Fred R., and J. C. Wachel. "Analysis of torsional vibrations in rotating machinery." *Proceedings of the Turbomachinery Symposium*. Vol. 22. 1993.
2. Vance, J. M., and R. S. French. "Measurement of torsional vibration in rotating machinery." *Journal of Mechanisms, Transmissions, and Automation in Design* 108.4 (1986): 565-577
3. Ginder J M, Nichols M E, Elie L D and Tardiff J L 1999 Magnetorheological elastomers: properties and applications Smart Structures and Materials 1999: Smart Materials Technologies, 131-138, 1999
4. Liao G J, Gong X L, Xuan S H, Kang C J and Zong L H 2012 Development of a real-time tunable stiffness and damping vibration isolator based on magnetorheological elastomer J. Intell. Mater. Syst. Struct.**23** 25-33
5. Jeong U C, Yoon J H, Yang I H, Jeong J E, Kim J S, Chung K H and Oh J E 2013 Magnetorheological elastomer with stiffness-variable characteristics based on induced current applied to differential mount of vehicles *Smart Mater. Struct.* **22** 115007
6. Jung H-J, Eem S-Hy, Jang D-D and Koo J-H 2011 Seismic Performance analysis of a smart base isolation system considering dynamics of MR elastomer *J. Intell. Mater. Syst. Struct.* **22** 1439-50
7. Behrooz M, Wang X and Gordaninejad F 2014 Performance of a new magnetorheological elastomer isolation system *Smart Mater. Struct.* 23 045014
8. Behrooz M, Wang X and Gordaninejad F 2014 Modeling of a new semi-active/passive magnetorheological elastomer isolator *Smart Mater. Struct.* 23 045013
9. Hegde, Sriharsha, Umanath R. Poojary, and K. V. Gangadharan. "Experimental investigation of effect of ingredient particle size on dynamic damping of RTV Silicone base Magnetorheological elastomers." *Procedia Materials Science* 5 (2014): 2301-2309
10. Hoang, N., et al. "Development of a torsional dynamic absorber using a magnetorheological elastomer for vibration reduction of a powertrain test rig." *Journal of Intelligent Material Systems and Structures* 24.16 (2013): 2036-2044
11. Hoang, N., N. Zhang, and H. Du. "A dual adaptive tunable vibration absorber using MREs for vehicle powertrain vibration control." SPIE, 2010
12. Kuang, Jao-Hwa, and Jing-Fung Lin. "Accelerometer Pair Measurements for Shaft Dynamic Parameters Analysis." PROCEEDINGS-SPIE THE INTERNATIONAL SOCIETY FOR OPTICAL ENGINEERING. SPIE INTERNATIONAL SOCIETY FOR OPTICAL, 1994
13. Yesilyurt, Isa, and Habibe Gursoy. "Estimation of elastic and modal parameters in composites using vibration analysis." *Journal of Vibration and Control* 21.3 (2015): 509-524
14. Alwan, V., et al. "Dynamic analysis of shafts of composite materials." *Journal of Reinforced Plastics and Composites* 29.22 (2010): 3364-3379
15. Poojary, Umanath R., Sriharsha Hegde, and K. V. Gangadharan. "Dynamic blocked transfer stiffness method of characterizing the magnetic field and frequency dependent dynamic viscoelastic properties of MRE." *Korea-Australia Rheology Journal* 28.4 (2016): 301-313
16. Jolly, Mark R., J. David Carlson, and Beth C. Munoz. "A model of the behaviour of magnetorheological materials." *Smart Materials and Structures*5.5 (1996): 607

17. Poojary, Umanath R., Sriharsha Hegde, and K. V. Gangadharan. "Dynamic deformation–dependent magnetic field–induced force transmissibility characteristics of magnetorheological elastomer." *Journal of Intelligent Material Systems and Structures* (2016): 1045389X16672730

Solid Modelling Interaction with Sensors for Virtual Reality Welding

Bharath V. G.[1*] and *Rajashekar Patil*[2]

[1] *School of Mechanical Engineering, REVA University, Bengaluru, Karnataka, India*
[2] *Department of Mechanical Engineering, Atria Institute of Technology, Bengaluru, Karnataka, India*

Abstract. The imperative part of strong displaying in virtual reality framework is to improve the movement of a user with definitive movement control in genuine intelligent condition. In a real environment motion characteristic is based on sensors and servomechanisms where as in virtual reality systems the motion fundamentals are kinematic in nature. To achieve physically correct interactivity suitable dynamic constraints, should be imposed which can be obtained by augmented reality interface. Beyond input and output hardware, the underlying software plays a very important role in virtual reality systems. It is responsible for the managing of input/output devices, analysing incoming data and generating proper feedback. This research paper focusses on movement displaying for connection between CAD models and Virtual Reality Models using 3ds Max 2017 and Unity 3d softwares. The paper portrays different equipment arrangement of sensors and Arduino for virtual reality welding.

Keywords: Welding, Virtual Reality, 3ds Max 2017, Unity 3d, Solid modelling, Arduino

1 Introduction

1.1 Virtual Reality

Virtual Manufacturing technique is a virtual reality technology to create virtual environment on the computer screen to simulate the physical world. The expertise and knowledge base gained from the work in the virtual environment enables the user to apply more meaningfully in the real-life situation. Banerjee and Tesic [1] has worked in Graphical User Interface, Virtual Reality advancements, auto translation.

1.2 Literature Survey

Balyliss [2] emphasized on VR methods with better 3D perception of a question and clarified hypothetical strong displaying strategies utilizing Virtual Reality Modelling Language (VRML) and 3ds Max. Bharath and Rajashekar Patil [3] created diverse parts of virtual welding stage and enlivened the stream of welding utilizing VR instruments. Leo

* Corresponding author: bharathvg@gmail.com

Louis [4] considered the working standard of Arduino, its equipment/programming elements and its favourable circumstances of building new gadgets of our own to make and execute creative items to worldwide market. Mendes et al. [5] clarified that the Arduino Mega 2560 is utilized for different sorts of utilization, for example, laser cutting, CNC and 3D Printing. Deepak et al. [6] has emphasized the significance of Arduino and its intelligence amongst machine and programming in 3D Printer.

2 Methodology

Motion modelling of virtual reality welding involves parameters such as modelling, simulation and integrity between virtual model and realistic model.

2.1 Methodology of modelling and simulation

3D modelling of virtual welding using 3DS Max software illustrates modelling of an entire virtual welding platform, simulation of a welding in virtual mode, algorithm.

2.2 Methodology of motion modelling

Unity 3d software is a game engine which illustrates the simulation of an entire model in a game mode and integrates model between virtual and real environment. Arduino is a microcontroller which integrates several sensors to a game engine by serial communication coding. Fig. 1 illustrates the methodology of Virtual Reality Welding.

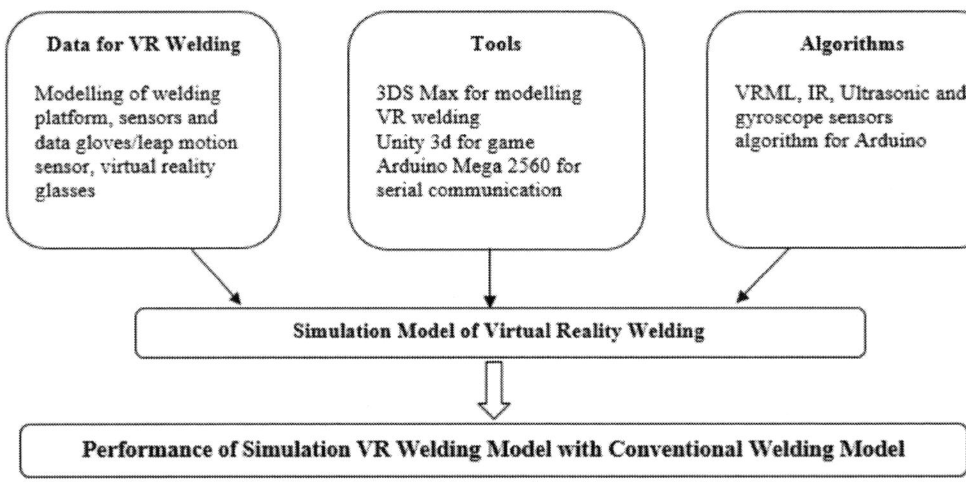

Fig. 1. Methodology of Virtual Reality Welding

3 Design

3.1 Arduino Mega 2560

The Arduino Mega 2560 is a microcontroller which has 54 advanced info/yield pins, 16 simple data sources, 4 Universal Asynchronous Receiver-Transmitter (UART) equipment serial ports for Transistor Logic (TTL) (5V) serial correspondence, a 16 MHz precious stone oscillator, a Universal Serial Bus (USB) association, a power jack, an In-Circuit

Serial Programming (ICSP) header, and a reset catch. It essentially associates with a PC with a USB link or power it with an AC-to-DC connector or battery to begin. The Mega is perfect with most shields intended for the Arduino. Fig. 2a demonstrates the arrangement of Arduino Mega 2560 with key marking and Fig. 2b gives a nitty gritty specialized drawing of Arduino Mega 2560 composed utilizing Solidedge ST9 programming.

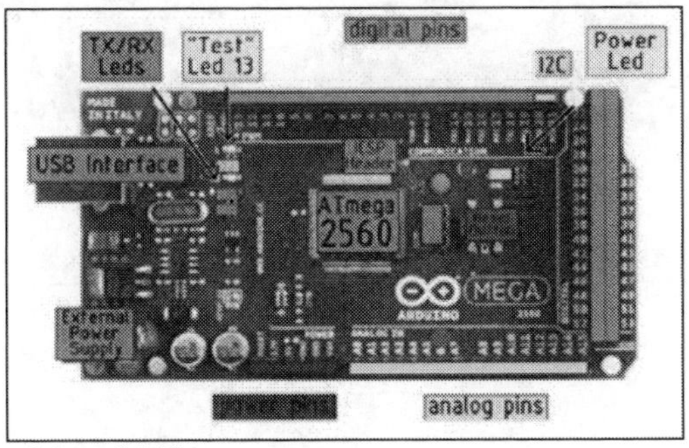

Fig. 2a. Top view of Arduino Mega 2560

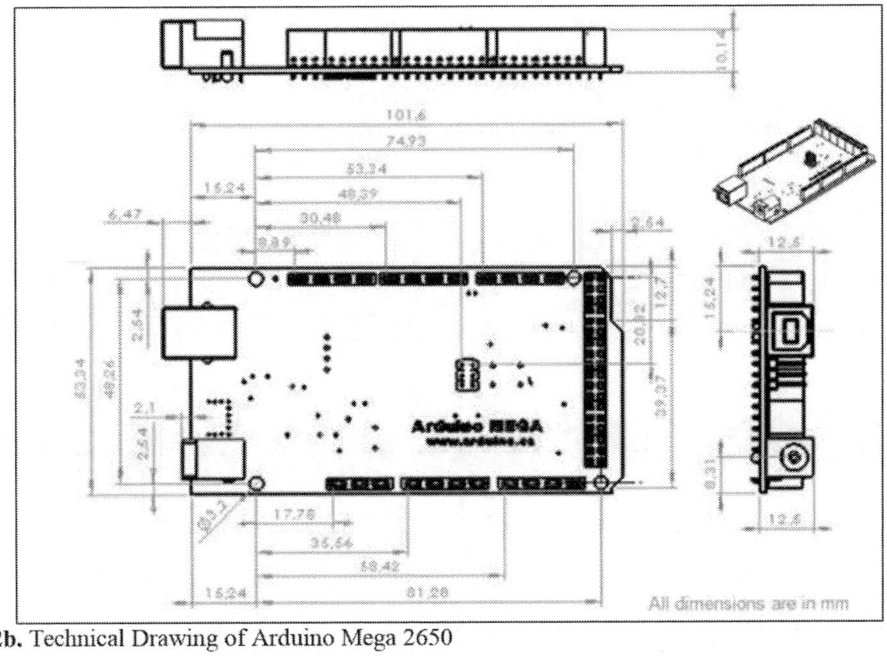

Fig. 2b. Technical Drawing of Arduino Mega 2650

3.2 Sensors and its coding for interactivity

Ultrasonic sensor transmits an ultrasound at 40,000 Hz which goes through the air and if there is a protest or impediment on its way, it will skip back to the module [7, 8]. Computation of the separation is acquired by considering the travel time and the speed of the sound. Ultrasonic sensor is utilized as a part of virtual welding to gauge the separation of a terminal development from one end to flip side of the workpiece [9, 10]. Fig. 3 demonstrates ultrasonic sensor and MPU 6050. The MPU-6050 gadgets consolidate a 3-hub spinner and a 3-hub accelerometer on a similar silicon bite the dust, together with an installed Digital Motion Processor (DMP), which forms complex 6-hub Motion Fusion calculations. These Motion Tracking gadgets are intended for measuring edge development of a terminal. Table 1 demonstrates the coding of Ultrasonic sensor and MPU 6050 sensor.

 https://doi.org/10.1051/mateconf/201814401008

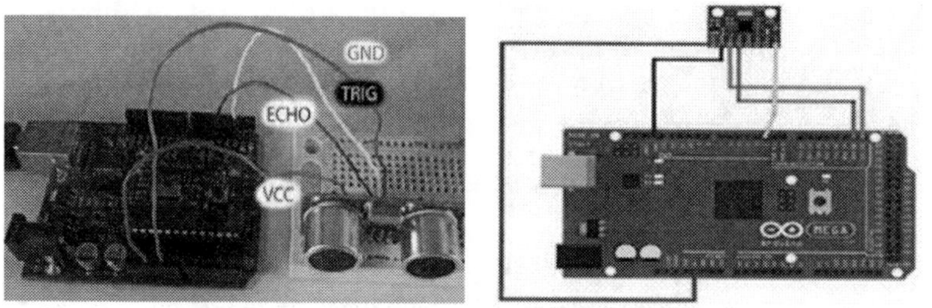

Fig. 3. Ultrasonic sensor and MPU 6050 Sensor

Table 1. Ultrasonic and MPU 6050 sensor coding.

```
const int trigPin = 2;
const int echoPin = 4;
void setup() {
  // initialize serial communication:
  Serial.begin(9600);
}
void loop()
{
   long duration, inches, cm;
   pinMode(trigPin, OUTPUT);
  digitalWrite(trigPin, LOW);
  delayMicroseconds(2);
  digitalWrite(trigPin, HIGH);
  delayMicroseconds(10);
  digitalWrite(trigPin, LOW);
   pinMode(echoPin, INPUT);
  duration = pulseIn(echoPin, HIGH);
  // convert the time into a distance
  inches = microsecondsToInches(duration);
  cm = microsecondsToCentimeters(duration);
   Serial.print(inches);
  Serial.print("in, ");
  Serial.print(cm);
  Serial.print("cm");
  Serial.println();
  delay(1);
}
```

```
#include<Wire.h>
const int MPU_addr=0x68;
int16_t AcX,AcY,AcZ,Tmp,GyX,GyY,GyZ;
int minVal=265;
int maxVal=402;
double x;
double y;
double z;
 void setup(){
  Wire.begin();
  Wire.beginTransmission(MPU_addr);
  Wire.write(0x6B);
  Wire.write(0);
  Wire.endTransmission(true);
  Serial.begin(9600);
}
void loop(){
  Wire.beginTransmission(MPU_addr);
  Wire.write(0x3B);
  Wire.endTransmission(false);
  Wire.requestFrom(MPU_addr,14,true);
  AcX=Wire.read()<<8|Wire.read();
  AcY=Wire.read()<<8|Wire.read();
  AcZ=Wire.read()<<8|Wire.read();
    int xAng = map(AcX,minVal,maxVal,-90,90);
    int yAng = map(AcY,minVal,maxVal,-90,90);
    int zAng = map(AcZ,minVal,maxVal,-90,90);
      x= RAD_TO_DEG * (atan2(-yAng, -zAng)+PI);
      y= RAD_TO_DEG * (atan2(-xAng, -zAng)+PI);
      z= RAD_TO_DEG * (atan2(-yAng, -xAng)+PI);
    Serial.print("AngleZ= ");
    Serial.println(z);
    Serial.println("-----------------------------------------");
    delay(400);
}
```

4 Virtual Welding Simulation Model

A simulation model of virtual welding is planned utilizing 3ds Max programming and a VRML content is produced for intuitiveness amongst sensors and simulation model. Unity 3d is a game engine used to impart between simulation display and practical condition in which representation is made utilizing virtual reality glasses and information gloves. Fig. 4 shows simulation model of virtual welding arc is generated by electrode contact.

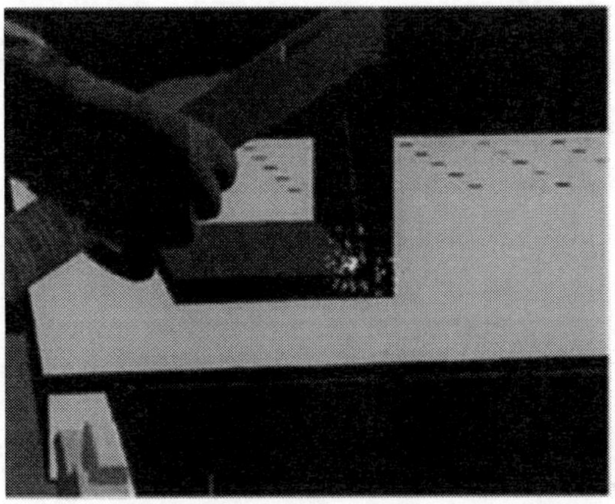

Fig. 4. Ultrasonic sensor and MPU 6050 Sensor

5 Conclusions

The present innovation is looking in front of virtual reality. As there is a shortage of materials step by step, what's to come will wind up noticeably virtual to expertise the works in welding. A virtual reality framework has been created to accomplish craft of welding in virtual environment with a prudent movement control gadgets of welding stage plan and resulting representation of circular segment welding. Different elements like camera scenes, VRML contents, serial correspondence coding are bound together to user for less demanding comprehension. In this present research work strong displaying of virtual welding is produced and created Arduino Mega code for ultrasonic and MPU sensors to communicate amongst virtual and genuine conditions. In future research scope haptic sensors which are costly, for example, Human Machine Interface and Programmable Logical Controller are utilized for better intelligence.

Acknowledgements

The authors thank Mr. Prashanth S. N. and Nuthan, student of Atria Institute of Technology for their support on analyzing and suggesting inputs in Arduino Mega 2650.

References

1. P. Banerjee and R. Tesic, Design of Virtual Objects for Exact Collision Detection in Virtual Reality Modeling of Manufacturing Processes, Proceeding of International Conference on Robotics and Automation, 18, 367-376 (1999).
2. G. M. Balyliss, A. Bowlyer, R. I. Talyor. and P. G. Willis, Theoretic Solid Modeling Techniques and Application Using the Virtual Manufacturing, Proceeding of CSG, 94 (1994)
3. V. G. Bharath and Rajashekar Patil, Designing Interactive 3D Modelling of Virtual Welding System from CAD to VRML, J of Control Theory and Applications, 10, 145-153 (2017)
4. Leo Louis, Working Principle of Arduino and using it as a Tool for Study and Research, J of Control Automation Communication and Systems, 1, 21-29, (2016)

5. Allen R. Mendes, Atur M. Mehta, Bhavya H. Gohil, Implementation of the Automatic and Interactive Chess Board, J of Electrical and Electronics Engineering, 9 01- 04 (2014)
6. Deepak D, Mohammed Murtaza, Prashanth S N, Harsha N, Rajashekar Patil and Bharath V G Economical 3d – Printer by Adopting FDM Technique, J of Mechanical Engineering and Technology, 8, 442- 447 (2017)
7. Manpreet Kaur, Distance Measurement of Object by Ultrasonic Sensor HC-SR04, J for Scientific Research & Development, 3, 503-505 (2015)
8. K. Shrivastava, Distance Measurement of an Object or Obstacle by Ultrasound Sensors using P89C51RD2, J of Computer Theory and Engineering, 2, 64-68 (2010)
9. Yi-Jen Mon, "The Gyroscope Sensor Test by Using Arduino Platform", J of Scientific & Technology Research, 4, 398-400 (2015)
10. Kyujung Lee and Choong Ho Lee, Effect of Gyro Sensor to the Position Error in Mobile Robot, J of Software Engineering and Its Applications, 9, 41-48 (2015)

MATEC Web of Conferences **144**, 01009 (2018) https://doi.org/10.1051/matecconf/201814401009
RiMES 2017

Comparative structural vibration analysis of machinery and GFRP with Al7075

K. Vinoth Kumar[1*], *T.G. Loganathan*[1], *A. Bharath*[1], *B. Shyam* Sundar[1] and *K. K. Abishek*[1]

[1]*Dept. of Mechanical Engineering, RMK College of Engineering and Technology, Puduvoyal, Thiruvallur, India.*

Abstract. The current scenario of the industries is that the major losses in efficiency of a machine are due to vibration and friction. To reduce the detrimental effects of vibration we need to decrease the frequency and amplitude of vibration or completely eliminate vibration. To do that one must quantify vibration that already occurs in machinery and structural components. Which is the aim of this paper. The intention of the paper is to obtain and characterize the vibration signature of equipment used in a company and composite material. We have designed a setup to vibrational properties composites, vibrational signature of industrial equipment .To study vibration properties, micro-electrical mechanical systems (MEMS) based accelerometers are used to measure acceleration of the material about the datum when displaced. The data obtained is processed in MATLAB using ARDUINO relayed to computer to convert the data to frequency spectra using Fast-Fourier transforms (FFT). We ultimately compared the vibrational properties of two lathes used at a metal fabrication plant operating at different Conditions and quantified the vibration results using Fast Fourier Transforms (FFT) algorithm. The vibration signatures of a composite is studied along with which various properties like Damping Coefficient, Free Vibration, GFRP, Natural Frequency applications are studied.

1 Introduction

Vibration is one type of a mechanical property where oscillations occur with or without repeated motion about a equilibrium position. The oscillations can be periodic, non periodic and many other types. But, vibration is not desired because energy is lost through vibration and noise is created. The studies of sound and vibration are closely related. Sound, or pressure waves, is generated by vibrating structures (e.g. vocal cords); these pressure waves can also induce the vibration of structures (e.g. ear drum). Hence, attempts to reduce noise are often related to issues of vibration. Design improvements can be done to reduce vibrations. Vibration and friction reduce the efficiency of a machine. Some of the reasons for vibrations are Imbalance, misalignment, wear and looseness and many more. Composite materials are widely used from aerospace industry to toys, and in such circumstances it is imperative to study vibrational properties of composite materials. Modern composite material has become viable alternatives to conventional substances as they have high strength to weight ratios in addition to other properties which render them to have a high

* Corresponding author: skvinothpec@gmail.com

structural integrity and stability. Fiber Reinforced Plastics, an essential composite used in modern manufacturing, can be classified as bidirectional FRP's, unidirectional FRP's, quad axial FRP's, twill FRP's etc. During vibrations, the resonance amplitude is greatly influenced by damping at each mode. During design of composite material, some properties of importance to be considered are static and dynamic properties in addition to damping factor. In testing, vibration frequencies show progressively increasing normal probability distribution function with rising amplitude, and the curve can be classified as dual peak. nonlinear vibration analysis can be performed using a C^0 assumed strain interpolated finite element plate model based on Reddy's third order theory, it was found that nonlinear frequencies show increasing non-Gaussian probability density function with increasing amplitude of vibration and show dual peaks at high amplitude ratios.[1]MEMS accelerometer is used to monitor the vibrations of the machine and the self test features. Impulse response of the accelerometer is measured with the help of output pin using an oscilloscope.[2] Vibration sensor is used convert mechanical load to an electrical signal and to enhance vibration sensor performance.[3] In future, piezo-electric sensors, actuators and the like can be integrated into 'intelligent structures' to achieve positive vibration control.[4] The vibration analysis systems and wear debris analysis can be integrated into industrial machinery, allowing for 'intelligent maintenance' programs to be implemented. The 'intelligent maintenance' programs allow for planning and assigning maintenance tasks to achieve a lean maintenance function in the firm, reducing waste (muda)- waste such as waiting due to equipment failure, maintenance inventory etc. can be avoided.[5]

2 Vibration Measurement Setup

The components used to capture vibrational properties of composite materials and industrial machinery using micro electrical mechanical systems based sensors are Arduino open source microcontroller and accelerometer and a stand has been designed to hold the specimen. The composite material has been fabricated according to ASTM standards for composite material vibration testing. The design of the stand has been shown in the diagram. The base of the stand is fabricated from M/S steel sheet of thickness 8mm.The base table has a railing mechanism attached to the top to hold a plunger. A plunger has been used to apply a uniform displacement and hence cause forced vibrations on the composite materials. A plunger has been explicitly used to apply a constant load. The Arduino Integrated development environment and MATLAB, which is a numerical computing environment that is built around an easy scripting language, which makes MATLAB perfect for quick testing and data analysis.

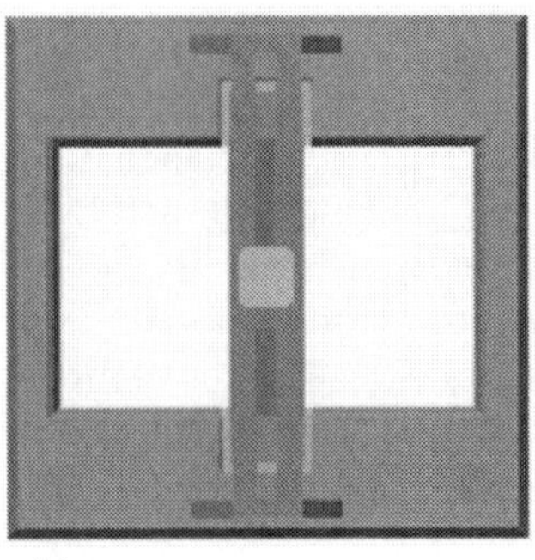

Fig. 1. Top view of the vibration machine

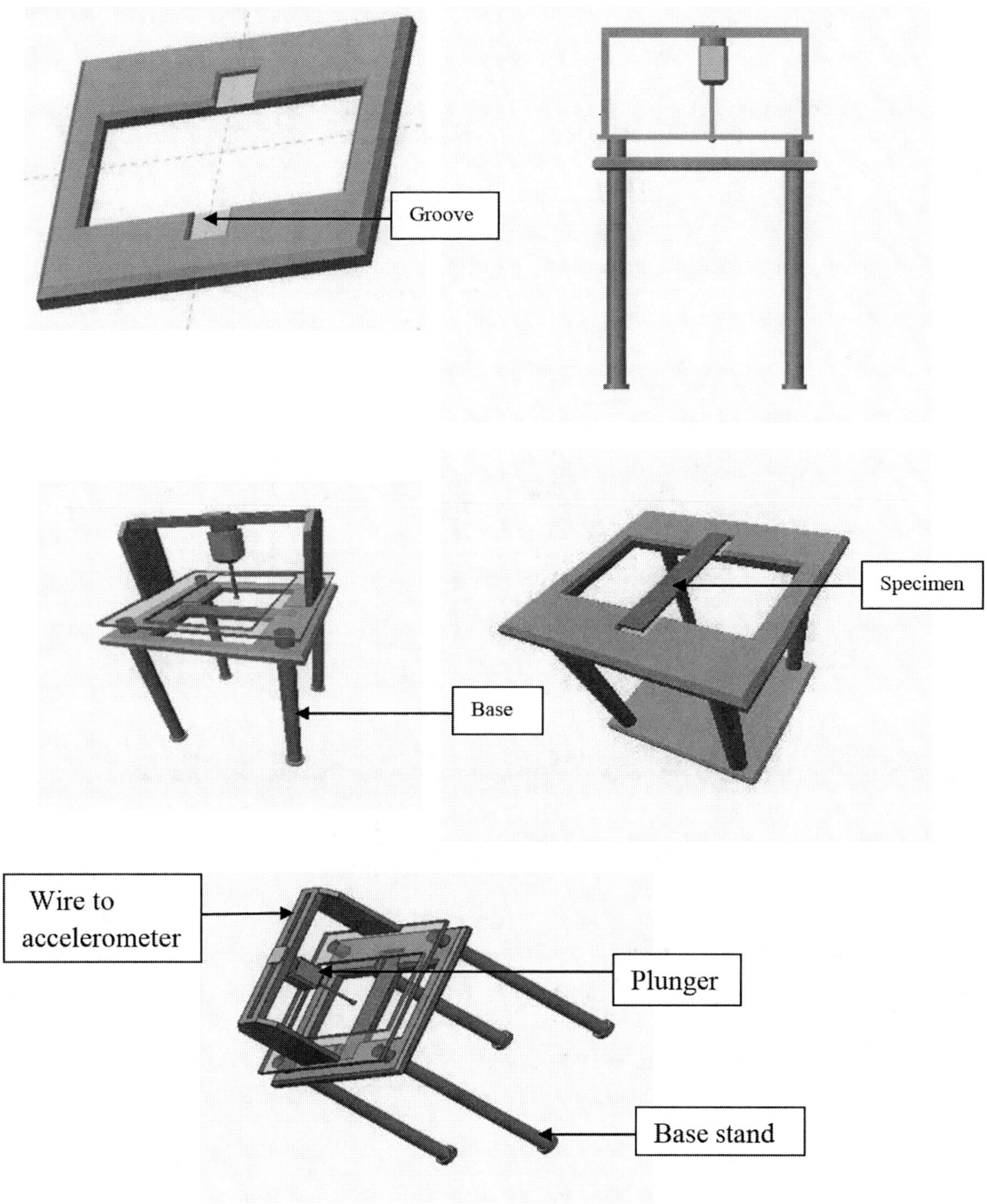

Fig. 2. (i) Specimen holding base with groove, (ii) Front view of vibration machine, (iii) 3-D view of setup, (iv) specimen placed in base, (v) different view of machine.

2.1 Design Calculation for plunger used

The force applied by the plunger is constant. Hence, the force applied has to be calculated. The dimension of diameter of the plunger tip is 5 mm. This is

$$pressure = \frac{force}{area} \gg 8\ bar = \frac{x}{0.000025} \gg x = 20\ newtons$$

Hence the plunger applies a load of 20 Newton force on the composite material. The plunger provides a displacement of 5mm on the composite material specimen.

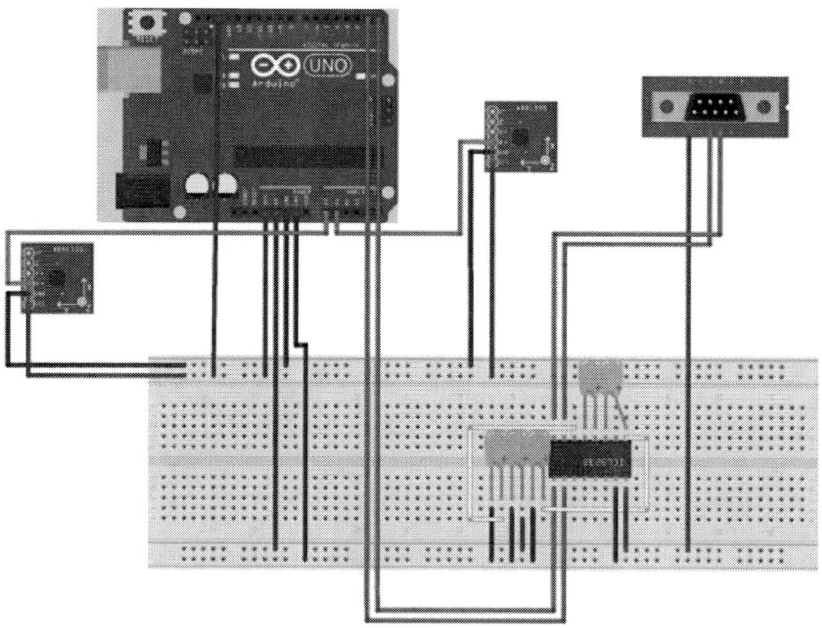

Fig. 3. Vibration Sensor Circuitry

2.2 Frequency Analysis

To test the vibration signatures of the machinery used in industries continuously to obtain a standard vibration signature. This standard vibration signature is used as reference. The copper block is attached to the chuck of the experimental lathe and the facing process is carried out at various test cutting speeds, at different feed and depth of cut; when the vibration signatures are captured by the vibration analysis systems. The so obtained vibration signatures are processed by MATLAB to perform Fast Fourier Transforms on the obtained data.

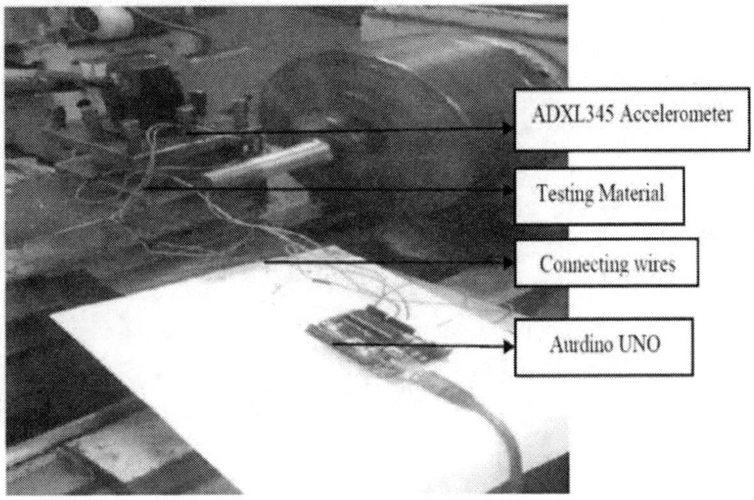

Fig. 4. Industrial Machinery Vibration Analysis

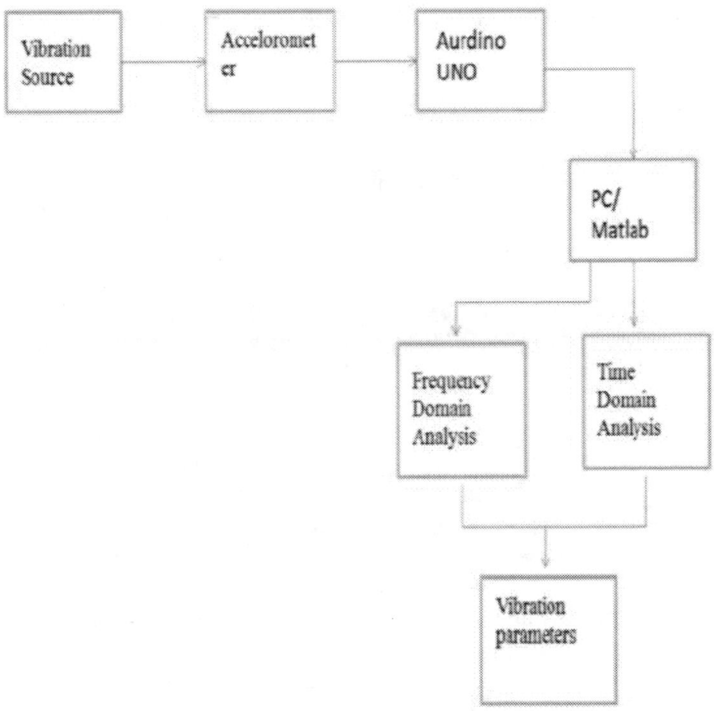

Fig. 5. Method to obtain the Frequency

The vibration analysis system is composed of accelerometers coupled with Arduino microprocessors to capture vibration signature and the data captured is analyzed by MATLAB software to record the data and serially input the data into the result file stored locally in the computer. The MATLAB program efficiently performs tune and frequency domain calculations; winch can be used for quantitative study and characterization of vibration. By using FFT, the data in time domain – in the form of a table with time stamp with the corresponding voltage output of the accelerometer – is converted to the frequency domain. The following MATLAB command is used:

$$V1\_fft*1 = abs(fft(V\_time) \quad (1)$$

The output of the above command is a matrix of voltage represented in the frequency Domain.

$$f = (0: N-1)/ (N*T) \quad (2)$$

Here, N represents the sampling rate in time domain.

3 Results and Discussions

The serial monitor program inbuilt in the Arduino package is used to capture and display the data output on the computer. This data is collected by the program called PUTTY to save the acceleration data of the vibrations in all three axis.

3.1 Vibration Analysis of Industrial Equipment

The sensor functionality to measure and quantize vibration of Computer Numeric Controlled (CNC) Lathe machine used at M/S Vasmo Technologies Pvt. Ltd. To obtain variation in results, brass and mild steel tools were used and the cutting speed has been altered during operation to obtain variety in results. The various cutting speeds selected were selected according to the operation manual of the lathe machines. While normal cutting speed for the tool piece was up to 1000 rpm the limit of the lathe machine was 800 rpm. The variety of 280,450 and 710 are preset speed on the machines were to act as minimum, moderate and maximum.

The vibration of equipment is studied for a period of five minutes and the collected data is formatted to produce fast Fourier transform, the end result of which is the frequency vs. amplitude graph and the amplitude vs. time graph. The time and frequency response of the vibration is also collected and presented.

The primary functionality performing fast Fourier transforms, plotting time waveforms etc. is to permit a trained engineer, researcher to perform the required vibration analysis in a rapid manner on the required specimen, be it a composite material on a piece of industrial machinery. For example, in industrial equipment has a specific vibration signature which will allow a technician to identify aberrant parts which might be failing or improperly configured, after which corrective action can be taken as appropriate.

3.1.1 Results Output

The output from the accelerometer placed gives vibration in terms of acceleration in all three axis. This data, whether for a composite material or for industrial equipment is processed in Microsoft Excel and then FFT transforms are applied. This is an example of data obtained from industrial equipment, which has been further been processed.

Table 1. Conversion of data from the accelerometer

Time Interval	Real Space A1	Real Space A2	Processed A1	Processed A2
1	-45	-119	45	119
2	-83	-142	83	142
3	-97	-180	97	180
4	-18	-111	18	111
5	-43	-124	43	124
6	-91	-147	91	147
7	-132	-136	132	136
8	-2	-80	2	80
9	-79	-101	79	101
10	-82	-99	82	99
11	-63	-89	63	89
12	-82	-109	82	109
13	-80	-124	80	124
14	-57	-138	57	138

3.1.2 Brass Tool time function, characteristics- 280 rpm

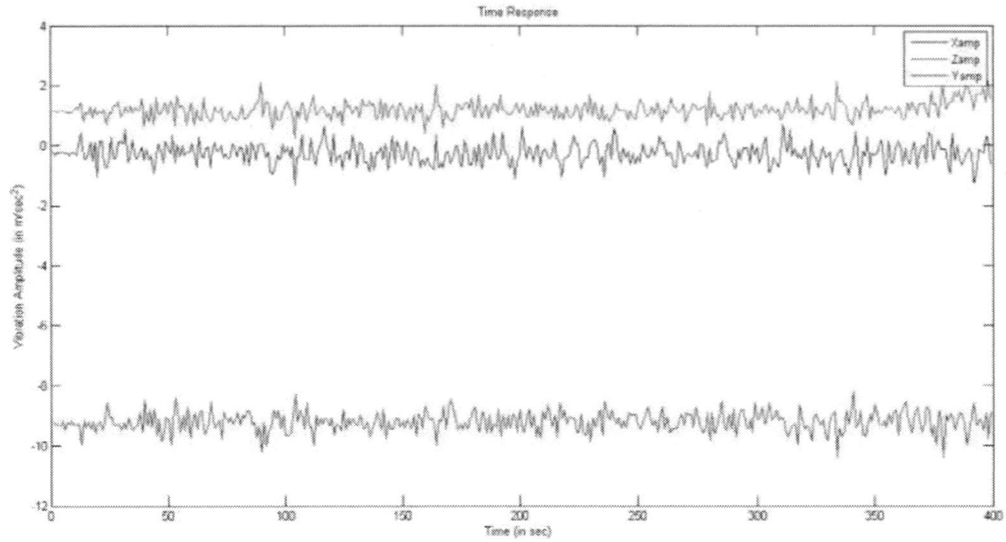

Fig. 6. Brass Tool time function, characteristics- 280rpm

3.1.3 Mild Steel Tool time function, characteristics- 280 rpm

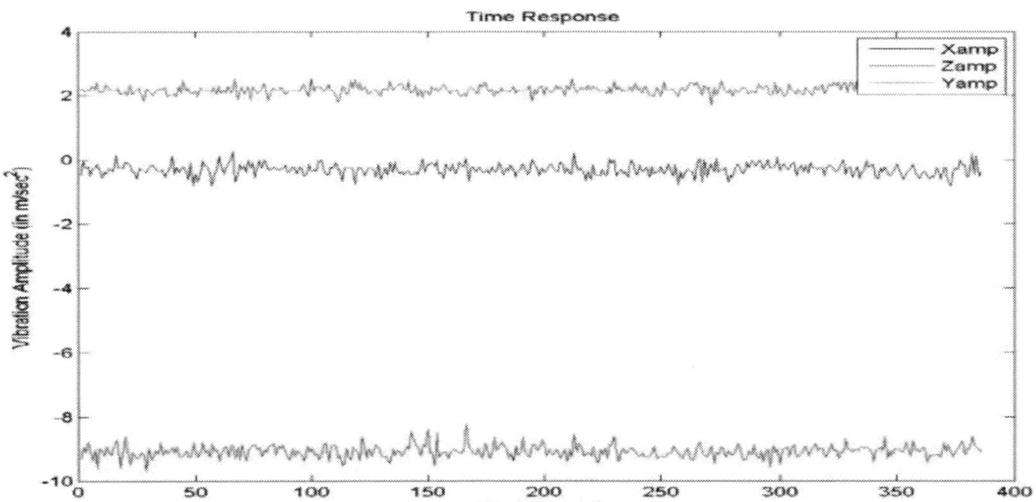

Fig. 7. Mild Steel Tool time function, characteristics- 280rpm

The amplitude of vibration and the frequency are cramped, indicating the proximity of operating equipment vibration with resonance frequency of the equipment. This increases stresses on the equipment and reduces the life of tool and the machinery. In addition to this, such vibrations cause flaw in work piece. Hence, at 280 rpm speed of operation mild steel is selected.

3.1.4 Brass Tool frequency function, characteristics- 450 rpm

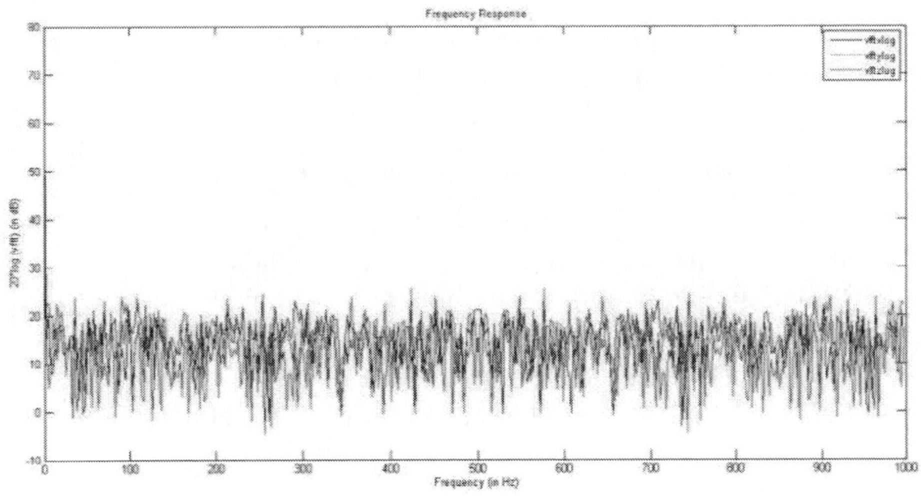

Fig. 8. Brass Tool frequency function, characteristics- 450 rpm

3.1.5 Mild Steel frequency function, characteristics- 450 rpm

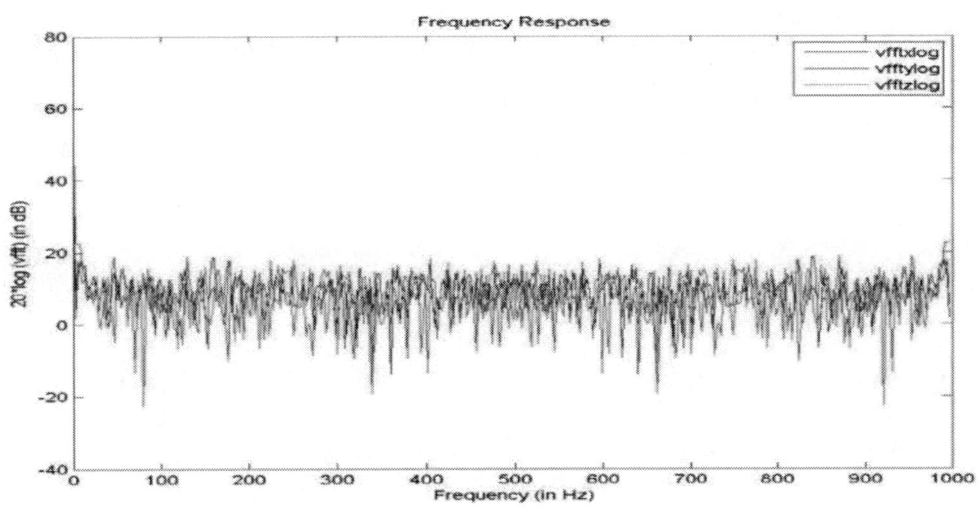

Fig. 9. Mild Steel frequency function, characteristics- 450rpm

The amplitude of vibration and the frequency are packed, indicating the closeness of operating equipment vibration with resonance frequency of the equipment. This increases stresses on the equipment and reduces the life of tool and the machinery. In addition to this, such vibrations cause flaw in work piece. Hence, at 450 rpm speed of operation mild steel is selected. However, this closeness in results can be due to the same material work piece being used.

3.1.6 Fast Fourier Transforms of Vibration in industrial Equipment:

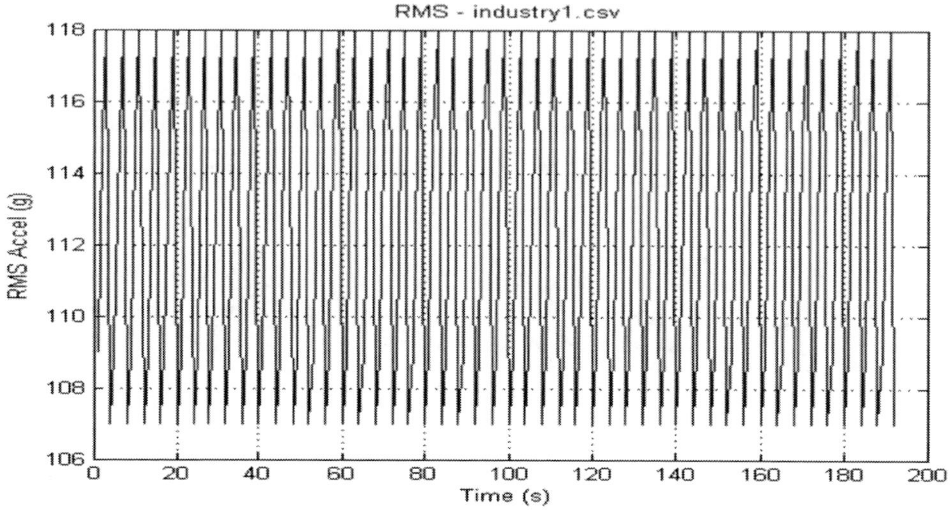

Fig. 10. Amplitude Time Graph of industrial Equipment.

The amplitude time graph shows that the equipment is operating in a stable condition with vibrations within prescribed limits. This graph can be used to detect if there is any fault in the equipment by analyzing outlying data. The above graph is calculated using MATLAB code. This data can be used to determine the limits of normal operation of the equipment and the maximum amplitude limits.

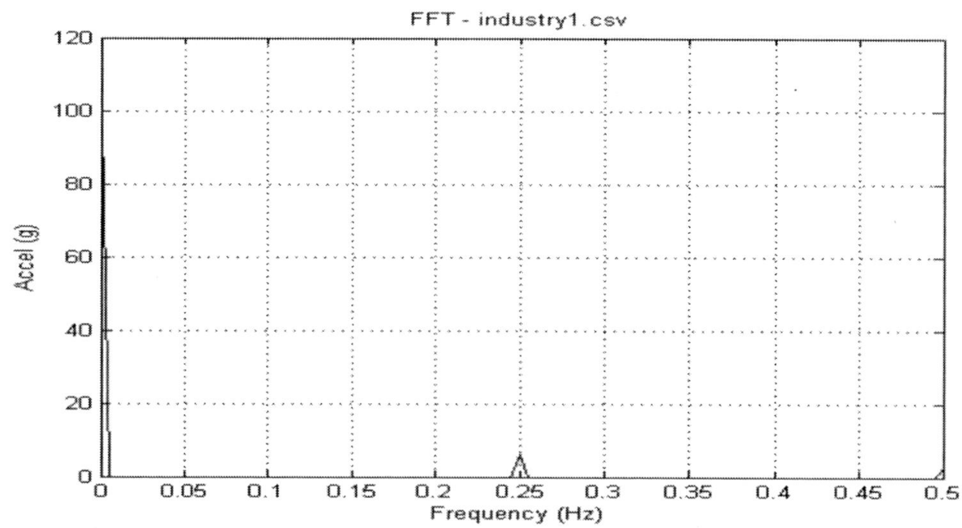

Fig. 11. Amplitude Frequency chart of the vibration of the industrial equipment.

The efficiency of fast Fourier transforms as an algorithm to calculate discrete Fourier transform in on display. The results of the fast Fourier transform in vibration analysis, which is frequency as a function of time to analyze the underlying cause of vibration. Another important advantage of FFT is the frequency resolution is not limited by computing power by the sampling rate of the equipment used and the duration for which the test is performed.

3.2 Vibration analysis of Glass fibre with Al7075 mesh reinforced with epoxy composite:

The composite material is fabricated according to ASME specifications. A uniform load is applied using a pneumatic plunger through a compressor. The composite material is mounted on the stand and then the load is applied.

The table 2 shows the raw data collected from the vibration produced form the composite material when a uniform load is applied.

Table 2. The raw data collected from the vibration

Time Interval	Axis 1
1	155
2	127
3	146
4	142
5	132
6	152
7	122
8	160
9	122
10	151
11	134
12	138
13	148
14	125
15	155

The above is the serial data obtained from the Arduino microprocessor. This data is then imported to the MATLAB software and then fast Fourier transforms and other such vibration analysis data is shows.

Failure analysis of composite material is more tedious compared to analysis of physical properties of composite materials as discussed so far. Failure mechanism of composite material, which is a function of temperature and load is difficult to simulate and is analogous to occurrences of micro fractures which have to be studied visually using an electron microscope. This vibration analysis is extremely important for fatigue testing of composite materials, to determine the maximum amplitude at which the failure of the composite material takes place.

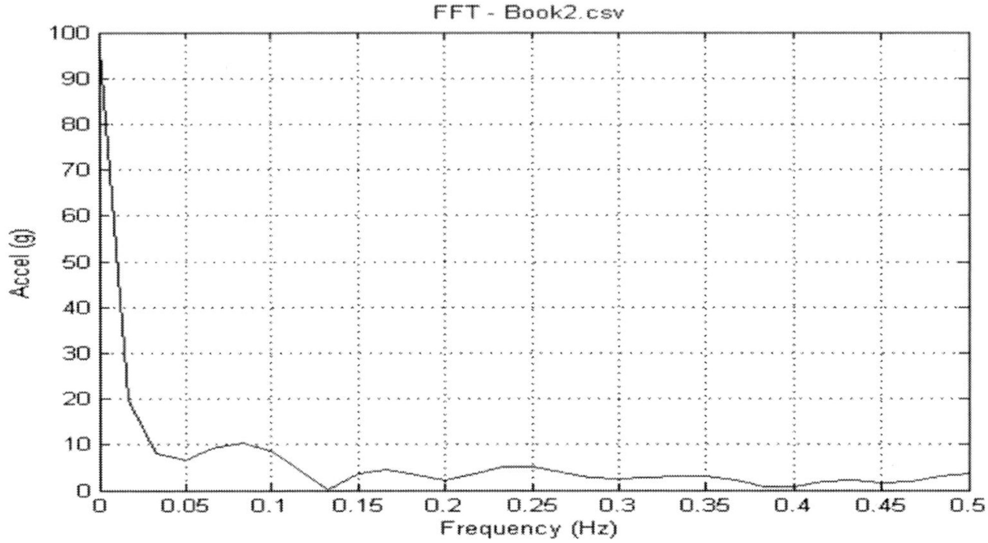

Fig. 12. FTT in Mat Lab Frequency vs. Acceleration

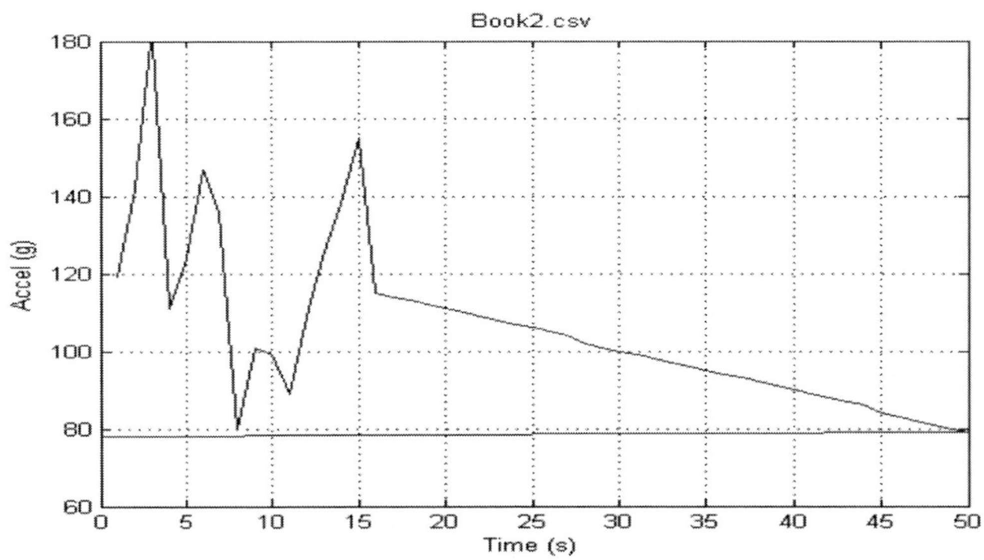

Fig. 13. FTT in Mat Lab Time vs. Acceleration

The above is the serial data obtained from the Arduino microprocessor. This data is then imported to the MATLAB software and then fast Fourier transforms and other such vibration analysis data is shows. This vibration analysis is extremely important for fatigue testing of composite materials, to determine the maximum amplitude at which the failure of the composite material takes place. The blue line shows the acceleration data corresponding to vibration about the vertical axis, i.e. the vibration about the y axis.

4 Conclusions

In this study we analyze the structural vibrational characteristics of brass, mild steel and GFRP with AL7075. The composite material is fabricated according to ASTM standards. The vibrational properties of composite materials and industrial machinery are captured using Arduino open source microcontroller and accelerometer and a stand has been designed to hold the specimen. The data is processed in MATLAB using ARDUINO relayed to computer to convert the data to frequency spectra using Fast-Fourier transforms (FFT).Vibrational properties of two lathes are compared at different conditions and quantified the vibration results using Fast Fourier Transforms (FFT) algorithm. In addition to which, the vibration signatures of a composite is studied. We also analyzed the vibration signature to identify the faults in industrial equipment.

References

1. M. Chandrashekhar, Ranjan Ganguli. Nonlinear vibration analysis of composite laminated and sandwich plates with random material properties. International, Volume 52, Issue 7, Pages 874-891. (July 2010)
2. Jonathan Young C. Vibration Analysis Using a MEMS Accelerometer. Thesis and dissertation
3. Zani Ghemari, Salah Saad. Development of Model and Enhancement of Measurement Precision of Sensor Vibration. IEEE Sensors Journal. Volume 12, Issue 12, Pages 3454 – 3459. (Dec. 2012)
4. GJ Sheu , SM Yang, W.L. Huang. Simulating Displacement and Velocity Signals by Piezoelectric Sensor in Vibration Control Applications. Smart Materials Research Volume 2012, Article ID 390873. (2012)
5. R.C.M. Yamp.W. Tsel. Lip. Tu,Intelligent Predictive Decision Support System for Condition-Based Maintenance Technology. The International Journal of Advanced Manufacturing Technology, Volume 17, Issue 5, Pages 383–391. (Feb.2001)
6. S Ebersbach, Z Peng, N J Kessissoglou .The investigation of the condition and faults of a spur gearbox using vibration and wear debris analysis techniques. WEAR, Volume 260, Issues 1–2, Pages 16-24.(Jan. 2006)
7. Han B, Sun S, Ding S, Zhang L, Yu X, Ou J. Review of nanocarbon-engineered multifunctional cementitious composites. Composites. Part A: Applied Science and Manufacturing. Volume 70, Pages 69-81. (Mar. 2015)
8. Abu Al-Rub RK, Ashour AI, Tyson BM. On the aspect ratio effect of multi-walled carbon nanotube reinforcements on the mechanical properties of cementitious nano composites. Construction and Building Materials. Volume 35, Pages 647-655. (Oct. 2012)
9. Konsta-Gdoutos MS, Aza CA. Self-sensing carbon nanotube (CNT) and nanofiber (CNF) cementitious composites for real time damage assessment in smart structures. Cement and Concrete Composites. Volume 53, Pages 162-169. (Oct. 2014)
10. Parveen S, Rana S, Fangueiro R.A Review on Nanomaterial Dispersion, Microstructure, and Mechanical Properties of Carbon Nanotube and Nanofiber

Reinforced Cementitious Composites. Journal of Nanomaterials. Volume 2013, Article ID 710175 (July. 2013)

Finite element simulation of low velocity impact loading on a sandwich composite

Vishwas M.[1*], *Sharnappa Joladarashi*[1] *and Satyabodh M. Kulkarni*[1]

[1]*Department of Mechanical Engineering, National Institute of Technology Karnataka, Surathkal 575025, Karnataka, India*

Abstract. Sandwich structure offer more advantage in bringing flexural stiffness and energy absorption capabilities in the application of automobile and aerospace components. This paper presents comparison study and analysis of two types of composite sandwich structures, one having Jute Epoxy skins with rubber core and the other having Glass Epoxy skins with rubber core subjected to low velocity normal impact loading. The behaviour of sandwich structure with various parameters such as energy absorption, peak load developed, deformation and von Mises stress and strain, are analyzed using commercially available analysis software. The results confirm that sandwich composite with jute epoxy skin absorbs approximately 20% more energy than glass epoxy skin. The contact force developed in jute epoxy skin is approximately 2.3 times less when compared to glass epoxy skin. von Mises stress developed is less in case of jute epoxy. The sandwich with jute epoxy skin deforms approximately 1.6 times more than that of same geometry of sandwich with glass epoxy skin. Thus exhibiting its elastic nature and making it potential candidate for low velocity impact application.

Keywords: Sandwich composite, Jute epoxy, Glass epoxy, Low velocity impact

1 Introduction

Composites materials are generally utilized as a part of designing applications, particularly in aeronautics in perspective of their light weight, high strength to weight ratio, stiffness, thermal and corrosion resistance. In any case, outline of composites is a great deal more perplexing contrasted with conventional materials because of their anisotropic behavior. Failure of composites are also complex and can happen because of splitting of fibers being part of whole, part/splitting of matrix, or fiber matrix de bonding or delamination at their interface and may also occur due to various other mechanisms of failure. Hence it needs cautious outline and investigation to be routed to precisely represent basic conduct, particularly when sandwich reaction may fundamentally vary with geometry, material, orientation, stalking sequence or loading condition of composite materials. A computational and demonstrating approach to be utilized for profitable research and plan for composites is particularly fundamental for such investigation by structural designers [1]. Sandwich

* Corresponding author: vishwasmahesh@gmail.com

structures are being utilized as a part of a few sections of aviation, and marine structures. For the most part these structures are subjected to effects, for example, dropping of the instrument during maintenance, strike due to solid precipitation or bird, and runway debris. Despite the fact that the sheets made of metal has higher strength than FRPs they possess large weight which results in increased cost [2]. Sandwich structures are obtained when two thin skins are appended to a core which is lighter in weight and thicker. These materials have high sp. strength and stiffness against bending [3]. Including transportation vehicles and packaging, sandwich composites are used as a part of basic/semi basic applications. Due to the geometrical advantages provided by sandwiches, they are broadly utilized as vitality retaining frameworks subjected to impact loads. During normal working, sandwich structures may experience low-speed impacts from shots being exceptionally sensible to such loads. In spite of extensive research on sandwich structures, their impact behavior is as yet not completely comprehended [4-5]. Lightweight composite sandwich boards, including fiber strengthened polymer confront sheets and polymeric core, are more generally utilized as a part of vehicles used in defense since their ability to bear the load per unit weight is high and less maintenance. In a few occasions, due to the flying debris, sandwiches may come under the influence of low velocity impact load [6]. PMCs are used widely in sandwiches as they delay the shot as it diminishes its KE [7]. Projectile impact contemplates on PMCs and a sandwich board has predominantly concentrated in the low velocity impact regime in light of its relationship with BVID [8-10]. Based on available literature, it was found that there are no studies available on sandwich composites having composite facings (Jute) with natural rubber core subjected to impact loading under low velocity. The investigations are carried out on impact parameters like Total Energy, contact force, deformation, von Mises stress and strain under low velocity impact condition and how they affect the performance of the components manufactured. The scope of the present work is limited to low velocity impact behavior of sandwich composite using commercially available software. Results pertaining to sandwich with jute epoxy and glass epoxy composite facing with rubber core are compared and an attempt has been made to explore the possibility of using them for low velocity impact applications.

2 Methodology

The sandwich model considered for the analysis is shown in fig1. The skin is modeled with a dimension of 150 mm × 150 mm × 6 mm and core is of thickness 30 mm which is five times that of skin thickness. Fig 2 shows the meshing of sandwich structure. The auto mesh option is selected to mesh the sandwich and tup. The type of mesh used is Quad with 24,696 nodes and 35,913 elements. The hemispherical tup impactor of radius 6.5 mm is considered to be made up of steel [11]. The explicit dynamics analysis type is selected to perform the low velocity impact test on sandwich structure. The boundary condition applied for the sandwich structure is fixed support on the edges of the sandwich structure as well as on the four side faces and velocity of 20m/s for the tup. The sandwich structure is defined as flexible material and tup as rigid material. Table 1 gives the properties of materials: glass epoxy and jute epoxy used during analysis [11-15]

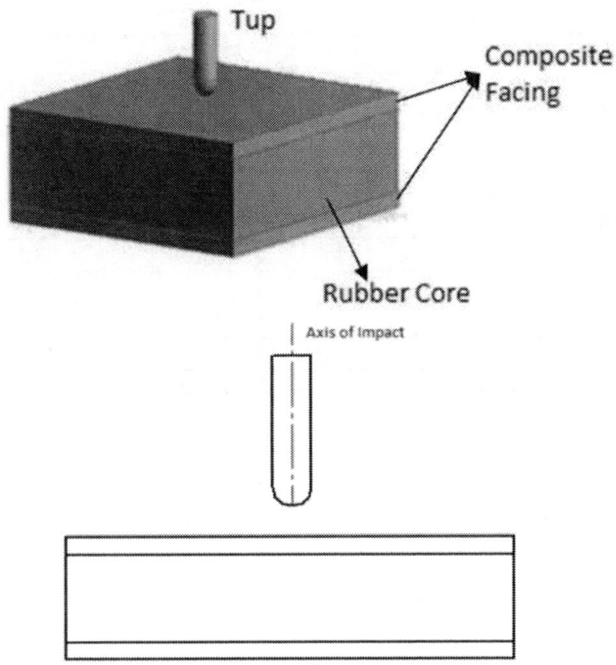

Fig. 1. Sandwich Composite subjected to Normal Impact

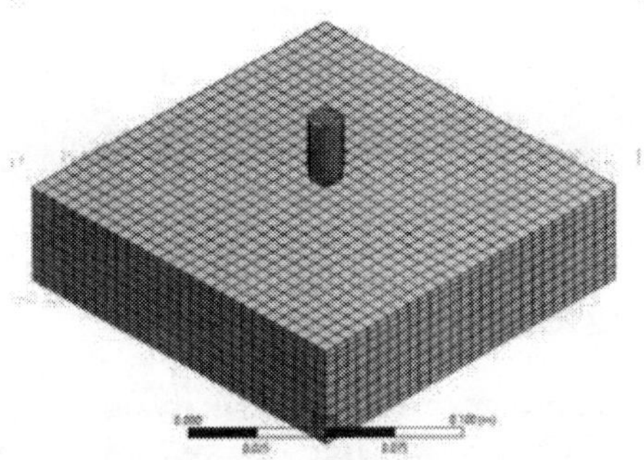

Fig. 2. Meshing of Sandwich Structure

Table 1. Properties of Materials: Glass Epoxy (GE) and Jute Epoxy (JE)

Parameter	**GE**	**JE**
Youngs Modulus (MPa)		
$E_{11} = E_{22}$	36,300	4,500
E_{33}	36,100	3,200
Shear Modulus (MPa)		
G_{12}	44,700	1,450
$G_{23} = G_{13}$	46,800	1,630
Poisson's ratio		
μ_{12}	0.173	0.24
$\mu_{13} = \mu_{23}$	0.279	0.27
Density (kg/m^3)	1,750	1,165
Tensile strength (MPa)		
$X_T = Y_T$	330	104
Z_T	35	11
Shear strength (MPa)		
$S_{12} = S_{13} = S_{23}$	35	23
Compressive strength (MPa)		
$X_C = Y_C$	320	95
Z_C	500	102

Table 2 gives the material property of hemispherical tup (structural steel) and core (natural rubber).

Table 2. Material Properties of Structural Steel and Rubber

	Density (kg/m^3)	**Modulus of Elasticity E (MPa)**	**Poisson's Ratio**	**Bulk Modulus (MPa)**	**Shear Modulus (MPa)**
Structural Steel	7,850	2,00,000	0.3	1,66,600	76.9×10^9
Rubber	1,000	1,00,000	0.5	1,09,000	300×10^3

3. Results and Discussion

Table 3 summarizes the results pertaining to impact analysis of sandwich composites with Jute Epoxy (top face sheet)-Rubber (Core) –Jute Epoxy (bottom face sheet)(JE-R-JE) and Glass Epoxy (top face sheet)-Rubber (Core) –Glass Epoxy (bottom face sheet)(GE-R-GE)

Table 3. Summary of Results

Sandwich Structure Specification	**Total Energy (mJ)**	**Contact Force (N)**	**Deformation (mm)**	**von Mises Strain**	**von Mises Stress (MPa)**
JE-R-JE	7,462.3	184.41	13.206	0.415	2.636
GE-R-GE	6,258.3	419.33	8.3085	0.101	17.73

3.1 Contact Force

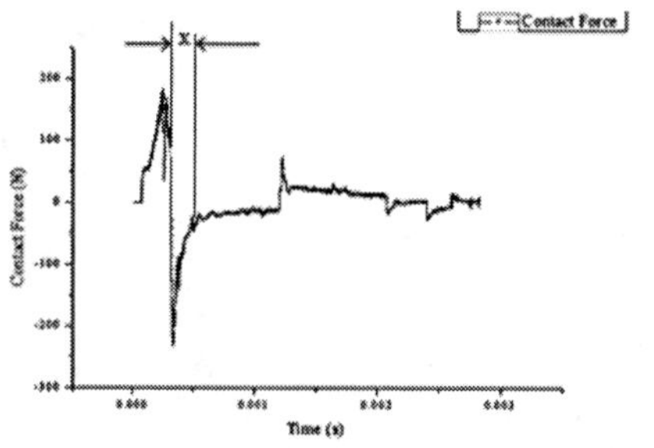

Fig. 3(a). Variation of contact force against tine for JE-R-JE sandwich

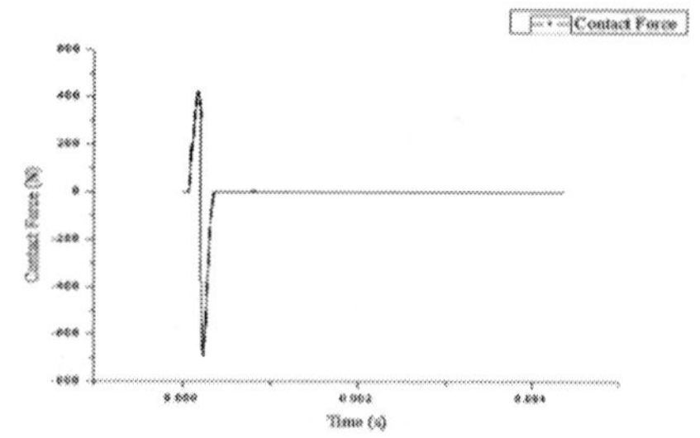

Fig. 3(b). Variation of contact force against tine for GE-R-GE sandwich

Variation of Contact force against time for JE-R-JE and GE-R-GE sandwich is shown in fig 3a and fig 3b respectively. The duration for which the impactor is in contact with sandwich is studied from the graph. Due to different types of fibers used in face sheet, a distinct period of contact with the impactor could be observed. Till the point of initiation of damage or peak load, the variation of force with time is linear. The point where the failure is just initiated on the graph is referred to as the maximum load carrying ability. Siowet al. [16] called it the incipient point of damage which is usually a matrix failure. The extent of damage is very little or no visible damage. Due to this there is a drop in the magnitude of force showing reduction in stiffness of material. The complete failure occurs at the second highest peak where the damage is due to fiber fracture, breakage or delamination and it is visible. This was described as classical mode of failure in composites by Banks et al. [17]. Penetration and perforation damages are the results of combination of such failures. The fig 3(a) shows contact force v/s time curve for jute epoxy skin and rubber core sandwich composite. The max value of force at contact is 184.41 N at 0.000248 s and min is -230.81 N at 0.00033256 s. For normal components of force and velocity, the Hertzian theory of impact is used where it is assumed that sticking and slipping regions are comprised at the contact area. At the beginning of impact, sliding could be generally observed. In order to reduce this, an opposing frictional force will be established. Sliding will allow rolling to occur if the impact velocity becomes zero which further makes frictional force to become zero. The impact will end once rolling is established.

Forces reverse its direction during impact loading which could be seen from the fig 3a and fig 3b. This is agreement with the continuum model established by Maw et al [18]. Full sliding may begin at some point toward the end of impact. At some point during the impact, coincident points on some central portion of the contact zone begin to stick together, while coincident points on the outer annulus of the contact zone may still have some relative slipping. The tangential force leaves the limiting friction envelope and reverses direction. Reverse sliding of the coincident points occurs toward the end of impact, with the tangential force equal to the negative limiting friction value, and persists until contact is lost.

'X' in the graph shows the time taken by the sandwich structure to stop the tup. Fig 3(b) shows contact force v/s time curve for glass epoxy skin and rubber core sandwich composite. The max value of force at contact is 419.33 N and min is at -691.05 N. The descending part of unloading is due to continuous loading beyond the peak point where there is a continuous progression of damage to the plate and thus reduction in the contact force. Therefore the major mode of failure for this impact loading scenario is due bending stress.

3.2 Total Energy

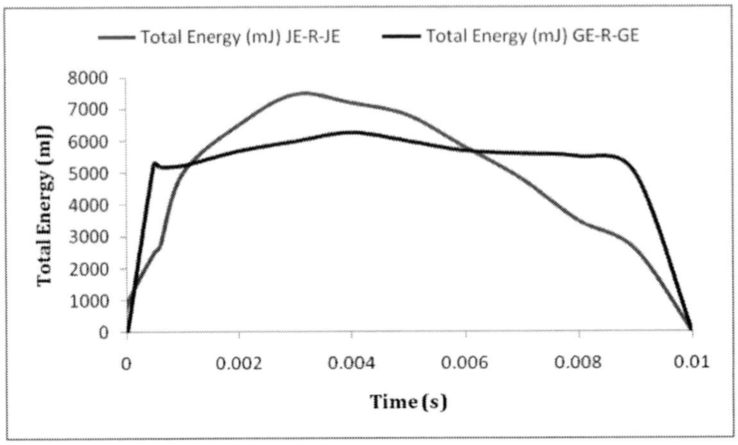

Fig. 4. Comparison of variation of total energy against time for JE-R-JE and GE-R-GE

Fig 4 shows the total energy v/s time graph for JE-R-JE and GE-R-GE. The total energy for JE-R-JE and GE-R-GE is maximum at 7462.3 mJ and 6258.3 mJ respectively. The energy absorbed increases gradually over a period of time and then attains zero at the end of impact event.

In case of normal impact loading, the plate gets displaced by the impact of the tup and velocity is constant which refers to the unchanging rate of displacement of an object in a certain period of time due to which the energy absorbed varies linearly. Depending on the variation of velocity of tup after coming in contact with the plate, the energy absorbed varies.

Energy absorption with time increases gradually. When the energy absorbed by the sandwich is small, the impactor bounces back as well as the damage area is also small. This indicates that the impact force can be reduced by increasing the impact time. Elastic deformation and failure mechanisms absorb the energy transferred from object to sandwich. In the graph, the region of linear profile represents the energy absorbed by elastic deformation and anything beyond this point by failure mechanisms.

3.3 Total Deformation

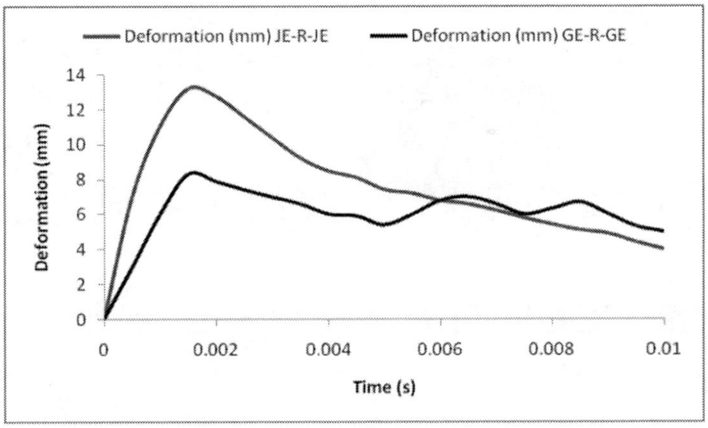

Fig. 5. Comparison of Deformation v/s Time for JE-R-JE and GE-R-GE

Fig 5 shows the deformation v/s time graph for JE-R-JE and GE-R-GE. Due to the impact at the velocity 20 m/s, JE-R-JE sandwich gets deformed to maximum extent of 13.206 mm on the top skin and GE-R-GE sandwich gets deformed by 8.3085 mm at the centre of the top skin and minimum deformation at the edges as the four side faces of sandwich are constrained. The maximum deflections of sandwich composite occur when the impact force becomes equal to zero. During the impact event, the travelling of the impacted surface is indicated by the displacement. Since, drop height of the tup is same in both cases, the amount of energy it delivered on the sandwich will be same [20]. The amount of energy dissipated by the sandwich is a function of the load and displacement of each sandwich, which means that the sandwich with the highest load resistance had the least displacement.

3.4 Equivalent Elastic Strain and Stress

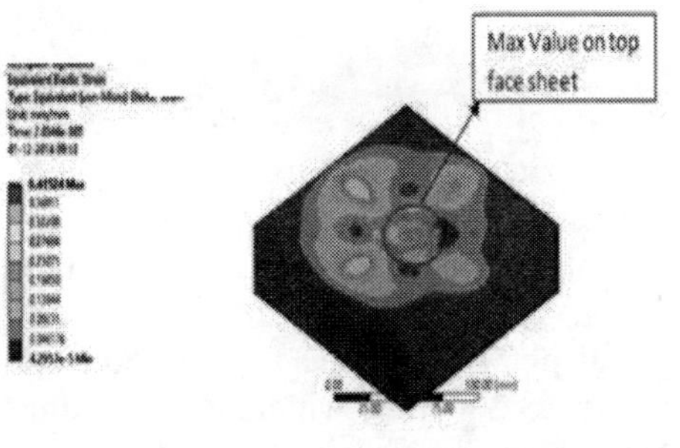

Fig. 6(a). Equivalent elastic strain: JE-R-JE

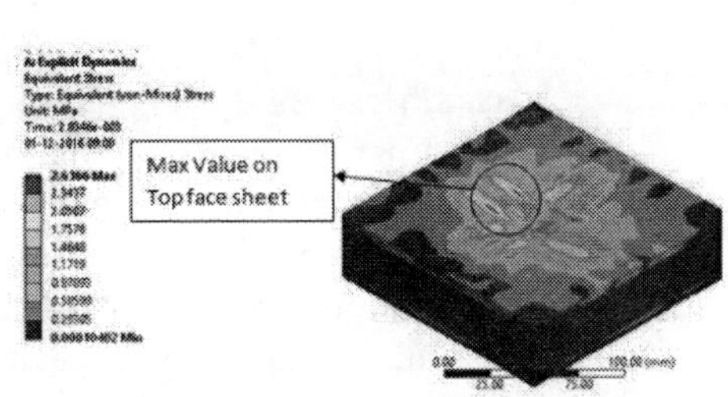

Fig. 6(b). Equivalent elastic stress: JE-R-JE

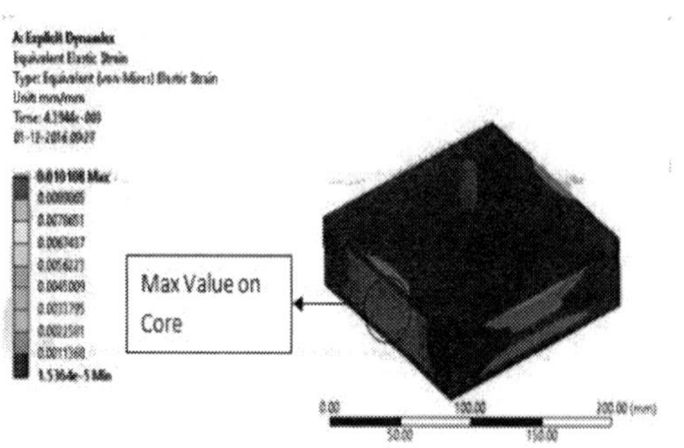

Fig 6(c). Equivalent elastic strain: GE-R-GE

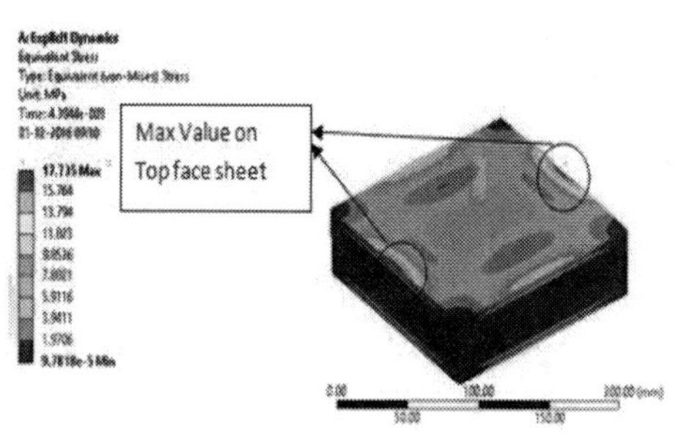

Fig 6(d). Equivalent elastic stress: GE-R-GE

Fig 6 (a-d) shows the sandwich is subjected to stress and strain due to the impact by the tup at the low velocity of 20 m/s. The equivalent elastic strain/von Mises strain obtained in case of JE-R-JE and GE -R-GE are 0.415 on top face sheet and 0.101 on core respectively. At the same time the von Mises stress obtained in case of JE-R-JE and GE-R-GE are 2.636 MPa and 17.73 MPa on top face sheet respectively. It can be found that the von Mises strain in case of JE-R-JE is more when compared with GE-R-GE which is due to its maximum deforming ability thereby absorbing more energy. Also the stress developed in JE-R-JE sandwich composite is less that of developed in GE-R-GE sandwich composite.

4 Conclusions

Low velocity impact behavior of two different sandwich composites (JE-R-JE and GE-R-GE) are investigated by using FE analysis. Some of the conclusions obtained from this study are as follows:

- Contact Force: The contact force developed in JE-R-JE sandwich is 2.3 times less when compared with GE-R-GE sandwich which shows that jute is less brittle than glass fiber.
- Total Energy: JE-R-JE sandwich absorbs 1.2 times more energy than GE-R-GE sandwich under low velocity normal impact loading. This exhibits the ability of JE-R-JE sandwich to absorb impact loading in low velocity regime.
- Deformation: The amount of energy dissipated by the sandwich is a function of the load and displacement of each sandwich, which means that the sandwich with the highest load resistance had the least displacement. The deformation of JE-R-JE sandwich under low velocity impact is 13.206 mm which is approximately 1.6 times more than that of GE-R-GE sandwich. This means that GE-R-GE sandwich has highest load resistance compared to JE-R-JE sandwich.
- Equivalent elastic strain and stress: von Mises strain in case of JE-R-JE is more (approximately 4.1 times) when compared with GE-R-GE which is due to its maximum deforming ability thereby absorbing more energy. Also the stress developed in JE-R-JE sandwich is less (approximately 6.7 times) that of developed in GE-R-GE sandwich.

The consequences of this examination shows that for low velocity impact application jute-epoxy has better impact resistance properties than glass epoxy making them suitable for structural/ semi structural applications. The use of natural rubber as core material enhances the energy absorbing ability of the sandwich structure. In spite of the fact that the composites have some benefits and disadvantages, the blend of the helpful properties of two unique materials, faster handling time, and lower producing cost make them as a flexible material in the field of designing and innovation.

References

1. Dinh Chi Pham and Sridhar Narayanaswamy, "An Optimization Tool for Impact Analysis of Composite Structures", Proceedia Eng., 75, pp3 – 8 (2014).
2. M. Sadighi and M. Saadati, "Unequal faces effect on fracture of composite sandwich beam with flexible core," Appl. Compos. Mater, 17(2), pp. 111–120 (2010).
3. Vinson, J.R. "The Behavior of Sandwich Structures of Isotropic and Composite Materials", Technomic Publishing, Lancaster, PA (1999)
4. Aktay, L., Johnson, A.F. and Holzapfel, M. 'Prediction of impact damage on sandwich composite panels', Comput. Mater.Sci., 32 (3–4), pp.252–260 (2005).
5. Brenda, L.B., Carlos, S., Sonia, S., Enrique, B. and Carlos, N. 'Modelling of composites and sandwich structures with honeycomb core subjected to high-velocity impact', Compos. Struct., 92(9), pp.2090–2096 (2010).
6. HooFatt, M.S. and Sirivolu, D. 'A wave propagation model for the high velocity impact response of a composite sandwich panel', Int. J. Impact Eng, 37(2), pp.117–130 (2010).
7. Morye, S.S., Hine, P.J., Duckett, R.A., Carr, D.J. and Ward, I.M. 'Modelling of the energy absorption by polymer composites upon ballistic impact', Compos. Sci. Technol, 60(14), pp.2631–2642 (2000).
8. Edgren, F., Soutis, C. and Asp, L.E. 'Damage tolerance analysis of NCF composites and sandwich panels', Comp. Sci. Tech., 68 (13), pp.2635–2645, (2008)
9. Karger, L., Baaran, J. and Tessmer, J. 'Rapid simulation of impacts on composite sandwich panels inducing barely visible damage', Compos. Struct, 79 (4), pp.527–534 (2007).
10. Meo, M., Vignjevic, R. and Marengo, G. 'The response of honeycomb sandwich panels under low-velocity impact loading', Int. J. Mech. Sci., 47 (9), pp.1301–1325 (2005)
11. Balasubramani V, Rajendra Boopathy S and Vasudevan R, "Numerical Analysis of Low Velocity Impact on Laminated Composite Plates", Procedia Eng., 64, pp.1089-1098 (2013)
12. Stuart M Lee, "Handbook of composite reinforcement", Wiley publications, (1992).
13. A. Mir, C. Aribi, B. Bezzazi, "Study of the Green Composite Jute/Epoxy", Int. J. Chem., Mol., Nuc., Mat. Met. Engg., 8 (2), (2014).
14. D Tabor, "The bulk modulus of rubber", Journal of Polymer, 35(13),pp. 2759-2763 (1994)
15. M. R. Hossain , M. A. Islam , A. V. Vuurea, and I. Verpoest, "Effect of Fiber Orientation on the Tensile Properties of Jute Epoxy Laminated Composite", J. Sci. Res. 5 (1), pp. 43-54 (2013).
16. Siow YP, Shim VPW. "An experimental study of low velocity impact damage in woven fibre composites" J.Comp. Mat., 32, (1998)
17. Banks WM, David-West OS, Alexander NV, Nash DH. Energy absorption and bending stiffness in CFRP laminates: The effect of 451 plies. Thin-walled Structures, 46, (2008)
18. Maw, N., Barber, J. R., and Fawcett, J. N., "The Oblique Impact of Elastic Spheres," Wear, 381, pp. 101–114, (1976)
19. Philip P. Garland and Robert J. Rogers, "An Experimental Study of Contact Forces during Oblique Elastic Impact" J. Appl. Mech., 76, (2009)

20. Alex M. Remennikov, Sih Ying Kong and Brian Uy, M.ASCE, "Response of Foam- and Concrete-Filled Square Steel Tubes under Low-Velocity Impact Loading" J. Perform. Constr. Fac, 373-381, (2011).

A Mathematical Model to Estimate the Position of Mobile Robot by Sensing Caster Wheel Motion

Jnana H. Amarendra[1], *Robins Mathew*[2] and *Somashekhar S. Hiremath*[2*]

[1] *Department of Electronics and Communication Engineering, Malnad College of Engineering, Hassan-573201, India*
[2] *Department of Mechanical Engineering, Indian Institute of Technology Madras, Chennai-600036, India*

Abstract. This paper describes the position estimation of mobile robot by sensing caster wheel motion. A mathematical model is developed to determine the position of mobile robot by sensing the angular velocity and heading angle of the caster wheel. Using the established equations, simulations were carried out using MATLAB version 8.6 to observe and verify the position coordinates of mobile robot and in turn obtain its trajectory. The simulation results show that the angular velocity of caster wheel and heading angle calculated from the sensor output readings with the help of inverse kinematics equations matches well with that of actual values given as input for simulation. Simulation result of tracking rectangular trajectory implies that the path traced by the mobile robot can also be determined from the sensor output readings. This concept can be implemented on a real mobile robot for estimation of its position.

Keywords: mobile robot, caster wheel, position estimation, kinematics, simulation.

1 Introduction

Creation of mobile robots has brought about prominent changes in various domains. In fact, with time mobile robots have substituted humans. They perform repetitive and dangerous tasks which humans prefer not to do or are unable to do because of inappropriate circumstances [1, 2]. Evolution in the field of robotics has made tougher tasks much simpler with appreciable efficiency and precision [3]. Mobile robots are capable of navigating an uncontrolled environment without the requirement of physical or electro-mechanical guidance devices. They are a major focus of current research and find their importance in industrial, military, medical and security related areas [4, 5].

On considering their importance in real time applications, estimating their position can prove to be advantageous in many sectors. Valet parking without the aid of a valet is one

* Corresponding author: somashekhar@iitm.ac.in

such example. The car is transported to the desired position and orientation with the aid of mobile robots and laser range scanners [6, 7]. With the current trend of automation in manufacturing technologies popularly known as Industry 4.0, autonomous mobile robots play a vital role. They provide flexible, efficient and integrated production by virtue of their locomotion feature [8]. In the search and rescue operation, mobile robots are operated to search victims on disaster scenes. These situations demand the implementation of teleoperated control and autonomous path following features on mobile robots for autonomous navigation [9]. Such autonomous mobile robots are not guided by physical path indicators and they are capable of dynamic route planning.

An appropriate hardware set up is required to sense the two dimensional motion of wheeled mobile robot to estimate its position. In this regard several techniques have been developed by many researchers. To mention a few, the Kalman Filter (KF) and the Extended Kalman Filter (EKF) was utilised for localisation of autonomous mobile robot based on Ackermann steering. It also states that the EKF is consistently better than the KF in estimating the robot position [10]. However, this technique lacks to represent ambiguities and in case of localisation failures, the ability to globally re-localize the robot is also absent. Another study involves installing the optical mouse sensors on mobile robots to detect its motion relative to the ground [11]. With the use of optical sensors, the position measurements are independent of kinematics of the mobile robot. Even though the optical sensors are cost effective and have high resolution sensing capability, they can be affected by errors. Placement and orientation of optical sensors on mobile robots must be accurate. In case of uneven surfaces, there are chances that the estimated position of robot is erroneous. In another research work, the Dynamic Markov Localization (DML) technique is applied to globally estimate the position of robot. This technique efficiently tracks its position and also recovers from localisation failures. The two approaches for localisation are local localisation and global localisation techniques. They can be distinguished based on the kind of localization problem they are designed to resolve. DML method combines the advantages of both the localisation approach for keeping track of the robot's position and global position estimation. This method uses probabilistic technique to achieve the efficiency and also determine when the robot lost track of its position [12]. Mobile robots can locate themselves and build three-dimensional model of their surrounding using scale-invariant visual landmarks. Simultaneous Localization And Map Building (SLAMB) is an important factor for successful navigation of mobile robots in this regard. Mobile robots can locate themselves without prior knowledge of their initial position with the help of vision based SLAMB algorithm [13]. The mobile robots serve their purpose only when they are capable enough to safely navigate in populated and dynamic environments. Extensions of Markov localization algorithm can be employed in position estimation of mobile robots in dynamic environmental conditions [14]. This simple mechanism can be applied so that they can operate autonomously in densely populated environments.

In the current work, a new approach is carried out for estimating the position of mobile robots by sensing caster wheel motion. Caster wheels unlike standard wheels are usually a part of mobile robots which are not driven. They are freely turning wheels having two degree of freedom and can rotate. Generally, these wheels are used for balancing the mobile robots. Since these wheels are already a built in part of wheeled mobile robots, this approach requires less hardware installation on mobile robots. The concept behind the mechanism of a mechanical computer mouse can be considered in this context. It is observed that with proper hardware arrangement, this concept can be applied to estimate the position of mobile robot by sensing the castor wheel motion. A mechanical computer mouse consists of a track ball with two freely rotating rollers, located 90^0 apart. The rotating rollers help in detection of forward–backward motion and left–right motion of the mouse. Likewise, the caster wheel is kept in contact with two perpendicularly arranged identical shafts connected to two sensors to sense caster wheel motion. Any translational motion causes movement in caster wheels which in turn can be used to estimate mobile robot's position. In this approach, slipping which is one of the common sources of measurement errors is taken care of.

With this hardware set up, an attempt can be made for estimating position of mobile robots by sensing motion of caster wheel. This technique provides a cost effective, simple hardware set up around caster wheels to estimate position and orientation of mobile robots. The results presented in this paper are of the pilot simulation study conducted to verify the concept for position estimation.

The current paper is organised as follows, the theoretical model for position estimation is deduced based on the concept of Kinematics and inverse Kinematics in Section II. Simulation results and discussions are presented in Section III followed by conclusions in Section IV.

2 Theoretical modelling

Consider caster wheel arrangement as shown in Fig. 1. Shafts of two sensors are assembled with caster wheel such that the two sensors connected to them are able to sense caster wheel motion in X and Y direction. The distance covered by caster wheel in one complete rotation, $s=r\theta$, where r is the radius of caster wheel.
Since $\theta=\omega t$,

$$s = r\omega t \tag{1}$$

Here t is the duration of rotation. Assuming that both the shafts connected to the sensors are identical, the distance covered by shafts in one complete rotation, s_1 is equal to the circumference of shafts,

$$s_1 = r_1\omega t \tag{2}$$

Here r_1 is the radius of shafts,

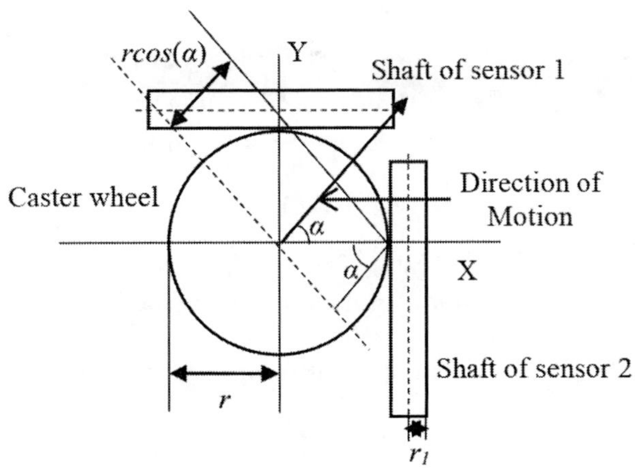

Fig. 1. Assembly of caster wheel along with two sensors

Consider direction of motion of caster wheel with an angle α with respect to positive x axis as shown in Fig. 1. Shaft 1 and shaft 2 measures X and Y components respectively. Number of rotations the shaft 2 covers if the caster wheel rotates with an angular velocity ω, is given by Eq. (3),

$$N_{r2} = \frac{\left(r\cos(\alpha)\right)\omega t}{2\pi r_1} \tag{3}$$

Number of rotations the shaft 1 covers, if the caster wheel rotates with an angular velocity ω is given by Eq. (4),

$$N_{r1} = \frac{(r\,sin(\alpha))\omega t}{2\pi r_1} \tag{4}$$

Sensor readings (X, Y) can be obtained from Eq. (5) and Eq. (6). Here, α is the heading angle, ω is the angular velocity of caster wheel and N_s is the resolution of sensor. The errors occur due to nature of surface over which caster wheel operates, slipping of wheels and effect of contamination on its surface were considered in the model. The errors were assumed to have Gaussian characteristics. Total effect of these errors was incorporated as random fluctuations that occur to the radius of the castor wheel. Maximum value of this error was assumed to be 10% of the actual dimension of the castor wheel. It was assumed that other errors such as misalignment in the axes of sensor shafts and caster wheel is constant and will be compensated during calibration.

$$X = \frac{((10 \cdot r + rand(-11) \cdot r)cos(\alpha\ ))\omega t N_s}{20\pi r_1} \tag{5}$$

$$Y = \frac{((10 \cdot r + rand(-1,1) \cdot r)sin(\alpha))\omega t N_s}{20\pi r_1} \tag{6}$$

Considering inverse kinematics, angular velocity (ω) of caster wheel and heading angle (α) can be estimated from outputs of two sensors. Values of for the same are calculated by solving the two simultaneous Eq. (5) and Eq. (6). On dividing Eq. (6) by Eq. (5), Eq. (7) is obtained.

$$\frac{Y}{X} = \frac{(r\,sin(\alpha))\omega t N_s}{2\pi r_1} \times \frac{2\pi r_1}{(r\,cos(\alpha))\omega t N_s} \tag{7}$$

Heading angle (α) is given by Eq. (8),

$$\alpha = tan^{-1}\left(\frac{Y}{X}\right) \tag{8}$$

Using trigonometric equation,

$$sin^2\,\alpha + cos^2\,\alpha = 1 \tag{9}$$

Eq. (5) and Eq. (6) are altered to get Eq. (10),

$$Y^2 \times \left(\frac{2\pi r_1}{r\omega t N_s}\right)^2 + X^2 \times \left(\frac{2\pi r_1}{r\omega t N_s}\right)^2 = 1 \tag{10}$$

From Eq. (10), Angular velocity (ω) of caster wheel can be obtained as given in Eq. (11),

$$\omega = \frac{2\pi r_1}{r\omega t N_s} \times \sqrt{X^2 + Y^2} \tag{11}$$

3 Results and discussion

The sensor output model developed in the previous section was simulated with the help of MATLAB version 8.6. Initial simulations were carried out to analyse output of sensors when the castor wheel was moving in predefined patterns. The effectiveness of this method was found out by estimating path tracked by robot with the help of sensor output. Simulations were conducted by considering diameter of caster wheel as 15 mm, diameter of shafts as 1 mm and resolution of sensor as 30.

3.1 Continuously varying heading angle

Simulations were conducted by continuously varying heading angle of castor wheel from 0 to 2π. The angular velocity of caster wheel was fixed at 4 rad/sec. The output of sensors during simulation run was calculated and plotted. Fig. 2. (i) and (ii) shows readings of sensor 1 and sensor 2 with respect to variation in heading angle for uniform angular velocity of caster wheel. It was observed from the graphs that the output of both sensors were varying sinusoidally. Output of sensor 2 was leading sensor 1 output by a phase angle of $\pi/2$. This is due to positioning of sensor 2 perpendicular to x axis of wheel. Since the wheel rolls towards x direction initially, maximum readings is obtained at sensor 2 and minimum at sensor 1. The readings will vary sinusoidally as direction of wheel motion varies. Thus the sensor output was varying as predicted.

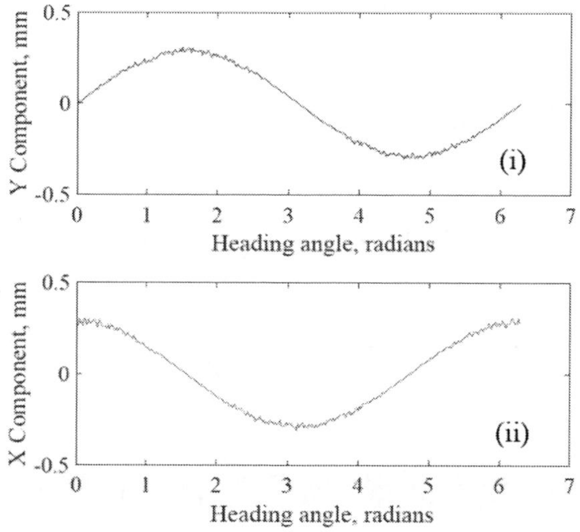

Fig. 2. Sensor output for varying heading angle (i) sensor 1 (ii) sensor 2

3.2 Continuously varying heading angle and angular velocity of caster wheel

In order to study variation pattern of sensor readings when both input parameters are varied periodically, another simulation was carried out. Values of heading angle and angular velocity of caster wheel was taken as $\alpha=sin(2t)$ and $\omega=sin(t)$. Fig. 3. depicts variations in (i) heading angle and (ii) angular velocity of caster wheel with respect to time. Based on the obtained periodic values of heading angle and angular velocity of caster wheel, sensor output was calculated. The plot of sensor output when heading angle and angular velocity of caster wheel are considered to be time dependent is as shown in Fig. 4. (i) and (ii). It can be clearly seen that sensor output readings vary proportionally as the time dependent input parameters fluctuates. Readings of sensor 1 and sensor 2 complement each other since the direction of wheel motion determines the output of sensors. Similar to the above case, it can be seen that the senor readings vary accordingly as the direction of wheel changes.

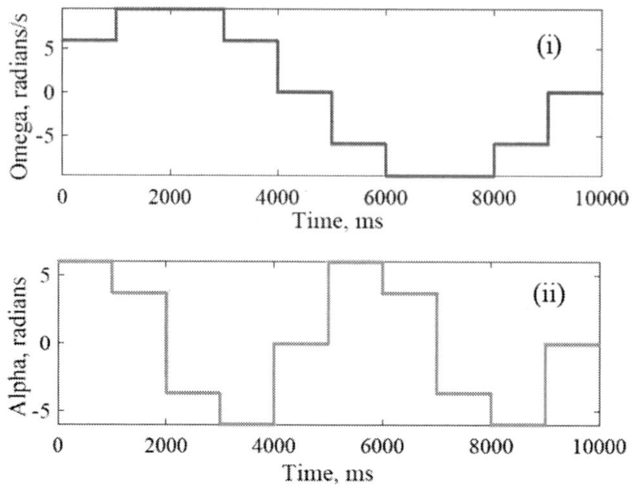

Fig. 3. Variations in (i) angular velocity,ω(ii) heading angle, α

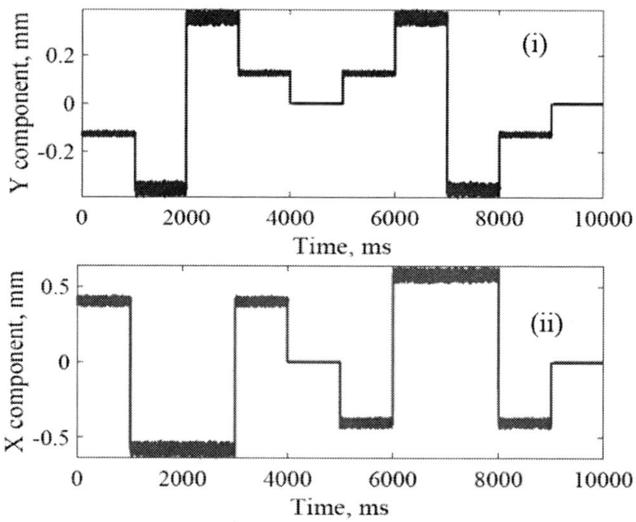

Fig. 4. Output responses when ω and α are time dependent quantities (i) sensor 1 (ii) sensor 2

3.3 Motion along a regular geometrical trajectory

Path traced by a mobile robot can be predicted with the help of sensor readings. For simplicity, consider path traced by mobile robot to be a rectangle of 150 mm by 75 mm as shown in Fig. 5. In this regard the heading angle and angular velocity of caster wheel was kept as given in Fig. 6 (i) and (ii). With input parameters as mentioned earlier, output of both sensors is obtained which is illustrated in Fig. 7 (i) and (ii). It is evident from the Fig. 5 that initially the output was only obtained from sensor 2. This is because initially heading angle was at 0 rad, which indicates that the robot motion was towards x direction. Similar to above discussed cases, the sensor readings vary according to the change in direction of wheel motion.

Concept of inverse kinematics was implemented to verify the correctness of sensor readings obtained. In addition to prove that the actual path tracked by robot can be retrieved from sensor readings. According to the concept, angular velocity of caster wheel and heading angle were estimated from sensor readings using established Eq. (5) and Eq. (6).

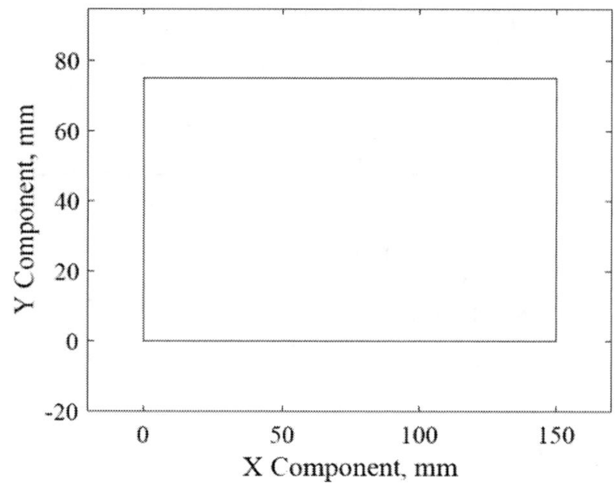

Fig. 5. Path traced by the mobile robot

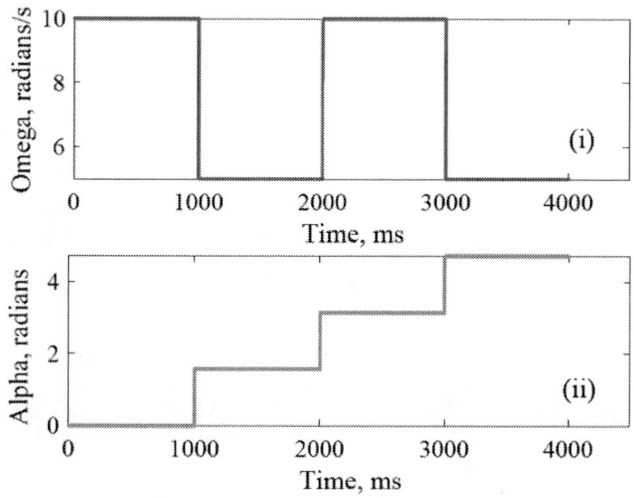

Fig. 6. Input parameters (i) angular velocity, ω (ii) heading angle, α

A comparison was made between actual and estimated values of input parameters (ω and α) for further analysis. Fig. 8 (i) and Fig. 8 (ii) depicts the plot of actual and estimated values of angular velocity of caster wheel and heading angle with respect to time respectively. From Fig. 8 (ii), it can be observed that the estimated value of angular velocity is affected by the disturbances and sources of error. The trend with which actual and estimated angular velocity varies is similar.

Similarly, in Fig. 8 (i), the estimated values of heading angle are in accordance with that of the actual values. The deviation in estimated heading angle from actual heading angle after 3 s is due to the inherent limitation of $\tan^{-1}$ trigonometric function in inverse kinematic equations to give value greater than π (i.e., $\pi + x = -\pi + x$). Thus this error in predicted heading angle will not contribute to any other errors. Therefore, it can be stated that the algorithm deduced in the previous section holds good for the purpose of position estimation of mobile robot.

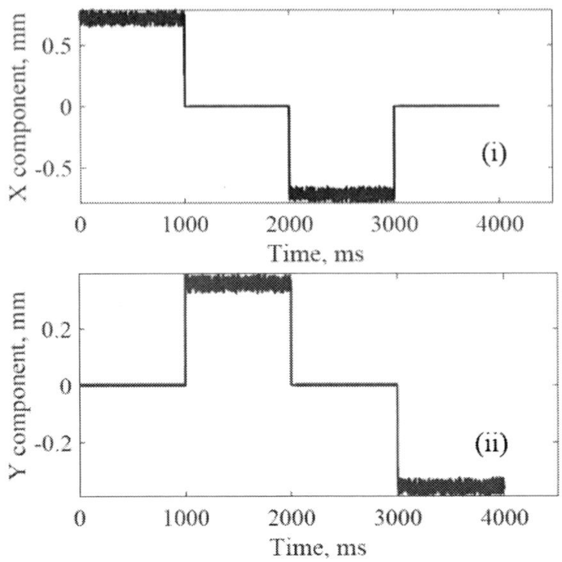

Fig. 7. Sensor readings (i) Sensor1 (ii) Sensor 2

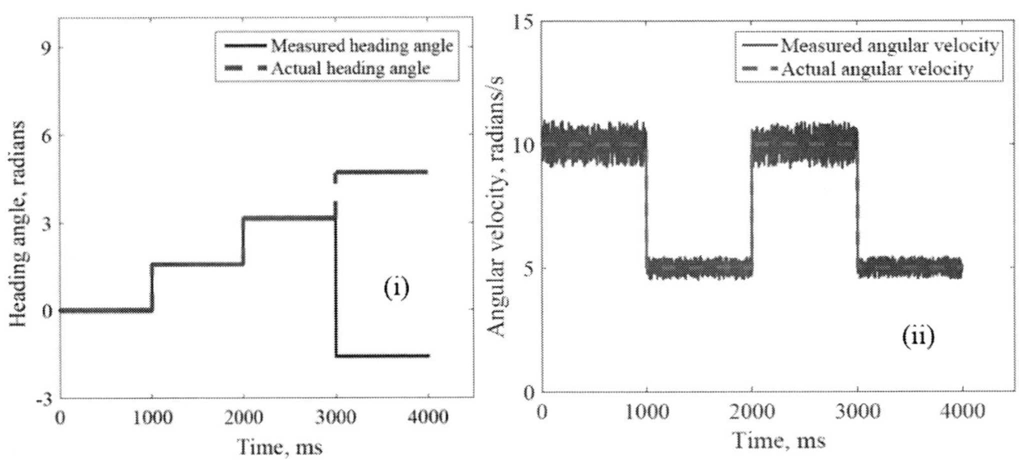

Fig. 8. Actual and estimated values (i) angular velocity of caster wheel (ii) heading angle

The verification is further continued by predicting path traced by mobile robot using estimated values of angular velocity of caster wheel and heading angle and established Eq. (5) and Eq. (6). This was conducted to analyse the path traced by mobile robot when estimated values of input parameters were considered. Fig. 9 makes a comparison between the path traced by mobile robot while considering the actual and estimated values of input parameters. It is worth to notice that the path tracked using actual and estimated values of input parameters are in good agreement. The effect of sensor errors is minimal on the estimated trajectory. This justifies that the position estimation of mobile robot by sensing caster wheel motion is possible with an appropriate hardware set-up.

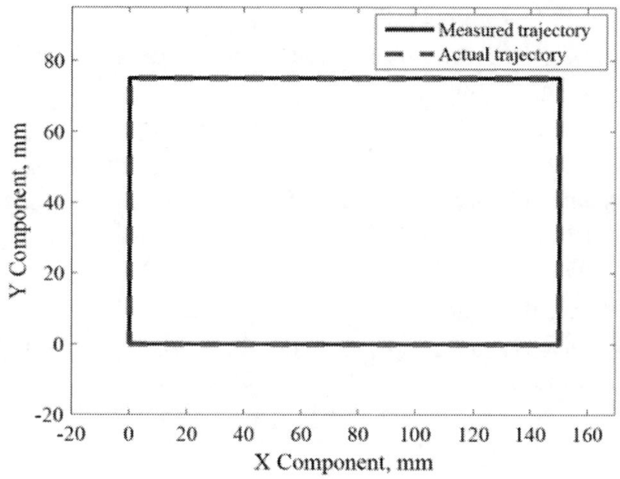

Fig. 9. Comparison between the path traced by the mobile robot while considering the actual and estimated values

4 Conclusions

The paper presented a mathematical model developed to estimate the position of mobile robot by sensing the motion of caster wheel. Different sources of errors that will occur during sensing were included in the model. A pilot simulation study was carried out to establish the effectiveness of method in detection of motion and position estimation. Sensor reading was calculated in different scenarios of castor wheel motion with different input parameters. The sensor readings are further used to recalculate the input parameters and compared with the actual input parameters. The sensor output when the robot was tracking a rectangular trajectory of 150 mm × 75 mm was also simulated. The robot was assumed to run at 150 mm/s in lengthy sides of the rectangular trajectory and 75 mm/s at short sides. Total travel time of 4 second was obtained. The results obtained from simulation show that the calculated and the actual values of input parameters, viz angular velocity of caster wheel and heading angle are on par. It is also observed from the simulation results that, the path traced by mobile robot can be determined from the output of the sensors although the sensors are exposed to external noises. On considering the results, it is seen that the concept can be experimented on a real mobile robot for estimation of its position.

References

1. M. Jamshidi, Robotics and Remote Systems for Hazardous Environments Prentice Hall, (1993)
2. S.G. Tzafestas, Introduction to Mobile Robot Control, Elsevier, (2014)
3. M. Endo, K. Hirose, Y. Hirata, Inter. Conf. on Intelligent Robots and Systems, 2795-2801, (2008)
4. J.M. Usher, R. McCool, L. Strawderman, Simulation Modelling Practice and Theory, 75, 96-112, (2017)
5. N. Chivarov, D. Chikurtev, K. Yovchev, IFAC-Papers On Line, 48, 24, 128-133, (2015)

6. N. Yonezawa, K. Kosuge, Y. Hirata, Proc. of the 2012 IEEE Int. Conf. on Robotics and Biomimetics, 1874-1880, (2012)
7. M. Endo, K. Hirose, Y. Hirata, Proc. of the 2009 IEEE Int. Conf. on Robotics and Automation, 2350-2355, (2009)
8. A. Lottermoser, C. Berger, S. Braunreuther, Procedia CIRP, 62, 594-599, (2017).
9. N. Ruangpayoongsak, H. Roth, J. Chudoba, Proc. of the 2005 IEEE Int. Workshop on Safety, Security and Rescue Robotics, 212-217, (2005)
10. C. Suliman, C. Cruceru, F. Moldoveanu, Scientific Bulletin of the Petru Maior University of TirguMures, (2009)
11. M. Cimino, P.R. Pagilla, Automatica, 47, 10, 2267-2272, (2011)
12. W. Burgard, A. Derr, D. Fox, A.B. Cremers, Proc. of the 1998 IEEE Int. Conf. on Intelligent Robots and Systems, 2, 730-735, (1998).
13. S. Se, D. Lowe, J, Little, Proc. of the 2001 IEEE/RSJ Int. Conf. on Intelligent Robots and Systems, 1, 414-420, (2001)
14. D. Fox, W. Burgard, S. Thrunz, Proc. of the Fifteenth National Conf. on Artificial Intelligence, 983-988, (1998)

Modified bug-1 algorithm based strategy for obstacle avoidance in multi robot system

Jom J. Kandathil[1], *Robins Mathew*[1] and *Somashekhar S. Hiremath*[1*]

[1]*Department of Mechanical Engineering, Indian Institute of Technology Madras, Chennai*

Abstract. One of the primary ability of an intelligent mobile robot system is obstacle avoidance. BUG algorithms are classic examples of the algorithms used for achieving obstacle avoidance. Unlike many other planning algorithms based on global knowledge, BUG algorithms assume only local knowledge of the environment and a global goal. Among the variations of the BUG algorithms that prevail, BUG-0, BUG-1 and BUG-2 are the more prominent versions. The exhaustive search algorithm present in BUG-1 makes it more reliable and safer for practical applications. Overall, this provides a more predictable and dependable performance. Hence, the essential focus in this paper is on implementing the BUG-1 algorithm across a group of robots to move them from a start location to a target location. The results are compared with the results from BUG-1 algorithm implemented on a single robot. The strategy developed in this work reduces the time involved in moving the robots from starting location to the target location. Further, the paper shows that the total distance covered by each robot in a multi robot-system is always lesser than or equal to that travelled by a single robot executing the same problem.

Keywords: BUG Algorithm, Obstacle Avoidance, Mobile Robot, Multi Robot System, Simulation.

1 Introduction

In recent years, mobile robots have become a more fundamental part of the industry. In general, obstacle avoidance has turned out to be pre-requisite to mobile robots and this has resulted in the introduction of a large variety of algorithms to support obstacle avoidance. Obstacle avoidance in unknown terrain is one of the key challenges in this regard. BUG algorithms are class of algorithms that deals with obstacle avoidance in unknown terrains. BUG algorithms were introduced as a methodology for non-heuristic motion planning in an uncertain environment where information about the environment is not fully pre-known [1]. The sensors attached to the robot are used to attain dynamic information about obstacles and the BUG algorithm uses this dynamic information for navigating the robot avoiding obstacles. Among the various preliminary BUG algorithms, BUG-1 is more reliable and safer for practical applications due to the exhaustive search algorithm present in it. Figure 1 show the trajectory of the robot when BUG-1 was used for obstacle avoidance. The robot

* Corresponding author: somashekhar@iitm.ac.in

initially moves towards target in a straight line. Upon detection of obstacle, the robots start circumnavigating the obstacle in either clockwise or anticlockwise direction. The point of shortest distance to the target is measured and stored during this circumnavigation. Once the robot completes the circumnavigation it will move towards the point of shortest distance through the shortest path. When it reaches the point of shortest distance, it will continue to move towards target in straight line.

There are many existing strategies, such as IBA [2] and K-Bug [3] that aims at improving the travel time of the BUG algorithm. More effective varieties of these strategies have been introduced to higher order problems such as military surveillance and video game designs. The working of various strategies that form the class of BUG algorithms and a comparison of their performances were provided by James et.al [4]. The potential for this branch of robotics is vast and hence this paper tries to address several issues of the basic variations of BUG algorithm with a new strategy using multiple robots in leader-follower formation. The concept of using a multi robot system to navigate around obstacles [5] is the basis of this new strategy.

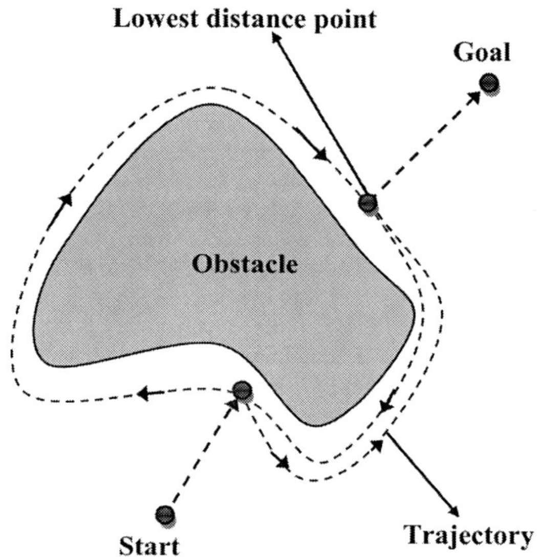

Fig .1. Bug-1 algorithm

Multi robot system in leader follower formation is a highly effective system in swarm robotics. The leadership exchange is the backbone of the leader-follower system [6], [7]. This was utilized in the present work to design an efficient multi robot obstacle avoidance strategy. This new strategy assumes that the robots in the system possess only basic sensing and movement capabilities thus allowing other robots to take over and complete the task in case of breakdown of the leader robot. The energy levels of individual robots were monitored in real time and the lower energy robots were made to conserve their energy by reassigning the task to other higher energy robots. These two features of the algorithm improve the success probability of the provided task. The concept of information exchange across multiple robots [8], [9] was also implemented to achieve better path navigation. Thus, this new strategy is a modified BUG-1 strategy which is implemented across multiple robots for improving travel time and energy expenditure of robots and providing a better success probability of task at hand.

The better travel time and energy efficiency of the proposed algorithm is described by comparing the algorithm with conventional algorithms. In principle, it has been shown that the strategy has better travel time than conventional BUG algorithms and provides more efficient and flexible energy distribution. Reminder of the paper is organized as follows. Section II explores the obstacle avoidance strategy in detail. Section III presents a comparison of this new strategy and BUG-1 algorithm with respect to travel time and energy efficiency. Preliminary experimental results are presented in section IV.

2 Multi robot obstacle avoidance

The multi robot bug algorithm, similar to conventional bug algorithm, navigates multiple robots across a planar surface with unknown obstacles of arbitrary shape provided the start and goal location are known. A conventional BUG-1 algorithm (hence will be referred to as BUG-1 algorithm) has memory element which enables it to store information about the coordinates tracked by the robot and hence can opt for the shortest path. The algorithm introduced in this paper is based on the exchange of this stored information across multiple robots. Since the robots can access information about the coordinates travelled by other robots, it enables them to make an optimal decision at a faster rate. The algorithm has three major parts: first two are concerning the motion of the robots while the third is regarding the formation. A pictorial representation of this strategy is shown in Fig.1.

Initially, two robots are assigned, a leader robot (at S_1) and a follower robot (at S_2). They move together towards the common goal (G). When these robots encounter an obstacle (O), they circumnavigate the obstacle in opposite direction and continue to move till they meet each other at a point (M). As the robots move around the obstacles, both the robots check the distance from its current position (P) to goal G and the point of minimum distance (D) is stored in the memory. From M, the robots take the shortest route to D and await the arrival of rest of the robots to D. At M, a new leader and a follower are assigned based on energy consideration. The direction of circumnavigation is chosen by taking the midpoint (C) of the line connecting start points of the leader and follower. An imaginary line is drawn from C to G as shown in Fig 2 and the robot that is on the left side of this line will be a right sensing robot (moves around the obstacle in the clockwise direction, sensing the obstacles on the right). The robot on the right side acts as follower robot and goes around in the other direction to that of the leader robot. This ensures that there is no collision between the robots before the completion of circumnavigation, which otherwise leads to overlapping and reduces efficiency of the system.

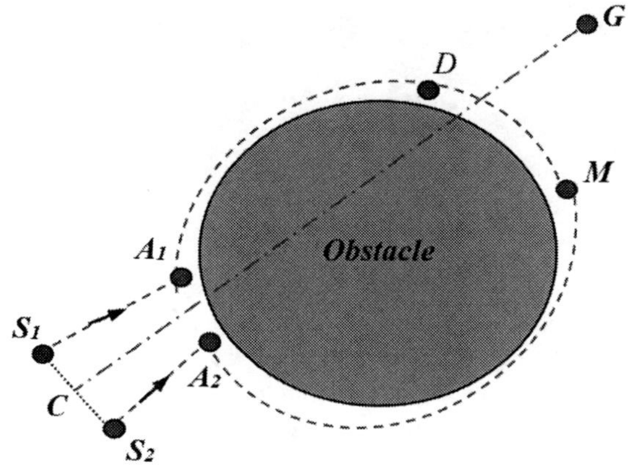

Fig. 2. Choice of direction of circumnavigation in multi robot system

The flow chart of the algorithm is shown in Fig. 3:

1) Leader and follower move towards target till one of the following events occur
   a) Target is attained- stop. Rest of the robots move to meet the leader and follower robots by taking the shortest path to reach the path tracked by leader and follower the path to reach target.
   b) Obstacle is encountered. Then move to step 2.
2) Circumnavigate the obstacle in opposite directions while recording the distance to target till the two robots meet. After meeting move to the point of minimum distance to target. Then move to step 3.
3) Rest of the robots move to meet the leader and follower robots by tracking the shortest distance path to reach the point of minimum distance. This ensures effective use of energy in reaching the goal. Then move to step 4.
4) The robot with highest leftover energy is assigned the role of leader and the robot with second highest leftover energy is assigned the role of follower robot. Then move to step 1.

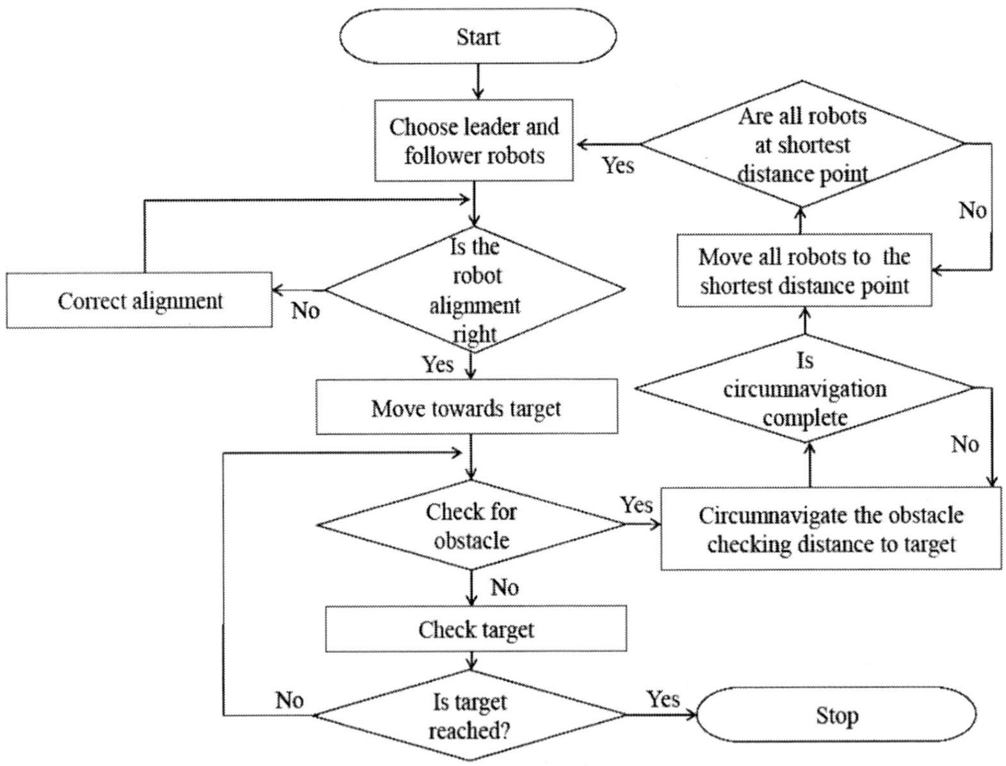

Fig. 3. Flow chart

There are many advantages for this approach in comparison to conventional algorithms especially when it comes to systems involving more than one robot. One of the most basic merits of the system is that it improves performance with increasing number of robots. Further, due to increased number of robots involved, the probability of success is also boosted considerably. Effective distribution of energy ensures that maximum number of robots reach the goal.

3 Algorithm analysis

3.1 Travel time

Without loss of generality, this paper considers an obstacle shape as shown in Fig.4. Tangents are drawn from S in Fig. 4 (i) and S_1, S_2 in Fig. 4 (ii) to the obstacle. These tangents provide the possible locations of goal since the obstacle does not obstruct a goal if these tangents do not enclose the goal. The worst case scenario is goal is along the tangent, or more generally, as the goal comes closer to the point of initial contact of robot and obstacle. The performance of the algorithm decrease in this scenario since the meeting point of the two robots will be away from the point of minimum distance.

Consider the arrangement as shown in Fig.4 (i) and (ii), S, S_1 and S_2 are the starting positions of robots while A, A_1 and A_2 are the initial points of contact of robots with obstacle and t is the time taken by the robots to reach A (or A_1 and A_2) from their respective starting positions (to avoid the effect of starting position on this time we have assumed a common time t for all cases). Let the time taken by a single robot to go around the obstacle and return to A be T_1 and the time taken for the leader and follower to reach M, the meeting point be T_2. If D is the point of minimum distance to goal, then time taken by the single robot to reach D from A will be T_3 and time taken by the leader-follower robots to reach D from M be T_4. Finally, let the time taken for the robots to move from M to goal be T_5.

The total time taken by a single robot to reach goal, T_{totalI} is given by Eq.1 and time $T_{total\text{-}M}$ taken by a multi robot system to accomplish the same is given by Eq.2. Equation 1 and Eq. 2 are compared in the following section to prove that $T_{totalI} \geq T_{total\text{-}M}$. For this comparison it is assumed that all the robots move with the same velocity. Further, to include the worst possible scenario, T_3 is taken as zero.

$$T_{total\text{-}I} = t + T_1 + T_3 + T_5 \quad (1)$$

$$T_{total\text{-}M} = t + T_2 + T_4 + T_5 \quad (2)$$

But,

$$T_3 = 0$$
$$T_4 \leq T_2$$
$$T_1 \geq 2T_2$$

Hence,

$$(t + T_1 + T_3 + T_5) \geq (t + T_2 + T_4 + T_5) \quad (3)$$

$$T_{total\text{-}I} \geq T_{total\text{-}M}$$

From above mathematical analysis, Eq.(3) states that the travel time associated with the multi robot system is always lesser than that of a single robot executing Bug-1 algorithm.

3.2 Energy distribution

Efficient distribution of energy in robots ensures task completion. The energy efficiency of the robots is analyzed using the concept of average energy of the robots and by calculating energy of individual robot. In the first section average energy of the robots involved in the multi robot system is analyzed. In the second section it is shown that after executing the task, energy of each robot in a multi-robot system will be higher than that of a single robot system.

The energy expenditure of a robot depends on several aspects such as motor, sensor elements, Transmission and reception of data and processing. Here it shall be assumed that all forms of energy expenditure are in direct relation to distance covered. This is a good enough approximation since we assume the robots to be idle when not in motion and almost all forms of energy expenditure mentioned above happen simultaneously when robot is in motion

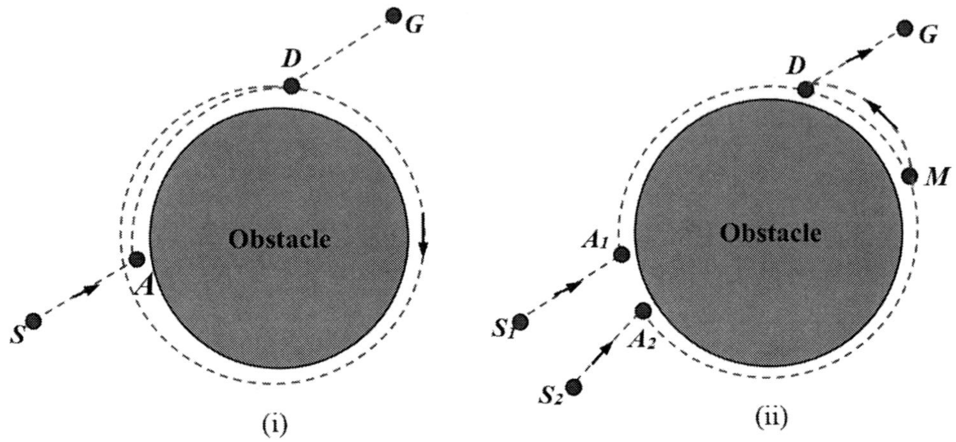

Fig. 4. Comparison of algorithms (i) Conventional Bug 1 algorithm (ii) Multi-robot Bug algorithm

3.2.1 Average energy

Total energy spend by a single robot, $E_{total\text{-}I}$ is compared with $E_{total\text{-}M}$ by using the same assumptions used to calculate travel time. To calculate the average energy from distance travelled, a constant k is used as the conversion factor as given in Eq.4 and Eq.5.

$$E_{total\text{-}I} = k(SA + curve(AM + MA + AD)) \tag{4}$$

$$E_{total\text{-}M} = k((SA_1 + SA_2) + curve(A_1M + A_2M + 2MD)) \tag{5}$$

From the analysis made in previous section, average distance covered by the leader-follower system is always lesser than the distance covered by a single robot. This implies that the average energy spent by a leader-follower system is lesser than that of a single robot executing the same task. i.e $E_{total\text{-}I} > E_{total\text{-}M}$. It also shows that the average energy expenditure shall always be lower in a multi-robot system irrespective of the number of robots involved.

3.2.2 Energy of an individual robot of multi-robot system

Consider one of the robots in the multi-robot system; if there is more than one obstacle and there are more than two robots, the initial leader-follower set of robots turn idle and another set of robots of higher energy levels take up the roles of leader and follower. Thus, assuming similar starting positions, the energy levels of individual robots of a system with more than two robots shall always remain higher compared with a two robot system. This inference arises from the fact that there is no other robot to take up the role of leader robot. So the worst case scenario in a multi robot system arises with only two robots in the multi-robot system. As in the case of average energy, distance covered multiplied by a conversion factor shall be considered as a measure of the energy expenditure, E_{total}. Equation 6 provides the total energy expenditure of a single robot. This is compared with $E_{total\text{-}M1}$ and $E_{total\text{-}M2}$ (Eq.(7) and Eq.(8)), the individual energy expenditure of the leader and follower robots in the two robot system.

$$E_{total\text{-}1} = k(SA + curve(AM + MA + AD)) \tag{6}$$

$$E_{total\text{-}M1} = k(S_1A_1 + curve(A_1M + MD)) \tag{7}$$

$$E_{total\text{-}M2} = k(S_2A_2 + curve(A_2M + MD)) \tag{8}$$

It can be deduced from these equations and from Fig.4 that even if curve (AD) = 0 in Eq.6, the distance covered by a single robot executing the task is more than the distance covered by both the individual robots in a multi-robot system.
The most important inference from the above two observations is that the final energy of individual robots in a multi-robot system is always more than that of a single robot executing the task with the help of bug-1 algorithm. This shows that the energy distribution in a multi-robot system provides better success rate and it ensures all the robots reaching the goal with sufficient energy.

4 Simulation results

The theoretical results in the previous section were examined with the help of a simulation study. The energy expenditure based on the distance covered by the robots and the total time taken for task completion is considered in this study. In this simulation study the distance is considered in meters and the time taken to complete the task is considered with clock cycle as the base. By counting the number of loops involved, an estimate of the total running time is obtained. Since the same base is used to measure the performance of all cases, the values provide are reliable source for comparison.

Path traveled by the robots during simulation for different number of robots is shown in Fig. 5.(i) to Fig.5.(vi). The distance covered by each robot, total cycles in simulation and the average distance involved are also calculated. In each of the simulation, the goal is assumed to be at coordinates (4,6) and the obstacles are placed at coordinates (2,2) and (3,4) with a radius of 0.5 units. The coordinates of the starting location of the robots are assigned as (1,1) for robot 1, (1.5,1) for robot 2, (1,1.5) for robot 3, (1.2,0.7) for robot 4, (1.25,1.35) for robot 5 and (1.3,0.9) for robot 6.

The results obtained from the simulations were plotted to draw inference regarding the travel time and average distance travelled by the multi robot system in comparison to a single robot executing the conventional Bug-1 algorithm. Figure 6 (i) displays the variation of average distance and Fig. 6 (ii) displays the variation in total travel time as the number of robot increases. The slight deviation from the decreasing trend occurs in the 5-robot and 6-robot system, which can be justified by the starting position that is further away from the goal which results in extra distance to cover thus taking more time to reach goal.

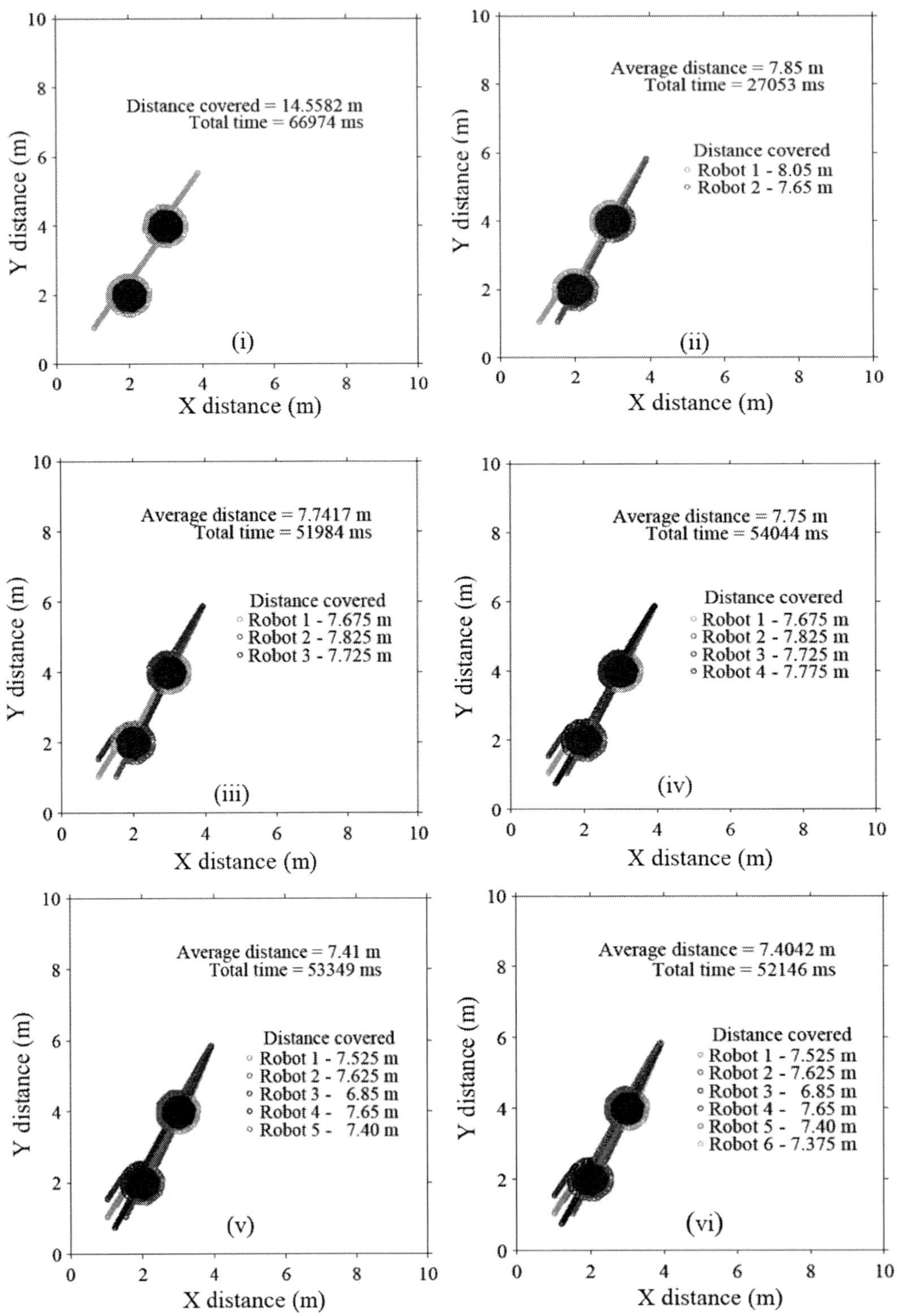

Fig. 5. Path tracked by robots (i) Single robot (ii) Dual robot system (iii) Three robot system (iv) Four robot system(v) Five robot system (vi) Six robot system

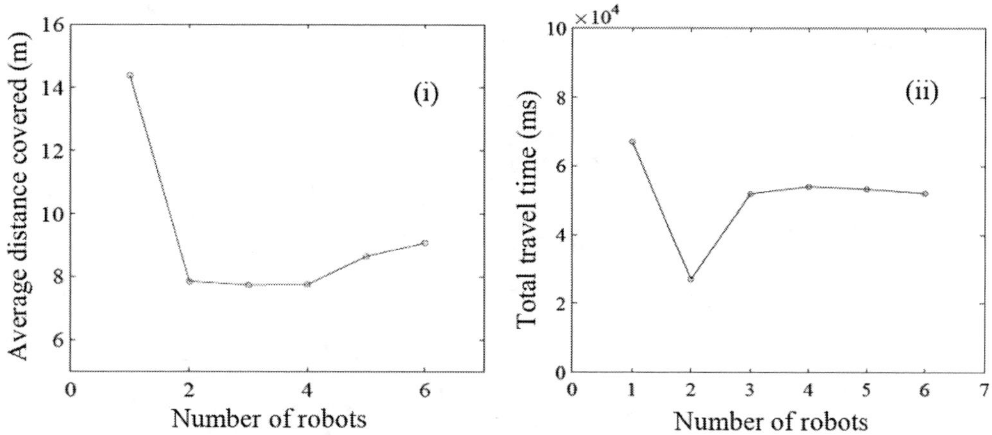

Fig. 6. Simulation results (i) Average distance plotted against number of robots in the system (ii) Total travel time plotted against number of robots in the system

The plots validates that the average distance travelled by the robots decreases as the number of robots involved increases. Another inference drawn is that the distance travelled by each robot is lower than that of a single robot executing Bug algorithm. The total time required for task completion reduces considerably when multiple robots are used instead of single robot. It was also observed that as the number of robots increases in multi robot system the total time to complete the task increases

5 Conclusions

This paper presented a more innovative form of Bug-1 algorithm for multi-robot system. The strategy makes use of exchange of information across a leader-follower robot system. The algorithm provides better performance in comparison to conventional Bug-1 algorithms. Theory behind this better performance was discussed in support. The travel time and energy distribution of the robots have been compared with that of a Bug-1 algorithm and it shows that the algorithm provided in the paper have a shorter travel time and better energy conservation strategy. A further argument for this strategy is that it also works better than conventional algorithms in many mazes where the conventional algorithm fails to deliver proper results. The simulation results suggests that the improved nature of this algorithm makes it a very effective in cases involving more than one robot dedicated to execution of a common task such as in swarm robotics

References

1. V. Lumelsky, A. Stepanov, IEEE Trans. Automat. Contr. **31**, 1058–1063(1986).
2. M. Zohaib, S.M Pasha, N. Javaid, J. Iqbal, *International multi topic conference*, 291-299 (2013)
3. R.A. Langer, L.S. Coelho, G.H Oliveira, *IEEE Inernational. Conference on control application*, 403-408 (2007)
4. J. Ng, T. Braunl, J. of Intelligent and Robotic Systems , ***50**,1*(2007)
5. T. Facchinetti, t-bots: a coordinated team of mobile units for searching and occupying a target area at unknown location

6. L. Consolini, F. Morbidi, D. Prattichizzo, M. Tosques, Automatica, **44**, 5, 1343-1349, (2008)
7. R.G. Lord, D.J Brown, S.J Freiberg, Organizational behavior and human decision processes, **78**, 3, 167-203, (1999)
8. J.Y.S. Luh, Y.F Zheng, The Int. J. of robotics res., **6**, 3, 60-70, (1987)
9. Y.F Zheng, J.Y.S Luh, *IEEE conference on decision and control* , 1761-1766, (1985)

MATEC Web of Conferences **144**, 01013 (2018) https://doi.org/10.1051/matecconf/201814401013
RiMES 2017

Foraging behavior analysis of swarm robotics system

E. Sakthivelmurugan[1*], *G. Senthilkumar*[1], *K. G. Prithiviraj*[1] and *K. R. Tinu Devraj*[1]

[1]*Department of Mechanical Engineering, Bannari Amman Institute of Technology, Sathyamangalam, Erode, Tamilnadu, India.*

Abstract. Swarm robotics is a number of small robots that are synchronically works together to accomplish a given task. Swarm robotics faces many problems in performing a given task. The problems are pattern formation, aggregation, Chain formation, self-assembly, coordinated movement, hole avoidance, foraging and self-deployment. Foraging is most essential part in swarm robotics. Foraging is the task to discover the item and get back into the shell. The researchers conducted foraging experiments with random-movement of robots and they have end up with unique solutions. Most of the researchers have conducted experiments using the circular arena. The shell is placed at the centre of the arena and environment boundary is well known. In this study, an attempt is made to different strategic movements like straight line approach, parallel line approach, divider approach, expanding square approach, and parallel sweep approach. All these approaches are to be simulated by using player/stage open-source simulation software based on C and C++ programming language in Linux operating system. Finally statistical comparison will be done with task completion time of all these strategies using ANOVA to identify the significant searching strategy.

Keywords: Foraging, Swarm Robotics, Foraging Algorithm, Item Searching Strategy, Item Detection Time.

1 Introduction

In current environment more people are subjected into hazardous work such as toxic waste clean-up, demining, search and rescue. In such a work the reliability of human life is relatively too low. In order to overcome that issue, the people are replaced with group of small robots. The group of small robots will accomplish the task at very faster and efficient manner. For this task accomplishment, various foraging behaviour is considered [1].

Many researchers have studied the behaviour of swarm robotics system. Also they have focused on these characteristics of swarm robotics system and applied the system to concrete problems from the engineering viewpoint. But most of them resort to study the qualitative aspect of swarm robotic system. And also they have studied swarm robotics by

* Corresponding author: sakthi.glen@gmail.com

adopting the swarm intelligence to mimic a swarm by lot of robots. In daily life social insects play some intelligent activities[3]. Also the researchers have concentrated only on random movements of robots with circular arena. The shell is placed in the centre of the circular arena. But they have not concentrated on various strategic movements of robots such as expanding square approach, expanding square with divider policy, parallel sweep approach and parallel sweep with divider policy. In this study, various strategic movements of robots and item inclusions are planned to be studied.

The objective of this study is to reduce the time consumption for searching through various strategic movements of robots with different strategies. And also to find the resource utilization, task completion time and efficiency of the system. This study will helpful to search the item in faster mode compare to random search. There is no possibility of collision and congestion in searching state and retrieving state. It will improve the system efficiency because of better utilization of resources. There is no interaction needed between robots because robots are placed at well defined arena.

2 Methodology

It explains about the various possible movements of robots from one stage to another stage through robot activity diagram. Also it explains about various assumptions to be followed in the foraging behaviour and various item searching techniques.

2.1 Robot Activity Diagram

In this swarm robotic system, various kinds of strategic movements are considered for the purpose of retrieving item to the shell. The possible robot activity was described in figure 1.

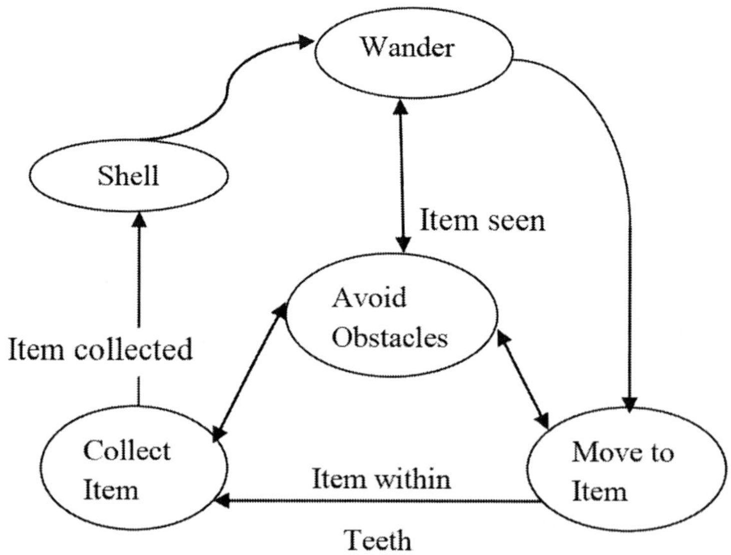

Fig. 1. Robot Activity Diagram

2.2 Assumptions

1. Limited number of robots is used.
2. Item is considered as a stationary object.
3. Environment boundary is well known.
4. Congestion and collisions are neglected.
5. Obstacles inclusion also considered.
6. Position of shell is well known.

2.3 Foraging Algorithm

Step 1: Start the program.

Step 2: Enable the port address.

Step 3: Calling the robots by the address R1, R2, R3, R4.

Step 4: Initialize variables Count=0, Delay time=2 Sec, Total number of item N=20, Bitmaps.

Step 5: Robots R1, R2, R3, R4 wandering.

Step 6: If any one of the robots R1, R2, R3, and R4 detects the item, it goes to step 7. Otherwise, it goes to step 5.

Step 7: Robot move to the item.

Step 8: If any other robots contact with the item means it enters into the ignore state. After delay of 2 Sec it goes to step 5. Otherwise, it goes to step 9.

Step 9: Robot makes physical connection with the item.

Step 10: Robot move to the shell.

Step 11: Robot release the item in the shell.

Step 12: If total count less than the total number of item means it goes to step 5. Otherwise it goes to step13.

Step 13: Print the total task completion time for a given arena.

Step 14: Stop the program.

2.4 Various Item Searching Strategies

Different types of item searching strategies are given below.

1. Straight line approach

2. Divider approach

3. Parallel sweep approach

4. Expanding square approach

5. Parallel line approach

2.4.1. Straight Line Approach

In this approach, the arena is divided into four sub elements. The shell is placed at bottom of the each sub elements boundary line. Also item is positioned throughout the environment. In this environment obstacles also placed. This environment is to be simulated and also total task completion time has to record.

2.4.2. Divider Approach

In this approach, the arena is divided into four sub elements. Here the shell is placed at each sub elements left corner side. The robot will search the item at allocated area only and do not interrupt with others. At last this environment is to be simulated and also total task completion time has to record.

2.4.3. Parallel Sweep Approach

In this approach, the arena is sweep in nature. Here, the shell is placed at bottom or top of the boundary line. Alternate search is possible in this approach. Finally this environment is to be simulated and also total task completion time has to record.

2.4.4. Expanding Square Approach

In this approach, the arena like square in nature. But square expands from one end to another end. Here also the shell is placed at either side of boundary line. Item and obstacles are placed throughout the environment. At end this environment is to be simulated and also total task completion time has to record.

2.4.5. Parallel Line Approach

This is similar to the straight line approach. In this case the arena is subdivided into eight elements. The shell is placed at top and bottom of the boundary line. Here the number of robots is high. This will applicable only, if the environment large in nature.

3. Simulation setup

For simulation Arena size, Number of robots, Strategy and Arena was described in table 1.

Table 1. Simulation Setup

S. No.	Type of Strategy	No. of Robots	Area (m^2)	Width (m)	Length (m)
1	Parallel sweep with divider	4	2.4 x 3	2.4	3.0
2	Parallel sweep with divider	4	3 x 3.5	3.0	3.5
3	Parallel sweep without divider	4	2.4 x 3	2.4	3.0
4	Parallel sweep without divider	4	3 x 3.5	3.0	3.5

3.1 Factorial design and strategy comparison

3.1.1 Introduction

When the number of factor increases, the number of experimental runs would be very large. Especially in the screening experiments conducted to develop new processes or new products, the number of factors considered will usually be large since the experimenter has no prior knowledge on the problem. Because resources are usually limited, the experimenter may restrict the number of replications. When only one replication is obtained, we will not have the experimental error for testing the effects.

3.2 Factor identification

In item searching mode a lot of factors that affects the robot performance. Here a strategy only plays a major role or any other factors are significantly contributed towards this robot performance. For this performance analysis, we considered the following factors namely Speed, Size of Arena, Number of Robots and Type of Strategy.

3.2.1. Factors Level Selection

Each factor was tested at two levels (low level and high level). The maximum allowable speed for the robot in the simulation is 1.0 m/sec. beyond that speed limit robot may collide with other robots or congestion will be possible. But the size of the arena, type of strategy and number of robots will be any according to their experimenter interest. Factors level is given in the table 2. In mentioned table, factors and levels are simulated by using player/ stage simulation software and responses are tabulated. With help of Yates algorithm, sum of squares and estimate of effect is calculated and tabulated in table 3.

Table 2. Various levels for each factor

S. No.	Factors	Low Level	High level
1	Speed of the Robot (A)	0.5 (m/sec)	1.0 (m/sec)
2	Size of the Arena (B)	20x30 (m^2)	50x30 (m^2)
3	Number of Robots (C)	2	4
4	Type of Strategy (D)	1	2

Table 3. Yates Algorithm for the 2^4 Design (Single Replicate)

S. No.	Treatment combination (s)	Response (Time)	(1)	(2)	(3)	(4)	Effect	Estimate of Effect	Sum of squares
1	(1)	105	232	802	1295	2617	I		
2	A	127	570	493	1322	-881	A	-110.13	48510.0
3	B	449	159	862	-459	1023	B	127.88	65408.0
4	ab	121	334	460	-422	-591	AB	-73.88	21830.0
5	C	102	264	-306	513	-711	C	-88.88	31595.0
6	ac	54	598	-153	510	379	AC	47.38	8977.56
7	bc	218	142	-324	-401	-321	BC	-40.13	6440.06
8	abc	116	318	-98	-190	429	ABC	53.63	11502.5
9	D	173	22	338	-309	27	D	3.38	45.56
10	ad	91	-328	175	-402	37	AD	4.63	85.56
11	bd	420	-51	334	153	-3	BD	-0.38	0.56
12	abd	178	-102	176	226	211	ABD	26.38	2782.56
13	cd	88	-82	-350	-163	-93	CD	-11.63	540.56
14	acd	58	-242	-51	-158	73	ACD	9.13	333.06
15	bcd	191	-34	-160	299	5	BCD	0.625	1.56
16	abcd	127	-64	-30	130	-169	ABCD	-21.13	1785.06

With help of analysis of variance, Degrees of freedom and percentage contribution of source of variation is computed and tabulated in table 4. From this table, we observed that source of variations such as A, B, C and AB are significantly contributed. The remaining source of variation is not significantly contributed and that is combined with error for conducting the F – Test. The F – Test is conducted and tabulated in table 5.

Table 4. Analysis of Variance for 2^4 design

Source of Variation	Sum of Squares	Degrees of Freedom	Contribution (%)
A	48510.06	1	24.27
B	65408.06	1	32.73
AB	21830.06	1	10.92
C	31595.06	1	15.81
AC	8977.56	1	4.49
BC	6440.06	1	3.22
ABC	11502.56	1	5.76
D	45.56	1	0.02
AD	85.56	1	0.04
BD	0.56	1	0.003
ABD	2782.56	1	1.40
CD	540.56	1	0.27
ACD	333.06	1	0.17
BCD	1.56	1	0.008
ABCD	1785.06	1	0.89
Total	199837.9	15	100

Table 5. Analysis of Variance after Pooling

Source of Variation	Sum of Squares	Degrees of Freedom	Mean Square	F_0
A	48510.06	1	48510.06	16.42 (S)
B	65408.06	1	65408.06	22.14 (S)
AB	21830.06	1	21830.06	7.39 (S)
C	31595.06	1	31595.06	10.70 (S)
Pooled error	32494.66	11	2954.06	

Table 6. Normal probability computations of the effects

Effect	Rank	Percentile	Rank-based z-score
-110.13	1	0.0333333	-1.833914636
-88.88	2	0.1	-1.281551566
-73.88	3	0.1666667	-0.967421566
-40.13	4	0.2333333	-0.727913291
-21.13	5	0.3	-0.524400513
-11.63	6	0.3666667	-0.340694827
-0.38	7	0.4333333	-0.167894005
0.63	8	0.5	-1.39214E-16

3.38	9	0.5666667	0.167894005
4.63	10	0.6333333	0.340694827
9.13	11	0.7	0.524400513
26.38	12	0.7666667	0.727913291
47.38	13	0.8333333	0.967421566
53.63	14	0.9	1.281551566
127.88	15	0.9666667	1.833914636

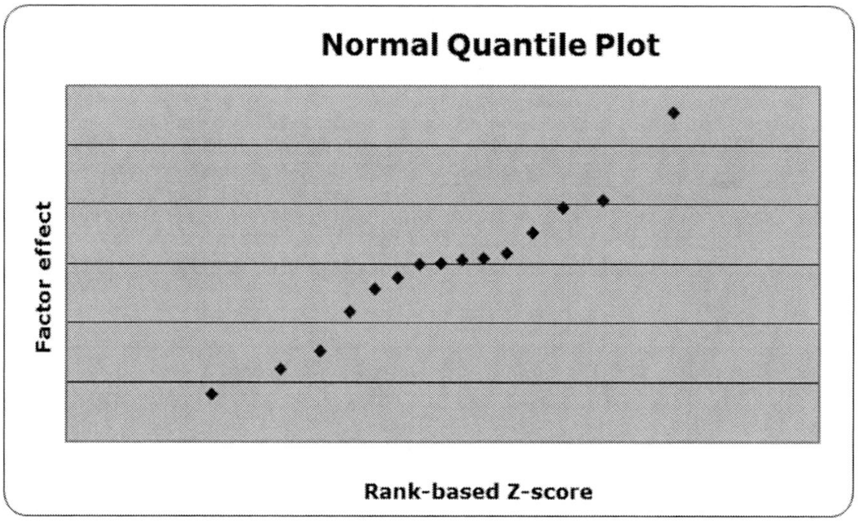

Fig. 2. Normal probability plot of factor effects for single replication

At 5% level of significance ($F_{1, 11}$=4.84), 2.5% level of significance ($F_{1, 11}$=6.72), main factors A,B, C and two factor interactions AB are significant as revealed by the normal probability plot.

3.3. Inference from factor identification

From this analysis, we noticed that in item searching mode apart from the strategies there are several factors plays a major role. But speed level, size of the arena and number of robots most significantly contributed towards this item searching mode. In order to avoid the collision and congestions, the experimenter must consider these three parameters for designing the experiment.

3.4. Various strategies item detection time comparison

It is very essential part in the simulation. Strategy that gives minimum item detection time based on the speed level, number of robots and size of the arena that should be the best strategy. For various strategies item detection time comparison is given in table 7.

Table 7. Various Strategies Item Detection Time Comparison

S. No.	Speed (m/sec)	Size of the Arena ($m^{2)}$	No of Robots	Item Detection Time (Sec)			
				1	2	3	4
1	0.5	20x30	2	105	173	204	158
2	0.5	50x30	4	218	191	208	198
3	1	20x30	2	127	91	152	124
4	1	50x30	4	116	127	132	122
5	0.5	20x30	4	105	88	120	102
6	1	20x30	4	54	52	68	60
7	0.5	50x30	2	449	424	428	430
8	1	50x30	2	121	178	142	164

3.4.1 Inference from Various Strategies Item Detection Time Comparison

From this various strategies Item detection time comparison, we noticed that the second strategy parallel sweep with divider policy gives minimum item detection time with various combinations of speed level, size of the arena and number of robots. According to the eight combinations of these three parameters Parallel sweep with divider policy contributed more compared with other three strategies. So finally, we concluded that parallel sweep with divider policy be the best strategy along with other three strategies

4 Conclusions

Swarm robotics is a number of small robots that are synchronically works together to accomplish a given task. It can be applied to many fields such as agriculture field, mining field, manufacturing industries and military purposes. Foraging is most essential part in swarm robotics. Foraging can be applied to a wide range of useful tasks such as toxic waste cleanup, search and rescue, demining, collection of terrain samples, deep sea and Polar Regions. The foraging behavior of multi-robot system is studied through the various literatures. From literatures various types of algorithms, mathematical model and computer simulations characteristics was studied. For this project state transition diagram, logical flow chart, algorithm, various types of item searching strategy and octagonal robot was developed. Various item searching strategies such as expanding square approach, expanding square with divider policy, parallel sweep approach and parallel sweep with divider policy are simulated by using player/stage software. From the Statistical analysis parallel sweep with divider policy be the best strategy along with other three strategies.

5. Scope for future work

- This paper mainly focused on foraging behaviour of swarm robotics system. Apart from foraging behaviour, there are many problem present in swarm robotics system such as pattern formation, aggregation, hole avoidance, self deployment and chain formation. It is possible to simulate all swarm robotics problem using player/ stage simulation software.
- In this work, few item searching strategies are developed and simulated. In future, the researcher can develop new strategies and it is possible to simulate with different simulation package.
- In this work, number of robots, arena size and position of item and shell are predefined. In future, the researcher can increase number of robots, arena size according to development of new strategies.

References

1. L. Bayindir and E. Sahin (2007) "A Review of Studies in Swarm Robotics", The Turkish Journal of Electrical Engineering and Computer Science Vol.15, No.2, pp. 115-147.
2. Campo and M. Dorigo, V. Longchamp, S. Garnier and C. Pinciroli (2010) "Artificial pheromone for path selection by a foraging swarm of robots", The Journal of Biological Cybernectics vol.103, pp. 339-352
3. Campo and M. Dorigo (2007) "Efficient Multi-foraging in Swarm Robotics", Book Advances in Artificial Life, 9th European Conference, ECAL, vol.4648, pp. 696-705, Lisbon, Portugal.
4. R. Hoff, A. Sagoff, J. Wood and R. Nagpal (2010) "Two Foraging Algorithms for Robot Swarms Using Only Local Communication", International Conference on Robotics and Biomimetics (ROBIO), pp. 123-130, Tianjin.
5. J. Lee, H. Kim and C. Ahn (2012) "Foraging Swarm Robots System Adopting Honey Bee Swarm for Improving Energy Efficiency", International Conference on Ubiquitous Information Management and Communication, ICUIMC '12, Kulalumpur, Malasia.
6. D. Lambrinos, R. Moller, R. Labhart, R. Pfeifer and R. Wehner (2000) "A mobile robot employing insect strategies for navigation", The Journal of robotics and autonomous systems, pp 39-64.
7. K. Sugawara, M. Sano, I. Yoshihara, K. Abe and T. Watanabe (1999) "Foraging Behavior of Multi-robot system and Emergence of Swarm Intelligence", International Conference on System, Man and Cybernetics, Vol.3,pp.257-262,Tokyo.
8. R. Sharpe and B. Webb (1998) "Simulated and situated models of chemical trail following in ants", International conference on simulation of adaptive behaviour, pp 195 -204.
9. K. Krishnaiah and P. Shahabudeen, "Applied Design of Experiments and Taguchi Methods", The Book ISBN-978-81-203-4527-0, pp. 85-153.

Design and Analysis of Split Fixture for Gear Hobbing Machine

Prajwal Shenoy[1], *Naik Nithesh*[2*], *Manjunath Shettar*[2], *Debarghya*[2] and *Wasim Abbas*[2]

[1]*Department of Mechatronics Engineering, Manipal Institute of Technology, Manipal Academy of Higher Education, Manipal, India.*
[2]*Department of Mechanical and Manufacturing Engineering, Manipal Institute of Technology, Manipal Academy of Higher Education, Manipal, India.*

Abstract. Compared to the conventional gear hobbing fixtures, split fixture can effectively reduce job set-up time during the manufacturing process. This paper investigates the behaviour and analysis of split fixture under varying static loading conditions. Design of the part was established by considering the ability of the split fixture to carry jobs of various diameters. In order to validate the design, Static structural analysis was carried out on two positional configurations of the split fixture. A load of 1 ton was applied on the resting face of the fixture to simulate the effect of holding the job. The analysis included a study of the Stress, Deformations, and Modal analysis at different resonating frequencies to check for failure of design. By applying varying loads similar to practical conditions, it was observed that the design successfully withstood the applied forces without failure and a factor of safety of 142 was achieved in a critical loading case. Investigating the effect of dynamic loads on the Split Fixture and including auxiliary assembly components in design analysis.

Keywords: Gear hobbing, Modal analysis, Split fixture, Static structural analysis

1 Introduction

Gear hobbing machine is used for machining and production of an external tooth form developed uniformly about a rotating center [1]. It is a relatively advanced form of machining. Specially designed fixtures are needed for gear hobbing. The design of fixture focuses on elimination of the movement of the workpiece undergoing machining process. Fixtures are generally designed for a specific operation and a given work piece individually. This increases the set up time, also adding to the labour cost and also wasting the company's money. To reduce the set up time without changing the fixtures for the different loads the design of split fixture was taken into consideration. Split fixture has a rotating screw which ensures the jaw expansion, thus reducing the setup time for the jobs of different diameter [2].

* Corresponding author: nithesh.naik@manipal.edu

2 Methodology

2.1 Design

The Split fixture is modelled using modelling software Solid Works®. The current design is derived from the combination of the fixture design and functionality of the currently used in gear hobbing machine which is widely used individually for different sized gears to be hobbed. The functional dimensions and attributes were considered to build the current prototype. The components of the developed design are listed as shown in Fig. 1:
1) Collar. 2) Job Resting section. 3) Ribs 4) Fixture structure plate. 5) Hand wheel.
6) Fixture base. 7) Extended slots 8) T Bar coupler 9) Sliding T bar. 10) Ball Screw Thread.

Fig. 1. Split Fixture Assembly

The split fixture we designed have certain manufacturing and dimensional considerations. It has a load carrying capacity of 1.5-2.5 tons. Ball screw thread is incorporated for easy alteration in the outside diameter of the fixture. [4] The prototype to be produced of Mild Steel 25HRC. The fixture is designed and analysed with two different center distances of 350mm and 550mm respectively. Also the height of the fixture from the fixture base to the job resting section is 360mm. There is a taper of 2 degrees of the fixture structure plate with the vertical. Sliding T bars are provided along worktable slots to slide the component manually and provided with a clearance of 1mm along its edges. [3]

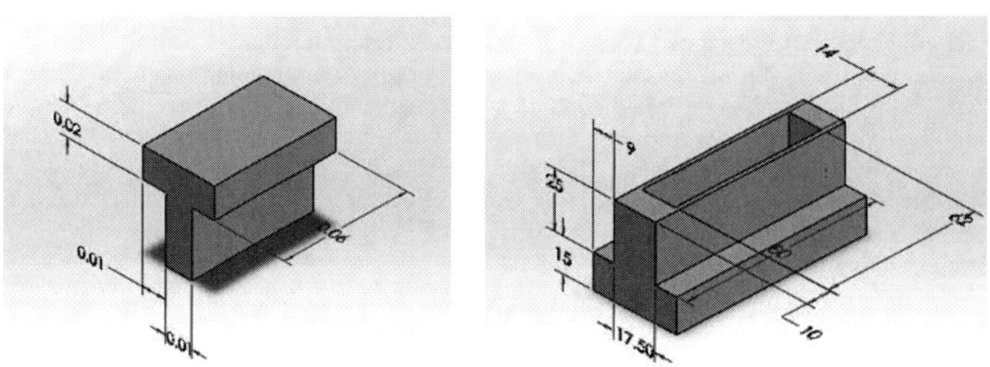

Fig. 2. T Bar and T Bar Coupler

The Fig. 3 shows the split fixture assembly with dimensions. Two models were prepared with different dimensions, Model 1 - Fixture A had CC*-350mm, Height-390mm and resting face-70mm, followed by Model 2 - Fixture C which had CC-550mm, Height of 450mm and the resting face of 75mm.

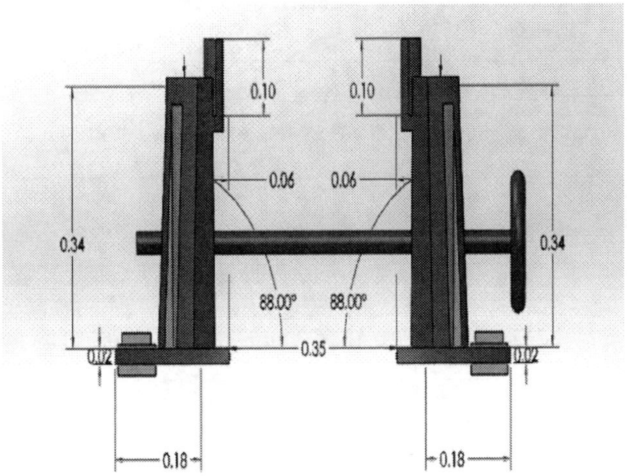

Fig. 3. Split Fixture A - Assembly Dimensions.

2.2 Analysis

The analysis of this split fixture was carried out using ANSYS Workbench Version 12.0. The meshing size is medium, and has a fine setting for proximity and Curvature. Analysis is carried out with a boundary condition of a fixed base (all DOF locked), whereas the topmost face, known as the Job resting face above has a pressure loading of 1tonne. The Modal analysis is being carried out considering the Fixture base as FIXED. The results being analysed involve the study of total deformations, Von Misses Stress, Modal analysis (first 5 frequencies). The underlying table gives a brief observation based on the analysis carried out. All the values enlisted in the table are the maximum values under the defined loading conditions. [5]

3 Results and Discussions

The Table 1 shows the results of total deformations, Von Mises stress and the five mode shapes of the Model 1 and Model 2 derived from modal analysis. The Model 1 and Model 2 with the centre distance of 350 mm and centre distance of 550 mm were selected based on the size of the gear to be hobbed. However the same fixture can be used for varying dimensions of centre distance over a range of 250 mm to 1000 mm.

As shown in the Fig. 4 and Fig. 5 the analysis of the 350mm configuration resulted in a maximum deformation of 0.001589m. The analysis of the 550mm configuration resulted in a maximum deformation of 0.001589m.

Table 1. Results of Total deformation, Stress and Modal Analysis of Model 1 and Model 2.

Analysed Parameter	Model 1 Centre Distance 550 mm Observations	Model 2 Centre Distance 350 mm Observations	Model 1 Centre Distance 550 mm Resonating frequency	Model 2 Centre Distance 350 mm Resonating frequency
Total Deformation	0.0015891 mm	0.0015891 mm	NIL	NIL
Von Mises Stress	1.6098e6	1.7611e6	NIL	NIL
Modal Analysis Mode Shape 1	13.307 mm	13.379 mm	490.64 Hz	490.66 Hz
Modal Analysis Mode Shape 1	13.31 mm	13.381 mm	490.65 Hz	490.67 Hz
Modal Analysis Mode Shape 1	12.096 mm	12.309 mm	1007.1 Hz	1007.3 Hz
Modal Analysis Mode Shape 1	12.096 mm	12.312 mm	1007.1 Hz	1007.4 Hz
Modal Analysis Mode Shape 1	38.748 mm	38.744 mm	1333.1 Hz	1333.1 Hz

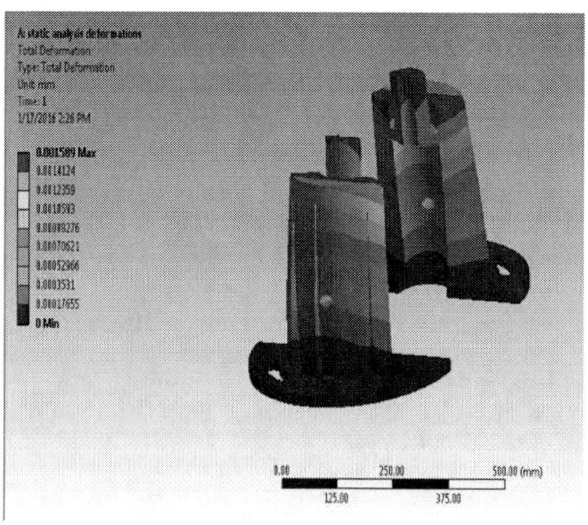

Fig. 4. Total Deformation (CC 350mm).

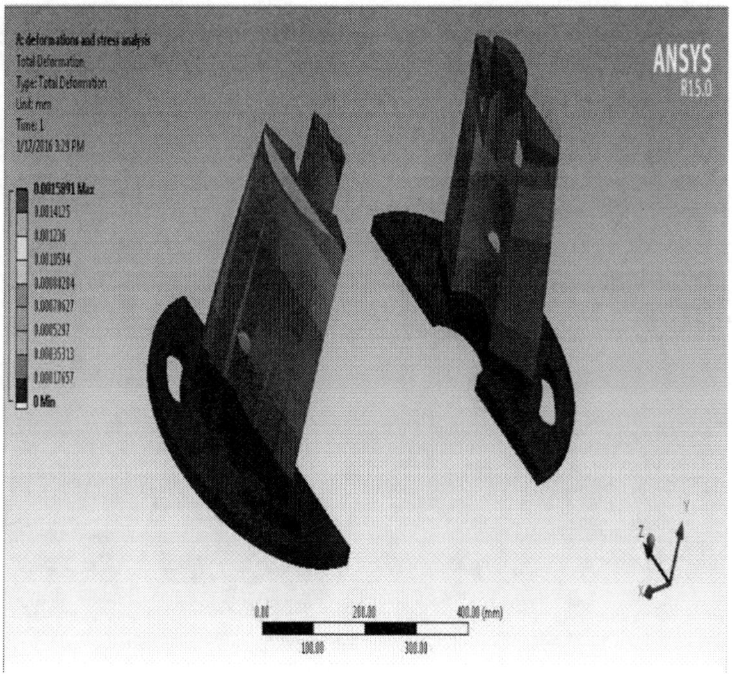

Fig. 5. Total Deformation (CC 550mm).

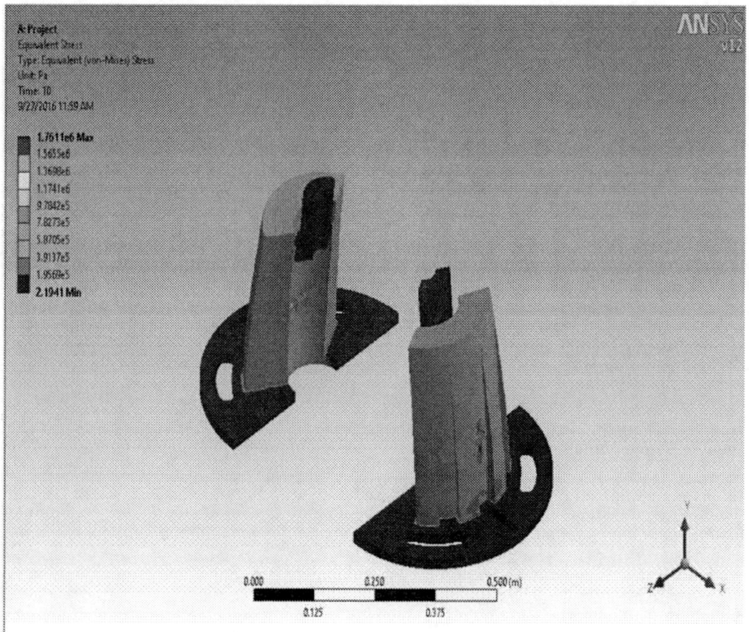

Fig. 6. Von-Misses Stress (CC 350mm).

The analysis of the 350mm configuration resulted in a maximum Von-Misses stress of 1.7611 MPa, which is less than 250 MPa; the yield strength of Static Structural Steel (25HRC).

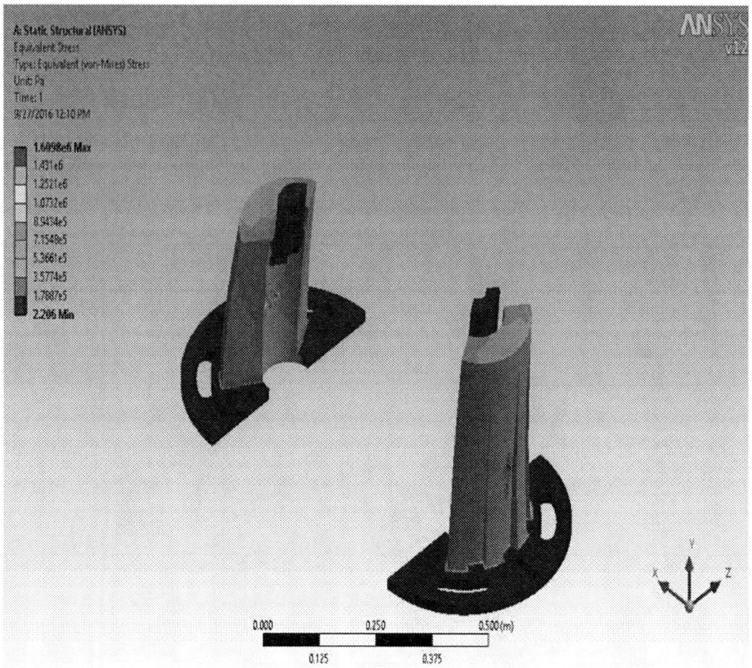

Fig. 7. Von-Misses Stress (CC 550mm).

The analysis of the 550mm configuration resulted in a maximum Von-Misses stress of 1.6098 MPa, which is less than 250 MPa; the yield strength of Static Structural Steel (25HRC).

The plot of force v/s displacement is given below (**Fig.** 8 and **Fig.** 9). It can be observed from the linear nature of the curve that the stress values on the fixture lie within elastic limits and hence below the yield stress of Structural Steel (25HRC). [6]

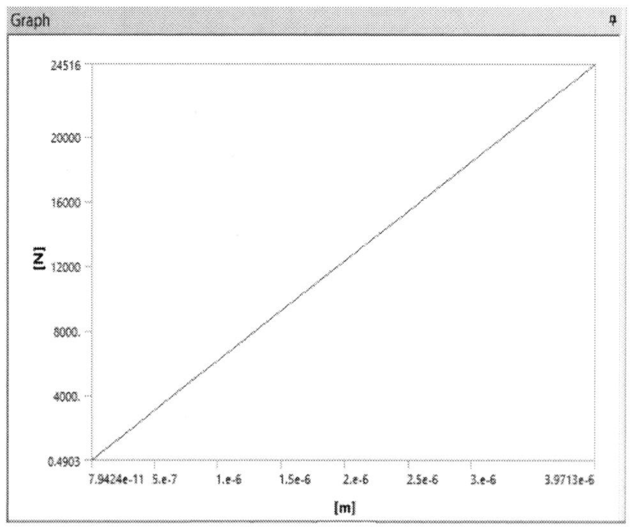

Fig. 8. Force v/s Deformation curve (CC 350mm).

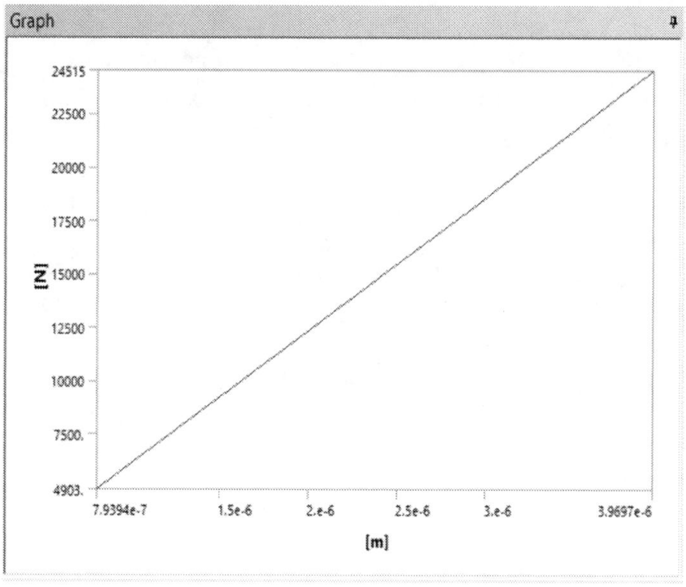

Fig. 9. Force v/s Deformation curve (CC 550mm).

4 Conclusions

The results derived through a simulation analysis considering the mechanical behaviour of the split fixture under uniformly distributed load of 1 Ton. The analysis was performed on two configurations of centre to center distance- (a) 350mm and (b) 550mm.

- The split fixture design built is suitable to cater heavy loads as heavy as uniformly distributed load of 1 Ton.
- The fixture designed will cater to the wide range of diameter of the gear whose OD lies between 200 mm to 1000 mm.
- The fixture designed will reduce the setup time hence reducing the overall manufacturing lead time of the gear production process.
- The above results show that designed structure lies within the acceptable limits of practical application and can be utilized for Industrial applications to reduce the set up time and increase the productivity.

The authors would like to be obliged to Department of Mechanical and Manufacturing. Engineering, Manipal Institute of Technology, Manipal Academy of Higher Education, for their extended support for providing lab and testing facility for the investigation

References

1. M. Yu Wang, P. Diana, CAD-based simulation of the hobbing process for the manufacturing of spur and helical gears, International Journal of Advanced Manufacturing and Technology, 347–357, (2009).
2. Karan Sharma, Naik Nithesh, Arun Prabhu, George Varghese, "Setup Time Reduction For CNC Hobbing Machine Implementing SMED And Design of "Split Fixture" International Journal of Mechanical And Production Engineering, ISSN: 2320-2092, Volume- 2, Issue-11, (2014).

3. Rong Y, Bai Y, Machining accuracy analysis for computer aided fixture design verification. Journal of Manufacturing Science Engineering, 118, 289–300, (1996)
4. Cai W, Hu SJ et al, A Variation method of robust fixture configuration design for 3-D workpieces. Journal of Manufacturing Science Engineering, 119, 593–602, (1997).
5. Djurdjanovic D, Ni J, Dimensional errors of fixtures, locating and measurement datum features in the stream of variation modelling in machining. Journal of Manufacturing Science Engineering, Transactions of ASME, 125(4), 716–730, (2003).
6. K. Kim, M.K. Kim, Volumetric accuracy analysis based on generalized geometric error model in multi-axis machine tools, Mech. Mach. Theory, 26(2), 207–219, (1991).

Power loss analysis in altered tooth-sum spur gearing

Sachidananda H. K.[1*], *Raghunandana K.*[2] and *Shivamurthy B.*[3]

[1]*Department of Mechanical Engineering, School of Engineering and IT, Manipal University, Dubai, UAE.*
[2]*Department of Mechatronics Engineering, Manipal Institute of Technology, Manipal Academy of Higher Education, Manipal-576104, India.*
[3]*Department of Mechanical and Manufacturing Engineering, Manipal Institute of Technology, Manipal Academy of Higher Education, Manipal-576104, India.*

Abstract. The main cause of power loss or dissipation of heat in case of meshed gears is due to friction existing between gear tooth mesh and is a major concern in low rotational speed gears, whereas in case of high operating speed the power loss taking place due to compression of air-lubricant mixture (churning losses) and windage losses due to aerodynamic trial of air lubricant mixture which controls the total efficiency needs to be considered. Therefore, in order to improve mechanical efficiency it is necessary for gear designer during gear tooth optimization to consider these energy losses. In this research paper the power loss analysis for a tooth-sum of 100 altered by ±4% operating between a specified center distance is considered. The results show that negative altered tooth-sum gearing performs better as compared to standard and positive altered tooth-sum gearing.

Keywords: Altered tooth-sum gearing, Power loss, Standard gearing, Churning losses, Windage losses.

1 Introduction

The windage power loss due to aerodynamic trial of air lubricant mixture is one of the important aspects which need to be considered in gear transmission systems at high speeds during optimization of gear tooth shape. In this context Fakher chaari et al., [1] have studied the windage power loss in spur gear sets and they concluded that windage power loss is a complex phenomenon and depends upon rotational speed, shape of tooth, number of tooth and module. The windage power loss using parametric modelling has been studied by satya seetharaman et al., [2]. According to them the windage power loss occurs due to interaction of gears with air-oil mixture and squeezing of the lubricant at gear mesh interface. The efficiency of the vehicle transmission system considering churning loss, bearing loss and windage loss on spur gear was studied by Wang et al., [3] using analytical and experimental results. They concluded that the losses estimated using analytical equations was in line with experimental results. The overall

* Corresponding author: sachidananda6@gmail.com

efficiency in multistage gear train has been studied by James kuria et al., [4]. They developed a mathematical model in order to determine efficiency considering lubricant, surface finish and operating conditions and concluded that sliding friction, rolling friction and windage losses are the main source of power loss in gearing.

While gear are in mesh the gears rolls and slides and the power loss taking place during this process is termed as sliding power loss. Chakri yenti et al., [5] have studied empirical formulae for friction coefficient considering gear module, pressure angle and gear ratio. They concluded using experimental analysis that sliding power loss are more in gears having larger module as compared to smaller module. Also, larger pressure angle have lower sliding losses as compared to smaller pressure angle. Franco concli et al., [6] have studied oil squeezing power loss in spur gears. According to them efficiency is more concerned in design of power transmission and they concluded that the power loss increases with gear rotational speed. The analytical expression of efficiency of standard and high contact ratio gears has been studied by Miguel pleguezuelos et al., [7]. According to them the friction coefficient varies along the length of path of contact due to sliding and the average values of friction coefficient needs to be considered during numerical calculations. Chang et al., [8] have studied meshing losses in high load and high speed involute gears considering gear module, pressure angle, addendum height, thermal conductivity and lubricant pressure and they concluded that these parameters have significant effect on gear meshing process. The theoretical gear power loss considering density, viscosity and speed have been studied and compared using experiment by vijay kumar et al., [9] and they concluded that these parameters influence power loss significantly.

The influence of pressure angle and number of teeth variation on mechanical efficiency during varying teeth pairs in contact have been studied by chaari et al., [10] considering internal loss coefficient during rolling and sliding of gears. The power loss due to trapping and squeezing of an incompressible fluid between meshing of gear teeth have been studied by Abdelilah et al., [11]. They concluded that the rotational speed has an influence on meshing time and the operating trapped fluid. Fernandes [12] have studied load distribution along the length of path of contact to estimate gear loss factor and its influence on power loss prediction and concluded that the classical formulas are accurate in some specific condition. Luis magalhaes et al., [13] have studied influence of tooth profile and oil formation on power loss. They concluded that the friction losses need to be reduced to improve lower energy consumption, lower operating temperature and lower risk of failures.

2 Methodology

2.1 Estimation of power loss.

The amount of power loss during gear tooth mesh depends upon type of gear tooth, characteristics of gear tooth engagement and coefficient of friction during mesh. The amount of power loss during transmission occurs due to sliding, rolling, formation of elasto-hydrodynamic film, windage and churning losses. During gear tooth mesh as the load is continuously changing the power loss due to sliding plays an important role. The power loss in case of gears is influenced by angular velocity, ratio of rolling velocities, point of contact, contact ratio and lubricant properties. Sliding friction depends upon sliding velocity and friction force which in turn depends upon tooth load and instantaneous coefficient of friction. In this research paper the sliding power loss and coefficient of friction along the length of path of contact has been taken into consideration. Also, the efficiency is calculated based on the power loss due to sliding. Benedict and Kelly [14, 15], developed a model considering coefficient of friction along the path of contact, tooth load, sliding velocity, tangential velocity and lubricant viscosity to find sliding power loss as shown in equation (1) and (2).

Coefficient of friction along the path of contact

$$FS\# = 0.0127\ log_{10}(29.66\ \left(\frac{W\#}{\vartheta}\right) V_s\#(V_{t\#}{}^2)) \quad (1)$$

Sliding power loss along the path of contact

$$PS\# = \frac{FS\# \times V_S\# \times W\# \times 10^{-3}}{cos\left(tan\frac{(T\#)}{R_b}\right)} \quad (2)$$

Efficiency considering sliding power loss

$$Efficiency = 1 - \left(\frac{Total\ power\ loss\ per\ mesh}{Power\ transmitted}\right) \times 100 \quad (3)$$

Where, FS is coefficient of friction, W is the tooth load, υ is the viscosity of the oil, V_S is sliding velocity, V_t is the tangential velocity, PS is sliding power loss, R_b is the base circle radius of pinion, # is different points along length of path of contact as A, B, D and E.

In this power loss analysis both the gears are made of steel C-40, υ=0.3, E=200 GPa, module of 2 mm, pressure angle of 20° (lower range), 25° (higher range) and tangential tooth load of 10 N per unit millimeter of face width is considered for computation. Based on the above model the power loss along length of path of contact due to sliding is computed for altered tooth-sum of 100 ± 4% at different points A, B, D and E along the length of path of contact.

3 Results and discussion

By using Benedict and Kelly models a C-program has been developed based on the methodology to estimate sliding power loss of the spur gear tooth for various values of profile shift for a fixed center distance for altered tooth-sum gearing (100 by ± 4%). Fig. 1(a) shows the plot of power loss versus profile shift for a 96 tooth-sum meshed gear system. It is observed from the plot that the power loss at point A, B, D and point E are 2.17 watts, 3.40 watts, 0.2 watts and 0.17 watts respectively for profile shift X_1=0.4 on pinion. The power loss at point A and B decreases continuously as the values of profile shift increases due to reduction in contact friction. Similarly, the power loss at point D and E increases continuously as the value of profile shift increases due to increased contact friction. However, at point A and B friction is more as compared to point D and E. Hence the power loss is more at these two points.

Similarly, it is observed that for a range of tooth-sums 100 ± 4%, the power loss decreases at point A and B and increases at point D and E for various values of profile shift coefficient on pinion (Fig. 1(a) to Fig. 1(c)). It is observed that the power loss for profile shift X_1=-0.1 at point E is zero watts (ideal case) for tooth-sum of 97 due to very small sliding distance (E to C) which in turn low sliding velocity. This leads to lower friction and lower power loss. From this it is evident that power loss depends on sliding distance, sliding velocity and friction force. Also, it is observed that sliding velocity depends upon normal tooth load and instantaneous coefficient of friction.

However, it is noticed that not much variation of power loss in case of negative alteration (96), standard (100) and positive alteration (104) in tooth-sum gearing. Also, the power loss at different points along the length of path of contact remains approximately the same for various values of profile shift on pinion.

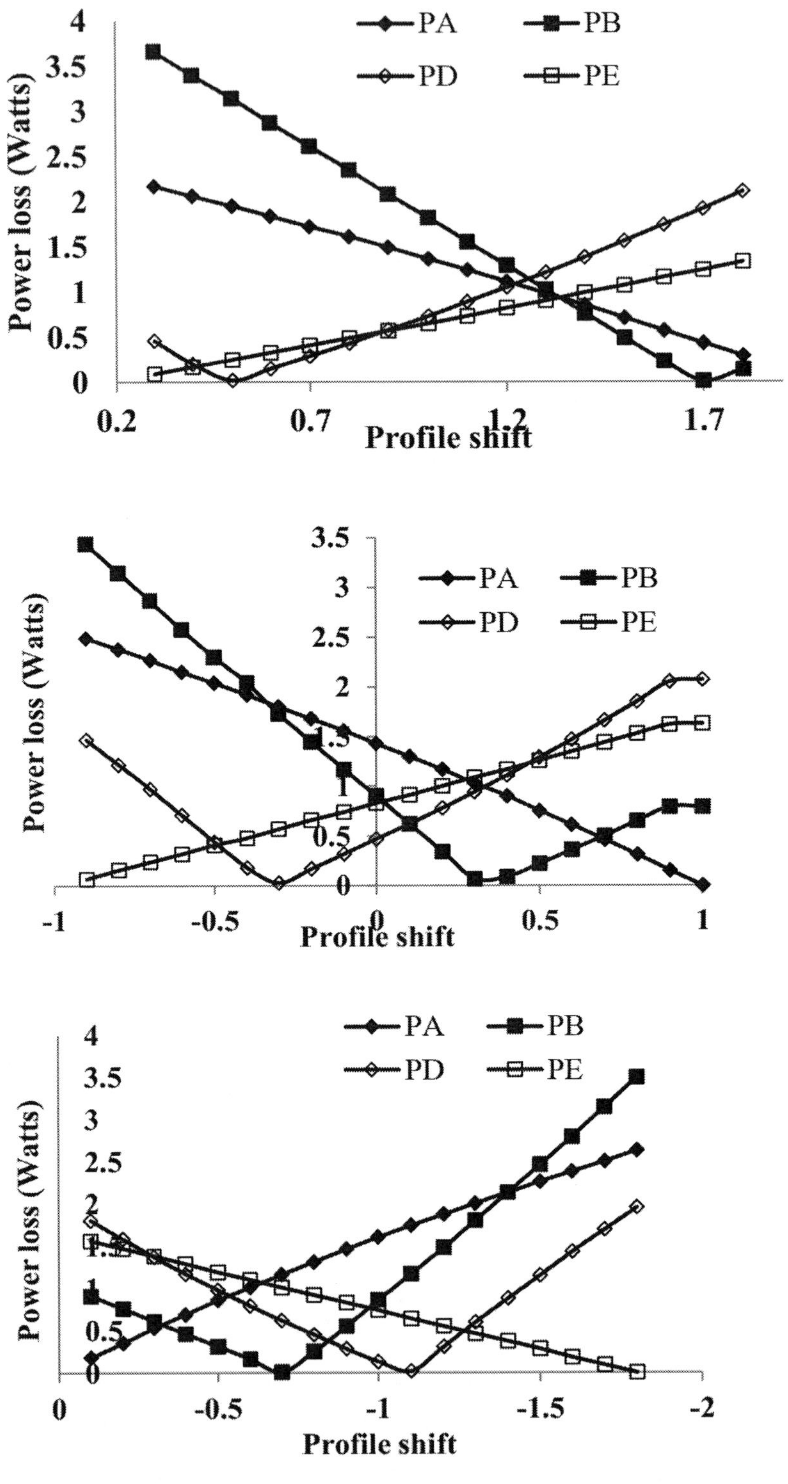

Fig. 1. Plot of power loss versus profile shift along the length of path of contact for higher range of pressure angle: a) 96 tooth-sum b) 100 tooth-sum c) 104 tooth-sum

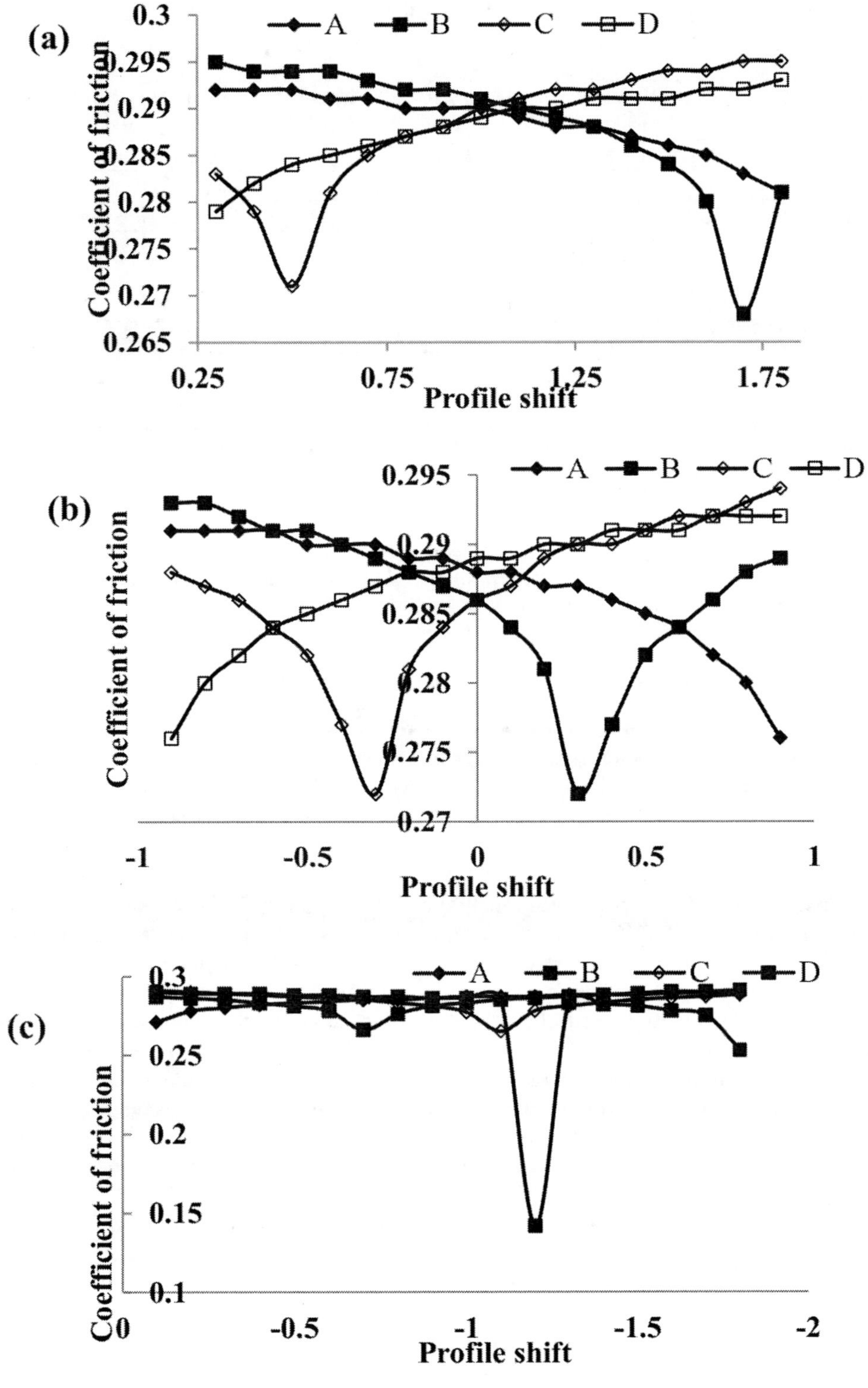

Fig. 2. Plot of coefficient of friction versus profile shift along the length of path of contact for higher range of pressure angle: a) 96 tooth-sum b) 100 tooth-sum c) 104 tooth-sum.

Table 1. Power loss and coefficient of friction for 96, 100 and 104 tooth-sum at point A, point B, point D and point E for higher range of pressure angle.

Profile shift	**96 tooth-sum**	**Point A**	**Point B**	**Point D**	**Point E**
	Power loss(Watts)	2.17	3.40	0.2	0.17
0.4	Friction coefficient	0.292	0.294	0.279	0.282
Profile shift	100 tooth-sum	Point A	Point B	Point D	Point E
	Power loss(Watts)	0.91	2.25	2.99	0.23
0.4	Friction coefficient	0.29	0.29	0.277	0.286
Profile shift	104 tooth-sum	Point A	Point B	Point D	Point E
	Power loss(Watts)	0.69	0.46	1.17	1.29
-0.4	Friction coefficient	0.282	0.283	0.288	0.289

Fig. 2 (a) shows the plot of coefficient of friction versus profile shift for 96 tooth-sum gearing (higher range of pressure angle). It is noticed that, the coefficient of friction decreases at point A and B and increases at point D and E for various profile shift. Similar trend was observed in 100 and 104 tooth-sum gearing (Fig. 2 (b) and Fig. 2 (c)). The power loss and coefficient of friction for 0.4 profile shift for altered tooth-sum gearing and standard gearing is tabulated in Table 1. From this Table 1 it is observed that for positive profile shift the total power loss is more as compared to negative profile shift and occurs due to pressure gradient difference [1]. It is also noticed from the tabulated values that the coefficient of friction is slightly higher for 96 tooth-sum as compared to 100 and 104 tooth-sum.

3.1 Power loss, film thickness and specific film thickness using KISSSOFT®

KISSsoft ® is the commonly used software by gear designers for strength analysis of gears which are used in construction equipment, cable car transmissions, mars rovers etc. The strength analysis of gears by KISSsoft ® is based on ISO, AGMA and DIN standards. In addition to the strength analysis, it also includes design and optimization. In order to compare the power loss, negative altered tooth-sum gearing (96 tooth-sum) and standard tooth-sum gearing (100 tooth-sum) have been considered. The analysis of power loss, film thickness and specific film thickness has been compared using KISSOSFT® for the above tooth-sum. Fig. 3 shows the plot of power loss versus angle of rotation for 96 tooth-sum for different points along length of path of contact for profile shift of 0.3. From this it is observed that the power loss at point A is 0.32 watts per unit mm of face width. Also, from the graph it is observed that the power loss at point C (pitch point) is zero and occurs due to pure rolling. The power loss at end point of contact is 0.04 watts per unit mm. From this it can be concluded that the power loss reduces from point B (HPTSC) continously upto pitch point C and it slightly increases from point C to end point of contact. Similarly, Fig. 4 shows the plot of film thickness versus angle of rotation for 96 tooth-sum. From this plot it is observed that the film thickness at initial point of contact A is 0.79 μm and reduces at point B (HPSTC) and slightly increases from point B to point D and again slightly increases from point D to end point of contact (point E). The film thickness at point B (HPSTC) and point D (LPSTC) approximately remains constant at these two points and the film thickness is lesser compared to other points of contact. These are the two critical points during gear teeth mesh as the load is carried by only one pairs of teeth during this region and more chances of crack propogation occurs at these points of contact.

Fig. 5 shows the specific film thickness versus angle of rotation for 96 tooth-sum at different points along the length of path of contact. Specific film thickness ratio is the ratio of lubrication oil film thickness to the combined roughness and is also termed as lambda ratio (λ). Using specific film thickness, it is possible to identify whether the lubrication condition occuring during gear teeth mesh can be full film or EHD, mixed wear lubrication and boundary lubrication. If the specific film thickness ratio lies above 3 then the lubrication occuring between the gear teeth is full film or EHD lubrication and the contact between the asperities is negligible which in turn friction and wear is less. Similarly, if the specific film thickness ratio is between 1.4 and 3 then the lubrication maintained between the gear teeth is mixed wear lubrication and under these circumstances some contact will takes place between the asperities and friction and wear will be more [16]. If the film thickness ratio is less than 1.4 then the lubrication maintained between the gear teeth is boundary lubrication and severe wear [17].

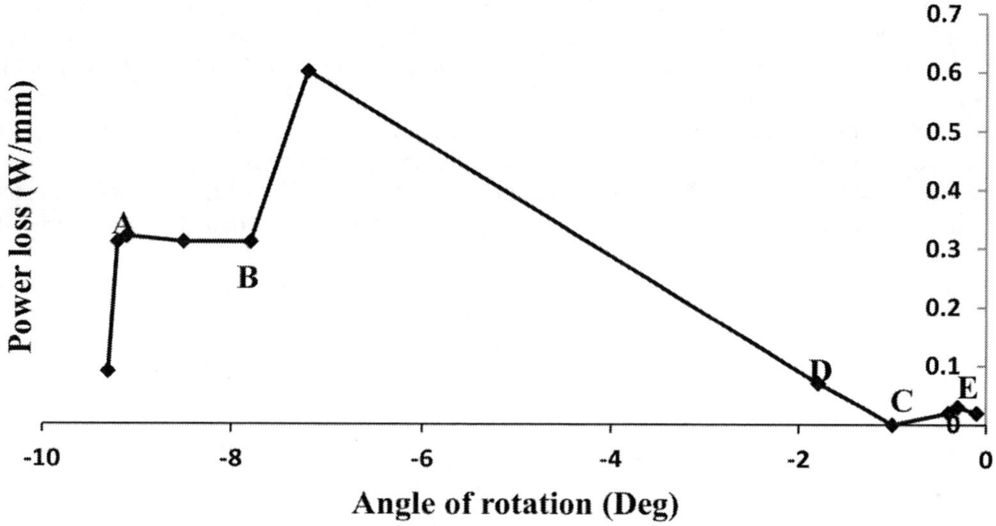

Fig. 3. Plot of power loss versus angle of rotation (Deg) for 96 tooth-sum

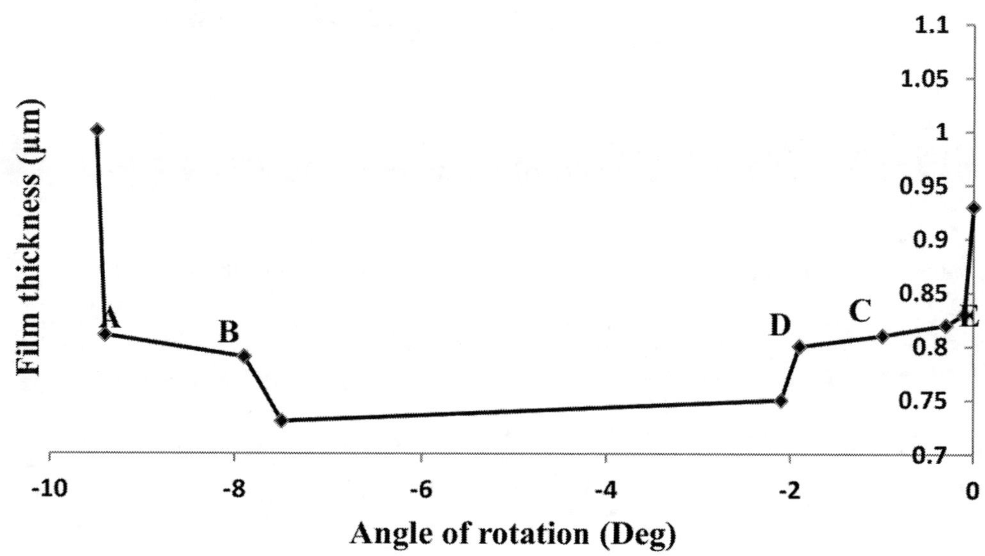

Fig. 4. Plot of film thickness versus angle of rotation (Deg) for 96 tooth-sum

Fig. 6 to Fig. 8 shows the plots of power loss, film thickness and specific film thickness for 100 tooth-sum for various angle of rotation of pinion. The analysis for power loss, film thickness and specific film thickness performs in the same way as discussed above for 96 tooth-sum.

As a power loss comparison for 96 tooth-sum and 100 tooth-sum from Fig. 3 and Fig. 6. It is observed that the angle of rotation in case of 96 tooth-sum will vary from -9.2 degrees to 0.1 degree whereas in case of 100 tooth-sum the angle of rotation will vary from -6.2 to 5.1 degrees. From Fig. 3, it is observed that the power loss continuously decreases from point B (LPSTC) to point D (HPSTC) for 96 tooth-sum whereas in case of 100 tooth-sum the power loss increases from point B (LPSTC) and then slightly decreases up to point D (HPSTC). From this it is observed that the power loss approximately decreases in case of 96 tooth-sum, whereas it behaves abruptly in case of 100 tooth-sum at different points along the length of path of contact. The power loss per unit mm in case of 96 tooth-sum is lesser as compared to that of 100 tooth-sum at initial point of contact and end of point of contact. Similarly, the film thickness in case of 96 tooth-sum is more as compared to that of 100 tooth-sum. From this it can be concluded that 96 tooth-sum have greater resistance to the damage to fatigue stresses and the process of damage is delayed as compared to 100 tooth-sum [18]. Also, it is observed that the specific film thickness ratio is more in case of 96 tooth-sum as compared to 100 tooth-sum. This shows that the service life of the gearing in case of 96 tooth-sum is more as compared to 100 tooth-sum.

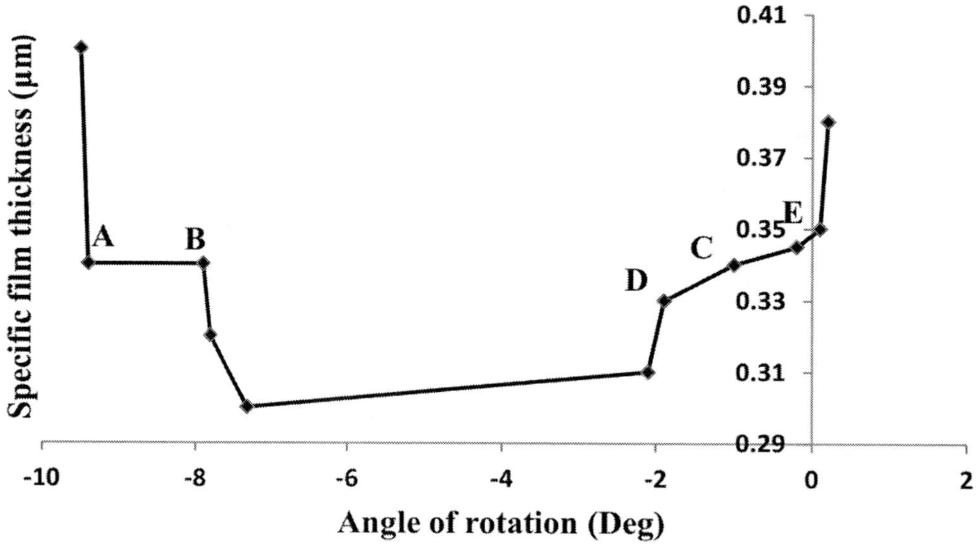

Fig. 5. Plot of specific film thickness versus angle of rotation (Deg) for 96 tooth-sum

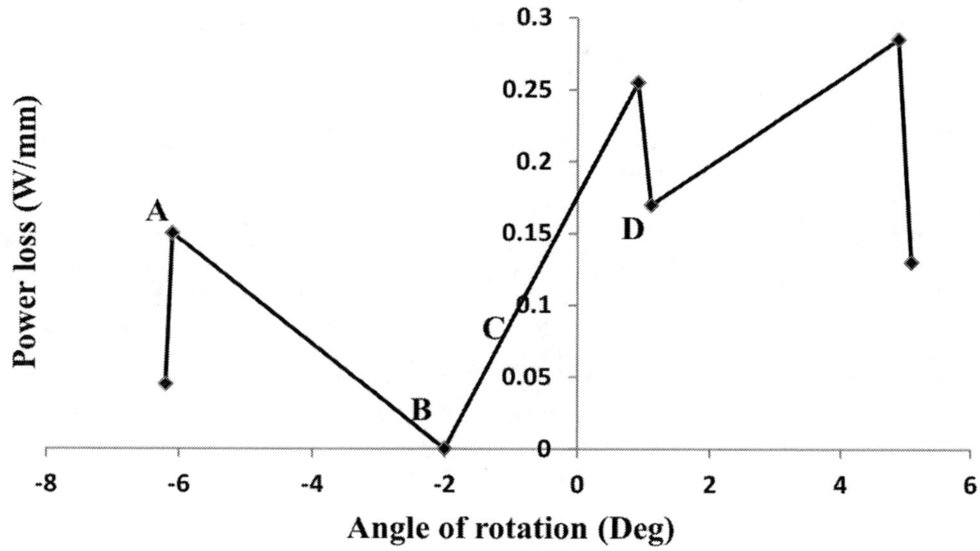

Fig. 6. Plot of power loss versus angle of rotation (Deg) for 100 tooth-sum

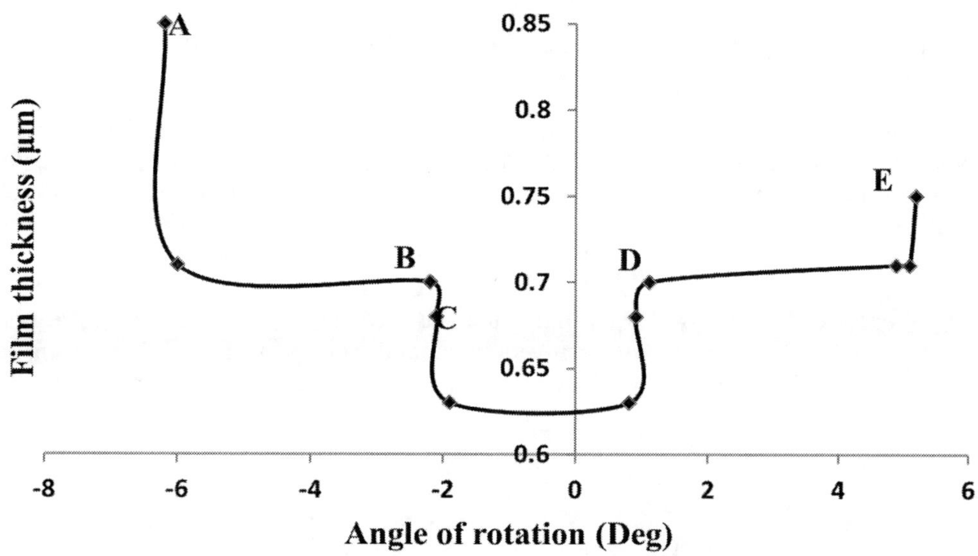

Fig. 7. Plot of film thickness versus angle of rotation (Deg) for 100 tooth-sum

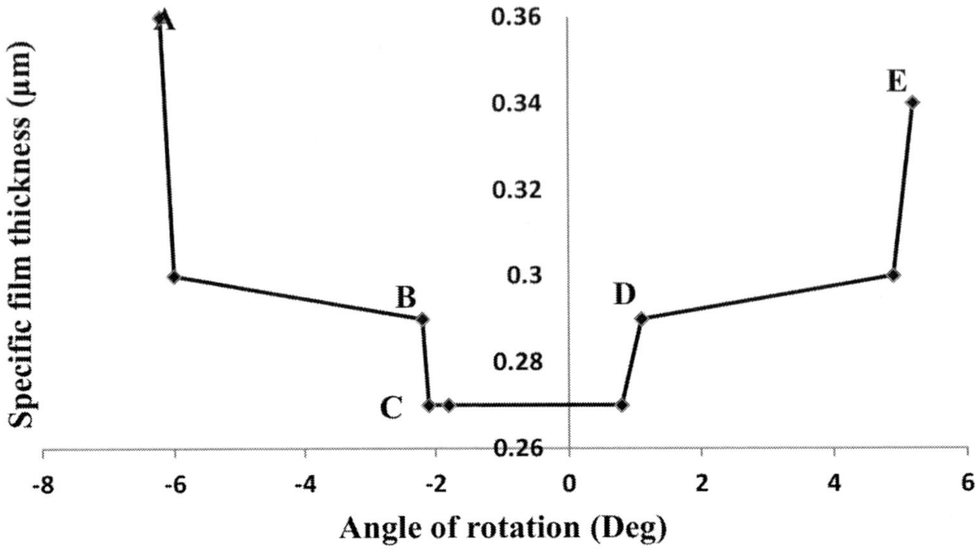

Fig. 8. Plot of specific film thickness versus angle of rotation (Deg) for 100 tooth-sum

3.2 Comparison of percentage change in power loss for higher and lower range of pressure angle.

In case of 96 tooth-sums the profile shift will vary from 0.3 to 1.8. But, in case of 100 tooth-sums the profile shift will vary from -0.9 to 0.9. An attempt is made to compare the power loss in common profile shift (between 0.3 to 1) of both altered tooth-sum gearing and standard gearing. It is noticed that the power loss is directly proportional to profile shift (Fig. 9 (a)). At point A negative alteration in tooth-sums has 2.73 times less power loss as compared to standard tooth-sum gearing. Similarly, negative altered tooth-sum gearing at different points B, D and E the power loss is 14.26, 11.28 and 4.01 times compared to standard tooth-sums gearing. The percentage changes in power loss of positive altered tooth-sum gearing at different points along A, B, D and E will be 2.29, 36.6, 0.25 and 3.17 times lesser as compared to standard tooth-sum gearing.

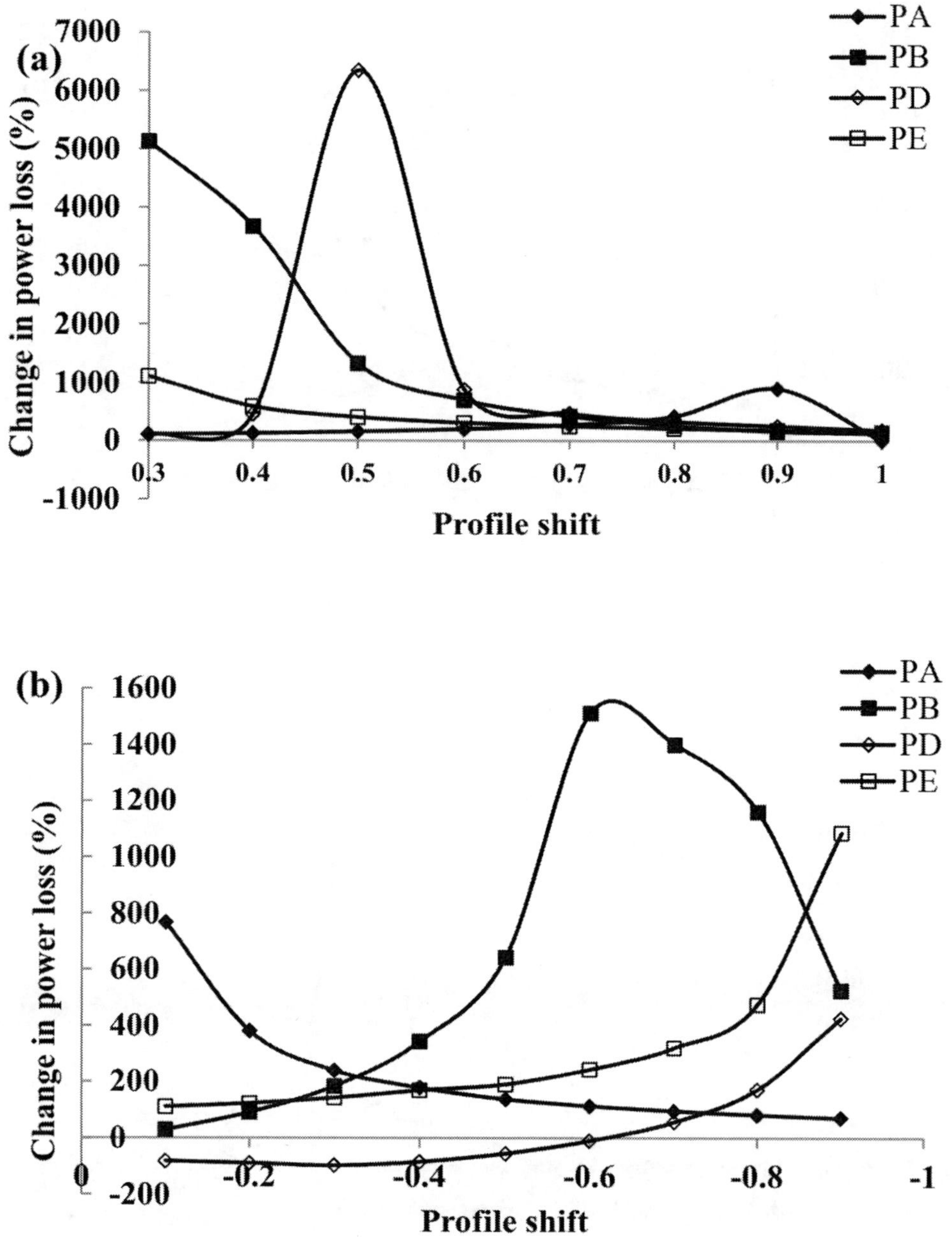

Fig. 9. Plot of percentage change in power loss along length of path of contact for higher range of pressure angle: a) 96 and 100 tooth-sum b) 100 and 104 tooth-sum.

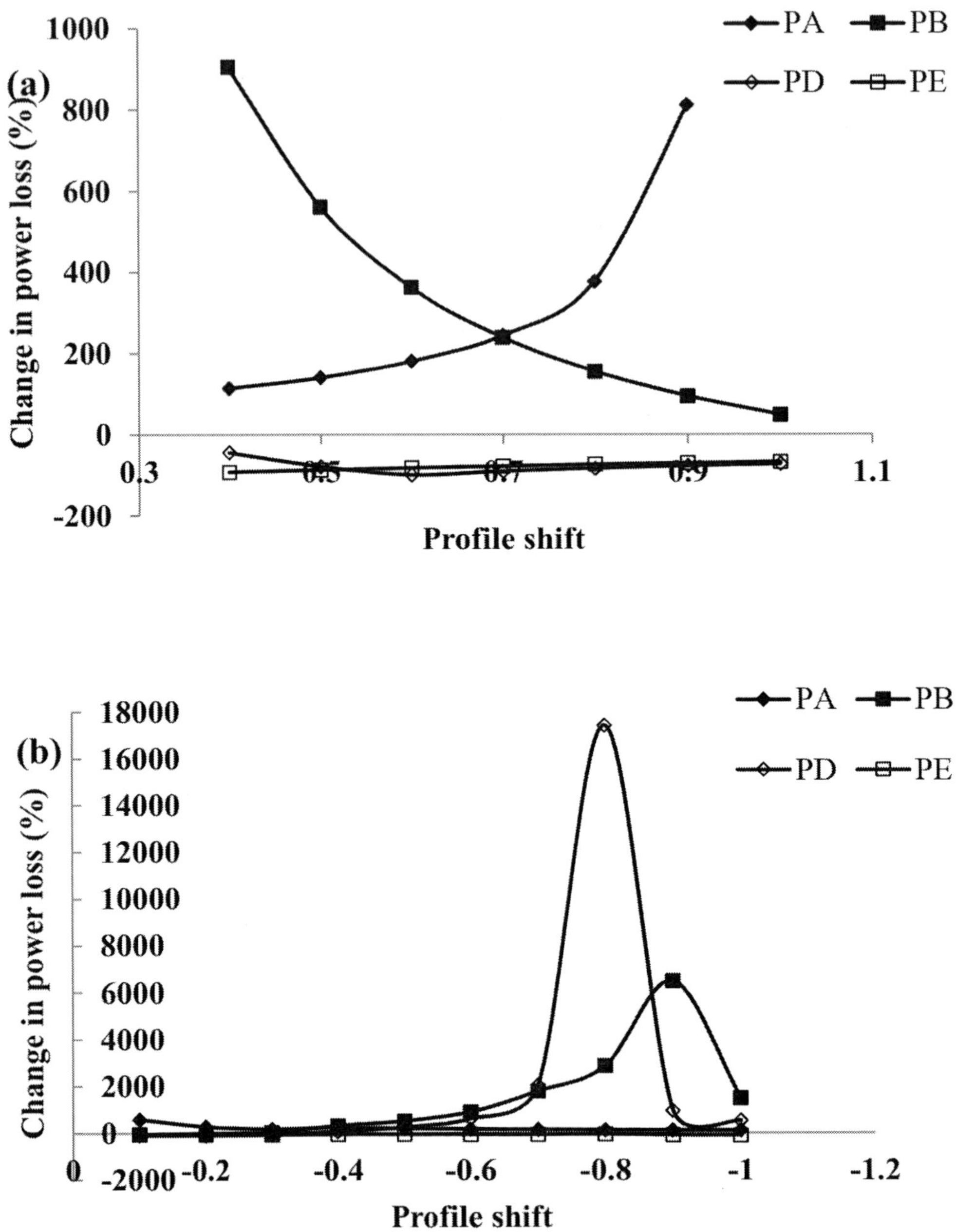

Fig. 10. Plot of percentage change in power loss along length of path of contact for lower range of pressure angle: a) 96 and 100 tooth-sum b) 100 and 104 tooth-sum.

Fig. 10(a) shows the percentage difference in power loss at different points along the length of path of contact for 96 tooth-sums and 100 tooth-sums for lower range of pressure angle. In case of 96 tooth-sums the profile shift will vary from 0.4 to 1.9 whereas for 100 tooth-sums the profile shift will vary between -1 to 1. The percentage change in power loss has been calculated by considering the range of profile shift from 0.4 to 1 which is common in both the cases. At point A negative alteration in tooth-sum has 2.66 times less power loss as compared to standard tooth-sum gearing. Similarly, negative altered tooth-sum gearing at different points B,

D and E the power loss is 3.38, -7.8 and -7.8 times compared to standard tooth-sum gearing. The percentage changes in power loss of positive altered tooth-sum gearing at different points along A, B, D and E will be 2.15, 14.3, 21.6 and -0.6 times lesser as compared to standard tooth-sum gearing.

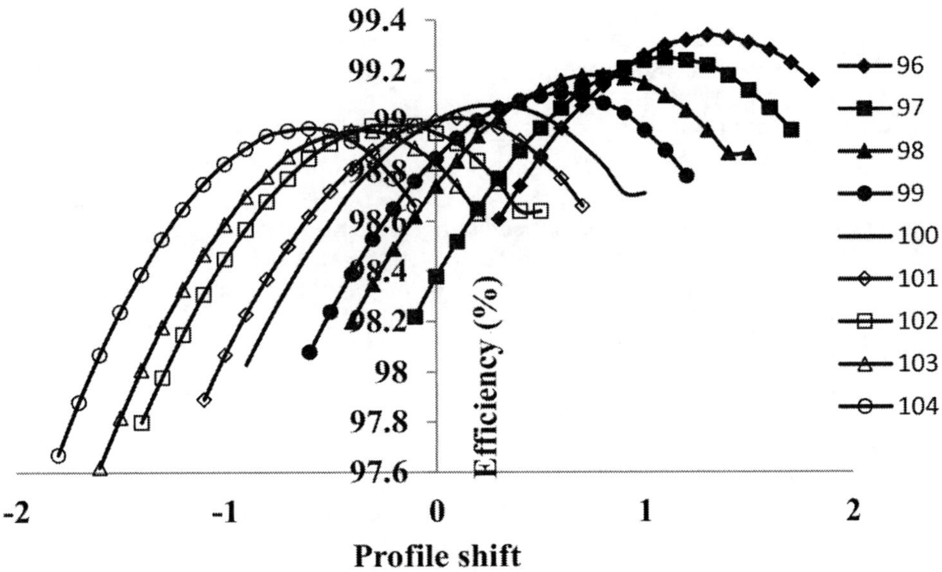

Fig. 11. Efficiency versus profile shift for a tooth-sum of 100 teeth altered by ±4 for higher range of pressure angle.

It is observed that the efficiency was more than 98% (Fig. 11) for both altered tooth-sum gearing and standard gearing. However, efficiency of negative alteration in tooth-sum is slightly more compared to standard and positive alteration in tooth-sum due to reduction in power loss. Also, it is noticed (Table 2) that the efficiency in case for higher range of pressure angle is more as compared to lower range of pressure angle.

Table 2. Comparison of maximum efficiency for higher and lower range of pressure angle for altered tooth-sum gearing.

Altered tooth-sum	For higher range of pressure angle		For lower range of pressure angle	
	Profile shift X_1	Efficiency %	Profile shift X_1	Efficiency %
96	1.3	99.34	1.4	99.31
100	0.2	99.06	0.3	98.79
104	-0.6	98.97	-0.5	98.48

4 Conclusions

The sliding power loss computed for altered tooth-sum operating between a specified center distance has been studied. From this analysis it is observed that the sliding power loss in case of negative altered tooth-sum gearing is less as compared to standard and positive altered tooth-sum gearing. Also, the efficiency observed in case of negative altered tooth-sum gearing is more as compared to standard and positive altered tooth-sum gearing. Efficiency observed in case of higher range of pressure angle is slightly more as compared to lower range of pressure angle.

References

1. Fakher, Chaari., Manel, ben, Romdhane., Walid, baccar., Taher, fakhfakh., and Mohamed, haddar., Windage power loss in spur gear sets, Wseas transactions on applied and theoretical mechanics, 7(2), 159-168, (2010).
2. Satya, Seetharaman., and Ahmet, Kahraman., A windage power loss model for spur gears, Tribology transactions, Society of tribology and lubrication engineers, Taylor and Francis, 54(4), 473-484, (2010).
3. Wang, B., Lu, Z., J, Chen, X. B., Wang, P., Luo, W., and Xu, h., Modelling and analysis of dynamic efficiency of manual transmission/reducer, Intl. J. of Auto. Tech. 16(3), 417-426, (2015).
4. James, Kuria., and John, Kihiu., Prediction of overall efficiency in multistage gear train, World academy of science, Engg and Tech, 5(2), 50-56, (2011).
5. Chakrit, yenti., surin, phongsupasamit., and chanat, ratanasumawong., Analytical and experimental investigation of parameters affecting sliding loss in a spur gear pair, Engg J. 17(1), 79-93, (2013).
6. Franco, Concli., and Carlo, Gorla., Analysis of the oil squeezing power losses of a spur gear pair by mean of CFD simulations. Proceedings of the 11th biennial conference on engineering systems design and analysis, ESDA2012, July 2-4, Nantes, France, 1-8, (2012).
7. Miguel, Pleguezuelos., Jose, I, Pedrero., and Miryam, B, Sanchez., Analytical expressions of the efficiency of standard and high contact ratio involute spur gears, Mathematical problems in engineering, Hindawi Publishing Corporation, 1-14, (2013).
8. Chang, l., yeau-ren, jeng., and pay-yau, huan., Modeling and analysis of the meshing losses of involute spur gears in high speed and high load conditions, J. of Tribo., ASME, 135(1), 1-11, (2013).
9. Vijay, Kumar, Karma., and Ram, Kumar., Comparison between experimental and theoretical investigation of power losses of single stage gear system, Engg, Indian J. of research, 2(8), 107-110, (2013).
10. Chaari, F., Kantchev, G., and Haddar, M., Modelling the mechanical efficiency of a spur gear system, proceedings of the ICE-Engineering and computational mechanics, 163(2), 83-90, (2010).
11. Abdelilah, Lasr., Lahcen, Belfal., Brahim, Najji., and Bernard, Mushirabwoba., Pressure estimation of the trapped and squeezed oil between teeth spaces of spur gears, Applied mathematical sciences, 8(107), 5317-5328 (2014).
12. Fernandes, C, M, C, G., Marques, P, M, T., Martins, R, C., and Seabra, J, H, O., (2015). Influence of gear loss factor on the power loss prediction, Mechanical sciences, 6 (2015), 81-88 (2015).
13. Luis, Magalhaes., Ramiro, martins., Cristiano, Locateli., and Jorge, Seabra., (2010). Influence of tooth profile and oil formulation on gear power loss, Tri. Intl, 43(10), 1861-1871.
14. Petra, Heingartner., and David, Mba., Determining power losses in helical gear mesh, Gear technology, 33-37, (2005).
15. Anderson, N, E., and S, H, Loewenthal., Spur gear system efficiency at part and full load, Technical report 79-46, NASA technical paper 1622 (1980).
16. Hutchings, I. M., Tribology: friction and wear of engineering materials, Ed. Butterworth-Heinemann, oxford, 58-76, (1992).
17. Krishnakumari, A., Elaya, perumal, A., and Panda, R. C., Prediction for condition monitoring of wear and film thickness in a gear box, Engg. Mechanics, 19(6), 393-405, (2012).

18. Muraro, M. A., Koda, F., Reisdorfer, Jr, U., and silva, C. H. D., The influence of contact stress distribution and specific film thickness on the wear of spur gears during pitting tests. J. of Brazilian soc. of mech. sciences and Engg, 34(2), 135-144, (2012).

Rational designing of the internal water supply system in reconstructed residential buildings of mass standard series

Evgeny Orlov[1*]

[1]*Moscow State University of Civil Engineering (National Research University), Department of water supply and wastewater treatment, 129337 Yaroslavskoe Shosse 26, Moscow, Russia*

Abstract. The issues of water supply system reconstruction in mass series buildings are reviewed with consideration of water- and resource saving. Principal points for location of plumbing cells in apartments, arrangement of water devices and wastewater receivers, selection of pipelines for reconstructed water line are described. Comparative analysis of design variants of inner water line before and following reconstruction are given. It was found that applying the developed system design approaches the head losses in the inner water supply line will be significantly decreased as well as the water mains length will be decreased with material and installation saving. Based on the data the conclusions on necessity to review standard arrangement solutions of water supply systems in the reconstructed buildings were made. Recommendations on water loss reduction in the system by installation of special water saving fittings on water devices and touchless faucets.

Keywords: water supply, sewage, housing, plumbing

1 Introduction

Today due to the came on economic depression as well as unsatisfactory state of old residential buildings the urgent issues are reconstructions of old and slum dwellings which can include buildings of the first standard series built mainly in 50-60th of the last century. These are different types of five-storeyed block and brick buildings equipped with outdated utilities (water mains, sewage etc.) [1-3].

In the work [4] of S.N. Bulgakov and V.V. Leontiev the main issues concerning reconstructions of these buildings considering architectural and space-planning solutions are reviewed but utilities reconstructions issues are not investigated at all.

Economic feasibility of reconstruction of these buildings is proved in the monograph [4]. It should be also noted that it is relevant not only for big megacities but also for small and medium cities. It will enable to attract attention of new buyers in such small localities proposing dwellings reconstructed according to new requirements but with better ecology and cheaper prices per square meter, e.g. cities located at 30-40 km from the capital.

* Corresponding author: jeks-2003@yandex.ru

2 Peculiarities of sanitary equipment application

A great number of five-storeyed buildings of 50-60th standard series (1-510, 1-511, 1-515 etc.) has one undeniable advantage - demolition of partition walls which are non-load-bearing. During building reconstruction, it enables to perform various replanning variants with area increase for reconstructed apartments improving quality of living for the future dwellers.

Concerning utilities of the inner water supply and water removal, they are in bad condition. The will be fully changed during building reconstruction. Problem of such old buildings is worsened by the fact that inner water supply calculation was performed for standard water devices (sink mixer, tap of toilet tank, single mixer above bath and wash stand). Nowadays a great number of dwellers install many devices for which the system was not accounted for (washers and dishwashers, jacuzzi tubs etc). Pressure drops were registered at consumers located on last floors (4-5th floors) which was caused by increased water consumption and use of a great number of water-retaining devices.

The disadvantages of reconstructed buildings are also very small areas of kitchens and utilities cores, availability of combined variants even in two-room apartments that is unacceptable in current terms. Thus upon reconstruction of such buildings it is required to considerably improve comfort of use of utilities e.g. water supply and sewage.

Considering the above problems, it is proposed to develop the reasonable methods for reconstruction of water supply and removal systems considering water and resource saving as well as energy efficiency as in the most cases during reconstruction these issues are almost not considered in the sphere of inner water supply and sewage.

Today the main problem arising during reconstruction of old and slum dwellings is that architectures engaged in reconstruction in most cases give attention to planning of dwelling premises in particular leaving in the second place the utilities of the building (water supply lines, sewage etc) and their position in the dwelling space more specifically. For example, utilities cores can be left with the same area that upon reconstruction so there is no question of full comfort of water line use.

During development, the sanitary utilities changed. Considering current requirements to comfort its area shall be increase as there shall be comfortably located also other water devices not designed previously in the system e.g. automatic washer or hydromassage bath. The Soviet concept that everything shall be minimized shall become history. The hygiene and sanitary level for the last 50-60 years has considerably increased which requires other attitude to this issue.

According to the author's opinion, the sanitary core size shall evolve in the same way as habitable rooms namely the area shall be increased not depending on a category (elite housing or economy class). Moreover, it is understood that the sanitary cores shall meet not only necessity in sanitary and hygienic procedures. It shall also provide for location of different fitness machines for relaxation in water and trainings also (e.g. bath-tubs for aquacycling - water cycle). Thus, the sanitary core can be a health compartment in an apartment of any consumer interested in maintenance of his/her physical state. This concept can be started to apply for elite housing smoothly introducing these ideas in buildings of business and economy classes.

The author proposes to design emergency chutes (dry hydroseals) for drain receipt in unforeseen leakage (it can be done in kitchen also) in all new sanitaryware cores (both in newly built buildings and in reconstructed premises). The existing leak protection systems work good in many case and cut off water supply in emergency in 3-5 seconds. But their installation is not regulated and required by any authorities which does not oblige habitants to install them. Flooding of neighbors living below can lead to deplorable consequences (equipment damage, wetting and excess dampening of building structures etc). It is feasible to install several receiving openings in the floor at once and rely on their emergency water receipt in case of leak.

It is more feasible to locate the utility core near the future kitchen. It enables to connect all water devices to one riser (cold water, sewage). Today there are different types of planning where kitchens are located at considerable distance from the utility core, which implies their connection to different risers. It results in excess expenditure of materials and makes the whole water supply and sewage system of the building more expensive. Also the consumer itself shall install several meters for the apartment. In most cases, it is related to absence of "resource saving designing" concept not only in architect thoughts but also in constructing engineer thoughts.

According to calculation, rules of the inner water line the pipeline length shall be minimum - head losses by length decreases and it is feasible to make as minimum turns in the system as possible - head losses to local resistance decreases. So it is required to proceed from the following. All water devices within an apartment shall be arranged on one wall avoiding pipeline bends (in-line device arrangement). As entry to the building is performed to the the part where the most devices are located so it will be rationally to located kitchens and utility cores on each floor as close to each other as possible. Thus, it is required to seek the minimum head losses, which in future may affect selection of pumping equipment. As an example two cases of arrangement solutions of utility cores and water supply systems before dwelling house reconstruction can be reviewed which will enable to draw the most objective conclusions.

Planning and routing of inner cold water supply system in a building before reconstruction (fig. 1, a) and basement plan (fig. 1, b) are given in fig. 1. It is seen that utility core sizes do not meet the comfort concept (they shall be increased in future, standard sizes are shown in the figure). The kitchen is located far from the utility core. The replanning variant considering up-to-date trends is shown in fig. 2. Previously not used area in single-room apartments is reconstructed in an additional room. Considering current trends, the kitchens are located near the utility cores removing a number of excessive risers in the system. All water feeders are arranged as close to each other as possible. Comparative data for two variants (before and after reconstruction) are given in table 1.

It is seen from table 1 that about 53 m of pipelines were saved (installation and materials), head losses were considerably decreases in comparison with the variant before reconstruction as well as the more efficient pump was selected which in future will enable to reduce electric energy costs.

In addition, it is feasible to make some additional activities following the reconstruction. It is feasible to install special water saving fittings [5] on each mixer to reduce water losses in the reconstructed building. These devices are screwed on any standard mixer instead of default aerators. The water saving fitting is equipped with a metal bar. Its operating principle is as follows. First, the consumer sets the required temperature and head (turn a faucet or lever). Then it presses on a metal bar and water begins to be supplied from the device. To stop water flowing it is required to press again on the bar. Thus the exact amount of water is spent which is required (time for setting temperature and head is absent; the consumer consumes water at once). In addition, there are water saving fittings equipped with timers available in the market. The consumer sets himself/herself that the metal bar itself (mechanically) stops water supply after the set time (as a rule within 3-10 seconds). It is relatively easy to design a mixer, which will be initially equipped with such water saving fitting.

In addition, to save water it is feasible to use touchless faucets actuating upon hand putting to them as per SP 30.13330.2012 "Internal water supply lines and sewages of buildings". This new type of water devices is widely described in the following literature [6-14]. However, today it costs rather expensive and benefit from its use is if it is installed on a wash stand in the utility core. In the kitchen, such device will create certain discomfort when filling large volumes (kettles, pots etc). Thus water saving will be similar to water saving fittings described above.

a)

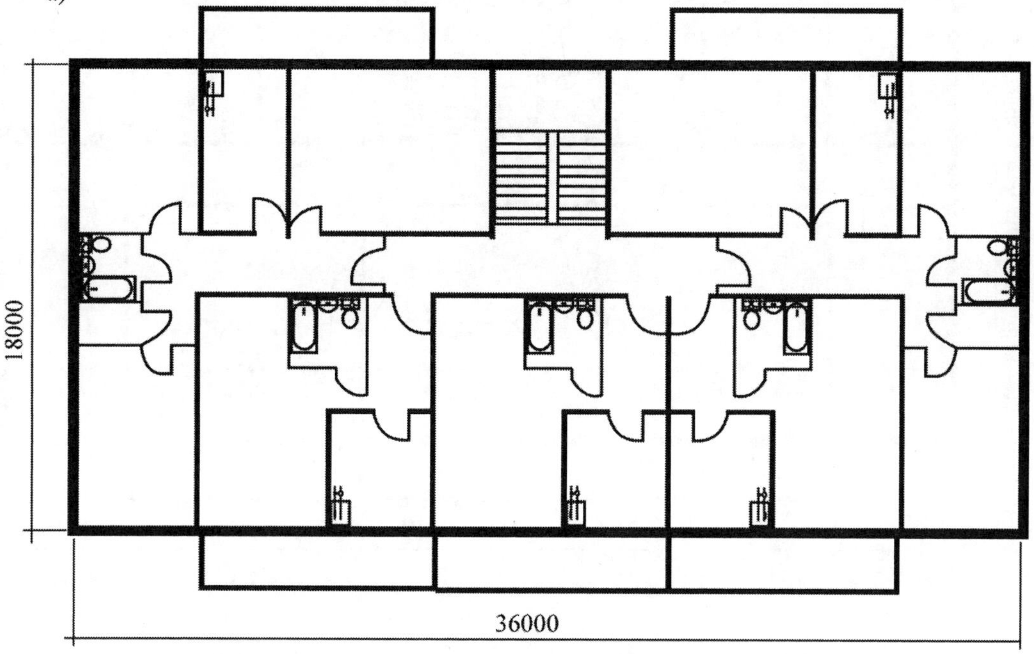

b)

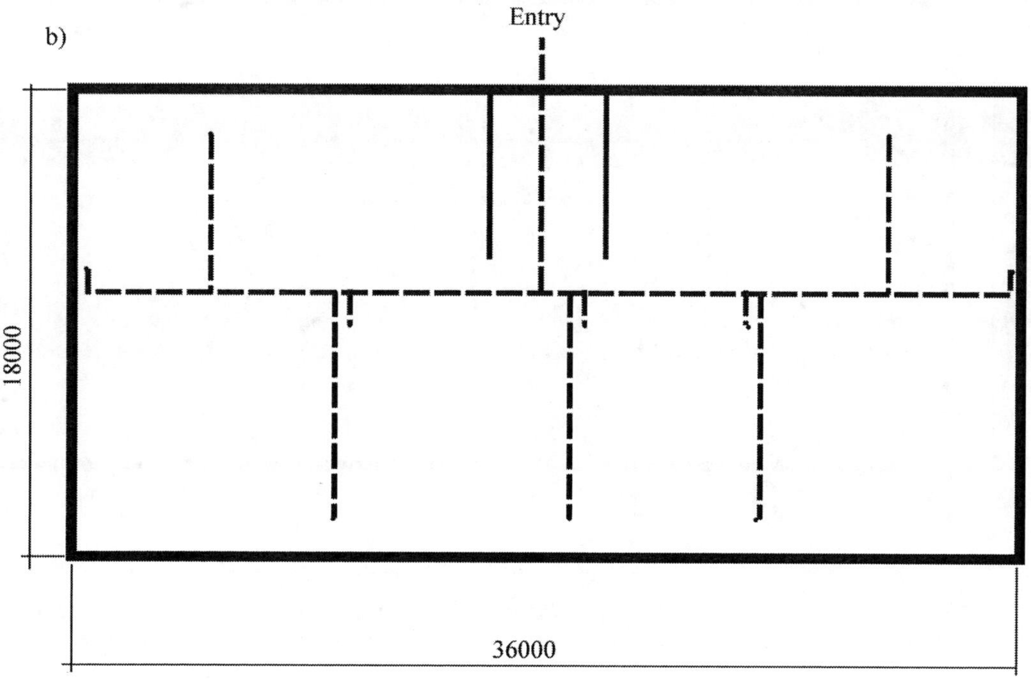

Fig. 1. Plan of standard floor of building (before reconstruction) with location of risers and sanitary ware (a) and basement plan (b) with location of pipe system and risers (dimensions in mm)

a)

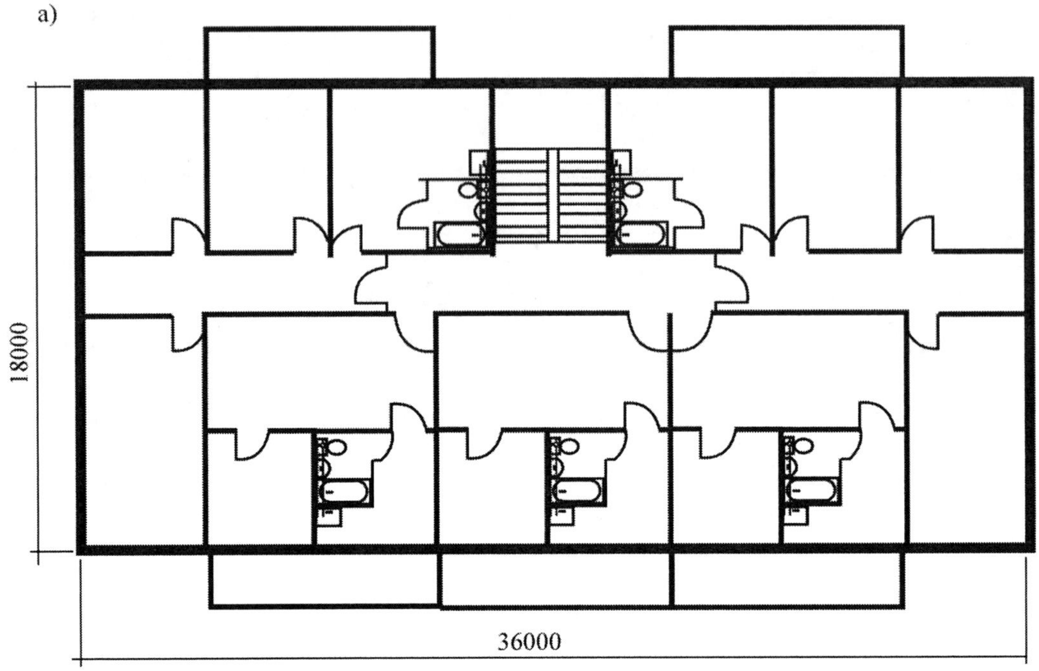

b)

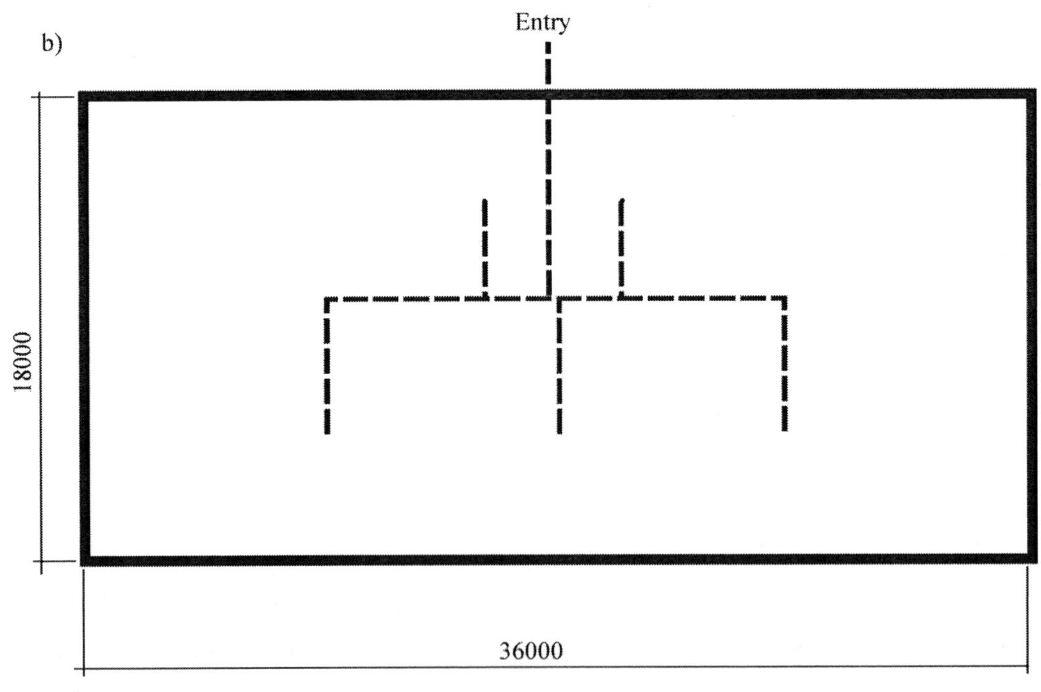

Fig. 2. Plan of standard floor of building (after reconstruction) with location of risers and sanitary ware (a) and basement plan (b) with location of pipe system and risers (dimensions in mm)

Table 1. Comparative data for two design variants before and after reconstruction

No.	Parameter	Before reconstruction	After reconstruction
1	Inner water supply main length in basement (without entry and connections to risers)	35.2 m	15.2 m
2	Length of the whole water supply main in basement	88.7 m	35.6 m
3	Quantity of water risers	10 nos.	5 nos.
4	Quantity of served water devices	20 nos.	20 nos.
5	Possibility to increase usable area of utility cores for creation of health cells	No, standard area	Yes, possibility to install additional water devices
6	Head losses by length	—	Decrease of head losses by length in 2.5 times
7	Head losses for local resistances	—	Decrease of head losses for local resistances in 2.5 times
8	Selection of new pumping equipment	—	Selection of more saving pumping equipment in comparison with the variant before reconstruction
9	Combined operation of pump and elevated water tank	No	Pressure leveling in the system in case of big water draw-off in peak hours
10	Inner sewage system	—	Decrease of outlet number in the system, pipeline number, less time for equipment installation

Upon building reconstruction of these standard series, the authors of the monograph [4] propose to add on several floors additionally increasing quantity of apartments in the building. In some cases, it is proposed to build one attic floor (added on on the top of the five-storeyed building) with planning which will not be same as the standard floor plan. However, in this issue it is proposed to use an attic floor (or its part) exactly for location of utilities e.g. elevated water tanks. It is particularly true for small localities where head is not sufficient for water supply to highly located consumers of water supply systems of top floors in particular hours, as a rule in the morning and in the evening upon large water draw-off.

3 Conclusions

Concerning pipeline selection for inner water supply systems then in accordance with SP 30.13330.2012 "Inner water supply lines and sewages of buildings. Revised edition of SNiP 2.04.01-95*" it is allowed to use the following pipe types: copper, bronze, brass, steel (with inner and external anticorrosion coating), polymeric pipes (polyethylene, polypropylene, polybutene etc.).

Use of copper pipes is topical now [15]. In foreign countries they are widely used due to their reliability (e.g. in high-rise construction) in comparison with steel pipelines. However, their high cost will not be acceptable for the customers who want to save money.

Upon selection of polymer pipelines it is required to review such parameters of particular polymers as operating conditions, service life, transported liquid temperature, low roughness coefficient etc. Preferences shall be given to polymers with high life ratios to which polybutene pipes are related as an example [16-19]. They obtain well flexibility, high resistance to cracking, well chemical resistance, and high resistance to chafing. Such pipelines are easily welded and connected using press-type and threaded fittings.

Thus it should be summarized that application of the above recommendations will enable not only to design efficiently the inner water supply and sewage system of reconstructed building but also to obtain required water and resource saving effect in future having reduced costs for residential building operation.

References

1. Yu.P. Grigoriev, Ind. and Civ. Const., 11, 3 (2003).
2. V.A. Anikin, V.V. Guriev, Ind. and Civ. Const., 11, 10 (2003).
3. S.N. Bulgakov, I.V. Rybalko, Ind. and Civ. Const., 2, 45 (2006).
4. S.N. Bulgakov, V.V. Leontiev, Justified Reconstruction of Residential Buildings of the First Mass Series (ASV publishing house, Moscow, 2016).
5. E.V. Orlov, Scientific review. 3, 32 (2016).
6. V.N. Isaev, A.M. Smirnovskaya, World Equip. and Tech.. 6, 33 (2015).
7. V.N. Isaev, V.A. Nechitaeva, Wat. Suppl. and Sew., 4, 88 (2010).
8. V.N. Isaev, V.A. Chukhin, A.V. Gerasimenko, Sanitaryware. 3, 14 (2011).
9. V.N. Isaev, V.A. Chukhin, A.V. Gerasimenko, Sanitaryware. 6, 16 (2010).
10. E.V. Orlov, World Tech.. 10. 35 (2012).
11. A.V. Chistyakova, V.A. Chukhin, A.P. Andrianov, Wat. Purif.,Wat. Treat., Wat. Suppl., 1, 48 (2016).
12. A.G. Pervov, A.P. Andianov, D.V. Spitsov, V.V. Kondratiev, Wat. Purif., 1, 38 (2008).
13. E.V. Orlov, Ecol. of Urban. Terr., 1, 43 (2015).
14. V.N. Isaev, N.V. Aksenova, M.Ya. Balavas, Nat. and Eng. Sci., 5, 200 (2015).
15. V.A. Nechitaeva, A.V. Eremeev, World Equip. and Techn., 10, 26 (2015).
16. A.A. Otstavnov, Water Supply and Water Removal of Public Buildings (AVOK-PRESS, Moscow, 2011).
17. A.A. Otstavnov, V.A. Kharkin, Sanitaryware, Heat., Air Cond.,1, 16 (2014).
18. N. Makisha, Procedia. Eng., 165, 1092-1097 (2016)
19. A. Kulakov, N. Makisha, MATEC Web of Conferences, 112, 10019 (2017)

Automated software for hydraulic simulation of pipeline operation

Roman Hurgin[1*]
[1]*Moscow State University of Civil Engineering, Yaroslavskoye shosse, 26, Moscow, Russia, 129337*

Abstract. Design of modern water supply systems of large cities as well as their management via renovation of hydraulic models poses time-consuming tasks to researchers, and coping with this task requires specific approaches. When tackling these tasks, water services companies come across a lot of information about various objects of water infrastructure, the majority of which is located underground. In those cases, modern computer-aided design systems containing various components come to help. These systems help to solve a wide array of problems using existing information regarding pipelines, analysis and optimization of their basic parameters. CAD software is becoming an integral part of water supply systems management in large cities, and its capabilities allow engineering and operating companies to not only collect all the necessary data concerning water supply systems in any given city, but also to conduct research aimed at improving various parameters of these systems, including optimization of their hydraulic properties which directly determine the quality of water. This paper contains the analysis of automated CAD software for hydraulic design and management of city water supply systems in order to provide safe and efficient operation of these water supply systems. Authors select the most suitable software that might be used to provide hydraulic compatibility of old and new sections of water supply ring mains after selective or continuous draw-in renovation and decrease in diameter of distribution networks against the background of water consumption decrease in the cities.

Keywords: water supply, design, CAD systems, software

1 Introduction

Reliable and efficient performance of modern water supply systems of large cities and management of power flow in extensive and long pipelines requires specific approaches based on the wide use of models [1].

Water services companies that have to tackle those issues often operate large sets of data regarding water infrastructure objects, such as water intake facilities, water treatment plants, pumping plants, pipelines etc. [2]. In these cases, special modern automated CAD systems with various subsystems, which help to solve a wide array of tasks, including systematization, evaluation, analysis and optimization of the most important parameters of any given city's water supply system, are of great help [3]. These automated CAD systems

*Corresponding author: hurgin@mail.ru

are becoming an integral tool for the management of utility systems, including water supply systems of large cities [4]. The capacities of this software allow operating companies to not only collect all relevant data regarding water supply systems, but also to conduct research aimed at improving certain parameters of these systems, including optimization of hydraulic features of pipelines, a key factor that determines the quality of water [5, 6].

2 CAD systems for design of water facilities

According to the results of analytical studies, it has been found that, among the CAD systems used to design study and analyze water supply and water discharge systems, the following ones are frequently utilized:

- Autocad Civil 3D, (Autodesk, USA);
- WaterCAD, WaterGEMS, HAMMER, SewerGEMS (Bentley Systems, USA);
- MIKE URBAN automated computer-aided design system (DHI Water & Environment, Denmark) [7];
- ZuluHydro (Politerm, USA) [8];
- HydroCalc, CityCom-Gidrograf solutions etc.

The approaches to designing, studying and analyzing of the city water supply systems are software specific. It has been found that Russian and foreign water services companies use CAD software designed by Bentley Systems and DHI Water & Environment more often [9].

3 Peculiarities of Bentley WaterGEMS application

Bentley WaterGEMS software can operate alone as well as in combination with ArcGis, Autocad or Microstation, the oldest and most important product developed by Bentley Systems [10].

This software gives its operator an opportunity to not only perform the hydraulic design of a water supply system and tackle water quality issues, but also to conduct cost benefit analysis and calculate energy consumption level.

Bentley WaterGEMS software allows for putting together a design model of a water supply system and for manual input of data of each element. Apart from designing a traditional ring pattern of a pipeline, it is also possible to install shutoff valves, control valves, retention tanks, expansion tanks as well as pumps with various features and operation modes. At the same time, the basic method of calculation might be specified according to one of three classic equations, such as Darcy-Weisbach equation, Hazen-Williams equation and Manning equation, depending on reference design conditions and features of analysis of a given water supply system [10].

WaterGEMS software gives its users an opportunity to align water supply systems at various hours of the day, with dynamically modified consumption levels during each hour. It is also possible to allocate various consumers for various nodes of a water supply system; each consumer can choose their own consumption mode. Systems with multiple water feeders (pump stations) may also be analyzed.

When using WaterGEMS software, it is possible to develop different operation scenarios for the water supply system. When utilizing various parameters of pipeline sections, the software determines the best possible option for design or redesign. All in all, Bentley WaterGEMS software is able to collect, analyze, and present large sets of data about water supply networks in large cities. After evaluating and analyzing a particular water supply network, the software provides output data regarding the city water supply system, for example, flow rate in pipelines, water loss, hydraulic and piezometric heads in water supply network nodes etc. Moreover, the direction and amount of flow rate in pipelines are determined. WaterGEMS software also provides users with a chance to design piezometric profiles for different directions of a water supply system as well as to plot dependency graphs for different values which change according to hours of the day in water supply network nodes. Besides, another feature has to be noted, which is the ability to plot a 2D diagram of pressure distribution right on various sections of a water supply system; values of pressure distribution at different sections of a water supply system might be represented graphically and then divided. Moreover, critical pressure points of the water network may be singled out.

4 Application of MIKE URBAN software

Apart from Bentley WaterGEMS, another CAD software based on the EPANET standard, MIKE URBAN (DHI Water & Environment, Denmark) [11], is widely used. This software is developed exclusively for calculation of water supply and distribution systems. MIKE URBAN is one of the most well-known and multi-functional software solutions worldwide, and it is used to analyze and study water supply and discharge systems of large cities, alongside products developed by Bentley Systems. MIKE URBAN software solution contains basic economic data, and, by knowing electricity rates, it may be possible to achieve efficient performance of pumps at required system parameters. This software also has a module which calculates hydraulic impact and takes into consideration the start-up and shutdown of pumps, power cuts, cavitation risks etc. Besides, MIKE URBAN software may be integrated with other software solutions, which allows for possible implementation of online operation system or decision-making system (dispatch aid).

High degree of detail of any water supply system element presented in the MIKE URBAN software provides users with an opportunity to input such parameters as the name of pipeline's main pipe, its inventory number, diameter, length, number of node wells, material and year of construction into a database [12].

With the use of models, MIKE URBAN software allows conducting dynamic analysis of water supply system operation modes, determining sections of a water supply system with low flow rates (these sections are subsequently included in flushing plans), organizing efficient control of water quality, especially during the summer period of low water consumption. Moreover, the software calculates the models of load flow when sections of a water supply system are shut down [13].

Therefore, two main tasks that can be solved by designing a hydraulic model of any city water supply system using the MIKE URBAN software may be singled out:

1. *Reduction in diameter.* The issue of reduction of pipeline diameter, their conservation or liquidation has recently become a major one for Russian water services companies due to the implementation of low water consumption policy in cities. With that in mind, in order to retain and increase sanitary security and efficiency of cities distribution networks, solutions regarding changing the diameters of existing pipelines, liquidation of specific waterways, water mains or duplicate sections of distribution networks are necessary. Taking into consideration the results of pipe diameter and water flow rate and volume and relying on roughness indexes of different materials, it is possible to make these solutions, for instance, decrease pipe diameter, reconstruct specific water mains, replace them with pipes made from different materials possessing different hydraulic features [14,15].

2. *Zoning.* Division of water supply systems into sections depends on three key factors: difference in drops at sections supplied with water, required free heads and maximum loss of head throughout the network. Pressure at certain sections is managed via simultaneous adjustment of network's pressure relief valves. Pressure in control area may be adjusted in real time. This system of pressure control enables the user to solve the following tasks: control pressure at waterworks and distribution networks of any given city, minimize water loss due to pipeline failures and prevent excess head in distribution networks. Thanks to this pressure control system, experts may obtain accurate data about water supply system of the entire city [16-18].

The results of the analysis show that MIKE URBAN software gives an opportunity to handle a wide array of tasks, including the analysis and management of water supply systems in large cities. The functions and options of this software solution are timely and completely coincide with the issues of and requirements for analysis and management of water supply systems of large Russian cities. This CAD solution allows for comprehensive analysis of various options of design and reconstruction as well as making relevant solutions regarding modernization of large water supply networks.

The most relevant software solution intended for analysis and management of a city water supply system should be selected based primarily on the results of the evaluation of requirements and objectives that engineering or operating companies face. Current CAD systems possess a wide range of basic functions and modules used to study various water supply systems. Nowadays, CAD modeling has reached tremendous progress, and the number of software solutions used in water supply systems management has increased drastically, compared to the 1970 s, 1980 s or 1990 s [19-21].

5 Conclusions

1. Automated CAD solutions like Bentley WaterGEMS and MIKE URBAN are considered the best software solutions for water supply systems management in any city. Both products operate on the EPANET industry standard developed exclusively to study and analyze hydraulic models of water supply and distribution systems. Both software products possess a wide range of functions, including design, representation and certification of components of a water supply system. They are also capable of conducting multivariate analysis based on operating scenarios of ring water supply networks, including their draw-in renovation and modernization using current efficient inside protective coatings with a wide variety of pipe wall roughness values, thus securing conditions for compatibility of old and new (or restored) sections of a water supply network.
2. importance of the aforementioned software solutions increases drastically when it comes to tackling the issues of management of water supply and distribution pipeline systems, provision of efficient performance of water supply networks under the conditions of lower water consumption, comparative evaluation of effects of pipeline renovation using various draw-in methods and performance analysis of a pipeline under reconstruction in the absence of hydraulic imbalance when alternative repair materials are used, as well as resources and energy saving when choosing various alternative inside protective coatings of pipelines.

References

1. N. Abramov, Water supply (1982)
2. M. Somov, M. Zhurba, Water supply. Vol. 1. Systems of water withdrawal, supply and distribution (2008)
3. Y. Galperin, WSAST, 6, 13–16 (1999)
5. S. Khramenkov, Strategies of modernization of water supply systems (2005)
6. V. Orlov, Pipelines. Automated maintenance of projects (2015)

7. N. Makisha, E3S Web of Conferences, 6, 01002 (2016)
8. ZuluHydr, hydraulic calculations for water supply systems. Politerm company. URL: www.politerm.com.
9. S. Govindan, CAD and graphics, 4, 36–38 (2009)
10. D. Borisov, CAD and graphics. 5, 64–68 (2009)
11. MIKE products by DHI Water & Environment. URL: www.mikebydhi.com. Accessed date: December 5, 2012
13. B. Rajani, Urban Water, 3,151-164 (2001)
14. M. Fattakhov, Design and use of plastic pipelines in oil industry (2009)
15. M. Najafi, Trenchless technology of piping: installation and inspection (2010)
16. S. Abulgafaov, Combined technology of draw-in repair of pipelines of irrigation systems (2004)
17. O. Davydenko, Increase of efficiency and reliability of draw-in repair of pipelines using toral engines (2013)
18. A. Zwierzchowska, Technologie bezwykopowej budowy sieci gazowych, wodociagowych i kanalizacyjnych (2006)
19. A. Kulakov, N. Makisha, MATEC Web of Conferences, 112, 10019 (2017)
20. GOST 52134–2003 Thermoplastic pressure pipes and their connecting elements for water supply and heating systems (2003)
21. N. Makisha, A. Kulakov, MATEC Web of Conferences, 112, 10018 (2017)

A new approach to hydraulic calculations of free-flow drainage systems

Elena Kitaytseva[1*]

[1]*Moscow state university of civil engineering, Yaroslavskoye shosse, 26, Moscow, Russia, 129337*

Abstract. The accuracy of hydraulic calculations is determined by the accuracy of the mathematical models inherent in their foundation. The results of calculations, obtained by different models, and differing from each other by 5-10% is considered identical. The present study examines classical ways of finding the Chezy formula that are used for calculations of free-flow pipelines. These formulas differ from each other not only in parameters but also in the form of the functional dependence. The equation, underlying the hydraulic calculations, is nonlinear. Its solution is possible only in the number of well-established methods. It offers numerical computing to find pipelines diameters and their filling, as well as the algorithm to realize it. It conducts numerical experiment to compare pipelines diameters and filling and flow velocities. It presents frequency characteristics to distribute the parameters under analysis. For a specific example, the difference of cost of pipes, the diameters of which was found using the various formulas for coefficient Chezy, was calculated.

1 Introduction

According to [1] "hydraulic calculation of sewerage free-flow pipelines should be carried out using charts, diagrams and nomograms". The A.A. Lukinykh and N.A. Lukinykh charts [2] that are widely accepted in engineering practice for calculations of free-flow drainage systems are based on N. N. Pavlovsky's function [3] and allow to do these tasks. There are online calculators automatizing use of these charts. When working out programs, the "hand counting" algorithm is used most frequently where the solution is sought by exhaustion [4].

2 Literature review

Hydraulic calculations of free-flow (gravity) pipelines are made if water flow should be stable and steady. These calculations are based on the assumption about the continuity of the flow:

$$q = 1000\omega v \tag{1}$$

and the Chezy formula:

$$v = C \cdot \sqrt{R \cdot i} \tag{2}$$

* Corresponding author: keh2@bk.ru

where q is flow rate, l/s; ω is water area, m^2; v is flow velocity, m/s; R is hydraulic radius, m; i is hydraulic slope; C is the Chezy coefficient, depending on hydraulic radius and roughness of wetted surface of pipelines, $m^{0,5}/s$.

When the pipe is filled partially the hydraulic radius *R,* water area ω and filling *h/D* can be stated in term of solid angle ϕ:

$$R=\frac{D}{4}\left(1-\frac{\sin\varphi}{\varphi}\right) \tag{3}$$

$$\omega=\frac{D^2}{8}(\varphi-\sin\varphi) \tag{4}$$

$$\frac{h}{D}=\frac{1}{2}\left[1+\sin\frac{180-\varphi}{2}\right] \tag{5}$$

where D, m, is inner pipe diameter.

The fundamental difficulty in making hydraulic calculations is to find the Chezy coefficient *C*. The N. N. Pavlovsky formula prevails:

$$C=\frac{R^y}{n_1} \tag{6}$$

$$y=2.5\sqrt{n_1}-0.13-0.75\sqrt{R}\left(\sqrt{n_1}-0.1\right) \tag{7}$$

where n_1 is relative roughness of pipe surface.

Alongside with this one, formulas developed by other authors are used, e.g.:

The S.V. Yakovlev formula:

$$C = 24.7\cdot\lg\left(\frac{10^3 Rv}{v\varepsilon+0.01}\right) \tag{8}$$

where ε, mm, is absolute surface roughness;

The A. D. Altshul formula:

$$C = 20\left(\frac{1000R}{k_e+\frac{0.004}{\sqrt{1000Ri}}}\right)^{\frac{1}{6}} \tag{9}$$

where k_e, mm is equivalent steady grained roughness;

The N. F. Fedorov formula:

$$C=\sqrt{\frac{8g}{\lambda}} \tag{10}$$

where λ is hydraulic friction coefficient that can be found with the help of the formula:

$$\frac{1}{\sqrt{\lambda}} = -2\lg\left(\frac{\Delta_e}{13680R}+\frac{a_2}{\mathrm{Re}}\right) \tag{11}$$

Re is the Reynolds number

$$\mathrm{Re}=10^4\frac{8q}{v\varphi D} \tag{12}$$

Δ_e, mm, is equivalent absolute roughness; a_2, is the coefficient which takes into account the characteristics of pipe surface roughness; v, m^2/s, is kinematic viscosity.

The coefficients from dependences (6-9, 11), characterize inner surface roughness in their own way. If the pipe material is given, they are defined with help of reference data [5]. For new materials the roughness values are found experimentally [6, 7]. In the research [8] the relation between coefficients n_1 and k_e is deduced.

Prescriptive literature [1] recommends using the N. N. Pavlovsky formula (6-7) and the N. F. Fedorov formula (10-12); in the meanwhile, it is underlined in [9] that formulas (6-7) are "right for areas of quite rough rubbing". At the same time a number of authors [10, 11] point to the opportunity for pipelines to work in the transition area where the Chezy formula depends on velocity (8) or the Reynolds number (11).

The formulas (6-12) are empirical or semiempirical dependencies and according to the authors, they correspond with experimental data well.

3 Methods and materials

Combining dependencies (1, 2) we get equation:

$$f(D,i,\varphi) = q - 1000\cdot\omega\cdot C\cdot\sqrt{R\cdot i} = 0 \tag{13}$$

When the flow rate is known, the equation (13) can be used to solve the following problems:

- to find diameter if pipe filling and slope are known;
- to find filling if pipe diameter and slope are known;
- to find slope if pipe diameter and filling are known.

The offered algorithm consists of the following steps:

1. To set the initial approximation to solid angle $\phi^{(0)}=3.55$;
2. To set the initial approximation to slope $i^{(0)}$: for off-nominal sectors - $i^{(0)}=0.007$, for nominal sectors $i^{(0)}=\min(0.00956q^{0.306};0.007)$;
3. To assign $i^{(k)}= i^{(0)}$;
4. To set initial approximation to $D^{(0)}=160$ (minimal permissible diameter for outer drainage systems);
5. To assign $D^{(k)}= D^{(0)}$;
6. To solve equation (13) for the unknown D;
7. To take as the answer a bigger diameter D which is nearest in pipe grade depending on pipe material;
8. To specify more exactly slope $i^{(k+1)}=\max(i^{(k)};1/D)$;
9. To specify initial approximation to solid angle $\phi^{(0)}$ according to the determined value of diameter D using the recommendations [2];
10. To assign $\phi^{(k)}= \phi^{(0)}$;
11. To solve equation (13) for the unknown ϕ;
12. To compute the filling h/d using the formula (5) and to compute the velocity v using the formula (2).

It is possible to solve the equation (13), which is nonlinear for the unknown D, only by means of numerical method [12, 13]. The Newton method was chosen as numerical method:

$$D^{(k+1)} = D^{(k)} - \frac{f\left(D^{(k)},i,\varphi\right)}{f'\left(D^{(k)},i,\varphi\right)} \tag{14}$$

where k = 0,1,2,... is iteration number;

$$f'\left(D^{(k)},i,\varphi\right)=\frac{\partial f}{\partial D}=1000\sqrt{i}\left[\frac{\partial\omega}{\partial D}\cdot C\cdot\sqrt{R}+\omega\left(\frac{\partial C}{\partial D}\sqrt{R}+\frac{C}{2\sqrt{R}}\frac{\partial R}{\partial D}\right)\right] \tag{15}$$

$$\frac{\partial\omega}{\partial D}=\frac{D}{4}(\varphi-\sin\varphi) \tag{16}$$

$$\frac{\partial R}{\partial D}=\frac{1}{4}\left(1-\frac{\sin\varphi}{\varphi}\right) \tag{17}$$

Iterations were over on the condition that:

$$\left|f\left(D^{(k)},i,\varphi\right)\right| < 0.001 \quad (19)$$

According to the model used, the derivative of the Chezy formula $\frac{\partial C}{\partial D}$ is calculated using the following formulas: The N. N. Pavlovsky model

$$\frac{\partial C}{\partial D} = C\left[\frac{\partial y}{\partial D}\cdot\ln R+\frac{y}{R}\cdot\frac{\partial R}{\partial D}\right], \quad (20)$$

where

$$\frac{\partial y}{\partial D} = -\frac{0.375\left(\sqrt{n_1}-0.1\right)}{\sqrt{R}}\frac{\partial R}{\partial D} \quad (21)$$

- the S. V. Yakovlev model

$$\frac{\partial C}{\partial D} = \frac{24.7}{\ln 10}\left[\frac{1}{R}\cdot\frac{\partial R}{\partial D}+\left(\frac{1}{v}-\frac{\varepsilon}{v\varepsilon+0.01}\right)\frac{\partial v}{\partial D}\right] \quad (22)$$

where

$$\frac{\partial v}{\partial D} = -\frac{q}{1000\omega^2}\frac{\partial \omega}{\partial D} \quad (23)$$

- the A. D. Altshul model

$$\frac{\partial C}{\partial D} = \frac{20}{6}\left(\frac{1000\cdot R}{k_e+\frac{0.004}{\sqrt{1000Ri}}}\right)^{-\frac{5}{6}}\cdot\frac{1000\cdot\left(k_e+\frac{0.004}{\sqrt{1000Ri}}\right)+\frac{2\sqrt{1000i}}{\sqrt{R}}}{\left[k_e+\frac{0.004}{\sqrt{1000Ri}}\right]^2}\cdot\frac{\partial R}{\partial D} \quad (24)$$

- the N. F. Fedorov model

$$\frac{\partial C}{\partial D} = \frac{4\sqrt{2g}}{\ln 10}\cdot\frac{\frac{\Delta_e}{13.68R^2}\cdot\frac{\partial R}{\partial D}+\frac{a_2}{\mathrm{Re}^2}\cdot\frac{\partial \mathrm{Re}}{\partial D}}{\frac{\Delta_e}{13.68R}+\frac{a_2}{\mathrm{Re}}} \quad (25)$$

where

$$\frac{\partial \mathrm{Re}}{\partial D}=-10^4\frac{8q}{v\varphi D^2} \quad (26)$$

As in most cases, the calculated diameter differs from the standard one, which is taken as the answer, it is necessary to specify the filling of pipelines h/D solving the equation (13) for the unknown solid angle ϕ. As this method the Newton method is used:

$$\varphi^{(k+1)} = \varphi^{(k)} - \frac{f\left(D,i,\varphi^{(k)}\right)}{f'\left(D,i,\varphi^{(k)}\right)} \quad (27)$$

where k = 0,1,2,... is iteration number;

$$f'\left(D,i,\varphi^{(k)}\right)=\frac{\partial f}{\partial \varphi}=1000\cdot\sqrt{i}\left[\frac{\partial \omega}{\partial \varphi}\cdot C\cdot\sqrt{R}+\omega\left(\frac{\partial C}{\partial \varphi}\sqrt{R}+\frac{C}{2\sqrt{R}}\frac{\partial R}{\partial \varphi}\right)\right] \quad (28)$$

$$\frac{\partial \omega}{\partial \varphi}=\frac{D^2}{8}(1-\cos\varphi) \quad (29)$$

$$\frac{\partial R}{\partial \varphi}=-\frac{D}{4}\cdot\frac{\varphi\cdot\cos\varphi-\sin\varphi}{\varphi^2} \tag{30}$$

Iterations are over on the condition that:

$$\left|f\left(D,i,\varphi^{(k)}\right)\right|<0.001 \tag{31}$$

According to the model, which is used, the derivative of the Chezy formula $\frac{\partial C}{\partial \varphi}$ is calculated using the formulas:

- the N. N. Pavlovsky model

$$\frac{\partial C}{\partial \varphi}=C\left[\frac{\partial y}{\partial \varphi}\cdot\ln R+\frac{y}{R}\cdot\frac{\partial R}{\partial \varphi}\right], \tag{32}$$

where

$$\frac{\partial y}{\partial \varphi}=-\frac{0.375\left(\sqrt{n_1}-1\right)}{\sqrt{R}}\frac{\partial R}{\partial \varphi} \tag{33}$$

- the S. V. Yakovlev model

$$\frac{\partial C}{\partial \varphi}=\frac{24.7}{\ln 10}\left[\frac{1}{R}\cdot\frac{\partial R}{\partial \varphi}+\left(\frac{1}{v}-\frac{\varepsilon}{v\varepsilon+0.01}\right)\frac{\partial v}{\partial \varphi}\right] \tag{34}$$

where

$$\frac{\partial v}{\partial \varphi}=-\frac{q}{\omega^2}\frac{\partial \omega}{\partial \varphi} \tag{35}$$

- the A. D. Altshul model

$$\frac{\partial C}{\partial \varphi}=\frac{20}{6}\left(\frac{1000\cdot R}{k_e+\frac{0.004}{\sqrt{1000Ri}}}\right)^{-\frac{5}{6}}\cdot\frac{1000\cdot\left(k_e+\frac{0.004}{\sqrt{1000Ri}}\right)+\frac{2\sqrt{1000i}}{\sqrt{R}}}{\left[k_э+\frac{0.004}{\sqrt{1000Ri}}\right]^2}\cdot\frac{\partial R}{\partial \varphi} \tag{36}$$

- the N. F. Fedorov model

$$\frac{\partial C}{\partial \varphi}=\frac{4\sqrt{2g}}{\ln 10}\cdot\frac{\frac{\Delta_e}{13.68R^2}\cdot\frac{\partial R}{\partial \varphi}+\frac{a_2}{\mathrm{Re}^2}\cdot\frac{\partial \mathrm{Re}}{\partial \varphi}}{\frac{\Delta_e}{13.68R}+\frac{a_2}{\mathrm{Re}}} \tag{37}$$

where

$$\frac{\partial \mathrm{Re}}{\partial \varphi}=-10^4\frac{8q}{v\varphi^2 D} \tag{38}$$

The developed algorithm was realized for Excel Microsoft Office 2007.

4 Research

The numerical experiment was carried out to compare the results which were received in numerical way using the N. N. Pavlovsky (P) model (6-7), the S. V. Yakovlev (Ya) model (8), the A. D. Altshul (A) model (9), the N. F. Fedorov (F) model (10-11) for the Chezy formula.

The scheme of the drainage system presented in Fig. 1 was used as research object. The scheme contained 126 pipelines, 35 of which are off-nominal (q<10). The flow rates in the pipelines varied within the limits 1.9 – 278.1 l/s. The lengths of the pipelines varied from 8 to 1615 m, the overall length of the whole system was 45.506 km.

The above algorithm was used to find the pipeline diameter and its filling for all the pipelines. With the assigned accuracy of 0.001 l/s the number of iterations in iteration processes (14) and (27) did not exceed 8.

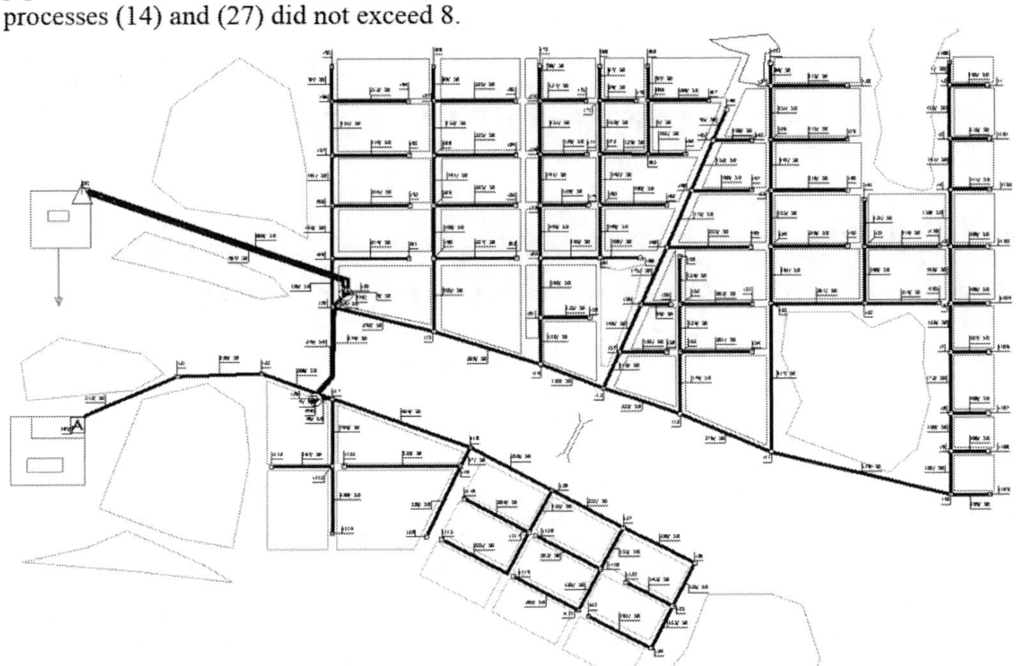

Fig. 1. Drainage System Scheme

Table 1. Combined parameters of comparison of hydraulic calculations results

Set of pipelines		Parameter								
		$(D_{in}-D_{in}^{(P)})/D_{in}^{(P)}*100,\%$			$(h/d-h/d^{(P)})/h/d^{(P)}*100,\%$			$(v-v^{(P)})/v^{(P)}*100,\%$		
		Model								
		Ya	A	F	Ya	A	F	Ya	A	F
All pipelines	min	-1.36	-4.75	-7.20	-2.17	-7.79	-11.53	-14.31	5.48	6.63
	max	8.24	-2.81	-3.55	11.54	-3.69	-4.42	2.80	10.41	16.27
	m	4.05	-3.37	-5.25	6.39	-5.25	-7.97	-7.49	7.12	11.21
	s	1.97	0.45	0.72	3.25	0.85	1.26	3.89	1.13	1.80
Nominal pipelines	min	-1.36	-4.75	-7.20	-2.17	-7.79	-11.53	-8.72	6.65	10.12
	max	4.70	-3.18	-4.72	7.56	-4.89	-7.22	2.80	10.41	16.27
	m	3.16	-3.51	-5.40	4.85	-5.63	-8.43	-5.64	7.56	11.76
	s	1.76	0.48	0.79	3.11	0.80	1.18	3.69	1.19	1.88

5 Results

The pipelines diameters D_{in}, their filling h/d, the flow velocity v were studied as analyzed parameters. For every pipeline the values D_{in}, h/d, v, received with the using of the S. V.

Yakovlev, A. D. Altshul, N. F. Fedorov models, were compared with the respective values $D_{in}^{(P)}$, $h/d^{(P)}$, $v^{(P)}$ received with the using of the N. N. Pavlovsky model.

Two sets were analyzed (chart 1): the first set included all the pipelines, the second one comprised only nominal pipelines for which q>10 l/s. The filling h/d and velocity v were compared only for those pipelines where D_{in} coincided with $D_{in}^{(\Pi)}$. For every sample and each parameter, the minimum, maximum, average values and standard divergence were evaluated. The removal of off-nominal sectors from sample 1 does not change minimum and average values of the studied parameters considerably, at the same time it decreases peak-to-peak value/ amplitude and standard divergence.

6 Discussion

The analysis that was conducted showed that the diameters which were calculated with the help of the S. V. Yakovlev model are by 3% in average bigger than the diameters calculated with the help of the N. N. Pavlovsky model whereas the diameters calculated with the help of the A. D. Altshul and the N. F. Fedorov models are by 3.5% and 5.4% respectively smaller than the diameters calculated using the N. N. Pavlovsky model (Fig. 2). The filling of the pipelines that was calculated using the S. V. Yakovlev model, is by 4.85% in average bigger than the filling calculated using the N.N. Pavlovsky model whereas the filling calculated using the A. D. Altshul and the N. F. Fedorov models is by 5.63% and 8.43% respectively smaller than the filling calculated with the help of the N. N. Pavlovsky model (Fig. 3). The flow velocity that was calculated with the help of the S. V. Yakovlev model, is by 5.64% in average smaller than the velocity calculated using the N. N. Pavlovsky model whereas the velocity calculated using the A. D. Altshul and N. F. Fedorov models is by 7.56% and 11.76% respectively bigger than the velocity calculated using the N. N. Pavlovsky model (Fig. 4).

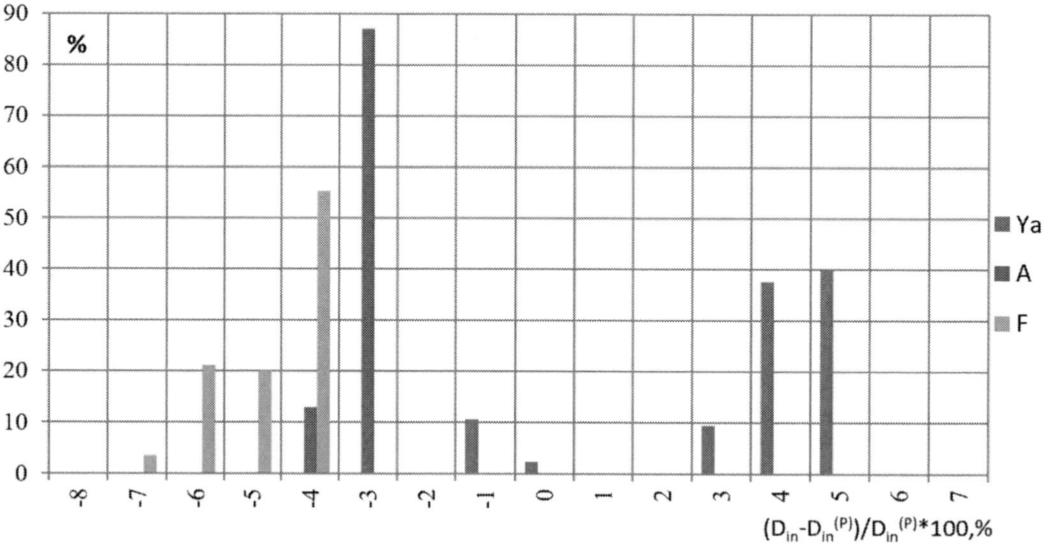

Fig. 2. Frequency characteristics of distribution of inner diameters calculated by different models

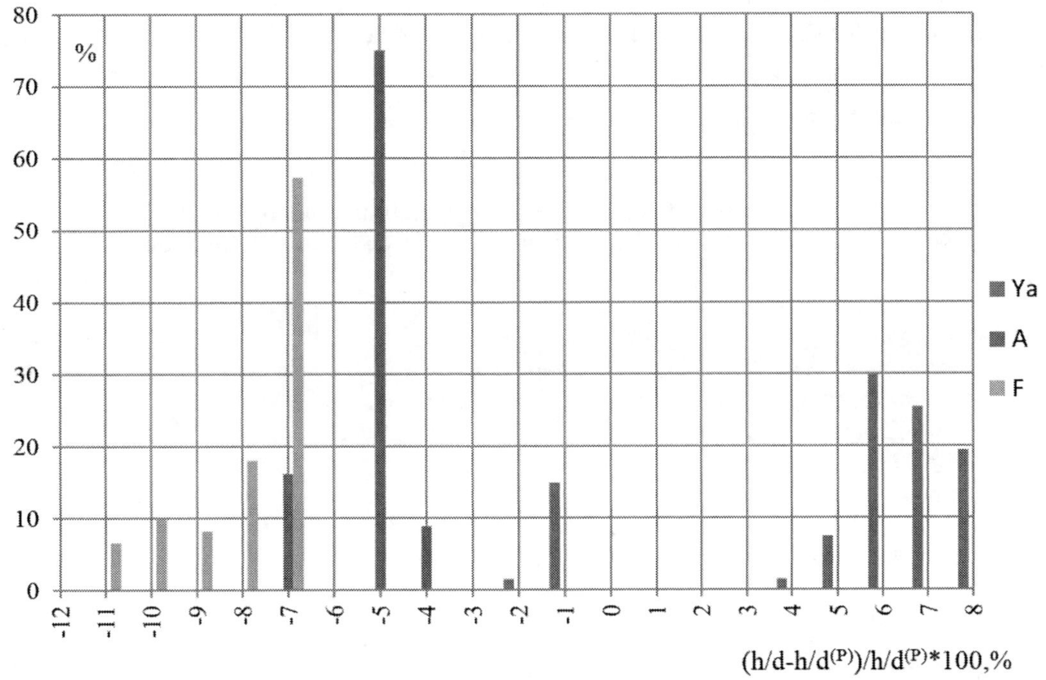

Fig. 3. Frequency characteristics of distribution of pipes filling calculated by different models

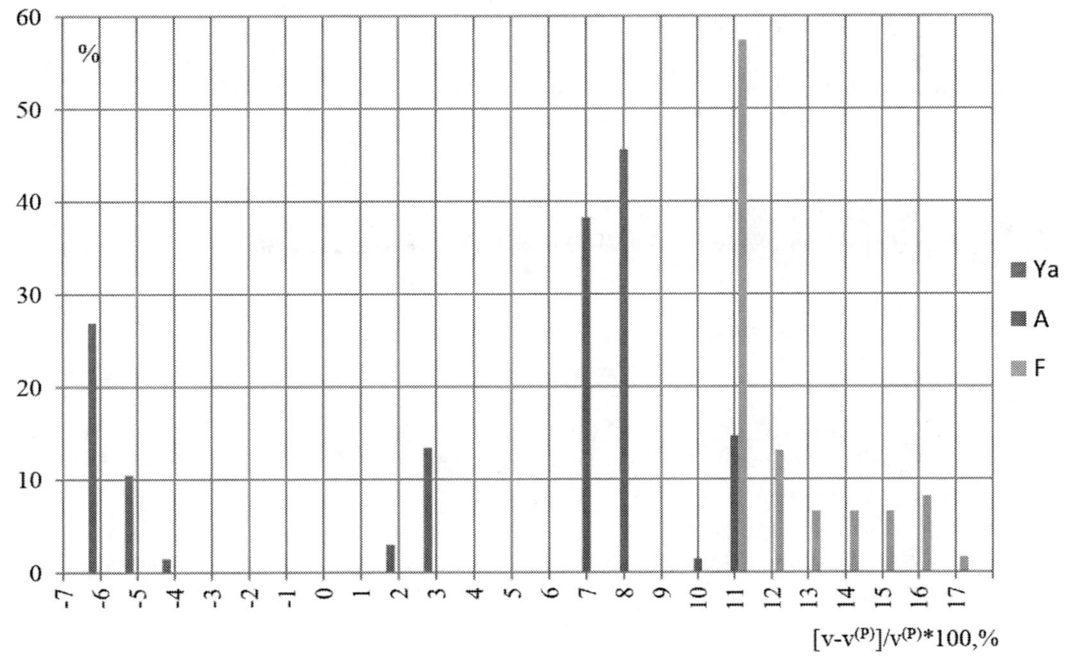

Fig. 4. Frequency characteristics of distribution of velocity calculated by different models

From the viewpoint of the accuracy of the results that were received, all the four models are acceptable for engineering calculations. However, matching the cost of pipes for the overall drainage system makes a stronger impression. The difference in diameters on the one hand allows saving 6.924 or 4.783 million roubles by means of using the N. F. Fedorov and the A. D. Altshul models respectively, on the other hand it increases expenses by 4.195 million roubles in case of using the S. V. Yakovlev model.

7 Conclusions

1. The recommended use of charts, diagrams and nomograms for hydraulic calculations does not satisfy the contemporary level of Information Technologies.
2. Numerical methods should be used in hydraulic calculations. Iteration methods have stable and rapid convergence.
3. The adequacy of the N. N. Pavlovsky, the A. D. Altshul, the N. F. Fedorov and S. V. Yakovlev models to the physics process can be verified by laboratory or natural experiment. The N. N. Pavlovsky model has the longest history of utilization, so the results received by this model, were chosen as the base set.
4. The fillings of the pipes and the speed of the streams received in all models keep within the admissible limits.
5. The N. F. Fedorov model (11) taking into account an important parameter, namely the Reynolds number Re, allows in certain cases to fit smaller diameters. It consequently allows decreasing expenses when purchasing pipes.

References

1. Svod pravil SP 32.13330.2012 "SNiP 2.04.03-85. Kanalisatsiya. Naruzhniye sety i soorugeniya" Aktualizirovannaya redaktsiya SNiP 2.04.03-85. (2012)
2. A.A. Lukinikh, N.A. Lukinikh, Tablitsy dlya guidravlicheskogo rascheta kanalizatsyonnyh setey po formule akad. Pavlovskogo N.N. (1974)
3. P.G. Kiselev, Spravochnik po gidravlicheskim raschetam (1972)
4. A.U. Khemich, A.V. Atkin, Internet-vesnik VolgGASU, 2 (4) (2007)
5. A.M. Kurganov, N.F. Fedorov, Guidravlicheskiye raschety system vodosnabgeniya i vodootvedeniya: Spravochnik (1986)
6. V.A. Orlov, R.E. Khurgin, Vesnik MGSU, 3, 118—122 (2010)
7. Orlov V.A., Zotkin S.P, Orlov E.V., Khurgin R.E., Maleeva A.V., Vestnik MGSU, 3, 205—210 (2012)
8. Rylova I.A., Borovkov V.S., Vestnik MGSU, 4, 181—187 (2013)
9. Kalitsyn V.I. Guidravlicheskiye soprotivleniya samotechnikh vodootvodjashikh setej (1986)
10. Bachir Achour , The Open Civil Eng. J., 9, 187-195 (2015)
11. Kalitsyn V.I. Vodootvodjashiye systemy I soorugeniya (1987)
12. Janet Bacro, Kenneth M. Wong, Michael K. Stenstrom// J. of Hydr. Eng., 134(4), 466-474 (2008)
13. Pushpa N. Rathie, Prabhata K. Swamee, Luan Carlos de S. M. Ozelim// Appl. Math. and Phys., 1 (4), 120-125 (2013)

Determination of the cavity mode in the water

Pavel Vikulin[1] and *Vera* Vikulina[1*]
[1]*Moscow State University of Civil Engineering, Yaroslavskoye shosse, 26, Moscow, Russia, 129337*

Abstract. Ultrasonic oscillations find their wide application in various sectors and industries: metallurgy, chemical and food industries, in engineering, in medicine. This is due to physical and chemical changes when exposed to the sound field. Cavitation in ultrasonic field causes the dispersion and emulsification of certain substances and promotes coagulation and degassing affects of crystallization and dissolution. It is known that ultrasonic vibrations also cause a variety of chemical transformations of substances such as oxidation, recovery, polymerization, and depolymerization. Researchers of cavitation action on the matter find the explanation of these phenomena: shock waves and, consequently, acoustic wind. The experiments were conducted in liquid environment (non-distilled water). The volume of the experimental sample amounted to 10 dm^3. Method of the magnetostriction was applied to receive ultrasonic vibration, the principle of which consists in transforming electrical oscillations into mechanical. The level of cavitation was checked in two stages to evaluate the cavitation mode according to the degree of erosion of the artificial barriers and to measure the intensity of cavitation noise in the volume.

1 Introduction

1.1 Nature of ultrasound

Ultrasound is elastic vibrations. The laws of non-linear acoustics can describe ultrasonic waves of high intensity. The distribution of ultrasonic waves in liquids is accompanied by acoustic flow, compression and rarefaction of the environment. Acoustic cavitation is among the important non-linear phenomena in ultrasonic field. Investigation of the influence of acoustic fields on substances is complicated due to various simultaneous processes that can influence each other in the ultrasonic field. Therefore, it is difficult to describe experimental results. At present, theoretical justification of physical methods is evaluated, in particular ultrasonic vibrations in water technologies. Application of ultrasonic field for the intensification of redox processes in the water environment, deposition of particulate impurities extends the scope of the possible use of this physical method [1, 2, 3, 4].

Sound as a physical phenomenon, is characterized by sound pressure, sound energy density, sound energy flux, the intensity level (power) of the sound. Fluid in a static state

* Corresponding author: nocviv@mgsu.ru

does not have the shear viscosity and is not able to withstand and transmit any shear stresses. Therefore, longitudinal waves only can distribute in liquids and gases in which the direction of the tangential movements of particles coincides with the direction of wave propagation. The speed of distribution depends on the density of the medium ρ and adiabatic compressibility βc calculated by the formula:

$$c = \sqrt{\frac{1}{\rho \cdot \beta_c}}. \quad (1)$$

Physical nature of elastic waves distribution can be described by the equation:

$$y = A \cdot \sin\omega \cdot t, \quad (2)$$

A – offset amplitude; ω – cycle frequency; *t* – time.

Acoustic vibrations in the environment creates additional pressure [5, 6, 7]. A sound wave passing through the fluid creates areas of compression and refraction, which change at every half period of the passing wave. When this occurs, the alternating pressure P, which can be determined by the formula:

$$P = \sqrt{r \cdot c \cdot I \cdot 4{,}6 \cdot 10^{-3}}. \quad (3)$$

Sound pressure is variable and changes periodically. At this point the environment during the period, the pressure P changes from a maximum to zero and then increases again to a maximum value, and corresponds to the description of harmonic oscillations:

$$P = P_{\max} \cdot \sin\omega \cdot t, \quad (4)$$

P_{max} – maximum sound pressure (pressure amplitude) defined by the formula:

$$P_{\max} = \omega \cdot c \cdot r \cdot A, \quad (5)$$

c – speed of sound.

Within the distribution of ultrasound in the environment, part of its energy is absorbed, heating of the environment. The absorption of acoustic energy happens due to the frequency of sound, viscosity, thermal conductivity and impedance of environment, i.e. the product of the density of the medium ρ and the speed of sound. The connection between sound pressure and impedance expresses the oscillatory speed. [8, 9]

The value of the medium wave resistance is determined by the ratio of sound pressure in a running flat wave to the particle velocity of the medium *V*

$$\frac{P}{V} = r \cdot c. \quad (6)$$

Vibrational velocity and sound pressure do not depend on the frequency, the amplitude is inversely proportional, and the acceleration is directly proportional to the frequency.

For moderate acoustic fields acoustic pressure does not typically exceed one MPa, however, the pressure gradient, especially at high frequencies, can reach large values. Amplitude accelerations of fluid particles in a field of ultrasonic waves may exceed the acceleration of free fall by several orders. Such acceleration is achieved only in special ultracentrifuge taking into account that such values of acceleration change sign twice during the period ultrasonic waves may be considered as a very powerful and distinctive physical factor influencing the substance.

1.2 Physical and chemical effects in the liquid medium under the action of ultrasonic cavitation

Basic physical and chemical effects that occur in fluid under the action of the acoustic field, are caused mainly to nonlinear effects, the most important of which is cavitation [10, 11, 12, 13]. One of the features of ultrasonic cavitation is that it is a unique and effective mechanism for the local concentration of a relatively low average energy of the acoustic field at very small volumes, resulting in extremely high energy densities.

The detailed mechanism of this effect remains unclear. Because of a great theoretical and practical interest in the basic physical-chemical cavitation problems in the following will be given significant attention.

The impedance of the environment is a very important characteristic that determines the conditions of radiation, absorption of acoustic waves, their reflection, refraction, etc.

The particles of an elastic medium, in which ultrasonic waves are distributed oscillate and therefore possess kinetic and potential energy.

The amount of energy carried the sound vibrations for 1 sec through an area of 1 cm^2 perpendicular to the direction of propagation, describes sound intensity and is determined by the formula:

$$I = E_1 \cdot c, \tag{7}$$

E_1 – density of energy.

The sound energy density at each point changes with time. The average value of the energy density at a given point is determined by the formula:

$$E_1 = \frac{1}{2} \cdot A^2 \cdot \omega^2 \cdot \rho. \tag{8}$$

Transforming equation (7) and equation (8) we get the expression (9):

$$I = \frac{1}{2} \cdot A^2 \cdot \omega^2 \cdot \rho \cdot c. \tag{9}$$

Within the distribution of sound waves in a liquid environment, the sound intensity I decreases with increasing distance from the radiation source according to the equation:

$$I = I_0 \cdot e^{2\alpha X}, \tag{10}$$

I_0 – sound intensity if $X = 0$; α – adsorption coefficient.

The absorption coefficient that depends on the physical properties of a substance is a characteristic of the substance and depends on external conditions (temperature, pressure) and frequency.

Although the physical nature of ultrasound and the basic laws describing the distribution are the same as for sound waves of any frequency range, it has a number of specific features. These features are due to the relatively high frequencies and, accordingly, the smallness of the wavelengths [14, 15, 16].

So, for high ultrasonic frequencies, the wave lengths:

$3{,}4 \cdot 10^{-3} - 3{,}4 \cdot 10^{-5}$cm In the air,

$1{,}5 \cdot 10^{-2} - 1{,}5 \cdot 10^{-4}$cm In the water,

$5 \cdot 10^{-2} - 5 \cdot 10^{-4}$cm In metals.

Ultrasonic waves attenuate much faster than waves of low frequency sound range, as the ratio of the "classical" sound absorption is proportional to the square of the frequency.

In the low frequency area, the coefficient of relaxation absorption also increases in proportion to the square of the frequency, however at higher frequencies this increase stops and the absorption coefficient tends to a constant value.

The relationship of the nature of ultrasound and, in particular, its high-frequency area is hypersonic, with the structure of matter and elementary excitations in it is one of the most important features of ultrasonic waves. It allows to judge about the structure of matter based on measurements of velocity and absorption in it, depending on the frequency, as well as some external factors such as temperature, pressure, etc.

Feature of the ultrasound in the high-hypersonic ranges is the possibility of using methods of quantum mechanics, since the wavelengths and sound frequencies in these bands be of the same order with the parameters and frequencies characterizing the structure of matter.

An elastic wave of a certain frequency in this case is mapped to a quasi-particle, the photon, or quantum of sound energy. Representation of quantum mechanics is convenient when considering the various interactions in solids. For example, scattering and absorption of sound vibrations of the crystal lattice can be considered as the interaction of coherent and thermal phonons.

In the ultrasonic field of high intensity acoustic flows are developed, the speed of which is usually small in comparison with the oscillatory velocity of the particles [17, 18, 19].

Flow appeared due to absorption of sound can occur in standing waves or in the boundary layer near obstacles of various types (Fig. 1)

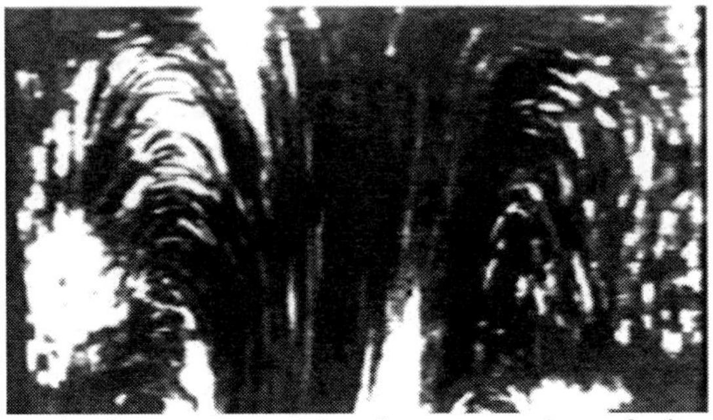

Fig. 1. Photo. The acoustic flow generated by the propagation of ultrasonic vibrations with a frequency of 5 MHz in benzene

The radiation pressure also increases with increasing frequency, because its magnitude is proportional to the intensity of sound in the ultrasonic frequency range. It is used in the practice of acoustic measurements for determination of sound intensity.

In order to determine various effects of sound field parameters sound intensity, sound pressure, vibration velocity, the radiation pressure has reached a noticeable size, with increasing frequency requires a smaller value of the amplitude of vibrational displacement.

The most important nonlinear effect of ultrasound is the cavitation that occurs in the liquid mass of pulsating bubbles filled with steam, gas or their mixture [20]. The complex motion of bubbles and their collapse, merge with each other, etc. generate in the fluid a pulse compression (microwaves), and therefore cause local heating of the environment and ionization. These effects have an impact on the substance: the destruction of solid particles in the liquid (cavitation erosion), there is a stirring of the fluid, different physical and chemical processes start and accelerate. By changing the conditions of occurrence of cavitation, it is possible to strengthen or weaken different cavitation effects, for example, with increasing frequency of the ultrasound increases the role of micro-flows and reduced cavitation erosion, with an increase in hydrostatic pressure in the liquid, the role of micro-impacts.

The increase in frequency oscillations typically leads to increased threshold intensity, corresponding to the onset of cavitation, which depends on the kind of liquid, its gas content, temperature, etc. For water in the low-frequency ultrasonic range at atmospheric pressure the intensity value is normally 0.3 – 1 W/cm^2 (Fig. 2).

Diverse applications of ultrasonic vibrations, using their various characteristics, can be divided into three areas. The first relates to obtaining information by means of ultrasonic waves, the second active impact on the substance and third – the processing and transmission of signals [21].

Fig. 2. Photo. The fountain of liquid that is formed when the radiation of the ultrasonic beam from the depth of liquid on its surface (the radiation frequency 1.5 MHz, intensity of 15 W/cm^2).

2 Aim and objective

The aim of this work is to confirm the presence of the cavitation and to identify the nature of the distribution of cavitation intensity in the liquid volume.

The objective of the work consists in the definition of cavitation according to the erosion (destruction) of the artificial barriers, and in obtaining a graphic image of the intensity of ultrasonic cavitation in the reactor volume of the experimental setup for fluid handling.

3 Materials and methods

There was ultrasonic reactor made of stainless steel assembled to estimate the influence of ultrasonic vibrations on water treatment with the introduction of acoustic vibrations from the bottom up, through the thickness of the liquid in the reactor volume at atmospheric pressure on the surface of the water.

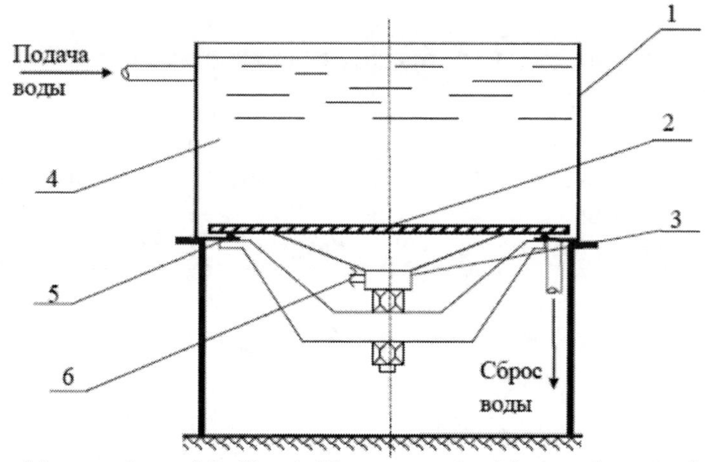

Fig. 3. Diagram of the experimental facility: 1. The reactor of stainless steel; 2. Membrane, emitting ultrasonic waves from the bottom to the top; 3. Magnetostrictive transducer. 4. The treated liquid; 5. Sealed (rubber) gaskets; 6. Coolant.

Ultrasonic reactor of rectangular shape was equipped with magnetostrictive transducer and had the dimensions LxBxH 400mm x 400mm x 300 mm (the size of the membrane 300 mm x 300 mm). Converter of acoustic vibrations is located in the lower part of the reactor under the liquid layer. The experimental setup of the processing liquid in the ultrasonic field is presented in Fig. 3.

Cavitation is the basis of the processes of ultrasonic influence on liquid environment. Therefore, the desired accurate registration of the occurrence of cavitation phenomena. It is advisable to carry out experimental research in two stages.

3.1 Stage 1 of the research

It is possible to assess the collapse of cavitation bubbles in the liquid, under the action of ultrasonic vibrations according to the degree of erosion of artificial barriers (indicator). An indicator is made of foil with a size of 400mm x 300mm, and immersed in fluid, in which there is a cavitation effect.

Formed in the ultrasonic field, the bubbles collapse and create implosions, shock waves, and micro flows. Bubbles, which are formed near the surface of the foil inevitably, lead to the erosion of foil itself. In the case of a strong influence of ultrasonic vibrations, the possibility of partial or complete destruction of the leaf foil.

The intensity of the cavitation depends on the chemical composition of the liquid, the intensity of the ultrasonic field and emission of ultrasonic vibrations. In our case, the emitter and the liquid remain constant; therefore, the chemical composition of the environment, exposure of ultrasonic influence and the temperature of the sample can be changed.

Pure water and industrial wastewater of the enterprise on primary processing of wool were chosen as environments for visual detection of the cavitation.

The leaf foil size 400mm•300mm, with a thickness of 15 microns in a special box that is placed in the reactor perpendicular to the radiating surface to the height of the sample.

The temperature of the source water samples was 13°C, and the duration of ultrasonic treatment on canvas – 10 min. To sewage, the exposure was changed from 10 sec to 10 min, and the temperature of the water was 38°C.

3.1 Stage 2 of the research

In the process of distribution of ultrasonic oscillations in the water, cavities begin to pulsate in phase with the current acoustic field. Registration of these sound effects can be done with a device that measures acoustic emission, filtering and removing the noise of the fundamental frequency. The studies determined that in the reactor under the action of ultrasonic field there are zones with different cavitation intensity and density. To identify areas of cavitation zones the device was used, allowing measuring the intensity of sound pressure when applying the ultrasonic field at any point in the reactor. This device is able to detect sound pressure, resulting from the collapse of cavitation cavities.

4 Results

In the first stage of experimental studies of the registration of cavitation according to the erosion of foil (the size of the canvas foil 300 x 400 mm.) are shown in figure 4 (a-e). The lower horizontal part of the foil was in close proximity to the emitter. Accordingly, the upper horizontal part of the fabric of the foil located on the free surface of the liquid.

Below are pictures of the actual images with traces of cavitation erosion on the canvas of foil. The resulting images presented in figure 4, confirm the presence of cavitation regime in the liquid.

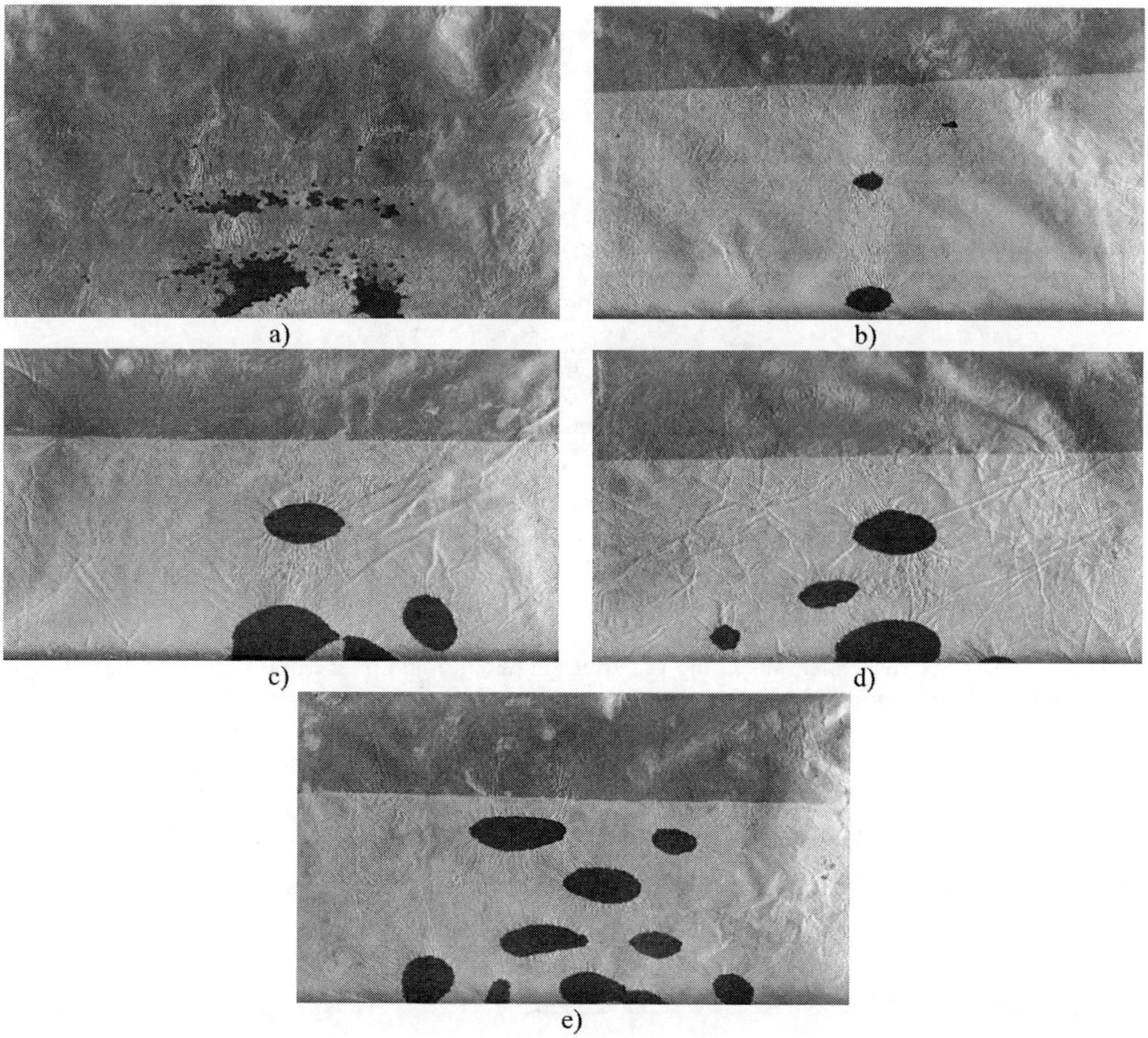

Fig. 4. Photos. The results of the first stage of the experiments: a) the Time of exposure to cavitation 10 min., drinking water quality, at t = 13°C; b) the Time of exposure to cavitation for 10 seconds. industrial waste water at t = 38°C; C) the Time of exposure to cavitation 1 min., industrial waste water, at t = 38°C; g) the exposure Time of cavitation 5 min., industrial waste water, at t = 38°C; d) the Time of exposure to cavitation 10 min., industrial waste water, at t = 38°C.

Cavitation destruction (erosion) was in the form of holes in the Central part of the foil. The degree of variation of physical and chemical properties of water affect the cavitation intensity in the ultrasonic field mode. Therefore, in the sample of wastewater with a high degree of variation of physical-chemical properties of water, the effect of cavitation action is superior to a similar action in water less variation of physical and chemical properties.

Thus, the resulting image on the canvas will help determine the location of foci of cavitation, and visually, with some certainty, to fix the mode of cavitation in the sample under study.

In the second stage, the obtained experimental data of the change of the sound pressure. The developed device allowed to record at selected points along the reactor cavitation pressure of the collapsing cavities. According to the obtained points to construct the plot of the isosurfaces of the distribution of sound pressure in the reactor (fig. 5).

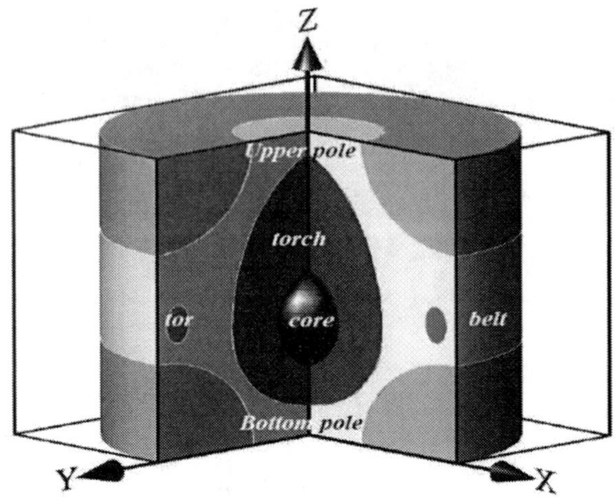

Fig. 5. Plot of the distribution of acoustic pressure in the ultrasonic reactor in the cavitation mode.

Perspective view allows witnessing the heterogeneity of areas of cavitation in the reactor volume. So, in the Central part of the plot allocated to the area of high cavitation, resembling in appearance the "torch" with a strong "core" the maximum value of acoustic pressure.

"Torch" is suspended above the radiating diaphragm and its Central part is not on the surface of the liquid medium.

The Central portion of the plot is very similar to "yule", and is characterized by an area of lower acoustic pressure than the Central part of the "torch" that corresponds to the physical parameters of the process: the length of the acoustic wave, the frequency and intensity of the radiation. However, in the Central part of the plot perimeter at the damage the "core", formed the area of the concentrated energy in the form of a "torus".

Power saturation region, the "torus" comparable area "torch", but significantly lower power of cavitation manifestations, than in the "core".

5 Conclusions

The first stage of experimental studies allowed obtaining real evidence of ultrasonic cavitation in the liquid medium of reactor volume.

In the second stage of experiments simultaneously installed and registered the points of maximum and minimum cavitation mode propagating in the reactor volume.

A statement of the phenomenon of cavitation in a liquid environment using the erosive characteristics of the artificial barriers is a simple and reliable method of recording the physical process. Based on a clear representation of areas of cavitation manifestations can justify the design parameters of the reactor depending on what effects you want to obtain in the liquid medium under the action of ultrasonic field in a cavitation mode.

References

1. I. P. Golyamina, Ultrasound. Little encyclopedia of Moscow (1979)
2. L. D. Rosenberg, Physics and technique of powerful ultrasound. vol. 2. (2012)
3. Vikulin, P. D., Physical and chemical manifestation of the acoustic field in the technologies of water conditioning (2004)
4. Flynn G., Physical acoustics (1967)

5. Kardashev G. A., Physical methods of intensification of processes of chemical technology (1990)
6. Raj Baldaev, Application of ultrasound (2006)
7. N. Makisha, E3S Web of Conferences, 6, 01002 (2016)
8. Elpiner I. E., Biophysics of ultrasound (1973)
9. Margulis M. A., Adv. in phys. Sci., 170, 3, 263-287 (2000)
10. Fridman, V. M., Ultrasonic chemical equipment (1967)
11. Mironov M. A., Acoust. J., 23 (5), 825-830 (2014)
12. Leighton, T. G., The Acoustic Bubble (1994)
13. V. N. Khmelev, A. V. Shalunov, A. V. Shalunova, Ultrasonic atomization of liquids (2010)
14. N. Makisha, Procedia engineering, 165, 1087-1092 (2016)
15. Shutilov V. A., Basic physics of ultrasound (1980)
16. Kundas S. P. and others, Ultrasonic processes manufacture of electronics products (2003)
17. N. Makisha, M.Yunchina, MATEC Web of Conferences, 106, 70016 (2017)
18. Khmelev, V. N. and others, Control of the parameters of cavitating liquid media subjected to ultrasonic influence (2012)
19. Fedotkin I. M., Guly I. S., Cavitation technique and technology, their use in the industry (theory, calculations and design cavitation machines). Part 1 (1997)
20. Dezhkunov N. B., Leighton, T. G., J. of eng. Phys., 77 (1), 45-51 (2004)
21. Vikulin, P. D., Development of processes of separation of highly concentrated Chertomlyk wastewater by ultrasound (1985)
22. N. Makisha, I. Panteleeva, MATEC Web of Conferences, 106, 70015 (2017)
23. Agranat B. A., Ultrasonic technology (1974)
24. Sirotyuk M. G., Acoustic cavitation (2008)
25. Zubrilov S. P., Physico-chemical aspects of ultrasonic activation of binding solutions (1975)

Hemodynamics of Idealized Carotid Artery

Chaitanya Puritipati[1], *Varun* Keri[1], and *Nitesh* Kumar[1*]

[1]*Dept. of Mechanical and Manufacturing Engg., Manipal Institute of Technology, Manipal Academy of Higher Education, Manipal, Karnataka, India*

Abstract. The behavior of blood during its flow in human arteries has always been a subject of keen interest. Flow parameters such as pressure, velocity and wall shear stress vary with changes mainly in the geometry of the artery. The work focusses on the variation of wall shear stress along the walls of an Idealized Carotid Artery. Simulations are carried out for both Newtonian and Non-Newtonian behavior of blood. In the Non-Newtonian flow, Carreau-Yasoda model is chosen for the calculations. Steady flow and Transient flow cases are also analyzed where the User Defined Function is interpreted for further calculations. It is observed that recirculation of flow occurs at a critical area in the geometry. The results obtained are compared with the numerical results of Eleyas Shaik. The Wall Shear Stress plots are found to be in close correspondence to those in the thesis.

Keywords: Carotid Artery, Wall Shear Stress(WSS), Newtonian Flow, Non-Newtonian Flow, Right Inner Wall (ROW), Right Outer Wall (ROW), Left Inner Wall (LIW), Left Outer Wall (LOW)

1 Introduction

The Carotid artery plays a significantly crucial role in the cardiovascular system of the human body. The important function of blood flow to the face, neck and the brain of the human body is performed by the Carotid artery [1]. There are 2 Carotid arteries, one on each side of the neck. Each Carotid Artery bifurcates into 2 divisions, namely the Internal Carotid Artery(ICA) and the External Carotid Artery(ECA). The ICA supplies blood to the brain while the ECA supplies blood to the face and the neck [1]. Hemodynamics of the Carotid artery gives us a better idea about the nature of the blood flowing through the artery and about the challenges faced during the flow in different locations inside the artery. Wall Shear Stress(WSS) is a major factor in analyzing the blood flow characteristics inside the artery [1].

* Corresponding author: nitesh.naik@manipal.edu

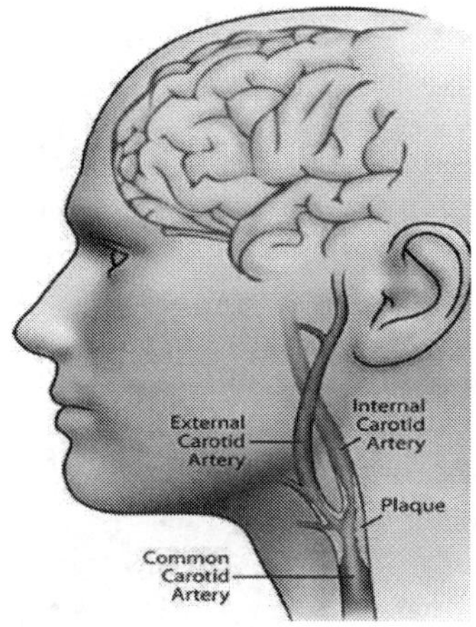

Fig. 1. Carotid Artery Structure

2 Methodology

3D idealized model of a carotid artery is modelled using Catia V6, and imported to Ansys CFX 17 to simulate the flow. The model is meshed using inbuilt mesh tool in Ansys workbench and tetrahedral elements are used in this simulation. An element size of 0.39 mm was selected after grid convergence study. Figure 5 shows the grid convergence of the elements used. Inflation of 4 layers with total thickness of 1mm is applied to capture the boundary layer and critical back flow near the wall of the artery. Figure 2 shows the model used in this study. A steady state analysis is performed to compare the results with those in the literature by Eleyas Shaik [1]. The boundary condition applied for steady flow analysis is an inlet velocity of 0.07m/s and an outlet pressure at both outlet surfaces of 0 Pa. Consequently, the boundary conditions for the transient flow analysis is the same outlet pressure of 0 Pa at both outlet surfaces while the inlet velocity is according to the user defined function. Further, analysis is also done considering different types of flow, i.e., Steady and Pulsatile flow and Newtonian and Non-Newtonian Flow. For the Non-Newtonian type of flow, Carreau-Yasuda model is adopted. The parameters required for it are mentioned in Table 1.

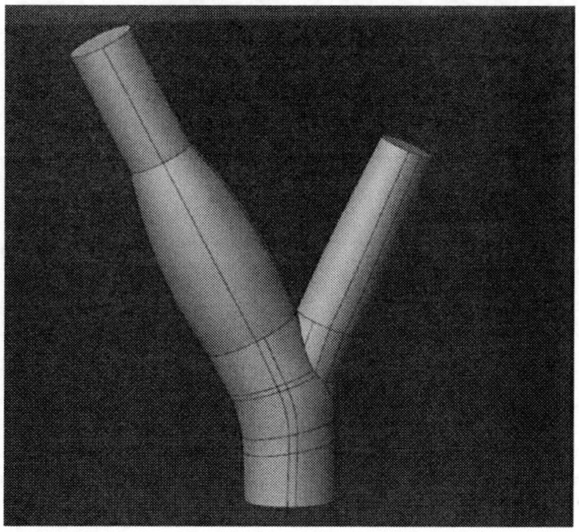

Fig. 2. CATIA model of Idealized Carotid Artery

2.1 Software Tools

The ANSYS Workbench platform upon which a geometric model is created, meshed and appropriate boundaries are identified and marked. With bi-directional CAD connectivity, powerful highly-automated meshing, a project-level update mechanism, pervasive parameter management and integrated optimization tools, the ANSYS Workbench platform delivers unprecedented productivity, enabling Simulation Driven Product Development. Ansys Fluent® is commercially available CFD software widely accepted for its high reliability. The software has different capabilities and functions. Both 2D and 3D models can be tested with laminar, inviscid, or turbulent flows. Steady and unsteady problems can also be handled using this software. Depending on the model and flow type the designer chooses the appropriate flow field type. Fluent also provides mesh flexibility including the ability to solve complex flow problems using unstructured meshes that can be generated about complex geometries with relative ease. Supported mesh types include 2D triangular or quadrilateral, 3D tetrahedral, hexahedral or polyhedral meshes and hybrid meshes. Fluent is flexible to allow us to refine or coarsen a grid based as required. Ansys Fluent® is written in the computer language C and makes use of the flexibility and variety offered by the language. User defined functions (UDF) written in C may also be appended to Fluent, thus increasing its versatility. Fluent uses finite volume discretization method in order to convert Partial Differential Equations of conservation of mass and momentum into non-linear algebraic equations [1]. CATIA platform is widely used for modelling various kinds of geometries. The solid part models created in CATIA can be exported to other softwares like ANSYS for further analysis. We have used all the above mentioned softwares for catering to our design and analysis needs.

2.2 Governing Equations

Continuous fluid flow is governed by the Navier-Stokes equations. The Navier Stokes equations consist of 3 equations, namely the Continuity equation, the Momentum equation and the Energy equation which is obtained by applying the conservation of the fluid flow. In this study, we assume isothermal conditions for the simulations, hence the energy equation is mentioned here for the sake of completeness. [1]

The principle of mass conservation applied to flowing fluid results in the continuity equation, which can be expressed as

$$\frac{d\rho}{dt} + \rho\nabla \cdot \vec{u} = 0$$

The momentum equation can be expressed as

$$\frac{\partial}{\partial t}(\rho \underline{v}) + div(\rho \underline{v}\underline{v}) = -\nabla p + div(\underline{\tau}) + \rho \underline{g}$$

$$\underline{\tau} = \mu\left[\left(grad(\underline{v}) + grad(\underline{v})^T\right) - \frac{2}{3}div(\underline{v})\underline{I}\right]$$

Where ρ is the density, $\underline{v}$ is the velocity, p is the pressure, $\underline{\tau}$ is the stress tensor and and $\underline{g}$ is the body force.

In case of pulsatile flow, the boundary condition at inlet is given in the form of the following waveform generated by the User defined function. [2]

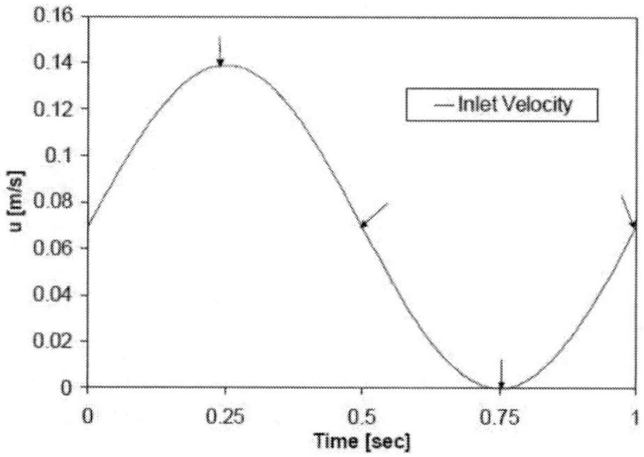

Fig. 3. Inlet Velocity Waveform for Pulsatile flow

2.3 Non-Newtonian Constitutive Equations

For the non-Newtonian case, the shear effect on viscosity must be considered. There are various models in several literatures to justify the non-Newtonian behavior of blood, e.g., the Casson model [4]. Yeleswarapu [5] provides a detailed report on these models. In the current investigation, the complex rheological behavior of blood is approximated using the shear-thinning Carreau-Yasuda model, where the apparent viscosity is expressed as a function of the shear rate, as shown in the equation below. [1]

$$\frac{\mu-\mu_\infty}{\mu_0-\mu_\infty}=\left[1+(\lambda\dot{\gamma})^a\right]^{(n-1)/a}$$

Table. 1 Parameters used for the Carreau-Yasuda Non-Newtonian Model

μ_∞ (Pa.s)	μ_0 (Pa.s)	a	n	λ (s)
2.2 x 10^{-3}	22 x 10^{-3}	0.644	0.392	0.110

2.4 Grid Convergence Test

Grid convergence test helps us in deciding the optimum value of the design parameters to be chosen for various calculations. A graph is plotted between one of the parameter/result and the number of elements. As the number of elements keep increasing, the value of the parameter becomes almost constant after a certain value. The number of elements after which the parameter becomes constant is considered as the optimum value which is used in the final calculations. The graph for the Grid Convergence test is shown in Figure 5. From the figure, it can be observed that at around 26,000 elements, the inlet pressure reaches a certain value and further increase in the number of elements do not affect the pressure value. Hence, the design conditions which gave us 26000 elements will be considered for all further calculations and results.

3 Results

3.1 Three-Dimensional Idealized Carotid Artery:

The 3-D model of the artery is developed using CATIA software. The model is then exported to ANSYS Workbench for meshing, calculation and further analysis [1].

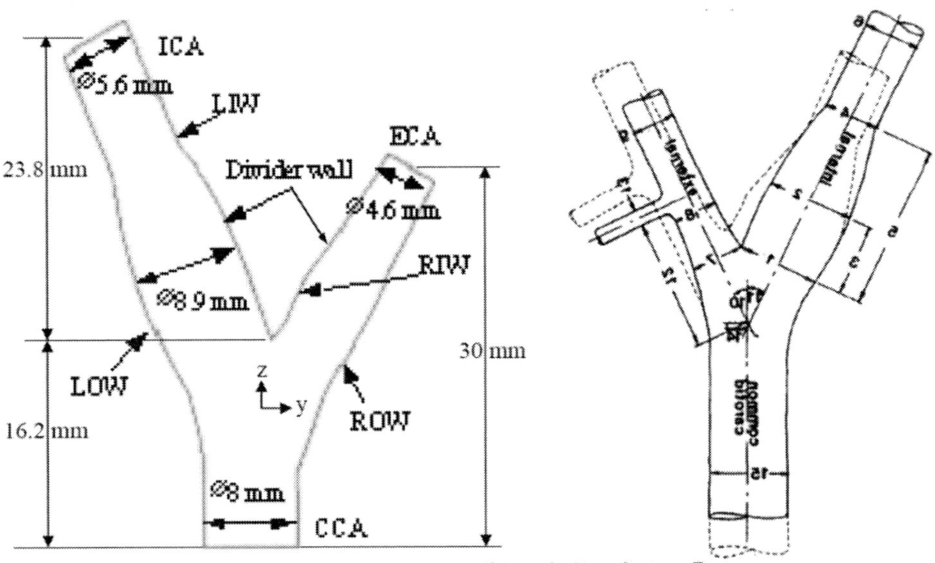

	Location in Fig. 1	Number of samples	(Dimensionless values) Mean	Standard deviation	True mean value (mm)
Internal carotid	1	40	1.04	0.21	8.3
	2	25	1.11	0.14	8.9
	3	11	0.91	0.31	7.3
	4	25	0.72	0.12	5.7
	5	13	2.14	0.45	17.1
	6	53	0.69	0.10	5.6
External carotid	7	36	0.69	0.17	5.6
	8	13	0.69	0.13	5.5
	9	40	0.58	0.10	4.6
Angles	10	23	25.1°	11.0°	25.1°
	11	23	25.4°	10.4°	25.4°
Other	12	11	1.86	0.81	14.9
	13	15	0.23	0.04	1.9
	14	51	−0.3	1.19	−0.2
	15	–	1.00	0.00	8.0

Fig. 4. Schematic Representation of 3D model along with dimensions [2]

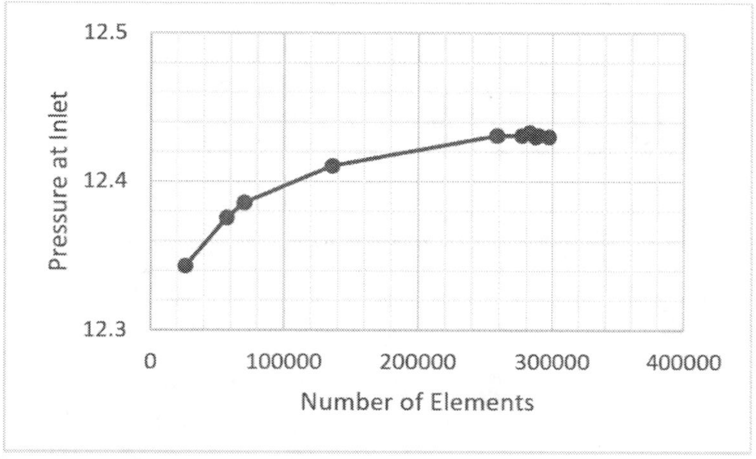

Fig. 5. Grid Convergence Test

Discussion:

From **Fig. 6**, we can observe that pressure contour on YZ plane is in close resemblance with that from the literature. Also from the results we can observe that the maximum wall shear stress occurs at the location of the bifurcation. Low wall shear is experienced in the bulging region (Carotid Sinus).

From the **Fig. 10**, we can observe low velocity in the Carotid Sinus region because of the bulge in the geometry.

Since the inlet surface has multiple contours, average pressure of all surrounding contours were calculated. It was observed that the average pressure at inlet is close to that of the literature value. The velocity vectors emerging out of different planes also were matching with the vectors from different planes in the literature.

Velocity and Pressure graphs in **Fig. 11** and **Fig. 12** are constructed by taking X axis as Y-distance length by creating lines at different locations such as inlet, ICA entrance, ECA entrance, ICA outlet, ECA outlet.

3.2 Newtonian Flow Results

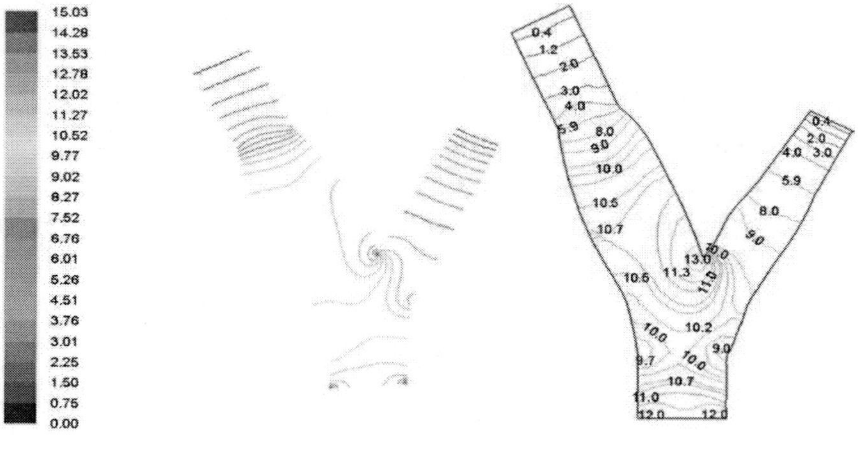

(a) (b)

Fig. 6. Pressure Contour on YZ plane in (a) Present study, (b) Literature [1]

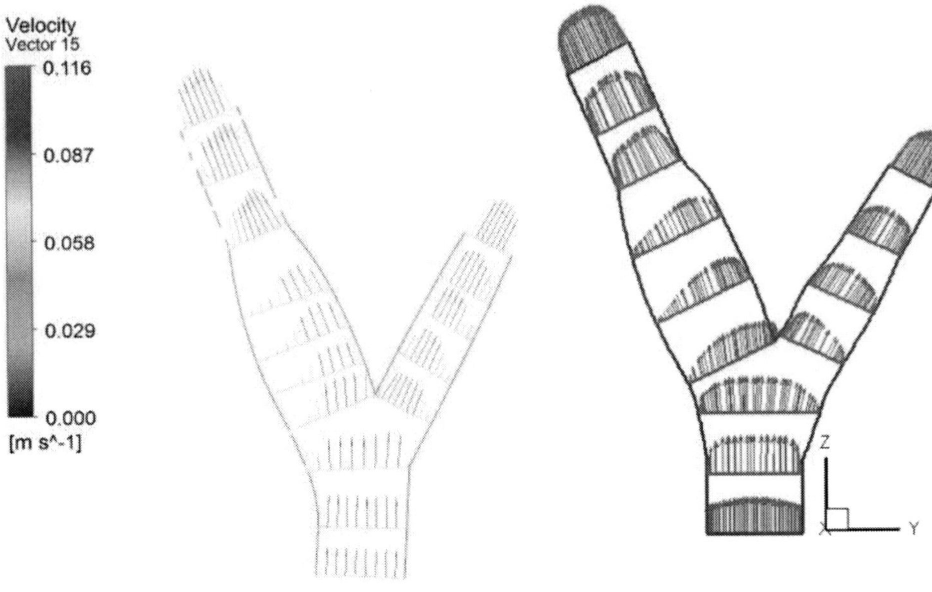

(a) (b)

Fig. 7. Velocity vectors on different XY plane in (a) present study (b) Literature [1]

Fig. 8. Streamlines of particles starting from inlet

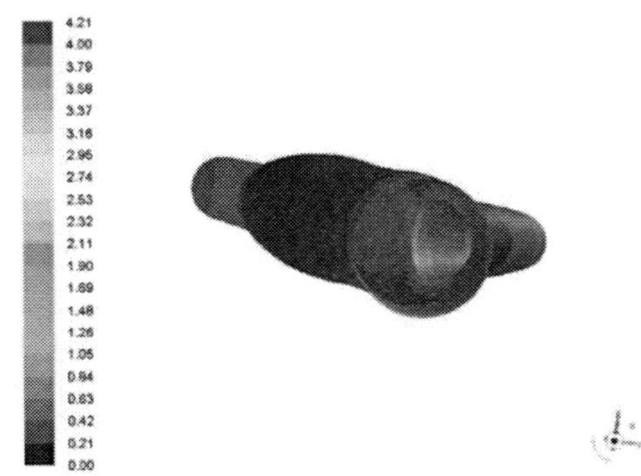

Fig. 9. Wall Shear Contour

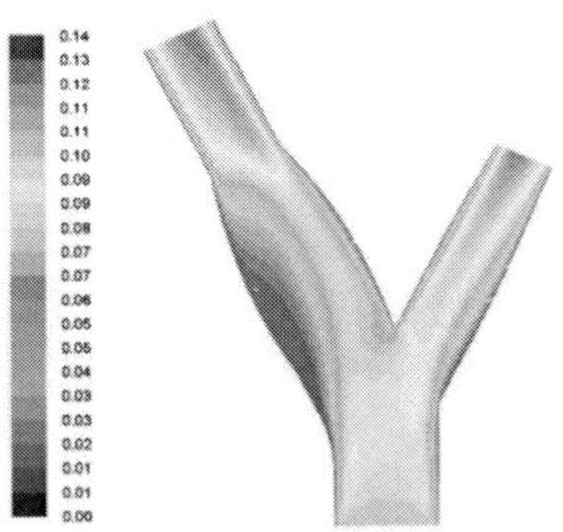

Fig. 10. Velocity contour on YZ plane

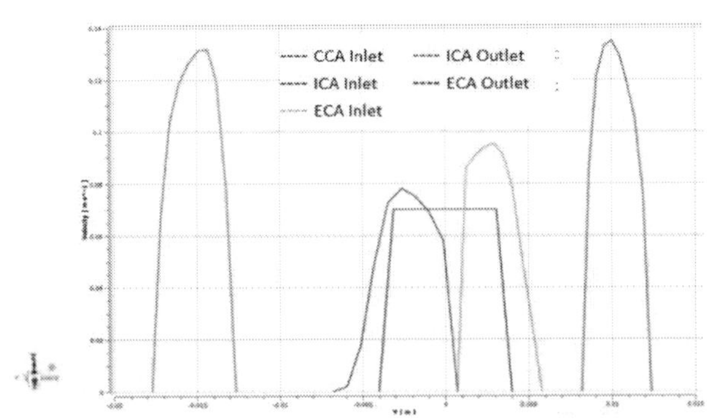

Fig. 11. Velocity vs Y distance at various locations

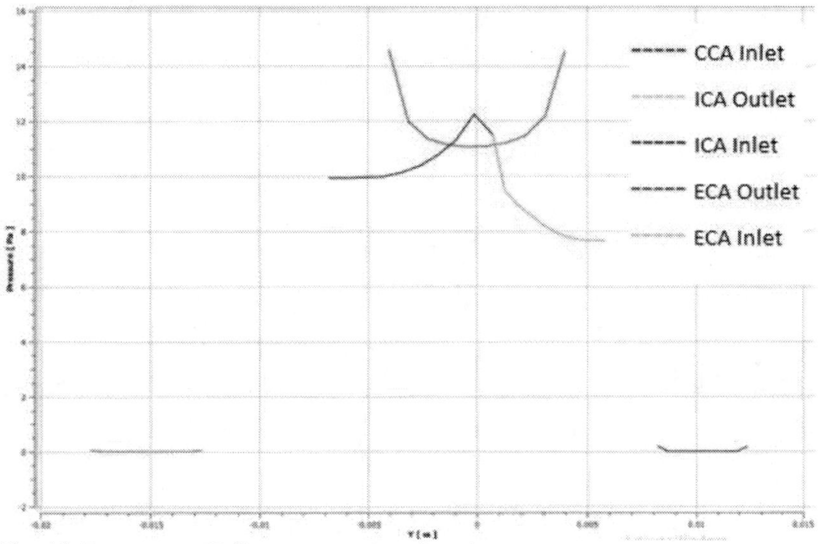

Fig. 12. Pressure vs Y distance at various locations

3.3 Non-Newtonian Flow Results

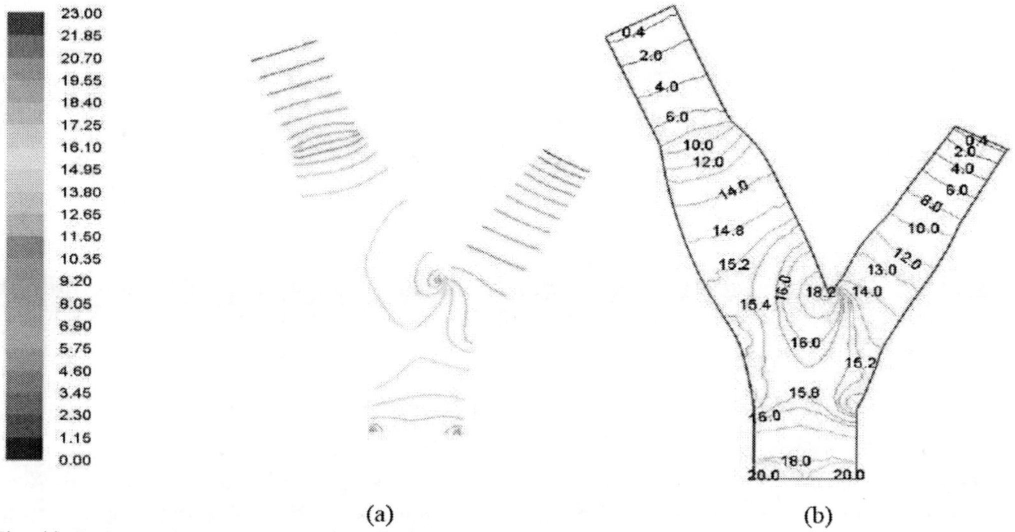

Fig. 13. Pressure Contour at YZ plane in (a) Present Study, (b) Literature [1]

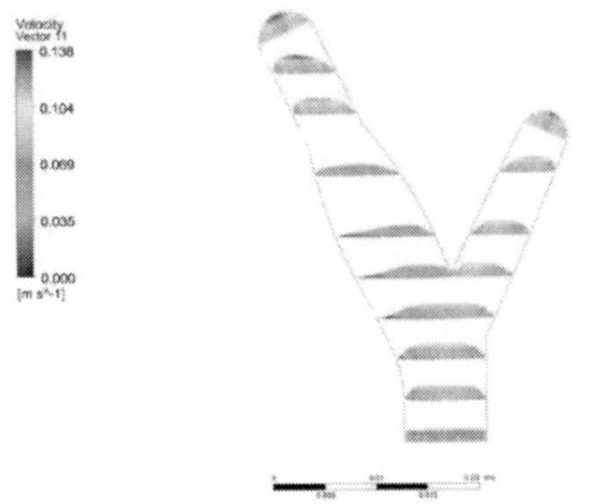

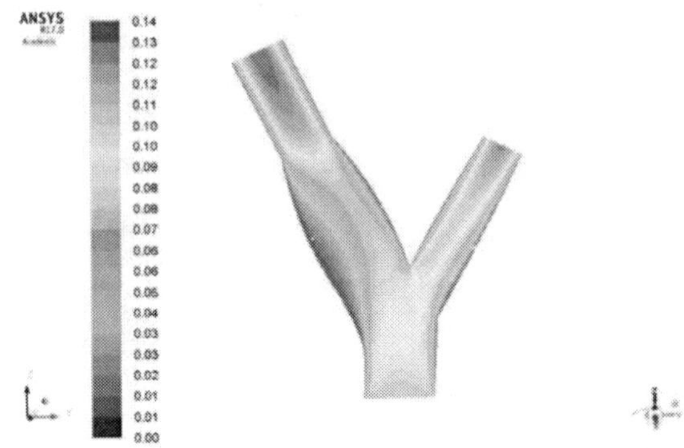

Fig. 14. Velocity contour at YZ plane

Fig. 15. Wall Shear Contour

3.4 Pulsatile Flow Results

In case of pulsatile flow, following changes are made before starting the calculations:

1) The necessary User Defined Function (UDF) for inlet velocity is interpreted
2) The inlet boundary condition accordingly from a constant value to the waveform generated from the UDF.
3) Number of time steps and number of iterations per time step are specified keeping in mind the total cycle time (1 second).

We have opted for 1000 time steps for our calculations.

Discussion:

A comparative study of the Wall Shear Stress Plots at the boundaries is presented as following:

For Inner Walls, comparison is made with the Wall Shear Stress (WSS) plots from the literature. For Outer Walls, the boundaries are shown to observe the variation of WSS along the respective boundary. From the results, it can be concluded that the WSS Plots obtained are in close correspondence with those from the literature.

As it can be seen from the comparative study in **Fig. 17**, the WSS is considerably high at the entry. The WSS value at the bulge is very low due to the recirculation of flow. The consequential increase in the WSS value after the bulge can be attributed to the fact that there is a significant decrease in the area which leads to increase in the velocity according to the continuity equation.

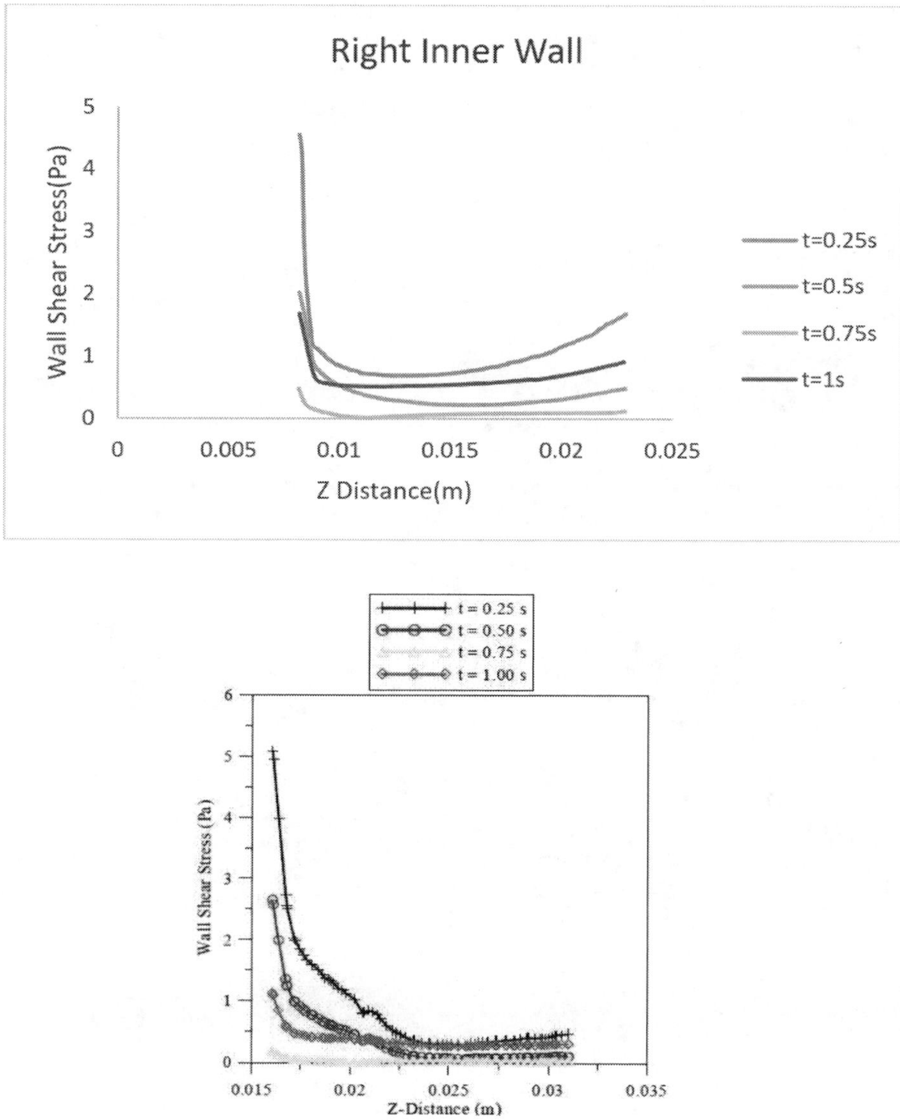

Fig. 16. Comparison of WSS along RIW

 https://doi.org/10.1051/matecconf/201814401020

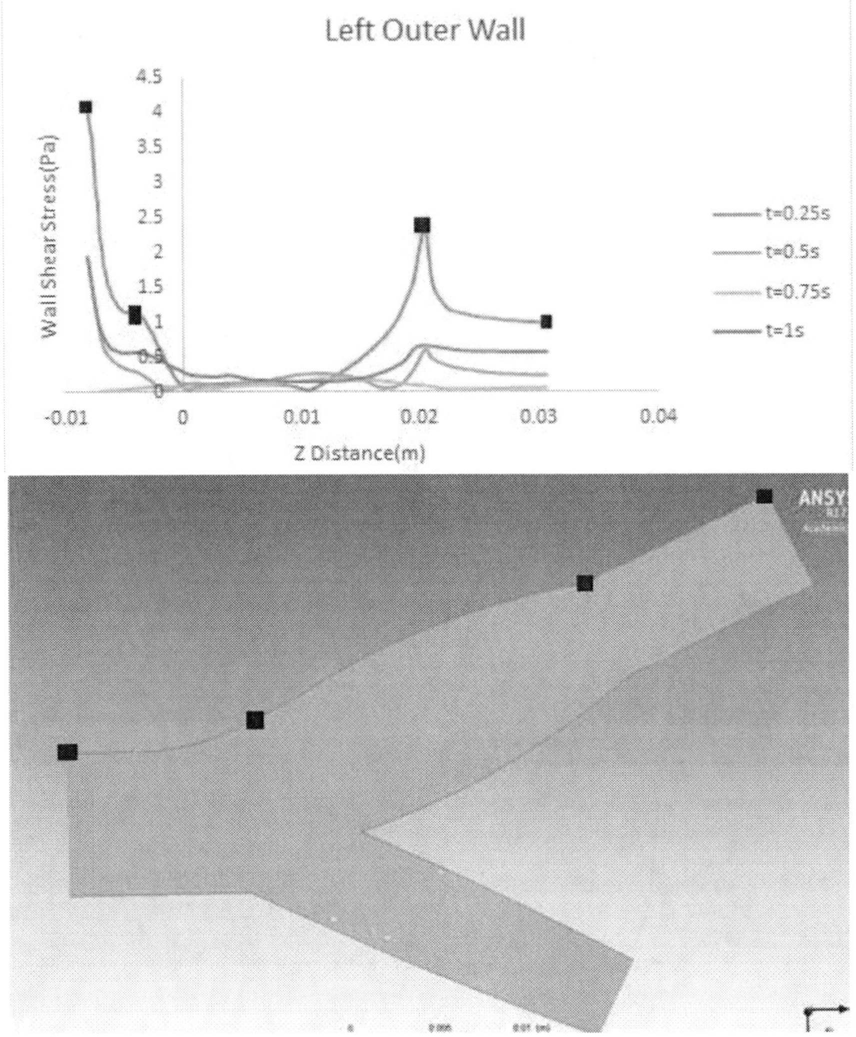

Fig. 17. WSS Plot along LOW

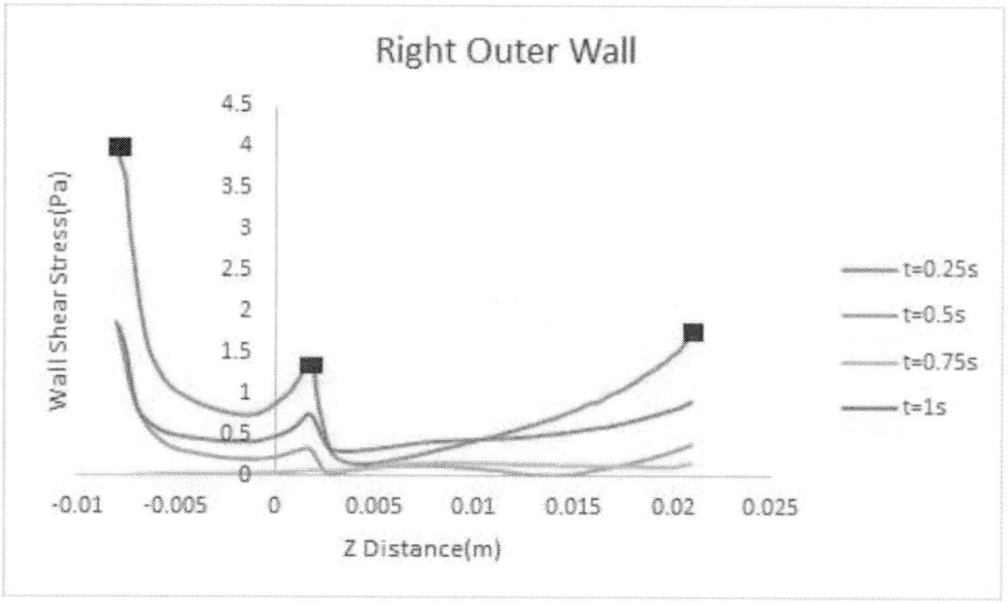

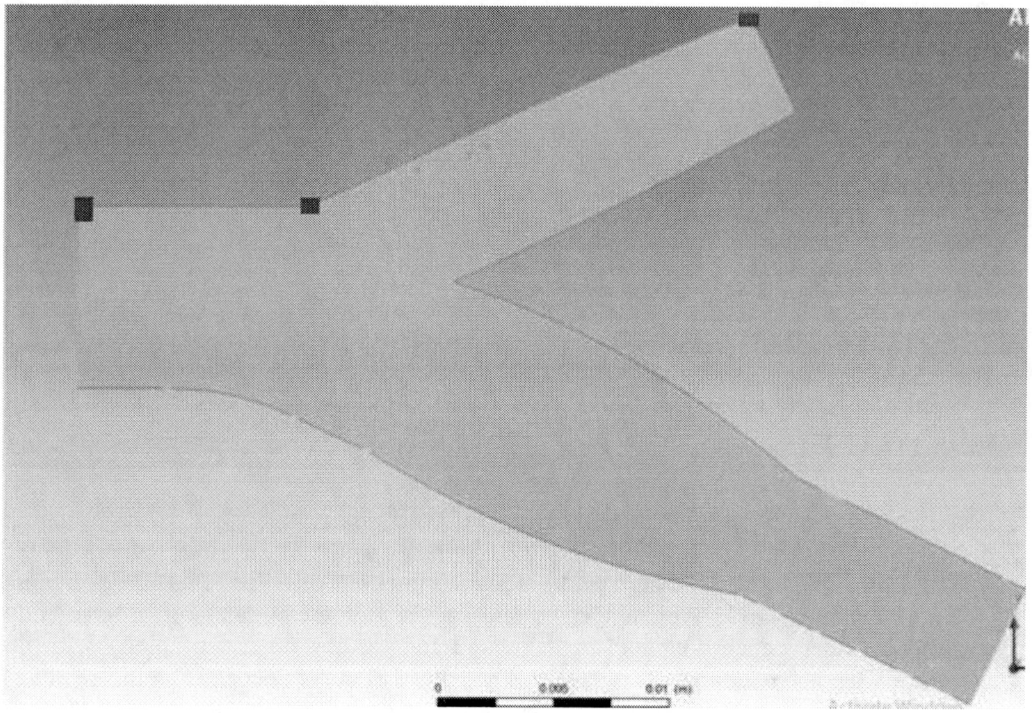

Fig. 18. WSS Plot along ROW

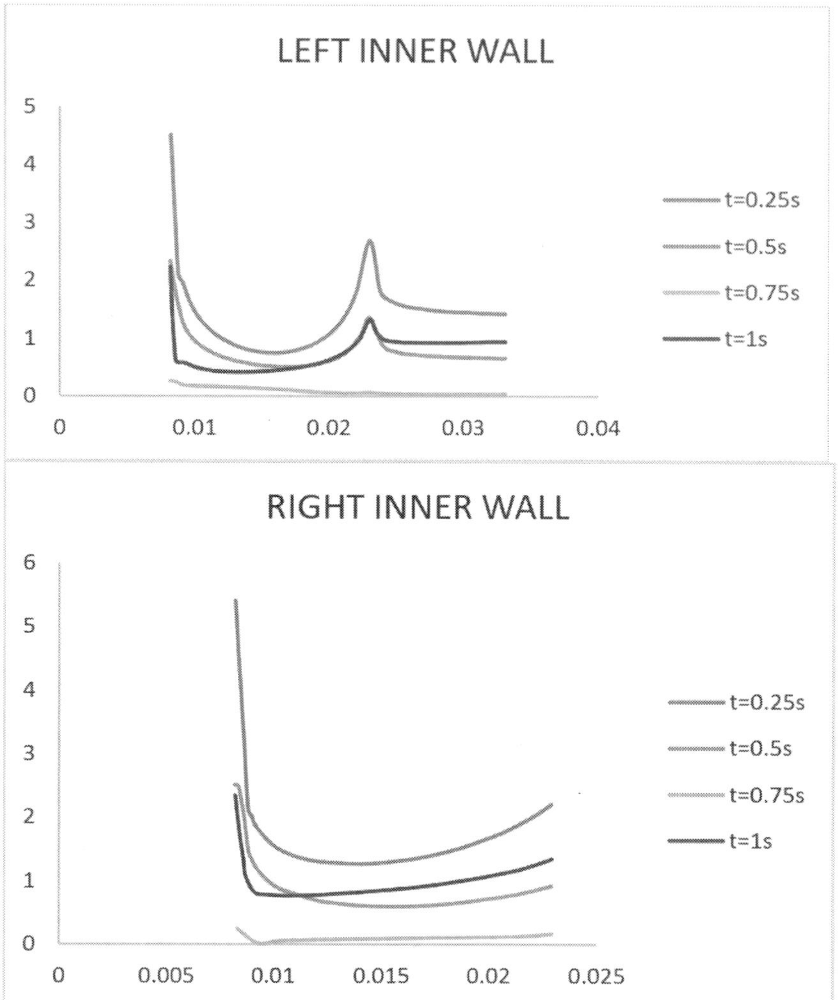

Fig. 19. WSS Plots along Inner Walls

Discussion:

The above graphs in Fig. 19 depict the behavior of Wall Shear Stress(WSS) with change in Z-distance in meters along both the Left and Right Inner walls. The various values of WSS on the curve are obtained by placing a probe at regular intervals on a line. This line is obtained by the intersection of the X-Y plane located midway through the body along with the body. The various values hence obtained are tabulated for various time steps along with the distances and the above curves are obtained. The nature of curves depends on the geometry of the corresponding wall. In case of the LIW, a steep increase can be observed at the point where the carotid sinus begins. The subsequent reduction in the WSS value can be attributed to the further change in geometry when the bulging portion ends. The following graphs in Fig. 20 depict the behavior of Wall Shear Stress(WSS) with change in Z-distance in meters along both the Left and Right Outer walls. The curves are obtained by following the same process as mentioned in this discussion. The curve of the Right Outer Wall is of considerable importance. The sharpness of the curve at the right outer wall causes the sudden increase in the WSS value due to the stress concentration in the corner. Hence the sudden rise in the WSS value can be observed.

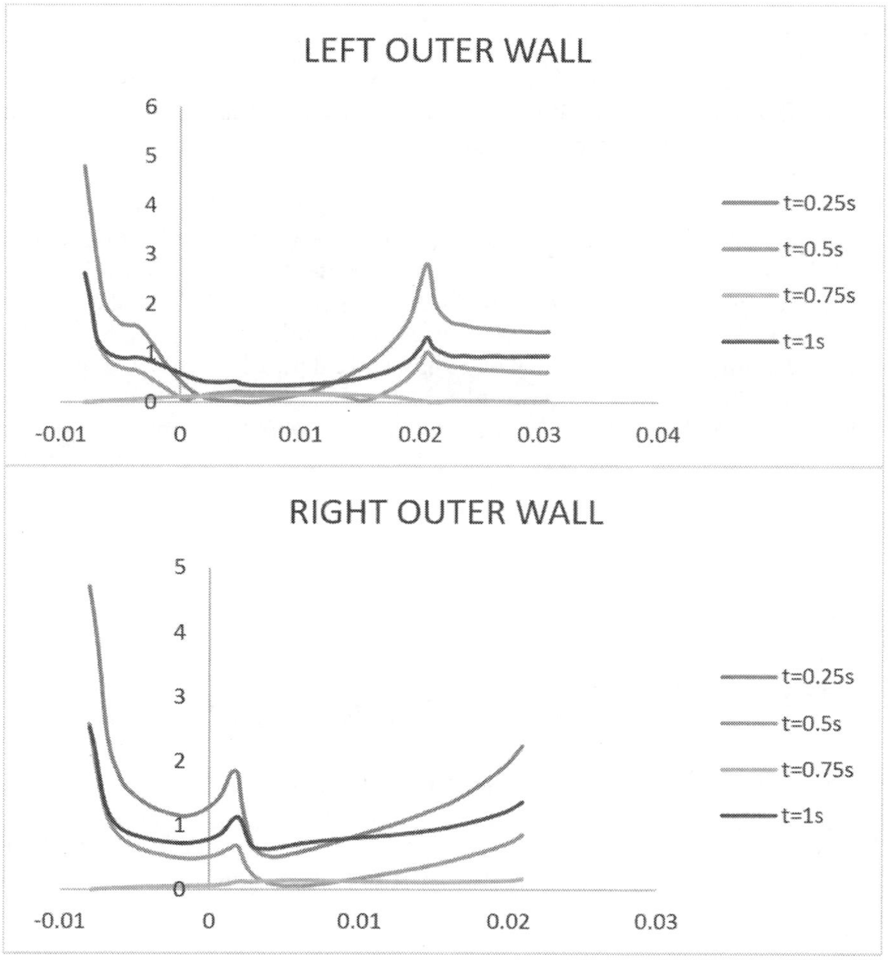

Fig. 20. WSS Plots along Outer Walls

4 Conclusions

1. According to the Idealized artery analysis we can conclude that the pressure reduces when there is sudden decrease in area due to bifurcation. This is because the velocity increases with decrease in area according to the continuity equation.
2. Due to the plaque, the heart needs to pump blood with increased pressure to maintain the constant flow rate to the artery.
3. More the constriction in the area of arteries, more will be the pressure at inlet thereby causing more the damage to the artery walls.
4. Flow separation occurs in stenosis because of the adverse pressure gradient.
5. In case of analysis for the Newtonian and Non-Newtonian models, the general trend of variation of the parameters remains almost same with slight deviation in the magnitude.
6. At the carotid sinus region, we observe low velocity and low wall shear stress values.
7. For Newtonian and Non-Newtonian type of flows, there is a change in the maximum wall shear stress value for the same dimensions of the geometry.
8. In case of pulsatile flow analysis, there is considerable variation in the wall shear stress value at different time steps as compared to the steady flow case.

4.1 Scope for Future Work

In our research we have considered the artery to be a rigid structure. But in real scenario, the artery is elastic in nature. Fluid- Structure interaction comes into picture where interface conditions can be specified. Structural deformations change the properties of the blood and also maximum wall shear value will be lesser than that in case of rigid wall. Also we have considered the waveform of the inlet flow to be sinusoidal in nature during the transient analysis. But in actual practice it is pulsatile in nature. The outlet pressure in reality is not zero Pascal because there exists some resistance for the blood flow in the arteries. Actual values of the pressure at the outlet can be specified taking into consideration the blood pressure in different cases. There are different ways to specify outlet pressure such as Resistance, Impedance and constant pressure models. Different user defined functions for different outlet pressure cases can be specified. As a part of further enhancements of analysis, the geometry of the idealized carotid artery can be modified with different percentage of stenosis. The results thus obtained can be compared with real time results from angiograms of patients.

5 References

1. "Numerical simulations of blood flow in arteries using fluid-structure interactions," no. July, 2007.
2. "MODELS IN A STENOSIS OF A CAROTID ARTERY," no. December, 2013.
3. R. F. Mabon and D. P. Giddens, "STEADY FLOW IN A MODEL OF THE H U M A N CAROTID BIFURCATION .
4. PART I - - F L O W VISUALIZATION *," vol. 15, no. 5, 1982.
5. Chien, S., Usami, S., and Richard, S., Blood Flow in Small Tubes. Handbook of Physiology, Section 2: The
6. Cardiovascular System, Microcirculation, Part I, Amer. Physio. Soc., Bethesda MD, Vol. IV, 1984, pp. 217-249
7. Yeleswarapu, K. K., "Evaluation of Continuum Models for Characterizing the Constitutive Behaviour of Blood," PhD.
8. Dissertation, Department of Mechanical Engineering, University of Pittsburgh, 1996

Vibration based condition monitoring of rotating machinery

Goutam Senapaty[1*] and *Sathish Rao U.*[1]

[1]*Department of Mechanical and Manufacturing Engineering, Manipal Institute of Technology, Manipal Academy of Higher Education, Manipal, India*

Abstract. This project looks at the different maintenance philosophies and the importance of vibration analysis in predictive maintenance. Since most industries and plants make use of rotational equipment, vibration analysis plays a major role in detecting machine defects and developing flaws before the equipment fails and potentially damages other related equipment and to avoid unwanted breakdowns and downtime. Vibration analysis can help increase the lifetime of equipment when the faults are diagnosed at the right time. Vibration analysis of a rotating table top model is also done to show that some faults might exist even though they are not visible to the naked eye.

1 Maintenance Philosophies and the role of Vibration Analysis

Plant maintenance is one of the most important aspects of running a plant or an industry, as the maintenance philosophy a plant follows will determine how much downtime the plant has to go through and the profit margin of the plant. Money saved in maintenance is a direct contribution to improved plant profitability. The main goal of maintenance is to minimize scheduled and unscheduled downtime, improve safety and maximize the availability of production equipment.

The common maintenance philosophies followed by plants and industries are as follows

- Reactive maintenance: This type of maintenance is also known as Break-down or On-failure maintenance. This type of maintenance is done during shutdowns after the failure of the equipment. Due to the nature of this philosophy, there is no warning of failure; there are uncontrolled plant outages and production losses or delays and a large inventory of spare parts I required.
- Preventive maintenance: This type of maintenance is also known as Fixed-time or time-based maintenance. Maintenance of equipment is based on a fixed time – calendar based, hours run based or number of equipment cycles based. This maintenance philosophy helps reduce failures, allows work, spare parts and labour to be planned. Despite its advantages, preventive maintenance is more costly and will cause frequent downtimes due to the nature of the maintenance, so it is only applicable to machines that undergo age related deterioration, otherwise it may lead to unnecessary replacement of part and components.

* Corresponding author: goutam.senapaty@gmail.com

- Proactive Maintenance: This type of maintenance is also known as predictive maintenance or condition based maintenance. This maintenance philosophy is very advantageous to the plant as machine faults can be detected before failure occurs, without doing regular scheduled maintenance. This maintenance philosophy requires regular and frequent monitoring of the machine health condition. Practices such vibration analysis, infrared thermography, ultrasound and various other forms of non-destructive testing can be used to detect deterioration of machine health and problems in the components before they fail and require the plant to be shut down. Usually critical plant equipment undergo this form of philosophy. It helps in reducing downtime, detecting root cause of the failure and managing resources such as labour and spare parts effectively.

Proactive maintenance philosophy is the best way to manage and maintain plant most the components and equipment in a plant or industry, but only if used effectively and in required places, otherwise it can get very expensive. It is important to find out defects with the equipment before they fail so that their failure doesn't affect the condition of equipment that are working with them and further cause more damage to the plant.

1.1 Why Vibration

Since the beginning of building machines, especially rotating machines, reduction and isolation of vibration has remained to be a key area of focus. Gradually as vibration reduction and isolation techniques have become more and more common, the need for accurate measurement of vibration has also become increasingly popular. Within the last 15-20 years, modern vibration measuring equipment have been developed for investigating high-load, high-speed machinery.

1.2 Vibration Characteristics

Amplitude – Helps in detecting the severity of the fault condition or defect.

Frequency – Helps in indicating the cause of the defect

Phase – Helps in determining the cause of the defect

Vibration amplitude can be measured in terms of displacement, velocity, acceleration. Displacement is measured peak to peak and usually in mils or microns. Velocity can be measured in peak or RMS and usually in inch/sec or mm/sec. Acceleration is expressed in peak and usually measured in g's. Phase indicates how a machine is moving to a reference of its part. The relationship of the movement of part of a machine to a reference – for example the position of both end of the shaft as it rotates.

This indicates the relationship of the movement between one or more points on a machine.

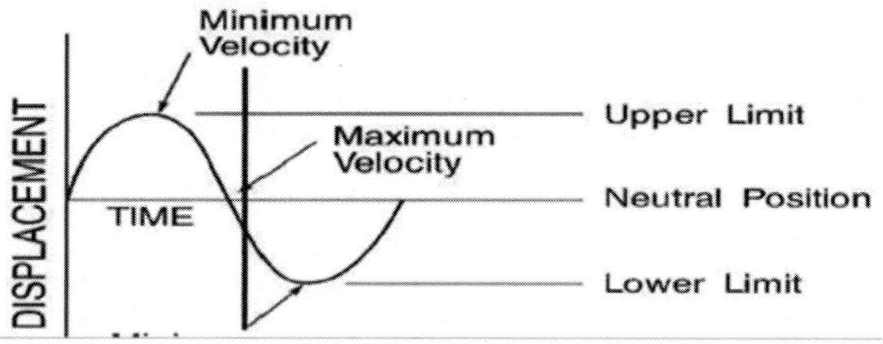

Fig. 1. Time waveform of a simple harmonic motion [1]

2 How to study and analyse vibration data

2.1 Quantifying the vibration level

The vibration amplitude, which describes the severity of the vibration, can be quantified in several ways. The Peak-to-Peak value indicates the maximum excursion of the wave, a useful quantity where, for example, the vibratory displacement of a machine part is critical for maximum stress or mechanical clearance considerations.

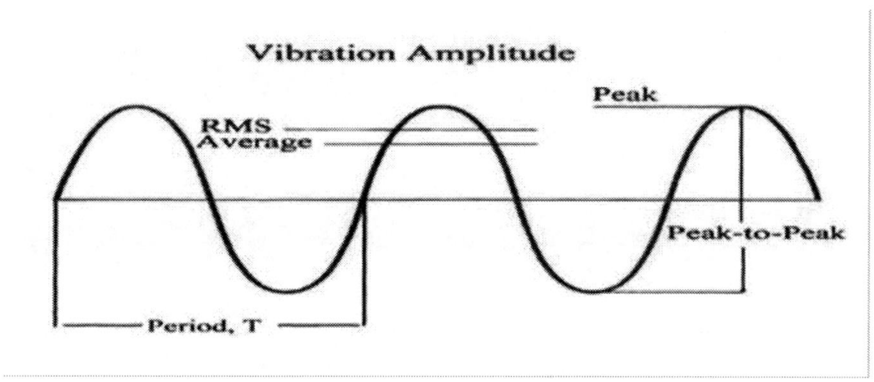

Fig. 2. Peak and RMS values [1]

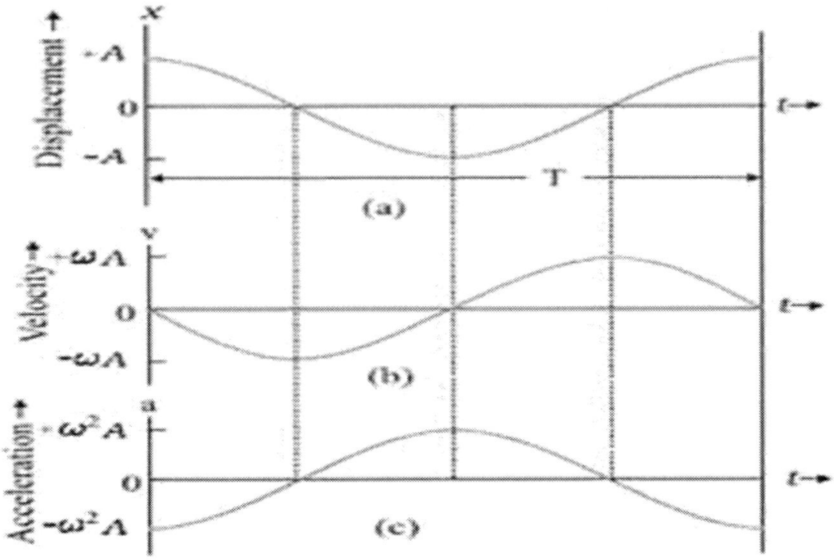

Fig. 3. Displacement, Velocity and Acceleration graphs [1]

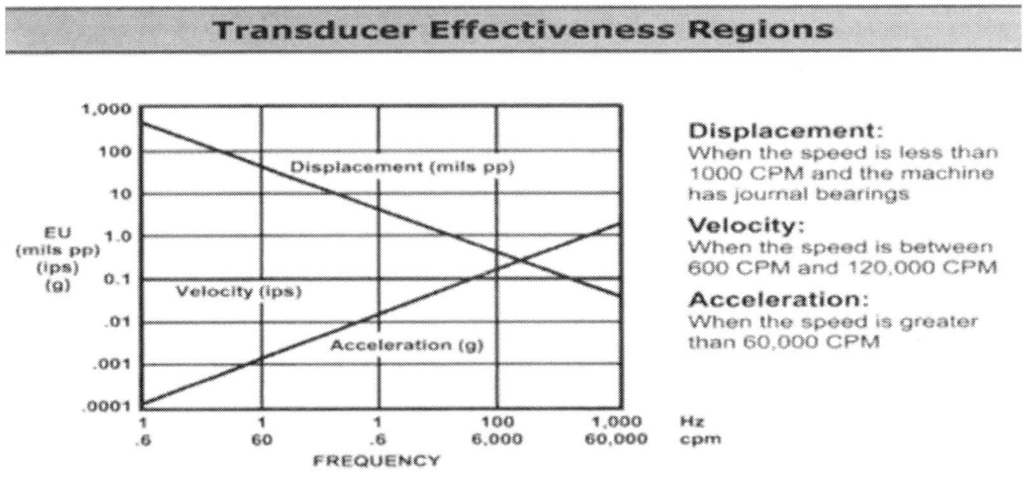

Fig. 4. Transducer effectiveness regions [1]

2.2 What are F_{max}, Lines and Averages ?

2.2.1 Guidelines for setting F_{max} (End Frequency)

- For machines with Rolling Element Bearings: F_{max} = 60 x RPM.
- For machines with Journal / Sleeve Bearings: F_{max} = 20 x RPM.
- For machines with Gear Box : F_{max} = 200 x RPM.

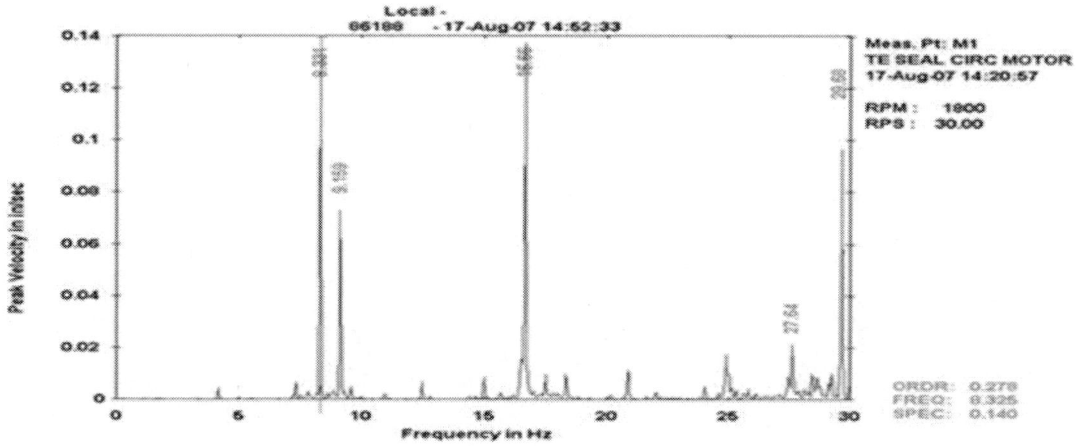

Fig. 5. Example of insufficient F_{max} [1]

2.2.2 Guidelines for setting Number of Lines (Resolution)

- Number of lines determines the Machine spectrum resolution.
- Band width (BW) = F_{max} / No. of Lines.
- More the lines, less is the bandwidth & hence higher the resolution & better the accuracy.
- Typically, 400, 800 or 1600 lines are used, as more the lines more the resolution.

2.2.4 Number of Signal Averaging

- Averaging is used to reduce random content in the vibration signal.
- Typically, 2-4 averages are used.
- Again, more the averages, more is the data collection time.

2.3 Time Waveform Analysis

Time waveform is the raw data generally acquired by the analyser and also plays a vital role in the analysis process. The amplitude units and the acceleration peak to peak values will be the initial focus for the analyst. The amplitude reflects the severity of the vibration against time domain. This is also known as the machine raw data, as it vibrates while running. The time waveform contains very important information. Although many analysts do not view the time wave form during data analysis, it should be noted that the waveform data is the source that is used to generate the spectral data and trend data that are commonly used for fault detection.

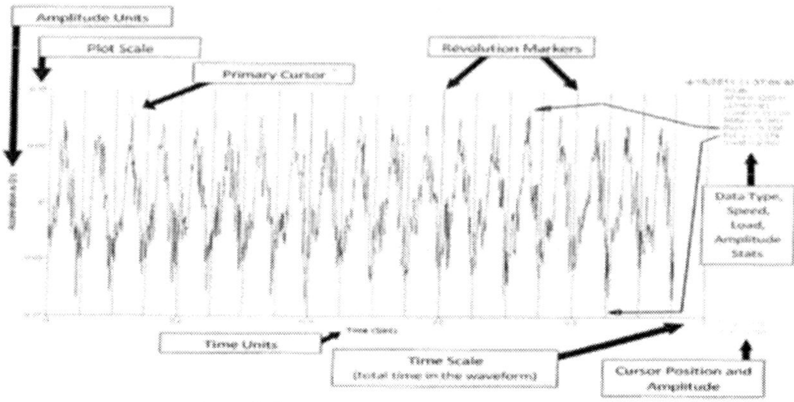

Fig. 6. Example of Time Waveform [1]

The length of the waveform should be long enough to include 10 to 15 shaft rotations.

2.9 What is FFT?

FFT stands for "Fast Fourier Transform". It is possible to take a vibration time waveform, whether simple or complex, and mathematically calculate the vibration frequencies present along with their amplitudes with the help of FFT.

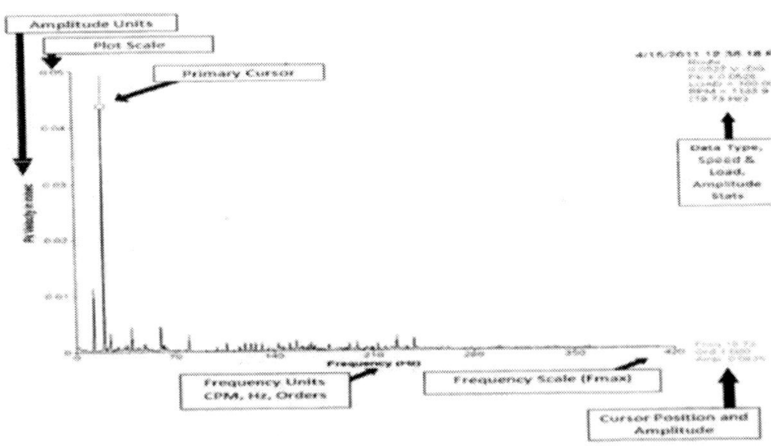

Fig. 7. Velocity Spectrum [1]

Spectrum analysis is the procedure of doing the transformation, and is invariably done with the help of FFT analyser. Signature term usually applied to the vibration frequency spectrum which is unique to a machine at a specific point in time, under specific machine operating conditions.

Normal Frequencies: Vibration peaks that appear in the spectrum as a normal condition of the function of the machine such as running speed, fan blade pass or pump vane pass etc.

Defect (fault) Frequencies: Calculated frequencies that appear in the spectrum only when there is a problem with the component such as a bearing defect frequency.

Fundamental Frequency: It is also called the first harmonic. Usually it's a pre-calculated frequency, such as turning speed of the machine or a frequency that is equal to a defect in the machine.

Dominant Peak: A peak in the spectrum with the highest amplitude value or a sharp maximum

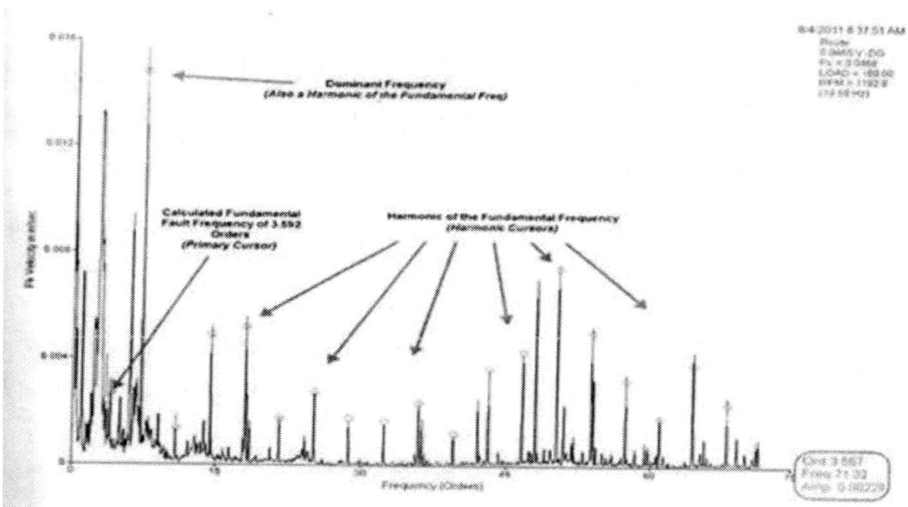

Fig. 8. Velocity Spectrum showing dominant peak [1]

Harmonics: Spectral peaks that occur at integer (whole number) multiples of any frequency. It is possible to have a spectrum where the fundamental is not visible but the harmonics will be spaced apart by the fundamental frequency.

Sub-harmonic: Frequency that is integral sub-multiple of a fundamental frequency.

Synchronous: These are spectral components which are integer multiples or harmonics of a fundamental frequency. They may be present as multiples of an integral fraction, which are called sub-harmonics.

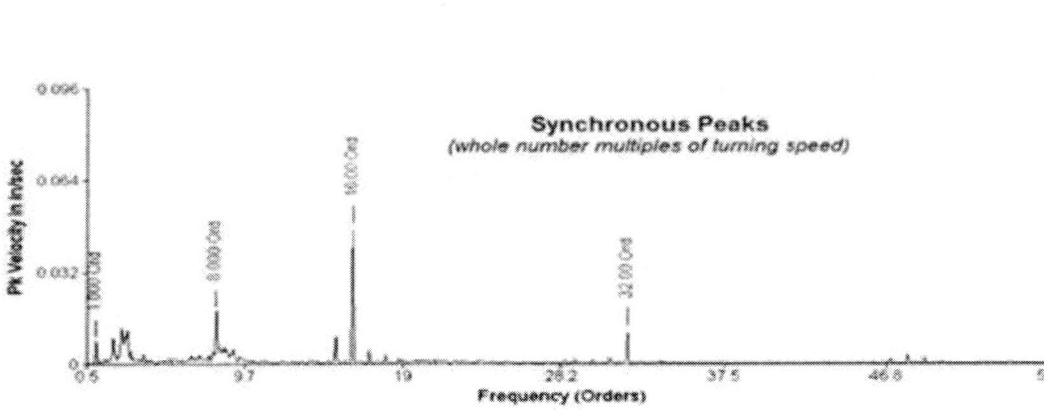

Fig. 9. Synchronous Peaks [1]

Sub-synchronous: Components of a vibration signal which has a frequency, less than the shaft rotation frequency

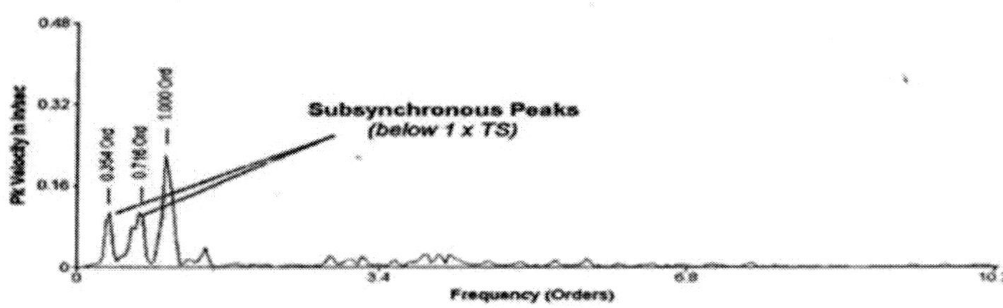

Fig. 10. Sub-Synchronous Peaks [1]

Nonsynchronous Spectra: Spectral peaks that are above turning speed, but are not whole number multiples of turning speed

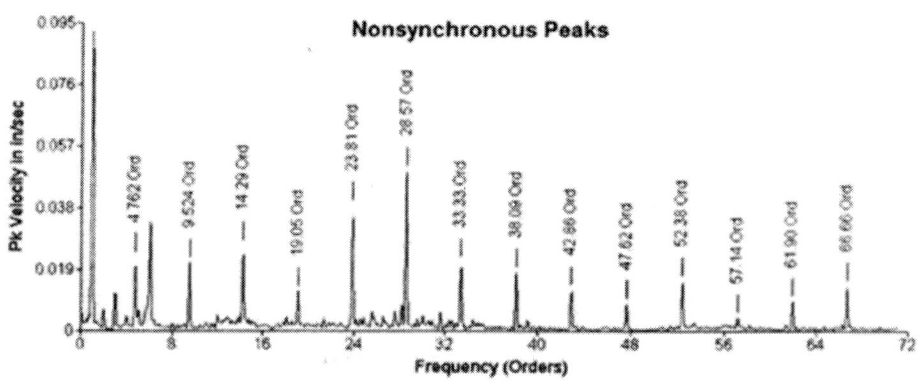

Fig. 11. Non-Synchronous Peaks [1]

Sidebands: A pattern of equally spaced peaks above and below a centre calculated frequency. Sidebands are caused by modulation of a carrier frequency.

Carrier (centre) Frequency: A vibration peak that is either changing in amplitude or frequency. The carrier frequency is the "centre" frequency that appears between the sideband peaks.

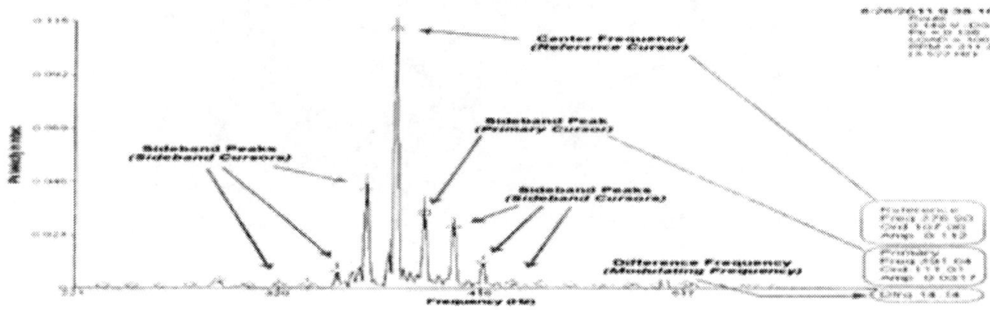

Fig. 12. Vibration spectra showing carrier frequency [1]

3 Performing vibration analysis on a table top model

A Table top model consisting of a variable speed motor (mainly used for a sewing machine) of 145 Watt was used to rotate the shaft holding a disc of weight of 500 grams, supported by two anti-friction bearings on either side of the disc. The input shaft is connected to the output shaft using a 3- jaw coupling.

Fig. 13. Table top rotating equipment

A laser tacho-probe was used to determine the speed of rotation and the phase. The tacho-probe is used to measure the speed of rotation with the help of reflective tape which was placed on the disc, it determines speed by recording the time elapsed between the first reflected laser beam and the second time when beam is reflected during the rotation.

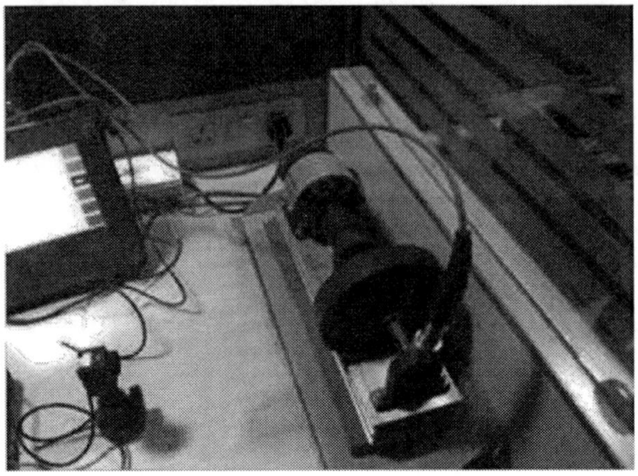

Fig. 14. Table top model with laser tacho-probe and accelerometer

Vibration of the model was recorded at 3 different spots, Motor outboard, Rotor inboard and Rotor outboard. Usually, 3 readings, namely axial, horizontal and vertical readings are taken at each of these positions, but it was not possible to take axial and horizontal reading on motor outboard and rotor inboard due to design limitation of the model and size limitations of the Accelerometer probe.

3.1 Methodology

1. First, speed is measured using the tacho-probe. The required cable attachments are made to the CSI 2140, the tacho-probe is placed parallel to the disc on which reflective tape is put, at a distance such that it is not too close, but not too far, such that the beam of laser is no wider than a point.

2. After speed is determined, the accelerometer cable is attached to the probe and to the machine health analyser and then placed on suitable respective points on the model, usually on the bearing housings.

3. Adjustments are made to the model to align it properly, then above steps are repeated again.

4. For single plane balancing,

- Create new job
- Set up job parameters (decide which plane to add/remove weights)
- Acquire reference run data
- Add trial weight
- Acquire trial run data
- View the balance correction

- o Remove trial weights and add/remove correction weight
- o Collect trim run data for corrected condition

3.2 Data collection and results obtained

1. First the monitoring locations (Six) for the Table top model was created in the computer interface of the analyser and then uploaded to the CSI 2140 analyser for collecting the data with the help of accelerometer for one of the locations.
2. All the collected data from the analyser was downloaded to the computer interface software and is viewed, showing the location Overall values and the spectra before and after the Alignment.
3. During Analysis in given state, model showed characteristics of having misaligned shaft, which was found out due to occurrence of 3x peaks in all the six locations
4. Readings after proper alignment of coupling was taken, and the shaft was known to be aligned as the 3x peaks had disappeared.

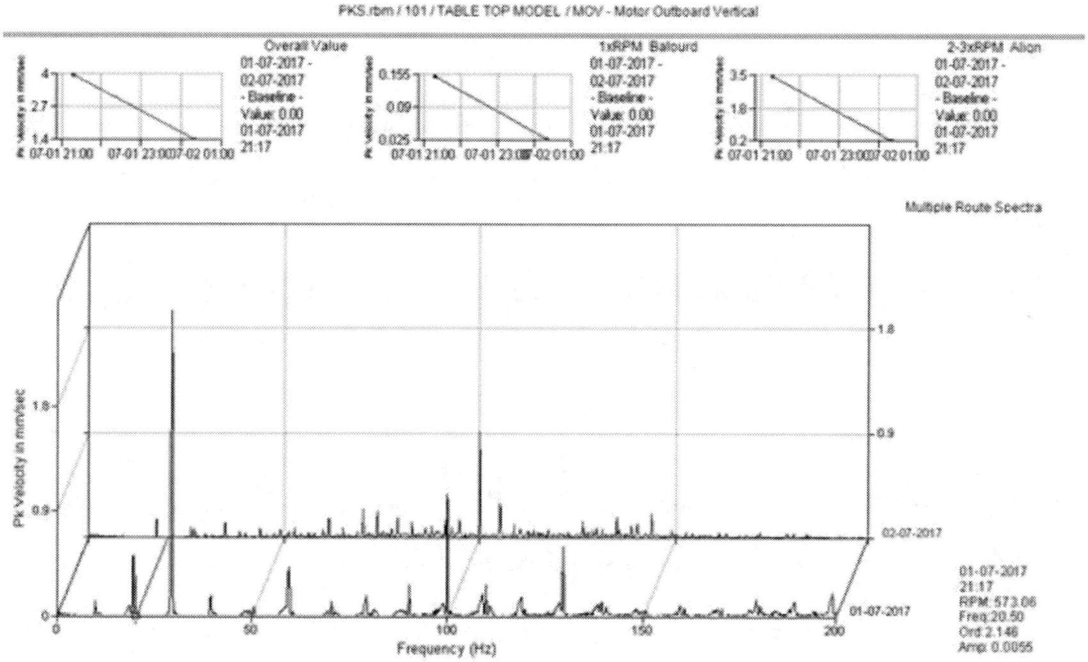

Fig. 15. Vibration spectra of motor outboard vertical

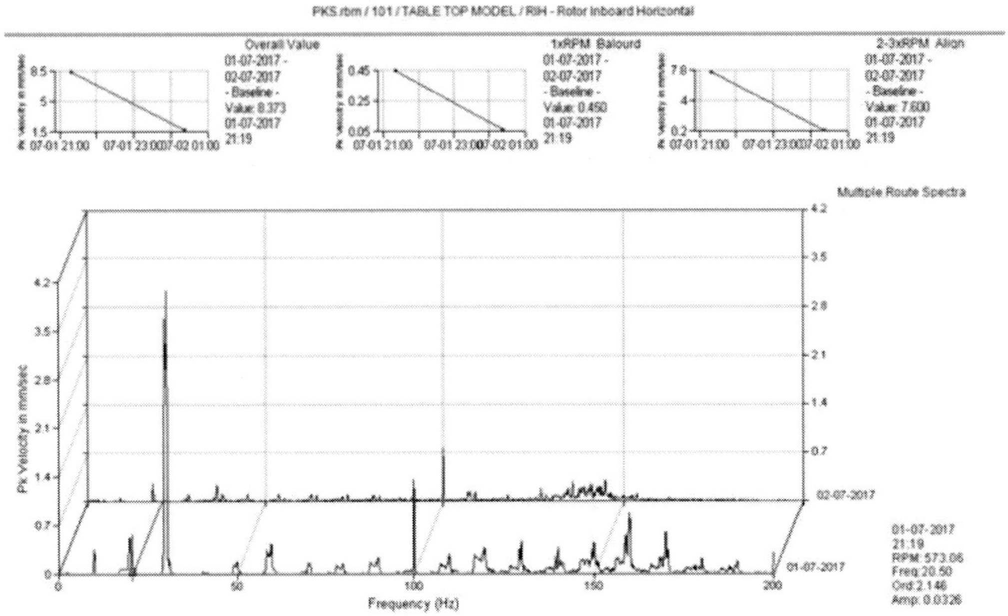

Fig. 16. Vibration spectra of rotor inboard horizontal

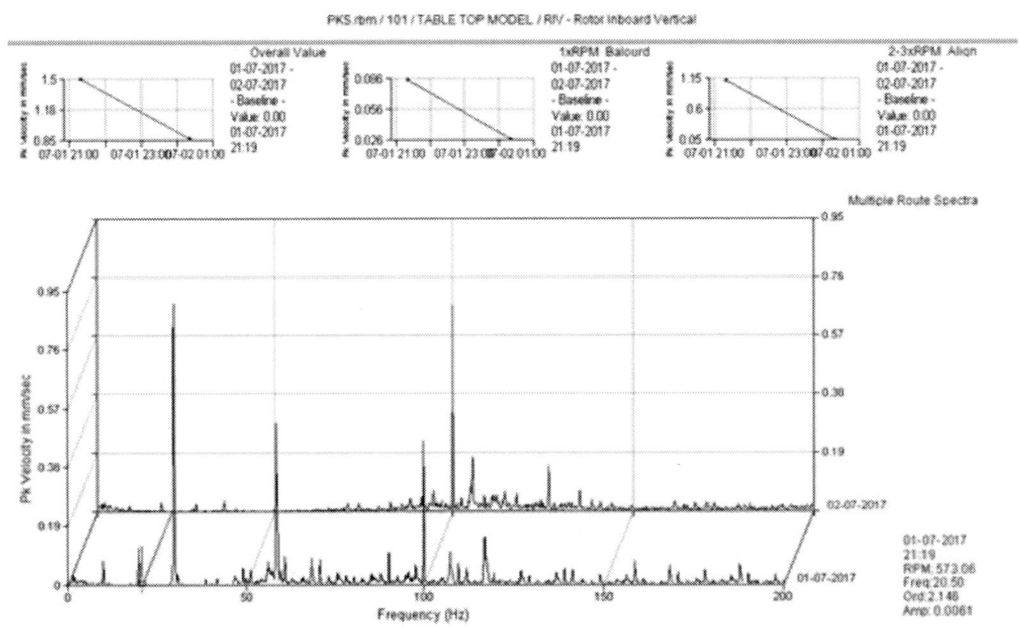

Fig. 17. Vibration spectra of rotor inboard vertical

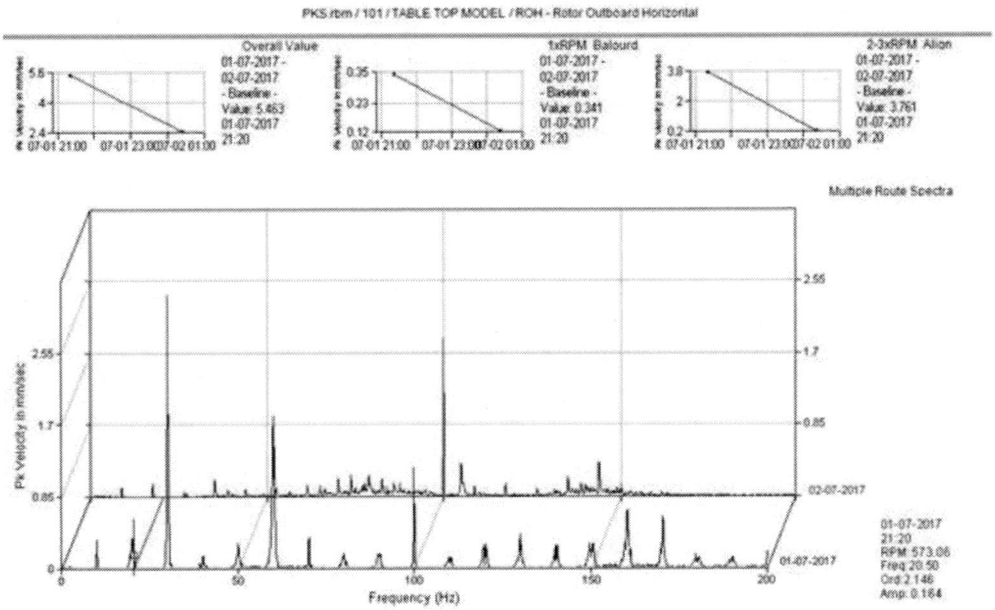

Fig. 18. Vibration spectra of rotor outboard horizontal

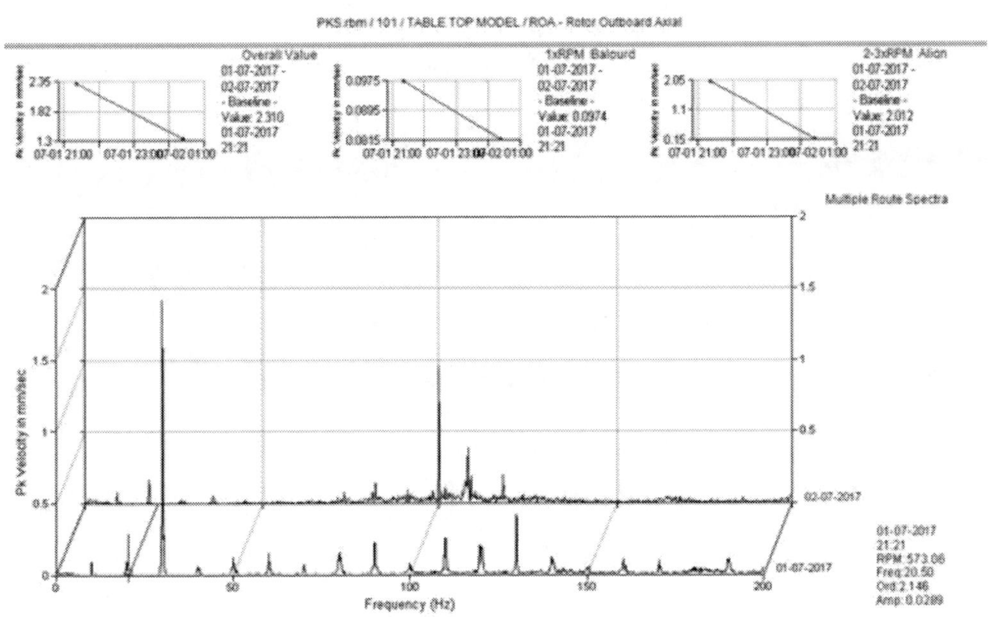

Fig. 19. Vibration spectra of rotor outboard axial

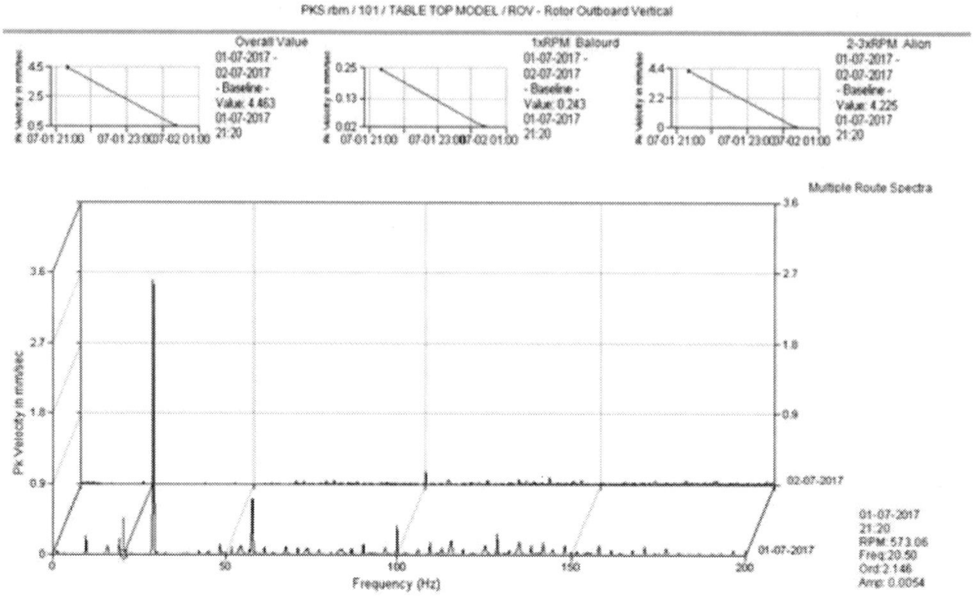

Fig. 20. Vibration spectra of rotor outboard vertical

5. As can be seen in above vibration spectra, 3x peaks were quite prominent before any alignment was made, which resulted in suspecting there might be some misalignment with the 3 jaw coupling.

6. After the three jaw coupling was aligned properly, we can see, as indicated with vibration spectra in the foreground, the 3x peaks completely disappear, which shows that the problem originated in the coupling.

For Balancing:

1) The table top model was used for studying how to balance a rotating machine. As it was a perfectly balanced model, first some weight in the form of Nut & bolt with washers were fitted to make the rotor unbalance. The same figure shows the reflective tape location to read the phase from the taco-probe.

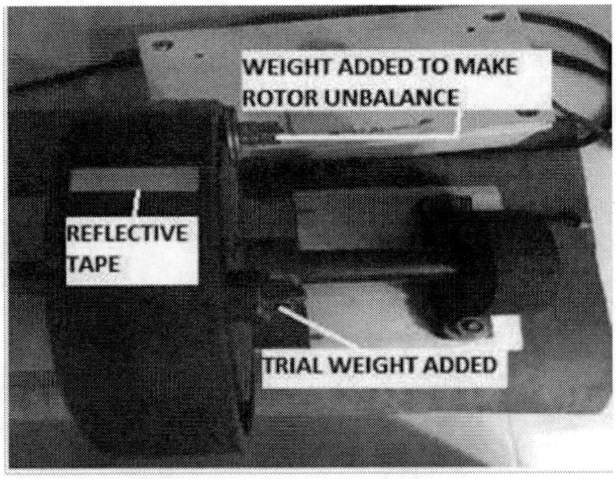

Fig. 21. Table top model to study unbalance

2) First a reference reading was taken with the CSI 2140 analyser in basic vibration mode, shows the reference run reading showing magnitude and phase was captured in the analyser.

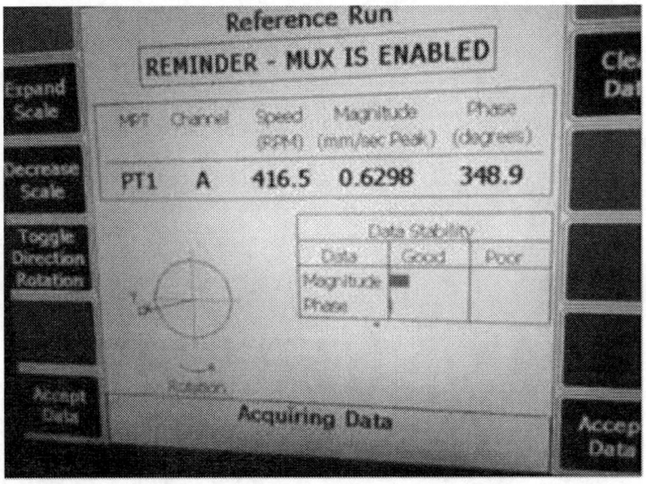

Fig. 22. Reference run taken for balancing

3) Next the trial weight was added to the rotor plane as shown in Fig 5.12 and another run was taken and data stored in the analyser

https://doi.org/10.1051/matecconf/201814401021

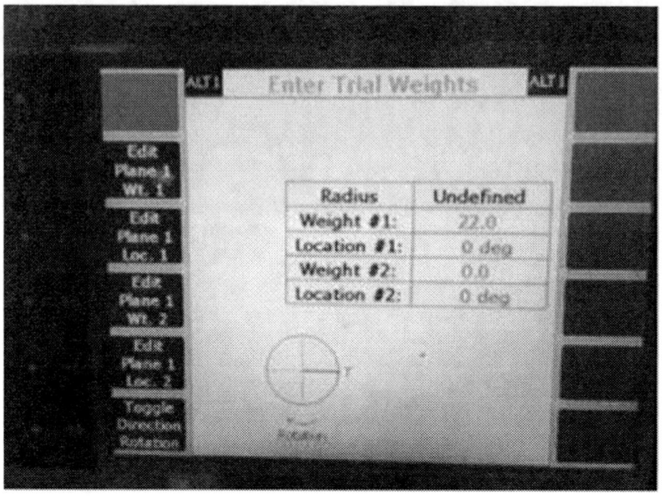

Fig. 23. Adding trial weight details to the analyser

4) Then the analyser with its in-built software calculates the amount of actual weight to be added or removed from the location to balance the machine.

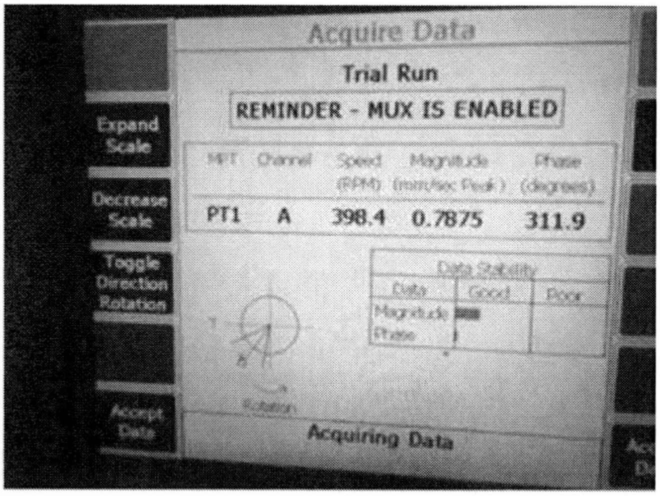

Fig. 24. Acquiring trial run data

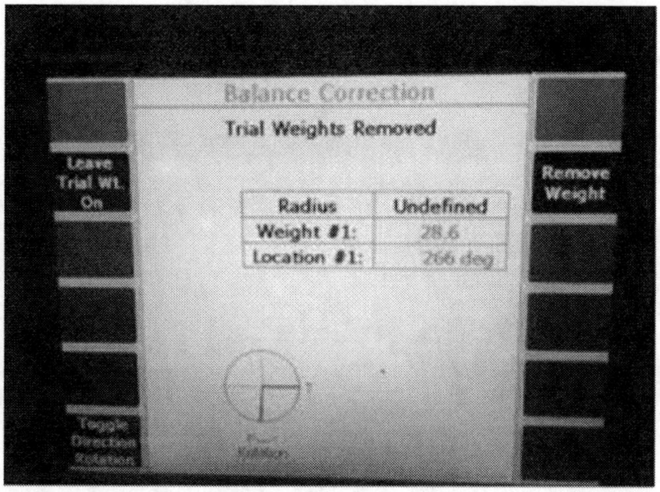

Fig. 25. Weight to be removed as calculated by the analyser

5) After removal of the weight as indicated by the analyser a run was taken, which shows the residual unbalance in the machine. To remove this further balancing has to be done, which is known as trim balancing with the repetition of same procedure.

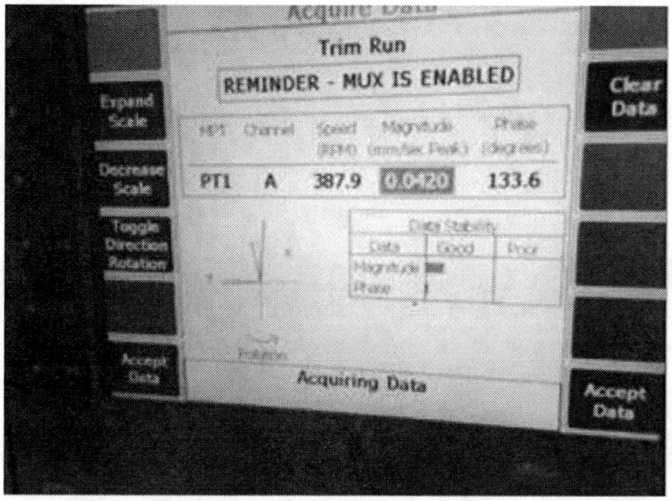

Fig. 26. Acquiring final trim run data

6) At the end the analyser indicates the balance quality, which checks whether the machine is balanced to its tolerance level or not. If not we can still proceed for trim balancing until it achieves the tolerance level for the particular machine.

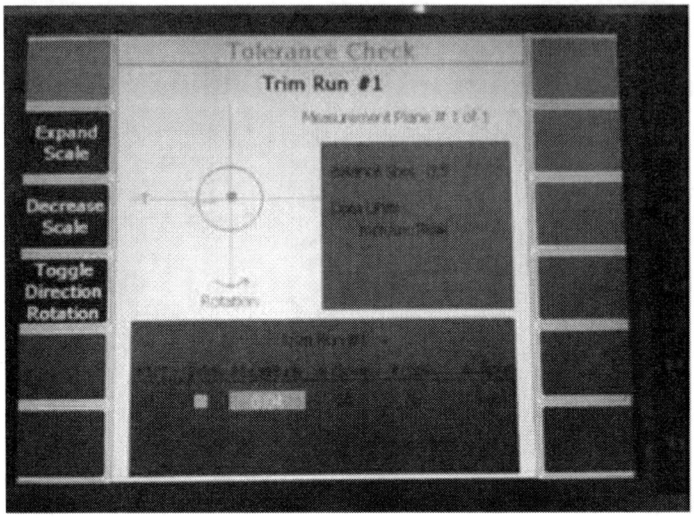

Fig. 27. Tolerance check for the balancing done

4 Conclusions

From the project, it can be concluded that, adapting the philosophy of predictive maintenance can be beneficial to all industries and plants. Using predictive maintenance, we can increase the life spans of various machinery and avoid unwanted shutdowns and downtime. Predicting a failure before it occurs can help analyze the root cause of the occurrence of machine fault and can prevent the failure of the machine in question which might in turn cause damage to other machines in the system. Vibration analysis plays a major role in predictive maintenance and as can be inferred from the experiments, defects that are not visible to the naked eye and that seemingly do not show any physical symptoms can be detected by this method, and thus is very important in the topic of plant maintenance.

References

1. Basic Vibration Analysis Course 2031, Educational Services, Machinery Health Management, EMERSON (2017)

2. Vibration training course book, Mobius (2012)

3. ISO 10816-3 Vibration Severity Chart.

Topological and fractal aspects of the main hydraulic parameters

Anton Lyapin[1*] and *Valery* Lyapin[2]

[1]*Moscow state university of civil engineering, Yaroslavskoye shosse, 26, Moscow, Russia, 129337*
[2]*Journal «Power Technology and Engineering», 2/1 Voroncovsky Pereulok, Moscow, Russia*

Abstract. The traditional concept of wetted cross-section, wetted perimeter and hydraulic radius seems fairly simple and clear, especially in the case of smooth boundaries of the flow. However, for the rough boundary surface covered with randomly arranged roughness the definition of the wetted cross-section as a normal to all elementary streams becomes notably problematic. More controversial seems the concept of wetted perimeter. This issue is of particular importance for the beds with high-dimensional roughness for which the geometric dimensions of the flow and the roughness elements are of the same order. The article shows that topological features of the boundary surfaces for the natural channels allow definitely refer them to fractals. Further study of the bed roughness based on its fractal nature can form a new approach to methodological soundness of main hydraulic parameters and open up new opportunities for their description.
Key words: hydraulics, hydraulic resistance, wetted cross-section, wetted perimeter, hydraulic radius.

1 Introduction

Hydraulic calculations of pipe and open channel flow in the practice of hydraulic engineering involve the following tasks:

- Definition of the size of the watercourse based on the specified discharge;

- Determining the capacity of the watercourse in terms of its size.

In engineering calculations based on one-dimensional flow model are used averaged flow characteristics: the average over the cross section of flow velocity V and the depth h. The calculation task is to determine the Chezy coefficient (C), or uniquely associated hydraulic resistance coefficient $\lambda = 8g / C^2$ (g - acceleration due to gravity). The relationship of these factors with the flow and channel elements called the hydraulic resistance law. To determine the coefficients C and λ there are a large number of empirical and semi-empirical dependences comprising the hydraulic radius R as the characteristic linear dimension of the watercourse. The exponential formula (Manning, Pavlovsky and

* Corresponding author: lyapinanton@gmail.com

others) with constant or variable exponent "y" are the most common type for the open channels [1-2]:

$$C = \frac{1}{n} R^y. \quad (1.1)$$

Obviously, the hydraulic radius R used in the hydraulic resistance laws should have a clear and understandable methodological substantiation. The classic definition is the following: "hydraulic radius is the ratio of the wetted cross-sectional area Ω to the wetted perimeter χ of the bed" [1, 2, 3]

$$R = \frac{\Omega}{\chi} . \quad (1.2)$$

The standard explanation of the R calculations often reduced to simple schemes, such as

$$\chi = b + 2h;$$

$$R = \frac{bh}{b + 2h}.$$

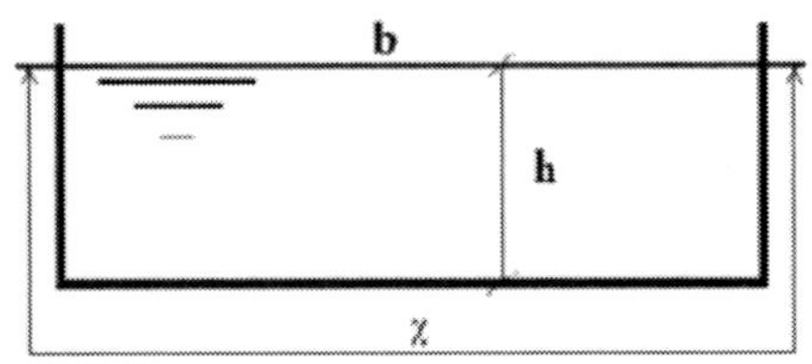

Fig.1. The open channel flow design scheme

Used herein the concepts of channel wetted cross-section, wetted perimeter and hydraulic radius, seem simple and clear especially in the case of "smooth" surface. The flow in this virtual "smooth" bed taken as consisting of elementary streams, which is a necessary justification for the using of the traditional interpretation of the wetted cross-section: wetted cross-section must be normal to all elementary stream tubes (or elementary streams) [2-3]. However, in practice we usually observe the rough surfaces often with a significant roughness for example, hydraulic tunnels traversed by drilling and blasting method. It is impossible to find a smooth contour, even in the case of usual water pipes (Fig. 2) provided with a corresponding increase.

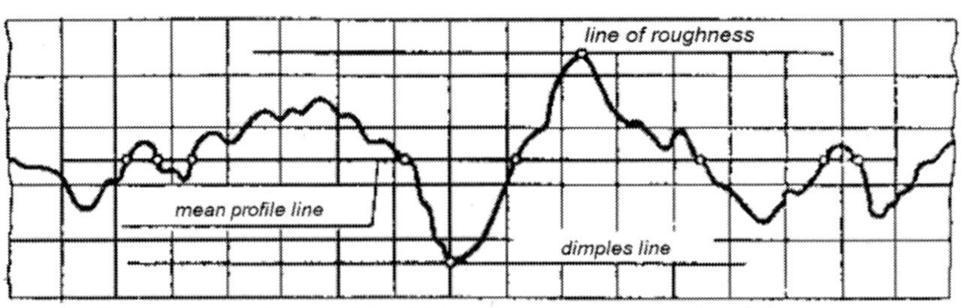

Fig. 2. The fragment of the inner surface profile of the concrete pressure pipe [4]

The eddy zones forming in the near bed region of rough channels [1], are the cause the specific flow around roughness. In these zones (zones of influence of the roughness), the flow cannot be considered as a flow consisting of elementary streams and there is uncertainty associated with the original concept of the wetted cross-section. This issue becomes important for the beds with increased roughness for which the geometric dimensions of the watercourse and the roughness elements are of the same order. Therefore, the definition of the wetted cross-section boundaries of the flow, which defined like the normal to all elementary streams, is very problematic.

More difficult is the case with the introduction of the concept of the wetted perimeter. Its current wording - "wetted perimeter is the part of the wetted cross-section perimeter along which the liquid contacts the bed of the channel" [1-3].

2 Application of fractal theory

Unfortunately, when the so ingrained concept was realized nothing had been known about fractals. The term "fractal" created by B.B.Mandelbrot from the Latin participle fractus [5]. The corresponding verb frangere means "break, crush, and create fragments of irregular shape". Under a stricter definition of a fractal fall sets for which dimension Hausdorff - Besicovitch is strictly greater than its topological dimension.

Consider as an example Fig. 2 and focus on the curve the envelope of the roughness elements. At first glance, this curve seems like each roughness element has a clearly defined contour at each point of which one we can draw a tangent. However, in subsequent approximation clearly revealed an increasing number of shape incorrectness and original clearness and the apparent smoothness of the curve would disappear. Boundary contour curve would include all the smaller irregularly shaped elements, which exclude the possibility of the existence of a tangent to the contour. In the absence of function's derivatives describing a continuous contour of the bed surface will not be surprising because "In nature all curves having no tangent could be considered as a rule, while the right curves - such as for example a circle - interesting but very special case» [5]. The length of the considered boundary curve will tend to an unlimited increase. This is because the curve has a very irregular shape and further approximation increases number of smaller parts, which in turn increases the overall length unlimitedly. Infinite length of the wetted perimeter of the channel confirms its fractal nature and raises questions about the lawfulness of the original concept of the wetted perimeter. The above-mentioned features of the boundary surface contour is also characteristic for natural riverbeds that certainly allows attributing them to the Mandelbrot fractal sets.

Obvious examples of fractals are also the coastlines of islands, mountain ranges forms and so on [5] (Fig. 3).

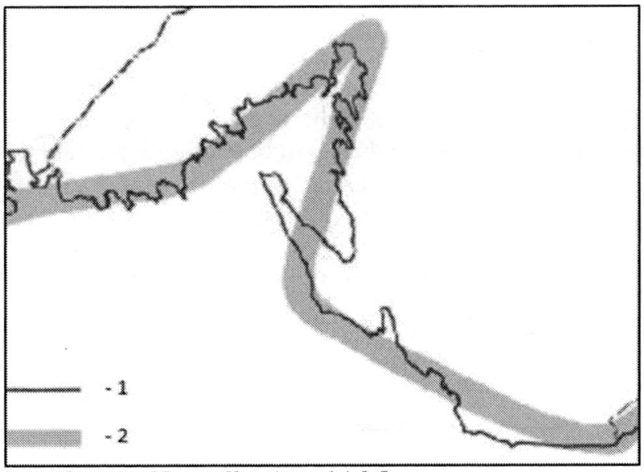

Fig. 3. The fragment of the Vyborg Gulf coastline (Russia) [6]
1- map scale of 1:500 000; 2- map scale of 1:22 000 000

Comparing Fig. 2 and 3 indicates the outward similarity of the contours in their respective degrees of approximation. These topologically similar lines including bed boundary contours nevertheless have different fractal dimension that promotes their further identification.

Fractal aspects consideration allows to move on to the next level of roughness description compared with a standard approach of Euclidean Geometry limited height, shape and mutual arrangement of the roughness elements. Considering that the development of a new approach to the evaluation of bed roughness given its fractal nature - the subject of future research we accept as a possible alternative the standard idealization in determining the length of the wetted perimeter of the rough bed to use channel with conditionally smooth bed.

I.Nikuradze [7] used this approach for the processing of experimental data received from pipes with sand roughness. He proposed to use as a flow model the flow in a pipe with smooth walls and the effect of roughness to consider in the form of the actual reduction of the wetted cross-section area. The diameter of the conventionally smooth pipe determined volumetrically. Thus, I.Nikuradze retained the ability to use the classical concepts of the wetted cross-section, wetted perimeter and hydraulic radius interpreting them for the virtual equivalent pipe with conditional smooth walls.

The transition to the model of "smooth" bed determined now as a choice of a method for calculating the conditional size (depth) of the flow. The lack of generally accepted methodology for calculating and obtaining experimental data greatly complicates the comparison of the numerous studies results. The most common is the equivalent roughness especially for pipes with natural or irregular roughness. However, the elementary analysis of different interpretations of this approach [7] reveals the absence of a harmonized concept of equivalent roughness. Most often the key conditions for transition from a given pipe to the equivalent pipe Nikuradze with sand roughness are:

- equality of the hydraulic resistance coefficients (λ) in quadratic resistance zone;
- equality of the head losses for a given pipe and the equivalent pipe with sand roughness.

Experience shows that the using of these two procedures does not lead to equality obtained through them results. Consequently, this approach needs further clarification and harmonization.

For open rough channels historically formed two basic approaches to the issue of what is meant by the line of contact of the flow with a rough bed and consequently the location of the zero reference plane:

- The study of topological (geometric) features of the boundary surfaces to determine the some "average" value of the height of roughness elements (roughness parameter). This approach is still be using in open channel hydraulics when determining roughness coefficient (Surface roughness in the hydraulic calculation of open channels is determined by the special tables linking its value to the characteristic features of the channel, including, as major factors linear dimensions and shape of the roughness).
- Search a virtual flow boundary near the rough bed using analytical (graphic) approach naturally specifying its form without taking into account the roughness contour (for example a circle, rectangle, trapezoid and the like). This approach uses a geometric factors (the volumetric method) as well as the combination of geometric and hydrodynamic parameters (determination the location of a zero reference plane using local velocity distribution, etc.) [8].

3 Conclusions

1. Unfortunately, the problem of choosing the location of zero reference plane is still far from being resolved. In particular, checking calculations performed for artificial roughness in the form of isosceles triangles [10] revealed the conventionality of the standard approach and appreciable divergences in the calculations of the hydraulic radius even in the case of infinitely small roughness elements.
2. This study marked the existence of gaps in the methodological foundations of the pipe and open channel hydraulics associated with the definition of the key design parameters.
3. The proposed approach to the evaluation of bed roughness based on its fractal nature eliminates the inconsistency of the original concepts of the wetted cross-section and wetted perimeter and opens up new possibilities for their description.
4. A complex combination of topological and fractal dimensions, obviously, allows to reach a higher quality of descriptions of roughness in comparison with Euclidean geometry limited by form, height and relative position of the roughness elements

References

1. V.T. Chow, Open channel hydraulics (Moscow, Stroyizdat, 1969)
2. A.D. Altschul, P.G. Kiselev, Hydraulics and aerodynamics (Moscow, Stroyizdat, 1987)
3. R.R. Chugaev, Hydraulics (Leningrad, Energoizdat, 1982)
4. GOST 12586.0-83. Reinforced concrete pressure pipes. Specifications. (Moscow, 1994)
5. Benoit B. Mandelbrot. Fractal geometry of nature (New York: W. H. Freeman, 1983)
6. K.S. Lazarevich, Geography, 31, 10-12 (2004)
7. J. Nikuradze, Forsch. Arb. Ing.-Wes., Heft 361, 2–22, 1931
8. V.U. Lyapin, Hydrotechnical Construction, 5, 42-44, 1994
9. Liapin V.U. Hydraulic resistance of gradually varied nonuniform and uniform flow in open channels (2004)
10. L.I. Vysotsky and others, Interuniversity scientific collection, Saratov, State Tech. Univ., 49-51 (2011)

Wastewater diffusive dilution and sedimentation of the fine contaminated particles for nonuniform flow in open channels

Anton Lyapin[1] and *Valery* Lyapin[2*]

[1]*Moscow State University of Civil Engineering, Yaroslavskoye shosse, 26, Moscow, Russia, 129337*
[2]*Journal «Power Technology and Engineering», 2/1 Voroncovsky Pereulok, Moscow, Russia*

Abstract. The influence of non-uniformity on mass transfer processes in open channels have been investigated under the action of urbanization factors. The study is related to the urgent problem of environmental degradation of water objects in urbanized areas. It is known that the water quality in the water objects depends on the manner in which the contaminants spread how they mix with the river water and diluted by it. The main results of the study consist of recommendations to incorporate non-uniformity factor to the calculation of diffusion dilution of wastewater and prediction of river processes. So the effect of the flow non-uniformity on the diffusion model of pollutants dilution and diffusion coefficient have been investigated. Formulas for the concentration profiles calculating and the average concentration of fine particulate matter in nonuniform gradually varied flow were presented. The deposition length of suspended contaminants were received, based on the hydraulic resistance laws of nonuniform gradually varied flow.

Keywords: nonuniform flow, diffusive dilution, hydraulic resistance, suspended contaminants

1 Introduction

The obvious result of intensive use of water objects in urban areas is the deterioration of water quality. Industrial and domestic wastewater from urbanized areas contain the large amount of different pollutants (oil, acid and alkali, suspended solids, salts, heavy metal ions, organics, solvents, surfactants, and etc.) in the form of soluble and suspended substances. Suspended solids are sandy, silty or clay particles of different sizes, which adsorb the surface contaminants. The subsequent deposition of suspended matter is one of the most important factors of soil pollution. In periods of most active riverbed processes is an active mixing of soils with pollutants of the riverbed. This leads to a more intense pollution of ground layers of substantial thickness (up to 1 m and more) [1]. Contaminated bed soils can be an independent source of water pollution for a significant period after termination of the wastewater discharge. The main sources of fine turbidity and organic contaminants are shown in Fig. 1.

* Corresponding author: lyapinanton@gmail.com

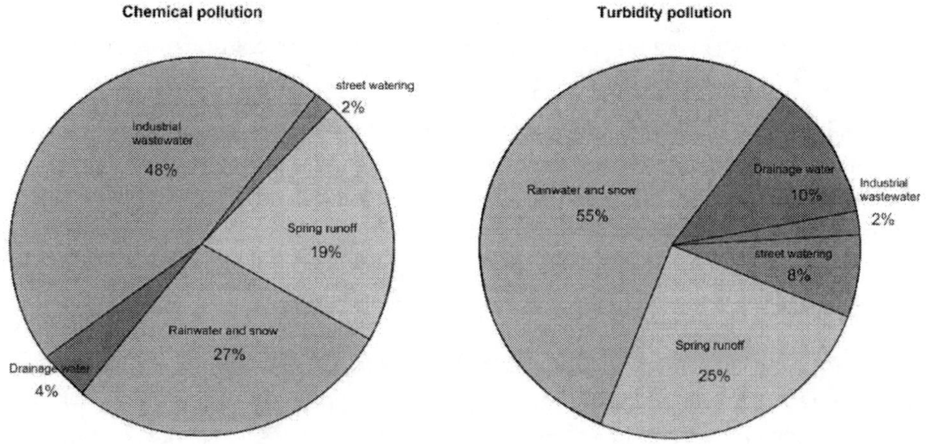

Fig. 1. Moscow river pollution sources according to MosvodokanalNIIproekt data

As can be seen from the figure the main source of turbidity pollution is runoff from urban areas. About half of all chemical contamination of organic nature constitute treated industrial and household wastewater. Long-term anthropogenic and technogenic impact on water flow leads to changes in water flow and sediments as well, as the channel geometry. The effect of these factors is reshaping and changing the river network, in some cases, negative and irreversible. Flow nonuniformity what is present in natural watercourses [2] plays important role in all above mentioned processes.

Dispersion of soluble pollutants in the aqueous medium refers to a category of non-equilibrium processes. These processes are due to varying concentrations of discharged pollutants in the river. Diffusion model of turbulent transport of conservative pollutants is described by Fick's law [3]

$$j = -D_T \frac{\partial c}{\partial y}, \tag{1}$$

j - diffusion flux, of which the dimension is amount of substance per unit area per unit time; D_T - diffusion coefficient; $\partial c / \partial y$ - gradient of concentration; y - vertical coordinate.

2 The diffusion coefficient for nonuniform flow

The diffusion coefficient characterizes the transfer of substance by chaotic turbulent motion and depends on the hydrodynamic flow characteristics. Since the flow of water in the natural watercourses is generally nonuniform let's investigate the influence of the factor of nonuniformity on the value of the coefficient of turbulent diffusion assuming the same depth values and the average flow velocities in the considered cross-sections of nonuniform and "equivalent" uniform flow. Especially that currently practiced approach to the calculation of the hydrodynamic characteristics of the middle section of gradually varied nonuniform flow needs, as shown by recent studies [2,4], in serious correction. Consider this question based on the identity determining of the diffusion coefficient D_T and the eddy viscosity ν_T [3].

Determine the coefficient of eddy viscosity in open nonuniform gradually varied flow using relationship [2]

$$\nu_T = \kappa U_* h\eta(1-\eta), \tag{2}$$

κ - von Kármán parameter; U_* - friction velocity; h - depth of flow; $\eta = y/h$ - dimensionless distance from the bottom.

In the case of an "equivalent" uniform flow, the Eq. (2) will differ by von Kármán parameter value (κ_p) and friction velocity value (U_{*p})

$$v_{Tp} = \kappa_p U_{*p} h\eta(1-\eta). \tag{3}$$

After dividing Eq. (2) by Eq. (3) and elementary mathematical transformations we have

$$\frac{v_T}{v_{Tp}} = \frac{\kappa}{\kappa_p}\frac{U_*}{U_{*p}}. \tag{4}$$

Thus, the influence of nonuniformity on the eddy viscosity (diffusion coefficient) can be identify analyzing the changes of parameters κ and U_* in a nonuniform flow.

Following F.Engelund [6], we can write the one-dimensional energy and momentum equations for nonuniform flow:

$$\frac{d}{dx}\left(h + \frac{\alpha q^2}{2gh^2}\right) = i - \frac{1}{2}\lambda\frac{q^2}{gh^3}, \tag{5}$$

$$\frac{dh}{dx} + \frac{1}{h}\frac{d}{dx}\left(\frac{\beta q^2}{gh}\right) = i - \frac{1}{2}C_f\frac{q^2}{gh^3}, \tag{6}$$

λ - hydraulic resistance coefficient of nonuniform flow defined by the hydraulic radius; a, β - kinetic energy and momentum correction factors; q - specific discharge; i - bottom slope; $C_f = 2(U_*/V)2$ – skin friction coefficient (V- mean velocity); x - longitudinal coordinate; g - acceleration due to gravity.

Subtracting Eq. (5) from Eq. (6), we will find after simple transformations

$$\lambda = C_f + 4m\frac{dh}{dx}, \tag{7}$$

there $m=3,12\,\lambda$ [6].

Eq. (7) may be rewritten as

$$C_f = \lambda\left(1 - 12{,}5\frac{dh}{dx}\right). \tag{8}$$

For decelerated flows ($dh/dx>0$) in a range $dh/dx \leq 0,02$ is valid relationship [4, 5]

$$\lambda = \lambda_p\left(1 + 21{,}2\frac{dh}{dx}\right), \tag{9}$$

λ_p - hydraulic resistance coefficient of "equivalent" uniform flow.

Substituting Eq. (9) into Eq. (8), we obtain

$$C_f = \lambda_p\left(1 - 12{,}5\frac{dh}{dx}\right)\left(1 + 21{,}2\frac{dh}{dx}\right). \tag{10}$$

Considering that $C_f = 2(U_*/V)^2$ and $\lambda_p = 2(U_*p/V)^2$ we can find

$$\frac{U_*}{U_{*p}} = \left(1 - 12{,}5\frac{dh}{dx}\right)\left(1 + 21{,}2\frac{dh}{dx}\right). \tag{11}$$

Opening brackets and neglecting small terms we have

$$\frac{U_*}{U_{*p}} \approx 1 + 8{,}7\left|\frac{dh}{dx}\right|. \tag{12}$$

For accelerated flows ($dh/dx <0$) $\lambda = \lambda_p$ [4, 5] and a similar dependence has the form

$$\frac{U_*}{U_{*p}} = \left(1 - 12{,}5\frac{dh}{dx}\right), or\ \frac{U_*}{U_{*p}} = \left(1 + 12{,}5\left|\frac{dh}{dx}\right|\right). \tag{13}$$

Processing of experimental data I. Nikuradze [7], F.Donhe [8] and A.G.Gurzhienko [9] for flat decelerated flows led [4] to the next Eq. for the von Kármán parameter

$$\frac{\kappa}{\kappa_p} = 1 + 76{,}5\frac{dh}{dx}. \tag{14}$$

After substituting Eq. (12) and Eq. (14) into Eq. (4) it is possible to find for decelerated flow

$$\frac{\kappa}{\kappa_p}\frac{U_*}{U_{*p}} = \frac{v_T}{v_{Tp}} = \left(1 + 76{,}5\frac{dh}{dx}\right)\left(1 + 8{,}7\frac{dh}{dx}\right),\ or\ \frac{v_T}{v_{Tp}} = 1 + 85.2\left|\frac{dh}{dx}\right|. \tag{15}$$

A similar dependence for accelerated flow can be obtained after the substitution Eq. (13) into Eq. (4)

$$\frac{\kappa}{\kappa_p}\frac{U_*}{U_{*p}} = \frac{v_T}{v_{Tp}} = \frac{\kappa}{\kappa_p}\left(1 + 12{,}5\left|\frac{dh}{dx}\right|\right). \tag{16}$$

In accordance with the well-known power-law dependence for velocity profile [10] the equality of the hydraulic resistance coefficients for accelerated and "equivalent" uniform flow simultaneously assumes the identity of the distributions of mean velocity in the comparable flows. Taking on this basis $\kappa = \kappa_p$ we find

$$\frac{\kappa}{\kappa_p}\frac{U_*}{U_{*p}} = \frac{\nu_T}{\nu_{Tp}} = 1 + 12{,}5\left|\frac{dh}{dx}\right|. \quad (17)$$

Based on the identity of determination the coefficient D_T and the eddy viscosity ν_T [3] we can turn from Eq. (15) and Eq. (17) to the analogous dependences for the diffusion coefficients D_T.

For decelerated flow

$$\frac{D_T}{D_{Tp}} = 1 + 85.2\left|\frac{dh}{dx}\right|. \quad (18)$$

For accelerated flow

$$\frac{D_T}{D_{Tp}} = 1 + 12{,}5\left|\frac{dh}{dx}\right|. \quad (19)$$

Comparing the values of diffusion coefficient for decelerated *(DTз)* and accelerated *(DTy)* flows in accordance with the Eq. (18) and Eq. (19) we obtain

$$\frac{D_{T3}}{D_{Ty}} = \frac{1+85.2\left|\frac{dh}{dx}\right|}{1+12{,}5\left|\frac{dh}{dx}\right|} \approx 6{,}8 - \frac{5{,}8}{1+12{,}5\left|\frac{dh}{dx}\right|}. \quad (20)$$

With a uniform flow *(dh/dx* = 0) the Eq. (20) is identically equal to one. Thus, the diffusion coefficients in decelerated flow can significantly exceed similar values for the accelerated flow (in several times) that explained by a significant increase in the intensity of turbulent pulsations in decelerated flow. Therefore, decelerated flows are preferable for the intensification of diffusion process for sewage dilution in the rivers.

3 Concentration profiles for uniform and nonuniform flow

In the case of suspended contaminants, small size of suspended particles contributes to their involvement in the turbulent flow. For a description of their movement, we used the diffusion model of turbulent transport in one-dimensional formulation [1]

$$\frac{\partial}{\partial y}\left(\nu_T\frac{dc}{dy}\right) + W\frac{dc}{dy} = 0, \quad (21)$$

c - volume concentration; W – fall velocity.

H.Rouse studies [11] showed that for a fine mixture the eddy viscosity ν_T doesn't depend on the concentration c and it is determined as well as for pure water. As a result of the integration of the Eq. (21) we find

$$\nu_T\frac{dc}{dy} + Wc = C. \quad (22)$$

Since the left-hand side of Eq. (22) represents a complete mass balance, passing through a unit area in the vertical direction, the constant C can be assumed equal to zero

$$\nu_T\frac{dc}{dy} + Wc = 0. \quad (23)$$

For the determination of the eddy viscosity, we use the Eq. (2). Substituting Eq. (2) into Eq. (23), we have

$$\frac{dc}{c} = -\frac{Wd\eta}{\kappa U_*\eta(1-\eta)}. \quad (24)$$

After integration of Eq. (24), we obtain

$$\ln c = \frac{W}{\kappa U_*}\ln\frac{1-\eta}{\eta} + C. \tag{25}$$

The integration constant C can be determined from the condition $\eta_0=0{,}05$ $c = c_0$. As a result, we find

$$\ln\frac{c}{c_0} = \frac{W}{\kappa U_*}\left(\ln\frac{1-\eta}{\eta} - \ln\frac{1-\eta_0}{\eta_0}\right). \tag{26}$$

Transform the Eq. (26) to the form

$$\frac{c}{c_0} = \left[\frac{\eta_0}{1-\eta_0}\left(\frac{1-\eta}{\eta}\right)\right]^{\frac{W}{\kappa U_*}}. \tag{27}$$

Substituting into Eq. (27) von Kármán parameter (κ_p) and friction velocity (U_{*p}), corresponding to the conditions of uniform flow, we arrive at a well-known profile Rouse-Velikanov [12]

$$\frac{c}{c_0} = \left[\frac{\eta_0}{1-\eta_0}\left(\frac{1-\eta}{\eta}\right)\right]^{\frac{W}{\kappa_p U_{*p}}}. \tag{28}$$

Consider possible changes in the concentration profiles in the case of non-uniform flow. To do this, multiply and divide the exponent in Eq. (28) on the same expression $\kappa_p U_{*p}$ different from zero

$$\frac{c}{c_0} = \left[\frac{\eta_0}{1-\eta_0}\left(\frac{1-\eta}{\eta}\right)\right]^{\frac{W}{\kappa_p U_{*p}}\frac{\kappa_p U_{*p}}{\kappa U_*}}. \tag{29}$$

For the ratio $\kappa U_*/\kappa_p U_{*p}$ previously were obtained Eq. (15) and Eq. (17). Substituting them into Eq. (29), we find for decelerated flow

$$\frac{c}{c_0} = \left[\frac{\eta_0}{1-\eta_0}\left(\frac{1-\eta}{\eta}\right)\right]^{\frac{W}{\kappa_p U_{*p}(1+85{,}2\frac{dh}{dx})}}. \tag{30}$$

The expression to accelerate the flow has the form

$$\frac{c}{c_0} = \left[\frac{\eta_0}{1-\eta_0}\left(\frac{1-\eta}{\eta}\right)\right]^{\frac{W}{\kappa_p U_{*p}(1-12{,}5\frac{dh}{dx})}}. \tag{31}$$

These profiles (30) and (31) differ from the Rouse-Velikanov concentration profile (28) by amendments to the exponent that take into account the degree of non-uniformity of flow. The graph (Fig. 2) shows that the resulting profiles are characterized by a more uniform distribution of the fine mixture along of the flow depth in non-uniform flow and close to the profiles of Rouse-Velikanov [12] and A.V.Karaushev [13].

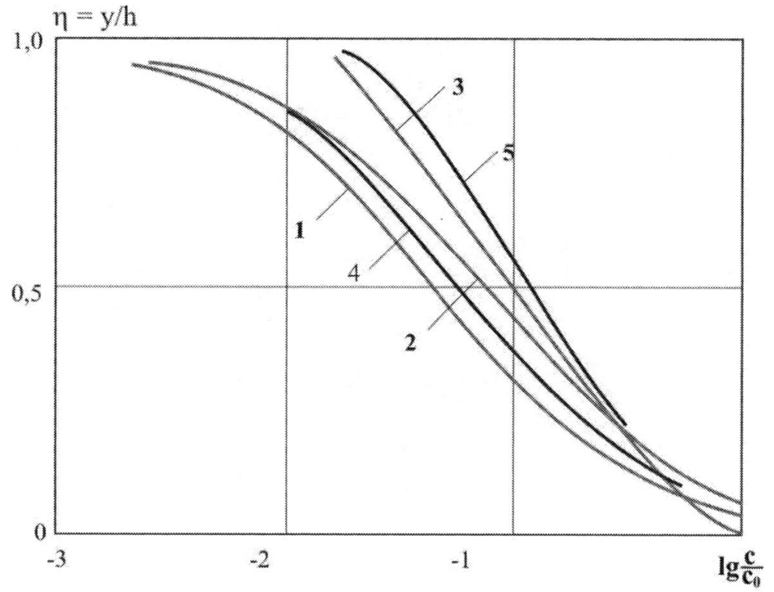

Fig. 2. Concentration profiles for unifom and nonuniform gradually varied flow:
1 – Rouse-Velikanov concentration profile, 2 – Data of V.Borovkov; 3 – data of A.Karaushev; 4 – accelerated flow; 5 – decelerated flow

The average concentration of fine mixture along of the flow depth can be determined by integrating of Eq. (27), after presenting it in the form

$$\frac{c}{c_0}=\left(\frac{\eta_0}{1-\eta_0}\right)^{\frac{W}{\kappa U_*}}\left(\frac{1-\eta}{\eta}\right)^{\frac{W}{\kappa U_*}}. \tag{32}$$

After decomposition the variable numerator of the second factor in the binomial set, we perform integration the Eq. (32) along of the flow depth, limiting to first two terms of the expansion

$$\frac{\bar{c}}{c_0}=\frac{\left(\frac{\eta_0}{1-\eta_0}\right)^{\frac{W}{\kappa U_*}}}{1-\frac{W}{\kappa U_*}}, \tag{33}$$

$\bar{c}$ - average concentration of suspended contaminants.

Substituting into Eq. (33) von Kármán parameter value (κ_p) and friction velocity (U_{*p}), corresponding to the conditions of uniform flow, we have for uniform flow [1]

$$\frac{\bar{c}}{c_0}=\frac{\left(\frac{\eta_0}{1-\eta_0}\right)^{\frac{W}{\kappa_p U_{*p}}}}{1-\frac{W}{\kappa_p U_{*p}}}. \tag{34}$$

Using the previously found Eq. (15) and Eq. (17), transform Eq. (33) for decelerated flow

$$\frac{\bar{c}}{c_0}=\frac{\left(\frac{\eta_0}{1-\eta_0}\right)^{\frac{W}{\kappa_p U_{*p}(1+85,2\frac{dh}{dx})}}}{1-\frac{W}{\kappa_p U_{*p}(1+85,2\frac{dh}{dx})}}. \quad (35)$$

The expression to accelerate the flow has the form

$$\frac{\bar{c}}{c_0}=\frac{\left(\frac{\eta_0}{1-\eta_0}\right)^{\frac{W}{\kappa_p U_{*p}(1-12,5\frac{dh}{dx})}}}{1-\frac{W}{\kappa_p U_{*p}(1-12,5\frac{dh}{dx})}}. \quad (36)$$

Curves built in accordance with (35) and (36) at $/dh/dx/= 0,005$, are represented on the graph (Fig. 3). On the same graph are plotted the experimental points obtained based on measurements on the Moscow River in the flood [1].

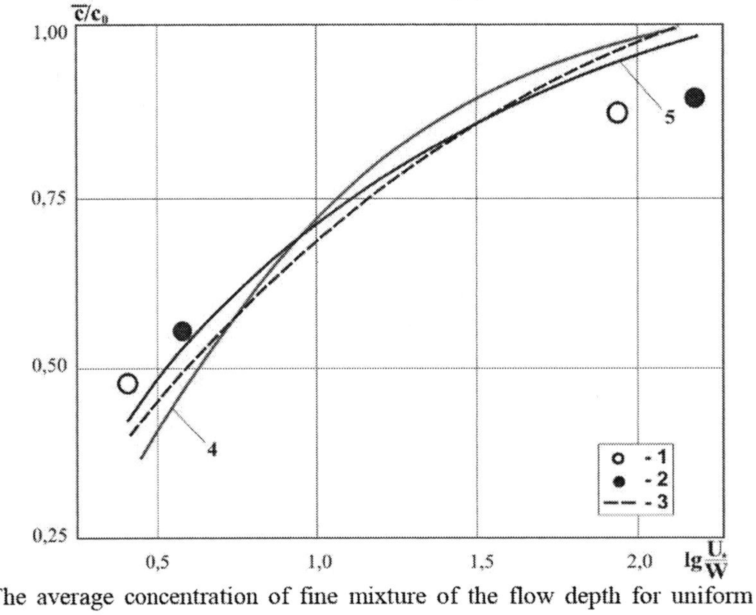

Fig. 3. The average concentration of fine mixture of the flow depth for uniform and nonuniform gradually varied flow when ldh/dxl=0,005:
1,2 – measurements on the Moskva-river in the flood; 2 – data of V. Borovkov; 4 – data of decelerated flow; 5 – data of accelerated flow.

The graph shows that the uneven distribution of suspended contaminants along the depth of gradually varied nonuniform flow increases considerably at minimum U_*/W values.

4 The deposition length for nonuniform flow

The accumulation of fine contaminating particles on the bottom sediments affects the bed erosion and sediment transport, worsens the sanitary condition of the river [13]. According to [1] let's use the fine particles deposition scheme supposing sediments balance for a unit length section of the flow in the absence of bed erosion

$$hdc = -cWdt, \tag{37}$$

c - volume concentration of suspended contaminants; t – time.

The solution of Eq. (20) with the initial conditions $t = 0$, $c = c_0$ has the form

$$c/c_0 = e^{-\frac{Wt}{h}}. \tag{38}$$

As follows from Eq. (21), the most intense sedimentation by fine suspended particles should be observed in the areas of flow with shallow depths (coastal zones). Considering the deposition process at the section of the flow from cross-section with the initial concentration c_0 to cross-section with the residual concentration of $c = 0{,}05c_0$ we find from Eq. (38)

$$Wt/h_c = 3,$$

h_c - average depth for the considered flow section.

Consequently, the time of suspension deposition

$$t = 3h_c/W. \tag{39}$$

Defining the time t using the mean flow velocity V for the flow section, we have

$$t = L/V, \tag{40}$$

L - length of the deposition.

Substituting Eq. (40) into Eq. (39), we have

$$L/h = 3V/W. \tag{41}$$

Calculate the mean velocity V using U_* and λ from the Eq. (8) for nonuniform gradually varied flow

$$2\left(\frac{U_*}{V}\right)^2 = C_f = \lambda\left(1 - 12{,}5\frac{dh}{dx}\right), \tag{42}$$

dh/dx – average slope of the free surface for the considered flow section; λ – hydraulic resistance coefficient for nonuniform flow.

Consequently,

$$V = \frac{\sqrt{2}\,U_*}{\sqrt{\lambda\left(1-12.5\frac{dh}{dx}\right)}}. \tag{43}$$

Substituting Eq. (43) into Eq. (41), we find after simple transformations

$$\frac{L}{h} = \frac{4{,}2}{\sqrt{\lambda\left(1-12{,}5\frac{dh}{dx}\right)}}\frac{U_*}{W}. \tag{44}$$

With a uniform flow dh/dx=0 and the Eq. (44) takes the form [3]

$$\frac{L}{h} = \frac{4{,}2}{\sqrt{\lambda_p}}\frac{U_*}{W}.$$

In the case of decelerated flow, we use the expression (9) for substitution into Eq. (44). After simple transformations we obtain

$$\frac{L}{h} = \frac{4{,}2}{\sqrt{\lambda_p\left(1+8{,}7\frac{dh}{dx}\right)}}\frac{U_*}{W}. \tag{45}$$

For accelerated flow we take $\lambda = \lambda_p$ [4] and according to Eq. (44) the resulting relation for hydraulic resistance coefficient has the form

$$\frac{L}{h} = \frac{4{,}2}{\sqrt{\lambda_p\left(1-12{,}5\frac{dh}{dx}\right)}}\frac{U_*}{W}. \tag{46}$$

We perform rationing for deposition lengths defined by Eq. (45) and Eq. (46) using the deposition length of equivalent uniform flow. For equal parameters U_*/W we obtain the relationships presented below.

For decelerated flow

$$\frac{L_3}{L_p} = \frac{1}{\sqrt{\left(1+8{,}7\frac{dh}{dx}\right)}}. \quad (47)$$

For accelerated flow

$$\frac{L_y}{L_p} = \frac{1}{\left(1-12{,}5\frac{dh}{dx}\right)}. \quad (48)$$

From the analysis of Eq. (47) and Eq. (48) follows that the deposition length for the fine particles in the decelerated streams at the same value of the hydraulic resistance coefficient λp and parameter U_*/W slightly less (within 10%) than the corresponding value for the uniform flow under other equal conditions. For accelerated flow, the deposition length is longer than the same length for uniform flow.

5 Conclusions

1. The effect of flow nonuniformity on the diffusion model of wastewater dilution was investigated.
2. A method was proposed for calculating the coefficient of turbulent diffusion for open channels flows.
3. The formulas for concentration profiles and the average concentration of fine suspensions were obtained.
4. The characteristic features of sedimentation in nonuniform open flows are revealed.

References

1. V.S. Borovkov, Channel processes and dynamics of river flows in urban areas (1986)
2. K.V. Grishanin, Hydraulic resistance of natural riverbeds (1992)
3. Bennet C.O., Myers J.E., Momentum, heat and mass transfer (1966)
4. V.Yu. Lyapin, Hydraulic resistance gradually varied nonuniform and uniform flow in open channels (2004)
5. V.Yu. Lyapin, Calculations of hydraulic structures (1982)
6. F. Engelund, Energy and Momentum Equation for Non-Uniform Flow (1981)
7. I. Nikuradze, Untersuchungen Uber die Stromungen des Wassers in Konvergenten und Divergenten Kanalen (1929)
8. F. Donch, Divergence und Convergente Turbulent Stromungen mit Kleinen Offnungswinkeln (1929)
9. G.A. Gurzhienko, On a steady turbulent flow in conical diffusers with a small expansion angles (1939)
10. A.D. Altschul, Hydraulic resistance (1982)
11. H. Rouse, Fluid Mechanics for hydraulic engineers (1967)
12. M.A. Velikanov, Riverbed process (1958)
13. A.V. Karaushev, Methodical bases of an estimation and regulation of human impact on surface water quality (1987)

Haemodynamic study of Flow in concentric and eccentric stenosed carotid Artery

S. M. Abdul Khader[1], *Raghuvir Pai*[2*], *Ganesh Kamath*[3], *Mohammed Zuber*[4] and *V. R. K. Rao*[5]

[1]*Department of Mechanical Engineering, School of Science and Engineering, Manipal International University, No.1, Persiaran MIU, Nilai-71800, Malaysia,*
[2]*Department of Mechanical and Manufacturing Engineering, Manipal Institute of Technology, Manipal Academy of Higher Education, Manipal, 576104, India*
[3]*Department of Cardiothoracic and Vascular Surgery, Manipal Academy of Higher Education, Manipal, 576104, India*
[4]*Department of Aeronautical and Automobile Engineering, Manipal Academy of Higher Education, Manipal, 576104, India*
[5]*Krishna Institute of Medical Sciences, Secunderabad, 500 003, India*

Abstract. Study of flow through arteries is gaining importance in the recent times due to onset of several cardio-vascular diseases like atherosclerosis (block in artery), aneurysms (bulging of artery), hypertension etc. The application of CFD will be useful in demonstrating the underlying mechanism of flow past the diseased arteries. In the present study, initially an approximate eccentric and concentric 50% stenosed carotid model is generated in ANSYS 12.0 based on patient data obtained from ultrasound Doppler scan. Later the percentage of stenosis is increased to 60%, 70%, 80% and 90% in both the cases. A transient analysis has been performed for several pulse cycles in ANSYS FLOTRAN for various percentage cases. The results obtained from ECN and CCN cases are compared to observe the changes in flow behavior in the downstream of stenosis as compared with normal case. The formation of vortices and flow separation zone in downstream in eccentric case is higher than in concentric case. Comparison of results concludes that, with the increase in severity of stenosis, the flow changes abruptly causing an increase in velocity and WSS at throat region. The simulation results obtained agree well with clinically observed data and available literature.

Keywords: CFD; Stenotic flow; Common Carotid Artery; ANSYS APDL

1 Introduction

The high level of maturity reached in many computational disciplines, prompts for the study of new multidisciplinary applications of great importance for engineering purposes. Such is the case of blood flow analysis through arteries. The flow analysis can help in

* Corresponding author: raghuvir.pai@manipal.edu

understanding the flow through elastic arteries. The CFD tools can be more promising in analysing such complex flow.

Besides theoretical work, experimental studies and clinical measurements have advanced the understanding in the past two centuries [1]. While experiments can be expensive or measurement data hard to attain, properly validated coupled computational simulations can hold the answer in characterizing conditions, such as flow and pressure and also to understand arterial flows [2]. This type of study can be extended in the applications such as disease research, where fluid mechanical conditions are correlated to regions prone to atherosclerosis, medical device design where the interactions between a device and the blood stream are modelled [3]. The study of stenosis, which is a constriction in the blood vessels, can be helpful in understanding the biomechanics of vascular diseases. The flow behavior in the stenosed artery is quite different than one in the normal arteries [4,5].

The concepts of fluid dynamics can be applied to study the interesting aspects of generation, detection and the treatment of stenosis [6]. The stenotic geometry is highly complex, but in the preliminary study it can be simplified as a constriction in a cylindrical tube. Generally the flow through the constricted tube is characterized by high velocity jet generated in constricted section and flow separation downstream to the stenosis [3,7]. The constriction can be broadly classified as symmetric-concentric and asymmetric-eccentric conditions. In case of concentric condition, a high speed jet of fluid extends downstream along the centerline of model, whereas in case of eccentric condition, the flow pattern is almost stagnant and forces the vorticity against the wall while flowing in downstream causing high flow separation [8].

An attempt has been made in the present study to demonstrate the significant changes for concentric and eccentric conditions with various stenosis percentages starting from 50% in steps of 10% till occlusion. This study focuses on detailed investigation of flow parameters like velocity, WSS, pressure contours. The comparison of results obtained reveals the high WSS zones on the far wall of downstream of the throat causing degradation of wall tissue [3,8]. Also shear layer dissipates more quickly due to formation of high eddies in downstream [9].

The flow behavior changes abruptly with the increase in velocity at throat and large recirculation in the downstream of stenosis. This type of study will be useful to predict the outcome of severity of stenosis, and aids the physicians with better insight into flow dynamics. Moreover it is also observed that severity less than 75% is considered as mild and anything greater than this is termed as critical [4,10]. This type of study will be useful to predict the outcome of severity of stenosis, and aids the physicians with better insight into flow dynamics.

2 Methodology

2.1 Governing Equations

In the present study, fluid model is assumed to be incompressible Newtonian fluid to minimize the complexity of analysis and modelled with a density of 1500 kg/m^3 and dynamic viscosity of 0.04 poise[5,6]. The dynamic flow is governed by the continuity and Navier-stokes equations [11, 12], respectively as

$$\nabla v = 0 \tag{1}$$

$$\rho\left(\frac{\partial v}{\partial t}+v\nabla v\right)=-\nabla p+\mu\nabla^{2}v \tag{2}$$

where v is the velocity vector, ρ is the density and P is the pressure.

2.2 Modelling

The present study is carried out considering the realistic physiological conditions. The basic parameters required for the modelling purpose is done using Doppler scan on a healthy young male volunteer [13]. The B-mode scan is used to obtain velocity spectrum and M-mode scan is obtain geometry details [14]. Based on the date obtained, an approximate eccentric model and concentric model having a constriction ratio of 50% and length of 18D is generated in ANSYS 12.0. Fig.1 shows the eccentric model where the stenosed region is encircled and similarly the concentric model is shown in Fig.2.

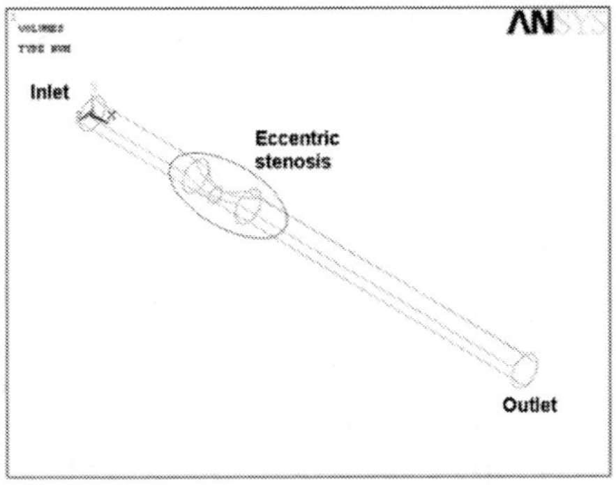

Fig. 1. Eccentric stenosed model

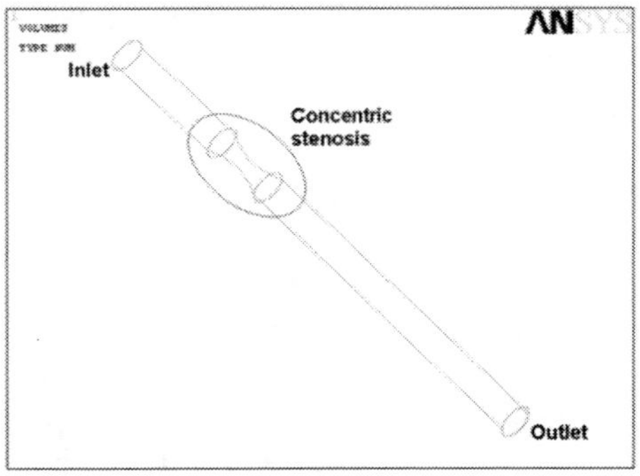

Fig.2. Concentric stenosed model

In order to analyse the effect of severity, the initial 50% ECN and CCN stenosis is increased and separate models are generated for 60%, 70%, 80% and 90% stenosis and the rest of the parameters remain same for all the later cases. This will yield useful simulation data when the stenotic severity progresses, which will be useful in quantifying haemodynamic factors that indicates plaque, rupture potential.

2.3 Analysis

An APDL is written in ANSYS to generate the model and mesh with 3D brick element which are discretised into approximately 150000 tetrahedral elements for both the cases [15]. The time varying velocity is applied at inlet as shown in fig.3, constant pressure boundary condition at outlet and no-slip condition is considered along the walls. The density and dynamic viscosity of the blood considered are 1050 kg/m3 and 0.004N-sec/m2 respectively [14, 15].

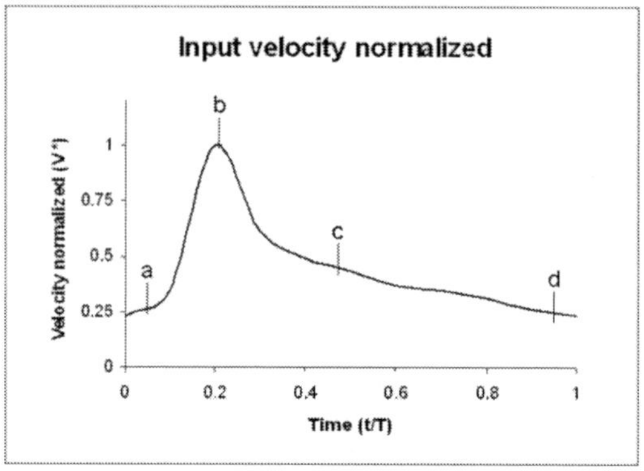

Fig.3. Concentric stenosed model

The analysis has been performed for 3 pulse cycles and each cycle is discritised into 80 steps. In the time varying velocity applied, the following regions of interest are highlighted. Region **a** - refers to early systole, Region **b** - refers to peak systole, Region **c** - refers to early diastole, Region **d** - refers to late diastole

3 Results and Discussion

A CFD study has been carried out to evaluate the effect of stenosis on the flow behavior. The results obtained from the various stenotic percentages for both eccentric and concentric models have been presented. The velocity contours, WSS and pressure have been considered in the present study for comparing the deviation in the flow. In the high grade stenosis, the flow resistance in arteries increases, which forces the body to raise the blood pressure to maintain the necessary blood supply [4]. Both the high pressure and the narrowing of blood vessel cause high flow velocity, high shear stress and low or negative pressure at the throat of the stenosis, low shear stress and flow separation at the distal side of the stenosis [5].

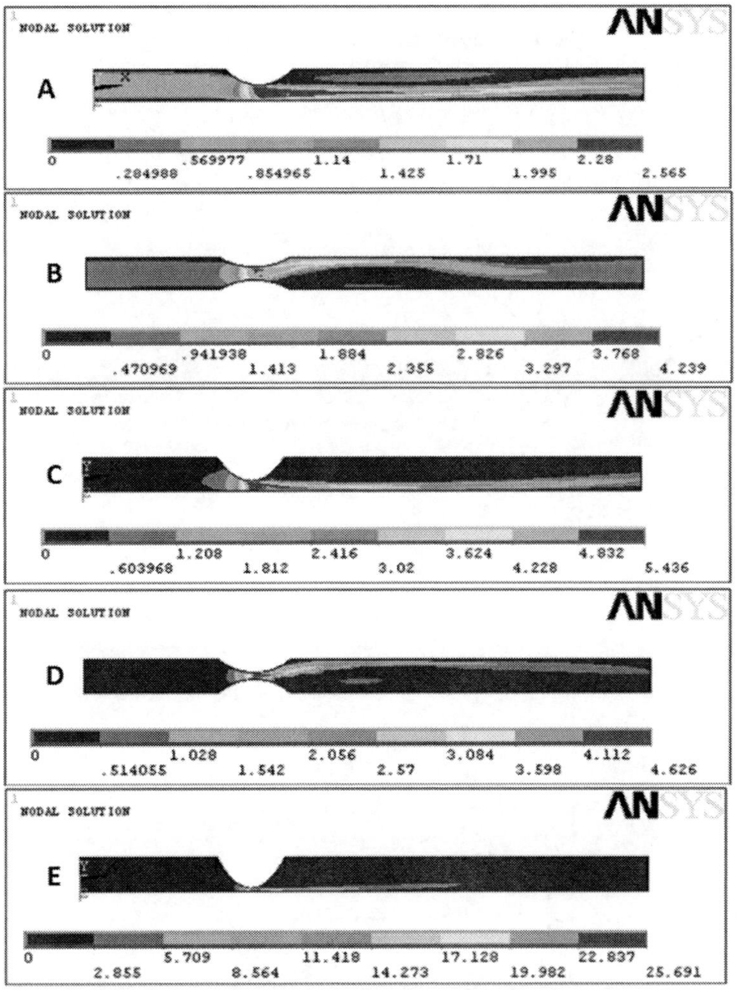

Fig.4. (a) 50 ECN peak systole (b) 60 CCN early diastole (c) 70 ECN early systole (d) 80 CCN end diastole (e) 90 ECN peak systole

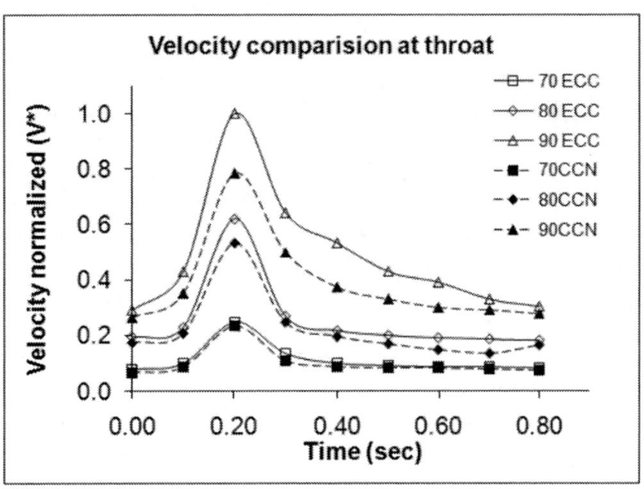

Fig.5. Velocity comparison at the throat

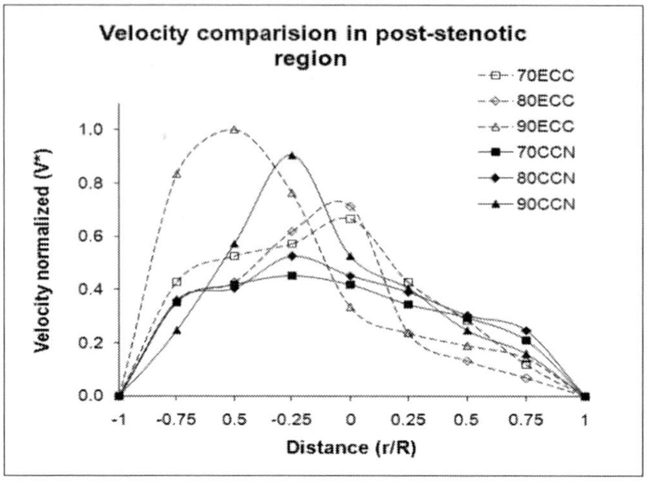

Fig.6. Velocity comparison in post-stenotic region

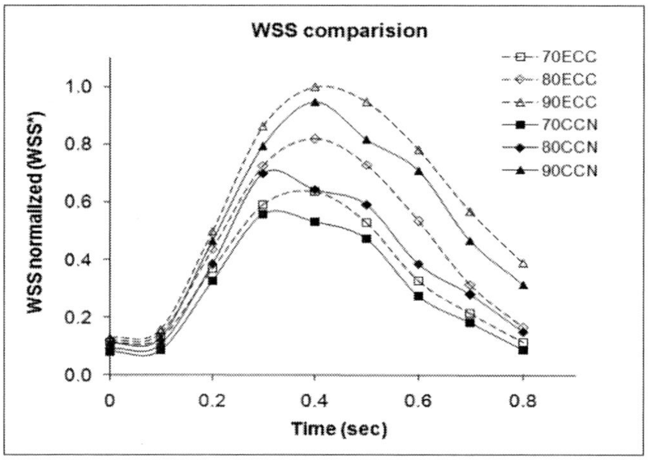

Fig.7. Wall shear stress comparison

Fig 4 shows the velocity profile for different values of stenosis during various instants of pulse cycle. The velocity contours depicts interesting flow features that varies throughout the pulse cycle. The flow behavior is almost similar in all the cases upstream of stenosis, whereas complex changes are observed in the distal side [3, 6]. Fig. 4(A) shows the velocity profile during peak systole for 50% of stenosis. Due to eccentric type, a flow jet is formed at the throat which extends along 3D times in downstream end. The flow separation is observed in the immediate vicinity in downstream side. Due to concentric type of stenosis as seen from Fig. 4(B) for 60%, the vortex formation is observed in the distal side extending for 2.5 D times during early diastole. The flow separation becomes more complex during early diastole due to flow decelaration.

It can be inferred from the fig. 4(C) and 4(E), that as the percentage of stenosis increases, the constriction of flow increases resulting in lesser flow rate on the downstream side. Similarly for CCN case as seen in Fig. 4(D) during end diastole, the formation of flow jet still persists in downstream end inspite of flow decelaration. Since the stenosis greater than 75% is considered critical, a comparison of velocity at throat for all the ECN and CCN percentages greater than 75% are compared as shown in the Fig.5. It is observed that for 90%, the velocity increases instantly at the ECN than at the CCN. The difference in velocity during peak systole between ECN and CCN is maximum when compared for less critical cases like 80% and 70%. During the later part of pulse cycle, in less critical cases, the difference in flow is less than those compared with more critical cases. The flow dynamics in stenosis is characterised by instant increase of velocity at throat and flow jet extending to certain distance in downstream side. The flow jet subsides after certain distance causing flow separation and formation of eddies in distal end. The flow behaviour is non-uniform throughout the pulse cycle in post-stenotic region as seen from Fig.6. In case of 90%, the flow jet in ECN still exists extending to more length than CCN. The flow orientation towards wall is more prominent in ECN than in CCN [6]. As percentage of stenosis decreases, the flow jet length subsides and flow orientation in CCN is more towards core, but in ECN it is still extending towards boundary.

As the severity of stenosis increases, the WSS also increases as observed from the comparison of WSS shown in Fig.7 for more critical cases. It can be observed that, during early pulse cycle, there are no significant changes in WSS pattern. Irrespective of the severity, the WSS is observed to be maximum during post peak phase in all ECN cases which is due to existence of flow jet after peak phase of cycle. In case of CCN, maximum WSS shifts before ECN condition and it immediately decreases because of sudden drop of flow jet in distal side. The result obtained agrees well with the clinical observations and also with available literature. An instantaneous increase in velocity is observed during the peak systole at throat region and with jet flow extending into the post-stenotic region. The flow behavior in case of ECN at the downstream of stenosis is characterized by flow separation leading to formation of eddies and in the case of CCN, the flow jet extends to larger length causing vortices. This behavior gets complex as the severity of stenosis increases [4, 5]. Further, the hypothetical cases of increased severity were used in prediction of changes in flow behavior as the percentage of stenosis increases. This can be beneficial in the prognosis of the clinical cases.

4 Conclusions

Present study demonstrates the major flow changes for eccentric and concentric stenotic condition. The flow is maximum during peak systole of the pulse cycle in concentric than in eccentric type. In the eccentric type, the formation of jet is more prominent for a longer length than in concentric type. Similarly, the flow separation is more intense in downstream side in former case than in later. But high vortex formation is observed to be more in concentric type than in eccentric type, causing severe turbulence. During flow deceleration, the flow gets more complicated due to back flow, causing severe vortex in CCN and complicated flow separation and eddy formation in ECN. Though a simple model is considered for the analysis, the change in flow behavior agrees well with available literature and clinical observations. The results demonstrate that the 3D stenotic CFD model is capable of predicting the changes in flow behavior for increased severity. Also this study can be further extended to various anatomically realistic models of stenosed carotid artery and predict a range of haemodynamic phenomena that are likely to have clinical significance.

Nomenclature

CCA	Common Carotid Artery
ECN	Eccentric type of stenosis
CCN	Concentric type of stenosis
WSS	Wall shear stress
V^*	Normalized velocity = V_t / V_{max}

V_t	Velocity at time, t
V_{max}	Maximum velocity in time period
WSS^*	Normalized WSS = WSS_t/WSS_{max}
WSS_t	Wall Shear Stress at time, t
WSS_{max}	Maximum Wall Shear Stress in time period

References

1. C. A. Taylor, T.J.R Hughes, Christopher K. Zarinsb, Finite element modeling of blood flow in arteries, Comput. Methods Appl. Mech. Engg., **158**, 155-196 (1998)
2. Wei Zhang, C. H, Effect of Surrounding Tissue on Vessel Fluid and Solid Mechanics, Transactions of the ASME, **126**, 760-769 (2004)
3. G. S. Tavakoli, M.M. Natarajan, Movahedi, Numerical Simulation of pulsatile flow with Newtonian and non-Newtonain behavior in arterial stenosis, Iranian Cardiovascular Research Journal, **1(3)**, 167-174 (2008)
4. G. Lorenzini and E. Casalena, CFD analysis of pulsatile blood flow in an atherosclerotic human artery with eccentric plaques, J. Biomech., **41**, 1862-1870 (2008)
5. Hun Jung, Jong Wook Choi, Chan Guk Park, Asymmetric flows of non-Newtonian fluids in symmetric stenosed artery, Korea-Australia Rheology Journal, **16(2)**, 101-108 (2004)
6. Khalifa A.M.A and D.P Giddens, Characterization and evolution of post-stenotic flow disturbances, J. Bio-mech, **14-5**, 279-296 (1981)
7. Mishra J.C and M.K. Patra, A non-Newtonian fluid model for blood flow through arteries under stenotic conditions, J. Bio-mech, **26**, 1129-1141 (1993)
8. S. M. Stanley, Transport Phenomena in the Cardiovascular System, *Wiley-Interscience Inc*,(1972)
9. W. W Nichols and M.F. O'Rourke, McDonald's Blood Flow in Arteries, *LEA & FEBIGER*, Philadelphia, London, UK, 81–85 (1998)
10. E.L. Seung, L. Sang-Wook, F. Paul, S. B. Hisham, L. Francis, "Direct numerical simulation of transitional flow in a stenose carotid bifurcation," J. Biomech., vol.41, 2008, pp 2551-2561.
11. Y.C Fung, Biodynamics- Circulation, *Springer Verlag*, New York Inc, (1984)
12. J. D. Humphrey and S. L. Delange, An Introduction to Biomechanics, *Springer Verlag*, New York Inc., (2004)
13. K. Perktold and G. Rappitsch, Computer simulation of local blood flow and vessel mechanics in a compliant carotid artery bifurcation model, J. Biomech., 28, 845–856 (1995)
14. S. Z. Zhao, X. Y. Xu and M. W. Collins, Blood flow and vessel mechanics in physiological realistic model of a human carotid arterial bifurcation, J. Biomech., 32, 975-984 (2000)

15. ANSYS Inc., ANSYS Release 12.0 Documentation, *ANSYS Company*, Pittsburgh, PA, (2008).

Fabrication of samarium strontium aluminate ceramic and deposition of thermal barrier coatings by air plasma spray process

Baskaran T [1*] and *Shashi Bhushan Arya*[1]

[1]*Department of Metallurgical and Materials Engineering, National Institute of Technology Karnataka, Surathkal, Mangaluru-575 025, India*

Abstract. Thermal barrier coatings (TBC) with the metallic NiCrAlY bond coat are often used in many aircraft engines to protect superalloy components from high-temperature corrosion thereby to improve the life of gas turbine components. The search for new TBC material has been intensified in recent years due to lack of thermo-physical properties of conventionally used Yttria stabilized Zirconia (YSZ) TBCs. Recently, the rare earth containing Samarium Strontium Aluminate (SSA) based ceramic was proposed as a new TBC material due to its matching thermo-physical properties with the substrate. The present work focused on the synthesis of SSA ceramics for TBCs application and its coatings development on Ni-based superalloy Inconel 718 substrate by air plasma spray process. The X-ray photoelectron spectroscopy (XPS) result confirmed the formation of single phase SSA ceramic after synthesis. The surface morphology of SSA TBCs is mainly composed of melted splats, semi and un-melted particles. The cross-sectional SEM micrographs did not show any spallation at the interface which indicated good mechanical interlocking between the bond coat and ceramic top coat. The Young's modulus and hardness of SSA TBCs were found to be 80 and 6.1 GPa, respectively. The load-depth curve of SSA TBC showed good elastic recovery about 47 %.

Keywords: Ceramic; Plasma spray; Superalloys; Thermal barrier coatings; Nano-indentation

1 Introduction

Thermal barrier coatings (TBC) plays an important role in providing low thermal conductivity, also act as a barrier to heat transfer between hot gas flow regions of the engine and coated superalloy substrate [1]. Presently Yttria stabilized zirconia (YSZ) has been used as a standard ceramic top coat material in TBCs applications due to its low thermal conductivity and high melting point [2]. However, due to the lack of sintering and

* Corresponding author: baskarannitk@gmail.com

phase transformation above 1200 °C intensify the search for new ceramic top coat materials [3].

In the past decade, the new ceramics such as zirconate perovskites (SrZrO3), zirconate pyrochlores (La2Zr2O7) and several hexa aluminates materials have been proposed to replace the currently used standard material YSZ [3, 4]. Among all the new materials, the zirconate based pyrochlores exhibited low thermal conductivity, high thermal expansion coefficients and good phase stability over YSZ. However, the poor thermo-chemical compatibility with thermally grown oxide at the interface reduced the thermal cyclic life of TBC at elevated temperatures [5]. Recently, the rare earth based strontium aluminate (Ln2SrAl2O7, Ln=Sm, Gd, Nd, La, etc.) have also been proposed as a thermal barrier coating materials more particularly Sm based aluminates showed low thermal conductivity, high melting point, phase stability and higher thermal expansion coefficient at higher temperature (>1200 °C) [6, 7]. Review of literature showed that most of the published work focused on the synthesis and thermal properties of ceramic but works on coating development and relevant studies are scarce. So, further research is required on the development of coatings of samarium strontium aluminate (Sm2SrAl2O7; SSA) ceramic for TBC applications.

In the present study, an attempt is made to synthesis the SSA ceramic in the laboratory and deposition of TBCs using air plasma spray process on nickel based super alloy substrate. The mechanical behavior of as deposited SSA ceramic, metallic bond coat and the substrate is investigated.

2 Experimental details

2.1 Synthesis of SSA powder by molten salt method

Samarium oxide (Sm_2O_3, 99.9%), Strontium carbonate ($SrCO_3$, 99.0%) and Aluminium oxide (Al_2O_3, 99.9%) were used as a starting precursor materials for the synthesis of SSA ceramic. These oxides were mixed with 0.5:0.5 eutectic compositions of potassium chloride (KCl, 99.5%) and sodium chloride (NaCl, 99.5%) for 30 min in an agate mortar. The mixture was placed in the alumina crucible and then heated up to 1100 °C for 24 h. Final products were continuously washed with hot deionized water to remove the salts. The obtained products were further dried in the hot air oven at 120 °C for 6 h.

2.2 Deposition of SSA ceramic by air plasma spray process

The as synthesized SSA ceramic was thermally sprayed on an Inconel 718 superalloy substrate using 55 kW F4-MB plasma gun. A metallic NiCrAlY was deposited on the substrate prior to the deposition of the ceramic top coat. The spray distance between nozzle and substrate approximately 100 mm. The feed rate of SSA ceramic powder is around 30 g/min. The thicknesses of the metallic bond coat and ceramic TBC top coat were about 50 and 100 μm respectively.

2.3 Characterization of SSA ceramic and as deposited coating

X-ray Photoelectron Spectroscopy (XPS) being used to examine stoichiometry of synthesized SSA (ESCALAB MkII, VG Scientific). Al Kα radiation (12 kV, 6 mA) was used to excite Sm 4d, Sr 3d, Al 2p, and O 1s electron from as-synthesized SSA. The obtained spectrum was calibrated with C 1s line at 285 eV.

The morphological analyses of as synthesized SSA ceramic and cross sectional studies of as deposited coatings were done using SEM (JEOL-JSM-6380LA, USA). The elemental analysis of ceramic powder and coatings were studied using Elemental Dispersive X-ray Spectroscope (EDS).

The mechanical properties such as Young's modulus, stiffness and hardness of substrate, metallic bond coat, and as deposited SSA TBCs were measured using Hysitron tribo-indenter fitted with Berkovich indenter.

3 Results and discussion

3.1 Characterization of synthesized SSA ceramic powder

A wide and high-resolution XPS spectrum of Sm, Sr, Al, and O are shown in Figs. 1 and 2 respectively. The major elements such as Sm, Sr, Al, and O were observed without any other impurities (Fig. 1). The binding energy (BE) at 284.6 eV corresponded to C 1s which is generally used as internal reference spectra (Fig. 1). The BE values at 976 eV and 1224 eV are more evident auger peaks of O and C respectively [8]. Fig. 2 (a-d) shows the high resolution XPS scan of Sm 3d, Sr 3d, Al 2p and O 1s. The Sm 3d peaks such as Sm $3d_{5/2}$ and Sm $3d_{3/2}$ were found at two distinct BE of 1084.3 eV and 1111.5 eV. The BE difference between these two peaks (27.2 eV) clearly indicated the +3 oxidation state of Sm ion (Table 1).

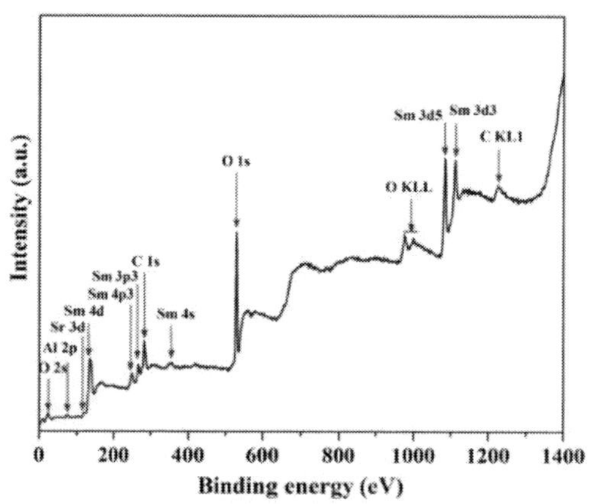

Fig. 1: XPS survey spectra of as-synthesized SSA ceramic powder.

The symmetry of Sm peak indicates that the presence of samarium ions in +3 valence state (Sm^{3+}). Two Gaussian peaks of Sr 3d peak were found at 134.06 eV and 135.8 eV corresponded to Sr $3d_{5/2}$ and Sr $3d_{3/2}$ lines, respectively in the SrO. The high resolution scan of Al (Al 2p) is shown in Fig. 2 (c) and this characteristic peak found to be at a BE of 73.03 eV. The BE values of O 1s peaks at 528.19, 529.50 and 530.10 eV revealed the presence of Al-O, Sm and /Sr-O bonds respectively. All noticed BE values of Sm_2O_3, SrO, and Al_2O_3 phases have shown good agreements with literature data's which are given in Table 1 [9-17].

Table 1: Binding energies of Sm 3d, Sr 3d, Al 2p and O 1s of synthesized SSA powder.

	Binding energy (eV)		Compound formation	
Electron level	Experimental value	Reference value	Experimentally observed	References
Sm $3d_{3/2}$	1111.5	1110.9	Sm_2O_3	(Ref 9-11)
Sm $3d_{5/2}$	1084.3	1083.4		
Sr $3d_{3/2}$	135.8	135.5	SrO	(Ref 12-13)
Sr $3d_{5/2}$	134.06	133.5		
Al 2p	73.03	74.50	Al_2O_3	(Ref 14-17)
O 1s	528.19	528.73	Al_2O_3	(Ref 15)
O 1s	529.50	530.40	Sm_2O_3	(Ref 11)
O1s	530.10	530.20	SrO	(Ref 13)

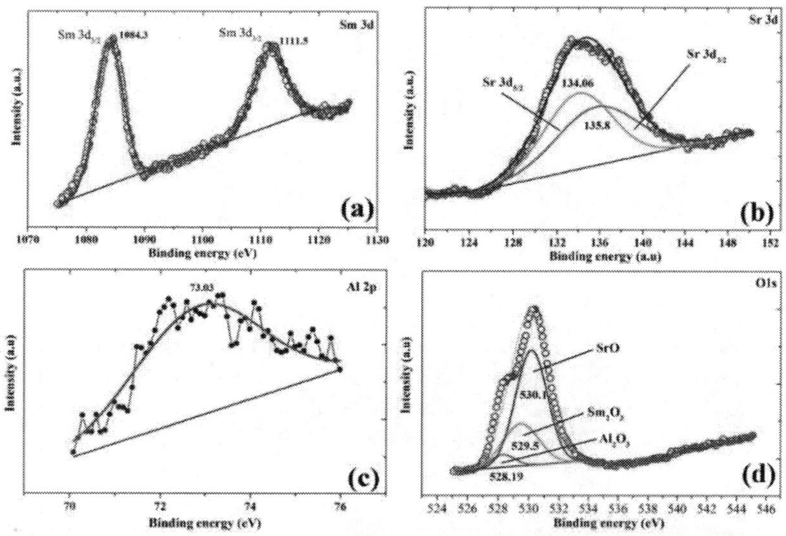

Fig. 2: Deconvoluted high resolution spectra of SSA ceramic powder: (a) Sm 3d; (b) Sr 3d; (c) Al 2p; and (d) O 1s.

3.2 Surface morphology of as synthesized SSA ceramic

Fig. 3 shows the surface morphology of as synthesized SSA ceramic and respective EDS analysis. The observed surface morphology of SSA particle was found to be cuboidal in shape (Fig. 3). In general cuboidal shape of particles would be observed in molten salt synthesis due to the mechanism of crystal growth in the chloride salts [18]. The elemental analysis of selected regions in Fig. 3 was done to find out the chemical composition of as synthesized SSA ceramic. The elemental compositions (at.%) of Sm, Sr, Al, and O were found to be 18.54, 8.08, 16.79 and 56.59 which is very similar to the theoretical stoichiometry of $Sm_2SrAl_2O_7$ (2:1:2:7) ceramic. So based on the XPS and EDS results confirm that the synthesized products correspond to single phase SSA without any impurities.

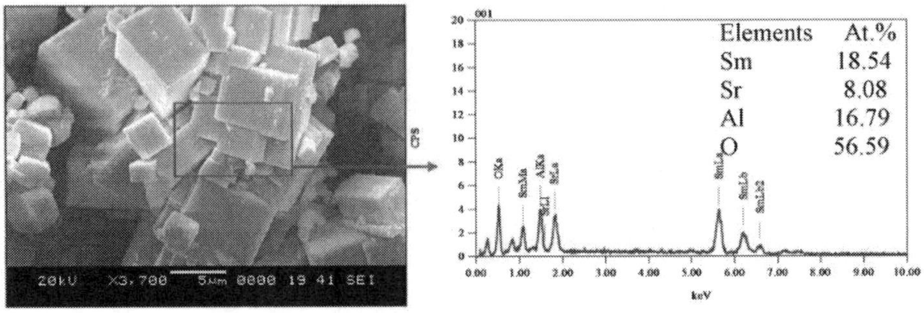

Fig. 3: Surface morphology of as synthesized SSA ceramic powder

3.3 Cross sectional micrograph of SSA after coatings

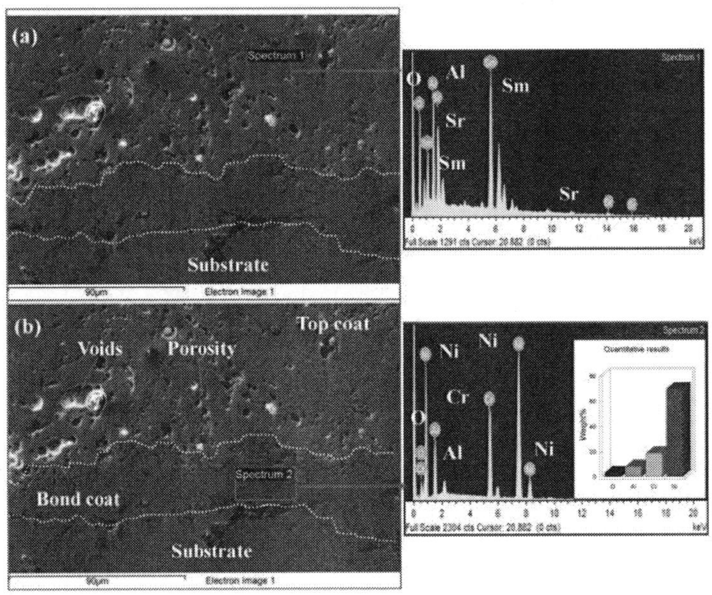

Fig. 4: (a and b) Cross sectional SEM microstructure of as deposited SSA TBCs. EDS analysis of region 1 in top coat (a); and region 2 in bond coat are also shown here.

Fig. 4 shows the cross sectional SEM micrographs and respective EDS analysis in the top coat (4a) and bond coat (4b) of as deposited SSA TBCs. The presence of voids and heterogeneous distributions of porosities were clearly seen in the Fig. 4 probably due to the presence of dissolved gases in the ceramic powder. The presence of such porosities and voids in the ceramic top coat TBC layer play a vital role in thermal shock resistance at higher temperature [19]. As compared to ceramic TBC (~1950 °C) layer the metallic bond coat seems to be dense due to completing melting of NiCrAlY (~1400 °C) particle in the plasma plume. The EDS analysis was conducted in the different places of as deposited coatings to know the chemical composition after plasma spraying process. The elemental analysis of selected region in Fig. 4 (a) shows the presence of Sm, Sr, Al and O (at.%) as 18.09, 7.95, 15.01 and 58.95 respectively. The obtained atomic percentage of Sm, Sr, Al and O which are similar to as synthesized SSA ceramic indicated that there is no compositional and any additional product formation during the plasma spraying process. The EDS analysis of selected region in the metallic bond coat shows the presence of Ni, Cr,

Al and O (69.75 wt.% Ni, 18.89 wt.% Cr, 8.16 wt.% Al and 3.20 wt.% O). A small amount of oxygen was also found in the metallic bond coat layer due to the oxidation of NiCrAlY during the deposition process.

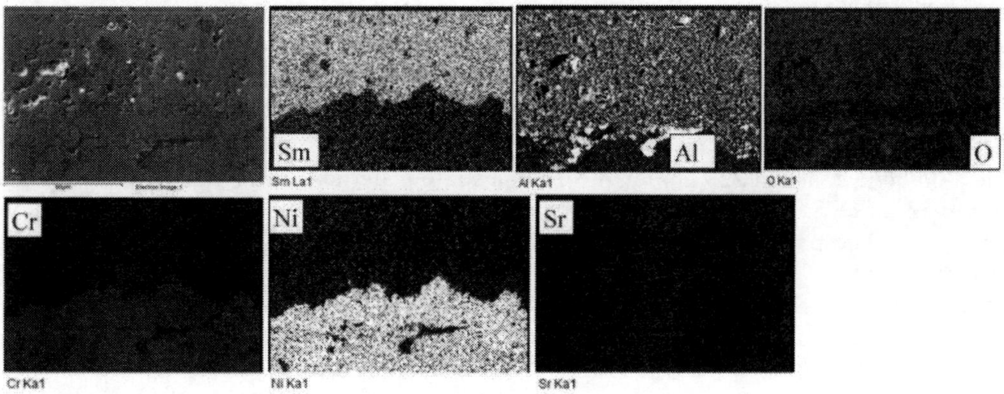

Fig. 5: Cross sectional EDS mapping analysis of as deposited SSA coatings

In order to understand the elemental interdiffusion between bond coat and top coat, the EDS elemental map scanning analysis was performed on the cross sectional sample. The map scanning images of Sm, Sr, Al, Ni, Cr, and O are shown in Fig. 5. It is observed that there is no interdiffusion of Sm from the top coat to bond coat and Ni or Cr from bond coat to ceramic top coat during the deposition process. Since the presence of Al in both top coat and bond coat, it is not able to distinguish the diffusion behavior by this analysis.

3.4 Mechanical properties of substrate, bond coat (NiCrAlY) and SSA TBCs

Fig. 6 shows the depth-time (6a) and load-depth (6b) curves of the substrate, metallic bond coat and as deposited SSA TBCs obtained from the polished sample surfaces (R_a<200 nm). The obtained mechanical properties such as stiffness, Young's modulus, and hardness are reported in Table 2. The load and depth variation of different layers of coatings were continuously measured during loading and unloading condition.

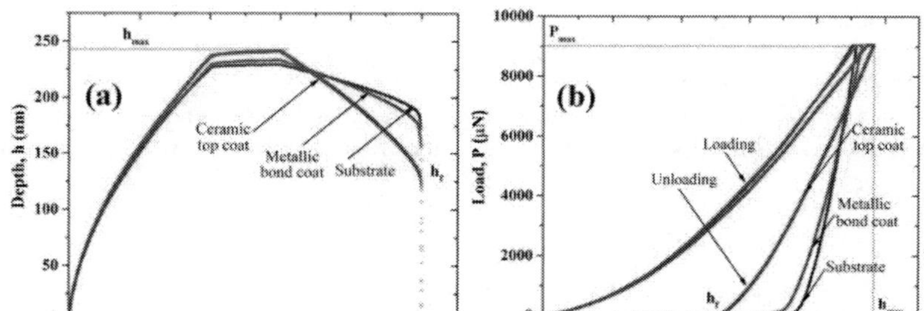

Fig. 6: Typical (a) depth-time; and (b) load-depth curves of substrate, metallic bond coat and as deposited SSA TBCs.

Table 2: Mechanical properties of substrate, metallic bond coat and as deposited SSA TBCs.

	Stiffness (μN/nm)	Young's modulus (GPa	Hardness (GPa)
Substrate	310±9	220±10	5.0±0.2
NiCrAlY bond coat	220±5	155±15	5.5±0.4
SSA TBC layer	110±3	80±0.3	6.1±0.5

The ceramic TBC showed maximum penetration depth of 267 nm followed by 235 nm for metallic bond coat and 223 nm for Inconel substrate material. The higher penetration depth observed for ceramic TBC layer may be due to the presence of defects such as micro pores and voids. The presence of such defects also decreased Young's modulus, hardness and stiffness of SSA TBC layer than the metallic bond coat and substrate material. The porosity of as deposited SSA TBCs was calculated to be about 65% using the theoretical relation between Young's modulus of solid and porous SSA ceramic. The load-depth curve of SSA TBCs also showed the highest elastic recovery of 49%, whereas for metallic NiCrAlY was about 25% and for Inconel substrate 19%. The highest elastic recovery of SSA TBCs was found due to the presence of higher porosity which can easily distribute the strain energy during loading and unloading condition. The lowest elastic recovery was obtained for the substrate which is due to dense nature; hence the maximum part of applied indentation load was utilized for the plastic deformation.

4 Conclusions

In the present study, the rare earth based samarium strontium aluminate ceramic was synthesized and coated on bond coated Inconel 718 substrate by air plasma spray process. There are no compositional changes of SSA ceramic was observed after the deposition process. The EDS mapping results showed that a negligible elemental interdiffusion between bond coat and top coat. The Young's modulus and hardness of SSA TBCs were found to be about 80 GPa and 6.1 GPa respectively. The highest elastic recovery was obtained for SSA ceramic top coat layer (49%) as compared to metallic bond coat (25%) and Inconel substrate (19%).

Acknowledgement

One of the authors (T. Baskaran) would like to thank the MHRD, Government of India, for a research scholarship. The authors acknowledge M/s. Metallizing Equipment Co. Pvt. Ltd, Jodhpur, India for providing plasma spray facility. The authors also thank Prof. U. Ramamurty (IISc, Bangalore) for providing the Nano-indentation facility.

References

1. R.A. Miller, Surf. Coat. Technol. **30**, 1-11, (1987).
2. G. Bertrand, P. Bertrand, P. Roy, C. Rio, and R. Mevrel, Surf. Coat. Technol. **202**, 1994-2001, (2008).
3. G. Moskal, L. Swadźba, M. Hetmańczyk, B. Witala, B. Mendala, J. Mendala, and P. Sosnowy, J. Eur. Ceram. Soc. **32**, 2035-2042, (2012).

4. H. Dong, Y. Ren, D. Wang, X. Li, Y. Bai, J. Wang, and W. Ma, J. Eur. Ceram. Soc. **34**, pp. 3917-3924, (2014).

5. W. Ma, S.K. Gong, H.B. Xu, and X.Q. Cao, Scripta Mater. **54**, 1505–1508, (2006).

6. C. Wan, T. D. Sparks, W. Pan, and D. R. Clarke, J. Amer. Ceram. Soc., **93**, 1457-1460, (2010).

7. J. Feng, B. Xiao, R. Zhou, W. Pan, and D. R. Clarke, Acta Mater. **60**, 3380-3392, (2012).

8. D.L. Hecht, K. Rohrmann, T. Stocker, and W. Thiel, Surf. Interface Anal. **39**, 845-851, (2007).

9. D. Petrov, B. Angelov, and V. Lovchinov, J. Rare Earths, **28**, 603-605, (2010).

10. Y. Uwamino, Y. Ishizuka, and H. Yamatera, J. Electron. Spectrosc. Relat. Phenom. **34**, 67-78, (1984).

11. C. D. Wagner, W.M. Riggs, L.E. Davis, J.F. Moulder, and G.E. Mullenberg, *Handbook of X-ray photoelectron spectroscopy*, Minnesota, (1979).

12. R. P. Vasquez, J. Electron. Spectrosc. Relat. Phenom. **56**, 217-240, (1991).

13. T. X. Wang, and W.W. Chen, Mater. Lett. **62**, 2865-2867, (2008).

14. B. R. Strohmeier, Surf. Sci. Spectra, **3**, 141-146, (1994).

15. N. Reddy, P. Bera, V.R. Reddy, N. Sridhara, A. Dey, C. Anandan, and A.K. Sharma, Ceram. Int. **40**, 11099-11107, (2014).

16. N.H. Turner, and A.M. Single, **15**, 215-222, (1990).

17. E. Gillet, and B. Ealet, Surf. Sci. **273**, 427-436, (1992).

18. C. Mao, G. Wang, X. Dong, Z. Zhou, and Y. Zhang, Mat. Chem. Phy. **106**, 164-167, (2007).

19. H. Guo, S. Kuroda, and H. Murakami, J. Am. Ceram. Technol. **89**, 1432-1439, (2006).

Experimental Studies on SiC and Rice Husk Ash Reinforced Al Alloy Composite

Y. M. Shivaprakash[1*], *Anunoy Sarkar*[1], *Sunit Jha*[1], *Muktesh Singh*[1] and *Shachin Bangera*[1]

[1]*Department of Mechanical and Manufacturing Engineering, Manipal Institute of Technology, Manipal Academy of Higher Education, Manipal-576 104, Karnataka, India.*

Abstract. In this research work Aluminium alloy with Cu (4.5%) as the major alloying element is used as the matrix in which SiC and Rice Husk Ash (RHA) are dispersed to develop a hybrid composite. The dispersion is done by the motorized stir casting arrangement. The composite is fabricated by varying the proportions of the reinforcements in the base alloy. The composite specimens were tested for density changes, hardness and the wear. The microstructure images showed a uniform dispersion of the reinforcements in the matrix and this resulted in higher strength to weight ratio. The increase in strength of the composite is probably attributed to the increase in the dislocation density. Also, the abrasive wear resistance of the produced composite is found to be superior as compared to the matrix alloy because of the hard-ceramic particles in the reinforcements.

1 Introduction

The research findings are required to meet the demands of industry maintaining the necessary property of materials and efficiency and cost saving in production. In meeting this requirement, a certain trend has been followed, the materials currently been utilized is tried for enhancement through known approaches of alloy blending, heat treatment, grain refinements, and the like. After the saturation, because of economic restrictions, difficulty in mass producing, or further enhancement is not possible so other line of approach emerged in further improving the properties or reducing cost and improving efficiency [1]. Among MMCs, Al-alloy-based composites were always extensively considered for research. Parallel areas of research had then emerged but after about two decades of research in various disciplines to further enhance the properties to satisfy the ever-increasing demand of the engineering sector, composites took a lead compared to the other processes when the cost and ease of fabrication were compared [2]. The other methods changed track and chose for themselves different areas of application and Al-based metal matrix composites remained as the most potential candidate to be researched on for making engineering components viable.

Worldwide research in Aluminum MMCs (AMMCs) established beyond doubt the advantages of Al-based metal matrix composites over the base alloy in the laboratory scale.

* Corresponding author: prakash.ym@manipal.edu

With time, the demands moved ahead and engineering components were demanded from AMMCs [3]. It can be easily concluded from the different results published from across the globe for two decades in AMMCs that the end use needs to be kept in mind before adopting a material and a process of fabrication. This is because the cost factors, ease of fabrication, shape, size, distribution, wettability between matrix alloy and second phase all must be considered for making a final choice depending on the component selected, mainly its working conditions for the engineering component to be fabricated.

In the past decade as research continued in this area yet the outcomes got were a just confirmation of the earlier obtained results, and a plateau in the possible enhancements has been reached [4]. Also attempts are being made to make engineering components from the AMMCs but still there is no mass utility of these composites. Some areas where research is still needed before these versatile materials see the light of the day as engineering components include machining, joining, repeatability of properties on mass production, and effect of secondary processing [5]. The difficulty in meeting the industry's demand can be easily seen from the lesser number of research publications in the last few years and old researchers were finding it difficult to answer questions related to the engineering potential of these composites. The material with better property needs to be converted to a component with better properties than presently available and here lies the challenge, like (i) cast AMMCs will seldom find use as engineering components, (ii) advantages accrued in cast composites do not always hold good when processed, (iii) only selected alloying elements can withstand the rigors and give meaningful results [6].

To date, the literature at least does not show a very promising picture regarding availability of engineering components been used which are made from AMMCs. However, a few components that have seen the light of the day from AMMCs include Swash plate, tennis racket, transfer bench, foot rest, and a few automobile components like brake drum, casings, head light combos, mining equipment, and a few components for structural applications, with an intention to replace either cast iron products or Al-alloy products presently used especially aimed at weight reduction, strength, stiffness, and energy efficiency [7]. The aerospace industries, automobile and railway sectors are keenly observing the composite research development.

The enormous amount of research and development that has gone into Al-based MMCs of every possible alloy with different reinforcements establishing beyond doubt the usefulness of making composites but a choice has to be made with both the base alloy selection and reinforcement size and volume percentage for making engineering components. Another point is that all engineering components will undergo some wear and damage related degradation with use even in an otherwise non-wear environment. The Al-Cu and Al-Zn-based alloys are commonly used for making engineering components used in environments like those mentioned above.

Earlier work on Al-Cu-based composites show that these materials have been broadly investigated both in terms of knowing the mechanism of formation and properties assessment of composites, the main reinforcements being SiC and Al_2O_3. Researches with variable quantity of dispersions have been done. The Al-Cu-based MMCs have been evaluated for their tensile strength, hardness and abrasive wear properties and in most cases, have reported improvement in the hardness and strength of composites over the control alloy [8]. But, the amount of enhancement relies on the test conditions and microstructure of the MMCs. MMCs with suitable reinforcements like SiC have been able to bring a reduction in the density of the composite as compared to the base alloy. A decrease in density and an increase in strength is a major factor in increasing the strength to weight ratio of the composites. Thus, MMCs have a large scope in the aerospace and automobile industries and it can very well be the most important material of these industries in the future [2].

In the present investigation, SiC and RHA are dispersed in Al alloy having 4.5% Cu to produce the hybrid composite. Two types of composites were produced with one composite having 2% SiC and 1% RHA and the other with 4% SiC and 1% RHA. The composites were then successively tested for the mechanical properties.

2 Experimental Details

2.1 Materials Used

2.1.1 Rice Husk Ash(RHA)

The Rice Husk Ash particles used were between 50-60 microns and are the reinforcements to disperse in the matrix alloy. The rice husk was burnt completely in the presence of air and the ash was collected 24 hours later. The ash was then conditioned by heating the ash at a temperature of 300°C for 90 minutes to reduce the carbonaceous and volatile constituents of the ash. The chemical components of the RHA used is presented in the Table 1.

Table 1. Chemical component of RHA reinforcement.

Element	Silicon dioxide	Carbon	Calcium oxide	Magnesium oxide	Potassium oxide	Hematite
% wt.	91.56	4.8	1.58	0.53	0.39	0.21

2.1.2 Silicon Carbide(SiC)

The SiC used in the present work is basically a hard-refractory material with particle size less than 20 microns. SiC has excellent thermal conductivity, low thermal expansion and it displays a good thermal shock resistance.

2.1.3 Aluminium Alloy

It is used as a matrix to hold the reinforcements. The aluminium alloy for the present investigation mainly consists of copper (4.5 % wt.) along with other elements as shown in the Table 2.

Table 2. Chemical component of aluminium alloy matrix

Element	Si	Mg	Cu	Mn	Fe	Ni	Ti	Zn	Pb	Sn	Al
%wt.	0.5	0.066	4.51	0.13	0.663	0.075	0.013	0.118	0.029	0.021	Rest

2.2 Production of Composite

The ingots of Al alloy were broken into pieces and 3 kg of it was weighed and loaded into the electrical resistance furnace of 6 kW capacity. The alloy that was charged into the graphite crucible is heated to 850^{0}C till the entire charge in the crucible was melted. The furnace is brick insulated. The reinforcement particles (RHA and SiC) were preheated by placing them on top of the cover of the furnace for one hour, before introducing them into the melt, so as to remove moisture. The moulds used for casting were made of cast iron. They are preheated to 500⁰C to obtain uniform solidification. The pre-heating was done using a kerosene fuel pump burner. After the molten metal was fully melted, degassing tablet (Hexachloro methane) was added to remove the slag, flux and gas in the molten metal. Scum Powder is also added along with the degassing tablet into the molten metal to remove the slag and flux.

A stirrer made up of HSS coated with ceramic powder was lowered into the melt slowly to stir the molten metal at a speed of 700 rpm. The speed of the stirrer was controlled by means of a regulator. The average grain size of SiC particles was 63 microns whereas RHA had a grain size of 210 microns. The preheated SiC/RHA particles were added into the molten metal at a constant rate during the stirring. Stirring was continued for 5–10 minutes even after the completion of particle feeding. The melt of Al/SiC and RHA is then maintained at 850^0 C for 10 minutes without stirring and then poured into the moulds allowing for solidification successively. The details of specimen prepared are shown in Table 3.

Table 3. Specimens of composite prepared

Specimen ID	RHA quantity	SiC quantity	Al alloy quantity(Kg)
#1	Nil	Nil	3
#2	30 gm (1% by wt.)	60 gm (2% by wt.)	3
#3	30 gm (1% by wt.)	120 gm (4% by wt.)	3

2.3 Testing of Composite

2.3.1 Density Test

The density variation of the composite with the increase in SiC content for the fixed RHA quantity are shown in the Table 4 and Figure 1. The density is calculated by the direct measurement of the mass and the volume of specimens.

Table 4. Density of specimens

Height (h) mm	Diameter (d) mm	Volume (V) m^3	Mass (M) gm	Density (ρ) kg/m^3	Average Density (kg/m^3)
Specimen #1					
20.12	15.16	3.63×10^{-6}	9.507	2617.0	2622.77
20.14	15.14	3.625×10^{-6}	9.508	2622.8	
20.11	15.14	3.62×10^{-6}	9.508	2626.5	
Specimen #2					
20.13	15.12	3.614×10^{-6}	9.315	2577.5	2576.10
20.13	15.12	3.6126×10^{-6}	9.313	2577.9	
20.17	15.12	3.6215×10^{-6}	9.314	2572.9	
Specimen #3					
20.22	15.16	3.649×10^{-6}	9.341	2559.8	2545.03
20.32	15.21	3.692×10^{-6}	9.341	2530.6	
20.27	15.20	3.678×10^{-6}	9.341	2545.2	

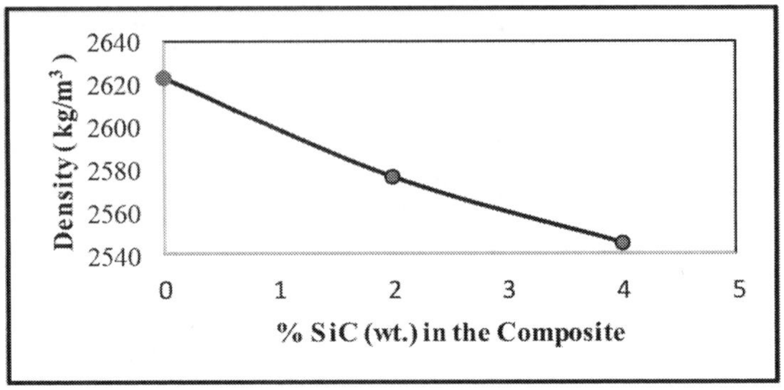

Fig. 1. Variation of density of specimens with quantity of SiC

2.3.2 Hardness Test

Hardness test was carried out using a Brinell Hardness Tester with an applied load of 500 kgf. A ball indenter was used to make the indentation on the surface of the Specimen. The diameter of the indentation was then measured and the Brinell Hardness Number(BHN) was calculated from the formula:

$$\mathrm{BHN} = \frac{2P}{\pi D\left(D - \sqrt{D^2 - d^2}\right)} \quad (1)$$

where, P= load applied=500 kgf

D=diameter of the indenter ball=10 mm

d=diameter of the indentation made on the specimen.

The results are tabulated in Table 5. And the comparison of hardness of specimens is shown in Figure 2.

Table 5. Hardness of specimens

Load (P) Kgf	Diameter (D) mm	Diameter (d) mm	BHN	Average BHN
Specimen #1				
500	10	3.5	50.32	
500	10	3.6	47.47	46.74
500	10	3.8	42.43	
Specimen #2				
500	10	3.5	50.32	
500	10	3.8	42.43	49.85
500	10	3.3	56.82	
Specimen #3				
500	10	3.4	53.43	
500	10	3.2	60.53	54.76
500	10	3.5	50.32	

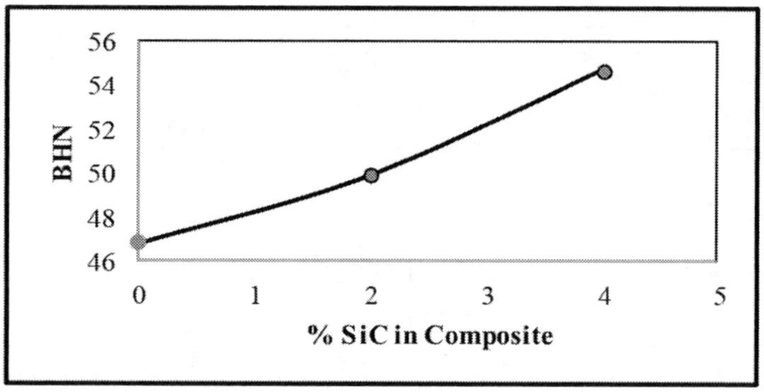

Fig. 2. Variation of BHN with increase in quantity of SiC

2.3.3 Wear Test

To know the wear behaviour of unreinforced alloy and the composite specimens in dry conditions, testing was carried out on a pin-on-disc type wear and friction monitor [DUCOM, India make; Model: TR-201CL] attached with data capturing system (Figure 3) The wear tests were conducted as per ASTM G-99 standards.

Fig. 3. Pin-On-Disc machine used for wear test

The test specimen used was a cylindrical pin (8 mm diameter and 30 mm length) which was held with its axis perpendicular to the surface of the disc, and one end of pin slid against the disc in a dry friction condition, under an axial load applied with a dead weight.

The test duration on each specimen was 10 min. A constant rpm was maintained throughout the experiment. Mass loss was noted at the end of ten minutes and the weight was increased for the next trial. Also, the frictional force acting on the surface of contact was noted at regular intervals. This value was used to calculate the average coefficient of friction of that specimen at that weight.

$$\text{Coefficient of friction } (\mu) = \text{Avg. Frictional Force in N/ Load applied in N} \quad (2)$$

The tabulated values as shown in the Table 6. This table show the wear test results for specimens. The comparison graph for the specimen is shown in Figure 4. The Variation of coefficient of friction, μ with Load applied at fixed rpm has been depicted in Figure 5.

Table 6. Results of wear test on specimen

Trial	Speed (rpm)	Load (N)	Track Diameter (mm)	Initial mass (gm)	Final mass (gm)	Friction force, F (N)	Friction Co-efficient, μ
Specimen #1							
1	500	25	60	3.755	3.752	2.84	0.116
2		35		3.752	3.747	7.06	0.206
3		45		3.747	3.741	8.08	0.1
Specimen #2							
1	500	25	60	3.392	3.390	7.2	0.293
2		35		3.386	3.383	12.7	0.369
3		45		3.383	3.380	12.4	0.280
Specimen #3							
1	500	25	60	3.963	3.962	13.26	0.540
2		35		3.962	3.960	14.68	0.427
3		45		3.955	3.953	18.32	0.415

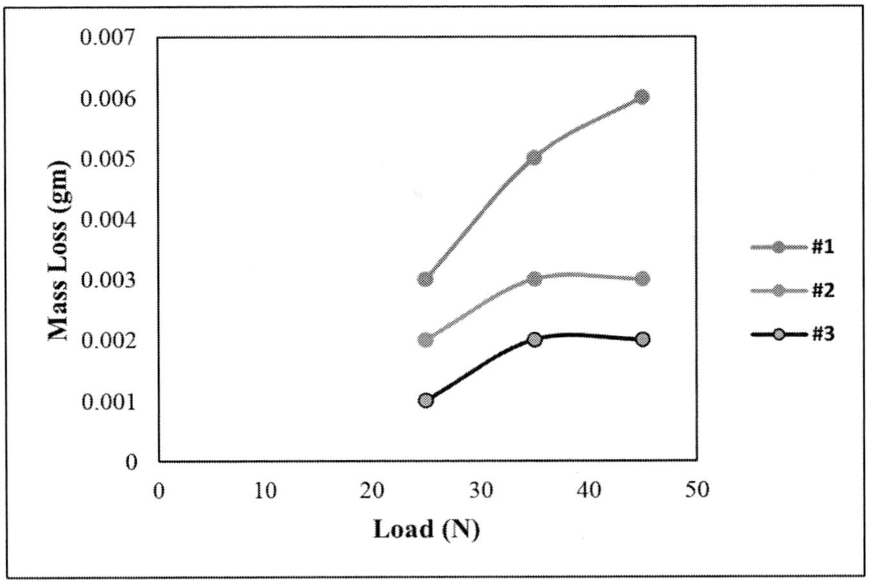

Fig. 4. Variation of mass loss with load at 500 rpm

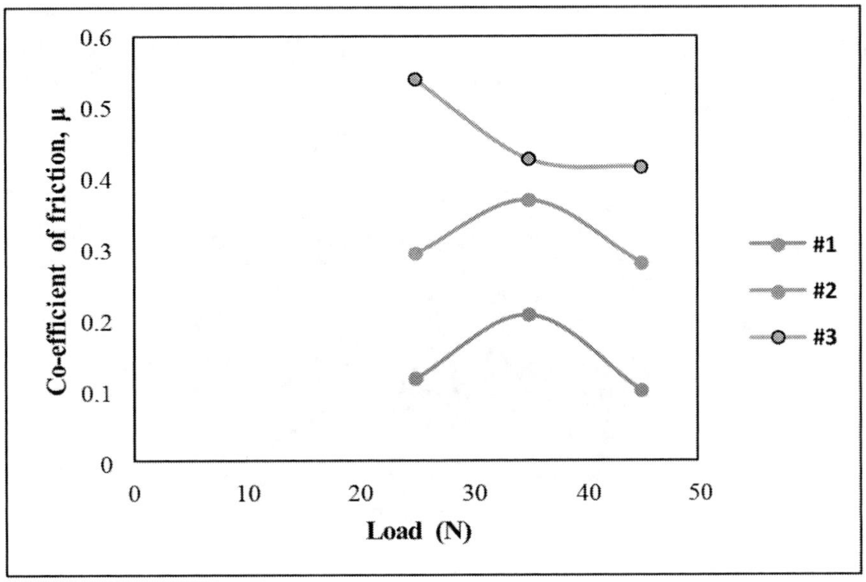

Fig. 5. Variation of Co-efficient of friction with load at 500 rpm

3 Results and Discussion

3.1 Microstructure

Figure 6 and 7 shows the optical micrographs of Al alloy and composite specimen #2, respectively. A uniform distribution of reinforcement particulates throughout the matrix alloy can be observed from these images with a very minimal porosity level. Also, there is good bonding between the matrix and the reinforcement particulates which aids for a better load transfer from the matrix to dispersed particulates. There was no agglomeration of reinforcements since sufficient care was taken to prevent agglomeration by preheating the reinforcement particulates. The micrograph shown in Figure 6 is for the pure aluminum alloy. The black spots represent $CuAl_2$ phase formed from the alloy of aluminum and copper in the matrix. They are mostly positioned along the grain boundary. Coarse grains suggest that the casting has been cooled slowly and at room temperature.

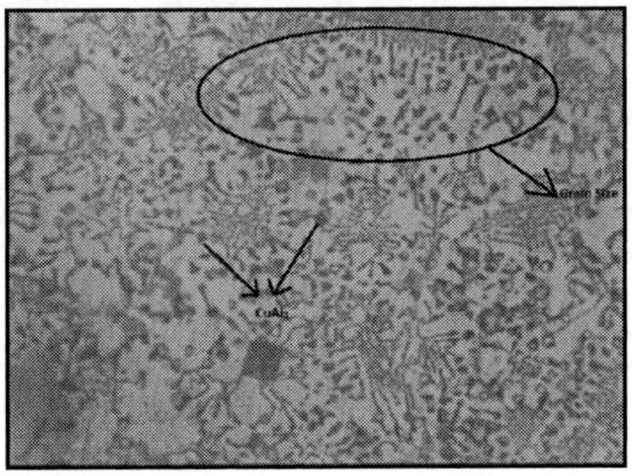

Fig. 6. Microstructure of Specimen #1(500X)

The micrograph shown in Figure 7 is for specimen #2. The SiC is black spots with a prismatic shape and sharp corners along the grain boundary. Mainly the slow cooling might have resulted in uniaxial grains.

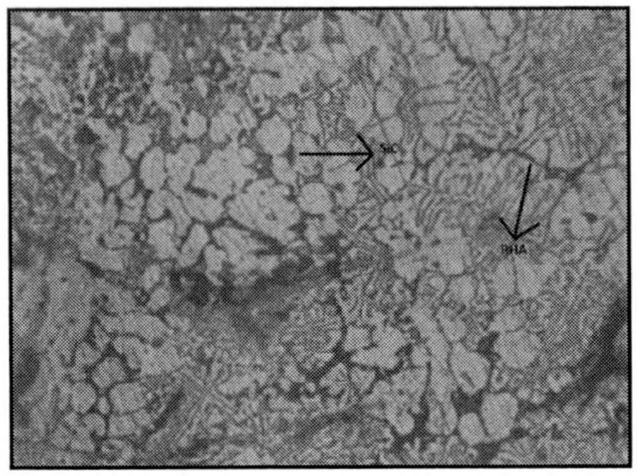

Fig. 7. Microstructure of Specimen #1(500X)

3.2 Density

As depicted in Figure 1 the density of composite decreases as the quantity of SiC is increased for the fixed amount of RHA. This trend is attributed to the increase in the level of porosity for higher amount of dispersions. The entrapment of gas while stirring, release of hydrogen, the shrinkage during solidification and existence of air pockets in the molten alloy would also decrease the density of composite [9]. Certain amount of porous structure is unavoidable during the fabrication composites. The similar nature of variation in density is noticed by some researchers in their work [10-16]. There will be always a tendency to increase density of composites due to ceramic reinforcing particles (SiC), but however there will be reduction in density because of dispersion of light weight reinforcements (RHA) in the hybrid composites.

3.3 Hardness

Hardness of the composite tends to increase with the addition of SiC over the alloy hardness. It may be because of hard SiC particles present in the matrix alloy. SiC particles along with $CuAl_2$ precipitates along the grain boundary will create more resistance for dislocation motion, and thus increasing the opposing ability to plastic deformation leading to higher hardness.

3.4 Wear

Higher amount of SiC in composite reduces the mass loss as compared to alloy without reinforcement. This is probably due to the hard nature of SiC particles that are exposed on to the rubbing contact surface preventing the amount of mass loss. Also, the coefficient of friction μ remains mostly constant and is maximum for the specimen with the highest percentage of reinforcements.

4 Conclusions

The conventional stir casting technique is successfully employed to produce the hybrid composite with uniform distribution of reinforcements. The density of the produced composite reduces as the quantity of reinforcement is increased. But the hardness increases with increase in SiC content. These particles along with $CuAl_2$ phase has contributed in opposing the dislocation motion. The SiC has significantly reduced the mass loss of composites as compared to the bare alloy which is because of hard ceramic nature of SiC and due to the exposure of these at the rubbing surface. Since the developed composites have shown better mechanical properties as compared to Al alloy without reinforcements the on-hand composite is suggested for use in the wear resistant applications.

References

1. S.D. Prasad, R.A.Krishna, Int J Adv Sci Technol. 33, 51-8 (2011)
2. H. Zuhailawati, P. Samayamutthirian, C.H. Mohd Haizu, J Phys Sci. 18, 47-55 (2007)
3. T.Senthilvelan, S. Gopalakannan, S. Vishnuvarthan, K.Keerthivaran, Adv Mater Res. 622-23, 1295-9 (2012)
4. T.V.Christy, N.Murugan, S.Kumar, J of Miner Mater Char Eng. 9(10, 57-65 (2010)
5. K.V.Mahendra, K. Radha Krishna, J Com Mater. (44)8, 989-1005 (2010)
6. S.Balasivanandha Prabhu, L.Karunamoorthy, S.Kathiresan, B.Mohan, J Mater Proce Technol. 171, 268-273 (2006)
7. M.K.Surappa, Sadhana. 281& 2, 319-34 (2003)
8. M.Adiamak, J Achevm Mater Manuf Eng.14, 43-7 (2006)
9. Jaswinder Singh and Amit Chauhan, J Mater Res Tech. 5(2), 159-169 (2016)
10. D.B.Miracle, Compos Sci Technol. 65, 526-40 (2005)
11. K.K.Alanemea, B.O.Ademilua, M.O.Bodunrin, Tribol Ind. 35(1), 25-35 (2013)
12. O.A.Olugbenga, A.A.Akinwole, IntJ Eng Technol. 2(4),212-9 (2010)
13. K.K.Alaneme, I.B.Akintundea, P.A.Olubambib, T.M.Adewalec, J Mater Res Technol. 2(1), 60-7 (2013)
14. M.Boopathi, K.P.Arulshri, N.Iyandurai, J Appl Sci. 10(3), 219-29 (2013)
15. Z.Gnjidi,D.Boi, M.Mitkov, Mater Charact. 147, 129-38 (2001)
16. J.B.Rao, D.B.Rao,N.R.M.R.Bhargava, Int J Eng Sci Technol. 2(11), 50-9 (2010)

Effect of retrogression and re-ageing heat treatment on microstructure and microhardness of aluminium 7010 alloy

Nandana M. S.[1], *Udaya Bhat K.*[1*] *and C. M. Manjunatha*[2]
[1]*Department of Metallurgical and Materials Engineering, National Institute of Technology Karnataka, India*
[2]*Structural Technologies Division, CSIR-National Aerospace Laboratories, Bangalore, India*

Abstract. Aluminium alloy 7010 is subjected to retrogression and re-ageing (RRA) heat treatment to study the influence of microstructural changes on hardness. Retrogression is performed at 190 °C for different time intervals ranging from 10 to 60 minutes. Optimum time for retrogression treatment is estimated based on the retrogression time that result with equivalent mechanical properties as that of peak aged (T6) condition. Retrogression performed for 30 minutes resulted with micro hardness of 203 HV, which is equivalent to that obtained by following T6 treatment. Microstructural characterization done with the help of transmission electron microscope (TEM) indicates RRA treatment results with the coarsened and discontinuous precipitates along the grain boundary which is similar to over aged (T7) condition, where as fine and densely populated precipitates in the matrix similar to T6 condition. Coarse and discontinuous grain boundary precipitates (GBP's) improves resistance to stress corrosion cracking. Fine and dense precipitates in the matrix ensures hardness equivalent to that of T6.

Keywords: RRA; Peak ageing; TEM; SCC; Microhardness

1 Introduction

Aluminium 7xxx series alloys is one of the high strength aluminium alloys used in aerospace structural applications mainly due to their high specific strength, damage tolerance and corrosion resistance properties [1]. They have Zn, Mg and Cu as alloying elements, these alloys are precipitation hardened alloys, with general precipitation sequence in the order of supersaturated solid solution (SSSS) $\rightarrow$ GP zones $\rightarrow \eta'$ $\rightarrow \eta$ ($MgZn_2$) [2]. The η' precipitates formed during the precipitation sequence is the metastable precipitate of $MgZn_2$, which is semicoherent with the matrix and is the major strengthening phase of the alloy. These alloys were earlier used in peak ageing (T6) condition, but during service, they are prone to stress corrosion cracking (SCC) due to the microstructural features like GBP's and coarse intermetallic particles [3]. Hence they are generally used in T7 condition which improves resistance to SCC. But the alloy during over ageing, looses its peak strength by about 10-15 % [4]. A three step heat treatment called retrogression and re-ageing is found out. It imparts the alloy with the strength characteristics of the T6 treated condition and SCC resistance of T7 treated condition [5].

* Corresponding author: udayabhatk@gmail.com

A substantial amount of research work has been put forward in developing a mechanism for improving the stress corrosion properties of 7xxx series alloys. The improvement in the SCC is related to the grain boundary characteristics of the alloys. The increase in the size of the GBP's during coarsening stage and more spacing of them are responsible for the improved SCC resistance [6,7]. The mechanism of SCC is related to the formation of corrosion pits near the grain boundary segregations. In the presence of stress, these corrosion pits act as stress raisers leading to intergranular cracking [8].

The high strength aluminium alloys are used as structural materials in aircrafts, such as bulk heads and lug joints [9]. Since aircrafts are frequently exposed to saline environment during service, the alloys used in T6 treated condition, are more prone to SCC. The alloys heat treated to RRA condition are more resistant towards SCC, retaining the strength of T6 alloy. RRA studies on AA7010 with retrogression temperature in the range of 180 - 230 °C has been reported [6]. But microstructure – mechanical property correlation for the RRA treated alloy is less studied. The aim of the present research work is to optimize the time of retrogression at 190 °C so as to obtain equivalent hardness as that of T6 treated sample and to study the microstructural changes during RRA.

2 Experimental Procedures

Material used for the current study is a high strength aluminium alloy, 7010 received in T7451 (over aged) condition. The alloy is composed of alloying elements Zn-6.3Mg2.21Cu1.65Zr0.124Fe-0.21Si-0.073 and rest is aluminium (wt %). The alloy is subjected to solutionizing at 490 °C for 6 h and quenched in water, subsequently aged at different tempering conditions. The alloy is peak aged by heat treating at 120 °C for 24 h. Retrogression is performed by re heating the peak aged sample at 190 °C for different time intervals from 10 – 60 min. The alloy is further re aged at 120 °C for 24 h. Optimized temperature of reversion is evaluated based on the equivalent hardness of RRA sample as that of T6 treated sample. The tempering conditions adopted during the heat treatment are listed in Table 1. Heat treatment of alloy is successfully performed in a salt bath furnace.

Mechanical characterization of the heat treated alloy samples are done by performing the microhardness tests using Vicker's microhardness instrument (make Shimadzu, model: G21). The indentation load is 500 gms and time of indentation maintained for 15 secs. Microstructural characterization of the heat treated samples is done with the help of transmission electron microscope (TEM make JEOL, model: JEM 2100) operating at 200 kV.

Table 1. Temper notations and corresponding heat treatment conditions

Temper notations	**Solutionization**	**Ageing treatment (error ± 2 °C)**
T6 (Peak ageing)	490 °C - 6 h	120 °C - 24 h
RRA (Retrogression and re-ageing	490 °C - 6 h + 120 °C - 24 h + Retrogression at 190 °C	Re-ageing at 120 °C - 24 h

3 Results and Discussion

The microstructure of the alloy sample in as received (T7451) condition is represented in Fig.1. The microstructure observed under light microscope reveals the grains oriented along the rolling direction with the average grain size of 12 μm. The evolution of hardening precipitates inside the matrix and along the grain boundaries is clearly evident from figure 1.

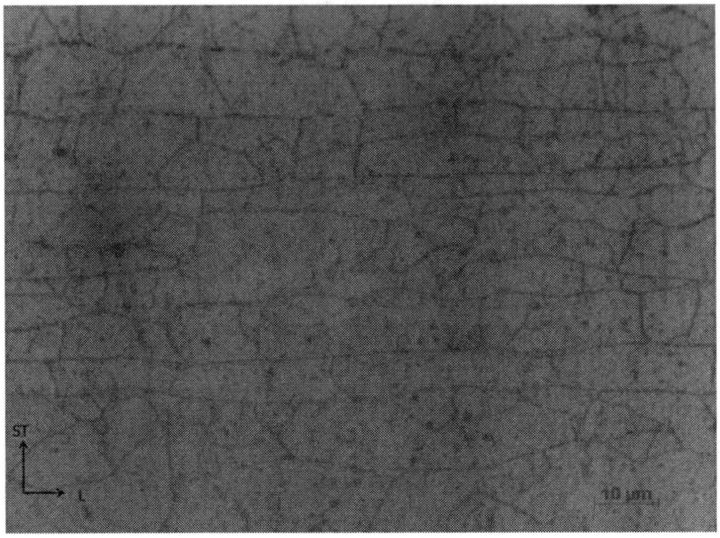

Fig. 1. Microstructure of AA7010 in T7451 (as received), etched with Kellers's reagent

The bright field TEM images corresponding to the matrix and the grain boundaries of T6 treated alloy is shown in Fig. 2. The matrix of the T6 treated sample is highly occupied with GP zones and fine scale metastable precipitates ή of size 3-5 nm. The GP zones are the clusters of solute atoms which are coherent with the matrix and ή are the major strengthening phase of the high strength aluminium alloys, which are semicoherent with the matrix [10]. The grain boundary is occupied with continuous precipitates. The PFZ (precipitate free zone) size of T6 treated sample is in the range of 25-30 nm.

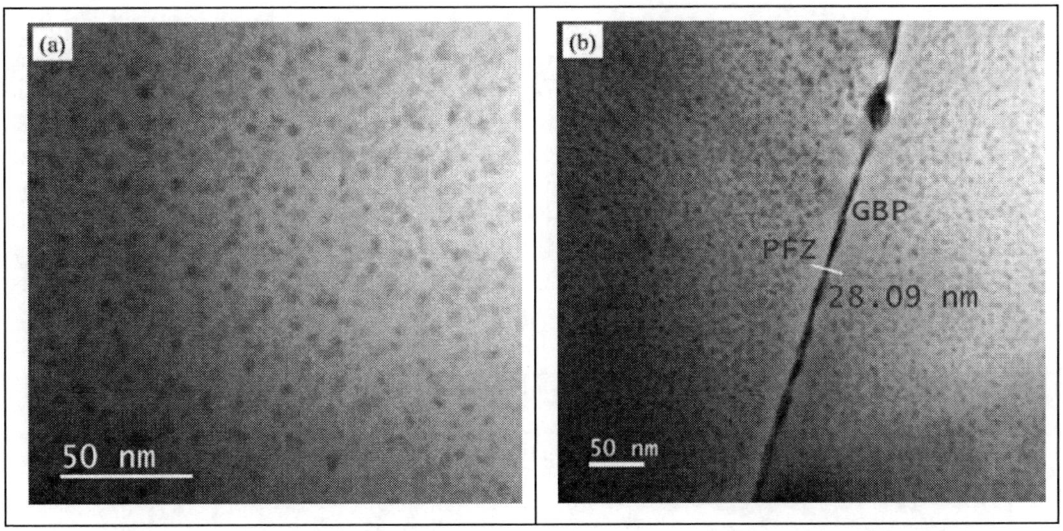

Fig. 2 TEM bright field micrographs of AA7010 in T6 condition: (a) matrix, (b) grain boundary

The microhardness of the T6 treated sample is 196 HV, represented by straight line in Fig. 3. Retrogression is performed at a temperature higher than the peak ageing temperature, at 190 °C, during which the GP zones gets completely dissolved in the matrix and fine scale ή precipitates are partially dissolved. This is evident from the valley in the hardness plot [11,6]. With the increase in the retrogression time, the hardness gets increased up to a limit from where coarsening and transformation of ή into stable η precipitates takes place [12,13]. It is evident from the development of secondary peak as noticed in the hardness plot in Fig. 3. Upon re-ageing the retrogressed sample, the dissolved solutes will re precipitate and grow. Presence of re nucleated fine scale precipitates in addition to the partially dissolved precipitates that remained during retrogression makes the matrix denser compared to the matrix of T6 treated sample as observed in Fig. 4(a).

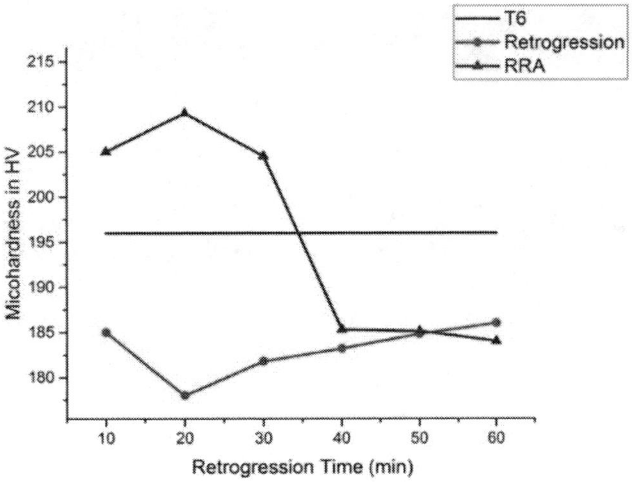

Fig. 3 Plot of microhardness vs retrogression time for AA7010 under different temper conditions.

The size of the matrix precipitates vary from 5-8 nm in RRA treated sample. During re-ageing, the existing continuous GBP's gets coarsened and get arranged in a discontinuous fashion [Fig. 4(b)]. The size of GBP's varies from 30-40 nm and PFZ size varies from 40-60 nm. Whereas, in T7451 (as received) condition the matrix is partly occupied with fine precipitates of size 7-10 nm and partly occupied with large and coarsened precipitates of 16-40 nm in size [Fig. 4(c,d)]. This observation is similar to the reported literature [14]. The coarsened precipitates are responsible for lowering the strength in over aged alloys. Apart from coarsening of GBP's the electrochemical potential of the GBP's gets increased by diffusion of more Cu into them, thereby delaying the formation of corrosion pits. This is expected to increase the resistance the alloy towards SCC [15,16].

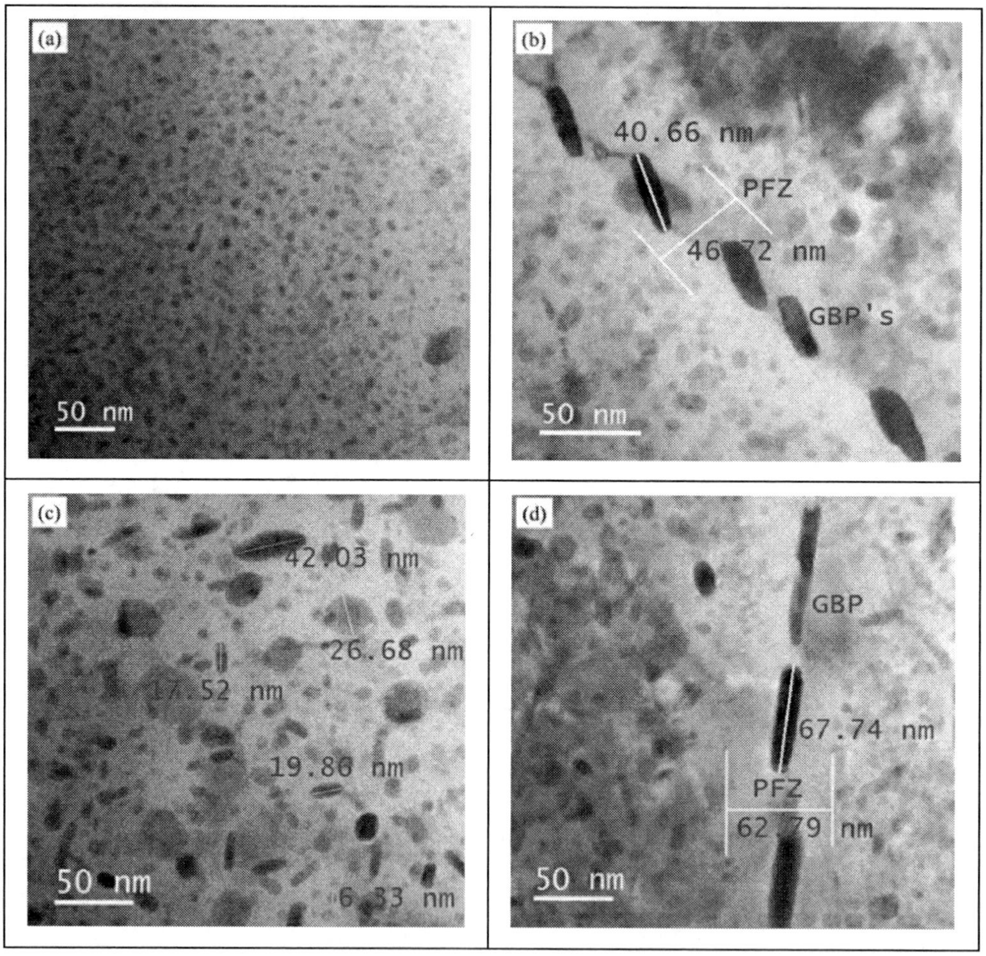

Fig. 4 TEM bright field micrographs of AA7010: (a,b) RRA (retrogressed at 190 °C for 30 min), (c,d) T7451 (as received)

4 Conclusions

1. The matrix of the RRA treated sample is more densely occupied with strengthening precipitates (ή) compared to T6 treated sample, hence hardness of the RRA treated samples is much higher than the standard peak aged sample.
2. The grain boundary precipitates of RRA treated sample is coarsened similar to that of the over aged sample which will influence in imparting better corrosion resistance to the alloy.
3. The isolated GBP's will enhance the resistance to crack propagation during SCC.

Acknowledgements

Authors are thankful to The Director, CSIR-NAL for facilitating to conduct this research work. One of the authors Mr. Nandana thanks the Director, NITK for the fellowship grant.

References

1. J. C. Williams, E. A. Starke, Acta Mater, **51**, 5775–5799 (2003)
2. G. Peng, K. Chen, S. Chen, H. Fang, Mater. Sci. Eng. A, **528**, 4014–4018 (2011).
3. M. B. Kannan, V. S. Raja, J mater Sci, **42**, 5458–5464 (2007).
4. M. Dixit, R. S. Mishra, K. K. Sankaran, Mater. Sci. Eng. A, **478**, 163–172 (2008).
5. C. Bora, US3856584 A, (1974)
6. F. Viana, A. M. P. Pinto, H. M. C. Santos, A. B. Lopes, J. Mater. Process. Technol, **92–93**, 54–59 (1999)
7. Y. P. Xiao, Q. L. Pan, W. B. Li, X. Y. Liu, Y. B He, Mater. Des, **32**, 2149–2156 (2011)
8. J. S. Robinson, Mater. Sci. Eng. Tech, **390**, 385–390 (2003)
9. J. Schijve, Int. J. Fatigue, **31**, 998–1011 (2009)
10. J. Buha, R. N. Lumley, A. G. Crosky, Mater. Sci. Eng. A, **492**, 1–10 (2008).
11. M. Kanno, I. Araki, Mater. Sci. Technol, **10**, 599–603 (1994).
12. N. C. Danh, K. Rajan, W. Wallace, Metall. Trans, **14**, 1843 (1983)
13. T. Marlaud, A. Deschamps, F. Bley, W. Lefebvre, B. Baroux, Acta Mater, **58**, 4814–4826 (2010)
14. P. Kumar Rout, M. Ghosh, K. Ghosh, Procedia Mater. Sci, **5**, 1214–1223 (2014).
15. M. Angappan, V. Sampath, B. Ashok, V. P. Deepkumar, Mater. Des, **32**, 4050–4053 (2011).
16. T. Marlaud, A. Deschamps, F. Bley, W. Lefebvre, B. Baroux, Acta Mater, **58**, 248–260 (2010).

A study of influence of material properties on magnetic flux density induced in magneto rheological damper through finite element analysis

T. M. Gurubasavaraju[1*], *Kumar Hemantha*[1] and *Mahalingam Arun*[1]

[1]*Department of Mechanical Engineering, National Institute of Technology Karnataka, Surathkal, India.*

Abstract. Magnetorheological fluids are smart materials, which are responsive to the external stimulus and changes their rheological properties. The damper performance (damping force) is dependent on the magnetic flux density induced at the annular gap. Magnetic flux density developed at fluid flow gap of MR damper due to external applied current is also dependent on materials properties of components of MR damper (such as piston head, outer cylinder and piston rod). The present paper discus about the influence of different materials selected for components of the MR damper on magnetic effect using magnetostatic analysis. Different materials such as magnetic and low carbon steels are considered for piston head of the MR damper and magnetic flux density induced at fluid flow gap (filled with MR fluid) is computed for different DC current applied to the electromagnetic coil. Developed magnetic flux is used for calculating the damper force using analytical method for each case. The low carbon steel has higher magnetic permeability hence maximum magnetic flux could pass through the piston head, which leads to higher value of magnetic effect induction at the annular gap. From the analysis results it is observed that the magnetic steel and low carbon steel piston head provided maximum magnetic flux density. Eventually the higher damping force can be observed for same case.

1 Introduction

The magnetorheological dampers are smart devices, whose damping force can be varied upon a change in the current. The MR damper force is highly influenced by the amount of magnetic flux induced in the MR damper due to the current supplied to the electromagnetic coil. Many investigations on computing the damping force through different approach have been investigated in the literature. Nguyen et al. (2008) [1] developed a novel optimization method based on finite element analysis to compute the optimal parameters. Zhang et al. (2006) [2] developed a method to design the MR damper constructed on the magnetic design through FE analysis. The MR damper design was experimentally verified. Tu et al.

* Corresponding author: gbrajtm@gmail.com

(2007) [3] conducted FE analysis using MR damper axisymmetric model in ANSYS to calculate the total magnetic field developed in the MR damper. Guo et al. (2003) [4] carried out optimal design of MR valve though FE analysis and this analysis was further coupled with a fluid plastic model to estimate the MR valve performance. Hu et al. (2015) [5] considered double electromagnetic coil MR damper for the FE based optimization to identify optimal geometric variables. Also, the effect of different piston configurations such as geometric features, shape and size on the damping force was investigated. Paul et al. (2014) [6] presented an investigation on characteristics of MR fluid and its effect on the MR damper through commercially available FE software and the analysis results obtained were validated with the experimental results. Gurubasavaraju et al. (2017) [7] evaluated the optimal geometric parameters of the monotube MR damper operating under shear mode through response surface and multi objective genetic algorithm optimisation technique. In the present paper, the variation in the damper performance due to usage of different materials for the piston head is investigated.

2 Finite Element Analyses

The FE analysis is employed particularly for calculating induced magnetic field in the flow gap of MR damper for various currents. The magnetostatic module in ANSYS Workbench is employed for the analysis. The dynamic characteristics of a damper is dependent on the total magnetic flux developed in the damper. There are still other parameters, which are having impact on generation the magnetic flux, such as materials used for the MR damper components, geometric dimensions etc. The materials used for piston head in the present FE analysis are given in table 1. The CAD model of the MR damper and boundary conditions for the analysis are depicted in figure 1.

There are totally 4330 tetrahedron elements generated in the meshed model. The electromagnetic coil incorporated in the piston head consists of 600 turns of copper wire. There are few assumptions are made in the analysis such as zero flux leakage from the cylinder and flux lines are parallel to the center plane. DC current in the range of 0 to 1.0 A is supplied to the electromagnetic coil with an increment of 0.25 A to compute the magnetic field in the fluid flow gap for the respective current.

Table 1. MR damper model details.

Sl.No	Model	Piston head material
1	'Model 1'	M36 Steel
2	'Model 2'	SA1020 Steel
3	'Model 3'	SST430 Steel

The materials chosen for different components of MR damper are shown the table 2. The material properties are also an important parameter, which influences the results of the analysis. In this study, all components of MR damper materials are kept unaltered except the piston head material. The material which is having high magnetic permeability could yield the higher value of magnetic flux. The change in the rheological properties of the MR fluid on application of magnetic flux is realized by calculating the shear yield stress. The magnetic field obtained from FEA is further used for estimating the field dependent yield stress using the equation 1 (Han et al. (2014) [8]). Calculated yield stress further employed to evaluate the damping force.

$$\tau_y(pa) = 39.71B_0^4 - 132.45B_0^3 + 119.2B_0^2 + 10.31B_0 + 0.11 \quad (1)$$

where, t_y is field dependent yield stress and B_0 is magnetic flux density (T).

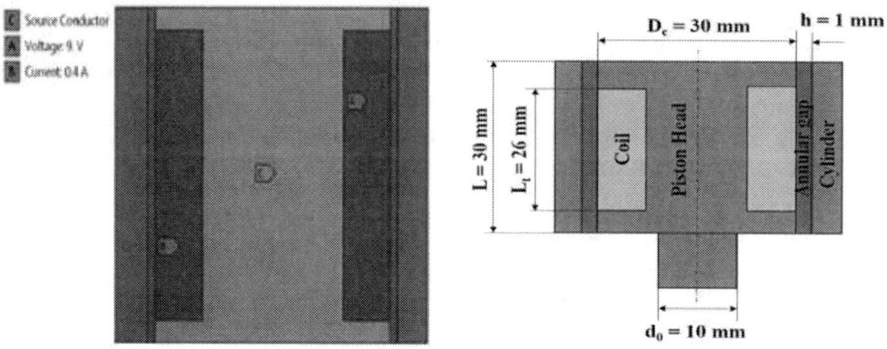

Fig. 1. MR Damper model dimensions

Table 1. MR damper model details.

Sl.No	Part name	Material	Relative Permeability
1	Piston Head	SA1020, M36, SST430	B-H Curve (Figure 2b, 2c, 2d)
2	Coil	Copper wire	1
3	Cylinder	Stainless Steel	2000

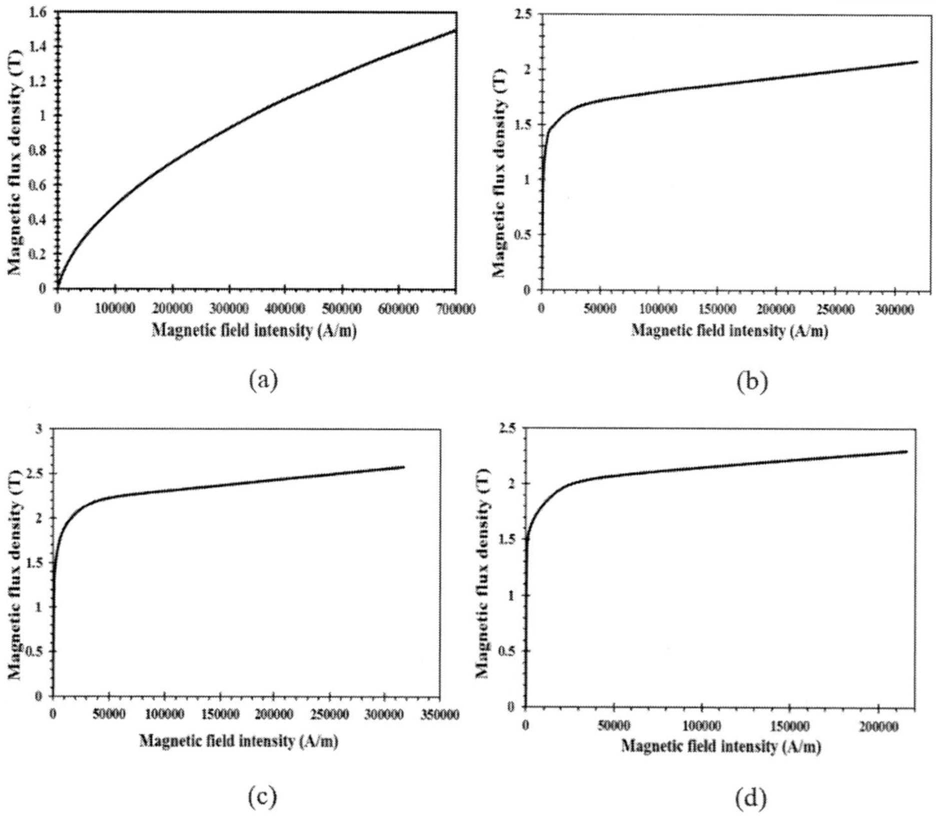

Fig. 2. B-H Curve of (a) MRF-132 DG (Lord Co.) [9] (b) SA1020 (c) SST430 and (d) M36

2.1 Results of FE analysis

The contour plots of magnetic flux density computed from the magnetostatic analysis are shown in figure 3. The variation of magnetic flux at the fluid flow gap is extracted and plotted with respect to the fluid flow gap length as shown in the figure 4. It can be observed from the results that; the magnetic flux increases with increase in DC current. The increase in the magnetic flux is almost linear up to 0.3 A and further it tends to saturation level with increase in current as depicted in figure 4 (d).

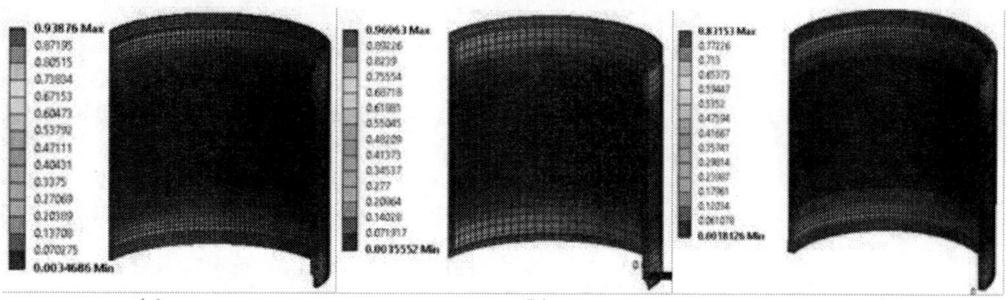

(a) (b) (c)

Fig. 3. Contour plots of magnetic flux density induced at the fluid flow gap of (a) 'Model 1', (b) 'Model 2' and (c) 'Model 3' MR dampers.

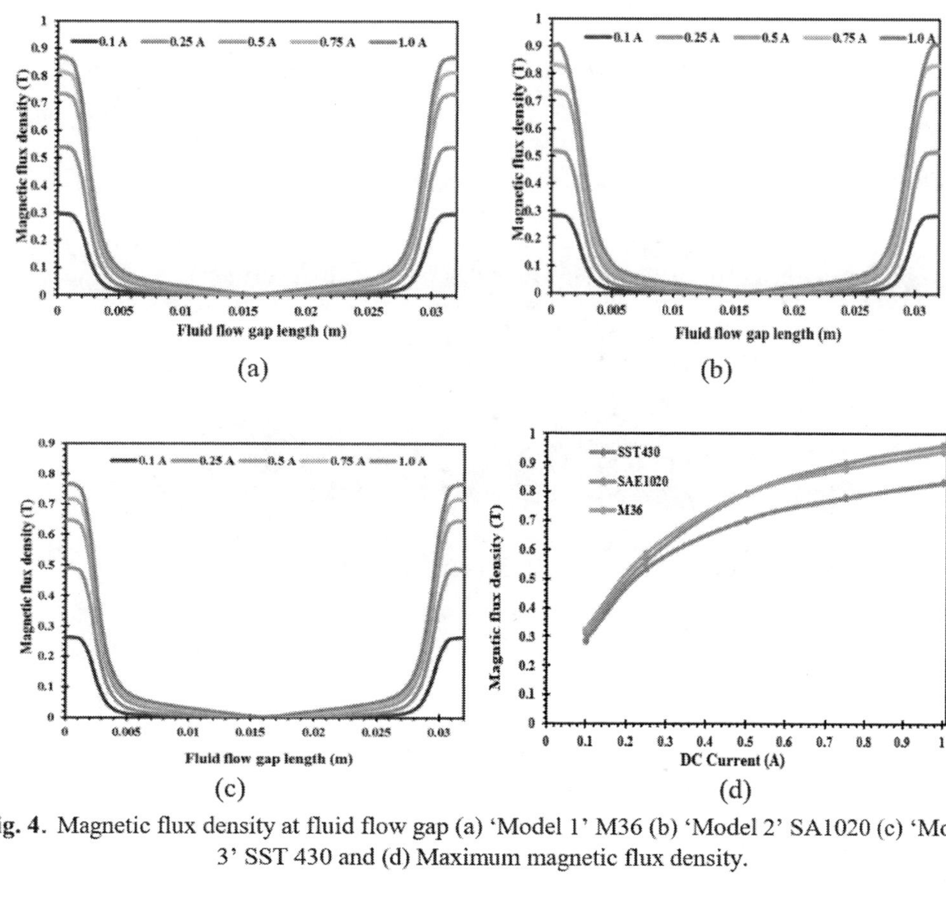

Fig. 4. Magnetic flux density at fluid flow gap (a) 'Model 1' M36 (b) 'Model 2' SA1020 (c) 'Model 3' SST 430 and (d) Maximum magnetic flux density.

From the figure 4, it can be observed that the value of magnetic flux of 'model 1' and 'model 2' are nearly same and there is considerable difference between the model 3 and other models. The maximum magnetic flux is obtained for 'model 2' MR damper, because magnetic permeability of SA 1020 is significantly higher than the other two materials. Hence, maximum value of magnetic flux can be obtained by employing SA1020 material as the core for the electromagnetic coil.

3 Estimation of damping force

Equations (2) to (6) are used for computing the damping force of MR damper. The characteristic of MR damper is represented by force versus velocity plots. The damper force is the combined effect of force due to shear, viscous and frictional effect. The force due to shear stress is referred as controllable force and force due viscosity and friction forces are referred as uncontrollable forces. (Xu et al. (2013) [10] and Ronge et al. (2011) [11]). Figure 5 shows schematic diagram of MR damper.

$$F = \left(2.07 + \frac{12Q\mu}{12Q\mu + 0.4wh^2\tau_y}\right)\frac{\tau_y L A_p}{h}\, sgn(\dot{u}) + \left(1 + \frac{wg\dot{u}}{2Q}\right)\frac{12\mu Q L A_p}{wh^3} \quad (2)$$

$$w = \pi\left(\frac{h}{2} + D + \frac{h}{2}\right) \quad (3)$$

$$A_p = \frac{\pi(D^2 - d_0^2)}{4} \quad (4)$$

$$\mu = 0.0006\dot{\gamma}^{-0.6091} \quad (5)$$

$$\dot{\gamma} = \frac{v}{2*h} \quad (6)$$

where,

$\dot{\gamma}$ = Shear rate, L = Effective pole length (mm), w = Circumference of fluid flow gap (mm), D = Piston head diameter (mm), h = Fluid flow gap (mm), μ = Viscosity (Pa-s), τ_y = Shear stress (Pa), d_0 = Piston rod diameter (mm), Q = Flow rate (mm^3/s), v = Velocity (mm/s), D_c= Cylinder diameter (mm) and A_p = Effective piston head area (mm^2)

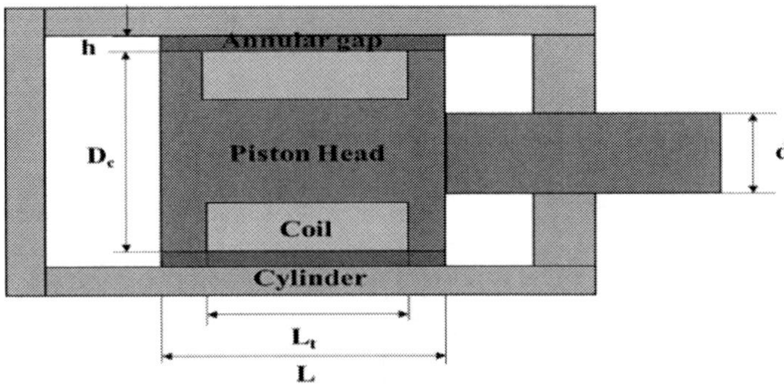

Fig. 5. Schematic diagram of MR damper

The force induced in the MR damper is calculated for 0.1 A, 0.25 A, 0.5 A, 0.75 A and 1.0 A current and for a piston velocity of 10 mm/s. Figure 6 illustrates the force versus velocity curves of three models. The maximum force developed in damper increases with current supplied to electromagnetic coil in each model of MR damper. The change in the damping force is significant due to increment in current. The maximum force obtained in case of ‘model 1’, ‘model 2’ and ‘model 3’ are 209N, 210 N and 192 N respectively for constant current of 1.0 A. Since magnetic flux density have direct influence on the damping force of MR damper, the damper which has higher value of magnetic flux yields higher damping force.

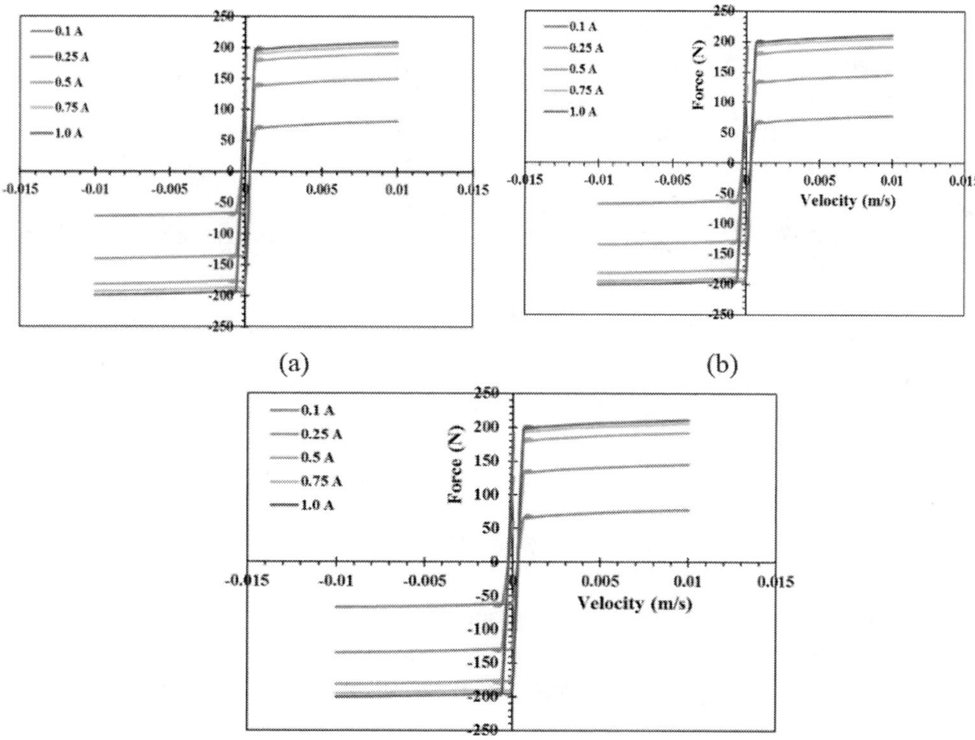

(c)

Fig. 5. Force versus velocity of MR damper for (a) ‘Model 1’ (M36) (b) ‘Model 2’ (SA1020) and (c) ‘Model 3’ (SST 430).

4 Conclusions

In the present work, piston head material influence on the damping force of the MR damper was evaluated through computational and analytical methods. Three different materials were considered as the material for piston head viz., M36, SA1020 and SST 430. Initially, the magnetic flux induced in the annular gap for different DC current input were computed through finite element analysis using ANSYS. The obtained magnetic flux was used for obtaining damper force versus velocity behavior through analytical method. It was observed that; higher value of magnetic flux can be obtained by employing SA1020 as the piston head material compared to other two materials. Consequently, the ‘model 2’ damper with SA1020 as the piston head material resulted in maximum damping force.

Acknowledgment
The authors acknowledge IMPRINT project No. IMPRINT/2016/7330 titled with “Development of Cost Effective Magneto-Rheological (MR) Fluid Damper in Two wheelers and Four Wheelers Automobile to Improve Ride Comfort and Stability” funded

by Ministry of Human Resource Development and Ministry of Road Transfer and Highways, Govt. of India.

References

1. Q. H. Nguyen, S. B. Choi, and N. M. Wereley, "Optimal design of magnetorheological valves via a finite element method considering control energy and a time constant," Smart Mater. Struct, 17(2), 025024, 2008.
2. H. H. Zhang, C. R. Liao, W. M. Chen, and S. L. Huang, "A magnetic design method of MR fluid dampers and FEM analysis on magnetic saturation", J. Ingt Mtrl. Sys and Struct, 17(8-9), 813-818, (2006).
3. S. H. Tu, and W. S. Jeng, "Finite Element Analysis of a Magnetorheological Fluid Damper." CSME conference, (2007).
4. N. Q. Guo, H. Du, and W. H. Li, "Finite element analysis and simulation evaluation of a magnetorheological valve," Int. J. adv. Manuf. Tech, 21(6), 438-445, (2003).
5. G. Hu, Z. Xie, W. Li, "Optimal design of a double coil magnetorheological fluid damper with various piston profiles" In11th World Congress on Struct. Multidis. Opt., pp. 1-6 Jun (2015).
6. D. Paul, A. Moinuddin, M.M.N. Islam, M. D. Paul, M. A. Moinuddin, and M. M. N. Islam, "Finite Element Analysis and Simulation of a Magneto-Rheological Damper," International Journal, 1, 12-19, (2014).
7. T. M. Gurubasavaraju, H. Kumar, and M. Arun, "Optimisation of monotube magnetorheological damper under shear mode," J. Braz. Soc. Mech. Sci. Eng, 39(6), 2225-2240, (2017).
8. S. B. Choi, Y. M. Han, "Magnetorheological fluid technology: applications in vehicle systems. CRC Press, (2014).
9. http://www.lordmrstore.com/_literature_231215/Data_Sheet_-_MRF-132DG_Magneto-Rheological_Fluid.
10. Z.D. Xu, L.F. Sha, X.C. Zhang, H.H. Ye, "Design, performance test and analysis on magnetorheological damper for earthquake mitigation". Sruct. Control. Hlth, 20, 6, pp. 956-70, Jun (2013).
11. B.P. Ronge, S.R. Kajale, and P.M. Pawar, "Theoretical and Experimental Analysis of MR Fluid Dampers," (LAP Lambert Academic publishing, 2011).

Effect of SiC Nano powder on Multiaxial Woven and Chopped Randomly Oriented Flax/Sisal Fiber Reinforced composites

Ganesh R. Kalagi[1*], *Rajashekar Patil*[1], *Sunil Kumar Shetty*[2] and *Madhukara Nayak*[3]

[1]*REVA University Bengaluru, Karnataka, India*
[2]*NMAMIT, Nitte, Karnataka, India*
[3]*SMVITM, Bantakal, Karnataka, India*

Abstract. A study has been carried out to investigate effect of SiC Nano powder on tensile and impact properties of Multiaxial layers of Flax and Sisal fiber reinforced composites and randomly oriented chopped Flax and Sisal fiber reinforced composites. It has been observed that tensile strength and impact strength were improved using 6% of SiC Nanopowder into Multiaxial layer (+45°/-45°, 0°/90°) of Flax and Sisal where as randomly oriented chopped Flax and Sisal fiber reinforced composites are improved in its stiffnes for the same composition of fiber, epoxy and SiC Nano powder. SEM Analysis are done to analyse the distribution of SiC in both Multiaxial layers of Flax and Sisal fiber reinforced composites and randomly oriented chopped Flax and Sisal fiber reinforced composites.

1 Introduction

Natural fiber reinforced composites are alternative materials for many applications such as small wind turbine blades, automobile applications, and aerospace applications [1]. Natural fibers shows superior properties such as tensile, flexural and impact than glass fiber [2]. Low cost, light weight, easy fabrication, ecofriendly, low manufacturing cost are the advantages of natural fibers [3, 4]. But high moisture absorption is major drawback of natural fiber [7]. European cars are already using Natural fiber reinforced composites due to more favorable economic, environmental and social aspects of vegetable fiber [9]. Also they are using them in trunk liners, instrumental panels, interior roofs, and door panels. Low level of volume fraction provides higher modulus of elastity and good mechanical strength under tensile and flexural loadings but also have values of apparent density, apparent porosity and water absorption [5,6]. The addition of SiC can improve the stiffness by maintaining its low density [8,]. The strength of short fiber composites depends on the type of fiber matrix, fiber orientation, fiber length, fiber concentration and the bonding between the fiber and matrix [10].

* Corresponding author: ganeshkalagi082@gmail.com

1.2 Literature Review

Mechanical properties of Sisal and Flax fiber reinforced composite are increses by selecting suitable composition of fiber and resin [11].

Sisal fiber gaves good potential as reinforcement in polymer composites. Low density and high specific properits of Sisal fiber, composites taking into account, these filaments may have great applications in the automotive and transportation industry [12].

A experimental investigation on effect of orientation of alkaline treated fiber reinforced composites. The charpy impact strength in the orientation 90° yielded the maximum impact strength. The orientation 90° shows the better mechanical properties compare to 0/90° and 45° [13].

Experimental investigation has been studied on the tensile properties of Sisal and coconut spathe reinforced composites. The distinctive rate of filaments is taken, because of the low density the composites can be viewed as a helpful light weight material [14].

The mechanical properties of Sisal and Flax fiber reinforced composites are incredibly affected by alkalization trématent. Alkaline trématent was found be effective in improvement of tensile and impact properties of natural fibres.

2 Fabrication

2.1 Materials used

The raw materials used in this work are, Sisal fiber, Flax fiber, SiC Nano powder, Epoxy and Hardener

2.2 Multiaxial Layers of Flax/Sisal Reinforced with Epoxy and SiC Nano powder

Flax and Sisal fiber reinforced composites are fabricated by hand layup process into dimensions of length and breadth of 300×200mm was used to prepare the specimen. The composite specimen consists of totally 6 multiaxial woven layers of Flax and Sisal at different angles of +45°/-45° and 0°/90° as shown in table 2.1. Specimens were fabricated using hang layup by taking 17% of Flax ,17% of Sisal and 6% of SiC Nano powder as shown in figure 2. 1..

After composite materials dried completely, unwanted surfaces are to be removed using secondary operations and also cut the materials as per the required ASTM standards.

Table 1. Details of multiaxial layers of Flax and Sisal

Layers	Angle	Fiber Specification
Layer 1	0°/90°	Sisal
Layer 2	+45°/-45°	Flax
Layer 3	0°/90°	Sisal
Layer 4	+45°/-45°	Flax
Layer 5	0°/90°	Sisal
Layer 6	+45°/-45°	Flax

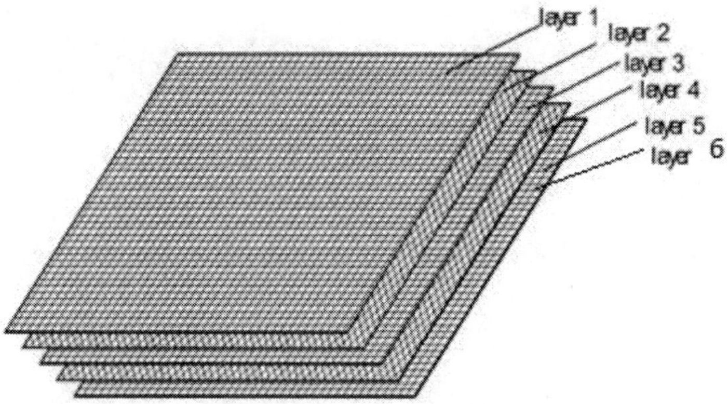

Fig. 1. Multiaxial layers of Flax and Sisal

2.3 Randomly oriented Chopped Flax/Sisal Reinforced with Epoxy and SiC Nano powder

Sisal and Flax were cut into a small length of 4 to 6 mm. Dimensions of length and breadth of 300×200mm was used to prepare the specimen. These chopped fibers are then fabricated in the ratio of 17% of Flax, 17% of Sisal and 6% of SiC Nano powder.

3 Mechanical Test

3.1. Tensile specimen

Tensile strength and young's modulus were determined by taking average value of three specimens and ASTM D3039 is used for testing tensile strength of composites. Specimens are fabricated according to ASTM Standard as shown in figure 2.

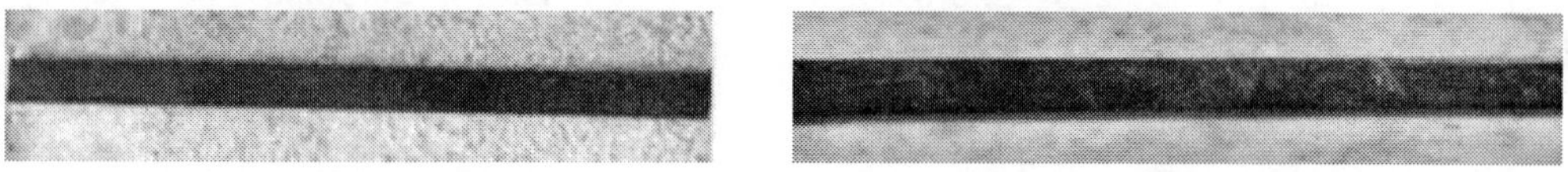

(a) (b)

Fig. 2. Fabricated Specimens for Tensile Test as per ASTM Standards
(a) Woven Flax and Sisal Fiber Reinforced Composite; (b) Randomly oriented Flax and Sisal fiber reinforced composite

3.2. Impact specimen

Impact strength of composite material is determined by by taking average value of three specimens carrying followed by ASTM Standards. ASTM E23-15b is used for testing impact strength of composites. Specimens are fabricated according to ASTM Standard as shown in figure 3.

(a) (b)

Figure 3: Fabricated Specimen for Impact Test as per ASTM Standards (a) Woven Flax and Sisal Fiber Reinforced Composite; (b) Randomly oriented Flax and Sisal fiber reinforced

4 Result and discussion

4.1. Tensile test

It is found that average Tensile strength of multiaxial woven Flax and Sisal fiber reinforced with epoxy and SiC is 28.13 N/mm², and average tensile strength of randomly oriented Flax and Sisal fiber reinforced composites is 13.183 N/mm². Tensile strength of composite materials increased by 53.14% for multiaxial layers of Flax and Sisal are increased.
Strain rate for multiaxial layers of Flax and Sial reinforced composites are 65% more than that of randomly oriented chopped fiber of Flax and Sisal. But stiffness of randomly oriented chopped fiber of Flax and Sisal reinforced composites are 36.12% better than multiaxial woven Flax and Sisal fiber reinforced composites.

4.3. Impact test

The impact strength of Flax and Multiaxial Woven Sisal fiber reinforced with SiC resulted in higher strength than the chopped fiber reinforced composites. Multiaxial layers of different angled woven fibers are ability to absorb more energy before its failure. Impact strength of composite material is improved by 62.47% by using multiaxial layers of Flax and Sisal Reinforced with SiC.

4.4 SEM Analysis

From the following Figure, Figure. a shows sample's surface topography of multiaxial Woven Flax/Sisal fiber reinforced with SiC Nano powder and figure.b for Chopped Flax and Sisal fiber reinforced composites.
The topography properties of the fabricated composite surface show the variation in mechanical properties through phase information of the fabricated composites specimens. From the SEM analysis on both composites (Woven and Chopped fiber), It is observed that distribution of fiber and resin is uniform in Woven cloth fiber composite which is resulted in improving impact strenth of composite materil. Where as in case of Chopped fiber composite the fiber and SiC are randmloy distributed and resulted in poor impact strenth but resukted in more stiffer than woven cloth fiber reinforced comosite.

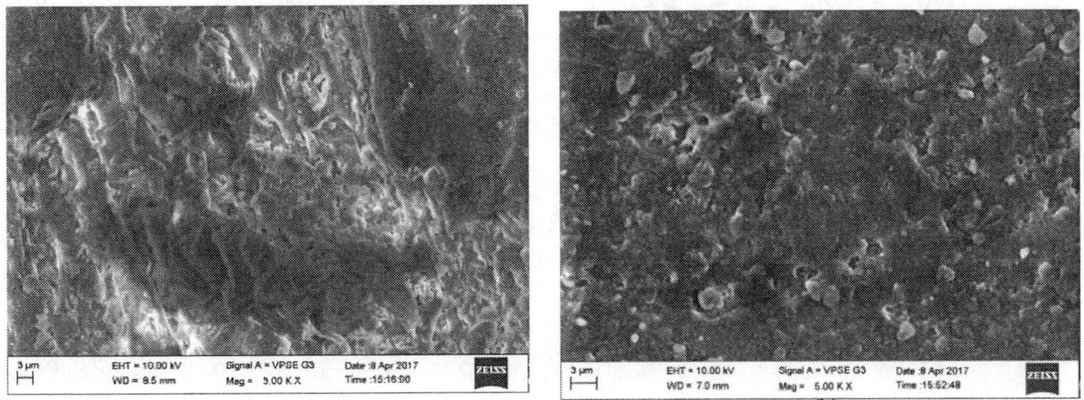

(a) (b)

Fig. 4. (a) SEM images of Woven cloth fiber composites (b) SEM images of Chopped fiber composites

5 Conclusions

Natural fibers are superior in mechanical properties and are also bio-degradable. Natural fiber are able to replace glass fiber and carbon for automobile and wind turbine blade applications. But it is very difficult by natural fiber alone. In order to increase the mechanical properties, it is very essential to use multiaxial layers of two or more than two types of natural fiber and Nano particle into it.

In this research, work has been carried out on multiaxial layers of Flax and Sisal fiber reinforced with SiC Nano powder and randomly chopped oriented Flax and Sisal fiber reinforced with SiC Nanopowder. It is found that uniform distribution of SiC into Flax/Sisal fiber improved impact strength and tensile strength by more than 50% than that of randomly oriented chopped fiber. But non uniform distribution of SiC into randomly oriented chopped fiber improved its stiffness by 36% more than that of woven cloth fiber reinforced composite.

References

1. Aakash Sirsshti, Senthiil Dr.P.V , Studies on Material and Mechanical Properties of Natural Fiber Reinforced Composites, The International Journal Of Engineering And Science (IJES), 2014, Volume- 3,Issue-11, pp-18-27.
2. Abdul Gafur, Jahangir A. Khan, Mubarak A. Khan, Rabiul Islam , Mechanical, Thermal and Interfacial Properties of Jute Fabric-Reinforced Polypropylene Composites: Effect of Potassium Dichromate, Materials Sciences and Applications, 2010, pp-350-357.
3. Ahmed F. Hamzah, Mohammed S. Hamzah, Najim A. Saad, Study of Fatigue Behavior of Composite Materials with the Basis of Polyphenylene Sulfide (PPS) Reinforced with Glass Fiber and Carbon, International Journal of Engineering and Technology, 2013, Volume-3 No. 4, pp-467-474.
4. Alagarraja1.K, Dhamodharan.A ,Gopinathan.K, Mathan Raj.R, Ram Kumar.K, Fabrication and Testing of Fiber Reinforced Polymer Composites Material, IOSR Journal of Mechanical and Civil Engineering (IOSR-JMCE), 2014, pp- 27-34.

5. Alberta Latteri, Gianluca Cicala, Giuseppe Cristaldi, Giuseppe Recca, University of Catania – Department of PhySiCal and Chemical Methodologies for Engineering, Composites Based on Natural Fiber Fabrics, 2010, volume-, pp-317-339.
6. Ashish Chauhan, Balbir Kaith, Priyanka Chauhan , Natural Fiber Reinforced Composite: A Concise Review Article, Chemical Engineering & Process Technology, 2012, Volume-3, Issue-2, pp-1-2.
7. Ashok Kumar.M, Dasaratha.H Fabrication and Performance of Natural Fibers: Sansevieria cylindrica, waste silk, jute and drumstick vegetable fibers (Moringa Oleifera) reinforced with rubber/polyester composites, International Journal of Fiber and Textile Research, 2011, pp-15-21.
8. Mr. Ganesh R Kalagi. Dr.Rajashekar Patil . Mr.Pradeepa shetty . Mr.Raviprasad . Mr.Sujith kumar . Mr.Vinod “"Mechanical Properties & SEM Analysis of Flax/Sisal/SiC Nano powder reinforced Hybrid Composites" Journal of Materials Science & Surface Engineering (JMSS) Vol. 5, Issue 5, 2017, pp 633-636.
9. Debiprasad Gon, Kousik Das, Palash Paul, SubhankarMaity, Jute Composites asWood Substitute, International Journal of Textile Science, 2012, Vol-6, pp-84-93.
10. De Rosa.M, Marra.F, Pulci.G, Santulli.C, Sarasini.F, Tirillo.J, Valente.M, Post-impact mechanical characterization of E-glass/basalt woven fabric interply hybrid laminates, express Polymer Letters, 2011, Vol-5, No.5, pp- 449–459.
11. Mr. Ganesh R Kalagi , Dr. Rajashekar Patil and Mr. Narayan Nayak, "Natural Fiber Reinforced polymer Composite Materials for Wind Turbine Blade Applications", International Journal of Science and Development Research Volume 1, Issue 9, September 2016, pp. 28-37.
12. Kuruvilla Joseph, Romildo Dias Tolêdo Filho, Beena James, Sabu Thomas & Laura Hecker de Carvalho,a study on a review on Sisal fiber reinforced polymer composites, v.3, n.3, p.367-379, 1999.
13. Kumaresan. M, Sathish. S and Karthi. N, a study on effect of fiber orientation on mechanical properties of Sisal fiber reinforced epoxy composites, Journal of Applied Science and Engineering, Vol. 18, No. 3, pp. 289_294 (2015).
14. Girisha.C, Sanjeevamurthy, Gunti Rangasrinivas, a study on tensile properties of natural fiber reinforced epoxy-hybrid composites, International Journal of Modern Engineering Research (IJMER) Vol.2, Issue.2, Mar-Apr 2012 pp-471-474.
15. Jinchun Zhu , Huijun Zhu, James Njuguna and Hrushikesh Abhyankar, a study on recent development of Flax fibers and their reinforced composites based on different polymeric matrices, materials2013,6,5171-5198;doi:10.3390/ma6115171, ISSN 19961944.

Study on Electrical Properties of PALF Reinforced Bisphenol-A Composite

Vinod B.[1*] and *L. J. Sudev*[1]

[1]*Department of Mechanical Engineering, VVCE, Mysuru, India*

Abstract. These days, composite materials successfully substitute the traditional materials due to its various significant applications. This paper examines the influence of fiber orientation and fiber length on electrical properties of PALF reinforced Bisphenol-A composite and explores the potential of using PALF as reinforcing material for electrical applications such as terminals, connectors, motor body cover, industrial and household plugs and circuit boards. The resistance and inductance of resin material is increased by 24.19% and 24.13% respectively after reinforcement of PALF in different orientations and its value increased by 37.93% and 37.81% respectively after reinforcement of PALF in different lengths. Similarly, Capacitance and Dielectric constant of resin material decreased by 19.44% and 19.39% respectively after reinforcement of PALF in different orientations and its value decreased by 27.67% and 27.50% respectively after reinforcement of PALF in different lengths. It can be inferred from this study that the fiber orientations and fiber lengths greatly effects the electrical properties of PALF composite.

Keywords: Pineapple leaf fiber, Bisphenol-A, Capacitance, Resistance, inductance, volume resistivity.

1 Introduction

Natural fiber reinforced polymer composite has attracted the interest of many of the researchers for developing competing material towards conventional material. These composite materials offer high strength to weight ratio, higher performing ability, easy in manufacturing, low cost and low density. Natural fibers are easily available, cheaper, environmental friendly and most of these fibers are wasted by burning out after extraction of fruits or vegetables, which lead to pollution of air. It is necessary to develop the newer material that is friendly towards environment, easily degradable, renewable and to be a substitution for depletion of conventional resources. Usually the natural fiber is from agricultural source, it is possible to modify the final properties of the composite material by selection of suitable fibers with a given chemical or morphological composition. In future, composites reinforced by natural fiber form a new class of materials and seem to have good potential in future as a standby for metals and wood based materials for structural applications of various small and large scale industries. [1, 2].

*Corresponding author: vinod@vvce.ac.in

India holds the second largest agricultural land in the world with diversified topography. Nearly 70% of population lead their livelihood from agriculture. The eatables that obtained from agriculture are basic need of human beings and animals and also a major source for various agro-based industries. The final products which are left out such as leaf, stem, roots etc. are burnt in most of the situations after extractions of useful products. One such commonly known fruit plant is pineapple. Even though being mechanically sound and environmentally excellent, pineapple leaves from which pineapple leaf fiber (PALF) are extracted are burnt or wasted after extraction of fruits. Though it is extensively used as textile material, it is the least studied fiber as reinforcement for composite material. Table 1 shows the chemical composition of PALF. The climatic condition, oldness of fibers, method of extraction and source of fibers are some of the factors responsible for varying composition of PALF. The most important of these is lignin, which acts as a cementing material by bonding different cells in plant fibers together. The lignin content in fiber obtained from plants impact on its structural properties and morphology. The cellulose fibrils embedded in lignin matrix of plant fibers is responsible for its high electrical resistance. This high electrical resistance property of the fiber can be utilized for manufacturing various industrial components such as switch board, PCB plates, terminals, connectors, switches, plugs and motor coverings etc. suitable for various electrical engineering applications by incorporating in polymer matrix material. The suitability of composite material for such electrical applications is determined by testing its properties in terms of resistance, inductance, capacitance and dielectric constant.

Table 1. Chemical composition of Pineapple Leaf Fiber (PALF)

Sl No.	Chemical Composition	Proportion
1.	Cellulose	67.12–82%
2.	Hemicelluloses	9.45–18.80%
3.	Lignin	4.4–15.4%
4.	Pectin	1.2–3%
5.	Fat and Wax	3.2–4.2%

2 Literature Survey

In most of the literature it is emphasized that natural fiber reinforced polymer composites shows better electrical resistance property and can be best suited for most of the electrical engineering applications. Navin Chand et al. [3] made investigation on the influence of fibre orientation on electrical properties of epoxy composites reinforced with sisal fiber. Sisal fibers are oriented parallel and perpendicular to the electric field. He concluded that sisal fibre reinforced polysulphide modified epoxy composites shows electrical anisotropic behaviour and Dielectric constant anisotropy decreases with frequency. P. A. Sreekumar et al. [4] made investigation on influence of short sisal fiber on electrical properties of polyester composites. The composites are fabricated by resin transfer moulding (RTM) technique with respect to percentage of fiber, frequency and temperature. The dielectric constant (ε'), loss factor (ε''), dissipation factor ($\tan \delta$) and conductivity increases with fiber content for the entire range of frequencies. The composites with 50% of fiber content shows better results due to the increment in orientation and interfacial polarizations resulting from the presence of polar groups of cellulose in natural fibers. Govardhan Goud et al. [5] made experimental studies on electrical properties of epoxy hybrid composite reinforced by Roystonea regia/glass fibre. He concludes that blending glass fiber with natural fibers leads to decrease in conductivity and dielectric constant at all frequency.

3 Materials and Methodology

3.1 Pine Apple Leaf Fiber (PALF)

PALF is extracted through anaerobic process, where pineapple leaf is soaked in water for 3 to 4 days in specially designed soaking tank connected with anaerobic reactor in which temperature between 20^0 to 37^0C and neutral pH of 6.7 to 8 is maintained. During the reaction, pectin the intercellular cementing matter present around the PALF fiber get degraded. This results in split-up of fibers from their surroundings. These fibers can be extracted easily without any damage when compared with other mechanical process for extraction of fibers [6].

3.2 Bisphenol-A

Bisphenol-A (BPA) is also called as Diglycidyl Ether of Bisphenol-A (DGEBA). It is an organic compound which belongs to the group of diphenylmethane derivatives and Bisphenol. BPA resin offers superior manufacturing flexibility, better adhering ability, excellent mechanical property, better resistance to fatigue and micro cracking, lower degradation from water ingress and improved resistance to osmosis. These properties made to choose BPA as matrix material [7].

3.3 Preparation of Sample

After extraction of fibers from pineapple leaf it is subjected for chemical treatment in order to improve adhering ability of the fibers by making its surface rough and to remove impurities. The treated fibers are ready to use as reinforcing material for preparation of composite laminates. The laminates are prepared for different fiber orientations like Uni-directional, Bi-directional and inclined at an angle of 45° as shown in Fig. 1. The manufacturing of composites by reinforcing with short fibers offers more flexibility than the long fibers. Hence laminates are prepared from fiber lengths ranging from 2 mm to 14 mm with step size of 2 mm in random orientations using hand layup method. Composite laminates are cut into test specimens of required size as per ASTM standards.

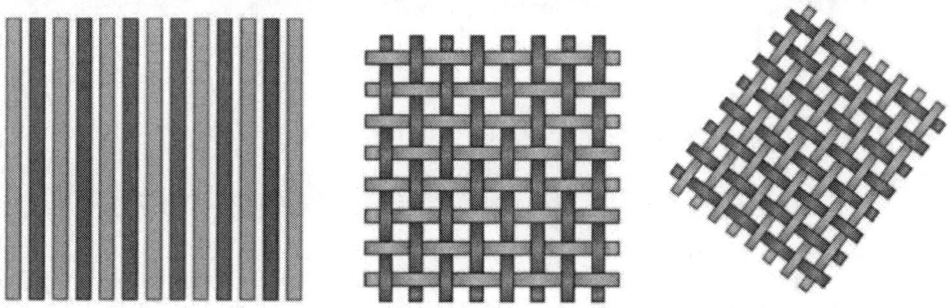

Fig. 1. (a) Uni-directional (b) Bi-directional (c) 45^0 oriented fibers

3.4 Experimentation

The electrical application of composite material makes it important to investigate electrical properties of composite. The resistance, inductance, capacitance and dielectric constant are some of the important electrical properties for different electrical applications of composite. Electrical resistance is a very important property of material for using it as electrical insulators. The resistance of the composite is calculated using an instrument called Megger.

Initially the instrument is peddled to energize the coil and peddling is continued at a constant speed till the needle showed a constant value of resistance. Using resistance value the other parameters like inductance, capacitance, dielectric constant, volume resistivity, dielectric dissipation factor are calculated using mathematical relationships as follows.

a) Inductance: It is a measure of an inductors 'resistance' to change in current flowing through the circuit. Good power factor (PF) plays a major role for correct operation of a power system. For very lower value of power factor an electrical system may suffer from excessive losses and poor voltage regulation. Also the inductors are used in filters, sensors, transformers, motors. Thus inductance plays a significant role in designing and varying the performance of the circuits and devices. The SI unit is Henry. Where R is resistance and L is inductance.

$$R = \frac{1}{\sqrt{R^2 + (2\pi fL)^2}} \tag{1}$$

b) Capacitance: It is an ability of a body to store an electrical Charge. It can be used to improve the power factor. The SI unit of capacitance is Farad. Where R is resistance and C is capacitance.

$$R = \frac{1}{\sqrt{R^2 + \frac{1}{(2\pi fC)^2}}} \tag{2}$$

c) Dielectric constant: The dielectric property decides the storage and dissipation of electric and magnetic energy in materials. Higher the dielectric constant is the greater the resistance it will impose upon the changing electric field. The various phenomenons in electronics, optics, and solid-state physics can be explained with the help of dielectrics property.

$$\varepsilon = \frac{Ct}{\varepsilon_0 A} \tag{3}$$

where, ε_o - Permittivity of air (8.85 × 10–12 Fm^{-1}), C - Capacitance (F), A - Area of cross-section of the sample (m^2) and t - Thickness of the sample (m) [8].

4 Results and Discussions

4.1 Electrical properties of PALF composite for different fiber orientations

Some of the important electrical properties such as resistance, inductance, capacitance and dielectric constant are determined for PALF reinforced composite as discussed above. Table 2 shows the experimental results of Bisphenol-A resin material alone without any reinforcement and for the resin material reinforced with PALF long fibers in different orientations. The values given in the table are the averaged values from 3 trials.

Table 2. Electrical properties of Resin and PALF composite in different fiber orientations

Sl no	Fiber orientation	Resistance R (MΩ)	Inductance L (kH)	Capacitance C (pF)	Dielectric constant ε
1	Bisphenol-A (Without any fibers)	145	461.55	21.91	7.07
Bisphenol-A resin after reinforcement of PALF in different orientations.					
2	Uni directional	148	466.27	21.47	6.931
3	Bi - directional	168	532.51	18.90	6.100
4	45° orientation	180	572.96	17.65	5.699

The variation of resistance of the unreinforced resin material and the resin reinforced with PALF in different orientations is shown in Fig. 2. Similarly, Fig. 3 shows variation of inductance. The change in capacitance and dielectric constant is depicted in Fig. 4 and 5 respectively.

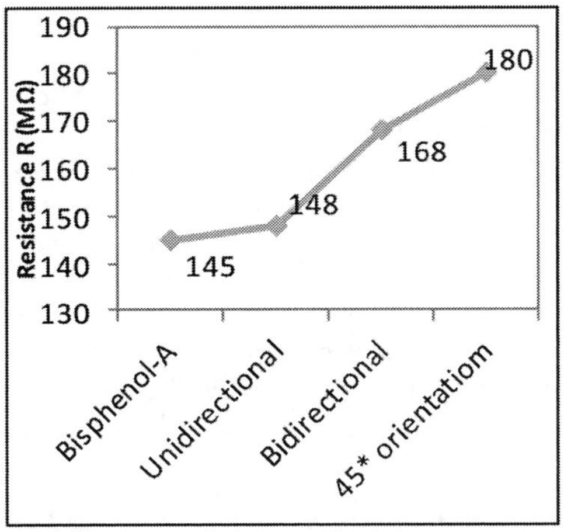

Fig. 2. Change in resistance for the resin & composite for different fiber orientation

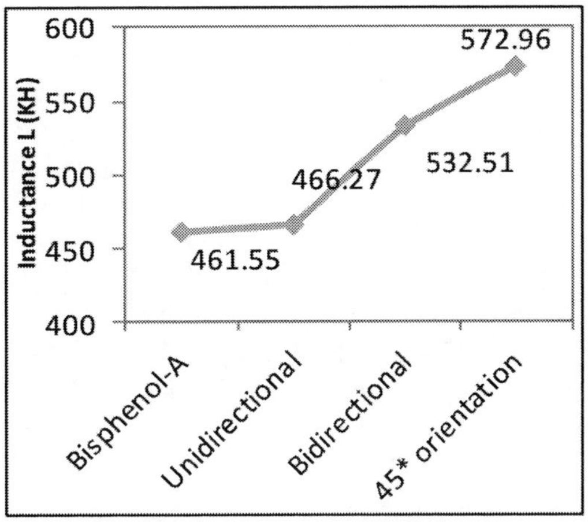

Fig. 3. Change in inductance for the resin & composite for different fiber orientation

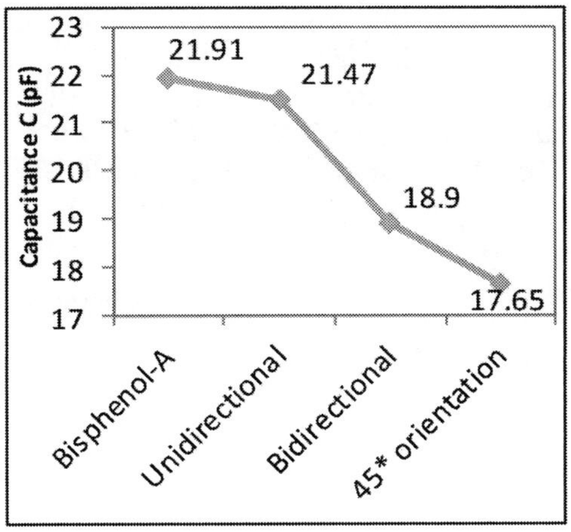

Fig. 4. Change in capacitance for the resin & composite for different fiber orientation

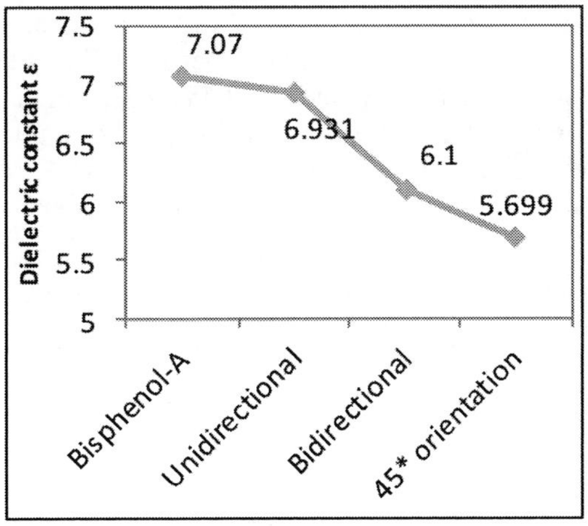

Fig. 5. Change in dielectric constant for the resin & composite for different fiber orientation

The electrical resistance and inductance of resin material is 145 MΩ and 461.55kH respectively and its value increases gradually after reinforcement of PALF in different orientations. The maximum value of 180MΩ and 572.96kH respectively is obtained for the composite with fiber orientation of 45^0 which is 24.13% and 24.23% greater than resin material without any reinforcement. The resistivity of the composites depends on percentage of water content in fiber, impurities, micro fibril angle, cellulosic structure etc. After extracting fibers from leaf, it is subjected to alkaline treatment in order to remove fatties and any other impurities that present in the fibers. Treated fibers are dried in oven for 1hr whose temperature is maintained at 60^0C. Hence finally obtained fibers are free from impurities and moisture content, which decreases the electrical conductance of fiber reinforced composite. Further, 45° oriented fiber arrangement brought the heterogeneity in to the composite. As a result of this, when electrical field is applied to the composite, OH-group presented in the cellulose of the PALF get polarized and trapped at the interfaces of heterogeneous system. Whereas, unidirectional fiber reinforced composite shows less heterogeneity.

The capacitance and Dielectric constant of resin material is 21.91pF and 7.07 respectively and its value decreases gradually after reinforcement of PALF in different orientations. The minimum value of 17.65pF and 5.699 respectively is obtained for the composite with fiber orientation of 45^0 which is 19.44% and 19.39% lesser than resin material without any reinforcement. It is found that dielectric constant increases with the presence of impurities, moisture content in fiber (which leads to flexibility for movement of water dipole) and with increase of temperature. As already discussed reinforced fibers are free from impurities and water content leads to decrease of dielectric constant. The value of dielectric constant of any composites is influenced by polarizability of that composite material. Higher the polarizability of particle, higher will be dielectric constant [9]. Hence, less polarization of polar groups due to alkali chemical treatment, which removes hydrophilic nature that is polar groups from the fiber surface, is also responsible for the decrease in dielectric constant [10]. The decrease in capacitance and dielectric constant after the PALF reinforcement indicates the decrease in the ability of a body to store an electrical charge.

4.2 Electrical properties of PALF composite for different fiber lengths

The electrical behavior of the composites for different fiber lengths is carried out similar to the experiments conducted for composites with different fiber orientations as discussed above. Table 3 shows the experimental results of Bisphenol-A resin material alone without any reinforcement and for the resin material reinforced with short PALF fibers in different lengths. The change in resistance of resin material and the resin reinforced with PALF with change in fiber lengths is shown in Fig. 6. Similarly, Fig. 7 shows variation of inductance. The change in capacitance and dielectric constant is shown in Fig. 8 and 9 respectively.

Table 3. Electrical properties of Resin and PALF composite in different fiber lengths

Fiber lengths	Resistance R (MΩ)	Inductance L (KH)	Capacitance C (pF)	Dielectric constant ε
Bisphenol- A (Without any fibers)	145	461.55	21.917	7.075
Bisphenol-A resin after reinforcement of PALF in different lengths.				
2	146	464.784	21.75	7.02
4	154	490.183	20.636	6.662
6	160	509.358	19.86	6.41
8	166	528.447	19.144	6.18
10	176	560.24	18.056	5.829
12	185	588.861	17.18	5.546
14	200	636.062	15.89	5.129

Electrical resistance and inductance increases gradually as fiber length increases. The maximum value of 200 MΩ and 636.062kH respectively is obtained for composite with fiber length of 14 mm, which is 37.93% and 37.81% respectively greater than unreinforced resin material. Similarly, Capacitance and Dielectric constant decreases gradually as fiber length increases. The minimum value of 15.89pF and 5.129 respectively is obtained for composite with fiber length of 14 mm, which is 27.67% and 27.50% respectively lesser than unreinforced resin material. As fiber length increases electrical resistance of composite material also increases due to collisions of adjacent atoms and electrons as there is a larger distance to travel along the fibers. The longer the length of fiber the more will be the collisions and hence higher the resistance and lower will be the rate of current change [11]. Hence, inductance increases with increase of fiber length. Resistance is inversely proportional to capacitance, capacitance decreases with increase of resistance and fiber lengths.

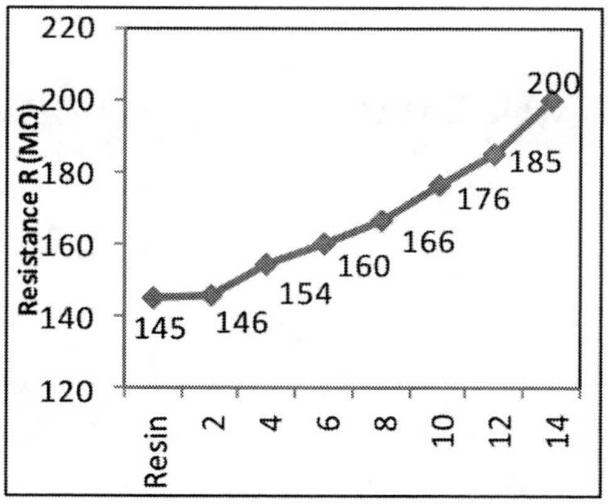

Fig. 6. Change in resistance for the resin & composite for different fiber length

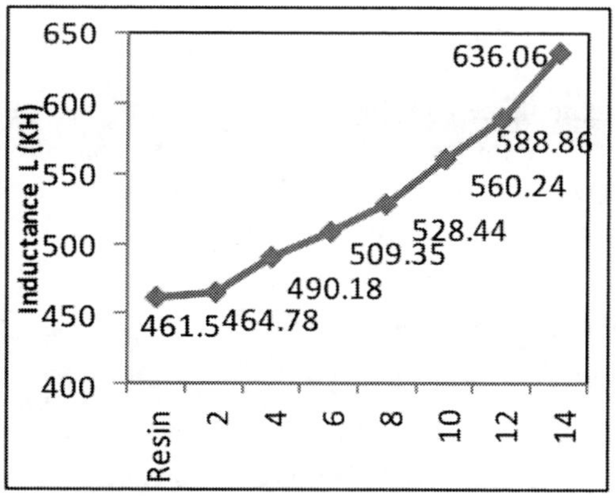

Fig. 7. Change in inductance for the resin & composite for different fiber length

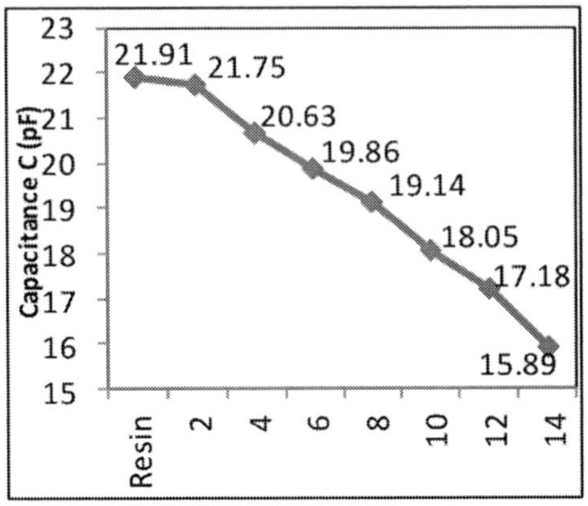

Fig. 8. Change in capacitance for the resin & composite for different fiber length

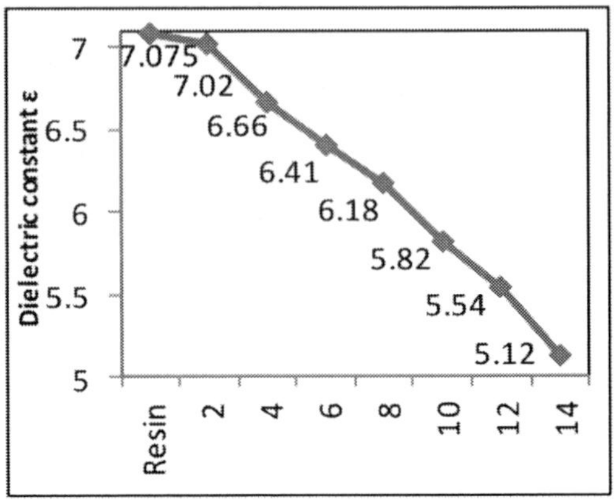

Fig. 9. Change in dielectric constant for the resin & composite for different fiber length

As the fiber length increases for constant loading, the number of interfaces and polar -OH groups decreases. Also alkaline treatment of the fibers after extraction from its leaf for removal of impurities and fatties is also responsible for decrease in orientation polarization. Sodium hydroxide (NaOH) in alkaline solution reduces the chance of interface of polar –OH groups of lignocellulosic fibers with that of water molecules, hence reduce the ability of fiber to absorb water and dielectric constant value [12].

5 Conclusion

From this present work, it is observed that fiber orientations and fiber lengths significantly influences the electrical properties of PALF reinforced Bisphenol-A composite. The composite with 45° orientation of fibers shows highest value of resistance and inductance and lowest values of capacitance and dielectric constant. As the fiber length increases resistance and inductance of the PALF composite increases gradually. Similarly, capacitance and dielectric constant decreases gradually. The composite with fiber length of 14 mm shows highest value of resistance and inductance and lowest values of capacitance and dielectric constant. Hence, PALF reinforced composite materials can be successfully used for several electrical applications.

References

1. Drzal, L.T., Mohanty, A.K., Burgueño, R. and Misra, M. "Biobased Structural Composite Materials for Housing and Infrastructure Applications: Opportunities and Challenges". Composite Science and Technology, 2003, PP.129-140.
2. Navin Chand, Deepak Jain. "Effect of Sisal Fibre Orientation on Electrical Properties of Sisal Fibre Reinforced Epoxy Composites". Elsevier Composites Part A 36, 16 August, 2004, pp. 594-602.
3. Navin Chand, Deepak Jain, Archana Nigrawal, "Investigations on gradient a.c. conductivity characteristics of bamboo (Dendrocalamus strictus)", Bull. Material Science, Volume. 29, Issue No. 2, April 2006, pp. 193–196.
4. P. A. Sreekumar, Jean Marc Saiter, Kuruvilla Joseph, G. Unnikrishnan, Sabu Thomas. "Electrical Properties of Short Sisal Fiber Reinforced Polyester Composites

Fabricated By Resin Transfer Molding". Elsevier Composites: Part A 43, Nov-2012, pp. 507–511.

5. Govardhan Goud and R N Rao. "Mechanical and electrical performance of Roystonea regia/glass fibre reinforced epoxy hybrid composites", Bull. Material Science, Volume 35, Issue No. 4, August 2012, pp. 595–599.
6. Uma Devi, L., Bhagawan, S.S. and Thomas, S. "Mechanical Properties Of Pineapple Leaf Fiber-Reinforced Polyester Composites". Journal of Applied Polymer Science. Nov-1996, pp. 1739-1748.
7. "Bisphenol-A: A safety and handling guide". Publication Number: AE-154. July 1997.
8. D. Pathania and D. Singh. "A review on electrical properties of fiber reinforced polymer Composites", International Journal of Theoretical & Applied Sciences, ISSN : 0975-1718, 2009, pp. 34-37
9. J. P. Siregar, S. M. Sapuan , M. Z. A. Rahman and H. M. D. K. Zaman. "The Effect of Alkali Treatment on the Mechanical Properties of Short Pineapple Leaf Fibre (PALF) Reinforced High Impact Polystyrene (HIPS) Composites". Journal of Food, Agriculture & Environment, Volume 8, Issue 2, 2010, pp. 1103 - 1108.
10. G. George, K. Joseph, E.R. Nagarajan, E. Tomlal Jose, K.C. George, "Dielectric behaviour of PP/jute yarn commingled composites: Effect of fibre content, chemical treatments, temperature and moisture" Composites Part A: Applied Science and Manufacturing, Volume 47, April 2013 pp: 12-21.
11. N.M. Mehta and P.H. Parsania, "Fabrication and evaluation of some mechanical and electrical properties of jute-biomass based hybrid composites" Journal of Applied Polymer Science, Volume 100, Issue 3, May 2006, pp: 1754-1758.
12. Elammaran Jayamani, Sinin Hamdan, Md Rezaur Rahman, and Muhammad Khusairy Bin Bakri, "Comparative Study of Dielectric Properties of Hybrid Natural Fiber Composites", 12th Global Congress on Manufacturing and Management, GCMM 2014, ScienceDirect, Procedia Engineering 97, 2014, pp: 536 – 544.
13. A.S. Singha, A.K. Rana, R.K. Jarial, "Mechanical, dielectric and thermal properties of Grewia optiva fibers reinforced unsaturated polyester matrix based composites" Materials & Design, Volume 51, 2013, pp: 924-934.
14. Z.N. Azwa, B.F. Yousif, A.C. Manalo, W. Karunasena, A review on the degradability of polymeric composites based on natural fibres, Materials & Design, Volume 47, 2013, pp: 424-442.

Mechanical characterization of precipitation hardened Al7075-grey cast iron powder reinforced metal matrix composites

Jamaluddin Hindi [1*], *Kini Achuta*[1] and *Murthy Amar*[1]

[1]*Department of Mechanical and Manufacturing Engineering, Manipal Institute of Technology, Manipal Academy of Higher Education, 576104, India*

Abstract. Al7075 alloy is the most commonly used by the aerospace industry. Al7075 alloy is characterized by its improved properties such as higher toughness, specific strength and hardness. The current work focuses on the preparation and characterization of age hardened Al7075-Grey cast iron composites. Two stage stir casting technique is used for the preparation of the composite. Age hardening treatment is imparted to enhance the mechanical characteristics. The variation of hardness and tensile strength with respect to aging temperature and percentage of reinforcement is analyzed. The composites exhibit higher hardness and tensile strength as the reinforcement percentage is increased at an aging temperature of 100^{0}C.

1 Introduction

Composite material is a combination of dissimilar materials which result in property enhancement than those of the individual elements. There will not be any change in the chemical, physical, and mechanical properties of individual elements i.e. matrix and reinforcement [1-4]. Composite materials are characterized by high strength to weight ratio, higher stiffness with low density resulting in considerable reduction in the weight of the component. Enhancement in the desirable properties is mainly due to the reinforcement which is generally harder, refractory, and stiffer than the base material [5]. The reinforcement phase can be continuous or discontinuous fiber or particulate. Metal matrix composites can be defined as a combination of two or more dissimilar materials, one of which is a metal, in which enhancement of desirable properties are achieved by systematic combinations of different constituents [6-8]. Very high specific strength and specific modulus can be achieved in MMCs consisting of continuous or discontinuous fibers, whiskers, or particles [9].

* Corresponding author: jamalhindi@gmail.com

Aluminum alloy composites are hardened by reinforcement of ceramic refractory particulates into the base matrix material. Due to their innate nature and creation of nucleation sites in the matrix material, reinforcements can enhance tensile related properties of the matrix material [10-12]. Particulate reinforced composites are characterized by economic manufacturing methods with finer grains. Selection of the type of reinforcement is based on the property enhancement and monetary investment [13,14]. Generally, continuous fiber reinforced MMCs have unidirectional properties in the direction of the fiber, but are costly. Chopped fibers can produce significant property enhancement in the two dimensionally in the direction of their orientation, at moderate cost. MMCs provide moderate but isotropic properties [15-19].

Grey cast iron is one of the types of cast irons which have unique microstructure containing carbon in free and combined form. Carbon is present as cementite and graphite. The weight percentage of these two constituents can be altered by the addition of graphite formers like Si. The fractured surface of the grey cast iron specimen is greyish in colour due to the presence of free carbon as graphite and combined carbon as cementite. The graphite also provides gray cast iron with an additional damping capacity for the alloy or matrix since it absorbs the energy [21-23].

2 Materials and methods

2.1 Al 7075 alloy

Al7075 alloy is the most commonly used by the aerospace industry. Al7075 alloy is characterized by its improved properties such as higher toughness, specific strength and hardness. The Al-Zn-Mg alloys [Al7075] are used in as cast and age hardened condition. As cast alloys are generally homogenized. Table 1 shows the chemical composition of Al7075 alloy [20].

Table 1. Chemical composition of Al7075 alloy.

Al	Cr	Cu	Fe	Mg	Zn	Si	Mn	Ti	Others
89.79	0.08	1.35	0.3	2.21	5.67	0.4	0.08	0.06	0.06

2.2. Grey cast iron

The chemical composition of grey cast iron is shown in table 2.

Table 2. Chemical composition of the grey and cast iron (wt.%).

Element in wt.%	C	S	P	Si	Mn	Fe
GCI	3.61	0.024	≤0.022	1.30	0.41	Bal

Grey cast iron has good potential as a reinforcement material due to following reasons:

- It is also an industrial waste which comes in the form of chips during machining operation. These chips can be re casted to be used as reinforcement.
- Grey cast iron is hard due to the presence of cementite but also has high self-lubricating property and machinability due to the presence of free graphite in the form of flakes.
- It serves as a hybrid reinforcement because microscopically there are three phases in GCI at room temperature i.e. cementite, ferrite and graphite.
- It will not react with Al7075 which will be used as a base matrix material.

2.3. Preparation of the reinforcement phases

Grey cast iron (GCI) rod is cast using chill casting technique with secondary operation i.e. turning operation to machine the casting to remove foreign material inclusions near the mould wall and slag at the top of the casting. Hardness values at three sample zones of casting is measured using Rockwell B scale to analyse the homogeneity of hardness values of the sample. To achieve homogenous chemical composition and hardness, the specimen is annealed at 800^0C for 10 hours [24-26] The casting is turned in lathe to collect grey cast iron dust (debris), washed and preheated to 300^0C for 1 hour to remove volatile substances present. Debris is ground in planetary ball mill as shown in figure 2 to reduce the size of particles upto 50 μm and sieving is performed to obtain uniform grade of particles. Sieving and grinding cycles is repeated till the required quantity of uniform grade of particles are obtained. The particles were seived through ASTM-140 sieve to obtain uniform grade of particles of average size 50 μm [18].

2.4 Stir casting of Al7075–GCI composites

Al7075-T6 rods of 1 inch diameter and 7 inch length were procured from the supplier. It was cut to 1 inch length and required number of pieces were placed in the crucible and melted in a furnace at 750^0C. After complete melting hexa chloroethane was added as a degasification agent. Alkaline powder was added to remove the slag. Permanent mold cavities for cylindrical and flat billets are cleaned using a wire brush and are coated with graphite powder and water emulsion. Assembled molds were preheated to a temperature of 450^0C for an hour in a separate furnace. Reinforcement prior to the addition to the molten metal was separately pre heated to 450^0C to remove volatile substances [9]. Three compositions of the composite were prepared by varying the weight percentage of the reinforcement. Stirring of the molten metal was carried using a mechanical stirrer placed on top of the Induction furnace. Required amount of reinforcement was added to the melt through a funnel placed in position. Vortex was created by rotating the stirrer at suitable speed, two step stirring was carried out to overcome heat loss and drop in furnace temperature. After the completion of stirring the molten metal with reinforcements was poured to a pre heated permanent mold and allowed to solidify at room temperature.

3 Results and discussion

3.1 Measurement of Peak hardness

Peak hardness values are noted for all the compositions. Age hardening curves were plotted for each specimen at half an hour interval. From the analysis of age hardening curves it is evident that lower the temperature of aging, higher will be the peak hardness value. As the aging temperature is increased peak hardness can be obtained at a faster rate but sacrificing on the value of peak hardness. At both lower and higher aging temperature, material exhibits higher peak hardness as the weight percentage of the reinforcement is increased. Figure 1 and 2 shows Age hardening curve at lower and higher aging temperature at 100 and 200°C. Figure 3 shows variation in peak hardness with respect to wt.% of GCI for as-cast and age hardened specimens.

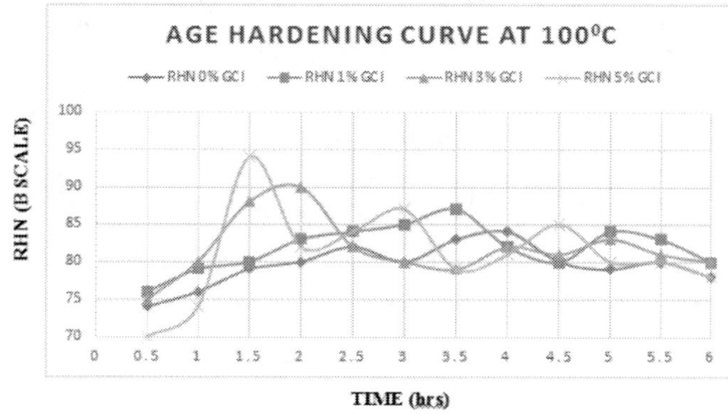

Fig. 1. Age hardening curve at lower aging temperature 100°C.

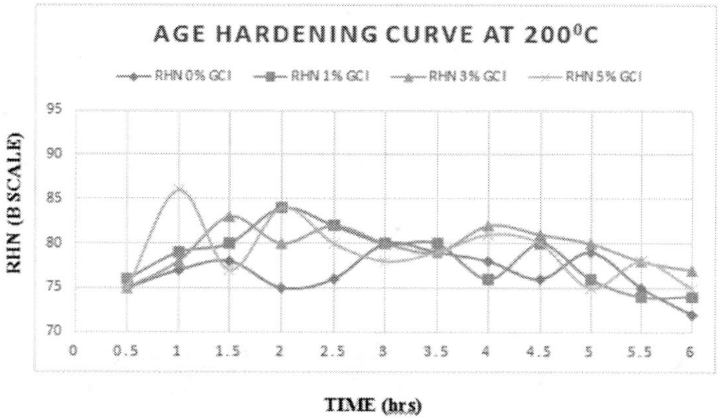

Fig. 2. Age hardening curve at higher aging temperature 200°C.

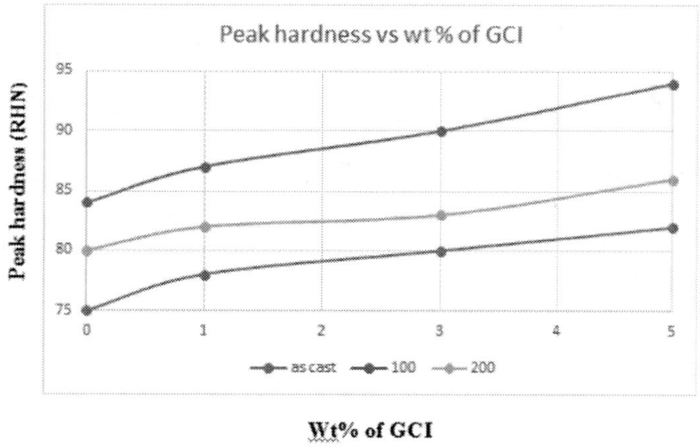

Fig. 3. Peak hardness vs wt.% of GCI for as-cast and age hardened specimens.

3.2 Measurement of tensile strength

From the analysis of the graph shown in figure, it is evident that lower the temperature of aging, higher will be the ultimate tensile strength value. At both lower and higher aging temperature, material exhibits higher ultimate tensile strength as the weight percentage of the reinforcement is increased. Figure 4 shows the variation in ultimate tensile strength with respect to wt.% of GCI for as-cast and age hardened specimens. The ductility decreases as the percentage of reinforcement is increased irrespective of the aging temperature as shown in figure 5.

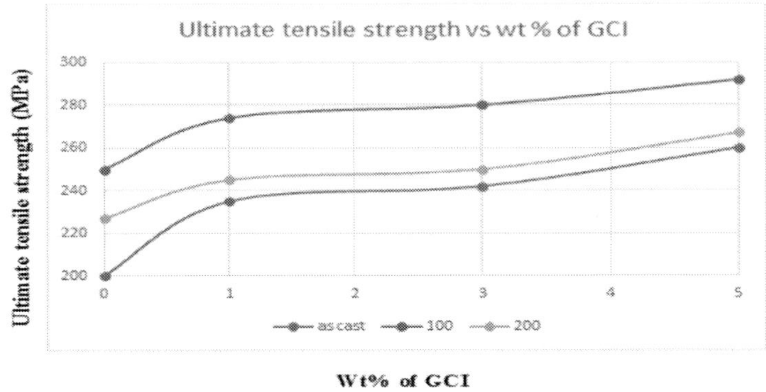

Fig. 4. Ultimate tensile strength vs wt.% of GCI for as-cast and age hardened specimens.

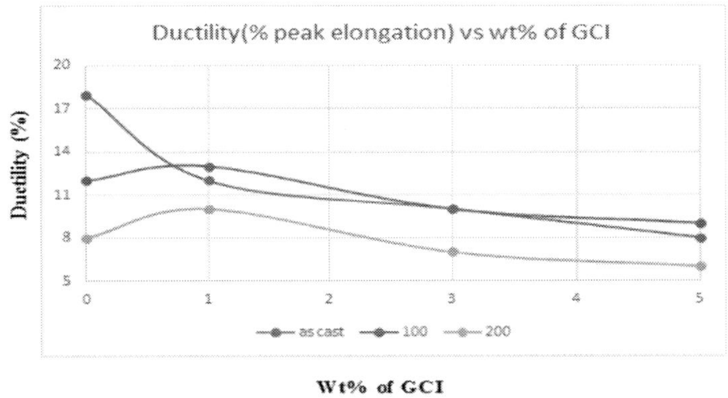

Fig. 5. Ductility vs wt.% of GCI for as-cast and age hardened specimens.

3.3 Microstructure analysis

The specimens are systematically polished with a series of silicon carbide embedded emery papers starting from coarser 100 microns to finer 600 microns in the steps of 100 microns. At every stage of polishing specimens are water washed and dried with acetone. Super finishing operation known as buffing is performed on disc polisher with wet diamond paste of 50 microns. Finally, the mirror like polished specimens is etched with etchant (Keller's reagent) [6]. Microstructures of all the samples are recorded in metallurgical microscope at 300X magnification. Figure 6 shows the microstructure of heat treated composites at 300X as recorded by ImageAnalyzer. Homogenised specimens show better dispersion of reinforcement without clustering.

Fig. 6. Microstructure of homogenized (a) AA7075 composite with 1% Grey cast iron powder (b) AA7075composite with 3% Grey cast iron powder (c) AA7075 composite with 5% Grey cast iron powder.

4 Conclusions

Al 7075 alloy can be successfully reinforced with grey cast iron powder using two step stir casting technique. Age hardening treatment significantly improves the mechanical properties of the composite material. Lower the aging temperature better is the peak hardness values in age hardening. Substantial improvement in the bulk hardness is observed in the composite with increase in the weight percentage of reinforcement. Substantial improvement in the ultimate tensile strength is observed in the composite with increase in the weight percentage of reinforcement. Peak hardness and maximum tensile strength was observed in Al 7075-5% GCI aged at 100^{0}C. The ductility decreases as the percentage of reinforcement is increased. The microstructure study reveals uniformity in the reinforcement distribution without any agglomeration.

References

1. Akhlaghi and Zare-Bidaki, "Influence of graphite content on the dry sliding and oil impregnated sliding wear behavior of Al2024–graphite composites produced by in situ powder metallurgy method", Wear, 266, 2009, 37–45.
2. Ali Kalkanli and Sencer Yılmaz, "Synthesis and characterization of aluminum alloy 7075 reinforced with silicon carbide particulates", Materials and Design, 29, 2008, 775–780.
3. Anand Pai, Sathya Shankar Sharma, Raynor Errol D'Silva and Nikhil, "Effect of graphite and granite dust particulates as micro-fillers on tribological performance of Al 6061-T6 hybrid composites", Tribology International, 92, 2015, 462–471.
4. Baradeswaran A and Elaya Perumal, "Study on mechanical and wear properties of Al 7075/Al2O3/graphite hybrid composites", Composites, 56, 2014, 464–471.
5. Baradeswaran A, Vettivel, Elaya Perumal, Selvakumar and Franklin Issac, "Experimental investigation on mechanical behaviour, modeling and optimization of wear parameters of B4C and graphite reinforced aluminium hybrid composites", Materials and Design, 63, 2014, 620–632.
6. Bayazida S.M, Farhangia, Asgharzadehb and Radanc "Effect of cyclic solution treatment on microstructure and mechanical properties of friction stir welded 7075 Al alloy", Materials Science and Engineering, 649, 2014, 293–300.
7. Campbell F.C, "Metal matrix composites" ASM international, 2010, 538-539.

8. Clark. R, Coughran, Traina, Hernandez. A and Scheck, "On the correlation of mechanical and physical properties of 7075-T6 Al alloy", Engineering Failure Analysis, 12, 2005, 520–526.
9. Deaquino Lar, Soltani, Bahrami, Gutiérrez Castañeda, and García-Sánchez, "Tribological characterization of Al7075–graphite composites fabricated by mechanical alloying and hot extrusion", Materials and Design, 67, 2015, 224–231.
10. Jin-feng, Zhuo-wei peng, Chao-xing li, and Zhi-qiang Jia, "Mechanical properties, corrosion behavior and microstructures of 7075 aluminium alloy with various aging treatments", Transactions of Nonferrous Metals Society of China, 18, 2008, 755–762.
11. Kiran T.S, Prasanna Kumar, Basavarajappa and Viswanatha, "Dry sliding wear behavior of heat treated hybrid metal matrix composite using Taguchi techniques", Materials and Design, 63, 2014, 294–304.
12. Mahathaninwonga N T, Plookphola, J, Wannasina, and Wisutmeth angoonb, "T6 heat treatment of rheocasting 7075 Al alloy", Materials Science and Engineering, 532, 2012, 91–99.
13. Michael Oluwatosin Bodunrin, Kenneth Kanayo, and Alaneme Lesley Heath Chown "Aluminium matrix hybrid composites: a review of reinforcement philosophies; mechanical, corrosion and tribological characteristics", Journal of Materials Research and Technology, 169, 2015, 169-181.
14. Murat Baydogan, Hüseyin Çimenoglu and Sabri Kayalı, "Study on sliding wear of a 7075 aluminum alloy", Wear, 257, 2004, 52–61.
15. Nikhilesh Chawla and Yu-Lin Shen, "Mechanical behavior of particle reinforced metal matrix composites", Advanced Engineering Materials, 3, 2001, 357-370.
16. Ravinder Kumar and Suresh Dhiman, "A study of sliding wear behaviors of Al7075 alloy and Al7075 hybrid composite by response surface methodology analysis", Materials and Design, 50, 2013, 351–359.
17. Reda Y, Abdel-Karim, and Elmahallawi, "Improvements in mechanical and stress corrosion cracking properties in Al7075 via retrogression and re aging", Materials Science and Engineering, 485, 2008, 468–475.
18. Rupa Dasgupta, "The stretch, limit and path forward for particle reinforced metal matrix composites of 7075 Al-alloys" Scientific Research Engineering, 2, 2010, 237-256.
19. Sahin Y, "Preparation and some properties of SiC particle reinforced aluminium alloy composites", Materials and Design, 24, 2003, 671–679.
20. Satish Kumar Thandalama, Subramanian Ramanathana and Shalini Sundarrajan, "Synthesis, microstructural and mechanical properties of ex situ zircon particles (ZrSiO4) reinforced metal matrix composites (MMCs): a review", Journal of Materials Research and Technology, 4, 2015, 333-347.
21. Siavash gholami, Esmaeil emadoddin, Mohammad tajally and Ehsan borhani, "Friction stir processing of 7075 Al alloy and subsequent aging treatment" Transactions of Nonferrous Metals Society of China, 25, 2015, 2847–2855.
22. Uvaraja V.C and Natarajan, "Tribological behavior of heat treated Al7075 aluminium metal matrix composites", Indian Journal of Engineering and Material Sciences, 22, 2015, 51-61.
23. Veeresh Kumar G.B, Rao, Selvaraj, and Bhagyashekar, "Studies on Al6061-SiC and Al7075-Al2O3 metal matrix composites", Journal of Minerals and Materials Characterization Engineering, 9, 2010, 43-55.

24. Warren H, Hunt, Darrell and Herling Text book, "Aluminium metal matrix composites", Advanced Materials and Processes, 2004.
25. Yoshiro Iwai, Hidetomo.Yonede, and Tomomi.Honda, "Sliding wear behaviour of SiC whisker reinforced aluminum composite", Wear, 181, 1995, 594-602.
26. Zeynep Taşliçukur, Gözde S.altuğ and Şeyda polat, "Characterization of microstructure and fracture behavior of GG20 and GG25 cast iron materials used in valves", Metals, 5, 2012, 23-25.
27. Zhang, Long S, and Flower, "Light alloy composite production by liquid metal infiltration", Composites, 25, 1994, 380–392.

Study on Spheroidization and Related Heat Treatments of Medium Carbon Alloy Steels

Harisha S. R.[1*], *Sathyashankara Sharma*[1], *U. Achutha Kini*[1] and *Gowri Shankar M. C.*[1]

[1]*Department of Mechanical and Manufacturing Engineering, Manipal Institute of Technology, Manipal Academy of Higher Education, Manipal, Karnataka 576104, India*

Abstract. The importance of medium carbon steels as engineering materials is reflected by the fact that out of the vast majority of engineering grade ferrous alloys available and used in the market today, a large proportion of them are from the family of medium carbon steels. Typically medium carbon steels have a carbon range of 0.25 to 0.65% by weight, and a manganese content ranging from 0.060 to 1.65% by weight. Medium carbon steels are more resistive to cutting, welding and forming as compared to low carbon steels. From the last two decades a number of research scholars reported the use of verity of heat treatments to tailor the properties of medium carbon steels. Spheroidizing is the novel industrial heat treatment employed to improve formability and machinability of medium/high carbon low alloy steels. This exclusive study covers procedure, the effects and possible outcomes of various heat treatments on medium carbon steels. In the present work, other related heat treatments like annealing and special treatments for property alterations which serve as pretreatments for spheroidizing are also reviewed. Medium carbon steels with property alterations by various heat treatment processes are finding increased responsiveness in transportation, aerospace, space, underwater along with other variegated fields. Improved tribological and mechanical properties consisting of impact resistance, stiffness, abrasion and strength are the main reasons for the increased attention of these steels in various industries. In the present scenario for the consolidation of important aspects of various heat treatments and effects on mechanical properties of medium carbons steel, a review of different research papers has been attempted. This review may be used as a guide to provide practical data for heat treatment industry, especially as a tool to enhance workability and tool life.

Keywords: Medium Carbon Steel, Heat Treatment, Spheroidizing, Microstructure, Annealing

1 Introduction

Medium-carbon steels have carbon content ranging from 0.25 to 0.65% by weight. These categories of steels respond intensely to heat treatment processes, with wide variation in the mechanical and thermal properties. Depending on their response to heat treatment, they are

* Corresponding author: harisha.sr@manipal.edu

divided into 2 categories. The first group containing carbon from 0.25 to 0.35% by weight, whereas the second group from 0.35 to 0.65% by weight. The second group response towards heat treatment is better, while of the first group is poor. That is why steels of the first group are suitable for moderately stressed components like, railway couplings, driving rings, hand tools, sockets and levers, tubes for bicycles, automobiles and aircrafts. Due to higher carbon content, 2nd group steels have more strength. Hence, they are used in high load applications, like spindles of machine tools, gears, bolts, cylinder liners, crank shafts, ball mill balls, etc. Typical medium carbon steels generally exhibit poor cold workability. It is widely known that their ductility can be improved significantly by spheroidization process. In 1970, a decisive major dynamic change was observed with the development of micro-alloyed medium carbon steels. Then, micro-alloyed steels containing Cr, Ni, V, W etc. as refractory metals proved to be highly cost effective by attaining superior properties with least heat treatment processing cost [1]. The excellent toughness with high tensile strength is possible due to the addition of Mn and Si followed by reduction in carbon content [2].

2 Spheroidization

2.1 Spheroidization of Medium Carbon Steels [3, 22, 28]

Spheroidization experiments on AISI 4037 steel were performed, using inter-critical and sub-critical annealing. The results show that in sub-critical heating cycle, the spheroidization process has occurred faster than inter-critical annealing. So shorter times are enough to produce high formability. The other result was that in the inter-critical cycle the hardness dropped faster. In the testing of AISI 4037 steel (0.37C, 0.80Mn, 0.22Si, 0.25Mo by wt.%) rods of 3/8 inches, the rate of spheroidization with inter-critical and sub-critical cycles were compared. In inter-critical cycle, specimen was heated to 748°C for 2 hours, slowly cooled to 704°C followed by isothermal holding for 10 hours and later cooled at 5°C/min approximately to room temperature. In the sub-critical cycle, the specimen was heated to 704°C for up to 10 hours and then cooled at approximately 5°C per minute.

To examine the microstructures developed, scanning electron microscopy (SEM) was used at each stage. After 4 and 12 hours of processing the surface was examined. They show that after 4 hours, in the sub-critical cycle, spheroidization was nearly complete whereas in the inter-critical cycle still un-spheroidized coarse pearlite was observed. It showed that the inter-critical cycle produces a much coarser spheroidised structure.

The more rapid rate of spheroidization during the subcritical process was attributable to the fact that a much finer pearlite was being spheroidized. With rapid spheroidization rate, high formability was the result of sub-critical route.

2.2 Spheroidization Cycles for Medium Carbon Steels [4, 29]

Two process cycles were considered in the spheroidization of medium carbon steels which are used in the temporary fastener industry. In the inter-critical cycle, the material was heated above the A_{C1} for 2 hours. The specimens were then allowed to cool below A_{C1} and were held there for various time intervals. This allowed the transformation of austenite and spheroidization of carbides. The second process was a sub-critical cycle. In this cycle the specimens were heated to temperature below the A_{C1} for different time periods. Two steels were studied in the process. Wires of AISI 1541 (high in manganese content and mild supporter of spheroidization) and AISI 4037 (prone to spheroidization, used extensively in fastener applications) were taken into consideration.

Similar drops in hardness were observed in both the cycles. But, 1 hour of the sub-critical cycle produced higher ductility than 32 hours of the inter-critical process. The formability test also showed that spheroidization was faster in the sub-critical cycle. It was concluded that degree of formability is directly proportional to degree of spheroidization but not with changes in hardness due to alloying elements. Less time was required for spheroidization with sub-critical process than the inter-critical process.

2.3 Sub-critical Spheroidization of Medium Carbon 50CrV4 Steel [5]

Tests on sub-critical spheroidization annealing of cold rolled specimens were carried out on 50CrV4 steel at 720°C. The chemical composition of 50CrV4 steel was 0.543 C, 0.827 Mn, 0.23 Si, 0.137 V, 1.012 Cr, 0.019 Al, 0.001 S and 0.011 P by wt.%. The A_{C1} of the steel is 740°C and the A_{C3} is 790°C. Steel sheets were cut into specimens with 300x200mm^2 rectangular dimensions. They were then annealed sub-critically at temperature of 720°C and furnace cooled at the rate of 10°C per minute to room temperature.

Tests showed that deformation with cold rolling accelerates the development of spheroidal cementite particles in the steel. If the deformation is high, the spheroidization occurs faster. Strength of the steel was decreased with respect to soaking time initially, but it had little effect on the hardness and the strength after 8 hours of spheroidization.

2.4 Effect of Si on the Low Temperature Thermo-Mechanical Deformation and Spheroidization of Medium Carbon Steels [6, 31]

Medium carbon steels SWRCH 45F, with composition 0.45 C, 0.35 Mn, 0.03 Cr and 0.2, 0.5, 0.8, 1.5 Si% by weight were subjected to heat treatment and deformed at elevated temperatures using Thermecmaster-Z. Rod specimens 8 mm in diameter and 12 mm in height were kept in the austenite region at a temperature of 950°C for 10 min and deformed under compression at 750, 700, 650, and 600°C, respectively, at a strain rate of 10 μm/s, followed by controlled cooling at a rate of 10°C/s. The specimens were subjected to spheroidization at 730°C for 30 min and 3 h, respectively, and subsequently underwent slow cooling with a cooling rate of 25°C/h down to 680°C, followed by air cooling. At the central region of the specimen microstructural characterization was performed as it is the region where the most severe plastic deformation is expected to take place. The average grain size and fraction of the ferrite were measured using a method specified in JIS standard G-0552 and an image analyzer, respectively. Electron microscopy was used to characterize the microstructures of the spheroidized specimens. The dilatometric change of the specimen was measured during heating (0.08°C/s) and cooling (5°C/s) using a Thermecmaster-Z to investigate the change in the transformation temperatures (Ac_3, Ac_1 for heating and Ar_3, Ar_1 for cooling) with varying Si content.

With increasing Si content the transformation temperatures increased. Average fraction of the ferrite and grain size after thermo-mechanical deformation was also observed to increase with high Si composition. Addition of Si in low temperature thermo-mechanically processed medium carbon steels led to the required extent of spheroidization. A modified thermomechanical deformation and spheroidization heat treatment process was proposed and found to be highly applicable to Si added medium carbon steels.

2.5 Influence of Spheroidizing Heat Treatment on Mechanical Properties of EN 47 steel [7,20-21]

The specific grade of medium carbon steel used to carry out the experiment was EN47 which is used widely in the formation of axles, gears etc. Heat treatment plan consisted of primary heating to a temperature of 815°C for a soaking time of 2 hours which is immediately followed by transfer of specimens to another furnace maintained at 750°C where the specimens are isothermally heated for different durations of time. The specimens are divided into three categories with 3 hours being the spheroidizing time for first set, 6 hours for second set and 9 hours for the third set of specimens. Mechanical tests carried out on the heat treated specimens were hardness, tensile, impact, wear and microstructure analysis. The microstructural analysis revealed the morphological change within the carbide phase with reference to the duration of processing. Spheroidal wearing was observed with heat treatment process increased beyond critical value. It was observed that with increase in treatment duration from 3 to 9 hours, the hardness of the material decreased continuously. With respect to the impact and tensile strength tests, it was observed that the strength showed an initial increase with increase in time of spheroidization, reached a peak and then decreased with further increase in processing time. Coefficient of friction values showed a decrease in their magnitude as the spheroidization processing time increased along with increase in wear resistance.

3 Annealing

3.1 A New Approach for Rapid Annealing of Medium Carbon Steels [8, 34]

The 42CrMo medium carbon steel was the primary material used in the test. 20 × 10 × 2 mm^3 dimensioned cuboidal specimens were obtained from the hot rolled rod, with the hot rolling direction being in parallel with the longitudinal direction. 15 T cryo-cooled superconducting magnet, 52 mm in bore size was used with a furnace set in it to perform magnetic annealing, where the central (zero magnetic force) region was used for positioning and placement of specimens with magnetic field direction made parallel to the direction of hot rolling. Specimens were heated to 880°C and soaked for 33 min with a subsequent 46°C/min cooling rate along with a 14T magnitude magnetic field. Conventional full annealing consisted of heating at 860°C with soaking time of 30 min in a chamber furnace of ordinary type followed by furnace cooling with a cooling rate of 1°C/min. Post and pre heat treated specimens were subjected to etching. An optical microscope equipped with image analyzer was used for microstructural examination. Vicker's micro-hardness test was performed with a 5 kgf load. In 42CrMo steel banded structure formation initiated due to preferential nucleation of pro-eutectoid ferrite within fine grained zones of hot rolled microstructure along the austenite grain boundaries. Uniform and fine microstructure of pearlitic colonies along with ferrite grains was produced using the dual strategy of 46°C/min high cooling rate and high magnetic field of magnitude 14T.

3.2 Formation of medium carbon TRIP steel microstructure during annealing in the inter-critical temperature range [9, 17-18]

Mechanically unstable retained austenite along with multiphase microstructure are the primary characteristics of Transformation Induced Plasticity (TRIP) steels. TRIP steel (41MnSi6-5) was subjected to dilatometric tests which comprised the initial stage of experiments. Through a heating rate of 3°C /min, 2×12 mm^2 sized samples were heated till 1200°C in dilatometer DT1000. Sample relative elongation L/Lo was recorded as a function of temperature T while heating. Predetermined inter-critical temperatures range of 730-870°C was used for a time duration of 30 min within which the samples of size 12×10×10 mm^3 were heated. Martensite was obtained through transformation from austenite because of water quenching. Metallographic micro-sections were subjected to X-ray quantitative phase analysis which was used to assess the retained austenite volume fraction. X-ray diffractometer TuR-M61 was used for these tests through the application of filtered cobalt radiation immediately after X-ray tests for the purpose of revealing structural components. 2% Nital was used to etch the samples for revealing the microstructure. Revealed microstructure characterization was conducted through scanning microscope FEI Nova Nano SEM 450 and Light microscope Axiovert 200MAT. Point-count and lattice variant quantitative metallography methods of manual nature were used on the visible microstructural components to obtain the volume fractions. Heat treated samples were subjected to Vickers hardness test at 30 kg load. At a load of 10 g the Knoop indenter was used for the 750°C annealed sample for verification of identified microstructural components. Largest volume fraction of retained austenite was obtained within the microstructure of samples annealed at low temperatures in the inter-critical range. It was observed that contrary to common notion of differential transformation of ferrite to low carbon austenite, the transformation actually runs simultaneously. For applications of final retained austenite volume fraction and increased carbon content in austenite in the TRIP steel microstructure, the strategy to be followed is to go for the modification which would allow to lower the annealing temperature range to Ac_1+ max. 30°C.

4 Unconventional Heat Treatments

4.1 Effect of Post Weld Heat Treatment on the Mechanical Characterization of Medium Carbon Steel Weld [10, 23-24, 30, 32]

Mechanical properties of weld in 0.36% C medium carbon steel was analyzed. Post weld heat treatment (stress relief) was done on welded samples at four different sub-critical temperatures 550, 600, 650 and 700°C followed by air cooling. The welding process used is SMAW with low hydrogen electrode (E7018) at a current of 130 A. Hardness and tensile tests were done on the samples later. The percentage loss in strength due to post weld heat treatment (PWHT) at the above temperatures followed by air cooling was 10.8, 12.2, 15.1 and 13.1% respectively. The best PWHT temperature for the property improvement of medium carbon steel is 650°C and the corresponding loss in strength was about 15.1%.

4.2 Some Laser Treatments Effects on Medium Carbon Steels [11]

Austenization and oil quenching were applied on plates 10 mm thick of 0.38 wt.% C-Ni-Cr-Mo low alloy (0.4 wt.% C) alloy steel. The plates were graphite coated to increase absorptivity and subjected to single pass laser irradiation. The beam translation speeds were 8.3 and 5.0 mm/s along with a beam size of 10 × 10 mm^2. Corresponding interaction times of laser material were 1.2 and 2.0 s and the incident power densities were 2.1 and 2.6 kW/cm^2 and along with energy densities of 25 and 52 MJ/m^2. The Mossbauer measurements were performed at room temperature using a 40 m Ci57co(Rh) source, and by detecting the re-emitted 6.4 keV X-rays and the 7.3 keV K-shell conversion electrons. Computer-controlled goniometer and CoKo radiation were used for X-ray diffraction analyses. Scanning electron microscopy and optical microscopy were utilized for the examination of the cross sections for the treated samples. A critical stage for laser treatments of steels is the control of the atmosphere composition. Irrespective of the use of a helium stream and graphite coating, all the treated samples underwent surface oxidation phenomena. Formation of low carbon phases is promoted under strongly oxidizing conditions and surface melting is hindered or prevented. Composition of the steel under laser processing conditions varying in the neighborhood of incipient melting has major influence on the extent to which melting takes place.

4.3 Influence of Heating Rate to Austenitic State on the Laser Surface hardening of a Medium Carbon Steel [12, 27, 33]

Specimens of 5 mm thickness medium carbon steel (SM 45C) was laser surface hardened at varying traverse speeds and beam spot diameters. Composition consisted of 0.26 Si, 0.73 Mn, 0.013 P, 0.005 S and 0.45 C by wt.%. For uniformity in conditions of absorption, before each hardening process surface of the specimen was cleaned and ground thoroughly. A 1 kW CO_2 laser system was used to perform surface hardening with varying traverse speeds of 0.5 -3 m/min along with beam spot diameters ranging from 1.6- 3.5 mm. Normal to the translation direction cutting was done of the processed specimens for metallographic examination. Through micro hardness testing and use of normal etching methods, a 100 gf load was used with a diamond pyramid indenter in the hardened zones. The theoretical result by Bramson's or Arata's equation and adopted for the FDM simulation to global absorptivity levels 15% less than that measured using a calorimeter. According to the distance of the work piece surface from the focal point for comparison of calculated and experimental results the critical beam spot diameter must be determined. The average diameter at which the beam average intensity falls to 5% of the maximum beam intensity was defined as the beam spot diameter. Through comparison of the burned zone widths produced on the acrylic plate under various defocused distances and traverse speeds, the relationship between the defocused distance and the beam spot diameter was determined. The criteria for choosing beam spot diameter were such that at the boundary of the burned zone the total incidence energy per unit area remained constant irrespective of irradiation conditions. Between the beam spot diameter D_b and the defocused distance f_d, the mathematical relationship could be expressed as

$$D_b = A \exp(Bf_d). \quad (1)$$

Where B and A are constants which are obtained using method of least square. To account for the shift in the austenitic transformation temperature the temperature dependent specific heat was modified, the specific heat in the solidification range was modified as follows taking latent heat into account

$$C_m = \frac{C_s + C_l}{2} + \frac{H}{\Delta T_{ls}}. \quad (2)$$

where C_l and C_s represent specific heats at the liquidus and solidus temperature respectively, ΔT_{ls} is the temperature difference between the solidus and liquidus temperature whereas H is the latent heat of fusion. Resistivity measurements or Dilatometer showed that heating rate strongly influences transformation behavior in steels.

For analysis of laser transformation hardening the finite difference method is pretty accurate. While determining isothermal lines for hardening the effect of forced convection was negligible. The calculated size and shapes of hardened zone correlated well with those obtained by experiment with the modified transformation temperatures. For accuracy in predicting the laser surface hardening zone, the relationship between the machine and beam spot diameter setting has to be analyzed. At the surface width of the hardened zone increases with increasing beam spot diameter at relatively small beam spot diameters. For a given laser power input and traverse speed there can be a range of beam spot diameters which without any surface melting produce a large width of the hardened zone.

5 Property Alterations in Heat Treatment Processes

5.1 The Microstructural Influence on the Mechanical Behavior of Medium Carbon Steel after Heat Treatment [13, 19]

Unalloyed carbon EN8 steel in the form of rolling bar was the test specimen used in this investigation. Tensile specimens were machined to standard dimensions and prepared using heat treatment experiments. Dimensions of ASTM standard were used to prepare tensile strength test specimens to be subjected to heat treatment cycles. Four sets of "A", "B", "C" and "D" groupings were made for tensile test. Microstructure characterization and phase identification of samples was obtained using Optical Electronic Microscope (OEM) and Scanning Electron Microscopy (SEM). Type, shape and size of granular structure were the principal properties to be examined of the test specimens. Hardness and tensile strength tests were primarily used for mechanical characterization of specimens. Four heat treatment plans were used for four specimens. The first one consisted of 950°C austenitizing soaked for 180 min with furnace cooling, second one consisted of 914°C austenitizing soaked for 10 min, with furnace cooling to 680°C, third heat treatment plan comprised of austenising at 914°C, held for 10 min, cooled in furnace and fourth heat treatment plan comprised of austenitizing at 914°C soaked for 3 min with furnace cooling to 715°C and subsequent oil quenching. Five different microstructures were obtained through the heat treatment cycles. Heavy impact on the characteristics of yielding was attributed to distribution and size of grain. For variable grain size distribution in a multiphase structure, high Von Mises stress was experienced in stronger phase while weaker phase was characterized by high plastic deformation. Distribution and grain size were profoundly impacted by holding time and annealing temperature. Increased formation of second phases (bainite and martensite) upon cooling and grain coarsening were promoted with increase in soaking time and annealing temperature. Material flow dependent manufacturing processes would be benefited through the "B" sample heat treatment because of coarse grains as yield strength increased linearly with decreasing grain size in conventional materials. "C" and "D" heat treatments would be appropriate for high hardness and high strength applications.

5.2 Effect of Thermal Treatment on the Bauschinger Effect for a Medium Carbon Steel [14, 25]

The steel used for this heat treatment process was DIN Ck60. The material was first hot-rolled, then normalized at 800°C for an hour and was cooled in the furnace. Then machined in the shape of button-head push-pull with a diameter of 14 mm. A 250 kN MTS testing machine was used to perform loading for all specimens. At room temperature, they were strained at 5×10^{-4} S^{-1} strain rate. The load versus displacement curves were obtained after measuring the strains using a longitudinal extensometer 25 mm in gauge length. In the initial condition, a pre-strain of 2.5% of total true strain in the mode of tension was applied on the specimens. The next process occurred over an hour where the specimen pairs were heated at temperatures from 373 -873 K. Then each pair was reloaded one in compression and the other in tension. Loading in compression was limited to 1.5% total true strain, to not cause buckling. The chemical composition of the steel used is of 0.26 Si, 0.55 Mn, 0.021 P, 0.032 S and 0.66 C by wt.%.

Heating medium carbon steel above $0.25T_m$ (where T_m is the melting point) reduces the Bauschinger effect. This is because of a recovery annealing assisted by short-range recoverable internal stresses. To determine the Bauschinger effect, the altered value of mechanical properties in the pre-straining direction should also be taken into consideration. It was observed that strain does not influence the Bauschinger effect.

5.3 Behavior of Medium Carbon Steel under Combined Fatigue and Wear [15, 26]

The choice of material for this experiment was AISI 1045. A cylindrical shape of 16 mm (5/8 in) in diameter was given to the material. To provide accurate shape to the testing region, a lathe tool of 12.7 mm radius was used. Emery paper was used to polish the necked area to a minimum diameter of 6.4 mm, followed by 6 μm particle size lapping compound embedded with diamond lap for finishing. A thin film of cadmium (0.1 μm) or nickel (0.05 μm) was used for plating on few specimens. A layer of gold (0.5 μm) was then applied on them. Then, lathe was used for combined fatigue wear tests and fatigue. The configuration of the load applied at the end of a cantilever rotating beam was adopted. Dead weights of 0.3 kg were applied as the normal load as well as fatigue load on the slider. A strain gage assembly with a slider attached to it was used to measure the force of friction. Argon atmosphere and air were used as the medium of tests. To have a controlled atmosphere to perform the tests, the slider and specimen were enclosed in a chamber, which was completely vacuumed out and then filled with argon gas. The tests were performed continuously until the specimen failed at an rpm of 1000. As the load on the slider was gradually increased, the rate of crack nucleation at the surface also enhanced and a doubling was observed in the actual number of fatigue loading cycles. The coating on the specimens reduced the wear rate, but it does not enhance the fatigue.

5.4 The Effect of Quenching on the Wear Rate of Medium Carbon Steel [16]

This experiment was conducted on a "ΦK" type fretting tester. The load was applied using a weighted lever and the contact pressure was calculated from the calculated contact area of the specimens. The conditions of operation for this machine were amplitude range of 20 to 1000 μm, frequency between 1.86 - 25 Hz and contact pressures of 5.3 - 110 MN m^{-2}.

Commercial cylindrical 45mm diameter carbon steel, with annular contact surfaces, was used for this experiment. A few of the specimen pieces were made to undergo quench hardening up to 587 HV. They were marked H45. Before starting the wear test, surfaces undergoing the test of every specimen was finished by grinding to get a rough state of grade 0.32 - 0.63 μm R_a,

Before the testing, the specimens were weighed to know their initial weight. After the test, they were weighed again, to know the weight loss. To obtain the cumulative reduction in mass, the waste material filaments were collected and weighed from time to time. Inert gas fusion oxygen determining machine was used to gather data about the percentage oxygen present in the shavings obtained after the test.

After that on the annular surfaces, a couple of cuts were made and along with that the NiCr-NiAl thermocouples of Φ0.5 mm were embedded in the specimen. A twin pen recorder was kept to note down the temperature rise from time to time. The entire experimental setup was put up at room temperature conditions. Quench hardening of carbon steel had a lot advantages. It enhances fatigue strength, tool life and hardness.

6 Summary

The extensive literature survey presented above portrays the effects of spheroidization, annealing and related heat treatment processes on various grades of medium carbon alloy steels under differing conditions of heat treatment process parameters. While it was observed that spheroidization was accompanied by good cold workability and formability characteristics and annealing showed a marked increase in the percentage of fine grained pearlitic structure; further in depth analysis has to be done to obtain better quantitative results. Along with the conventional heat treatment processes like annealing and spheroidization, this paper also delves into the realm of certain unconventional processes like post weld and laser heat treatments which have not been adopted at an industrial scale and thus holds great research and commercial potential in the near future. Material science researchers at all levels will find this review on medium carbon steels helpful in their own scientific pursuits.

References

1. T. Gladman, D. Dulieu, I. D. McIvor, Microalloying 75 symposium (Washington DC, 1977)
2. S. Sankaran, V. S. Sarma, K. A. Padmanabhan, Mater Sci. Engg., A345, 328 (2003)
3. J. M. O'Brien, W.F. Hosford, J. of Mat. Eng. Per. 6, (1997)
4. J. M. O'Brien, W.F. Hosford, Met. Mat. Tran A 33A, (2002)
5. W. Y. Guo, J. Li, X. F. Jiang, J. of Mat. Eng. Per. 21, (2012)
6. Y. S. Choi, I.M. Park, J.H. Kim, K.M. Cho, D.L. Lee, Met. Mat. Int. 11, 327 (2005)
7. S.R Harisha, S.S. Sharma, U. A. Kini, Mat. Sci. For. 880, 136 (2016)
8. Y. Zhang, C. He, X. Zhao, C. Esling, L. Zuo, Adv. Eng. Mat. 6, 5 (2004)
9. A. Kokosza, J. Pacyna, Arch. Met. and Mat. 59, 3 (2014).
10. S. M. Manladan, B. O. Onyekpe, App. Mech. Mat. 315, 6 (2013)
11. M. Carbucicchio, G. Sambogna, G. Palombarini, Hy. Int. 28, 1021 (1986)
12. S. J. NA, Y. S. Yang, Sur. Coat. Tech. 34, 319 (1988)
13. K.O. Sansui, C.M. Moleejane, O.L. Ayodele, G.J. Oliver, Tran. Eng. Tech. 525 (2015)
14. Aran, M. Demirkol, Mat. Sci. Eng. 47, 89 (1981)
15. T. Nagao , J.J. Pamies-Teixeira, Wear 44, 101 (1977)
16. D. X. lin, Wear 123, 77 (1988)
17. A. Bhateja, A. Verma, A. Kashyap, B. Singh, I. J. of Eng. Sci. 1, 253 (2012)
18. A. Saha, D. K. Mondal, J. Maity, Mat. Sci. Eng. A, 537, 4001 (2010)
19. Sarwar M, Priestner R., J. of Mat. Sci. 31, 2091–2095. (1996),
20. A. Saha, D. K. Mondal, J. Maity, Mat. Sci. Eng. A, 537, 4001 (2010)
21. J. Y. Li, S. Suguyama, J. Yanagimoto, Y. L. Chen, G. W. Fan, J. Mat. Pro. Tech. 208, 165 (2007)
22. S. Y. Oh, I.H. Son, K.H. Jung, D.K. Kim, D.L. Lee, Y.T. Im, Mat. Sci. Eng. 528, 5833 (2011)
23. Y. W. Lee, Y. Son, S. J. Lee, Mat. Sci. Eng. A, 585, 94 (2013)
24. A. J. Abdalla, L. R. O. Hein, M. S. Pereira, T. M. Hashimoto, Mat. Sci. Tech. 15, 1167 (1999)

25. S. Sun, M. Pugh, Mat. Sci. Eng, 335, 298 (2002)
26. M. Erdogan, R. Priestner, J. Mat. Sci. Tech. 18, 369 (2002)
27. H. Goto, Y. Amamoto, Wear 254, 125 6(2003)
28. D. H. L. Ng, K. S. Cho, M. L. Wong, S. L. I. Chan, Mat. Sci. Eng 358, 186 (2003)
29. P. K. Mitra, S. Paul, S. Chatterje, J. Mat. Mat, 85, 78 (2004)
30. Y. Zia, W. Liu, Q. Xue, Wear, 38, 508 (2005).
31. S. M. Anijdan, H. Vahdan, Mat. Let. 59, 1828 (2005)
32. P. Larour, P. Verleysen, W. Bleck, J. Phy. 134, 1085 (2006)
33. P. L. Menezes, Kishore, S. V. Kailas, Wear 261,578 (2006)
34. A. P. Modi, Tri. Int. 40, 490 (2007)
35. M. Sarwar, E. Ahmad, K. A. Qureshi, Manzoor, Mat. Des, 28, 335 (2007)

Experimental and numerical investigation of mode II failure behavior evaluation using three point bend, end notched flexure test

R. Nikhil[1*], *S. Shivakumar*[1], *Kallol Anupama*[1] *and Shettar Manjunath*[2]

[1]*Department of Mechanical Engineering, KLS Gogte Institute of Technology, Belagavi, India*

[2]*Department of Mechanical Engineering, Manipal Institute of Technology, Manipal, India*

Abstract. In the present paper the primary task is the study involving calculation of elastic properties of the composite from the individual properties of the E-glass fiber (650 GSM) and the properties of resin LY 556 with Hardener HY951. The properties of varying volumetric ratio of fiber are obtained from calculation of the properties by using rule of mixtures. Experimentally validating the theoretical and numerical approaches by comparing the load-displacement response and crack paths observed in large scale bridged crack propagation in laminated fiber-reinforced composites specimens. An effort is being made to develop a numerical framework for cohesive crack propagation and demonstrating its effectiveness by simulating failure through crack propagation in materials with complex microstructure like fiber reinforced composites. Experimentally validating the theoretical and numerical approaches by comparing the load-displacement response and crack paths observed in large scale bridged crack propagation in laminated fiber-reinforced composites specimens.

1 Introduction

The introduction of fiber-reinforced composites has been a major step in the evolution of structural frames. Limited attention has been paid to modeling the fracture of crack-like flaws in pultruded composite systems, fracture mechanics has previously been advocated and applied to thin-section laminated composites. Sih et al.[1] used a complex variable approach to derive the general equations of crack-tip stress fields in anisotropic bodies. Finite element analysis using the virtual crack closure technique [2-3] has been used to determine the strain energy release rate and relate it to the SIF. In an attempt to use existing stress intensity factor (SIF) solutions derived for isotropic conditions, Mandell et al. [4] used hybrid FE analysis to show that the SIF can be dependent on the degree of material anisotropy for various specimen geometries. The SIF change between isotropic and anisotropic cases was relatively constant for varying crack lengths in a given geometry. The material constants were found to have an effect on the crack-tip stresses. Boa et al. [5-6]

[*] Corresponding author: nikhilr@git.edu

used an orthotropy rescaling technique to fit formulae into numerical calibrations to quantify the effect of material orthotropy on notched bars, delaminated beams, and hybrid sandwiches.

Stress based failure criteria were proposed to determine the strength of notched composites. Some of these criteria were evaluated point-wise while others were determined using average variables or characteristic lengths. Whitney and Nussimer [7] proposed a failure criterion using the average stress distribution calculated from the edge of the hole to a given characteristic distance. The distance from the edge was considered as a material property since it did not depend on the size of the notch. A second related criterion was proposed where failure is assumed to occur when the stress, at a distance away from the notch, reaches the strength of the unnotched material. The proposed criteria were applied to graphite / epoxy and glass / epoxy laminated composites with straight cracks and were used to explain the reduction in strength of composites with larger hole sizes. Sih et al. [8] proposed a strain energy density failure criterion to predict fracture in unidirectional composites subject to off-axis loading. Despite limitations, such as the violation of self-similar crack growth in various laminates, Linear Fracture Mechanics (LFM) has been applied to study failure in different composite systems [9-11] used a stress intensity factor (SIF) calculated based on amodified crack size. The larger effective crack length was determined by equating the intense energy region ahead of the crack to a material dependent characteristic length.

The paper starts with general theory regarding the determining of crack behaviour due to propagation and lifetime estimation. A numerical solution is conducted in ANSYS, and a process of programming due to obtain generals numerical solutions for crack trajectories and lifetime estimations is accomplished. In order to verify the numerical solutions from ANSYS, experimental results are taken into account for benchmarking. The methodology intended to be followed includes:

Phase 1: Overview of fractography and stress concentration factor

Phase 2: Development of elasto-plastic FEA model

Phase 3: Experimental studies

Phase 4: Comparison between FE and Experimental studies.

2 Experimental study

The experimental study involves calculation of elastic properties of the composite from the individual properties of the E-glass fiber (650 GSM) and the properties of resin LY 556 with Hardener HY951. The properties of varying volumetric ratio of fiber are obtained from calculation of the properties by using rule of mixtures.

The three points bending flexural test gives values for the modulus of elasticity in bending Ef, flexural stress σ_f, flexural strain ε_f and the flexural stress-strain response of the material. The main advantage of a three point flexural test is the ease of the specimen preparation and testing. However, this method has also some disadvantages: the results of the testing method are sensitive to specimen and loading geometry and strain rate.

Fig. 2.1 shows the experimental setup of the specimen in UTM. The testing procedure for conducting the test usually involves a specified test fixture that can be movable to adjust according to the required support span needed on a universal testing machine where loading is done uniformly and at constant rate with respect to time. Details of the specimen preparation, conditioning, and conduct affect the test results as it could lead to improper binding of laminates thus three specimens of same volumetric ratio are taken and their average peak load is taken. The sample is placed on two supporting pins a set distance apart and a third loading pin is lowered from above at a constant rate until sample failure.

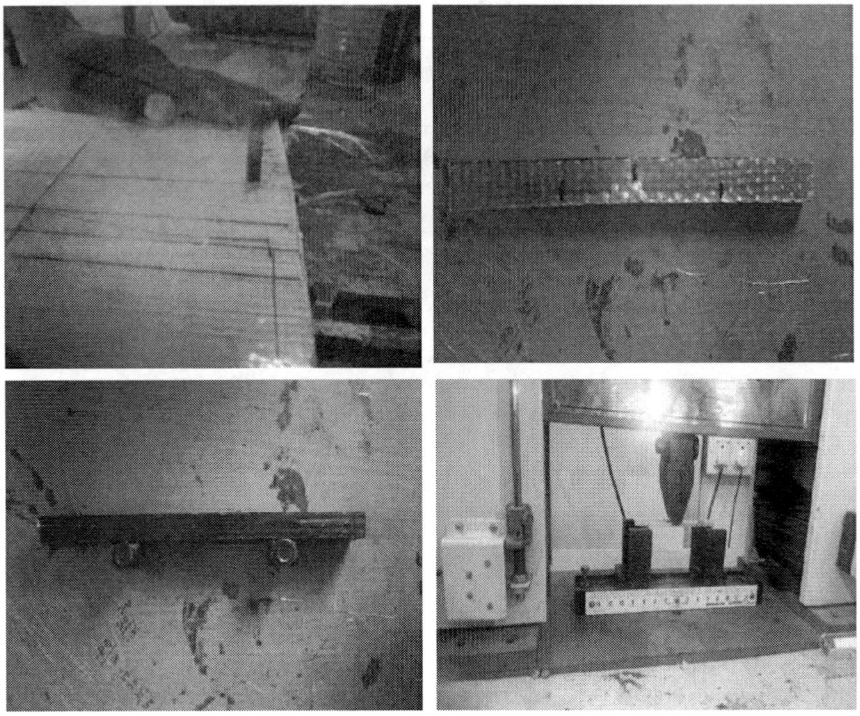

Fig. 2.1 Specimen preparation & experimental setup for fabrication and testing on UTM

Flexural stress given in following equation and is calculated by

$$\sigma_f = \frac{3FL}{2bd^2}$$

The experimental three point bending test by compliance calibration method is conducted on the ENF specimens to obtain peak critical load Pc at which the specimen fails and mode II crack propagates accompanied by sudden decrease in load in the load vs deflection graph. Then experiment is conducted for all specimens of varying volumetric ratio of fiber-resin and also on uncracked specimens.

For these values for 0.4, 0.5, 0.6 fiber volume ratio the G_{IIC} is calculated

$$G_{IIC} = \frac{9P^2Ca^2}{2w(2L^3 + 3a^3)}$$

The compliance (C) can be found by Equation

$$C = \frac{2L^3 + 3a^3}{8Ewh^3}$$

3 Cohesive zone modelling (CZM)

The ENF specimen is analyzed by CZM method to obtain SERR By using the peak load from the experimental values. These numerical values are compared with experimental values. Fig. 3.1 shows that CZM analysis is done on ENF specimen and Von Mises stress is obtained for different volumetric ratio of fiber, with peak load from experimental values taken and input as point load at the centre of specimen in direction being perpendicular to fibers along the length of specimen.

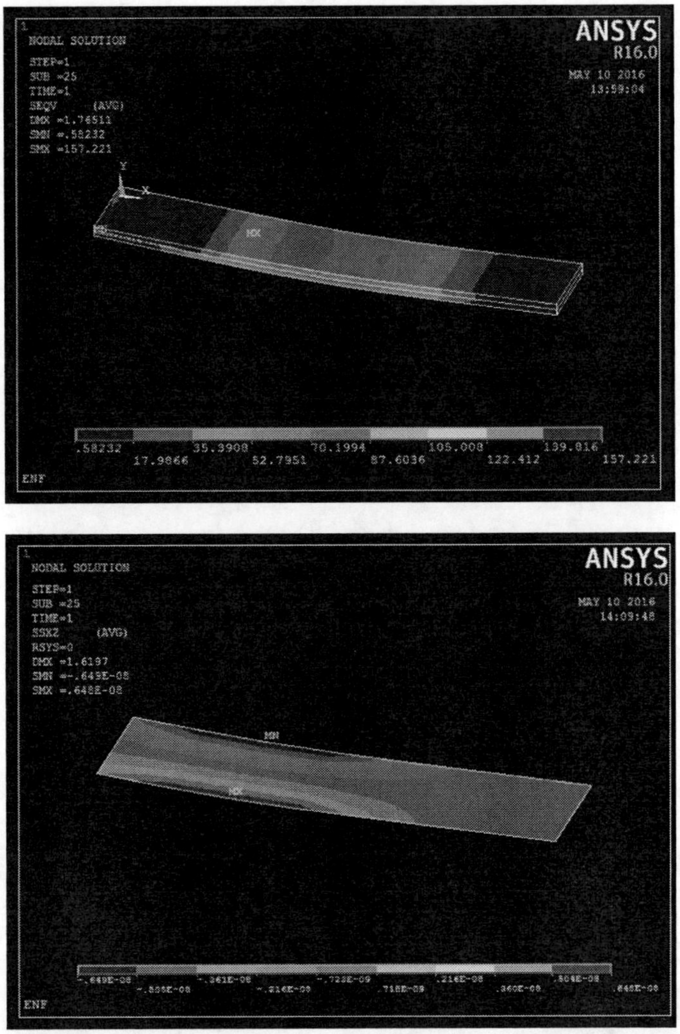

Fig. 3.1 Von- Misses Stress and displacement by CZM approach

Now, comparing the results from experimental and CZM approach shows that the values of SERR obtained from CZM has slightly higher values than experimental values of SERR (Fig. 3.2) and the difference of SERR for 0.5 fiber volume fraction is slightly less when compared to other values.

Table 3.1 Strain Energy Release Rate (SERR) of Experimental and CZM

Volume fraction of fiber	Average peak load with crack (N)	Average peak load without crack (N)	Strain Energy Release Rate - Experimental Method (J/mm2)	Strain Energy Release Rate - CZM Analysis (J/mm2)
0.4	2324.7	3379	0.74	1.15
0.5	2617.3	3547	1.59	1.78
0.6	3172.2	3726	2.10	2.37

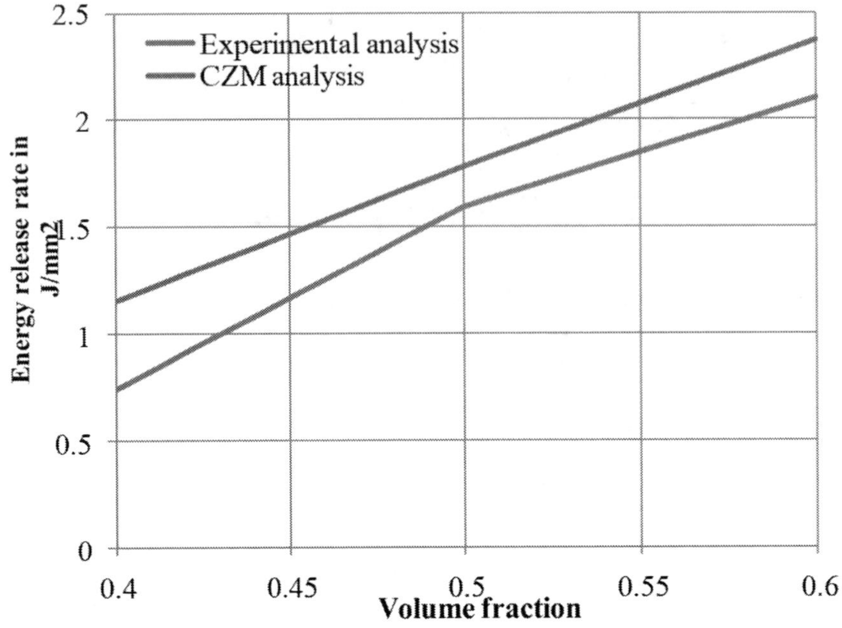

Fig. 3.2 Comparision of energy release rate of experimental and CZM approach

3.1 Fracture surface

Fig. 3.3 (a) and 3.3 (b-c) shows the fracture surfaces of un-notched epoxy/ glass composites and notched epoxy /glass composites respectively. Un-notched composites show lesser degree of fibre pullout (Fig. 3.3(a)), which is due to the delay in crack initiation and higher interfacial strength. The notched composites showed high fibre breakage due to initial damage and lesser interfacial strength. The resin-fibre bonds have broken and lumps of resin are observed in amorphous form, Fig. 3.3 (b-d)

Fig. 3.3: Tensile fracture surfaces of a) unmodified glass laminates and b) modified glass laminates

4 Summary of results

- From the experimental analysis performed it shows that the delamination method of crack propagation is mainly dominated by inter-laminar delamination between the layers and the crack propagates exactly along the middle layer where the initial crack was initiated, intra-laminar delamination, buckling of fibers and fiber effects like hygrothermal, moisture absorption and temperature effects do not contribute significantly to contribute (i.e. its effects are neglected during testing).
- Delaminations also occur due to improper binding, non uniform distribution of resin, presence of foreign particles ,absorption of moisture, non uniform compaction during pressing of laminate and curing of resin along the layers.
- During experimentation there is variation and load fluctuation during peak loads due buckling of fibers and after the peak load accompanied by loud sound, the load carrying capacity decreases indicated by steep fall in curve (load vs displacement).
- For varying volumetric composition the increasing content of fiber in ENF specimens have more load carrying capacity thereby having greater SERR and thus offers more resistance to crack propagation as compared to lower percentage of fibers. But too high fiber percentage (>75 %) means more brittle nature and less bonding due low resin content thereby crack propagates easily.
- Comparison of numerical results with available experimental ones showed that the CZM method, in spite of simplicity, cannot predict the propagation behaviour in unidirectional ENF specimen with fiber bridging very well. Hence a method such as Virtual Crack Closure Technique (VCCT) can also be used to study mode II failure behavior evaluation using three point bend, end notched flexure test

References

1. Sih. G. C., Chen E. P., Huang S. L., and Mc.Quillen E. J., J. Comp. Mate., vol. 9, pp. 167- (1975).
2. Raju I. S., Engg. Frac. Mech., vol. 28, pp. 251-274, (1987).
3. Krueger R., NASA/CR-2002-211628, ICASE Report no. 2002-10, 2002.
4. Mandell J. F., McGarry F. J., Wang S. S., and Im J., J of Comp. Mate., vol. 8, pp. 106-116, (1974).
5. Boa G., Ho S., Suo Z., and Fan B., Int. J. of Solids & Struc., vol. 29, no. 9, pp. 1105-1116, (1992).
6. Boa G., Ho S., Suo Z., and Fan B., Corrigenda, Int. J. Solids & Struc., vol. 29, no. 16, pp. 2115, (1992).
7. Whitney J. M., Pagano, NJ editors. Interlaminar response of composite materials. Amsterdam, Elsevier, pp. 161-250, (1989).
8. Sih G. C., Paris, P. C., Irwin, G. R., Int. J. of Frac. Mech., vol. 1, no. 3, pp. 189-203, (1965).
9. Konish Jr., H. J., Swedlow, J. L. Cruse, T. A., AIAA Journal, vol. 11, no. 3, pp.40-43 (1973).
10. Saxena, Ashok, Fibre Sci. and Tech., vol. 12, no. 2, pp. 111-128, (1979).
11. Parhizgar S., Zachary L.W., and Sun C. T., Int. J. Frac., vol. 20, pp. 3-15, (1982).

Sliding wear studies of microwave clad versus unclad surface of stainless steel 304

Akshata M. K.[1*], *Ajit M. Hebbale*[1], and *Srinath M. S.*[1]

[1]*Department of Mechanical Engineering, NMAM Institute of Technology, Nitte. 574110, India*

Abstract. Small and large scale (gas power plant, hydro power plant), automobile industries are suffering by failure of component. Sometimes, it is also observed that the component which was failed due to these reasons are very much costly and replacement of those also very difficult due to the complex geometry. By using Microwave hybrid heating, WC-12Co based clads were developed on austenitic stainless steel (SS304). Microwave clads were developed by introducing the preplaced, preheated powder for a duration of 15 min to microwave radiation at 2.45GHz frequency and 900 W power in domestic microwave applicator. By using optical microscope and scanning electron microscope (SEM), the developed clads were characterized. By using pin-on-disk, wear performance of the WC-12Co based clads and unclad samples were tested. It is observed that developed clad samples performed superior wear resistance than unclad samples.

Keywords: Microwave, Cladding, SEM, Wear, Austenitic Stainless Steel

1 Introduction

Austenitic Stainless steel has superior wear and erosion unsusceptible in several environments and has been broadly used in variety of sectors of industries. Generally, components of stainless steel fail more frequently due to erode instead of corrosion related problems. Hence, it is very much essential to modify the surface of the components to improve the service life of the component. Wear, corrosion and tear of the component surface are the main reason for the failure of the engineering materials [1-4]. It is comprehensively not verbalized in potential wear applications. However, to reach an increased functional surface of the steel can be a conceivable arrangement in many conditions. Bloyce [5] reported surface alloying is the best solution to decrease the wear rate. Now days various methods like cyaniding, carburizing, nitriding, and cladding techniques are available to improve the surface properties. Cladding, generally a popularly used technique to enhance the surface property of the intend material by fully changing the phase from solid to liquid of clad powder particles along with partial melting of the substrate. In laser cladding the laser beam melts the clad particles as well substrate and forms enough engineering thickness with fine metallurgical bond with substrate [6].

The accomplished review of material development through microwave energy and sintering of ceramics includes aluminum, zirconium oxide, barium titanium, and WC-Co based ceramic metal composites are reported[7]. Oghbaei et al. [8] reported the theory of microwaves and usage in different areas of manufacturing system. Authors extended his work to composites, polymers, and metallic powders. The detailed investigation of sliding wear studies was done on WC10Co2Ni . For lower sliding speed these has shown better

* Corresponding author: akshatha_213@nitte.edu.in

wear resistance [9]. Sharma et al. [10] described the joining of material through domestic microwave oven. Srinath et al. [11] Researches bonding of copper in mass form within short duration through microwave energy and it was noted that consistent microstructure secured the uniform volumetric heating all around the weld region. An organized layer of SS-316 and MS was formed by the authors when implemented microwave joining of different metals with Ni based powder [12]. Badiger et al. reported [13] Inconel-625 alloy was weld through microwaves using Ni based powder. Cobalt based clad are created through microwave vitality on martensitic stainless steel (AISI-420). The developed clad shown excellent metallurgical bond with 132% of higher wear resistance than the unclad surface. The cobalt based clad are also formed to be having superior erosion wear resistant [14, 15].Very recently, many researcher explored the cladding process through microwave energy. The innovation of this microwave cladding has confirmed that the developed clad has least amount of porosity and least defect as well crack free structures. This crack free structure leads to perform developed mechanical properties of material surface which is under analysis. It was discovered that the enhancement of clad surface through microwave energy has immense impact on material development cost and also involves minimal time and energy [16-18].

The current study investigate, the possibility of using domestic microwave oven for development of WC-12Co based clad on steel SS304. Dry sliding wear studies were carried out on developed clads and finally analyzed through SEM.

2. Experimental Details

2.1. Materials and Microwave Cladding process

A domestic microwave oven having microwave energy was used in this work are the well familiar commercially available hard facing alloying of WC-12Co based powder having particle size of 35 μm has been used to improve the surface SS304 steel. Typical SEM image of substrate SS304 is depicted in Fig.1.

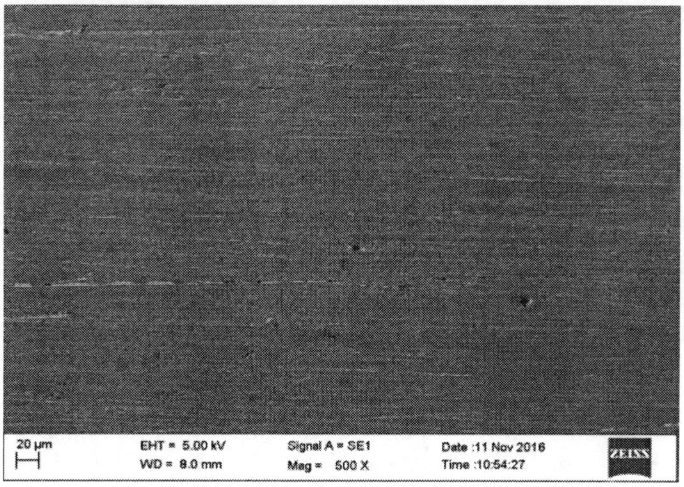

Fig. 1. Morphology structure of SS304.

The desirable samples are sliced to a size of 12×12×5 mm^3 from SS304 steel plate. Prior to the operation, samples are polished to remove any possible impurities such as grit and oil. Further, in a microwave oven the clad powder was heated in advance at approximately 100^0-120^0 C for 1h . By conserving a mean thickness of about 1mm, the clad powder was preplaced physically on the substrate. The preplaced clad powder was melted and that leads to strong metallic bond between the substrate. In the current work charcoal powder was used to gain the range of temperature powder particles. To avoid the adulteration of clad powder with charcoal powder, a precise flat graphite sheet is used as a barrier. Inside a refractory insulating box, the entire arrangement was kept as appeared in the Fig.2. It is noticed that the microwave cladding process effectively depends on the subjected time of microwaves.

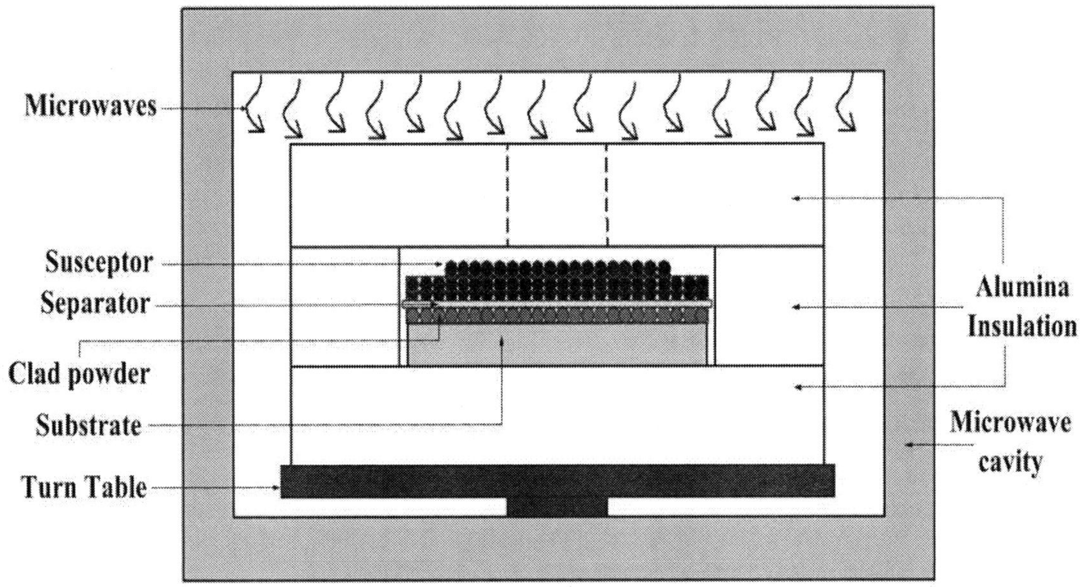

Fig. 2. Schematic of experimental setup of microwave cladding process [21].

2.2. Dry sliding wear studies

Pin on disc wear test is done to calculate the amount of wear rate on the surface of developed clad and unclad surface of SS304. A dry sliding wear execution of the created clad surface was assessed through pin on dics tribometer (Make: Ducom India) under different loading conditions. The sample size of 12x12x5 mm^3 were machined of developed clad surface. The 10mm diameter steel pins were welded to the opposite side of the developed surface to make easy to fix to the fixture. Finally the developed samples were polished through #1000 grit Sic paper to preserve a consistent surface contact. To gauge the wear rate of the pins before the test and after the test a microelectronic weight balance was used. Finally, the typical load was applied on the created clad surface through lever, string and pulley setup. The schematic of exploratory setup utilized as a part of the present work is appeared in Fig. 3.

Fig. 3. Schematic Pin on disc arrangement.

3 Results and Discussions

The wear studies of WC-12Co based developed clad surface was analysed through SEM to determine the sliding effect on the clad surface.

3.1. Microscopic inspection of worn surface

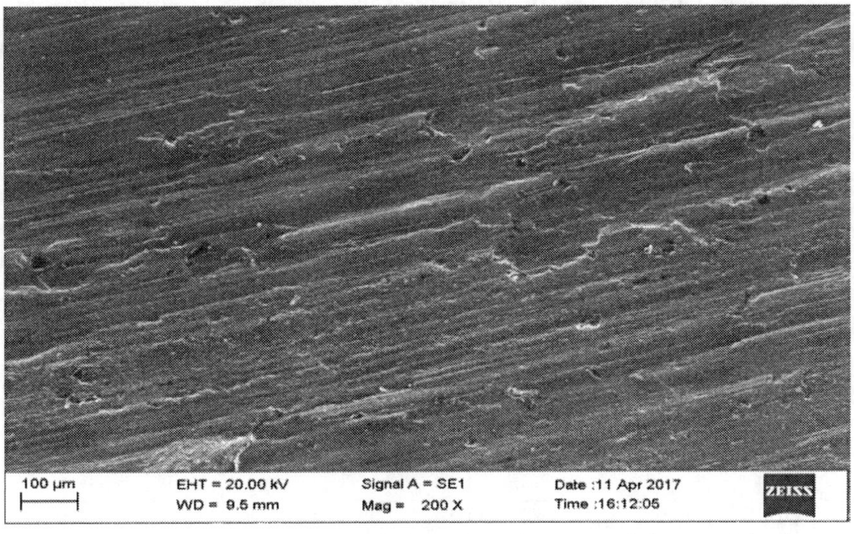

Fig 4. Typical SEM analysis of the worn surface of WC-12Co microwave clad at 20 N loads.

The WC-12Co based developed clads surface are subjected to dry sliding wear studies as per L9 Taguchi orthogonal array. The orthogonal L9 array of both the worn rate of clad and unclad surface is tabulated in Table 1. The worn surface was observed through scanning electronic microscope. At low loads a imperceptible wear in the form of abrasion and adhesion was noticed. However, comparatively smooth wear tracks and small amount material loss (wear rate) from the contact surface could be observed from Table.1.

Figure 4 shows shallow wear grooves, these are clear indication for the low wear rate on the developed surface than the unclad surface. At higher load (40 N), the worn surface of exhibits the confirmation of carbide pullouts. These carbide pullouts are established by the occurrence of indentation on the worn surface is shown in Fig.5.

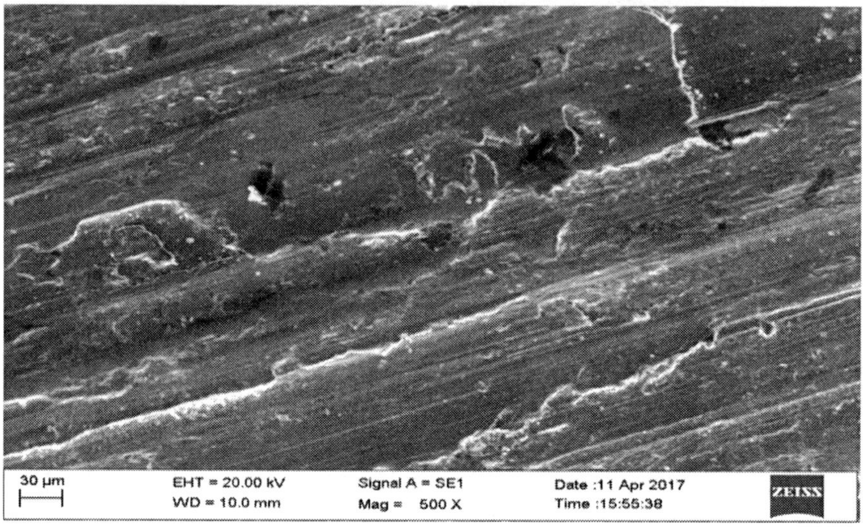

Fig. 5. Typical SEM analysis of the uneven surface of WC-12Co microwave clad at 30 N loads.

It was discovered that the material removed from unclad surface was more than the developed clad surface. The wear rate of clad surface is minimal when contrasted with unclad surface. The wear rate of clad surface is 0.001 g where as unclad surface is 0.043 g as listed in Table.1. At higher load (40 N) the transfer of material occurs between the clad surfaces to the counter face. The maximum wear took place for higher load 0.156 g on unclad surface as compared to developed clad surface.

Table 1. Dry sliding wear results of clad and unclad surface.

SNO	Sliding Distance(m)	Sliding Speed (m/s)	Load (N)	Wear rate of unclad surface (g)	Wear rate of clad surface (g)
1	1000	4	30	0.043	0.001
2	1000	6	40	0.081	0.073
3	2000	2	30	0.131	0.088
4	2000	4	40	0.181	0.15

 https://doi.org/10.1051/matecconf/201814402010

5	2000	6	20	0.036	0.031
6	3000	2	40	0.272	0.256
7	3000	4	20	0.216	0.191
8	3000	6	30	0.156	0.153

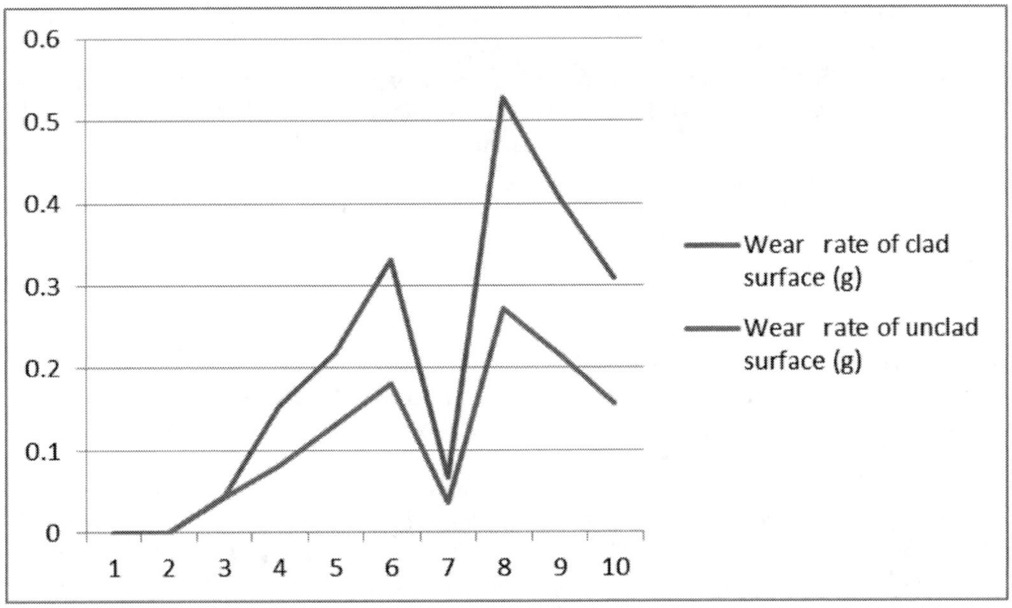

Fig. 6. Analysis of eear rate of clad versus unclad

4 Conclusions

Material processing through microwave energy on the metallic substrate has been demonstrated successfully. The current work highlights the improvement of clad surface developed through microwave energy. Metallurgical and mechanical characterization were carried out for developed clads. The few general observations are listed as follows.

- WC-12Co based clads are developed on SS-304 through microwave energy.
- The created clad shows great metallurgical fusion between substrate.
- As received SS304 samples and WC-12Co based developed clads surface are subjected to pin on disc wear studies.
- Taguchi orthogonal array is successfully used to conduct the wear studies.
- The worn surface was observed through SEM. At lower loads a imperceptible wear in the form of abrasion and adhesion was noticed.
- The wear rate of clad surface (0.001g) is less as compared to unclad surface (0.043g).
- At higher load (40N) the transfer of material occurs between the clad surfaces to the counter face. The maximum wear took place for higher load 0.156 g on unclad surface as compared to developed clad surface.

•The observation of experimental study proved that the developed clad shown better wear resistant than the as received substrate.

References

1. Jiang, W.H.; Yao, X.D.; H R Guan, H.R.; Hu, Z.Q. *Secondary Carbide Precipitation in a Directionally Solidification Cobalt-base Super alloy.* (Metallurgical and Materials Transactions 1999), 30A, 513-520.
2. Shuangjie Chu.; Jianguo Li.; Zhongyuan Liu .*On Morphologies, Micro segregation and Mechanical behavior cooling Rate.* (Metallurgical and Materials Transactions 1994), 25A, 637-642.
3. Selvan J.S.; Subramanian, K.; Nath, A.K. *Effect of laser surface hardening on En18 (AISI 5135) steel.* Journal of Material Process Technology (1999), 91, 29–36.
4. Sun, Y.; Bell, T.; *Dry sliding wear resistance of low temperature plasma carburized austenitic stainless steel.* Wear 2002, 253,689–693.
5. Bloyce A. *Surface engineering of titanium alloys for wear protection.* Proceeding of IMechE Journal, Journal of Engineering Tribology (1998), 212,467–476.
6. Lin, W.C.; Chen, C. *Characteristic of thin surface layer of cobalt-based alloy deposited by laser cladding. Surface and coating technology* (2006), 200, 4557-4563.
7. Agrawal, D. *Latest global developments in microwave materials processing.* (Materials Research Innovations 2010), **14**, 3–8.
8. Oghbaei, M.; Mirzaee, O. *Microwave versus conventional sintering: a review of fundamentals, advantages and applications. Journal of Alloys and Compounds* (2010), 494, 175–189.
9. Gupta, D.; Sharma, A.K. *Investigation on sliding wear performance of WC10Co2Ni cladding developed through microwave irradiation. Wear* (2011), 271, 1642– 1650.
10. Sharma, A.K.; Srinath, M.S.; *Pradeep, K. Microwave Joining of Metallic Materials. Indian Patent* (2009). Application No. 1994/Del/2009.
11. Srinath, M.S.; Sharma, A.K.; Kumar, p. *A new approach to joining of bulk copper using microwave energy.* (Materials and Design 2011), 32, 2685–2694.
12. Srinath. M.S.; Sharma. A. K.; Kumar. P. *Investigation on microstructural and mechanical properties of microwave processed dissimilar joints.* (Journal of Manufacturing Process 2011), 13,141–146.
13. Ravindra I. Badigera.; Narendranath. S.; Srinath. M.S. *Joining of Inconel-625 alloy through microwave hybrid heating and its characterization.* (Journal of Manufacturing Process 2015), 18,117-123.
14. A M Hebbale.; Srinath M S.; *Microstructural Studies of Cobalt Based Microwave Clad Developed on Martensitic Stainless Steel* (AISI-420). Trans Indian Inst Met DOI 10.1007/s12666-017-1206-7.
15. A M Hebbale.; Srinath M S.; *Taguchi analysis on erosive wear behaviour of cobalt based microwave cladding on stainless steel AISI-420.* Measurement 99 (2017) 98-107.
16. Srinath M S.; A M Hebbale.; *Microstructural investigation of nickel based cladding developed on austenitic SS304 through microwave irradiation. Journal of materials research and technology.* 5(4) (2016) 293–301.
17. A M Hebbale.; Srinath M S.; *Microstructure and experimental design analysis of nickel based clad developed through microwave energy.* Journal of Perspectives in Science. **8** (2016) 257-259.
18. Srinath M S.; A M Hebbale.; *Fuzzy prediction of slurry erosive behavior of cobalt based clad developed through microwave energy.* Materials Today: Proceedings 4 (2017) 1804–1811.

Influence of post curing on GFRP hybrid composite

Vithal Rao Chavan[1*], *K. R. Dinesh*[2], *K. Veeresh*[1], *Veerabhadrappa Algur*[3] and *Manjunath Shettar*[4]

[1]*Department of Mechanical Engineering, Rao Bahadur Y. Mahabaleshwarappa Engineering College, Bellary, India.*

[2]*Department of Mechanical Engineering, Government Engineering College, Raichur, India.*

[3]*Department of Industrial and Production Engineering, Rao Bahadur Y. Mahabaleshwarappa Engineering College, Bellary, India.*

[4]*Department of Mechanical and Manufacturing Engineering, Manipal Institute of Technology, Manipal Academy of Higher Education, Manipal, India.*

Abstract. Composite materials for the most part depicted as the mixes of two or more materials that outcome in the unmistakable properties than that of guard materials. Fibre strengthened plastics have been all around utilized for get-together flying machine and transport key parts as a delayed consequence of their specific mechanical and physical properties, for example, high particular quality and high particular robustness. Another pertinent application for fibre maintained polymeric composites (particularly glass fibre strengthened plastics) is in the electronic business, in which they are utilized for passing on printed wiring sheets. The utilization of polymer composite materials is winding up being powerfully essential. The present work delineates the change and mechanical portrayal of new polymer composites including glass fibre fortress, epoxy and maple cellulose fibre. The starting late made composites are delineated for their mechanical properties. The composite spreads were set up by utilizing hand layup framework. The experiments were conducted on and studied the effect of post curing on hybrid composites. The result reveals that the samples only with natural fibre have more promising results compared with synthetic fibre. The synthetic fibres get wrinkled due to post curing were as no such visuals in the natural fibres.

Key words: Post curing, GFRP, Cellulose fibre, Tensile, Impact, Hardness.

1 Introduction

Composite materials are defined as "a combination of two or more materials that results in better properties than those of the individual components used alone". These materials have high strength and stiffness tailored with low density, as compared to the conservative

* Corresponding author: nrvithalchauhan@gmail.com

metals. Polymer matrix composites (PMCs) are viscoelastic materials. Mechanical properties of PMCs are significantly influenced by heat i.e., variation in temperature. The PMCs are mostly used in the area like aerospace, automobiles and marine applications [1].

Curing is an irreversible reaction where chemical covalent cross-links are formed whichare thermally and mechanically stable. The curing process plays a vital role to achieve the end mechanical properties and chemical resistance of PMCs. For the fabrication of PMCs, resin and curing agents (hardener) with proper ratio are used. During crosslinking, the stateof matrix changes from liquid to gel and then transforms into solid. Curing can be done at room temperature as well as at elevated temperature [2].

2 Literature review

Cagri Uzay et al., worked on the post cure heat treatment on the impact toughness and tensile properties of fibre reinforced composites. The laminates were prepared by utilizing the hand layup vacuum bag moulding of woven carbon and glass fibre fabric plies. At three distinct temperature, the post curing was done for one hour i.e. 25, 62.5 and 100°C. The impact (charpy) and tensile test were conducted along with the statistical analysis to know the effect of post curing and different type hybrid plies. The results reviled that post curing has positive effect on energy absorption capacity and tensile properties of FRP [3].

As the operating temperature varies the characteristics of FRP got altered, especially impact behaviour, under various operating temperature ranges Viz., -70 to 180 °C [4], -75 to 75 °C [5] and -75 to 25 °C [6]. The result stated that FRP made up of carbon fibre were highly responsive to the impact loads due to its behaviour of brittleness, therefore both toughness and impact was directly proportional to the temperature.

Gonzalez et al., investigated the impact behaviour of PEEK and its short carbon fibre reinforced composite (SCFR PEEK) at low temperatures. The brittleness of PEEK and SCFR, PEEK composites at low temperatures could restrict the application of this composite in aeronautical applications [6].

Aktaş and Karakuzu, investigated on the properties such as tensile, compression and shear, of unidirectional glass/epoxy composite plies under distinct temperature from 20 to 100°C and the results showed that, the fibre direction influenced decrease in tensile strength and modulus of elasticity (after 60 and 80°C). As the temperature increased the rigidity and strength of composites were decreased [7].

A similar study was made by Cao et al. [8] and tested the carbon FRP from 16 to 200°C. The result showed that, the tensile strength decreased significantly with increasing temperature and remained almost stable after the polymer exceeds its glass transition temperature. On the other side, at cryogenic temperatures, it was shown that CFRP composites had better tensile properties [9]. The tensile properties of FRP composites have also been affected by curing temperature used during fabrication of composite materials. Generally, the specimens have been cured at room temperature. However, the increase in curing temperature improves the mechanical properties compared to curing at room temperature [10, 11]. As the post-curing process improves mechanical properties of the composites and enhances their chemical durability and heat resistance [12], post-curing heat treatment processes have been studied for composites and its adhesives [13].

Hiremath et al. determined the optimum post-curing temperature for epoxy/alumina polymer nanocomposite in order to obtain better viscoelastic and flexural properties [14]. Similarly, Kumar et al. determined both optimum post-curing temperature and time for interlaminar shear strength (ILSS) and the corresponding glass temperature (Tg) of glass

FRP composites. While the curing temperature during fabrication and the effect of various operating temperature conditions have already been investigated, and discussed, the knowledge about post-curing effect on impact and tensile strength of FRP composites is limited and further studies are needed. As it is stated in the literature, without post-curing heat treatment process, the polymer matrix is not strong enough and the failure of composites under loading mostly occurs due to the crushing of polymer matrix [15].

3 Methodology

Design and modification of polymer matrix composite:

Based on the composite design the specimen has to be fabricated by using hand lay-up technique. The proper volume fraction of fibres, epoxy, and cellulose fibres is to be controlled [16]. The laminates are cured for a period of 12hr. The sample is cut to a required size for the tests [17]. And post cured the samples for 1hr by varying the temperature.

3.1Specimen Preparation

The materials used for the specimen preparation were E-glass fibre plain woven mat (200gsm) and epoxy (LAPOX L-12) with hardener (K-6), purchased from ATUL India Ltd, Gujarat, India. The hand layup technique (figure 1) was adopted for specimen preparation. In connection of volume fragment, the estimations were made for 60-40 (60% Resin– 40% Fibre). One filler material i.e. cellulose fibre were added to various mixes by keeping Epoxy rate as continuing (60%). Table 1 shows the proportionate of three materials used for laminate preparation.

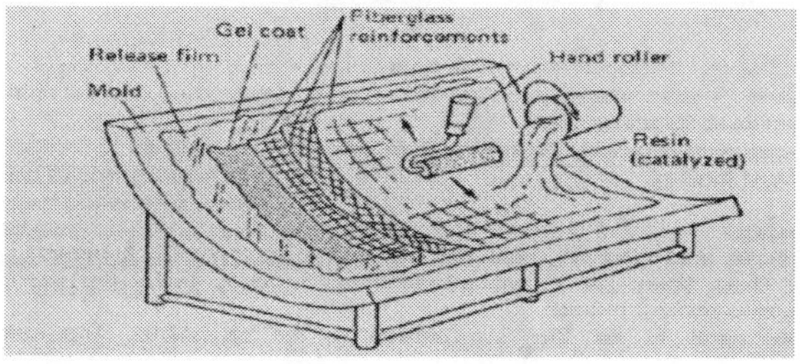

Fig. 1. Hand layup technique

Table 1. Details of Glass/Epoxy and Cellulose fibre Proportions

Laminate No	Glass fibre(Wt %)	Cellulose fibre(Wt %)	Matrix material (Wt %)
1	35%	5%	60%
2	30%	10%	60%
3	25%	15%	60%
4	20%	20%	60%
5	5%	35%	60%
6	0%	40%	60%

3.2 Testing of mechanical properties

- Tensile test is to be conducted as per ASTM D3039 standards.
- Impact test (charpy) ASTM D2344 standards.
- Hardness test ASTM E20-00 standard

The ultimate tensile strength and percentage of elongation of the composite was studied with varying percentage of reinforcement and filler material and varying the temperature of post curing.

4 Result and Discussion

In this study, the tensile, impact test and hardness test were carried out on glass epoxy-cellulose fibre filled composites, to study the effect of fibre proportionate and post curing of those composites. In this section, experimental result obtained for tensile, impact and hardness tests are widely discussed. The results were analyzed by plotting bar graphs for materials that were having different proportions of cellulose fibre and varying temperature materials.

4.1 Tensile Test

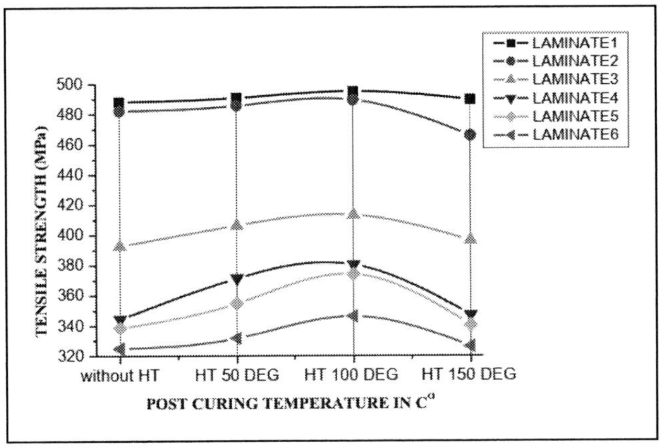

Fig. 2. Comparison of tensile strength

Tensile test was carried out for the different fibre weight percentage of laminates with different post curing temperature viz., 50, 100 and 100°C. Figure 2 shows comparison of tensile strength for the different laminates at different post curing temperature. The result reveals that laminate 1 without post curing i.e. at 5 wt% of cellulose fibre, 35 wt% of E-glass fibre and 60 wt % of epoxy resin, showed the higher tensile strength compared to the other samples. The tensile fracture is due to delamination of the fibres. Delamination of the fibre may be found due to weak bonding between the fibres and resin. The fracture due to delamination of the fibre decreased as the percentage of cellulose fibre is increased because cellulose fibre is laid up along with epoxy as the percentage of cellulose fibre increases E-glass fibre percentage decreases this is done to maintain constant volume of 300mm × 300mm × 3mm. However, increase in post curing temperature decreased the strength. The cellulose fibre is having more tendency of getting adhere with resin, as temperature is increased the viscosity got decreased and most of the resin is been consumed with cellulose fibre, hence there is decrease in strength.

4.2 Impact Test

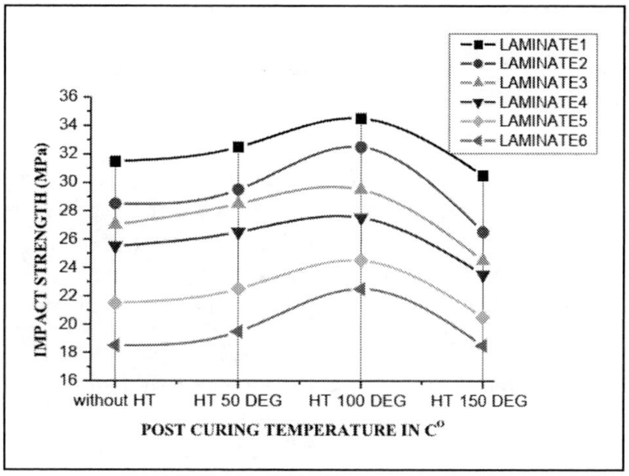

Fig. 3. Comparison of impact strength

As shown in figure 3, laminate 1 with post curing at 100° C had higher impact strength compared to the other samples. The test reveals that, the increase in the impact strength is due to the proper wettability of the fibre with resin. The decrease in viscosity of resin and overheating at higher temperature leads to reduction in its impact strength. All other samples are follows the same pattern as shown figure i.e., increase the impact strength up to 100°C of post curing, but drastically reduces at 150°C of post curing. This is because the fracture process of fibres is generally accompanied by the damage of chemical bond, intermolecular slip and the weakening of van der Waals force, etc. Heating will undoubtedly weaken the chemical bond and van der Waals force and then lower the strength of fibres. Also, increase the in cellulose fibre percentage reduces the impact strength, it might be due to agglomeration of fibre.

4.3 Hardness Test

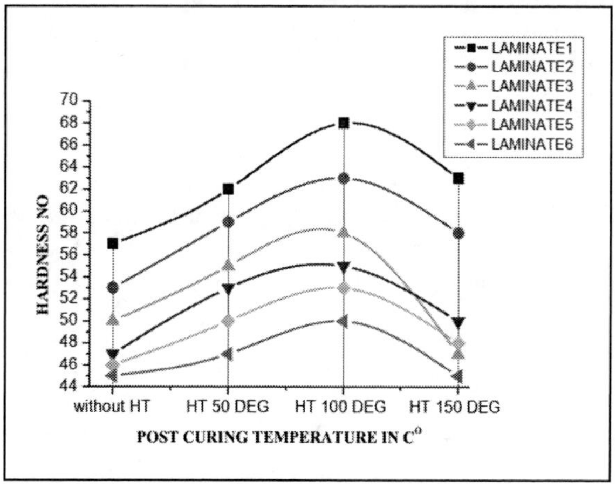

Fig. 4. Comparison of hardness number

As shown in figure 4, all laminates, at 100°C post curing exhibits higher hardness value compared to other samples. The post curing results in increase in the bonding between fibres and resin, due to increase in cross-linking and stacking, which reduces the movement of polymer molecules and making it to become more resistant to the penetration of indenter. The overheating at 150°C resulted in decrease in hardness value due to degradation the matrix materials, warp, etc.

5 Conclusions

All composite with filler material exhibited satisfactory mechanical properties than unfilled composites. Composites showed failure due to delamination as the filler percentage increases this may be because weak bonding between the fibres. Hence to strengthen the bonding between the synthetic and natural fibres the post curing is adopted. The post cured samples of both synthetic and natural fibres give the better results compared with the after and before post curing.

- The Tensile Strength of post cured samples at 100°C for 1 hour of 40 Wt% of cellulose fiber and 60 Wt% of resin sample had increased with 6.74%,35 Wt% E-glass fiber ,5Wt% of Cellulose fiber and 60% of resin sample had increased with 1.38% , 20 Wt% E-glass fiber ,20Wt% of Cellulose fiber and 60% of resin sample had increased with 9.42%, 5Wt% E-glass fiber,35Wt% of Cellulose fiber and 60% of resin sample had increased with 7.28%, when compared with, samples without post curing.
- The Impact Strength of post cured samples at 100°C for 1 hour of 40 Wt% of cellulose fiber and 60 Wt% of resin sample had increased with 10.81%,35 Wt% E-glass fiber , 5Wt% of Cellulose fiber and 60% of resin sample had increased with 8.69%, 20 Wt% E-glass fiber,20Wt% of Cellulose fiber and 60% of resin sample had increased with 7.27%,5Wt% E-glass fiber,35Wt% of Cellulose fiber and 60% of resin sample had increased with 14.63%, when compared with, samples without post curing.
- The Hardness of post cured samples at 100°C for 1 hour of 40 Wt% of cellulose fiber and 60 Wt% of resin sample had increased with 10.86% ,35 Wt% E-glass fiber ,5Wt% of Cellulose fiber and 60% of resin sample had increased with 16.17% , 20 Wt% E-glass fiber,20Wt% of Cellulose fiber and 60% of resin sample had increased with 14.54%, 5Wt% E-glass fiber,35Wt% of Cellulose fiber and 60% of resin sample had increased with 12.22%, when compared with, samples without post curing.

Hence the post curing of the samples only with natural fibre has more promising results compared with synthetic fibre. The synthetic fibres get wrinkled due to post curing were as no such visuals in the natural fibres.

References

1. Ray B. C. and Rathore D. " Durability and integrity studies of environmentally conditioned interfaces in fibrous polymeric composites"Critical concepts and comments Adv. Colloid Interface Sci. 209 68–83(2014).
2. Aruniit A, Kers J., Krumme A, Poltimäe T. and Tall K. "Preliminary Study of the Influence of Post Curing Parameters to the Particle Reinforced Composite's Mechanical and Physical Properties"Mater. Sci. 18 256–61(2012).
3. Cagri Uzay "Effect of post-curing heat treatment on mechanical properties of fibrereinforced polymer (FRP) composites" Materials Testing downloaded from www.hanser-elibrary.com by Hanser - Library on April 4, (2017).
4. J. R. Tarpani, O. Maluf, M. Cristina, A. Gatti "Charpy impact toughness of conventional and advanced composite laminates for aircraft construction", Materials Research 12, No. 4, pp. 395-403 DOI:10.1590/S1516-14392009000400004.(2009)

5. S. W. Hong, S. S. Ahn, H. Li, J. K. Kim, S. J. Ko, J. M. Koo, C. S. Seok "Charpy impact fracture characteristics of CFRP composite materials according to variations of fibrearray direction and temperature", International Journal of Precision Engineering and Manufacturing 14, No. 2, pp. 253-258 DOI:10.1007/s12541-013-0035-9.(2013)
6. D. G. Gonzalez, M. R. Millan, A. Rusinek, A. Arias: "Low temperature effect on impact energy absorption capability of PEEK composites",Composite Structures 134, pp. 440-449 DOI:10.1016/j.compstruct.2015.08.090.(2015)
7. M. Aktaş, R. Karakuzu "Determination of mechanical properties of glass-epoxy composites in high temperatures", Polymer Composites 30, pp. 1437-1441 DOI:10.1002/pc.20708.(2009)
8. S. Cao, Z. Wu, X. Wang "Tensile properties of CFRP and hybrid FRP composites at elevated temperatures", Journal of Composite Materials43, No. 4, pp. 315-330 DOI:10.1177/0021998308099224.(2009)
9. M. G. Kim, S. G. Kang, C. G. Kim, C. W. Kong "Tensile properties of carbon fibre composites with different resin compositions at cryogenic temperatures", Advanced Composite Materials19, pp. 63-77 DOI:10.1163/156855109X434838.(2010)
10. Y. Cao, J. Cameron "The effect of curing conditions on the properties of silica modified glass fibre reinforced epoxy composite", Journal of Reinforced Plastics and Composites26, No. 1, pp. 41-50 DOI:10.1177/0731684407069950.(2007)
11. S. C. Joshi, Y. C. Lam, U. W. Tun "Improved cure optimization in pultrusion with pre-heating and die-cooler temperature", Composites: Part A 34, pp. 1151-1159 DOI:10.1016/j.compositesa.2003.08.003.(2003)
12. A. Aruniit, J. Kers, A. Krumme, T. Poltimae, K. Tall "Preliminary study of the influence of post curing parameters to the particle reinforced composite's mechanical and physical properties", Materials Science (Medžiagotyra) 18, No. 3, pp. 256-261 DOI:10.5755/j01.ms.18.3.2435.(2012)
13. V. A. A. Chetti, R. L. Macchi, M. E. Iglesias "Effect of post-curing treatment on mechanical properties of composite resins", Acta Odontologica Latinoamericana 27, No. 2, pp. 72-76 DOI:10.1590/S1852-48342014000200005.(2014)
14. V. Hiremath, M. Singha, D. K. Shukla "Effect of post curing temperature on viscoelastic and flexural properties of epoxy/alumina polymer nanocomposites", Procedia Engineering97, pp. 479-487 DOI:10.1016/j.proeng.2014.12.272.(2014)
15. D. S. Kumar, M. J. Shukla, K. K. Mahato, D. K. Rathore, R. K. Prusty, B. C. Ray, "Effect of post-curing on thermal and mechanical behavior of GFRP composites", Materials Science and Engineering 75, pp. 1-6 DOI:10.1088/1757-899X/75/1/012012.(2015)
16. Manjunath Shettar, Vithal Rao Chauhan "Investigation On Mechanical Properties Of E- Glass, Epoxy Resin With AsbestosFilled /Hybrid Composites" Global Journal Of Engineering Science And Research Management Issn: 2349-4506 February, (2015).
17. Vithal Rao Chavan "Effect Of The Tribological Behaviour Of Hematite Filled Hybrid Composites – A Taguchi Approach" International Journal of Engineering Research and Advanced Technology Special Volume. 02 Issue.01, ISSN: 2454-6135 May (2016).

Study on fracture behaviour of Hybrid Aluminium Composite

Niranjan D. B.[1*], *G. S. Shiva Shankar*[1] and *Latha B. Shankar*[2]

[1]*Department of Mechanical, Siddaganga Institute of Technology, Tumakuru, India*
[2]*Department of IEM, Siddaganga Institute of Technology, Tumakuru, India*

Abstract. This work makes an attempt to study of fracture behaviour of hybrid Composite reinforced with hard ceramic Boron carbide as primary phase and secondary phase as soft solid lubricant Molybdenum Disulphide with base matrix Al2219 by two stage stir casting method is incorporated. Compact tension type specimen was utilized for test and dimensions are conformity to ASTM E-647.Fatigue crack growth and fracture toughness test conducted as per ASTM Standard E-399 and E-1820. This study discloses that reinforcing the hard ceramic Boron Carbide of constant 3wt% and soft solid lubricant Molybdenum Disulphide with increasing wt% of 3%, 4%, 5% to Al2219, results increase in energy required to open the crack front to create two new crack surfaces during mode-I type failure and also significantly variation of fracture toughness. Surface of fractured specimens were study under scanning electron microscope and observes that pulled regions tiny dimples results in balance between ductility and strength of prepared hybrid composite.

Keywords: Stir casting, Al2219, Fracture Toughness

1 Introduction

Aluminium is the most abundantly available material in earth crust. It exhibits numerous properties as conventional alloy however still concern about like strength and stability. In 1990s hard ceramic particles are reinforced with Al, starts new era of aluminium in Metal matrix composites (MMCs), exhibits superior properties like high strength to low weight ratio, resistance to corrosion, thermal stability these properties makes to replace conventional Al alloy[1-3]. The multifunctional property of aluminium metal matrix has seen utilized major material in space research sector, aerospace technology, substrates of solar panel, electronic heat sinks, antenna reflectors, explosion engine components, fins of drive shaft in automotive and others. Enhancement of material property from aluminium base MMC have been derived to be highly dependent on (i) Composition of base alloy of Al (ii) utilization of weight and volume fraction of reinforcing material (iii) Processing techniques incorporated for fabrication of composites. Thus, most study on composites described in literature has attempted to determine how these causes influences to properties and enhancement of performance of Aluminium Metal matrix composites (AMMCs). Most of study stated in literature has been dedicated to Al composite like A359, 6061, 6063,

* Corresponding author: contactniranjandb@gmail.com

7075, A357 [4-6]. Research work on Al2219 as hybrid composite is less. However, Al2219 made its name in applications like contact parts of airframe and fuel tank of rockets [7-9]. The primary objective of this study is developing hybrid aluminium metal matrix composite using Al(2219) reinforced with Boron Carbide(B_4C) and Molybdenum Disulphide(MoS_2). The compact tension (CT) specimen were prepared from as-cast and hybrid composite (Al2219) for fracture toughness test of mode-I failure criteria.

2 Materials and methods

2.1 Materials

For this study Al2219 alloy used as base material as received from in the form of slabs having chemical composition included copper, magnesium as higher percentage and small percentage rare earth material zirconium along with small percentage of vanadium and titanium. Reinforcing particles used are Boron carbide (B_4C) of 90μm and Molybdenum Disulphide of 1.8μm. B_4C acting as hard ceramic particle and MoS_2 acting as soft solid lubricant. Along with reinforcing particles Potassium titanium fluoride (K_2TiF_6) halide salt is used for improve of wettability of Al-B_4C-MoS_2 during melting.

2.2 Method

2.2.1 Stir casting

The two stage stir casting technique is involved in this study accordance to K alaneme [1] and Kumar et al [2]. Standardized calculations follows while charging the Al(2219) billets and reinforcement of B_4C and MoS_2 to produce the combinations of Al2219-3%B_4C-3%MoS_2, Al2219-3%B_4C-4%MoS_2, Al22193%B_4C-5%MoS_2. K_2TiF_6 was incorporated as wetting agent along with composite to improve wettability of base matrix and reinforcement. For the degassing process Hexachloroethane (C_2Cl_6) was used. Preheated reinforcements are introduced to process by novel two stage and stir casting process carried to form vortex at speed of 200-250 rpm about 5-8min, for this zirconia coated stainless steel used. Casting was then accomplished on metallic moulds at poring temperature of 720°C.

2.2.2 Fracture toughness, K_{IC}

The prepared composites are fabricated to Compact tension specimen as dimensions are conformity to E-647 ASTM standard for the determination of K_{IC} of composites. Figure 2.2.2 shows the line diagram of CT specimen. Fatigue crack growth is accordance of E-1820 ASTM standard follows the equation (1) for fatigue loading. Notch preparation accordance with E-399 ASTM standard.

$$P_f = \frac{0.4\, B\, b_0^2 \sigma_y}{(2W + a_o)} \quad (1)$$

where,

$$\sigma_y = \frac{\sigma_{YS} + \sigma_{TS}}{2} \quad (2)$$

$$b_o = W - a_o \quad (3)$$

where a_o is initial crack length, W is the length from centre of hole grove to initial notch length as shown in line diagram 2, B is thickness of material. σ_y and b_o are obtained as shown in equation (2) and (3).

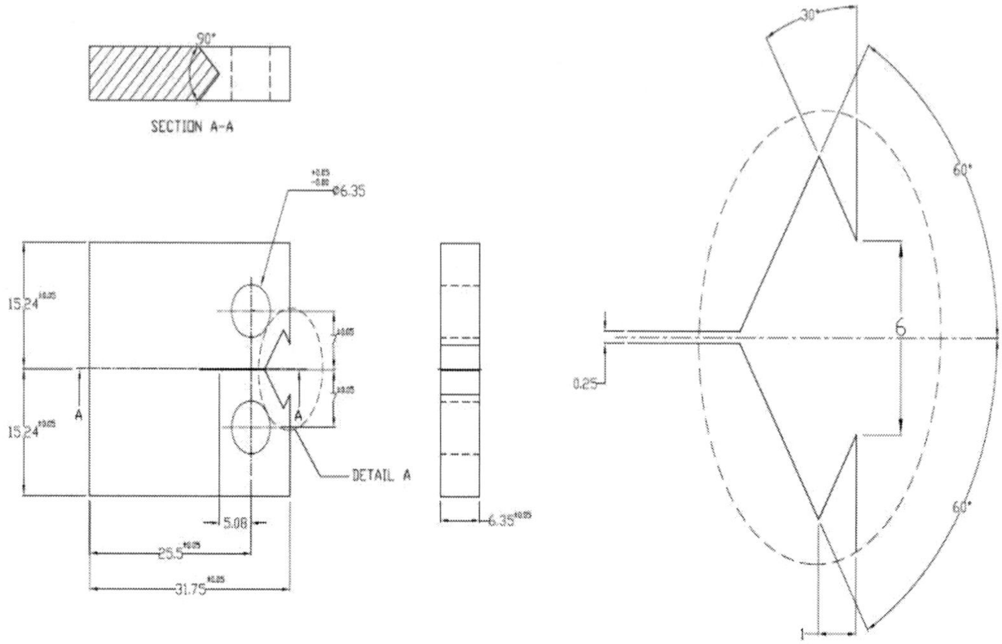

Figure 2: C(T)SPECIMEN FOR 25.4W-6.35 THK

2.2.3 Microstructure

The intention of microstructure study is to know distribution of reinforcement particles to base matrix. Initially small diameter of specimen cut down from prepared composites and polished from emery papers, grit size ranging from 200 to 1200. Characterization of specimen performed by field emission scanning electron microscope.

3 Results and Discussion

3.1 Characterization

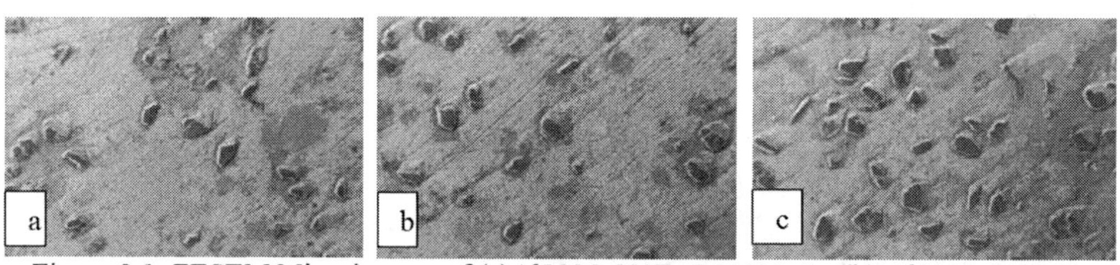

Figure 3.1: FESEM Microimages of (a)Al2219-3%B_4C-3%MoS_2, (b) Al2219-3%B_4C-4%MoS_2, (c) Al2219-3%B_4C-5%MoS_2

Figure 3.1 shows the particles distribution in the matrix. It clearly observes that from figure 3.1 (a) shows for (a)Al2219-3%B_4C-3%MoS_2, (b) Al2219-3%B_4C-4%MoS_2, (c) Al2219-3%B_4C-5%MoS_2, Characterization of specimens are made by field emission scanning electron microscope and this clearly visible of particles of reinforcement B_4C and MoS_2 makes even distribution in Al(2219) hybrid composite.

3.2 Fatigue Crack growth

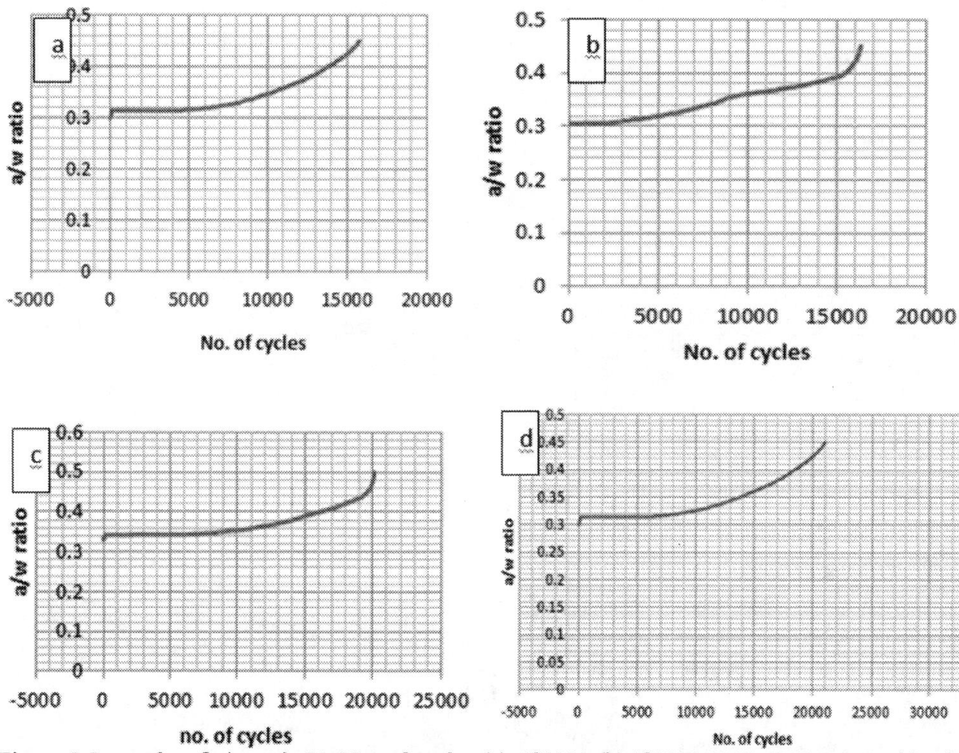

Figure 3.2: graphs of a/w ratio Vs No. of cycles (a) Al2219 (b)Al2219-3%B_4C-3%MoS_2, (c) Al2219-3%B_4C-4%MoS_2, (d) Al2219-3%B_4C-5%MoS_2

In this study fatigue crack growth procedure followed by ASTM-E399.Specimen were loaded BiSS axial UTM for tensile loading condition, variable amplitude fatigue crack propagation test was initiated for applying fatigue load. Initially specimens were prepared for a/w ratio of 0.3 however by VAFCP test a/w ratio increased to 0.45 and test stops automatically after achieving a/w ratio between 0.45 to 0.55. For the purpose of obtaining sharp notch at the tip, fatigue load was given. Figure 3.2 shows required number of cycles to achieve sharp notch at tip and standardized a/w ratio.

3.3 Fracture Toughness

The significant deviation of fracture toughness is shown in Table 3.3 for as cast and hybrid composite. It observe that addition of hard ceramic B_4C and soft solid lubricant MoS_2 makes increase significantly in the fracture toughness of prepared composites. Reason for improvement is addition of reinforcements makes minimizes in interparticle arrangement and strong obstacle to open the crack front. However only increase in MoS_2 makes decrease in fracture toughness as observed in table 3.3. The prepared composites are balance between strength and ductility and Al2219-3%B_4C-5% MoS_2 combination of hybrid composite exhibits maximum load bearing capacity of 5.565 KN to create new crack surface and also upto 20% increase in fracture toughness compare to base alloy.

Table 3.3: Validation K_{IC}

Composite	Pmax	Pmin	P_{max}/P_Q	K_{IC}(Mpa m^0.5)	%increase in K_{IC}
Al 2219	2.052	1.959	1.047	21.485	-
Al2219-3%B_4C-3% MoS_2	2.147	2.054	1.045	23.002	7%
Al2219-3%B_4C-4% MoS_2	4.236	3.863	1.091	27.326	27%
Al2219-3%B_4C-5% MoS_2	5.565	5.122	1.086	25.894	20%

3.4 Fractography

Figure 3.4: SEM microimages of fractured surface: (a)Al2219-3%B_4C-3%MoS_2, (b) Al2219-3%B_4C-4%MoS_2, (c) Al2219-3%B_4C-5%MoS_2

Scanning electron microscope is incorporated for study the new crack surface obtained by fracture test. Fig3.4 (a) shows fractured surface of Al2219-3%B_4C-3%MoS_2 fig 3.4(b) shows fractured surface of Al2219-3%B_4C-4%MoS_2 Fig3.4(c) shows fractured surface of Al2219-3%B_4C-5%MoS_2 after conduction of test. Study on surface reveals that the fracture occurred in both intragranular and Trans granular type fracture, combination of cracked particles and pulled regions and tiny plastic dimples, results in mixed mode fracture, which possesses high strength before fracture.

4 Conclusions

The present study makes an attempt to study fracture behaviour which involves study of crack tip opening condition and determination of fracture toughness of Al hybrid composite by utilization of Compact Tension specimen with Mode-I failure criteria. Fatigue crack growth and fracture toughness test of as cast alloy and hybrid composite influenced by ASTM E-399 and ASTM E1820 standard.

- It observed that maximum load bearing to open the crack surface of prepared hybrid composite significantly increases.

- Effect of addition of reinforcement comprise the interparticle arrangement, makes strong obstacle to open the crack front, which results in increase in fracture toughness upto 27% maximum for Al2219-3%B_4C-4%MoS_2 hybrid composite.

References

1. K. K. Alaneme, "Frature toughess and tensile properties of as cast and age hardened aluminium (6063)-silicon carbide particulate composite", Scientia Iranica A (9(4), 992-996 (2012).
2. Kumar N. G., "Mechanical and wear behaviour of aluminium metal hybrid composites", Procedia Materials science 5, 908-917 (2014)
3. Jayasheel Harti, "Hardness and Tensile Behavior of Al2219-TiC Metal Matrix Composites", Journal of Mechanical Engineering and Automation 2016, 6(5A): 8-12 DOI: 10.5923/c.jmea.201601.02. (2016)
4. Miracle, D. B., ''Metal matrix composites-from science to technological significance'', Compos. Sci. Technol., 65 (15–16), pp. 2526–2540 (2005).
5. Garcia-Romero A. M., Egizabal P. and Irisarri A. M., ''Fracture and fatigue behaviour of aluminium matrix composite automotive piston'', Appl.Compos. Mater., 17, pp. 15–30 (2010).
6. Adiamak M. ''Selected properties of aluminium base composites reinforced with intermetallic particles'', J. Achiev. Matls Manufac. Tech., 14(1–2), pp. 43–47 (2006).
7. Surappa M. K. ''Aluminium matrix composites: challenges and opportunities'', Sadhana, 28 (1–2), pp. 319–334 (2003).
8. Soboyejo W. O., "Mechanical Properties of Engineered Materials", Marcel Dekker Inc., New York, USA pp. 1–608 (2003).
9. B. D. Dunn, "The corrosion properties of Spacelab structural alloy aluminium 2219 - T851", Published by: ESA Scientific and Technical Publications Branch, ESTEC, Noordwijk (NL) Printed in The Netherlands ISSN 0379 – 4067
10. Nikhil Gupta, "An Experimental and Computational Investigation of Crack Growth Initiation in Compact Tension (CT) Specimen", International Journal of Scientific and Research Publications, Volume 2, Issue 8, August 2012 1 ISSN 2250-3153 (2012).
11. Y. Lei, "Finite element crack closure analysis of a compact tension specimen", International Journal of Fatigue 30 (2008) 21–31 (2008).
12. Shreyas Pawar, "Experimental Investigation of Mechanical Characterisation of Al6061 Reinforced with Molybdenum Disulphide (MOS2)", Volume 4 Issue IV, April 2016 IC Value: 13.98 ISSN: 2321-9653 International Journal for Research in Applied Science & Engineering Technology (IJRASET)(2016).

Microstructure and wear behavior of austempered high carbon high silicon steel

Palaksha Acharya[1]*, *Ajit Kumar*[1] *and Ravishankar Bhat*[1]

[1]*Department of Metallurgical and Materials Engineering, National Institute of Technology Karnataka, Surathkal, Mangaluru 575025, India.*

Abstract. In the present investigation, the influence of austempering temperature and time on the microstructure and dry sliding wear behavior of high silicon steel was studied. The test specimens were initially austenitised at 900°C for 30 minutes, thereafter austempered at various temperatures 280°C, 360°C and 400°C, for varying duration from 30 to 120 minutes. These samples after austempering heat treatment were subsequently air cooled to room temperature, to generate typical ausferritic microstructures and then correlated with the wear property. The test outcomes demonstrate the slight increase in specific wear rate with increase in both austempering temperature and time. Specific wear rate was found to be minimum at an austempering temperature of 280°C, that exhibits lower bainite microstructure with high hardness, on the other hand specific wear rate was found to be slightly high at increased austempering temperatures at 360°C and 400°C, due to the upper bainite structure that offered lower hardness to the matrix. The sample austempered at 280°C for 30 minutes offered superior wear resistance when compared to other austempering conditions, mainly due to the presence of fine acicular bainitic ferrite along with stabilized retained austenite and also some martensite in the microstructure.

1 Introduction

Steels with an ausferritic structure (i.e., a blend of ferrite and carbon rich austenite) have a few points of interest including high strength, great toughness, superior wear resistance relying on the transformation temperature and time [1]. The extraordinary blend of these properties with that of the low expenses of the raw material and heat treatment contrasted with customary steels discover their utilization in various fields like automotive, railway components and cutting tools applications [2]. Manganese content in steel underneath 2% favours hardenability, which adds to solute strengthening [3]. The deleterious effect of cementite is overcome by smothering their precipitation by adding adequate quantity of silicon (atleast 1.5 wt. %) [4], due to insolubility of silicon in cementite. Thus carbon remains in and stabilizes austenite [5].

*Corresponding author: palaksha.pa@gmail.com

The present investigation was focused to evaluate the effect of varying austempering heat treatment parameters on the microstructure and wear behavior of AISI 9260 high silicon steel.

2 Experimental procedures

The test steel used for the present research work is AISI 9260 steel in the form of cylindrical rolled bars of 13 mm diameter and 1000 mm long. The as-received steel was checked for its chemical composition using optical emission spectrometer, with composition details shown in table 1.

Table 1. Chemical composition of the as-received AISI 9260 steel (weight %)

C	Si	Mn	Cr	Ni	Mo	Cu	V	S	P
0.6	2.06	0.84	0.13	0.019	0.007	0.009	0.002	0.005	0.013

Wear test specimens were machined from the as-received AISI 9260 steel bars, with dimensions 8 mm diameter and 25 mm length, which is as per ASTM G99 standard [6].

To perform austempering heat treatment, the machined specimens were subjected to full austenitisation at 900°C for about 30 minutes in a muffle furnace and there after quenched in a molten salt bath, (mixture of 45% $NaNO_3$ and 55% KNO_3) maintained at various temperatures (280°C, 360°C and 400°C) and varying duration of 30, 60, 90 and 120 minutes for each temperatures.

For microstructural analysis, the austempered specimens were metallographically polished and etched with 2% nital (98 ml concentrated HNO_3 and 2 ml ethanol), then observed under JEOL JSM-6380LA scanning electron microscope (SEM). SEM was also used to characterize the worn out surfaces of the samples in order to understand the wear mechanisms. X-ray diffractometer (XRD) analysis was carried out using JEOL-JDX-8P XRD (operated with Cu-Kα radiation at 30KV and 20mA), scanned in the 2θ range of 40° to 50° and at a scan speed of one degree per minute.

The hardness tests were carried out for all the austempered samples using Shimadzu HMV-G 20ST Vickers hardness tester (1kg load for dwell time of 15 seconds). The average of at least five readings was reported as hardness values.

Pin-on-disc apparatus was used to evaluate the dry sliding wear behavior of the austempered specimens, under a fixed load of 2 kg, wear track diameter set to 120 mm and disc speed of 1000 rpm, Wear test was performed for duration of three hours in a cumulative way i.e. the worn specimens were weighed every half an hour. Specific wear rate (SWR) was registered toward the finish of three hours, using the equation (1).

$$SWR(mm^3 / N-m) = \frac{\Delta w}{\rho \times L \times d} \qquad (1)$$

Δw = weight loss (grams), ρ= Density of the test steel (gram/mm^3), L= load (Newton), d= sliding distance (meters).

3 Results and discussion

3.1 Microstructure

(Fig. 1(a-f)) demonstrates the microstructural investigations of the various austempered specimens that exhibits bainitic structure that comprises of ferrite, retained austenite and presence of some martensite as well.

Austempering at lower temperature i.e. at 280°C for 30 minutes reveals acicular structured bainitic ferrite, retained austenite and martensite (Fig. 1(a)), which is because of high nucleation rates of ferrite and furthermore expanded level of supercooling of bainitic transformation and lower carbon diffusion rates.

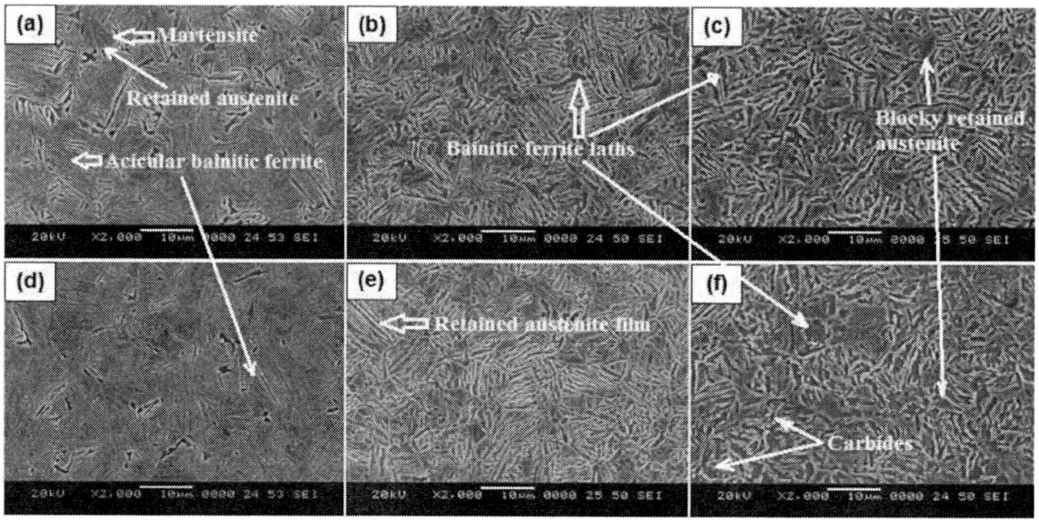

Fig. 1. Scanning electron micrographs of the specimens austempered at various temperatures and duration (a) 280°C- 30 min. (b) 360°C- 30 min. (c) 400°C- 30 min. (d) 280°C- 120 min. (e) 360°C- 120 min. (f) 400°C- 120 min.

(Fig. 1(b & c)) exhibits the microstructures of specimens austempered at 360°C and 400°C for a duration of 30 minutes, where impressive measure of retained austenite and bainitic ferrite with enlarged laths (upper bainite) was noticed, due to the extremely high carbon diffusion rates from bainitic ferrite to the adjoining retained austenite, which stimulates the growth rate of bainitic ferrite which thus causes augmenting of ferrite laths.

Referring to the Fig. 1(d, e & f) which shows the impact of increment in austempering time (120 minutes for temperatures 280°C, 360°C and 400°C) where fineness in the microstructure was seen, due to the adequate availability of time for further nucleation of bainitic ferrite needles, bringing about lessened retained austenite content at 280°C (Fig. 1(d)). However at elevated austempering temperature i.e., at 360°C adequate availability of time for growth of bainitic ferrite laths and finally at 400°C, isothermal holding for longer duration actuates the second stage response of bainitic transformation, where carbides get precipitated in retained austenite (Fig. 1(f)), which reduces the retained austenite content and were confirmed from XRD analysis displayed in the figure 2(c).

3.2 X-ray diffractometry

XRD studies were carried out to estimate retained austenite and bainitic ferrite quantitatively by direct comparison method suggested by B. D. Cullity [7].

Figure 2(c) demonstrates that for all austempering time, the retained austenite content increases from 280°C to 360°C and after that drops at 400 °C (except for 30 minutes). In addition to this decrease in the retained austenite content was observed for all temperatures with extending austempering time, the conceivable reason is for isothermal duration past 30 minutes, the bainitic transformation proceeds on austenite grain interface, bringing about nucleation and growth of individual bainitic ferrite plates.

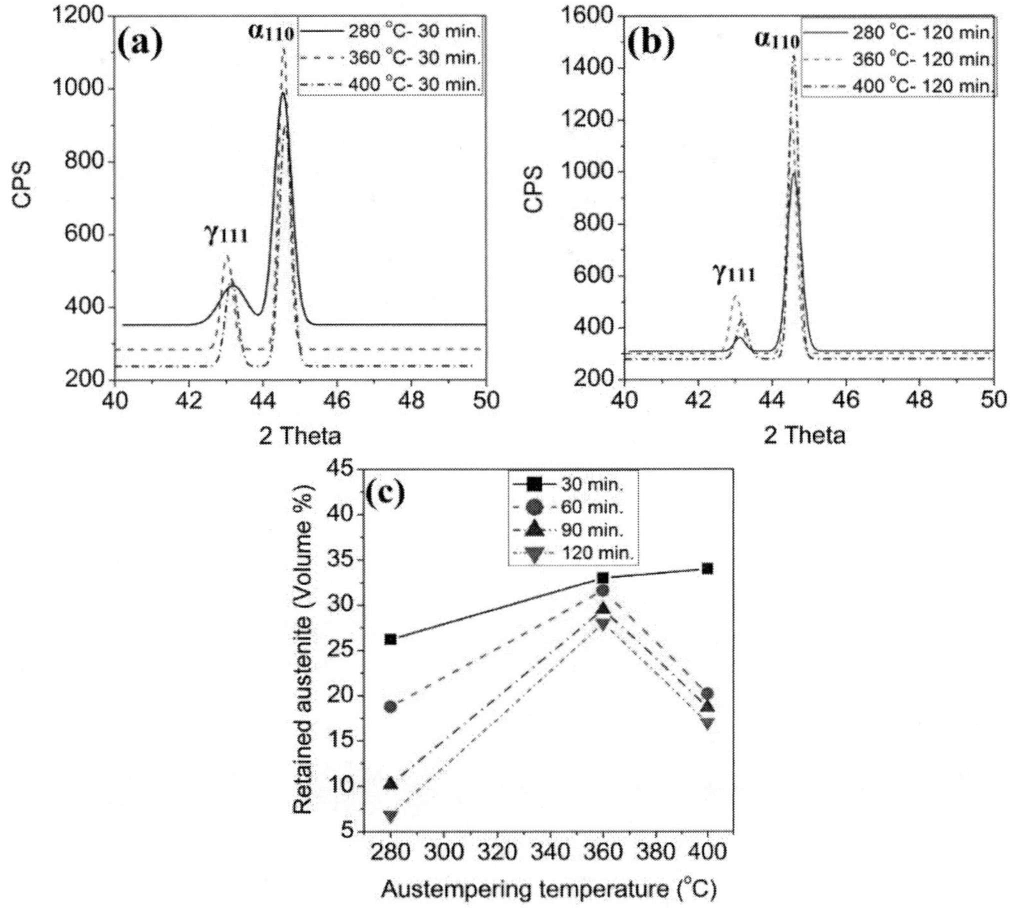

Fig. 2. (a) XRD patterns for samples austempered at 280, 360 and 400°C for 30 minutes.
(b) XRD patterns for samples austempered at 280, 360 and 400°C for 120 minutes.
(c) Retained austenite (volume %) as a function of austempering temperature and time.

The blend of ferrite laths inciting the course of action of a greater bainitic ferrite lath enables a more prominent measure of austenite to be as blocks rather than film morphology. Consequently bainitic ferrite content increments and hence the retained austenite content decreases for extended isothermal duration [8].

3.3 Vickers hardness

Fig. 3 shows variation of hardness as a function of varying austempering parameters.
The hardness of the material decreases with increase in austempering temperature for all austempering time (particular case for 120 minutes 595 HV to 378 HV). At lower

 https://doi.org/10.1051/matecconf/201814402013

austempering temperature of 280°C, the microstructure comprising of acicular bainitic ferrite, retained austenite and martensite contributes for high hardness.

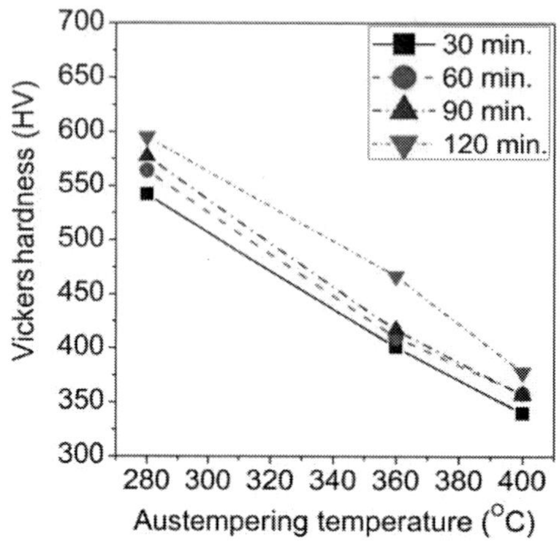

Fig. 3. Vickers hardness as a function of austempering temperatures and duration.

At elevated austempering temperatures at 360°C and 400°C, the increased coarseness of bainitic ferrite and retained austenite results in lower hardness to the matrix.

3.4 Wear property

Specific wear rate (SWR) reveals very small increase with extending austempering temperature and time (Fig. 4(a)). This was ascribed to the way that while extending the austempering time, retained austenite content decreases, accordingly SWR shows marginal increase (Fig. 4(b)), which implies higher retained austenite content favors to enhance wear resistance.

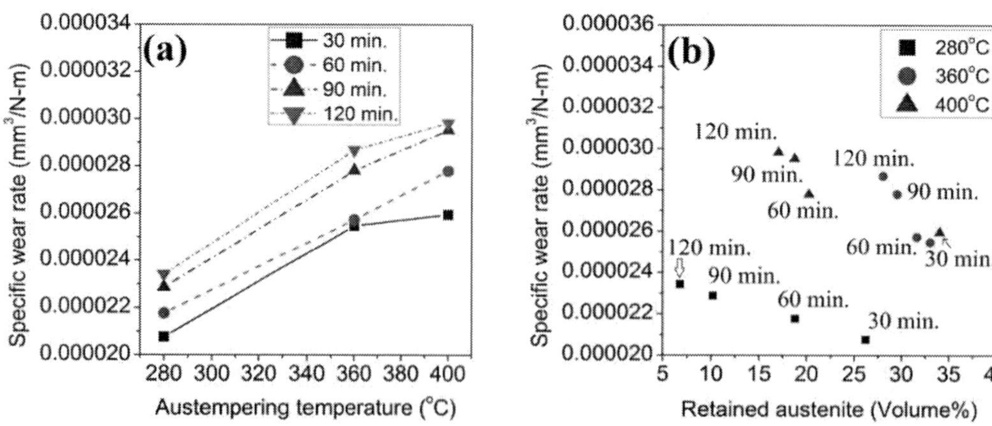

Fig. 4. (a) Specific wear rate as a function of austempering temperatures and duration.
(b) Specific wear rate as a function of retained austenite content.

The sample austempered at 280°C for 30 minutes, here the growth of fine acicular ferrite along with stabilized austenite and martensite adds to the lower SWR, despite the fact that the hardness value of 280°C for 30 minutes sample is lesser contrasted with those of examples austempered at higher duration (Fig. 3), the fundamental explanation for this is hardening amid wear, were the specimens with low hardness tends to harden amid deformation i.e., a greater amount of the retained austenite content experiences transformation to martensite [9] during wear bringing about high surface hardness, which was confirmed by microhardness measurements (load = 200 grams and dwell time = 30 seconds) on the light and dark regions of the worn surface (Fig. 5(a)), light regions showed a maximum hardness value of 860 HV which implies transformation from retained austenite to martensite and dark areas are oxidized debris that exhibits high hardness in the range 950 to 1160 HV. Hence the material removal decreases due to adhesive and abrasive forces, in this manner enhances the wear resistance. Wear resistance is obtained by a blend of hardness and toughness of materials, by temperance of sliding contact, hardness assumes an imperative part in wear resistance, inferable from the little impact in the midst of sliding friction. Additionally, increase in austempering temperature, hardness of austempered samples reduces, bringing about higher SWR.

3.4.1 Wear mechanism

After completion of wear test, the worn surfaces of the austempered specimens were characterized using SEM in order to understand the wear mechanisms (Fig. 5). Initial stages sliding wear was the predominant mechanism of material removal. The wear debris is pushed into the contact and is smeared by the acting forces. The main damage mechanisms observed in different austempered high silicon steels, due to sliding contact were adhesion, indentation and surface fatigue.

Fig. 5. Scanning electron micrographs of worn out samples austempered at (a) 280°C- 30 minutes (b) 360°C- 120 minutes (c) 400°C- 120 minutes

During the wear process formation of micro cracks, voids, crater, oxides, plastic flow of material and delamination takes place. The voids are the initiator of the crack formation; therefore crack will form near to voids. The linkage of voids enhance delamination of worn surface or in other words the sub surfaces present below the worn out surface propagates and comes out at the surface and shows delamination. Further running-in time the test specimen undergoes plastic shearing and indentation damage takes place due to hard wear debris [10].

4 Conclusions

The following conclusions were drawn from the present study on microstructural aspects and wear behaviour of austempered AISI 9260 high silicon steel.

- Austempering heat treatment of this steel over a temperature range 280 to 400°C generates ausferritic microstructure consisting of bainitic ferrite and retained austenite.

- Retained austenite content was found to be maximum at 360°C. Decrease in retained austenite content was observed with increasing austempering time for all temperatures.
- Hardness decreases as the austempering temperature increases, mainly due to the increase in coarseness of bainitic ferrite and austenite. Increase in austempering time showed moderate increase in hardness for all austempering temperatures.
- Specific wear rate shows marginal increase with the increase in austempering temperature and time. However specimen austempered at 280°C for 30 minutes offered superior wear resistance i.e. least specific wear rate.
- Formation of voids, micro cracks, oxide debris and surface delamination was observed in almost all worn out samples.

Acknowledgement

The authors gratefully acknowledge Steelmart, Mumbai, India for providing us this specific grade of steel as per our requirement. We are also thankful to National Institute of Technology Karnataka, Surathkal, India for providing us the financial support and facilities to carry out this investigation.

References

1. S. K. Putatunda, V. S. Arjun, T. Ronald and L. Gavin, Material Science and Engineering A **513-514**, 329 (2009)
2. S.K. Putatunda, Materials and Design **24**, 435 (2003)
3. D. Mandal, M. Ghosh, J. Pal, P.K. De, G. S. Chowdhury, S. K. Das, G. Das and S. Ghosh, Journal of Material Science **44**, 1069 (2009)
4. S. Sharma, S. Sangal and K. Mondal, Metallurgical and Materials transactions A **42A**, 3921 (2011)
5. L. C. Chang, Wear **258**, 730 (2005)
6. *A. S. T. M. G99 Standard*, ASTM International, (West Conshohocken, PA 2008)
7. B. D. Cullity, *Elements of X-Ray Diffraction* (Addison-Wisley, Reading, MA, 1974)
8. C. Liu, Z. Zhao and S. D. Bhole, Materials Science and Engineering A **434**, 289 (2006)
9. J. Yang, T. S. Wang, B. Zhang and F.C. Zhang, Scripta Materialia **66**, 363 (2012)
10. A. Leiro, A. Kankanala, E. Vuorinen and B. Prakash, Wear **273**, 2 (2011)

Mechanical Characterization and Finite Element Analysis of Jute-Epoxy Composite

Rajole Sangamesh[1*], *Naveen Kumar*[1], *K. S. Ravishankar*[1] and *S. M. Kulkarni*[2]

[1]*Department of Metallurgical and Materials Engineering, National Institute of Technology Karnataka, India*
[2]*Department of Mechanical Engineering, National Institute of Technology Karnataka, India*

Abstract. Natural fiber composite materials are such an appropriate material, that replaces synthetic composite materials for many of practical applications where we need high strength and low density. Natural fiber composites combine the technological, ecological and economical aspects. This leads to discovering its vast applications in the aeronautics, automotive, marine and sporting sectors. This paper deals with the study on mechanical characterization (Tensile, Compression and Flexural) of jute/epoxy (JE) polymer composite. The flexural properties of composites are experimentally tested and are simulated in commercially available FEA software. Flexural tested results are in good agreement with FEA results. Scanning electron microscopy (SEM) analysis of the failed samples reveals the matrix dominated failure.

Keywords: Jute/epoxy composite, Mechanical characterization, FE analysis.

1 Introduction

Throughout the 20th and 21st century natural fibers are replacing the existing synthetic fiber because of their better mechanical properties such as stiffness, modulus, and flexibility with respect to glass fibers in the fields of automobile, locomotives, marine structure and commercial application [1,2]. The outstanding preferences of natural fibers are low cost, low density, abundantly available and simple recyclability [3-8]. Jute/epoxy composite is one of the fiber reinforced composites which holds the all the above properties.

Basically, mechanical properties of composite materials depend on the many factors such as the type of matrix and fiber, fiber-length, direction, orientation, weaving patterns and number of layers used in composite [9-12]. Flexural test gives the information about the mechanical properties such as yield stress, ultimate stress, strain, maximum load capability [13].

In recent decades, research progress has occurred based on mechanical properties of natural fiber composite and it is noted that improvement was because of fiber selection, extraction, and treatment [12, 14, 15]. There is scope for modeling of these natural fiber

* Corresponding author: rajolesangmesh@gmail.com

composites for a various mechanical testing. From the experiments, it's difficult to analyze the crack initiation and its development in the material which can be easily performed using Finite element analysis software.

The present study addresses the mechanical characterization (tensile, compression and flexural) and flexural simulation of jute/epoxy (JE) polymer composite. Flexural tested and FEA simulated results are compared.

2 Materials, method and testing

The Jute/Epoxy composite has been fabricated using woven jute fabric and epoxy (Lapox L-12 with 10% K-6 hardener). The jute fabric and epoxy resin are used in the ratio of 55% and 45% volume fraction respectively. The appropriate quantity of epoxy resin and hardener are mixed in a glass jar using stirrer for a period of 5 minutes. Jute fabric is soaked in the epoxy resin-hardener mixture in order wet the fabric completely. The soacked layers are placed in a square shape steel mold (300 × 300 × 5 mm). Then it covered by a removable plate at above it and that whole system is placed under a compression molding machine with 25kg/cm^2 constant pressure for more than or equal to 24 hours to cure. The specimen was cut from the manufactured plate using a computerized cutting machine. The geometry of specimens was based on ASTM 3039 for tensile, ASTM 695-02a for compression and ASTM D7264 standards for flexural testing. Specimens were tested in each condition to obtain the properties. The test was carried out using Shimadzu AGX plus 100 KN (Japan) UTM which is shown in Fig. 1(a) for tensile, 1(b) for compression and 1(c) for flexural with a typical arrangement.

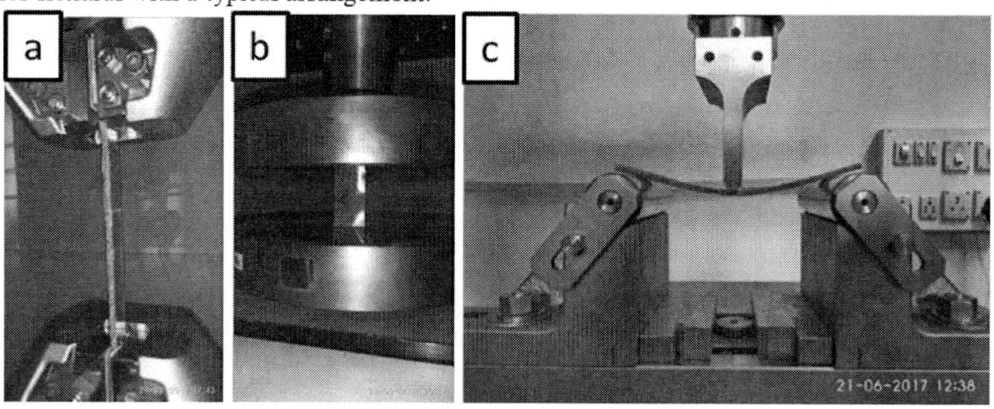

Fig. 1. Testing of samples with fixture for (a) Tensile loading (b) compression loading (c) Flexural loading

These mechanical characterizations have been performed to evaluate the basic properties required for the modeling of these composites so as to validate.

3 Modeling of jute-epoxy composite for flexural analysis

FE analysis has been performed to simulate the flexural properties of JE composite which goes as follows:

In CAE "Part Module" JE plate is modeled as Three-dimensional (3D) deformable solid element whereas the support roller and loading rollers are modeled as a 3D discrete rigid solid element. Density and mechanical properties are defined in CAE "Property Module". These properties are shown in Table 1. JE plate, support roller, and loading rollers are assembled in the "Assembly Module" to perform a flexural simulation as shown in Figure.2

(a). In the "Loading Module," the boundary conditions are fixed such that loading roller moves downside (- ve Y direction) and support rollers are fixed in all directions. The meshing of all the parts created using 8 noded linear brick 3D solid element C3D8R in "Mesh Module" which is shown in Fig. 2 (b) and the job is submitted and results are extracted.

Table1 Material properties of Jute/epoxy composite [16]

JE Composite	Density (ρ) kg/m^3	Modulus of elasticity (E) GPa	Poisons ratio (ν)	Shear Modulus (G) GPa
	1337.5#	12.57#	0.3395*	2.452*

Experiment *Rule of mixture

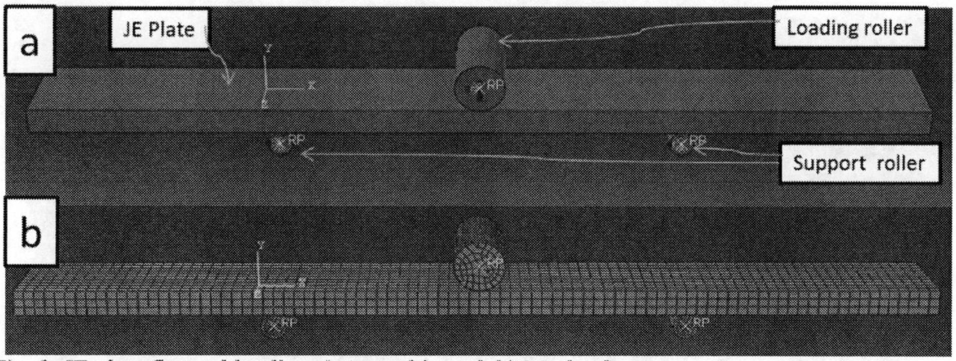

Fig. 2. JE plate flexural loading a) Assembly and b) Mesh of parts

4. Results and discussion

The typical experimental stress-strain graph for the jute/epoxy composite is shown in Figure.3 which proves that JE composites have higher flexural strength than tensile and compression. The tensile strength of the JE composite was found to be 55 MPa which is almost 20% higher than the woven jute-glass-epoxy hybrid composite characterized by Soma Dalbehera and S. K. Acharya [17]. The compression strength of JE composite was 62 MPa with failure strain was 0.09. Whereas the flexural strength of JE was 76 MPa with very less failure strain of 0.015.

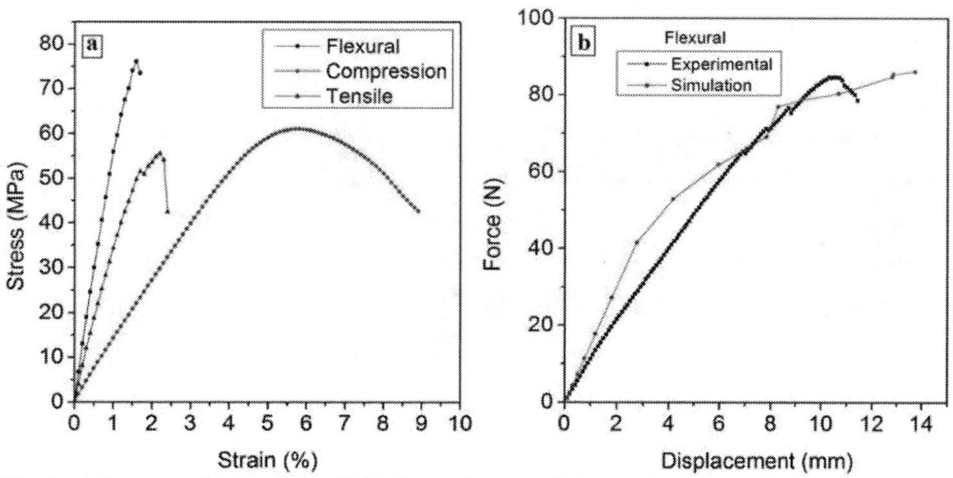

Fig. 3. a) Stress-strain Graph for JE b) Flexural comparison graph of force-displacement curve for JE composite.

The very similar trend of the force-displacement curve obtained for experimental and simulation is shown in Fig. 3(b). Fig. 4(a) shows simulation at zero seconds and Fig. 4(b) reveals the stress development in the material due to bending thereby very small crack appears on the JE composite which was matrix crack formation. As the bending force increases the development of the crack takes place opposite side of the loading that is bottom side of the plate which can be seen from Fig. 4(c) due to the pulling of the fibers in the opposite direction. On further increasing the load the crack propagates towards the upper side of the plate due to the failure of the jute fibers and finally, failure of component takes place which is seen from the Fig.4 (d). The Fig. 5(a) indicating the matrix dominated failure. Fiber pulls out due to tension at the bottom side of the plate and compression at the top side of the plate. Failed samples which are no more capable of taking the bending load are shown in Fig. 5(b)

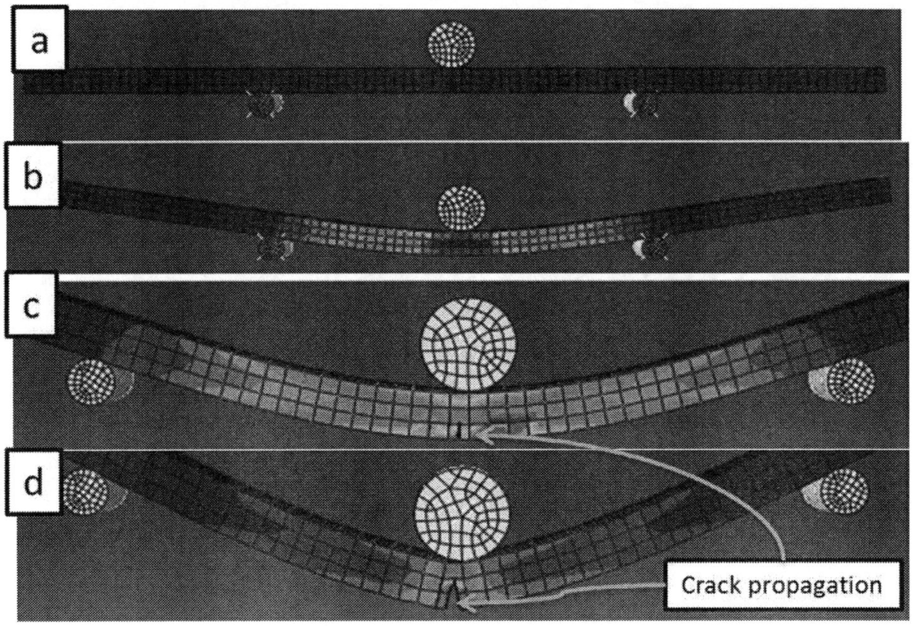

Fig. 4. Flexural simulated stress variation with time for a) 0 s b) 50 s c) 70 s d) 110 s failed sample

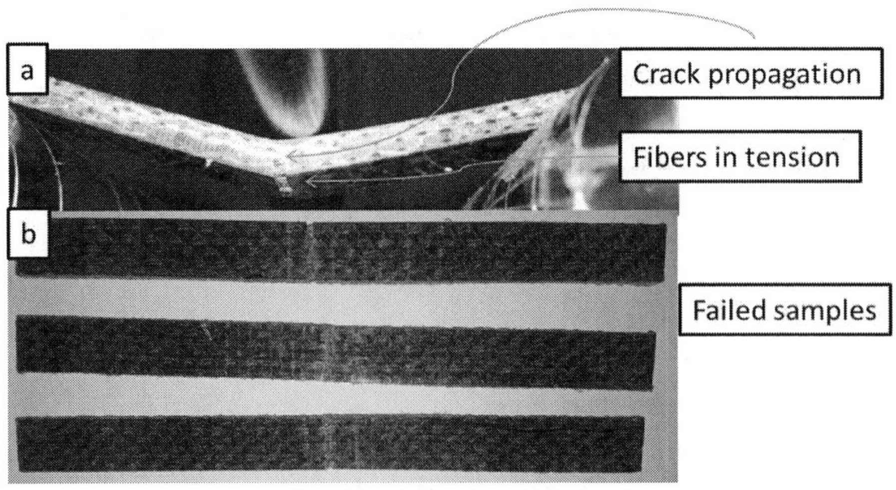

Fig. 5. Experimental JE composite a) Flexural crack propagation and b) Failed sample

A very similar kind of crack propagation and failure of the bending plate takes place which can be seen from the FE simulation and experimentation plots of Fig. 4(d) and Fig. 5(a) respectively.

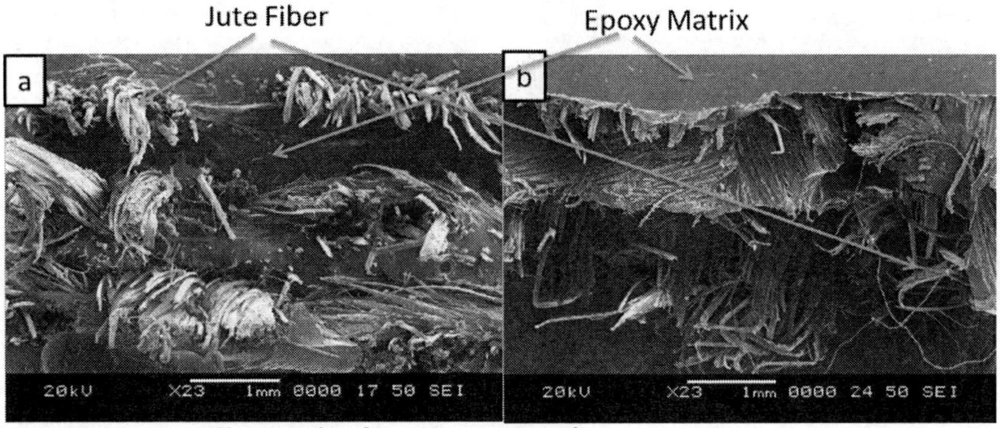

Fig. 6. SEM Fracture micro-graphs of Jute-Epoxy composite

Fig. 6 shows the fracture micro-graph for the jute-epoxy composite flexural tested mixed mode failed specimen, where matrix cracked at the very first with large pulling of fibers, appears on the lower side of the load applied which is very similar to that obtained from the simulation.

5. Conclusion

Jute-epoxy composites were fabricated successfully by compression molding technique. Flexural strength of fabricated samples comparatively higher than tensile and compression strength.

The flexural analysis of Jute-epoxy composite has been performed and crack initiation and propagation predicted was very similar to that are obtained from experimental testing.

SEM fractography reveals matrix dominated failure which was also true from the FEA simulation.

References

1. S. V. Joshi, L. T. Drzal, A. K. Mohanty, & S. Arora, Composites Part A: Applied science and manufacturing, **35**(3), 371-376 (2004)
2. C. W. Chin, & B. F. Yousif, Wear, **267**(9), 1550-1557 (2009)
3. B. F. Yousif, & H. Ku, Materials & Design (1980-2015), **36**, 847-853 (2012)
4. Z. Leman S. M. Sapuan, M. Azwan, M. M. H. M. Ahmad, & M. A. Maleque, Polymer-Plastics Technology and Engineering, **47**(6), 606-612 (2008)
5. U. M. K. Anwar, M. T. Paridah, H. Hamdan, S. M. Sapuan, & E. S. Bakar, industrial crops and products, **29**(1), 214-219 (2009).
6. S. M. Sapuan, M. Harimiand, & M. A. Maleque, Arabian Journal for Science and Engineering, **28**(2), 171-182 (2003)
7. A. A. A. Rashdi, S. M. Sapuan, M. M. H. M. Ahmad, & A. Khalina, International Journal of Mechanical and Materials Engineering, **5**(1) (1970)

8. M. Jawaid, H.A. Khalil, & A. A. Bakar, Materials Science, and Engineering: A, **528**(15), 5190-5195 (2011)
9. G. Cantero, A. Arbelaiz, R. Llano-Ponte, &, I. Mondragon, Composites science and technology, **63**(9), 1247-1254 (2003)
10. P. Saha, S. Manna, S. R. Chowdhury, R. Sen, D. Roy, & B. Adhikari, Bioresource technology, **101**(9), 3182-3187 (2010)
11. S. Shibata, Y. Cao, & I. Fukumoto, Composites part A: Applied science and manufacturing, **39**(4), 640-646 (2008)
12. V. Vilay, M. Mariatti, R. M. Taib, & M. Todo, Composites Science and Technology, **68**(3), 631-638 (2008)
13. S. Jayabal, & U. Natarajan, The International Journal of Advanced Manufacturing Technology, **54**(5), 639-648 (2011)
14. G. L. S. R. Prasad, M. S. Kumar, & G. Rajesh, International Journal of Engineering Science Invention, ISSN, 2319-6734 (2014)
15. K. Begum, & M. Islam, Research J Eng Sci, 2278, 9472. (2013)
16. K. S. Ahmed, & S. Vijayarangan Indian journal of engineering and material science **13** 435-442 (2006)
17. S. D. S. Acharya, Advances in Polymer Science and Technology, **4**(1). (2014)

Investigation of tensile properties of RTV Silicone based Isotropic Magnetorheological Elastomers.

Bhaktha Sandesh[1], *Hegde Sriharsha*[2*], *Rao Sathish U.*[1] and *Gandhi Nikhil*[2]

[1]*Department of Mechanical and Manufacturing Engineering, Manipal Institute of Technology, Manipal Academy of Higher Education, Manipal- 576104, India*
[2]*Department of Aeronautical and Automobile Engineering, Manipal Institute of Technology, Manipal Academy of Higher Education, Manipal- 576104, India*

Abstract. Magnetorheological elastomer (MRE) consists of an elastomer matrix and a Ferro-magnetic ingredient. The mechanical properties of MR elastomers can be reversibly controlled by applying a magnetic field of suitable intensity. The current work focusses on the enhancement of tensile property of RTV (Room Temperature Vulcanization) silicone based elastomer. The influence of Carbonyl iron powder (CIP) content and magnetic field were experimentally investigated. Addition of CIP increases the tensile modulus but it reduces the percentage elongation and tensile strength making it brittle. Under the influence of magnetic field, the enhancement of tensile properties up to 20% content was linear. The behavior above 20% is observed to be non-linear. The onset of non-linear stress-strain behavior is investigated. Regression equation is developed from the experimental data relating percentage content with the mechanical properties of MRE. The developed equation predicted the behavior of 27% MRE with an error of less than 8%. Hyperelastic model developed by Yeoh was fitted to the stress-strain behavior of MRE with minimal error.

Keywords: Magnetorheological elastomer, RTV Silicone, Tensile testing, Regression Analysis, Hyperelastic modelling.

1 Introduction

Magnetorheological materials include MR fluids, MR elastomers and MR foams belong to a class of smart materials whose rheological properties can be controlled instantly and reversibly with the externally applied magnetic field. MR materials consists of Ferro-magnetic particles suspended in a non- magnetic matrix such as liquid carrier fluid in case of MR fluid, elastomeric matrix in MR elastomer and a foam like matrix in MR foams [1, 2]. MR fluids and MR foams have a field dependent yield stress and MR Elastomers have a field dependent modulus [1, 2]. The response time of MR fluids are fast but, fluids require a separate container for storage [2]. The sedimentation of the ferromagnetic ingredients over a period of time reduces the efficiency [3]. These limitations are overcome by employing magnetorheological elastomers.

* Corresponding author: sriharsha.hegde@manipal.edu

Magnetorheological elastomers consists of micron sized Ferro-magnetic particles dispersed in a non-magnetic polymeric matrix. The particle are mixed before the curing process. When curing is done under a magnetic field, the interaction between the adjacent particles resulting in induced dipoles forming columnar structures parallel to the applied field. This imparts anisotropic nature to MRE [1, 4]. These are also referred as structured MRE. Curing process without the application of magnetic field results in uniformly distributed particles within the matrix causing isotropic effect [5]. Both these type of MREs show significant enhancements in properties when tested under magnetic field [1, 4, 6]. Anisotropic MRE show better MR effect than isotropic samples [4, 5]. Lot of matrix materials are available for the processing of MRE [3]. RTV silicone rubber is the most widely used matrix by various researchers due to the ease of working in the laboratory and economical availability [2, 4]. The working temperature of silicone elastomer is also wider than most matrix materials. Carbonyl iron powder is the obvious choice as ferromagnetic ingredient due to low coercivity and high saturation magnetization. The stiffness and the damping properties of the MRE is a function of the strength of the magnetic field [4, 7, 8, 9], size [10, 11], shape [12] and percentage content of carbonyl iron particles [7, 8, 13,14]. The aforementioned properties make MRE as semi – active /active suspension devices in various applications such as adaptive tuned vibration absorber (ATVA) [15], variable stiffness suspension bushing [16] etc. To successfully commercialize MRE, the mechanical characterization is necessary to ascertain its feasibility at different operating conditions.

Structured MRE display better stress-strain relationship than isotropic MRE. The maximum stress increases with increase in field strength [17]. Addition of Ferro-magnetic particles to the silicone rubber increases the tensile strength, stiffness and hardness. Anisotropic MREs aligned in the longitudinal direction have higher tensile strength and strain compared to transversely aligned MRE and isotropic MRE (27% volume fraction) [18]. The addition of 20% CIP enhanced the tensile stress by 95%. The elongation at break increased by 111% [19]. The tensile strength at break remains approximately constant up to 30% volume of CIP. Further increase in CIP content deteriorates the mechanical properties [20].

In the present work, RTV-Si based MRE samples were prepared by varying the percentage content of CIP. The stress-strain behavior of the samples were tested under varying magnetic fields. Hyper-elastic model (Yeoh model) was fitted to the experimental stress-strain data. Regression equations relating the Young's modulus, tensile strength and the percentage elongation as a function of %CIP content was developed. The relation was used to predict the properties of 27% CIP content MRE which was experimentally validated. The error was found to be within acceptable limits.

2 Methodology

2.1 Sample preparation

A two part Silicone RTV rubber from Dow Corning Corporation, India (Silastic 3483 base and Silastic 83 curing agent) was used as matrix. CIP of average diameter 6.23μm (Chengdu Nuclear 857 New materials Ltd, China) was employed as Ferro-magnetic ingredient. In the first step, the CIP was mixed with the elastomer base and stirred thoroughly to ensure uniform distribution. The curing agent was mixed then it was stirred again and the mixture was kept inside a vacuum chamber to remove air bubbles. The mixture is poured in to acrylic mould (dimensions as per ASTM D412 – Die C) and allowed to cure for a period of 24 hours. Samples with 0%, 5%, 10%, 15%, 20%, 25% and 30% CIP were prepared. One of the test samples is shown in figure below (Fig. 1).

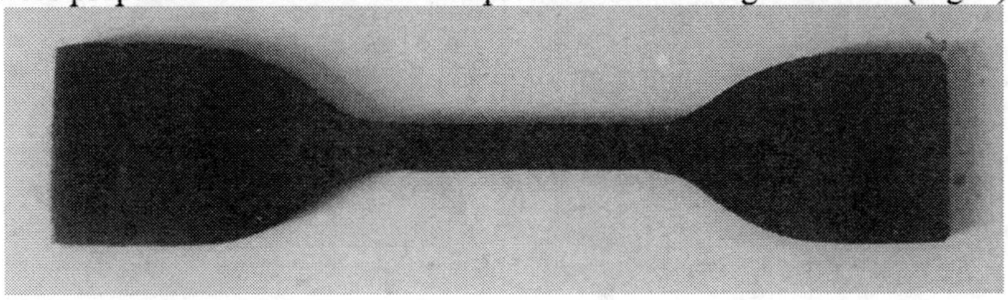

Fig. 1. Tensile test sample

2. 2 Magnetic field analysis

Rare earth Neodymium permanent magnets were used in the form of Halbach array to produce magnetic field (Fig. 2). An FEMM (Finite Element Method Magnetics) analysis of the same was conducted to ensure uniform distribution of magnetic field (Fig. 3a). The variation of the field with respect to the gap between the poles shown in figure 3b indicates that approximately 0.5T field is produced at the center. This was validated using a Gaussmeter (SES instruments, Roorkee).

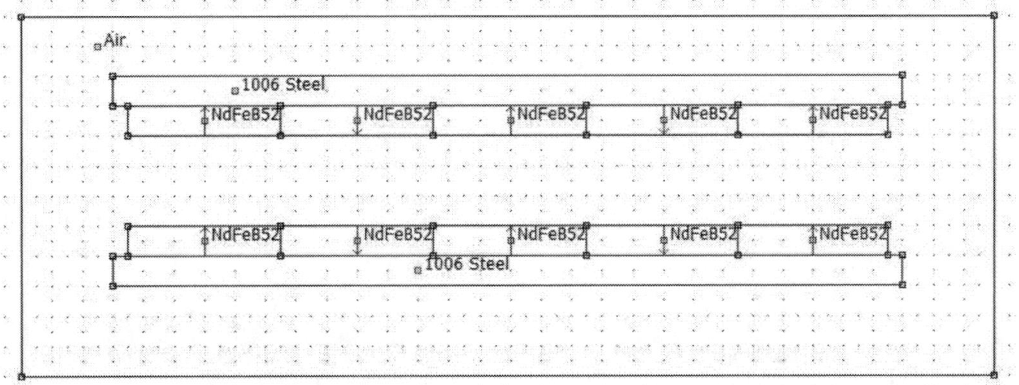

Fig. 2. Arrangement of permanent magnets

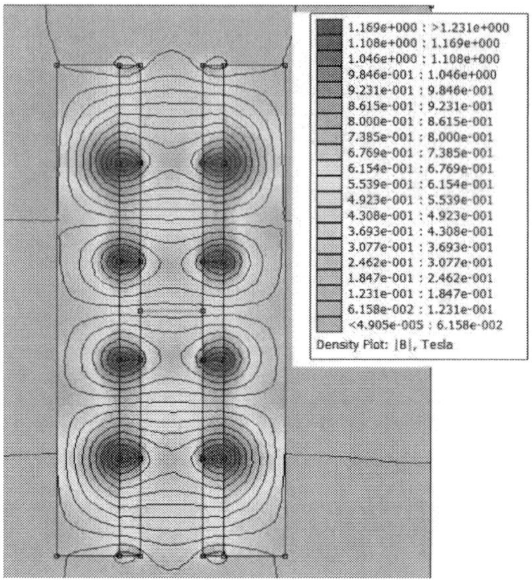

Fig.3a. The flux density of Halbach array

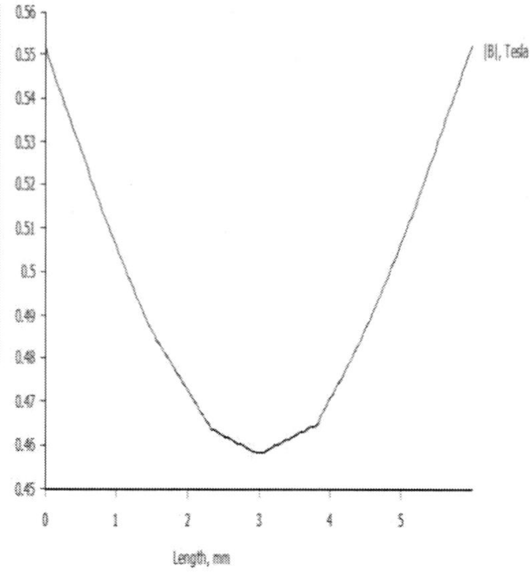

Fig.3b. Flux density v/s distance

2.3 Tensile test

Tensile test was performed on a Universal testing machine (INSTRON 3366) both in the presence and absence of a magnetic field. The magnetic field was applied using Halbach array along the gauge length (33mm) of the test specimen. The samples were tested at a cross head speed of 500mm/min as per standard ASTM *D*412 [21] at room temperature. The photograph of the test set up is depicted in Fig. 4a and 4b.

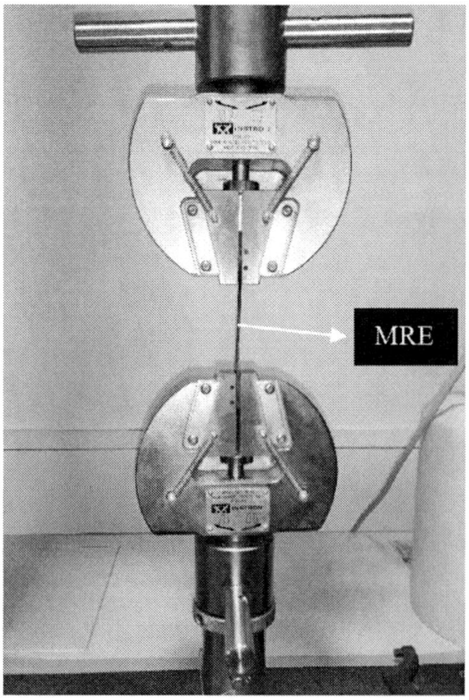

Fig.4a. Tensile testing without magnets

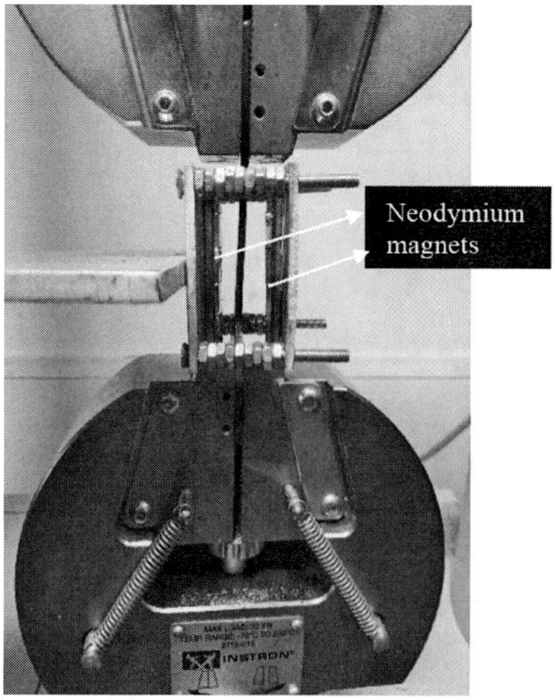

Fig.4b. Tensile test under magnetic field

2.4 Hardness measurement

The variation of hardness of the test samples with and without the application of the magnetic field was measured by a Shore (A) durometer as per ASTM *D*2240 [22]. The hardness is measured at 5 different locations and the averaged valued is considered for analysis.

2.5 Hyperelastic modelling

Hyperelastic material models are used to explain the nonlinear elastic behavior of rubber like materials under the assumptions of material isotropy and incompressibility. These constitutive models mainly rely on the definition of strain energy functions which is governed by the values of strain invariants. Silicone rubber used in the current study exhibit hyperelastic characteristics [23, 24] and this behavior is modelled using the constitutive equations. The Yeoh model [25] is used in this study since the quality of the fit was observed to be good. The quality of fit was ensured by coefficient of determination (R_2) whose value is required to be close to 1 which meant that the difference between the experimental value and model's predicted value are small. The equation showing the relationship between the strain energy and the strain invariants for this model could be represented as

$$\psi = \sum_{i=1}^{N} C_{i0}(I_1 - 3)^i \tag{1}$$

where
ψ =Strain energy density
I_1 = First variant of C
C = Right Cauchy-Green def tensor
C_{i0} = Stress like parameters defining stiffness of material
N = Number of terms

Stress-strain data of all the samples obtained from experiments was imported to Hyperfit freeware and curve fitting was done for Yeoh model (N=3). Stress-strain curves were generated based on the values of the strain variants obtained with averaged R_2 value being 0.998.

3 Results and discussions

3.1 Stress-strain behavior

The Fig. 5 shows the variation of stress-strain (average data of 3 samples each obtained from force-displacement graph) behavior of the samples tested in the absence of the magnetic field. From the stress-strain graph it is observed that the pure silicone rubber sample exhibited good ductility where the stress increased with the increase in the strain. As the percentage CIP content was increased, the reinforcing effect increases the stiffness of the matrix. When the ingredients are absent, the cross-linking of the polymer chain is more continuous rendering it more elastic. With the increase in percentage content the continuity of the polymer chain breaks making the elastomer more brittle resulting in a decrease in elongation at break. The addition of Carbonyl iron particles to the matrix decreased the tensile strength (2.25 MPa for pure silicone and 1.37 MPa for 30% MRE) due to the increase in the brittleness of the MRE. However, the tensile modulus increases with the increase in the percentage CIP in a nonlinear manner which indicates that the stiffness of the MRE increases. The increase in the stiffness is theoretically quantified by hyperelastic modelling. The variation of tensile properties with respect to the change in the percentage carbonyl iron powder content have been shown in the Table 1.

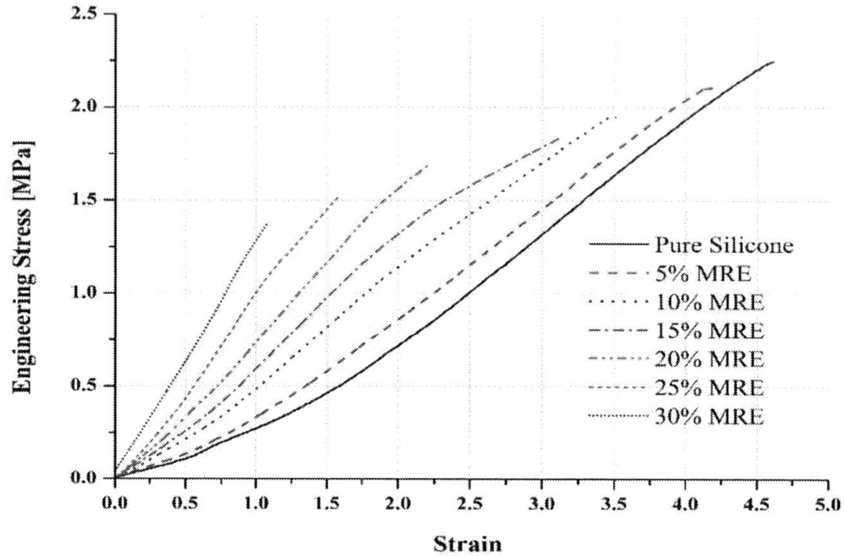

Fig. 5. Stress v/s Strain behavior of isotropic MREs at 0T.

Table1. Tensile Properties of MRE

Percentage volume of CIP	Tensile strength, s (MPa)	Tensile modulus, E (MPa)	% Elongation
0	2.25	0.62	461.1
5	2.11	0.75	417.9
10	1.95	0.84	350.72
15	1.83	0.94	314.2
20	1.69	1.03	223.1
25	1.59	1.17	152.7
30	1.37	1.39	107.6

3.2 Stress-strain behavior under the influence of magnetic field

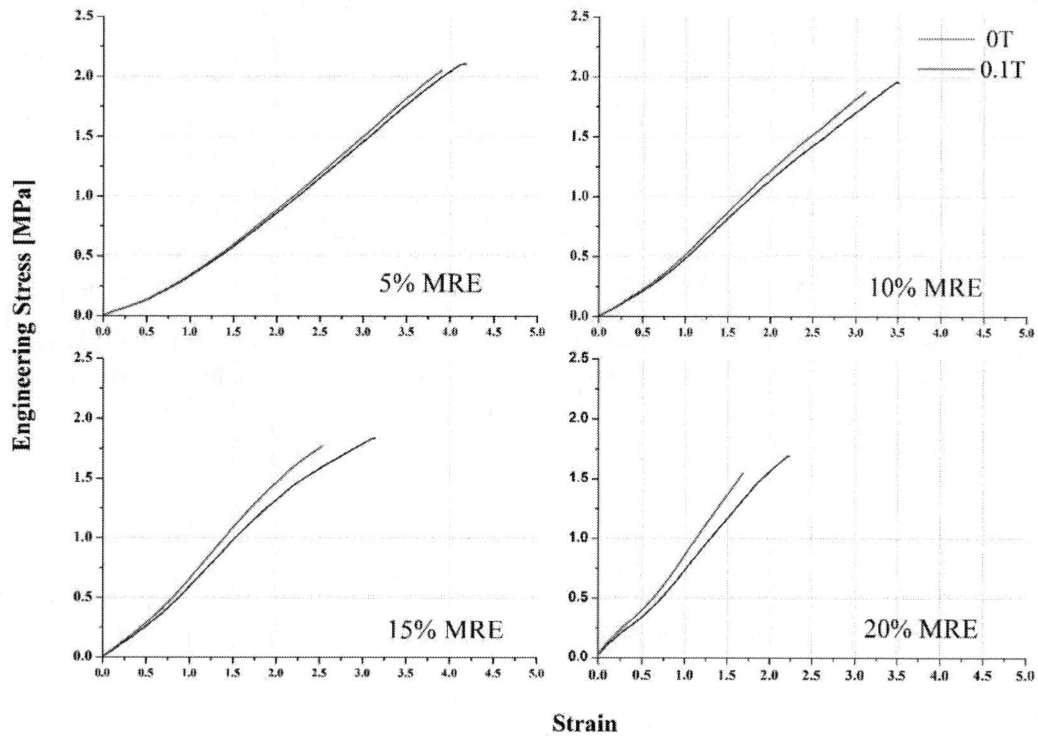

Fig. 6. Stress-strain behavior of MRE with magnetic field

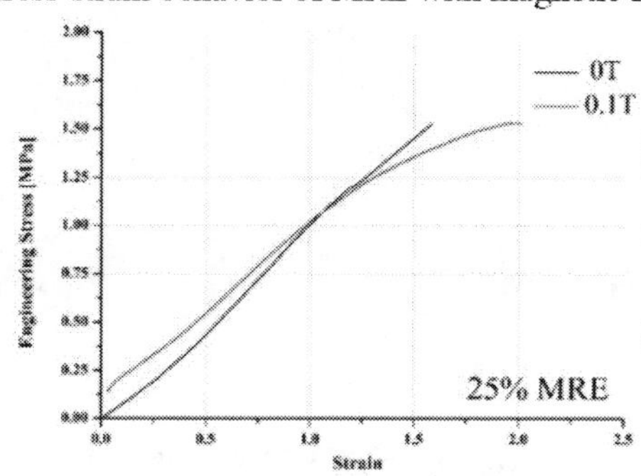

Fig.7a. Stress strain behavior 25% MRE

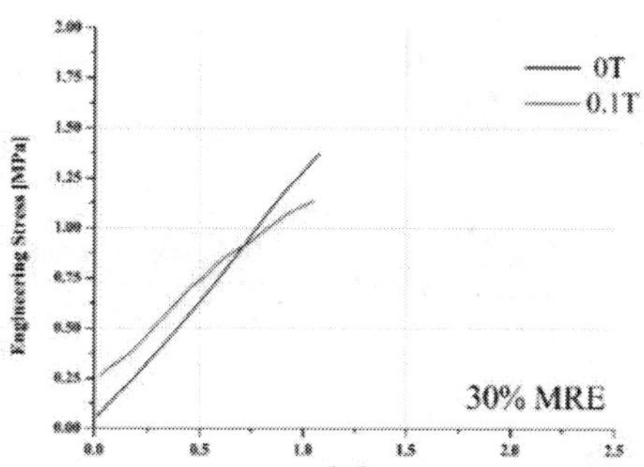

Fig.7b. Stress strain behavior 30% MRE

The elongation of pure sample at break is about 460% which is more than MRE samples. The addition of CIP is resulting in improved Young's modulus, but the elongation at break is reducing. The trend is uniform across all samples under the absence of magnetic field. When the field is applied, the behavior is similar, but with an improved field induced modulus resulting from hardness increase up to 20% CIP concentration. The ductility is higher at lower CIP content and the distance between the adjacent particles is also relatively higher. As the strain is increased, the distance between the adjacent particle increases. This behavior is linear if the matrix is relatively more ductile. For 25% and 30% CIP samples, the field induced behavior is slightly different resulting from non- linearity. The 25% sample starts to show non-linear behavior at about 100% strain at 0.1T (figure7a). This is lower than the samples with lower CIP content. Even though, the elongation is slightly improved when compared to 0T the linearity is compromised. The onset of non-linearity for 30% CIP sample is further reduced at approximately at 75% strain (Fig.7b).

The phenomenon of on-set of non- linearity can be explained by considering a unit cell of MRE with 4 CIP particles for lower percentage content of CIP (0-20% CIP samples) (Fig. 8) and 6 CIP particles for higher percentage (Fig. 9) content of CIP (25% and 30% CIP samples) . When the MRE is subjected tensile force in the longitudinal direction, equal and opposite force acts at the fixed end which resists the tensile loading increasing the modulus. During tensile loading the CIP particles in the longitudinal direction tend to move away from each other. However, particles in the lateral direction come closer which prevents further elongation in the longitudinal direction. With the application of magnetic field, the distance between the particles decreases in the lateral direction due to the dipole. The dipole attraction between the adjacent particles in the lateral direction results in increased resistance to tensile force increasing the tensile modulus. For lower percentage of CIP samples, the increase is not enough to cause non-linearity. For higher percentage of CIP samples the gap between the particles is lower which further prevents the elongation in the longitudinal direction making it more brittle and reducing the tensile strength. Further, under the influence of magnetic field, the force of attraction between the particles in the lateral direction is higher causing increased resistance to the loading which accounts for increase in tensile modulus for lower values of strain. However, at higher values of strain, since the distance between the adjacent particles is considerably small, the resistance to the tensile force exerted by the sample decreases. The sample is highly brittle such that the effect of the magnetic field is arrested reducing the tensile modulus. This results in nonlinear behavior.

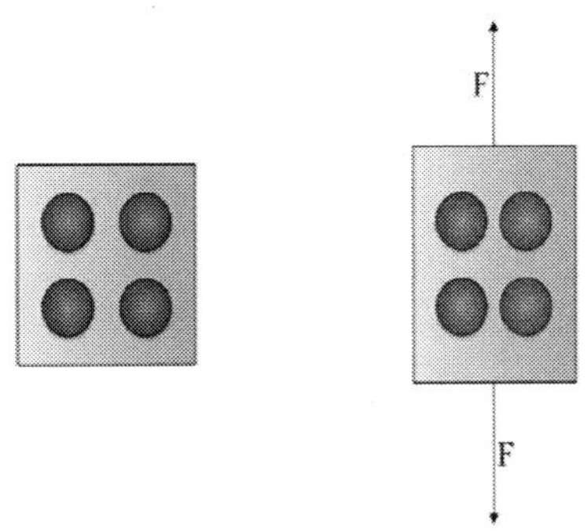

Fig. 8a. Behaviour under tensile force

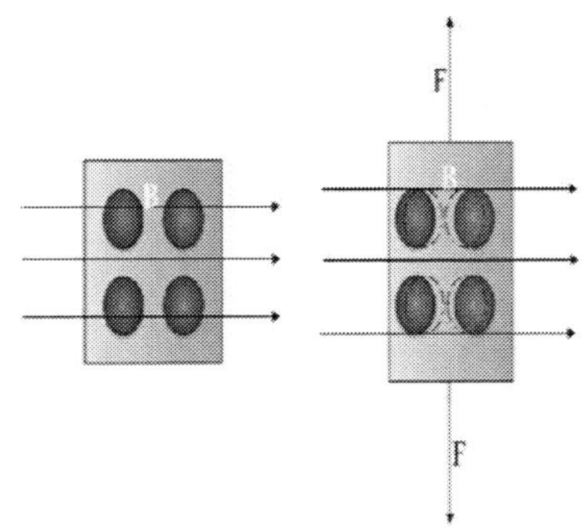

Fig. 8b. Behaviour under magnetic field

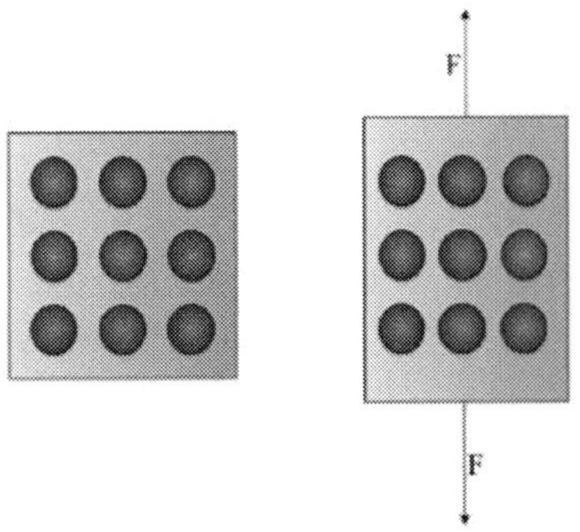

Fig. 9a. Behaviour under tensile force

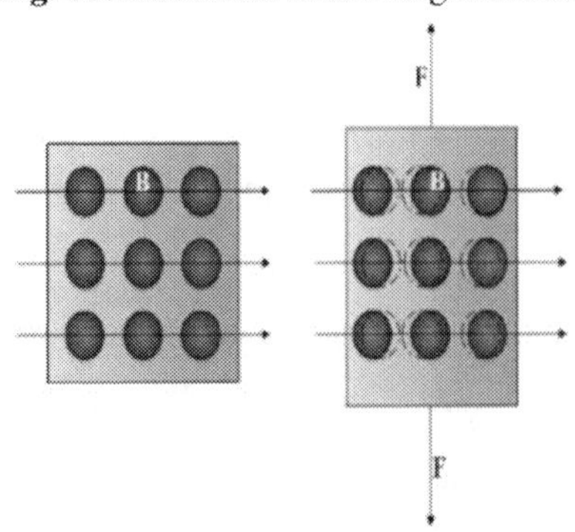

Fig. 9b. Behaviour under magnetic field

3.3 Hyperelastic modelling

The material constants determined for pure silicone and MRE samples are listed in Table 2. It is observed from the table that with the increase in percentage content of CIP, the value of C_{10}, C_{20} and C_{30} increases. The greater the value of C_{i0}, higher is the stiffness of the sample. Silicone reinforced with 30% CIP has the highest value of C_{i0}, which indicates that this sample is the hardest among all the samples. Pure silicone rubber has the lowest value of C_{i0}, which indicates that the percent elongation is highest among all samples. It can be concluded that the addition of the CIP has an effect on the stiffness of the samples which in turn effects the tensile properties.

Table 2. Parameters of Yeoh model

Yeoh Model (N=3)			
% CIP	**C_{10}**	**C_{20}**	**C_{30}**
0	0.06	0.005	0
5	0.078	0.006	0
10	0.115	0.008	0
15	0.125	0.012	0
20	0.148	0.018	-0.001
25	0.163	0.041	-0.004
30	0.256	0.044	-0.005

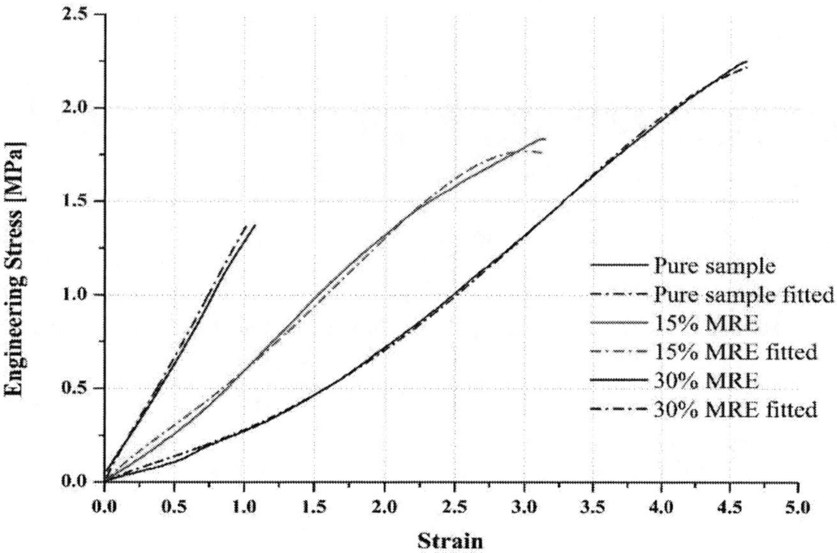

Fig.10. Yeoh model (N=3) fitted for samples having 0, 15 and 30% CIP

Fig. 10 shows the Stress-strain characteristics of MREs having 0, 15 and 30% CIP and their corresponding Yeoh Model (N = 3) fitted plots. Selected graphs are shown here for the clarity and other samples also show a similar trend.

3.4 Hardness

From the Table 3, it is observed that, the addition of CIP enhances the hardness due to reinforcing effect. The hardness of the MRE samples increase with the magnetic field due to the alignment of the CIP along the flux lines. This trend is consistent for all the MRE samples.

Table 3. Effect of addition of CIP and application of magnetic field on the hardness of the MREs.

Percentage volume of CIP	**Hardness (Shore A)**	
	Without magnetic field	**With magnetic field (0.1T)**
0	9	9
5	10	12
10	12	14
15	16	20
20	22	26
25	29	32
30	33	36

3.5 Regression equation

Regression analysis is a statistical tool used to determine the relationship between two variables. It is used to determine the effect of the input parameters on the output. In this study, the effect of percentage volume of CIP on the tensile strength, tensile modulus and % elongation is analytically determined using regression equations. The regression equations are generated based on the experimental results. A second order polynomial is used in this study since the correlation coefficient (r^2) was observed to be close to 1. Table 4 shows the Regression equation to predict Tensile strength, Tensile modulus and percentage elongation. Using these equations, the tensile parameters are calculated for 27% volume of CIP. The stress-strain behavior of 25%, 27% and 30% MRE samples is depicted in Fig. 11.

Table 4. Regression equation and Correlation coefficient for tensile parameters

Response	**Regression Equation**	**Correlation coefficient (r^2)**
Tensile Strength	$R_{TS} = (2.24095-(0.02614*X)-(6.67*10^{-5}*X^2)$	0.99
Tensile Modulus	$R_{TM} = (0.64609+(0.01392*X)+(3.3981*10\text{-}4*X^2)$	0.98
% Elongation	$R_E = (465.72-(10.35*X)-(0.0641*X^2)$	0.98

where X = Percentage volume of CIP

Table 5 shows the percentage error between the experimental values and predicted value generated using the regression equations. The average error for the model is observed to be less than 6% which showed a good correlation between the experimental values and predicted values.

Table 5. Calculated and Experimental value comparison for 27% MRE

Parameters	Predicted values	Experimental Values				% Error
		Trial 1	Trial 2	Trial 3	Average	
Tensile Strength (MPa)	1.48	1.54	1.57	1.56	1.55	4.7
Tensile Modulus (MPa)	1.27	1.30	1.32	1.33	1.31	3.14
% Elongation	139.54	147.37	152.99	149.91	149.9	7.42

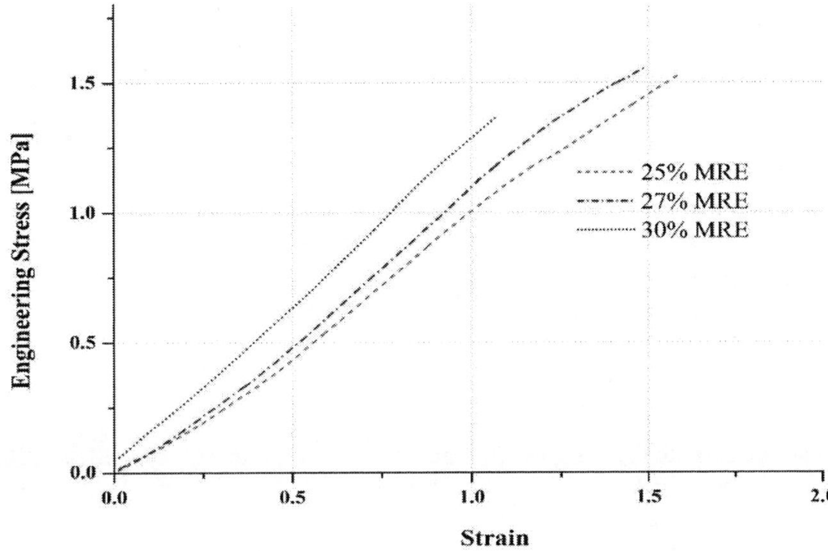

Fig.11. Stress v/s Strain behavior of 27% MRE

4 Conclusions

RTV silicone based MRE samples were prepared with different CIP percentage content. Tensile and hardness tests were conducted at varying magnetic fields, which resulted in the following conclusions.

1. The addition of CIP improves modulus of the MRE compared to pure silicone sample. The ductility of the samples reduces with the percentage increase in the CIP content. The elongation at break for pure sample is 460% which reduces to 107% for 30% MRE. Under the influence of magnetic field, the stress strain behavior is improved and up to 20% CIP, the behavior is linear. Higher percentage CIP samples showed non-linear behavior under the influence of the field.

2. The addition of CIP increases the hardness due to reinforcement effect. The application of magnetic field further increases the hardness.
3. Stress-Strain behavior of MRE was employed to Yeoh hyper-elastic modelling. The model was fitted to experimental data with negligible errors.
4. The regression equation was developed which can predict the modulus, tensile strength and percentage elongation as a function of CIP percentage content. The equation predicted the behavior of 27% content CIP with good effect.

References

1. J. D. Carlson and M. R. Jolly, Mechatronics, 10, 555-569 (2008).
2. Weihua Li and X. Zhang, Recent Patents on Mechanical Engineering, 1,161-166 (2008).
3. Y. Li, J. Li, W. Li and H. Du, Smart Mater. Struct, 23, 1-24, (2014).
4. J. M. Ginder, M. E. Nichols, L. D Elie, and J. L. Tardiff, Smart Mater. Struct., 131-138 (1991).
5. X. L. Gong, X. Z. Zhang, and P. Q. Zhang, Polym Test., 24, 669-676 (2005).
6. M. Lokander and B. Stenberg:, Polym Test., 22, 677-680 (2003).
7. L. Chen, X. L. Gong, and W. H. Li, Chin. J. Chem. Phys, 15, 271-274(2008).
8. M. R. Jolly, J. D. Carlson, and B. C. Munoz, J. Intel. Mat. Syst. Str., 7, 612-622 (1996).
9. J. F. Li and X. L. Gong, J. Cent. South Univ., 15, 261-265 (2008).
10. S. Hegde, U. R. Poojary, and K. V. Gangadharan, International Conference on Advances in Manufacturing and material engineering. Mangalore, India, 2301-2309 (2014).
11. Y.C. Fan, X. Gong, S. Xuan, W. Zhang, J. Zheng, and W. Jiang, Smart Mater. Struct., 20, 1-8 (2011).
12. H. J. Song, O. Padalka, M. Werely and R. C. Bell, 50th AIAA/ASME/ ASCE/ AHS/ ASC Strucutres, Structural Dynamics, and materials conference, Palm springs, USA, AIAA, 2009-2118 (2009).
13. A. Boczkowska and S. F. Awietjan, J. Mater. Sci., 44, 4104-4111 (2009).
14. L. C. Davis, J. App. Phys., 85, 3348-3351(1999).
15. H. X. Deng and and X. L. Gong, J. Intel. Mat. Syst. Str., 18, 1205-1210 (2007).
16. J. R. Watson, Patent US5609353, 1997.
17. C. Bellan and G. Bossis, Int. J. Modern Phys., 16, 2447-2453 (2002).
18. M. Farshad and Andre Benine, Polym. Test., 23,347-353 (2004).
19. I.A.P Martinez, L.M.P Pineda, L.M.L Sanchez, O.S Romero, J.G.P. Cordova and A.E Zuniga, Polym. Test., 57, 78-86 (2017).
20. Lokander M and B Stenberg, Polym Test., 22, 245-251(2003).
21. ASTM D 412 –15a, 2013, Standard test methods for Vulcanized rubber and Thermoplastic Elastomers - Tension, American society for Testing and Materials, United States
22. ASTM D 2240 – 15, 2017, Standard Test method for rubber property – Durometer Hardness, American society for Testing and Materials, United States

23. V. Vahapoglu, S. karadeniz, and I Yazici, Experimental techniques, 35, 17-23(2011).
24. M. Noor, S.N. Azizzat, and J. Mahmud, Advanced materials research, 915, 858-866 (2014).
25. OH. Yeoh, Rubber chemistry and Technology, 66, 754-771(1993).

Case study on water based cellulose couplant in contact type ultrasonic testing

B. Shivamurthy[1*], *Krishnamurthy*[1] and *H. K. Sachidanada*[2]

[1]*Department of Mechanical and Manufacturing Engineering, Manipal Academy of Higher Education Manipal, India*
[2]*School of Engineering and IT, Manipal University, G04, PO. Box: 345050, DIAC, Dubai, UAE*

Abstract. In this paper the ultrasonic test performance of water based Methylcellulose couplant on steel is reported. The couplant used in the ultrasonic testing has an influence on the amplitude of the pulse echo signal. The ultrasonic energy which is transmitted from the transducer to the test material, through the couplant which is present in the interface. The amplitudes of back wall echo increases with increasing acoustic impedance of the couplant. In this research work couplant was prepared by using 0 to 50g of methylcellulose in steps of 10g in 250 ml of water. All five types of water based methylcellulose couplant were used for testing a standard stainless steel calibration U2 block. The effect on the amplitudes of back wall echoes was studied and best possible combination of water and methylcellulose are reported.

Key words: Couplant, Amplitude, Ultrasonic Testing, Methylcellulose

1 Introduction

The contact pulse echo type ultrasonic testing has a wide range of applications such as dimensional measurements, flaw detection, medical ultrasound, sonography etc. Recently researchers given attention on influence of various parameters on test performance of ultrasonic test. The main parameters which influences the results are; instrument and transducer performance, couplant used, material variations and defect variations.

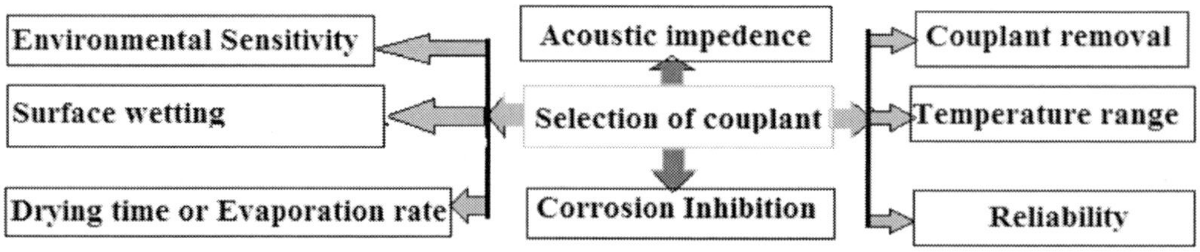

Fig. 1. Parameters to be considered for selection of couplants

* Corresponding author: b_shivamurthy@yahoo.co.in

A couplant is a material which is applied in between the probe and the test specimen to form a medium through which the sound waves transmit to and fro. Depending on the requirements of testing there are many suppliers who have specific couplants for specific materials. The materials were tested under controlled conditions and suggested for industrial use to get the optimum results in ultrasonic testing. The selection of couplant for ultrasonic testing is based on various parameters. Figure 1 shows the parameters to be considered while selecting the couplant. High acoustic impedance couplant reduces the surface noises and improves coupling on rough and/or curved metal surfaces. Couplant selected must be non-corrosive in nature. Higher viscosity provides good coupling and reduces surface noise. The temperature of couplant is also an important factor. The test specimen to be checked for its temperature, the area of the test specimen surface, the time required for inspection as the couplant needs to be present during the entire course of inspection. After applying couplant if the time taken for the inspection is long then the drying time is a crucial consideration. Once the inspection is completed the applied couplant must be easily removable. Couplant should be non-toxic, easy to dispose and environmental friendly. Good wetting character and the consistency of the couplant is also a desired factor. In this context few researchers worked and developed different materials for couplants and transducers. Kim Young H., et al. (2001) reported in their research about couplant effects during contact ultrasonic testing. They prepared different mixture proportions of water and glycerin as couplant and conducted ultrasonic test on carbon steel test specimen. The experiments conducted are showing different back wall echo's on the CRT screen of the ultrasonic flaw detector. They discuss also about the attenuation coefficient which is influenced by the reflection coefficient factor. Also,Vincent (1987) reported that, the amplitude of ultrasonic test signals are affected by the type of couplant used in the testing technique. Bindal (1992) reported that the petroleum products such as oils and grease as couplants over the test specimen like refractory bricks, timber, stone and civil structures leaves some marks which may be difficult to remove it later. Hence, an alternative liquid couplant is required. Ashok Kumar (1996) reported thickness measurement using ultrasonic technique. He concluded that the quality of results depends on the correction factor of the couplant being used. Bindal (2000) presented in their research that, the thicker oils, greases and viscous liquid is preferred to be applied on vertical and inclined surfaces and provide good coupling characteristics. Young H., et al. (2001) described in his research that, if the layer thickness of the couplant is applied too thin it may cause an error in the measurement of the velocity and attenuation. Kumar.et al. (2007) has discussed in his research about various coupling conditions on ultrasonic echo parameters which may have significant effects on the couplant used and their amplitude during the ultrasonic testing. They discussed various characteristics of the couplants correction factors with various test specimen thickness. Mubuza et al., (2012) reported that the reflection and transmission coefficients at the interface of transducer, couplant and specimen surface. The specimen they tried on steel and aluminum. They concluded that couplant (Honey) with high acoustic impedance and high transmission coefficients are most effective for detection of defects. Also, aluminum as low acoustic impedance as compared to steel results in back wall echoes of high amplitudes. Based on above research output in this paper a mixture of water and methylcellulose has been developed as couplants for ultrasonic testing and results are reported.

2. Materials and methods

2.1 Materials

In this work Methylcellulose was used to prepare the couplants. Methylcellulose is easily available and generally used for adhesive purposes. Due to its adhesive nature a good coupling characteristics has been achieved. Also, it is very low environmental impact and after application it can be removed easily without leaving any marks on the test specimen. A V2-calibration block was used as a specimen to conduct ultrasonic test to evaluate effect of coupling on amplitude of back wall echoes. An ultrasonic testing machine Einstein-II TFT was used to conduct the test. A Normal probe diameter of 12 mm was used for generating pulse with 5 MHz frequency.

2.2 Couplant Preparation

Methylcellulose is a chemical compound derived from cellulose. It is a hydrophilic white powder in pure form and dissolves in water. Methyl cellulose has lower critical solution temperature, it is between 40 °C and 50 °C. At temperatures below this it is readily soluble in water. Hence, 10 g of Methyl cellulose was mixed with 250 ml of water at 50°C by using magnetic stirrer and cooled the mixture while mixing to a lower temperature of 40°C .Due to this the particles are dispersed uniformly and rapid dissolution was taken place. This method of mixing is repeated for different combination as shown in Table 1. Then each type of combination kept in ink filler bottle for easy application on surface of specimen.

Table 1. Composition ratio of water and Methylcellulose

Trial No.	Quantity of water (ml)	Quantity of Methylcellulose (gm)
1	**250**	**0**
2	**250**	**10**
3	**250**	**20**
4	**250**	**30**
5	**250**	**40**
6	**250**	**50**

2.3 Experiments and results and discussion

The densities and the ultrasonic velocities of the prepared couplant by above mentioned method are measured. The characteristic acoustic impedance of those couplants was calculated using the measured densities and ultrasonic velocities and tabulated as shown in Table 2. It is found that the characteristic acoustic impedance increased with increased content of Methylcellulose in the water. This is because of increased viscosity and adhesive nature of the Methylcellulose in the couplant.

Table 2. Acoustic Impedance of water based Methylcellulose couplant at different proportion.

Couplant Code	Combination of Couplant (ml. of Water +g. of Methyl Cellulose)	Density $(kg.m^{-3}) \times 10^3$	Sound Velocity $(m.s^{-1})$	Acoustic Impedance $(kg.m^{-2}.s^{-1}) \times 10^6$
C_1	250 + 0	0.998	1490	1.49
C_2	250 + 10	1.019	1426	1.45
C_3	250 + 20	1.042	1474	1.46
C_4	250 + 30	1.067	1523	1.55
C_5	250 + 40	1.092	1567	1.64
C_6	250 + 50	1.119	1616	1.74

Note: Acoustic Impedance of steel (Specimen V2 Block) = Z_3=45.7×10^6 kg.m^{-2}.s^{-1}

The probe makes direct contact with the surface of the specimen using the above mentioned coupling. The couplant thickness was maintained constant using the support of aluminum foil. An ultrasonic testing machine – Einstein –II TFT was used to excite the probe to transmit and receive the pulse echo signals on specimen. The measured signal is passed on to the flaw detectors CRT screen. The screen shots were captured and mentioned as shown in figure 3.

Considering the first back wall echo it is found that, when using only water (without methylcellulose) i.e., C_1 as a couplant (refer to figure 3(a)) we have 10 dB gain when compared to the couplant mixture C_6 i.e. 250ml of water and 50 gm of methyl cellulose as the couplant as shown in Figure 3(f). In the Figure 3(a), since the couplant is only water the coupling nature is very good and the back wall echo is observed as 97%. The amplitude of the back wall echo is much large when only water is used as a couplant. Further it is observed decrease amplitude when different proportion of methyl cellulose has been added to water and the blend was used as the couplant. The intensity of the first echoes increases with the transmission coefficient and the transmission coefficient decreased with the following consecutive echoes decrease.

In the Figure 3(b), considering the first back wall echo for analysis using couplant C_2 (250 ml of water and 10g of methyl cellulose). It is observed a gain of 2.5 dB when compared to the couplant mixture C_3. The couplant used is a little more viscous compared to the first couplant and the amplitude of the back wall echo observed here is 95% of the screen.

In the Figure 3(c), considering the first back wall echo for analysis using couplant C_3 (250ml of water and 20g methyl cellulose). It is observed a gain of 1.5 dB when compared to the couplant mixture C_4. The couplant used is a little more viscous compared to the first couplant and the amplitude of the back wall echo observed here is 95% on the screen.

In the Figure 3(d), considered the first back wall echo for analysis when using couplant C_4 (250 ml of water and 30g methyl cellulose). It is found that a gain of 2 dB when compared to the couplant mixture C_5. The couplant used is a little more viscous compared to the first couplant and the amplitude of the back wall echo observed here is 90% on the screen.

In the Figure 3(e), considered the first back wall echo for analysis using couplant C_5 (250 ml of water and 40g of methyl cellulose). It is observed a gain of 2 dB when compared to the couplant mixture C_6. The couplant used is a little more viscous compared to the first couplant and the amplitude of the back wall echo observed here is 76% on the screen.

For the analysis of the first back wall echo we consider 250 ml of water and 50 methyl cellulose as a couplant as seen in figure 3(f). The amplitude of the back wall echo is observed to be 75% on the screen.

Table 3. Calculated transmission coefficients for interfaces of the various combination of water based cellulose couplant with PZT transducer element and steel block

Couplant Code	T_1	T_2	Gain (Db)	Amplitude of back wall echo (%)
C_1	0.0864	1.936	46	97
C_2	0.0842	1.938	48	95
C_3	0.0847	1.938	50.5	95
C_4	0.0897	1.934	52	90
C_5	0.0947	1.931	54	76
C_6	0.1002	1.927	56	75

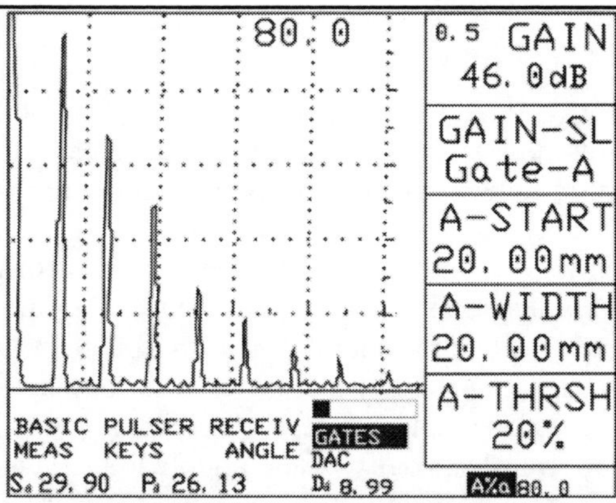

(a) Back wall echo using couplant C_1 (water)

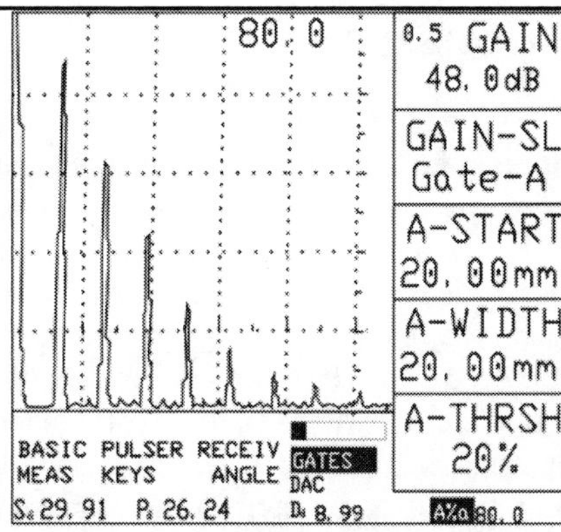

(b) Back wall echo using couplant C_2 (250 ml. of Water + 10 g. of Methyl cellulose)

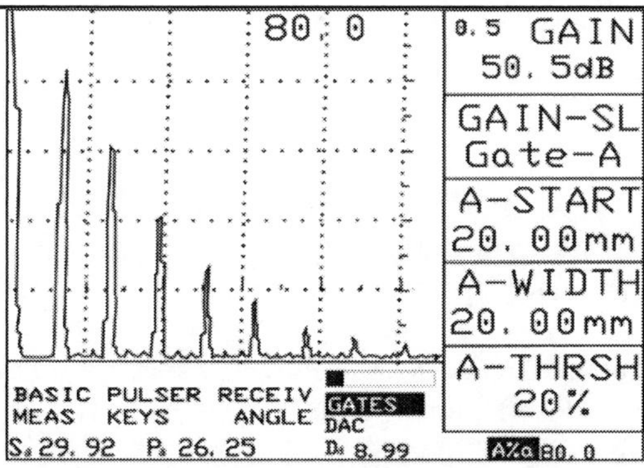

(c) Back wall echo using couplant C_3 (250 ml. of Water + 20 g. of Methylcellulose).

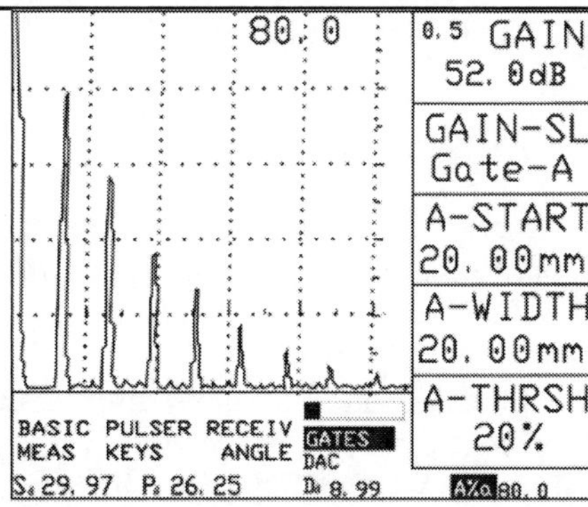

(d) Back wall echo using couplant C_3 (250 ml. of Water + 30 g. of Methylcellulose)

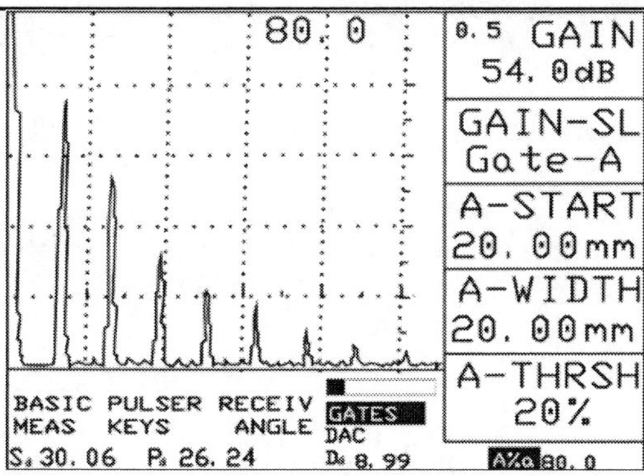

(e) Back wall echo using couplant C_3 (250 ml. of Water + 40 g. of Methylcellulose).

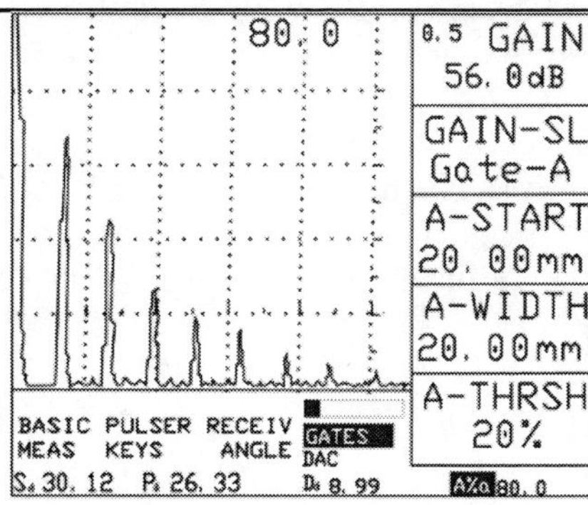

(f) Back wall echo using couplant C_3 (250 ml. of Water + 50 g. of Methylcellulose).

Fig. 3. Back wall echo analysis for different couplants

3 Conclusions

By using the different proportions of the couplants prepared by water and methyl cellulose the ultrasonic testing is conducted by applying the couplant and their back wall echoes are measured. It is found that the amplitudes of the first back echoes and the attenuation coefficients are increased with the characteristic acoustic impedance of the couplants used. The research shows the larger the transmission coefficient and the acoustic impedance results in a higher sensitivity of the detection of the flaw in the contact pulse echo technique. Here the amplitudes of the back wall echoes are observed and their results are noted for further analysis or research when using methyl cellulose as a couplant. In industrial practice only water cannot be used as a couplant due to various factors like viscosity, corrosion etc. So there must some optimum viscosity of the couplant such that it can be applied which can withstand high temperatures, should not leave marks on the test specimen, chemically stable and should not create any impact to the environment. So with the above six couplants being used the couplant mixture C2 has better results having gain of decibels when compared to the couplant mixture C6 and also the amplitude of the back wall echo is observed to be 90% on the screen.

References

1. A, Kumar; N, Gupta; R, Gupta; & Y, Kumar; Effect of coupling conditions on ultrasonic echo parameters, J. Pure Applied Ultrasonic. 2007, 27, 70-79.
2. Ashok, Kumar; Correction factor due to couplant in ultrasonic thickness measurement, Insight, UK, 1996, 38, 336-337.
3. Bindal, Arvind; A new high gain couplant for ultrasonic NDT inspection, Paper presented at National Seminar on (NDE-92). Tiruchirappalli, 1992, 3-5 Dec.
4. Bindal, V. N; Water-based couplants for general purpose use for ultrasonic NDT applications. Journal of scientific and industrial research. 2000, 59 (11), 935-939.
5. Kim, Young, H; and Song, Sung-Jin; A study on the couplant effects in contact ultrasonic testing, 10th Asia-Pacific Conference on Nondestructive Testing, Brisbane, Australia, 2001,17th -21st September.
6. Mabuza, B, R; and N; Netshidavhini; Inspection of flaws in steel using ultrasonic testing and fracture mechanics approach. Emerging Technologies in Non-Destructive Testing V 2012 (225).
7. Vincent, A; Influence of wear plate and coupling layer thickness on ultrasonic velocity measurement, Ultrasonic. 1987, 25, 237-243.
8. Young, H, Kim; Sung-Jin, Song; Sung-Sik, Lee; Jeong-Ki, Lee; Soon-Shin, Hong; A study on the couplant effects in contact ultrasonic testing, 10th APCNDT Brisbane, 2001.

Hydraulic characteristics of internal protective coatings used for rehabilitation of pipelines of water supply and sanitation systems

Vladimir Orlov[1*]
[1]*Moscow State University of Civil Engineering, 26, Yaroslavskoe Shosse, Moscow, 129337 Russia*

Abstract. The purpose of this article is to provide a comparative analysis of hydraulic characteristics of some modern organic coatings used for trenchless rehabilitation of engineering pipelines. Scotchkote 169HB pipeline inner protective coating manufactured by «3M» production company (USA) and Subcote FLP coating by «Group POLYPLASTIC» Ltd (RF) were the objects of this investigation. The article presents general characteristics of alternative protective coatings. A methodical approach to determination of the protective coating hydraulic characteristics is described by the tests at a special patented small-sized bench through estimation of the coatings' hydrophoby with the terms of hydraulic characteristics. Based on the obtained results the conclusions are made concerning the feasibility of application of organic coatings compared to non-organic ones (for example, cement-sand coatings). This will enable power savings due to a lower hydraulic resistance during the water transportation. Based on the experimental studies in static and dynamic conditions on a small-sized testing bench it was proved, that the Subcote FLP protective coating has hydrophilic properties. There are shown results of an automated processing of experimental data and comparison of the values of the coefficients of hydraulic friction, resistivity and roughness, which have been obtained at the large-sized and small-sized benches, as well as at a roughness indicator. Provision has been made of a possible and reasonable use of small-sized benches instead of large-sized ones for determination of hydrophoby and hydraulic parameters, which will considerably facilitate the traditional method of comprehensive estimation of the hydraulic performance of various types of protective coatings.

1 Introduction

By the XXI century, the RF cities have faced the poor state of underground pipeline systems of water supply and sanitation, where more than 60% of the pipelines need an urgent repair and modernization with a wide use of trenchless technologies [1]. The trenchless rehabilitation of the pipeline transportation is subject to different technologies, which include pulling of round and end-section deformed polymer pipes [2], use of

* Corresponding author: orlov950@yandex.ru

polymeric sleeves and application of inner spiral wrapping of infinite polymeric ribbons at the inner surface of pipelines [3, 4]. Protective coatings, which are applied to the inner surface of the pipelines by the centrifugal spraying, are one of the perspective methods of the trenchless pipeline rehabilitation. These coatings include non-organic materials on the basis of construction cement-sand mortars, which have been practiced in the pipeline rehabilitation for nearly 100 years, as well as new types of coatings, which are classified as organic ones [5]. Organic coatings are much more advantageous than non-organic ones due to some higher strength characteristics and lower hydraulic resistances, which increase the power saving effect during the water flowing [6, 7].

Protective coatings, which are used in the trenchless rehabilitation works, shall provide solution of the resource saving problems in the flow transportation pipelines both of natural and sewage water, i.e. enable creation of the conditions of accident-free mode of operation of engineering networks through efficient and, to the extent possible, quick repair of all types of defects on pipelines.

The actual construction market proposes a wide range of the pipeline protective coatings. However, before a proposal is made of application of such or another material for rehabilitation of the pipelines, the proposed coating shall be tested to estimate its strength and hydraulic properties [8].

2 Literature review

The following information is given herein as a brief characteristic of the above-mentioned coatings. Over the last decade in European countries and the United States the internal protective coating Scotchkote 169HB [9, 10] is widely applied in water supply systems for pipeline rehabilitation. It has high strength characteristics, but the technical literature does not provide data on its hydraulic performance, or these data are contradictory [11]. The material of the coating is 100% insoluble aliphatic isocyanate polyurethane. The coating is applied to the inner surface of the pipelines by centrifugal spraying in the range of 100-1000 mm diameters. The coating thickness for a single application can be up to 5 mm (typically, a layer is 3 mm thick). Gelation time (at 25 °C) is 1 minute; the formation of a film takes 2 minutes and polymerization on the inner surface of the pipe - about 10 minutes.

To the next generation of protective coatings company «3M» should be attributed Scotchkote 2400, which is a rapidly solidifying polyurea. The coating is applied in the process of centrifugal spraying. The coating composition is a two-component one, where the basic component is a white thixotropic liquid and the thixotropic black liquid is an activator. The maximum thickness of the coating when applied to a single layer is 2.75 mm. When such a coating thickness is in the range of 100-1000 mm diameters, it is able to fill in the corrosion holes (fistulas) and crevices (cracks), annular gaps (for example, stretched joints, annular tears) and other similar defects up to 5 mm size. The internal surface roughness of the restored pipeline is reduced, which is reflected in the flow rate profiles [12]. The gelatinization time is 60 seconds, the formation of the film makes 3 minutes, and the polymerization time takes about 60 minutes.

A relatively new type of protective coatings is a material called Subcote FLP, developed by «Group POLYPLASTIC» Ltd [13, 14]. It is a quickly solidifying polyurethane resin, which is sprayed on the inner surface of steel, iron, concrete and asbestos cement pipes with a nominal diameter from 100 to 1200 mm. The composition of Subcote FLP coating is a two-component one: the basic component is a pale blue liquid; the activator is a brown liquid. The maximum thickness of the coating when applied to a single layer is 4 mm. When the thickness of the protective coating is 3 mm it is able to block the through holes (fistula) with a diameter up to 8 mm, as well as circumferential and longitudinal cracks. The gelatinization time is 32 seconds, formation of a stable film makes 6 minutes, and the polymerization time with the possibility of CCTV inspection is 60 minutes.

3 Materials and methods

Experimental study of the hydraulic properties of the materials of Scotchkote 169HB, Scotchkote 2400 and Subcote FLP coatings were made in the laboratory of the Department "Water supply and Sewerage" of NIU MGSU in the years 2016-2017. The aim of the protective coating investigation was to obtain the hydraulic characteristics of protective coatings, which may be used by a designer, in particular: the coefficient of equivalent roughness k, of the hydraulic friction coefficient λ and the ratio of the resistivity A. The hydraulic studies were made on two benches and were compared with results of experiments aimed at determining the roughness, got by a TR 1000 roughness indicator. The data, which have been got at the roughness indicator, were used in the automated software in the simulation series aimed at determining hydro-mechanical values of protective coatings.

To determine the hydraulic characteristics provision was made both of the traditional methods and the improved ones based on the use of a patented test equipment for investigation of the protective coating hydrophoby. For the general view of the experimental bench see Fig. 1, the diagram of the equipment for traditional hydraulic tests is given in Fig. 2.

Fig. 1. General view of the hydraulic bench with pipes made of different materials

The experimental hydraulic test bench has the following main elements: a storage tank with a capacity of 4 m^3; the trestle with 15 m long pipelines which have a protective coating and are rigidly fixed on it, a pumping unit on the base of a multistage pump Grundfos CRE 90-2-2, with a capacity of 90 m^3/h, 30 m pressure and a motor of 11 kW power and 360 – 3,600 revolutions per minute; ultrasonic flow meter US-800. As an apparatus for measuring the pressure of water in the pipeline use has been made of the roughness indicators spaced at the distance of 10 m (the pipeline operation part) between them. The above described bench has been used for the hydraulic experiments with Scotchkote 169HB, Scotchkote 2400 coatings.

The experiments with Subcote FLP coating were carried out both at the aforementioned large-sized bench, and a special compact test stand for determination of the hydrophoby degree and the hydraulic performance indices [15]. The target of the parallel experiments was a comparison of the received hydraulic characteristics of the coating to examine possible refuse from large-sized installations and possible future works on compact benches, when it is necessary to carry out hydraulic experiments with new protective coatings.

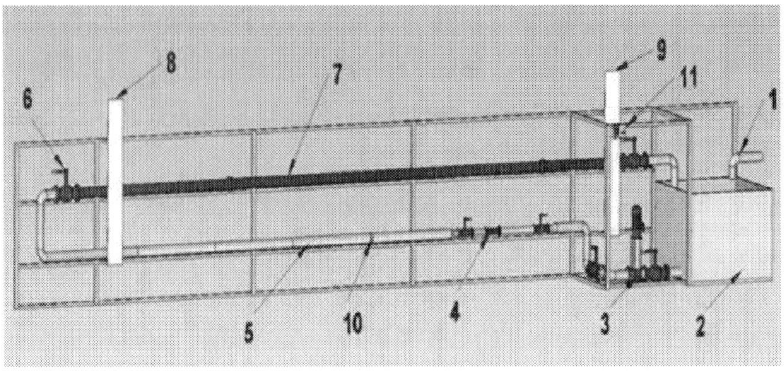

Fig. 2. General scheme of the equipment for hydraulic tests of sprayed protective coatings: 1 – water supply; 2 – storage tank; 3 – pump; 4 – sensor of a flowmeter; 5 – feed pipe; 6 – valve; 7 – pipeline coated with a protective coating; 8, 9 – bench with roughness indicators; 10 – mounting frame for the pipeline fixation; 11 – ultrasonic flow meter

The proposed method of the study was to investigate aspects of the interaction of the liquid with the surface of the protective coating, namely:

-behavior of water drops (no more than 0.55 ml) in statics on the surface of the protective coating;

-behavior of the flowing stream in dynamics, if it flows by a sloped surface with the protective coating in the form of a chute (Fig. 3 and Fig. 4).

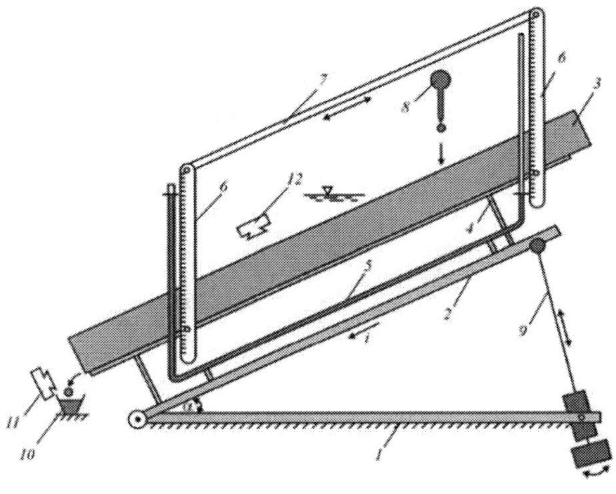

Fig. 3. Schematic drawing of the experimental equipment at the maximal slope of a 1 m long chute: 1 – support frame; 2- rack; 3 – an open chute (tray); 4 – posts; 5 – the system of communicating vessels; 6 –measuring rules; 7 – the control bar; 8 – pipette with water; 9 – rod mechanical Jack; 10 – a collection of drops; 11, 12 -respectively, the camera of frontal and coaxial front shooting of the flow

The Fig. 5 shows a design in the form of an open chute made of polyvinyl chloride of 130 mm rigidly attached Subcote FLP coating with a thickness of 1.1 mm. The drinking water was used as a liquid for formation of droplets and a mini-flow (or mini-strim) in the chute to simulate operation of diameter with a liquid transfer gravity pipeline.

As the instrumentation for photo and filming of the drop configuration and the liquid mini-flow use has been made of a digital SLR camera Sony 550, the lens is DT 1.8/50 SAM in complex with extension tubes Kenko Extension tube with continuous multi-frame and a digital video camera Sony model HDR-CX250E.

The methods of static studies of the protective coating Subcote FLP on a small-sized bench were aimed at determining the degree of wettability of the protective coating, namely, the contact angle α of the spreaded drop and the mini-flow. This indicator is a criterion for a classification of the protective coating to the category of hydrophobic ($\alpha > 90°$) or hydrophilic ($\alpha < 90°$) one [16].

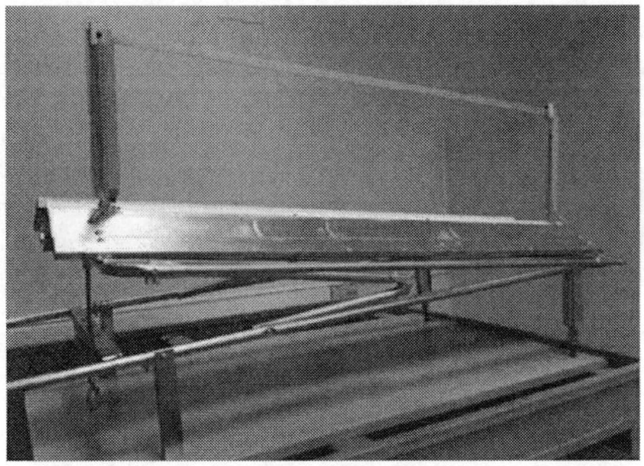

Fig. 4. Experimental equipment for the study of water hydrophoby in nature

Fig. 5. Fragment of the experimental equipment with an open chute, and the applied coating Subcote FLP

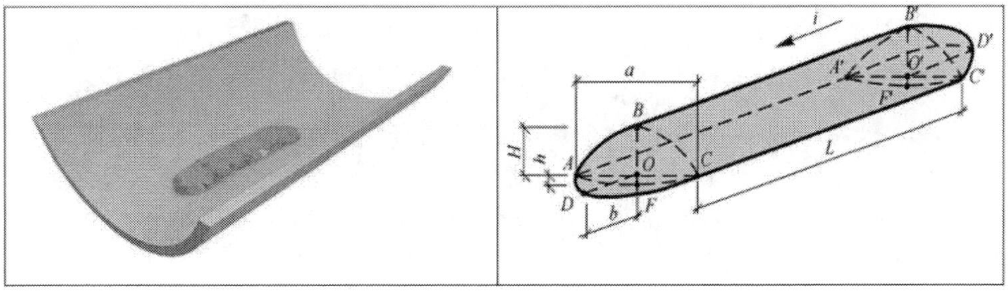

Fig. 6. Axonometric diagram of the mini-flow generated due to the flow of the critical volume of the liquid along a sloped chute with the protective coating Subcote FLP in the form of a tray of 1 m length: a, L ,respectively, - average values of the width and length of the compact part of mini-flow; H, h , respectively, the heights of the upper and lower lenses; b - height of the half of the ball segment of the mini-flow head part; AFC, AA `C`C respectively- the wetted perimeter and the surface of the compact part of mini-flow; i – slope of the chute.

In turn, the technique of dynamic research was to study the formation and spontaneous movement of the liquid mini-flow in the chute due to its appropriate slope and volume of the liquid critical mass. General view of the mini-flow at a slope I of the working surface is shown in Fig. 6. The head and the tail part of the mini-flow are presented in the form of a three-dimensional structure (half of a ball segment), respectively *ABCDA* and *A'B'C'D'A'*. Their projections on the vertical plane can be represented in the form of a lens (the circumference) by analogy with front projection of the drops.

Determination of the geometric dimensions of the mini-flow was carried out by the front and coaxial shooting, a mathematical description of the mini-flow element shape: the heights of the top lens H and the bottom lens h, the equation of the top of the lens and a tangent thereto at the point of intersection with the horizontal plane, the boundary angles α, the areas of the projections of the lenses to the vertical plane (for example, to the top of the lens $Sv=2aH/3$) and other parameters, allowing to estimate the degree of hydrophoby of the material and if possible to identify its relationship with geometric figures of the mini-flow.

4 Results

Mathematical processing of results of tests of the structure «steel pipe + protective coating Subcote FLP» on a long bench with an inner diameter of 0.094 m and a layer thickness of 2.5 mm showed that the average value of the hydraulic friction coefficient is λFLP=0.01772, the structure «steel pipe + protective coating Scotchkote 169HB» with an inner diameter of 0.0952 m and thickness 2.4 mm - λ169HB=0.01925 and the structure «steel pipe + protective coating Scotchkote 2400» with an inner diameter of 0.0952 m and thickness 2.4 mm - λ12400=0.01959. Fig. 7 shows a photograph of drops on the surface of the protective coating Subcote FLP under static conditions.

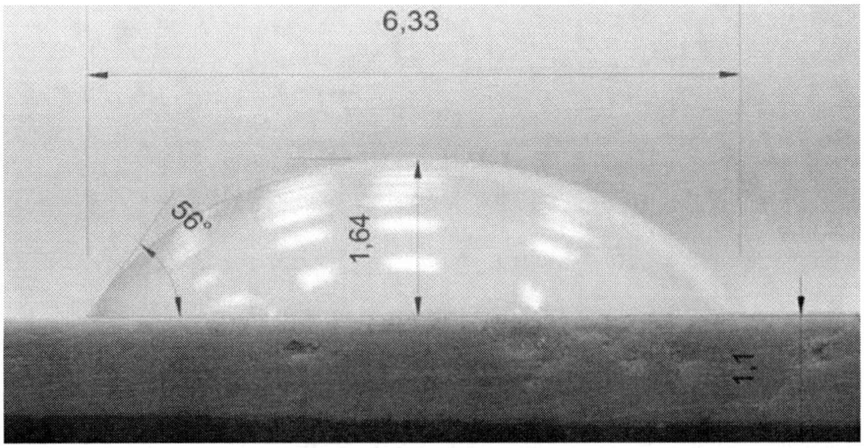

Fig. 7. Photograph of a drop on the horizontal surface of the protective material Subcote FLP with the edge angle 56°

The geometric dimensions of the diameter size of the spreading of the drop on the surface (6.63 mm) and the maximum height of the drop (1.64 mm) were determined from the photographs of frontal projection of drops on the protective coating. The results of the experiments the contact angle was 56° (< 90 degrees), indicating the hydrophilic nature of the protective coating Subcote FLP.

The fragments of the frontal and coaxial filming of the mini-flow in a chute with Subcote FLP protective coating under dynamic conditions are presented in Fig. 8.

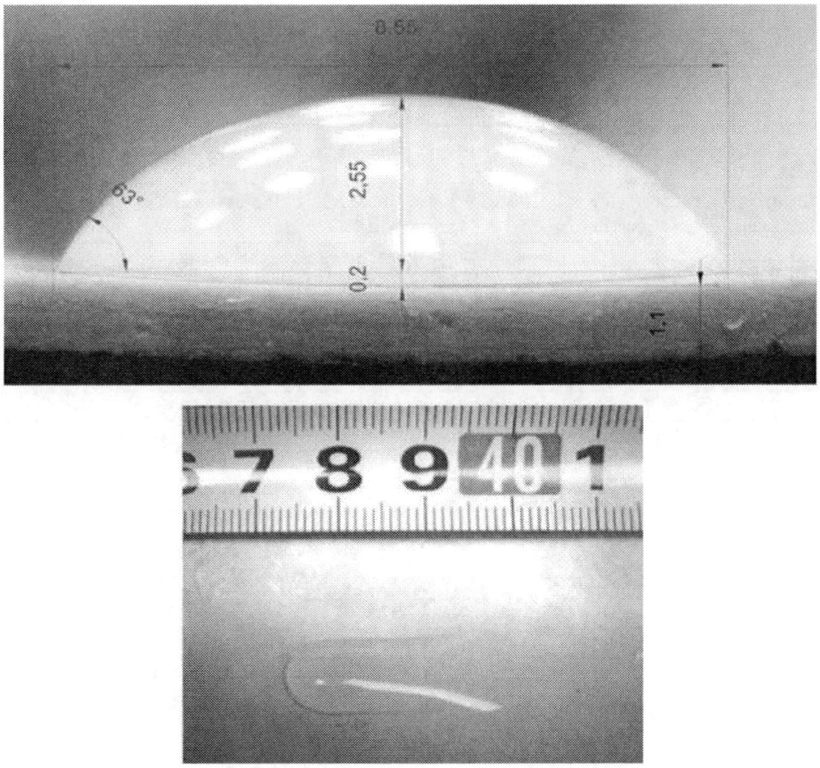

Fig. 8. Selective fragments of frontal (top) and coaxial (bottom) photos of the flow

According to the results of the performed experiments on the TR 100 roughness indicator, the following data were obtained on the average roughness of the protective coatings:

-for Scotchkote 169HB: k_{169} =*3.50 mkm = 0.0385 mm*

-for Scotchkote 2400: k_{2400}=*mkm of 45.57 = 0.04557 mm*

-for Subcote FLP: k_{FLP} =*36.80 mkm = 0.0368 mm*

5 Discussion

5.1 Interpretation of results of hydraulic experiments on a large-size bench

To determine the values of the *A* specific resistances obtained on the hydraulic bench the data were processed using a computer program which supports the hydraulic study of pressure pipelines. The program allows the calculation of the hydraulic parameters and performance of hydro-mechanical modeling of pipelines with the change of their diameters d in obtaining the dependence $A = f(d)$. According to the results of the computer processing of the experiments' results on the hydraulic stand the mathematical dependence of $A = f(d)$ type has been got, as well as the values of specific resistances *A* for a wide range of diameters of the pipelines with applied protective coatings (in the range of 0.1 – 1.2 m). The results of the processing are presented in the Table 1. It also includes values of the specific resistance of the pipes coated with cement-sand coating.

Table 1. Calculated values of specific resistance of pipelines of various diameters c applied protective coatings

No	Inner diameter of the pipe d, m	Subcote FLP	Scotchkote 169HB	Scotchkote 2400	Cement-sand coating
		A_{FLP}=0.0008d$^{-5.188}$	A_{169}=0.0008d$^{-5.19}$	A_{2400}=0.00095d$^{-5.19}$	
		k $_{FLP}$ = 0.0368 mm	k $_{169}$= 0.0385 mm	k $_{2400}$= 0.04557 mm	
		Specific resistance values A, s^6/m^2			
1	0.1	123.336036	123.905330	147.137579	322.91
2	0.15	15.049724	15.106935	17.939485	33.49
3	0.2	3.383346	3.394254	4.030676	6.048
4	0.25	1.063108	1.066059	1.265945	1.841
5	0.3	0.412843	0.413838	0.491433	0.7374
6	0.35	0.185552	0.185942	0.220806	0.3192
7	0.4	0.092812	0.092982	0.110416	0.1657
8	0.45	0.050376	0.050456	0.059917	0.8533
9	0.5	0.029163	0.029204	0.034679	0.04939
10	0.6	0.011325	0.011337	0.013462	0.01951
11	0.7	0.005090	0.005094	0.006049	0.009424
12	0.8	0.002546	0.002547	0.003025	0.004784
13	0.9	0.001382	0.001382	0.001641	0.002596
14	1.0	0.000800	0.000800	0.000950	0.001503
15	1.2	0.000311	0.000311	0.000369	0.0005833

As follows from the Table 1 the resistivity values for three types of the coatings under investigation are approximately equal, whereas for cement-sand coating the resistivity is much higher (in the range of 2.2 -1.55 times). The got data may attest to the fact that for the resource and power efficiency during the water transportation by pressure pipelines it is more advantageous to apply organic coatings as they have a lower hydraulic resistance. Similar conclusions can be also made according to the values of the coefficients of the equivalent roughness k, where the Subcote FLP coating prevails.

The practical significance of the mathematical dependences $A = f(d)$, which have been got during the experiments, consists in their application in the hydraulic calculations during designing of the pressure pipeline systems. In particular, to determine the H_d pressure loss in the pressure pipes of a l_d given length coated on their inner surface by Subcote FLP, if the pipe flow pass q, the calculation may be made according to the formula (1):

$$H_d=Al_dq2 \tag{1}$$

where H_d is the pressure loss, m of water column; A – coefficient of resistivity, s^2/m^6; l_d – the length of the pipeline part, m; q – flow rate, m^3/s.

Using the got empirical dependence $A=f(d)$, the formula (1) takes the form (2):

$$H_d=0.0008d^{-5.188}l_dq^2 \ [\mathrm{m}] \tag{2}$$

where d is the inner diameter of the pipeline part, m.

For example, when the inner diameter of the pipeline is 0.3768 m (including the thickness of the coating layer), the length makes 100 m and the flow rate is 0.3 m^3/s, the head losses H_d will be:

$$H_d=0.0008d^{-5.188}l_dq^2=0.0008\cdot 0.3768^{-5.188}\cdot 100\cdot (0.3)^2=1,14\ m$$

5.2 Interpretation of the results of experiments on small-sized bench for the study of hydrophoby

The main conclusion according to the results of the static studies (see Fig. 7) is that the studied surface of the protective coating can be classified as hydrophilic one. Analysis of the photo in Fig. 8 shows that the dynamics show a similar trend, confirming the hydrophilic nature of the protective coating Subcote FLP (the contact angle α is 63°). Assessment of the degree of hydrophobicity in the dynamics was carried out by determining the relative coefficient of hydrophobicity K_o (3):

$$K_o = S_v/(S \cdot i) \quad (3)$$

where S_v is the area of the projection of the upper lens (*ABCOA* in Fig. 6), mm^2; S – area of the wetting surface of the chute by the compact part of L mini-flow by the width of the arc (the wetted perimeter) *AFC, mm*^{2}; i is the slope of the chute.

Using the geometrical dimensions and the mini-flow average rates, which have been got during the experiments, whole range of necessary hydraulic parameters of the investigated protective coatings has been determined followed by statement of the roughness coefficient of the appropriate protective coating «*n*». The algorithm for the hydraulic parameter analysis consists in determination of: the coefficient of hydrophoby K_o, the mini-flow filling (*h/d*) and the wetted perimeter (the length of the arc *AFC*); the hydraulic radius R in terms of filling and using the wetted perimeter and their comparison with the subsequent adjustment of the formula corresponding to the design filling; the mini-flow rate V, as the quotient of division of the path U which has been passed by the mini-flow head for the time t recorded by a stopwatch; the coefficient Chezie C $(C=V/(Ri)^{0,5})$, which depends on the hydraulic radius R, the rate V and the slope i; the coefficient of relative roughness «*n*» according to Manning $C=\frac{1}{n}\cdot R^{1/6}$

The processing of the experimental data on the study of the hydrophoby / hydrophily of the Subcote FLP coating has been made using a patented automated program «Hydrophoby» [17].

When performing the automated calculation 12 experimental and reference data were taken into account. It is conducive to a complex calculation of parameters of hydrophoby and determination of the hydraulic parameters with the subsequent determination of the coefficient of roughness «*n*» and the hydraulic friction coefficient λ. Below is the printout of the results of the automated calculation program «Hydrophoby» for Subcote FLP coating.

1. The total volume of the mini-flow W_1, ml: $W_1 = 0{,}0568N + 0{,}4943 = 2.1983$
2. The mini-stream total volume W, mm^3: $W = 10^3 W_1 = 2198.3$
3. Top lens area S_v, mm^2: $S_v = 2aH/3 = 14.535$
4. Bottom lens area S_n, mm^2: $S_n = 2ah/3 = 1.14$
5. Small chord length m, mm: $m = [(a/2)^2 + h^2]^{1/2} = 4.2797$
6. Wetted perimeter p, mm: $p = 2m + (2m\text{-}a)/3 = 8.5625$
7. Total area of the top and bottom lenses S_s, mm^2: $S_s = S_n + S_v = 15.675$
8. Theoretical volume of the compact part of the mini-stream W_t, mm^3: $W_t = S_s L = 423.225$
9. The wetted surface of the compact part of the mini-stream S, mm^2: $S = pL = 231.1867$
10. Relative index of hydrophobicity K_o: $K_o = S_v/(S\ i) = 0.4191$
11. Edge angle (angle of wetting) α, deg.: $\alpha = 50.05$
12. The estimated volume of the mini-stream W_r, mm^3: $W_r = W\text{-}\ (1/12)[H_g(H_g^2+0.75a^2)+H_k(H_k^2+0.75a^2)] = 423.225$
13. Estimated length of the mini-stream compact part L_r, mm: $L_r = W_r/S_s = 136.4464$
14. The approximation between the projected length L_r and the average length L of the mini-stream P_L, %:

$P_L = |100 - L \cdot 100/L_r| = 8.212$

15. Tray filling (h/d): $(h/d) = h/d = 0.0015385$
16. Hydraulic radius R_1 (through filling), mm: $R_1 = 62.2(h/d)^{0.8833} = 0.20377$
17. Hydraulic radius R_2 (through the wetted perimeter and live section), mm: $R_2 = S_n /p = 0.13314$
18. The absolute approximation value P_R between R_1 and R_2, %: $P_R = |100 - R_1 \cdot 100/R_2| = 53.0517$
19. The adjusted exponent «*x*» in the formula to determine the hydraulic radius $R_2 = 62.2(h/d)^x$: $x = 0.94901$
20. The general corrected formula for determining the hydraulic radius R and its value, mm: $R = 62.2(h/d)^X = 62.2(h/d)^{0.94901} = 0.13314$
21. Average mini-stream speed V, mm / s: $V = U/t = 9.3069$
22. The Chezie's Coefficient C: $C = V/(R\ i)^{1/2} = 65.85791$
23. Manning's relative roughness coefficient n: $n = (1/C)R^{1/6} = 0.0109$
24. Hydraulic friction coefficient λ_{FLP}^*: $= 8g/C^2 = 8\ 9.81/C^2 = 0.0181$

According to the results of the experiments which have been performed on the pipe study hydraulic bench and the plant for study of the hydrophoby degree of materials it shall be noted, that the convergence of the results is very high as for definition of the average value of the hydraulic friction coefficient λ.

The error in determining values for λ was:

$$|(\lambda_{FLP} - \lambda_{FLP}^*)/\lambda_{FLP}^*| = |(0{,}01772 - 0{,}0181)/0{,}01772| = 0{,}0214\ (2{,}14\ \%)$$

Thus, the results obtained by high convergence of λ values confirmed possible refuse from a large-sized equipment and transition to small-sized benches for performance of the hydraulic experiments with protective coatings.

6 Conclusions

1. Hydraulic parameters of alternative organic protective coatings which can be used during trenchless rehabilitation of pressure pipelines to achieve the effect of energy saving have been determined and analyzed on special benches.
2. The methodology of field experiments to determine the degree of hydrophoby and hydraulic parameters has been developed for a small-sized testing bench under dynamic conditions by studying the behavior of a liquid mini-flow at a wide range of the sliding surface slopes.
3. There is given description of a patented bench model for determination of the hydrophoby degree of the materials and their hydraulic characteristics, as well as the results of an automated calculation of the Subcote FLP coating. The results of the experiments and the automated calculation data enabled comparison of the coefficients of hydraulic friction λ, which had been obtained on large- and small-sized benches and showed a high convergence of the results.
4. The possibility of application of small-sized benches instead of large-sized ones was proved for determination of hydrophoby and hydraulic parameters. In the future, it will greatly simplify the traditional method for comprehensive assessment of hydraulic parameters of various types of protective coatings and reduce time and financial costs of the experiments to be performed.

References

1. S. V. Khramenkov, O. G. Primin, Rekonstruktsiya truboprovodnyh system (2008)
2. A. Kuliczkowski, Rury kanalizacyjne (2004)

3. Janssen A., 30 NO-DIG International Conference and Exhibition, 012287, 1-10 (2012)
4. Combeer G., Melville S. 31 NO-DIG International Conference and Exhibition, 1.19, 1-6 (2013)
5. D-H. Koo, C-H. Bae, J-H. Kim, T-J. Kweon 31 NO-DIG International Conference and Exhibition, 2013, 2.8, 1-6 (2013)
6. A. Kuliczkowski, E. Kuliczkowska, A. Zwierzchowska Technologie beswykopowe w inzeynierii srodowiska (2010)
7. Rabmer-Koller U. 29 NO-DIG International Conference and Exhibition, NO-DIG Berlin 2011, Paper 2C-1, pp. 1-10
8. Roscher H., J. Wasserwirt.-Wassertechn., 1-2, 36-40, 2007
9. G. N. Greaves, A. L. Greer, R. S. Lakes & T. Rouxel, J. Nature Materials, 10, 823-837 (2010)
10. Material safety data sheet 3M Scotchkote Rapid Setting Polymeric Lining 169HB // 3M Company, UK, 2012.
11. M. Rameil, Handbook of pipe bursting practice (2007)
12. S. Grossmann, D. Lohse, C, Eur. Phys. J. E., 40, 16 (2017)
13. M. I. Gorilovsky, J. Communal complex of Russia, 8 (122), 34-35 (2012)
14. M. I. Gorilovsky, Buyanovskii, V. M., S. J. Pliskin, A. N. Menshov, A. B., Branzburg, J. Plasticheskie MASSY, 8, 1-8 (2011)
15. V.A. Orlov, I.S. Dezhina, E.V. Orlov, I.A. Averkeev, Ispytatelniy stend po opredeleniyu stepeni hydrofobnosti materialov diya izgotovleniyq trub i remonta truboprovodov, Patent no.157695 (Russian Federation) (2015)
16. Sakai, M.; Song, J. H.; Yoshida, N.; Suzuki, S.; Kameshima, Y.; Nakajima, A. Langmuir 22, 4906-4909 (2006)
17. V.A. Orlov, S.P. Zotkin, I.S. Dezhina, A.A. Pelipenko, Programma rascheta stepeni hydrofobnosti i hydravlitcheskih parametrov trub i zaschitnyh pokrytiy, (2017)

Sedimentation of suspended solids in ultrasound field

Vera Vikulina[1*] and *Pavel* Vikulin[1]
[1]*Moscow State University of Civil Engineering, Yaroslavskoye shosse, 26, Moscow, Russia, 129337*

Abstract. Physical and chemical effects of aquatic environment that occur in an ultrasonic field change the sedimentation rate of coagulated suspension. This might only happen in case of cavitation of ultrasonic filed that causes a change of potentials of the medium. Research of the influence of ultrasonic vibrations on coagulation of suspended solids within water purification allows expanding their scope of implementation. The objective of the research is to estimate the effect of ultrasound on the sedimentation of the suspended solids, to determine of the efficiency of the process in relation to the dose of the coagulant, and to calculate the numerical values of the constants in the theoretical equation. The experiment condition was held in the water with the clay substances before the introduction of the coagulant. The method of magnetostriction ultrasonic generator was applied to receive ultrasonic vibration. Estimate of concentration of clay particles in water was performed using photometry. As a result of the research, the obtained data allow determining the increase in efficiency of suspended particles sedimentation related to the dose of coagulant, depending on time of ultrasonic treatment. The experiments confirmed the connection between the effect of sedimentation in the coagulation process, the coagulant dose and the time of scoring. Studies have shown that the increase in the duration of ultrasonic treatment causes a decrease of administered doses of coagulant.

Keywords: sedimentation, ultrasound treatment, coagulant

1 Introduction

Various physical and chemical effects occur in the ultrasonic field, may be associated with the transformation of energy of elastic vibrations. In this case, we can say that they are one of the types of mechanic and chemical reactions [1, 2]. These effects are accompanied by the phenomenon of cavitation [2]. On the other hand, elementary processes of sound and chemical reactions that are similar to radiolysis, reactions in electric discharge, and photolysis [3, 4].

Napiers and Nolting put forward the theory of high temperatures occurrence by adiabatic compression of a cavitation bubble [4]. According to their calculations, within continuous increase the velocity of the motion of bubble – liquid boundary temperature in the collapsing bubble can reach 10000°K, and the pressure in the shock wave up to 10^4 bar. [5]. Such high temperatures occurring in the gas-filled cavity of the bubble, can determine

* Corresponding author: nocviv@mgsu.ru

the occurence in it of electric charges of dissociated and ionized molecules, atoms, free radicals [6]. Thus, the this cavity can serve as a source of products with a high reactivity and launch physical and chemical actions in the ultrasonic field.

Soviet physicist Y. I. Frenkel suggested the occurrence on the walls of the formed lenticular cavity electric charges of opposite sign. In the initial stage of formation of a cavity in it there are electric charges. Y. I. Frenkel believes that the break of continuity of the environment in the areas of rarefaction are formed not spherical cavity and lenticular, molecular sizes. For such a gap requires the amplitude of the pressure, of the order of 10^3 – 10^4 bar.

The transition from lenticular to spherical cavity occurs due to penetration of dissolved gases or vapors of surrounding liquid. The schematic structure of a cavity is shown in Fig. 1 [7].

According to the theory of Ya. Frenkel, in conditions of an electric discharge energy-rich particles, the ionized molecules and the ions, free radicals arise in the cavity [7].
Considering that the charges formed as a result of the uneven distribution of ions on the walls of the bubble in the fluid gap, Ya. I. Frenkel determined the field strength within the cavity at the time of its formation:

$$E_п = \frac{4\bar{e}}{r_п}\sqrt{N_п \delta_п}\,, \tag{1}$$

$E_п$ – field intensity, [V/cm];
$\delta_п$ – gap between liquid layers, [*Å*];
$N_п$ – the number of dissociated molecules per unit volume;
$r_п$ – the radius of a cavity, [cm];
$\bar{e}$ – electron charge, [Cl].

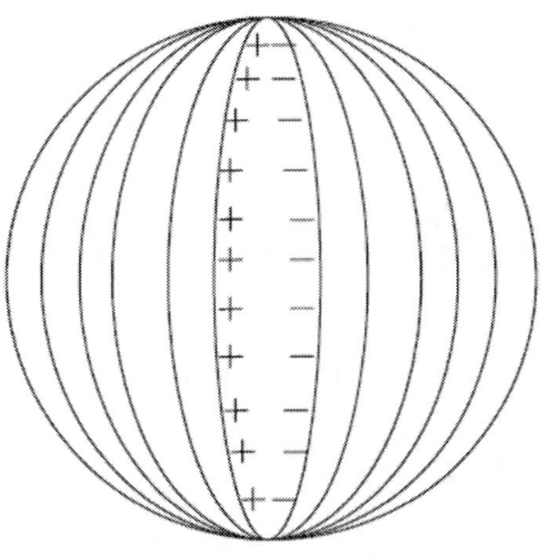

Fig. 1. Cavity structure

R. Parker [8] carried out experimental study of coagulation under the action of ultrasonic vibrations. Experiments were done on the smoke of magnesium oxide in the air. It was observed that in the absence of acoustic field normal coagulation and sedimentation process was negligible. There are examples of use of the coagulating action of ultrasound during the deposition of suspended solids.

At intensities of the order of 0.3 – 0.5 W/cm^2 ultrasound promotes coagulation of suspended particles. Scoring of a suspension of calcium carbonate (particle size of 47 μm), quartz dust (23 μm) and anthracite (76 – 152 μm) leads to an acceleration of the process of settling.

Conducted research the possibility of coagulation of high-ash coal slurries with polymers and ultrasound. For experimental studies have used ultrasound generator high pressure (USHP) – 2 frequency range from 0.1 to 10.0 MHz [9, 10].

Research conducted in the all-Union scientific research Institute VODGEO at ultrasonic coagulation of finely dispersed systems L – S and L – L (Hydrosol of iron, activated sludge, polystyrene latex, concentrate copper ore, water, coal tar, etc.) showed that exposure to ultrasound coagulates finely dispersed system with a particle size of from 1 to 5 microns; for the system of L – S the best results occur when the frequency of 8 – to 18 kHz, and for systems L – L – at a frequency of 400 – 800 kHz. With the increase of intensity of ultrasonic vibrations increases the effect of coagulation [1]. Observed coagulation liquid systems, for example, an aqueous suspension of Al_2O_3 under the action of ultrasound. The experiments showed that already after a few minutes there is a sharp coagulation of the slurry, whereas in normal conditions it is completely absent [11, 12, 13]. Analyzing the above, we can say that the ultrasonic vibrations in the water system can have a significant impact on the processes associated with the oxidation of impurities, the phenomenon of crystallization and coagulation of suspended solids.

It must be emphasized that the desired effect for the above-mentioned process in the ultrasonic field can be achieved only in the presence of cavitation regime in volume. The process by which the ultrasound can have an effect is the coagulation of fine suspended matter in the water [14].

2 Research aim

The aim of this work is to conduct research on the sedimentation of the suspended solids in the ultrasonic field and to receive data to verify the theoretical equation [15, 16, 17, 18]

$$\frac{\Delta G}{G \cdot q} = a + b \cdot t \tag{2}$$

a и *b* – constants;
G – Clay particles per volume unit [mg/l];
ΔG – suspended clay particles [mg/l];
q – Coagulant dose, [mg/l of Al_2O_3];
t – Sound duration, [min].

The objective of the work is to obtain numerical values of constants a and b for theoretical equations (2) and the relative standard deviation (dispersion) [19].

3 Materials and methods

The main experiment was the deposition of clay particles with a concentration of 120 mg/l when added to water coagulant – aluminum salts with a dose of Al_2O_3 – 2.5, 5.0, and 10.0 mg/l [20, 21].

Carried out pre-treatment of water with ultrasound with subsequent introduction of the coagulant. Cavitation mode of ultrasonic field in water was maintained by an ultrasonic generator (ultrasonic generator)–2–4 and a magnetostrictive transducer (ICP) 6-22 (frequency and intensity for excitation in water cavitation f = 18 – 22 kHz, I = 2.0 W/cm2).

Pre-treatment of water with suspended clay particles was carried out with ultrasound for 0.1; 0.5; 1.0; 2.0 and 5.0 minutes. Estimation of concentration of clay particles in water was carried out according to standard techniques, the photometric method by comparing samples of the investigated water with standard suspension.

4 Results

To verify the theoretical equation (2) the decrease in the concentration of clay particles suspended in water was determined, as the difference between initial and final concentration (the deposition process is completed). As a value that characterizes the deposition process, the ratio of the decrease in the concentration of ΔG to its final value of G_2 taken, related to the dose of the coagulant was taken. Experimental data on the effect of ultrasound on the coagulation process placed in the Table 1.

Table 1. Research data

No	Initial SS concentration G [mg/l]	Final SS concentration G_2, [mg/l]	SS concentration decline ΔG, [mg/l]	Sound duration t, [min]	Coagulant dose q, [mg/l]	$\frac{\Delta G}{G_2 q}$
1	120	91	29	0.5	2.5	0.1
2	120	83	37	1.0	2.5	0.12
3	120	71	49	2.0	2.5	0.16
4	120	51	69	5.0	2.5	0.23
5	120	69	51	0.17	5.0	0.09
6	120	66	54	0.5	5.0	0.09
7	120	57	63	1.0	5.0	0.11
8	120	54	66	2.0	5.0	0.11
9	120	34	86	5.0	5.0	0.14
10	120	114	6	0	2.5	0.023
11	120	86	34	0	5.0	0.079
12	120	31	89	0	10.0	0.287

Table 1 shows that, the maximum deposition of clay particles was observed at the dose of coagulant on Al_2O_3 equal to 10 mg/l without ultrasonic treatment, which is MAX level.

When measuring the concentration of suspended solids by the constant value of the instrument was determined, allowing it to switch from the scale to the weight concentration of clay particles in water. It is installed for stationary uniform distribution of particles in a given volume of water, so the colorimeter measured the concentration of particles in the water just before the experiment (initial concentration) and after all flakes settled, and the water was the final (residual) concentration of only one clay particles. All intermediate readings are unable to fix the bleaching process as such, but only indicate the duration of the dynamics of the process.

5 Results processing

Statistical analysis of experimental data performed with the goal of finding the coefficients a and b in equation (2) and the calculated relative standard deviation (dispersion) values of the experiments [19].

The experimental data on the deposition of suspended solids using the cavitation regime in the ultrasonic field is revealed in table 2. To perform the calculation adopted the following notation: time scoring t = x_k, the ratio of the concentration of deposited clay particles to a final concentration of suspended solids is related to the dose of coagulant $\frac{\Delta G}{G_2 q} = y_k$. In the calculation, we used the deviations from the arithmetic mean values k – x results of the experiments.

Table 2. Experiment and calculated data

No	$t = x_k$	$\frac{\Delta G}{G_2 q} = y_k$	$x = \bar{x} - x_k$	$y = \bar{y} - y_k$	$x{\cdot}y$	x^2	y^2
1	2	3	4	5	6	7	8
1	0.5	0.1	+1.41	+0.028	+0.039	1.990	0.0008
2	1.0	0.12	+0.91	+0.008	+0.007	0.828	0.00006
3	2.0	0.16	-0.09	-0.032	+0.003	0.0081	0.001
4	5.0	0.23	-3.09	-0.102	+0.315	9.548	0.0104
5	0.17	0.09	+1.74	+0.038	+0.0066	3.027	0.0014
6	0.5	0.09	+1.41	+0.038	+0.054	1.988	0.0014
7	1.0	0.11	+0.91	+0.018	+0.016	0.828	0.00032
8	2.0	0.11	-0.09	+0.018	-0.002	0.0081	0.00032
9	5.0	0.14	-3.09	-0.012	+0.037	9.548	0.00014
∑	17.17	1.15			0.535	27.77	0.0164
Av. value	1.91	0.128			0.0594		

The calculation was made under the formula:

$$K_{xy} = \sum_{k=1}^{n} x_k y_k - \overline{xy}\,, \quad (3)$$

At the same time the coordinates of all experimental points, subtract the coordinates of a point located in the middle region [19]. The correlation coefficient is determined by the formula:

$$r = \frac{\overline{xy}}{\sigma_x \sigma_y}, \quad (4)$$

σ_x and σ_y – the standard deviation.

For calculating the dispersion the formula was used:

$$\sigma_x^2 = \frac{1}{n}\sum_{k=1}^{n} x_k - \bar{x}^{\,2} \text{ и } \sigma_y^2 = \frac{1}{n}\sum_{k=1}^{n} y_k - \bar{y}^{\,2} \quad (5)$$

,

We substitute the experimental data and calculated values for x:

$$\sigma_x^2 = \frac{27.77}{9} = 3.09, \quad \sigma_x = \sqrt{3.09} = 1.76; \quad (6)$$

And for *y* as well:

$$\sigma_y^2 = \frac{0.0164}{9} = 0.00182, \quad \sigma_y = \sqrt{0.00182} = 0.0425; \quad (7)$$

Then we find the correlation coefficient

$$r = \frac{0.0594}{1.76 \cdot 0.0426} = 0.79 \quad (8)$$

We use the general expression (9) to estimate the coefficients for equation (2)

$$y - \bar{y} = \frac{\sum xy}{x^2}\; x - \bar{x}\,. \quad (9)$$

Made for the y (ordinate), and it was postponed the value of the expression $\frac{\Delta G}{G_2 \cdot q}$ and on the x-axis (abscissa) expression of the form x = t, we can write

$$\frac{\Delta G}{G_2 \cdot q} - \bar{y} = \frac{\Sigma xy}{x^2} \cdot t - x\,. \quad (10)$$

By substituting the obtained numerical values of the expression (10) takes the form

$$\frac{\Delta G}{G_2 \cdot q} - 0.128 = \frac{0.535}{27.7} \cdot t - 1.91 \tag{11}$$

The solution of equation (11) we get equation (12):

$$\frac{\Delta G}{G_2 \cdot q} = 0.03 \cdot t + 0.071 \tag{12}$$

The resulting equation (12) substituted the values of t and calculated ratio $\frac{\Delta G}{G_2 q}$, the results of which are shown in Table 3.

Table 3. The results of calculations of the duration of ultrasonic treatment from 0 to 5 minutes

t [min]	0	1	2	3	5
$\frac{\Delta G}{G_2 \cdot q}$	0.071	0.101	0.131	0.161	0.221

The graphic description of the data obtained (Table 1, Table 2 and Table 3) is shown in Fig. 2: experimental and theoretical values; theoretical line; MAX level; area of maximum dispersion, equal to 3σ.

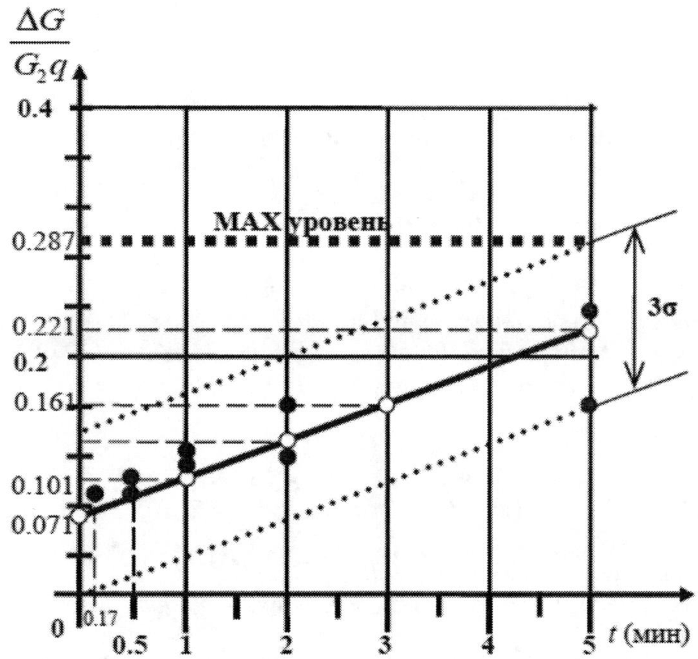

Fig. 2. Diagram of change of the relative content of suspended solids in the water treated with the coagulant, duration of ultrasound exposure: ○ – theoretical points, ● – experiment points

On the chart of Fig. 2 values $\frac{\Delta G}{G_2 q}$ plotted depending on the time of the t sound. Treatment of water containing suspended solids, ultrasound, in all cases, remained provisional, i.e. until the introduction of coagulant, as it was supposed earlier about the possible effect of ultrasonic vibrations on suspended solids.

There is a horizontal dotted line plotted on the graph passing through the value

$\frac{\Delta G}{G_2 q} = 0.287$, which shows the maximum level at the dose of coagulant of 10 mg/l on Al_2O_3 (without sound, Table 1).

As can be seen from the graph of Fig. 2, the experimental points lie in the area of possible spread, equal to 3σ. Thus, confirmed the assumption of the description of phenomena of coagulation using ultrasound to equation (2).

6 Conclusions

1) Experimental studies have confirmed the assumptions about the changing potentials of the suspended particles in the water, because of the action of ultrasonic field.
2) Change of potential under the action of ultrasound leads to an increase of the effect of deposition.
3) Determined the numerical values of the constants a and b of the theoretical equation (2).
4) Statistical processing of experimental data allowed obtaining the area variation equal to 3σ.
5) The experiments confirmed the relationship between values:
   – The duration of the sound;
   – The dose of coagulant;
   - Effect of deposition in the coagulation process.
6) Increase the duration of ultrasonic treatment leads to a reduction of administered doses of coagulant.
7) Experimental studies and calculations have shown that the effect of deposition of suspended solids in water, time-dependent scoring and coagulant dosage increased three times.

References

1. P. D Vikulin. *Physical and chemical manifestation of the acoustic field in the technologies of water conditioning.* (2004).
2. I. P. Golyamina, *Ultrasound. Little encyclopedia of Moscow* (1979).
3. V. Vikulina, P. Vikulin.. *Water treatment coagulation under the action of ultrasonic field* (2016)
4. M. G. Sirotyuk, *Acoustic cavitation* (2008).
5. M. A. Margulis, Adv. In Phys. Sci. **170** (3), 263-287. (2000).
6. K. Hill, J. Bamber, H. Ter Haar, *Ultrasound in medicine. The physical basis for the use* (2008).
7. N. Makisha, Pr. Eng., **165**, 1087-1092 (2016)
8. Y. Zhou, L. Zhai, L., R. Simmons, P. Zhong, *J. Acoust. Soc. Am.* **2 (120)**, 676-685 (2006)
9. N. Makisha, I. Panteleeva, MATEC Web of Conferences, **106**, 70015 (2017)
10. L. R. Gavrilov, *Focussed high intensity ultrasound in medicine* (2013)
11. V. V. Zavyalov, *Study of electrocoagulation purification of drinking waters* (2004)
12. T. G. Leighton, *The Acoustic Bubble* (1994)
13. Edited by Y. Mason, *Physical acoustics. part 1* (1967)
14. V. B. Vikulina, *The use of ultrasonic vibrations for the processes of softening of water and deposition of suspended matter* (1980)
15. E. V. Alekseev and others, *Fundamentals of water systems modeling* (2015)

16. P. D. Vikulin, V. B. Vikulina, *Hydraulics of water Supply and Sanitation. Second edition* (2015)
17. N. Makisha, Pr. Eng., **165**, 1092-1097 (2016)
18. V. B. Vikulina, P. D. Vikulin, *Water treatment coagulation under the action of ultrasonic field* (2016)
19. E. S. Wentzel, *The theory of probability. Tenth edition*, (2005)
20. V. B. Vikulina, P. D. Vikulin, The *use of ultrasound in coagulation* (2016)
21. B. D. Summ, *Foundations of colloid chemistry* (2007)

A Study on Tensile and Shear behaviour of Untreated and Cryogenically Treated Al-SiC and Al-Gr Metal Matrix Composite

Arunkumar K. N.[1*], *G. B. Krishnappa*[1] and *Mohammed Salman Pasha*[1]

[1]*Department of Mechanical Engineering, VVCE, Mysuru - 570002, INDIA,*

Abstract. The present study aims at evaluating the shear and tensile behaviour of untreated and cryogenically treated aluminium-silicon carbide and aluminium graphite metal matrix composite. In this composite Al6061 is used as matrix material with varying Silicon carbide quantity from 2.5wt% to 10wt% and also varying the quantity of graphite from 1wt% to 4wt%. The composites used for this study were fabricated using stir casting technique. Electron dispersive X-ray spectroscopy and X-ray diffraction tests were carried out to know the composition and phase identification of the composite sample. Microstructure analysis was carried out to study the structure of the composites with and without cryogenic treatment.

Keywords: Al6061+SiC, Al6061+Gr, Cryogenic treatment.

1 Introduction

1.1 Composites

In recent decades, composite materials have effectively submitted the customary materials in a few light weight and high quality applicants. A *composite* material is a material system composed of two or more physically distinct phases, the properties of a composite as a whole are enhanced as compared with the properties its components. Composite material consists of two phases, primary phase and secondary phase. Matrix forms the primary phase within which secondary phase is imbedded. The embedment is also known as reinforcement or secondary phase.

1.2 Metal Matrix Composite

Metal Matrix composite is metal reinforced with other metal, particulate or fibres of ceramics materials. These are advanced materials giving properties that have not accomplished by customary materials. MMCs are generally prepared by dispersing the reinforcements in metal matrix. The alloys of aluminium, magnesium, titanium, and copper,

* Corresponding author: arunkn.10@vvce.ac.in

are chosen as matrix material. The reinforcement is in the form of both continuous and discontinuous fibres, whiskers, and particulates.

1.3 Aluminium Metal Matrix Composites

Aluminium metal matrix composites are light weight and high performance aluminium centric material system. Aluminium metal matrix composites generally utilized in high-tech structural and functional applications including defence, aerospace, automotive, and thermal management areas and sports. AMCs are anticipated to replace monolithic materials incorporating aluminium alloys, ferrous alloys, titanium alloys and polymer based composites in a number of appliances.

1.4 Fabrication of Metal Matrix Composites

There is an extensive choice of creation methods in favour of composite materials yet here emphasis will be on the creation of MMCs and particularly particulate reinforced composites.

On the account of creation of MMC's, there are two stages existing. They are: matrix phase and reinforcement phase. Metal or an alloy of a base metal constitute matrix phase and the ceramic generally stand for reinforcement phase. Mechanically combined matrix and reinforcement is called MMCs. If it is not mixed mechanically, the product will be an alloy, not a composite material.

1.5 Cryogenics

The utilization of thermal treatments to enhance mechanical properties of metal components is an antiquated craftsmanship from the previous eras and is used till date. The treatments that are applied to metals to enhance mechanical properties are in a choice of temperature higher than the room temperature. Since few decades, attention has been shown in the cause of sub-zero treatment on the performance of metals. Sub-zero treatment is generally classified as either cold treatment down to dry ice temperature 193k or deep cryogenic treatment at near liquid nitrogen temperature 77k.

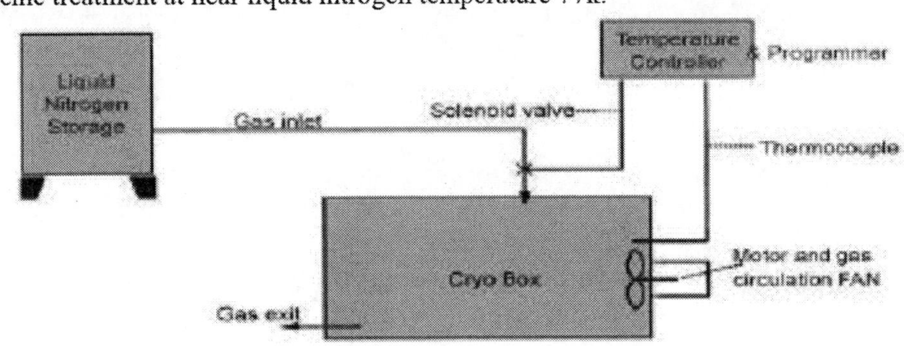

Fig. 1. Cryogenic Treatment Equipment

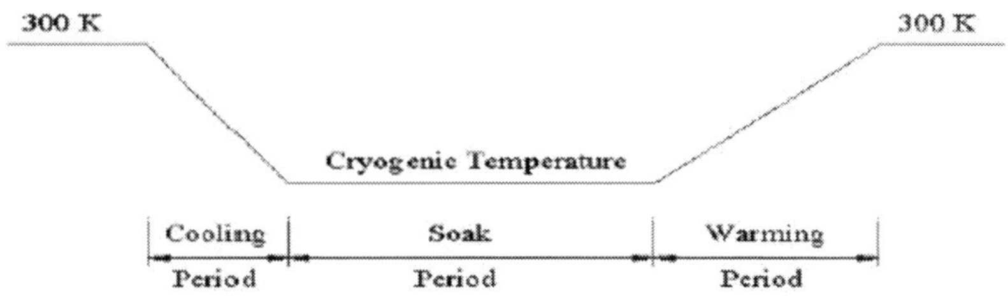

Fig. 2. Cryogenic Treatment Cycle

Cryogenic treatment imparts no visible changes in the metal after treatment. Despite the fact that it is unrealistic to observe external physical changes in the metal, a more uniform and refined microstructure with greater density is observed through metallurgical microscopes.

Typically, cryogenic treatment cycle comprises of cooling the test specimens from room temperature to cryogenic temperature in around 4 hours followed by soaking it at cryogenic temperature (normally liquid nitrogen temperature) for around 24hours and then warming it to room temperature for about 9hours. The soak temperature and duration are to a great extent in charge of changes in the properties after cryogenic treatment.

2 Materials Used

2.1 Materials

The raw materials as well as reinforcement materials used in synthesis of AMMCs are briefly described below.

2.1.1 Al6061

The aluminium Al6061 grade was utilized as the matrix material. Still some compelling studies are being conducted on aluminium alloys. It has low density, which gives additional positive circumstances in a couple of utilizations. Cast iron and Bronze have been supplanted by aluminium alloys to make wear resistance parts.

Table 1. Chemical Composition of Al6061 by Wt%

Elements	Percentage
Al	97.01
Mg	0.88
Si	0.71
Fe	0.64
Cu	0.23
Cr	0.15
Ti	0.04
Zn	0.09
Mn	0.11

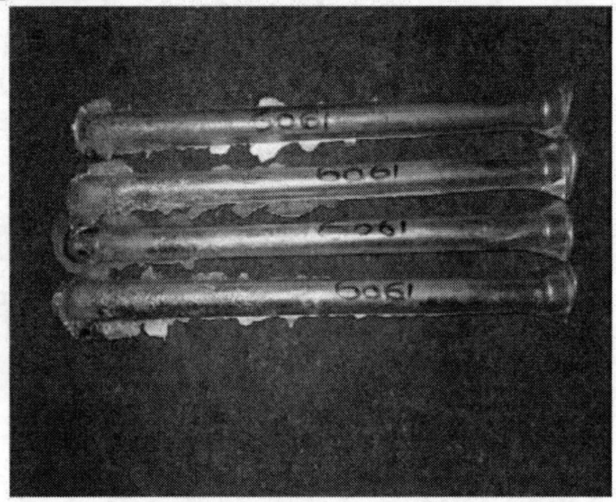

Fig. 3. Aluminium 6061 Ingot

2.1.2 Silicon Carbide

Silicon carbide (Sic) can be utilized as a reinforcement in the form of particulars, whiskers or fibres to enhance the properties of the composite. SiC improves the overall strength of the composites along with very high wear resistance and corrosion.The wear resistance of carbides is very high, therefore the wear resistance of material obtained is also high. Also, the hardness of the composites will increase. SiC are harder than Aluminium metals. If we add SiC in aluminium then it increases the stiffness of the material. After adding hard reinforcement, it resulted in improvement in resistance to wear and abrasion of the material. Content of SiC in material must be carefully added to avoid problem of grain boundaries, embrittlement and cracks formations.

Table 2. Properties of Silicon Carbide (SiC)

Properties	Values
Elastic Modulus(Gpa)	410
Density(g/cc)	3.1
Poisson's ratio	0.14
Hardness(HB500)	2800
Compression Strength(Mpa)	3900
Melting temperature(^{O}C)	3100

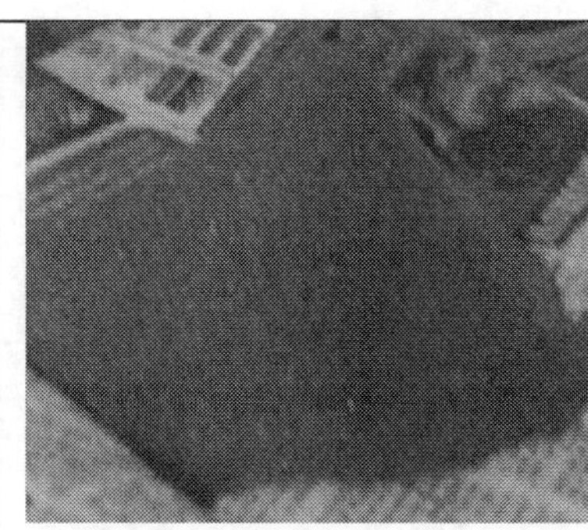

Fig. 4. Silicon Carbide Powder

2.1.3 Graphite

Graphite is a crystalline form of carbon having a layered structure with sheets of closely arranged carbon atoms. Consequently, graphite is fragile when sheared along the layers. This characteristic in turn gives graphite its low frictional properties as a solid lubricant. Its frictional properties are low only in an air or moisture. Unlike other materials, strength and stiffness of graphite increases with temperature.

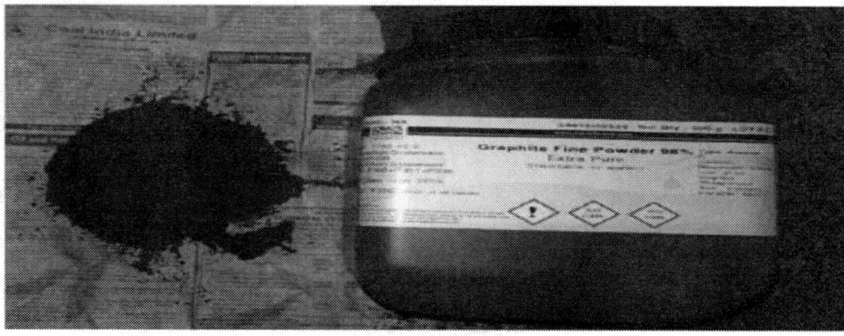

Fig. 5. Graphite Powder

3 Methodology Flow

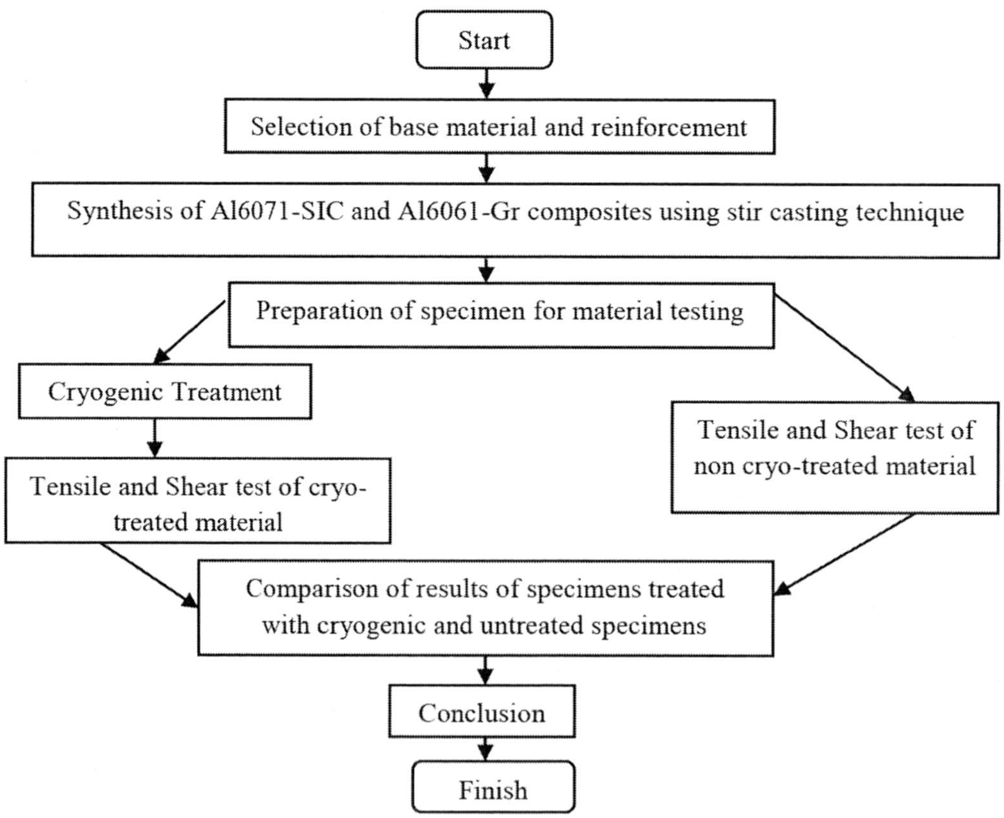

Fig. 6. Method Flow Chart

4 Preparation of Composites

Stir casting is utilized to fabricate the AMMCs specimens with multiple weight fractions of SiC and Gr particles as shown in table 4.1. Al6061 alloy is taken as matrix material. SiC and Gr are used as reinforcements. Aluminium alloy is reduced to molten state in the furnace and then preheated reinforcement particles are added. The composites are prepared by stir casting technique. The furnace used for fabrication is shown in fig 6. The required composition was initially obtained by raw ingots and those ingots were melted at 800°C for almost 120 minutes in stir casting machine.

Table 3. Weight fraction of SiC and Gr in Al6061

SL No.	Aluminium	Silicon Carbide	Graphite
1.	Al6061	2.5% SiC	1%Gr
2.	Al6061	5% SiC	2%Gr
3.	Al6061	7.5% SiC	3%Gr
4.	Al6061	10% SiC	4%Gr

Fig. 7. Furnace

5 Tensile and Shear Test

Tensile test and shear test were carried out on the AMMCs separately. For tensile test ASTM E8 standard was followed. Similarly for the shear test specimen size of 85mm length and 10mm diameter was used.

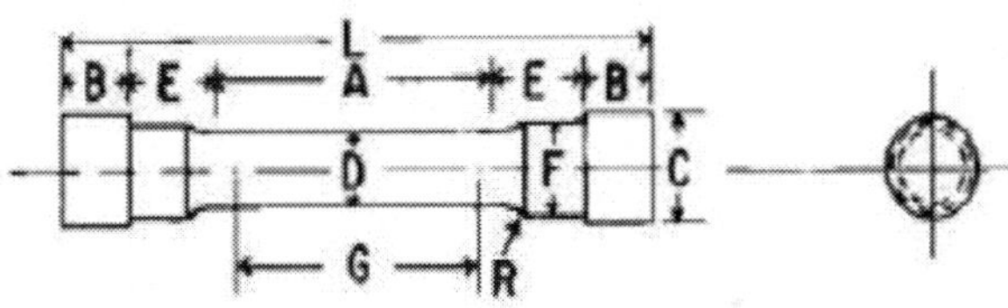

Fig. 8. ASTM E8 standard

Fig. 9. Shear box

6 Results and Discussion

6.1 Scanning Electron Microstructure

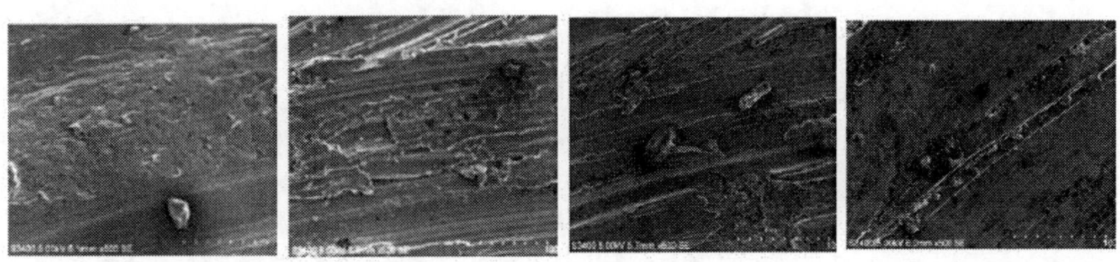

(a) **(b)** **(c)** **(d)**

Fig. 10. SEM of Untreated (a) Al6061+2% Gr (b) Al6061+4% Gr (c) Al6061+5% SiC (d) Al6061+10% SiC

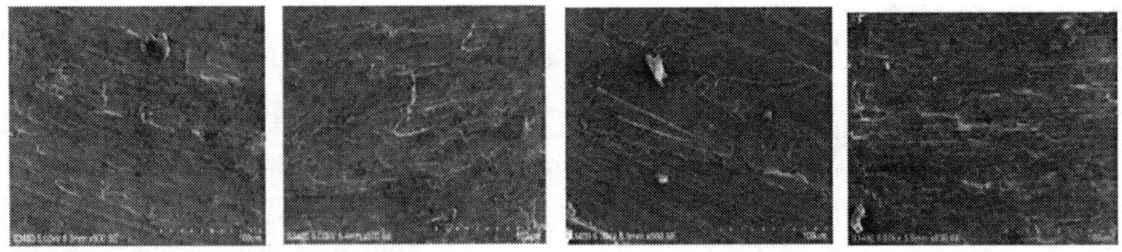

(a) **(b)** **(c)** **(d)**

Fig. 11. SEM of Cryo-treated (a) Al6061+2% Gr (b) Al6061+4% Gr (c) Al6061+5% SiC (d) Al6061+10%SiC

6.2 Shear test

Shear test has been carried out using universal testing mahine(UTM) with the specimen of dimension 85mm length, 10mm diameter and cross sectional area of 78.53mm^2

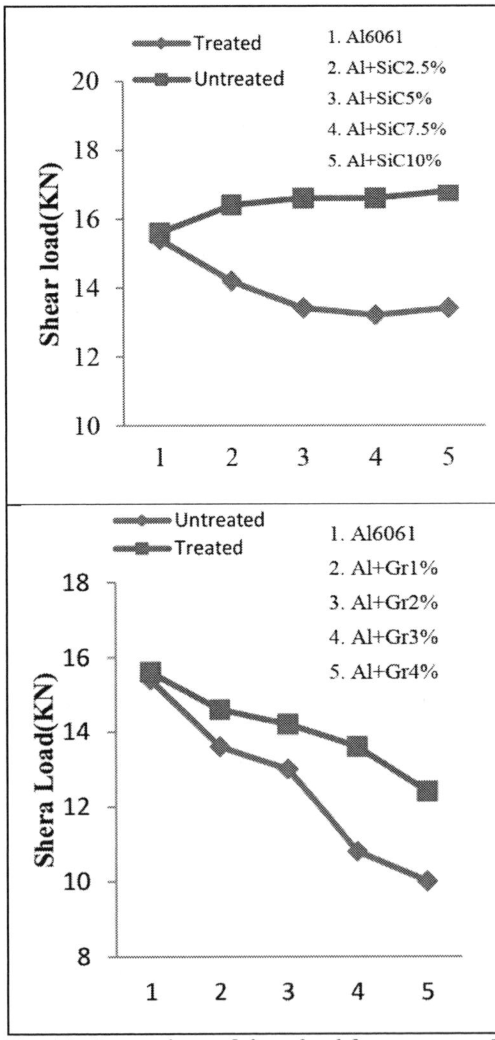

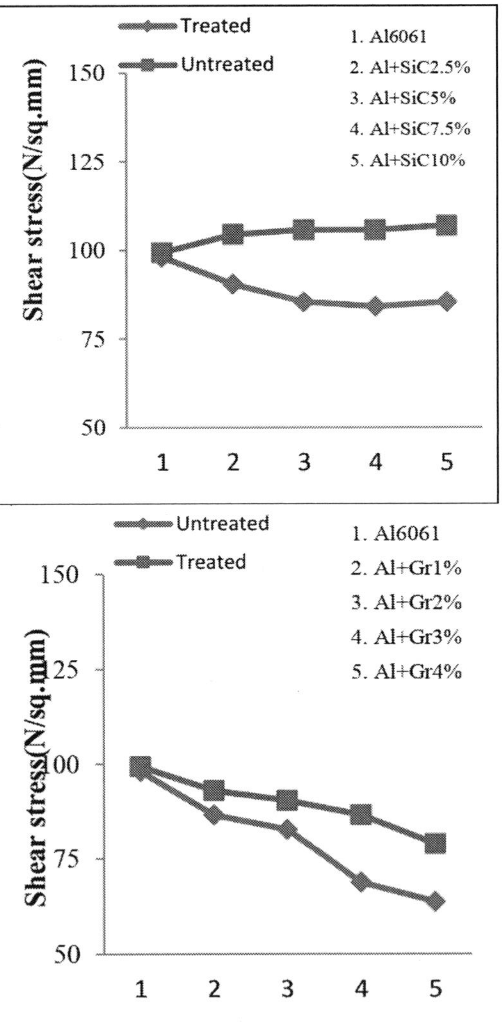

Fig. 12. Comparison of shear load for Al+SiC and Al+Gr AMMCs

Fig. 13. Comparison of Shear stress for Al+SiC and Al+Gr AMMCs

6.3 Energy Dispersive Spectroscopy(EDS) Analysis:

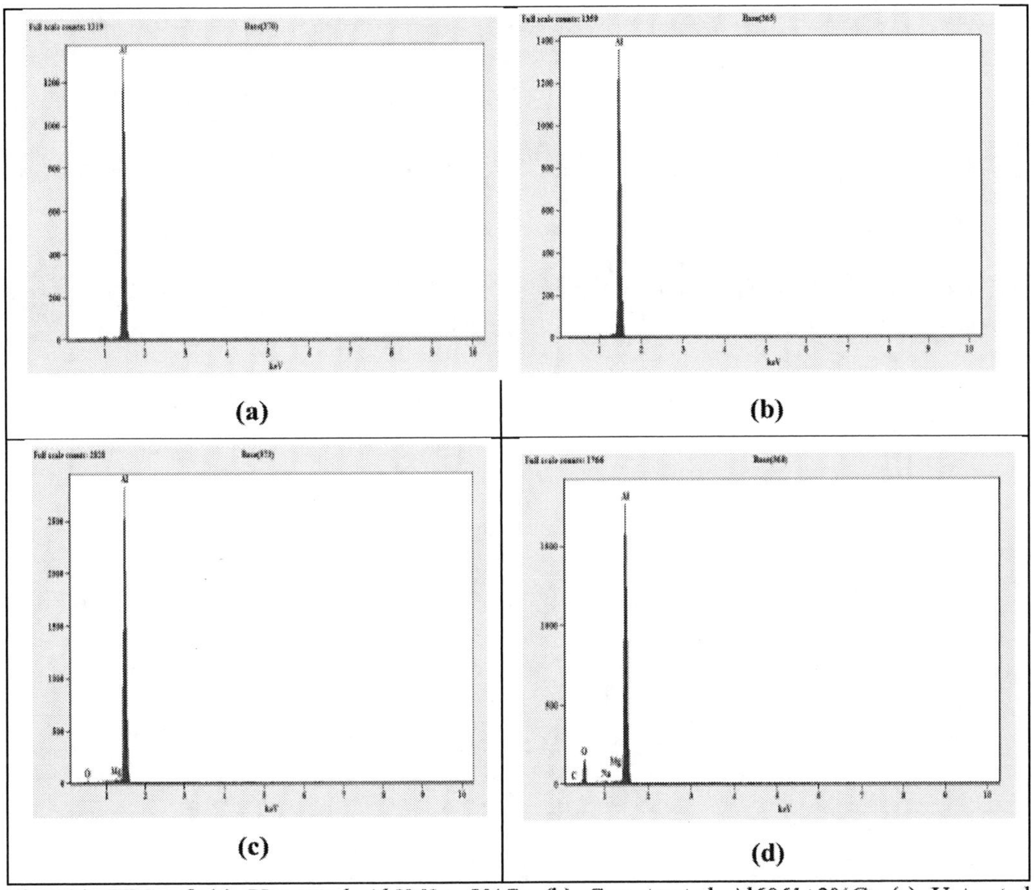

Fig. 14. EDS of (a) Untreated Al6061+ 2%Gr (b) Cryo-treated Al6061+2%Gr (c) Untreated Al6061+5%SiC (d) Cryo-treated Al6061+5%SiC

Energy dispersive spectroscopy analysis of the Al6061 reinforced with Gr and SiC with weight fraction of 2% and 5% respectively. It is shown that Al, Si, O and C peaks the graph and EDS analysis depicts the presence of AL6061, SiC, Gr particles. The Mg_2Si (magnesium silicate) and SiO_2 (silicon oxide) are the products of reaction which shown as Mg and Si peaks in EDS graph after cryogenic treatment.

6.4 Tensile Test

The Peak load recorded for untreated Al6061 specimen was 1092.6 N. The highest peak load recorded for cryo-treated Al6061 was 1176.8N. Al6061+7.5%SiC was found to best among the combinations.

Similarly, the highest break load recorded for untreated for unwanted specimen was 690.6 N. The highest break load for cryo-treated specimen was found to be 784.6N. Compared to untreated specimen, cryo-treated specimen has higher value of break load which indicates that cryo treatment improves strength of the specimen. Al6061+7.5%SiC was found to be best among the combinations.

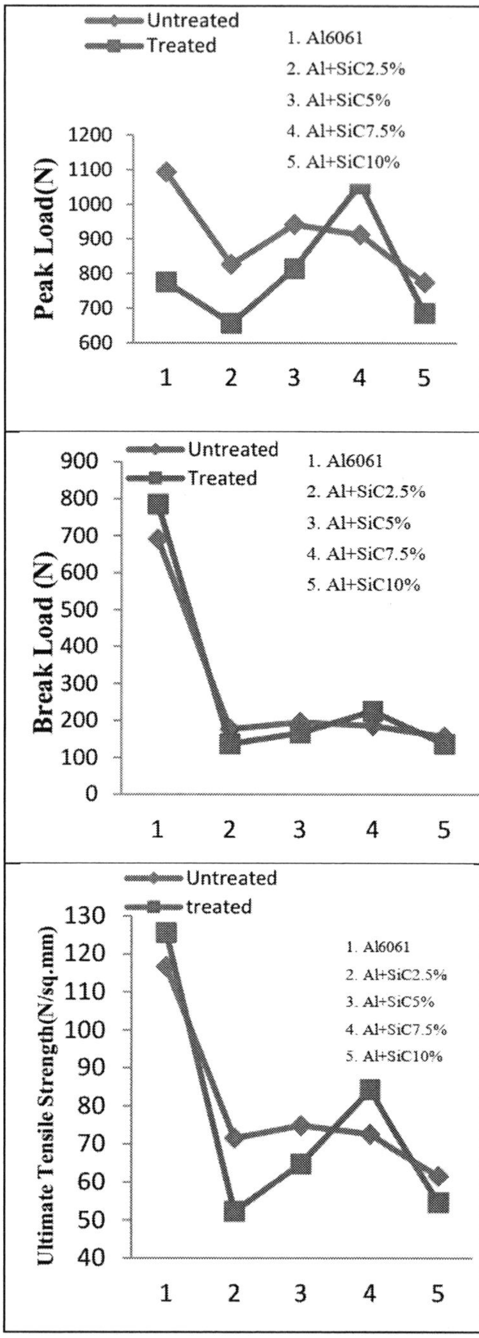

Fig. 15. Comparison of peak load, Break load and Ultimate Tensile Strength of Al+SiC AMMCs

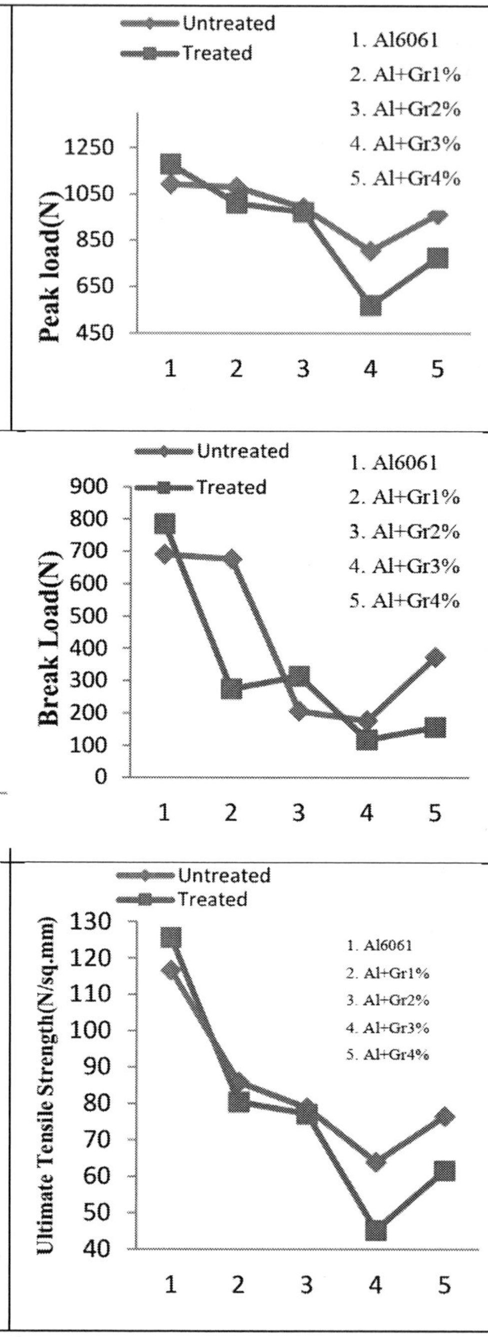

Fig. 16. Comparison of peak load, Break load and Ultimate Tensile Strength of Al+Gr AMMCs

7 Conclusions

1. Tensile properties improvement has shown for cryo-treated and cryo-treated Al6061+2%Gr and Al6061+7.5% SiC has optimum value. But as Gr percentage increases strength decreases.
2. Shear test shows 5-6% improvement after cryogenic treatment
3. SEM images show a uniform distribution of reinforcement in the material after cryogenic treatment.
4. By adding SiC as reinforcement shows improved characteristics in cryogenic treatment but by adding Gr as reinforcement does not show any improvement in cryogenic treatment.

References

1. G. Chigal, G. saini, "Mechanical testing of Al6061/silicon carbide metal matrix composites", IJREAS Volume 2, Issue 2 February 2012 pp 221-238
2. J T. Black, Ronald A. Kohser "Materials and process in manufacturing", Text book tenth edition (2007), pg 351.
3. Seymour. G Epstein, "Aluminium and its alloys", the Aluminium association, Inc.(2001)
4. Surappa.M.K "Aluminium Matrix Composites Challenges and Opportunities", Sadhan, Vol.28, Parts 1 and 2, (2003), pp319-334
5. P. Baldissera and C. Delprete "Deep Cryogenic Treatment: A Bibliography Review" The Open Mechanical Engineering Journal,(2008), 2, 1-11
6. Ajit Behera , S.C Mishra "Comparative Study of Cryo-Treated Steel" IJSTR research volume 1, issue 7, august (2012)
7. F. Farhani, K. S. Niaki "Cryogenic treatment of metals-a review of experimental studies, underlying mechanism and process equipment" proceedings of the 4th BSME-ASME ICTE 27-29 December, (2008), Dhaka, Bangladesh
8. Khalid Mahmood Ghaur, Liaqat Ali, AkhlaqAhamed, Rafiq Ahmad "Synthesis and Characterization of Al/SiC Composite Made by Stir Casting Method" J. Engg & Appl. Sci. Vol. 12, Jan, (2013) p.102-110

A Study on Hardness and Wear behaviour of Untreated and Cryogenically Treated Al-SiC and Al-Gr Metal Matrix Composite

G. B. Krishnappa[1], *Arunkumar K. N.*[1 *] and *Mohammed Salman Pasha*[1]

[1]*Department of Mechanical Engineering, VVCE, Mysuru - 570002, INDIA*

Abstract. The present study aims at evaluating the hardness and wear behaviour of untreated and cryogenically treated aluminium-silicon carbide and aluminium graphite metal matrix composite. In this composite Al6061 is used as matrix material with varying Silicon carbide quantity from 2.5wt% to 10wt% and also varying the quantity of graphite from 1wt% to 4wt%. The composites used for this study were fabricated using stir casting technique. Electron dispersive X-ray spectroscopy and X-ray diffraction tests were carried out to know the composition and phase identification of the composite sample. Microstructure analysis was carried out to study the structure of the composites with and without cryogenic treatment. Hardness behaviour of a metal plays a vital role in understanding the resistance offered by a particular metal to the various shape changing forces and the ability of the metal to resist deformation. Similarly, wear is erosion of material from its original position on a solid surface performed by the action of any other material surface. It is very important to know the wear property of any given material in order to gain an understanding of the wear process, to determine the effects of variables, to characterize materials, and to select materials for specific applications.

Keywords: Al6061+SiC, Al6061+Gr , Cryogenic treatment

1 Introduction

1.1 Composites

In recent decades, composite materials have effectively submitted the customary materials in a few light weight and high quality applicants. A composite material is a material which is made from two or more constituent materials with different properties that, when combined together, produce a material with unique characteristics. Composite material consists of two phases, primary phase and secondary phase. Matrix forms the primary phase within which secondary phase is imbedded. The embedment is also known as reinforcement or secondary phase. Metal Matrix composite is composite material which consists of at least two constituent parts, one being a metal and other may be a different metal or any material, such as polymer or organic compound. These are advanced materials giving

* Corresponding author: arunkn.10@vvce.ac.in

properties that have not accomplished by customary materials. MMCs are generally prepared by dispersing the reinforcements in metal matrix. The alloys of aluminium, magnesium, titanium, and copper, are chosen as matrix material. The reinforcement is in the form of both continuous and discontinuous fibres, whiskers, and particulates. Aluminium metal matrix composites are light weight and high performance aluminium centric material system. Aluminium metal matrix composites generally utilized in high-tech structural and functional applications including defence, aerospace, automotive, and thermal management areas and sports. AMCs are anticipated to replace monolithic materials incorporating aluminium alloys, ferrous alloys, titanium alloys and polymer based composites in a number of appliances.

1.2 Fabrication of metal matrix composites

There is an extensive choice of creation methods in favour of composite materials yet here emphasis will be on the creation of MMCs and particularly particulate reinforced composites.

On the account of creation of MMC's, there are two stages existing. They are: matrix phase and reinforcement phase. Metal or an alloy of a base metal constitute matrix phase and the ceramic generally stand for reinforcement phase. Mechanically combined matrix and reinforcement is called MMCs. If it is not mixed mechanically, the product will be an alloy, not a composite material. In our study, aluminium metal matrix composites were synthesized using stir casting method.

1.3 Cryogenics

The utilization of thermal treatments to enhance mechanical properties of metal components is an antiquated craftsmanship from the previous eras and is used till date. The treatments that are applied to metals to enhance mechanical properties are in a choice of temperature higher than the room temperature. Since few decades, attention has been shown in the cause of sub-zero treatment on the performance of metals. Sub-zero treatment is generally classified as either cold treatment down to dry ice temperature 193k or deep cryogenic treatment at near liquid nitrogen temperature 77k.

Fig. 1. Cryogenic Treatment Equipment

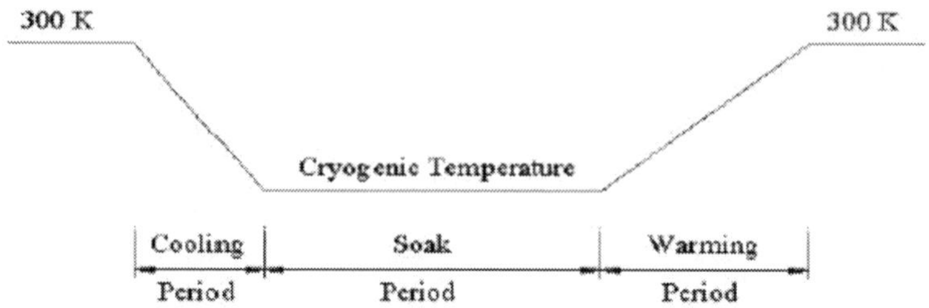

Fig. 2. Cryogenic Treatment Cycle

Cryogenic treatment imparts no visible changes in the metal after treatment. Despite the fact that it is unrealistic to observe external physical changes in the metal, a more uniform and refined microstructure with greater density is observed through metallurgical microscopes.

Typically, cryogenic treatment cycle comprises of cooling the test specimens from room temperature to cryogenic temperature in around 4 hours followed by soaking it at cryogenic temperature (normally liquid nitrogen temperature) for around 24hours and then warming it to room temperature for about 9hours. The soak temperature and duration are to a great extent in charge of changes in the properties after cryogenic treatment.

2. Materials used

2.1 Materials

The raw materials as well as reinforcement materials used in synthesis of AMMCs are briefly described below.

2.1.1 Al6061

The aluminium Al6061 grade was utilized as the matrix material. Still some compelling studies are being conducted on aluminium alloys. It has low density, which gives additional positive circumstances in a couple of utilizations. Cast iron and Bronze have been supplanted by aluminium alloys to make wear resistance parts.

Table 2. Chemical Composition of Al6061 by Wt%

Elements	Percentage
Al	97.01
Mg	0.88
Si	0.71
Fe	0.64
Cu	0.23
Cr	0.15
Ti	0.04
Zn	0.09
Mn	0.11

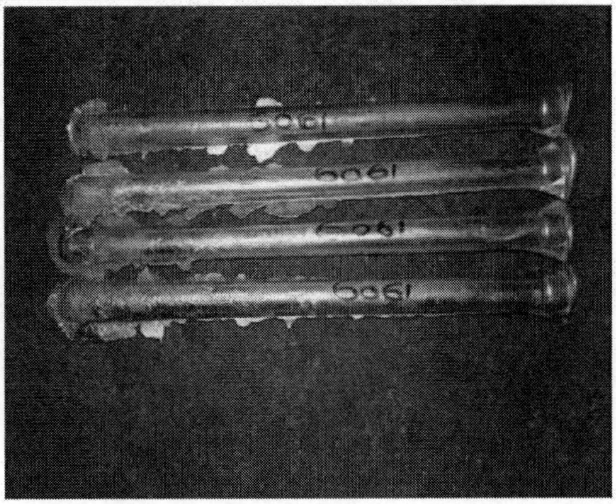

Fig. 3. Aluminium 6061 Ingot

2.1.2 Silicon Carbide

Silicon carbide (Sic) can be utilized as a reinforcement in the form of particulars, whiskers or fibres to enhance the properties of the composite. SiC improves the overall strength of the composites along with very high wear resistance and corrosion. The wear resistance of carbides is very high, therefore the wear resistance of material obtained is also high. Also, the hardness of the composites will increase. SiC are harder than Aluminium metals. If we add SiC in aluminium then it increases the stiffness of the material. After adding hard reinforcement, it resulted in improvement in resistance to wear and abrasion of the material. Content of SiC in material must be carefully added to avoid problem of grain boundaries, embrittlement and cracks formations.

Table 3. Properties of Silicon Carbide (SiC)

Properties	Values
Elastic Modulus(Gpa)	410
Density(g/cc)	3.1
Poisson's ratio	0.14
Hardness(HB500)	2800
Compression Strength(Mpa)	3900
Melting temperature(^{o}C)	3100

Fig. 4. Silicon Carbide Powder

2.1.3 Graphite

Graphite is a crystalline form of carbon having a layered structure with sheets of closely arranged carbon atoms. Consequently, graphite is fragile when sheared along the layers. This characteristic in turn gives graphite its low frictional properties as a solid lubricant. Its frictional properties are low only in an air or moisture. Unlike other materials, strength and stiffness of graphite increases with temperature.

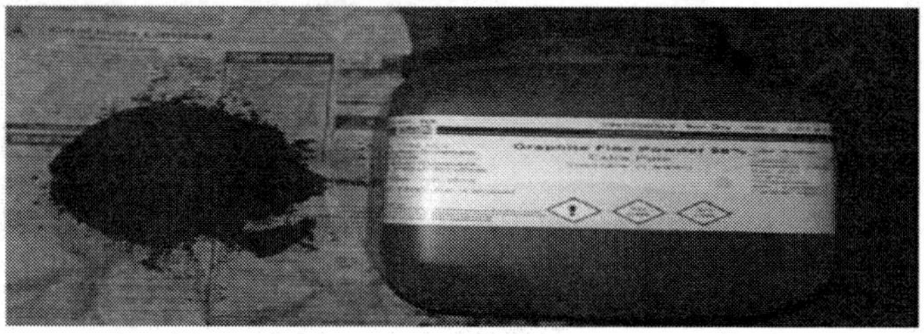

Fig. 5. Graphite Powder

3. Methodology

In order to carry out study on hardness and wear behaviour of both cryotreated and untreated AMMCs following methodology was followed. (i) Selection of material: This phase was carried out to select the best material available to reinforce with aluminium. By its application it has decide to SiC and graphite as reinforcement. Because SiC acts as binding agent and give good strength and graphite acts as good lubrication agent. (ii) ***Synthesis of specimen***: In this study, stir casting method was used to synthesize Al6061-SiC and Al6061-Gr composites. (iii) ***Cryogenic treatment***: Specimens obtained by synthesizing were made to go through cryogenic treatment in order to manipulate the physical properties. (iv) ***Tensile and Shear test***: Both cryotreated and untreated specimens with SiC weight fraction of 2.5wt%, 5wt%,7.5wt% and 10wt% and Gr weight fraction of 1wt%,2wt%,3wt% and 4wt% were made to undergo various tensile test and also shear test was carried out on the specimens and results were recorded. (v) ***Comparison of results***: After carrying out various tests on both cryotreated and untreated specimens, results were recorded and compared with each other.

4. Wear and Hardness test

Both treated and untreated specimens of AMMCs are subjected to wear and hardness test. Wear test is conducted using a tribometer which is an instrument that measures the tribological quantities like co-efficient of friction, friction force and wear volume between two surfaces in contact. Similarly, to carryout hardness test "Rockwell Hardness Testing Equipment" is used. Readings are recorded for both treated and untreated specimens and compared

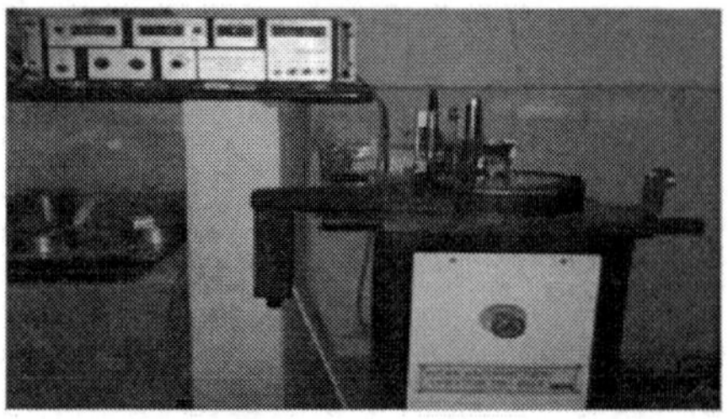

Fig. 6. Wear testing Setup

Fig. 7. Rockwell Hardness Testing Equipment

5. Results and discussion

5.1 Scanning Electron Microstructure

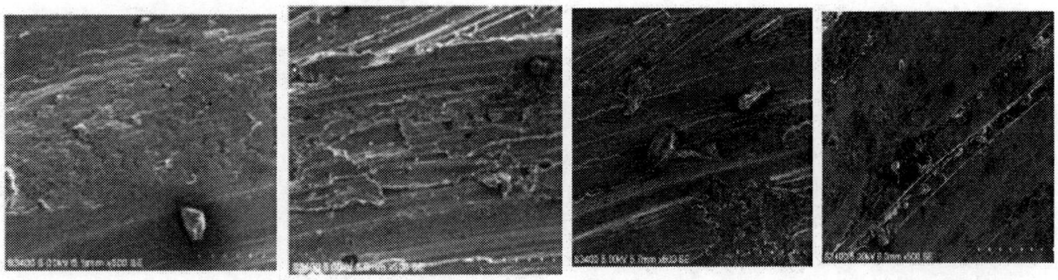

(a) (b) (c) (d)

Fig. 8. SEM of Untreated (a) Al6061+2% Gr (b) Al6061+4% Gr (c) Al6061+5% SiC (d) Al6061+10% SiC

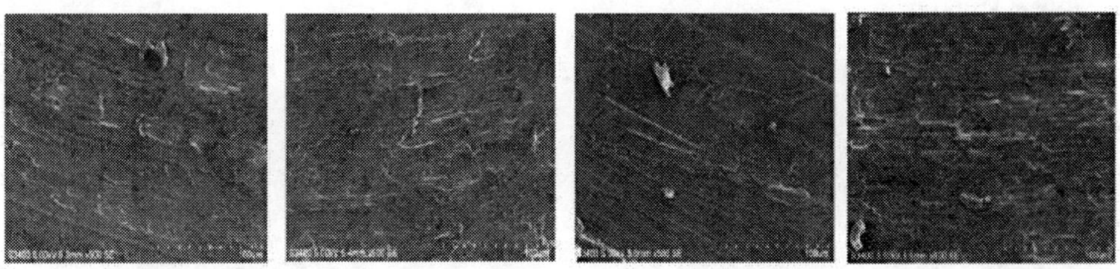

(a) (b) (c) (d)

Fig. 9. SEM of Cryo-treated (a) Al6061+2% Gr (b) Al6061+4% Gr (c) Al6061+5% SiC (d) Al6061+10% SiC

5.2 Wear and Co-efficient of friction:

The tribological behaviour of metal matrix composite was studied using a pi-on-disc tribometer developed according to ASTM G99. The surface of 25mm Length and 10mm diameter comes in contact with the hardened steel disc of hardness 62 HRC. The test was conducted on a track of 120mm diameter, velocity 1 ms-1, time 15 min and load 20N.

Wear in the case of Al+Gr, showed a varying reading for cryotreated and untreated Al+Gr. Whereas in the case of Al+SiC, wear values for cryotreated increased when compared to untreated Al+SiC. Among different weight fraction of Gr in Al+Gr, wear property of untreated Al+Gr3% was increased significantly compared to cryotreated Al+Gr3%. Similarly, In the case of Al+SiC, Wear values of Al+SiC5% showed a significant change from 58 micrometres to 248 micrometres. Coefficient of friction values of Cryotreated and untreated Al+SiC and Al+Gr were recorded. Among all the weight fractions, Al+SiC7.5% showed appreciative change in coefficient of friction.

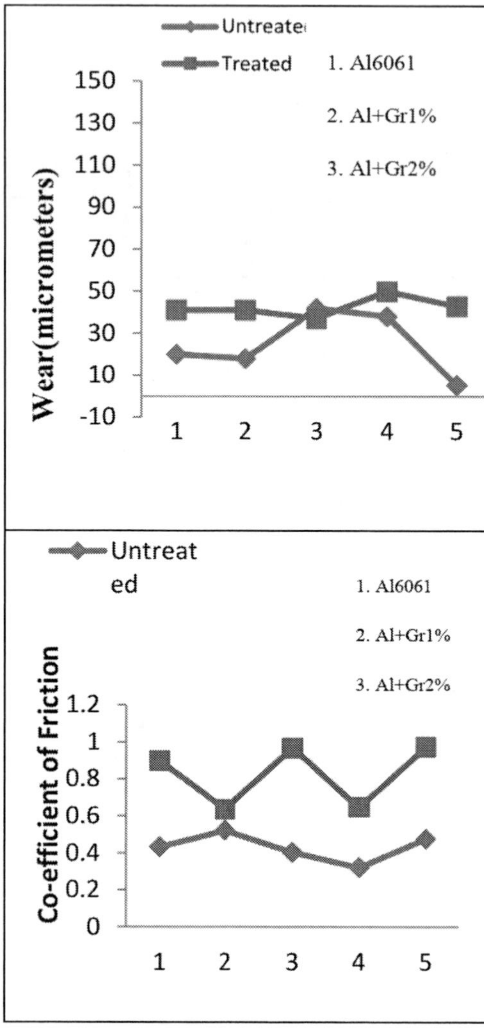

Fig. 10. Wear and co-efficient of friction plot of Al+Gr AMMCs

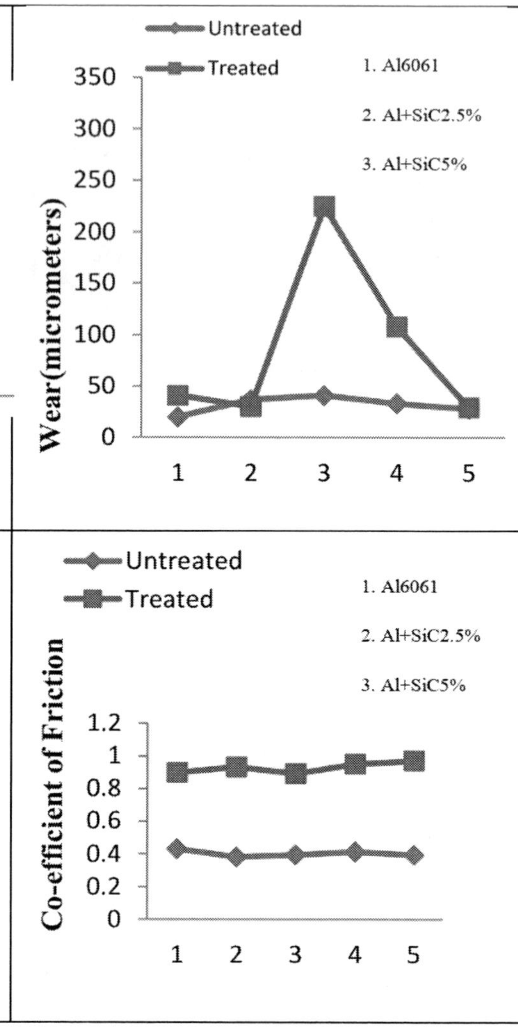

Fig. 11 Wear and Co-efficient of friction plot of Al+SiC AMMCs

5.3 Energy Dispersive Spectroscopy(EDS) Analysis:

Energy dispersive spectroscopy analysis of the Al6061 reinforced with Gr and SiC with weight fraction of 2% and 5% respectively. It is shown that Al, Si, O and C peaks the graph and EDS analysis depicts the presence of AL6061, SiC, Gr particles. The Mg2Si (magnesium silicate) and SiO2 (silicon oxide) are the products of reaction which shown as Mg and Si peaks in EDS graph after cryogenic treatment.

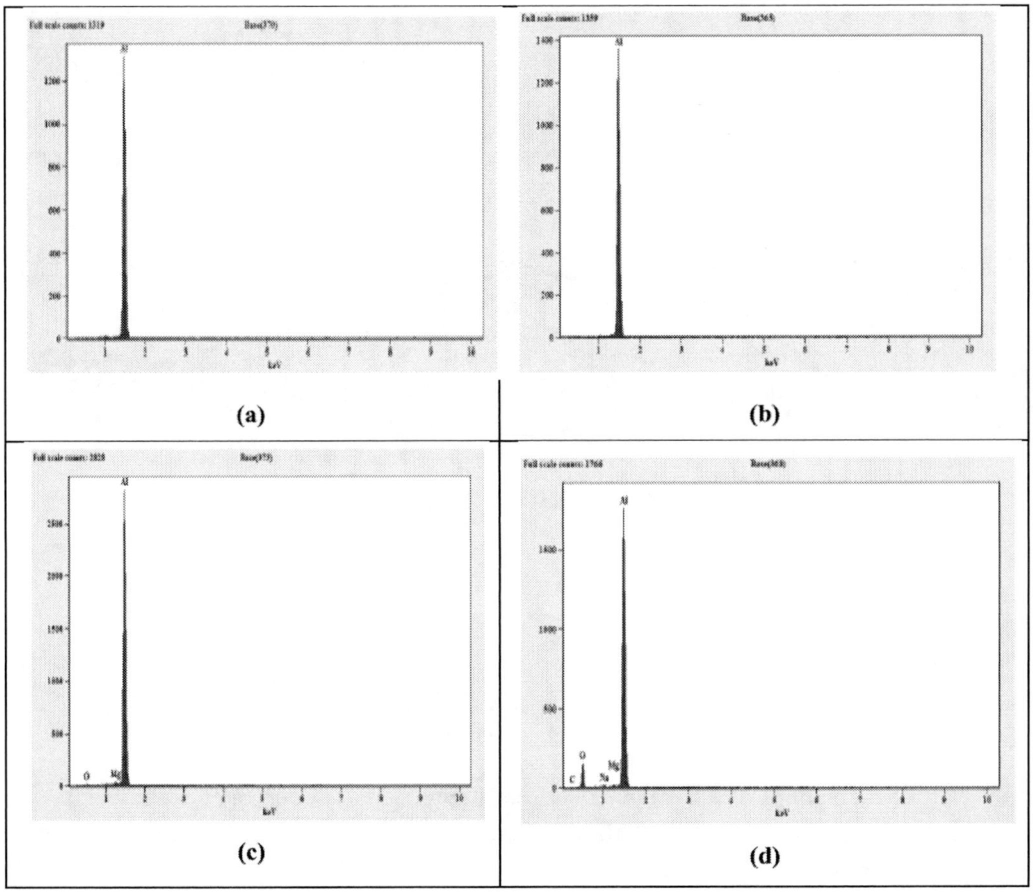

Fig. 12. EDS of (a) Untreated Al6061+ 2%Gr (b) Cryo-treated Al6061+2%Gr (c) Untreated Al6061+5%SiC (d) Cryo-treated Al6061+5%SiC

5.4 Hardness Test

Hardness measurement is conducted by using Rockwell hardness testing machine. The surface being tested generally requires a metallographic finish and it was done by using 600 and 100 grit size emery paper. Load used on Rockwell`s hardness tester was 100kg at dwell time of 10 seconds for each sample. It was observed that hardness of cryotreated Al+SiC was increased. Similarly, hardness values for both cryotreated and untreated Al+Gr was less deviating and was found me almost near to each other. Among all the AMMCs, Al+SiC7.5% showed the significant change in hardness value from 190 to 212.

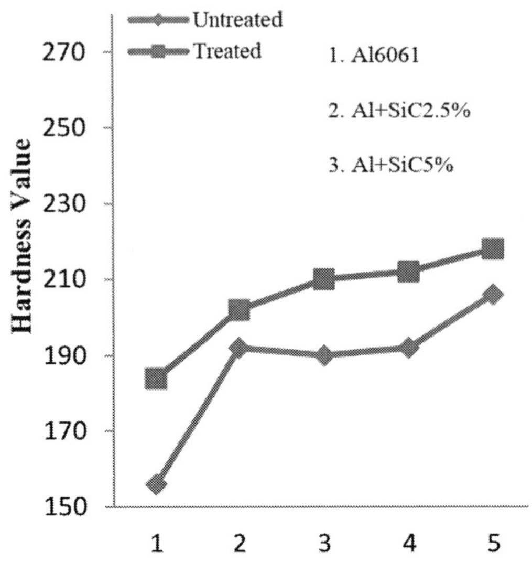

Fig. 13: Hardness Value of Al+SiC AMMCs

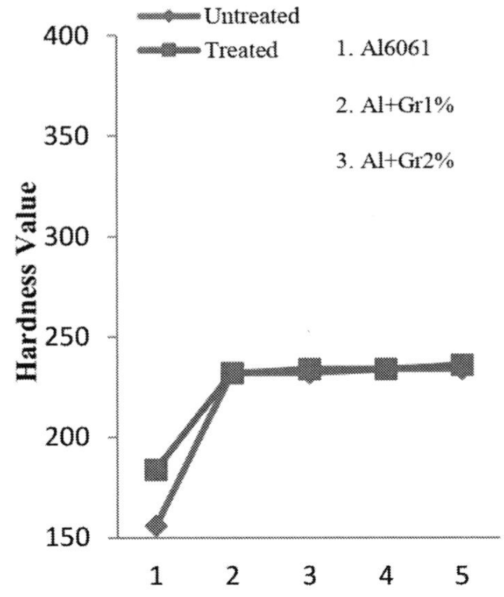

Fig. 14. Hardness Value of Al+Gr AMMCs

6 Conclusions

1. Hardness of composites have been increased upto 2-3% after cryogenic treatment
2. In wear test, the weight loss of alloy and composites material has been decreased after cryogenic treatment
3. SEM images shows a uniform distribution of reinforcement in the material after cryogenic treatment
4. Finally, cryogenic treatment shows an improvement in the microstructure, wear and hardness of AMMCs

References

1. G. Chigal, G. saini, "Mechanical testing of Al6061/silicon carbide metal matrix composites", IJREAS Volume 2, Issue 2 February (2012) pp 221-238
2. J T. Black, Ronald A. Kohser "Materials and process in manufacturing", Text book tenth edition (2007), pg 351.
3. Seymour. G Epstein, "Aluminium and its alloys", the Aluminium association, (Inc.2001)
4. Surappa.M.K "Aluminium Matrix Composites Challenges and Opportunities", Sadhan, Vol.28, Parts 1 and 2, (2003), pp319-334
5. Prashant S N, Madev Nagaral, V Auradi, "Preparation and evaluation of mechanical and wear properties of Al6061 reinforced with graphite and SiC particulate metal metal matrix composites", Vol. 1, No. 3, October (2012) IJMERR, pp 106-112

6. Rajender singh, Yogesh Sharma, Vikas Sharma, "Examine of Mechanical Properties of microstructure, mechanical and fracture properties of Al2O3 metal matrix composites" IJRET Vol. 1, Issue 2, July (2013), pp193-198

A Review on Removal of Heavy Metal Ions from Waste Water using Natural/ Modified Bentonite

Prathiksha P. Prabhu[1*], and *Balakrishna* Prabhu[1]

[1]*Department of Chemical Engineering, Manipal Institute of Technology, Manipal, Manipal Academy of Higher Education, India.*

Abstract. The pollution of water due to various toxic components especially heavy metals are causing an adverse effect on the flora and fauna and also have a potential effect on the human well-being. Due to this there is an immediate requirement to find various techniques for the removal of lethal toxins in wastewater. Amongst all of the existing methods, adsorption is one of the most effective and efficient technique for the elimination of contaminants such as heavy metals from wastewater. The increasing number of research publications on adsorption of heavy metals using modified bentonite leads to the fact that there has been a surging importance in the production of a variety of economical adsorbents for water treatment. Outcome from the latest advances in using bentonites and modified bentonite shows the adaptable nature of the clay and its environment friendly nature. The present review shows the removal of heavy metals using modified bentonite. The main purpose of this review is to describe the flexible way of natural bentonite and modified bentonite and its ability to absorb array of inorganic pollutants, which are present in the waste water. It is apparent from the review that modified bentonite have displayed high removal potential for certain kind of inorganic contaminants from wastewater.

Keywords: Modified Bentonite, Adsorption, Heavy metals, Inorganic pollutants

1 Introduction

Industries have been evolving rapidly in the last few decades which have resulted in increase in all kinds of pollution one of which is water pollution. The accessibility of water supply with regard to both quality and quantity is important to human survival. The environmental concern due to globalization and rapid industrialization are becoming a hindrance for human being. Hence efficient and constructive methods are required specifically for process industries. Heavy metals accounted in wastewaters and industrial discharges are one of the major concerns of environmental pollution. Heavy metals are normally contemplated as those whose density exceeds 5 g/cm^3. Most of the components belonging to this grouping are extremely water soluble, popularly-known toxins and carcinogenic agents.

* Corresponding author: prathikshaprabhu25@gmail.com

Heavy metals are contemplated to be the following elements: Lead, Copper, Zinc, Cadmium, Chromium, Nickel, Arsenic, Tin, Iron, Selenium, Molybdenum, Silver, Gold, Mercury, Cobalt, Aluminium and Manganese. These symbolize serious threats to the human populace and the animals and plants of the receiving water bodies. These can be soaked up and assembled in human body and cause consequential health issues like organ damage, cancer, nervous system damage, and in extreme cases, death. It's also observed to decrease growth and evolution. Heavy metals such as Lead, Copper, Zinc, Cadmium, Chromium, Nickel, Arsenic, Tin, Iron, Selenium, Molybdenum, Silver, Gold, Mercury, Cobalt, and Aluminum are produced in textile industry, leather tanning industry, zinc refineries, metallurgical industries, automobile emissions, incineration of metals, mining and refining of heavy metals, cement and asbestos industries, cadmium nickel batteries, fungicides, insecticides etc. Substantial amount of heavy metal wastes like lead, Tin, and nickel result from building construction waste, storage batteries, paints, cable coatings, and prosthetics containing Ni [1].

Chromated copper-arsenate wood processing industry generates selenium that contains wastes and inorganic pigment manufacturing industries generates pigments which contains cadmium sulfide and chromium compounds; petroleum refining produces conversion catalysts contaminated with vanadium, chromium and nickel; and photographic operations producing film with high concentrations of Ferro cyanide and silver. The adverse effect of heavy metal may manifest in various kinds of disorders like retardation of growth, decrease in longitivity, detrimental changes in reproductivity cycle, morbidity, formation of tumors, symptoms of chronic diseases, pathological changes, target organs affected by heavy metal pollutants are liver, brain, lungs, kidney, bones and teeth. These consequences are reduced by limiting the variety and concentration of heavy metals that may be present in the discharged wastewater. Hence it is a requirement to process and treat the metal polluted wastewater previous to its disposal to the surrounding environment. Heavy metal removal from inorganic effluent can be accomplished by frequently used treatment processes. Removal of heavy metals from industrial effluents can be achieved through various treatment techniques, including filtration or floatation, centrifugation, sedimentation, chemical precipitation, membrane processes such as reverse osmosis, Electrodialysis, ultrafiltration, ion exchange, coagulation/flocculation, adsorption [1].

2 Conventional methods for removal of heavy metal

2.1 Chemical precipitation

Chemical precipitation process refers to a process which includes the input of chemical components (reagents), continued with the segregation of the precipitated particles from the purified water. Removal of metals was brought about by the input of coagulants such as hydrated lime, alum, ferric chloride, chlorinated copperas and other organic polymers. Researchers found that 70% removal of Cd, Cu, and Pb from electroplating, textile and leather tanning industry wastewaters was possible by using coupled system of clariflocculator and hydroxide precipitation [2].

2.2 Electro dialysis

Electro Dialysis (ED) is an electro-chemical reduction of dissolved/colloidal species to elemental metal, transferred between two liquids through semi-permeable ion-exchange membranes under the influence of electric potential. The membranes either let positive or negatively charged ions i.e. either cations or anions to flow through them. An electromotive force brings about the charge separation of metal ions according to the charge. Removal of Cu, Zn, and Ni takes place from this process [2, 3].

2.3 Coagulation/ flocculation

Coagulation is the most economical and effective means to remove impurities. The drinking or waste water are treated with certain chemicals in order to remove finely divided suspended solids and colloidal particles by forming "flocs". The chemicals used for coagulation are well mixed with baffled channels followed by mechanical stirring where the finely divided particles coalesce into larger particles and settle out. Pre-aeration for 5mins before sedimentation helps in removal of entrained particles and improves flocculation. Various studies have been done to determine the effect of pH, turbidity and quantity of coagulants on removals of impurities by this process [2, 3].

2.4 Ultrafiltration

Ultrafiltration is a type of membrane filtration in which the hydrostatic pressure forces a liquid against the membrane with pore sizes in the range of 0.1-0.001µ. This process is used for the removal of suspended solids, colloids, organic and inorganic polymeric molecules. The advantage of ultrafiltration compared to conventional processes such as clarifiers and rotary filters are its high resistance to feed water quality. They are commercially available in hollow fiber, tubular plate and frame and spiral wound configurations. The typical feed pressure is 4-100 psig [2, 3].

2.5 Reverse osmosis

This is a technique in which only the water molecules are allowed to pass through semi-permeable membrane and does not let metal ions to pass through it. In this process the membrane is supported on a filter cloth and pressure is applied for the separation of concentrated side (side containing impurities) and dilute side (side containing pure water). This process finds application in various industrial applications such as removal of heavy metals or organics from industrial waste water from industries such as textile industry, pulp and paper industry, electroplating industry, fertilizer industry, leather tanning industry, rubber industry, detergent industry etc. [2, 3].

2.6 Adsorption

In this process, the gaseous and liquid effluents/solutes accumulate on porous solid adsorbents forming a thin film (adsorbate). The organic or inorganic solid/liquid or gaseous particles adhere onto the surface either by physical or chemical adsorption. The effectiveness of the process depends on various factors such as the surface area/unit weight of adsorbent, physical or chemical properties of the adsorbent and the nature of the particle being adhered on the surface of the adsorbent. This process is being used most often in comparison to the other techniques due to reasons such as cost effectiveness and easy mechanism. There are various adsorbents like silica gel, commercial zeolites, porous pellets of NaF, iron oxide, pulverized lime stone or dolomite, alkalized alumina, bauxite, xanthate, chitosan, activated carbon, peat moss, bone gelatin beads, bentonite, lignin etc. The specific and selective nature of adsorbents renders the adsorbents effective only for few certain applications. The most commonly used adsorbent is activated carbon due to its highly porous structure, low cost and easy availability. It has the advantage of removing wide variety of small quantities of toxic metals. In certain cases industrial by-product or wastes are used as low cost adsorbents since they are locally available in enormous quantities due to which they are inexpensive. This process is widely used in water purification for removal of toxic metals, traces of organics and color [2, 3].

3 Bentonite

Bentonite is classed as natural clay of layered phyllosilicates, composed of alternating octahedral and tetrahedral sheets in the ratio of 2:1. The tetrahedral and octahedral sheets have a loose negative charge Si^{4+} and Al^{3+} in the crystal which is balanced out by reacting it with a lower valency cation such as sodium, calcium etc. The structure, chemical composition, exchangeable ion type and small crystal size contribute to the unique property of bentonite including a large chemically active surface area, high cation exchange capacity and high porosity. The porous structure is classified into macro pores, mesopores and microporores. The adsorption capacity depends on presence of micro and mesopores. Also, bentonite is easily available all over the world.

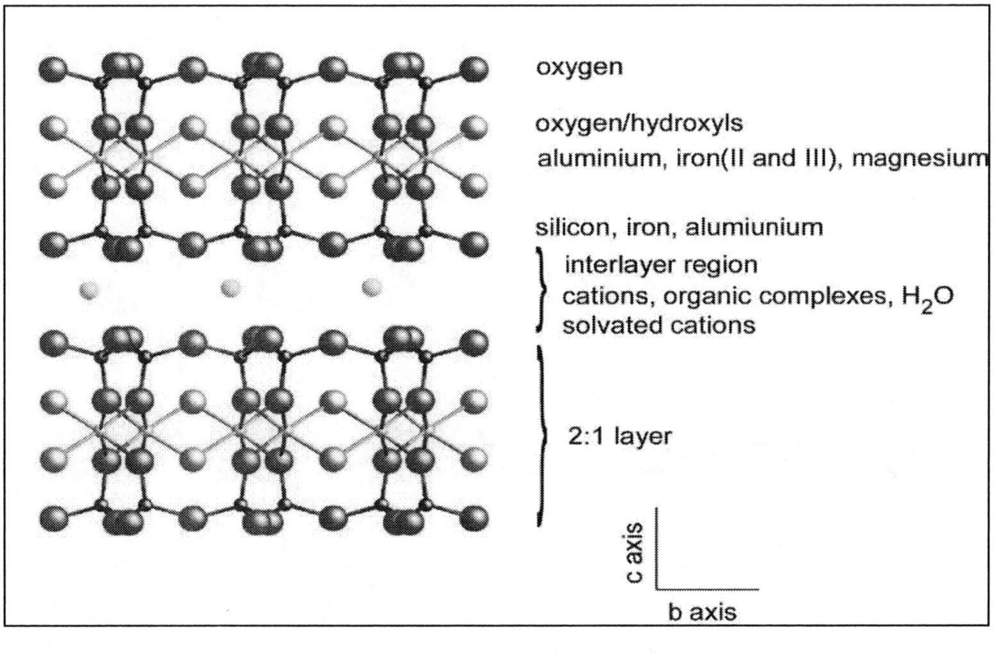

Fig. 1. Structure of Bentonite [1]

Bentonite finds various commercial application in syrups and wines, water purification, clarification and purification of sugar solutions, sewage and effluent treatment, pharmaceutical and therapeutic preparations, carrier for catalysts, refining and bleaching of glyceride oils carrier for fungicides, drilling fluids, fertilizer sprays, wall support for boreholes, formulation of ceramic glazes, non-drip paints, formulation of mortars, putties, adhesives, pelletizing animal feedstuffs, building agents for special moulding sands. Bentonite has advantages such as high surface area for adsorption, high porosity, and high cation exchange capacity and low cost while it also has the disadvantage that it requires modification for adsorption of anions. Modifying the structure of clay helped in increasing the surface area and sorption site thereby enhancing the adsorption capacity. Altering the structure of bentonite is required for improving its applicability for different processes. The composition of the bentonite clay varies depending on the type of operation it is being used for, adsorption capacity, from where it is obtained and the original layer charge.

Table 1. Surface and pore parameters of natural and modified bentonite.

Sl. no.	Adsorbent	Pore diameter	Surface area(m^2/g)	Ref
1.	Natural Bentonite (B)	9.8 nm	27.6	[4]
2.	Natrified Bentonite (Na-B)	7.7 nm	48.2	[4]
3.	Quartz sand bentonite mixture(Mn-M)	9.9 nm	56.3	[4]
4.	Manganese Oxide onto Sodium bentonite (Mn-NaB)	12. 9nm	25.9	[4]
5.	Sodium activated bentonite (Na-AB)		150.3	[5]
6.	Bentonite supported nanoscale zero valent Iron(B-nZVI)	14.23 nm	39.94	[6]
7.	P4V4-Bentonite	250μm	27	[7]
8.	GMZ-Bentonite	53 μm	29.5	[8]
9.	Humic acid immobilized amine modified Polyacrylamide bentonite(HA-Am-PAA-B)		336.3	[9]
10.	8-Hydroxy quinine bentonite (HQ-B)	63 μm	45.69	[10]
11.	Polyacrylamide composite bentonite (BENT-PAAm)	53 μm	26.14	[11]
12.	Activated Bentonite (ABn)	4.6 nm	29.1	[12]
13.	Manganese oxide coated bentonite(MCB)	7.82 μm	30	[13]
14.	Amine modified Polyacrylamide bentonite(Am-PAA-B)		203.5	[9]

There are different methods by which the natural clays can be activated, few of which are acid activation, thermal activation, treatment with cationic surfactants, clay rubber composites, polymer addition, pillaring by different types of poly (hydrxo metal) cations, intraparticle and interparticle polymerization, dehydroxylation and calcination, delamination and reaggregation of smeclites, lyophillization, ultrasound and plasma, adsorption and ion exchange with inorganic and organic cations, binding of organic and inorganic anions and grafting of organic compounds [1]. The bentonite can be modified by addition of polymer in the interlayer of the clay spaces with the help of any simple process like physical adsorption or chemical grafting. The modification of bentonite with polymer improves the chemical and physical properties without actually altering its structure and nature of the clay. But has a disadvantage that the bond between the clay molecules and polymer is not very strong. The bentonite can also be modified by treating it with strong acids such as HCl and H_2SO_4 under favourable conditions of high temperature. The modification by acid activation helps alter the surface area, increases the sorption sites, enhances thermal stability and enhances adsorption capacity [1].

Table 2. Adsorption capacities of natural and modified bentonite

Sl. no.	Adsorbent	Metal ion removed	Adsorption capacity (mg/g)	Ref
1.	Natural Bentonite (B)	Cd(II)	63.29	[4]
2.	Natrified Bentonite (Na-B)	Cd(II)	72.99	[4]
3.	Quartz sand bentonite mixture(Mn-M)	Cd(II)	116.28	[4]
4.	Manganese Oxide onto Sodium bentonite (Mn-NaB)	Cd(II)	108.69	[4]
5.	Sodic Bentonite(B-Na)	Cd(II)	29.4	[14]
6.	Bentonite Oxine complex(B-Oxine)	Cd(II)	61.4	[14]
7.	Sodium activated bentonite (Na-AB)	Pb(II)	232.75	[5]
8.	Sodium activated bentonite (Na-AB)	Cd(II)	28	[5]
9.	Sodium activated bentonite (Na-AB)	Ni(II)	16	[5]
10.	Bentonite supported nanoscale zero valent Iron(B-nZVI)	Cr(VI)	81.69	[6]
11.	P4V4-Bentonite	Cr(VI)	45.25	[7]
12.	GMZ-Bentonite	Cu(IV)	44.25	[8]
13.	Humic acid immobilized amine modified Polyacrylamide bentonite(HA-Am-PAA-B)	Cu(II)	112.5	[9]
14.	8-Hydroxy quinine bentonite (HQ-B)	Cu(II)	56.55	[10]

15.	Sodium activated bentonite (Na-AB)	Cu(II)	30	[5]
16.	Polyacrylamide composite bentonite (BENT-PAAm)	Cu(II)	33	[11]
17.	4-methybenzo-15-crown-5 bentonite(MB15C5-B)	Pb(II)	47.5	[15]
18.	4-methybenzo-15-crown-5 bentonite(MB15C5-B)	Cu(II)	34.2	[15]
19.	Manganese oxide coated bentonite(MCB)	Cu(II)	58.88	[13]
20.	Acid activated bentonite (AAB)	Pb(II)	8.92	[12]
21.	8-Hydroxy quinine bentonite (HQ-B)	Pb(II)	142.94	[16]
22.	Manganese oxide coated bentonite(MCB)	Pb(II)	55.88	[13]

4 Characterization of modified bentonite

The surface properties of the bentonite on modification can be determined by various methods such as Brannuer-Emmet-Teller (BET) method, Fourier transform infra-red spectroscopy (FTIR), scanning electron microscopy (SEM) and X-Ray diffraction (XRD) method.

4.1 Evaluation of pore size and surface area

The specific surface areas of bentonite modified by acid activation, chemical grafting and physical adsorption are estimated using BET method. The pore width was determined from nitrogen adsorption and surface area analyzer. For homogenous and regular particles the specific surface area can be found using the following formulae:
SSA = (Surface area/mass)

4.2 Fourier transform infra-red spectroscopy

FTIR spectroscopy is carried out in order to determine the change in structure and bonding between the clay molecules by the effect of acid activation, chemical grafting and physical adsorption. An IR spectrum is obtained post the modification by above mentioned processes. The sample should be powdered and dried completely prior to the analysis in order to prevent error while analysing the structure of the clay.

4.3 Scanning electron microscopy

Bentonite samples are evaluated under scanning electron microscopy (SEM) to determine the morphological and elemental characteristics. The evaluation of the composition of the bentonite is done over a range of temperature using the sample of modified bentonite [17, 18].

5 Modified bentonite for waste water treatment

Bentonite is finding wide application due to its enhanced surface properties, low cost and improved sorption capacity. There are a number of clay minerals such as montmorillonite, TBA-montmorillonite, natural bentonite, poly (hydroxyl) zirconium-modified montmorillonite, clinoptilolite, Kaolinite, siderite, Ballclay, bentonite, coffee residue binding with clay, natural Jordanian clay and wollastonite that are widely used for the cost effective removal of chemical pollutants from wastewater. Among all of the above mentioned natural clays bentonite is finding wide application due to its easy availability and good adsorption characteristics [19].

5.1 Modified Bentonite for removal of inorganic contaminants

Heavy metals such as Lead Pb (II), Cadmium Cd (II), Chromium Cr (VI) and Copper Cu (II) &Cu (IV) have a toxic effect on the living being and environment. Hence their removal is very vital for the survival of life.

5.1.1 Cadmium Cd (II)

The potential of removal of cadmium using diverse modification of bentonite was studied by various researchers. The adsorption capacity of quartz sand modified bentonite mixture (Mn-M)[1] was better than manganese oxide modified sodium bentonite (Mn-NaB), natrified bentonite(Na-B), bentonite oxine complex (B-Oxine) [12], sodic bentonite (B-Na) and sodium bentonite (Na-AB)[6]. The effect of various parameters such as pH, temperature, contact time, dose of adsorbent for the extraction of cadmium from aqueous solutions was studied. An optimum pH range of 5-6 was observed for cadmium removal using (Mn-M) & (Mn-NaB) whereas a pH of 4 using (B-Na) and pH of 2-7 using (Na-AB). The optimum temperature range of 293-323K was observed for using (B-Oxine) and a range of 290-328K using (Na-AB). The most suitable isotherm and kinetic model that best fits the experimental data was found. A 3-step mechanism was proposed for removal using (B-Na) which consisted of ion exchange step, complexation step and precipitation step [4, 5, 14].

5.1.2 Chromium Cr (VI)

The adsorption potential of various adsorbents such as bentonite supported nanoscale zero valent iron (B-nZVI) and (P4V4-bentonite) for removal of cadmium from aqueous solutions was studied by researchers [5, 6]. 99% Cr (VI) removal was achieved at 35°C using bentonite supported on nZVI whereas in comparison only 66% Cr (VI) removal was observed using unsupported bentonite. For Cr (VI) removal at 25°C and pH of 1 using P4V4-bentonite showed clearly that the pyridinium's positive charge on polymer makes it efficient in chromium extraction. Amount of chromium adsorbed was calculated using

$$Q = (C_o - C_t)\, m / V \qquad (1)$$

The researchers came to a conclusion that (B-nZVI) adsorbed better than (P4V4-bentonite). The adsorption of chromium was observed to increase with increase pH, temperature, contact time with adsorbent and dosage of adsorbent [6, 7].

5.1.3 Copper Cu (II) &Cu (IV)

The adsorption potential of various adsorbents such as GMZ Bentonite [8], Humic acid immobilized amine modified Polyacrylamide bentonite(HA-Am-PAA-B) [9], 8-Hydroxy quinoline immobilized bentonite (HQ-Bentonite) [10], Polyacrylamide complex of Bentonite (PAAm-Bentonite) [11], 4-methylbenzo-15-crown-5 Bentonite(MB15C5-Bentonite) [15], Manganese oxide coated bentonite (MCB)[13] for removal of copper (II) & Cu (IV) was studied. The optimum pH range for each of the above mentioned adsorbents was found to be 6.5, 5-6, 1.5-5.5, 2.4-7, 3.5-6 and 3-6.5 respectively. The (Am-PAA-Bentonite) on modification with humic acid was found to have higher thermal stability than just (Am-PAA-Bentonite). Desorption of Cu (II) and easy regeneration characteristic of (HA-Am-PAA-B) acts as a better adsorbent than others for copper removal. The thermodynamic parameters such as ΔG^o, ΔH^o and ΔS^o were studied for each of the above adsorbents. The negative value of ΔG^o indicated that the adsorption was spontaneous under the applied conditions. The decrease in ΔG^o with increasing temperature signified more efficient sorption at higher temperature. The positive value of ΔH^o signified that the sorption was endothermic and that the increase in ΔH^o increased the difficulty of sorption of Cu (II) ion on the adsorbent. The positive value of ΔS^o showed affinity of metal ions on adsorbent and that Cu (II) ions are less hydrated in bentonite layers than in aqueous solutions. The positive value also indicated the randomness in the system with increase in hydration of adsorbing Cu (II) ions [8-13, 15, 23].

Table 3. Thermodynamic parameters for various adsorbents

Sl. no.	Adsorbent	Temp. (k)	δg^o ($kjmol^{-1}$)	δh^o ($kjmol^{-1}$)	δs^o ($jmol^{-1}k^{-1}$)	Ref
1.	GMZ Bentonite	293.15	-16.26	1.24	59.7	[8]
		313.15	-17.76	0.935		
2.	HA-Am-PAA-B	293	88.1	143.7	-251.7	[9]
		303	90.6			
		313	93.2			
		323	95.6			
3.	HQ-Bentonite	293	-24.14	17.18	140.69	[10]
		303	-25.45			
		313	-26.12			
		323	-28.46			
4.	PAAm-Bentonite	293.15	-22.68	28.46	174.9	[11]
		313.15	-27.56	27.78		
		323.15	-30.18	28.53		
5.	Bentonite	293.15	-20.37	20.62	139.82	[12]
		318.15	-23.86			
6.	MCB	303	-9.50	43.81± 5.22	175.85± 238	[13]
		313	-11.25			

Table 4. Parameters comparison between various modified bentonite

Sl no.	Adsorbent used	Metal ion removed	Ph	Contact time	Rpm	Dose of adsorbent (mg/l)	Temp (k)	Characterization	Isotherm	Ref
1	Na-bentonite	Pb(II), Cd(II) Ni(II)	2-7	1,3,5, 10,15, 20,30, 40,50, 60mins	50, 100, 150, 200	0.125, 0.250, 0.375, 0.5	290, 298, 313, 328	FT-IR,XRF, XRD	Freundlich, Langmuir	[5]
2	Na-bentonite	Pb(II)	3-7	180-1440 mins	5000-9000	-	298, 318, 338	FT-IR, XRD	D-R isotherm, Freundlich, Langmuir	[21]
3	Mn immobilized bentonite & bentonite Quartz	Cd(II)	5-6	30-1500 mins	700	10-700	298 (+ or -)10	XRD, SEM analysis	Langmuir	[4]
4	Hydroxy 8-quinolene Intercalated bentonite	Cd(II)	5.6-11	15-180 mins	3000	4-100	293-323	XRD, SEM analysis	Langmuir	[5]
5	Poly(4 vinyl Pridinium) bentonite	Cr(IV)	1-11	0-1440 mins	Ten thou-sand	25-200	2298-333	FT-IR, XRD	Freundlich, Langmuir	[22]
6	Natural Bentonite	Cu(II), Cd(II)	-	5-60 mins	1500	50-300	298, 308, 318, 328, 348	FT-IR, SEM analysis	Freundlich, Langmuir	[22]
7	Nanoscale Zero valent Iron (n-ZVI) Bentonite	Cr(VI)	3,7, 9,11	0-30mins	150	0.25, 0.5, 0.75, 1.0, 1.5	298 (+ or -)10	XRD, SEM analysis, XRF	D-R isotherm, Langmuir	[6]
8	GMZ Bentonite	Cu(II)	5-6.5	0-1440 mins	9000	1-3	293, 313, 333	XRD, FT-IR	D-R isotherm, Langmuir	[8]
9	Humic-acid Immobilized Bentonite	Cu(II)	4-7	0-300mins	200	25-100	283-333	FT-IR	D-R isotherm, Freundlich, Langmuir	[9]

10	8-hydroxy Quinoline Immobiliz-ed Bentonite	Cu(II)	1.5-5.5	10-180 mins	-	1-2	293, 303, 313, 323	XRD, SEM analysis, XRS, FT-IR	D-R isotherm, Freundlich, Langmuir	[10]
11	Activated bentonite	Cu(II), Pb(II)	2-9	0-720mins	400	1-20	275-343	XRD-IRS, FT-IR, SEM analysis	Langmuir	[12]
12	Sintered bentonite	Pb(II), Zn(II)	3-5	0-1800 mins	120 (+ or –)10	0-200	298 (+ or -)10	XRF, XRD, SEM analysis	Langmuir	[21]

5.1.4 Lead Pb (II)

The adsorption capacity of various adsorbents such as manganese oxide coated bentonite (MCB) [20], acid activated bentonite (ABn) [12], 8-Hydroxy quinoline immobilized bentonite (HQ-Bentonite) [16] and sodium activated bentonite (Na-AB) for removal of lead was studied. Adsorption of lead on MCB is influenced by pH which was found to be optimum around 6; ionic strength increased from 0.01 to 0.1 and temperature was found to be optimum in the range 303 to 338K. Bentonite was activated using 6N H_2SO_4 to give AAB, it was found that specific surface area and pore volume increased by 3.3 times the raw bentonite (RB). The most suitable isotherm and kinetic model for removal of metal ion was determined. A comparative study of the adsorption capacity for various adsorbents was done and is shown in the table below.

Table 5. Adsorption capacity of various clay-based adsorbents for removal of lead (II) ions [16]

SL. NO.	Adsorbent	Adsorption capacity (mgg^{-1})
1.	Montmorillonite	31.05
2.	Poly(hydroxyl) zirconium-modified montmorillonite	31.44
3.	TBA-montmorillonite	30.67
4.	Natural bentonite	20.00
5.	Clinoptilolite	62.00
6.	Kaolinite	19.27
7.	Siderite	14.06
8.	Ballclay	3.52
9.	Coffee residue binding with clay	19.50
10.	Natural Jordanian clay	66.24
11.	Wollastonite	1.68
12.	8-Hydroxy quinoline-immobilized bentonite	142.94
13.	Illite	4.29
14.	Acid activated bentonite	13.05

6 Conclusions

Removal of heavy metals has been extensively studied using various adsorbents and still research is being continued on which technique among the various available conventional methods is most efficient. In this review paper adsorption was used as technique for removal using different types of modified bentonite and their capacities to effectively remove heavy metals from waste water was compared and following are some findings:

1. Most of the modified bentonite has good potential for adsorption of heavy metals.
2. Initially for higher metal ion concentration in waste water, the uptake of metal ion by the adsorbent was high and then slowly reached equilibrium.
3. Adsorption isotherms of two parameter and three parameter was used for studies and it was observed that sorption by various adsorbents has been found to fit well by Freundlich and/or Langmuir isotherm models.
4. Adsorption kinetics for the adsorption processes was found to fit well with the pseudo-second-order kinetic model rather than pseudo-first-order kinetic model and intra-particle diffusion.
5. The effect was various parameters such as the pH, temperature, dose of the adsorbent, contact time; agitation speed and adsorption capacity etc. was compared for different modified bentonite.

References

1. Manjot Kaur Toor, "Enhancing adsorption capacity of bentonite for dye removal: physicochemical modification and characterization", Clays Clay Miner 77-120 (2010)
2. Robert W. Peters and Linda Shem, "Separation of heavy metals: removal from industrial wastewaters and contaminated soil", Desalination 244 91-96 (2011)
3. Debasree Purkayasthha, Umesh Mishra, Swarup Biswas, "A comprehensive review on Cd (II) removal from aqueous solution", J. Water Process Eng. 2 105-128 (2014)
4. Dimple Lakherwal, "Adsorption of heavy metals: A review", Int. J. Environ. Res. Dev. 1 41-48 (2014)
5. TomášSchütz, Silvia Dolinská, PavolHudec, Annamária Mockovčiaková, Ingrid Znamenáčkova, "Cadmium adsorption on manganese modified bentonite and bentonite–quartz sand blend", Int. J. Miner. Process. 150 32–38 (2016)
6. A.A. Taha, Mohamed A. Shreadah, A.M. Ahmed, Hany Fathy Heiba, "Multi-component adsorption of Pb (II), Cd (II), and Ni (II) onto Egyptian Na-activated bentonite; equilibrium, kinetics, thermodynamics, and application for seawater desalination", J. Environ. Chem. Eng. 4 1166–1180 (2016)
7. Li-na Shi , Yu-Man Lin, Xin Zhang , Zu-liang Chen, "Synthesis, characterization and kinetics of bentonite supported nZVI for the removal of Cr(VI) from aqueous solution", Chem. Eng. J. 171 612–617 (2011)
8. A. Mansria, K.I. Benabadji, J. Desbrières, J. François, "Chromium removal using modified poly(4-vinylpyridinium) bentonite salts" , Desalination 245 95–107 (2009)
9. Jiaxing Li, Jun Hu, Guodong Sheng, Guixia Zhao, Qing Huang, "Effect of pH, ionic strength, foreign ions and temperature on the adsorption of Cu(II) from aqueous solution to GMZ bentonite" , Colloids Surf., A 349 195–201 (2009)
10. T.S. Anirudhan, P.S. Suchithra, "Humic acid-immobilized polymer/bentonite composite as an adsorbent for the removal of copper (II) ions from aqueous solutions and electroplating Industry wastewater", J. Ind. Eng. Chem. 16 130–139 (2010)
11. Ozer G, Adnan Ozcan, Bilge Erdem, A. SafaOzcan, "Prediction of the kinetics, equilibrium and thermodynamic parameters of adsorption of copper (II) ions onto 8-hydroxy quinoline Immobilized bentonite", Colloids Surf., A 317 174–185 (2008)

12. GuixiaZhaoa, HongxiaZhang, QiaohuiFan, XuemeiRen, Jiaxing Li, Yixue Chena, Xiangke Wang, "Sorption of copper (II) onto super-adsorbent of bentonite–Polyacrylamide Composites", J. Hazard. Mater. 173 661–668 (2010)
13. Radheshyam R. Pawar, Lalhmunsiama, Hari C. Bajaj, Seung-Mok Lee, "Activated bentonite as a low-cost adsorbent for the removal of Cu (II) and Pb (II) from aqueous solutions: Batch and column studies", J. Ind. Eng. Chem. 34 213–223 (2016)
14. E. Eren, A. Tabak, B. Eren, "Performance of magnesium oxide-coated bentonite in removal process of copper ions from aqueous solution", Desalination 257 163–169 (2010)
15. A. Bentouami, M.S. Ouali, "Cadmium removal from aqueous solutions by hydroxy-8 quinoleine intercalated bentonite", J. Colloid Interface Sci. 293 270–277 (2006)
16. Liu Yun, Shen Xing, Xian Qiming, Chen Haidong, Zou Huixian, Gao Shixiang, "Adsorption of copper and lead in aqueous solution onto bentonite modified by 4 - methylbenzo-15-crown-5", J. Hazard. Mater. B137 1149–1155 (2006)
17. Stella Triantafyllou, E. Irini Christodoulou, and Paraskevi Neou-Syngouna, "Removal Of Nickel And Cobalt From Aqueous Solutions By Na-Activated Bentonite", Clays Clay Miner. 5 567-572 (1999)
18. A. Mellah and S. Chegrouche, "The Removal of Zinc from Aqueous Solutions by Natural Bentonite", War. Res. 3 621-629 (1997)
19. C. Femina Carolina, P. Senthil Kumara, A. Saravanan, G. Janet Joshiba, Mu. Naushad, "Efficient techniques for the removal of toxic heavy metals from aquatic environment: A review", J. Environ. Chem. Eng. 5 2782-2799 (2017)
20. A. Safa Ozcan, Ozer G, Adnan Ozcan, "Adsorption of lead (II) ions onto 8-hydroxy quinoline-immobilized bentonite", J. Hazard. Mater. 161 499–509 (2009)
21. E. Eren, B. Afsin, Y. Ona, "Removal of lead ions by acid activated and manganese oxide coated bentonite J. Hazard. Mater. 161 677–685 (2009)
22. Shitong Yanga, DonglinZhao, Hui Zhang, Songs hengLud, Lei Chen, Xianjin Yu, "Impact of environmental conditions on the sorption behavior of Pb (II) in Na-bentonite suspensions", J. Hazard. Mater. 183 632–640 (2010)
23. N. Karapinar, R. Donat, "Adsorption behaviour of Cu2+ and Cd2+ onto natural bentonite", Desalination 249 123–129 (2009)

A Review on Adsorption of Cationic Dyes using Activated Carbon

Nikita Chrishel Corda[1*] and *M.* Srinivas Kini[1]

[1]*Department of Chemical Engineering, Manipal Institute of Technology, Manipal, Manipal Academy Of Higher Education, India.*

Abstract. In this article efficiency of activated carbon as a potent adsorbent of cationic dyes has been reviewed. Non-biodegradable nature of pollutants and their removal in the present generation is a great challenge. Therefore, extensive study on adsorption of these classes of pollutants from water bodies is being carried out. Methylene blue (majorly) a dye seen in the effluent streams of textile, printing, paper industries along with some of the commonly used cationic dyes in process industries and their sorption on activated carbon are reviewed here. High cost of commercially activated carbon which is a limitation to its extensive use have paved way for study of adsorption by naturally obtained and extracted activated carbon from agricultural wastes and various other sources. The purpose of this review paper is to summarize the available information on the removal of cationic dyes using naturally extracted and commercially obtained activated carbon. Various parameters such as temperature, initial dye concentration, pH, contact time, adsorbent dosage, particle size, stirring, agitation etc. were studied and the optimum parameters were determined based on the experimental outcomes. Equilibrium data was examined using Langmuir, Freundlich, Temkin and Dubinin–Radushkevich and few other isotherm models. Kinetic studies also have been carried out to find the most suitable way of expressing the adsorption process.

Keywords: Adsorption, Cationic, Langmuir, Isotherm, Activated carbon.

1 Introduction

Urbanization has been taking place rapidly in recent years which has resulted in increase of all forms of pollution one of which is water pollution. Exhaust and waste water streams being let out from industries carry huge amount of various toxic substances. Dyes are one of the main pollutants of textile industries. They are composed of two main components: chromophone one which imparts color and the other is auxophone this not only behaves as a supplement to the former component but also gives the water-soluble property to the dye and enhances the affinity of the dye to the cloth material. There are different classes of dye depending upon various factors. One of the classification of dyes is based on the charge the molecule carries. Some of the dyes are cationic in nature (e.g., methylene blue) and some are anionic in nature (e.g., methyl orange). Dyes present in effluent streams that are emptied in water bodies can have adverse effects on human and marine life.

---

* Corresponding author: nikkicorda24@gmail.com

Presence of methylene blue reduces the diffusion of light and thereby photosynthesis [1]. Concentration of dyes higher than 1mg/l in water streams make it unfit for consumption and therefore quality testing and treatment of water becomes important [2]. But overall there are about 100,000 commercial dyes which are produced and account for massive amounts ranging from 7×10^5-1×10^6 tons annually of which 65-70% of dyes are azo dyes [3]. Its low biodegradability [1] makes it a challenging task to eliminate it from water bodies. There are various methods that are used to treat dye contaminated water, some of them are osmosis, reverse osmosis, filtration, membrane filtration etc. In any process certain factors are to be taken to consideration such as economic feasibility, time required, availability of materials, design of the process. Adsorption is one of the recent methods that has gained attention for treatment of waste water streams as it is efficient compared to other existing methods. A good adsorbent is one that is highly porous and has a large surface area. Depending on the availability, its properties and ease of usage, activated carbon can be classified as a good adsorbent. Activated carbon can be produced or extracted by various waste materials such as organic wastes (coconut husk, rice husk, pomegranate peels, olive stones, apricot stones etc.) Activated carbon is known for its high adsorptivity.

Researches have been conducted to study the effect of activated carbon on adsorption of various dyes depending upon the source from which activated carbon is obtained. Some research have been carried out by using activated carbon from olive stones [4], tea waste [2], sugarcane bagasse, apricot stones [5], date stones [6], banana peel [7], tamarind seeds [7], orange peels [6], etc. For studies on adsorption and its efficiency there are some basic requirements: rate of adsorption and equilibrium data [1]. Parameters such as concentration of adsorbent, pH, agitation temperature at which experiments are carried out can be varied to study the effect of adsorption on dyes [1].

1.1 Cationic dyes

As mentioned in the introduction dyes can be classified as cationic and anionic in nature depending upon the charge they carry. Cationic dyes possess a positive charge on its molecule and are also called as basic dyes. They are usually found along with a $ZnCl_2$ or HCl complexes [2] Some of these dyes on which extensive studies have been carried out are methylene blue, malachite green [8], rhodamine 6G[9] basic yellow[10]. All these dyes are known to be toxic, capable of causing mutation, carcinogenic in nature. These dyes are capable of imparting color to water and are soluble with it [6]. Prolonged exposure to methylene blue can cause eye irritation [6], tachycardia [2], nausea, vomiting, slower the mental growth or retardation. Malachite green a cationic dye widely used in clothing industries, fish industries and at a large scale used as disinfectants in medical sector [8]. This pollutant too is known for having carcinogenic properties and cause mutation. Rhodamine 6G a dye which gives a dark reddish-brown appearance is from the rhodamine family known for stunting the growth in plants due to its adherence to lipid and phosphate molecules present in the cell membrane of plants.

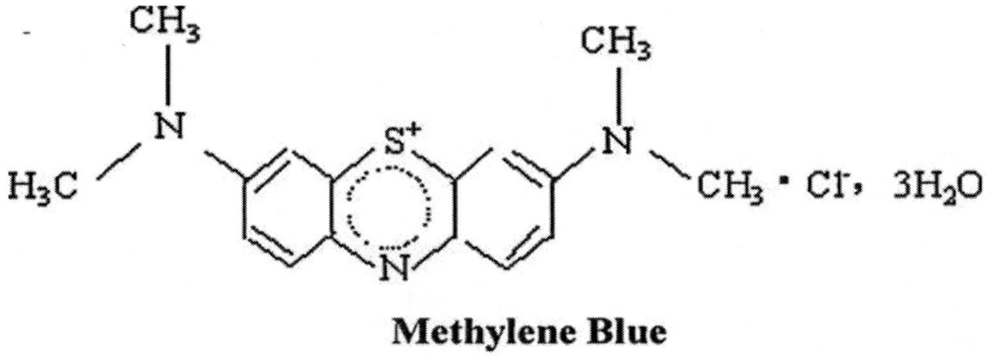

Fig. 1. Methylene Blue structure [11]

1.2 Different methods of dye removal

Few of the common techniques used for treatment of waste water streams polluted by dye are discussed below:

1.2.1 Filtration

There are different types of filtration process such as ultra-filtration, micro filtration, reverse osmosis, nano-filtration etc. Each of them are used for different purposes. Nano and ultra-filtration are the two types of processes that are generally used for treatment of waste water stream for removal of dyes. Drawback of this method is clogging of the pores which makes it an inefficient process. [3]

1.2.2 Chemical treatment

In this method coagulating and flocculating agents are used for dye removal. Ions such as Aluminium (Al^{3+}), Calcium (Ca^{2+}) etc. induces flocculation. Though the process is very feasible, the disadvantage of this method is the production of concentrated sludge due to which disposal becomes a problem. [3]

1.2.3 Oxidation process

Oxidizing agent are used in treatment of effluent streams. Chemical oxidation and UV assisted oxidation are the two common types in this process. These use low quantities of agent and short time for reaction. But the drawback of this process is formation of sludge [3], and (H_2O_2) agent need activation for efficient performance [2].

1.2.4 Adsorption using activated carbon

Of the various methods explained and enlisted above adsorption is one of the most commonly used methods for removal of cationic dyes from industrial effluents. In this process the adsorbate or dye molecules adhere to surface of adsorbent. Adsorption is widely applicable for removal of wide range of dyes. The disadvantage of adsorption is the cost of adsorbent which makes it economically infeasible. In this paper we are reviewing different sources of agricultural waste from which activated carbon is prepared and can be used as adsorbent. By this method the process can be made economical.

Table 1. Studies on adsorption of cationic dyes using activated carbon from different sources

Adsorbents	Dye	Source
F400 commercial activated carbon	Methylene blue	[12]
Pinewood (AC)	Methylene blue	[2]
Activated carbon from textile sludge	MB	[13]
CAC	MG	[8]
Rice husk activated carbon	MG	[2]
Carbon from jute fibers	MG	[8]
Acorn AC	MB	[1]
Oil palm wood AC	MB	[1]
A C peanut shell	MG	[3]
A C from pinedust	MG	[3]
A C plum kernel	Basic red 22	[3]
A C bagasse	Basic red 22	[3]
A C corncob	MB	[6]
A C rice husk	MB	[6]
A C date stones	MB	[6]
A C orange peels	MB	[6]

1.3 Activated carbon

Activated carbon (AC), as the name suggests, is conventionally prepared from carbon from its various physical forms such as coal, coconut shell, wood etc., using physical or chemical activation methods and for all its virtues is the oldest known adsorbent to mankind. Physical and chemical activations [14], as discussed below, have their own merits and drawbacks. On one hand physical activation is more dependent or inclined towards use of elevated temperature and prolonged activation time and on the other hand chemical activation needs a rigorous and exhaustive cleaning of substrate because of exposure to chemical reagents [12].

1.3.1 Physical activation:

Initially, the substrate or source material is subjected to hot gases to be further developed into activated carbon. After the hot air treatment, the source material is then required to be burnt out of gases, giving way to a graded, concealed and dustless form of activated carbon. To achieve such an outcome from a relatively complicated process, a combination of techniques is used that are discussed below:

- *Carbonization*: It is the term for the conversion of an organic substance into carbon or a carbon-containing residue through destructive distillation, commonly known as pyrolysis, at 600–900 °C, conventionally with inert gases such as argon or nitrogen.

- *Activation/Oxidation*: Material obtained from carbonization is subjected to oxidization (oxygen or steam) with temperatures ranging well above 300°C, is in the vicinity of 600-900°C

1.3.2 Chemical activation: It acts as a precursor to carbonization. Here, the raw material is typically injected with an acid, base or salt. After this process, the previous step of carbonization is repeated but this time at a slightly lower temperature range of 400-900°C. The raw material is then carbonized at lower temperatures (450–900 °C). Although there is no certain proof for it but scientists widely agree that the activation step proceeds parallel to chemical activation.

Activated charcoal has been used for over 100 years for the treatment of toxic ingestions, and continues to be the most common form of gastrointestinal decontamination for the poisoned patient [15]. Though commercially available activated carbon (CAC) are usually derived from natural materials such as biomass, lignite or coal, but almost any carbonaceous materials may be used as precursor for the preparation of carbon adsorbents because of its commercial availability and cheapness, coal is the most widely used precursor to produced activated charcoal [10,16].

1.3.3 Properties [17]

1 gram of activated carbon has surface area ranging from 500 m^2 to 1500 m^2 which is available commercially. Carbon aerogels, which have higher surface areas are expensive, and are used in specific applications. Electron microscope, reveal the structure of high surface area of activated carbon. Particles that are extremely complicated exhibit innumerable kinds of porosity; but there can be regions where flat planes of graphite-like material run parallel to each other, which may be disconnected by few nanometers. These micro-pores provide excellent conditions for adsorption such that adsorbing material can interact with various surfaces concurrently. Testing of adsorption behavior are generally carried out with nitrogen gas at 7 K under high vacuum condition. On a general basis activated carbon is perfectly capable of producing its equivalent, by adsorption from its environment, liquid water from steam at 100°C (212 °F) and a pressure of 1/10,000 of an atmosphere.

1.3.4 Applications of activated carbon

The unique properties of activated carbon find its way to following applications:

1. Water purification
2. Gas purification
3. Gold and precious metal recovery
4. Pharmaceuticals and drug industry
5. Thermal and nuclear power stations
6. Waste water treatment
7. Petroleum

Coal is a mixture of carbonaceous materials, formed by the degradation of plants and formation of fossils over ages. The adsorption properties of each coal are determined by the manner by which the plants cultivated and the extent of the physical–chemical changes occurring after deposition. Coal adsorption capacities are given below.

Table 2. Adsorption capacities of commercially activated carbon for MB [18-20]

Adsorbents	**Adsorption capacity (mg/g)**
Commercial activated carbon	980.3
Activated carbon produced from New Zealand coal	588
Filtrasorb 400	476
Activated carbon	400
Activated carbon produced from Venezuelan bituminous coal	380
Peat	324
Coal	323.68
Filtrsorb 400	299
Norit	276
Picacarb	246
Filtrasorb 300	240
Coal	230
Commercial activated carbon	200
Bituminous Coal	176
Charcoal	62.7

1.3.5 Classification

Depending upon size of particle, methods of preparation and industrial applications activated carbon can be classified as follows:

1. Powdered activated carbon (R1, PAC)
Activated carbon is generally prepared in particulate form as fine granules which have a size of 1mm having an average diameter ranging from 0.15 to 0.25 mm. With this structure achieving higher surface to volume ratio and small diffusion distance becomes easier. Activated carbon (R1) are those particles of activated carbon which retain on a 50-mesh sieve (0.297mm).

2. Granular activated carbon (GAC)
Granular activated carbon, these comparatively have larger particle size than powdered activated carbon and hence represent a smaller external surface. GAC are classified depending on their dimensions where particles having sizes 8×20, 20×40, or 8×30 are used for liquid phase applications and 4×6, 4×8 or 4×10 are used for vapor phase applications.

3. Bead activated carbon (BAC)
Petroleum pitch is used for the production of bead activated carbon and are available in diameter ranging between 0.35mm to 0.8mm. It is also known for its low ΔP, high mechanical strength and low dust content, with a smaller pellet size. It is widely used in fluidized bed applications such as water filtration which could be credited to its spherical structure.

2 Effect of parameters

2.1 Effect of temperature

Temperature is an important parameter in adsorption studies. For any process it is essential to know to what extent does temperature bear an effect on the efficiency of adsorbent. Experiments are conducted at different temperatures ranging from 293K-333K by keeping all other parameters constant [1]. The nature of the process can be understood depending upon the way process reacts with variation in applied temperature. A process or a reaction is said to be endothermic in nature if the sorption of dye on the adsorbent sees a rise due to increase in temperature and exothermic if there is a fall in the adsorption capacity with increase in temperature. Adsorption of methylene blue dye on commercially activated carbon is an endothermic process. Two main effects of temperature on this process are increase in the rate of diffusion of the dye molecules on the outer surface of AC and the other increased movement of molecules thereby leading to reduced viscosity [1]. The same was seen in adsorption of MB using olive stone and olive stone activated carbon (OSAC) [4]. Increasing the temperature beyond a certain limit to achieve better adsorptivity is also not advisable since it has an effect on the operating cost [5]. It is seen that for most of the adsorbent used for uptake of MB (whatever the source of activated carbon used) had an endothermic natured process [11, 2, 21-23].

2.2 Effect of contact time

Contact time refers to the amount of time the adsorbent is in contact with the dye solution. Amount of MB that was adsorbed onto OSAC increased with the rise in contact while keeping the other parameters constant [4]. For a dye solution of 50mg/l a contact time of 15min was sufficient whereas for a dye of about 120mg/l concentration the time to attain equilibrium was about 30 minutes [4]. With increase in dye concentration the contact time required to reach equilibrium gets higher. The adsorption of MG was rapid till 120 min of contact and thereafter plateau is observed until equilibrium. And no further dye could be adsorbed [8]. The uptake of MG-5 increased continuously until 30 min of contact time after which removal rate decreased slightly and plateaued after 120 min. The quick adsorption from 50.1 to 72.8% within 1 min suggests strong affinity for the cationic dye molecule [24].The same trend was followed when C-PAN (*Persean Americana* nuts)was introduced in basic yellow 28 dye solution where during the first 15 min the adsorbing capacity was high followed by low adsorption rate until equilibrium was established at 30 min. depending upon the type of adsorbent being used and the removal % expected from the adsorbent the optimal contact time can range from few min to few hours[25].

2.3 Effect of pH

pH refers to the acidity or basicity of a solution. Solution having pH value below 7 are termed as acids/acidic in nature and pH above 7 are alkaline in nature where neutrality is at pH 7. pH gives an account of degree of ionization of acidic or basic compounds in adsorption process [1]. pH values are generally altered by using NaOH and HCl [5]. pH value can alter the surface charge of A C and functional group of the dye [23]. In the experiment of MB uptake using olive stone AC it was seen that there was increase in the amount adsorbed from 6.4-8.3 mg/g but optimum value was found at pH 5 having 9.5 mg/g uptake. There is no significant change in the adsorption capacity of adsorbent with variation in pH [22]. The same is in accordance with adsorption of MB dye using CAC by varying pH between 2-11[1], but there was a significant change in the surface charge the drop in the pH value shows reduced hydrolysis of the groups, simultaneous increase in the stability of $(MB)^+$ was seen. In adsorption process of MB using ASAC there was an evident rise in the efficiency of adsorption with increase of pH credited to the protonation dye in the acidic medium and presence of excess H^+ ions showing optimum value at pH 5[5]. Whereas in adsorption of MG using CO_2 activated carbon it was seen that max adsorption (35.5%-93.2%) took place at pH 2-8 but further increase from 8-12 showed no such effect and was almost negligible impact due to the increased negative charges on CO_2 activated carbon surface at this pH [8].

2.4 Effect of initial dye concentration

Initial dye concentration studies and its effects on adsorption can be understood by treating different concentration dye with adsorbent dose maintained constant. For experiment of AC treatment of MB dye by varying the initial dye concentration, it was seen that achieving equilibrium with lower concentration of dye was much easier compared to higher concentration levels. It is to be mentioned that the rate at which dye is adsorbed is faster at high concentration initially and then steadily attains equilibrium [1]. The same was seen in textile sludge activated carbon used as adsorbent also indicating the faster rate is due to high driving force initially which gradually reduces [13]. In most of the cases it's seen that the initial dye concentration and removal % are inversely related to each other as active sites are saturated in the presence of higher initial concentration [2].

Table 3. Adsorption capacity on varying initial dye concentration

Adsorbents	Dye	Initial dye concentration (mg/l)	% removal	Adsorption Capacity (mg/g)	References
Sawdust Carbon	MB	50-250	99%-82.2%	12.49-51.4	[2]
Apricot Stone AC (ASAC)	MB	5-100	99%-67%	----	[5]
Olive Stones AC	MB	50-150	------	4.8-12.4	[4]
Ficus Carica Bast AC (FCBAC)	MB	40-90	80-50	30-45	[26]
Coconut shell AC	MB	10-50	87.8-39.6	--	[21]

2.5 Effect of adsorbent dosage and particle size

Optimal amount of adsorbent to be used for adsorption can be studied extensively by carrying out experiments where the adsorbent dosage is varied when other conditions are maintained constants. In an experiment of removal of MB dye using CAC it was seen that a dosage of 0.1 g was sufficient to get maximum removal % where initial dye concentration was 10mg/l with agitation speed of 200rpm at 298K were maintained [5]. This was in correspondence with removal of MB using FCBAC when the adsorbent dosage was varied from 0.1/100ml to 0.5/100ml [26]. Same was seen that with an increase in dosage from 2-12g/l the removal % varied from35-90% for CP13[27]. Generally, it is seen that with rise in the adsorbent dosage the removal % or its efficiency gets better as the number of vacant sites available for uptake of dye are more [2]. But to achieve better adsorption we cannot keep increasing the adsorbent dose as cost becomes a limiting factor and therefore finding an optimal dosage for any process of dye adsorption has to be considered.

The CAC used for removal of MB dye had the following physical parameters [1]

Table 4. Physical properties of commercially activated carbon

BET surface area (m^2)	900
Particle size (μm)	150
Mesopore vol. (cm^3/g)	0.058
Micropore vol. (cm^3/g)	0.294

It is seen in an adsorption process study of MB uptake using PAC2 the adsorption capacity of the adsorbent did rise with smaller particle size of PAC2. The solution of $600mg/dm^3$ showed an increase in adsorption from 253-325mg/g as the particle size of adsorbent was varied between 106-180μm when compared to particle size which was maintained below 106μm [10].

2.6 Effect of stirring /agitation speed

Adsorption of MB using CAC as adsorbent was studied at different mixing rates experiments were carried out at 293K by varying stirring speed from 90-150rpm for a solution of 120mg/l concentration. From the experiment, it was concluded that this parameter did not have any significant effect on removal % of dye [1]. Depending upon the agitation speed of a solution it can be understood that the rate controlling step could either be film diffusion or pore diffusion. On agitating MB solution between 50-300rpm with initial dye concentration of 10mg/l adding 0.01g of adsorbent at 298K following were concluded [5].

Table 5. Optimum stirring speed for MB adsorption

Adsorbent	Optimum Stirring speed (rpm)	Efficiency/ removal %
CAC	200	97.4
ASAC	100	99.2

3 Adsorption isotherms

There are different types of adsorption isotherms such as: Freundlich, Langmuir, and Temkin, Harkins-jura, R-D isotherms among which the former two are widely used. Depending on the type of data available any of them could be used for best fitting of the data. Adsorption isotherm in general refers to a graphical representation giving a relation between amount of adsorbate adsorbed on to surface of an adsorbent. There are basically two types of isotherm depending on the parameters

1. Two parameter isotherms: Langmuir, Freundlich, Temkin, R-D isotherms, hill isotherms.
2. Three parameter isotherms: this include Redlich-Peterson, sips model, khan isotherm, Radke–Prausnitz isotherm [28].

3.1 Two Parameter Models [4]

3.1.1 Langmuir isotherm

Langmuir isotherm assumes that a single layer of adsorption takes place on the surface of adsorbent having certain number of vacant sites for adsorption of uniform strategies of adsorption. The expression for Langmuir isotherm is given as follows

$$\frac{C_e}{q_e} = \frac{1}{Q_O K_L} + \frac{C_e}{Q_O} \tag{1}$$

Where C_e refers to concentration of adsorbate at equilibrium.
q_e refers to adsorbate that has been adsorbed /unit mass of adsorbent
Q_O, K_L are the adsorptivity rate constant of adsorption process
A dimensionless parameter R_L gives the essential characteristics of Langmuir isotherm and can be given as

$$R_L = \frac{1}{1+K_L C_e} \tag{2}$$

3.1.2 Freundlich isotherm

In this isotherm heterogeneous/non-homogenous surface with a non-uniform distribution of heat of sorption over surface is assumed. This is given by

$$\log q_e = \log K_F + \frac{1}{n} \log C_e \tag{3}$$

K_F, n are Freundlich isotherm constants where n values generally range from 1-10. Increased value of n shows better adsorption

3.1.3 Temkin isotherm

Here it is assumed that heat of adsorption of all molecules in the layer reduces linearly with coverage. This is due to the interactions that exist between the dye and activated carbon molecules in this case. It is given by:

$$q_e = B_T \ln K_T + B_T \ln C_e \tag{4}$$

K_T is equilibrium binding constant,

$$B_T = RT/b$$

R is the gas constant, T absolute temperature.

3.1.4 Dubinin-Radushkevic: The model is given as follows:

$$\ln q_e = \ln Q_s - B\dot{\varepsilon}^2 \tag{5}$$

Where $\dot{\varepsilon}$ is known as Polanyi potential

$$\dot{\varepsilon} = RT \ln (1+\frac{1}{C_e}) \tag{6}$$

The relation between mean energy for sorption and B is given as

$$E = \frac{1}{\sqrt{2B}} \tag{7}$$

The E value gives relation of physical and chemical nature of adsorption

3.1.5 The Harkins-Jura isotherm [4]

$$\frac{1}{qe^2} = \frac{B_2}{A} - \frac{1}{A}\log C_e \tag{8}$$

Where q_e -adsorption at equilibrium and C_e is the concentration at equilibrium
A is Harkins-Jura isotherm parameter and B_2 is the isotherm constant.
In this isotherm it is assumed that multilayer adsorption takes place in an heterogenous pore distribution.

3.2 Three Parameter Models [28]

3.2.1 Redlich-Peterson isotherm

$$q_e = \frac{K_R C_e}{1+a_R C_e^g} \tag{9}$$

A combination of both Langmuir and Freundlich isotherms, which three parameter isotherm. The isotherm is linearly dependent on concentration in the numerator and an exponential function in the denominator to represent adsorption equilibria. This could be done over a wide concentration range in both homogeneous or heterogeneous systems due to its versatility.

3.2.2 Sips isotherm

Just like Redlich-Peterson isotherm this also is a combination of both Langmuir and Freundlich isotherm which is generally applicable to heterogeneous adsorption systems. This isotherm is influenced by the working conditions of the adsorption process such as pH, temperature and concentrations.

$$q_e = \frac{K_s C_e^{\beta s}}{1+ C_e^{\beta s} a_s} \tag{10}$$

Following is a table of various adsorbent and adsorbate combination and their relation with adsorption isotherms.

Table 6. Adsorption isotherms for dyes using different adsorbents

Adsorbent	Dye/adsorbate	Isotherm	Constants	Source
CAC	MB	Langmuir	q_m=146.46;K=1.11; R^2=0.998(293K)	[1]
Olive stone activated carbon	MB	Temkin	R^2=0.997	[4]
Olive stone	MB	Temkin	R^2=0.995	[4]
Textile sludge AC	MB	Freundlich	K_F=0.165; n=1.280	[13]
Apricot stones AC	MB	Langmuir	q_m= 36.68 ;K_L=23.3 ;R^2=0.961	[5]
CAC	MB	Langmuir	q_m=199.60 ; K_L=2.97; R^2=0.964	[5]
Pinewood AC	MB	Langmuir	q_m=5.56	[2]
AC from pinedust	MG	Langmuir	q_m= 370.37	[3]
A C peanut shell	MG	Langmuir	q_m =222.22	[3]
Mesoporous A C	MB	Langmuir	q_m=336.3,K_L=1.004; T=293K	[23]
AC rice husk	MB	Freundlich	T=303K; q_L=9.8mg/g	[6]
AC sunflowerhulls	MB	Langmuir	T=303K; q_L=473.4mg/g	[6]
AC walnut shells	MB	Redlich-Peterson	T=298K; q_L=355.5mg/g	[6]
AC Date stones	MB	Sips	T=303K ;q_L=369.4mg/g	[6]

4 Adsorption kinetics

The rate of adsorption is given by adsorption kinetics in terms of order of the rate constant [2]. This adsorption rate is essential to decide the type or the best suited adsorbent for a specific process and this is governed by two main properties of adsorbent. One is the adsorption capacity and the other is the rate of adsorption. Most of the adsorption kinetics fall under one of the two types:

1. Pseudo- first order
2. Pseudo-second order

The adsorption process is then evaluated and is checked for the best kinetics that is applicable.

4.1 Pseudo- first order (PFO)

The adsorption rate for this kinetics is given by Lagergren; [4]

$$\ln (q_e - q) = \ln q_e - k_1 t \tag{11}$$

where q_e refers to the dye adsorbed (mg/g) at equilibrium and q is the amount of dye adsorbed at any time t

k_1 is the rate constant for pseudo first order kinetics (min^{-1})

when the adsorption is characterized by diffusion through a boundary the following equation applies when n=1 [2].

$$-r_A = C_A^n$$

4.2 Pseudo-second order model [2]

$$\frac{t}{q_t} = \frac{1}{K_2 q_e^2} + \frac{t}{q_e} \tag{12}$$

4.3 Intraparticle diffusion model [4]

The adsorption process has three steps: Film diffusion, Pore diffusion, intra-particle transport.
The slowest step among the three is called as the rate determining step. Intraparticle diffusion is given by

$$q_t = k_{diff}\, t^{0.5} + C \tag{13}$$

where k_{diffi} is the intraparticle diffusion rate constant (mg/g min) and C is the intercept (mg/l) indicating that the line does not pass through the origin which is attributed to the difference in the initial and final mass transfer rates. The intercept gives an idea about the thickness of the boundary layer higher the value of which greater is the thickness.

Where q_t and q_e are the amount of dye adsorbed at time t and at equilibrium respectively.
k_2 is the rate constant for pseudo second order kinetics (g/mg min).
This model shows or follows a linear path. Depending upon the linear regression coefficient R^2 the best model for process can be inferred.
Generally, for cationic dye adsorption pseudo second order represents the system better [2].

For adsorption of MB from macore fruit shells revealed that the R^2 value was max for pseudo second order kinetics [22]. The same was seen in adsorption of the MB dye using OSAC and OS as adsorbent suggesting that overall rate was controlled by chemical adsorption. But in contrast the MB dye adsorption onto surface of ASAC should better results for intraparticle diffusion [2]. An experiment of malachite green adsorption using CO_2 activated carbon gave the best result for intraparticle diffusion model [8].

5 Regeneration of activated carbon

Regeneration of the adsorbent is essential as excess amounts untreated can lead to problems in disposal. The adsorbent used can be reused for future applications. It is seen that regeneration of adsorbents from different agricultural sources show good amount of adsorption. The different methods of regeneration are biosorbents [25], carbonization of activated carbon, thermal regeneration [8]. It is seen that as we increase the regeneration cycles the adsorption capacity decreases.

Table 7. Adsorption capacity of adsorbents on regeneration

Adsorbent	Dye	Adsorption Capacity (mg/g)	References
C-PAN	MB	235-37.3	[25]
A C from fuel oil	MB	37-28	[9]
A C from fuel oil	Rhodium 6G	27-18	[9]
CAC	MB	210.18-148.51	[8]

6 Conclusions

Adsorption has been extensively studied and still research is being continued on efficient methods of dye removal.in this review paper various works have been compared and following are some of the findings

- Most of the agriculture based AC have good potential for adsorption of cationic dyes
- On uptake of MB dye varying pH over certain range and stirring did not show any significant changes.
- Initially for higher dye concentration the uptake of dye by the adsorbent is high and then slowly reaches equilibrium
- Depending upon the agitation and the process response towards it one can identify the rate limiting step between film and pore diffusion
- Adsorption isotherms of two and three parameter were used for studies and it was observed that most of them gave good results for Langmuir isotherms.
- Adsorption kinetics for the adsorption processes were not very satisfying for pseudo-first order kinetics when compared to the second order or intraparticle diffusion.

7 References

1. S.Karaca, A. Gurses, M.Acikyildiz, M.Ejder "Adsorption of cationic dye from aqueous solutions by activated carbon" Microporous Mesoporous Mater 115 376-382 (2008).
2. Mohammad Amran Mohd Salleh, Dalia Khalid Mahmood, Wanazlina wan abdul karim, Azniidris "Cationic and anionic dye adsorption by agricultural solid wastes: A comprehensive review" Desalination (2011).
3. V.K. Gupta, Suhas "Application of low cost adsorbents for dye removal-A Review" J. Environ. Manage.90 2313-2342 (2009).
4. RihamHazza, Mohammed Hussein "Adsorption of Cationic dye from aqueous solution onto activated carbon prepared from olive stones" Environmental Technology and Innovation 4 36-51 (2015).
5. Chahrazed Djilani, Rachida Zaghdoudi, Faycal Djazi, Bachir Bouchekima, Abdelaziz Lallam, Ali Modarressi, Marek Rogalski "Adsorption of dyes on activated carbon prepared from apricot stones and commercial activated carbon" J Taiwan Inst Chem Eng 53 112-121 (2015).

6. J Muthanna Ahmed "Applications of agricultural based activated carbons by microwave and conventional activations for basic dye adsorption:Review" J Environ Chem Eng 4 89-99 (2016).
7. V.Tharaneedhar, P.Senthilkumar, A.Saravannan, C.Ravikumar, V.Jaikumar "Prediction and interpretation of adsorption paameters for the sequestration of methylene blue dye from aqueous solution using microwave assisted corncob activated carbon" SM&T 11 1-11 (2017).
8. Miao Yu, Yingyinghan, Jianli, Lijuan Wang "CO2-Activated porous carbon derived from cattail biomass for removal of malachite green dye and application as super capacitors" Chem. Eng. J 317 493-502 (2017).
9. Nabeel Jarrah "Competitive adsorption isotherms of rhodium 6G and methylene blue on activated carbon prepared from residual fuel oil" J Environ Chem Eng S2213-3437 30390-1 (2017).
10. N.Emad El Qada, J.Stephen Allen, M.Gavin Walker "Adsorption of basic dyes from aqueous solution onto activated carbons" Chem. Eng. J 135 174-184 (2008).
11. Wenhong Li, Qinyan Yue, Peng Tu, Zuohao Ma, Baoyu Gao, Jinze Li, Xing Xu "Adsorption characteristics of dyes in columns of activated carbon prepared from paper mill sewage sludge" Chem. Eng. J 178 197-203 (2011).
12. Mohd.Rafatullah, Othman Sulaiman, Rokiah Hashim, AneesAhmed "Adsorption of Methylene blue on low cost adsorbents: A Review" J. Hazard. Mater. 177 70-80 (2010).
13. Syieluing Wong, Nurul Atiqah Najlaa Yac'cob, Nozrita Ngadi, Onn Hassan, M.Ibrahim Inuwa "From Pollutant to solution of wastewater pollution: Synthesis of activated carbon from textile sludge for dyes adsorption" S1004-9541 (17) 30748-6
14. J. Romanos, M. Beckner, T. Rash, L. Firlej, B. Kuchta, P. Yu, G. Suppes, C. Wexler, P. Pfeifer. "Nanospace engineering of KOH activated carbon". Nanotechnology. 23 (1): 015401. doi:10.1088/0957-4484/23/1/015401 (2012).
15. American Academy of Clinical Toxicology Position Paper: "Single-Dose Activated Charcoal. Clin Toxic".
16. G.G. Stavropoulos, A.A. Zabaniotou, "Production and characterization of activated carbons from olive-seed waste residue", Microporous Mesoporous Mater
17. P.J. Paul "Value Added Products from Gasification – Activated Carbon". The Combustion, Gasification and Propulsion Laboratory (CGPL) at the Indian Institute of Science (IISc).
18. N. Kannan, M.M. Sundaram, "Kinetics and mechanism of removal of methylene blue by adsorption on various carbons—a comparative study", Dyes Pigm.
19. A.N. Fernandes, C.A.P. Almeida, C.T.B. Menezes, N.A. Debacher, M.M.D. Sierra, "Removal of methylene blue from aqueous solution by peat", J. Hazard. Mater.
20. K.V. Kumar, S. Sivanesan, "Equilibrium data, isotherm parameters and process design for partial and complete isotherm of methylene blue onto activated carbon", J. Hazard. Mater
21. A.Umar Isah, Giwa Abdulraheem, Salisu Bala, Sallahudeen Muhammed, "Kinetic, equilibrium and thermodynamics studies of C.J Reactive Blue 19 dye adsorption of coconut shell based activated carbon" Int Biodeterior Biodegradation 102 265-273 (2015).
22. Kouassi Narcisse Aboua, Yao Augustin Yobouet, Kouassi Benjamin Yao, Droh Lancine Gone, Albert Trokourey "Investigation of dye adsorption onto activated carbon from the shells of Macore fruit" J. Environ. Manage 156 10-14 (2015).

23. Lei Yu, Yong-Ming Luo "The adsorption mechanism of anionic and cationic dyes by Jerusalem artichoke stalk-based mesoporous activated carbon" J Environ Chem Eng 2 220-229 (2014).
24. Hai Nnguyen Tran, Sheng-Jie You, Huan-Ping Chao, "Fast and efficient adsorption of methylene green 5 on activated carbon prepared from new chemical activation method", J. Environ. Manage, 188, 322-336 (2017).
25. Abdelmajid Regti, My Rachid Laamari, Salah-Eddine Stiriba Mohammadine, El Haddad "Potential use of activated carbon derived from Persea species under alkaline conditions for removing cationic dye from wastewaters" J. Assoc. Arab Univ. Basic Appl. Sci (2017)
26. Deepak Pathani, Shikha Sharma, Pardeep Singh, "Removal of methylene blue by adsorption onto activated carbon developed from Ficus carica bast", Arab. J. Chem, 10, S1445-S1451 (2017).
27. F. Asaad Hassan, Hassan Elhadidly, "Production of activated carbons from waste carpets and its applications in methylene blue adsorption: Kinetic and thermodynamic studies", J. Chem. Eng, 5 (2017), 955-963
28. K.Y. Foo, B.H. Hameed, "Insights into modeling of adsorption isotherm systems", Chem. Eng. J, 156 (2010), 2-10

Current Status of Methods Used In Degradation of Polymers: A Review

Aishwarya Kulkarni[1] and *Harshini* Dasari[1*]

[1]*Department of Chemical Engineering, Manipal Institute of Technology, Manipal, Manipal Academy of Higher Education, India.*

Abstract. Degradation of different polymers now a day is the most crucial thing to carry out. It possesses threats to human health as well as to the environment. Different polymers like PVA, PVC, and PP with high density and low density are one of the most consumed by population and also their degradation is a bit difficult. For this many people have started working on effective methods of degradation of these polymers. This can be done by thermal degradation and pyrolysis which requires high temperature, bio degradation using starch, bacteria etc and photo degradation. Traditional gravimetric and respirometric techniques are the methods currently used in testing. They fit readily for degradable polymeric materials usually. Also they are well suited for biodegradable components with polymer blends. But the recent polymer generation is comparatively resistant to bio degradation of polymers hence they cannot be used here. The polymer matrices are readily present in the plasticizers boosting the strength of polymeric material hence in addition; there is the mechanism of degradation. The information on various methods discussed in this review is planned to illustrate a best fit of methods for those who are interested in testing the degradation of polymers under different environmental conditions and selection of appropriate technique for specific combination of mixture of polymer and catalysts which helps to degrade the polymeric material.

Keywords- Polymers, PVA, PP, PVC, thermal degradation, biodegradation, photo degradation, gravimetric.

1 Introduction

Chemical engineering involves new technologies employing highly novel materials whose unusual response at the molecular level endows them with unique properties. In recent years, polymers have gained an increased presence in the chemical industry with their highly defined properties and uses. Natural and synthetic polymers find a vital and pervasive use in day to day life [1] due to their versatile traits [2]. Polymers are the molecule whose structure contains multiple repeating units, which results into a compound of high relative molecular mass and associated properties[3].Polymers are widely used in petrochemical industries, oil

* Corresponding author: kulkarniaishwarya66@gmail.com

& gas companies, steel & aluminium industries, plastic industries which involves bags, boxes etc. Also they find their adequacy in packaging of food items, agricultural produce, and electrical supplies and so on. Central Pollution Control Board stated that in 2017 plastic usage in India itself is 8,500,000 tonnes.

However, some of their most useful features, the chemical, physical and biological inertness, durability, low density, low cost has made their production efficient and easy, resulting in their accumulation in the environment if not recycled. The management of wastes generated by polymers is difficult whereas recycling can be feasible, environmentally and technically. For example waste generated by PE if sunken into yards, it causes adverse effect on soil which can lower the water drainage and agricultural produce will be reduced. Unfortunately, the accumulation of polymers has become a serious problem for all countries in the world. Therefore the need of hour is to use polymer that do not harm the environment or to find easy and energy efficient methods which helps to degrade them. Polymer degradation is the term used for process which depletes chemical and physical properties of polymer including their outward appearance. Various types of polymer degradation method including thermal, photo oxidative, mechanical, biological, by irradiation and chemical have been used traditionally. Ionic liquids have immersed as chemicals and solvents due to their unique properties and structural interactions for degradation. General mechanism for polymer degradation can be described briefly as follows.

1.2 Mechanism of polymer degradation

The major factors causing degradation are heat, mechanical energy, radiation and ozone leading to various types such as thermal, thermo oxidative, photo degradation, chemical, biological, hydrolytic and degradation by irradiation.

In a depolymerisation reaction, a reversal of polymerisation occurs which includes Initiation at chain ends, Depropogation, Termination.

1.2.1 Depolymerization

In depolymerisation process low activity free radical takes place at the end of polymer chain. The change in molecular chain takes lot of time as the monomers are lost by polymeric chain one by one as per the mechanism of chain reaction [4].

1.2.2 Random chain scission

Decrease in the molecular weight occurs rapidly as the backbone breaks down randomly which occurs at any site of backbone. Here two reactions namely disproportion termination and intermolecular chain transfer occurs resulting into formation of new free radicals having high reactivity.

1.2.3 Side –group elimination

Bonds connecting the backbone are much stronger than the bonds which connect the side groups to the backbone chain. When depolymerisation occurs side groups are removed from the backbone before any further breakage.

For depolymerisation of PVC:

$$CH_2\,(Cl)\,CHCH_2CH\,(Cl) \rightarrow CH{=}CH{-}CH{=}CH + 2HCl$$

Depolymerisation of polymers is favoured when initiation by main chain depletion is possible or intermediates formed during the degradation reaction are stable or the activation energy required for the degradation reaction is very less. Therefore,
Activation energy of depropogation = (activation energy of polymerization propagation) + (heat of polymerization ΔH_p)

2 Thermal degradation methods

Polymer degradation and recycling is intended to prevent the environmental pollution caused by their production as they are often disposed in landfills or water. The long chain backbone of the polymer begins to separate and react at relatively elevated temperature.

Polymers have their own mechanism of degradation due to heat, light and chemicals. A few polymers are perceptive to oxidation and UV radiation. Whereas some are receptive to hydrolysis, acids and alkalis. Generally degradation reactions due to heat or thermal energy occur by depolymerization (main chain reactions) and substituent reactions. Depolymerization reaction occurs by breaking of main chain polymer which gives intermediate stage product which is similar to the parent polymer but the monomer units are distinguishable. This is called as unzipping. In substituent reactions the substituent's chemical nature is changed where as the chain structure remains the same [5]. Whereas by chemical energy which involves two techniques namely hydrolytic depolymerization which is considered to be an effective technique for chemical recycling of polymers and recovery of monomers and another as microwave irradiation which shortens the time required for degradation[6].

Tamer Karayildrum carried out a degradation of Polyvinylchloride, Polyethylene by stepwise pyrolysis using polymer combination with presence of HCl [7]. With the help of dynamic measurements, kinetic data for isothermal thermal decomposition of Polyethylene was done by Bockhorn et.al. Which proposed that pyrolysis with step by step process is possible if the sample of single polymer can be separated at particular temperature [8,9]. This degradation experiment was conducted in a cycled shape reactor as this reactor acts as a heat transfer agent. The extruder was kept at 190°C and polymer mixtures with any one of the polymer and catalyst was fed to reactor. Temperature was maintained at 330°C. After the dechlorination, temperature was increased and then the mixture was held at 440°C for 60 min. Residue was collected from bottom. Then analysis was conducted like TGA and liquid analysis which showed that $CaCO_3$ was most suitable for HCl fixation which produced liquid product. And RM was the most effective for HCl fixation and reduction.

A study for LDPE waste by thermal degradation was done by Ali Karaduman and Emir H. Simsek [10]. They suggested that this reaction is pseudo first order reaction and the reaction was completed in30 min at 400°C and 425 °C. Different models were generated for this experiment.

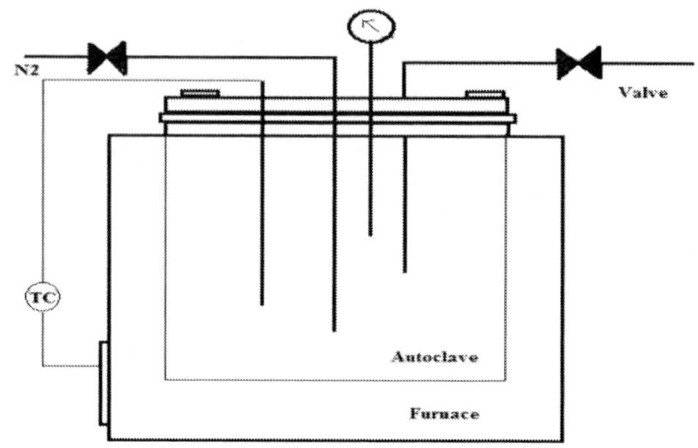

Fig. 1. Experimental set up.

Another method for thermal degradation was proposed by V.A.Kraikin et al [11]. They estimated that PTP, a heat resistant polyheteroarylene prone to carbonization and degradation of PTP begins at 480°C. This process took place by decomposition of phthalide and evolution of CO and CO_2. And CH_4 and H_2 and carbonization of polymer were at 500°C. Degradation done by CO_2 was about 12%. Correlation between chemical structure and experimental data of molecular chains of polymer was studied by O.V.Okatova and N.N.Ul'yanova in 2007[12]. Polymer Poly (napthylimides) was thermally degraded with 96%H_2SO_4 in PNI solutions which are observed to be stable and degradation occurred was 6.4-9.6%. R.W.J.Westerhout proposed a method which is performed at elevated temperature typically from 300°C to 1000°C [13], for this high pressure equipment and solvents with high boiling point are required. To solve this issue, pyrolysis of polymers or supercritical water has been used in some cases. These type of reactions are usually performed in such apparatus which with stand the high pressure condition. Similarly in another experimental procedure developed by Akio Kamimuna, a 6-nylon chip having a molecular weight of 22,000 was reacted with a solvent at 300°C for one hour. Same conditions were provided under vacuum. The monomerized material was collected and checked for purity by GC-MS analysis. Depolymerization did not happen. 43% yield was gained after ionic liquid showed some preference [14].

However, polymers can be degraded by using ionic liquids as they offer specificity required to break the bonds at comparatively low temperatures. The most general and efficient methodology that may fulfill this requirement is phase transfer catalysis [PTC], studied by G.R. Gupta et al. Various reviews have been published on both engineering and chemistry view points of phase transfer catalyst in the last few decades. Cryptands, solvents, Onium salts, crown ethers, and co-solvents have been widely used as PTC [15]. Considering compatibility and relatively extensive solubility of ionic liquids with many inorganic and organic functional polymers it can also be used for dissolution of wood which comprise of cellulose, lignin, hemicelluloses and extractives. The complete acetylation of wood is done by ionic liquid solutions. Traditional ionic liquids were used to prepare by the metathesis reaction. Ionic liquids were used as catalytic solvents in dimeraization of 2-methylpropene to octane. Similarly for the depolymerization of PET such as aminolysis, hydrolysis, and alcoholysis have been put forward which in the absent of catalyst were very slow. The catalyst used were titanium phosphate, metal acetates, metal oxide, and solid super acids, sulphates, carbonates which were difficult to separate from the depolymerized products and disturbs the product analysis. Therefore it necessitated use of ionic liquids as studied by Martin J.Earle and Kenneth R.Seddon [16].

The products of all the reaction are insoluble in the ionic liquid and can be separated by decantation, leaving the ionic liquid and catalyst behind; whence the catalyst and solvent can be reused [17]. Use of ionic liquid for depolymerization of PVC stabilizes it by preventing the liberation of HCl. While the PVC degrades rapidly therefore the solubilization and degradation depends upon structural characteristics of polymeric molecules [18]. Thermal degradation technique surely gives the desired output if correct parameters are chosen. But thermal degradation includes energy usage with very high demand. And for every industry it is not feasible to generate a degradation plant with high demand of energy and also it may involve evolution of gases.

3 Biodegradation methods

Four different mechanisms for study of biodegradation- charge formation, solubility, hydrolysis, enzymatic catalyzed degradation [19]. One of the major polymer, Polyethylene consumed is always thrown into environment. Until the olefins having high molecular weight were not found, polyethylene was assumed to be a non degradable by means of biodegradation. Vimala P.P. and Dr. Lea Mathew [20] in 2016 proposed a method which involves biodegradation of plastic. For this experiment they collected LDPE (18μ) and HDPE (19μ). Some sheets of polyethylene of specific thickness were exposed to UV treatment for some time. *Bacillus subtilis* which produces bio surfactant were also collected. Conical flasks were poured with bacterial species combined with polymer films, microbes and biosufactants. This was kept at room temperature till 30 days at 180rpm at about 32°C. A weight loss of 9.36% was observed in 18μ PE films. Formation of ketones, aldehydes, alcohols and carboxylic acids was observed. Weight loss after the UV treatment is listed as follows.

Table 1. Gravimetric analysis showing weight loss [20]

Polymer Type	Before U V (g)	After U V (g)	Weight Loss (g)
PE (18μ)	3.000	2.999	0.001
PE(41μ)	3.000	3.000	0.000

A similar approach was done in order to degrade polythene with the help of marine microorganisms mainly *Bacillus pumilus, Bacillus subtilis* and *Kocuriapalustris*. Polyethylene samples including bacterial strain which was incubated and another one which was non incubated. Based on properties like hydrobhobicity, FTIR analysis etc, most suitable bacteria was found for biodegradation. Properties showed the growth on surface of polymer with their metabolic activity giving rise to effective degradation and the most suitable bacteria found was *Bacillus subtilis*.

Another biodegradation methodology was put forward by K.S. Veethahavya, Rajath B.S. et al [21]. Biodegradation is the method in which the cost is very less as compared to any of the methods like thermal degradation or photocatalyatic degradation. The method proposed by them included degradation of low density polymer, polyethylene in liquid media. Starch content and polymer samples with different ratios were taken and a mixture containing aqueous media, ammonium nitrate, disodium hydrogen, and bacterial cells were subjected to a rotary shaker at room temperature. The results showed that with the proceeding number of days, starch content was also raised depending upon the combination used with polymer. Weight loss was observed from 1.53% up to 54.33% for pure low density polymer. For higher LDPE starch content, degradation achieved was higher and at lower levels, degradation was constant. CO_2 was also evaluated from the bioreactor. The level was also stepped up with time.

From this experiment it is stated that at higher starch content, degradation was higher which was done by microorganisms as observed from HPLC analysis. Also, when there was degradation the peaks were vanished.

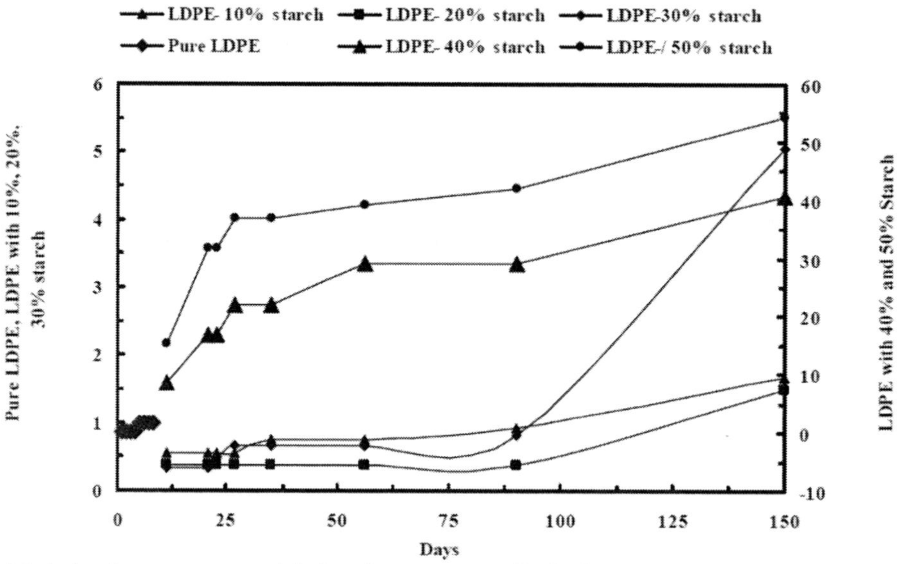

Fig. 3. Variation in percentage weight loss in aqueous media [21]

S. Schusser, M. Backer et al [22] suggested a technique for biodegradation using a field effect sensor base of which was semiconductor. This experiment involved a system which was monitored in a medium including *Rhizomucormiehei* which an enzyme lipase for degradation of poly (D,L-lactic acid). This polymer after spin coating gave a thickness of 500nm when solved with methyl ethyl ketone. Conductivity was maintained with KCl addition. Two different types of samples were prepared, one which was with lipase *Rhizomucormiehei* and another one with a buffer solution (pure) with pH 7. The results showed that polymer which was showing better degradation was suspected with buffer solution and the time required for this was about 12 days. For the mixture containing enzyme, sensor signal was reduced.

This same team proposed another system which was included with real time and also in-situ process degradation in 2015.It consisted of buffer solutions having different pH values,

Rhizomucormiehei and a medium of cell culture. This time the lipase showed more rate of degradation than buffer solutions with pH 11 and pH 7. But the buffer solution after addition of NaOH showed tremendous change in behaviour and degradation was favoured.

These experiments thus showed that the polymer degradation using sensor based system shows a great potntial indicating in situ and real time monitoring. Biodegradation methods are very econimic and they do not cause huge pollution as they involve treatment of polymeric mateials with micro-organisms. Hence this method is environmentally effective as there are no harmful effects. However it is bit difficult to get desired output as the microorganisums tend to affect slowly on the degradation process.

4 Photo degaradation methods

Photochemical degradation is another type of degradation which is used to degrade the polymer chains with UV rays and combinations used with different chemicals by means of thermal degradation. In the field of paints, art work poly vinyl acetate is generally used.For this experiment done by Shuya Wei, Valentina Pintus et al [23] different pigments were used. The samples to be treated were exposed to UV rays with H2 filter supported with a specific wavelengh range and ample radiation with temperature at 48.8˚C. For the analysis FTIR-ATR was used. It showed that the range obtained was abot 370 and 4000/cm. Gas chromatography/ mass spectroscopy was used and different pyrolysis temperatures were attended. With increased amount of detactable acetic acid aged and uaged poly vinyl acetate paint samples were found and differentiated. The aged samples can be used for degradation as they can be used as effective indicators.

Another major polymer which is commonly found in textile waste water is poly vinyl alchohol(PVA) which is the major reason for incresed organic carbon content. For the degradation of this polymer, Diana Hamad, Ramdhane Dhib et al. [24] experimented photochemical degradation with contineous UV/H_2O_2 process. Three aqueous solutions of PVA was prepared and mixed with a flask having distilled water. The solution was poured into a photo reactor. Hydrogen peroxide was added in intervals. Then sample was taken out after some specific time interval and molecular weight was tested along with total carbon content. Simultaneous exposure of feed to continuous flow of UV and H_2O_2 resulted in significant weight loss of polymer and total carbon content. The most significant factor was interaction between PVA and hydrogen peroxide affecting PVA degradation within the operating range of condition.

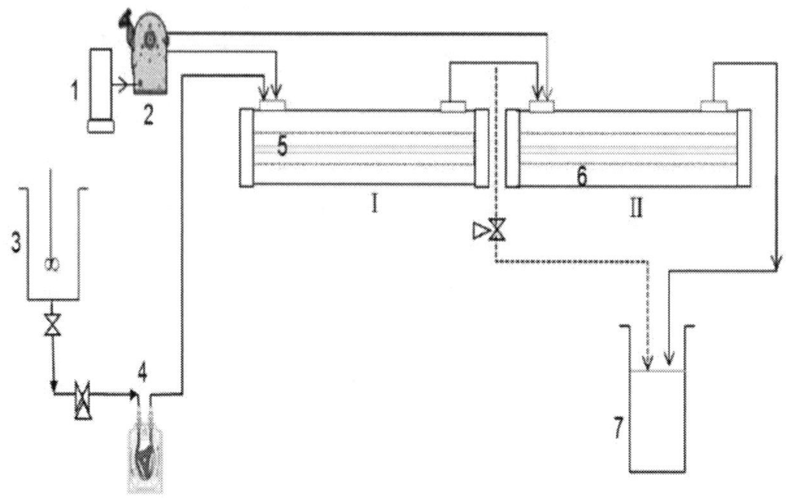

Fig. 5. Experimental setup for continuousUV/H_2O_2 process [24].

Saba Sadaqat Ali, Ishtiaq A. Quazi et al [25] worked on degradation of LDPE (low density polymer) by using titania nanotubes with the help of photo catalytic degradation. For this degradation, PE was used which now a day's very hard to dispose and this is becoming a major issue. Photocatalyatic degradation in visible light is another effective method to use now days. Photocatalyst used here is titania nanotubes. The experimental setup included exposure of LDPE and composite films to UV light for 15 days and at room temperature. Again the experiment was carried out for 45 days. Each sample was weighted at regular time interval.

This study demonstrated photocatalytic degradation process of BG sensitized TNT composites PE films under visible light exposure. Dye sensitization of nanotubes has increased the degradation rate of LDPE films under visible light and those films can degrade under sunlight with good degradation rate hence it is more economical. Volatile organic product generation was found by FTIR, SEM and tensile strength procedures. This new LDPE-TNT material is environment friendly polymer material which could be degraded when exposed to the outdoor environment.

Telmo Ojeda, Ana Freitas et al. Studied the degradation of linear polyfins under natural weathering [26]. High density polyethylene, isotactic polypropylene and low density linear polythene having antioxidant additives at lower or zero levels were excluded and moulded as films by blowing. Mechanical properties and structural properties were studied by exposing these samples to natural weather conditions for one year. Weight loss of HDPE and LDPE was significant but the results were seen very slowly in a few months. Presence of pro oxidant additive and frequency of tertiary carbon atoms in polymer chain were the most influencing factors. Decline in degree of crystallinity after thermal destruction during exposure to natural weathering resulted by PP whereas increase in crystallinity was shown by HDPE/LLDPE oxo-biodegradable blend.

Behaviour of three linear polymer samples poly methyl methacrylate (PMMA) having different molecular weights in a continuous ink jet printer is investigated by Joseph S.R. Wheeler, Stuart W. Reynolds et al [27]. The experimental setup is shown below in the figure. In this experiment the main source of polymer degradation is the continuous recycling of ink through pump resulting in mono chemical polymer degradation. Random scission achieves the molecular weight degradation and it found at lower shear rates. The delivery of high molecular weight polymers and materials sensitive to mechano-chemical degradation has the significance in total ink jet setup.

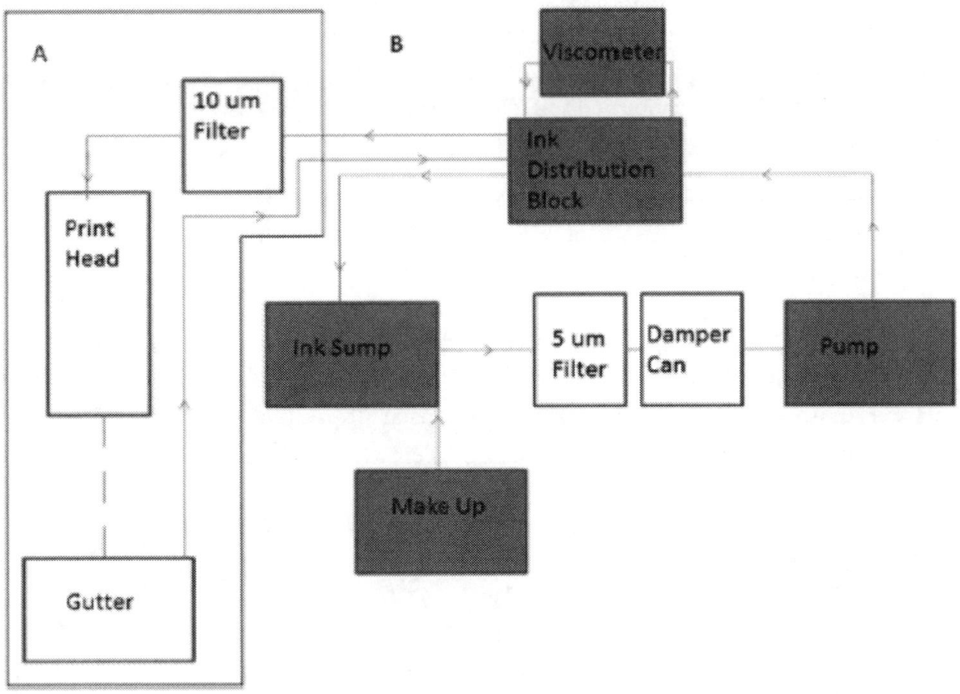

Fig. 6. Diagrammatic representation of ink jet system [27].

Photo degradation of polymers is one of the primary means by which the polymeric bonds can be broken. Generally in this method polymers tend to absorb energy radiations with high amount and it is not economical. But with certain limitations and proper selection of parameters, this method is useful.

5 Conclusions

In this study, different degradation models were proposed for degradation of polymers. Thermal degradation took place at very high temperatures and the time required for process was comparatively less than any other processes. The average temperature required for degradation with pyrolysis was around 400°C. The degradation was done but the cost required for this and the heat economy was not suitable for all industries. Another method was to degrade polymers with microorganisms like *Bacillus subtilis* which was the most promising bacteria in order to degrade the polymers economically. But all the polymers can be degraded by this method as the environmental conditions are not suitable and also the time span is also large. Polymer degradation with photocatalytic and photo chemical degradation was studied which produced better degradation quality by using natural weathering and ink jet printers.The information will be helpful in control strategies in the extension of polymers' service life and formulation to recover it for better use.

6. References

1. N.G. McCrum, C.P. Buckley, C.B. Bucknall, Principles of polymer engineering. Oxford; New York: Oxford University Press. p.1.I SBN 0-19-856526-7(1997).
2. Paul C Painter, Michael M Coleman, Fundamentals of polymer science: An introductory text. Lancaster, Pa: Technomic Pub. Co. p.1.ISBN 1-56676-559-5. (1997).
3. http://goldbook.Iupac.org/Mo3667.html
4. http://nvl.nist.gov/pub/nistpubs/sp958-lide/344-346.pdf
5. A. Dharaskar Swapnil, Res.j.chem.sci. 2(8).80-85,August(2012)
6. Ahmad Adlie and DzulkeflyKuandAbdullah,"Ionic liquids: Preparations and limitations", makara,sains,14, no.2, 101-106,(2010).
7. Tamer Karayildirim, JaleYanik, MithatYuksel, Mehmet Sagkam, and Michael Haussmann, J.Polym.Environ.13, No.4, DOI: 10.1007/s10924-005-5531-2 (2005).
8. H. Backhorn, A.Hornung and U. Hornung, J.Anal.Appl.Pyrol.46,1-13
9. H. Backhorn, A.Hornung, U. Hornung,S. TeepeJ.Weichmann, Combust.sci.Technol 116/117, 129-151.
10. Ali Karaduman and Emir H. Simsek, , J.Polym.Environ.9, No.2,(2001).
11. V. A. Kraikin, Z.N.Musina, E.G.Galkin, S.I. Kuznetsov,A.E.Egorov, and S.N.Salzkin, J.Polym.Sci. A, 48,No.12,pp 1287-1303.(2016)
12. O.V.Okatova,N.N. Ul'yanova,E.G.Bilycheva,L.B.Elshia,A.L.Rusanov, and P.N. Lavrenko, J.Polym.Sci. A,49,No.6,pp 716-721.(2007)
13. R.W.J.Westerhout,J.Waanders,J.A.M.Kuipers,W.P.vanSwaaij.Ind.Eng.Chem.Res.37,2 993-2300(1998).
14. AkioKamimuna, ShigehiroYamamoto,Organic Letters.9,No.13,2553-2535. (2007).
15. G.R.Gupta,G.R.Choudhary,PreetiTomar,G.P.Waghulade,Kesharsingh J. Patil, "Synthesis of ionic liquids and derivatives using molten salt reaction media" Asian J.Chem.(2012)
16. Martyn J.Earle and Kenneth R.Seddon, Pure Appli. Chem.,72,No. 7,pp.1391-1398,(2000).
17. Q.F.Yue, C.X.Wang, L.N.Zhang, Y.Ni.Y.X.Jin, Polym Degrad Stab 96 399-403. (2011).
18. Preeti A.Tomar, S.M.Yadav and Gurav R.Gupta, "The thermal gravimetric studies for polymer samples of PVA and PVC obtained with treatment with ionic liquid [bmim]Br-" ,Polym. Bull, ISSN 0170-0839.
19. Gilding D.K., Biodegradable polymers. In:Williams, D.F.(Ed), Biocompatibility of clinic implant materials, 2 .CRC Prss, Boca Raton, FL.chapter 9.
20. P.P.Vimala and Lea Mathew, Proc. Technol.,24.232-239. (2016).

21. K.S. Veethahavya, Rajath B.S., Sabike Noobia, B. Manoj Kumar, Procedia Environ Sci,35.709-713. (2016)
22. S. Schusser, M.Krischer,D.G.M.Molin,N.M.S.van den Akker, M.Backer, A. Poghossian, M.J.Schoning, Procedia Eng. 120, 948-951. (2015).
23. ShuyabWei, Valentina Pintus, Manfred Schreiner, Journal of Analytical and Applied Pyrolysis 97,158-163. (2012)
24. Dina Hamad, RamdhaneDhib,MehrabMehrwar, J Polym Environ,24:72-83,(2016).
25. SabaSadquat Ali, Ishiaq A. Quazi, Muhammad Arshad, Zahiruddin Khan, Thomas C. Voice,Ch. Tahir Mehmood, ENMM, 5, page no.44-53,(2016).
26. Telmao Ojeda, Ana Fretias, Katia Birk, EmileneDalmolin, Rodrigo Jacques, Fatima Bento, Flavio Camargo, Polym Degrad Stab,96,page no.703-707, (2011).
27. Joseph S.R.Wheeler, Stuart W. Reynolds, Steven Lancaster, Veronica Sanchez Romanguera, Stephen . Yeates, Polym Degrad Stab.105,page no.116-121, (2015).

Influence of Sawdust Bio-filler on the Tensile, Flexural, and Impact Properties of Mangifera Indica Leaf Stalk Fibre Reinforced Polyester Composites

Srinivas Shenoy Heckadka[1], *Suhas Yeshwant Nayak*[1*], *Gouthaman P. V.*[1], *Abhishek Talwar*[1], *Ravishankar V. A.*[1], *Linto George Thomas*[1] and *Ankur Mathur*[1]

[1]*Department of Mechanical and Manufacturing Engineering, Manipal Institute of Technology, Manipal Academy of Higher Education, Manipal, India*

Abstract. The need to have biodegradable composites is aloft in today's market as they are environment friendly and are also easy to fabricate. In this study, mangifera indica leaf stalk fibres were used as reinforcement along with saw dust as bio-filler material. Unsaturated isophthalic polyester resin was used as the matrix. The fibres were treated with 6 % vol. NaOH and neutralized with 3 % vol. of dilute HCl. Treatment of sawdust fillers was done by using 2% vol. NaOH solution. Hand layup method and compression moulding technique was used to fabricate the composite laminates. Specimens for evaluating the mechanical properties were prepared by using water jet machining. The results indicated an increase in tensile, flexural and impact strength of composites with addition of sawdust upto 3%. Further addition of the bio-filler resulted in decrease of mechanical properties.

Keywords: Mangifera indica leaf stalk fibre, Sawdust, Polyester, Handlayup, Compression Moulding, Mechanical Properties

1 Introduction

In the present scenario, wide varieties of composite products are manufactured from thermosetting polymers reinforced with synthetic fibres which are expensive, have health hazards, and are not nature friendly [1, 2]. Natural fibres are thus gaining lot of attention and extensive research is going on, as these fibres offer better strength and stiffness, low density, reduced embodied energy and possesses excellent biodegradability [3]. The abundant availability of natural fibres and flexibility in bringing about improvements in their properties by surface treatments make these fibres an excellent alternative to traditionally used synthetic reinforcements. Such composites find applications in automobile and construction industry [2, 4, 5]. Roe and Ansell [6] analysed the characteristics of polyester composites reinforced with jute fibre. The results obtained indicate that, the specific modulus shown by jute fibre is far better to the glass fibre along with the modulus per cost. Also, the specific strength per unit cost exhibited by jute fibre

[*] Corresponding author: synayak@gmail.com

are comparable to that shown by the glass fibre. Rachchh et al. [7] examined the properties of rattan fibre polyester composites. Fibre addition improved the flexural and tensile properties up to 12.5% weight percentage after which a drastic drop in the mechanical properties was observed due to insufficient resin content. They observed reduction in density with addition of fibres indicating that such composites can be used for low strength applications.

For a proper matrix-fibre adhesion, surface of natural fibres should be chemically modified. Consequently, the hydrophilic nature of the natural fibre reduces due to partial elimination of hemicellulose content. Hill and Khalil [8] studied the effect of chemical treatment and mechanical characteristics of oil palm and coir fibre reinforced polyester composites. Acetylation and usage of silane and titanate coupling agents were considered for the study. The tensile properties increased due to stronger interfacial bonding and better wetting of the fibres. A minor improvement in impact strength was also noticed with chemically modified fibres. Naguib et al. [9] studied the physical properties of green bagasse polyester composite. Fibre treatment was performed with sodium hydroxide solution and then diluted with sulfuric acid. The authors reported an improvement in flexural strength and Young's modulus of the treated fibres. The study suggested that bagasse polyester composite can be used for various industrial applications that require bearing higher loads under mild operating conditions. Pothan et al. [10] studied the mechanical characteristics of polyester composites reinforced with short banana fibres. Silane treated banana fibres exhibited better tensile strength. The morphological images clearly showed a good adhesion between polyester and banana fibre. The impact strength showed an increasing trend with fibre loading. Kini et al. [11] investigated the effect of alkali treatment on borassus and tamarind fruit fibres. They observed that chemical treatment removes the surface impurities of the raw fibre and makes the fibre suitable for the better bonding with the matrix.

Dubey and Angnihotri [12] analysed the usage of midrib of coconut palm leaves (MCL) as reinforcement in polyester composites. Flexural and impact properties showed a significant increase with the addition of fibre reinforcement in the composite. They observed that the MCL/polyester composite was found to be superior to most of the polyester composites with regards to flexural and impact properties. Shanmugam and Thiruchitrambalam [13] examined the influence of alkali treatment on the mechanical properties (flexural and tensile strength) of palmyra palm leaf stalk fibre (PPLSF)/jute fibre polyester hybrid composites. The study confirmed that the fibre surface roughness increases with alkali treatment thereby providing more surface area leading to better fibre matrix adhesion.

To enhance the thermal and mechanical properties, fillers are added to polymer matrices. Organic fillers have added benefits over inorganic fillers such as less weight and reduced wear of the processing machinery. Marcovich et al. [14] carried out a study on composites made from sawdust and unsaturated polyester. The bending modulus showed an increasing trend up to a certain extent with the addition of filler. They concluded that saw dust composites showed satisfactorily comparable specific mechanical properties with respect to inorganic filler composites. Suarez et al. [15] analysed the tensile properties of polypropylene and malleated polypropylene (PP) coated sawdust composites. The results obtained indicated that the tensile strength of PP/MAPP sawdust composites was much superior to PP composite. Also, the micrographs revealed that the fibre pulled out is coated with the matrix and filler material indicating an effective interaction between the filler and matrix.

Many studies have been conducted to explore the use of sawdust as fillers in thermoset composites, but its usage as a filler with mangifera indica leaf stalk fibre as reinforcement in polyester has not been explored. Hence the present work aims at studying the mechanical properties of such composites finding applications in automotive and household interiors.

2 Experimental Details

2.1 Materials

Mangifera Indica leaf stalk fibres and saw dust filler were used as the reinforcements. Mangifera Indica leaves were collected from plantations within the campus in Manipal. Sawdust bio-filler was obtained from a basic workshop at our institute. Unsaturated polyester resin, its accelerator cobalt naphthalene and catalyst methyl ethyl ketone peroxide were supplied by M/s Sri Mookambika Poly Products, Udupi, Karnataka. Chemicals such as NaOH and HCl were procured from Nice Chemicals Pvt. Ltd, Kerala. Hand layup technique with compression moulding was used for fabricating the composites. The matrix properties are shown in Table 1.

Table 1. Properties of Unsaturated isophthalic polyester resin [16].

Properties	Remarks
Appearance	Pale Yellow
Density (g/cm^3)	1.15
Tensile strength (MPa)	17-21
Flexural strength (MPa)	39- 42
Viscosity at 25°C (cps)	650

2.2 Extraction and chemical treatment of the fibres

Fibres were extracted from the leaf stalk of the Mangifera Indica tree. It was sun dried for 2 days so that the sticky material on the outer surfaces loosens up. The dried leaf stalk fibres were then washed thoroughly with distilled water to remove the sticky material and greasy layers present on its surface. The fibres were later dried at room temperature for 120 h. To ensure better chemical bonding between the fibre and matrix, the fibres were subjected to mercerization treatment. The dried fibres were treated with 6 % vol. NaOH solution for 3 h at room temperature and neutralized with 3 % vol. dilute HCl. The chemically modified fibres were placed in a hot air oven at 50°c for 4 h to dry it thoroughly. The leaf stalk and fibers being treated are shown in Fig. 1.

2.3 Preparation of sawdust bio-filler

Saw dust is a by-product formed because of machining operations carried out on wood. In the present study saw dust of teak wood was used. Approximately two kilograms of sawdust bio-filler was collected which was then sieved to size of 200 microns [17]. Sieved sawdust fillers were then treated with 2% vol. NaOH solution at room temperature for 1 h, to enhance the interfacial bonding between the polyester matrices. Post chemical treatment, the fillers were rinsed with distilled water and dried in a hot air oven for 8 h at 50°c. The fillers were placed in a desiccator to prevent moisture absorption.

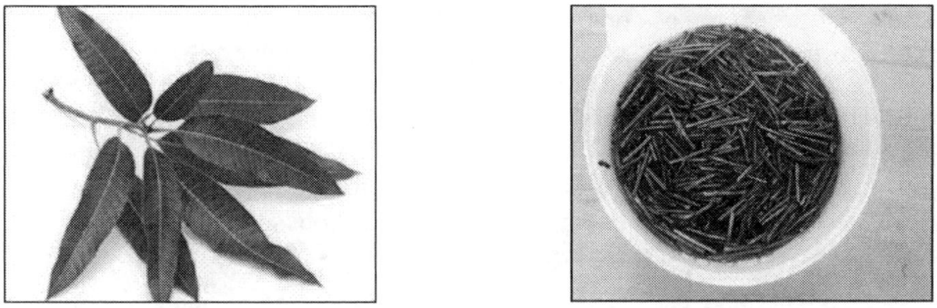

(a) (b)

Fig. 1. Mangifera Indica (a) leaf stalk (b) leaf stalk fibres undergoing chemical treatment

2.4 Preparation of composite

Chemically treated mangifera indica leaf fibre was used as reinforcement and unsaturated polyester resin used as the matrix with sawdust as filler. Composites were fabricated using hand layup followed with compression moulding technique. To ensure good quality of surface and for easy demoulding, white wax was applied over the mould surface. Fibre content of 30% by weight was kept constant while the filler content was varied from 3% to 9% by weight. A peel ply film is placed on the top surface of the mould to achieve a smooth surface before applying compressive load. Post curing was done at 50° C for duration of 3 h. The size of all the composite panels was 300 mm × 300 mm with a thickness of 5 mm. Fig. 2. shows the cured panels.

(a) (b)

Fig. 2. Cured Composite panels with (a) 3% filler (b) 6% filler

 https://doi.org/10.1051/matecconf/201814402024

2.5 Testing of mechanical properties of composites

The mechanical properties of composites like tensile, flexural and impact strengths were evaluated. Tensile and flexural tests were performed on a Zwick Roell Universal Testing Machine at a cross head speed of 2 mm/min and 1 mm/min respectively, while impact testing was done on a Zwick Roell Impact tester where the specimen was subjected to an impact energy of 7.5 J. Specimens for tensile, flexural and impact tests were prepared using water jet cutting. Details of standards and specimen dimensions are shown in Table 2. Five specimens were considered for each of the above mentioned tests and average value was arrived at. Test setups are shown in Fig. 3.

Table 2. Standards and specimen dimensions [18-20].

Properties	Standard Used	Dimensions
Tensile Strength	ASTM D3039	250 mm × 25 mm
Flexural Strength	ASTM D7264	Overall Length: 120 mm; Span length: 100 mm
Impact Strength	ISO 179-1/1eu	Length 80 mm; Span length 64 mm

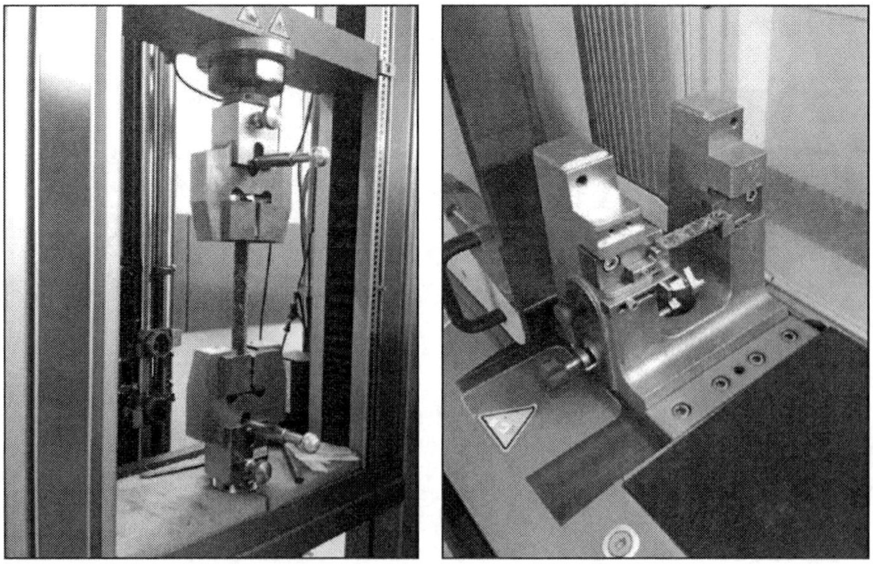

(a) (b)

(c)

Fig. 3. Test setups (a) Tensile (b) Impact (c) Flexural

3 Results and Discussions

Fig. 4 shows variation in tensile strength with the addition of sawdust to the composites. Treatment of fibres and fillers make them more hydrophobic resulting in improved bonding between the various constituents. When compared to neat polymer, mangifera indica leaf stalk fibre composite with no filler showed an increase of about 31% in its tensile strength. Highest tensile strength of 28 MPa was obtained for composites with 3% filler by weight, which is an increase of about 12% over composites without any filler. Further addition of saw dust to the composites resulted in reduction of tensile strength.

 https://doi.org/10.1051/matecconf/201814402024

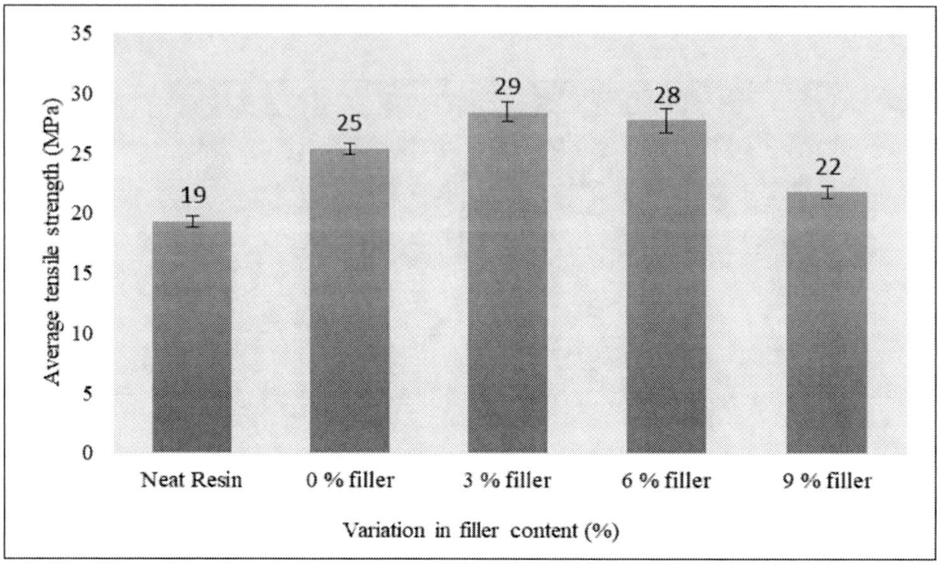

Fig. 4. Variation in tensile strength

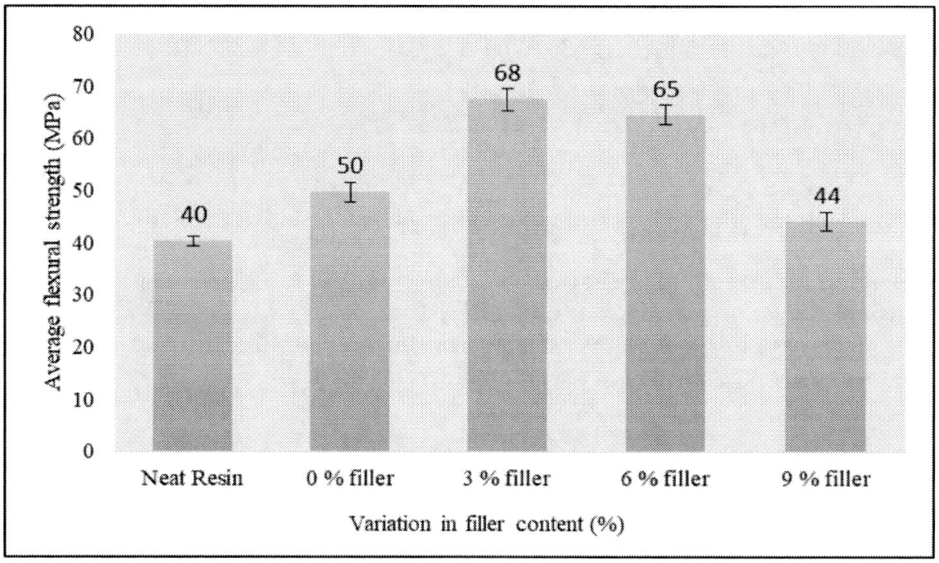

Fig. 5. Variation in flexural strength

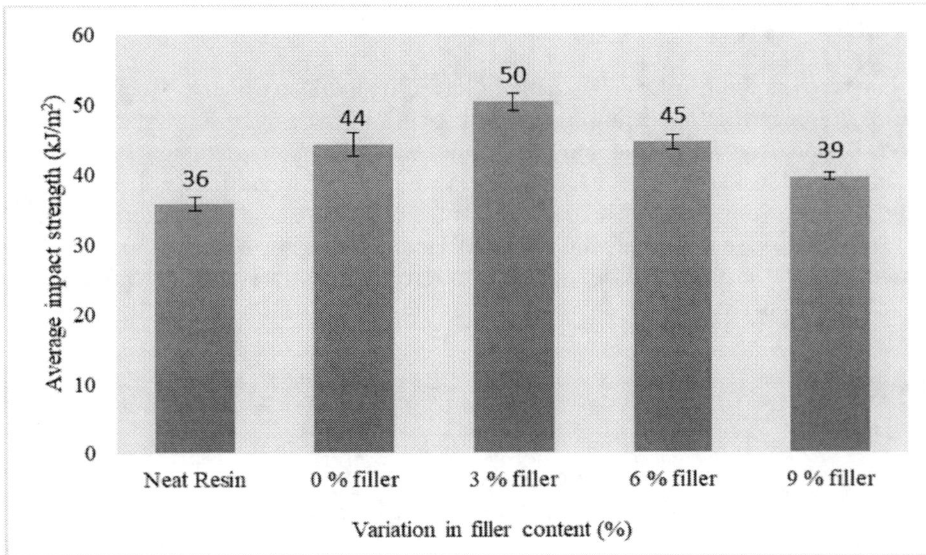

Fig. 6. Variation in impact strength

Flexural strength of the composites is presented in the Fig. 5. Addition of sawdust increased the flexural strength of the composites by about 68% when compared to neat polymer where the strength was observed to be 68 MPa which was the highest among all composites. Increase in saw dust filler to 6% and 9% resulted in lower flexural strength and the lowest strength of 44 MPa was obtained in composites with 9% saw dust filler.

The variation of impact strength is presented in Fig. 6. As seen in case of tensile and flexural tests, presence of sawdust resulted in improved energy absorption when compared to composites with only mangifera indica leaf stalk fibres as reinforcement. Increased resistance to impact loading is mainly because of the resistance provided by the fillers to the propagation of cracks [21]. Highest impact resistance of 50 kJ/m^2 is seen in composites with 3% weight filler. But with increase in filler content further reduced the ability of the composite to absorb impact load. Reduction in mechanical properties of composites beyond 3% filler content is due to reduced resin content which results in ineffective stress transfer between the various phases in composite material [7].

4 Conclusions

Composites were fabricated by varying saw dust filler content in mangifera indica leaf stalk fibre reinforced polyester matrix. Mechanical properties such as tensile, flexural and impact strengths were evaluated and compared. Significant inferences are, weight fraction of sawdust fillers have greater influence on the mechanical properties of the composites. Composites with 3% saw dust filler content showed improved mechanical properties. Increasing the filler content beyond 3% led to decrease in strength of the composites. Decrease in strength with increase in filler content is attributed to ineffective load transfer between the matrix and reinforcements. Results indicate that saw dust has a potential as filler in polymer based composites.

Acknowledgements

The authors are thankful to Dr. K Jagannath, Head of the Department, Mechanical and Manufacturing Engineering for permitting us to make use of the Advanced Material Testing and Research Laboratory. The authors are also indebted to Dr. Satish Shenoy B, Professor and Head, and Dr. Dayanand Pai, Professor, Department of Aeronautical and Automobile Engineering for allowing us to use their Advanced Composite and Material Testing Laboratory. The authors thank Manipal College of Dental Sciences, Manipal for permitting us to use their material testing facilities. The authors would also like to thank M/s. Konkan Speciality Polyproducts Pvt. Ltd., Mangalore for providing us their testing facility.

References

1. B. V. Ramnath, S. J. Kokan, R. N. Raja, R. Sathyanarayanan, C. Elanchezhian, A. R. Prasad, and V. M. Manickavasagam, Evaluation of mechanical properties of abaca–jute–glass fibre reinforced epoxy composite. Materials & Design, **51**, 357-366 (2013)
2. S.C. Venkateshappa, B. Beerhall, M. G. Kenchappa, and R. P. G. Ranganagowda, Flexural behaviour of areca fibres composites. BioResources, **5**(3), 1846-1858 (2010)
3. S. Mukhopadhyay, R. Fangueiro, Y. Arpaç, and Ü. Şentürk, Banana Fibres--Variability and Fracture Behaviour. Journal of Engineered Fabrics & Fibres (JEFF), **3**(2) (2008)
4. K. S. Kumar, I. Siva, N. Rajini, P. Jeyaraj, and J. W. Jappes, Tensile, impact, and vibration properties of coconut sheath/sisal hybrid composites: Effect of stacking sequence. Journal of Reinforced Plastics and Composites, **33**(19), 1802-1812(2014).
5. A. Atiqah, M. Jawaid, M. R. Ishak, and S. M. Sapuan, Effect of Alkali and Silane Treatments on Mechanical and Interfacial Bonding Strength of Sugar Palm Fibres with Thermoplastic Polyurethane. Journal of Natural Fibres, 1-11 (2017)
6. P. J. Roe, and M. P. Ansell, Jute-reinforced polyester composites. Journal of Materials Science, **20**(11), 4015-4020 (1985)
7. N. V. Rachchh, P. S. Ujeniya, and R. K. Misra, Mechanical characterization of rattan fibre polyester composite. Procedia Materials Science, **6**, 1396-1404 (2014)
8. C. A. S. Hill, and H. P. S. Abdul Khalil, Effect of fibre treatments on mechanical properties of coir or oil palm fibre reinforced polyester composites. Journal of Applied Polymer Science, **78**(9), 1685-1697 (2000)
9. H. M. Naguib, U. F. Kandil, A. I. Hashem, and Y. M. Boghdadi, Effect of fibre loading on the mechanical and physical properties of "green" bagasse–polyester composite. Journal of Radiation Research and Applied Sciences, **8**(4), 544-548 (2015)
10. L. A. Pothan, S. Thomas, and N. R. Neelakantan, Short banana fibre reinforced polyester composites: mechanical, failure and aging characteristics. Journal of Reinforced Plastics and Composites, **16**(8), 744-765 (1997)
11. U. A. Kini, S. Y. Nayak, S. S. Heckadka, L. G. Thomas, S. P. Adarsh, and S. Gupta, Borassus and Tamarind Fruit Fibres as Reinforcement in Cashew Nut Shell Liquid-Epoxy Composites. Journal of Natural Fibres, 1-15 (2017)
12. N. Dubey, and G. Agnihotri, Flexural and Impact Properties of Midrib of Coconut Palm Leaves Reinforced Polyester. Materials Today: Proceedings, **4**(2), 3422-3430 (2017).
13. D. Shanmugam, and M. Thiruchitrambalam, Influence of alkali treatment and layering pattern on the tensile and flexural properties of Palmyra palm leaf stalk fibre (PPLSF)/jute fibre polyester hybrid composites. Composite Interfaces, **21**(1), 3-12 (2014)
14. N. E. Marcovich, M. M. Reboredo, and M. I. Aranguren, Composites from sawdust and unsaturated polyester. Journal of applied polymer science, **61**(1), 119-124 (1996).

15. J. C. M. Suarez, F. M. Coutinho, and T. H. Sydenstricker, SEM studies of tensile fracture surfaces of polypropylene—sawdust composites. Polymer testing, **22**(7), 819-824 (2003)
16. P. A. Sreekumar, S. P. Thomas, J. M. Saiter, K. Joseph, G. Unnikrishnan, and S. Thomas, Effect of fiber surface modification on the mechanical and water absorption characteristics of sisal/polyester composites fabricated by resin transfer moulding. Composites Part A: Applied Science and Manufacturing, **40**(11), 1777-1784 (2009)
17. N. Sombatsompop, C. Yotinwattanakumtorn, and C. Thongpin, Influence of type and concentration of maleic anhydride grafted polypropylene and impact modifiers on mechanical properties of PP/wood sawdust composites. Journal of Applied Polymer Science, **97**(2), 475-484 (2005)
18. ASTM D3039 / D3039M-14, Standard Test Method for Tensile Properties of Polymer Matrix Composite Materials, ASTM International, West Conshohocken, PA, (2014) www.astm.org
19. ASTM D7264 / D7264M-15, Standard Test Method for Flexural Properties of Polymer Matrix Composite Materials, ASTM International, West Conshohocken, PA, (2015) www.astm.org
20. ISO 179-1: 2010, Plastics -- Determination of Charpy impact properties -- Part 1: Non-instrumented impact test, International Organization for Standardization, Vernier, Geneva, Switzerland, (2010)
21. S. Y. Nayak, S. S. Heckadka, U. A. Kini, L. G. Thomas, and I. Gupta, Pistachio shell flakes and flax fibres as reinforcements in polyester based composites, International Conference on Engineering and Information Technology, 17-24 (2017)

Tensile and Flexural Properties of Chopped Strand E-glass Fibre Mat Reinforced CNSL-Epoxy Composites.

Suhas Yeshwant Nayak[1], *Srinivas Shenoy Heckadka*[1*], *Linto George Thomas*[1], and *Anil Baby*[1]

[1]*Department of Mechanical and Manufacturing Engineering, Manipal Institute of Technology, Manipal Academy of Higher Education, Manipal, India*

Abstract. Glass fibres have the principal advantages such as high tensile strength, high chemical resistance, low cost, and excellent insulating properties which makes them an important constituent in fibre reinforced plastic and composite industry. In this study, E-glass fibre in the form of Chopped Strand Mat (CSM) with different weight fractions such as 15%, 30%, and 45% were used as reinforcement in CNSL-epoxy resin composites. Fabrication of the composites was done by hand layup technique. Micro-hardness, tensile and flexural properties were investigated for all the composite panels of different compositions. The results clearly indicated an improvement in micro hardness, tensile and flexural properties with the increase in fibre content.

Keywords: E-glass fibre, Cashew nut liquid shell resin, Hand layup, Tensile strength and Flexural strength

1 Introduction

Glass fibre composites are being used in manufacture of numerous parts/products such as automobile parts, construction equipment parts, sports goods like surfboard, tennis rackets, fishing rods, golf clubs, archery bows and many other household applications due to reasonably high specific strength, tensile strength and hardness. Also, glass fibres are less expensive; possess good resistance to chemical attacks and insulating properties [1-3]. Utilization of glass fibre with appropriate orientation and proportion can yield composites with strength which are equivalent to steel, superior to aluminum in stiffness and one fourth of the steel in terms of specific gravity [4]. Mechanical and tribological properties of the composites can be enhanced by the usage of randomly oriented chopped fibre, chopped strand mat and woven mats [5]. Tungjitpornkull et al. [6] carried out the study on chopped strand E-glass fibre reinforced wood/PVC composites. They observed that tensile, flexural moduli and impact strength increased with the increasing content of the glass fibre due to its high stiffness. Gupta and Siddhartha [7] did a comparative study between bi-directional and chopped e-glass fibre epoxy reinforced composite materials. Bi-directional e-glass fibres exhibited better flexural and tensile modulus, wear characteristics and impact

[*]Corresponding author: srinivas.shenoy.h@gmail.com

properties due to improved interface strength and superior hardness when compared to randomly orient chopped fibre composites. Bhaskar and Kolla [8] evaluated the properties of composites made of chopped strand mat and woven roving E-glass fibre. The results showed that chopped strand mat composites have better mechanical properties than woven roving E-glass fibre composite. Khan et al. [9] investigated the mechanical and interfacial properties of jute and E-glass fibre reinforced composites. The studies revealed that E-glass fibre composites have twice the mechanical strength when compared to jute fibre reinforced composites.

Thermosetting polymers are an excellent substitute for most of the conventional materials. Epoxy resins are one of the extensively used thermosetting resins. It is mostly used in the form of coatings and structural adhesives for many engineering applications in addition to its usage as matrices for fibre reinforced composites due to its superior thermal and mechanical properties, excellent corrosion and chemical resistance [10]. Hameed et al. [11] conducted a study on the utilisation of the modified epoxy resin as matrix in glass fibre composites. The results indicated that the thermoplastic modification is an excellent way to enhance the mechanical, thermal and dynamic properties of composites. Composites fabricated using modified matrix indicated better storage modulus, increased glass transition temperature and enhanced thermal stability. Muhammad and Ahmad [12] examined the mechanical and thermal properties of glass fibre reinforced epoxy composite with matrix modification using liquid epoxidized natural rubber. Results indicated an increase in mechanical properties due to matrix modification.

Cashew nut shell liquid (CNSL) is a heavy liquid with a lingering scent, produced from the shell of Anacardium occidentale (cashew nut tree). The raw CNSL contains 5% of cardonal, 18% cardol and 70% anacardic acid and remaining polymeric materials. This raw CNSL is processed by heating to get modified CNSL. Composition of modified CNSL is 10% cardol, 52% cardanol, 30% polymeric material [13, 14]. Resins which are developed from cashew nut liquid are hydrophobic in nature, due to which they have better resistance over water, acids, and alkalis. These resins also possess superior properties against other synthetic or conventional resins such as better friction resistant properties, corrosion resistance, are an effective thermosetting binder, and provide protection against heat and fire. CNSL resin modified with formaldehyde has found applications as automotive disk-brake pads, preparation of medium density particulate board, and composite wood [15]. Kim [16] explored the potential of CNSL as a green adhesive. Cashew nut shell liquid formaldehyde resin (CF) and Polyvinyl acetate resin (PVAc) was used. Better bonding over commercially available tannin adhesive, reduced formaldehyde and VOC emissions were reported. Bisanda and Ansell [13] analysed the thermal properties of sisal reinforced cashew nut shell liquid composite. CNSL polymerised with formaldehyde and sodium hydroxide as a catalyst was used to produce resol type polymer matrix for the sisal fibres. Priya et al. [17] studied the tensile behavior and chemical resistance of chopped glass strand mat reinforced epoxy composites toughened with epoxy phenol CNSL. Authors observed improved tensile, interfacial bonding and chemical resistance over regular epoxy composites.

From the literature survey conducted it is evident that CNSL has a potential in modifying traditional thermoset resins to obtain improved mechanical, thermal and chemical properties. CNSL is readily available as a by-product during processing of cashew nuts and hence is less expensive. Use of CNSL as a resin either as a whole or as a part in some other resin will help in bringing down the fabrication cost. Thus, the present study aims at determining the micro hardness, tensile and flexural properties of chopped E-glass CSM reinforced CNSL-epoxy matrix for engineering applications.

2 Experimental details

2.1 Materials

E-glass CSM with areal density of 290 gsm was used as the reinforcement. The fabric was purchased from S & S Polymers, Bangalore, India. The matrix is a blend of CNSL and epoxy resins. CNSL resin was procured from Shivam Cashew Industry Ahmedabad, India, and an unmodified medium viscosity Epoxy Bisphenol-12 resin (LAPOX L-12) with curing hardener (K6) were supplied by Atul Private Limited, Bangalore, India. The physical properties of the resins are shown in Table 1.

Table 1. Physical properties of matrices [18, 19]

Properties	Resin-1	Resin-2	Hardener
Trade Name	CNSL-resin	Epoxy-resin	K-6
Density	0.970 kg/m^3	1162 kg/m^3	954 kg/m^3

2.2 Fabrication of composite

Composites were prepared by hand layup technique. E-glass CSM was used in three different weight fractions viz. 15%, 30% and 45% with CNSL-epoxy polymer matrix. The matrix was prepared by blending epoxy with CNSL in the ratio of 85:15 by weight. The blend was continuously stirred at room temperature for 30 min.

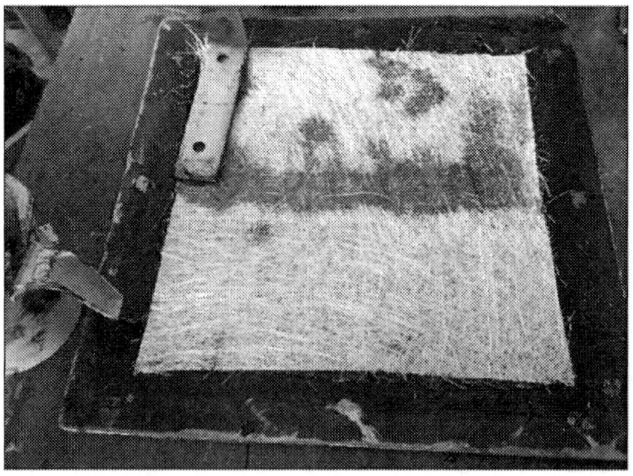

Fig. 1. Fabrication of composite laminates

A closed mould of size 300 mm x 300 mm was used to fabricate the panels. To begin with a layer of CNSL-epoxy resin was uniformly spread over the gel coated mould surface. This was followed by laying up of alternate layers of CSM and resin mix. Peel ply film was placed on the outside to get a smooth surface finish. Fig. 1 shows the fabrication of composite panels with the first layer of reinforcement being laid. Total of 6 plies of CSM was used in the fabrication. The mould was closed and kept under a load of 6 kg for duration of 24 h at room temperature for curing. Post curing for 4 hours at 50°c was carried out in a hot air oven so that the crosslinking of the resin is complete. The thickness of the all cured laminates was 3 ± 0.5 mm. Test specimens were cut as per ASTM standards using water jet cutting at Om Waterjet Machine Pvt. Ltd, Bangalore, India. Table 2 presents the designation of composite laminates.

Table 2. Designation of composite laminates

Composition	Designation
CNSL-epoxy polymer resin	C1
15% E-glass fibre + 85% CNSL-epoxy polymer resin	C2
30% E-glass fibre + 70% CNSL-epoxy polymer resin	C3
45% E-glass fibre + 55% CNSL-epoxy polymer resin	C4

2.3 Mechanical characterization of composites

Micro hardness was determined by Matsuzawa micro-hardness tester (Model: MMT-X7A; Matsuzawa Co., Ltd. Japan). For each sample, six indentations were performed with an indentation load of 100 g and a dwell time of 15 s for indentation. A diamond indenter is forced into the material under a load F, which leaves an indentation of square shape on the material surface. The lengths of the two diagonals S and T of the indentation are noted. The following equation calculates Vickers micro hardness.

$$V_h = 0.1889\,\frac{F}{L^2} \tag{1}$$

$$L = \frac{S+T}{2} \tag{2}$$

where L is the mean diagonal of square impression (mm), F is the applied load (N), S (mm) and T (mm) are the horizontal and vertical diagonal length respectively. Tensile and flexural strength of the composite laminates were evaluated on a Zwick Roell universal testing machine. The samples for the tensile tests were cut to a dimension of 250 mm × 25 mm based on the ASTM D3039 [20] while the flexural samples were cut to an overall length of 120 mm with span length of 96 mm according to ASTM D7264 [21]. Cross head speeds were set at 2mm/min and 1mm/min for tensile and flexural tests respectively. Five specimens from each composition were tested for and the average values were reported.

3 Results and discussion

The variation in average micro-hardness of E-glass fibre reinforced CNSL-epoxy composites with variation in the fibre weight fraction is presented in Fig. 2. Hardness improved with increase in fibre content, an increase of 68% in hardness was observed in C4 composites having fibre content of 45%, when compared to neat polymer (C1). This is an indication of increased resistance to deformation and abrasion under indenter load. Reinforcements like glass fibres which are the discontinuous phase in any composite material are also the hardest phase unless it's a filled composite. Also, when tested for hardness, concentrated compressive loads are applied on the surface, which try to deform the material. These forces are resisted by the reinforcements thus increasing the hardness and results in minimum plastic deformation.

Fig. 3 presents the variation in the tensile strength of composites. It is evident that increase in fibre content improves the tensile strength of the composites. Highest tensile strength was obtained for composites with fibre content of 45% which is about 10 times more than the strength of neat polymers. For composites with fibre content of 15% and 30%, the tensile strength obtained was 116 MPa and 160 MPa respectively. A huge improvement in tensile properties of the composites over neat polymers is mainly due to the load bearing capacity of E-glass fibres. Though E-glass fibres are known for its high strength and modulus, ineffective interfacial bonding between the various constituents would lead to unsound composites. Thus, it can be concluded that modification of epoxy with CNSL was successful.

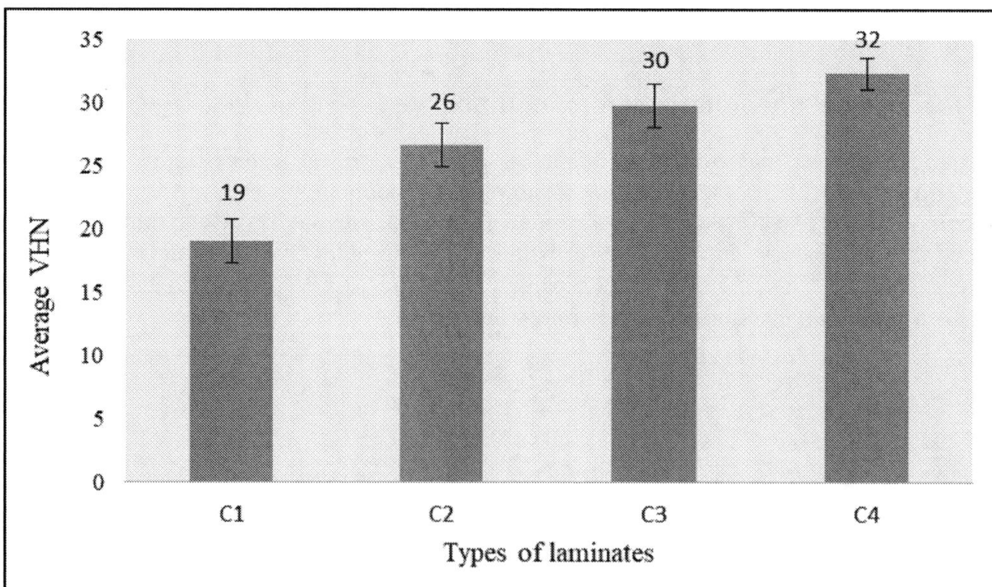

Fig. 2. Variation in average micro-hardness

Flexural strength (Fig. 4) of the chopped E-glass CSM reinforced CNSL-epoxy composite materials increased significantly with the fibre loading. This is an indication of good dispersion of fibres in the CNSL-epoxy matrix. The fibres in the mat are randomly oriented and are held together which increases it's through thickness strength thereby providing good resistance to bending forces. The maximum flexural strength of 166 MPa was obtained for C4 composites with 45% fibre weight fraction. Similar observations were reported by researchers in the past [7, 12] wherein they established that increase in the fibre weight fraction induces an increase in flexural strength property of the epoxy composites.

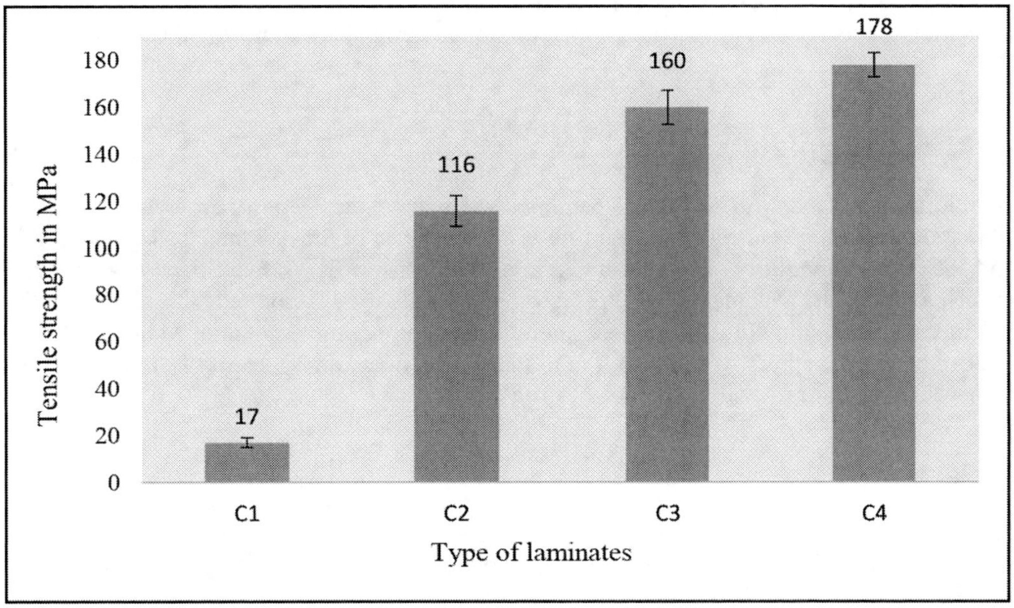

Fig. 3. Variation in average tensile strength

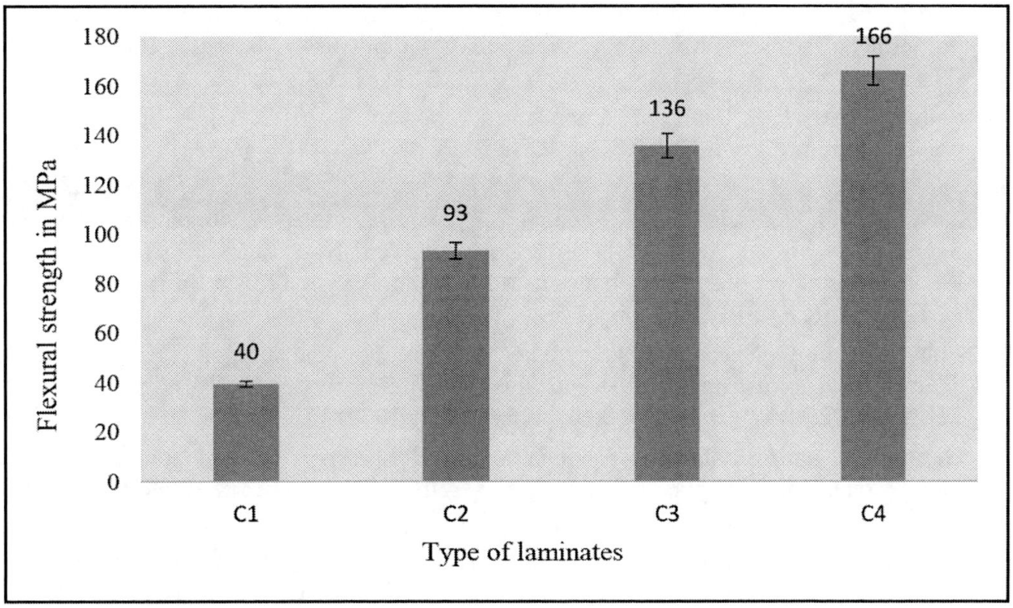

Fig. 4. Variation in average flexural strength

4 Conclusions

Composites fabricated with E-glass CSM in epoxy-CNSL resin showed improvement in micro hardness, tensile strength, and flexural strength over neat polymers. Identical trends were observed for all tests where increased fibre content resulted in improvement in properties. Improvement in properties is due to the inherent capability of the glass fibres to take up higher static loads before failure and also due to effective interfacial bonding between the constituents of the composites. This study shows that by adding CNSL to epoxy resin, mechanical properties of the composite laminates can be enhanced.

Acknowledgment

The authors are thankful to Dr. K. Jagannath, Head of the Department, Mechanical and Manufacturing Engineering for permitting us to make use of the Advanced Material Testing and Research Laboratory. The authors are also indebted to Dr. Satish Shenoy B, Professor and Head, and Dr. Dayanand Pai, Professor, Department of Aeronautical and Automobile Engineering for allowing us to use their Advanced Composite and Material Testing Laboratory. The authors would also like to thank M/s. Konkan Speciality Polyproducts Pvt. Ltd., Mangalore for providing us their testing facility.

References

1. Suresha, B., Chandramohan, G. and Renukappa, N. M., "Mechanical and tribological properties of glass–epoxy composites with and without graphite particulate filler". J APPL POLYM SCI, 103(4), pp.2472-2480. (2007)
2. Mallick, P. K., "Fiber-reinforced composites: materials, manufacturing, and design". CRC press. (2007)
3. Dadfar, M. R. and Ghadami, F., "Effect of rubber modification on fracture toughness properties of glass reinforced hot cured epoxy composites". Mater. Des, 47, pp.16-20. (2013)
4. Sathishkumar, T. P., Satheeshkumar, S. and Naveen, J., 2014. "Glass fiber-reinforced polymer composites–a review". J REINF PLAST COMP, 33(13), pp.1258-1275.(2013)
5. Nayak, S. Y., Heckadka, S. S., Sadanand, R. V., Bharadwaj, K., Pokharna, H. M. and Sanjeev, A. R.,"2D woven/3D orthogonal woven Non-crimp E-glass Fabric as Reinforcement in Epoxy Composites using Vacuum Assisted Resin Infusion Molding". J ENG FIBER FABR, 12(2), pp.12-19.(2017)
6. Tungjitpornkull, S., Chaochanchaikul, K. and Sombatsompop, N.,"Mechanical characterization of E-chopped strand glass fibre reinforced wood/PVC composites". J THERMOPLAST COMPOS, 20(6), pp.535-550.(2007)
7. Gupta, K., and Siddhartha, "Mechanical and abrasive wear characterization of bidirectional and chopped E-glass fibre reinforced composite materials". Mater. Des, 35, pp.467-479.(2012)
8. Bhaskar, V. V. and Srinivas, K., "Mechanical characterization of glass fiber (woven roving/chopped strand mat E-glass fibre) reinforced polyester composites". AIP CONF PROC, 1859, No. 1, p. 020108. AIP Publishing.(2017)
9. Khan, R. A., Khan, M. A., Zaman, H. U., Pervin, S., Khan, N., Sultana, S., Saha, M. and Mustafa, A. I., "Comparative studies of mechanical and interfacial properties between jute and E-glass fibre-reinforced polypropylene composites". J REINF PLAST COMP, 29(7), pp.1078-1088.(2010)

10. Shivamurthy, B., Bhat, K. U. and Anandhan, S., "Mechanical and sliding wear properties of multi-layered laminates from glass fabric/graphite/epoxy composites". Mater. Des, 44, pp.136-143.(2013)
11. Hameed, N., Sreekumar, P. A., Francis, B., Yang, W. and Thomas, S., "Morphology, dynamic mechanical and thermal studies on poly (styrene-co-acrylonitrile) modified epoxy resin/glass fibre composites". Compos Part A Appl Sci Manuf., 38(12), pp.2422-2432.(2007)
12. Muhammad, Y. H. and Ahmad, S., "Mechanical and thermal properties of glass fibre-reinforced epoxy composite with matrix modification using liquid epoxidized natural rubber". J REINF PLAST COMP, 32(9), pp.612-618(2013)
13. Bisanda, E. T. N. and Ansell, M. P., "Properties of sisal-CNSL composites". J MATER SCI, 27(6), pp.1690-1700.(1992)
14. Akinhanmi, T. F., Atasie, V. N. and Akintokun, P. O. "Chemical composition and physicochemical properties of cashew nut (Anacardium occidentale) oil and cashew nut shell liquid". J AGR FOOD ENVIR SCI, 2(1), pp.1-10.(2008)
15. Telascrêa, M., Leão, A. L., Ferreira, M. Z., Pupo, H. F. D. F., Cherian, B. M. and Narine, S. "Use of a Cashew Nut Shell Liquid Resin as a Potential Replacement for Phenolic Resins in the Preparation of Panels–A Review". Mol. Cryst. Liq. Cryst., 604(1), pp.222-232. (2014)
16. Kim, S."The reduction of formaldehyde and VOCs emission from wood-based flooring by green adhesive using cashew nut shell liquid (CNSL)". J HAZARD MATER, 182(1), pp.919-922.(2010)
17. Priya, S. P., Ramakrishna, H. V., Rai, S. K. and Rajulu, A. V. "Studies on tensile properties and chemical resistance of glass mat-epoxy toughened with epoxy phenol CNSL composites". J REINF PLAST COMP, 25(2), pp.141-147.(2006)
18. Lubi, M. C. and Thachil, E. T. "Cashew nut shell liquid (CNSL)-a versatile monomer for polymer synthesis". Des. Monomers Polym., 3(2), pp.123-153.(2000)
19. Heckadka, S. S., Nayak, S. Y., Narang, K. and Vardhan Pant, K. "Chopped Strand/Plain Weave E-Glass as Reinforcement in Vacuum Bagged Epoxy Composites". J MATER, 2015, (2015)
20. ASTM D3039 / D3039M-14, Standard Test Method for Tensile Properties of Polymer Matrix Composite Materials, ASTM International, West Conshohocken, PA, 2014, www.astm.org
21. ASTM D7264 / D7264M-07, Standard Test Method for Flexural Properties of Polymer Matrix Composite Materials, ASTM International, West Conshohocken, PA, 2007, www.astm.org

Hygrothermal Studies on GFRP Composites: A Review

Manjunath Shettar[1], *Aakarshit Chaudhary*[1*], *Zaid Hussain*[1], *U. Achutha Kini*[1] and *Sathyashankara Sharma*[1]

[1]*Department of Mechanical and Manufacturing Engineering, Manipal Institute of Technology, Manipal Academy of Higher Education, Manipal, Karnataka - 576104, India*

Abstract. The objective of this paper is to review the hygrothermal environment effects on Glass Fiber Reinforced Polymer (GFRP) composites. A brief summary of the hygrothermal phenomenon and its mechanisms of GFRP is followed by detailed review of hygrothermal effects on the GFRP. The review also includes the different hygrothermal aging tests viz., cold soaking, boiling soaking, thermal shocks and use of environmental chamber, procedures and significance. At the end, hygrothermal effects on the individual constituents of GFRP viz., fiber, matrix and the fiber-matrix interface, are discussed.

Keywords: Hygrothermal Aging, GFRP, Accelerated Aging

1 Introduction

In general, "composite" means "a material prepared by two or more different constituents" or "a combination of two or more different constituents, which are mixed at macroscopic level to get the best of their properties" [1, 2].

Glass Fiber Reinforced Polymer (GFRP) composites have gained an extensive application in marine, aerospace, automobile and construction industries. Composite materials are not only used to produce new products, but also are used to repair and rehabilitate of existing parts [3]. GFRP composites are also widely used in architectural applications viz., roofing and partition walls [4]. The physical and mechanical properties of GFRP composites could be affected by humidity, elevated temperatures, water, alkalis and other harsh environments in service life [5, 6]. The reliability and performance of composite materials over long time period is the concern to be addressed. There are several ways to investigate hygrothermal effects including some empirical methods for accelerated aging or real exposure tests. The purpose of this paper is to review the different tests available for hygrothermal aging conditions and summarize the procedure and significance of each test.

*Corresponding author: c.aakarshit21@gmail.com

1.1 Hygrothermal Diffusion

Thermal and moisture gradients are the major contributors for hygrothermal diffusion. In most of the cases absorption of water or moisture obeys Fick's law. Diffusion is determined by the moisture concentration gradient between the surrounding environment and material, producing continuous absorption till saturation point reaches. The rate of diffusion rises swiftly with increase in temperature [7, 8].

1.2 Theory of moisture absorption

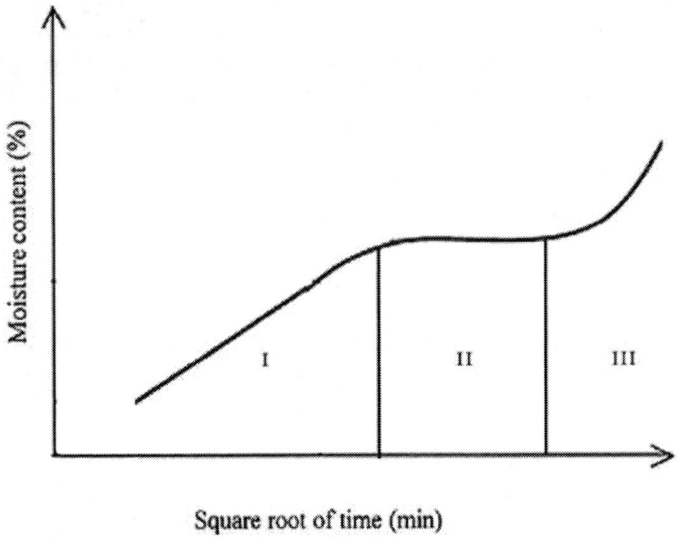

Fig. 1. Moisture absorption curve

Fig. 1 shows three stages of moisture absorption [9],

I - Moisture absorption is Fickian, means, the curve follows linearity and observed in the rubbery state of polymers.

II - Once saturation point reaches, the curve deviate from linearity.

III - Moisture absorption is non-Fickian pattern, sudden rise in absorption due to microcracks or delamination or voids in the matrix or poor boding at fiber matrix interface.

2 Literature Review

S. Larbi et al. [10] investigated the effects of hygrothermal conditioning on mechanical properties of glass fiber reinforced polymer (GFRP) laminates. The laminates were kept in sea water at room temperatures for different time periods. The weight of the samples was noted down on regular intervals. During each weighing, samples were dried using an absorbent to remove water on the outer surfaces. The results show that the absorption kinetics is Fickian at the beginning of the process; but in the long term it is non-Fickian. This could be due to an irreversible degradation of material of lixiviation or hydrolysis type. Also, there was weight loss of samples in long term due to the harmful nature of the high pH of seawater that caused degradation of the matrix. Tests were conducted for composite properties after oven drying. Three-point bending characterization of materials aged hygrothermally had a negative impact. It was particularly illustrated with a loss of tensile strength.

B. C. Ray [11] worked on unidirectional and woven carbon fiber composite specimens at room temperature and at varied higher temperature to assess the environmentally induced damage in composites. At elevated temperature, faster is hygrothermal ageing, rises the moisture uptake rate. The interfacial bonding of fibers and matrix was more affected by aging at elevated temperature due to longer time of exposure. The effect of swelling and thermal stress is greater at elevated temperature. The non Fickian behavior could be correlated to the creation of supplementary free volume in the composite by viscoelastic effects or hydrolysis or by a combination of both. The hygrothermally created weaker and porous interface might lead to capillarity to absorb moisture molecules into the composites at elevated temperature.

S. Shivakumar and Shivarudraiah [12] studied hygrothermic behavior of GFRP, by exposing the composite specimen in water at 30, 50 and 70°C for a period of 200 days. It was observed that the moisture uptake rate and moisture content were increased with increase in immersion time. The moisture uptake percentage of composites with respect to immersion time was far greater in the case of elevated temperature compared to the room temperature. The debonding between fiber and matrix at interface due to swelling, creates voids which act as a reservoir for moisture. Also, the results revealed that hygrothermal aging conditions for longer duration (200 days) and at higher temperature (70°C) decreased the flexural strength by 30% as compared to virgin samples. But at the same time, thickness of the samples increased notably.

A. Rashdi et al. [13] investigated the effects of water absorption on the mechanical properties of kenaf fiber-polyester composites. Specimens were prepared with different weight percentages of fiber. The specimens were immersed for 4 months at 25°C in distilled water. Maximum specimens grasped the saturation point of moisture content in about 1920 – 2400 hours. Increase in the percentage of weight of the fiber leads to an improvement of the tensile properties of the dry specimen, but a decline in the tensile properties of the immersed specimen due to the moisture absorption.

H. N. Dhakal et al. [14] worked on the water absorption effects on the mechanical properties of the hemp fiber-polyester resin composites. The specimens were soaked in de-ionized water at room and boiling temperatures for varied time intervals. The pattern for water absorption is found to follow Fickian behavior at room temperature, but it follows non-Fickian pattern at the boiling temperature. It was observed that, there is rise in weight of the samples (23%) in boiling water compared to samples soaked in water at room temperature. The faster and higher weight gain of the samples aged in boiling water, leads to interfacial cracks because of degradation at the fiber matrix interface. The tensile strength of aged samples is higher than that for unaged samples. This might be due to high content of water that causes swelling in the fibers, which could have filled the gaps (voids) between the fiber and matrix and ultimately might have led to increase in the tensile properties of the composites.

H. S. Almeida et al. [15] investigated the effect of hygrothermal aging effects on tensile, compressive, in-plane and interlaminar shear properties of carbon fiber/epoxy laminates. The specimens were subjected to an accelerated hygrothermal aging at 80°C and 90 % relative humidity for 60 days as per ASTM D5229M standard. It was observed that there was 0.42% of weight of specimens increase due to moisture absorption and reached saturation level in around 33 days. Tensile strength reduced by 15% for aged laminates.

The SEM analysis indicated the degradation of the epoxy matrix, which led to fall in interfacial bonding, followed by decrease in tensile strength.

R. Zenasni and A. S. Bachir [16] investigated the effect of hygrothermo-mechanical aging on the inter-laminar fracture behavior of GFRP. Hygrothermo-mechanical aging characterization was carried out to know the loss in crack propagation resistance, on cantilever beam and notched flexural inter-laminar fracture tests.

Z. Sereir and N. Boualem [17] worked on the effects of long term hygrothermal loading on the hybrid composites. They also analyzed inter laminar stresses for the specimen by 3 point bending test, that resulted in change in the properties.

S. Pilli et al. [18] worked on the mechanical properties and long-term durability of composites under the influence of moisture. The test technique was developed to accelerate humidity by rising the pressure inside the test chamber for moisture ingression.

Y. I. Tsai et al. [19] worked on water absorption and mechanical property tests. The specimens were examined after immersion in water for 32 weeks at varied temperatures. Later specimens were weighed to calculate the moisture uptake percentage. The short beam strength was decreased while there was increase in moisture uptake, with temperature and immersion time.

R. M. Rowell et al. [20] studied the aspen fiber polyester resin with micro filler materials for cyclic humidity and water soaking conditions. Specimens have shown increase in moisture absorption. The addition of micro fillers decreases the moisture absorption compared to composites without any fillers.

3 Hygrothermal Aging

There are several ways to investigate hygrothermal effects including some empirical methods for accelerated aging or real exposure tests. Each has its own advantages and limitations. Real exposure tests consume lot of time to perform [5, 6, 7].
Different hygrothermal aging tests are as follows;

1. Cold soaking
2. Boiling soaking
3. Thermal shocks
4. Use of environmental chamber

3.1 Cold Soaking

The composite specimens are soaked in water at room temperature (Fig. 2). During testing, specimens are removed from water weekly once, dried with paper napkin and weighed by an electrical balance with least count of 1 mg. The weight and time periods are noted down to analyze uptake moisture. Similarly, thickness of specimens are measured for any swelling.

Fig. 2. Specimens immersed in water at room temperature

Significance

The water absorption is not typically in the form of liquid. It comprises of molecular group of hydrogen bonds linked to the polymer molecule chains. Adding to that, water could be absorbed by "capillary action" along the microcrack present in the specimens [11, 21, 22].

3.2 Boiling Soaking

The specimens are dried in an oven between 100 to 110 °C for a period of 12 hours. Later the specimens are weighed and measured for thickness. The specimens are boiled for 2 hours (Fig. 3) and instantly weight and thickness are measured. Specimens are re-dried in oven for 24 hours and again measured for weight and thickness. The process is repeated 10 times for a total of 20 hours of boiling [14].

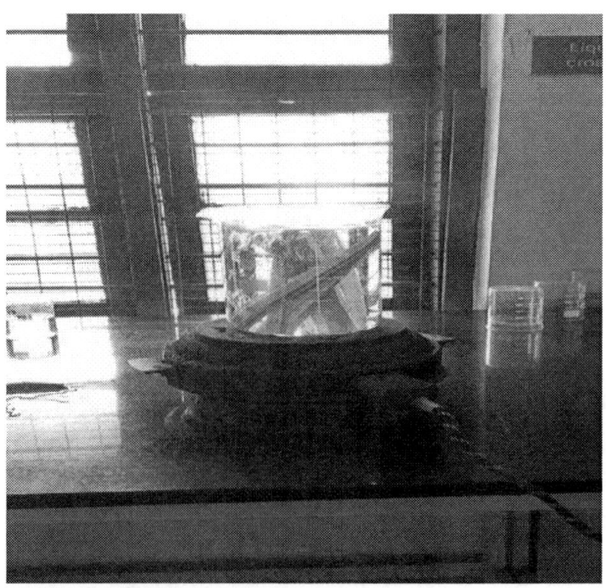

Fig. 3. Specimens in Boiling Environment

Significance
Water absorption increases rapidly with rise in temperature. The water molecules migrate from higher concentration to the lower concertation region in the composite materials.

3.3 Thermal Shocks

In order to identify the effect of sudden changes in environment, specimens are placed in an oven maintained at 70 ^{0}C for a period of 2 hours (Fig. 4) and then immediately placed in freezing temperature of -10 ^{0}C for another 2 hours. The process is repeated for 5 cycles.

Fig. 4. Specimens in oven

Significance
Sudden change in temperature from hot to cold region or vice versa has impact in residual stresses, due to differences in thermal expansion coefficients of the constituents. Residual stresses could result in initiation and propagation of microcracks. Under continued thermal cycling microcracks endure to grow in number and size, which could result in degradation of stiffness and other matrix-dominated properties.

3.4 Use of Environmental Chamber

The specimens are kept in environmental chamber at known relative humidity (RH), temperature and time period. The specimens are withdrawn at consistent intervals, wiped dry to take away the surface moisture and then weighed using an electronic balance to monitor the mass change behaviour or moisture diffusion [12, 15].

For the study of the behaviour of the composite under high humidity and temperature conditions one should,

1. know the temperature and humidity ranges for which the composite is exposed to during its designed life.
2. know what properties of the composite are most important for the further technological performance of the material.
3. study the change of these properties by increasing wet mass of standard probes at a given temperature for different relative humidity (RH).
4. repeat the study in step 3 for different temperatures.

5. correlate the results of the studies proposed in steps 3 and 4.

Significance

FRP composites exposed to humid environment with elevated temperature absorb moisture by instantaneous surface absorption and diffusion. An elevated temperature enacts like an activator for moisture diffusion throughout the composite.

4 Hygrothermal Effects

Hygrothermal effects on GFRP composites depend on the effects on the individual constituents viz., fiber, matrix and the fiber matrix interface. But the overall effects depend on the hygrothermal effects on the dominant component.

4.1 Effects on Glass Fiber

The effect of water on glass fibers could be established from the following chemical equations. A sequence of reactions lead to cleavage of silicon-oxygen bonds and to their conversion to hydroxysilane.

$$Si - ONa + H_2O \longrightarrow Si - OH + NaOH$$
$$Si - O - Si + OH^- \longrightarrow Si - OH + Si - O^-$$
$$Si - O^- + H_2O \longrightarrow Si - OH + OH^-$$

The overall chemical reaction could result in degradation and micro flaw formation on the glass fiber surface, due to the rise in the pH level, since the glass fibers are covered and protected by the matrix material. But, microcracks or voids present in composite materials might allow water to diffuse through the matrix. The matrix and fiber matrix interface degrade at a higher rate compared to glass fiber.

4.2 Effects on Polymer Matrix

Moisture absorption into a polymer matrix affects both physical and mechanical properties through different mechanisms viz., swelling, hydrolysis, crazing and plasticization. Hygrothermal environments have robust effects on the polymer properties. These effects could be named as reversible and irreversible. For example, at the point when water diffuses into the polymer matrix, it plasticizes and swells the polymer molecular network, diminishes the surface free energy and enhance the free volume of the polymer, consequently actuating a decrease in the glass transition temperature (Tg). This process is of reversible effects, so the property will recuperate after drying the polymer material.

The higher exposure span to moisture results in irreversible damage (hydrolysis and microcracking). Hydrolysis might be the prime reason for the polymer weight loss after long term soaking. On the other hand, microcracking in the matrix might contribute to decrease in mechanical properties.

4.3 Effects on the Fiber / Matrix Interface

The interface is regarded as the non-homogenous region that lies between the matrix and the fibers. Usually matrix is bulk material, which absorbs more moisture compared to the fibers. The mismatch in the moisture absorption leads to localized stress and strain field because of nonuniform volumetric expansion between the fibers and matrix. The localized stress and strain field between fibers and matrix will lead to initiation and propagation of crack throughout the boundary layer of the fiber matrix interface.

5 Concluding Remarks

The hygrothermal environment effects on Glass Fiber Reinforced Polymer (GFRP) composites was reviewed in this paper. The fundamental points can be summarised as follows:

1. Moisture absorption into a polymer matrix affects both physical and mechanical properties through different mechanisms viz., swelling, hydrolysis, crazing and plasticization. Hygrothermal environments have robust effects on the polymer properties.
2. Thermal and moisture gradients are the major contributors for hygrothermal diffusion. In most of the cases absorption of water or moisture obeys Fick's law. The results show that the absorption kinetics is Fickian for shorter duration; but in the long term it is non-Fickian.
3. Also, the results revealed that hygrothermal aging conditions for longer duration and at higher temperature had negative impact on physical and mechanical properties of GFRP composites.

References

1. K. K. Chawla, *Composite Materials Science and Engineering* (Springer Science & Business Media, New York, 2012)
2. R. M. Jones, *Mechanics of Composite Materials* (CRC Press, Taylor & Francis Group, Philadelphia 1998)
3. F. C. Campbell, *Manufacturing Processes for Advanced Composites* (Elsevier, London, 2003)
4. A. K. Kaw, *Mechanics of Composite Materials* (CRC press, Taylor & Francis Group, Philadelphia 2005)
5. L. H. Hihara, R. P. Adler, R. M. Latanision, *Environmental Degradation of Advanced and Traditional Engineering Materials* (CRC Press, 2013)
6. R. Martin, *Aging of Composites* (CRC Press, Washington 2008)
7. V. K. Thakur, *Green composites from natural resources* (CRC Press, Washington 2013)
8. Y. Zhong, S. C. Joshi, Mat. Des. **92**, 866 (2016)
9. K. E. Pickering, *Properties and performance of natural-fibre composites* (CRC Press, Washington 2008)
10. S. Larbi, S. Bensaada, A. Bilek, S. Djebali, *AIP Conference Proceedings*, 1653 (2015)
11. B. C. Ray, J. of Col. Int. Sci. **298**, 111 (2006)
12. S. Shivakumar, Shivarudraiah, I. J. Adv. Eng. Tech. **1**, 225 (2010)
13. A. Rashdi, M. S. Salit, K. Abadan, M. M. H. Megat, Pert. J. Sci. Tech. **18**, 433 (2010)
14. H. N. Dhakal, Z. Y. Zhang, M. O. W. Richardson, Comp. Sci. Tech. **67**, 1674 (2007)
15. J. H. S. Almeida, S. D. Souza, E. C. Botelho, S. C. Amico, J. Mat. Sci. **51**, 4697 (2016)
16. R. Znasni, A. S. Bachir, J. Ther. Comp. Mat. **19**, 385 (2006)
17. Z. Sereir, N. Boualem, J. Theo. App. Fra. Mech. **47**, 145 (2007)

18. S. Pilli, P. Simmons, L. Kevin, J. Comp. Part A, **40**, 1501 (2009)
19. Y. I. Tsai, E. J. Bosze, E. Barjastech, T. S. R. Nut, J. Comp. Sci. Tech. **69**, 432 (2009)
20. R. M. Rowell, S. E. Lange, R. E. Jacobson, *Proceedings of the 7th Pacific Rim Bio-based Composites Symposium* (2004)
21. E. Osman, M. W. Rashid, M. E. A. Manaf, E. Mohamad, M.R. Salleh, T. Ito, T. Moriga T. J. Adv. Man. Tech. (2016)
22. E. C. Botelho, L.C Pardini, M. C. Rezende, J.Mat. Sci. **41,** 7111 (2006)

Influence of plasma modification on mechanical and thermal properties of Polypropylene/ Nano-Calcium Silicate Composites

Ajeesh. G[1*], *Aditya Raman*[1], *Adarsh Parameswaran*[1], *Sundhara Pandiyan*[1], *Parvathy S.*[1] and *Philip George*[2]

[1]*Department of Aerospace Engineering, Amrita School of Engineering, Amrita Vishwa Vidyapeetham, Coimbatore-641112, India*
[2]*Vijnan Institute of Science and Technology, Elanji, Ernakula, Kerala, India*

Abstract. The aim of the research is to study the influence of plasma modification on nano calcium silicate/polypropylene composites. Polypropylene (PP) is considered for this study as it possesses high impact strength, toughness and availability. Calcium silicate is considered as reinforcement because of its high temperature resistance, high flexural strength and high strength to mass ratio. Fourier transform infrared spectroscopy (FTIR) results show that there is a change in the functional group on the surface of calcium silicate after modification. Thermo-Gravimetric Analysis (TGA), Differential Scanning Calorimetry (DSC) results show that the decomposition temperature increased with increasing amount of filler particles. It is also observed that the modification has produced a marginal increase in the decomposition and glass transition temperature. Tensile test results also show a gradual increase in the tensile properties of composites when high ratio is reinforcement. Tensile test results also show that there is a marginal increase in the tensile strength when reinforced with modified calcium silicate when compared to non-modified. Scanning Electron Microscopy (SEM) reveals that there is a enhanced dispersion of nano particles on modification. Based on the findings it can be concluded that plasma modification enhances the thermal and mechanical property marginally.

Keywords: Polymeric nano composites, Plasma modification, Scanning Electron Microscopy, Differential scanning calorimeter, Superior dispersion.

1 Introduction

From the start of human existence, materials have played a significant role in the development of civilization. In the late 19th century Composites gained importance because of its high strength, low weight, superior thermal properties and design flexibility. The first composite made by man was having mud as matrix and straw as reinforcement. Among metal matrix, ceramic matrix and polymer matrix composites, polymer composites have a

[*] Corresponding author: g_ajeesh@cb.amrita.edu

high resistance to chemical corrosion and scratching. They provide easy fabrication of very complex parts with low tooling cost. Thermoplastic polymers are chosen over thermoset due to their advantages such as high toughness, shorter manufacturing cycles, reprocessing possibilities and they also do not require refrigeration storage. Carbon short fiber or glass short fiber reinforced in thermoplastic matrices such as polyetherimide (PEI), polyetheretherketone (PEEK) and polyphenylene sulfide (PPS) have been studied considering their excellent mechanical and thermal properties. The major drawback of these polymers are that they are costly and difficult to process when compared to thermosetting composites. Considering the cost and ease of manufacturing, polypropylene (PP) was considered as matrix for this research [1]. Polypropylene is a semi-crystalline thermoplastic and is known for its superior thermal and chemical properties.
Polypropylene can be used in automobiles appliances in which creep resistance needs to be considered in addition to stiffness and toughness. But non reinforced PP resin has low mechanical and thermal properties. To improve its strength and thermal properties, various types of fillers (such as fibres, clay, silica, graphene and CNT) have been incorporated. Calcium silicate is considered for this study as it possesses high thermal properties. Research has been performed on modification of nano particles and it has been noted that modification of nano particles enhances the mechanical and thermal properties of the composites. It is also observed that the filler show better dispersion on modification [2, 3].
The modification of nano calcium silicate has not been performed in the past and this research would help in understanding the influence of plasma modification on nano calcium silicate- polymer composites.

2 Experimental

2.1 Materials

Isotactic polypropylene of grade H110MA supplied by Repol, India was used. Nano-calcium silicate was procured from Sigma Aldrich, Switzerland.

2.2 Plasma modification

Plasma treatment is a surface modification technique that uses the fourth state of matter plasma. In order to attain the plasma state, high energy is provided to a gas [Ar, O_2, N_2, etc] which ionizes it. The process is carried out under vacuum conditions. Oxygen plasma removes organic contaminants by reacting chemically with highly reactive oxygen radicals. Oxygen plasma surface treatment promotes surface oxidation and hydroxylation (O-H groups) which increases the surface wettability. The figure of plasma modification of nano calcium silicate is as shown in the Fig. 1.

Fig. 1. Plasma Modification of nano-calcium silicate

2.3 Preparation of PP/ (unmodified/ modified) nano calcium silicate composites

The compounding was done using twin-screw extruder. The compounded mixture was then fed into injection molding for preparation of the specimens. Weight percentage of 0.5, 1 and 1.5 were the mass ratios taken initially for both modified and unmodified calcium silicate. The extruder screw rpm was 71 and the die temperature was 232^{0}C. The outcome was in the form of strands which was then cut to smaller pieces using a shear cutter and then fed into the injection molding machine. The mold temperature was set as 250^{0}C. Polypropylene containing more than 5% filler had to be manufactured using Twin screw extruder. These pieces were then injection molded to obtain the specimens.

3 Characterization

3.1 FTIR spectroscopy

The effect of modification in the nano filler was observed using a Nicolet iS10 machine manufactured by Thermo Scientific, USA. Few grams of the filler was dried and grounded along with 200 mg of KBr. They were then made into transparent pellets and the interferogram was obtained which was converted into a spectrum using Fourier transform.

3.2 Tensile test

Tensile test was done using a Universal Testing Machine. The machine used was Zwick/ Roell Z010. The gauge length of the specimen was 50 mm and the cross-head speed was 1mm/ min according to ASTM D-638.

3.3 Thermo-gravimetric analysis

Thermo Gravimetric Analysis is performed to study the influence of fillers on the decomposition temperature of composites. SDT Q600 V20.9 Build 20 instrument manufactured by TA Instrument Inc., USA, was used to measure the thermal stability of the sample.

3.4 Differential scanning Calorimetry

Differential scanning Calorimetry (DSC) is performed to understand the influence of nano calcium silicate on the glass transition temperature of composites. DSC Q20 V24.10 Build 122 instrument manufactured by TA Instrument Inc., USA, was used. Few grams of samples was measured and kept in the aluminium DSC pan provided and heated up to 300°C in nitrogen atmosphere. The heat flow curves were obtained as a function of temperature from which the melting temperature was determined.

3.5 Scanning Electron Microscopy

SEM is performed to study the effect of modification on the dispersion of the nano fillers in the composites.

4 Results and Discussion

4.1 FTIR spectroscopy of the unmodified and the plasma modified calcium silicate fillers

Comparing the FTIR spectrum of the unmodified nano calcium silicate as shown in Fig. 2, and the spectrum of the plasma-modified filler as shown in Fig. 3, it is observed that the transmittance value of the hydroxyl group (O-H) [corresponding to the wavenumber range 3200-3600] has increased by 18% (from 50% to 68%); alkene group (C=C) [corresponding to wavenumber 1620 – 1680] and aromatic group (C=C) [corresponding to wavenumber 1400-1600] by 12% (from 68% to 80%) and another alkene group (=C-H) by 14% (70% to 84%) for the modified filler. The absorption band around 1640 cm^{-1} is commonly assigned to O–H bending vibration but can occur also from vibrations of the SiO_2 network and is often hidden by water O–H band.

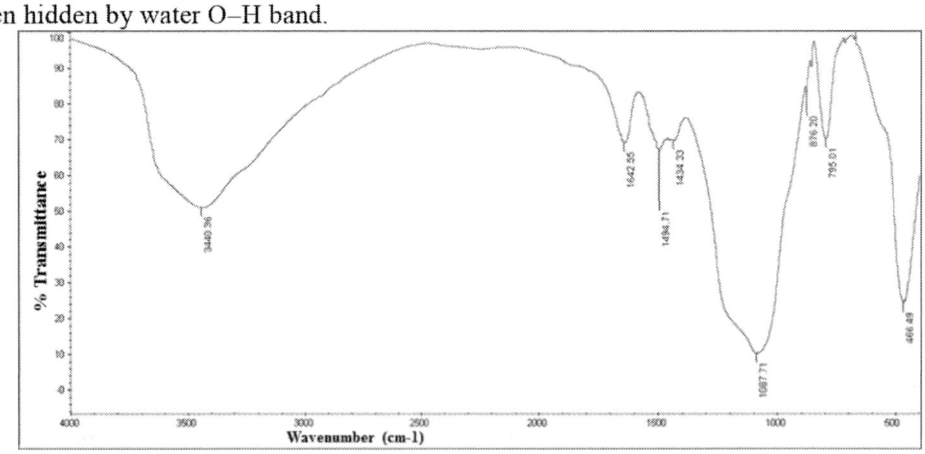

Fig. 2. FTIR plot of untreated Calcium Silicate

The FTIR spectra (Figure 2) reveal an absorption band at 1090 cm^{-1} due to C-O stretched vibration and a characteristic very narrow band for the spectra are dominated by a broad signal in the 750–1300 cm^{-1} spectral range, where the stretching vibrations of the Si–O–Si and Si–O–Ca bridges occur [4].

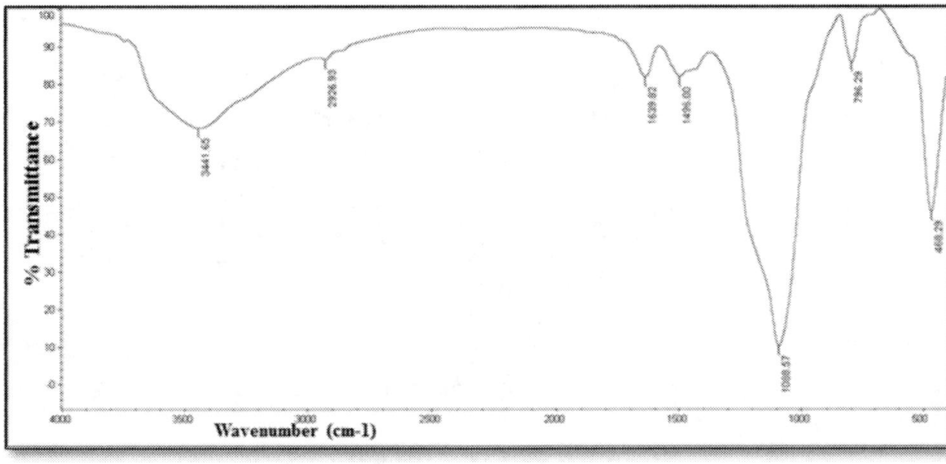

Fig. 3. FTIR plot of treated Calcium Silicate

4.2 Tensile Tests

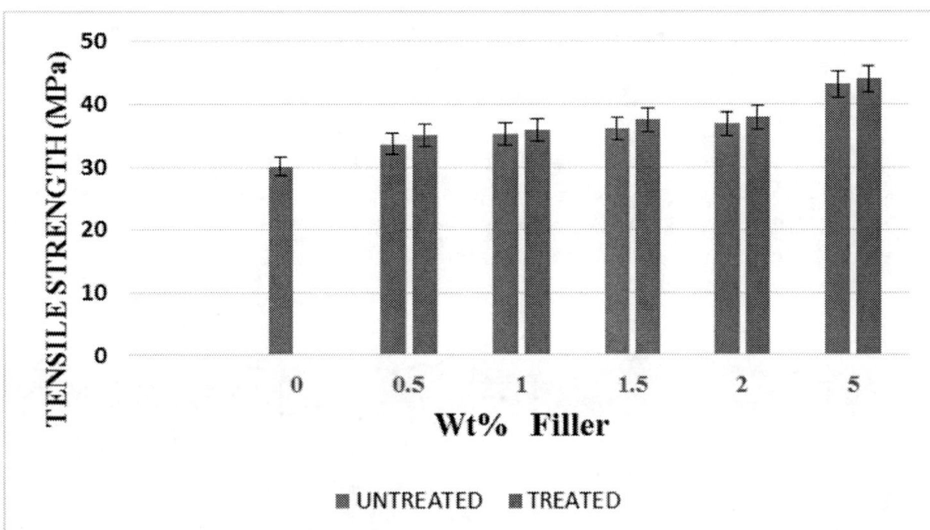

Fig. 4. Tensile strength of composites

It is observed from Fig. 4 and Fig. 5 that the tensile strength and the Young's modulus increases significantly with increasing mass ratio for the composite containing unmodified calcium silicate. The increase is because the filler absorbs energy [5]. The nano filler have higher surface area and hence significant increase is seen. On the contrary it is observed that on modification, there has been a marginal increase compared to untreated nano calcium silicate. This may be attributed to better dispersion.

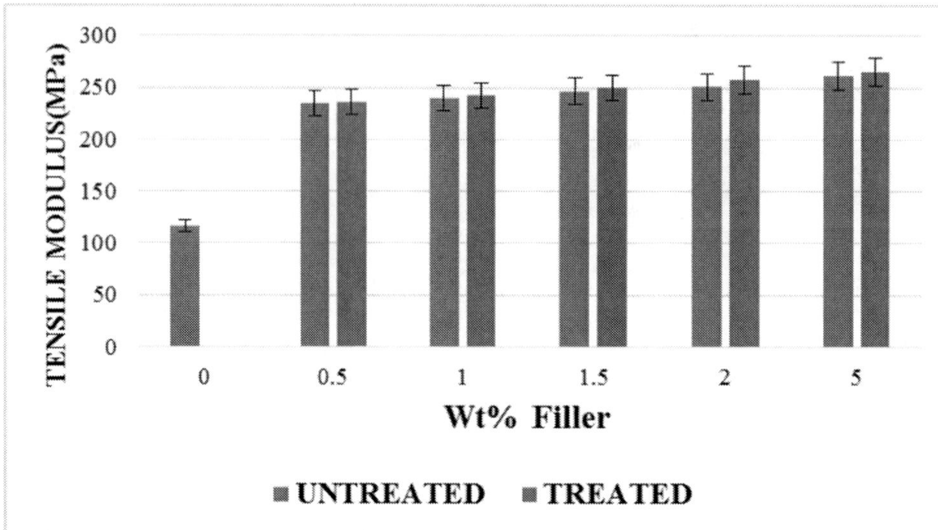

Fig. 5. Tensile modulus of the composites

4.3 TGA

It is observed from Fig. 6 that the addition of the fillers increased the decomposition temperature of the composite. Since smaller mass ratios were taken, the increase was not prominent. Both the unmodified and the modified filler showed the same pattern of increase. The thermal stability of the PP matrix increased with the introduction of small amounts of nano calcium silicate. The high specific area of nano calcium silicate also contributed to the greater interaction between the nano filler and the PP matrix which led to the enhancement of both interfacial bonding and thermal stability.

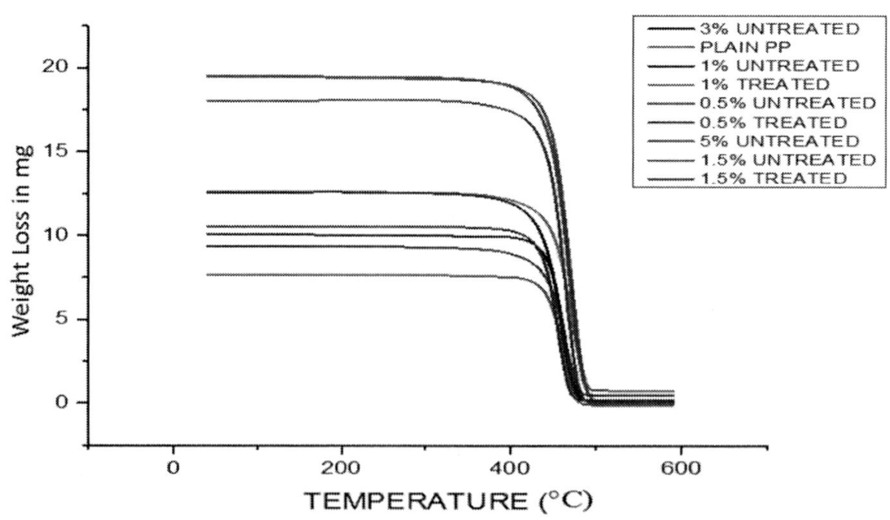

Fig. 6. TGA Plots of composites

This result also indicated that the well-dispersed nano-calcium silicate particles successfully acted as effective nucleation agents in the PP matrix composites. The increase in the thermal stability for both the unmodified and the modified fillers may be attributed to the presence of filler loading.

4.4 DSC

It is observed from the heating curve (Fig. 7) that the melting transition temperature (defined by the maximum at the endothermic peak) is not influenced by the presence of nano-fillers. The slight fluctuations may be due to the impurities in the crucible [6, 7]. The broadening of the melt endotherm has been associated with the formation of less perfect crystals because of the spatial hindrance of the filler [8, 9].

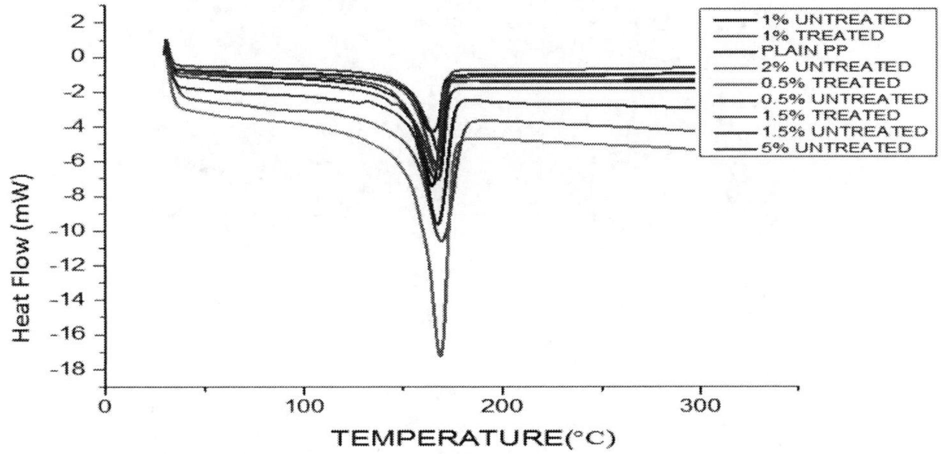

Fig. 7. DSC plot of composites

4.5 SEM

SEM was carried out to study the dispersion of nano filler in the composites. Fig. 8 shows the composite figures which does not contain nano filler. Fig. 9 and Fig. 10 shows the composites with untreated and treated nano fillers.

Fig. 8. SEM image of non-reinforced composites

 https://doi.org/10.1051/matecconf/201814402027

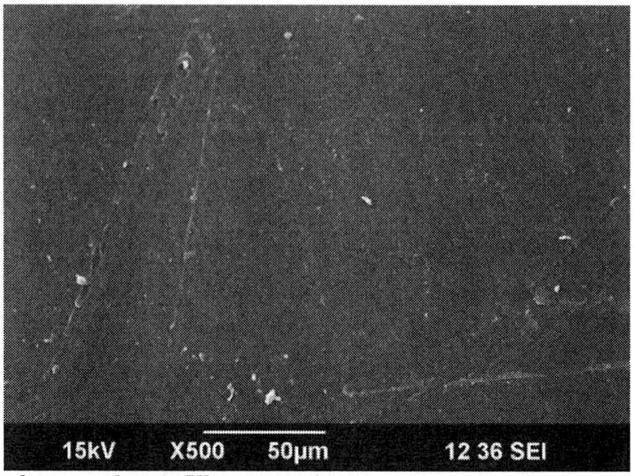

Fig. 9. SEM image of untreated nano filler composites

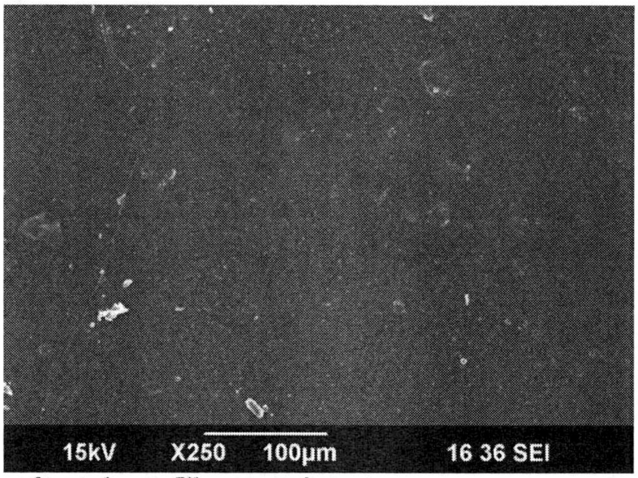

Fig. 10. SEM image of treated nano filler composites

Fig.11 shows a more magnified image of plasma modified calcium silicate composite. This shows dispersion of nano filler within the polymer.

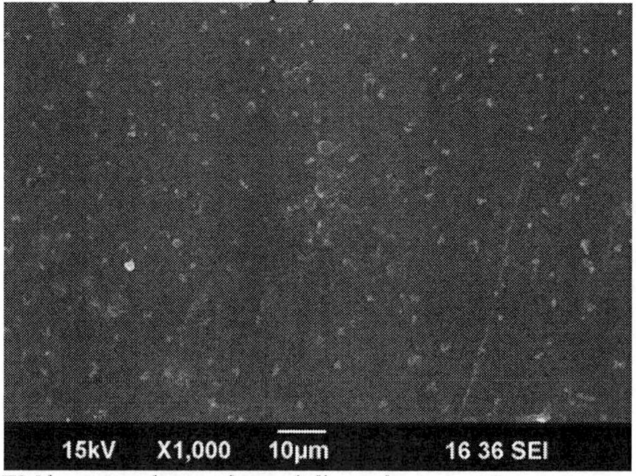

Fig. 11. Magnified SEM image to show enhanced dispersion

Based on these findings it is evident that plasma modified nano filler shows better dispersion.

5 Conclusions

The study revealed the following conclusions:

- The mechanical properties of the composite containing unmodified filler had increased with increasing mass ratio.
- The decomposition temperature of the composite containing unmodified filler increased with increasing mass ratio.
- There was no much change observed in the melting temperature values. But there were marginal fluctuations. This may be because the ratio of filler used was not high.
- The oxygen plasma modification done on the filler increased its surface energy. Hence there was proper bonding between the filler and the polymer. Therefore the composites containing modified calcium silicate gives marginally better results with regard to tensile and dispersion properties.

References

1. M. Gabr, W. Okumura, H. Ueda, W. Kuriyama, K. Uzawa, I. Kimpara. Composites Part B: Engineering, 69, 94-100 (2015).
2. G. Ajeesh, S. Bhowmik, V. Sivakumar, L. Varshney, Journal of Composite Materials, 51(8), 1057-1072 (2017).
3. G. Ajeesh, Shantanu Bhowmik, Sivakumar Venugopal, Lalit Varshney, Abrar Baluch, Yurim Park, Son Gilsang and Chun Gon Kim. High Performance Polymers, 28(5), 570-580 (2016).
4. Lin, Jia-Horng, Chien-Lin Huang, Chi-Fan Liu, Chih-Kuang Chen, Zheng-Ian Lin, and Ching-Wen Lou. Materials 8(12), 8279-8291(2015).
5. George Philip, Shantanu Bhowmik, Mathew Abraham, P. K. Sriram, Mohan kumar Pitchan, and G. Ajeesh. Proceedings of the Institution of Mechanical Engineers, Part L: Journal of Materials: Design and Applications 1464420716660874 (2016).
6. Mingliang G Demin, Journal of Reinforced Plastics and Composites. 28(1):5-16 (2008).
7. D. Pérez-Rocha, A. Morales-Cepeda , F. Navarro-Pardo , T. Lozano-Ramírez , P. LaFleur . International Journal of Polymer Science. 1-8 (2015).
8. A. Petrov Saakov A. Technology and equipment for plasma surface hardening of heavy-duty parts. Materials and Manufacturing Processes. 17(3):363-378 (2002).
9. N. Phong, M. Gabr , K. Okubo , B. Chuong , T. Fujii, Materials & Design.,47:624-632 (2013).

Machinability study of Al-TiC metal matrix composite

Siddappa P. N.[1*], *B. P. Shivakumar*[1], *Yogesha K. B.*[1], *Mruthunjaya M.*[1] *and Hanamantraygouda M. B.*[2]

[1] *Department of Mechanical Engineering, JSS Academy of Technical Education, Bengaluru-560060, India,*

[2] *Department of Mechanical Engineering, Sir M. Visvesvaraya Institute of Technology, Bengaluru-560057, India*

Abstract. Aluminum Metal Matrix Composites have emerged as an advanced class of structural materials have a combination of different, superior properties compared to an unreinforced matrix, which can result in a number of service benefits such as increased strength, higher elastic moduli, higher service temperature, low CTE, improved wear resistance, high toughness, etc. The excellent mechanical properties of these materials together with weight saving makes them very attractive for a variety of engineering applications in aerospace, automotive, electronic industries, etc. Hence, these materials provide as alternative substitutes for conventional engineering materials when specific mechanical properties necessary for required applications. In this work an attempt is made to study the machining parameters of Al6061/TiC MMC. The composite is developed by reinforcing TiC particles in varying proportions of 3, 6, 9 and 12 % weight fractions to the Al6061 matric alloy through stir casting technique. Cutting forces were measured by varying cutting speed and feed rate with constant depth of cut for different % weight fractions. The results showed that the cutting force increases with the increase of feed rate and decreases with the increase of cutting speed for all the weight fractions. Cutting parameters were optimized using Taguchi technique.

Keywords: Metal matrix composite; Weight fraction; Stir casting; Machinability.

1 Introduction

Machining is a secondary manufacturing process and it is described as an elimination of excess material from the metal component using power-driven machine. Machining of composites is more difficult and tedious task compared to conventional metals due to their complex nature. The composites require superior machining characteristics in terms of cutting tool material, tool geometry, tool design, cutting speed, feed, depth of cut, etc. Thus, in today's world, non-traditional methods of manufacturing or higher grades of cutting tools are preferred for the machining composites. Machinability can be compare in terms of tool life, cutting force, and surface finish under similar conditions. Tool life criterion is

[*] Corresponding author: siddappa.pn@gmail.com

most important from the point of view of machining economics. Machinability rating of test material is given by MR = (V60t /V60s)×100. The specific cutting pressure and specific cutting power are often considered as index of machinability of the given work material. The larger the specific cutting energy, the smaller the machinability index.

2. Literature review

Composites have found application in many areas of daily life for quite some time and in engineering applications. Extensive development in the enhancement of composites have been achieve in recent days, so that they could be used for variety of applications. Pushpendra Kumar et al. [1] investigated on machining of Aluminum Metal Matrix Composites (AMMCs) by controlling the process parameters. The literature concluded that, by suitably selecting the machining parameters, AMMC could be machined economically. PCD tool is the most preferred. The process is still challenging due to non-homogenous and anisotropic nature of composites. V. Songemene et al. [2] studied machinability of aluminium alloy. The factors like cutting force, cutting speed, feed rate and chip formation were optimized while machining Al7075-T6. It was found that, thrust force is increases with feed rate at different cutting speeds. Cutting force is sensitive to the variation of feed and feed determines the chip thickness. Rabindra Behra et al. [3] conducted the machinability test on SiC reinforced AMMC by varying factors like cutting force, cutting speed, feed rate, cutting depth and surface finish. It was found that cutting force decreases with the increase of cutting speed. It was also noticed that cutting force increases with the increase of feed rate at constant depth of cut. It is found that at constant feed rate as the cutting speed increases, surface roughness decreases and the size of chip formation decreases with increasing weight fraction. Anand Kumar et al. [4] dealt with cutting parameters of Al-4.5 % Cu-TiC metal matrix composites using coated ceramic tools. Machining factors like cutting speed, cutting force and depth of cut were varied their effects on machining were studied. It is concluded that, tensile strength and hardness increases ductility reduces with the increase of wt.% of TiC. Cutting force increased up to medium cutting speed of 80 m/min and then decreased when cutting speed increased from medium to high speed (120 m/min) wherein good surface finish was observed. Nassim Samer et al. [5] studied the microstructure and mechanical properties of Al-TiC metal matrix composite by reactive synthesis method. Aluminium composite prepared by nanometre sized TiC particles. The study concluded that, the composite Al-TiC MMC showed enhancing mechanical properties for a high volume fraction of reinforcement. The distribution of reinforcement particles was attribute to reactive synthesis route. Himanshu Kala et al. [6] investigated on the mechanical properties of AMMC contain single and multiple reinforcements. Reinforcement of aluminium and its alloys with ceramic particles showed considerable increase in mechanical properties. Suraya Sulaiman et al. [7] observed the process of determining the mechanical properties of Al-11.8% Si/TiCp composite. The study revealed that, addition of reinforcement particles, considerably improves the value of axial tensile strength of the MMC. The result shows that hardness and tensile strength decreases after 10 wt. % of TiC reinforcements. P. Dhanasekaran et al. [8] found the tensile and tear behaviour of TiC reinforced with Al 6063 alloy and SEM witnessed the dispersal of particulates in metal matrix. It was observed that, by incorporating TiC particles into matrix, dry wear properties, hardness and fluidity was enhanced. The wear of composite specimen decreases with increase in reinforcement fraction, which improves hardness. Venkatesan et al. [9] did the comparative analysis of Al MMC reinforced with TiC and TiB2 with different weight percentages. This literature also analysed the mechanical and wear characteristics. The study revealed that increasing the percentage of TiC directly increases the ultimate tensile strength, yield strength, elasticity of modulus and decreases the percentage of elongation. Increasing the percentage of titanium boride increases the tensile strength and co-efficient of friction.

3. Experimental Work

3.1 Material properties of Matrix and reinforcement

Table 1 gives the chemical composition of Al6061 matrix and Table 2 provides the details of material properties of both matrix and reinforcement at room temperature.

Table 1. Chemical composition of Al6061 alloy

Elements	Mg	Si	Fe	Cu	Ti	Cr	Zn	Mn	Al
Percentage	1.50	0.90	0.70	0.40	0.15	0.35	0.25	0.20	Balance

Table 2. Material properties of Al6061 alloy and TiC particles

Properties	Al 6061	TiC
Density	2.7 g/cm3	4.93 g/cm3
Melting Point	5800 C	31400 C
Modulus of Elasticity	70-80 GPa	400 GPa
Hardness	95-97 BHN	3400 VHN
Particle size	-	18μm

3.2. Composite preparation

Preheated TiC particles were poured in to the molten Al 6061 alloy at 800°C in the graphite crucible, which is placed in the induction furnace and stirred for 15 minutes in a rotating graphite stirrer at 350 rpm. Composite mixture was degassed using commercially available hexachlro ethane degassing tablet and then poured in to preheated metallic moulds as per AFS standards as shown in Fig. 1 for 3%, 6%, 9% and 12% weight fractions. The obtained casts were machined to the required size of 30 mm diameter and 300 mm length as shown in Fig. 2.

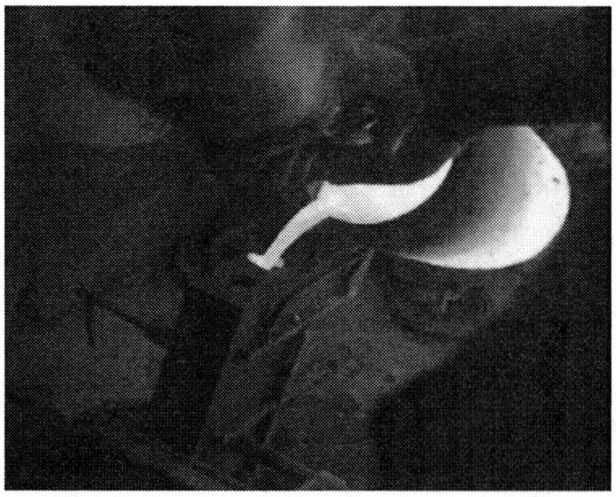

Fig.1. Pouring of molten metal in to the metal mould

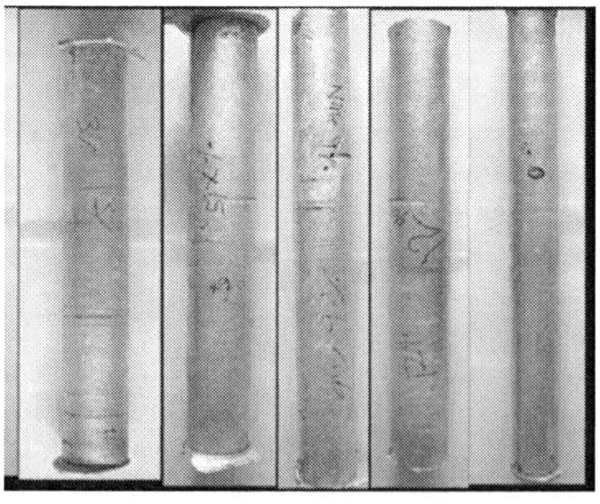

Fig.2. Cast Specimens for different weight fractions

3.3 Experimental details

Experiments were conducted on turning using turnmaster-35 lathe. IEICOS Model 621A lathe tool dynamometer was used to measure the cutting force, the details of lathe tool dynamometer is as shown in Table 3. In this study, three feed rates 0.2, 0.28 and 0.355 mm/rev. and three speeds 26.39, 42.41 and 66.91 m/min is considered for the measurement of cutting force at constant depth of cut of 1 mm for different weight fractions. All the turning experiments were carried out in dry condition [12, 13].

Table 3. Specifications of lathe tool dynamometer

Sl. No.	Specification	Values
1	Model	621A
2	Range of force	200 kgf
3	Directions	X - Horizontal; Y-Vertical/Tangential; Z –Thrust
4	Sensor type	Strain gauge
5	Sensor resistance	350 Ω
6	Bridge voltage	8V DC Max
7	Operating temperature range	Ambient temperature

Fig.3. Experimental setup - Lathe and lathe tool dynamometer

4. Results and Discussions

4.1. Microstructure study

To ascertain the uniform distribution of reinforcement particles in the Al6061 matrix alloy, microstructural study was carried for all the samples. SEM photographs were obtained for all the test specimens, which reveals that reinforcement particles are uniformly distributed throughout the composite material as shown in figures from 4 to 6. Fig. 7 to 9 shows the EDX spectrums of the composite for 3%, 6%, 9% and 12% weight fractions respectively. In all the spectrums it is observed that the presence of TiC particles along with other elements of Al6061 alloy [8].

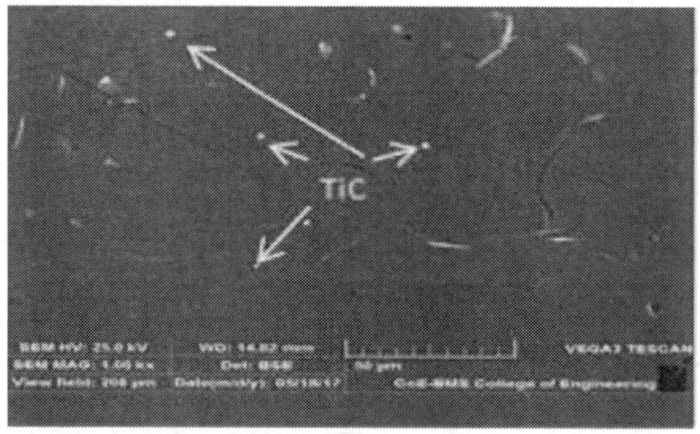

Fig. 4. SEM Micrograph depicting the presence of TiC particles in the composite for 3% reinforcement

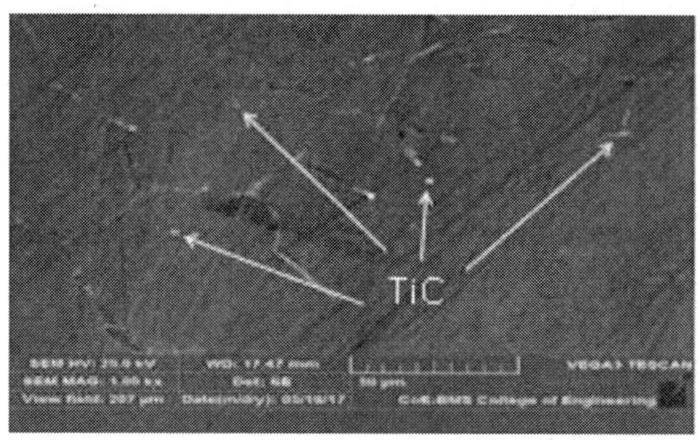

Fig. 5. SEM Micrograph depicting the presence of TiC particles in the composite for 6% reinforcement

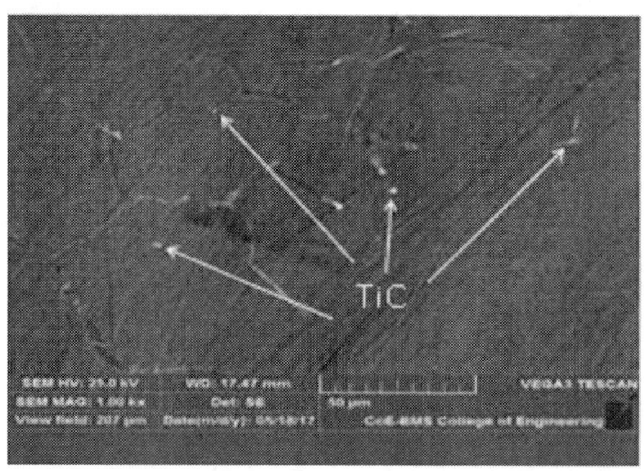

Fig. 6. SEM Micrograph depicting presence of TiC particles in the composite for 12% reinforcement

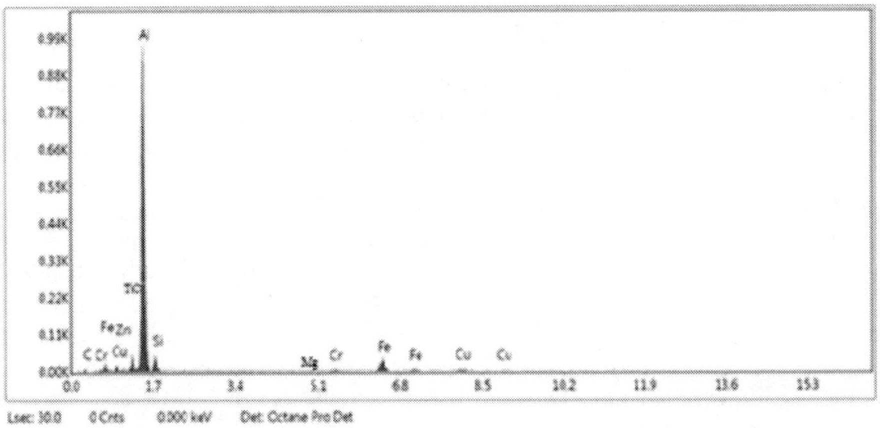

Fig. 7. EDX Pattern of Al-3 % TiC

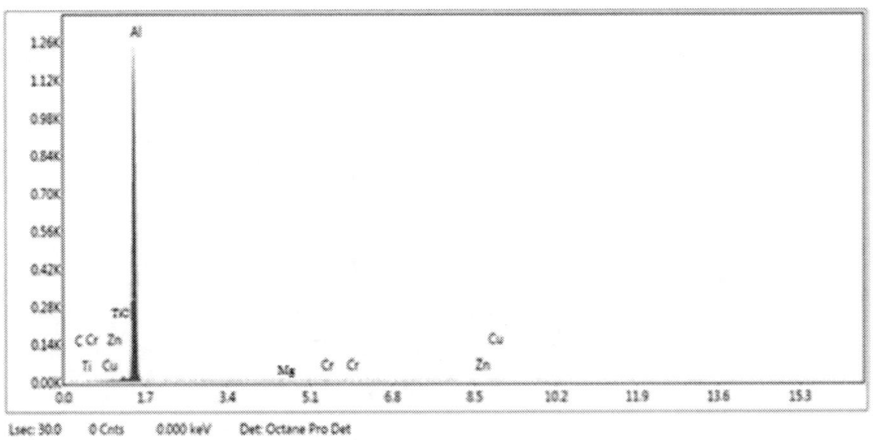

Fig. 8. EDX spectrum of Al6061/6 % TiC MMC

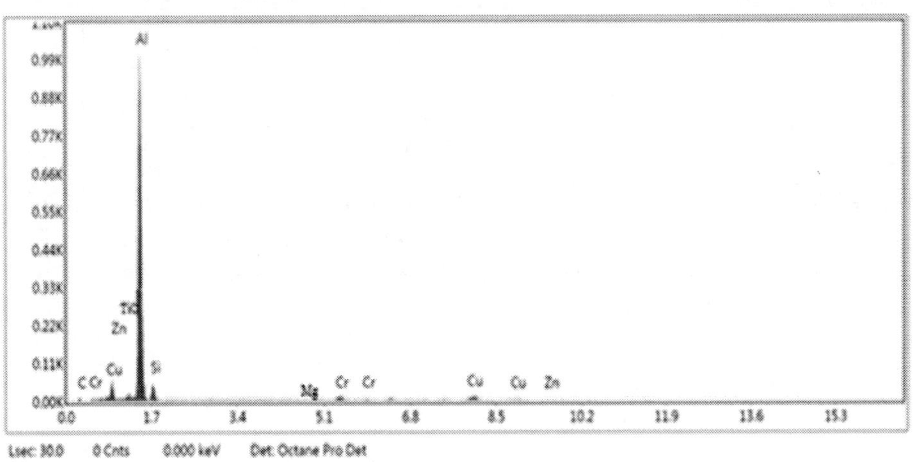

Fig. 9. EDX spectrum of Al6061/12 % TiC MMC

4.2. Measurement of cutting force

Machining of test specimens was carried out at a constant speed of 26.39 m/min for three different feed rates, and cutting forces were measured for all the weight fractions. The same procedure was repeated for 42.41 m/min and 66.91 m/min speeds and the results were tabulated as shown in Table 4. The variation of measured cutting force against cutting speed and feed rates were plotted for all the weight fractions as shown in Fig. 10 to 15.

It is observed that, the cutting force increases with the increase of feed rate for all the constant speeds and also for all weight fractions. It is due to the increase of brittleness and also more friction between chip interfaces. In some cases, the cutting force decreases with the increase of feed rate, probably due to the delamination of TiC particles from the matrix and also due to slight reduction of brittleness of the composite because of increasing the temperature at higher feed rate. On the other hand, the cutting force decreases with the increase of cutting speed for all the constant feed rates and weight fractions, this may be due to reduction of friction at higher speeds [11].

Table 4. Cutting force values for different weight fractions

Trial No.	% of TiC particles	Cutting speed m/min	Feed rate (mm/rev)	Cutting force (N)
1	0	26.39	0.2	49
2			0.28	88
3			0.355	98
4		42.41	0.2	49
5			0.28	59
6			0.355	108
7		66.91	0.2	39
8			0.28	98
9			0.355	118
1	3	26.39	0.2	59
2			0.28	69
3			0.355	78
4		42.41	0.2	39
5			0.28	69
6			0.355	98
7		66.91	0.2	59
8			0.28	49
9			0.355	118
1	6	26.39	0.2	29
2			0.28	118
3			0.355	118
4		42.41	0.2	20
5			0.28	20
6			0.355	29
7		66.91	0.2	39
8			0.28	49
9			0.355	98
1	9	26.39	0.2	59
2			0.28	78
3			0.355	118
4		42.41	0.2	78
5			0.28	118
6			0.355	137
7		66.91	0.2	108
8			0.28	118

9			0.355	147
1	12	26.39	0.2	69
2			0.28	78
3			0.355	78
4		42.41	0.2	49
5			0.28	59
6			0.355	128
7		66.91	0.2	78
8			0.28	128
9			0.355	137

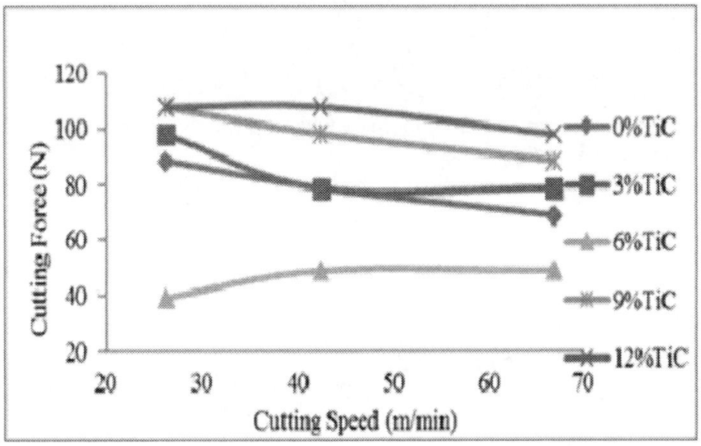

Fig. 10. Variation of cutting Force with the increase of cutting speed for different weight fractions at feed rate of 0.2 mm/rev & 1 mm depth of cut

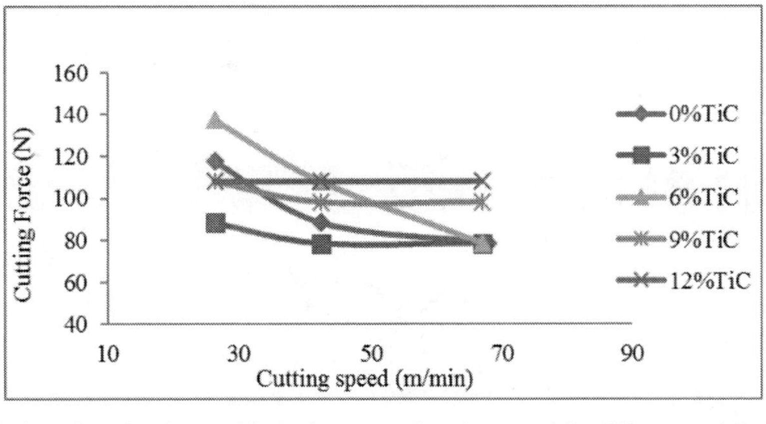

Fig. 11. Variation of cutting force with the increase of cutting speed for different weight fractions at feed rate of 0.28 mm/rev & 1 mm depth

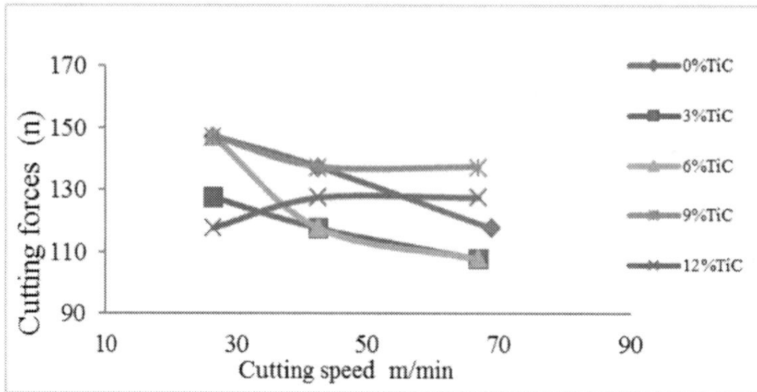

Fig. 12. Variation of cutting force with the increase of cutting speed for different weight fractions at feed rate of 0.355 mm/rev & 1 mm depth of cut

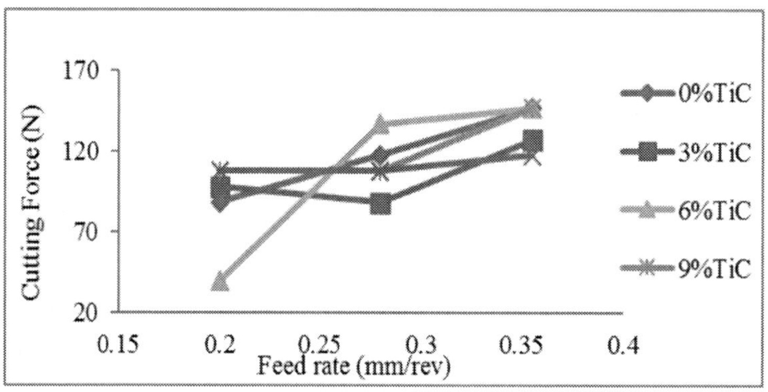

Fig. 13. Variation of cutting force with the increase of feed rate for different weight Fractions at constant speed of 26.4m/min & 1 mm depth of cut

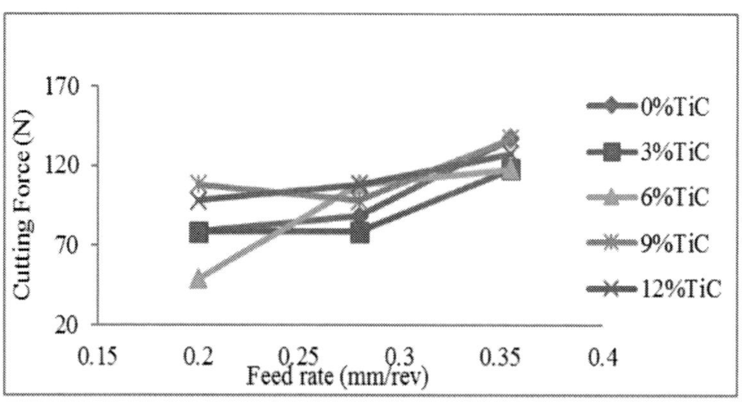

Fig. 14. Variation of cutting force with the increase of feed rate for different weight fractions at constant speed of 42.41 m/min & 1 mm depth of cut

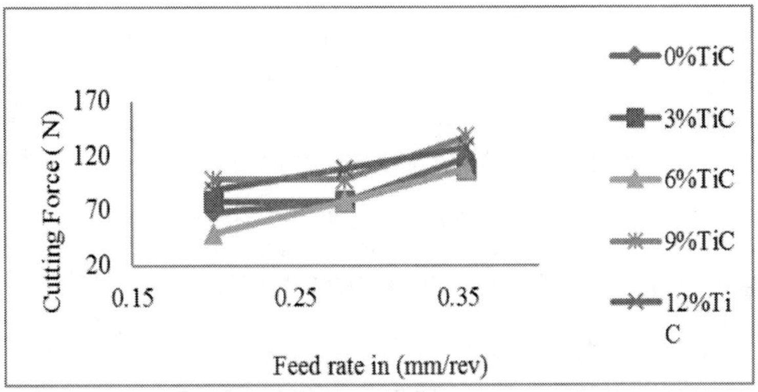

Fig. 15. Variation of cutting force with the increase of feed rate for different weight fractions at constant speed of 66.92 m/min & 1 mm depth of cut

4.3. Hardness

Hardness test was carried out on model KB3000H Brinell hardness tester. Three trials were conducted for each composition, average values were calculated, and the results were tabulated as shown in Table 9. It is observed from that, hardness of the composite increases with the increase of wt.% of TiC particles because of increasing the brittleness of the composite.

Table 5. Hardness test results

Sl. No.	Material	Force applied in N	Indentor diameter, in mm	Indentation diameter, in mm	Brinell Hardness Number (BHN)
1	Al-6061 + 0% TiC	500	10	3.7	44.88
2	Al-6061 + 3% TiC	500	10	3.6	47.46
3	Al-6061 + 6% TiC	500	10	3.5	50.31
4	Al-6061 + 9% TiC	500	10	3.3	56.81
5	Al-6061 + 12% TiC	500	10	3.1	64.60

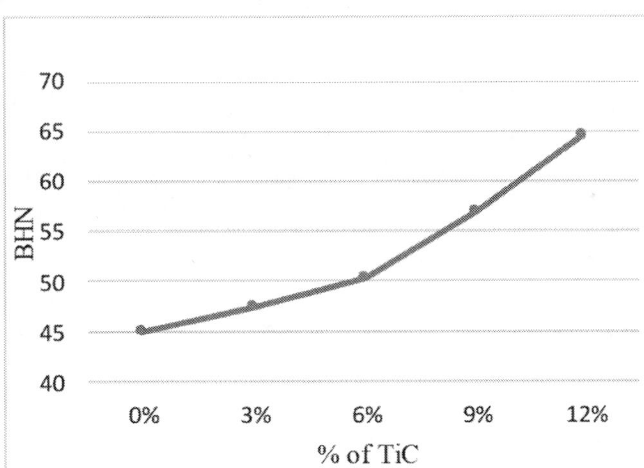

Fig. 16. Variation of hardness with the increase of reinforcement in the composite

4.4 Design of experiments (DOE)

Taguchi technique was adopted to determine the best probable grouping of process parameters to compute minimum cutting force [11]. Taguchi techniques like orthogonal arrays and signal to noise (S/N) ratio are the major tools used for minimizing of machining parameters. Orthogonal array provide several design parameters instantaneously and design of experiments with array of orthogonal declares stable evaluation and interaction of design factors. The representations for the columns of the orthogonal array with experimental process factors were optimized and the rows represent the levels of each parameter.

Table 6. Process parameters and their levels

Process parameter	Level 1	Level 2	Level 3
Cutting speed (m/min)	26.39	42.41	66.91
Feed rate (mm/rev)	0.2	0.28	0.355
Depth of cut (mm)	1	1	1

S/N ratio was calculated using Taguchi method in Minitab software (version: 17). The experiments were carried out using parameters of the L9 orthogonal array. Table 6 shows the experimental plan based on L9 orthogonal array.

4.5 Analysis of S/N Ratio

In the present study, minimum cutting force was considered as quality characteristic for turning of Al 6061/TiC MMC. Lower of cutting force is desirable, thus "Lower is better" concept was adopted for optimizing the process parameters using Taguchi technique.

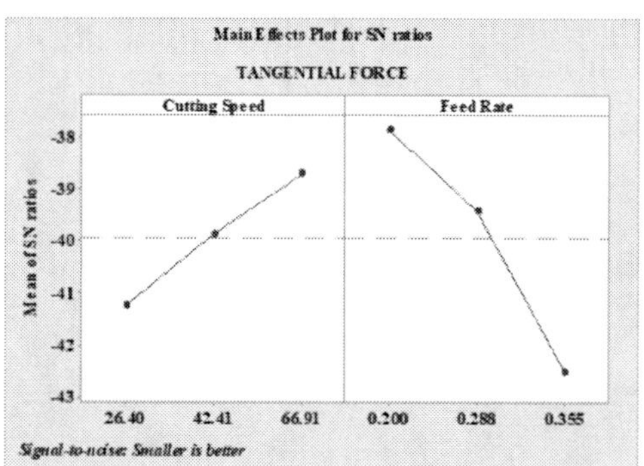

Fig. 16. Plots for SN ratio of Al6061- Alloy TiC

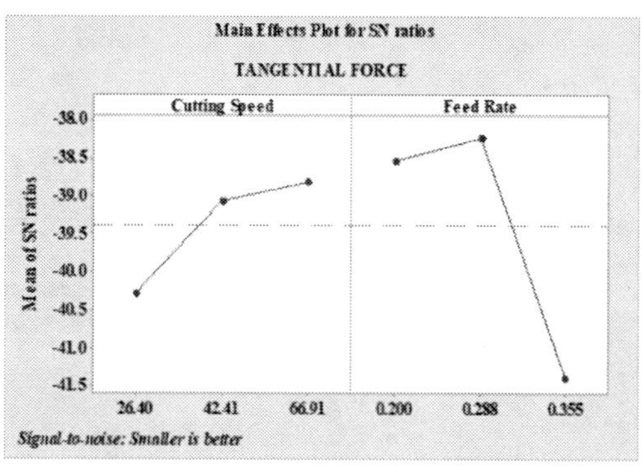

Fig. 17. Plots for SN ratio of Al-6061 with 3%

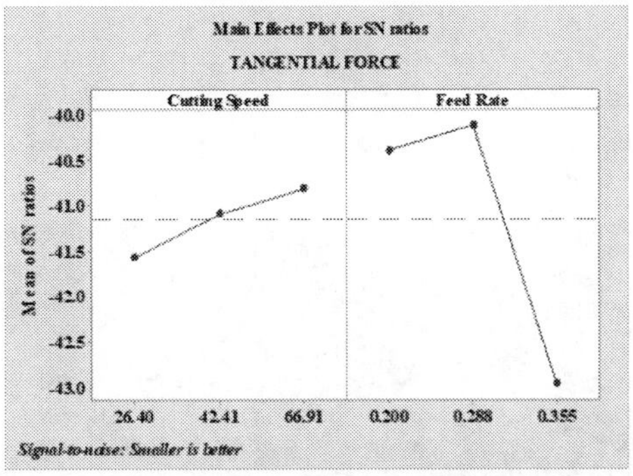

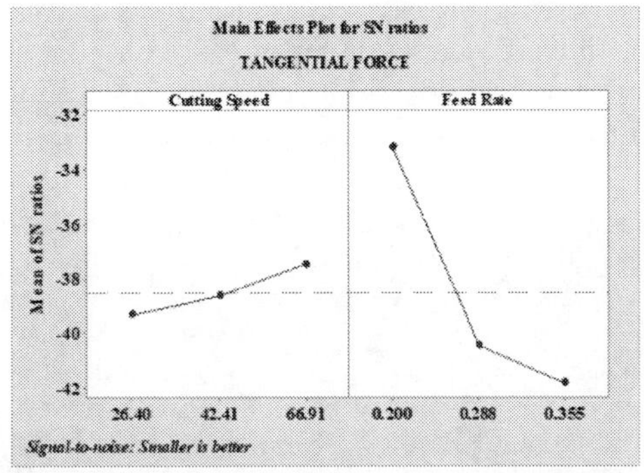

Fig. 18. Plots for SN ratio of Al-6061 with 6% TiC **Fig. 19**. Plots for SN ratio of Al-6061 with 9% TiC

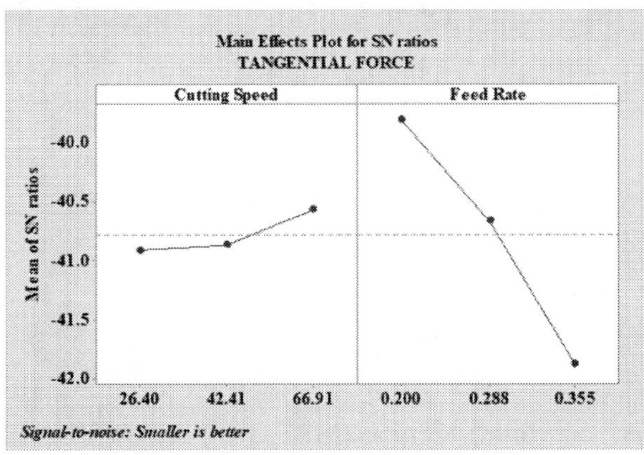

Fig. 20. Plots for SN ratio of Al-6061 with 12% TiC

Fig. 16 to 20 shows the plots for SN ratio to decide the ideal set of machining parameters for this experimental design. The control factors used are cutting speed and feed rate. Table 7 shows the optimum parameters and their level, which gives the minimum cutting force [11]. Also, there are no conflicts happening in determining the various parameters while the standard parameters of the lowest response and uppermost S/N ratio.

Table 7. Process parameters and their levels

Wt % of TiC in 6061Al	Optimum Cutting Speed	Optimum Feed Rate
0	Level 3 (66.91m/min)	Level 1 (0.2m/rev)
3	Level 3 (66.91m/min)	Level 2 (0.28m/rev)
6	Level 3 (66.91m/min)	Level 1 (0.2m/rev)
9	Level 3 (66.91m/min)	Level 2 (0.28m/rev)
12	Level 3 (66.91m/min)	Level 1 (0.2m/rev)

4.6 Chip Formation

Chip formation during machining was complemented by drastic plastic deformation at shear zone. TiC particulate added into Al alloy was initiated to decrease in ductility and produces a discontinuous type of chips. It is witnessed that the size of chip decreases on increasing the wt % of TiC. Fig. 22 (a, b, & c) shows the photographs of chip shapes produced by machining of cast composites. The size of the chips formed during machining of pure Al alloy is the chips produced by composites. Further the size of the chip formed during machining of Al reinforced with 3% TiC is longer compared to Al reinforced with 12% TiC. This shows that reinforcement causes reduction in the size of chips.

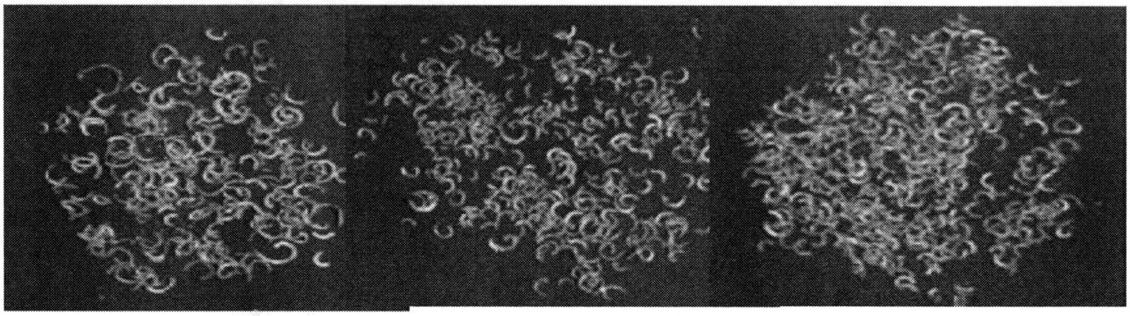

Fig. 22a. Shape of chips produced in Al alloy at 710 rpm & 0.355 mm/rev.

Fig. 22b. Shape of chips produced in Al/ 3% TiC MMC at 710 rpm & 0.355 mm/rev.

Fig. 24c. Shape of chips produced in Al/ 12% TiC MMC at 710 rpm & 0.355 mm/rev.

5 Conclusions

- SEM micrograph shows uniform distribution of TiC particles with minimum agglomeration.
- EDX tests depicted the presence of TiC particles in the composites.
- It is observed that, the cutting force increases with the increase of feed rate for all the constant speeds and for all weight fractions. In some cases, the cutting force decreases with the increase of feed rate, probably due to the delamination of TiC particles from the matrix and due to slight reduction of brittleness of the composite because of increasing the temperature at higher feed rate.
- Cutting force decreases with the increase of cutting speed for all the constant feed rates and weight fractions. This may be due to reduction of friction at higher speeds.
- Hardness of the composite is increases as the wt. % of TiC particles increases for all wt. % fractions.
- The optimum process parameters for achieving minimum cutting force for the material 0%,6% and 12 % TiC composite at cutting speed of 66.92 m/min and feed rate is 0.2 mm/rev compare with remaining Al- TiC composite materials.

References

1. Pushpendra kumar Jain, S C. Soni and Prashanth V.Baredar, Review on Machining of Aluminium Metal Matrix Composite, Material Science Research India, **14(2),** 114-120, (2014).
2. V. Songemene, R. Khettabi, J. Kouam, A. Dejebra, Machining and Machinability of Aluminium Alloys, Intech, pp 378 - pp 400.
3. Rabindra Behra, S.Kayal, N.R. Mohanta, G.Sutradhar, Study on Machinability of Aluminium Silicon Carbide MMC, Indian Foundry Congress, (2013)

4. Anand Kumar, M.M .Mahapatra, P.K.Jha, Effect of Machining Parameters on Cutting Force and Surface Roughness of In-Situ Al-4.5%Cu/TiC MMC, Elsevier,Measurment,**48**,325-332, (2014)
5. Nassim Samer, Bruno Gardiola, Nikhil Karnatak, Olivier Martin, Microstructure and Mechanical Properties of Al-TiC by Reactive synthesis, Elsevier, Composites Part A,**72**,50-57 (2015)
6. Himanshu Kala, K.K.S Mer, Sandeep Kumar, a Review on Mechanical and Tribological Behaviour of Stir Cast AMMC, Elsevier, Procedia Materials Science, **6**, 1951-1960, (2014).
7. Suraya Sulaiman, Shamsuddin Sulaiman, Noor Jaffer, Studies On Tensile Properties Of TiC Particulate Composites, Universiti of Malaysia Pahang, (2014)
8. P. Danashekaran, M. Ravishankar, S.Sathishkumar, N. Velavan, Tensile Test and Wear Behaviour of Titanium Carbide Reinforced With Al6063 MMC, International Research of Innovative Research in Science, (2015)
9. S. Venkatesan, M.Anthony Xavior, Comparative Analysis Of Aluminium MMC Reinforced With TiC And TiB2 Using Stir Casting Process, IJSRES,**4**, (2015)
10. C.R. Prakash Rao, Bhaghyashekar M.S, Narendra Vishwanath, Machining Behaviour of Al6061-Fly Ash Composites, Elsevier,**5**,1593-1602, (2014)
11. D. Saichaitanya Kishore, K. Prahalad Rao, A. Mahamani, Investigation of Cutting Force, Surface Roughness and Flank Wear in Turning of In-Situ Al6061-TiC MMC.Elsevier,**6**,1040-1050, (2014)
12. Dr. R.R Malagi, Rajesh BC, Factors Influencing In Turning And Development Of Software To Estimate Cutting Forces In Turning, IJEIT,2,37-43, (2012)
13. Dr. Thella Babu Rao, *Theory of Metal Cutting and Machining Tools*, GITAM University.

Friction stir welding of Aluminium matrix composites – A Review

Prabhu Subramanya[1*], *Murthy Amar*[2], *Shettigar Arun*[2], *Herbert Mervin*[1] *and Rao Shrikantha*[1]

[1]*NITK Surathkal, Karnataka, India*
[2]*Department of Mechanical and Manufacturing Engineering, Manipal Institute of Technology, Manipal Academy of Higher Education, Manipal, Karnataka, India*

Abstract: Friction stir welding (FSW) is established as one of the prominent welding techniques to join aluminium matrix composites (AMCs). It is a solid state welding process, takes place well below the melting temperature of the material, eliminates the detrimental effects of conventional fusion welding process. Although the process is capable to join AMCs, challenges are still open that need to be fulfill to widen its applications. This paper gives the outline of the friction stir welding technique used to join AMCs. Effect of process variables on the microstructure and mechanical properties of the joints, behavior of reinforcing materials during welding, effect of tool profiles on the joint strength are discussed in detail. Few improvements and direction for future research are also proposed.

1 Introduction

Aluminium Matrix composites reinforced with particulates are considered as an ideal material for aerospace automotive, marine application as it displays better mechanical properties compared to aluminium and its alloys. AMC possess high strength to weight ratio, higher formability and higher resistance to wear and corrosion [1, 2]. However, the potential application of AMCs are limited if a proper welding process is not developed [3, 4]. AMCs joined through conventional fusion welding methods unveils several short comes in terms of quality of joints. Fusion welding leads to the formation of harmful phases in the joint region due to excessive heat, cracking, porosity, distortion and settling of hard particles at the bottom of the weld region due to the difference in densities [5-7]. A solid state welding process seems a favorable joining method for AMCs.

Friction stir welding is a solid state welding process, was developed and patented by The Welding Institute (TWI) in 1991, to weld aluminium alloys [8]. It was then efficaciously tried to join Cu, Mg, Ti alloys, steel and aluminium, magnesium, steel based metal matrix composites [9, 10]. A prolific literature shows that FSW process can be successfully employed to weld difficulty-to-weld materials with improved mechanical

* Corresponding author: subramanya.prabhu@gmail.com

properties, without weld defects [11]. But the presence of reinforced materials in the AMCs limits the range of welding parameters in comparison with the aluminium alloys.

Several review papers were available in the field of FSW process. Threadgill [9] reviewed the significance of FSW method used to join aluminium alloys. Mishra et al. [12] gave a meticulous review on friction stir welding and friction stir processing. Thomas et al. [13] and Zang et al. [14] discussed thoroughly about different FSW tools and their effect on FSW processes. Wang et al. [15] and Salih et al. [16] gave a detailed review on FSW of aluminium based metal matrix composites. Significance of FSW process variables on microstructure and mechanical properties of the welded joints are discussed in detail in their reviews. Fenoel et al. [17] systematically reviewed FSW of Metal matrix composites with various kinds of matrix material. They discussed modification occurs in the microstructure and mechanical properties of the friction stir welded zones in detail. The present paper reviews the latest development took place in FSW of AMCs. After a brief description of FSW method, comprehensive assessments of issues in the FSW of AMCs are addressed. Effects of FSW process on mechanical properties, micro and macrostructure of AMCs, welding defects and wear of tools are explained in the subsequent sections. The paper is ended with Conclusion and future research directions.

2 FSW Process:

FSW is a solid state, environment friendly welding process, where a non- consuming specially designed rotating tool is plunged into a pair of plates which are tightly held together on a back plate in a rigid FSW fixture (Fig. 1). The tool consists of shoulder and prolonged pin whose length is slightly less than the thickness of the welding plate. Rotating tool is inserted into the plates till the shoulder touches the top surface of the plate. The frictional heat is produced between the tool shoulder and the plate interface resulting in plasticization of the material below the shoulder. Rotating tool remains in the plunged position for a short duration of time. The tool made to traverse along the joint weld line after a short delay time. As the tool traverses, the plasticized material is get solidified behind the rotating tool. Plates are being welded due to both forging and extrusion effects. The tool is lifted at the end if the process, leaves a key hole of a size of rotating pin in the plates.

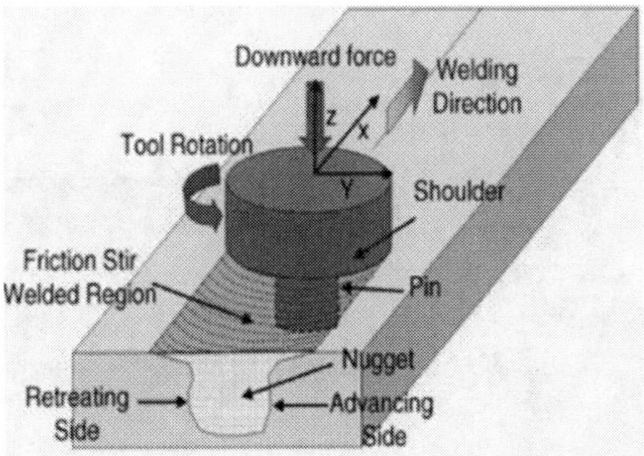

Fig. 1. Schematic illustration of FSW process [12].

The weld line divides the work piece into two parts, i.e. advancing side and retracting side [12]. The advancing side is a side where tool rotation and tool movement is in the same direction and in the retracting side the tool rotates opposite to that of tool movement. As the tool moves along the weld line, the plasticized softened material is swept from advancing side to the retracting side and gets solidified behind the rotating tool. The welding institute initially developed FSW process to join non weld-able aluminium alloy series, and then FSW technique is extended to join AMCs. Plates with dissimilar compositions or different metals can be joined by this technique. Composition compatibility can be ignored in FSW process which is a major concern in conventional fusion welding which leads to solidification cracking [18].

Various welding parameters affect the material flow around the tool pin which decides the quality and strength of the joint. Rotation speed of the tool, tool traverse speed, tool geometry, tool material, axial force and tool tilt angles are the major parameters influence the quality of the weld [8, 12].

3 Grain Structure of a FSW Joint:

Frictional heat generates at the rotating tool and work piece interface leads to rigorous plastic deformation of the material at the weld zone. The amount of heat induced due to the rotation of tool will be varied for different rotation speed of the tool. Defects are formed if the amount of heat induced in the weld zone either excessive or insufficient. The grain structure, mechanical strength and the quality of the joints depends upon the quantity of heat produced in the weld zone. Hence proper process parameters to be chosen to get good quality welds. By evaluating the macro and micro structure of the weld zone, the quality of the FSW can be assessed [12, 18, 19].

3.1 Macrostructure evaluation of Friction stir weld zone:

Due to the variation in the thermo mechanical effect, four distinct zones are visible in the weld region. The grain size, orientation and distribution of particles are different at these four zones [12]. The zones are classified as Nugget zone (NZ) or stirred zone (SZ), Thermo mechanically affected zone (TMAZ), Heat affected zone (HAZ) and the parent or base material (PM or BM) as shown in the Fig. 2. The central zone of the weld area is called nugget or stirred zone which is substantially deformed by the rotating tool pin. Severe stirring action of the tool pin, heats up this area leads to the plastic deformation of the material. Fine equi-axed grains are visible in this area arranged in the form of onion shape [20] as depicted in Fig. 3. Softened material flow and dislocation density differences during the welding process are the main reason to form onion shaped structure in the NZ [21, 22].

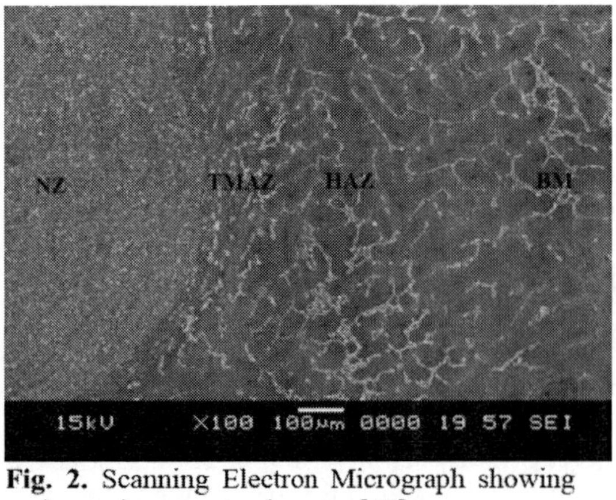

Fig. 2. Scanning Electron Micrograph showing various microstructural zones. [19]

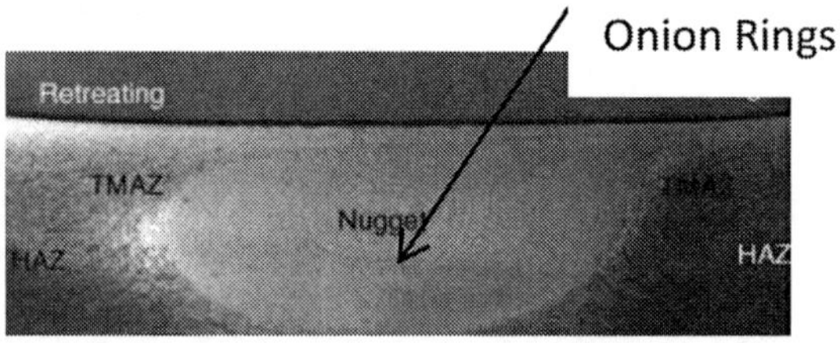

Fig. 3. Onion Ring structure in Nugget Zone [12]

Next to nugget zone is TMAZ whose peak temperature is slightly lesser than that of NZ. The grains are elongated and oriented in the inclined direction due to the thermo mechanical effect of rotating tool. Heat affected zone which is in between Base material and TMAZ, experiences only heating effect. Rotating pin does not mechanically deform this area. Grains are coarsened due to the thermal effect. The peak temperature attains in HAZ is much less than that of TMAZ. The base material or parent material is unaffected zone where grain structure is not changed due to FSW.

Welding defects such as tunnel holes (Fig. 4) are visible during macrostructure assessment of the FS welded joint. Tunnel defect is formed mainly due to the inadequate heat input and inapposite flow of plasticized softened material in the weld zone. Nami et al. [27] pointed out that tunnel defects are formed at both low and high rotating speed. They showed that inappropriate material flow in the stirred zone due to either less heat input at low rotation speed of 700-900 rpm or excessive heat at the higher rotation speed of 1300 - 1400 rpm, results in tunnel defects. Similarly Cioffi et al. [23] observed the formation of tunnel defects at low rotation speed and defect is disappeared as the rotation speed is increased to 800 rpm.

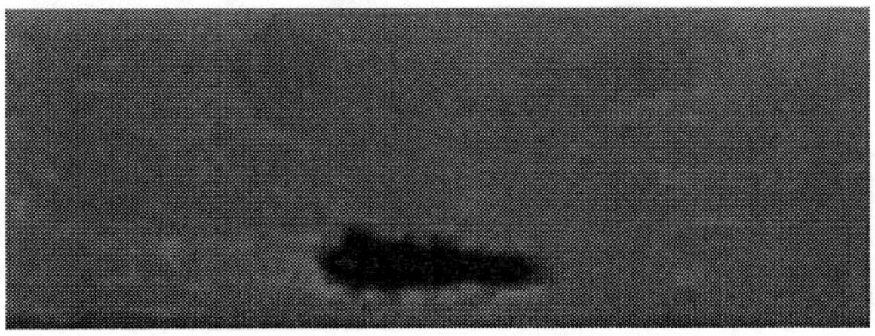

Fig. 4. Tunnel hole in cross section of an AA6061-4.5 Cu- 5 Sic joint welded at 1500 rpm and 50mm/min [19]

3.2 Microstructure of FSW zone:

During FSW process, the friction between the rotating tool and work piece increases the peak temperature in the weld zone up to 0.75 to 0.8 times of the melting temperature of the base material. Redistribution of reinforced particles, reorientation, grain refinement, recrystallization and grain growth will takes place due to the rise in the temperature at the stirred zone. During production of AMCs either by stir casting or powder metallurgy process, it is common that the reinforced particles are clustered, non-uniformly distributed in the matrix. This will affects the strength characteristics of the AMCs, which is different at different location, results in anisotropy mechanical properties. Stirring action and plastic deformation caused by the rotating pin, separates the clustered particles and redistributes the particles uniformly in the stirred zone [20, 23, 25-27]. Due to the stirring action, the edges of the reinforced particles are rounded off and the size of the particles decreases compared to the parent material. This occurs due to the abrasion and mixing effect between work piece and the circumference of rotating pin, shoulder face [5, 20, 28, 29]. Size of the particles is closer to the shoulder is smaller than the tool pin tip [30]. However there is no change in the particle size after FSW process, if the particle size is already smaller in the base material as reported by Storjohann et al. [28]. They show that particles were reoriented and bigger particles were exposed to grain refinement.

Periyasamy et al. [24] showed that lower heat input at the nugget zone results in faster cooling of plasticized material causes development of coarse grains due to the lack of complete recrystallization in this zone. On the contrary higher heat input at the nugget zone due to the high rotational speed, results in finer grain structure and eliminates dendritic structure in this zone. Fine equi axed grains are observed in stirred zone which is smaller than the base material grain size, indicates nucleation of new grains in this zone [32]. Wang et al. [33] observed high energy point on the broken surface of reinforced material (Sic) is the area of the nucleation and growth of the grains.

In conclusion, to attain superior mechanical properties in the AMCs, the reinforced materials are to be uniformly distributed in the matrix. Apposite FSW process parameters produces essential heat input results in proper material flow and uniform distribution of reinforced particles and improved joint quality. The sharp edges of reinforcements are rounded off and bigger particles are split into smaller one. FSW changes the design of the AMCs at the weld region by changing the spacing, distribution, morphology and size of the reinforced particles. This modification is caused by the dynamic recrystallization in the nugget zone. The reformed grain structure decides the properties of the AMCs at the weld zone.

4 Mechanical properties of FS welded joints:

Mechanical properties such as hardness and tensile strength of friction stir welded joints are depending on the process parameter. Optimization of process parameters, that yields the higher joint efficiency, is done by evaluating these properties.

4.1 Micro hardness of joints:

Micro hardness across the weld zone indicates the distribution of reinforced particles and various phases. In the weld zone two different micro hardness curve profiles are observed. One type of graph shows highest value of hardness in the stirred zone and its goes on decreasing along TMAZ and HAZ, and shows minimum value in the base material region as shown in Fig. 5 [24, 26, 27, 31, 33-35]. Refinement of grain caused by dynamic recrystallization and proper distribution of particles due to the severe stirring action of FSW tool, results in higher hardness at the stirred zone. Decrease in heat input at the nugget zone leads to the incomplete recrystallization, develops coarser grain and there by reduces the hardness (Fig. 5).

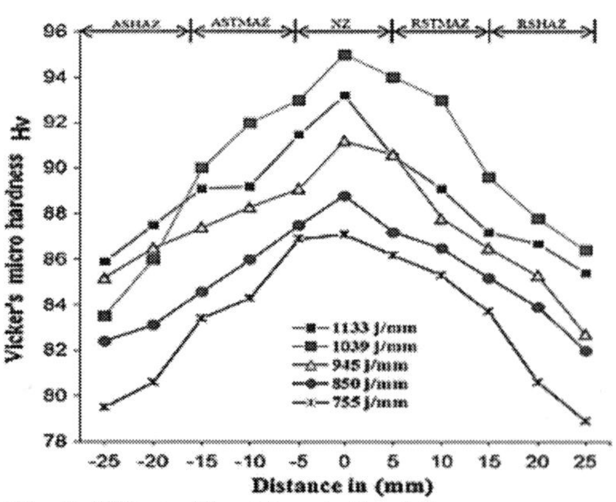

Fig. 5. Effect of heat input of welding process on micro hardness profile across the weld region of AA6061-10 SiC [24]

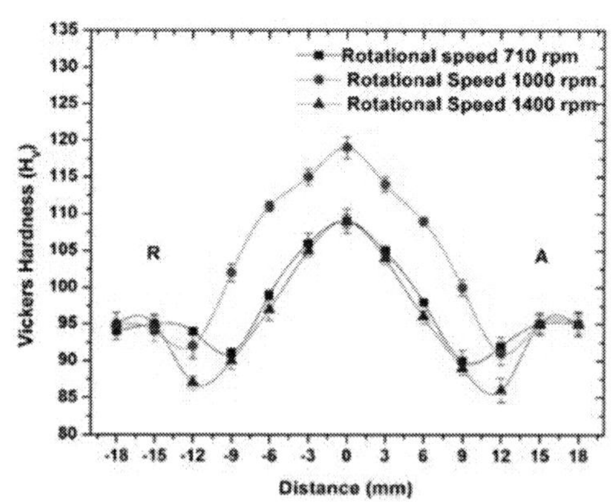

Fig 6. Micro hardness profile distribution across the weld zone of AA6061-4.5Cu-10SiC [19]

Second type of graph shows the lowest hardness at the HAZ, and highest value in the NZ as shown in the Fig. 6 [20, 23, 25, 36]. In the microstructural analysis it was observed that the microstructure of TMAZ show the transition of base metal into nugget zone. TMAZ is characterized by the elongated grains, where the material is plastically deformed and thermally affected. Temperature attained in HAZ is sufficient to cause dissolution of fine secondary precipitate. The dissolution of fine strengthening precipitate in turn promotes coarsening of aluminium grains which led to reduction in hardness value at HAZ. Hence W shaped hardness distribution across the joint profile is observed.

4.2 Tensile Properties of FS Welded Joints:

Compared with conventional fusion welding, FSW yields higher joint efficiency. Joint efficiency is defined as the ratio of tensile strength of friction stir welded AMC to that of Base material. Various factors affect the joint efficiency such as process variables, tool profile and development of intermetallic compounds. Table 1 summarizes the joint efficiencies of AMCs welded by FSW.

4.2.1 Effect of Process parameters:

FSW process parameters such as rotation speed of the tool, tool traverse speed and axial force affects the quantity of heat generated at the tool – work piece interface, stirring of softened material and finally solidification of the material behind the rotating tool and forming the joint. To obtain superior quality joints, it is necessary to choose optimum process parameters. The joint efficiency is normally in the range of 60 to 96 % of those of the base material. UTS of welded joints can be increased by increasing the rotational speed up to a certain limit. Several researchers [20, 24, 26, 27, 37] are reported that maximum UTS is attained by using a rotational speed in the range of 1000 to 1200 rpm depending on the type of AMCs. Rotational speed less than 1000 rpm results in inadequate heat input, affect the plasticization and proper mixing of material in the weld zone. Similarly rotational speed higher than 1200 rpm results in excessive heat input, leads to turbulent material flow in the nugget zone, thereby deteriorates the weld quality and strength.

Tool traverse speed is another important factor which affects the strength of the joint. The amount of heat transfers to the work piece, which is generated at the tool work piece interface and the rate at which material cools and solidifies, is depended on the tool traverse speed.

Several researchers [36, 37-39] are pointed out that there is no proportional relation between tool traverse speed and the UTS of the joint. Increase in tool traverse speed increases the UTS of the joint up to a certain value then it start decreasing as the traverse speed further increases. The quantity of heat induced at the weld zone depends on the tool rotation speed. Whereas the amount of heat transferred to the work piece and rate at which work piece cools behind the moving tool is influenced by the tool traverse speed [19]. For the given rotational speed of the tool, lower tool traverse speed increases the amount of heat transferred to the work piece and reduces the cooling rate, results in formation of coarser grains. Higher tool traverse speed results in lesser heat transferred to the work piece and increases cooling rate. This leads to softened material in the welding zone experiences improper stirring action results in low quality welds. Both high and low tool traverse speed results in tunnel defect. Therefore lesser UTS is observed at both low and high tool traverse speed. Wang et al. [36] showed that joint efficiency increases as the traverse speed increases and attains maximum value of 97 % at 800mm/min. At lower traverse speed, the work piece fails at HAZ (a low hardness zone) due to the dissolution and growth of precipitates. This will reduces strength of the joint. On the contrary, higher traverse speed increases the cooling rate results in partially dissolving clusters in the matrix. This results in increase in the hardness of HAZ and failure of joint is shifted from HAZ to NZ. Thus joint efficiency is increases as the traverse speed increases.

Table 1. FSW process variables and joint efficiency of AMC joints.

Base Material	Tool Material	Tool Geometry	Rotation Speed RPM	Traverse Speed mm/min	Axial force (kN)	Joint Efficiency	Ref.
AA6061/4.5Cu / 5SiC	Tool Steel (60 HRC)	Combined Square and Threaded 15-6-5.7	710-1400	50-80	-	90 %	19
AA 2124/ 25SiC	-	Cylinder thread	1120	40	-	81	20
6061/20 Al_2O_3 7005/10 Al_2O_3	-	-	800	56	-	83.87	21
AA 2124/ 25SiC	H13 Steel 48 HRC	Cylinder thread 20-7.95 to 6.35-7.62	550	75	8.5	86.5	23
AA 6061/10 SiC	High Speed steel	Cylinder thread 18-6-5.7	1100	45	6	74	24
AA2009/17 SiC	Cermet	Cylindrical with Triangular tip 14-5-2.7	1000	50	-	77	25
AA6061/12 B_4C	High carbon high chromium steel (62 HRC)	Square 18-6-5.7	997	78	9.3	96.83	26
Al/Mg_2Si	H13 Steel	Tapered (7°) Cylindrical Threaded 18-6-5.7	710-1400	125	-	~100	27
AA 6061/ 20Al_2O_3	Ultra-hard material	Cylindrical 20-8-?	700	300	-	70	28
AA 6061/ 22 Al_2O_3	Wear resistant steel 64 HRC	Cylindrical 15-4-?	880	260	-	99	29
Al/4.5Cu/TiB_2	M2 Steel(62 HRC)	Cylindrical Threaded 18-6-5	710-1400	63	-	-	30
AA 2009/ 15SiC	Steel	Cylindrical Threaded 24-8-?	600	5	-	95	31
AA 2009/ 15SiC	Ultra hard Cermet	Conical 20-8-5.8	800	100		82	32
LM25/ 5SiC	HSS Coated with TiAlN	19.5-6/4-5.7 Tapered	1200-1800	20-60	6/7/8		33
AA6061/ZrB_2	High carbon high chromium steel (62 HRC)	Square 18-6-5.7	1150	50	6	94-95	34
Al/4.5Cu/ 10TiC	Steel Shoulder Titanium pin	Threaded 18-7-4.8	500	20	6	89	35
AA6061-T6/ AlNp	High carbon high chromium steel (62 HRC)	Square 18-6-5.7	1000-1400	25-85	3-7	93.42	37
AA6061/4.5Cu / 10SiC	Tool Steel (60 HRC)	Combined Square and Threaded 15-6-5.7	710-1400	50-80	-	90 %	38
AA6061/AlNp	High carbon high chromium steel (62 HRC)	Square 18-6-5.7	1200	55	5	93.42	39
Al/12Si/10 TiC	Steel Shoulder Titanium pin	Threaded 18/20/22-7-5.8	710-100	20-40	-	-	40
AA6061/12B_4C	high carbon high chromium steel (62 HRC)	Square 18-6-5.7	997	78	9.3	96.83	41
Al/10 TiB_2	High carbon high chromium steel	Various tool pin profile	2000	30	19.6	99.47	42

Tool geometry information includes shoulder dia- pin dia (x/y with x at root and y at tip if conical or pin side if square- pin length. All dimensions are in mm.

Few studies [20, 35, 40] showed that the effect of traverse speed on the joint efficiency is not much significant as that of other process variables such as rotation speed, axial force and tool geometry. They found marginally inverse relation of traverse speed with UTS of joint. As the traverse speed increases, the UTS of joint decreased. Traverse speed of 20 mm/min and 40 mm/min are the best suited for FSW of SiC reinforced and TiC reinforced composite respectively.

Another significant parameter in FSW is applied axial force. The friction coefficient quantified by the applied axial force, governs the quantity of heat induced due to the friction between tool and the work piece. Higher axial force results in increased heat input at the weld zone, results in proper material flow. Defect less, high quality and better joint strength can be attained with appropriate axial force.

Several researchers analyzed the effect of axial force on the FS welded joint strength. Lower axial force, less than 5 kN, results in lesser heat generation leads to the tunnel defect in FSW of AA6060/AlNp [37]. On the contrary axial force more than 5 kN causes thinning of stirred zone and worm hole formation. Thus axial force beyond the optimum value results in drop in joint efficiency.

Similar results were shown by Dinaharan and Murugan [38]. Joint strength of FS welded AA6061/ZrB2, shows higher value for the axial force of 6kN. Excessive flash and thinning of nugget zone is observed for forces larger than 6 and formation of micro voids are visible for the axial force less than 6 kN. Kalaiselvan and Murugan [41] were pointed out similar finding in their work of FSW of B_4C reinforced AA6061 matrix with an ideal force of 10 KN.

In conclusion, UTS of FSW joint greatly affected by the Process variables including tool traverse speed, rotational speed and axil force. In comparison with other two factors, rotational speed has the major effect on UTS of the joint. However there is no fixed relation between process parameter and the joint strength, for all types of AMCs. Therefore each AMCs needs its own analysis to accomplish its maximum joint strength.

4.2.2 Effect of tool Profile:

Different shoulder profiles (concave, convex, flat) and pin geometries (Cylindrical, Square, triangular, hexagonal, straight tapered etc.) are used by researchers to carry out FSW.

Kumar et al. [35] studied the effect of shoulder surface (2 mm flat, 1 mm flat and fully flat surface) with 7° concave angle on the tensile properties and suggested shoulder with 2mm flat surface gives higher tensile strength. Higher contact surface between tool and work piece induces higher heat, results in proper mixing of softened material in the nugget zone. Yigezu et al. [40) studied the effect of shoulder diameter on tensile properties with three different diameters (18, 20 and 22mm). They reported that to obtain maximum efficiency, shoulder with 20 mm diameter is preferred.

Vijay and Murugan [42] showed that straight square profiled pin delivers highest tensile strength with a joint efficiency 99.47% in comparison with all other tool profiles like cylindrical hexagonal and orthogonal, both tapered and straight. Higher volume ratio (ratio of swept volume during rotation to the static volume) of the tool results in higher joint efficiency. Volume ratio of square, hexagon and octagon pin geometries are 1.56, 1.21 and1.11 respectively Wang et al. [36] and Hassan et al. [43] showed that conical threaded pin at high welding speed, gives higher joint efficiency compared to straight cylinder pin. This is due to the enhancement in the flow ability of softened material.

4.2.3 Effect of heat treatment and strain rate:

Post weld heat treatment, development of intermetallic compounds and strain rate also influences the joint strength. Feng et al. [31] explored the effect of post weld treatment on FS welded AA2009/SiC/15p composite. Heat treatment enhanced the joint efficiency both in longitudinal and traverse direction and reached 82 % and 95 % respectively. They reported that in the stirred zone the development of secondary phase caused by interaction of base material and the tool worn out material reduces the joint efficiency. Formation of the secondary phases can be eliminated by using ultra hard tool material. Wang et al. [36] achieved nearly 100 % joint efficiency after T4-treatment. Ni et al. [44] investigated the strain rate effect on strain hardening and mechanical properties of SiC reinforced AA2009 composite. Minor reduction in the tensile strength is observed as the strain rate increased and most of the specimens are fractured in the HAZ of retreating Side.

5 Tool wear in FSW of AMCs:

FSW tools made of standard materials have a negligible wear rate when used to weld softer materials like aluminium and its alloys. But they experience higher wear in the welding of harder materials like aluminium matrix composites and steels. Presence of hard reinforcing materials resulting in abrasion of tool material leads to the tool wear as shown in Fig. 7. If the hardness of tool material is more than the work piece, nearly no tool wear is observed. However the lower hardness of tool compare to the work piece leads to the higher tool wear. [45]. Reduction of the pin length or, erosion of tool profile due to the tool wear hampers the material flow, which increases the welding defect [46]. Root flaw defect is formed due to the lack of consolidated material at the base of the joint due to the reduction in the pin length. The tool wear mechanism is depends upon tool geometry interaction between the tool and work material and welding parameters. The wear in polycrystalline cubic boron nitride (PCBN) tools is mainly due to adhesive wear at low rotation speed while its due to abrasive wear at high tool rotational speed [12, 47]. Liu et al. [48] observed that diminution in the diameter of the pin with increasing weld length. Severe wear is noticed in the radial direction whereas negligible changes are seen in the shoulder size and pin length. Radial wear of the pin mainly depends upon the welding speed. Higher wear rate is observed during low welding speed and its maximum in the initial welding. Prater [45] reported the volumetric wear of the tool to the process parameters, i.e., rotation speed (ω), traverse speed (v) and distance welded (l). He has shown existence of direct proportionality between wear and linear weld length in his studies.

Fernandez and Murr [49] explained that as the tool rotation speed increases, tool wear also increases, whereas tool wear decreases as the welding speed increases. They experimentally proved that, In the FSW process, tool wear was shear phenomenon than drag phenomenon. Prado et al. [50] examined the tool wear rate in FSW of Al6061 + Al2O3 composite and showed that severe wear is observed during FSW of composite whereas no apparent wear for FSW of Al6061. They showed that the tool wear is increases as the tool rotation speed increases and it reaches maximum for the speed of 1000rpm and then it decreases as the speed increases beyond 1000 rpm due to the increases fluid like behavior and turbulent material flow. They also noticed that enough tool material is removed at the beginning of the welding. However as the welding continues the wear rate decreases and tool attain a self-optimized shape, where no more tool wear is observed. Tool will lose its initial shape due to the tool wear but still produces good quality weld after attaining self-optimized shape.

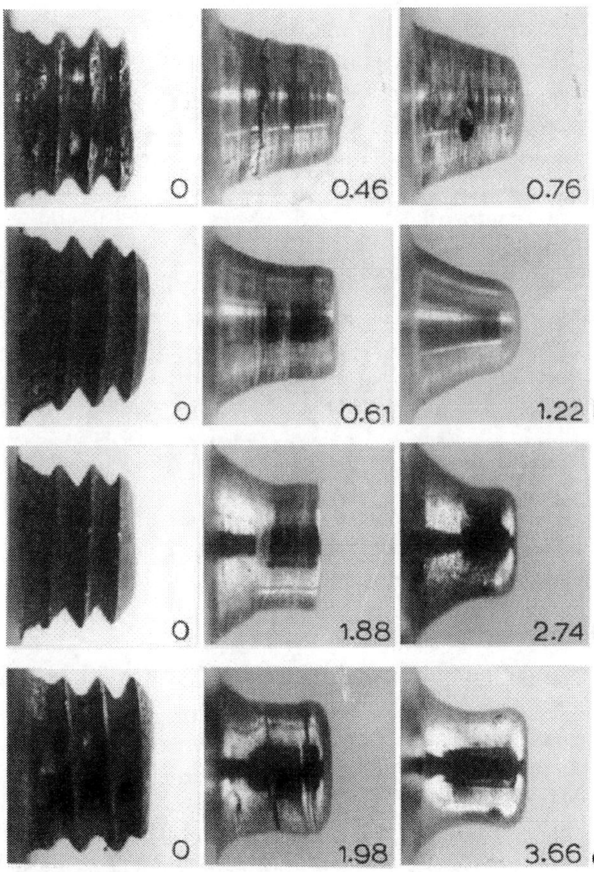

Fig. 7. Tool pin sequences showing wear features for constant rotation speed of 1000 rpm for specific traverse speeds. (a) *T*= 60mm/min; (b) *T*=180mm/min; (c) *T*=360 mm/min; (d) *T*=540mmimin. Traverse distance showed in meters [50].

6 Summary

Various aspects of FSW process is reviewed including process parameters, their effect on mechanical properties, modifications in the microstructure of weld zone and tool wear. As a solid state joining technique, FSW eliminates the defects of conventional fusion welding in AMCs such as non-uniform distribution of reinforcements, formation of deleterious phases, porosity etc. Superior joints with better mechanical properties could be attained by FSW. However research in FSW is still in its early stage and potential application of this technique to join AMCs in industry is yet to be implemented.

The joint strength and weld quality of AMCs joint are mainly dependent on composition of AMCs and the process parameters. Hence the performance of the AMCs joined by FSW should be assessed by considering both composition and process variables. Lot of research is going on to study the effect of process parameters on the joint strength. However meager studies are done to understand the effect of type and percentage of reinforcement on joint efficiency. Future studies should emphasize on the preparation of those composite with required percentage of reinforcement which gives better mechanical properties.

Selection of apposite process variables plays a key role in obtaining high quality joint with better properties. To have better material flow, material mixing and thereby to get sound joint, a large amount of heat to be induced during FSW. High heat input can be achieved by increasing the tool rotational speed and reducing the tool traverse speed. However this leads to tool wear. Generally a tradeoff between tool wear and higher heat input is found in FSW. Range of process variables used in FSW of AMCs is narrower compared to the aluminium and its alloys. Also each AMCs acts differently for given value of process parameters. Future study should focus on creating a FSW window for each AMCs for optimized mechanical performance. Also there is need for analyzing the fracture toughness and fatigue strength of FS welded AMCs to explore the potential application FSW of AMCs in industries.

Key issue in FSW of composites, reinforced with hard materials is severe wear of tool, and it is major limiting factor for the large scale industrial application of FSW. Tools with surface coating, tool made of wear resistant material, heat treatment of tools are some of the doable solution to overcome the tool wear.

References:

1. N. Chawla, K.K. Chawla, *Metal Matrix Composites*, (Springer, 2006).
2. M. Haghshenas, *Metal–Matrix Composites, In Reference Module in Materials Science and Materials Engineering*, (Elsevier, 2016).
3. M.B.D. Ellis, Joining of aluminium based metal matrix composites, Int. Mater. Rev., **41(2),** 41–58, (1996).
4. Pal, T.K., Joining of aluminium metal matrix composites, Mater. Manuf. Process., **20**, 717–726, (2005).
5. D. Storjohann, O.M. Barabash, S.S. Babu, S.A. David, P.S. Sklad, E.E. Bloom, Fusion and friction stir welding of aluminum metal–matrix composites, Metall. Mater. Trans. A, **36A**, 3237–3247, (2005).
6. K. Peng, H.C. Cui, F.G. Lu, X. M. Wu, X.H. Tang, S. Yao, S.N. Lou, Mechanical properties and wear resistance of aluminum composites welded by electron beam, Trans. Nonferrous Metals Soc. China, **21**, 1925–1931, (2011).
7. J. Niu, L. Pan, M.Wang, C. Fu, X. Meng, Research on laser welding of aluminium matrix composite SiCw/6061, Vacuum, **80**, 1396–1399, (2006).
8. W. Thomas, E. Nicholas, J. Needham, M. Murch, P. Temple-Smith, and C. Dawes, Friction Stir Butt Welding, International Patent No. PCT/GB92/02203, GB Patent No. 9125978.8, 1991, U.S. Patent No. 5,460,317, 1995. (1991).
9. P.L. Threadgill, A.J. Leonard, H.R. Shercliff, P.J. Withers, Friction stir welding of aluminium alloys, Int. Mater. Rev., **54(2),** 49–93, (2009).
10. G. Cam, Friction stir welded structural materials: beyond Al-alloys, Int. Mater. Rev., **56(1)**, 1–47, (2011).
11. R. Nandan, T. DebRoy, H.K.D.H. Bhadeshia, Recent advances in friction stir welding - process, weldment structure and properties, Prog. Mater. Sci., **53**, 980–1023, (2008).
12. R.S. Mishra, Z.Y. Ma, Friction stir welding and processing, Mater Sci Eng R, **50,** 1–78, (2005).
13. W.M. Thomas, D.G. Staines, I.M. Norris, R. de Frias, Friction stir welding tools and developments, Weld. World, **47**, 10–17, (2013).
14. Y.N. Zhang, X. Cao, S. Larose, P.Wanjara, Review of tools for friction stir welding and processing, Can. Metall. Q., **51(3)**, 250–261, (2012).
15. D. Wang, B.L. Xiao, D.R. Ni, Z.Y. Ma, Friction stir welding of discontinuously reinforced aluminum matrix composites: a review, Acta Metall Sin, **27**, 816–824, (2014).
16. O.S. Salih, H. Ou,W. Sun,M.C. DG, A review of friction stir welding of aluminium matrix composites, Mater. Des., **86**, 61–71,(2015).

17. M. A. Fènoël, A. Simar, A review about Friction Stir Welding of metal matrix composites, Materials Characterization, **120**, 1-17, (2016).
18. R.S. Mishra, P.S. De, N. Kumar, *Friction Stir Welding and Processing, Science and Engineering*, (Springer, London, 2014).
19. S. Prabhu, A. K. Shettigar. K. Rao, S. Rao and M. Herbert, Influence of Welding Process Parameters on Microstructure and Mechanical Properties of Friction Stir Welded Aluminium Matrix Composite, Materials Science Forum, **880**, 50-53, (2017).
20. Y. Bozkurt, H. Uzun, S. Salman, Microstructure and mechanical properties of friction stir welded particulate reinforced AA2124/SiC/25p-T4 composite, J. Compos. Mater., **45(21)**, 2237–2245, (2011).
21. P. Cavaliere, E. Cerri, L. Marzoli, J. Dos Santos, Friction stir welding of ceramic particle reinforced aluminium based metal matrix composites, Appl. Compos. Mater., **11**, 247–258, (2004).
22. L. Ceschini, I. Boromei, G. Minak, A. Morri, F. Tarterini, Microstructure, tensile and fatigue properties of AA6061/20 vol.% Al_2O_3p friction stir welded joints, Compos. A: Appl. Sci. Manuf., vol. 38, no.4, pp. 1200–1210, 2007.
23. F. Cioffi, R. Fernández, D. Gesto, P. Rey, D. Verdera, G. González-Doncel, Friction stir welding of thick plates of aluminum alloy matrix composite with a high volume fraction of ceramic reinforcement, Compos. A: Appl. Sci.Manuf, **54**, 117–123, (2013).
24. P. Periyasamy, B. Mohan, V. Balasubramanian, Effect of heat input on mechanical and metallurgical properties of friction stir welded AA6061-10% SiCp MMCs, J. Mater. Eng. Perform, **21(11)**, 2417–2428, (2012).
25. D.R. Ni, D.L. Chen, D.Wang, B.L. Xiao, Z.Y. Ma, Influence of microstructural evolution on tensile properties of friction stir welded joint of rolled SiCp/AA2009-T351 sheet, Mater. Des., **51**,199–205, (2013).
26. K. Kalaiselvan, I. Dinaharan, N. Murugan, Characterization of friction stir welded boron carbide particulate reinforced AA6061 aluminum alloy stir cast composite, Mater. Des., **55**, 176–182, (2014).
27. H. Nami, H. Adgi, M. Sharifitabar, H. Shamabadi, Microstructure and mechanical properties of friction stir welded Al/Mg2Si metal matrix cast composite, Mater. Des., **32(2)**, 976–983, (2011).
28. L.M. Marzoli, A.V. Strombeck, J.F. Dos Santos, C. Gambaro, L.M. Volpone, Friction stir welding of an AA6061/Al_2O_3/20p reinforced alloy, Compos. Sci. Technol., **66(2)**, 363–371, (2006).
29. G. Minak, L. Ceschini, I. Boromei, M. Ponte, Fatigue properties of friction stir welded particulate reinforced aluminium matrix composites, Int. J. Fatigue, **32(1)**, 218–226, (2010).
30. S. Peddavarapu, S. Raghuraman, R. J. Bharathi, G. V. Sunil and Manikanta, Micro Structural Investigation On Friction Stir Welded Al--4.5Cu--5TiB_2 Composite, Transactions of the Indian Institute of Metals, **70(3)**, 703—708, (2017).
31. Feng, B. Xiao, Z. Ma, Effect of microstructural evolution on mechanical properties of friction stir welded AA2009/SiCp composite, Compos. Sci. Technol., **68(9)**, 2141–2148, (2008).
32. D. Wang, B.L. Xiao, Q.Z.Wang, Z.Y. Ma, Evolution of the microstructure and strength in the nugget zone of friction stir welded SiCp/Al–Cu–Mg composite:, J. Mater. Sci. Technol., **30(1)**, 54–60, (2014).
33. C. Devanathan, A.S. Babu, Friction stir welding of metal matrix composite using coated tool, Proc Mater Sci, **6**, 1470–1475, (2014).
34. Dinaharan, N. Murugan, Effect of friction stir welding on microstructure, mechanical and wear properties of AA6061/ZrB_2 in situ cast composites, Mater. Sci. Eng. A, **543**, 257–266, (2012).

35. A. Kumar, M.M. Mahapatra, P.K. Jha, N.R. Mandal, V. Devuri, Influence of tool geometries and process variables on friction stir butt welding of Al–4.5%Cu/TiC in situ metal matrix composites, Mater. Des., **59**, 406–414, (2014).
36. D. Wang, Q.Z. Wang, B.L. Xiao, Z.Y. Ma, Achieving friction stir welded SiCp/Al–Cu– Mg composite joint of nearly equal strength to base material at high welding speed, Mater. Sci. Eng. A, **589**, 271–274, (2014).
37. N. Murugan, B. Ashok Kumar, Prediction of tensile strength of friction stir welded stir cast AA6061-T6/AlNp composite, Mater. Des., **51**, 998–1007, (2013).
38. A. K. Shettigar, S. Prabhu, R. Malghan, S. Rao and M. Herbert, Application of Neural Network for the Prediction of Tensile Properties of Friction Stir Welded Composites, Materials Science Forum, **880**, 128-131, (2017).
39. B. Ashok Kumar, N. Murugan, Optimization of friction stir welding process parameters to maximize tensile strength of stir cast AA6061-T6/AlNp composite, Mater. Des., **57**, 383–393, (2014).
40. B.S. Yigezu, D. Venkateswarlu, M.M.Mahapatra, P.K. Jha, N.R.Mandal, On friction stir butt welding of Al + 12Si/10 wt% TiC in situ composite, Mater. Des., **54**, 1019–1027, (2014).
41. K. Kalaiselvan, N. Murugan, Role of friction stir welding parameters on tensile strength of AA6061–B_4C composite joints, Trans. Nonferrous Metals Soc. China, **23(3)**, 616–624, (2013).
42. S.J. Vijay, N.Murugan, Influence of tool pin profile on the metallurgical and mechanical properties of friction stir welded Al–10 wt.% TiB_2 metal matrix composite, Mater. Des., **31(7)**, 3585–3589, (2010).
43. A.M. Hassan, T. Qasim, A. Ghaithan, Effect of pin profile on friction stir welded aluminum matrix composites, Mater. Manuf. Process, **27(12)**, 1397–1401, (2012).
44. D.R. Ni, D.L. Chen, D. Wang, B.L. Xiao, Z.Y. Ma, Tensile properties and strain hardening behavior of friction stir welded SiCp/AA2009 composite joints, Mater. Sci. Eng. A, **608**, 1–10, (2014).
45. T.Prater, M. S. Alvin, E. C. George, T. G. Brian, D. C. Chase, A comparative evaluation of the wear resistance of various tool materials in friction stir welding of metal matrix composites, Journal of Materials Engineering and Performance, **22**, 1807–1813, (2013).
46. A. Bist, S. J. Saini, B. Sharma, A review of tool wear prediction during friction stir welding of aluminium matrix composite, Trans Nonferrous Met Soc China, **26**, 2003–2018, (2016).
47. M. Collier, R. Steel, T. Nelson, C. Sorensen and S. Packer, Grade development of polycrystalline cubic boron nitride for friction stir processing of ferrous alloys, Materials Science Forum, **426–432**, 3011–3016, (2003).
48. F. J. Liu, J. C. Feng, H. Fujii and K. Nogi, Wear characteristics of a WC–Co tool in friction stir welding of AC4A+30% vol. SiCp composite, International Journal of Machine Tools and Manufacture, **45**, 1635–1639, (2005).
49. G. J. Fernandez, L. E. Murr, Characterization of tool wear and weld optimization in the friction-stir welding of cast aluminum 359+20% SiC metal matrix composite, Materials Characterization, **52**, 65–75, (2004).
50. R. A. Prado, L. E. Murr, K. F. Soto, J. C. Mcclure, Self-optimization in tool wear for friction-stir welding of Al 6061+20%Al_2O_3, Materials Science and Engineering A, **349**, 156–165, (2003).

Performance improvement studies for cutting tools with perforated surface in turning of titanium alloy

Rao Charitha[1]*, *Rao Shrikantha*[1] *and Herbert Mervin*[1]

[1]*Department of Mechanical Engineering, NITK, Surathkal- 575025, India*

Abstract: In turning process, the cutting tool is essential for shaping materials. The cutting tools with various perforated surfaces help to increase the cutting tool life. Also, advances in CNC machining technologies have enhanced the productivity of machining process. One of the best or futuristic approaches in modern manufacturing engineering is the use of FEM Simulation for the metal cutting process. FEM simulation helps in understanding the metal deformation process and also helps in the reduction of experiments. The simulation helps the researchers to predict the major influencing cutting variable values without carrying out any experiment which is time-consuming and expensive. This research presents the simulation study of the performance of micro-hole patterned Polycrystalline Diamond cutting insert in machining Titanium alloy (Ti-6Al-4V). Micro-holes are drilled using Electrical Discharge Wire Drilling machine on the rake face of Polycrystalline Diamond (PCD) cutting inserts. FEM analysis is carried out to evaluate the effect of perforations on the mechanical integrity of insert. The micro-hole patterned insert is modeled in PRO-E modeler and simulated using DEFORM-3D software. The effective stress, strain, and temperature distribution are analyzed and the results are compared with the normal insert.

1 Introduction

Titanium alloys are categorized as difficult to cut materials due to their low density, chemical reactivity at high temperatures and poor thermal conductivity. These Titanium alloys have various applications in the field of aerospace, commercial, as well as military and medical applications. The excellent strength-to-weight ratio of Titanium alloy provides a decrease of aircraft parts weight and therefore there is a decrease in fuel consumption and emissions. In aircraft engines, Titanium alloy is used in high and low-pressure compressors and also in the components which are subjected to high centrifugal loads such as disks and blades which have reduced the flow diameters and also components which are worked under severe fatigue conditions [1].

The unique characteristics of Titanium alloy result in high cutting temperature and high level of tool vibration at the tool-chip interface. Usually crater wear is common in the material removal process and is difficult to eliminate. Cutting fluids are applied often during metal removal to improve the wear and friction conditions. The purpose of using

* Corresponding author: charithamrao@gmail.com

cutting fluid is for cooling and lubrication. Moreover, it is believed that cutting tool inserts with micro-holes or micro-dimples filled with lubricants are more significant in wear and friction fall at the chip-tool interface [2]. These micro-holes or micro-dimples with lubricants reduce the chip-tool contact area and the lubricant may be pushed out of holes due to the thermal expansion at the chip-tool interface, [3]. The cutting tool with different textures such as perpendicular, cross patterned and parallel geometries, which were used in the machining of aluminium alloy with various groove sizes ranging from microscale to nanoscale have been studied. As a result, coefficient of friction and cutting forces were decreased with a perpendicular textured tool when compared with another type of textured tools, [4]. Micro-textured tools with different geometries such as elliptical, parallel and perpendicular were used to machine steel (#45 grade) using MoS_2 solid lubricant. It was noticed that elliptical grooves had better performance than the parallel and perpendicular textures, [5]. Similar kinds of investigations were performed on different materials with different texture patterns by various authors and they finally concluded that for any given machining parameter the textured cutting tools had better results when compared with the normal insert or non-textured cutting tool[6-9].

The machining process is very complex and cost effective due to the strange properties of a material and different textured pattern on the surface of the cutting insert. Finite Element simulations are boon to the manufacturing industries of machining process to its skills in optimizing the production systems. The FEM tools have assisted in cost reduction of change in design of tools which helps in the improvement of product quality by reducing the manufacturing [10].

Numerical models are important in machining process to predict or enhance the machining factors and leads to the reduction of cost in the experimental process. The FEM has predicted the machining output parameters such as stress-strain, cutting temperatures and cutting force components during the machining of Titanium alloy.

The Johnson-Cook (JC) model is used extensively in deformation process due to its flexible nature of data in flow stress is suitable for FEM simulation code. But there is a requirement to estimate the different material models based on the applicability in the context of deforming simulations. The Johnson-Cook material model selected for this analysis is Lesuer (2000) [11].

The thermal and mechanical properties of the tool and workpiece, friction consideration, and fracture conditions are put up in the FEM model. Flow stress data for the model is found out for the range of effective Stress, temperature distribution, and strain. The simulations are being performed based on the machining conditions of Titanium alloy (Ti-6Al-4V). A relative study on the enactment of different designs of Micro-holes on the rake and flank face of the cutting insert is done.

2 Johnson-Cook Model (JC model)

The constitutive material model, for FEM analysis, is essential to relate stress, strain, and temperature. To calculate the flow stress for Titanium alloy, the eq. (1) is used to model, and was proposed by Johnson Cook. This constitutive Johnson-Cook model describes the effective strain, stress, and temperature distribution effects. The main disadvantage of using Johnson-Cook model is that the deformations could not be predicted below the room temperature owing to lack of thermal tempering [12].

$$\sigma = [P + Q\,\varepsilon^{n}]\,[1 + R\ln(\varepsilon'/\varepsilon'_{o})]\,[1 - \{(T - T_{room})/(T_{melt} - T_{room})\}^{m}] \quad (1)$$

where, T is the temperature of work material, T_{room} is the room temperature; T_{melt} is melting temperature of workpiece, ε^{n} is plastic strain, ε'_{o} is reference plastic strain rate, ε' is plastic strain rate, P is yield stress, Q is modulus of hardening, R is coefficient of strain rate sensitivity, m is thermal softening coefficient and n is co-efficient of hardening. The strain rate ε' is normalized with reference strain rate ε'_o.

The constants [11] were calculated with the help of least square technique which is fitted with an equation of power law. The other parameters are preferred based on the experimentation and the flow stress data to compute for JC material model. Table 1 shows the material properties and JC material model for the research work.

Table 1. JC model constants for Titanium alloy (Ti-6Al-4V) [13]

Young's modulus	380GPa
Thermal conductivity	4.63 W/mK
Poisson's ratio	0.23
Heat Capacity	565 J/KgK
Thermal Expansion	$6.5x10^{-6}$
Emissivity	0.07
Yield Strength (P)	896 MPa
Modulus of hardening (Q)	656 MPa
Strain rate sensitivity Coefficient (R)	0.128
Thermal softening Coefficient (m)	0.8
Hardening Coefficient (n)	0.5
Plastic Strain (ε_o)	$1s^{-1}$
Melting Temperature of work-piece	1650

3 Orthogonal Machining of Titanium alloy

The orthogonal machining experiments were performed on Maxturn CNC lathe with a billet of Titanium alloy (Ti-6Al-4V) as work-material under dry condition and Minimum Quantity Lubrication (MQL) system. The cutting speeds were 100,125,150,175 and 200 m/min with a constant feed rate and depth of cut of 0.05mm/rev and 0.5mm [14-15]. An advanced Polycrystalline Diamond (PCD) tool was used as a Cutting insert. The cutting insert was modified with the micro-holes on the rake face of the insert which help in the micro-pool lubrication in the reduction of heat generation at the tool-chip interface. The cutting force is determined based on the power consumption during the cutting operation. The chip morphology based on the chips collected during the experimentation was scrutinized under Scanning Electron Microscope (SEM), which calibrates and captures the chips. The chip thickness and the tool wear are observed under Tool Maker's Microscope. Natural image of serrated chip formation of Titanium alloy is measured based on the pitch, Valley, average peak and length of the chip. Table 2 shows the experimental details for the machining of Titanium alloy (Ti-6Al-4V). Fig. 1 shows the experimental set-up.

Table 2. Machining Conditions of Titanium alloy (Ti-6Al-4V)

Work Material	Titanium alloy (Ti-6Al-4V)
Dimensions of workpiece	100*25 (Length * Diameter) (mm)
Tool Material and specification	Polycrystalline Diamond, DNMG110408
Rake angel of tool	-5°
Clearance angel	+5°
Cutting conditions	Dry and MQL
Cutting speed	100,125,150,175 and 200 (m/min)
Feed rate	0.15 (mm/rev)
Depth of cut	0.5 (mm)

Fig. 1. Experimental setup

3.1 Micro-holes on Polycrystalline Diamond Insert

Polycrystalline Diamond (PCD) is used as a cutting tool in this research work. Micro holes are made on the rake face and flank face of the cutting tool insert, very close to chip-tool contact area to effect the heat reduction through the micro-pool lubrication along the cutting edge. In designing the holes pattern, circular holes are selected because this shape is easy to be drilled using drill bits. Two different types of designs are modeled and simulated to select the best designs for further experimentation. The holes diameters are varied from 0.4mm to 0.8mm and the depth of the hole from 1mm to 3mm respectively with the equidistance of 1mm and 2mm. Table 3 shows the different designs with varied depth and diameter of the holes. The main goal is to drill deep holes so that they retain more lubricants to reduce the temperature. Fig. 2(a) and 2(b) shows the two designs of the Micro-hole pattern on the rake and flank face of the PCD insert.

Table 3. Different designs with varied depth and diameter of the holes

Design Pattern	Dimensions
Design 1	0.4mm Dia. and 1mm equidistance
Design 2	0.6mm Dia. and 1mm equidistance
Design 3	0.6mm Dia. and 2mm equidistance
Design 4	0.4mm Dia. and 2mm equidistance
Design 5	0.4mm Dia. 0.6mm Through Hole, 2mm equidistance
Design 6	0.6mm Dia. 0.6mm Through Hole, 2mm equidistance
Design 7	0.6mm Dia. 0.8mm Through Hole, 2mm equidistance

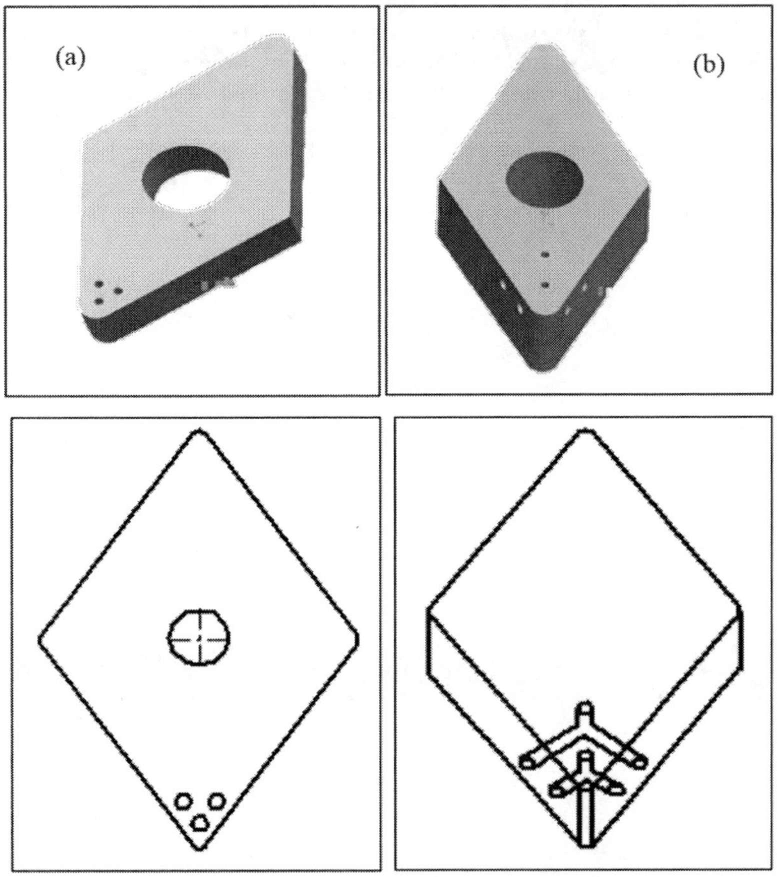

Fig. 2. CAD model of different type of perforation designs on rake and flank face of PCD insert

4 FE Modeling and Simulation

To estimate the effect of micro-holes on the mechanical strength of the cutting tool when in use [3], Finite Element modeling is performed using DEFORM-3D™ 10.2 contingents upon an updated Lagrangian formulation. This contemplates the mesh to be involved with a workpiece during the machining and as a resulting chip is formed which is a function of deformation process for which the cutting conditions and material constants need not be preset. The workpiece material is fully constrained and the tool material movement is allowed in the Y-axis [10]. The thermo-mechanical and physical properties of the workpiece and tool materials of Titanium alloy (Ti-6Al-4V) and PCD were calculated from material properties and are merged into the FEM model. The workpiece is modeled as a plastic material and the tool insert as a rigid material. Fourteen thousand tetrahedral elements were taken for the workpiece material. PRO-E Creo 5 software is used to model the micro-holes on cutting tools that are drilled along the preferred orientations on insert DNMG110408 tool geometry and micro-hole pattern as shown in Table 3 are considered.

3D tool model with micro-holes in different orientations and related tool material properties are imported into the DEFORM-3D™ software package during the formulation of machining simulation. The cutting tool is defined by 16,000 elements and was remeshed with 1,40,000 elements for the superior scattering of the output parameters like cutting force, Stress distribution, Strain distribution and temperature distribution.

The 3D FEM simulation deformation process is carried out with varying cutting speed and constant feed rate and depth of cut. The simulation is performed with the micro-hole textured tool and the normal insert. The results after completion of simulation such as cutting force, stress distribution, strain distribution and temperature distribution are validated with experimental results. Fig. 3 shows the different FE model for different design pattern.

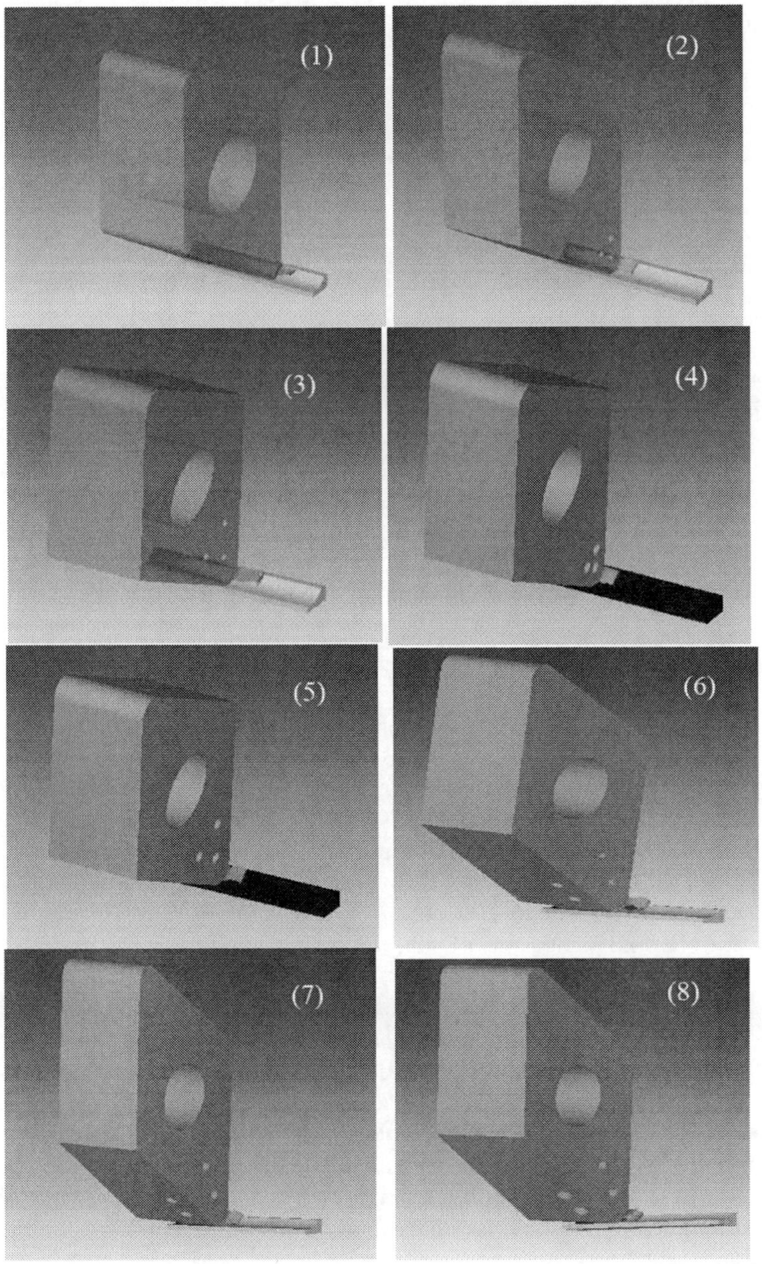

Fig. 3. FE model (1) Normal Insert (2) 0.4mm Dia. and 1mm equidistance (3) 0.4mm Dia. and 2mm equidistance (4) 0.6mm Dia. and 1mm equidistance (5) 0.6mm Dia. and 2mm equidistance (6) 0.4mm Dia. 0.6mm Through Hole, 2mm equidistance (7) 0.6mm Dia. 0.6mm Through Hole, 2mm equidistance (8) 0.6mm Dia. 0.8mm Through Hole, 2mm equidistance

4.1 Friction Modeling

DEFORM-3D™ 10.2 software uses a coefficient of friction as defined by the Coulomb friction in the following Eq. (2) for FEM code to model the frictional demeanor of Titanium alloy (Ti-6Al-4V) deformation. This friction law is used from the time when it has been verified that friction coefficient is more important in friction modeling than the law which is based.

As far as force is concerned the good predictive value for the frictional coefficient ranges from 0.8 - 0.9 [16].

$$\tau = \mu\sigma_n \tag{2}$$

τ =shear stress
μ = friction coefficient
σ_n = normal stress
Shear stress is the product of Coulomb friction coefficient and the normal stress

5 Results and Discussions

The results of Finite Element Analysis for cutting force, temperature distribution, strain, and stress, with the material model input for different designs, are shown here. The analysis is performed with varying cutting velocities ranging from 100m/min to 200m/min and constant feed and depth of cut [17]. The Finite Element output is measured around at near stable state circumstances to ensure the comparisons with normal insert.

5.1 FEM based Prediction of Cutting Force

Fig. 4 plots the predictions of cutting force by FEM Simulation model comparing with the results of normal insert and the designed insert. The predicted cutting force increased with the cutting velocity ranging from 100m/min to 125m/min and showed the marginal decrease and increase from 125m/min to 175m/min and thereafter steadily decreased for cutting velocity between 175m/min to 200m/min. The increase in the cutting force is as a result of the strain hardening effect which modifies the inclination slightly and then decreases between 175m/min to 200 m/min is owing to increase in cutting temperature resulting in the thermal tempering of the workpiece in the normal insert. The predicted values of cutting force are shown in Table 5.

Finally, there is a flawless decrease in cutting force between the normal insert and the micro-hole textured inserts. Compared to that normal insert, a reduction in cutting force by 14% to 22% is noticed when machining steel [18] with micro-hole textured insert as shown in Table 4. The cutting forces are also impacted by the coefficient of friction and also the law of friction. The higher the friction values, the cutting force also increases. In this research work, the JC-model is employed with μ = 0.3 Coulomb friction constant in friction modeling [16]. Several simulations were performed with the same circumstances to study the repeatability of the results since the experimentation is costlier. All the simulations showed similar results.

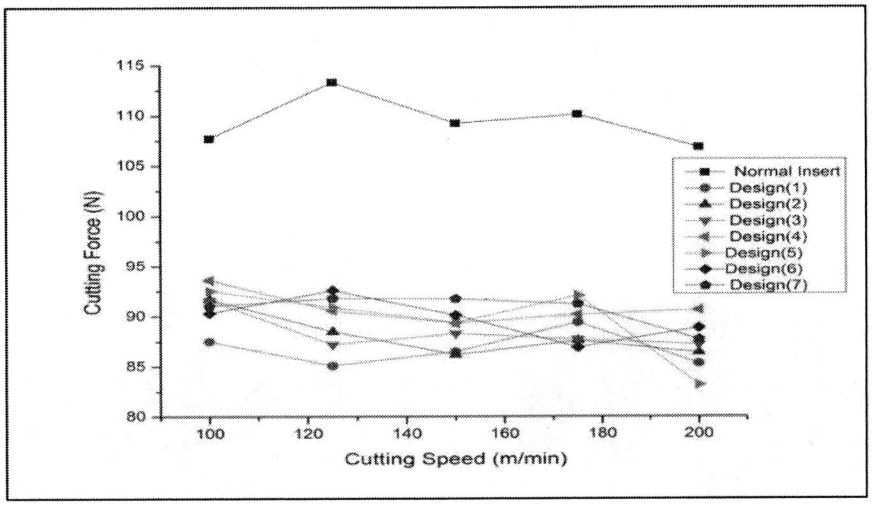

Fig. 4. Comparison of cutting force with normal and designed inserts

Table 4. Error or Percentage decrease in Cutting force for FEM simulated Model

Comparison of Cutting Force with Error %					
	100m/min	125m/min	150m/min	175m/min	200m/min
Normal Insert	107.71	113.32	109.26	110.12	106.83
Design 1 (Error %)	87.50 (-18.76)	85.10 (-24.90)	86.5 (-20.83)	89.4 (-18.81)	85.3 (-20.15)
Design 2 (Error %)	91.7 (-14.86)	88.5 (-21.90)	86.2 (-21.10)	87.6 (-20.45)	86.4 (-19.12)
Design 3 (Error %)	91.6 (-14.95)	87.2 (-23.04)	88.31 (-19.17)	87.7 (-20.35)	87.2 (-18.37)
Design 4 (Error %)	93.6 (-13.09)	90.57 (-20.07)	89.34 (-18.23)	90.18 (-18.10)	90.64 (-15.15)
Design 5 (Error %)	92.52 (-14.10)	90.91 (-19.77)	89.39 (-18.18)	92.07 (-16.39)	83.19 (-22.12)
Design 6 (Error %)	90.32 (-16.14)	92.59 (-18.29)	90.12 (-17.51)	86.92 (-21.06)	88.86 (-16.82)
Design 7 (Error %)	91.04 (-15.47)	91.80 (-18.99)	91.75 (-16.02)	91.23 (-17.15)	87.70 (-17.90)

5.2 FEM based Prediction of Effective Stress

Fig. 5 predicts the effective stress variations in FE models of the normal insert and different designed patterns of inserts. The effective stress plot for all design patterns reveals that maximum stresses are usually found in the primary deformation region where there is a contact between tool and workpiece. The stress distribution at the deformed surface is residual in nature and decreases at the uncut surface and at chip deformed surface. The seven designs prove good stability in modeling, proposing that JC law is noble when evaluating overall output stress.

All seven designs predict the similar ranges (1738-1860 MPa) for effective stress, proposing that material constants of the model combine to produce the similar stress pattern in the course of machining when compared with the normal insert as shown in Table 5. The stress concentrations are less near the segments in the designed pattern at their deformation zones. The normal insert shows irregular stress distribution at primary and secondary zones when compared. It should be acclaimed that there is no specific stress concentration due to the perforations on the rake face and flank face of the insert.

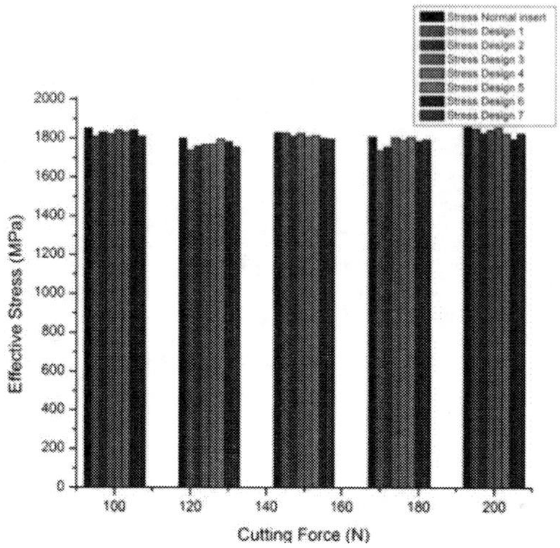

Fig. 5. Predicted Stress distribution graph of Normal Insert and Designed Inserts

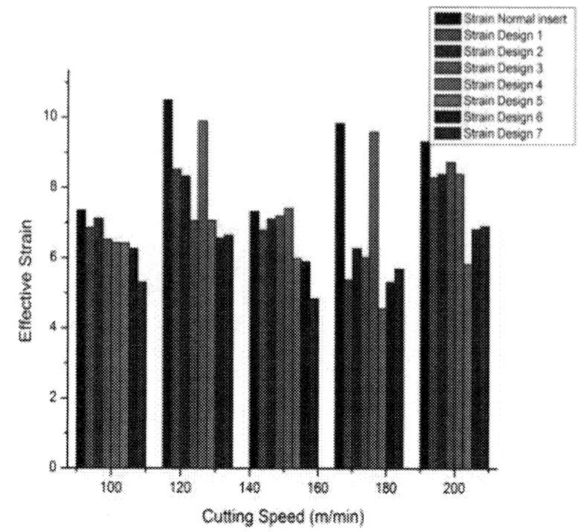

Fig. 6. Predicted Strain distribution graph of Normal Insert and Designed Inserts

5.3 FEM based Prediction of effective Strain

Fig. 6 shows the effective strain variation of predicted values with FE models of the normal insert and different designed patterns of inserts. The strain distribution in the plastic region is greater at the primary deformation region followed by secondary region and very less at the end of the chip. The formation of chip in titanium alloy suggests a convoluted formation mode resulting in segments and serrations even at lower cutting speeds. In other material, this type of formation occurs only during high-speed cutting. The simulated chip formation for different designs shows similar strain patterns. The strain distribution is less in designed inserts when compared with the normal insert, as the values of which are shown in Table 5. Minor variations can be observed in the secondary and primary deformation zones.

5.4 FEM based Prediction of Temperature

Fig. 7 shows the temperature predictions with FE models for normal insert and different designed patterns of inserts. The heat transfer takes place in the primary region where the deformation occurs. The plastic work material is converted to heat in shear zone and frictional heat is generated at the chip-tool interface. The Titanium alloy has the low thermal conductivity property which ensures less heat dissipation which leads to rapid tool wear of cutting tool insert which reduces the tool life. Hence, lubrication is necessary with cutting fluids or the cryogenic coolants for the removal of heat generation. In this FEM simulation model, the workpiece is evaluated as plastic and the cutting tool as rigid material to understand the heat transfer during the deformation of titanium alloy in the machining process. The experimental temperature will usually be more near chip-tool interface followed by shear plane and least at the uncut surface.

https://doi.org/10.1051/matecconf/201814403003

Table 5 shows the predicted temperature distributions of machining Titanium alloy. The difference in temperature is less for micro-holed textured inserts when compared with normal inserts. This can prove that micro-hole textured pattern helps in the decrease of tool wear which finally increases tool life. The minimum value can be attributed to the discrepancy in the value of 'm', the thermal softening coefficient in the model. The cutting speed can also affect the temperature distribution. Generally, the temperature distribution may be less than normal value on account of a small cutting process in the simulation.

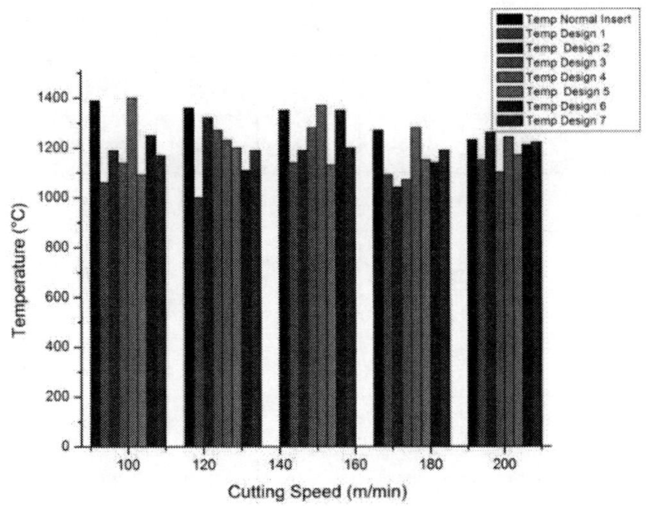

Fig. 7. Predicted Temperature distribution graph of Normal insert and Designed inserts.

Table 5. Simulation values of Cutting Force, Stress, Strain and Temperature distributions for Normal insert and designed insert

Cutting Speed (m/min)	Simulation Output parameters	Normal Insert	Design Insert 1	Design Insert 2	Design Insert 3	Design Insert 4	Design Insert 5	Design Insert 6	Design Insert 7
100	Cutting Force (N)	107.71	87.50	91.70	91.6	93.60	92.52	90.32	91.04
	Stress (MPa)	1850	1810	1830	1820	1840	1830	1840	1810
	Strain	7.35	6.86	7.12	6.53	6.43	6.42	6.26	5.30
	Temperature (°C)	1230	1060	1190	1140	1400	1090	1250	1170
125	Cutting Force (N)	113.32	85.10	88.50	87.20	9.57	90.91	92.59	91.80
	Stress (MPa)	1800	1740	1759	1768	1769	1795	1782	1756
	Strain	10.5	8.52	8.33	7.06	9.89	7.07	6.57	6.65
	Temperature (°C)	1360	1000	1320	1270	1230	1200	1110	1190
150	Cutting Force (N)	109.26	86.50	86.20	88.31	89.34	89.39	90.12	91.75
	Stress (MPa)	1830	1827	1812	1825	1808	1815	1801	1796
	Strain	7.33	6.80	7.11	7.20	7.42	5.99	5.89	4.84
	Temperature (°C)	1350	1140	1190	1280	1370	1130	1350	1200
175	Cutting Force (N)	110.12	89.40	87.60	87.70	90.18	92.07	86.92	91.23
	Stress (MPa)	1810	1738	1757	1805	1793	1807	1786	1794
	Strain	9.84	5.37	6.28	6.03	9.59	4.55	5.31	5.69
	Temperature (°C)	1270	1090	1040	1070	1280	1150	1140	1190
200	Cutting Force (N)	106.83	85.3	86.40	87.20	90.64	83.19	88.86	87.70
	Stress (MPa)	1860	1850	1827	1842	1856	1824	1796	1825
	Strain	9.32	8.30	8.40	8.74	8.40	5.82	6.83	6.92
	Temperature (°C)	1390	1150	1260	1100	1240	1170	1210	1220

6 Conclusions

The research work aims at scrutinizing the effect of different micro-hole textured PCD inserts of material constitutive model parameters of the Johnson-Cook model in foreseeing the cutting force, effective stress, strain, and temperature distribution in deformation of Titanium alloy (Ti-6Al-4V). The material models flow stress data is assigned to FEM code and the model is built to show the replica of machining process of Titanium alloy (Ti-6Al-4V). The cutting force, effective strain, stress, and temperature were investigated for all the seven different micro-textured inserts and compared with the normal insert. The following conclusions are drawn:

1. The micro-hole textured insert is very efficient in dropping cutting force, consequently cutting energy and chip-tool contact length compared with normal insert.
2. Design 2, 4, 6 and 7 showed the better enactment in terms of Cutting Force, Stress, Strain and Temperature distribution when compared with other designs. Out of simulation results obtained for other designs it was deduced to that, the designs 2,4, 6 and 7 give better results
3. All the designs predicted the global phenomena of stress, strain, and temperature with worthy correspondence to each other. This shows that JC constitutive equation is robust in modeling the deformation behavior from the global point of view.
4. The FE model assumption of material properties is found to be appropriate to model the machining of titanium alloy and appraises the pros and cons employed in the model parameter.

References

1. E. O. Ezugwu, J. Bonney, & Y. Yamane, An overview of the machinability of aeroengine alloys, *Journal of materials processing technology*, **134(2)**, 233-253, (2003).
2. L. E. Sanchez, D. A. Palma, G. L. Marinescu, I. Modolo, D. L. Nalon, L. J. & A. E. Santos, Effect of different methods of cutting fluid application on turning of a difficult-to-machine steel (SAE EV-8), *Proceedings of the Institution of Mechanical Engineers, Part B: Journal of Engineering Manufacture*, **227(2)**, 220-234, (2013).
3. Shuting. Lei, S. Devarajan, & Z. Chang, A study of micropool lubricated cutting tool in machining of mild steel, *Journal of materials processing technology*, **209(3)**, 1612-1620, (2009).
4. N. Kawasegi, H. Sugimori, H. Morimoto, N. Morita, & I. Hori, Development of cutting tools with microscale and nanoscale textures to improve frictional behavior, *Precision Engineering*, **33(3)**, 248-254, (2009).
5. D. Jianxin, W. Ze, L. Yunsong, Q. Ting, & C. Jie, Performance of carbide tools with textured rake-face filled with solid lubricants in dry cutting processes, *International Journal of Refractory Metals and Hard Materials*, **30(1)**, 164-172, (2012).
6. T. Sugihara, & T. Enomoto, Development of a cutting tool with a nano/micro-textured surface—Improvement of anti-adhesive effect by considering the texture patterns, *Precision Engineering*, **33(4)**, 425-429, (2009).
7. P. Koshy, & J. Tovey, Performance of electrical discharge textured cutting tools, *CIRP Annals-Manufacturing Technology*, **60(1)**, 153-156, (2011).
8. C. Ma, & H. Zhu, An optimum design model for textured surface with elliptical-shape dimples under hydrodynamic lubrication, *Tribology International*, **44(9)**, 987-995, (2011).
9. W. M. Da Silva, M. P. Suarez, A. R. Machado, & H. L. Costa, Effect of laser surface modification on the micro-abrasive wear resistance of coated cemented carbide tools, *Wear*, **302(1)**, 1230-1240, (2011).
10. K. S. Vijay Sekar, & M. Pradeep Kumar, Finite element simulations of Ti6Al4V titanium alloy machining to assess material model parameters of the Johnson-Cook

constitutive equation, *Journal of the Brazilian Society of Mechanical Sciences and Engineering*, **33(2)**, 203-211, (2011).

11. Donald. Lesuer, Experimental investigation of material models for Ti-6Al-4V and 2024-T3, *Livermore: University of California, Lawrence Livermore National Laboratory*, 1-36, (1999).
12. M. Calamaz, D. Coupard, & F. Girot, A new material model for 2D numerical simulation of serrated chip formation when machining titanium alloy Ti–6Al–4V, *International Journal of Machine Tools and Manufacture*, **48(3)**, 275-288, (2008).
13. K. Satyanarayana, A. V. Gopal, & P. B. Babu, Finite Element Simulation of Cutting Forces in Turning Ti6Al4V Using DEFORM 3D, *In ASME 2013 International Mechanical Engineering Congress and Exposition, American Society of Mechanical Engineers*, **volume 2A**, V02AT02A083, (2013).
14. S. Thamizhmanii, R. Ahmad, & S. Hasan, Performance of CBN and PCBN Tools on the Machining of Hard AISI 440C Martensitic Stainless Steel. *In Advanced Materials Research*, **264**, 1137-1147, (2011).
15. M. Fahad, P. T. Mativenga, & M. A. Sheikh, A comparative study of multilayer and functionally graded coated tools in high-speed machining, *The International Journal of Advanced Manufacturing Technology*, **62(1-4),** 43-57, (2011).
16. L. Filice, F. Micari, S. Rizzuti, & D. Umbrello, A critical analysis on the friction modelling in orthogonal machining, *International Journal of Machine Tools and Manufacture*, **47(3)**, 709-714, (2011).
17. B. Ozcelik, E. Kuram, & B. T. Simsek, Comparison of dry and wet end milling of AISI 316 stainless steel, *Materials and Manufacturing Processes*, **26(8)**, 1041-1049, (2011).
18. D. Arulkirubakaran, V. Senthilkumar, & V. Kumawat, Effect of micro-textured tools on machining of Ti–6Al–4V alloy: an experimental and numerical approach, *International Journal of Refractory Metals and Hard Materials*, **54**, 165-177, (2016).

Cryogenic Deflashing for Rubber Products

Abhilash M.[1], *G. V. Naveen Prakash*[2*] and *S. Kasthurirengan*[3]

[1]*Dept. of Mech. Engg., Maharaja Institute of Technology, Mysore, Karnataka, India*
[2]*Dept. of Mech. Engg., Vidyavardhaka College of Engineering, Mysuru, Karnataka, India*
[3]*CCT, Indian Institute of Science, Bengaluru, Karnataka, India*

Abstract. Deflashing is the process of removal of excess flashes from the rubber products. Initially deflashing was a manual operation where dozen of workers, seated at small work stations would take each part and trim the excess rubber off with scissors, knives or by grinding. Still the same method is employed in most of the rubber industry. The drawbacks of this method are demand inconsistent and repeatable quality. Work done by hand is often inconsistent. There are commercially available cryogenic deflashing machine but they are too expensive hence cost effectiveness is also a prime factor. The objective of this paper is to develop a technique, to identify the media through which the flashes can be removed easily and effectively. Based on the test results obtained from testing of five different types of media, ABCUT Steel media gave best results. The testing of the ABCUT Steel media on rubber samples like O-rings, grommet tail door, bottom bush etc. shows good results.

Keywords: Cryogenic deflashing, media, rubber products

1 Introduction

"FLASH" is a surplus rubber that oozes out between the two halves of mold during the moulding of rubber parts. In whatever technique employed to manufacture rubber products flashes cannot be avoided. The flashes can also be caused from worn mold cavities which cannot be fit together. The other reason may be due to the type of the material being molded, and its attendant viscosity in the liquid form which leads to the formation of unwanted mold flash. The one area where all the manufacturers constantly trying to improvise there gain in productivity is the region of deflashing. Earlier to the final inspection and packing, the vast majority of molded rubber parts must undergo a finishing process to remove excess flashes and this process is termed as Deflashing. [1] There have been a number of ways employed in order to remove this excess rubber flash, all of which mainly revolves around mechanical methods.

* Corresponding author: gvnp@vvce.ac.in

A. **Manual Deflashing:** Initially deflashing was a manual operation (still many manufacturers in India employ the same). Many workers seated at small work stations would take each part and trim the excess rubber off with scissors, knives, grinders, buffing or Machine stamping. The hand deflashing has several disadvantages. The quality of work done by hand is inconsistent, generally slow and expensive due to high labour costs and also often given a poor quality finish leading to high rejection rates.

B. **Vibrators and Tumbling:** Vibration is the most delicate way of removing flash. This process is used where the cross sections of moulding is very thin. Tumbling of rubber components in their embrittled state is the major technique employed over the last 20 years. These are effective on a broad range of parts. They are limited, due to the reason that they are less effective on inner dimensional (ID) and on the parts which has complex configuration [7].

C. **Cryogenic Shot Blast Deflashing:** In this process, the parts to be deflashed slowly tumbles in a barrel (or) basket and are simultaneously blasted with media propelled by a throwing wheel. Despite its productivity the machine had relatively high cost (US $75,000 to US $1, 70,000) and large space requirement put it out of reach for many molders.

Hanson Li, et al. [5] in their work compares the different deflashing methods. It gives knowledge about the advantages and disadvantages of the different deflashing methods. Maxi-Blast [6] made a review which includes an overall review on the cryogenic deflashing. It covers the growth of cryogenic deflashing starting from manual deflashing to cryogenic deflashing (the second generation). It also gave knowledge about the operational parameters which are included in the deflashing equipment such as the cycle time, wheel speed, media and the temperature.

From the literature survey found that much of the work is not carried out in the area of cryogenic deflashing in India and the commercially available cryogenic deflashing machine is too expensive. The process involved in its construction is a complex one. Hence an attempt is made to develop a cryogenic deflashing system using a simple technique and to reduce the overall cost of the system.

2 Methodology

Fig. 1 shows the schematic diagram of the cryogenic deflashing system. It consists of a bowl, rubber gasket, top enclosure, and a shaft carrying two cantilever beams which are placed at certain angle with respect to each other, which is connected to a motor with the help of a pulley. Fig. 2 shows the complete set up of cryogenic deflashing machine, which includes the top enclosure which has been clamped perfectly and the liquid nitrogen supply tank.

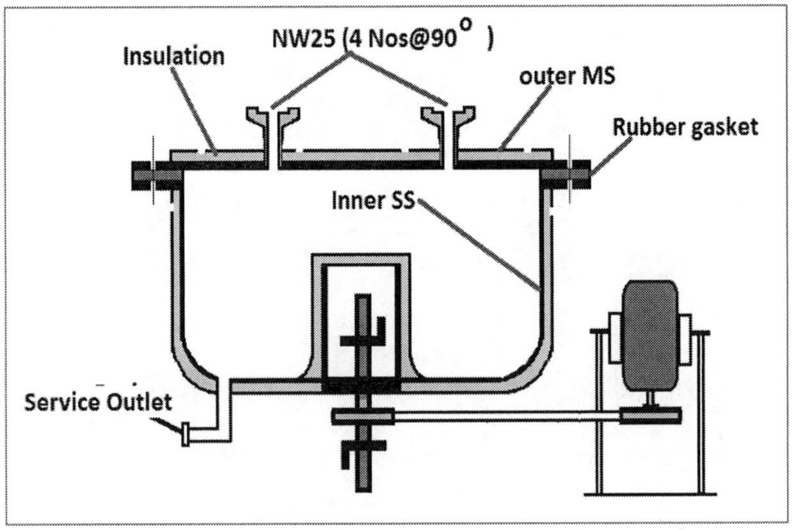

Fig. 1. Diagram of cryogenic deflashing machine

Fig. 2. Cryogenic deflashing machine

Theoretical calculation is made in order to know how much amount of liquid nitrogen is required approximately for the system in order to make a trial run.

$$\text{Amount of heat transfer } (Q) = m \times C_P \times \Delta T \tag{1}$$

Assumption: Material used is Stainless steel

where, m = mass (kg) = 1kg
C_p (steel) = specific heat (kJ/kg.K) = 530kJ/kg.K
ΔT = temperature difference (K)

$$Q = 1\times 530\times (300-77)$$

$$Q = 118.19 \times 10^3 \text{ J}$$

Latent heat of liquid nitrogen (h_{fg}) = 200 kJ/kg.

$$Q = m \times h_{fg} \quad (2)$$

$$m = \frac{Q}{hfg} = \frac{118.19 \times 10^3}{200 \times 10^3}$$

$$m = 0.59095 \text{ kg}$$

Density of liquid nitrogen = 0.807 g/ml = 807 kg/m^3

$$\text{Specific volume} = \frac{1}{Density} = \frac{1}{807} = 1.23911 \times 10^{-3} \text{ m}^3/\text{kg}$$

$$\text{Volume} = \text{m} \times \text{specific volume} = 0.59095 \times 1.23911 \times 10^{-3}$$

The amount (volume) of liquid nitrogen required per kg of steel = 7.3218×10^{-4} m^3.

Volume (per kg) = 0.732 liters.

Media place a crucial role in the deflashing process. During the time of embrittlement of the rubber, impact from the media will cause the excess rubber to deflash. There are wide ranges of media available. Selection of the particular one is a tedious task. Several tests are carried out for identifying the media to be used in the system.

3 Results & Discussions

Fig. 3 shows the different types of media which are available and tests were carried out on each of the media.Table1 shows the results obtained from testing various types of media available.

Based on the results obtained ABCUT media was most suitable for deflashing process. Since the amplitude obtained using this media is high we made use of the same media for the cryogenic deflashing process.

The results obtained from testing of various rubber samples with ABCUT media are as shown in Table 2. Fig. 4 shows the original and deflashed samples. Tests are also carried out on other types of samples. The results are equally good in all the types of samples tested.

4 Conclusions

This section describes the overall conclusions drawn from the work carried out and the results obtained. A cryogenic deflashing technique through which the flashes can be removed easily and effectively was identified. Based on the test results obtained from testing of five different types of media, ABCUT Steel media gave best results. The testing of the ABCUT Steel media on rubber samples like O-rings, grommet tail door, bottom bush, etc., showed good results.

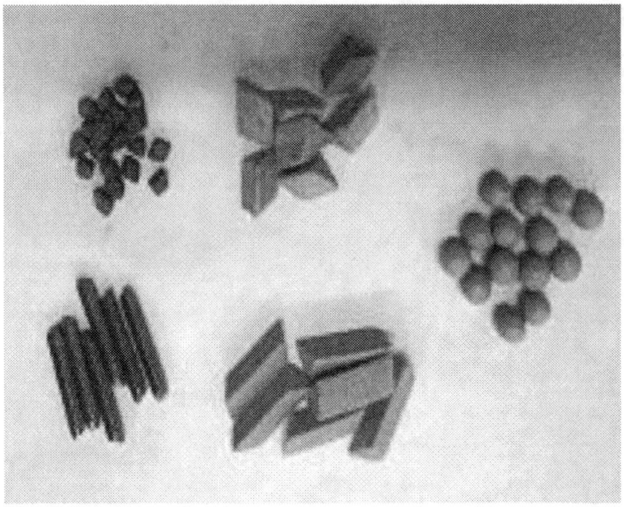

Fig. 3. Different types of media

Table 1. Results from different types of media

Media	Type of Sample	No. of samples	No. of accepted samples	Amplitude (mm)
Alumina Deburring 4X4	O-Ring(55mm)	10	5	1.5-2
Ceramic Ball	O-Ring(55mm)	10	1	Less than 1.5
T8X20 Grey cut	O-Ring(55mm)	10	4	2-2.5
Tri star	O-Ring(55mm)	10	8	3-3.5
ABCUT	O-Ring(55mm)	10	10	3.5-4

Table 2. Results from different samples tested

Type of sample	No. of Sample	Total time (minutes)	No. of okay samples
O-RINGS	150	90	140
Grommet Tail Door	50	90	42
Bottom bush	50	90	44

Fig. 4. Results from different samples tested

References

1. D. Spooner, J Mech. Engg., **10**, 209 (1984)
2. R. M. Leliaert, S Bend, *U.S. Patent,* 2,996,846, (1959)
3. R. A. Rhodes, *www.cryogenicsociety.org*, (2010)
4. R. Wieland, *Cold Facts,* (1997)
5. H. Li, S. Zhu, M. Dubois, *Technical bulletin*, Solvay (2015)
6. CRYOGENIC DEFLASHING: A REVIEW, Maxi- Blast, Inc.
7. R. A. Potterfield, T. Morrow, *U.S. Patent,* 7,883,396 (2011)
8. P. Warrier, *http://rubbermachineryworld.com,* (2015)

Optimize the Rolling Process Parameters for Material AA1100 using Metal Forming Simulation

Hemanth T. S.[1]*, *Y. Arunkumar*[1] *and Srinath. M. S.*[1]
[1]*Affiliated to Visvesvaraya Technological University, Belagavi, Department of Industrial and Production Engineering, Malnad College of Engineering, Hassan, Karnataka, India*

Abstract. Metal forming plays a very important role in the manufacturing. Simulation of manufacturing process aids in the improvement of quality, reduce energy and resource consumption and helps in visualization of the process. The design of experiment helps in optimization of the parameters in any processes. In this paper, Taguchi optimization technique is used to predict the best results for the given inputs such as roller diameter, friction value, velocity of the rollers and percentage reduction to the forming process and get the optimized values for spread, hardness, effective stress, power required, strain rate and torque using the manufacturing simulation software. It is found that the important parameter is percentage reduction affecting the effective stress. Optimal parameters with desirability value of 0.87 have been obtained.

Keywords: Metal Rolling, DOE, AFDEX, Optimization, Aluminum

1 Introduction

The metal forming process is most important processing technique. One of the bulk metal forming processes is rolling where the metal is deformed by passing it through the two rollers which are rotating in opposite directions. Metals, when passed through the rolls, are reduced from initial thickness to predefined thickness. The roll gaps define the final thickness of the metal and because of the volume constancy of material after and before rolling; the reduction amount plays a major role in governing the elongation in length. There is not the only reduction in the thickness in the rolling process, but also there is a significant change in materials properties such as hardness, strength, microstructures and other properties [1]. Many research has been done in this area to improve the quality of metal rolling.

Finite element Method (FEM) is among one of the established and most widely used techniques for obtaining solutions for various complex engineering problems of many fields of engineering. Because of the advent of faster processors and high computation,

* Corresponding author: hemanththulasi@gmail.com

complex engineering problems can be solved using a lesser amount of time with the greater amount of accuracy. Use of FEM in metal forming is gaining considerable attention because of the various advantages like less cost of material, a large number of trials in shorter duration and more accuracy. There are many commercially available applications to predict the deformation and other process parameters in metal forming like DEFORM 3D, ANSYS, AFDEX, FORGE, STATISTICA, FEM, Monte Carlo method and many other methods[2]–[6]. There are many factors present in the rolling process which can be simulated and optimized. The main purpose of this paper is to determine the optimal parameters that influence the quality of AA1100 alloy during the rolling process using simulation software called AFDEX. To determine the optimal parameters through individual simulations can be time-consuming. The design of experiment technique is employed to optimize the process parameter by minimum simulation runs. There are lot of research conducted to optimize the process parameters for various applications and manufacturing processes.

In rolling operation, two main changes elongation and spread occur. The main challenge here is to minimize spread. Many authors have predicted spread using either introducing new formulae or by giving modifications to experimental methods [7].The precision of the rolled products depends on the rolling force. The rolling torque is computed using the rolling force. This rolling force is important to predict as this parameter depends on the design of the rolling mill and it is the driving force if there is any breakdown of the mill [8]. Roller speed plays a major role in determining the quality of the rolled product. There are several works done to find the effect of the rolling speed on other parameters of rolling. Strain rate, flow stress, heat of deformation, roll force and coefficient of heat transfer are all controlled by rolling speed directly. Therefore, rolling speed can be considered as one of the important parameters for rolling [9]. Strain rate is another parameter which plays a vital role in deciding the quality of roll. The strain rate can be interrelated to the speed of the roller. It can be suggested that the strain distribution is more homogeneous at high-speed rolling [10]. Shahani et al. studied the AA5083 aluminum alloy hot rolling process using simulation. The parameters such as the geometry of slab, thickness reduction, rolling speed, frictional coefficient and load are used in developing Artifical Neural Network (ANN) using the output of FE simulation by training the network. This network predicts the workpiece behavior during the rolling process [11]. In the present work, the effect of rolling parameters and optimisation of AA1100 for the billet of size 40mm × 40mm × 100mm length is considered. The parameters such as stress, strain rate, torque, hardness and power required are optimised for the given input parameters reduction percentage, roller speed, roller diameter and coefficient of friction.

2 Simulation Procedure

The basic step for simulating any process is to define the model and the required tools for computation. In this analysis, the necessary modeling is done using the NX software which enables easy conversion of the model to any format for analysis. For the rolling process, the rollers and billet are the geometric parameter and depending on the requirement it is modeled in the NX environment. The modeled process in this analysis is then converted to .STL (Stereo lithographic format) so that it would be easy to manipulate the model in the simulation environment. In the NX environment, the required model is oriented towards Y-axis so that when the model is imported to the simulation, the billet moves perpendicular to the Y-axis as it happens in real-world. The software used for simulation is AFDEX (Advisor as friend for Forging Design Experts) which is a metal forming simulation software. The finalized CAD model is imported to simulation environment, further minor changes in the positions are made and finally all the necessary inputs parameters, properties are applied to the model for analysis. The parameters used in the simulation environment is given in Table 1. The material selected for the present analysis is AA1100 alloy. This aluminum is low strength commercial grade which is useful in pyrotechnic powder, chemical plant equipment, architectural flashing, cable sheathing and lamp reflectors. The chemical composition of available commercial AA1100 is Fe 0.57%, Cu 0.12%, Si 0.13%, Ti 0.03%, Mg is 0.02%, Mn is 0.013, Ni is 0.017, Zn is 0.01% and remaining is aluminum [12]. The mechanical property of the selected material are Modulus of Elasticity is 70-80 GPa, Poisson's ratio is 0.33, density is 2.71 g/cc, tensile strength is 110 MPa and yield strength is 105 MPa [12].

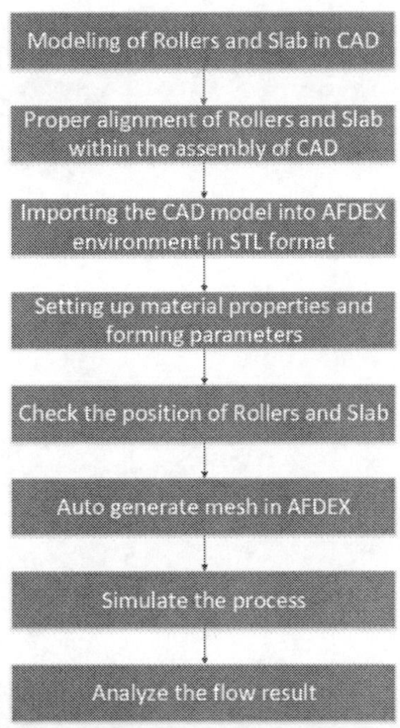

Fig. 1. Flow chart of Rolling Process Simulation

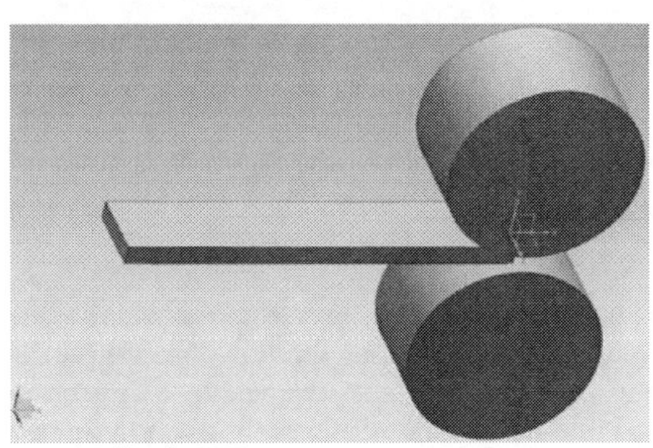

Fig. 2. Modeling of Rollers and Billet in NX

Table 1. Parameters selected from simulation

Type of Forming	Cold Rolling
Type of Simulation	3D without flash
Type of analysis	Flow analysis
Deformation	Rigid Plastic
Billet material	AA1100
Translation Velocity	Depends on the simulation trial
Lubrication	Depends on the simulation trial

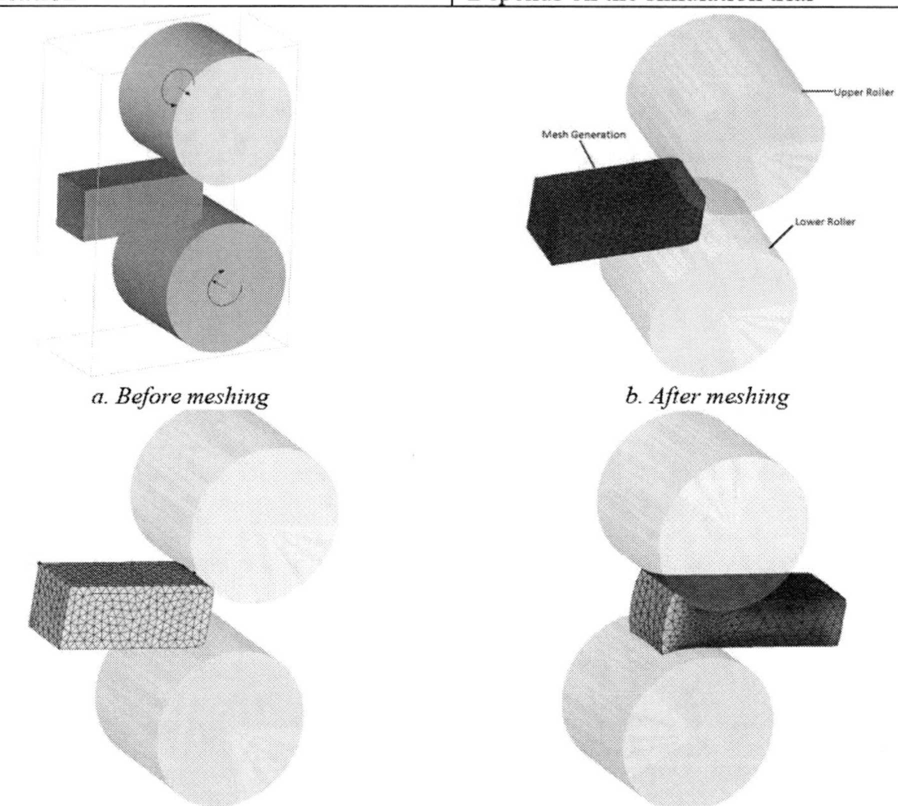

a. Before meshing *b. After meshing*

a. Before Rolling *b. After Rolling*

Fig. 4. Meshing before and after Rolling Process

2.2 Plan for Simulation

The Taguchi method is used in the present investigation; this is one of the powerful design of experiment tools available. Taguchi method is the simple, systematic and efficient approach in determining optimal forming parameters. Conventional experiment methods are costly, complex and time-consuming. Relatively many numbers of experiments needed to be carried out to study the entire process, whereas Taguchi method employs an orthogonal array for studying the process with a lesser number of experiments. The traditional experiment required experimenting by changing one parameter and keeping other parameters constant. This method fails to understand the possible interactions between parameters. It would be unlikely to understand the main effects in an experiment involving the different change in parameters. Taguchi method overcomes all these disadvantages, and it can be used for optimization of process parameters[13].

The steps involved in Taguchi Techniques are [14]

1. Identify the process parameters and response functions.

2. Determine the possible interactions and number of levels between the parameters.

3. Select the suitable orthogonal array.

4. Select the optimal level of process parameter through ANOVA analysis.

5. Perform the confirmation experiment for verification of optimal process parameter.

The chosen input parameters and their levels for the simulations are as shown in Table 2. For the selection of the values of the input process parameters, trial runs were made by varying one parameter and keeping other variables constant. By measuring the possible reduction without any defects and with equal intervals, the working ranges for the parameters were selected. The range and number of levels for the design parameters is as shown in Table 2. The rolling operations were performed as per the conditions governed by the design matrix developed by the Taguchi techniques.

Table 2. Parameters and their levels

Parameters	Level 1	Level 2	Level 3
Roller diameter (mm)	100	200	300
Speed (RPM)	60	120	240
Reduction (%)	25	30	40
Frictional Coefficient	0.3	0.25	0.15

3 Selection of Design matrix

For the present work, an L9 orthogonal array with four columns and nine rows is used. This array is applied for three level process parameters. Only nine experiments can be conducted to study the rolling process for this parameters. The layout of experiments need to be carried out using L9 orthogonal array is as shown in Table 3. The value 1represents level 1; value 2 represents level 2 and value 3 represents level 3.

Table 3. Simulation layout using L9 array

Run	Control factors and levels			
	Roller Diameter(mm)	**Speed, RPM**	**Reduction %**	**Friction**
1	1	1	1	1
2	1	2	2	2
3	1	3	3	3
4	2	1	2	3
5	2	2	3	1
6	2	3	1	2
7	3	1	3	2
8	3	2	1	3
9	3	3	2	1

4 Analysis of simulation results

The simulation results are analyzed, to understand the main effect of the parameters, difference between different levels of the chosen parameters and effects of levels 1 on other parameters and vice versa on the out parameters. In the present work, as it is simulation only one run is performed for all the nine sets of the experiment. Some of the output parameters are given higher importance than other parameters as they play important role in quality. The parameters such as effective stress, spread, hardness are givens higher preference than power and torque. For optimization, effective stress, spread, power and torque should be minimum whereas hardness and strain should be more. The objective of this work is to obtain the optimal value for the given set of the parameter with a minimum number of trials to save the time. However, while conducting the actual experiment using this parameters and levels are performed using the Taguchi analysis through the design expert software. The values of each trial were taken by selecting the average at six different points and for all the trails, the values were taken from the same region that is at the lower die middle region. The results from the simulation for the L9 orthogonal array is as given in Table 4.

Table 4. L9 generated Orthogonal array with results obtained from AFDEX

Exp.No.	Roller diameter (mm)	Speed (rpm)	Reduction%	Coefficient of Friction	Effective Stress (N/mm^2)	Spread (mm)	Hard-ness (BHN)	Strain Rate	Power,(Watt)	Torque(N-m)
1	100	60	25	0.3	32.42	4.66	229.5	79.69	0.2017	0.0305
2	100	120	30	0.25	34.3	5.23	235.5	86.49	0.3838	0.0283
3	100	240	40	0.15	38.27	7.94	245.6	108.3	0.32	0.0165
4	200	60	30	0.15	36.66	8.09	249.4	76.15	0.394	0.0622
5	200	120	40	0.3	57.73	8.3	252.4	119.9	1.143	0.0914
6	200	240	25	0.25	72.34	8.48	253.4	120	0.2477	0.0985
7	300	60	40	0.25	66.11	7.07	251.1	121	0.898	0.0936
8	300	120	25	0.15	56.41	4.57	192.8	124	1.13	0.0891
9	300	240	30	0.3	60.37	4.82	215.5	140	0.3362	0.0933

The obtained Desirability is 0.87 for these set of experiments which is within the acceptable limit. The optimized value obtained is as shown in Table 5.

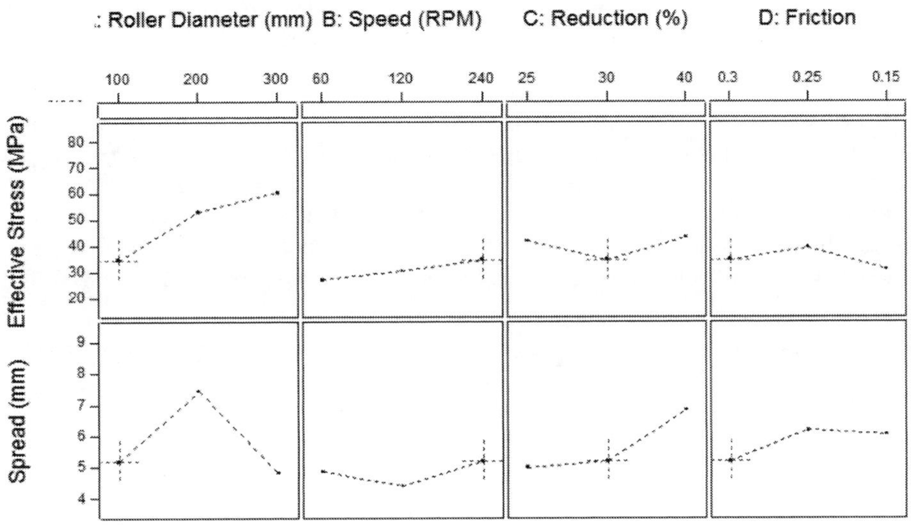

Fig. 5. Graph between Effective stress and Spread vs. the input parameters

From the Fig. 5, it can be seen that the Effective stress increases with increase in Roller diameter and speed but it decreases and increases with reduction values. Effective stress is increased and again decreased to the lowest point in case of decreased friction. Therefore, speed and Roller diameter are the important factors. Spread should be minimized and under control, from the Fig. 5 it can be observed that reduction plays a vital role for spread.

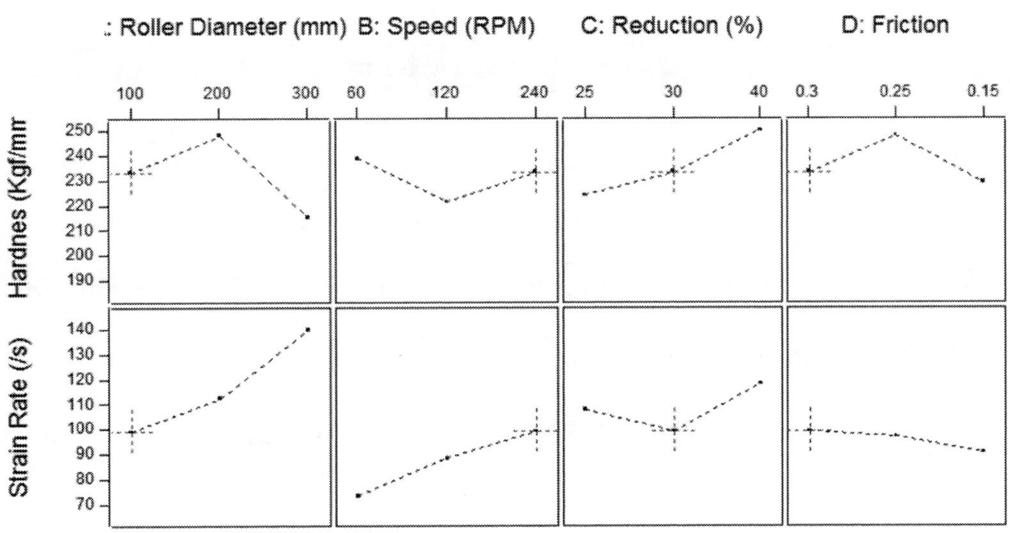

Fig. 6. Graph between Hardness and Strain Rate vs. the input parameters

From Fig. 6, it can be seen that the reduction plays an important role in the hardness and maximum hardness can be obtained by increasing the reduction and minimum hardness can be obtained by increasing the roller diameter. Strain rate is directly proportional to the Roller diameter, speed and the friction but higher strain rates are found with increased Roller diameter.

Similarly, the power and torque relation with the input variables has been shown in Fig. 7. Both power and torque have same response concerning the input responses. There is an increase in both power and torque with an increase in Roller diameter. These responses are computed using the design of experiment software to find the optimal values for the output. There was 23 output were given by the software but the optimal values for the given input are obtained as shown in Table 5 with the desirability value was found to be 0.87 for the given value.

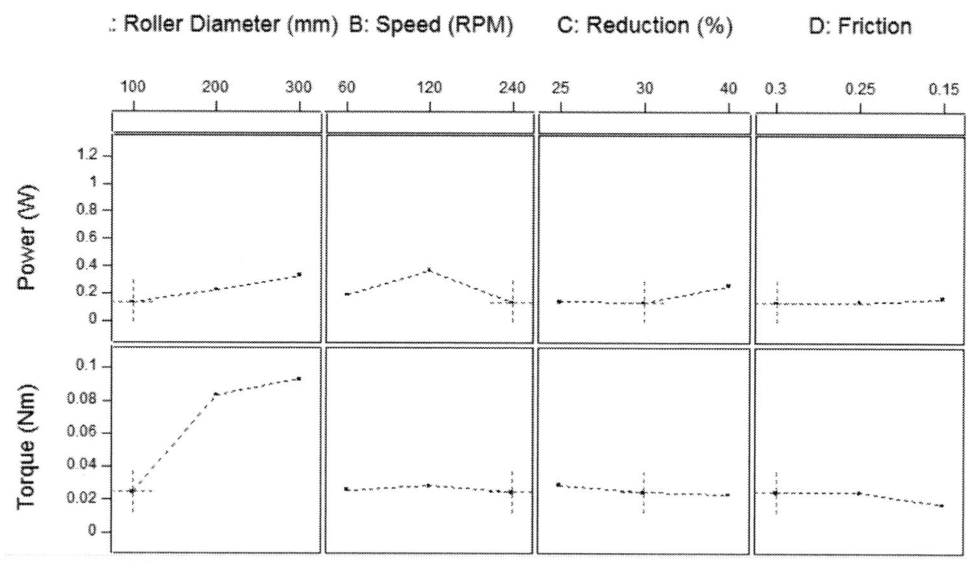

Fig. 7. Graph between Power and Torque vs. the input parameters

Table 5. Optimized levels for the factors

Roller diameter (mm)	Speed (rpm)	Reduction %	Coefficient of Friction	Effective Stress (N/mm2)	Spread (mm)	Hardness (BHN)	Strain Rate	Power,(Watt)	Torque(N-m)
100	240	30	0.3	34.65	5.18	233.51	99.16	0.140	0.025

5 Confirmation Simulation

A confirmation simulation is needed to determine the optimum conditions and to compare the results with the expected conditions [15]. Using the optimal data, confirmation simulation was done to validate the generated solution. For the roller diameter of 100mm, the speed of 240 rpm, reduction of 30%, and coefficient of friction of 0.3, the result obtained were as given in the table. These are the condition for getting the effective values of the rolling parameters. The parameters like effective stress, spread, power, and torque are lower values where as the strain rate and hardness are at optimal levels. The optimal value obtained by simulation and prediction are similar with some slight deviation. The deviations are lesser than 5% in most of the cases except for the hardness. The hardness and strain rate have more deviations compared with others. This is because it is not possible to predict the exact hardness and strain rates through simulation as it depends on many factors like microstructures and work hardening. These two parameters should be isolated and using other mathematical model it can be simulated to get the actual value to the experiment.

Table 6. Comparison of Simulated and Predicted value

	Roller diameter (mm)	Speed (rpm)	Reduction %	Coefficient of Friction	Effective Stress (N/mm²)	Spread (mm)	Hardness (BHN)	Strain Rate	Power,(Watt)	Torque(N-m)
Simulation value	100	240	30	0.3	35.9	5.5	217.2	110	0.15	0.029
Predicted value	100	240	30	0.3	34.65	5.182	233.5	99.2	0.140	0.025

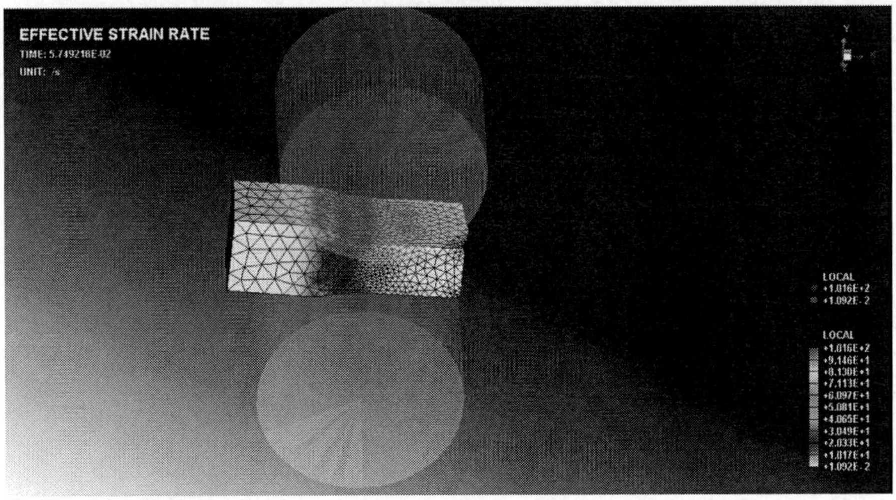

Fig. 5. Rolling trial for optimal value

6 Conclusions

Using the Taguchi method, the process parameters that affect the rolling process are studied. The variables such as roller diameter, speed, reduction and coefficient of friction affect the output parameters such as effective stress, strain rate, hardness, power torque and spread. The following are the observation made

- The optimum rolling conditions are said to be at roller diameter of 100mm, the speed of 240 rpm, reduction of 30% and coefficient of friction of 0.3.
- Effective stress increases with increase in roller diameter and speed.
- Spread increases linearly with increase in reduction.
- Strain rate increases significantly with increase in roller diameter.
- Friction has no significant influence on the power.
- Speed and seduction cause fewer variations in the torque value.

Acknowledgment

Authors would like to acknowledge for providing research grants from TEQIP-II, World Bank grants, Govt. of India and DHIO, Research and Engineering Pvt. Ltd. Bangalore for their research support.

References

[1] R. Jain, K. Kumari, R. K. Kesharwani, S. Kumar, S. K. Pal, and S. B. Singh, *Modern Manufacturing Engineering*. 2015.

[2] G. Banaszek, S. Berski, H. Dyja, and A. Kawałek, "Theoretical modelling of metallurgical defect closing-up processes during forming a forging," *J. Iron Steel Res. Int.*, vol. 20, no. 9, pp. 111–116, 2013.

[3] T. Kroib, U. Engel, and M. Merklein, "Comprehensive approach for process modeling and optimization in cold forging considering interactions between process, tool and press," *J. Mater. Process. Technol.*, vol. 213, no. 7, pp. 1118–1127, 2013.

[4] T. Gangopadhyay, D. K. Pratihar, and I. Basak, "Expert system to predict forging load and axial stress," *Appl. Soft Comput. J.*, vol. 11, no. 1, pp. 744–753, 2011.

[5] M. R. Doddamani and M. Uday, "Simulation of Closed die forging for Stud Bolt and Castle Nut using AFDEX," vol. 1, no. 3, pp. 16–22, 2012.

[6] M. Sanjari, P. Saidi, A. Karimi Taheri, and M. Hossein-Zadeh, "Determination of strain field and heterogeneity in radial forging of tube using finite element method and microhardness test," *Mater. Des.*, vol. 38, pp. 147–153, Jun. 2012.

[7] T. Sheppard and D. X, "A new spread formula for hot flat rolling of aluminium," *Model. Simul. Mater. Sci. Eng.*, vol. 10, pp. 597–610, 2002.

[8] A. Pesin, A. Korchunov, and D. Pustovoytov, "Finite Element Simulation of Shear Strain in various Asymmetic cold Rolling Processes," *Издается с марта 2003 года*, pp. 32–40, 2014.

[9] M. Bagheripoor and H. Bisadi, "Effects of rolling parameters on temperature distribution in the hot rolling of aluminum strips," *Appl. Therm. Eng.*, vol. 31, no. 10, pp. 1556–1565, 2011.

[10] E. I. Poliak and J. J. Jonas, "Critical Strain for Dynamic Recrystallization in Variable Strain Rate Hot Deformation," *ISIJ Int.*, vol. 43, no. 5, pp. 692–700, 2003.

[11] A. R. Shahani, S. Setayeshi, S. A. Nodamaie, M. A. Asadi, and S. Rezaie, "Prediction of influence parameters on the hot rolling process using finite element method and neural network," *J. Mater. Process. Technol.*, vol. 209, no. 4, pp. 1920–1935, 2009.

[12] H. Pirgazi, A. Akbarzadeh, R. Petrov, and L. Kestens, "Microstructure evolution and mechanical properties of AA1100 aluminum sheet processed by accumulative roll bonding," *Mater. Sci. Eng. A*, vol. 497, no. 1–2, pp. 132–138, 2008.

[13] Y. S. Tarang, "Optimization of the submerged arc welding process," *Mater. Manuf. Process*, vol. 13, no. 3, pp. 455–467, 1998.

[14] P. M. George, B. K. Raghunath, L. M. Manocha, and A. M. Warrier, "EDM machining of carbon-carbon composite - A Taguchi approach," *J. Mater. Process. Technol.*, vol. 145, no. 1, pp. 66–71, 2004.

[15] Ö. Savaş and R. Kayikci, "Application of Taguchi's methods to investigate some factors affecting microporosity formation in A360 aluminum alloy casting," *Mater. Des.*, vol. 28, no. 7, pp. 2224–2228, 2007.

Surface roughness optimization in machining of AZ31 magnesium alloy using ABC algorithm

Abhijith[1], *Pai Srinivasa*[1], *D'Mello Grynal*[1] *and Hebbar Gautama*[1*]

[1]*NMAM Institute of technology, Nitte-574110, Karnataka, India*

Abstract: Magnesium alloys serve as excellent substitutes for materials traditionally used for engine block heads in automobiles and gear housings in aircraft industries. AZ31 is a magnesium alloy finds its applications in orthopedic implants and cardiovascular stents. Surface roughness is an important parameter in the present manufacturing sector. In this work optimization techniques namely firefly algorithm (FA), particle swarm optimization (PSO) and artificial bee colony algorithm (ABC) which are based on swarm intelligence techniques, have been implemented to optimize the machining parameters namely cutting speed, feed rate and depth of cut in order to achieve minimum surface roughness. The parameter Ra has been considered for evaluating the surface roughness. Comparing the performance of ABC algorithm with FA and PSO algorithm, which is a widely used optimization algorithm in machining studies, the results conclude that ABC produces better optimization when compared to FA and PSO for optimizing surface roughness of AZ 31.

Keywords: Surface roughness, optimization, Artificial Bee colony (ABC), Firefly algorithm (FA), Particle swarm optimization (PSO).

1 Introduction

The applications of Magnesium alloys are very wide. They are used in electronics, automobile, and aerospace fields for its properties such as low density and high strength-weight ratio. Magnesium alloy, with the density of 1700 kg/m^3 is regarded as one of the lightest among the metallic materials because aluminum alloys are 35 % heavier than Mg alloys [3]. AZ31 is a magnesium alloy which consists of 96 % magnesium, 3 % aluminium and 1% zinc [5]. Dry milling is widely used in industrial manufacturing as a high performance cutting technology. Milling operation is required to produce finished products [6]. Surface integrity in machined products is emerging as the new focus in machining research [6]. Guo et al. (2010) analysed the surface integrity by high speed dry milling of Mg alloy. The findings of this study are high-speed dry milling can be performed using PCD inserts safely with slight flank build-up and surface integrity characterized by low roughness [2]. Kaining Shi et al. 2015) conducted dry face milling on magnesium alloy.

* Corresponding author: gouthamahebbar@nitte.edu.in

Taguchi with grey relational analysis of experimental data was carried out to determine an optimum combination of machining parameters for better surface roughness. Feed rate was regarded as the most dominant factor influencing surface integrity which was found out by the analysis of variance of grey relational grade. The optimum combination of process parameters were validated by validation experiments and the validations revealed that the Taguchi with grey relational analysis is an efficient method to determine available cutting parameters for a desired surface integrity during milling of magnesium alloy. [6]. Grigoraş et al. (2015) analysed the influence of milling parameters on surface roughness of AZ61A, a magnesium-aluminium alloy. The results of ANOVA analysis revealed that only speed and feed have a significant impact [7]. Zagórski et al conducted end milling experiments on AZ31 and AZ91 alloy and they concluded that the best surface quality was achieved for a tool with a PCD cutting edge and the worst for a coated solid carbide tool and slightly higher roughness parameters were obtained for AZ31 than for AZ91 [8]. Yusup et al. (2012) presented an overview and comparison of the research work carried out during 2007-2011, that used evolutionary optimization techniques to optimize machining process parameters. Five techniques namely genetic algorithm, simulated annealing, particle swarm optimization, ant colony optimization and artificial bee colony (ABC) were considered. And among the various machining parameters that were optimized, surface roughness was the main parameter, followed by machining costs and material removal rate [9]. There have been some efforts with regard to use of optimization techniques in machining magnesium alloys to improve surface integrity / surface roughness. There have been very less reported literature regarding study of optimization techniques like Genetic Algorithm and the newer evolutionary optimization techniques for the optimization of surface roughness in machining of magnesium alloys. In this study the focus is on the use of evolutionary optimization techniques like PSO, Firefly algorithm (FA) and ABC for optimizing surface roughness in terms of parameter R_a during milling of magnesium alloy AZ31. The cutting speed, feed rate and depth of cut are to be optimized.

2 Experimentation

The work piece used in this study is AZ31 magnesium alloy of width 60 mm and length 110 mm. CNC vertical machining center (BMW 45 T 20 BFW make) which is having 6000 rpm as the maximum spindle speed was utilized for conducting experiments. Face milling experiments were carried out for a length of 110 mm. Milling experiments were conducted with cutter diameter of 50 mm and single uncoated carbide insert for different cutting conditions using three level full factorial design. The specification of the milling cutter is R217.69-2020.3-12-2AN (SECO make) and insert specification is XOEX120404FR-E06, H15 (SECO make) with corner radius of 0.8 mm. After each machining pass, the tool insert is removed from the tool holder and inspected for wear using Mitutoyo make Tool makers' microscope (TM505/510). No tool wear was observed for all the experiments carried out. Dry machining was carried out with no use of coolant.

Table 1. Machining parameters

Cutting speed (m/min)	Feed rate (mm/ tooth)	Depth of cut (mm)
200	0.4	0.15
350	0.8	0.25
500	1.2	0.35

Fig. 1. Machining setup

Fig. 2. Surface roughness measurement

Surface roughness has been measured after each machining pass using Taylor Hobson Taly Surf 50 which is a contact type stylus instrument. The traverse length of stylus is set for 2.5 mm which is considered while measuring the surface roughness parameter R_a (arithmetic average surface roughness). The surface roughness parameter R_a is measured at three different locations along the mid portion of the milled surface and mean value has been calculated. Fig. 1 shows the machining setup and Fig. 2 shows the surface roughness measurement process.

3 Optimization techniques

Determining the optimum machining conditions are of great concern in manufacturing environments, where economy of machining operation plays an important role to survive in the market. In highly competitive manufacturing industries nowadays, the final goal of a manufacturer is to produce high quality products. Surface roughness is one of the output parameter which influences the quality of the machined products. Surface roughness depends upon process parameters. In optimizing the machining process parameters, the selection of these parameters are important. It can be done based on human judgment and experience, but this will not always result in optimal results. Machining process is very complex and there are several factors, which influence their performance [9]. Optimization techniques are useful to find the optimum solution. An optimization algorithm is an iterative procedure executed by comparing different solutions till a satisfactory solution is obtained. The method selected for a particular case is dependent on the objective function, the nature of constraints, number of independent and dependent variables [10]. Currently a number of optimization techniques are being used to optimize machining process parameters and include evolutionary or metaheuristic techniques, most of which are inspired by the nature of animal behavior and include genetic algorithm (GA), PSO, ABC, ACO etc [9]. In this research work, three optimization techniques namely PSO, FA and ABC algorithm have been applied for optimizing surface roughness during milling magnesium based alloys.

3.1 Artificial Bee Colony algorithm

The algorithm consists of colony of artificial bees contain bees of three categories: employed bees, onlookers and scouts. Initial half of the bee colony consists of the artificial bees and the latter half includes the onlookers. One employed bee is utilized for each food source. The employed bee without a food source becomes a scout. The food search carried out by the artificial bees is as follows:

- Food source is identified by employed bees within the neighbourhood of the food source in their memory.
- Onlookers gain information about food sources and select one food source from employed bees inside the hive.
- A food source is selected by onlooker within the boundary of food sources chosen by themselves.
- A scout is an employed bee without a source will search for a new food source randomly.

The four main steps are:

Initialize

REPEAT

(1) Employed bees are moved onto their food sources and determine their nectar amounts.

(2) Circulate the onlookers onto the various food sources and determine their nectar amounts.

(3) Scouts are being moved to search for fresh food sources.

(4) Memorize the best possible food source detected so far.

The steps are repeated until requirements are met. [15]

3.2 Objective function development

For any non-conventional optimization problem, it is very much essential to define the objective function. There are several methods to obtain the objective function when studying a predictive task. Artificial Neural Network (ANN) based objective functions have been used in all the optimization techniques studied in this work. Puneet N. P. et al. (2015) in a study related to surface roughness optimization in high speed turning of mild steel have established that ANN based objective function is better than Response Surface Methodology (RSM) based objective function in the use of both genetic algorithm (GA) and PSO techniques [10].

The primary objective of this work is to optimize (minimize) surface roughness parameter R_a. The objective function is of the form,

$$R_a = [V_c f d V_c^2 f^2 d^2 V_c * f V_c * d f * d][C] \quad (3)$$

[C] = matrix containing coefficients of terms in the equation

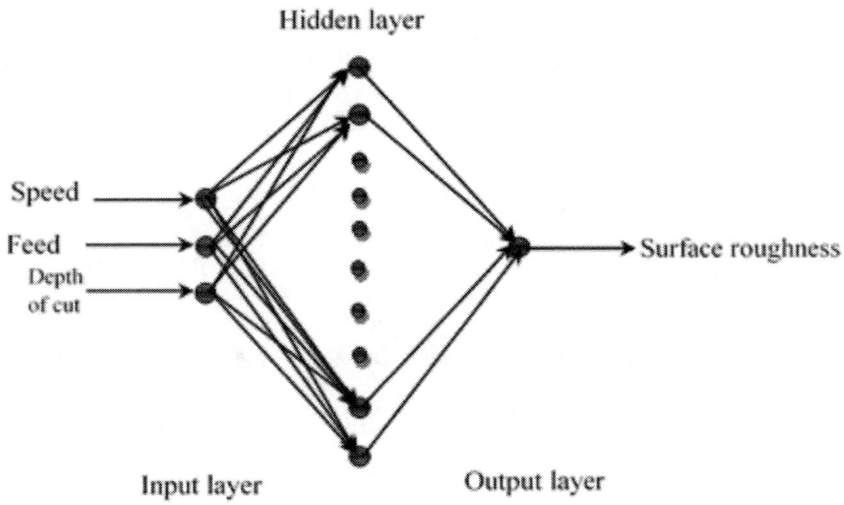

Fig. 3. ANN Model for R_a

A Multi-Layer Perceptron (MLP) model has been developed using the neural network toolbox in MATLAB [13]. The network has been trained using tansig and logsig transfer along with different training algorithms namely *trainlm*, *trainbfg*, *trainscg* and *traincgf*. The neurons in the hidden layer has been changed from 5 to 25 and the model which gave the highest R^2 (correlation coefficient) has been selected. The highest R^2 of 0.998 is achieved using *tansig* transfer function with *trainlm* (Levenberg-Marquardt) training algorithm for 13 hidden neurons with an error of $5.64*10^{-5}$ in 1000 epochs.

3.3 Implementation of Algorithms

3.3.1 Firefly algorithm

The implementation of FA is based on the MATLAB code developed by Xing-She Yang at MATHWORKS file exchange website [12]. The FA algorithm is run for 100 iterations. The input parameters for FA include population of fireflies which have been varied between 15 and 100, randomization is varied between 0.3 and 0.8, light absorption coefficient is set a value of unity, the attractiveness coefficient is set to 0.2.

The range of input parameters are:
Range of milling speed is 200 m/min – 500 m/min
Range of feed rate is 0.4 mm/revolution – 1.2 mm/revolution
Range of DOC is 0.15 mm – 0.35 mm

3.3.2 PSO algorithm

The implementation of PSO algorithm has been done in MATLAB based on the work done by S.B.Raja and N. Bhaskar (2011) [14]. The PSO algorithm is run for 100 iterations. The particle dimension is set to 5, with population size varied between 20 and 40. The learning factors are given the values 1 and 3. The range of input parameters are same as described in previous section.

3.3.3 ABC algorithm

The population size is varied between 15 and 25.The setting of different parameters are as follows: Number of iterations = 40, Dimensions of particle = 3. The range of input parameters are same as described previously.

4 Results and discussion

The goal of this work is to minimize surface roughness in terms of parameter (R_a). The values of the simulation parameters for all the algorithms have been set by trial and error. The minimum value of the surface roughness parameters obtained using all the algorithms is given in Table 3 along with the experimental values.

Table 2. Performance of Optimization algorithms

Roughness parameter	Algorithm	V_c	f	DOC	R_a μm
R_a	PSO	499.871	0.4	0.347	0.4606
	FA	500	0.471	0.265	0.374
	ABC	462.5	0.45	0.27	0.36
	Experimental values	500	0.4	0.35	0.4146

It is recommended to use high cutting speed, low feed rate and high depth of cut for obtaining minimum R_a.

Table 3. Comparison of optimization algorithms

Comparable techniques	Parameter	Techniques	Surface Roughness parameter R_a	Better technique	% Improvement
ABC and FA	R_a	ABC FA	0.36 0.374	ABC over FA	3.74%
ABC and PSO	R_a	ABC PSO	0.36 0.4606	ABC over PSO	22.8%

The focus of the present work is to study the use of ABC algorithm for optimization of R_a of AZ31 magnesium alloy during milling. The desired surface roughness value is in the range of R_a (0.31 to 0.4 μm). The focus in optimization studies is to identify the minimum value of R_a, which can be achieved practically based on the experimental studies carried out for the given machining operation, alloy machined and cutting conditions used. Table 4 gives the comparison in performance between ABC and other two algorithms. The first comparison is between ABC and FA. It is found that ABC is better than FA in minimization of surface roughness and the improvement is 3.74 %. The next comparison is between ABC and PSO and the improvement is 22.8%. Compared to FA and PSO, ABC is better and its performance is much better than PSO when compared to FA. ABC algorithm is very effective in these kinds of applications. There is no much difference in the execution times of all the three algorithms.

4.1 Validation experimentation

The results obtained have been validated by conducting an experiment, by setting the optimized cutting conditions obtained from ABC namely cutting speed, feed rate and depth of cut as 500 m/min , 0.4 mm/tooth and 0.374 mm respectively. The surface roughness value obtained is 0.34995 μm (average value). A comparison with the values obtained using ABC has resulted in an error of 2.792 %. This establishes the effectiveness of ABC for optimization of surface in milling of magnesium alloy.

5 Conclusions

In the present work, the main objective is to obtain minimum surface roughness during milling of magnesium alloy AZ31.To achieve this, optimal machining parameters have been obtained using different optimization algorithms like firefly algorithm, ABC algorithm and a widely used algorithm PSO. The focus is to compare ABC algorithm with firefly and PSO. The objective function for all algorithms has been obtained from output of an ANN model. The following important conclusions can be drawn from the study.

1. ABC algorithm is found to be better compared to firefly and PSO in terms of minimization of R_a.
2. From the results it is preferred to use high cutting speed, low feed rate and high depth of cut for obtaining minimum R_a.
3. The validation results has resulted in an error of only 2.792 % for minimization of R_a.

References

1. Xue-Nan GU et al., "Magnesium based degradable biomaterials", Front. Mater. Sci., Vol.8, No.3, (2014), pp 200-218.
2. Y.B. Guo and M. Salahshoor, "Process mechanics and surface integrity by high speed dry milling of biodegradable magnesium-calcium implant alloys", CIRP Annals – Manufacturing Technology, Vol. 59, (2010), pp 151-154.
3. Kaining Shi, Dinghua Zhang and Junxue Ren, "Optimization of process parameters for surface roughness and microhardness in dry milling of magnesium alloy using Taguchi with grey relational analysis", Int. J. Adv. Manuf. Technol., Vol.81, (2015), pp 645-651.
4. Yongjun Chen et al., "Recent advances on the development of magnesium alloys for biodegradable implants" Acta Biomaterialia 10 (2014) 4561–4573.
5. Shaoxiang Zhang et al., "Research on an Mg–Zn alloy as a degradable biomaterial", Acta Biomaterialia 6 (2010) 626–640.
6. Kaining Shi et al., "Effect of cutting parameters on machinability characteristics in milling of magnesium alloy with carbide tool", Advances in Mechanical Engineering (2016), Vol. 8(1) pp 1–9.
7. Grigoraş, Cosmin-Constantin, and Gheorghe Brabie. "The influence of milling parameters on the surface roughness in the case of magnesium alloy AZ61A." Elastic 3, no. 20 (2015) pp 41-45.

8. Zagórski, Ireneusz, and Józef Kuczmaszewski. "Selected problems of effectiveness of processing of magnesium alloys with a Kordell design toll." Management and Production Engineering Review, Vol. 4, (2011), pp 90-96.
9. Norfadzlan Yusup, Azlan Mohd. Zain and Siti Zaiton Mohd. Hashim, "Evolutionary techniques in optimizing machining parameters: Review and recent applications (2007-2011)", Expert System with Applications, Vol.39, (2012), pp 9909-9927.
10. Puneet N.P., Srinivasa Pai P., and Grynal D'mello, "Optimization techniques in Turning – A review", Manufacturing Technology Today, A CMTI Publication, May 2015, pp 3 – 21.
11. Grynal D'Mello, Srinivasa Pai P. and Puneet N. P., "Optimization studies in High Speed turning of Ti-6Al-4V", Applied Soft Computing, Vol. 51, (2017), pp 105-115.
12. X.-S. Yang, Firefly Algorithm, Mathworks File Exchange Website, December2010 http://www.mathworks.com/matlabcentral/fileexchange/29693-firefly-algorithm.
13. MATLAB and STATISTICAL Tool box, R 2012A, The Mathworks Inc., Natick, MA, 2013.
14. S.B.Raja and N.Baskar, "Particle swarm optimization technique for determining optimal machining parameters of different workpiece materials in turning operation", Int. J. Adv. Manuf. Technol., Vol.54, No. 5-8, (2011), pp 445-463.
15. D. Karaboga, B. Basturk, "On the performance of artificial bee colony (ABC) algorithm", Applied Soft Computing, Vol. 8, (2008), pp 687–697

Study of mould hardness for alpha set type resin bonded moulding sand system using Taguchi approach for metal casting applications

Deshpande Anand[1], *Gharge Aishwarya*[2], *Deshpande Pratik*[2], *Patil Pratik*[2], *Vaidya Tejashree*[2] and *Chate Ganesh*[1*]

[1]*Dept. of Mechanical Engineering, KLS Gogte Institute of Technology, Belagavi, Affiliated to Visvesvaraya Technological University, Belagavi, Karnataka, India*
[2]*Dept. of Industrial and Production Engineering, KLS Gogte Institute of Technology, Belagavi, Karnataka, India, Affiliated to Visvesvaraya Technological University, Belagavi, Karnataka, India*

Abstract. The quality of the castings are depended on many properties of sand mould like mould hardness, green compression strength, shear strength etc. and these properties in turn dependent on factors such as grain fineness number of sand, setting/curing time, amount of resin, amount of hardener, moisture content etc. Mould hardness is very important while transferring mould from moulding station to pouring station when the weight of the mould is above 100kg and may lead to defects. In the present work, an effort is made to study the mould hardness of a moulding sand specimen using Taguchi technique.L_9 orthogonal array is used and experiments are conducted randomly. The main factors such as the amount of resin, amount of hardener and setting time were considered. Factors were selected based on Literature review and brainstorming session with foundries. The chemical (resin and hardener) used is alphas set type two part system. From experiments, it was observed that the Amount of resin and setting time are significant for mould hardness. Confirmation tests were conducted to validate the obtained results.

1 Introduction

Casting is the most basic manufacturing process to manufacture the engineering products made up of metals. There are many types of castings like sand casting, shell moulding, injection moulding etc. But sand casting is the widest method adopted to make cast products. Based on the type of sand and binding agents used, sand casting is again divided into the green sand moulding, chemicals and moulding, dry sand moulding, no-bake sand moulding, cold box moulding. Many researchers worked in the field of sand moulding and also tried to optimize the moulding sand system used for making moulds using different techniques. Rasik et al. used Taguchi technique to optimize process parameters in sand casting [1].Danko et al. investigated the properties of the reclaimed sand for making no-bake sand moulds using alpha set binding system by using newly reclamation

* Corresponding author: ganeshchate@git.edu

apparatus[2].Himanshu et al. used oil urethane no-bake binder system and used Taguchi technique to optimize the process, shrinkage of the specimen is taken as the response[3].Raghwendra et al. used Taguchi technique for green sand moulding sand to get effective and economical castings[4].Parappagoudar et al., used full factorial design to obtain a linear model for green sand moulding system[5].Dietert et al. emphasized that density of sand mould increases with mould hardness as a result of wide range of grain size [6].Sand drop defects in casting are influenced by mould hardness and also shear strength, which in turn is dependent on by grain fineness number (GFN), the quantity of binder (resin and hardener), curing time, the degree of ramming pouring velocity and so on. Barlow et al. studied the influence of hardness to prevent the mould wall movement in moulding[7]. Frost et al. analyzed pressure and hardness distributions in sand moulds using theoretical and experimental methods. The authors observed that frictional interface between moulding sand and pattern to be the governing factors that significantly affect mould hardness[8].From the above literature, it is understood that not much work is being done on Alpha set type sand mould systems and also mould hardness is highly influenced by the process variables. Proper control of process parameters and setting these process parameter is essential to get good quality castings i.e., castings with less sand drop defect. The present work focused on determining the optimum levels of process parameters to better mould hardness. Not much work reported yet to obtain optimum levels for a process. Taguchi method has been reported to reduce the process variation through a robust DOE to gain high-quality products at low costs. Taguchi method uses orthogonal arrays to decide the design matrix for experiments, determine the optimal process parameter setting for each response, and estimate the percentage of contribution of each individual factor towards the response [9].Taguchi method has been employed to determine the optimum input parameter levels on different quality characteristics (that is, outputs) of squeeze casting process[10], drilling[11], injection moulding process[12], cold box core shooter machine[13].From the above literature, it was understood that Taguchi technique reduces experimentation cost and time. Hence Taguchi technique was used for this study.

From the literature survey and also brainstorming session with foundries, the problem was defined. Most of the industries add resin and hardener to the sand by trial and error basis, hence at the beginning of the shift moulds start braking and gradually the workers increase the hardener and resin percentage. But too much of hardener and resin leads to difficulties in shakeout operation and also the gases evolved after pouring the molten metal will be entrapped because of the reduction in permeability. Hence it was decided to set the process parameters for mould hardness for the alpha-set type no-bake sand system.

The objectives of the work are

1. To set the levels of process parameters that affect the mould hardness.
2. To identify the factors which are significant for mould hardness.
3. To study the effect of hardener, resin, curing time and interaction between these parameters using an L9 orthogonal array on alpha set type no base moulding sand system.

2 Materials and equipment used

The work deals with chemically bonded moulding sand system. Sand is the main component in the system, a fresh silica sand with grain fineness number (GFN) number 55-60 was taken, that is the medium sand. If coarse sand was taken, dimensional accuracy and surface finish of the casting will be compromised. If too fine sand was taken, it will reduce the permeability of mould and leads to blowholes and pin holes. The fresh sand is heated in sun rays in order to remove the moisture traces in the sand. The alpha set type no-bake two-part system resin and hardener were taken for the study. An Ester Cured Alkaline Resin binder (Commonly known as Alpha-set Resin).Alpha set chemicals are alkaline Phenolic systems and are eco-friendly, and gives good Casting Finish. As per industrial standards, the amount of resin (part A) to be added is from 1.8% to 2.2% weight of sand and amount of hardener(part B) is 20 to 22% weight of the resin. The fishbone diagram for mould hardness is as shown in fig 1. There are many factors that affect the mould hardness. The factors which can be practically controlled easily, effectively and economically were considered. The factors and levels considered for experimentation are shown in Table 1.The values of levels and also the ranges are chosen based on the consultation with industry experts and also literature available.

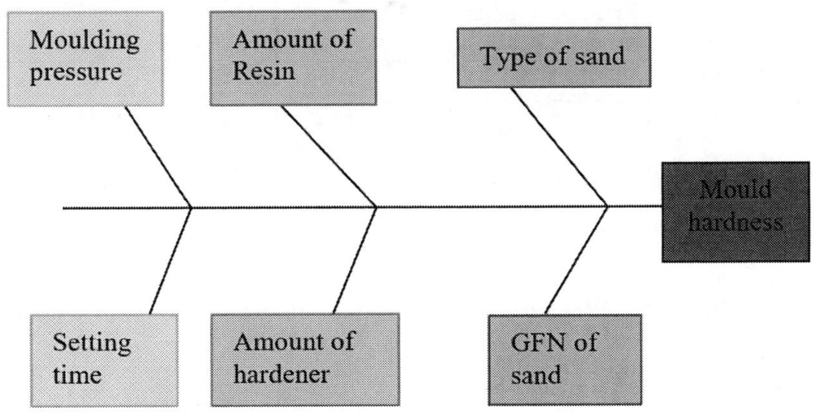

Fig. 1. Fishbone diagram for mould hardness

The measured quantity of sand is mixed with metered quantity resin and hardener using sand muller.2 minutes of mixing (mulling) is done. The experiments were conducted using an L9 orthogonal array, as there were 3 factors and at 3 levels.

Table 1. Process parameters (factors) and their levels

S.No	Parameters	Notations	LEVELS		
			1	2	3
1	Resin (% weight of Hardener)	A	1.8	2	2.2
2	Hardener (% weight of Resin)	B	20	22.5	25
3	Curing time(Mins)	C	60	180	300

The GFN of the sand to be used is confirmed by using sieve shaker, then the silica sand is dried by keeping it in sun rays for about 1hour.1kg of sand is weighed and added in sand muller along with the resin and hardener. The quantity of resin and hardener are varied as these are considered to be factors. The variation is done on the basis of experimental design matrix as shown in Table 2.

Table 2. Standard L9 orthogonal array and variable settings

	DESIGN MATRIX			VARIABLE SETTINGS		
S.No	A	B	C	A	B	D
1	1	1	1	1.8	20	60
2	1	2	2	1.8	22.5	180
3	1	3	3	1.8	25	300
4	2	1	2	2	20	60
5	2	2	3	2	22.5	180
6	2	3	1	2	25	300
7	3	1	3	2.2	20	60
8	3	2	1	2.2	22.5	180
9	3	3	2	2.2	25	300

The mixture is then taken out from the muller and standard specimens are prepared using standard rammer. The specimens are allowed to settle for the predefined time (settling time) as per the experimental design matrix. The mould hardness is then measured using mould hardness tester. The step by step procedure is shown in Fig. 2.

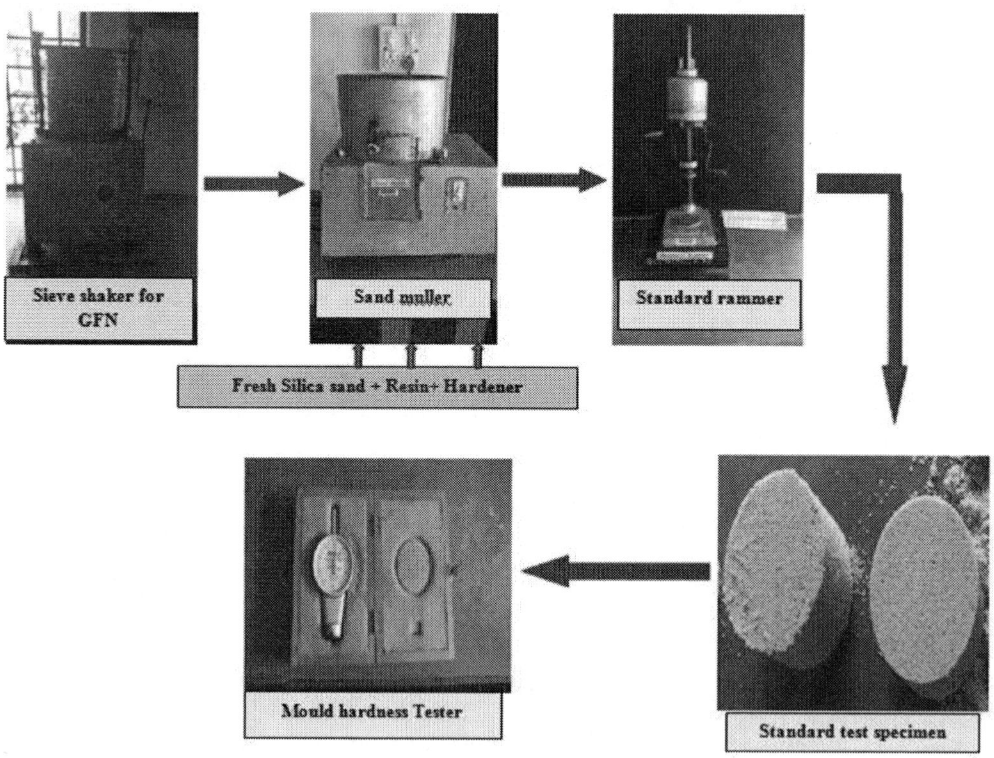

Fig. 2. Stepwise représentation of expérimentation

3 Results and Discussion

The mould hardness (MH) is measured for 3 replicates. The average or mean value of all 3 replicates along with the signal to noise ratio are shown in Table 3.Eqution 1 is used to find a signal to noise ratio. Larger the better condition is used to find a signal to noise ratio, as mould hardness should be maximum (objective is to maximize).

$$\text{Signal to noise ratio(Larger is better)} = -10\log_{10}\frac{\left(\Sigma\frac{1}{Y_i^2}\right)}{n} \quad (1)$$

Table 3. Experimental design matrix, mean value along with signal to noise ratio

Resin (%wt.of sand)	Hardener (%wt.of resin)	Setting time (hrs)	MH1	MH2	MH3	Signal to noise ratio(S/N)	Mean
1.8	20	1	58	59	59	35.36	58.66
1.8	22.5	3	60	62	64	35.83	62
1.8	25	5	67	67	66	36.47	66.66
2	20	3	65	67	65	36.34	65.66
2	22.5	5	71	70	72	37.02	71
2	25	1	60	60	60	35.56	60
2.2	20	5	72	73	73	37.22	72.66
2.2	22.5	1	62	63	61	35.84	62
2.2	25	3	66	67	66	36.43	66.33

ANOVA test for the signal to noise ratio of mould hardness is given in Table 4 represents the signal to noise ratio. From the table, it can be inferred that resin and setting time are significant factors for mould hardness. Even though hardener acts as a catalyst, it does not play a significant role for mould hardness. The error in the ANOVA table indicates the interaction among the factors.

Table 4. Analysis of Variance for SN ratio

Source	Degrees of Freedom	Sum of squares	Adjusted Mean Squares	F-test	P-test	Percentage contribution
Resin (%wt.of sand)	2	0.578	0.28933	80.36	0.012	17.91
Hardener (%wt.of resin)	2	0.035	0.01785	4.96	0.168	1.11
Setting time (hrs)	2	2.606	1.30322	361.96	0.003	80.86
Residual Error	2	0.007	0.00360			0.22
Total	8	3.22801				100

The interaction of the factors is also not significant as per the ANOVA table. The highest percentage contributing factor is setting time followed by the amount of resin.Fig3.shows the fitted values that are the theoretical values and residuals(i.e., the difference between experimental values and fitted values).It also shows the variation in the residuals in all experiments and also the upper and lower range fitted values from zero residual. The residual is the difference between theoretical value and experimental value. The residual values fall +0.4 below and -0.2 below fitted (theoretical value). The residual for all the experiments are also shown in Fig. 3.

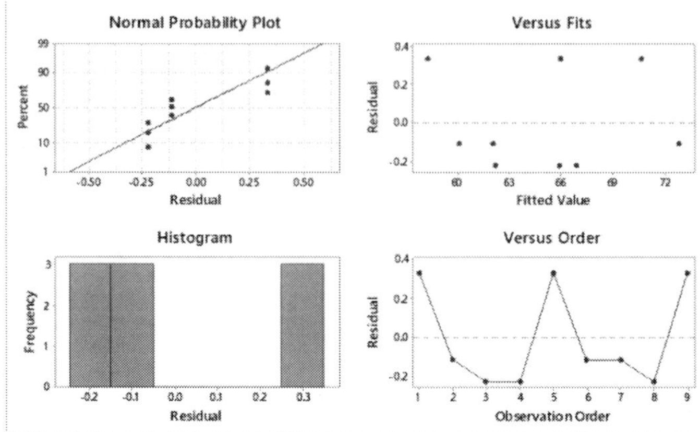

Fig. 3. Fitted values and residual plots

Fig. 4 is a graph of signal to noise ratio v/s factors. This graph shows the variation in signal to noise ratio with respect to different levels of the factors (process parameters).The highest point from the graph is considered to be the optimum levels of factors.

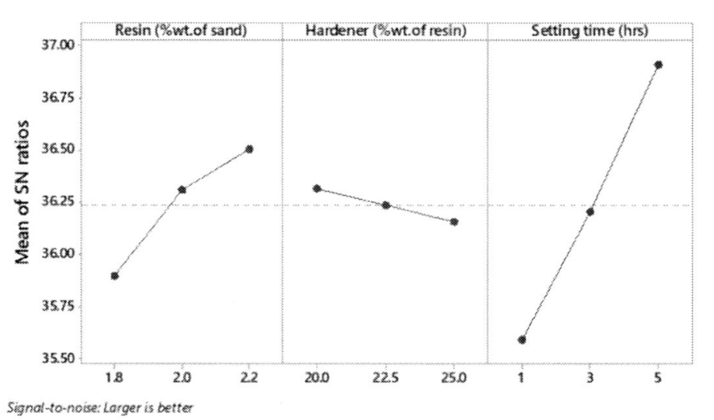

Fig. 4. Main effect plots for SN ratios v/s factors

The optimum process parameters can also be set by using a graph of mean v/s factors as shown in Fig. 5. As a number of resin increases mould hardness increases, it is because the resin is the binding agent. As the amount of hardener increases the mould hardness decreases, it is because too much hardener quantity will decrease the binding strength between resin and sand grains just like dilution effect (hardener acts as a catalyst). As setting time increases, mould hardness increases.

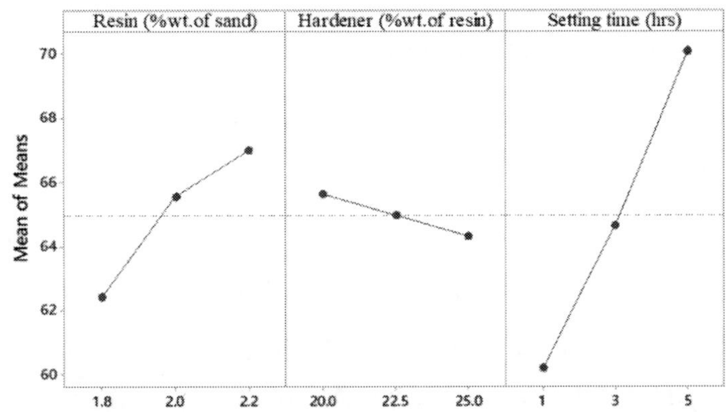

Fig. 5. Main effect plots for Means v/s factors

A surface plot of average mould hardness v/s significant factors (setting time and amount of resin) is shown in Fig. 6. As the setting time increases there will be a gradual increase in the mould hardness (In figure average value of mould hardness for 3 replicates is indicated as mean).For the same setting time, if resin content is increased within the range, there will be an increase in mould hardness. If the amount of resin is increased beyond experimental range then specimens will become wet thus getting irregularly shaped specimens. Similarly, when setting time is increased within the range with a constant level of resin, there will be an increase in mould hardness. If setting time is increased beyond the experimental range, then specimen will absorb moisture. At 2.2% of resin, there exists a point (between 4.0 hrs to 5.0 hrs) where there will be a slight reduction in the average mould hardness value. Hence the relationship between the significant factors and average mould hardness value (mean1 in the graph) is non-linear.

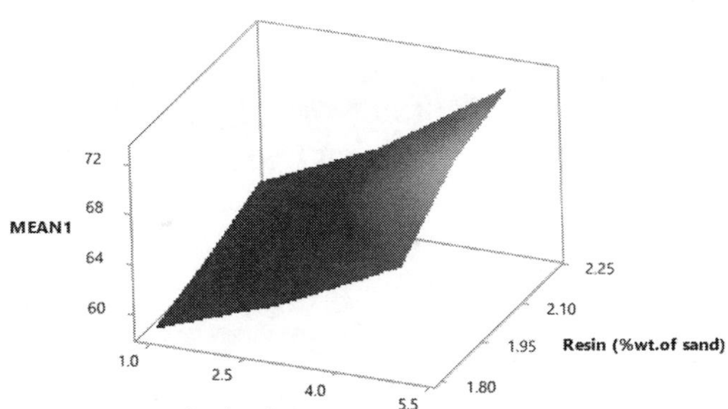

Fig. 6. Surface plot of mean response (mould hardness) v/s significant factors

4 Conclusions

The experimental results obtained are used to set process parameters to get maximum mould hardness in no-bake moulds to get quality castings. From the result, it can be concluded that setting time and resin are significant factors for mould hardness. The process parameters to be set to get maximum mould hardness is resin at 2.2% weight of sand, hardener at 20% weight of resin and setting time of 5hrs.These values hold good for alpha set type no-bake sand moulds only and temperature condition of 250-400 Celsius.

References

1.RasikUpadhye ,Dr.Ishwar P Keswani, IJERT, **1**, 1(2012).

2.R. Danko, J. Kaminska,Arch of foundry engg, **9**, 33, (2009)

3.Himanshu Khandelwal,.B,Jr. of Manuf. Proc, **22**,127, (2016)

4.Raghwendra Banchhor, S.K. Ganguly, IJAET,**V**,25,(2014).

5.M. B. Parappagoudar, D. K. Pratihar,G. L. Datta, , IJCMR,**20**,1, 2007

6.Dietert, H.W., Fairfield, H.H. & Hasty, E.J,AFS Transactions, **55**, 175, (1947).

7.Barlow, T.E, AFS Transactions, **74**,70, (1966).

8. Frost J. Hillier M.J. AFS Transactions, **75**, 126, (1967).

9.Montgomery D. C, DOE- John Wiley & Sons, New York, (2008).

10.Patel G. M, Krishna P, and Parappagoudar M.B. Procedia Tec., **14**,157 (2014).

11.GaitondeV.N, Karnik S. R, Achyutha, B.T, and Siddeswarappa, B. IJAMT,**34**,1 (2007).

12. Ozcelik B, and Erzurumlu T, Jr.of Materials Proc. Tech., **171(3)**,437, (2006).

13 .Chate G.R., Bhat R.P, & Chate U.N. Procedia Materials Sci.. **5**, 1976, (2014).

Vision system in quality control automation

Ravi Kiran[1], *Amarendra H. J.* [1] and *Shashank Lingappa*[1*]

[1]*Dept. of Mechanical Engineering, Malnad College of Engineering, Hassan 573202, Karnataka, India*

Abstract. Measurement of surface roughness is one of the quality control processes, usually carried out off line. Contact type surface roughness measurement method is commonly used in quality control. The processes consume lot of time with human interaction. In order to reduce or to eliminate non value added time, effective quality inspection tool and automation of the processes has to be utilized. An attempt has been made to automate the process with integration of vision camera in capturing the image of the component surface. The image process technique has the advantage of analyzing the single captured image for multiple area measurement. Hence, the in-line quality control of each component surface roughness measurement is ensured. The automation process involves component movement, image capturing, image processing, and decision making, using sensors, actuators and microcontroller. The proposed in-line quality control of surface roughness with vision system has been successfully developed. The designed automated system has fulfilled the objectives in respect of the scope of the present work.

Keywords: Quality Control, Surface Roughness, Image Processing, Intensity Factor,

1 Introduction

Over the past decades, the need of the mankind has demanded increase in productivity with the improved quality of the products. This has led to innovations, and these innovations have transformed the traditional manufacturing to advanced manufacturing. In view of the increase in productivity, non-value added manufacturing activities (NVAs) need to be eliminated through thorough study. In the manufacturing process, the quality control is one of the important stages. There are several methods of quality checking. One such activity in the manufacturing process is the quality inspection of components' flat surface roughness. Some of the contact type surface roughness measurement methods are Stylus Profilometry, Atomic Force Microscopy etc. Non-contact methods are Laser Measurement Technique, Binary Speckle Image Analysis, Image Processing Techniques etc. Each method has its own advantages and disadvantages. Contact type equipment (Stylus Profilometer) consumes more time for quality check than non-contact type [1]. Atomic Force Microscopy (AFM) has better speed of operation and accuracy. Laser technique can take multiple

* Corresponding author: shashanklingappam@gmail.com

readings over a desired region. The disadvantage with laser technique is limited working distance [2]. All these surface roughness quality checking methods require an atmosphere free from vibrations, dust, and heat.

The conventional quality checking of a component is made by taking one sample out of a lot to ensure the quality of the particular lot. Such quality checking method may lead to rejection of the whole lot or even acceptance of defective parts. In order to ensure the quality of the product, in the present competitive world, each component has to undergo quality check, which raises the need of in-process inspection. In-process inspection ensures successful control over the quality of the component, reduces the quality check time, ensure inspection of each component, ensure reliability and efficiency of the system. In-process inspection of surface roughness can provide a high-speed quality control check, to reduce skilled labor cost and thereby increasing productivity in modern manufacturing environments.

The image processing technique can be used to analyze quality of the component's surface roughness by capturing the surface image of the work piece through digital camera [3, 4]. Image processing technique can be incorporated along the manufacturing line. Such system has the advantages over other techniques. The integration of vision based inspection system reduces production lead-time, increases inspection activity efficiency and quality by eliminating the human interaction in the inspection process.

In the present study, an attempt has been made to incorporate the image processing technique in the manufacturing system for flat surface roughness measurement along with component acceptance/rejection automated system based on the analyzed data collected through image processing. The system is developed to demonstrate the concept with the following; a system with a conveyer belt system, metal detection system, component gripping and component positioning system, a camera positioning system, a carriage rotation system, an indicative grinding system along with the delivery conveyor belt system. The Graphical User Interface (GUI) is used for image processing and data handling, and microcontroller is used for controlling the overall process.

2 System development and component arrangement

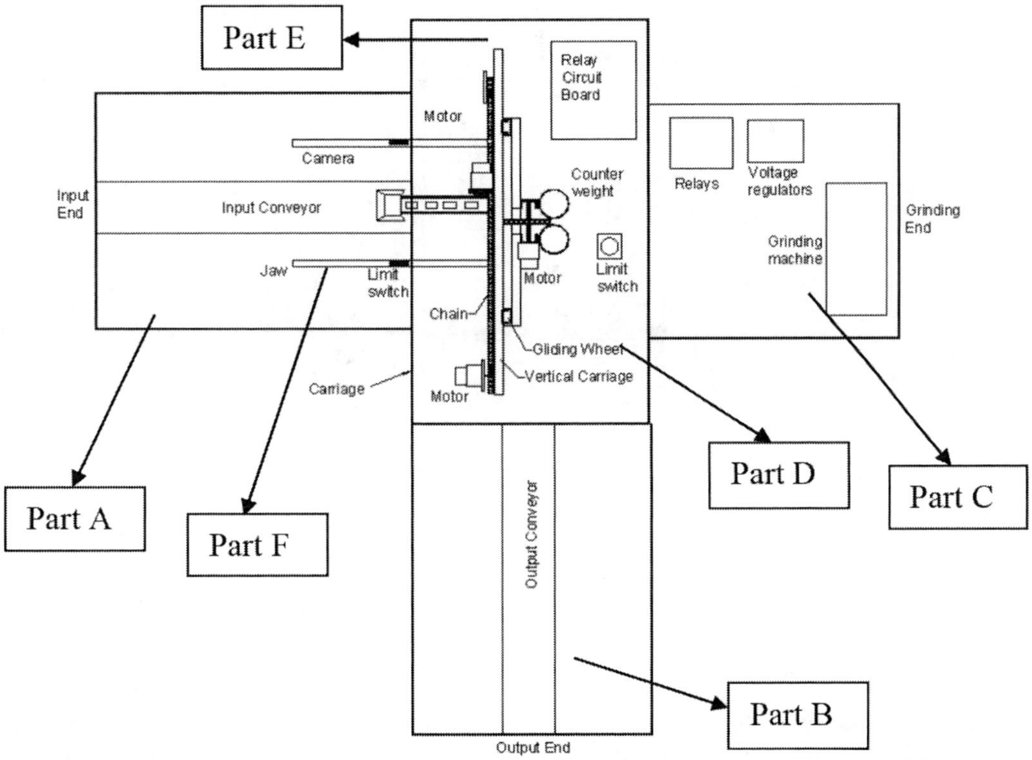

Fig. 1. Schematic diagram of the proposed system (top view) Part A-Input end, Part B- Output end, Part C – Grinding Assembly, Part D –Glider support, Part E – Vertical carriage, Part F- Jaw.

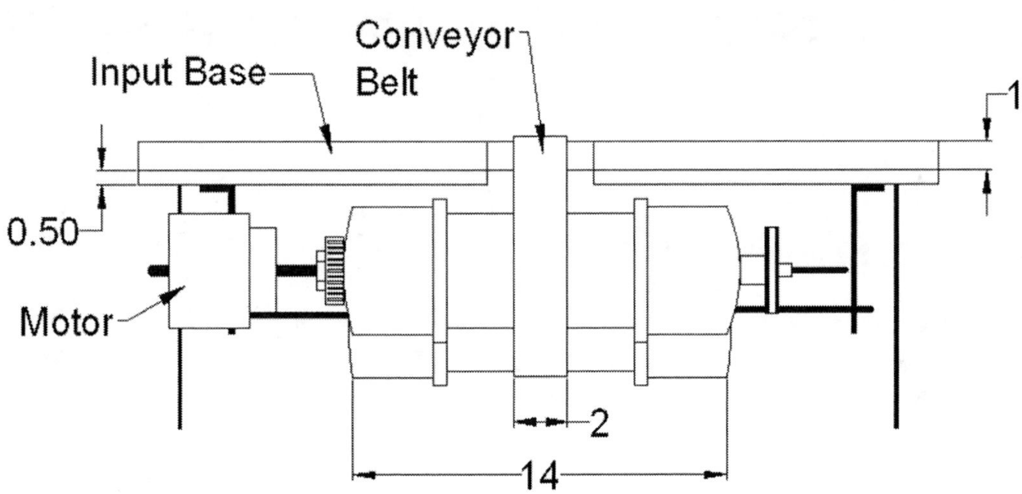

Fig. 2. Schematic diagram of input end

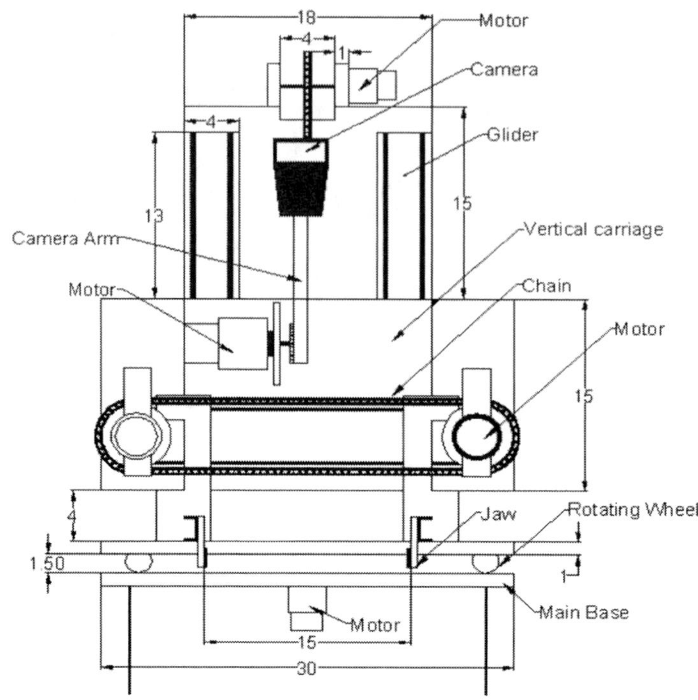

Fig. 3. Front view of the system

Fig. 3 shows the front view from the input end or the component supply end of the system. The camera arm is designed to rotate with an angle of 120°. The camera is connected to the tip of the arm at an angle of 30° to reach to the required position to capture the image of the component surface. The image captured is processed using GUI program to calculate the surface roughness over a specified region. The system is designed to capture the image from the region of interest before processing the image. This feature facilitate for the better and detailed quality check.

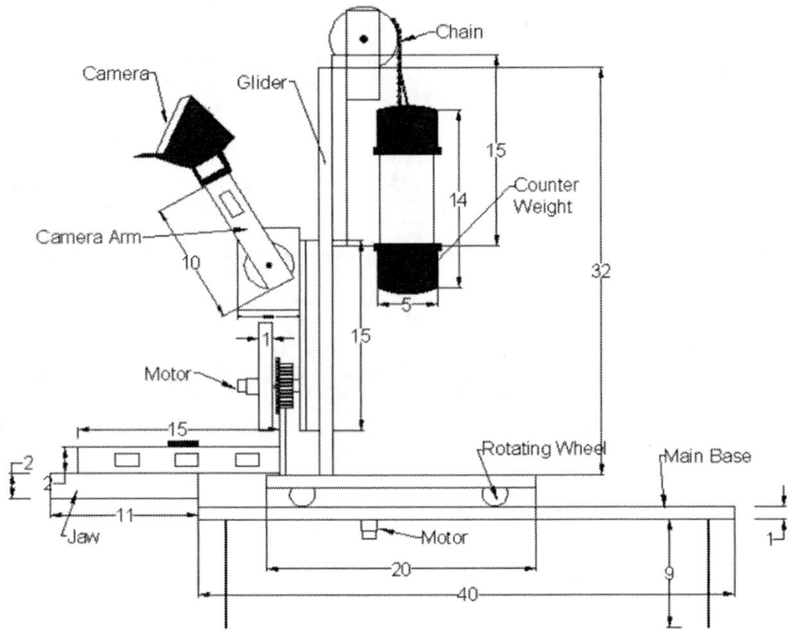

Fig. 4. Right side view of the system

Fig. 4 shows the schematic diagram of the right side view of the system from the delivery end. Once the image is captured by the camera and the surface roughness of the component is measured by the GUI software, the written program logic verifies the result to accept or to reject the component. The instructions to microcontroller are fed according to the surface roughness value obtained through GUI software. Further, the camera arm retracts to its initial position. The jaw closes to hold the component firmly and with a delay of 5 s, the vertical carriage moves up by 10 cm. The carriage then rotates to delivery end if the component surface roughness value falls within the acceptance limit, else to place it on the grinding end as rejected and to continue the grinding process.

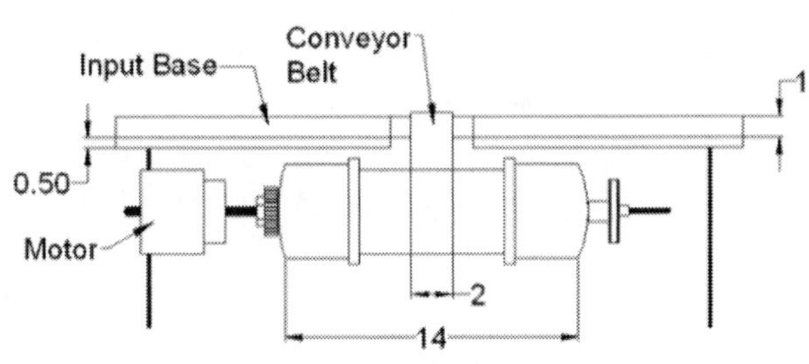

Fig. 5. Schematic diagram of the output end

Schematic diagram of the output end or the delivery end of the developed system is shown in Fig. 5. The delivery program executes to place the component on the delivery end conveyor. The program is designed to start the conveyor motor with 5 s delay. The delivery end is designed to operate for a load of 200 g and to operate at a speed of 0.04 m/s to cover a distance of 40 cm within 10 s. The program is designed to move the vertical carriage immediately after the component reaches the delivery end. The vertical carriage moves up by 10 cm and the carriage rotates by an angle 90° in clockwise direction to the home position for the next operation.

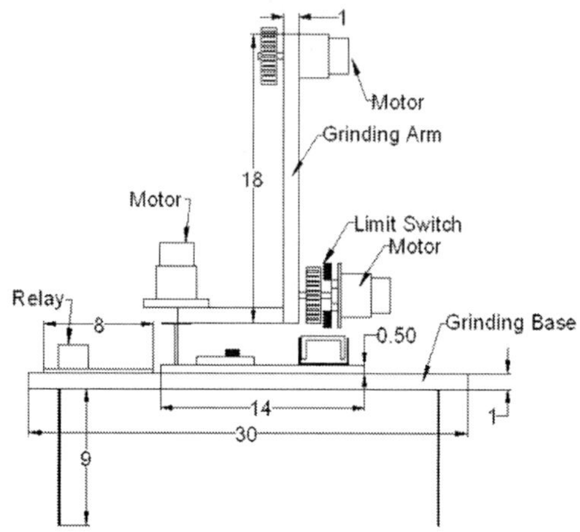

Fig. 6. Schematic diagram of the grinding assembly

Schematic of the grinding process setup is shown in Fig. 6. If the component is rejected by the GUI, the component from input end is moved to the grinding end. The carriage rotates by an angle of 180° to reach the grinding end. Initially, the grinder carriage moves forward and further the grinder arm moves down to carryout indicative grinding process. After the grinding process, grinder arm and grinder carriage moves to their home position. The camera arm move towards the component to capture the image and to analyze the surface roughness. If the component fulfils the required surface roughness, it is placed in the delivery end by rotating by an angle of 90° in a clockwise direction of the carriage, else the grinding process is repeated.

3 Camera arm kinematics

Kinematics refers to the analytical study of the motion of a robot manipulator [5]. Camera arm kinematics is calculated by applying the forward kinematics. Fig. 7 shows the schematic diagram of a camera arm.

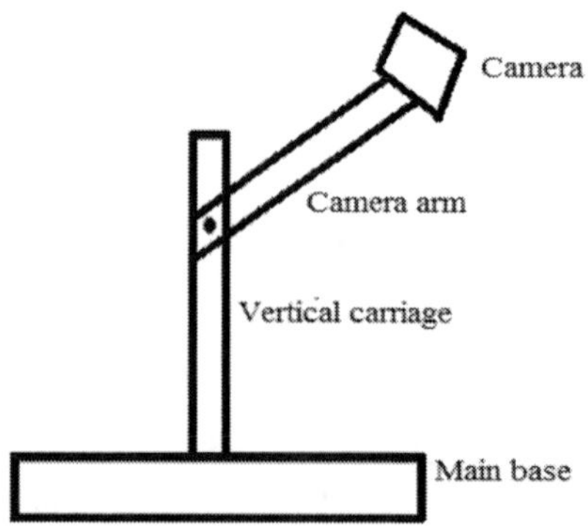

Fig. 7. Schematic diagram of a camera arm

Fig. 8 shows the link co-ordinates of the camera arm. Vertical carriage has a linear movement of 10 cm either in upward or downward position in Z-axis. Camera arm has an angular movement of 120° in Y-axis and camera is connected to the tip of the camera arm at an angle of 30°.

$$T=\begin{bmatrix} -0.1 & 0 & 0.3247 & 0 \\ 0 & -1 & 0 & 0 \\ 0 & 0 & 0.0625 & -10 \\ 0 & 0 & 0 & 1 \end{bmatrix}$$

The matrix T shows the overall transformation from the initial position to the final position of the camera arm. The matrix obtained is called the arm equation in kinematics. The value '-10' gives the vertical carriage is in the ' –Z' direction along the Z-axis when it comes to the initial position from the final position.

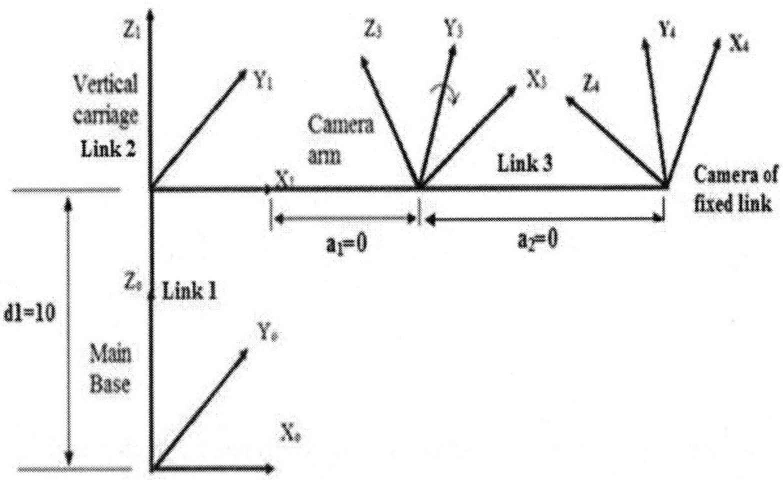

Fig. 8. Link co-ordinates

4 Image Processing Techniques

Image processing is a technique to enhance and analyze the raw images received from cameras. Image processing is the first step to prepare the image that has been captured for analysis [6]. Besides that an image is a representative of a real scene, this image is presented in such a way in an array or a matrix of square pixels (picture elements) arranged in columns and rows. Therefore, it may be defined as a 2-dimensional function, $f(x, y)$, where x and y are spatial coordinates.

Edges are often used in image analysis for finding region boundaries. They are the regions where pixel brightness changes abruptly. An edge essentially distinguishes between two distinctly different regions or in short, an edge is the border between two different regions [7]. Edge detection of an image reduces significantly the amount of data and filters out information that may be regarded as less relevant, preserving the important structural properties of an image. Most edge detection methods work on the assumption that an edge occurs where there is a discontinuity in the intensity function or a very steep intensity gradient in the image. Using this assumption, the derivative of the intensity values across the image is taken and finding the points where the derivative is a maximum, the edge will have to be marked.

The image representation technique is mainly used in the present work to acquire the maximum and minimum pixel values of the image obtained from the work piece. A coloured image is captured by the camera. It consists of all the 3colours i.e. Red, Green and Blue. It is very difficult to process a color image. In the present work, only red color of the image is extracted. It is nn 8-bit grey scale image. The image can now be accessed as a two-dimension array of data, where each data point refers to a pixel for digital images. It consists of values from 0 to 255 only i.e. 256 different pixel values. 0 being pitch black and 255 being white.

The image of the component is captured and stored as bitmap image in the installed drive of the visual basic software. Maximum and minimum pixel of the image is extracted in the particular area of interest and is displayed in the visual basic interface. After getting the pixel values, the intensity factor of the component is obtained using a specified formula. With the knowledge database, surface roughness value is picked corresponding to intensity factor value.

5 Graphical User Interface

A graphical-based language allows the user to work directly with graphics. A project is a collection of files that makes the user program. They may consist of form, modules, and active x controls [8].

The visual basic software version 6 is used as a graphical user interface. Firstly when the software is opened it asks for new project. Standard EXE is opted and option open is clicked. A form window is opened. In the form window, text boxes, labels, command buttons, picture boxes are created. Properties of all the created icons are set according to the requirement. Double click on the particular icon allow us to write the code. For this project, area of the co-ordinates is specified to find the surface roughness. In the particular area of the co-ordinates, maximum and minimum pixel values are extracted. Minimum pixel value is assigned to the text11.text and maximum value is assigned to the text12.text. These maximum and minimum values are used to find the intensity factor of the component. Text1.text is assigned to the intensity factor formula. If the text1.text value is less than the 0.5 then the PC give the message to microcontroller to deliver the component in the delivery end. If the text1.text value is greater than the 0.5 then the PC give the message to microcontroller to deliver the component in the grinding end. At the same time surface roughness of the component is picked corresponding to intensity factor value and is displayed in the GUI. The Graph of intensity factor versus surface roughness is also displayed.

6 Experimental setup of the developed setup

Input end: Component moves on the conveyor and is stopped by the proximity sensor. The proximity sensor is connected at the bottom end of the input conveyor. Fig. 9 shows the top view of the setup.

Grinding end: An indicative grinding machine is used to show how the grinding machine works. The machine grinds the component after the component has been placed in the grinding end by the jaw.

Output end: Once the component has met the quality standards it will be placed on the output conveyor by the jaw. The component moves till the end of the output conveyor.

Carriage: It contains camera, vertical carriage with gripper. The carriage moves both in clockwise and anti-clockwise direction. A camera is used to capture the component image. Jaw is used to hold the component. Chain and sprocket is used to move the vertical carriage up and down.

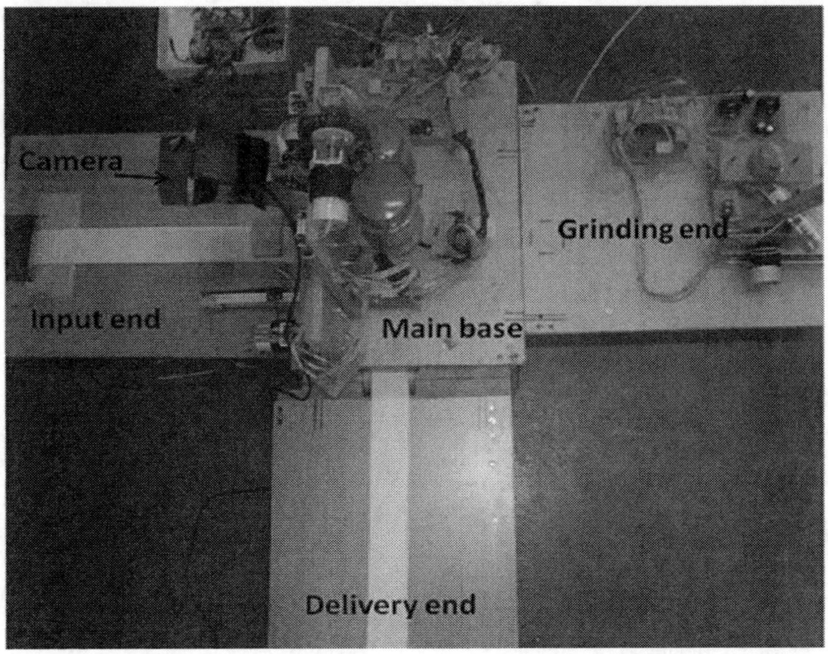

Fig. 9. Top view of the setup

7 Results

The design intention is achieved successfully and below shows the developed GUI window for acceptance and rejection process, time taken for the test are shown in table 1 and the red box shown in GUI can be moved to any point within the sample snapshot boundary.

7.1 Acceptance and rejection process

Output 1: Rejected component with IF>0.5 and R_a*=8.75 μm*

The image captured by the camera has to be processed for finding the surface roughness value. Code has been written in such a way that it has to extract maximum and minimum pixel values of red colour over a specified area for finding the intensity factor value. With the help of knowledge database, surface roughness value is picked corresponding to intensity factor value. Variation of R_a value v IF is also showed in the GUI. The 1st component has not met the quality standard where the IF is greater than 0.5 and that component needs to be grinded to meet the standard by rotating an angle of 180° in anticlockwise direction by the carriage for placing the component in the grinding end. (Fig. 10)

Output 2: Accepted component with IF<0.5 and R_a=3.75 μm

After the grinding process, the camera captures the image of the 2nd component and the same procedure is followed for finding the surface roughness of the component. If the component has met the quality standard where the IF is less than 0.5, the component is moved to the delivery end by rotating an angle of 90° in clockwise direction. The component on the conveyor moved a distance of 40 cm for 10 s at a speed of 0.04 m/s. Carriage comes to the initial position to do the next operation. The conducted experiment shows that the design intention is fulfilled. (Fig. 11)

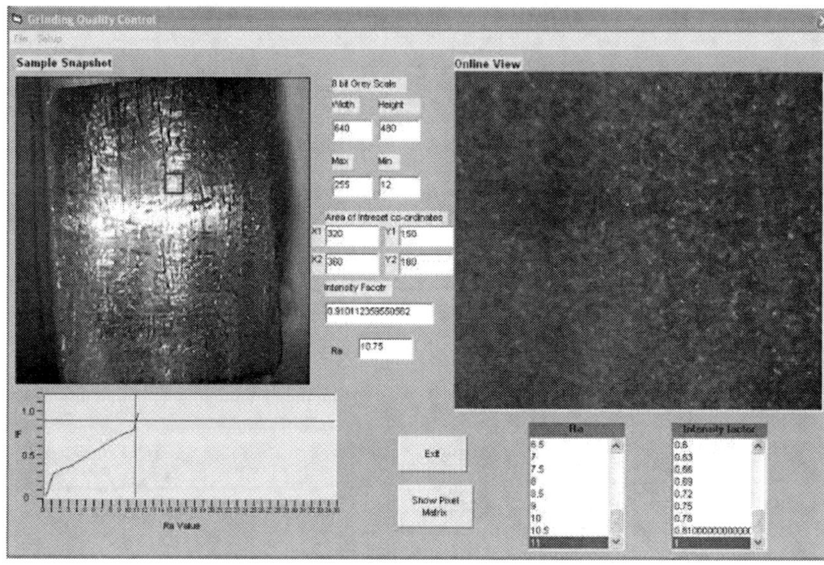

Fig. 10. Graphical User Interface with an intensity factor > 0.5 and R_a =10.75 μm

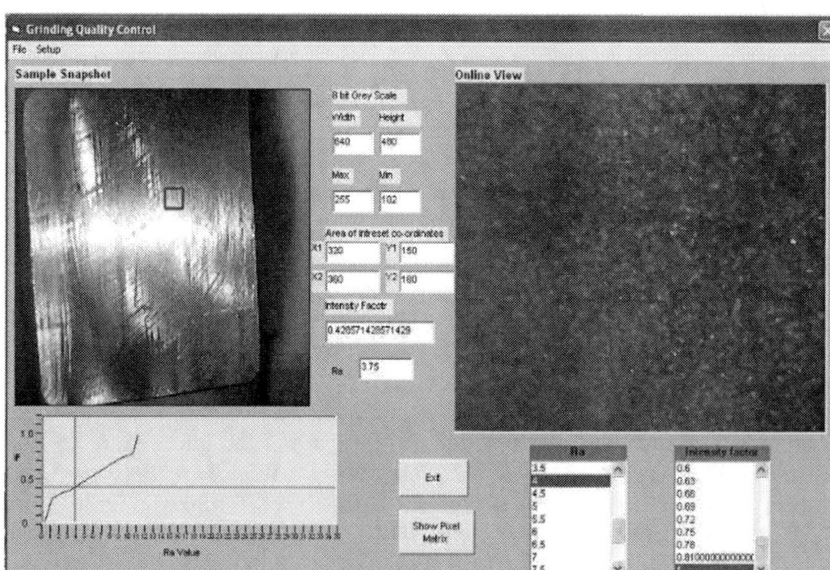

Fig. 11. Graphical User Interface with an intensity Factor < 0.5 and R_a=3.75 μm

7.2 Time taken for the test procedure

In the present work, time taken by the designed automation of the testing process is evaluated. The time taken for the acceptance, rejection and for the consecutive operation is shown in the Table 1. The time consumed for both acceptance and rejection is within ~90 s which include testing of the component for its quality and movement of the component to the delivery end. The integration of image processing technique in to the manufacturing line ensure not only the quality of the component by checking each component but also saves time in terms of movement of the component, as well as the time consumed for the measurement process.

Table 1. Time taken for the test

Sl. No	Test	Time (s)
1.	Acceptance	75
2.	Rejection	89
3.	Next operation	24

8 Conclusions

Present work describes the design, development and implementation of automating the vision based system for surface roughness measurement by incorporating the image processing technique in quality control. This system provides fast and flexible quality checking of every component, thereby eliminating the need for separate quality department and ensures 100% defect free production. The designed GUI system also provides facility for the measurement of the whole surface or desired area in the component. Due to the elimination of non value added time viz. movement of component to quality department, the proposed work drastically reduces the lead time and in turn increases productivity and reduces the cost of production.

References

1. Peter J. Blau, Rebecca L. Martin, Laura Riester, "A comparison of several finish measurement methods as applied to ground ceramic and metal surfaces", Propulsion system materials program office of transportation technologies energy efficiency and renewable energy, U.S. Department of energy, Jan 1996.
2. Ersin Kayahan, Hasan Oktem, Fikret Hacizade, Humbat Nasibov, Ozcan Gundugdu, "Measurement of surface roughness of metals using binary speckle image analysis", Tribol Int, 43, 307–311 (2010).
3. Frank Y. Shih, "Image Processing and Pattern Recognition", Institute of E & E Eng Inc. (2010) ISBN: 978-0-470-40461-4.
4. B. Y. Lee, Y. S. Tarng, "Surface roughness inspection by computer vision in turning operations", International Journal of Machine tools and Manufacture, 41, 1251-1263 (2001).
5. Robert J. Schilling, "Fundamentals of robotics, analysis and control", Prentice Hall of India Private Limited (2005) ISBN: 81-203-1047-0.

6. Rafael C. Gonzalez, "Digital image processing", Prentice Hall, 2nd Edition (2002) ISBN: 0201180758.
7. Nurul Hakeem, "Development of visual basic software using visual basic 6.0", Faculty of Electrical and Electronics Engineering, Universiti Malaysia Pahang (2008)
8. Nisreen S. Mohammed, "Visual Basic 6.0", Building and Construction Department (2012) ISBN: 1576102831.

The effect of Process Parameters on Material Removal Rate in Drilling of Al 6061-SiC composite

Deepak D.[1*] and *Rajendra B.*[2]

[1]*Department of Mechanical and Manufacturing Engineering, Manipal Institute of Technology, Manipal Academy of Higher Education, Manipal – Karnataka-576104, INDIA*
[2]*Centre for Applied Sciences, Manipal Institute of Technology, Manipal Academy of Higher Education, Manipal – Karnataka-576104, INDIA*

Abstract. This study investigates the effect of process parameters such as diameter of drill, cutting velocity, feed rate and coolant supply on material removal rate while drilling of Al-SiC composite. Sample specimens are prepared by stir-casting method and experiments are designed using Taguchi's L9 orthogonal array. Response data is analysed using S/N ratio and significant process parameters are identified using Analysis of Variance (ANOVA). Study shows that drill diameter, followed by feed rate are found to be the most significant factors which influence MRR.

Keywords: Taguchi method, ANOVA, S/N ratio, Al6061-SiC

1 Introduction

Metal matrix composites (MMC) possess superior quality compared to metals. The strength-to-weight ratio and the strength exhibited by such materials make them suitable for wide range of applications like cylinder liners, structural components of landing gear of aircraft and radio frequency quadrupoles etc. Aluminium MMCs are used for various applications in aircraft, aerospace, automobile industries. Often these composite materials are generally reinforced with particles of Silicon Carbide (SiC) and Aluminium Oxide (Al2O3), etc. Tamer Ozben et al. [1] studied the mechanical properties of 5 % - 15 wt.% silicon carbide particle (SiC-p) reinforced aluminium MMC. It was observed that increase of reinforcement element produced better impact strength and hardness. Drilling is the most commonly performed machining operation during fabrication and assembly process. The quality of hole produced depends upon diameter of the drill, speed, feed rate and coolant supply. The quality of hole is measured by dimensional accuracy and surface finish. In order to obtain the best drill quality, selection of appropriate process settings is important. There are cited literatures pertaining to machining and mechanical properties of various MMC. Habibur Rahman et al. [2] found that that the hardness, tensile strength and wear resistance of Al-SiC composite increased with increase in composition of SiC. The influence of the feed, speed and point angle on tool wear, surface finish for various samples of Al-SiC composites was investigated by Shrinivas Jadhav et al. [3]. It was observed that

* Corresponding author: raj.beedu@manipal.edu

higher percentage of SiC resulted in faster tool wear. The experimental evaluation of thrust force on drilling of Al-2219/B4C/Gr composite was made by Ravindranath et al.[4]. The feed rate was found to drastically increase the thrust force for mono-alloy and the addition of graphite resulted in reducing the thrust force and surface roughness. The mechanism of failure of a composite material is complex in nature which require rich set of experiments. Hence statistical tools are generally used to reduce number of experiments in selection of optimum machining parameters. Priyank et al.[5] used grey relational analysis to optimise the drilling parameters in which feed rate was found to be the most significant factor influencing the responses. Similar results were also observed by Kumar Abhishek et al. [6] using principle component analysis. Reinforcement of B4C into Al-7XXX composite showed that composition and feed rate were significant parameters which affected the trust force and torque [7]. Genetic Algorithm method was adopted by Dhawamani et al. [8] to study the effect of drilling parameters on the tool wear and surface finish of work piece. Metin Kok et al. [9] studied the effect of drilling Al/Al2O3 composite on surface finish and tool life and optimised the process parameters such as composition, drill speed, drill diameters for obtaining higher material removal rate. The present work investigates the effect of various drilling parameters on Al-6061 composite with various percentage of SiC particles on MRR.

2 Materials and methods

2.1 Preparation of Al/SiCp MMC

The work-piece is prepared by stir-casting method. 40 grams of SiC particles having mesh size 170 mesh is thoroughly mixed with 20 grams of fly ash. This mixture is pre-heated to 650°C. The addition of fly-ash prevents the formation of Al_4C_3. Al-6061 (960 gram) material is melted in a crucible which was maintained at a temperature of 850°C. The reinforcements in the form of powder is mixed with molten metal and distributed uniformly by stir casting. The molten metal is poured into pre-heated (at 200°C) metal mould and allowed for cooling to room temperature. Figure 1 shows the specimen preparation process.

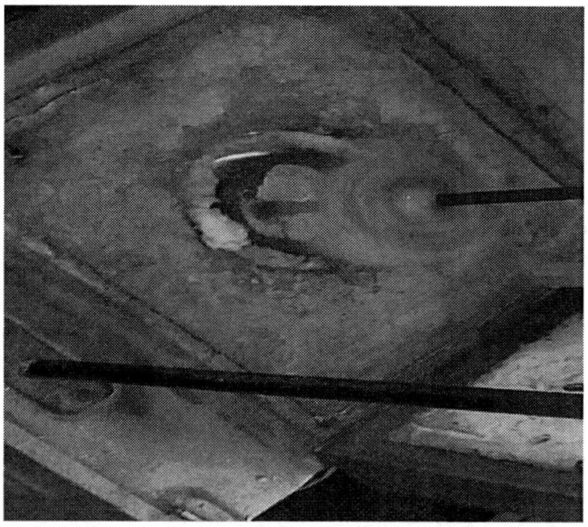

Fig. 1. Specimen preparation process

2.2 Experimental set up

Experimental work is carried out using vertical machining centre (Make: AMS-SPARK). The machine is fitted with disc type tool changing system for 6 tools with table movement of 300x250x250 mm in X, Y and Z directions respectively. Figure 2 shows the experimental set-up used in the present work. The cutting tool used is high speed steel.

Fig. 2. Experimental setup

2.3 Experimental design

The experimental study involves 4 control factors of each 3 levels. Since, full factorial experimental design requires 81 runs for the study, Taguchi's L_9 orthogonal array is chosen with 2 replications. The response parameters of the study are material removal rate which is calculated based on the weight loss method during experiments. Table 1 shows the response data obtained corresponding to various experimental conditions. Table 2 shows the parameter and their levels chosen for the study.

Table 1. Experimental design and corresponding MRR

Run	Parameters				0 % SiC		2 % SiC		4 % SiC	
	A	B	C	D	Trial-1	Trial-2	Trial-1	Trial-2	Trial-1	Trial-2
1	1	1	1	1	0.022	0.023	0.023	0.025	0.026	0.022
2	1	2	2	2	0.037	0.034	0.033	0.03	0.040	0.042
3	1	3	3	3	0.057	0.059	0.059	0.061	0.058	0.06
4	2	1	2	3	0.070	0.072	0.058	0.062	0.067	0.07
5	2	2	3	1	0.104	0.101	0.095	0.100	0.105	0.103
6	2	3	1	2	0.038	0.035	0.037	0.035	0.040	0.043
7	3	1	3	2	0.165	0.160	0.132	0.128	0.176	0.178
8	3	2	1	3	0.060	0.057	0.061	0.059	0.066	0.06
9	3	3	2	1	0.093	0.095	0.120	0.125	0.120	0.116

Table 2. Process parameters and their levels

Parameter	Code	Level 1	Level 2	Level 3
Drill Diameter mm	A	06	08	10
Cutting speed m/min	B	50	80	100
Feed rate m/min	C	0.04	0.08	0.12
Coolant flow rate ml/s	D	0	62	312

3 Results and discussion

Table 2 shows the MRR obtained for different experimental conditions which are replicated twice. The response data thus obtained is subjected to S/N ratio analysis with higher the better criteria. The figure 3, 4 and 5 shows the average S/N ratio of samples with various compositions of SiC. The effect of operating parameters is discussed in the following sections.

3.1 The effect of drill diameter

Figure 3-5 shows the effect of drill diameter on MRR obtained for work-piece with 0, 2 and 4 % SiC respectively. It is observed that, the mean S/N ratio of MRR is linearly increasing with increase in diameter of the drilled hole. This is due to the fact that, tool having larger diameter removes more material per revolution i.e., the volume of material removal is proportional to proportional to $\pi D^2 f/4$ mm^3/min, where D is the diameter of drill bit (mm) and f is the feed rate (mm/min). Hence it is seen that, increase in drill diameter increases the MRR. ANOVA values shown in table 3-5 indicate that the drill diameter is the most significant factor influencing MRR with percentage contribution 49 % to 54% for different samples.

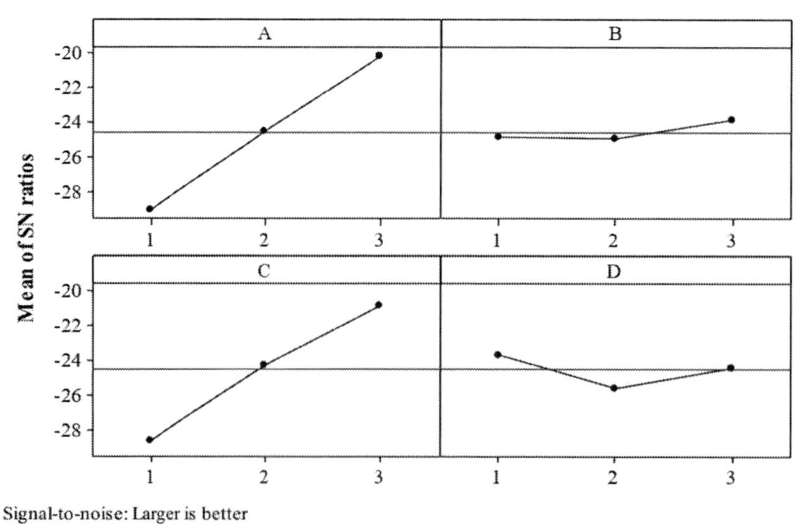

Fig. 3. The effect of process parameters on MRR for Al-6061 without SiC

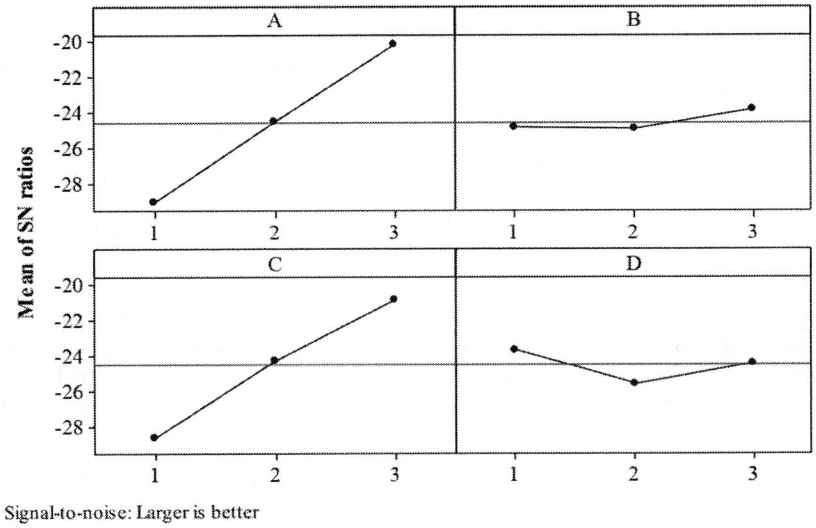

Fig. 4. The effect of process parameters on MRR for Al-6061 with 2 % SiC

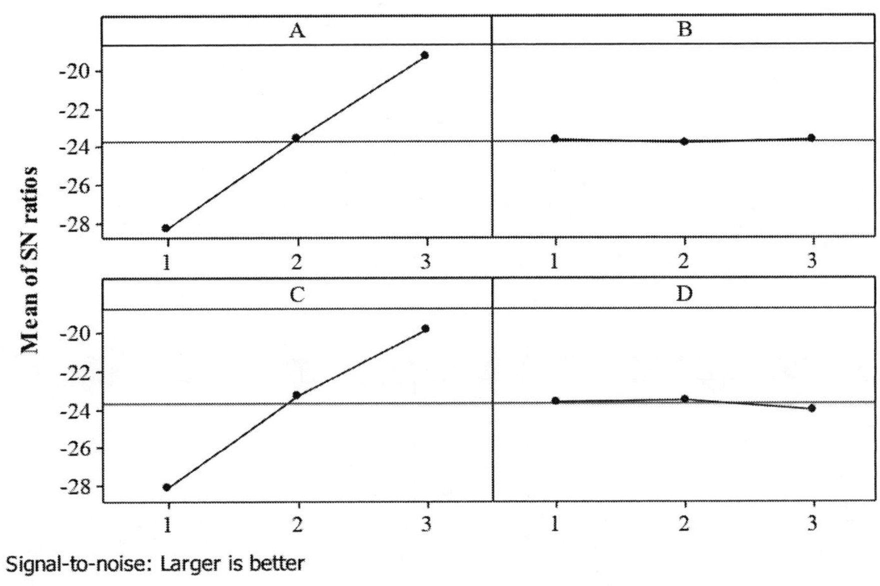

Fig. 5. The effect of process parameters on MRR for Al-6061 with 4 % SiC

3.2 The effect of cutting velocity

Figure 3, 4 & 5 shows the effect of cutting velocity on MRR obtained for work-piece with 0, 2 and 4 % SiC respectively. A slight increase in MRR is observed due to increase cutting speed. However, the percentage contribution observed by the ANOVA from table 3-5 indicates that the effect of this factor is insignificant. Hence, increasing the cutting velocity did not improve the material removal rate.

3.3 The effect of Feed rate

From the figure 3-5, it is observed that increase in feed rate increases the MRR. As mentioned earlier, volume swept by the cutting tool is proportional to feed rate. Hence increase in feed rate increases the MRR. ANOVA values shown in tables 3-5 indicate that, feed rate is also most prominent parameter which influences the MRR. The contribution of this factor varies from 42 % to 50 % for work-piece with different composition of SiC particles.

3.4 The effect of coolant flow rate

The effect of coolant flow rate on MRR is shown in figure 3-5. It is observed that the coolant flow has marginal influence on MRR. The function of coolant is to cool the drill bit and work-piece as well as to flush-out the chip formed. Use of coolant usually results in better the dimensional accuracy during machining and also improve the surface finish, but the present study shows its influence on rate of material removal is negligible. The influence of coolant is shown by its contribution (0.11% -2.5 %) as mentioned in tables 3-5.

Table 3. ANOVA for the response data - 0 % SiC

Source	DF	Seq SS	Adj MS	% Contribution
Drill diameter	2	111.49	55.74	49.25
Cutting velocity	2	0.976	0.488	0.43
Feed rate	2	113.64	56.819	50.19
Coolant supply	2	0.263	0.1313	0.11

Table 4. ANOVA for the response data - 2 % SiC

Source	DF	Seq SS	Adj MS	% Contribution
Drill diameter	2	117.175	58.5875	54.20
Cutting velocity	2	2.133	1.0666	0.98
Feed rate	2	91.250	45.6250	42.21
Coolant supply	2	5.616	2.8082	2.59

Table 5. ANOVA for the response data - 4 % SiC

Source	DF	Seq SS	Adj MS	% Contribution
Drill diameter	2	123.171	61.5855	54.02
Cutting velocity	2	0.091	0.0457	0.04
Feed rate	2	104.314	52.1572	45.75
Coolant supply	2	0.411	0.2056	0.18

3.5 Optimisation of process parameters

Based on the mean MRR is obtained at different levels of drill diameter, cutting velocity, feed rate and coolant supply and is shown in table 6, 7 & 8. The maximum change in response for change in levels of the process parameter is shown as delta values in the same tables. Using delta values, the process parameters are ranked based on their order of influence on the MRR as mentioned. Further, it is observed from these tables that the maximum MRR is observed at A_3-B_1-C_3-D_2, i.e., at diameter drill-bit : 10 mm, cutting velocity: 50 m/s, feed rate : 120 mm/minute and coolant : 62 ml/s.

Table 6. Mean response - 0 % SiC

Level	A	B	C	D
1	0.03867	0.08533	0.03917	0.07300
2	0.07000	0.06550	0.06683	0.07817
3	0.10500	0.06283	0.10767	0.06250
Delta	0.06633	0.02250	0.06850	0.01567
Rank	1	4	2	3

Table 7. Mean response - 2 % SiC

Level	A	B	C	D
1	0.03850	0.07133	0.04000	0.08133
2	0.06450	0.06300	0.07133	0.06583
3	0.10417	0.07283	0.09583	0.06000
Delta	0.06567	0.00983	0.05583	0.02133
Rank	1	4	2	3

Table 8. Mean response - 4 % SiC

Level	A	B	C	D
1	0.04133	0.08983	0.04283	0.08200
2	0.07133	0.06933	0.07583	0.08650
3	0.11933	0.07283	0.11333	0.06350
Delta	0.07800	0.02050	0.07050	0.02300
Rank	1	4	2	3

4 Conclusions

Following conclusion are drawn from present work.

- The influence addition of SiC (up-to 4%) particles into Al-6061 on MRR is negligible.
- The most significant factor influencing MRR while drilling Al-6061 SiC composite using HSS tool is the drill diameter, followed by feed rate.
- The effect of coolant on MRR is negligible.
- Optimum settings for maximum MRR is $\mathbf{A_3}$-$\mathbf{B_1}$-$\mathbf{C_3}$-$\mathbf{D_2}$, i.e., at diameter drill bit : 10 mm, cutting velocity: 50 m/s, feed rate : 120 mm/minute and coolant : 62 ml/s for all compositions of work material.

The author wishes to express their sincere thanks to Manipal University for providing laboratory support for this research work.

References

1. Tamer Ozben, Erol Kilickap and Orhan Cakir, Investigation of mechanical and machinability properties of SiC particle reinforced Al-MMC, Materials processing technology 198 (2008)
2. Md. Habibur Rahman, H. M. Mamun Al Rashed, Characterization of silicon carbide reinforced aluminum matrix composites, 10th International Conference on Mechanical Engineering, ICME 2013, Procedia Engineering pp. 103 – 109, 90 (2014)
3. Shrinivas Jadhav, Jaywant Sankpal, Mukesh Chaudhari and Nilesh Patil, A review on drilling of metal matrix composites, International Journal of Current Engineering and Technology, pp. 1-7 (2016)
4. V M Ravindranath, G S Shiva Shankar, S Basavarajappa, R Suresh, Effect of cutting parameters on thrust force and surface roughness in drilling of Al-2219/B4C/Gr metal matrix composites, IOP Conf. Series: Materials Science and Engineering149 (2016)
5. D. Priyank, K. Arun, P. Venkata Ramaiah, Optimization of process parameters in drilling of hybrid aluminium metal matrix composite", International Journal of Innovative Research in Science, Engineering and Technology , 5 (2016)
6. Kumar Abhishek, Vikas Sonkar, Saurav Datta, Siba Sankar Mahapatra, "Optimization in drilling of MMC composites- A case research", Journal of Basic and Applied Engineering Research, 1(2014)
7. Ahmet Taþkesen, Kenan Kütükde, Optimisation of the drilling parameters for the cutting process in B4C reinforced Al-7XXX series alloys based on Taguchi method, Materials and technology, pp. 169–176, 47 (2013)
8. C Dhawamani and T Alwarsami, Optimisation of machining parameters for Al/ SiC composite using genetic algorithm, International Conference on Modelling Optimisation and computing, Procedia Engineering 38 (2012)
9. Metin Kok, A study on the machinability of Al2O3 particle reinforced aluminium alloy composite, 11th Int. Inorganic Bonded Fibre Composite Conference , Mandrid (2008)

Effect of variable thermal conductivity on entropy generation in a plate with internal energy generation

T. K. Favas[1*] and *G. Jilani*[1]

[1]*Department of Mechanical Engineering, National Institute of Technology Calicut, Kerala, India*

Abstract. The current numerical investigation aims at analyzing the effect of variable thermal conductivity on local and global entropy generation rates in an energy generating plate dissipating heat by conjugate conduction-forced convection heat transfer. In order to fulfill this objective, the physical model of the plate dissipating heat into surrounding coolant is transformed into a mathematical model governing the temperature field in the plate as well as flow and thermal fields in the fluid. The resulting mathematical model, being a set of coupled and non linear partial differential equations, is solved by adopting stream function-vorticity formulation and by employing Alternating direction implicit scheme. Keeping Prandtl number of the fluid, temperature of the free stream coolant and maximum permissible plate temperature as fixed, numerical predictions are obtained for wide range of values of aspect ratio, conduction-convection parameter, energy generation parameter and flow Reynolds number. It is concluded that unrealistic constant thermal conductivity assumption leads to underestimation of entropy generation rates. It is also found that an increase in energy generation parameter results in significant increase in underestimation of global entropy generation rate.

1 Introduction

Plates with volumetric energy source find application in nuclear reactors [1], printed circuit boards [2], thermal insulation, metal casting, heat transfer gauges [3], hardening of structural materials [4], wall heating systems [5], laser irradiation [6], blade shaped electrical resistance heater [7], curing of concrete slabs [8] etc. For example, in printed circuit boards and packages, the highest temperature called as ‘hot spot’ must not exceed a specified ceiling value. If the temperature rises above this ceiling value, the error-free operation of the electronic device is threatened and hence heat dissipation to the surrounding cooling medium is highly essential [9]. Nuclear reactor is another application where the energy generated as a result of fission process in a nuclear fuel element is dissipated to surrounding stream of coolant by combined conduction-convection process so as to prevent from detrimental changes in the fuel element material by maintaining the

* Corresponding author: favastk@gmail.com

maximum temperature well below certain limit [10]. These heat transfers, occurring in electronic boards and nuclear fuel elements, through finite temperature difference will definitely lead to entropy generation [11]. As entropy generation is a measure of lost available work [11], it has to be minimized for an optimum design of the concerned thermal system. Due to the preceding facts, a number of investigators have paid their attention to studies on entropy generation in energy generating plates in the recent past. A brief review of the literature which is closely related to the present study is illustrated below.

Ruocco [12] numerically predicted entropy generation rates in a plate mounted with discrete energy sources dissipating heat into a laminar impinging jet. Shuja et al. [13] numerically predicted entropy generation rates arising out of cooling of an energy generating rectangular block having constant thermal conductivity. Ibanez et al. [14] analytically predicted entropy generation rates associated with unidirectional conduction in an energy generating plate with constant thermal conductivity by imposing average convective heat transfer co-efficient over its surfaces. Bautista et al. [15] analytically studied the problem of entropy generation associated with unsteady state unidirectional conduction in a constant thermal conductivity slab having uniform internal energy source. El Haj Assad [6] analytically predicted entropy generation rates associated with unidirectional conduction in a non-uniformly energy generating slab with constant thermal conductivity by imposing average heat transfer co-efficient over its surfaces. Aziz and Khan [16] analytically and numerically investigated the problem of entropy generation associated with steady unidirectional conduction in an energy generating plate by considering dependence of thermal conductivity on temperature. Sahin [17] analytically studied the problem of entropy generation arising out of heat transfer by conduction in a wall. Assuming uniform wall temperature boundary conditions, the consequences of variable thermal conductivity and internal energy source on global entropy generation rate were investigated. Aziz and Khan [18] analytically obtained entropy generation rates in an energy generating slab having constant thermal conductivity by assuming steady one-dimensional conduction within the slab and average heat transfer co-efficient over its surfaces. Torabi and Zhang [19] analytically investigated the problem of entropy generation associated with conduction heat transfer in a plane wall with variable thermal conductivity and internal energy source by assuming convective along with radiative conditions over the heat dissipating surfaces.

The review of the literature pertinent to entropy generation studies in plates with volumetric energy sources undoubtedly discloses that except Aziz and Khan [16], Sahin [17] and Torabi and Zhang [19], no one has paid their concentration to the study of entropy generation associated with cooling of a plate with volumetric energy source by taking into consideration of variation of thermal conductivity with temperature of the plate. Nonetheless, these studies are dealing with idealistic assumption of unidirectional conduction within the plate. Besides, idealistic situations of either uniform wall temperature or uniform convective heat transfer co-efficient are imposed on the heat dissipating surfaces. The shortfalls of the earlier studies are the inspiration for the current study. Thus, the main objective of this study is analyzing the outcome of thermal conductivity variation with temperature on local and global entropy generation rates in an energy generating plate by considering more realistic two-dimensional conjugate conduction-forced convection heat transfer analysis.

2 Mathematical formulation

Fig. 1 depicts a plate with uniform volumetric energy generation having dimensions H and $2W$ dissipating heat into an upcoming stream of coolant having free stream velocity and temperature U_∞ and T_∞ respectively. The Prandtl number of the coolant is Pr. The energy generated within the plate is first transferred to its heat dissipating surfaces by heat conduction process resulting in entropy generation and finally dissipated by forced convection to the surrounding fluid stream. Fig. 2 illustrates the computational model with respect to the physical domain explained above. Obviously, temperature distribution within the plate and flow and thermal fields within the fluid will be symmetric about the centre line of the plate. Thus, in the present study, the right half of the physical domain is taken as the computational domain as depicted in Fig. 2.

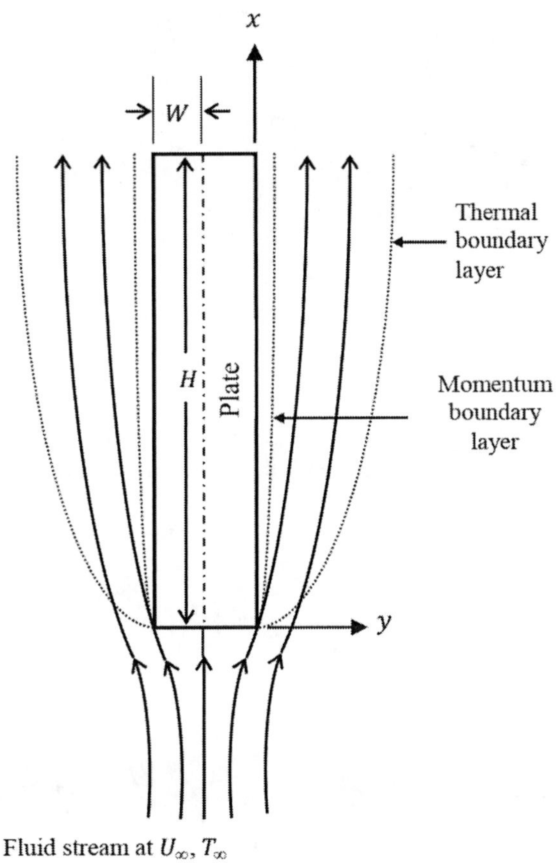

Fig. 1. Physical model.

With the intention of numerically simulating the problem depicted above, the following supplementary assumptions are introduced for obtaining the most appropriate mathematical model.

(i) The plate material is isotropic and homogenous.
(ii) The heat conduction in the plate is two-dimensional.
(iii) The flow of the coolant is incompressible, laminar, and two-dimensional.
(iv) Coolant is viscous and Newtonian.
(v) Effect of viscous dissipation is negligibly small.
(vi) The effect of thermal radiation on heat dissipation rate is negligible.

Taking into account the assumptions stated, dimensionless equations governing temperature field in the plate as well as thermal and flow fields in the fluid domain are deduced as follows:

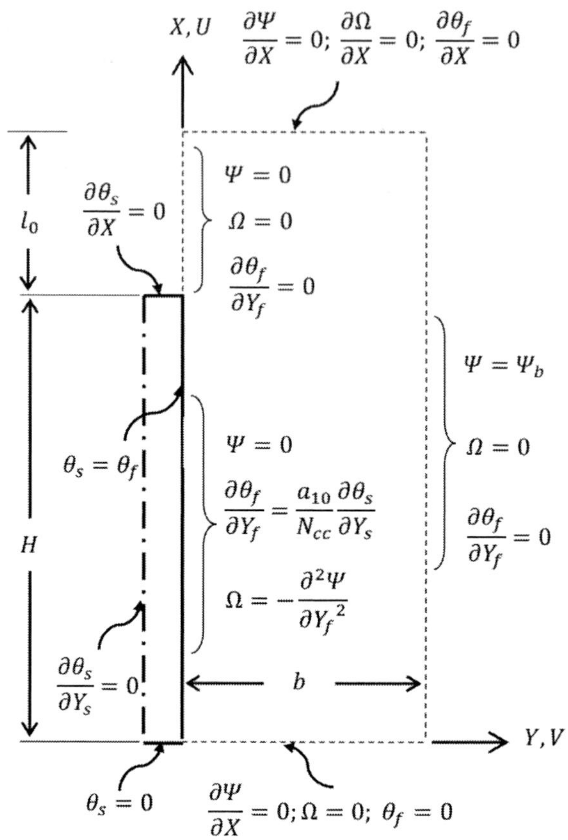

Fig. 2. Computational domain.

Solid domain:

$$\frac{\partial^2 \theta}{\partial X^2} + C\frac{\partial^2 \theta}{\partial Y_s^{\ 2}} - \frac{a_3}{(a_4 + a_3\theta)}\left(\frac{\partial \theta}{\partial X}\right)^2 - \frac{a_5}{(a_4 + a_3\theta)}\left(\frac{\partial \theta}{\partial Y_s}\right)^2 + (a_6\theta + C)Q = 0 \qquad (1)$$

Where, θ is the dimensionless temperature in the plate and Q is the energy generation parameter. The constants in Eqn. (1) are defined as

$$C = 4A_r^2 \quad a_3 = a_2\,(T_0 - T_\infty) \quad a_4 = a_1 + a_2 T_\infty \quad a_5 = Ca_3 \quad a_6 = \frac{a_5}{a_4}$$

Where, A_r is the plate aspect ratio and T_0 is maximum allowable temperature in the plate. The presence of constants a_1 and a_2 in the expressions of the constants a_3 and a_4 is due to dependence of thermal conductivity k_s on plate temperature T_s which is expressed for Uranium Dioxide, a commonly used nuclear fuel element material as [10].

$$k_s = \frac{1}{a_1 + a_2 T_s} \qquad (2)$$

Eqn. (1) is a highly non-linear partial differential equation and therefore it has to be linearized before seeking its numerical solution. The linearized form of this equation can be obtained as

$$\frac{\partial^2 \theta}{\partial X^2} + C\frac{\partial^2 \theta}{\partial Y_s^{\ 2}} - a_7\frac{\partial \theta}{\partial X} - a_8\frac{\partial \theta}{\partial Y_s} + a_9 Q = 0 \qquad (3)$$

Where,

$$a_7 = \frac{a_3}{(a_4 + a_3\theta)}\left(\frac{\partial\theta}{\partial X}\right) \qquad a_8 = \frac{a_5}{(a_4 + a_3\theta)}\left(\frac{\partial\theta}{\partial Y_s}\right) \qquad a_9 = a_6\theta + C$$

Fluid domain:

$$\frac{\partial^2\Psi}{\partial X^2} + \frac{\partial^2\Psi}{\partial {Y_f}^2} = -\Omega \tag{4}$$

$$U\frac{\partial\Omega}{\partial X} + V\frac{\partial\Omega}{\partial Y_f} = \frac{1}{Re_H}\left(\frac{\partial^2\Omega}{\partial X^2} + \frac{\partial^2\Omega}{\partial {Y_f}^2}\right) \tag{5}$$

$$U\frac{\partial\theta}{\partial X} + V\frac{\partial\theta}{\partial Y_f} = \frac{1}{Re_H Pr}\left(\frac{\partial^2\theta}{\partial X^2} + \frac{\partial^2\theta}{\partial {Y_f}^2}\right) \tag{6}$$

Where, Ψ and Ω represents dimensionless stream function and vorticity respectively.

Once plate temperature field is found, the dimensionless local and global entropy generation rates S_l and S_g can be computed using the following equations:

$$S_l = \frac{a_{10}}{(\theta + \theta_\infty)^2}\left(\left(\frac{\partial\theta}{\partial Y_s}\right)^2 + \frac{1}{C}\left(\frac{\partial\theta}{\partial X}\right)^2\right) \tag{7}$$

$$S_g = 2\int_0^{-1}\int_0^{1} S_l dX dY_s \tag{8}$$

Where,

$$a_{10} = \frac{a_4}{(a_4 + a_3\theta)}$$

The dimensionless parameters and variables used in this work are expressed as

$$X = \frac{x}{H} \qquad Y_s = \frac{y}{W} \qquad Y_f = \frac{y}{H} \qquad U = \frac{u}{U_\infty} \qquad V = \frac{v}{U_\infty} \qquad \theta = \frac{T - T_\infty}{T_0 - T_\infty}$$

$$A_r = \frac{H}{2W} \qquad N_{cc} = \frac{k_f}{k_{s\infty}}\left(\frac{W}{H}\right) \qquad Pr = \frac{\mu_f c_p}{k_f} \qquad Q = \frac{q''' W^2}{k_{s\infty}(T_0 - T_\infty)}$$

$$Re_H = \frac{\rho_f U_\infty H}{\mu_f} \qquad S_l = \frac{S'''_{gen} W^2}{k_{s\infty}} \qquad S_g = \frac{S_{gen} W}{k_{s\infty} H} \qquad \theta_\infty = \frac{T_\infty}{T_0 - T_\infty}$$

3 Numerical solution

Eqns. (3) - (6) are coupled partial differential equations of non-linear nature and therefore, these equations have to be solved numerically in an iterative manner. While Eqns. (3) and (4) are discretized using central difference schemes and obtained algebraic equations are solved numerically by employing Line-by-Line Gauss-Seidel iterative method, Eqns. (5) and (6) in their pseudo-transient form are discretized using ADI scheme and obtained algebraic equations are solved iteratively by employing Thomas Algorithm. Once the converged values of plate temperature is obtained, local and global rates of entropy generation are computed by means of Eqn. (7) and Eqn. (8), respectively.

4 Results and discussion

The current numerical investigation deals with the outcome of variable thermal conductivity k_s on entropy generation arising out of conjugate conduction-forced convection in a plate having uniform volumetric heat sources. Keeping maximum permissible plate temperature T_0, temperature of the free stream coolant T_∞ and Prandtl number of the fluid Pr as constant at 2354 K, 673 K and 0.005 respectively, results are exhibited in the form of transverse and axial local entropy generation rate S_l profiles and variation of global entropy generation rate S_g with plate aspect ratio A_r, conduction-convection parameter N_{cc}, flow Reynolds number Re_H and energy generation parameter Q for temperature dependent and temperature independent thermal conductivity cases.

Fig. 3 depicts the consequence of variable k_s on transverse S_l profiles in the plate while keeping A_r = 10, N_{cc} = 0.50, Q = 0.50 and Re_H = 2500 as fixed. It is very much clear from this figure that S_l takes its minimum value at Y_s = -1, and it keeps on increasing to its maximum value at Y_s = 0, for both constant and variable thermal conductivity cases. It is also evident that constant thermal conductivity assumption results in significant underestimation of S_l except in the region close to the centre line of the plate. Further, it is worth noticing that an underestimation of 31.41% in S_l can be noticed at Y_s = 0, i.e, along the solid-fluid interface.

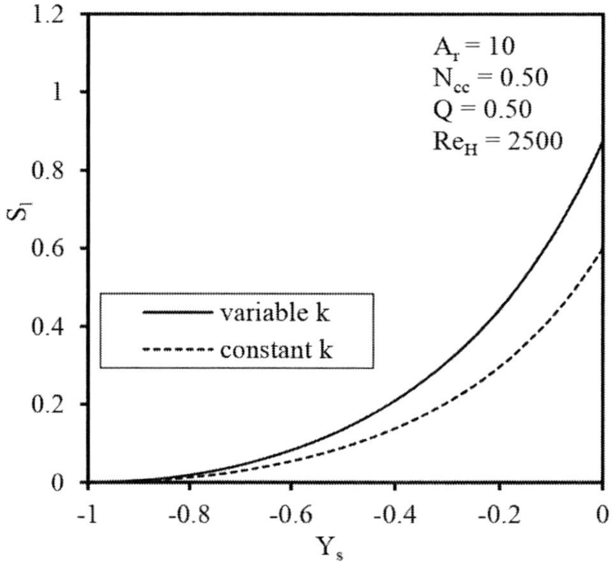

Fig. 3. Transverse S_l profiles for constant and variable k_s studies at X = 0.50

Fig. 4 presents the result of variable k_s on S_l profile along solid-fluid interface while keeping the values of A_r = 10, N_{cc} = 0.50, Q = 0.50 and Re_H = 2500 as constant. It is clearly evident that S_l takes its minimum value near the leading edge of the plate, it keeps on increasing abruptly to maximum value at X = 0.125, and thereafter gradually diminishes towards the trailing edge. Further, as discussed in Fig. 3, it is very much evident that the idealistic constant thermal conductivity assumption results in underestimation of S_l. Furthermore, it is worth noticing that error in prediction of S_l increases to maximum value near trailing edge of the plate. Precisely, error in prediction of S_l is as high as 38.17% at X = 1, which represents the trailing edge of the plate.

Fig. 5 exhibits the consequence of variable thermal conductivity on the variation of S_g in the plate with aspect ratio A_r, while N_{cc} , Q and Re_H are kept constant. It is abundantly clear that for constant and variable k_s , the value of S_g is almost invariant with A_r. It can be also noticed that constant thermal conductivity assumption results in considerable underestimation of S_g. Accurately, error in calculation of S_g increases from 28.29% to 30.49% as A_r varies from 5 to 15.

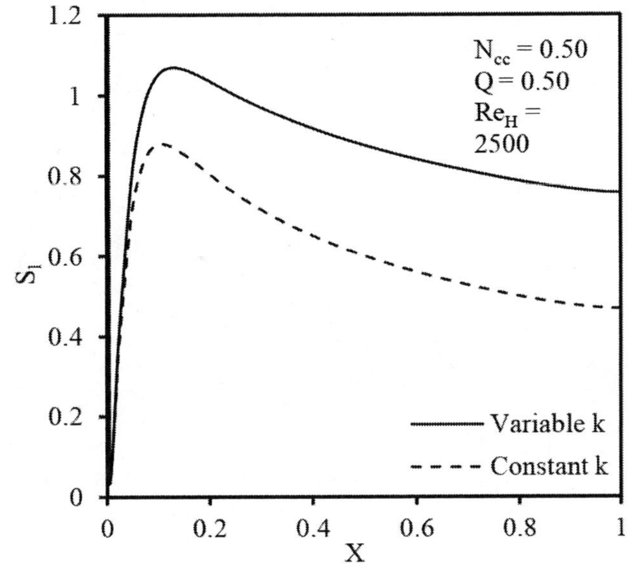

Fig. 4. Axial S_l profiles at solid-fluid interface for temperature dependent and constant k_s studies

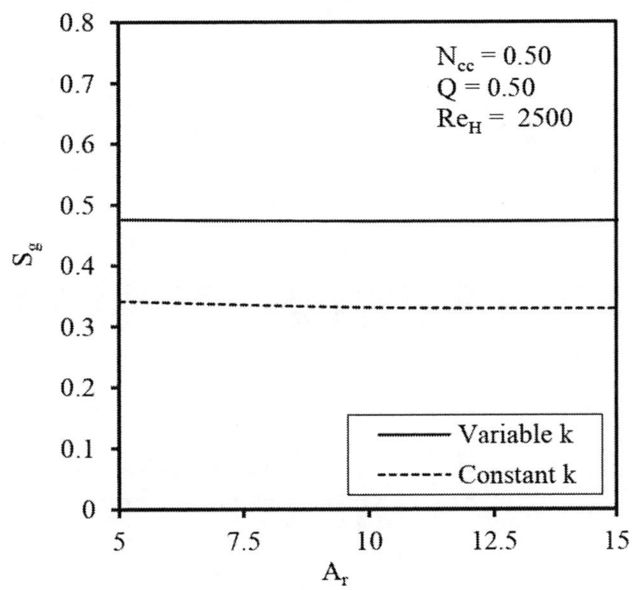

Fig. 5. Variation of S_g with A_r for both variable and constant k_s studies

Fig. 6 depicts the outcome of temperature dependent k_s on the variation of S_g with N_{cc} while other parameters are being kept fixed. It is exclusively evident that for constant and variable k_s, S_g increases considerably as N_{cc} increases. Further, it is noticeable from this figure that S_g increases almost in the same manner for both cases. Furthermore, it is important to emphasize here that the idealistic constant k_s assumption results in erronious prediction of S_g which remains almost consistent with increase in N_{cc} , but the percentage error diminishes from 32.19% to 25.97% as N_{cc} varies from 0.40 to 0.70

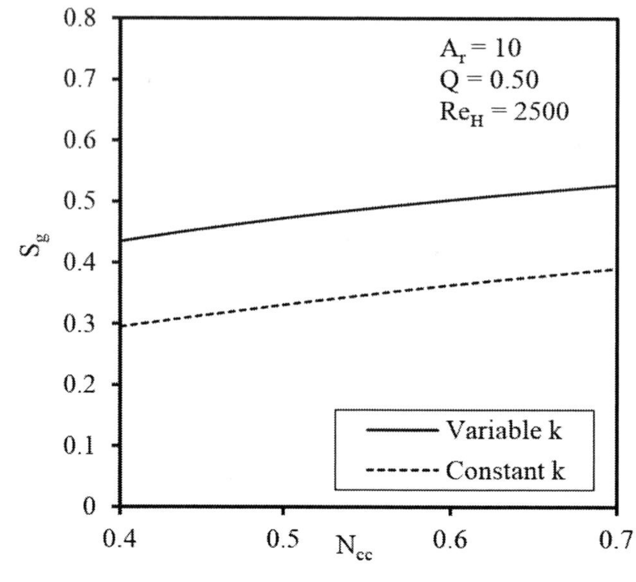

Fig. 6. Variation of S_g with N_{cc} for both constant and variable k_s studies

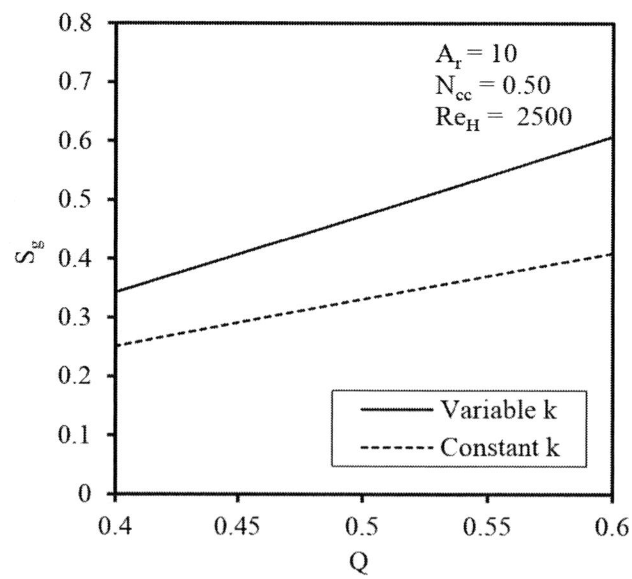

Fig.7. Variation of S_g with Q for both constant and variable k_s studies

Fig. 7 illustrates the result of temperature dependent k_s on the variation of S_g with Q for fixed values of A_r , N_{cc} and Re_H. It is clear from this figure that for both cases, S_g increases significantly with increase in Q. It is also evident that the rate of increase in S_g with Q is higher to some extent for variable k_s case. Further, it is apparent that error in calculation of S_g resulting from the idealistic constant k_s assumption amplifies extensively while Q takes its larger values. Furthermore, the under prediction in S_g varies from 26.71% to 32.74% as Q varies from 0.35 to 0.60.

Fig. 8 shows the result of temperature dependent $k_s$on the variation of S_g with Re_H for fixed values of A_r, N_{cc} and Q. It is quite visible that for constant and variable k_s, a raise in Re_H results in considerable raise in S_g. It is also clear that the rate of increase in S_g with Re_H remains more or less same for both cases. Further, it is noticeable that the disparity in prediction of S_g resulting from constant k_s assumption remains more or less same with increase in Re_H, but the percentage error decreases from 31.70% to 27.92% as Re_H increases from 1500 to 3500.

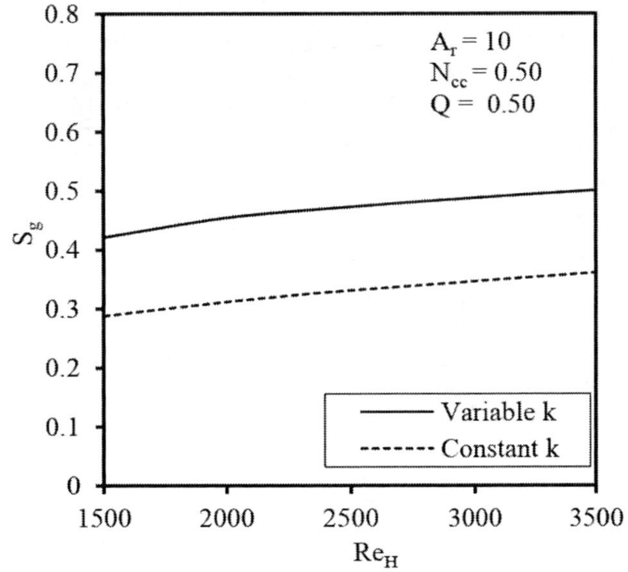

Fig. 8. Variation of S_g with Re_H for both constant and temperature dependent k_s studies

4 Conclusions

The foremost aim of the current investigation is to inspect the consequence of thermal conductivity variation with temperature on local and global rates of entropy generation in a plate with volumetric energy source being cooled by forced convection. Employing finite difference method, the partial differential equations governing the plate temperature field, fluid flow and coolant thermal fields are solved at the same time in an iterative manner. Keeping Prandtl number of the fluid, temperature of the free stream coolant and maximum permissible plate temperature as fixed, results are expressed in terms of local and global entropy generation rates for different values of plate aspect ratio, conduction-convection parameter, flow Reynolds number and energy generation parameter. It is concluded that unrealistic constant thermal conductivity assumption in the plate results in significant under prediction of entropy generation rates. Further, it is concluded that an increase in energy generation parameter leads to significant increase in the underestimation of global entropy generation rate.

References

1. G. Kim, D.C. Davis, Nucl. Eng. Des. **158**, 1-17 (1995)
2. S.M. Sawant, C. Gururaja Rao, Heat. Mass. Transf. **44**, 1485-1495 (2008)

3. A. Aziz, O.D. Makinde, J. Thermophys. Heat. Tr. **24(2)**, 438-444 (2010)
4. O.D. Makinde, A. Aziz, Heat. Mass. Transf. **47**, 1407-1415 (2011)
5. M. Torabi, K. Zhang, Energ. Convers. Manage. **89**, 12-23 (2015)
6. M. El Haj Assad, Int. J. Exergy. **9(3)**, 355-369 (2011)
7. A. Bejan, A.M. Morega, S.W. Lee, S.J. Kim, Int. J. Heat. Fluid. Fl. **14(2)**, 170-176 (1993)
8. A.J. Chapman, *Heat Transfer* (Macmillan Publishing Company, 1989)
9. S.J. Kim, S.W. Lee, *Air Cooling Technology for Electronic Equipment* (CRC Press, 1996)
10. R.H.S. Winterton, *Thermal Design of Nuclear Reactors* (Pergamon Press, 1981)
11. A. Bejan, *Entropy generation through heat and fluid flow* (John Wiley & Sons, 1982)
12. G. Ruocco, Int. Commun. Heat. Mass. Transf. **24(2)**, 201-210 (1997)
13. S.Z. Shuja, B.S. Yilbas, M.O. Budair, I.S. Hussaini, Int. J. Energy Res. **23**, 1133-1142 (1999)
14. G. Ibanez, S. Cuevas, M. Lopez de Haro, Int. J. Heat. Mass. Transf. **46**, 1321-1328 (2003)
15. O. Bautista, F. Mendez, J.L. Martinez-Meyer, Int. J. Therm. Sci. **44**, 570-576 (2005)
16. A. Aziz, W.A. Khan, Energy. **36**, 6195-6207 (2011)
17. A.Z. Sahin, Int. J. Phys. Sci. **6(12)**, 2826-2831 (2011)
18. A. Aziz, W.A. Khan, Heat. Transf. Asian. Res. **4**, 260-271 (2012)
19. M. Torabi, K. Zhang, Energy. **65**, 387-397 (2014)

Design and Simulation of a Vapour Compression Refrigeration System Using Phase Change Material

Raju Siddharth[1*], *Korody Jagannath*[2], *Kini Giridhar P.*[3] and *K. Kedlaya Vishnumurthy*[4]

[1] *Department of Electronics and Communications Engineering, Manipal Institute of Technology, Manipal Academy of Higher Education, Manipal, India*
[2] *Department of Mechanical and Manufacturing Engineering, Manipal Institute of Technology, Manipal Academy of Higher Education, Manipal, India*
[3] *Department of Electrical and Electronics Engineering, Manipal Institute of Technology, Manipal Academy of Higher Education, Manipal, India*
[4] *Department of Electronics and Communications Engineering, Manipal Institute of Technology, Manipal Academy of Higher Education, Manipal, India*

Abstract. The paper details the design and simulation of a solar powered vapour compression refrigeration system. The effect of a phase change material, in this case ice, on a vapour compression refrigeration system powered by solar panels is discussed. The battery and solar panels were sized to allow the system to function as an autonomous unit for a minimum of 12 hours. It was concluded that the presence of a phase change material in the refrigeration system caused a considerable increase in both the on and off time of the compressor. The ratio by which the on time increased was greater than the ratio by which the off time was increased. There was a 219% increase in the on time, a 139% increase in the compressor off time and a 3.5% increase in compressor work accompanied by a 5.5% reduction in COP. Thus, under conditions where there is enough load in the system to cause the initial on and off times of the compressor to be comparable, the presence of a phase change material may result in a greater on period than an off period for the compressor.

1 Introduction

On average nearly 300 days a year in India are sunny[1], with about 1,500–2,000 sunshine hours per year depending upon location. The daily average solar energy incident over India varies between 4 to 7 kWh/m^2[1]. This provides huge potential for projects that use solar energy to be implemented. Although there has been considerable progress in this field, a lot of scope remains to rework and improve existing ways of utilizing solar energy in a viable manner. At the same time, there is also room for creation of new ways to efficiently engage with solar power.

* Corresponding author: siddharth.raju94@gmail.com

Many vaccines such as polio vaccines are temperature sensitive and must be stored in storage units with controlled temperatures. In regions that experience irregular supply of electricity, the project may have potential to store vaccines at the desired temperatures. The project aims to describe the design and simulation of a solar powered vapour compression refrigeration system employing a phase change material. The constraints of this system were defined by its application, the storage of vaccines. The components of the system included solar panels, a maximum power point tracking unit, a battery bank, vapour compression refrigeration system. The system was simulated using Matlab/Simulink. Each component was sized and configured appropriately in order to optimise the performance of the system. The battery and solar panels where sized to allow the system to function as an autonomous unit for a minimum of 12 hours. The characteristic curves of SSI150W solar panels by Solar India were simulated and studied.

2 Literature Review

Mba E.F. et al. [2] (2012) developed a mathematical model of a solar vapour compression system and simulated the photovoltaic modules using MATLAB. The designed system consisted of a vapour compression refrigerator, solar panels, solar charge controller, a DC inverter and a lead acid battery. The relationship between solar radiation intensity and temperature attained in the refrigerator was studied. Different photovoltaic models were analysed and their characteristic curves were studied. No power was produced by the panel under short circuit and open circuit conditions. The open circuit voltage remained constant with an increase in solar irradiation while the short circuit current increased linearly. It was also noted that the power produced by the panels had a unique operating point at which power produced was maximum. It was concluded that with an increase in load current, there was a non-linear decrease in the panel voltage.

Kalbande and Deshmukh [3,4] (2015,2016) constructed a PV based vapour compression refrigeration system for vaccine preservation. The desired temperature to be maintained in the storage unit was 2°C to 8°C.The model was implemented using a DC vapour compression refrigerator of 25 litre capacity, two 80 W PV panels connected in parallel with 35° tilt angle, a 12 V – 150Ah sealed lead acid battery and a charge controller. It was observed that the average PV conversion efficiency was 8.5% and PV exergy efficiency was 11% in both no load and full load conditions. This indicated that load conditions of the refrigeration system do not affect the PV system.

Del Pero et al. [5] (2015) conducted a feasibility study for a solar refrigeration kit for remote areas in developing countries. A system consisting of PV panels, a refrigeration unit and a controller was designed with an additional aim that the thermally insulated envelope equipped with an energy storage system could be designed on site using local materials. An energy model was defined and simulation was carried out to identify the optimal size of the refrigerated volume, energy storage and the PV section. It was observed that there was a surplus of energy generated by the panel when compared with compressor power consumption. However, the excess power produced remained unutilized. It was also estimated that the cost of the system would be 500 Euro.

Alkelani and Kanyarusoke [6] (2016) designed and constructed an inverter-less solar assisted refrigeration system for the storage of fruits and vegetables. The system consisted of 12-volt PV panels, a battery bank, charge controller and a DC vapour compression refrigerator. The desired temperature to be maintained was between 5°C and 15°C. The PV system was sized according to the refrigeration system requirements. The solar radiation on the PV panels was estimated using the Perez model, as the panels were tilted. It was observed that the total solar irradiation on an inclined PV panel was greater than the solar irradiation on a horizontal surface. The size of the refrigerator components was computed on the heat loads of the system. The system was tested without load and then with 20kg of fruits. It was concluded that the system was able to maintain the desired temperature, with the overall COP based on the input to the panels being 2.8.

It can be concluded from the literature review, that a solar powered refrigeration system should consist of the following components: PV panels, charge controller, battery bank and a vapour compression system. If an AC compressor is used in the system an inverter will be required to convert the direct current produced by panels to alternating current that will be fed to the refrigerator. This can be avoided by the use of a DC compressor. The sizing of the battery bank and the PV array depends on the expected time the refrigerator has to run however it is independent of the load present in the refrigerator at a given time.

3 Methodology

3.1 Refrigeration System

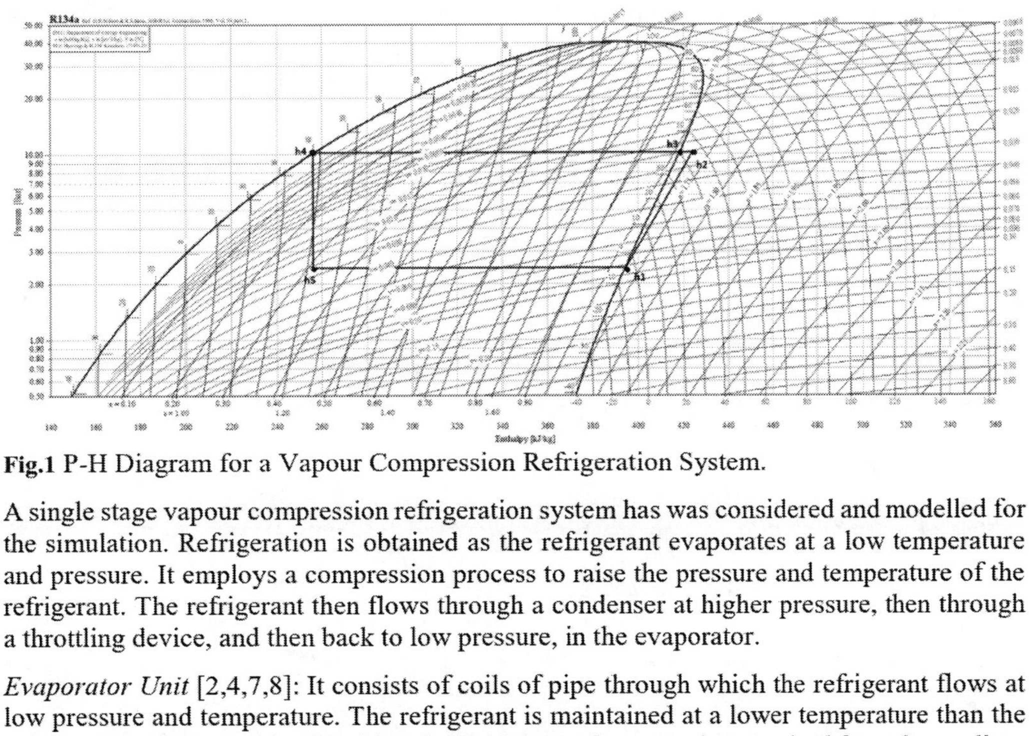

Fig.1 P-H Diagram for a Vapour Compression Refrigeration System.

A single stage vapour compression refrigeration system has was considered and modelled for the simulation. Refrigeration is obtained as the refrigerant evaporates at a low temperature and pressure. It employs a compression process to raise the pressure and temperature of the refrigerant. The refrigerant then flows through a condenser at higher pressure, then through a throttling device, and then back to low pressure, in the evaporator.

Evaporator Unit [2,4,7,8]: It consists of coils of pipe through which the refrigerant flows at low pressure and temperature. The refrigerant is maintained at a lower temperature than the surrounding medium, thus absorbing the latent heat of evaporation required from the medium to be cooled. Heat transfer rate at evaporator $\dot{Q}_e$is given by:

$$\dot{Q}_e = \dot{m}\mathrm{R} \tag{1}$$

$$\dot{Q}_e = \dot{m}(\mathrm{h}_1 - \mathrm{h}_5) \tag{2}$$

Where:
$\dot{m}$: Mass flow rate of refrigerant (kg/s)
R: The refrigeration effect, which is equal to the heat transferred at the evaporator per kilogram of refrigerant.
h_1: Specific enthalpy at the exit of the evaporator (kJ/kg)
h_7: Specific enthalpy at the inlet of the evaporator (kJ/kg)

Compressor Unit [2,4,7,8]: It maintains the desired evaporator pressure corresponding to the requirement of low temperature. It continuously draws the refrigerant vapour from the evaporator, allowing low pressure and temperature to be maintained in the evaporator. The compressor also raises the pressure and temperature of the refrigerant so that it can reject heat to the external environment in the condenser. Power input to the compressor $\dot{W}_c$, is given by:

$$\dot{W}_c = \dot{m}W \tag{3}$$

$$\dot{W}_c = \dot{m}(h_2 - h_1) \tag{4}$$

where:
W: The work done by the compressor, which is equal to the work input to the compressor per kilogram of refrigerant.
h_2: Specific enthalpy at the outlet of the compressor (kJ/kg)

Condenser unit [2,4,7,8]: In this unit, the heat absorbed in the evaporator and the heat added in the compressor to the refrigerant is rejected to the external environment, resulting in the condensation of the refrigerant. Heat transfer rate at condenser $\dot{Q}_c$, is given by

$$\dot{Q}_c = \dot{m}(h_3 - h_4) \tag{5}$$

where
h_3: Specific enthalpy at the inlet of the condenser (kJ/kg)
h_4: Specific enthalpy at the outlet of the condenser (kJ/kg)

Expansion Device [2,4,7,8]: It restricts the flow of the refrigerant leading to a pressure drop, resulting in a throttling process. Thus reducing the pressure of the refrigerant. For the isenthalpic expansion process:

$$h_4 = h_5 \tag{6}$$

Coefficient of Performance [6-8]: The COP of the system is given by:

$$\text{COP} = \frac{\dot{Q}_e}{\dot{W}_c} = \frac{\dot{m}(h_1 - h_5)}{\dot{m}(h_2 - h_1)} = \frac{(h_1 - h_5)}{(h_2 - h_1)} \tag{7}$$

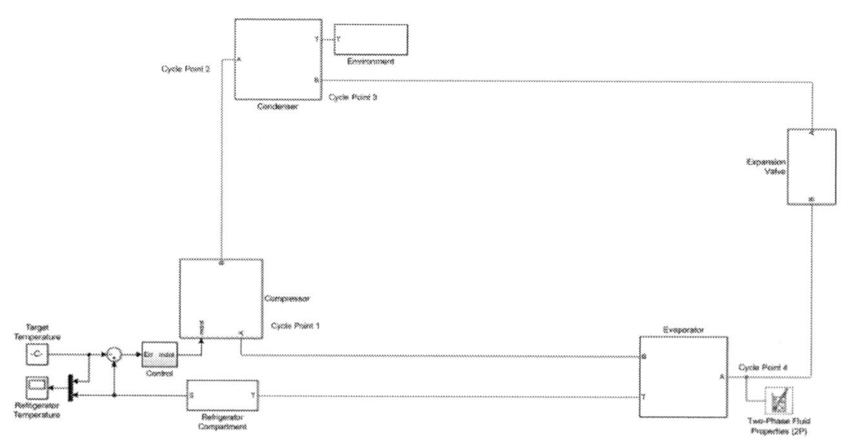

Fig. 2. Modelling the Refrigeration System.

3.2 Heat Transfer in Cooling Chamber

The sensible heat Q present in the cooling chamber can be estimated by [5-8]:

$$Q=mc_p\Delta T \tag{8}$$

Where
m = Mass of vaccines (kg)
c_p = Specific heat of medium (kJ/kgK)
ΔT = Change in temperature of cooling chamber (K)

Assuming the walls of the chamber are composite walls, the heat transfer can be estimated by [6-8]:

$$\frac{Q}{A}=\frac{\Delta T}{\left(\frac{L_1}{K_1}+\frac{L_2}{K_2}+\frac{L_3}{K_3}\right)} \tag{9}$$

Where:
Q: Heat transfer (W)
A: Surface area of refrigerator (m^2)
ΔT = Difference in temperature between external environment and cooling chamber (K)
L: Length of walls (m)
K: Thermal conductivity (W/mk)

The compressor ON and OFF time can be determined using the following equations [9]:

$$Q_i + Q_1 = Q_f \tag{10}$$

During compressor OFF time:

$$Q_1 = t_{off} * W_1 \tag{11}$$

Where
Q_i: Energy content of the chamber at the end of the ON duration (kJ)
Q_f: Energy content of the chamber at the end of the OFF duration (kJ)
t_{off}: Compressor OFF time duration/cycle (s)
W_1: Average heat transfer rate from external environment to cooling chamber (kJ/s)

$$mc_p\Delta T= Q_f - Q_i \tag{12}$$

$$mc_p\Delta T=t_{off} * W_1 \tag{13}$$

During compressor ON time assuming W_1 remains constant for both ON and OFF time

$$Q_e = (Q_f - Q_i) + Q_2 \tag{14}$$

$$Q_e = t_{on} * W_e \tag{15}$$

$$Q_2= t_{on} * W_1 \tag{16}$$

Where
W_e: Evaporator capacity (kJ)
t_{on}: Compressor ON time duration/cycle (s)

$$(Q_f - Q_i) = Q_e - Q_2 \tag{17}$$

$$mc_p \Delta T = t_{on}(W_e - W_1) \tag{18}$$

3.3 Phase Change Material

Most temperature sensitive vaccines are required to be stored at temperatures between 2°C to 8°C. Under these constraints the desired PCM should ideally have a melting point, within the desired temperature range. Thus, paraffin wax consisting of mixtures of n-Tetradecane was considered as they have a melting point near 6°C. However, based on cost and ease of availability, ice was chosen as the desired phase change material. It was desired that the phase change material provides the same cooling effect as the evaporator coil for an estimated amount of time. The mass of ice required by the system can be estimated by the following equation [7]:

$$Q_{ice} = Q_e = m_{ice}(C_s \int_{T_1}^{0} dT + L + C_f \int_{0}^{T_2} dT) \tag{19}$$

Where
Q_e: Heat removed by the evaporator (kJ)
m_{ice}: Mass of ice (Kg)
C_s: Specific heat of ice (kJ/kgK)
C_f: Specific heat of water (kJ/kgK)
L: latent heat of fusion (kJ/kg)
T_1: Initial temperature of ice (K)
T_2: Final temperature of ice (K)

3.4 Sizing of Battery and PV module

Having calculated the power being consumed by the compressor, the capacity C of the battery in ampere hours can be estimated using the following equation [8,10,11]:

$$C = \frac{Daily\ power\ consumed * Days\ of\ autonomity * Battery\ operation\ period}{Nominal\ Voltage * Depth\ of\ Discharge * Corrective\ Factor} \tag{20}$$

Similarly, PV module size can also be estimated by:

$$P = \frac{Daily\ power\ consumed * Battery\ operation\ period}{Sunlight\ hours * Depth\ of\ Discharge * Corrective\ Factor} \tag{21}$$

3.5 Photovoltaic Cell

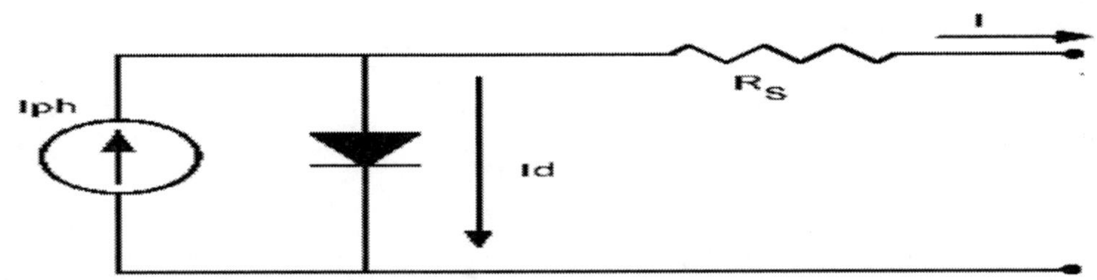

Fig.3. Equivalent PV cell circuit

Table. 1. Solar Panel SSI150W

Power	150 W
Nominal Voltage	12 V
Open Circuit Voltage	21.6 V
Short Circuit Current	9.21 A
Maximum Power Point Voltage	17.3 V
Maximum Power Point Current	8.67 A
Number of Cells	36

The panels have been modelled on the basis of a single diode, series resistance model of a solar panel. The panel selected for the purpose of the project were Solar India 150 W 12 V Polycrystalline Solar Panel SSI150W. The model was simulated on Matlab/Simulink, under conditions of constant solar radiation intensity and varying module temperature. Then under conditions of constant temperature and varying solar radiation intensity. The configuration of the PV cell can be mathematically by [12-18]:

$$\mathrm{I} = I_{ph} - I_d \tag{22}$$

$$I_d = I_0\left[e^{\frac{V+I.R_S}{a}} - 1\right] \tag{23}$$

$$I_{ph} = \frac{G}{G_{ref}}\left(I_{ph,ref} + \mu_{sc} * \Delta T\right) \tag{24}$$

$$I_{0,ref} = I_{sc,ref}\left[e^{\frac{-V_{oc}}{a_{ref}}}\right] \tag{25}$$

The final output current for N_p parallel cells present in the array, is determined by:

$$I = N_p{*}I_{ph} - N_p{*}I_d\{e^{[\frac{q}{K*N_S*A*T_{op}}*(V+I*R_S)]} - 1\} \tag{26}$$

Where:
V= k*T_c/q

a= $(N_s * A * k * T_c)/q$
I_{ph}: the photocurrent (A)
I_d: the diode current (A)
I_0: The reverse saturation or leakage current of the diode (A)
T_c: The operating temperature of the cell (K)
k: Boltzmann constant= $1.381 \cdot 10^{-23}\, J/K$
q: Electron charge = $1.602 \cdot 10^{-19}$ C
N_s: Number of PV cells connected in series.
A: the ideality factor
G: Irradiance (W/m^2)
G_{ref}: Irradiance at STC (W/m^2)
μ_{sc}: Coefficient temperature of short circuit current (A/K)
ΔT: $T_c - T_{c,ref}$ (K)
$T_{c,ref}$: Cell temperature at STC (K)
D: Diode diffusion factor
ε_G: Material band gap energy (eV)

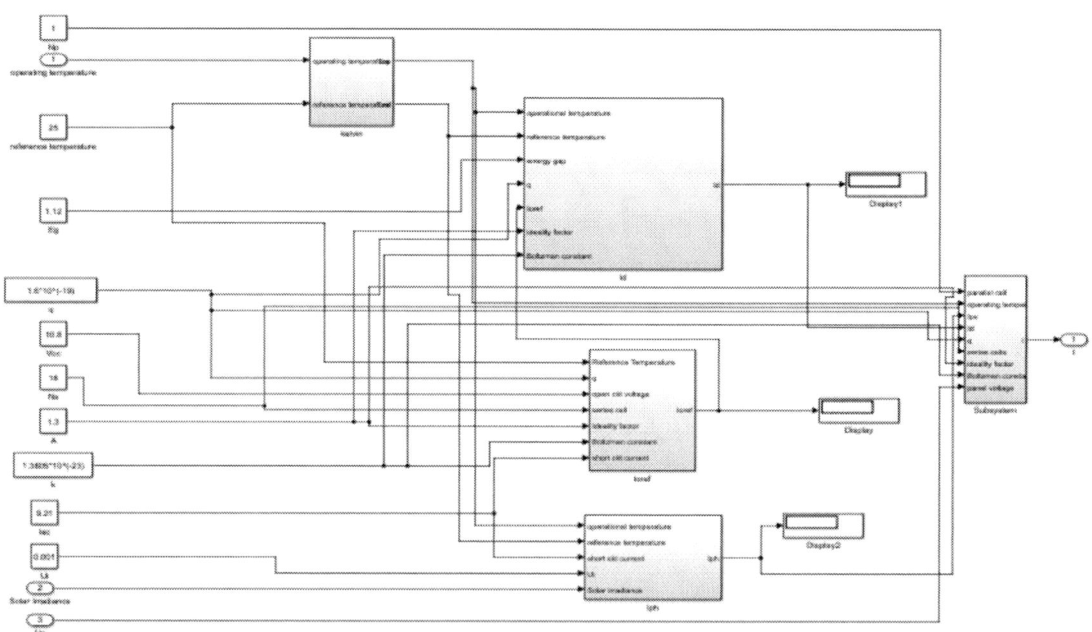

Fig. 4. Modelling of PV Subsystem.

3.6 MPPT using Perturb and Observe Algorithm

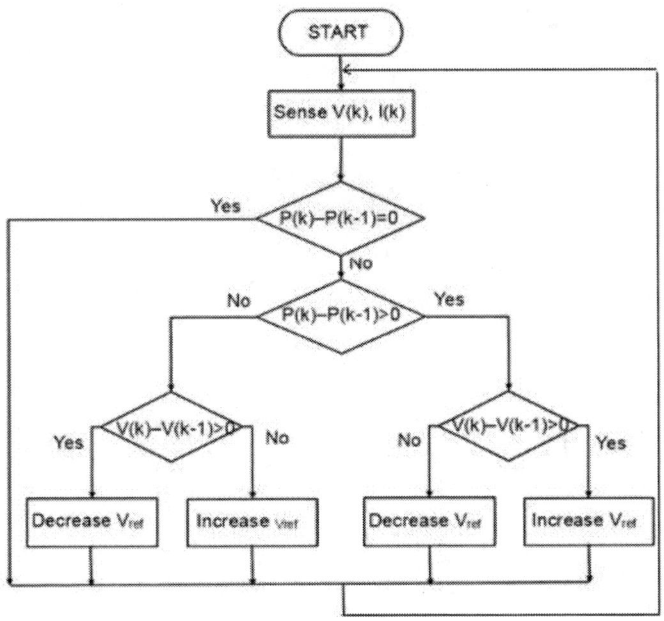

Fig. 5. P&O Algorithm.

A Maximum Power Point tracking unit [19-22] allows optimum power to be transferred from the PV module to the load under given atmospheric conditions. The converter acts as a load to the PV module. Altering the duty cycle results in the variation of impedance of the converter circuit. By the maximum power transfer theorem when the impedance of the converter will be similar to the internal impedance of the cell under given conditions, maximum power will be transferred. Based on the output of the PV module and the nominal voltage of the battery the duty cycle of the converter is varied using MPPT algorithms. The Perturb and Observe (P&O) algorithm [19-22] perturbs the operating voltage to ensure maximum power. The P&O algorithm reads the voltage and current produced by the solar photovoltaic array. It then calculates the power produced by the array. As can be seen from the characteristic PV curve of the solar module, the slope of the curve is positive to the left of the maximum power point and negative to the right. The power calculated at the instant 'k' is compared with the power calculated at the instant 'k-1' similarly the voltages at both these instances are also compared. If $dP/dV>0$, then it is known that the operating point is on the left side of the MPP. The P&O algorithm would increase the solar PV panel reference voltage to move the operating point towards the MPP. Alternatively, if the output voltage of the solar PV panels is perturbed and $dP/dV<0$, then it is known that the operating point is on the right side of the MPP. The P&O algorithm would decrease the solar PV panel reference voltage to move the operating point towards the maximum power point. When the steady state condition is touched the P&O technique oscillates around the maximum power point.

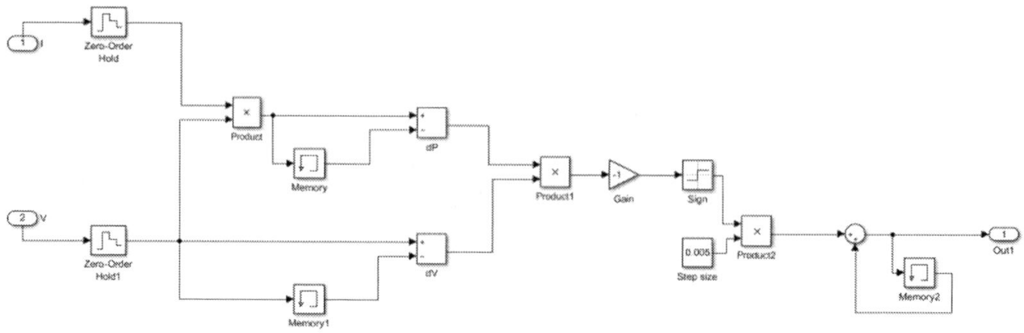

Fig. 6. Modelling the P&O Algorithm.

3.7 Buck-Boost Converter

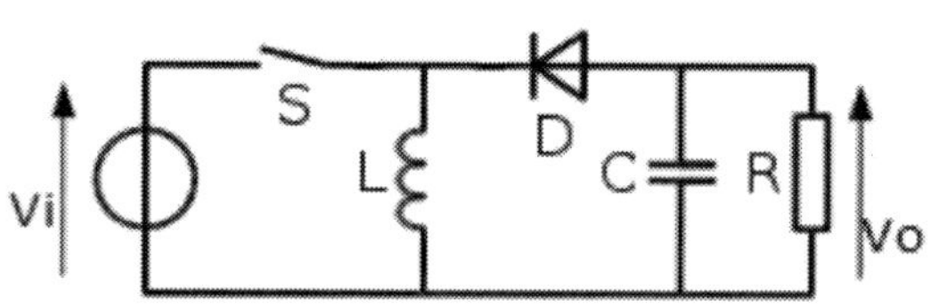

Fig. 7. Configuration of the Buck-Boost Converter

A buck boost converter [23] is a dc-dc converter that allows the output voltage to be either greater or lesser than the input voltage. When switch S is turned on, the diode D becomes reversed biased. Thus, the current flows through inductor L, charging it. When switch S is off the diode D becomes forward biased, resulting in the inductor discharging through capacitor C to the load. In this state, energy stored in the inductor is transferred to the load and the inductor current continues to fall till the switch S is turned on. The input and output voltage of the converter are related to the duty cycle of the switch by the following equation:

$$V_o = -\frac{DV_{in}}{(1-D)} \tag{27}$$

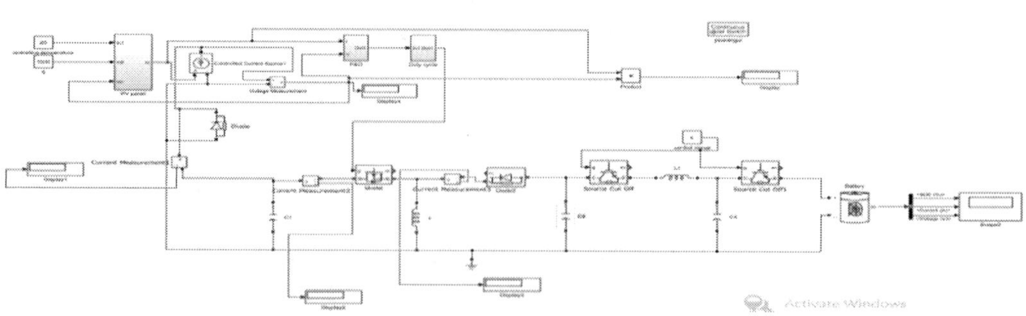

Fig. 8. MPPT using Buck-Boost Converter with Battery

 https://doi.org/10.1051/matecconf/201814404002

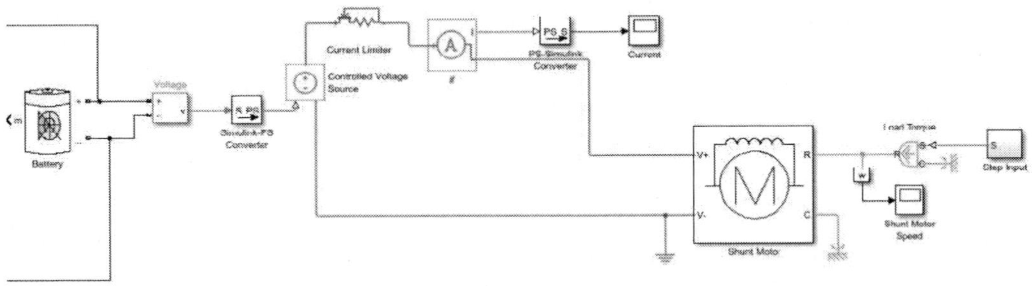

Fig. 9. Modelling of DC Shunt Motor Circuit.

3.8 Modelling of Systems

The subsystems have been designed and simulated in Matlab. The mechanical system is based on the example "Two-Phase Fluid Refrigeration" [24] available in the software. It has been designed using the thermodynamics library in Simscape. The PV module, MPPT unit and battery were simulated in Simulink. In addition, a shunt motor [25] was modelled in Simscape to verify that the solar panels could generate enough power to run the refrigeration system.

4 Results

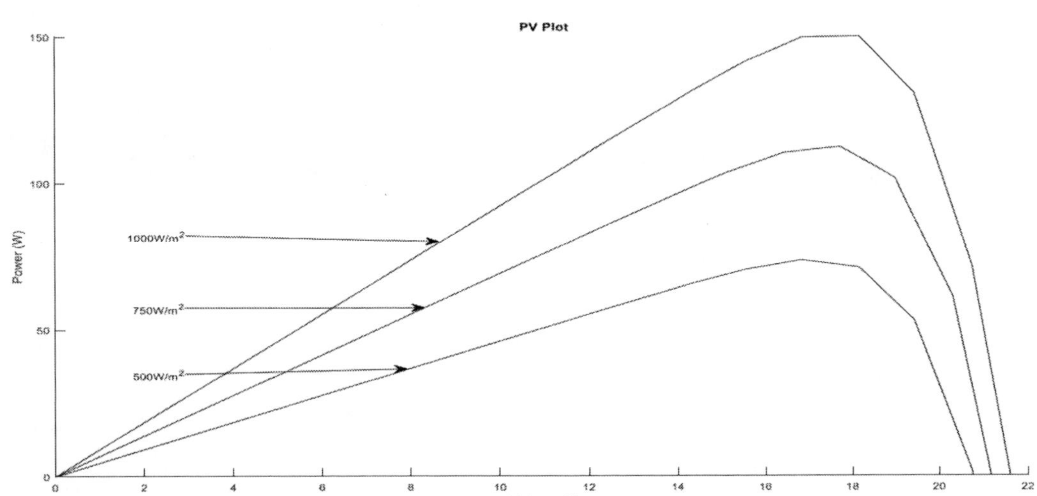

Fig. 10. Variation of Solar Irradiance at constant Temperature on P-V Characteristics of Solar Panel.

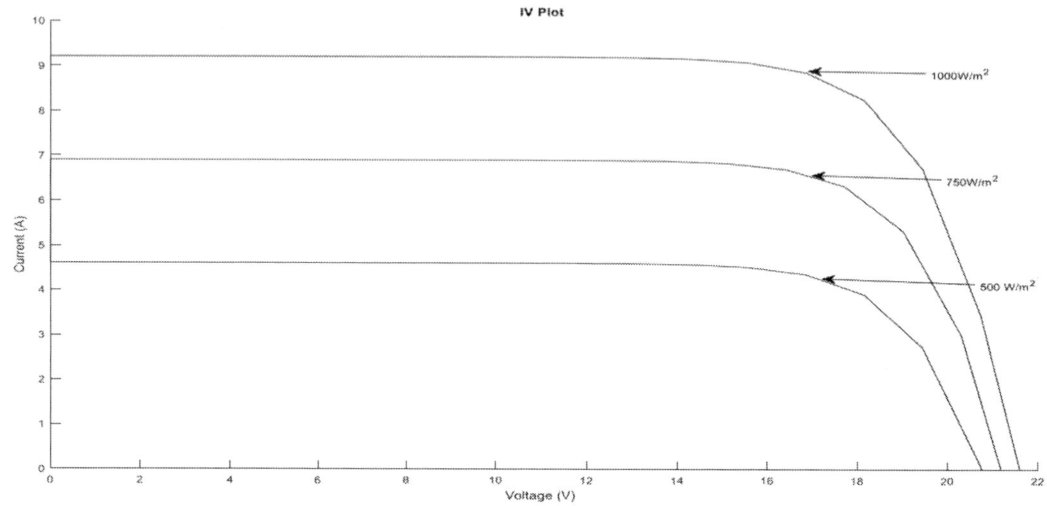

Fig. 11. Variation of Solar Irradiance at constant Temperature on I-V Characteristics of Solar Panel.

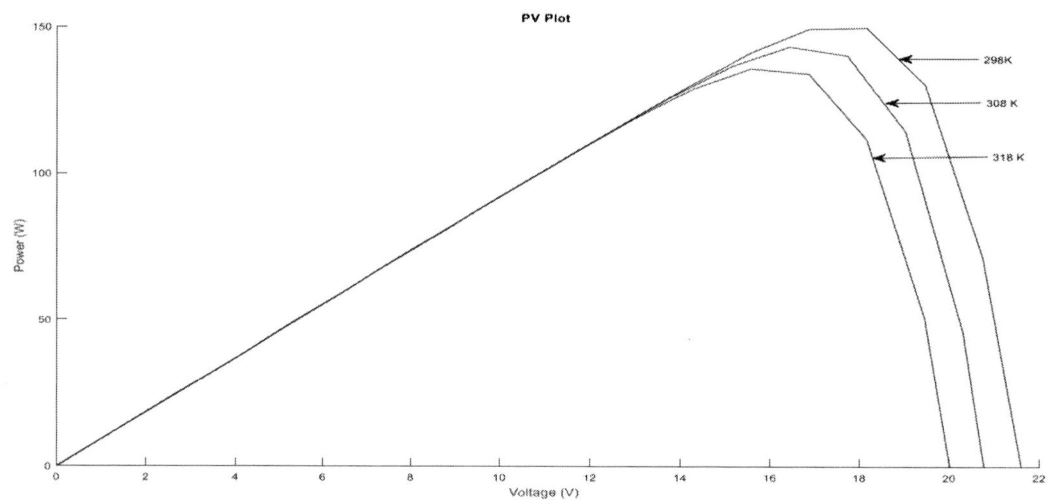

Fig. 12. Variation of Temperature at constant Solar Irradiation on P-V Characteristics of Solar Panel.

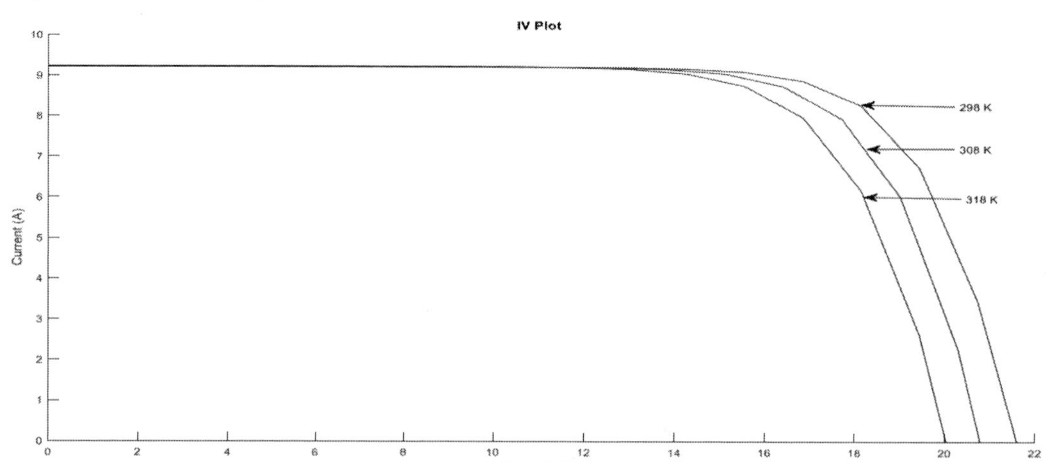

Fig. 13. Variation of Temperature at constant Solar Irradiation on P-V Characteristics of Solar Panel.

4.1 Simulation of Solar Panels

It was observed from the simulation of the solar panels that:

- Under conditions of varying irradiance at a constant temperature, there was a significant rise in module current with the rise in irradiance while the voltage increased at a much slower rate. There was an increase in module power with increase in solar irradiation.
- Under conditions of varying temperature at constant irradiance, with decreasing temperature, module current decreased slightly but PV voltage decreased significantly. There was an evident decrease in the output power of the PV cell with an increase in temperature.

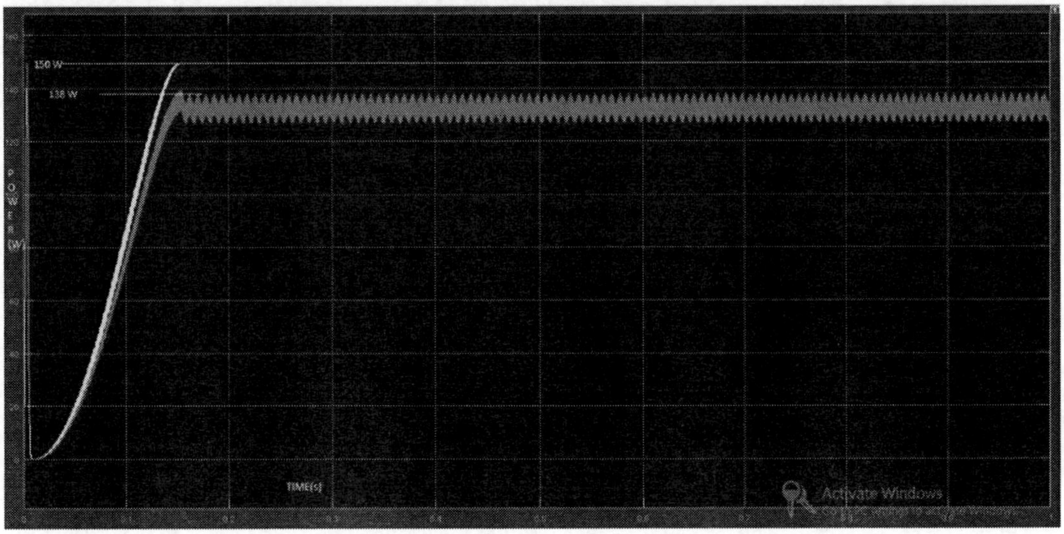

Fig. 14. Output Power of PV Module Compared with Output at DC-DC Converter.

4.2 Simulation of MPPT

It was observed during the simulation of the MPPT that:

- The P&O algorithm was followed and the unit tracked the output of the module. The output of the buck boost converter increased along with the panel power and then oscillated once steady state was achieved.
- While the module generated 150 W of power, there was loss of power at the output of the buck boost converter owing to switching losses, resistance of the components and forward voltage of diode.

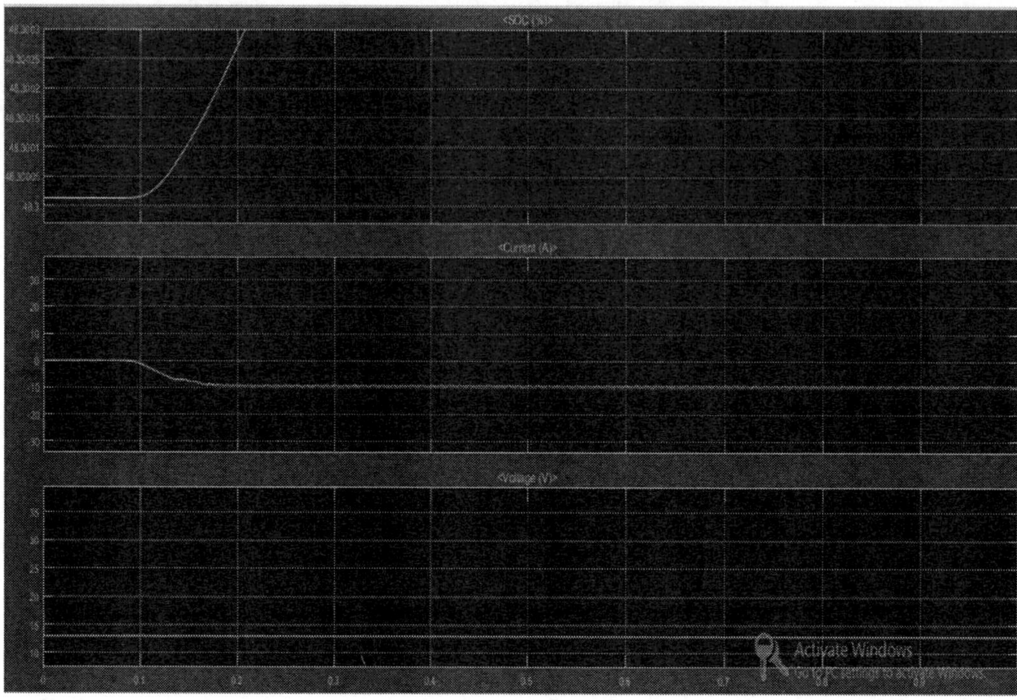

Fig. 15. Battery Parameters

4.3 Simulation of Battery Parameters

It was observed that with the presence of an MPPT unit most of the power generated by the panels was channelled to charge the battery. However, at the interface of the battery and the converter heavy interference was noted. Thus, an additional EMI filter was applied in order to reduce the noise. From the above simulation it was observed that approximately 129.5 W was transferred to the battery. This was in the form of a charging voltage of 12.95V with 10A of charging current.

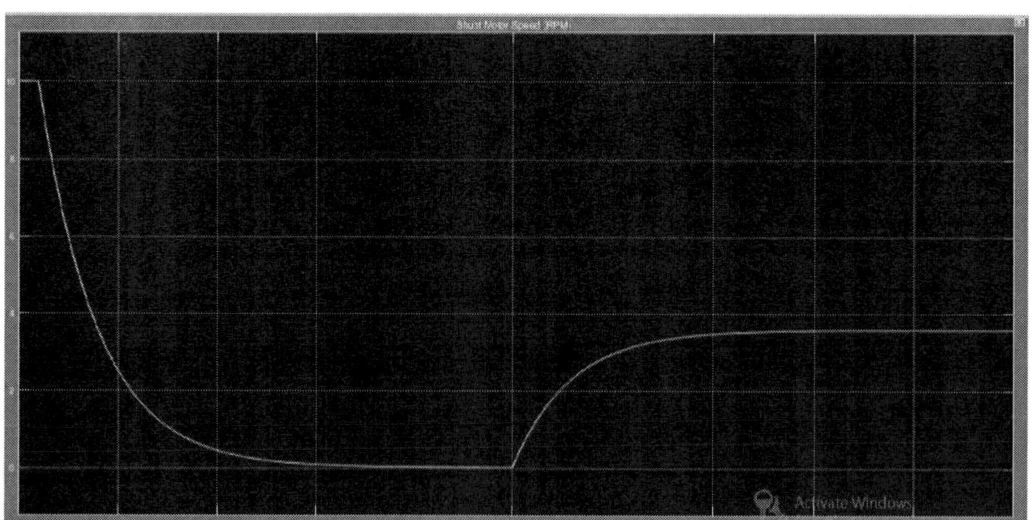

Fig. 16. DC Motor Current.

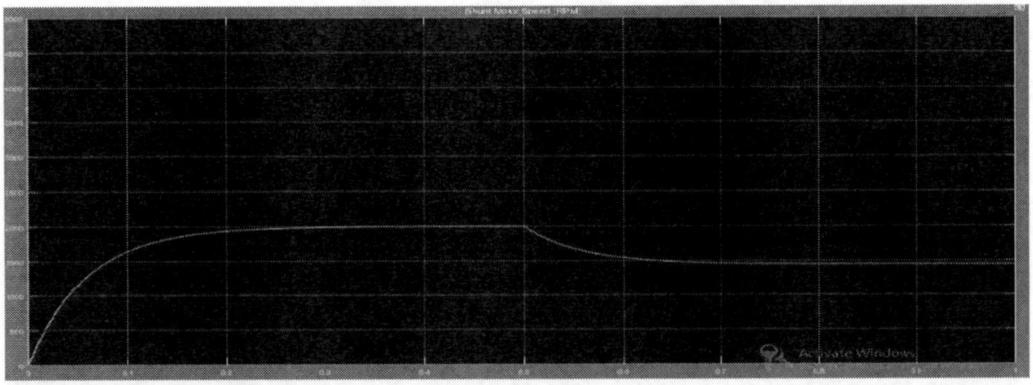

Fig. 17. DC Motor Speed.

4.4 Simulation of DC Motor

It was observed during the simulation of the DC motor that:

- With the aid of the current limiter the surge current was limited to 10 A. After the initial surge current was drawn, under zero load torque conditions the current drawn tended to zero. It is also noted that motor speed is at its highest under these conditions at 2000 rpm.
- As the load torque was increased there is an increase in the current drawn to 3.62 A, resulting in a speed of 1450 rpm and a power consumption of 43 W which was in line with the calculations for the maximum power drawn by the compressor.

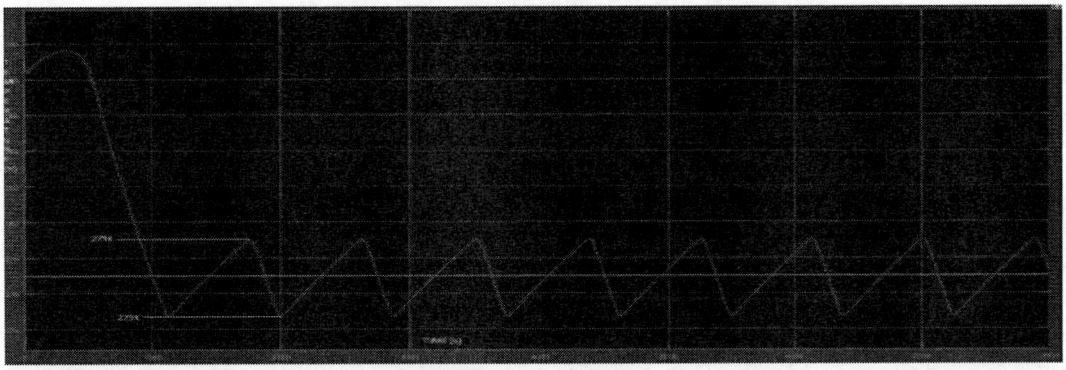

Fig. 18. Refrigerator Compartment Temperature without PCM.

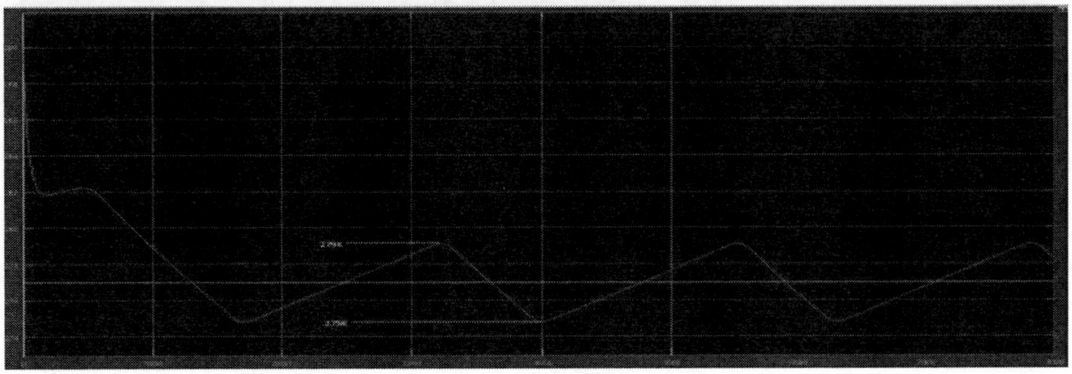

Fig. 19. Refrigerator compartment temperature with PCM.

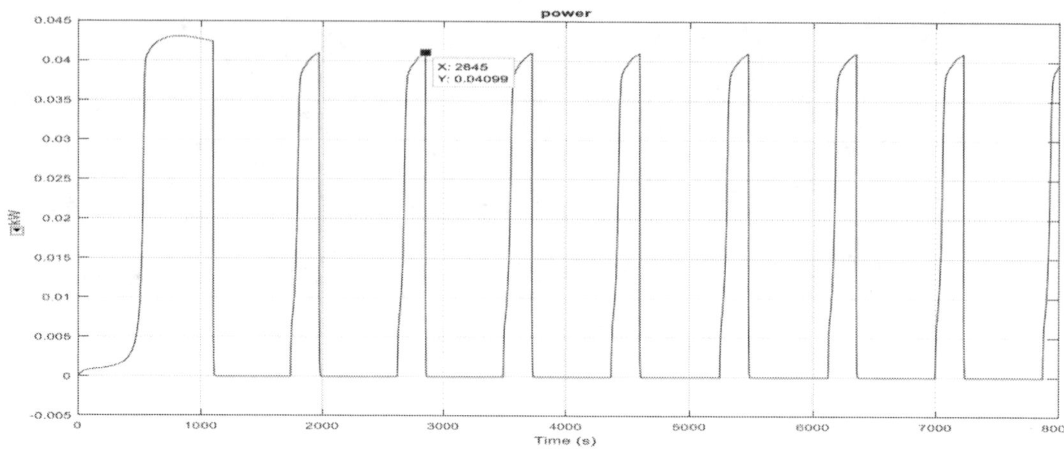

Fig. 20. Compressor Power Consumption Without PCM

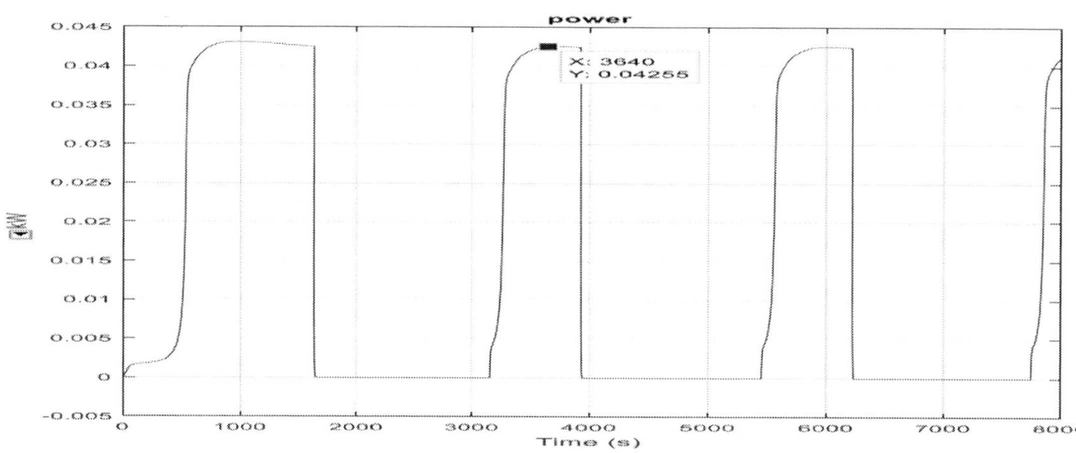

Fig. 21. Compressor Power Consumption With PCM.

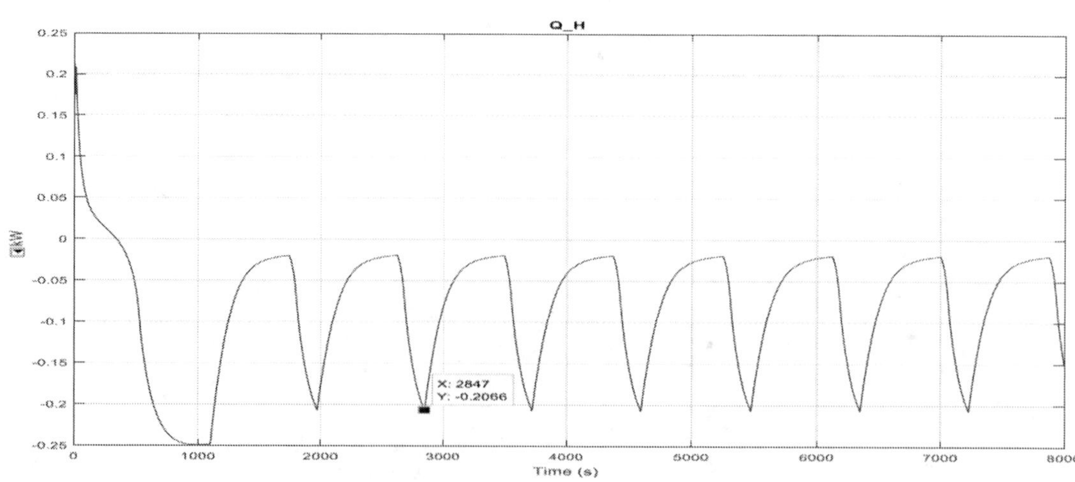

Fig. 22. Condenser Coil Capacity Without PCM

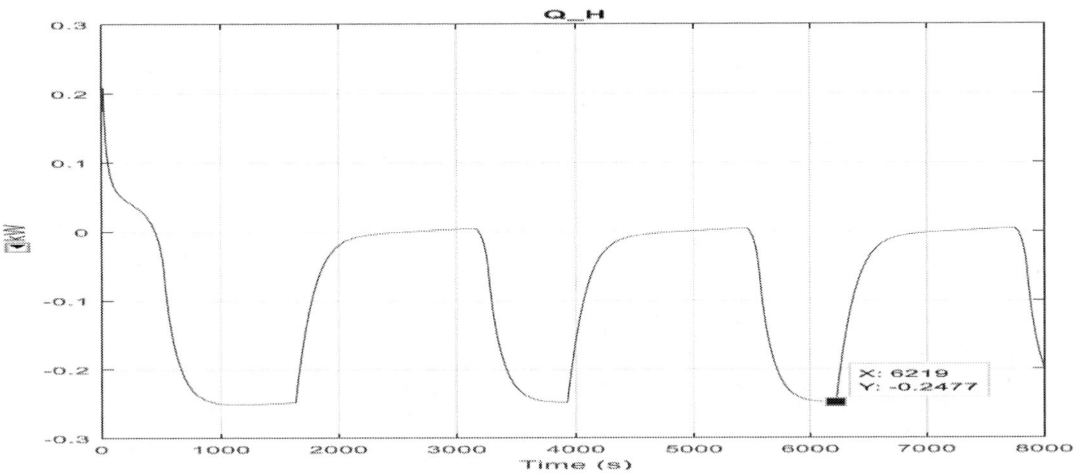

Fig. 23. Condenser Coil Capacity With PCM.

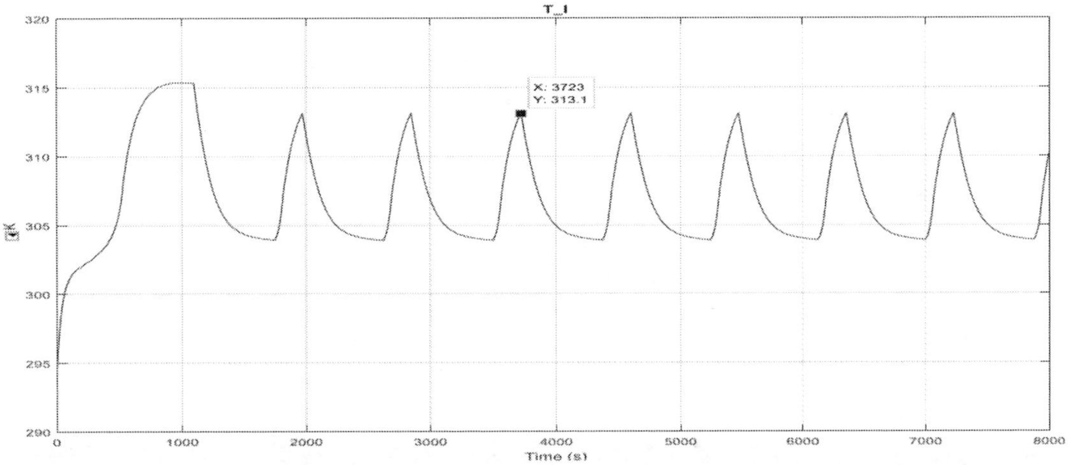

Fig. 24. Condenser Coil Temperature With PCM.

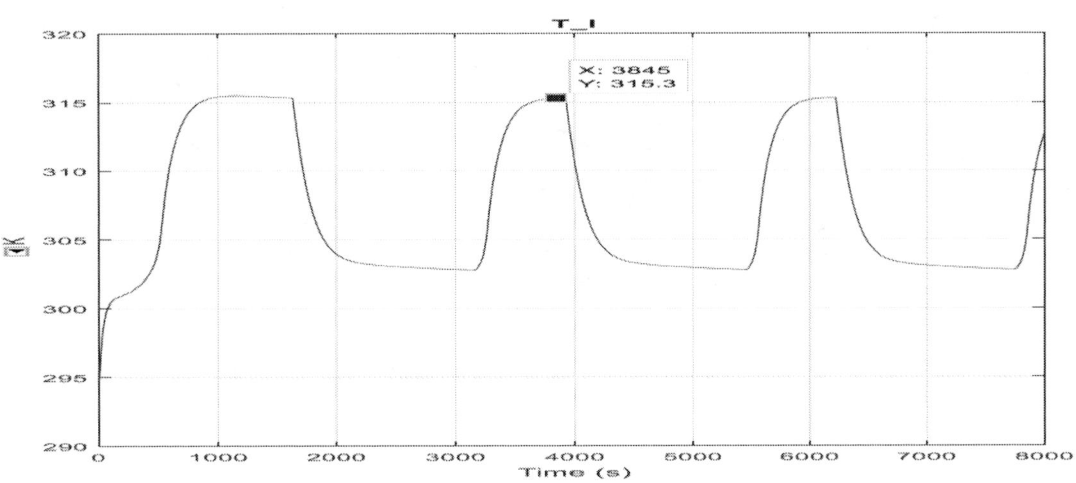

Fig. 25. Condenser Coil Temperature With PCM.

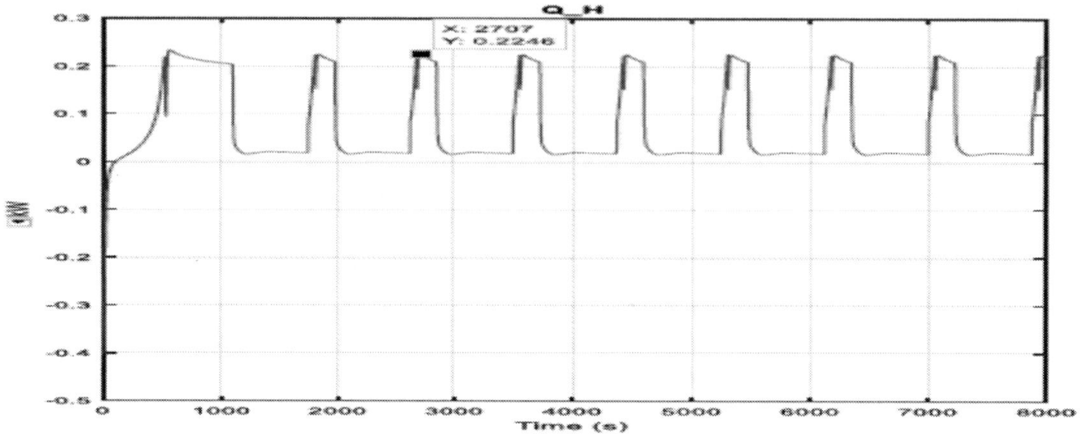

Fig. 26. Evaporator Coil Capacity Without PCM.

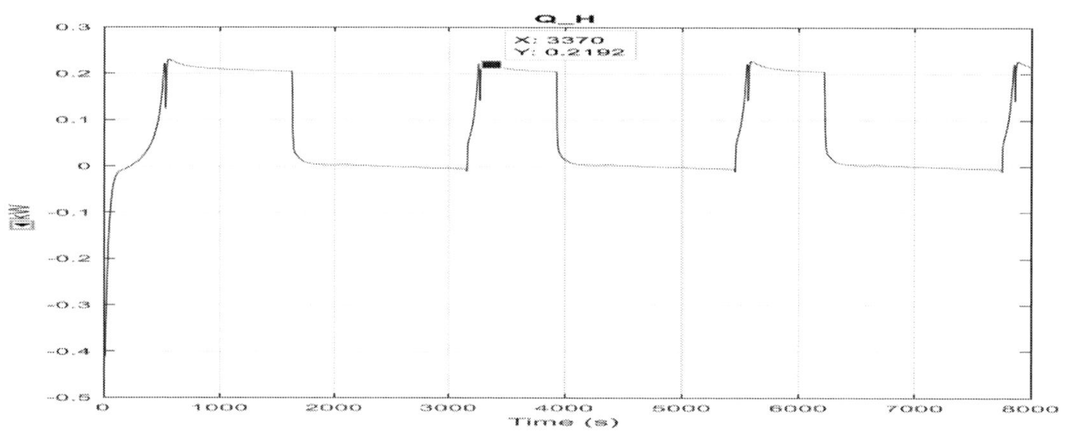

Fig. 27. Evaporator Coil Capacity With PCM.

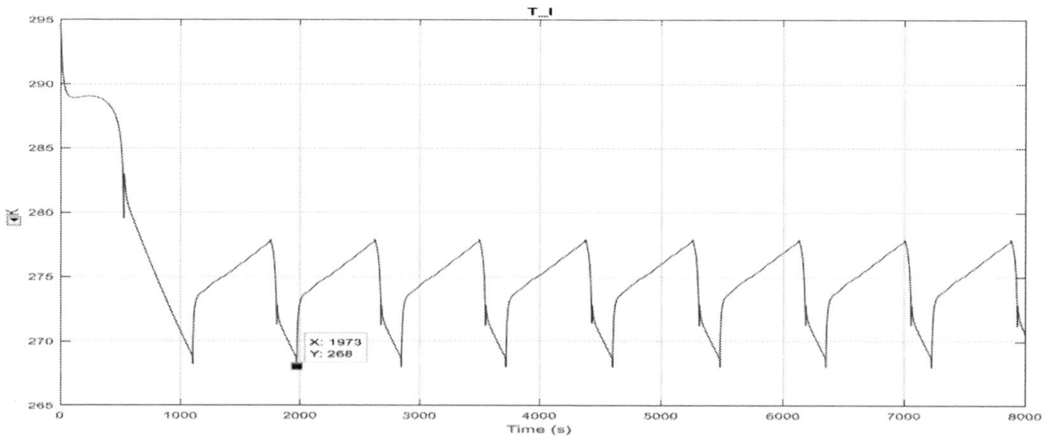

Fig. 28. Evaporator Coil Temperature Without PCM.

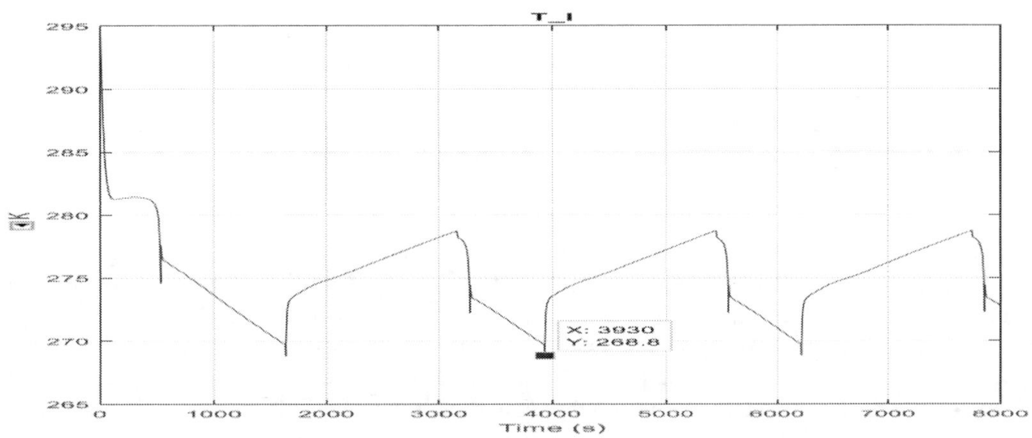

Fig. 29. Evaporator Coil Temperature With PCM

4.5 Simulation of Refrigeration Unit

Table. 2. Simulation of Refrigeration Unit

PCM	Power (Watt)	T_{on} (sec)	T_{off} (sec)	Evaporator Temp (K)	Evaporator Capacity (Watt)	Condenser Temp (K)	Condenser Capacity (Watt)	COP
Absent	41	249	627	268	224.6	313.1	207	5.48
Present	42.5	794	1500	268.8	219.2	315.3	248	5.16

It was noted in the simulation results of the refrigeration system using R134a refrigerant that:

- The presence of the phase change material increased both the on and off time of the compressor in a disproportionate manner as predicted by the theoretical calculations. There was a 219% increase in the on time, but only a 139% increase in the off time. There was a 5.8% fall in COP and a 3.5% increase in compressor work.
- The presence of the phase change material increased the minimum temperature to be maintained in the evaporator coil by 0.8 K. However, raised the temperature maintained in the condenser by 2.2 K.

6 Conclusions

The presented work aimed at designing and simulating a solar powered vapour refrigeration system using a phase change material. The project aimed at designing the system for the purpose of storing vaccines. The results of project have led to the following conclusions being drawn.

A system consisting of 150W panels with nominal voltage 12V, with the presence of a battery bank rated at 75Ah at nominal voltage 12V can successfully run a refrigeration system with rated at power consumption at 50W. As per theoretical calculations the system should be able to run as an autonomous unit for 12 hours at least, under the conservative estimate that only 6 hours of sunlight are present during the day. The battery bank provides enough charge for the system to run for 6 hours during which 50W is constantly consumed.

A single diode with series resistance model of a photovoltaic cell was implemented to validate the performance characteristics of solar panel SSI150W by Solar India. It was

observed with increasing PV module temperature, open circuit voltage decreased but short circuit current slightly increased, it was also observed that maximum power decreased with an increase in temperature. It was further observed that the open circuit voltage as well as the short circuit current increased with an increase in solar irradiation.
A maximum power point tracking unit was implemented using the Perturb and Observe algorithm and a buck boost converter. It was observed that the algorithm was able to track the maximum power point by altering the duty cycle of the buck boost converter. The output of the converter was observed and it was noted that the power oscillated after the maximum power point was found owing to the nature of the algorithm that requires constant perturbation of the voltage. It was also noted that the presence of a maximum power point tracking unit allowed greater charging current to be available for the battery.
It was concluded that the presence of a phase change material in the refrigeration system causes a considerable increase in both the on and off time of the compressor. This will aid in smooth running of the compressor without subjecting it to high temperature. However, the ratio by which the on time is increased is greater than the ratio by which the off time is increased. Numerically there is a 219% increase in the on time, a 139% increase in the compressor off time, a 5.8% fall in COP and a 3.5% increase in compressor work. Thus, under conditions where there is enough load in the system to cause the initial on and off times of the compressor to be comparable, the presence of a phase change material may result in a greater on period than an off period for the compressor. In terms of absolute value however the presence of ice in the refrigeration system does result in a significant increase in the off time. This implies that with the presence of ice as a PCM in the refrigeration system, a smaller battery bank would be required to operate the system. This increases the viability of using lower wattage solar panels to operate the same refrigeration system. Thus, reducing the overall cost of the system.

7 Future Work

There is scope to expand upon and further improve the presented work by:

- Employing alternate phase change materials, as the melting point of ice is 0°C and the maximum acceptable temperature for the storage of vaccines is 8°C. A phase change material, in particular, mixtures of n-Tetradecane which have a melting point near 6°C may perhaps provide better results.
- Varying the quantity of vaccine present in the cooling chamber would allow for a better understanding of how the ratio of PCM and load present in the refrigeration chamber affects the on and off time of the compressor.
- A major drawback arising from the current system is its inability to perform desired function in the prolonged absence of sunlight. In India this becomes relevant generally in the context of the monsoon season. While the occurrence of absolute absence of electricity is unlikely, it is possible to overcome such a situation by using hybrid systems to generate power.

References

1. Bala Bhaskar, *Energy Security and Economic Development in India: a Holistic Approach*, (TERI, 2013)
2. Mba E. F., Chukwuneke J. L., Achebe C. H. and Okolie P. C., JIEA, **2(10),** (2012)
3. S. R. Kalbande, Sneha Deshmukh, UJES, **3(2),** 17 (2015)

4. S. R. Kalbande, Sneha Deshmukh and V. P. Khambalkar, IJRANSS, **4(12),** 87 (2016)
5. C. Del Pero, F.M. Butera, M. Buffoli, L. Piegari, L. Capolongo and M. Fattore, ICCEP, 701 (2015)
6. Fouad M. Alkelani and Kant E. Kanyarusoke, ICUE, 178 (2016)
7. S. C. Arora & S. Domkundwar, *Refrigeration and Air-Conditioning*, (Dhanpat Rai Publication, 1992)
8. Armand Noël Ngueche Chedop, Noël Djongyang and Zaatri Abdelouahab, IJSR **3(11),** 1066 (2014)
9. Pradip Subramaniam, Chetan Tulapurkar,Ramasamy Thiyagarajan, G Thangamani, *Proceedings International Refrigeration and Air Conditioning Conference at Purdue*, (2010).
10. Sizing of Battery and PV module, Available from http://blog.su-kam.com, (9/3/2017)
11. Sizing of Battery and PV module, Available from http://www.leonics.com, (9/3/2017)
12. Ami Shukla, Manju Khare and K N Shukla, IJIRSET, **4(1),** 18516 (2015)
13. J. A. Ramos-Hernanz, J.J. Campayo, J. Larranaga, E. Zulueta, O. Barambones, J. Motrico, U. Fernandez Gamiz, I. Zamora, IJTPE, **4(10),** 45 (2012)
14. Habbati Bellia, Ramdani Youcef, Moulay Fatima, NRIAG, 53 (2014)
15. Jay Patel, Gaurag Sharma, IJRET, **2(3),** 225(2013)
16. Marcelo Gradella Villalva, Jonas Rafael Gazoli, and Ernesto Ruppert Filho, *IEEE Transactions on Power Electronics*, **24(5),** 1198 (2009)
17. M. G. Villalva, J. R. Gazoli, E. Ruppert F., SOBRAEP, **14(1),** 35(2009)
18. Chen Qi, Zhu Ming, *Physics Procedia*, **24,** 94 (2012)
19. Duy C. Huynh, Thu A. T. Nguyen, Matthew W. Dunnigan and Markus A. Mueller, ICIEA, 864 (2013)
20. A. Pradeep Kumar Yadav, S. Thirumaliah and G. Haritha, IJAREEIE, **1(1),** 18 (2012)
21. Aurobinda Panda, M. K. Pathak and S. P. Srivastava, IJETP, **1(2),** 18 (2011)
22. MPPT Algorithm, Available from https://www.mathworks.com, (13/4/2017)
23. Daniel W. *Hart, Power Electronics*, (Tata McGraw-Hill, 2011)
24. Two-Phase Fluid Refrigeration, Available from https://www.mathworks.com, (1/5/2017)
25. Shunt Motor, Available from https://www.mathworks.com, (18/5/2017)

Thermoacoustic Refrigerator for High Temperature Gradient

Abhishek Mishra[1], *Ankur* Kumar Choudhary[1], *Taniya* Tomar[1] and *Jagannath* Korody[1*]

[1]*Manipal Institute of Technology, Manipal Academy of Higher Education, Manipal, Karnataka, India*

Abstract: Leakage of HFCs, thermal shocks to sensitive devices, moving mechanical components which in turn require more maintenance, are some of the cons of a conventional vapor compression system .The newest field in refrigeration overcoming such flaws is Thermoacoustic refrigerator (TAR), where the sound energy is used to generate a temperature difference which can be used for refrigeration. TAR are environment friendly, do not use any kind of refrigerants, and as the compressor and expander are not used, hence, its life is more than a conventional VCR. The objective is to obtain a substantial temperature gradient using a Thermoacoustic Refrigerator. The methodology is to choose the suitable material and design the prototype. A large part of this project is devoted to explore the different designs for thermo acoustic refrigerators, and select the optimum one for maximum utility. With the present system, it was possible to achieve a temperature difference across the stack of 14 K.

1 Introduction

The basic working of the Heat engines and refrigerators can be explained by the laws of thermodynamics. A commonly used Vapor Compression Refrigeration Cycle which consists of 4 units; Compressor, Condenser, Expander or a Throttle Valve (A small capillary tube is generally used) and an Evaporator. The problem with VCRs is the power required by the Compressor is high, wear & tear of the compressor as it has moving parts and also the leakage of refrigerant can be hazardous. Therefore there is a need to go for alternative refrigeration concepts. One of the concepts that can be used is Acoustics. Thermoacoustic Refrigerator is such a concept which uses sound to create a refrigeration effect. So the compressor and the expander processes of a conventional VCR are replaced by the sound sources which have compression & rarefaction zones. In the compression zones the temperature is high and in the rarefaction zone the temperature is low. Thus, this concept is used to create a refrigeration effect. Thermoacoustic Refrigerators are environment friendly as the working medium chosen is air.

1.1 Literature Review

The performance of Thermoacoustic Refrigerator depends upon the different stack geometry i.e. spiral stack, parallel plate stack or pin array stack. It also depends upon the working medium inside the resonating tube i.e. He, Air, He-Xe mixture, hydrogen, etc. The selection

* Corresponding author: jagan.korody@manipal.edu

of materials for the resonating tube and the stack plays a vital role in the performance of Thermoacoustic Refrigerator. Some of the materials suitable are Kapton, Mylar, Polymer Film, Plexi glass, PVC glass, and acrylic fiber. One of the literatures discuss about the behavior of Standing wave TAR regarding the COP. The COP of the TAR is generally low. The Thermoacoustic Refrigeration has a linear relationship with frequency and thus high operating frequency is desired. A higher operating frequency results in a small stack plate spacing but makes the construction difficult. The position of the stack from the driver end plays a vital role in the performance of Thermoacoustic Refrigerators hence it is desired to find an optimum position for the stack. It also depends upon the porosity of the stack. According to various literatures it is desirable to use quarter wavelength resonator over half wavelength resonator because it dissipates only half the energy compared to half wavelength resonator. Hence the minimization of the energy loss can be used to make the device compact and efficient.

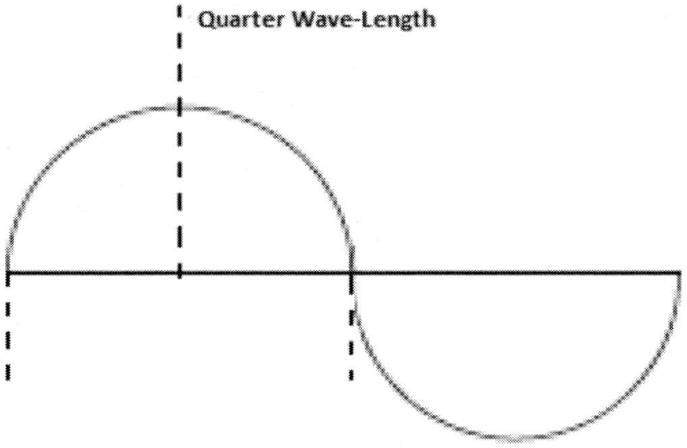

Fig. 1. Quarter Wave Length Representation

2 Working of the system

A Thermoacoustic refrigerator consists of generally a Speaker, an Amplifier, a Resonating Tube, a Stack, and a Speaker Casing. The speaker is used to create sound waves which can be monitored at a certain frequency. The amplifier is used to vary the amplitude of the sound waves. Resonating tube (Resonator) consists of the working medium and encases the stack. It is the resonator in which the whole working takes place. A stack is the heart of the Thermoacoustic refrigeration which is one of the most crucial components for having the cooling effect. A sound wave compresses and expands at different regions in the resonator. A stack is located at a certain position inside the resonator. The sound wave absorbs heat from the cooler region and pumps it to the stack. The next sound parcel takes the heat from that layer of the stack and passes it to the next adjacent layer of the stack. This step continues and finally the heat is transferred from the cooler region to hot region and thus the cooler region becomes cooler and the hotter region becomes hotter. A heat exchanger can also be used to aid the performance of the Thermoacoustic refrigerator.

The Thermoacoustic refrigerator demonstration is of a standing wave type and consists of a quarter wavelength resonator driven by a speaker. In the apparatus demonstration an 860 W speaker is used. The primary purpose of the refrigerator is to demonstrate the Thermoacoustic refrigeration effect and hence the efficiency was not our primary concern.

The amplifier used was Flyball IPL-108 of low capacity to change the amplitude of the sound waves. The setup is shown in the figure below

Fig. 2. Experimental setup for the TAR

The resonator for this refrigerator was a 22 cm length of glass tube with inner diameter 2.2 cm. The length defines the resonant frequency of the system as 379 Hz for our apparatus. Frequency generator was used to vary the frequency. A glass casing of glass thickness 4 mm was made with dimension 10 X 10 X 5 inches. A One inch hole was cut on the top face of the glass casing, and one end of the resonating tube was placed on top of the hole with central axis aligned. The glass tube was adhered to the glass box and sealed properly with silicon resin as the adhesive. Two O-ring of 9 inch diameter were chosen to avoid leakage and seal properly the speaker to the surface of the glass casing top surface from the inner side. A plug made of mild steel was used to seal the resonator from the top end and a hole was drilled at the center of the plug with the help of a 5.5 mm drill bit on a vertical drilling machine. Then the facing operation of the plug was done on an engine lathe to obtain high surface finish. This hole was drilled to pass the thermocouple wires through it.

Stack material should have a low thermal conductivity and a higher thermal capacity than the working medium. The stack was made from a 35 mm photographic film. The stack geometry used was of Spiral (rolled) type. The spacing between the adjacent layers of the stack is filled by the nylon fishing line (0.50 mm thick). Fishing lines were stuck onto the surface of the sheet with the help of a polymer tape and then spirally rolled, being aligned parallel to the length of the resonator. The spacing between the linings has to be optimum; if the layers are close together viscous effects hamper the motion of the gas particles and if the layers are too wide apart the heat transfer rate would not be much due to leakage of air across the gaps.

Thermocouple of K-type was used. The wires of the thermocouple passed through the hole of the plug and measured the temperature at the two ends of the stack. The important factor while placing the thermocouple wires was that they should not touch the stack surface but they should be located as closely as possible to the stack. Two displays for the thermocouples were used to indicate the temperature. Sealing of the whole system was done with silicon resin and M-seal.

Thermal penetration gap is given by the expression

$$\delta = \sqrt{(2K/\rho c\omega)} \qquad (1)$$

K = Gas thermal conductivity
ρ = Density
c = Isobaric Specific Heat
ω = Operating frequency in radians per second

Therefore thermal penetration gap is calculated to be 1.30×10^{-5} m or 0.013 mm.

3 Results and Discussion

Testing of the apparatus was done at various frequencies ranging from 345-385 Hz. Also the position of the stack was varied to obtain various results.

Initially the testing was done with a frequency of 385 Hz and stack position of 5 cm from the top (plug surface). The setup ran for 12 minutes and a temperature gradient of 2 Kelvin was noted along with 1 Kelvin of cooling. With a frequency of 379 Hz and stack position of 4 cm from the top, the trial resulted in a gradient of 4 Kelvin and cooling of 1 Kelvin. With a frequency of 374 Hz and a stack position of 3 cm from the top, the gradient obtained was 5 Kelvin with 2 Kelvin of cooling. With a frequency of 369 Hz and a stack position 2 cm from the top, a gradient of 8 Kelvin with 3 Kelvin of cooling was observed.

Finally a frequency of 364 Hz showed the best result with the stack being placed at 1 cm from the top which resulted in a thermal gradient of 14 Kelvin with 5 Kelvin of cooling in just 33 minutes, i.e. the hot side of the stack rose up by 9K and the cooler side cooled down by 5K. It can be noted from the Fig.2 that the variation of the temperature is not linear with the time.

Figure 4 shows that after some period of time (in minutes) the temperature-time behaviors lies near to the straight line where as in during the initial minutes there is a deviation from the straight line. Also the temperature gradient is very rapid during the initial few minutes of the run of the apparatus but very slow afterwards.

Table 1. Variation of Temperature for f = 364 Hz

Time(min)	Temperature Gradient	T_1 °C (K)	T_2 °C (K)
0	0	30(303)	30(303)
1	4	32(305)	28(301)
2	6	34(307)	28(301)
3	7	34(307)	27(300)
4	8	35(308)	27(300)
8	10	37(310)	27(300)
15	11	38(311)	27(300)
22	12	38(311)	26(299)
27	13	39(312)	26(299)
33	14	39(312)	25(298)

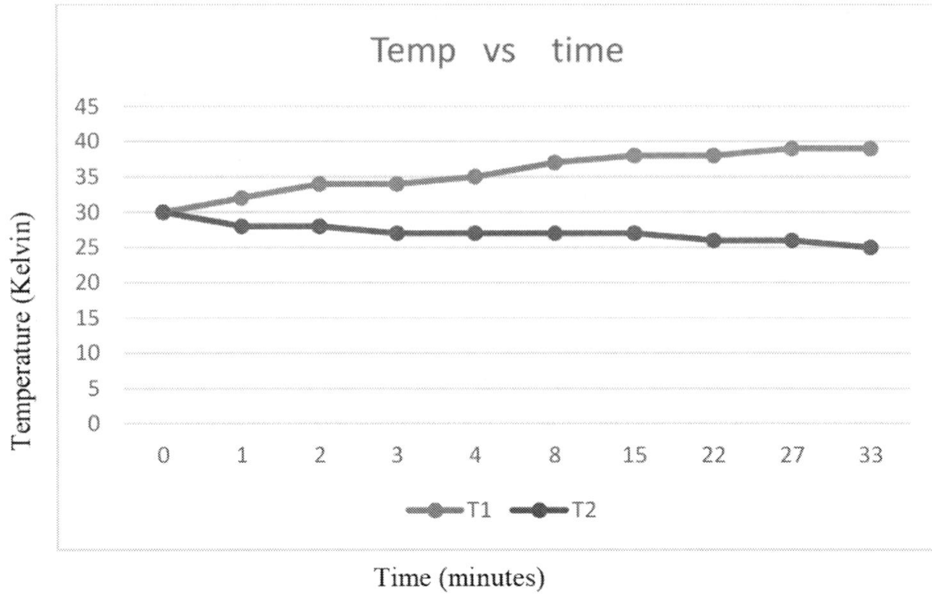

Fig. 3. Plot of variation of Hot and Cold Region across a stack with respect to time

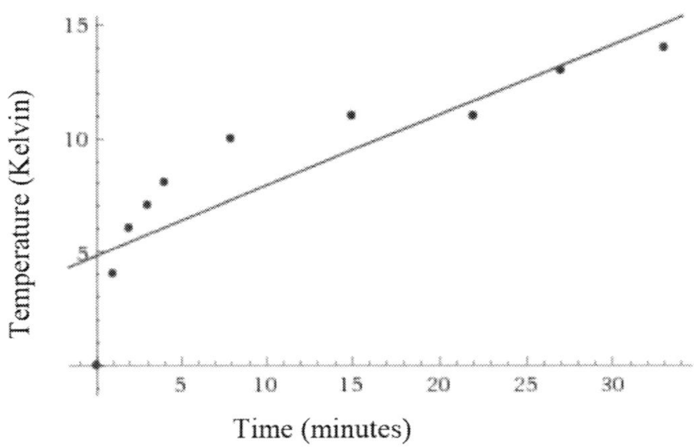

Fig. 4. Scatter plot for the Temperature Gradient

4 Conclusions

The results from the experiment show that a significantly high amount of Temperature gradient can only be achieved when all the significant parameters are well taken care of. Hence we conclude from the trials that:

i. The maximum temperature gradient obtained was 14K, with the maximum drop in colder region temperature being 5K, when the stack was placed at its most optimum position and the setup was operated with the most optimum frequency along with proper sealing which is quite good for a basic Thermoacoustic refrigerator model. The testing setup was started at an initial temperature of 303K. The hotter thermocouple went on to reach a temperature of 312K, i.e. a rise of 9K, whereas the cold thermocouple showed a drop in temperature of 5K and thus read 298K after the setup was run for about 33 minutes.

ii. The most efficient stack position was 1 cm from the top (distance measured from the bottom of the Mild Steel Plug). The optimum stack position is very critical in obtaining the best possible or the highest achievable thermal gradient which depends on the geometry of the resonator and also on the frequency of the sound waves. A temperature difference of 14K was obtained when the stack was placed 1 cm from the top of the resonator tube.

iii. The most optimum frequency was found to be 364 Hz. The calculations according to the length of the resonating tube and stack position resulted in a frequency of 369 Hz, but by hit and trial it came to 364 during the testing phase. We hence concluded that there are few unaccounted losses in transmission of sound waves which resulted in the deviation from the ideal calculations, owing to friction due to glass surface, viscous damping effects of air, and also maybe because of the thermal penetration depth of the stack layers were not perfectly uniform.

iv. Very less or almost zero temperature gradient will be obtained even if there is a slight leak in the setup, and hence for maximum efficiency of the apparatus it has to be hermetically sealed.

v. Stack being the most important part of the apparatus has to be rolled very carefully. If it is rolled too tight then viscous damping effects dominate which lead to loss in energy transfer and if it is rolled too loosely then the air particles tend to escape through the gaps without any heat exchange with the stack surfaces.

vi. Thermocouples are used to measure the air temperature and not the stack surface temperature.

vii. Use of heat exchangers can significantly increase the thermal gradient as it will accelerate the heat exchange process across the stack.

viii. Higher the power rating of the speaker, higher is the maximum amplitude of the sound wave, and hence higher the amount of thermal gradient that can be achieved.

ix. Use of pressurized air or lighter gases like helium can give better results but involve costlier setups and more complicated designs.

x. Stack material should have low thermal conductivity and higher thermal capacity than the working medium.

xi. The power density of the Thermoacoustic refrigerator is a linear function of the frequency, and hence high value of frequency is desirable. But high frequency results in a small stack plate spacing and that makes construction difficult. Hence the selection of the optimum frequency is required, which is obtained using a suitable loudspeaker.

References

1. Timothy S.Ryan B.S in Mechanical engineering, "Design and Control of a Standing-Wave Thermoacoustic Refrigerator", University of Pittsburg, (2006)
2. M.E.H. Tijani, J.C.H Zeegers, A.T.A.M de waele "Design of Thermoacoustic refrigerators", – Elsevier Science, (2002).
3. Ratish Savant, Sandesh Rasal, Gaurav Yadav, Arunkumar Yadav, Darshan Kadam , "Design & Fabrication of Thermoacoustic Refrigeration", International Journal of Research in Advent Technology , **Vol.3 , No.4**,April (2015)
4. Ratish Savant, Sandesh Rasal, Gaurav Yadav, Arunkumar Yadav, Darshan Kadam , "Design & Fabrication of Thermoacoustic Refrigeration", International Journal of Research in Advent Technology , **Vol.3 , No.4**,April (2015)

5. Jinshah B S, Ajith Krishnan R, Sandeep V S ,"Study on A Standing Wave Thermoacoustic Refrigerator Made of Readily Available Materials" International Journal of Scientific and Research Publication , **Volume 3, Issue**, July (2013)
6. B.G Prasantha, M.S.Govinde Gowda, S. Sreetharamu , "Effect of mean operating pressure of the performance of Stack Based Thermoacoustic Refrigerator" International Journal of Thermal & Environment Engineering, **Volume 5**, (2013)

Performance analysis of PV panel under varying surface temperature

Tripathi Abhishek Kumar[1*], Murthy Ch. S. N[1] and Aruna *Mangalpady*[1]

[1]*Department of Mining Engineering, National Institute of Technology Karnataka, Surathkal, India*

Abstract. The surface temperature of PV panel has an adverse impact on its performance. The several electrical parameters of PV panel, such as open circuit voltage, short circuit current, power output and fill factor depends on the surface temperature of PV panel. In the present study, an experimental work was carried out to investigate the influence of PV panel surface temperature on its electrical parameters. The results obtained from this experimental study show a significant reduction in the performance of PV panel with an increase in panel surface temperature. A 5W PV panel experienced a 0.4% decrease in open circuit voltage for every 1°C increase in panel surface temperature. Similarly, there was 0.6% and 0.32% decrease in maximum power output and in fill factor, respectively, for every 1°C increase in panel surface temperature. On the other hand, the short circuit current increases with the increase in surface temperature at the rate of 0.09%/°C.

Keywords: Surface temperature, open circuit voltage, short circuit current, maximum power output and fill factor.

1 Introduction

The use of energy plays a very vital role in one's life. Thus, it's supply and generation should be secure and sustainable. At the same time, it should be environmentally friendly, socially and economically acceptable. The present scenario of energy generations is neither secure nor sustainable and depends on fossil fuels, nuclear and renewable energy sources. Among these, the renewable energy could be considered as a source of secure and sustainable energy source [1]. There are different types of renewable energy sources, such as hydro, geothermal, biomass, wind and solar. Among all available renewable energy sources, solar energy experienced a rapid growth and popularity in the last one decade. Therefore, solar energy is the most promising and vital energy source to produce electricity in a present scenario [2-4].

In solar energy, photovoltaic (PV) panel is a device that converts sun radiation (solar energy) into electrical energy with the help of photovoltaic effect [5]. Photovoltaic effect is a process in which PV panel generates electric power, when it is exposed to sunlight. Sun light consists of the bundle photons, which are simply absorbed by a PV panel and when a suitable wavelength of photon incident on the panel surface, the energy of the photon is transferred to the electron of the PV material. This causes the electron to move towards a

* Corresponding author: abhinitrkl12@gmail.com

higher energy state (i.e. conduction band) from a lower energy state (i.e. valence band). This movement of electron is the main cause of generation of an electric current in the PV device [6-8].

The performance of PV panel is highly depended on the environmental parameters, such as solar radiation, wind speed, humidity, dust and atmospheric temperature. The panel surface temperature is a key environmental parameters that influence the performance of PV panel by changing its electrical parameters, such as open circuit voltage (VOC), short circuit current (ISC), maximum power output (PM) and fill factor (FF) [9-10].

In general, PV panels are made of silicon semiconductor material and like all semiconductor material, PV panels are also temperature sensitive. Hence, temperature of PV panel is an important environmental parameter, which affects its performance. The panel surface temperature depends on the encapsulating material, solar radiation, atmospheric temperature, humidity and wind speed [11]. As the temperature of PV panel increases its VOC, PM and FF decreases. Moreover, the ISC of PV panel experiences a very small increments with the rise in its surface temperature. As a result of this, the overall performance of PV panel reduces due to increase in its surface temperature [12]. The rise in panel temperature for c-Si PV panel decreases its VOC at the rate of -0.45%/K [13]. Similarly, the decrease in power output and fill factor of the PV panel to increase in panel surface temperature are at the rate of -0.65%/K and -0.2%/K, respectively [14]. A study has shown that the 9% increments in panel efficiency due to 20% reduction in the panel surface temperature [15]. Similarly, one more study reported that with the 22°C reduction in the panel surface temperature, its power output experienced a gain of 10.3% [16]. The available literature shows that the surface temperature of PV panel is a vital parameter that affect its performance. Therefore, in this paper, an attempt was made to understand the influence of surface temperature on PV panel performance.

2 Experimental set-up and methodology

A 5W polycrystalline PV panel was used to carry out the present investigation. The detail technical specifications of PV panel considered are given in Table 1. The PV panel was mounted on the flat frame with zero inclination and subjected to a constant solar radiation of 1182W/m^2. This range of solar radiation was generated by using a set of solar simulators and measured by TM-207 solar power meter. The experimental setup used in this study is presented in Figure 1. The PV panel was connected to a variable rheostat of 320Ω rating through the ammeter (connected in series) and a voltmeter (connected in parallel). A Digital Multimeter Fluke 178+ and DT830B were used as a voltmeter and ammeter in the circuit, which is given in Figure 2. To increase the surface temperature of PV panel a set of infrared dryer bulbs were used and a digital pyrometer was also used to measure the panel temperature of PV panel. The panel temperature was increased from 35°C to 65°C and the readings of electrical parameters (i.e. current and voltage) were taken at every increase of 10°C of panel surface temperature. Based on the obtained reading the other electrical parameters of PV panel, such as power output and fill factor were calculated. The equation used for calculating the fill factor (FF) of PV panel is given in equation 1.

$$FF = \frac{V_m \times I_m}{V_{OC} \times I_{SC}} \quad (1)$$

where,

V_{OC} = open circuit voltage (volt)

I_{SC} = Short circuit current (amp)

I_m = Maximum current (amp)
V_m = maximum voltage (volt)

Table 1. Technical Specification of 5W PV panel.

Sl. No.	Specification	Rating
1	Maximum power	5 watt
2	Maximum voltage	9.64 volt
3	Maximum current	0.52 amp
4	Open circuit voltage	11.57 volt
5	Short circuit current	0.57 amp
6	NOCT	(47±2)°C
7	Tolerance level	0 to 3%

(a) Heating of PV panel (b) Electric measurement of PV panel

Fig. 1. Experimental set-up.

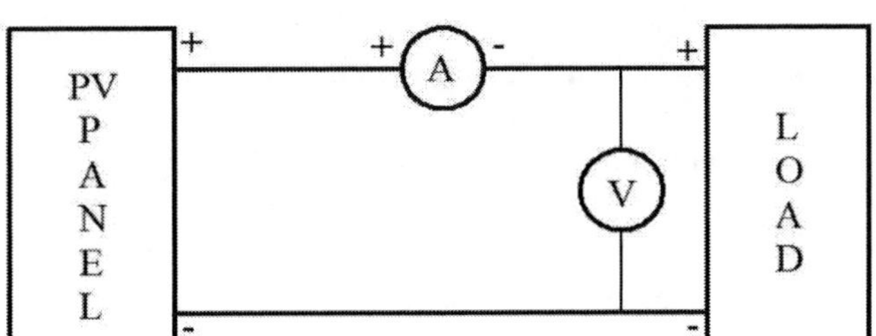

Fig. 2. Circuit diagram.

3 Error analyses

The uncertainty in the measurement may be taken as the possible error of the experimental data. The uncertainty in the measurements may arise from the various experimental errors, such as instrumental error, calibration error, observation error and human error. The uncertainty in the measured values can be explained by the error analysis [17].

Consider, a set of measurements, such as x_1, x_2, x_3.......x_n, during N number of trials. Suppose, the possible error E_1, E_2, E_3........E_n are introduced in measured results. Then, the possible error in the measured result is given by equation (2). The measured error in the various measurements like, open circuit voltage, maximum power output and fill factor is given in Table 2.

$$E= \{ [(x_1-x_{avg})^2+ (x_2-x_{avg})^2............(x_n-x_{avg})^2]/N\}^{1/2} \quad (2)$$

where,

E = measured error

N = number of trials

x_{avg}= average values of the n trials

Table 2. Percentage errors introduced in the measurements

Surface temperature of the panel (°C)	Percentage error in the measured parameters (%)		
	V_{OC}	P_M	FF
35	4.40	4.76	2.39
45	3.80	8.20	2.49
55	2.02	5.13	2.08
65	1.21	4.98	0.9

4 Results and Discussions

The recorded electrical parameters of PV panel, such as V_{OC}, I_{SC}, P_M and FF under varying panel surface temperature are given in Table 3. Based on the recorded electrical parameters the current (I)-voltage (V) and power (P)-voltage (V) characteristics of PV panel are plotted at four different surface temperatures (i.e., 35°C, 45°C, 55°C and 65°C), which are shown in Figure 3 and Figure 4. As depicted in Figure 3 and Figure 4, surface temperature of PV panel has a significant impact on its I-V and P-V characteristic. The current is almost same in lower voltage range but as the voltage increases a fast reduction in the current is observed for the all four defined surface temperature conditions. As presented in Figure 3 the deformation in the I-V characteristic gets increases with the panel surface temperature. There is a significant reduction in the V_{OC} of the PV panel with increases in the panel surface temperature, whereas a very small increment in I_{SC} is observed. Due to this significant reduction in V_{OC} the other electrical parameters (i.e., maximum power output and fill factor) of PV panel also reduces. Therefore, the overall performance of PV panel reduces with the rise in its surface temperature. Moreover, as presented in Figure 4 the maximum power point (MPP) in the PV curve moves towards the lower output voltage with the increase in panel surface temperature. Due to this, the maximum power point tracker algorithm would not be able to operate at a particular maximum power point, which reduces the performance of solar charge controller. Hence, the shifting of MPP in the PV curve directly affects the effective operation of the whole PV systems (i.e., PV panel and charge controller).

Table 3. Recorded electrical parameters of PV panel.

Panel surface temperature (°C)	**V_{OC} (volt)**	**I_{SC} (amp)**	**P_M (watt)**	**FF**
35	9.70	0.36	2.475	0.708
45	9.30	0.36	2.325	0.694
55	8.90	0.365	2.175	0.669
65	8.52	0.37	2.025	0.640

As indicated in Table 3, the reduction in V_{OC} is 12.16% with the increase in panel surface temperature from 35°C to 65°C. Similarly, the reduction in P_M and FF are 18.18% and 9.60%, respectively, due to increase in panel surface temperature. The P_M and FF are found to decrease with increase in panel surface temperature due to change in respective current and voltage. The reduction in P_M and FF with the increase in panel surface temperature are presented in Figure 5 and Figure 6. The maximum power output and fill factor of PV panel decreases linearly with the increase in panel surface temperature. From the Table 2, the temperature coefficient for V_{OC}, I_{SC}, P_M and FF are calculated with the help of Equations (3), (4), (5) and (6). The temperature coefficient for V_{OC}, I_{SC}, P_M and FF are 0.4%/°C, 0.09 %/°C, -0.6%/°C and -0.32%/°C, respectively.

$$\text{Temperature Coefficient of } V_{OC} = \frac{\Delta V_{OC}}{V_{OC}\times\Delta T} \quad (\%/^{\circ}C) \tag{3}$$

$$\text{Temperature Coefficient of } V_{OC} = \frac{\Delta I_{SC}}{I_{SC}\times\Delta T} \quad (\%/^{\circ}C) \tag{4}$$

$$\text{Temperature Coefficient of } P_M = \frac{\Delta P_M}{P_M\times\Delta T} \quad (\%/^{\circ}C) \tag{5}$$

$$\text{Temperature Coefficient of FF} = \frac{\Delta FF}{\Delta FF\times\Delta T} \quad (\%/^{\circ}C) \tag{6}$$

where,

ΔT = change in panel surface temperature (°C)

V_{OC} = open circuit voltage of PV panel (volt)

ΔV_{OC} = change in open circuit voltage (volt)

P_M = maximum power output of PV panel (watt)

ΔP_M = change in maximum power output of PV panel (watt)

FF = fill factor of PV panel

ΔFF = change in fill factor

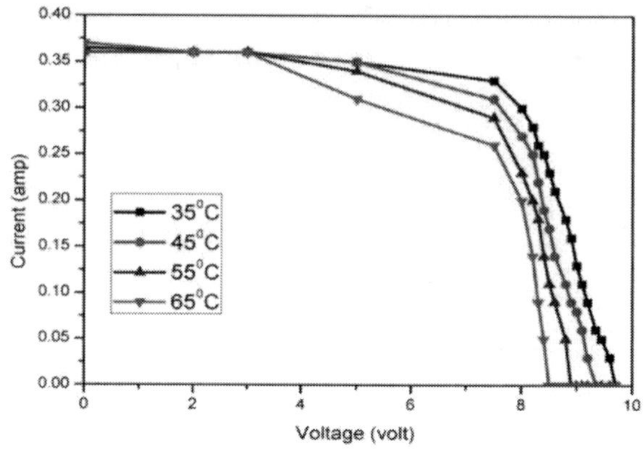

Fig. 3. I-V characteristic of PV panel.

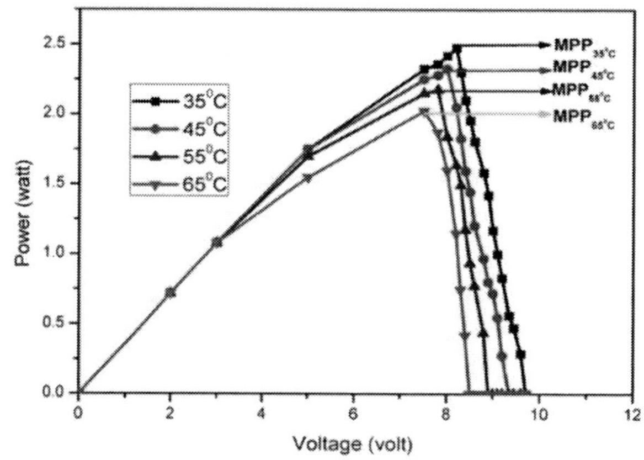

Fig. 4. P-V characteristic of PV panel.

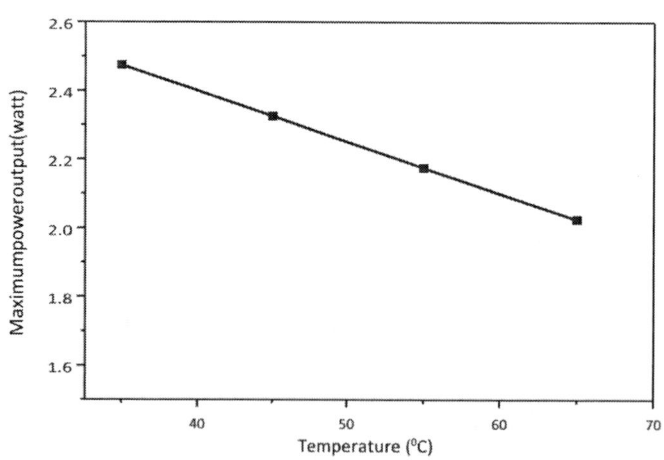

Fig. 5. Variation of maximum power output with panel surface temperature.

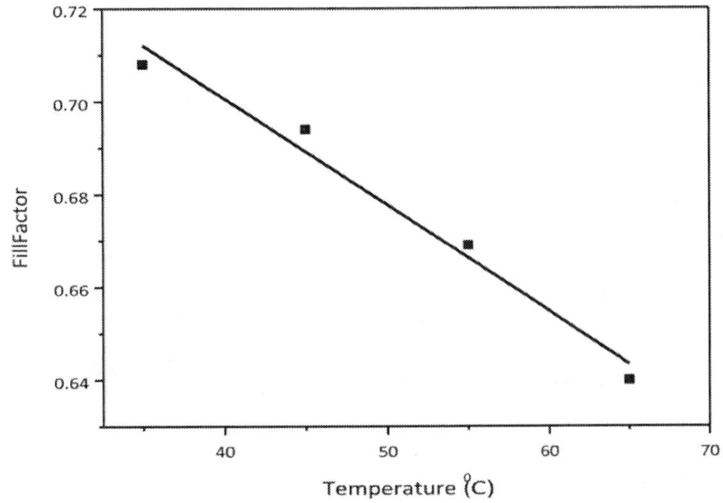

Fig. 6. Variation in fill factor of PV panel with panel surface temperature.

5 Conclusions

The performance of PV panel depends on the temperature at which the panel is operating. In this study, the performance analysis of PV panel was evaluated under its varying surface temperature condition. The performance of PV panel was calculated in terms of its electrical parameters, such as V_{OC}, I_{SC}, P_M and FF. The results showed that the P_M and FF of PV panel have a negative linear relation with the panel surface temperature. According to obtained results, the reduction in V_{OC}, P_M and FF are respectively 12.16%, 18.18% and 9.60% of the increase in panel surface temperature from 35°C to 65°C (i.e. 30°C increase in panel surface temperature). On the other hand, short circuit current of PV panel experiences very slight increments of 2.7% with the 30 °C increase in panel surface temperature. Over and above, this study also demonstrates that the maximum power point (MPP) in the P-V curve moves towards the lower output voltage as the surface temperature of PV panel increases. This movement of MPP disturbs the operation of solar charge controller, which reduces the overall performance of whole PV systems.

Abbreviations

V_m	Maximum voltage
I_m	Maximum current
P_M	Maximum power output
V_{OC}	Open circuit voltage
I_{SC}	Short circuit current
PV	Photovoltaic
FF	Fill factor
ΔV_{OC}	change in open circuit voltage
ΔP_M	change in maximum power output of PV panel
ΔFF	change in fill factor

References

1. Z. Graham and E. Economics., Peak People: The Interrelationship between Population Growth and Energy Resources, *Energy Bulletin* 20 (2009).
2. B.R. Angel, A.O. Bielsa and A.Martínez-Gracia, Photovoltaics on flat roofs: energy considerations, *Energy* 36, 4, 1996-2010 (2011).
3. M. Oliver, and T. Jackson, Energy and economic evaluation of building-integrated photovoltaics, *Energy* **26**,4, 431-439 (2001).
4. C.Fernández, J. Francisco and B.R. Ángel, Power supply for pumping systems in northern Chile: photovoltaics as alternative to grid extension and diesel engines, *Energy* **35**,7, 2909-2921 (2010).
5. G. Adolf, C. Hebling and H.W. Schock, Photovoltaic materials, history, status and outlook, *Materials Science and Engineering: R: Reports* **40,**1, 1-46 (2003).
6. P. Bhubaneswari, S. Iniyan and R. Goic, A review of solar photovoltaic technologies, *Renewable and sustainable energy reviews* **15,** 3, 1625-1636 (2011).
7. G. K. Singh, Solar power generation by PV (photovoltaic) technology: A review, *Energy* **53,** 1-13 (2013).
8. El. Chaar and N. El. Zein, Review of photovoltaic technologies, *Renewable and sustainable energy reviews* **15,** 5, 2165-2175 (2011).
9. V. Silvano, G. Acciani and Ottavio Falcone, A finite-element approach to analyze the thermal effect of defects on silicon-based PV cells, *IEEE Transactions on Industrial Electronics* **59,** 10, 3860-3867 (2012).
10. S. Chander, A. Purohit, A. Sharma, S.P. Nehra and M.S. Dhaka, Impact of temperature on performance of series and parallel connected mono-crystalline silicon solar cells, *Energy Reports* **1,** 175-180 (2015).
11. M. A. García and J. L. Balenzategui, Estimation of photovoltaic module yearly temperature and performance based on nominal operation cell temperature calculations, *Renewable energy* **29,** 12, 1997-2010 (2004).
12. M.G. Angee, Study the operating temperature of a PV module, *Project Report* (2013).
13. A. Virtuani, D. Pavanello, and G. Friesen, Overview of temperature coefficients of different thin film photovoltaic technologies, *5th World Conference on Photovoltaic Energy Conversion*. 2010.
14. E. Radziemska, The effect of temperature on the power drop in crystalline silicon solar cells, *Renewable Energy* **28,** 1, 1-12 (2003).
15. H. Bahaidarah, A. Subhan, P. Gandhidasan and S. Rehman, Performance evaluation of a PV (photovoltaic) module by back surface water cooling for hot climatic conditions, *Energy* **59,** 445-453 (2013).
16. K. Stefan, Increased electrical yield via water flow over the front of photovoltaic panels, *Solar energy materials and solar cells* **82,** 1, 131-137 (2004).
17. J.P. Olman, Experimental techniques for engineers, 7th ed, New Delhi, Tata McGraw Hill, 2004.

Performance Testing of Diesel Engine using Cardanol-Kerosene oil blend

Ravindra[1*], *Aruna* M[1] and *Vardhan* Harsha[1]

[1]*Department of Mining Engineering, National Institute of Technology Karnataka, Surathkal, 575025, India*

Abstract. Awareness of environmental pollution and fossil fuel depletion has necessitated the use of biofuels in engines which have a relatively cleaner emissions. Cardanol is a biofuel, abundantly available in India, which is a by-product of cashew processing industries. In this study performance of raw Cardanol blended with kerosene has been tested in diesel engine. Volumetric blend BK30 (30% kerosene and 70% Cardanol) has been used for the test. The properties like flash point, viscosity and calorific value of the blend have been determined. The test was carried out in four stroke diesel engine connected with an eddy current dynamometer. Performance of the engine has been analysed by finding the brake specific fuel consumption (BSFC) and brake thermal efficiency (BTE). The results showed that the brake thermal efficiency of the blend is 29.87%, with less CO and smoke emission compared to diesel. The results were also compared with the performance of Cardanol diesel blend and Cardanol camphor oil blend, which were already tested in diesel engines by other researchers. Earlier research work reveals that the blend of 30% camphor oil and 70% Cardanol performs very closer to diesel fuel with a thermal efficiency of 29.1%. Similarly, higher brake thermal efficiency was obtained for 20% Cardanol and 80% diesel blend.

Keywords: Biofuel; Cardanol; Kerosene; Emission; Alternative fuel

1 Introduction

The reserves of fossil fuels were depleting due to their increased usage. The inventors in the field of fuels have to think about an alternative to fossil fuel, which is renewable in nature and green to environment. Biofuels were one such source of fuel originated from plants and animals [1]. The oils extracted from seeds such as coconuts, cotton seed, soya bean, sunflower, jatropha, peanuts, mustard seed, linseed, palm and cashew nuts are converted to biodiesel and tested as a substitute fuel for CI engines [2, 3].

The Cardanol is a by-product of cashew industry, extracted from CNSL. It is abundantly available at low cost in India. Cardanol has been tested in diesel engine as a fuel by many researchers. Cardanol was blended with diesel and its performance and emissions were studied in a double cylinder CI engine. One such study reported that up to 20% Cardanol

* Corresponding author: ravindra@nitte.edu.in

blending with diesel the emissions were lower with improved efficiency [4]. To reduce the viscosity of Cardanol it was preheated up to 80°C before blending with diesel. The diesel engines operated with preheated Cardanol diesel blend reported 20% higher BTE than unheated oil [5]. Cardanol was mixed with 10% methanol as an additive and blended with diesel. The B20M10 blend (20% Cardanol, 70% diesel and 10% methanol) showed lower emissions and improved BTE in CI engines [6]. Diesel engines operated with B20M10 blend with extra intake oxygen also reported improvement in BTE with less emissions [7]. In another study Cardanol was blended with camphor oil to reduce its viscosity and the blend has been tested in CI engine [8]. The performance of CMPRO30 (30% camphor oil and 70% Cardanol) blend was very close to diesel fuel. Some of the researchers have blended the biodiesel with kerosene and tested the possibility of usage of the blend as a fuel in diesel engines. They reported improvement in performance with lower emissions [9, 10]. In the present study Cardanol has been blended with kerosene to reduce the viscosity and the blend has been tested in diesel engine. The results were compared with the performance of diesel Cardanol blend and camphor oil Cardanol blend.

2 Materials and Methods

The test fuel Cardanol has been purchased from a local cashew industry. The kerosene and diesel has been purchased from local market. The test blend BK30 (30% kerosene and 70% cardanol) was prepared by mixing 30% kerosene and 70% Cardanol on volume basis. The properties, like flash point, calorific value and kinematic viscosity of the blend and diesel fuel has been determined as per ASTM standards. The kinematic viscosity was determined as per ASTM standard D445 using Cannon-Fenske Viscometer. The flash point was determined using Pensky-Martens closed cup apparatus as per ASTM Standard D93. The calorific value of the fuels were found using bomb calorimeter as per ASTM Standard D240. The results of all these tests were tabulated in Table 1.

Table 1. Properties of diesel and BK30 blend

Properties	**ASTM code**	**Diesel**	**BK30**
Kinematic viscosity @40°C (in cSt)	D445	3.17	5.9
Flash point(in °C)	D93	51	65
Calorific value (in kJ/kg)	D240	43580	41331

The experiments have been conducted in a four stroke water cooled CI engine running at 1500 rpm (Fig. 1).The engine technical specifications were given in Table 2. The load on the engine was applied by a water cooled eddy current dynamometer, which was directly connected to the engine. Thermocouples were mounted on the exhaust gas calorimeter for temperature measurement. A burette was fitted in the fuel line for the measurement of fuel consumption. The exhaust emissions was measured using exhaust gas analyser (Netel exhaust gas analyser, Model: NPM-MGA-1) and the smoke opacity was measured by using smoke meter (Netel smoke meter model, NPM-SM-111B).

The initial settings of the engine were checked as per manufacturer specification. The water flow rate through engine and calorimeter were adjusted to 250 litres/hour and 100 litres/hour, respectively. Diesel and BK30 blend were filled separately in the dual fuel tank. The air lock on the fuel line, if any, was cleared by loosening the set screw at the fuel pump. Start the engine with diesel fuel by cranking and warmed up for 15 minutes. Load was applied and fuel consumption rate was measured. The exhaust gas temperature,

Exhaust emissions and smoke values were noted. The trail was repeated three times to get repeated values and average value was considered. Then the engine was operated with the BK30 blend by changing the position of the fuel tank valve and the same procedure was repeated. The brake specific fuel consumption and BTE can be calculated using the equations given below.

BSFC= mass flow rate of fuel / Brake power
BTE=1/BSFC x lower calorific value

Table 2. Engine specifications

Number of cylinders	1
Number of strokes	4
Power	3.5 kW
Bore and stroke	87.5 mm x 110mm
Rated speed	1500 RPM
Compression ratio	12 to18:1(18:1)
Dynamometer arm length	184 mm

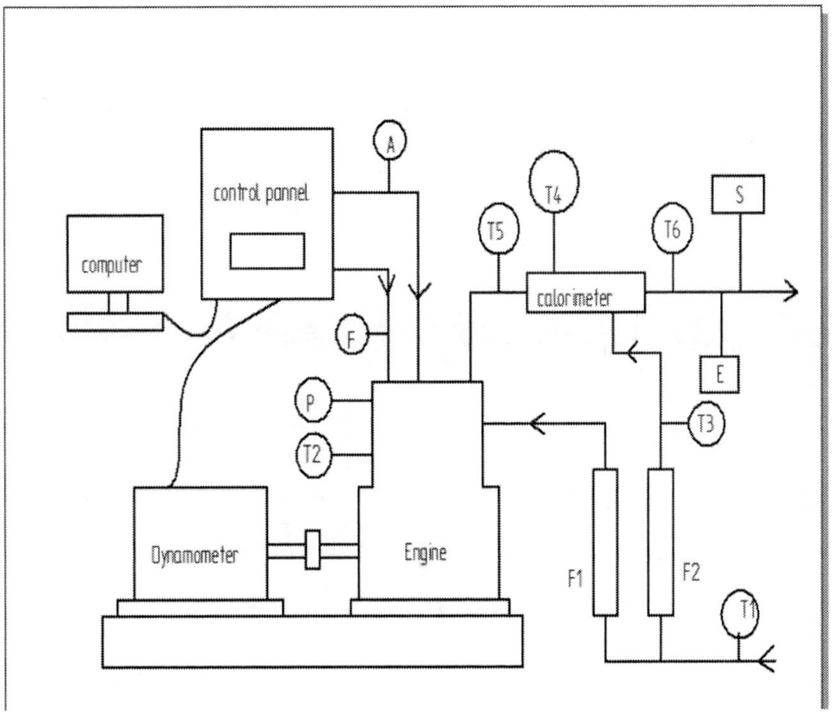

Fig. 1. Experimental Set up

P: Pressure sensor
A: Air flow rate measurement
F: Fuel flow rate measurement
F1: Water flow rate through engine
F2: Water flow rate through calorimeter

S: Smoke meter
E: Exhaust gas analyser
T1 and T2: Engine inlet and outlet jacket water temperature
T3 and T4: Calorimeter inlet and outlet water temperature
T5 and T6: Exhaust gas temperature at inlet and outlet of calorimeter

3 Results and Discussions

The raw Cardanol was blended with 30% kerosene and the blend was tested in diesel engine to find the feasibility of using it as a fuel. Fig. 2 shows the variation of BTE for BK30 blend and diesel at various loading conditions. At peak load the BTE of BK30 blend was 29.87% and for diesel it was 30.36%. The engine was operating smoothly with the blend at all the loads similar to diesel fuel. By blending kerosene with Cardanol the combustion was fast because of vigorous burning of kerosene, so the BTE was only 1.6% less than diesel fuel. At lower load, due to low combustion chamber temperature the combustion was slow. More BK30 fuel was injected compared to diesel (higher BSFC) at lower loads. So more kerosene particles available which are more volatile than diesel. Due to higher volatility of kerosene molecules the surrounding biofuel molecules will burn fast. So at lower load good combustion compared to diesel fuel.

BSFC variations with load is shown in Fig. 3. For BK30 blend a higher BSFC was recorded than diesel fuel, which is due to poor calorific value. The BSFC of the blend was 6.8% higher than diesel fuel. A higher BSFC for biodiesel blends were already reported [11].

Exhaust gas temperature variations has been represented in Fig. 4. At peak load for BK30 blend EGT was lower than diesel fuel, which shows the fast combustion of the kerosene Cardanol blend. At lower loads due to lower temperature, the combustion of biofuel extended for longer duration which leads to higher exhaust gas temperature.

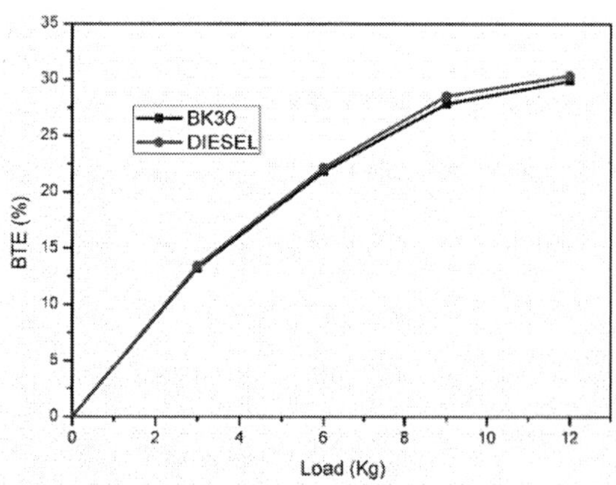

Fig. 2. BTE vs load

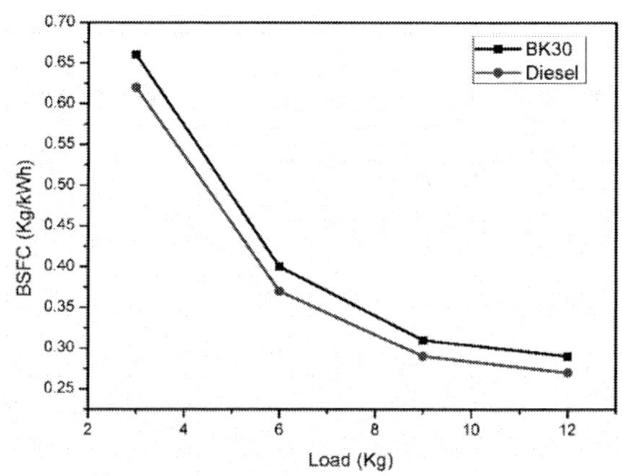

Fig. 3. BSFC vs load

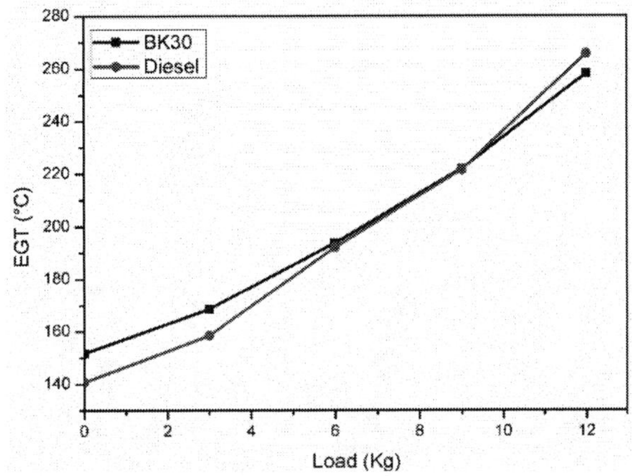

Fig. 4. Exhaust gas temperature vs load

The emission of CO was due to incomplete combustion. The CO emission for various loading condition has been presented in Fig. 5. From the figure it was observed that for the Cardanol kerosene blend CO emission reducing with increase in the load and for diesel fuel it was increasing. Similar observations were recorded by other investigators on diesel fuel [5, 12, and 13].

The emissions of smoke for different loading conditions was shown in Fig. 6. As shown in Fig. 6 when the load on the engine was increasing the smoke emission was also increasing [14]. At lower loads the Cardanol kerosene blend emits more smoke than diesel fuel. But the smoke was lower for the blend at peak load compared to diesel. At lower loads due to lower temperature the poor combustion of some of the biofuel particles leads to higher smoke emission.

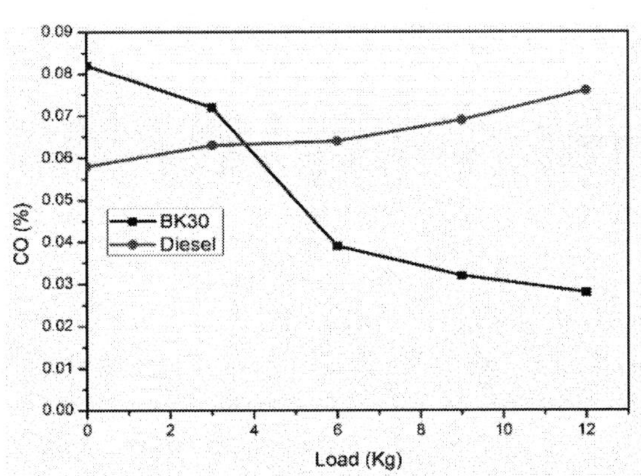

Fig. 5. CO vs load

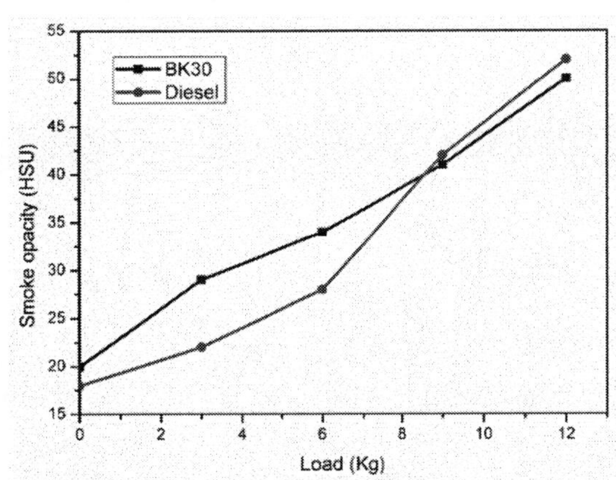

Fig. 6. Smoke vs load

Cardanol has been already used as diesel engine fuel in different ways by many researchers. Table 3 gives comparison of performance of these kinds of blends.

Table 3. Comparison of different blends of Cardanol

Particulars	Cardanol Kerosene blend (BK30)	Cardanol camphor oil blend [8] (CMPRO30)	Cardanol diesel blend [14] (20CNSO)
BTE	29.87%	29.1%	~28%
EGT	258 °C (531K)	686K	224 °C(497K)
CO emission	0.028%	0.3%	0.62%
Smoke emission	50HSU	3.91BSU	~32 BSU

From Table 3 it can be observed that brake thermal efficiency of Cardanol kerosene blend and Cardanol camphor oil blend were higher than Cardanol diesel blend with lower emissions of smoke and CO.

4 Conclusions

The Cardanol can be blended with kerosene and the blend can be a fuel for CI engine. The performance of BK30 blend is very nearer to diesel fuel with respect to BTE and BSFC. Emissions of smoke and CO with BK30 blend were lower than diesel at peak load. Therefore Cardanol kerosene blend can be replaced for diesel without much modifications in the engine with minimum pollution.

Nomenclature

BTE	Brake thermal efficiency
CO	Carbon monoxide
BSFC	Brake specific fuel consumption
CI	Compression ignition
EGT	Exhaust gas temperature
CNSL	Cashew nut shell liquid

References

1. R. Behcet, R.Yumrutas, H. Oktay, Energy, 71, 645-655 (2014)
2. G. Sakthivel, G. Nagarajan, M. Ilangkumaran, A. B. Gaikwad, Fuel, 132, 116-124 (2014)
3. A. Sanjid, H. H. Masjuki, M. A. Kalam, S. M. A. Rahman, M. J. Abedin, S. M. Palash, Journal of Cleaner Production, 65, 295-303 (2014)
4. D. N. Mallikappa, R. P. Reddy, Ch. S. N. Murthy, Renewable Energy, 38, 150-154 (2012)
5. S. Vedharaj, R. Vallinayagam, W. M. Yang, S.K. Chou, K. J. E. Chua, P. S. Lee, International Journal of Green Energy, 12, 359-367(2015)
6. C. Jagadish, B. Patil, P. Mohanan, International Journal of Engineering Research and Technology, 1, 9, 1-6 (2012)
7. P. Dinesha, P. Mohanan, Distributed Generation and Alternative Energy Journal, 30, 1, 6-14 (2015)
8. G. Kasiraman, B. Nagalingam, M. Balakrishnan, Energy, 47, 116-124, (2012)

9. M. M. Roy, W. Wang, M. Alawi, Energy Conversion and Management, 84, 164-173 (2014)
10. H. Aydin, H. Bayindir , C. Ilkilic, Energy Sources, Part A: Recovery, Utilization and Environmental Effects, 33, 130-137 (2011)
11. S. Godiganur, Ch. S. Murthy, R. P. Reddy, Renewable Energy, 35, 355-359 (2010)
12. J. Devaraj, Y. Robinson, P. Ganapathi, Energy, 85, 304-309 (2015)
13. N. K. Natesan, Energy Sources, Part A: Recovery, Utilization and Environmental Effects, 35, 1648-1656 (2013)
14. S. Santhanakrishnan, B. K. M. Ramani, International journal of Ambient Energy, 38, 1, 104-107 (2017)

Development of Powerhouse Using Fresnel lens

Nawar Saif Al-Dohani[1], *Nagaraj* S. Nayak[1*], *Anarghya* A.[2] and *Abhishek* V. N.[3]

[1]*Caledonian College of Engineering, Seeb, Sultanate of Oman*
[2]*National Institute of Technology Karnataka, Surathkal, India*
[3]*Toyota Boshoku Automotive India Pvt. Ltd., Bangalore, India*

Abstract. Solar energy is an alternative source of renewable energy. Sultanate of Oman government showed initiation on utilization of solar energy for domestic and industrial applications. Fresnel lens is one of the methods to collect maximum energy by gathering heat of the sun in the concentrated form (using solar collectors). Earlier research work discloses that Fresnel lens gave better result in terms of power output and produces lower heat loss as compared to linear –parabolic solar collectors. In this work, development of a proto Fresnel lens power house was made to generate electricity. The focused heat from Fresnel lens was used to heat the molten salt in a heat exchanger to produce the steam. The generated steam was used to rotate the steam engine coupled to a generator. In the current work, a maximum power of 30 W was produced. In addition, comparative study was carried out regarding solar salts and heat exchanger materials to understand the Fresnel powerhouse performance. Overall the present study gave valuable information regarding usage of Fresnel lens for electricity generation in Oman.

Keywords: Solar energy, Fresnel lens, solar salt, steam engine, electricity

1 Introduction

Renewable energy is a type of energy which can be acquired from natural resources that can be constantly recharged. Sultanate of Oman geographical location helps to develop the hot climatic conditions throughout the year and every city in Oman can make it an ideal location for exploitation of solar energy to produce electricity or used for water heating systems. However, recent study demonstrated Oman is not using solar energy effectively. In this context, Oman government has initiated on effective utilization of solar energy in residential and industrial areas. Research work is underway to collect the solar energy in effective manner. Normally, solar collector is used to collect the solar heat and there are different types of collectors, namely power tower, dish sterling, parabolic trough and Fresnel reflector. These approaches were executed in different parts of the world and generated energy in terms of MW. Facao and Oliveira (2010) studied the replica of a linier Fresnel solar collector and work demonstrated the optical and thermal performances of a new trapezoidal cavity for a small linear Fresnel receiver. In their work, the Ray –tracing simulation and computational fluid dynamics (CFD) approach was used. In addition,

* Corresponding author: nagarajsnayak@gmail.com

Mahmoud and Mohamed (2011) carried out humidification and dehumidification experiments on water desalination with Fresnel lens. Lin et al (2013) carried out theoretical and experimental analysis on linear Fresnel lens reflector with V-shaped cavity receiver. The study showed linear Fresnel lens reflector technology is viewed as a most promising approach to focus the solar power. Zhao et al (2014) studied the thermal performance of the Fresnel lens using different receivers and the study demonstrated that Fresnel lens solar collector could be extensively used in many fields for energy generation. Kumar and Shunmugalatha (2014) developed a proto model of Fresnel power house with solar salt and their study gave good information on implementation of Fresnel power plant in India. In addition, Meraj Rajaee and Seyed Mahamad (2015) proposed a unique photovoltaic solar cell system that consists of semi-Fresnel lens convergent structure and a novel two axis sun tracking module to enhance the efficiency of solar cell by using less cell area and energy losses. Udawant R R et al (2016) studied the performance of Fresnel lens solar concentrator with continuous fluid flow and steam flow test methods. The study revealed that Fresnel lens Solar Concentrator system has a very good potential in generation of low pressure steam.

Thus, from the literature reviews it is confirmed that Fresnel lens collectors are important to extract the sun energy. Accordingly, in this work a prototype Fresnel lens powerhouse was developed and its performance with regard to solar salts and heat exchanger materials was studied. The outcome of the present work could be useful to implement large scale Fresnel power house in Sultanate of Oman to overcome the energy crisis.

2 Fresnel lens powerhouse

2.1 Linear Fresnel lens

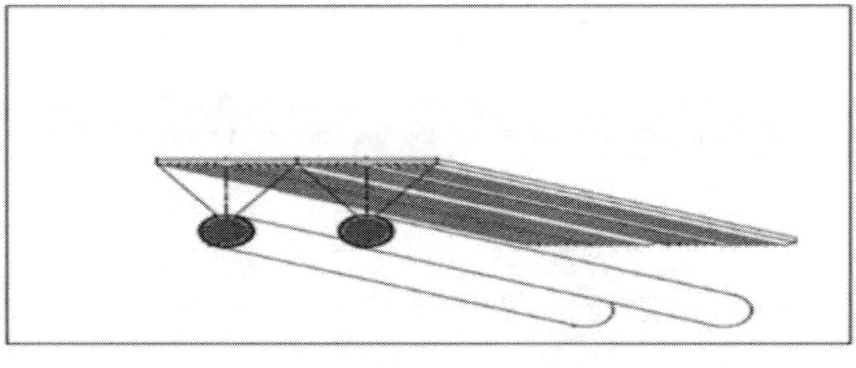

Fig. 1. Solar rays focused on tube from Linear Fresnel lens (Kumar, 2014)

Linear Fresnel lens solar collector is a non-imaging lens and is designed to forward all the solar rays on absorber. Fresnel lens is typically designed with a very short focal length, which is considered as simplified form of conventional lens. Mass and volume of material required for fabricating a Fresnel lens will be less than that the conventional lens. Hence, the cost of Fresnel lens is generally less than a conventional lens.

2.2 Power of focal length and spot intensity of solar radiation

In general, the focal length, f, of the lens is defined as distance between the lens and focal point on the object. The focal length is calculated from

$$f = \frac{s_1}{1-\frac{1}{m}} \dots\dots (1)$$

Where s1and m are the distance between the Fresnel lens and receiver, and magnification factor of the lens respectively.

The power of focal length determines the effectiveness of lens and is calculated on the basis of focal length and diameter of lens. i.e., power of focal length, F is

$$F = \frac{f}{D_{lens}} \dots\dots (2)$$

Where, D lens is the diameter of the Fresnel lens.

In addition, heat received from the Fresnel lens is calculated from solar spot intensity and area of Fresnel lens.

$$Q = A.I_{spot} \dots\dots (3)$$

The solar spot intensity is directly related intensity of solar radiation and power of focal length. Equation 4 is used to calculate the solar spot intensity.

$$F^2.I_{spot} = \left(\frac{L}{R_{sun}}\right)^2.I_0 \dots\dots (4)$$

Where F is power of the focal length in Watt, Ispot is the spot intensity in W/m2, L is the length between sun to earth in m, I0 is the solar intensity in W/m2 and Rsun is the sun radius, m.

2.3 Solar Salt Mixture

Solar thermal power plants are the key technology for generating electricity. In this context, thermal energy storage (TES) materials will able to meet the intermediate power capacity. Hence, molten salts are attractive storage materials because they have advantages in terms of high heat capacity, high density, high thermal stability, relatively low cost and low vapour pressure. One major problem with molten salts is it freezes during the operation. The freezing of molten salt is prevented from heat exchanger or storage tanks approach via auxiliary heating systems. Nowadays, salt mixtures with low melting temperature is developed. However, at high temperatures, the salt stabilities and corrosion aspects play a major role.

Nowadays, non-eutectic salt mixture of 60% sodium nitrate and 40% potassium nitrate is typically used in thermal power plants and this mixture is commonly called Solar Salt.

2.4 Heat exchanger

In the present work, heat exchanger was developed to overcome the freezing of eutectic salt mixture, figure 2. The concept of heat exchange such that tube side water is flowing and shell side is filled with solar salt. The solar heat is transferred to the heat exchanger from the focal point of Fresnel lens and it melts the molten salt and transfers heat to the water. Thus, steam is generated and is used to rotate the steam engine.

The essential quantity of mixer salt required to generate a steam from the heat exchanger is calculated from

$$Q = m_o \times hfg, \dots\dots (5)$$

Where Q is amount of heat received from Fresnel lens, m is mass of salt and hfg is the latent heat of fusion for solar salt.

With the help of energy balance, the required quantity of heat needed to generate the steam is evaluated i.e., the heat taken by water is equal to the heat lost by salt. Similarly, the size of heat exchanger is determined from equation 6.

$$Q = U\,A\,\Delta T \dots\dots (6)$$

Where U and ΔT are the overall heat transfer coefficient in W/m2K, and the change of temperature across the heat exchanger respectively.

In addition, in the current work external auxiliary heating unit is coupled to heat exchanger to avoid the freezing of solar salt.

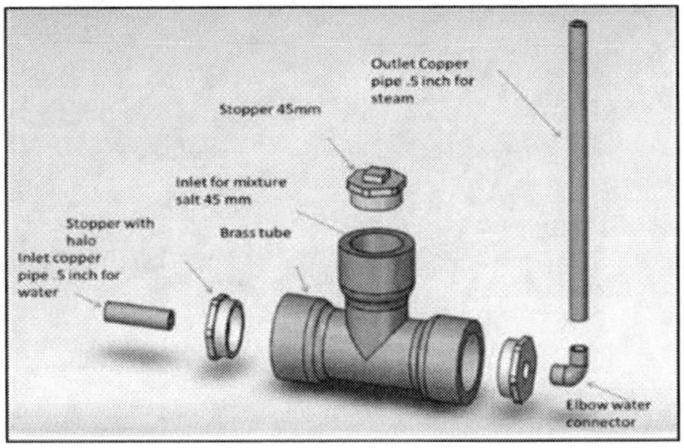

Fig. 2. Parts of eutectic salt heat exchanger

3 Experimental methodology

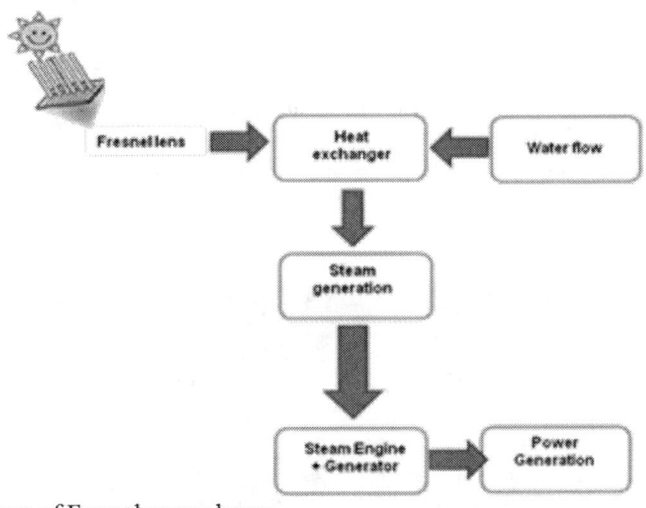

Fig. 3. Block diagram of Fresnel power house

Fig. 3 represent the block diagram of Fresnel powerhouse. Fresnel lens focused the solar rays on the heat exchanger and transfers the heat to the solar salt. Solar salt stores the thermal energy and releases this energy to covert water to steam. The generated steam passes through steam engine coupled with generator. The generated DC current is used to glow the bulb. Fig. 4 explains the experimental set-up of proto model of developed Fresnel powerhouse.

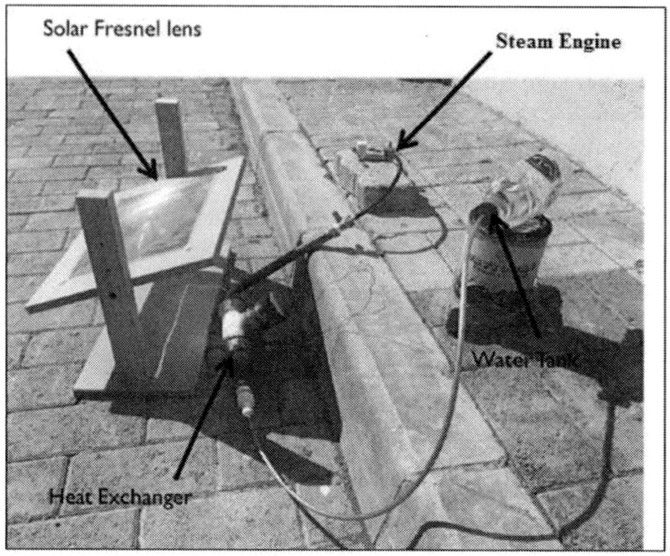

Fig. 4. Experimental set up of proto Fresnel lens powerhouse

4 Results and discussion

4.1 Solar intensity

In the beginning, calculation of solar intensity is made to find the intensity of the solar radiation in Oman. Theoretical calculations showed an average solar intensity of 650-750 W/m2in Oman in the month of November. To verify the theoretical results, solar intensity data were recorded from November 22-25, 2016. The experimental results indicated a maximum solar intensity of 725 W/m2 at 12-1PM, see Fig. 5.

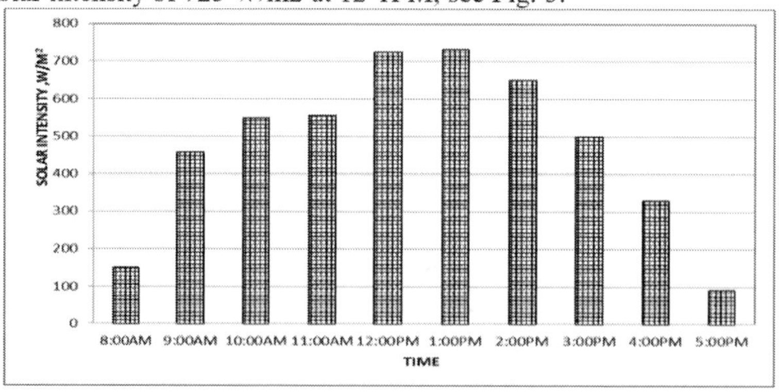

Fig. 5. Solar intensity Dated 22nd November 2016

4.2 Performance study with KNO3 and Mixtures of KNO3 + NaNO3 salt

At the beginning, the study has been carried out with the use of KNO3 salt and later with the mixtures of KNO3+NaNO3 salt in the heat exchanger. For both the cases, power developed by the steam engine is calculated from voltage and current relation. Figure 6 indicates the power output of the generator using KNO3 salt and mixtures solar KNO3 +NaNO3 salt. Similarly, Fig. 7 explains the steam temperature variation with time for

KNO3 salt and mixtures of KNO3 +NaNO3 salt. From the result it confirms that mixture salt gives best performance of output power developed 0.8 W. This value is very small due to the low solar intensity and less flow of water inside the pipe and the quantity of steam generated is less. From the study it is confirmed that mixing 60% of potassium nitrate and 40% of sodium nitrate gave high temperature for the steam as compared to single salt (KNO3). The technical reason is that freezing point of the mixture salt is below 140oC, and it helps to increase the thermal storage capacity of the salt for longer duration. In addition, it improves the heat transfer and minimizes the salt volume. Hence, in the later part of experiments mixture salt of KNO3 +NaNO3 is used.

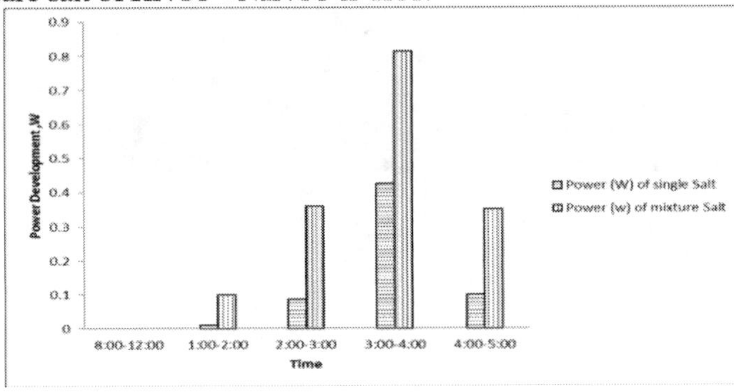

Fig. 6. Comparison of power development with KNO3 salt and mixture salt (KNO_3+$NaNO_3$)

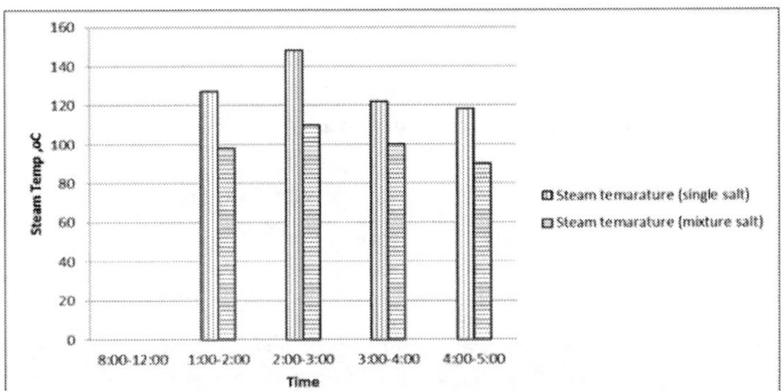

Fig. 7. Comparison of power development with KNO3 salt and mixture salt (KNO3+NaNO3)

4.3 Significance of engine speed on power development

The significance of engine speed on the power development is discussed in Fig. 8. The speed of the turbine at the beginning is 60 rpm and then increased to 160 rpm. The observed maximum power across generator is 0.8 W. From the result, it is confirmed that output power is very low to switch ON the bulb due to low solar intensity on the test day and it led to low steam generation to drive the turbine.

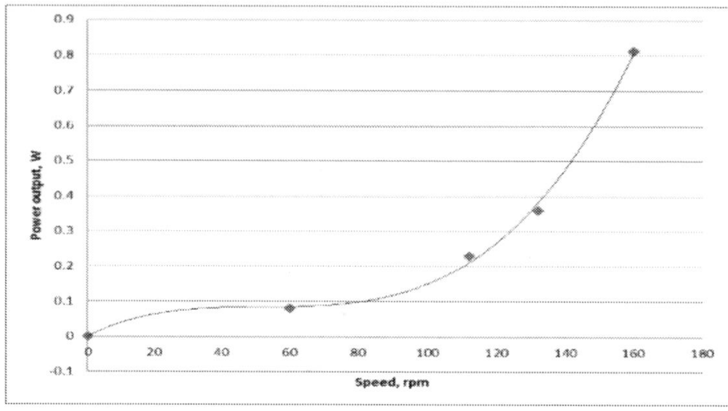

Fig. 8. Power variation with turbine speed -Solar mixture salt (KNO3+NaNO3)

To improve the steam generation quantity, heat exchanger design is modified and is discussed in the next section.

4.4 Study with heat exchanger

In the current work, shell and tube side dimensions are modified to improve the performance of heat exchanger. Also, the performance study has been carried out with brass and cast iron materials at the shell side. Due to low solar intensity, induction heater is coupled to the heat exchanger to melt the solar thermal salt. Fig. 9 indicates the power output comparison of the Fresnel powerhouse for two types of heat exchangers (shell made with cast iron and brass materials). Similarly, variation of steam temperature with cast iron and brass heat exchangers is showed in Fig. 10. From the results, it is confirmed that brass material gave best performance of 30 W output power and the steam outlet temperature of 160oC as against the cast iron heat exchanger developed 134o C with the same input conditions. So, the above study results revealed that corrosion resistant good quality brass material would be an ideal choice at shell side of the heat exchanger and corrosion resistant copper material at the tube side of the heat exchanger to enhance the performance of Fresnel lens powerhouse.

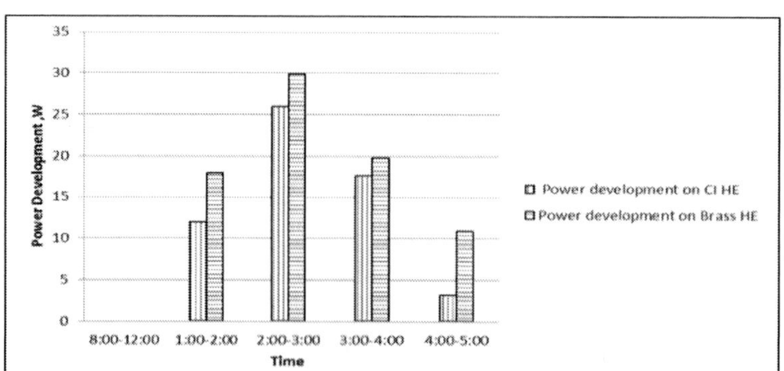

Fig. 9. Power variations with Time (Cast iron and Brass as a material for the heat exchanger)

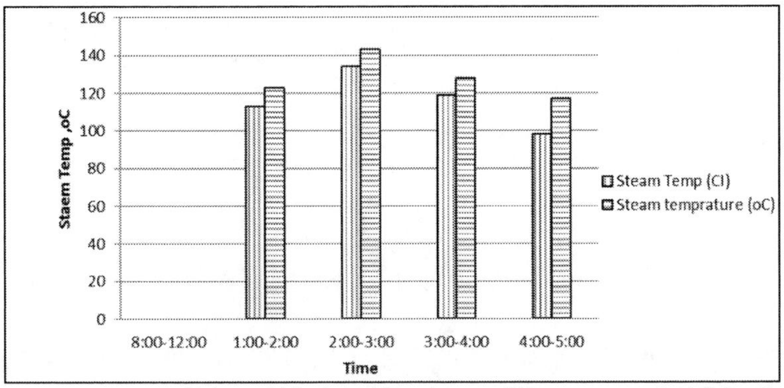

Fig. 10. Steam temperature variations with time (Cast iron and Brass as a material for the heat exchanger)

In addition, complete assembly of proto Fresnel powerhouse is studied under high intensity solar radiation. Figure 11highlights the significance of turbine speed on the developed power. A maximum power output value of 30 W is achieved in the afternoon session and the corresponding turbine speed is 1832 rpm. From the graph, it is confirmed that the trend of the curve is in line with fan law relationship of power (varies polynomial with an exponent of three for rotational speed). Overall, the present study gave valuable thoughts on usage of solar thermal salt in Fresnel powerhouse development. It minimized the size of heat exchanger with improved heat transfer and higher thermal energy storage capacity.

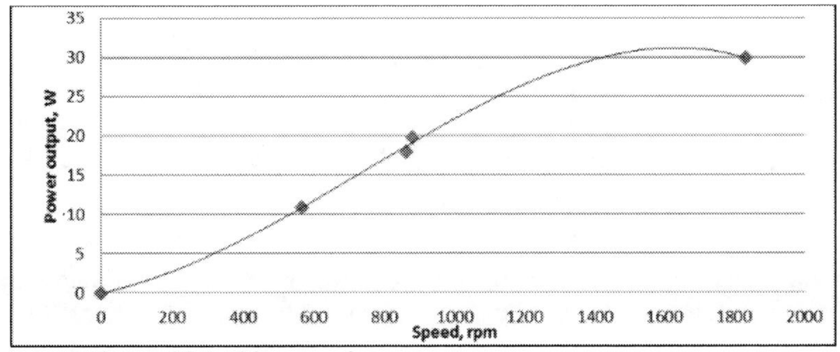

Fig. 11. Power variations with turbine speed

5 Conclusions

In the present work, detailed study was carried out for the development of proto Fresnel lens power house. Overall this work gave valuable information on usage of Fresnel power house in Oman to overcome the energy crisis. The following are the outcomes of the study:

- The proto model Fresnel lens powerhouse gave the maximum power output of 30W at 700 W/m2 solar intensity and it is reasonable in comparison to the size of Fresnel lens and quantity of steam generated.
- Solar thermal salt would be the better media for steam generation in the heat exchanger. In the present work, a mixture of KNO_3 and $NaNO_3$ solar salts gave better results compared to the use of the single salt, KNO_3.
- Study on heat exchanger indicated that corrosion resistant brass material at the shell side and copper material at the tube side would be a better choice to generate more power and high steam temperature

- Increase in turbine speed gave more power and it is in line with Fan power law relationship.

References

1. A. Alvarez, Finite element modelling of a solar collector, Renewable Energies and Power Quality, 6, pp.4-15 (2010).
2. J. Incropera & D. DeWitt, Fundamentals of Heat and Mass Transfer, 2nd edition New York: John Wiley and Sons (1990).
3. F. Jorge & O. Armando, Numerical simulation of a trapezoidal cavity, Journal of Renerwable energy , 36, pp.90-96 (2011).
4. P. Kumar, Development of Fresnel powerhouse prototype with thermal energy storage concentrated solar power, International Journal For Technological Research In Engineering .5 (1), pp. 2347-4718 (2014)
5. H. Petcu & A. Badea, Theoretical and experimental assessment of stationary and dynamic running of one flat plate solar collector, International Journal of Low-Carbon Technologies, 72(2), pp. 48-55 (2010).
6. A. Rebel, Active Solar Collectors and Their Applications. [e-book] Oxford University Press, (1985).
7. I. Soriga & C. Neaga, Thermal analysis of a linear Fresnel lens solar collector with black body cavity receiver, International Journal of Low-Carbon Technologies, 74 (4) pp. 106-116 (2012).
8. B. Vick, Thermal solar track, American Solar Energy Society Conferences, San Francisco, California, Sunday 6th to Thursday 10th July2014, California:PubM.pp.2-7 (2014).
9. J.A. Duffie, &W.A.B Eckman, Solar engineering of thermal processes. 2nd edition New York: John Wiley and Sons (1980).
10. M. Rajaee, & Syed M B G., Experimental measurements of a prototype high-concentration Fresnel lens and sun-tracking method for photovoltaic panel's efficiency enhancement, Journal of Theoretical and Applied Physics, 9(4) pp. 251-259 (2015).
11. R.R Udawant, K.C Mohite & M.G Takwale, Study on performance of Fresnel lens solar concentrator, International Journal of Energy Engineering, 6(1) pp. 14-22 (2016).

Performance computation of window air conditioner with very low GWP near azeotropic refrigerant mixtures as a drop in Substitutes to R22

Shaik Sharmas Vali[1*], *Talanki Puttaranga Setty*[1] *Ashok Babu*[1]

[1]*National Institute of Technology Karnataka Surathkal, Mechanical Engineering Department, Mangalore-575025, INDIA.*

Abstract. The principal objective of the present study is to compute the thermodynamic performance of window air conditioner based on standard vapour compression refrigeration cycle using R22, R407C and nineteen refrigerant mixtures. In this work nineteen R290/R1270 blends at different compositions are developed. A MATLAB code is developed to compute the thermodynamic performance parameters of all the studied refrigerants at condensing and evaporating temperatures of 54.4^{0}C and 7.2^{0}C respectively. The performance parameters are cooling effect, compressor work, COP, compressor discharge temperature, power per ton of refrigeration and volumetric cooling capacity respectively. Analytical results revealed that COP of new binary mixture R290/R1270 (90/10 by mass %) is 2.82% higher among R22, R407C and nineteen studied refrigerants. Energy required by the compressor per ton of refrigeration for R290/R1270 (90/10 by mass %) is 2.73% lower among R22, R407C and nineteen studied fluids. The discharge temperature of the compressor for all the nineteen investigated blends are reduced by 6.0-8.9oC compared to R22. Overall thermodynamic performance of window air conditioner with R290/R1270 (90/10 by mass %) is better than R22 with significant savings in energy consumption and hence it is an energy efficient ecofriendly refrigerant mixture as a drop in substitute to R22.

Keywords: COP; Discharge temperature; GWP; Ecofriendly: Energy savings; R290/R1270 blends

1 Introduction

Hydrochlorofluorocarbon (HCFC) refrigerant R22 was extensivenly used in air conditioning as well as heat pump industries from past several years. Since R22 has better thermophysical and thermodynamic properties. But R22 contains ozone layer desctruction substance called chlorine. Therefore R22 was eventually phase out in all the developed nations by the year 2030 as per Monetreal protocal [1-3]. Till now no pure refrigerant is avialble to replace R22. Therefore it is essential to develop new ecofriendly refrigerant blends. Earlier several theoretical and expermintal investigations were carried to find an appropriate fluid to replace R22. From the past several years refrigerants R407C and R410A were the foremost substitues to R22 [4]. Theorotical results reported that R444B was a viable refrigerant to replace R22 [5]. Experimental results reported that R407C was a suitable retrofit candidate to R22 [6]. Experimental test was conducted in an air conditioner with R290 as a drop in refrigerant. Test results exhibited that performance of R290 was 2.8-

* Corresponding author: sharmasvali.nitk@gmail.com

7.9% higher than R22 [7]. Experimental tests revealed that the ternary refrigerant mixture R32/R125/R161 (15/34/51 by weight percentage) was suitable choice to replace R407C. Since COP of above mixture was better than R407C [8]. Experimental studies reported that performance of binary mixture R170/R290 (4/96 by mass %) was similar to that of performance of R22 [9]. A detailed review paper suggested that hydrocarbons and its blends were most suitable candidates to air conditioning and refrigeration sectors [10]. Experimental studies were conducted in a heat pump apparatus with R22 and binary blend R431A (R290/R152a 71/29 by mass %) used for heat pump and air conditioning applications. Test results exhibited that COP of R431A was 3.5 – 3.8 % superior than R22 [11]. Performance tests were done in a heat pump equipment with R22 and binary blend R432A (R1270/RE170 80/20 by mass %) used for heat pump and air conditioning applications. Experimental results showed that performance of R432A was 8.5-8.7% superior than the R22 [12]. Experimental investigations were carried out in a heat pump device with R22 and binary blend R433A (R1270/R290 30/70 by mass %) used for heat pump and air conditioning applications. Test results revealed that performance of R433A was 4.9-7.6% superior than R22 [13]. The present study mainly focuses on finding an alternative refrigerant to replace R22. For this thermodynamic performance computation of nineteen refrigerant blends (R290/R1270) is carried out based on standard vapour compression refrigeration system.

2 Properties of refrigerants

In the present study apart from R22 and R407C, total nineteen refrigerant blends (R290/R1270) of various compositions are developed. All the nineteen binary refrigerant blends R290/R1270 (Refrigerants 2 to 20 as given in table 2) are closer to azeotropic in nature. Since their temperature glide is less than 0.5oC. A MATLAB code is developed to compute the thermodynamic properties of refrigerants like R22, R407C and R290/R1270 blends by using Martin-Hou equation of state [14]. Since thermodynamic properties of considered refrigerants are useful for the computation of thermodynamic performance of an air conditioner.

$$P = \frac{RT}{v-b} + \frac{a_2 + b_2T + c_2 e^{\frac{-5.475T}{T_C}}}{(v-b)^2} + \frac{a_3 + b_3T + c_3 e^{\frac{-5.475T}{T_C}}}{(v-b)^3} + \frac{a_4}{(v-b)^4} + \frac{b_5T}{(v-b)^5} \qquad (1)$$

The computed properties of R22 and R407C shows good agreement with the experimental reported properties available in ASHRAE fundamental data hand book. The deviation of computed properties of R22 and R407C from that of ASHRAE is within 2 to 4% for the given operating conditions. Therefore the developed MATLAB code is reliable and hence it can be used for the development of thermodynamic properties of other new refrigerant blends. The properties of new refrigerants blends (R290/R1270) are not available in literature and hence they are not compared. However properties of refrigerant R290/R1270 blends are compared with REFPROP. The deviation of properties from that of REFPROP is within 1.5 to 4 % for the given working conditions. Hence the same developed thermodynamic properties are used in the thermodynamic analysis. The code followed for the refrigerants is given in table 1 and the basic physical and critical properties of R22, R407C and nineteen refrigerant blends (R290/R1270) taken from REFPROP is given below in table 2 respectively[15].

Table 1. Code of the refrigerants

Refrigerant code	Composition (by mass %)
1 (R22)	Pure fluid
2 (R290/R1270)	5/95
3 (R290/R1270)	10/90
4 (R290/R1270)	15/85
5 (R290/R1270)	20/80
6 (R290/R1270)	25/75
7 (R290/R1270)	30/70
8 (R290/R1270)	35/65
9 (R290/R1270)	40/60
10 (R290/R1270)	45/55
11 (R290/R1270)	50/50
12 (R290/R1270)	55/45
13 (R290/R1270)	60/40
14 (R290/R1270)	65/35
15 (R290/R1270)	70/30
16 (R290/R1270)	75/25
17 (R290/R1270)	80/20
18 (R290/R1270)	85/15
19 (R290/R1270)	90/10
20 (R290/R1270)	95/5
21 (R407C R32/R125/R134a)	23/25/52

Table 2. Basic physical and environmental properties of investigated refrigerants

Refrigerant code	M (kg/kmol)	T_b (°C)	P_c (Mpa)	T_c (K)	ODP	GWP (100 years)	T_g (°C)
1	86.5	-40.81	4.99	369.3	0.055	1760	0
2	42.176	-47.45	4.538	364.3	0	<3	0.03
3	42.273	-47.27	4.521	364.4	0	<3	0.05
4	42.37	-47.07	4.505	364.5	0	<3	0.09
5	42.468	-46.85	4.490	364.7	0	<3	0.13
6	42.566	-46.62	4.475	364.9	0	<3	0.16
7	42.665	-46.37	4.460	365.1	0	<3	0.21
8	42.764	-46.11	4.445	365.3	0	<3	0.25
9	42.864	-45.84	4.431	365.6	0	<3	0.29
10	42.964	-45.56	4.417	365.9	0	<3	0.32
11	43.064	-45.27	4.403	366.2	0	<3	0.35
12	43.165	-44.97	4.388	366.5	0	<3	0.37
13	43.267	-44.67	4.374	366.8	0	<3	0.38
14	43.368	-44.36	4.360	367.2	0	<3	0.39
15	43.471	-44.04	4.345	367.5	0	<3	0.38
16	43.574	-43.73	4.330	367.9	0	<3	0.35
17	43.677	-43.41	4.315	368.3	0	<3	0.31
18	43.781	-43.08	4.299	368.7	0	<3	0.27
19	43.885	-42.76	4.284	369.1	0	<3	0.19
20	43.99	-42.44	4.267	369.4	0	<3	0.10
21	86.204	-43.63	4.631	359.3	0	1774	7.0

3 Thermodynamic performance methodology

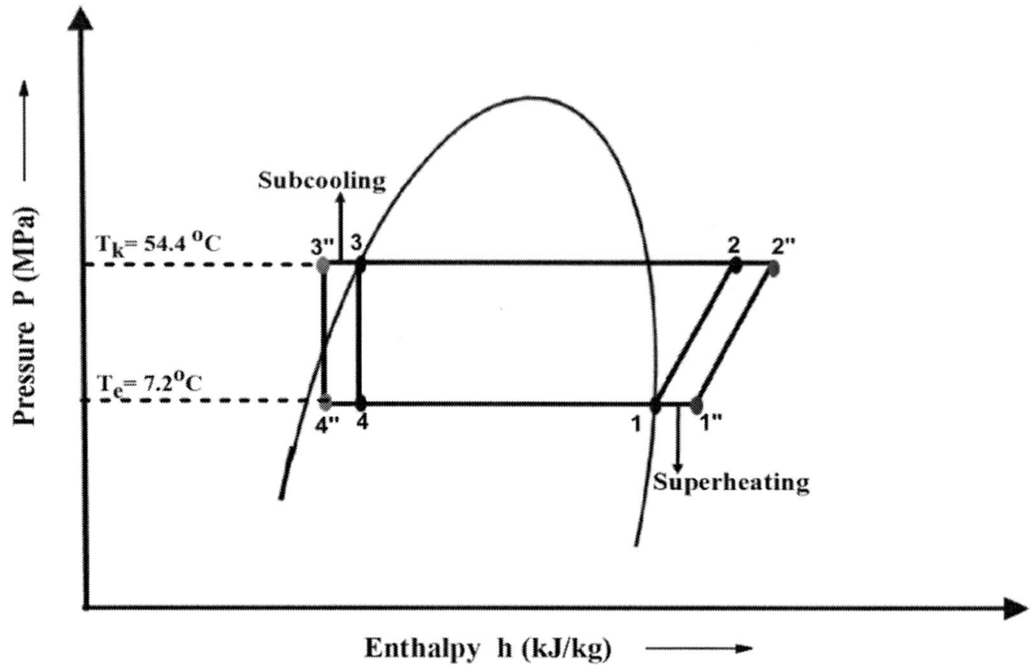

Fig 1. P-h diagram of standard vapour compression refrigeration cycle.

Computational thermodynamic analysis of R22, R407C and nineteen refrigerant blends (R290/R1270) is done based on standard vapour compression refrigeration cycle. The p-h diagram of corresponding cycle is shown in figure 1 [16-17]. In this methodology the various pressure losses and heat losses or heat gains occurred at the various components of the system are ignored. The degree of superheating and subcooling can be taken as 5°C respectively. In this study a MATLAB code is developed to compute the thermodynamic performance parameters of refrigerants at ARI conditions (T_k=54.4°C and T_e=7.2°C) respectively. The various performance parameters are refrigeration effect, compressor work, coefficient of performance, compressor discharge temperature, power per ton of refrigeration and volumetric cooling capacity.

The capacity and operating conditions of air conditioner considered in the present study for R22, R407C and nineteen binary refrigerant blends are Q=1.5 TR, T_k=54.4°C and T_e=7.2°C respectively. The summary of results for the considered refrigerants are given in table 3.

4. Mathematical computations

The mathematical computations involved in the thermodynamic analysis of for the considered refrigerants are given below.

Refrigerant mass flow rate is computed by

$$\dot{m} = Q/RE \qquad (2)$$

Isentropic compressor work is computed as

$$W_c = h_{2''} - h_{1''} \quad (3)$$

Refrigerating effect (Cooling effect) is calculated as

$$RE = h_{1''} - h_{4''} \quad (4)$$

Coefficient of performance (COP) is calculated as

$$COP = RE/W_c \quad (5)$$

Energy (Power) consumption by the compressor per ton of refrigeration is computed by

$$PPTR = 3.5167/COP \quad (6)$$

Volumetric cooling capacity is calculated as

$$VCC = \rho_{1''} \times RE \quad (7)$$

The compressor discharge temperature (T_d) is computed using superheated property tables and interpolating for the degree of superheat corresponding to the entropy difference.

Table 3. Summary of results for the various considered refrigerant blends

Refrigerant code	$\dot{m}$ (kg/min)	RE (kJ/kg)	W_c (kJ/kg)	COP	T_d (°C)	PPTR (kW/TR)	VCC (kJ/m³)
1	2.261	139.944	34.958	4.003	78.09	0.878	3646
2	1.309	241.723	64.187	3.765	69.34	0.933	3560
3	1.297	243.930	64.454	3.784	69.26	0.929	3553
4	1.286	246.007	67.343	3.653	71.87	0.962	3543
5	1.276	247.954	64.951	3.817	69.17	0.921	3524
6	1.266	249.833	65.139	3.835	69.15	0.916	3508
7	1.257	251.599	67.751	3.713	71.59	0.946	3494
8	1.249	253.257	67.845	3.732	71.54	0.942	3479
9	1.241	254.858	67.913	3.752	71.51	0.937	3464
10	1.234	256.374	67.941	3.773	71.49	0.931	3451
11	1.227	257.792	67.952	3.793	71.5	0.926	3437
12	1.221	259.152	67.891	3.817	71.5	0.921	3426
13	1.215	260.418	65.583	3.970	69.3	0.885	3414
14	1.209	261.590	67.731	3.862	71.55	0.910	3405
15	1.204	262.718	65.422	4.015	69.41	0.875	3394
16	1.199	263.755	65.265	4.041	69.47	0.870	3385
17	1.195	264.704	67.297	3.933	71.74	0.894	3376
18	1.191	265.612	67.086	3.959	71.82	0.888	3367
19	1.187	266.433	64.718	4.116	69.74	0.854	3356
20	1.184	267.162	66.712	4.004	72.07	0.878	3338
21	2.296	137.816	42.645	3.231	78.07	1.088	3172

5 Results and discussions

The results and discussions of various performance parameters of R22, R407C and nineteen refrigerant blends (R290/R1270) are given below.

5.1 Refrigeration effect

Figure 2 shows the cooling effect (refrigeration effect) of various refrigerants blends. Referring to figure 2 it is noticed that refrigeration effect of nineteen refrigerant blends R290/R1270 (2 to 20) are higher than the R22 and R407C. This is due to high latent heat of vaporization of refrigerants when compared to R22 and R407C.

5.2 Compressor work

Figure 3 shows the work input required by the compressor for the various refrigerant blends. From the figure 3 it is observed that the compressor work of all the nineteen investigated refrigerant mixtures (2 to 20) are higher than the R22 and R407C. This is due to high vapour enthalpy of the refrigerant blends compared to R22 and R407C.

5.3 Coefficient of performance

Figure 4 shows the coefficient of performance of various studied refrigerants. COP is measured as an index of energy efficiency of the device charged with specific refrigerant. COP is the ratio of refrigeration effect to the compressor work. From the table 3 it is noticed that both cooling (refrigeration effect) effect and compressor work increase for all the nineteen investigated refrigerant blends. Therefore combined effect of cooling effect and compressor work of these refrigerants on coefficient of performance may be either increases or decreases or remains same which depends on the type and composition of the refrigerant blend and also on the working conditions of the device. From the figure 4 and table 3 it is clear that the COP of new binary refrigerant blend R290/R1270 (90/10 by mass %) is 2.82 % higher among R22, R407C and nineteen studied refrigerants.

5.4 Compressor discharge temperature

While introducing the new refrigerant blend into the equipment the consistency and lifespan of compressor motor should be studied. This can be achieved by computing the compressor discharge temperature of new refrigerants. Referring to figure 5 it is noticed that compressor discharge temperature of all the nineteen investigated refrigerants (2 to 20) are reduced by 6.0-8.9oC when compared to R22 and R407C. This is due to lower adiabatic index of refrigerant blends. Hence all the nineteen studied fluids exhibits better reliability and lifespan of compressor motor.

5.5 Power per ton of refrigeration

Figure 6 shows the power (energy) spent by the compressor per ton of refrigeration for various studied refrigerants. It denotes the electrical energy required to the compressor to produce per ton of refrigeration. Referring to figure 6 it is observed that electrical energy spent by the compressor per ton of refrigeration for the refrigerant blend R290/R1270 (90/10 by mass %) is 2.73 % lower among R22, R407C and nineteen investigated refrigerants. This is due to type and composition of refrigerants and also due to operating conditions of the given device.

5.6 Volumetric cooling capacity

Figure 7 shows the volumetric cooling capacity of various refrigerant blends. Volumetric capacity depends on density of vapour refrigerant and cooling effect of the given fluid. It indicates the cooling capacity per unit volume of refrigerant vapour at the evaporator outlet. And also it denotes the volume of refrigerant handled by the compressor for the given fluid. From the figure 7 it is evident that volumetric cooling capacity of all the nineteen investigated refrigerants are closer to that of volumetric capacity of R22. Hence same size of compressor can be used for all the nineteen fluids as that of R22 with little modifications.

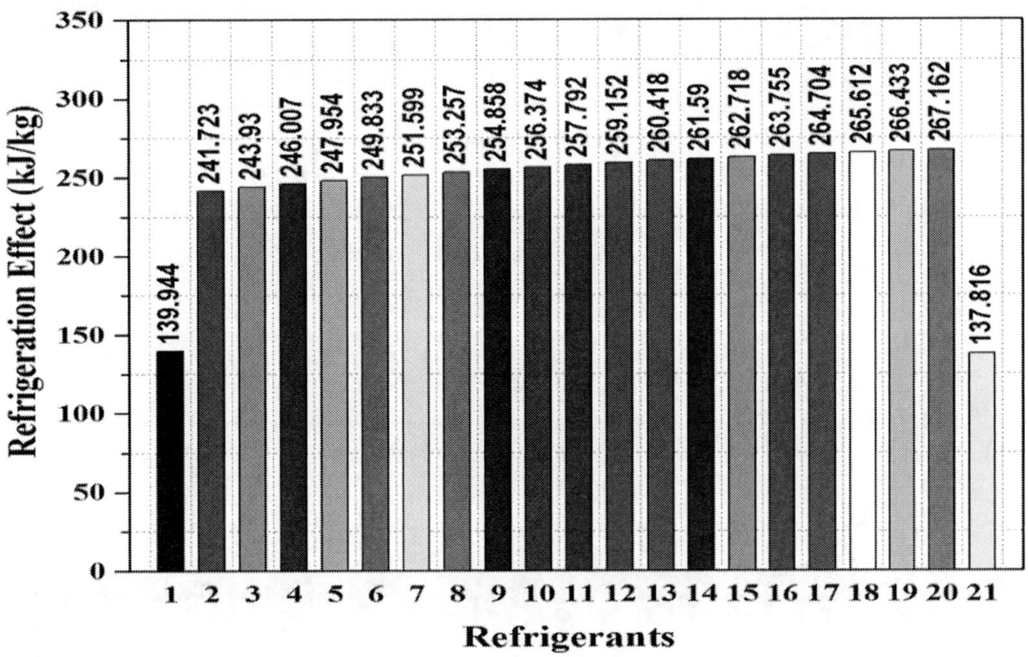

Fig 2. Refrigeration effect of various R22 alternative blends

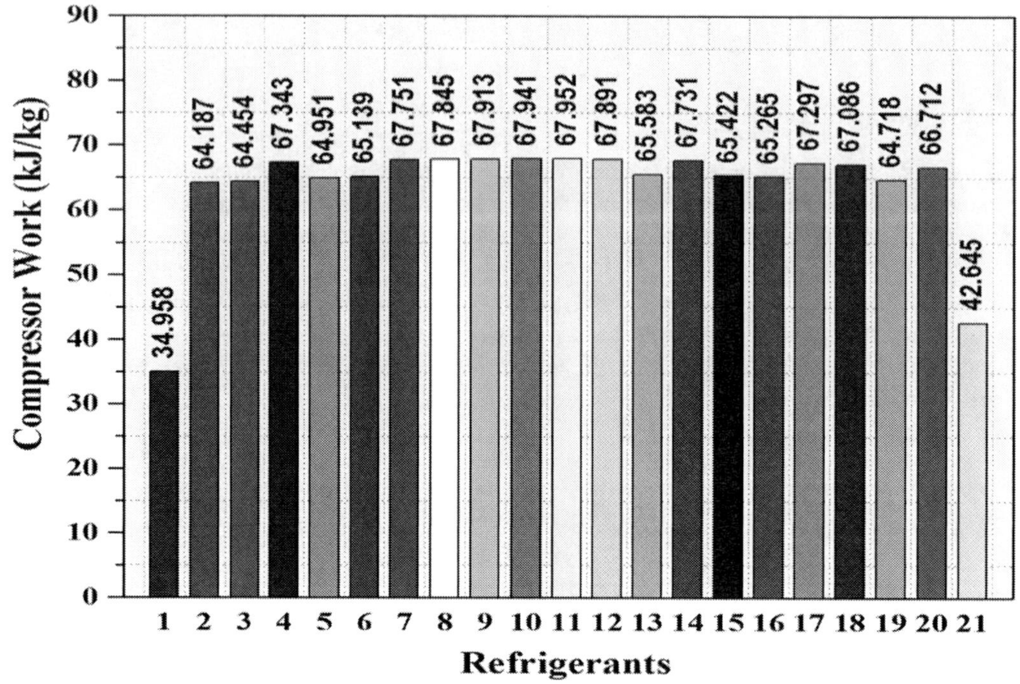

Fig 3. Compressor work of various R22 alternative blends

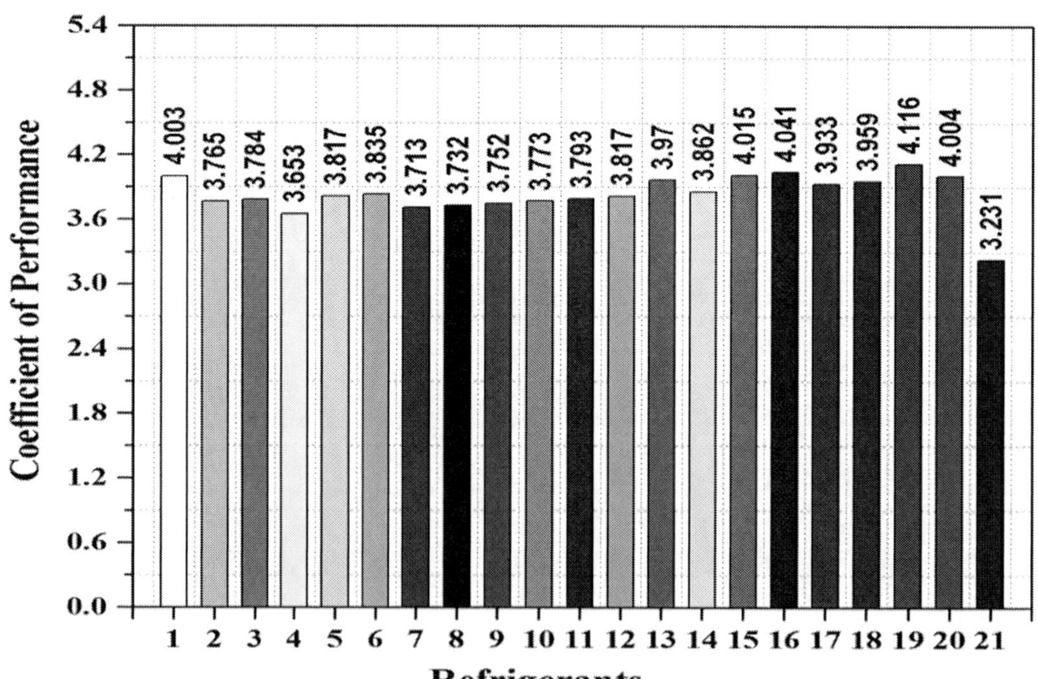

Fig 4. COP of various R22 alternative blends

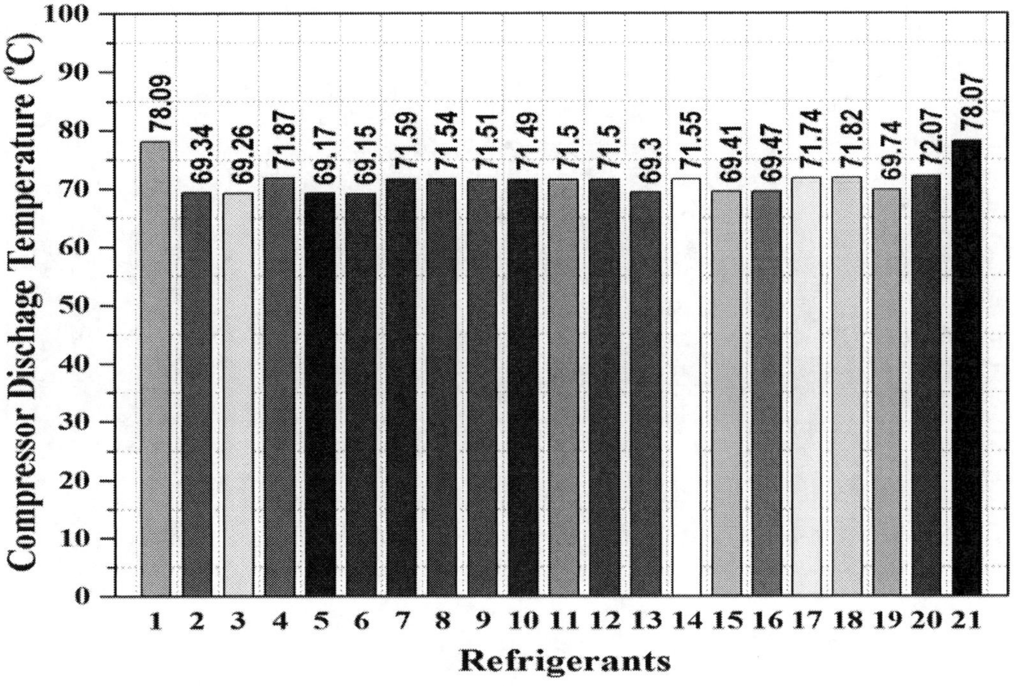

Fig 5. Compressor discharge temperature of various R22 alternative blends

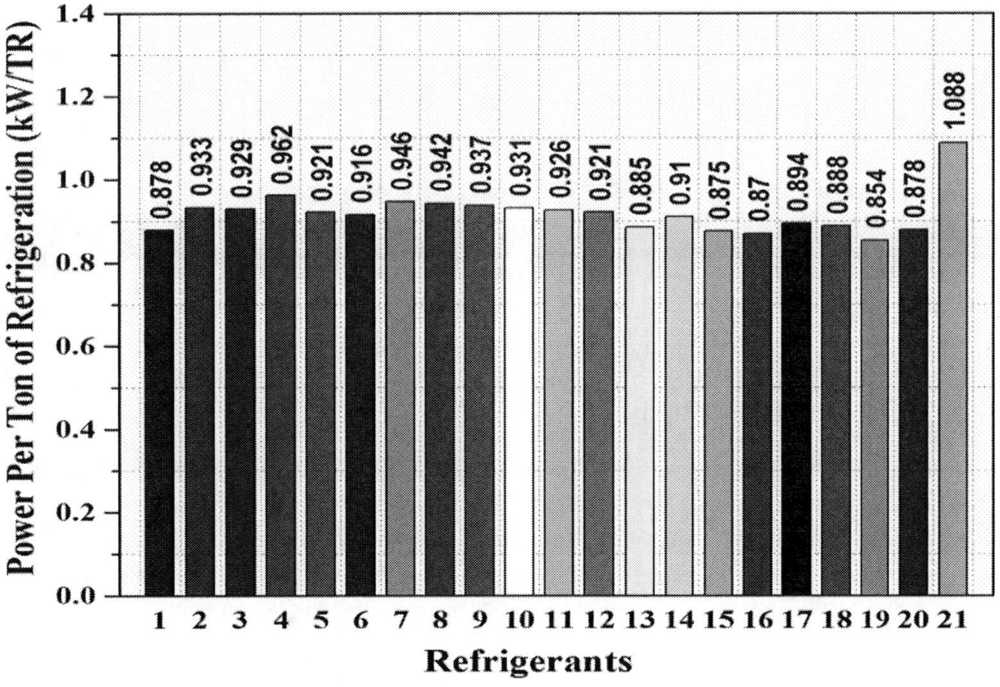

Fig 6. Power per ton of refrigeration of various R22 alternative blends

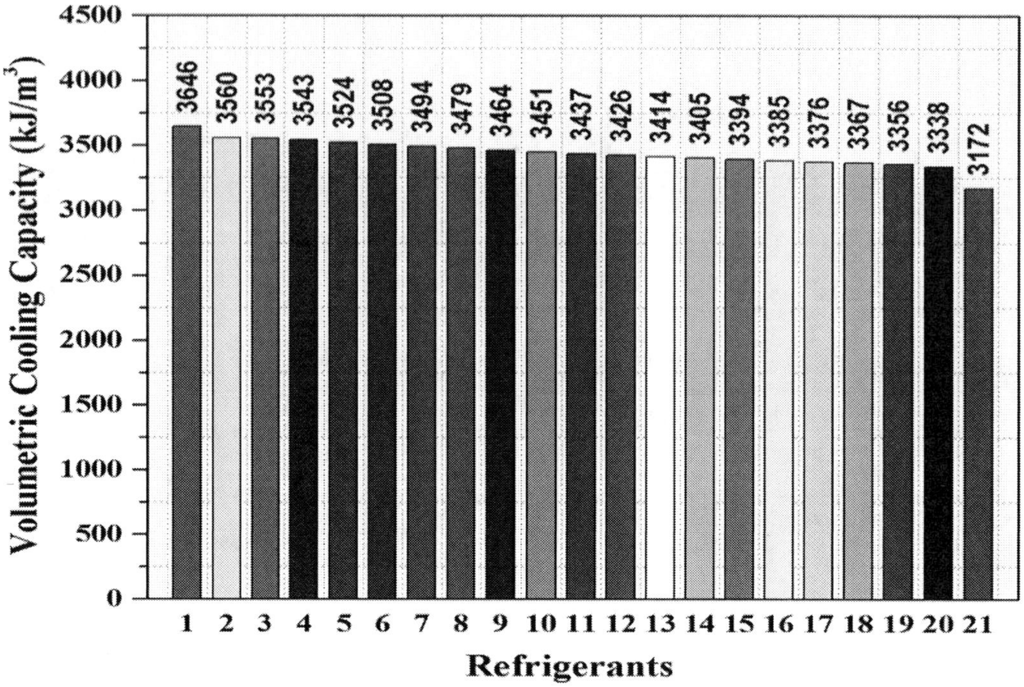

Fig 7. Volumetric cooling capacity of various R22 alternative blends

6 Conclusions

From the performance computation of various refrigerant blends the following conclusions can be drawn.

1. COP of new binary refrigerant blend R290/R1270 (90/10 by mass %) was 2.82 % higher among R22, R407C and nineteen studied refrigerants.
2. Compressor discharge temperature of all the nineteen refrigerant blends were reduced by 6.0-8.9°C compared to R22 and R407C. Therefore all the nineteen studied fluids would be advantageous from the standpoint of lifespan of compressor motor.
3. Energy (Power) spent by the compressor per ton of refrigeration for the refrigerant R290/R1270 (90/10 by mass %) was 2.73% lower among R22, R407C and nineteen studied refrigerants.
4. Volumetric cooling capacity of all the nineteen studied fluids were closer to that of volumetric capacity of R22. Hence same size of compressor can be used for all the nineteen refrigerants as that of R22 with very little modifications.
5. Overall thermodynamic performance of window air conditioner with R290/R1270 (90/10 by mass %) was better than R22 with significant savings in energy consumption and hence it is an energy efficient sustainable refrigerant mixture as a drop in substitute to R22.

Nomenclature

ARI	Air conditioning and refrigeration institute
GWP	Global warming potential
ODP	Ozone depletion potential
RE	Refrigeration effect (kJ/kg)
TR	Ton of refrigeration (kW)
P_c	Critical pressure (MPa)
Q	Refrigeration capacity (kW)
$\mathcal{R}$	Universal gas constant (J/molK)
T_b	Boiling point (°C)
T_c	Critical temperature (K)
T_d	Discharge temperature (°C)

$h_{1'}$	Enthalpy at the compressor entry (kJ/kg)	T_e	Evaporating temperature (°C)
$h_{2'}$	Enthalpy at the compressor exit (kJ/kg)	T_g	Temperature glide (°C)
$h_{4'}$	Enthalpy at the evaporator inlet (kJ/kg)	T_k	Condensing temperature (°C)
$\dot{m}$	Mass flow rate of refrigerant (kg/min)	v	Specific volume (m^3/kg)
M	Molecular weight (kg/kmol)	W_c	Compressor work (kJ/kg)

References

1. United Nations Environmental Programme, Montreal Protocol on substances that deplete the ozone layer, Final act. New York: United Nations; 1987.
2. RL. Powell, J. Fluorine Chem, 114, 237-250 (2002)
3. M. Mohanraj, S. Jayaraj, C. Muraleedharan, Int. J. Greenh Gas Control 3, 108-119 (2009)
4. J.M. Calm, P.A. Domanski, ASHRAE J, 46, 29-39, (2004)
5. Atilla Gencer Devecioğlu, Vedat Oruc, j.egypro, 75, 1452-1457 (2015)
6. S. Devotta, A.V. Waghmare, N.N. Sawant, B.M. Domkundwar, Appl. Therm. Eng 21, 703-715 (2001)
7. S. Devotta, A.S. Padalkar, N.K. Sane, Int. J. Refrigeration, 28, 594-604, 2005.
8. X.H. Han, Q. Wang, Z.W. Zhu, G.M. Chen, Appl. Therm. Eng , 27, 2559-2565, (2007)
9. Ki-Jung Park, Dongsoo Jung, Appl. Energy, 86, 2598-2603, (2009)
10. B.O. Bolaji, Z. Huan, Renew. Sustainable Energy Rev, 18, 49-54, (2013)
11. Ki-Jung Park, Yun-Bo Shim, Dongsoo Jung, Energy Convers. Manage 50, 1671-1675, (2009)
12. Ki-Jung Park, Yun-Bo Shim, Dongsoo Jung, Appl. Therm. Eng, 29, 597-600, (2009)
13. Ki-Jung Park, Yun-Bo Shim, Dongsoo Jung, Appl. Energy, 85, 896-900, (2008)
14. Martin Joseph J, Yu-Chun Hou, AlChE J, 1, 142-151, (1955)
15. EW Lemmon, ML Huber, MO McLinden, "NIST Standard Reference Database 23, Version 9.1 Reference Fluid Thermodynamic and Transport Properties-REFPROP, National Institute of Standards and Technology, Standard Reference Data Program, Gaithersburg, (2013).
16. C P. Arora, Refrigeration and Air conditioning (Tata McGraw-Hill, New Delhi, 2009)
17. R.C. Arora, Refrigeration and Air conditioning (PHI learning Private Limited, New Delhi, 2010)

Effect of Various External Shading Devices on Windows for Minimum Heat Gain and Adequate Day lighting into Buildings of Hot and Dry Climatic Zone in India

Gorantla Kirankumar[1]*, *Shaik Saboor*[2], *Putta Ranga Talanki Setty*[1] and *Ashok Babu*[1]

[1]*Mechanical Engineering Department, National Institute of Technology Karnataka, Surathkal, Mangalore-575025, Karnataka, India*
[2]*Department of Thermal and Energy Engineering, VIT University, Vellore-632014, Tamilnadu, India.*

Abstract. Glass is the major component of the building envelope to provide visual comfort to inside the buildings. In général clear and bronze glass was used as a main building envelope for both residential and commercial buildings to provide better day lighting into the buildings. If we use more glass area as a building envelope more radiation allows into the buildings. So that it is necessary to reduce more solar radiation and provide sufficient daylight factor inside the building's through glass windows with the help of external devices called shading devices. In this work four shading devices was tried on bronze glass window to find the heat gain and daylighting into buildings. This paper presents the experimental measurement of spectral characteristics of bronze glass which include transmission and reflection in entire solar spectrum region (300nm-2500nm) based on ASTM standards. A MATLAB code was developed to compute visible and solar optical properties as per the British standards. A building model was designed by design builder software tool. 40% window to wall ratio was considered for building models, thermal and day lighting analysis of buildings through windows was carried out in Energy plus software tool for hot and dry climatic zone of India.

Keywords: Bronze glass window, spectral characteristics, spectrophotometer, design builder, adequate day light factor and external shading devices

1 Introduction

Now a day's glass is used as main building envelope material to construct the buildings for providing visual comfort from inside to outside the occupants. Especially most of the commercial buildings are used different glasses like clear and bronze as main building envelopes. Due to usage of more quantity of glasses to allow direct solar radiation and high percentage of day lighting so that more heat gain and comfortable conditions will not occur into buildings because due to their visible and solar optical properties. So that we need

* Corresponding author: gorantla.kirankumar@gmail.com

external devices to allow minimum solar radiation and as well as sufficient day lighting into buildings and provide thermal comfort to the occupants, such devices call external shading devices. External devices are classified as mainly four types overhang, louvers with different orientations, egg-crate shading and (egg-crate + louver) shading devices.

Previous authors have worked on different external shading devices on various glass materials to reduce the direct solar radiation into buildings. To study the various shading devices vertical fins, overhangs and egg-crate windows find the best shading device to Jordan University of science and technology administration office buildings in south-west façade orientation to control the direct solar radiation and improve the visual comfort and protect glare by using various simulation software's IES/Sun-Cast and Radiance to analyzed annually and compared with experimental results in detail [1]. By arranging different slant lengths, tilts of fixed horizontal louver shading devices on window glasses and analyzing the thermal performance of these buildings by using TRNSYS simulation of four climatic regions in Italy it was found that 70% decrease of solar radiation in summer and 40% increase in winter seasons has been reported annually [2]. To decrease the direct solar radiation through various louver shading areas arranged on windows to various climatic zones of different countries in European continent by considering horizontal and vertical louver shadow geometry were analyzed by energy equation solver to save the power consumption by using TRNSYS simulation software annually were reported in literature [3]. Detailed study of roller blinds and venetian blinds of shading device models developed and numerically found the solar and visible properties of total fenestration systems, finally suggests that venetian blinds results better than roller blinds when experimented to reduce the solar radiation and improved the day light illuminance levels in both clear and overcast sky conditions of Lisbon latitude climatic region in Portugal country [4]. By using building energy performance simulations software between past and present on shading devices and comparing to various selected solar position on different shading devices resulted an error in present has been described in detail at Algeciras latitude climatic region [5]. To decreases heat losses from the windows in winter season Poznan latitude climatic region of Poland country to analyze the air temperature, wind effect on double glazed windows with an air gap 20mm and also tried with internal and external roller blinds save the power nearly 29% to 44% in Central European continent studied in detail [6]. To find out the interior day light illuminanace levels from the double clear glass window at different points in a room and analyze day lighting performance assessment algorithms radiosity and split-flux methods with DE light and Energy simulation programs comparing and validated experimentally in detail [7].

To find the inside day light illuminance levels of building model with south oriented clear glass window and analyzed on other orientations such as north, east and west comparative study of two day lighting analysis methods with regard to window orientation and interior all reflectance [8]. To study of various double glazing windows like clear, bronze and low E-glasses to keep the windows as building envelopes and find the best suitable window reduce the electric energy consumption Dhahran latitude hot climatic region in Saudi Arabia was done by [9]. Find the total solar energy transmittance of glazing with venetian blind shading of different geometries to keep the laboratory applied physics measure the global solar radiation experimentally and numerically validated to near Zurich latitude climatic region Switzerland country in detail [10]. A numerical and analytical method was developed to keep the egg-crate shading devices on vertical windows of arbitrary orientation to reduce solar radiation Venice and Bologna latitudes climatic regions to Italy country was studied in detail [11].

To determine the simplified procedure i.e. using nomographs and waldram diagrams establish and described to find indoor daylight illuminance using daylight coefficient concept to university of building and construction of university of Hong Kong was reported in detail [12].

To find the day light illuminance levels on horizontal plane of an office building room with a controlled shading device of double glazing window with internal blinds and control glare levels and provide adequate daylight factor inside the buildings from different points from window was reported by [13]. Daylight factor model was developed under clear sky conditions in east oriented window and experimentally validated to New Delhi climatic region and find out the daylight illuminanace levels of windows high and bottom to keep the building walls in south and north oriented windows at different sky conditions was reported in [14]. The main objective of the present work is reduce the most of the solar radiation by keeping various external shading devices on bronze glass windows and also provide adequate daylight factor inside the buildings of hot & dry climatic region (Jodhpur) in India.

2 Experimental procedure to measure the spectral characteristics of bronze glass window

Perkin Elmer Lambda 950 spectrophotometer is an UV-VIS-NIR instrument used to measure the spectral characteristics of glass windows. For measuring the spectral characteristics of glass sample is keeping vertical i.e. normal angle of incidence (90^0 from horizontal) wavelength range from (300nm-2500nm) based on ASTM Standards [15]. To find the visible properties to measure wavelength range from (380nm-780nm) and solar optical properties to measure from wavelength range 300nm to2500nm which is obtained from the measured spectral characteristics data as shown in Figure 1. Solar heat gain coefficient factor is an important factor to find the solar radiation passing through window glass. A MATLAB code was developed to compute the solar optical properties like transmittance, reflectance in entire solar spectrum region (300nm-2500nm) by using equations 1, 2 and 3 based on British standards [16,17].

$$T_{Sol} = \sum_{\lambda=300}^{\lambda=2500} S_\lambda \tau(\lambda)\Delta\lambda \Big/ \sum_{\lambda=300}^{\lambda=2500} S_\lambda \Delta\lambda \qquad (1)$$

$$R_{Sol} = \sum_{\lambda=300}^{\lambda=2500} S_\lambda \rho(\lambda)\Delta\lambda \Big/ \sum_{\lambda=300}^{\lambda=2500} S_\lambda \Delta\lambda \qquad (2)$$

Solar absorbance of the glass window can be found by knowing the solar transmittance and reflectance

$$A_{Sol} = 100 - T_{Sol} - R_{Sol} \qquad (3)$$

To find the day light factor inside the buildings through windows it is necessary to measure the spectral data of glass windows from wavelength range 380nm to780nm at normal angle of incidence. A MATLAB code was developed to compute the visible properties like transmittance, reflectance in the visible spectrum region (380nm-780nm) by using equations 5, 6 and 7 based on British standards. Daylight factor is defined as the ratio of indoor illumination to outdoor illumination available in outside the building. Units of day light factor is in (%) or in (Lux).

$$DF = \frac{Ei}{Eo} \times 100\% \qquad (4)$$

The day lighting strategies not only reduce lighting energy use but they are also responsible for an increase in cooling/heating loads. Table 1 shows the visible, solar optical properties and solar heat gain coefficient of bronze glass window.

$$T_{Vis} = \sum_{\lambda=380}^{\lambda=780} D_\lambda \tau(\lambda) V(\lambda) \Delta\lambda \Big/ \sum_{\lambda=380}^{\lambda=780} D_\lambda V(\lambda) \Delta\lambda \quad (5)$$

$$R_{Vis} = \sum_{\lambda=380}^{\lambda=780} D_\lambda \rho(\lambda) V(\lambda) \Delta\lambda \Big/ \sum_{\lambda=380}^{\lambda=780} D_\lambda V(\lambda) \Delta\lambda \quad (6)$$

Visible absorbance of the glass window can be found by knowing the visible transmittance and reflectance of the glass window.

Table 1. Visible, solar optical properties and solar heat gain coefficient of bronze window glass of thickness 6mm.

	Visible spectrum (380nm-780nm)			Total solar spectrum (300nm-2500nm)			
Window glass	T_{Vis} (%)	R_{Vis} (%)	A_{Vis} (%)	T_{Sol} (%)	R_{Sol} (%)	A_{Sol} (%)	SHGC(%)
Bronze glass window	50	6	44	49	6	45	62

$$A_{Vis} = 100 - T_{Vis} - R_{Vis} \quad (7)$$

Figure 1 (a) shows the spectral characteristics of bronze glass window in UV-VIS spectrum region and figure 1 (b) shows the spectral characteristics of bronze glass window in entire solar spectrum region. These properties are very important to find the solar optical properties of glass windows like transmittance and reflectance to calculate the solar heat gain coefficient of the window glass and also to find the total solar radiation through glass windows into the buildings.

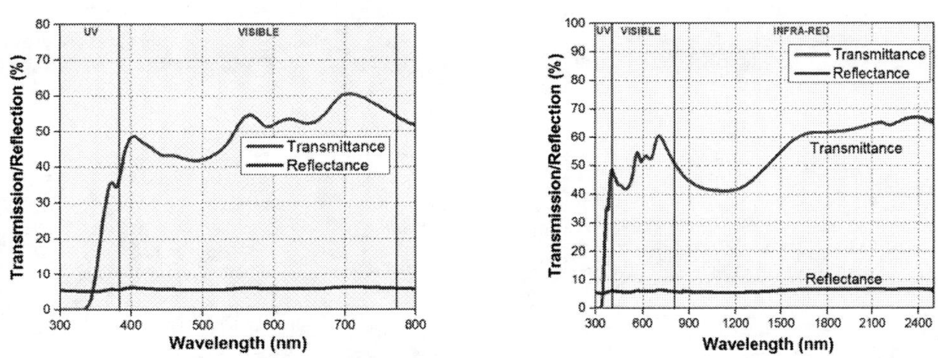

(a) (b)

Fig. 1. Spectral transmission and reflection of bronze glass window (a) UV-VIS spectrum region (b) Solar spectrum region

3 Design and simulation methodology

In this work reading room building model was designed with 4m×4m×4m dimensions by using design builder 4.3.0.039 version tool. Building external walls designed by laterite wall of thickness 0.2m and cement plaster was added on both sides of the wall thickness

0.0125m. Roof of the building material was designed by reinforced cement concrete of thickness 0.150m and cement plaster was added on top and bottom sides of the roof of thickness 0.0125m. Floors of the building model were designed by dense concrete of thickness 0.150m and cement plaster was added on top side of the floor of thickness 0.0125m. 40% window to wall ratio was considered as per Indian standards [18], window of dimensions 3.2mX2m is keeping in all orientations like east, west, north and south and also keeping various shading devices like 0.5m overhang, 0.5m louver thickness 0.020m of five quantity, 0.5m Egg-crate window and (0.5m Egg-crate +0.5m Louver) on windows of thickness is 5mm and quantity 5 keep spacing between plates is 0.400m, of four type was keeping one by one like that total sixteen building models were analyzed in four orientations. In this work find the solar radiation gain into the buildings through windows to hot and dry climatic region Jodhpur at summer solstice and also at winter solstice was done as per Indian standards [19-20]. As per Indian standards for reading room recommended daylight factor is in between 1.9% to 3.75% [21]. To find the average daylight factor levels inside the building were computed at a height of 0.75m from the floor by using split-flux algorithm as per International Commission on Illumination standards (CIE Standards) in both summer and winter seasons by using Energy plus 8.1.0.009 version tool from morning 9:00 hrs to 16:00 hrs. Figure 2 shows the building model of dimensions 4mx4mx4m and window of 3.2mX2m window size four shading devices namely overhang 0.5m, 0.5m Louvers, 0.5m Egg-crate and (0.5m Egg-crate+0.5m Louvers) arranged.

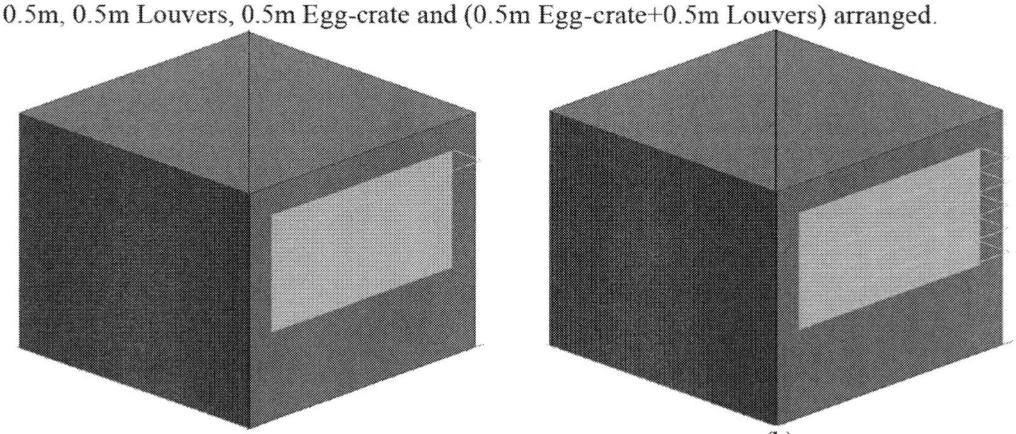

(a) (b)

(c) (d)

Fig. 2. Building model of dimensions 4m×4m×4m with external shading devices (a) Overhang shading device; (b) Louver shading device; (c) Egg-crate shading device; (d) (Egg-crate 0.5m +0.5m Louver) shading device.

4 Results and discussions

4.1 Solar radiation gain into buildings through various shading devices on bronze glass window to Jodhpur (Hot & Dry) climatic region in summer and winter season

Figure 3 (a) and Figure 3 (b) shows the solar radiation passing through different shading devices on bronze glass windows from four directions like east, west, north and south on windows to Jodhpur (26.30^{0}N, 77.58^{0}E) climatic region at summer solstice and at winter solstice as per Indian standards. From the graphs it is clearly observed that the solar radiation gain into the buildings through all four types of shading devices on windows from west orientation is more and less in south orientation in summer season, and also clearly observed that the solar radiation gain into the buildings through all four types of shading devices on windows from south orientation is more and less in north orientation in winter season. In summer season at peak summer day June 21st in south direction solar radiation gain into buildings with (0.5m Egg-crate +0.5m Louver) shading device is gaining less heat 2.79 kWh when compared to all other shading devices. In winter season at peak summer day June 21st in south direction solar radiation gain into buildings with 0.5m shading device is gaining more heat 11.01 kWh, when compared to all other shading devices. For this work Energy plus 8.1.0.009 version tool used to find the day lighting, this software gives average daylight factor inside the buildings in percentages at a distance from 0.75m from the floor.

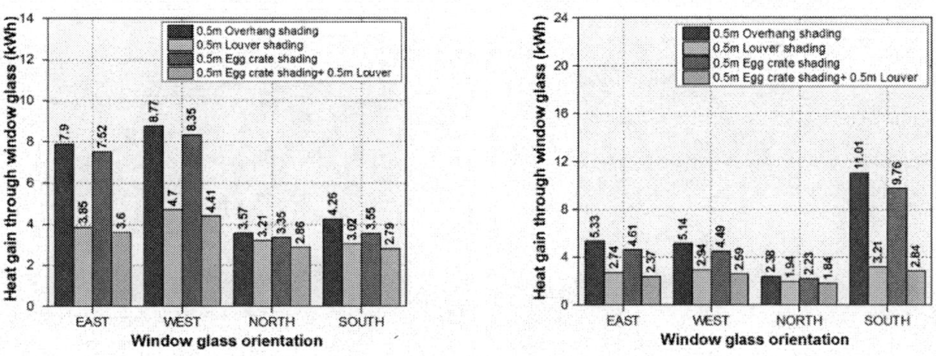

(a) (b)

Fig. 3. Heat gain into buildings through different external shading devices to Jodhpur climatic region (a) Summer season; (b) Winter season.

4.1.1 Effect of 0.5m overhang shading device on windows for average daylight factor into buildings in both summer and winter seasons of Jodhpur (Hot & Dry) climatic region

From Figure 4 (a) to Figure 4 (d) shows the Average daylight factor in an laboratory building at a height of 0.75m from the floor with bronze glazing window placed in East, West, North and South orientation at peak summer day (June 21st) and peak winter day (December 21st) of Jodhpur city (Hot and dry climatic zone) as per CIE standards from morning 9:00 hrs to 16:00 hrs. From Figures, it is observed that daylight factor in buildings in east orientation window is decreasing with time in both summer and winter seasons. It is also noticed that the daylight factor in buildings in west orientation window increases with the increase in time during both summer and winter seasons. Average daylight factor in summer season is dropped during 13:00 hrs in all orientations i.e. because sun is at zenith position and sun moves from zenith to west orientation and also sky illumination (Clear sky) available at that time. Average daylight factor in winter season is coming curve shape i.e. because sky illumination (Overcast sky) available at that time and day. In north orientation, average daylight factor is periodic during summer as shown in Figure 4 (c) that is due to sun travels from northeast to northwest orientation and also depends on sky illumination available outside at that particular day and time. In winter, average daylight factor is look like parabola as shown in Figure 4 (c) that is due to that is due to sun travels from southeast to southwest orientation and also sky illumination is available outside at that particular day and time. In south orientation, the average daylight factor is periodic during summer as shown in Figure 4 (c) that is due to sun travels from northeast to northwest orientation and also depends on sky illumination available outside at that particular day and time as shown in Figure 4 (d) during summer season as shown in Figure 4 (d). In winter, average daylight factor is look like parabola as shown in Figure 4 (c) that is due to that is due to sun travels from southeast to southwest orientation and also sky illumination is available outside at that particular day and time.

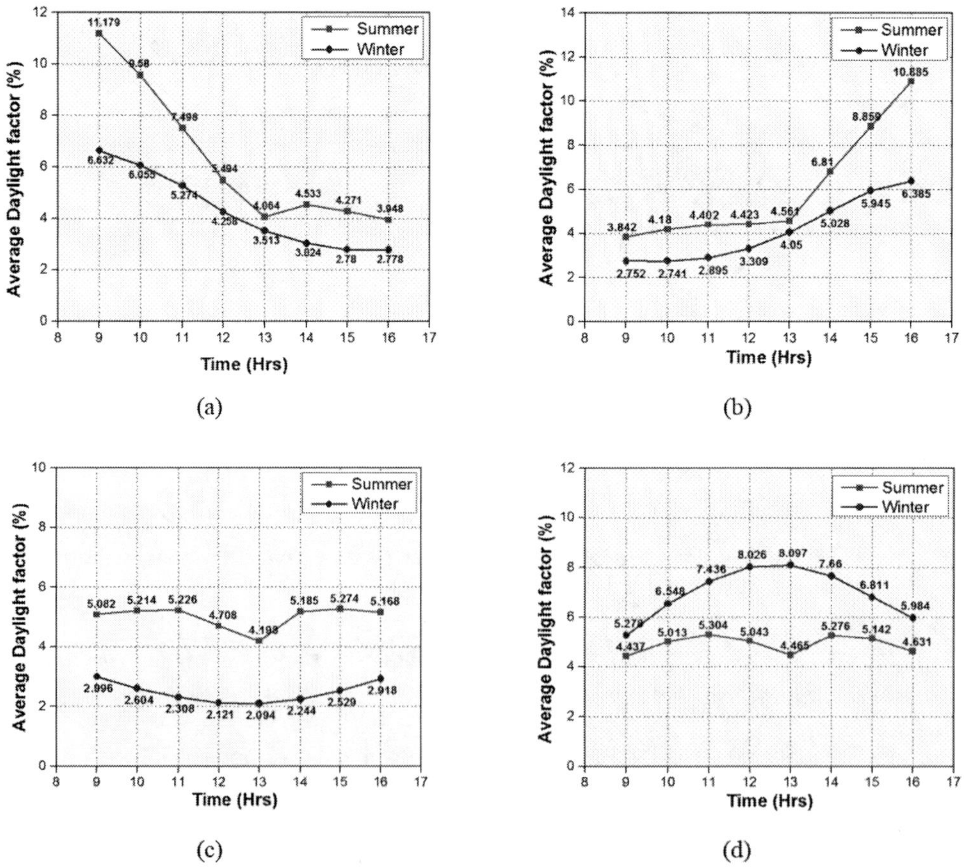

Fig. 4. Average daylight factor into buildings through 0.5m overhang shading device on bronze glass window (a) East direction (b) West direction (c) North direction and (d) South direction.

4.1.2 Effect of 0.5m louver shading device on windows for average daylight into buildings in both summer and winter season to Jodhpur (Hot & Dry) climatic region

From Figure 5 (a) to Figure 5 (d) shows the Average daylight factor in an laboratory building at a height of 0.75m from the floor with bronze glazing window placed in East, West, North and South orientation at peak summer day (June 21st) and peak winter day (December 21st) of Jodhpur city (Hot and dry climatic zone) as per CIE standards from morning 9:00 hrs to 16:00 hrs. From Figures, it is observed that daylight factor in buildings in east orientation window is decreasing with time in both summer and winter seasons. It is also noticed that the daylight factor in buildings in west orientation window increases with increase time during both summer and winter seasons. In north orientation, average daylight factor is periodic during summer as shown in Figure 5 (c) that is due to sun travels from northeast to northwest orientation and also depends on sky illumination available outside at that particular day and time. In winter, average daylight factor is look like parabola as shown in Figure 5 (c) that is due to that is due to sun travels from southeast to southwest orientation and also sky illumination is available outside at that particular day and time. In south orientation, the average daylight factor is periodic during summer as shown in Figure 5 (c) that is due to sun travels from northeast to northwest orientation and also depends on sky illumination available outside at that particular day and time as shown in Figure 5 (d) during summer season as shown in Figure 5 (d). In winter, average daylight factor is look like parabola as shown in Figure 5 (c) that is due to that is due to sun travels from southeast to southwest orientation and also sky illumination is available outside at that particular day and time.

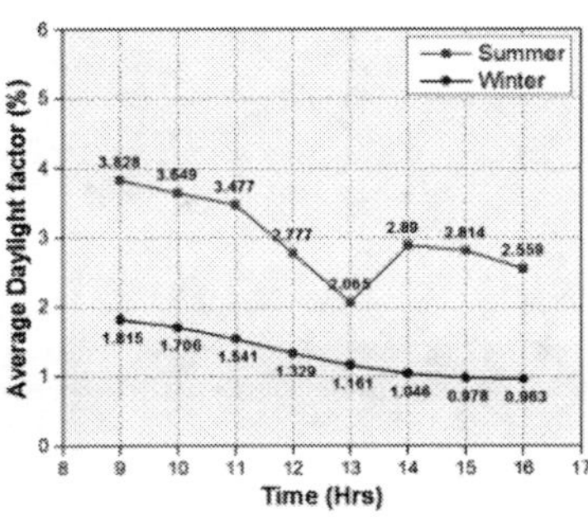

(a)

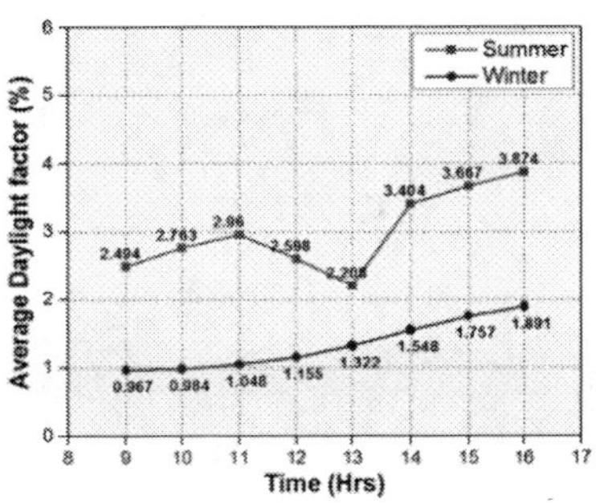

(b)

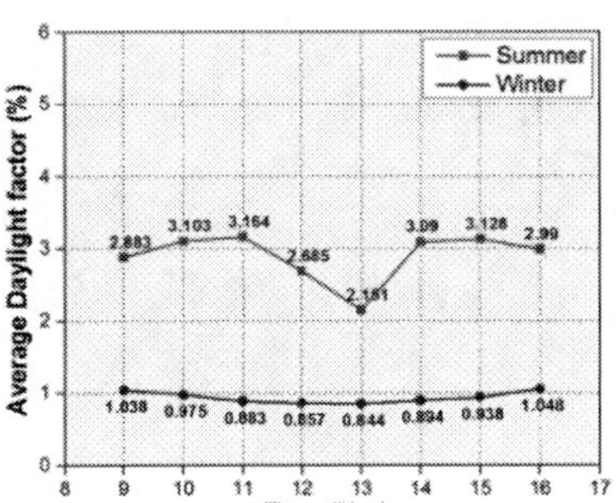

(c)

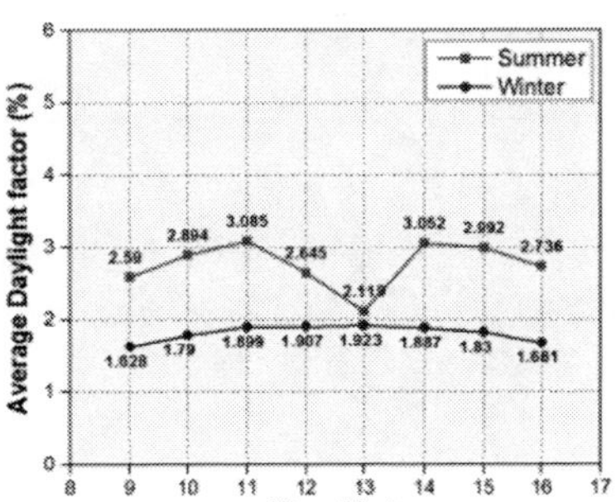

(d)

Fig. 5. Average daylight factor into buildings through 0.5m louver shading device on bronze glass window (a) East direction (b) West direction (c) North direction and (d) South direction.

4.1.3 Effect of 0.5m Egg-crate shading device on windows for average daylight into buildings in both summer and winter season to Jodhpur (Hot & Dry) climatic region

From Figure 6 (a) to Figure 6 (d) shows the Average daylight factor in an laboratory building at a height of 0.75m from the floor with bronze glazing window placed in East, West, North and South orientation at peak summer day (June 21st) and peak winter day (December 21st) of Jodhpur city (Hot and dry climatic zone) as per CIE standards from morning 9:00 hrs to 16:00 hrs. From Figures, it is observed that daylight factor in buildings in east orientation window is decreasing with time in both summer and winter seasons. It is also noticed that the daylight factor in buildings in west orientation window increases with increase time during both summer and winter seasons. In north orientation, average daylight factor is periodic during summer as shown in Figure 6 (c) that is due to sun travels from northeast to northwest orientation and also depends on sky illumination available outside at that particular day and time. In winter, average daylight factor is look like parabola as shown in Figure 6 (c) that is due to that is due to sun travels from southeast to southwest orientation and also sky illumination is available outside at that particular day and time. In south orientation, the average daylight factor is periodic during summer as shown in Figure 6 (c) that is due to sun travels from northeast to northwest orientation and also depends on sky illumination available outside at that particular day and time as shown in Figure 6 (d) during summer season as shown in Figure 6 (d). In winter, average daylight factor is look like parabola as shown in Figure 6 (c) that is due to that is due to sun travels from southeast to southwest orientation and also sky illumination is available outside at that particular day and time.

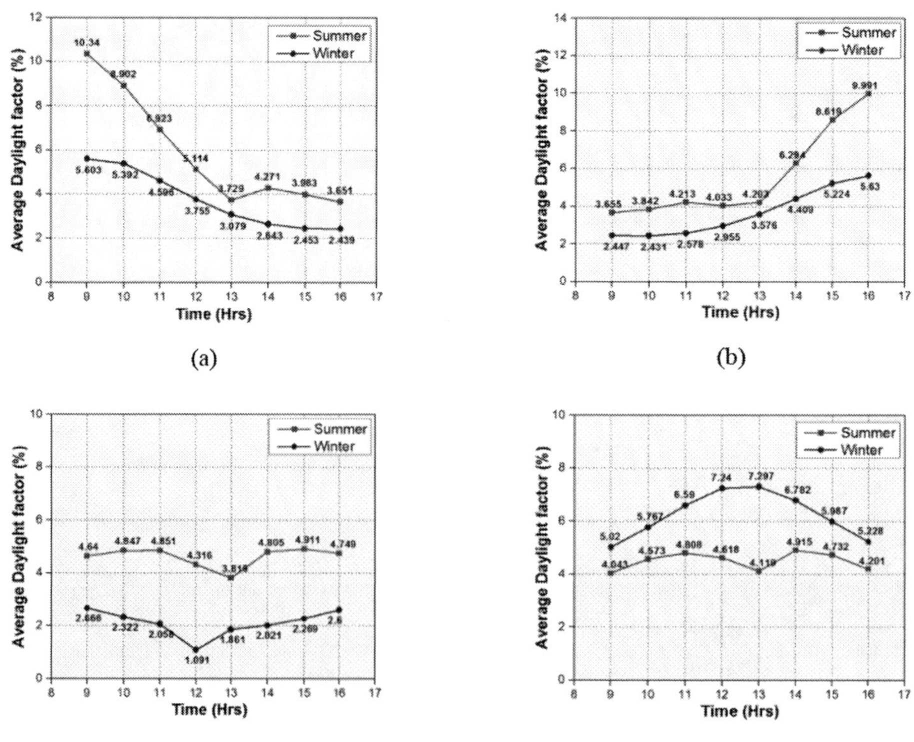

Fig. 6. Average daylight factor into buildings through 0.5m Egg-crate shading device on bronze glass window (a) East direction (b) West direction (c) North direction and (d) South direction.

4.1.4 Effect of (0.5m Egg-crate + 0.5m Louvers) shading device on windows for average daylight into buildings in both summer and winter season to Jodhpur (Hot & Dry) climatic region

From Figure 7 (a) to Figure 7 (d) shows the Average daylight factor in an laboratory building at a height of 0.75m from the floor with bronze glazing window placed in East, West, North and South orientation at peak summer day (June 21st) and peak winter day (December 21st) of Jodhpur city (Hot and dry climatic zone) as per CIE standards from morning 9:00 hrs to 16:00 hrs. From Figures, it is observed that daylight factor in buildings in east orientation window is decreasing with time in both summer and winter seasons. It is also noticed that the daylight factor in buildings in west orientation window increases with increase time during both summer and winter seasons.

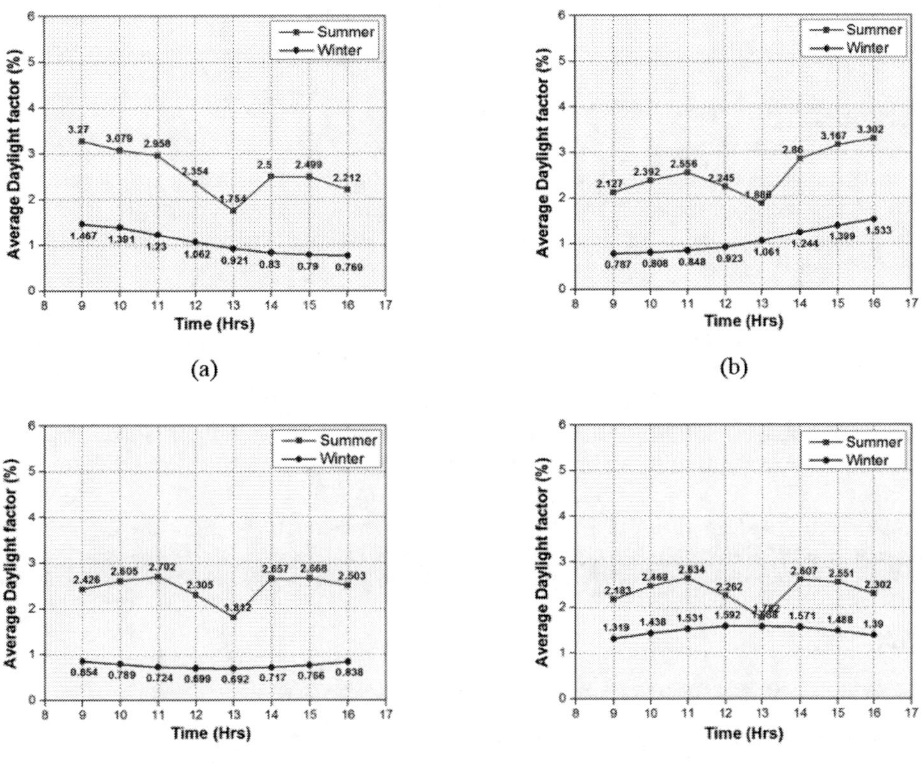

Fig. 7. Average daylight factor into buildings through (0.5m Egg-crate + 0.5mSide fins) shading device on bronze glass window (a) East direction; (b) West direction; (c) North direction and (d) South direction.

In north orientation, average daylight factor is periodic during summer as shown in Figure 7 (c) that is due to sun travels from northeast to northwest orientation and also depends on sky illumination available outside at that particular day and time. In winter, average daylight factor is look like parabola as shown in Figure 7 (c) that is due to that is due to sun travels from southeast to southwest orientation and also sky illumination is available outside at that particular day and time. In south orientation, the average daylight factor is periodic during summer as shown in Figure 7 (c) that is due to sun travels from northeast to northwest orientation and also depends on sky illumination available outside at that particular day and time as shown in Figure 7 (d) during summer season as shown in Figure

7 (d). In winter, average daylight factor is look like parabola as shown in Figure 7 (c) that is due to that is due to sun travels from southeast to southwest orientation and also sky illumination is available outside at that particular day and time.

Nomenclature			
		$\Delta\lambda$	Wavelength interval (nm)
A_{Vis}	Visible absorbance (%)	D_λ	Relative spectral distribution of illuminant D65 (Dimensionless)
A_{Sol}	Solar absorbance (%)	$V(\lambda)$	Spectral luminous efficiency for photopic vision defining the standard observer for photometry (Dimensionless)
R_{Vis}	Visible reflectance (%)	$\tau(\lambda)$	Spectral transmission (%)
R_{Sol}	Solar reflectance (%)	$\rho(\lambda)$	Spectral reflection (%)
T_{Vis}	Visible transmittance (%)	$\alpha(\lambda)$	Spectral absorption (%)
T_{Sol}	Solar transmittance (%)	S_λ	Relative spectral distribution of the solar radiation (W/m^2)
Greek letters			
λ	Wavelength (nm)		

5 Conclusions

The present work presents the reduce solar radiation passing through different shading devices on bronze glass window as well as provide adequate day light factor into buildings in four orientations of hot and dry climatic region (Jodhpur), in both summer and winter seasons.

- In Summer season south orientation window having (0.5m Egg-crate+0.5m Louver) shading device, heat gain inside the building is less (2.84kWh) compared to all other directions and shading devices and also provide adequate day light factor into reading room (2.183% to 2.302%) of hot & dry climatic zone in India.
- In winter season south orientation window having 0.5m overhang shading device on window, heat gain inside the building is more (11.01kWh) compared to all other directions and shading devices, as far as daylight is concerned adequate day light factor into reading room is (2.996% to 2.918%) in north orientation is preferable to hot & dry climatic zone of India.

References

1. A. Ahmed, Y. Freewan, Sol. Energy, 102, 14-30, (2014)
2. D. Gouri, Renew. Energy, 23, 497-507, (2001)
3. I. Ana, M. Palmero, CO. Armando, Appl. Energy, 87, 2040-2049, (2010)
4. M. Gloria Gomes, A.J. Santos, A. Moret Rodrigues, Build. Environ. 71, 47-59, (2014)
5. R.M. Ismael, F.B. Juan Luis, G.G. Francisco Javier, R.C. Paloma, Energ. Build. 101, 144-152, (2015)
6. OP. Czeslaw, S. Marta, Energ. Build. 73, 48-58, (2014)
7. B.Y. Yeo, R.J. Woo, H.L. Kwang, Energies, 7, 2362-2376, (2014)
8. B.Y. Yeo, M. Rashmi, and H.L. Kwang. Energies, 7, 2362-2376, (2014)
9. A.F. Mohammed and M.B. Ismail, Energ. Build. 108, 307-316, (2015)
10. S. Hans and B. Bruno, Build. Environ. 43, 197-204, (2008)
11. S, Giovanni and Barozzi, Solar Energy, 39(4), 329-342 (1987)

12. H.W.Li, Danny, H.W.C, Gary and C.S.L. Chris. Build. Environ. 41, 578-589 (2006)
13. A.K. Athienitis and A. Tzempelikos, Sol. Energy, 72(4), 271-281, (2002)
14. Su. Madhu, G.N. Tiwari, and I.M. Al-Helal. Sol. Energy, 115, 379-389, (2015)
15. ASTM E424 Test for Solar energy Transmittance and Reflectance (terrestrial) of sheet materials Washington DC, USA. (1971)
16. BS EN 410 Glass in Building-Determination of luminous and solar characteristics of the glazing British Standards, (1998)
17. ISO 9050:2003(E), Glass in building-Determination of light transmittance, solar direct transmittance, total solar energy transmittance, ultraviolet transmittance and related glazing factors (2003)
18. ECBC, Energy Conservation Building Code. Bureau of Energy Efficiency, New Delhi, India (2009)
19. NBC National Building Code of India 2005, Section 1 Building and Services Lighting and Ventilation Part-I, Bureau of Indian Standards, New Delhi, India. (2005)
20. A. Mani, Solar radiation over India Allied Publishers private limited, India. (1982)
21. SP: 41. (S&T) Handbook on functional Requirement of Buildings other than industrial buildings. Bureau of Indian Standards, India. (1987)

Analytical computation of thermodynamic performance parameters of actual vapour compression refrigeration system with R22, R32, R134a, R152a, R290 and R1270

Shaik Sharmas Vali [1*], *Talanki Puttaranga Setty*[1] and *Ashok Babu*[1]

[1]*National Institute of Technology Karnataka Surathkal, Mechanical Engineering Department, Mangalore-575025, India.*

Abstract. The present work focuses on analytical computation of thermodynamic performance of actual vapour compression refrigeration system by using six pure refrigerants. The refrigerants are namely R22, R32, R134a, R152a, R290 and R1270 respectively. A MATLAB code is developed to compute the thermodynamic performance parameters of actual vapour compression system such as refrigeration effect, compressor work, COP, power per ton of refrigeration, compressor discharge temperature and volumetric refrigeration capacity at condensing and evaporating temperatures of 54.4oC and 7.2oC respectively. Analytical results exhibited that COP of both R32 and R134a are 15.95% and 11.71% higher among the six investigated refrigerants. However R32 and R134a cannot be replaced directly into R22 system. This is due to their higher compressor discharge temperature and poor volumetric capacity respectively. The discharge temperature of both R1270 and R290 are lower than R22 by 20-26oC. Volumetric refrigeration capacity of R1270 (3197 kJ/m3) is very close to that of volumetric capacity of R22 (3251 kJ/m3). Both R1270 and R290 shows good miscibility with R22 mineral oil. Overall R1270 would be a suitable ecofriendly refrigerant to replace R22 from the stand point of ODP, GWP, volumetric capacity, discharge temperature and miscibility with mineral oil although its COP is lower.

Keywords: Discharge temperature; Ecofriendly; Miscibility; R1270; Volumetric capacity

1 Introduction

Chlorofluorocarbons (CFCs) and Hydrochlorofluorocarbons (HCFCs) was extensivenly used in vapour compression based refrigeration industries from past several years due to their better thermophysical and thermodynamic properties [1]. However CFCs and HCFCs contains ODP (ozone depletion potential) and high GWP (global warming potential) and hence CFCs and HCFCs were eventually phase out in all the developing nations by the year 2010 and in all the developed nations by the year 2030 respectively in accordance to Monetreal protocal [2-3]. Therefore it is essential to develop new ecofriendly refrigerants.

* Corresponding author: sharmasvali.nitk@gmail.com

Formerly several theoretical and expermintal studies were carried to find an appropriate fluids to replace R22 and R134a respectively. Experimental peformance tests were carried out in a vapour compression plant with R134a, R152a, R290, R600a, R1234yf and R1234ze(E) respectively [4]. Tests results revealed that refrigerants R152a and R1234yf were suitable drop in replacements to R134a. Expermimental studies were done in a vapour compression test setup with R513A and R134a [5]. Results reported that superior peformance of R513A would be obtained only with the adjustment of thermostatic expansion valve. Peformance tests were conducted in a heat pump tester under various test conditions with hydrocarbons like R290 and R1270 [6]. Experimental results showed that performance of both R1270 and R290 were better than the R22. Particulary R290 exhibited 6-12% higher COP under various test conditions compared to R22. Therotical thermodynamic analyis was carried out in a standard vapour compression system with RE170, R290, R1270, R600a, R32, R134 and R152a respectively [7]. Reults exhibited that peformance of RE170 was higher among other studied refriferants and hence it was a viable alternative to R134a. Theoretical and expermintal investiagtions were done in an air conditioner under different operating conditions with R161 and R22 [8]. Both theoretical and exprimental results exhibited that energy efficiency ratio of R161 was superior than the R22. A thorough review study reported that hydrcarbons and its mixtures were viable refrigerants to refrigeration related industries [9]. Therortical performance studies were carried out in a simple saturation vapour compression system with R1234yf, R161, R152a and R134a respectively [10]. Results revealed that R152a would be a better choice to replace R134a from the stand point of COP, power savings, volumetric capacity and GWP. The present work emphasis on performance computation of six pure refrigerants based on actual vapour compression cycle in order to find an appropriate candidate to replace R22.

2 Properties of investigated refrigerants

In this study six pure refrigerants like R22, R32, R134a, R152a, R290 and R1270 are considered and thermodynamic properties of all the six pure refrigerants are must require to do the thermodynamic analysis of vapour compression refrigeration system. Hence a computer MATLAB program is written to develop the thermodynamic properties of six pure refrigerants by using Martin-Hou equation of state [11].

$$P = \frac{RT}{v-b} + \frac{a_2 + b_2 T + c_2 e^{\frac{-5.475T}{T_C}}}{(v-b)^2} + \frac{a_3 + b_3 T + c_3 e^{\frac{-5.475T}{T_C}}}{(v-b)^3} + \frac{a_4}{(v-b)^4} + \frac{b_5 T}{(v-b)^5} \quad (1)$$

The developed properties of all the six studied refrigerants shows good agreement with the experimental reported properties available in ASHRAE hand book [12]. The deviation of computed properties of all the six pure refrigerants from that of ASHRAE are within 1.5 to 3% for the given operating conditions. Hence the developed MATLAB code is reliable and therefore the same developed thermodynamic properties are used in this study. The basic physical and critical characteristics of six pure refrigerants are taken from the ASHRAE hand book and it is given below in table 1.

https://doi.org/10.1051/matecconf/201814404009

Table 1. Basic properties of pure refrigerants

Refrigerant	M (kg/kmol)	T_b (°C)	P_c (MPa)	T_c (K)	ODP	GWP (100 years)
R22	86.5	-40.81	4.99	369.3	0.055	1760
R32	52.024	-51.651	5.782	351.25	0	677
R134a	102.03	-26.074	4.0593	374.21	0	1300
R152a	66.051	-24.023	4.5168	386.41	0	140
R290	44.096	-42.11	4.2512	369.89	0	3
R1270	42.08	-47.62	4.5548	364.21	0	3

3 Thermodynamic performance analysis

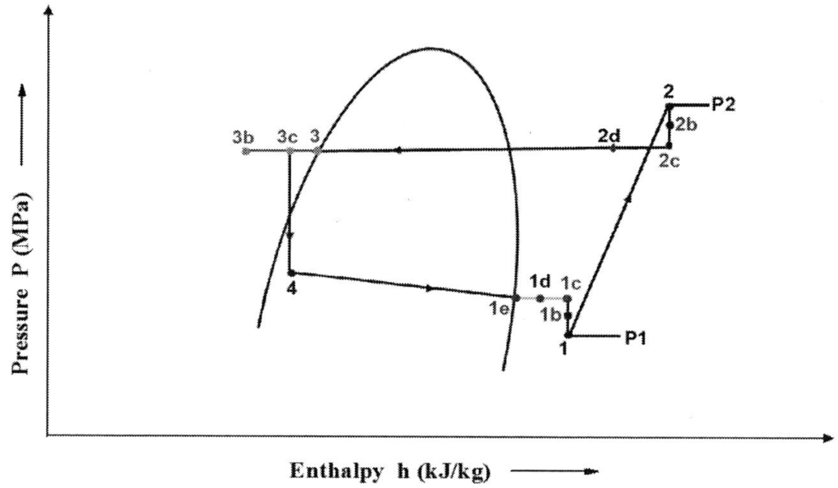

Fig 1. P-h diagram of actual vapour compression refrigeration cycle.

In the present methodology analytical computation of thermodynamic performance of six pure refrigerants (R22, R32, R134a, R152a, R290 and R1270) is done based on actual vapour compression refrigeration cycle. In this cycle superheating, subcooling, various pressure and heat losses or heat gain occurred at different components of the system are considered and they are given in table 2 and 3 respectively [13-14].The p-h diagram of actual vapour compression refrigeration cycle is shown in figure 1 [15-16]. A MATLAB code is developed to compute the thermodynamic performance parameters of six pure refrigerants at ARI conditions (Tk=54.4oC and Te=7.2oC) respectively. The various performance parameters are refrigeration effect, compressor work, coefficient of performance, compressor discharge temperature, power per ton of refrigeration and volumetric refrigeration capacity. The capacity and operating conditions of the system considered for all the six refrigerants are Q=1.5 TR, Tk=54.4oC and Te=7.2oC respectively.

Table 2. Explanation of different state points of the cycle

State points of the cycle	Explanation
4-1e	Pressure drop in the evaporator
1e-1d	Superheating of refrigerant vapour occurred in the evaporator
1d-1c	Heat gain and refrigerant superheating in the suction pipe line
1c-1b	Pressure drop through suction line
1b-1	Pressure drop across suction valve
1-2	Compressor work
2-2b	Pressure drop across discharge valve
2b-2c	Pressure drop in the discharge line
2c-2d	Heat loss and refrigerant desuperheating in the discharge pipe line
2c-3	Pressure drop in the condenser
3-3b	Subcooling
3b-3c	Heat gain occurred in the liquid pipe line

Table 3. Description of various losses considered in the cycle

S. No.	Description	Values
1	Pressure loss across suction valve	0.2 bar
2	Pressure loss across discharge valve	0.4 bar
3	Pressure drop occurred in the suction line	0.1 bar
4	Pressure loss occurred in the discharge line	0.1 bar
5	Pressure drop in the evaporator	0.1 bar
6	Rise in temperature due to heat gain at the inlet of compressor	5°C
7	Temperature drop due to heat loss at the outlet of compressor	5°C
8	Degree of superheating	10°C
9	Degree of subcooling	5°C

4 Mathematical computations

The mathematical computations involved in the thermodynamic analysis of investigated refrigerants are given below.

Refrigerant mass flow rate is computed by

$$\dot{m} = Q/RE \quad (2)$$

Isentropic compressor work is computed as

$$W_c = h_2 - h_1 \quad (3)$$

Refrigerating effect (Cooling effect) is calculated as

$$RE = h_{1d} - h_4 \quad (4)$$

Coefficient of performance (COP) is calculated as

$$COP = RE/W_c \quad (5)$$

Energy (Power) consumption by the compressor per ton of refrigeration is computed by

$$PPTR = 3.5167/COP \quad (6)$$

Volumetric cooling capacity is calculated as

$$VRC = \rho_{1d} \times RE \quad (7)$$

The compressor discharge temperature (T_d) is computed using superheated property tables and interpolating for the degree of superheat corresponding to the entropy difference.

Table 4. Summary of results for the pure refrigerants

Refrigerant code	$\dot{m}$ (kg/min)	RE (kJ/kg)	W_c (kJ/kg)	COP	T_d (°C)	PPTR (kW/TR)	VRC (kJ/m³)
R22	2.236	141.502	38.642	3.661	91.40	0.960	3251
R32	1.302	242.923	57.220	4.245	110.18	0.828	5929
R134a	2.230	141.915	34.693	4.090	76.08	0.859	2103
R152a	1.447	218.678	61.441	3.559	92.22	0.988	1903
R290	1.334	237.232	71.110	3.336	65.19	1.054	2582
R1270	1.300	243.373	71.260	3.415	70.42	1.029	3196

5 Analytical validation

In the present study a MATLAB code was developed and it is used to compute the thermodynamic performance of R22 and its alternatives. To validate the MATLAB code, results are compared with the previously published data [7, 17]. Dalkilic AS and Wongwises S has computed the thermodynamic analysis of simple saturation vapour compression system with R22 at Tk = 50°C and Te=-10°C whereas Baskaran A and Koshy Mathews P has computed the thermodynamic performance of standard vapour compression system with R134a at Tk = 50°C and Te=-10°C by considering degree of superheating and subcooling as 10°C and 5°C respectively. For the validation of present MATLAB code, the same refrigerants R22, R134a and operating conditions are used as that of the previous authors. The results of thermodynamic performance parameters of refrigerants R22 and R134a obtained from the MATLAB code are compared with published data results and they are given in table 5 and 6 respectively. The deviation of MATLAB code results with Dalkilic AS, Wongwises S, Baskaran A, and Koshy Mathews P for both the R22 and R134a is less than 1%. Therefore the developed MATLAB code is reliable and hence it can be used for the computation of thermodynamic performance of any refrigerants.

Table 5. Comparison of thermodynamic performance parameters of MATLAB code results with Dalkilic AS and Wongwises S results for the refrigerant R22

S.NO	Performance parameters	Dalkilic AS and Wongwises S results for R22	MATLAB code results for R22	Deviation (%)
1	Evaporator pressure (MPa)	1.943	1.9427	-0.015
2	Condenser pressure (MPa)	0.355	0.35481	-0.053
3	Pressure ratio	5.476	5.4753	-0.012
4	Refrigeration effect (kJ/kg)	138	137.9490	-0.036
5	Compressor work (kJ/kg)	43.40	43.7740	0.861
6	COP	3.180	3.1514	-0.899
7	Power per ton of Refrigeration (kW/TR)	1.101	1.1106	0.871
8	Volumetric cooling capacity (kJ/m^3)	1273.912	1280.525	0.5191

Table 6. Comparison of thermodynamic performance parameters of MATLAB code results with Baskaran A and Koshy Mathews P results for the refrigerant R134a

S.NO	Performance parameters	Baskaran A and Koshy Mathews P results for R134a	MATLAB code results for R134a	Deviation (%)
1	Evaporator pressure (kPa)	200.6	200.41	-0.094
2	Condenser pressure (kPa)	1317.9	1317.0	-0.068
3	Pressure ratio	6.57	6.5715	0.022
4	Refrigeration effect (kJ/kg)	137.28	138.2360	0.696
5	Compressor work (kJ/kg)	41.42	41.5229	0.248
6	COP	3.315	3.3292	0.428
7	Power per ton of Refrigeration (kW/TR)	1.057	1.0563	-0.066
8	Volumetric cooling capacity (kJ/m^3)	1314	1317.8	0.289

6 Results and discussions

The results and discussions of various thermodynamic performance parameters of six pure studied refrigerants are given below.

6.1 Refrigeration effect

Figure 2 shows the cooling effect (refrigeration effect) of various refrigerants. From the figure 2 it is noticed that refrigeration effect of refrigerants like R32, R152a, R290 and R1270 are higher than the R22. This is due to their high latent heat of vaporization when compared to R22.

6.2 Compressor work

Figure 3 shows the compressor work of six studied refrigerants. Referring to figure 3 it is observed that the compressor work of both R290and R1270 refrigerants are higher than the R22. This is due to high vapour enthalpies of refrigerants (R290 and R1270) compared to R22.

6.3 Coefficient of performance

Figure 4 shows shows the coefficient of performance of various considered refrigerants. COP is measured as an energy efficiency index of the device charged with particular refrigerant. From the table 4 it is seen that both cooling (refrigeration effect) effect and compressor work increase for the refrigerants R32, R152a, R290 and R1270 respectively. Therefore net effect of refrigeration effect and compressor work of these refrigerants on coefficient of performance may be either increases or decreases or remains same which depends on the type of the refrigerant and also on the operating conditions of the device. From the figure 4 it is clear that the COP of R32 and R134a is 15.95% and 11.71% higher than R22 respectively.

6.4 Compressor discharge temperature

Before introducing the new refrigerant into the device, it is essential to study the steadiness and lifetime of compressor. This can be done by computing the compressor discharge temperature of all the studied refrigerants. Referring to figure 5 it is noticed that compressor discharge temperature of both R1270 and R290 are lower than R22 by 20-26°C. This is due to lower adiabatic index of refrigerants. The lower compressor discharge temperature is advantageous from the stand point of durability of the compressor life. Hence both R290 and R1270 fluids exhibits better reliability and lifespan of compressor motor. However R32 is not preferable from the stand point of compressor life because its discharge temperature is very high.

6.5 Power per ton of refrigeration

Figure 6 shows shows the power (energy) per ton of refrigeration for various studied refrigerants. It indicates the electrical work input required to the compressor to produce per ton of refrigeration. Referring to figure 6 it is observed that energy spent by the compressor per ton of refrigeration for the refrigerant R32 is 13.75 % lower among six investigated refrigerants. This is due to refrigerant type and also due to operating conditions of the given device.

6.6 Volumetric cooling capacity

Figure 7 shows shows the volumetric refrigeration capacity of various refrigerants. Volumetric refrigeration capacity depends on density of refrigerant vapour and refrigeration effect of the given fluid. It denotes the cooling capacity per unit volume of vapour refrigerant at the outlet of evaporator .And also it denotes the volume of refrigerant handled by the compressor for the given fluid. From the figure 7 it is evident that volumetric refrigeration capacity of R134a and R152a are lower than R22 and hence these refrigerants requires larger size of compressor whereas volumetric capacity of R32 is very high among six studied refrigerants. Therefore R32 requires redesign or smaller size of compressor.

Volumetric cooling capacity of refrigerant R1270 is very close to R22 and hence same size of compressor can be used for R1270 as that of R22. Volumetric capacity of R290 is in between R22 and R1270.

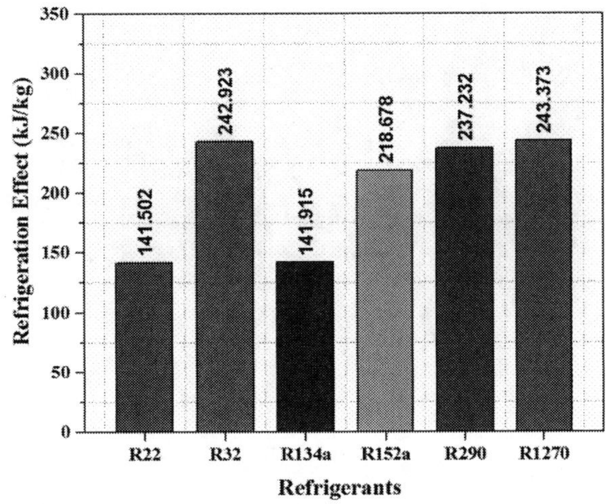

Fig 2. Refrigeration effect of various refrigerants

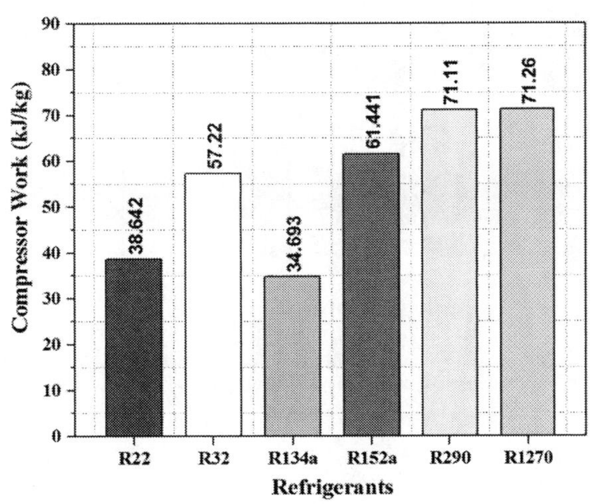

Fig 3. Compressor work of various refrigerants

 https://doi.org/10.1051/matecconf/201814404009

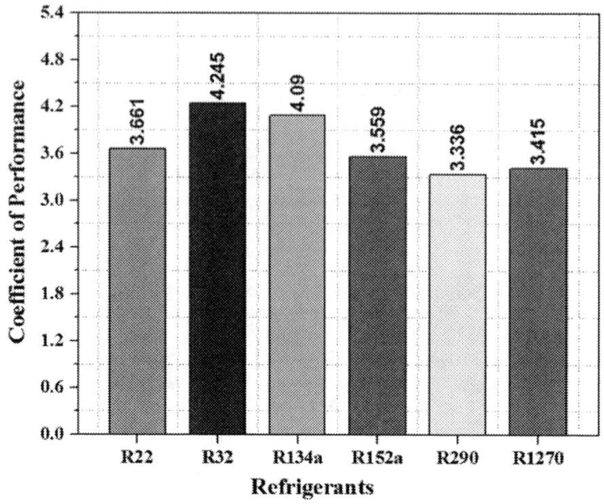

Fig 4. COP of various refrigerants

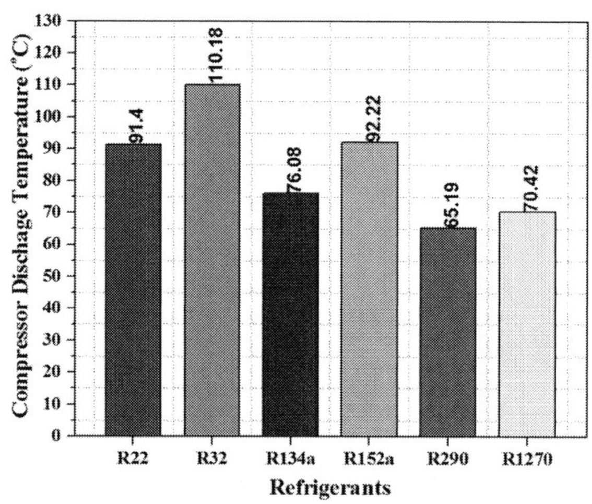

Fig 5. Compressor discharge temperature of various refrigerants

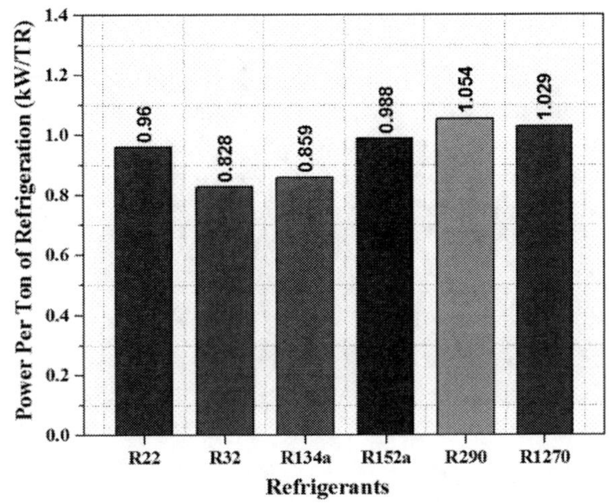

Fig 6. Power per ton of refrigeration of various refrigerants

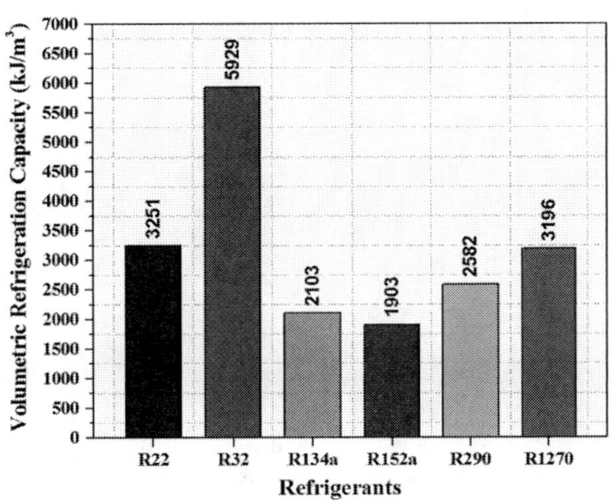

Fig 7. Volumetric refrigeration capacity of various refrigerants

7 Conclusions

From the thermodynamic performance computation of six pure refrigerants the following conclusions can be drawn.

- Eventhough COP of R32 and R134a were 15.95% and 11.71% higher among the six investigated refrigerants, they cannot be replaced directly into R22 system. This is due to their higher compressor discharge temperature and poor volumetric capacity respectively.
- Among the six investigated refrigerants the discharge temperature of both R1270 and R290 were lower than R22 by 20-26°C. Hence both R290 and R1270

exhibits better reliability and lifespan of compressor motor. Refrigerants R290 and R1270 shows better miscibility with R22 mineral oil.

- Volumetric refrigeration capacity of R32 (5929 kJ/m3) was 82.3% higher than the R22 (3251 kJ/m3). Therefore R32 requires redesign or smaller size compressor compared to R22 whereas volumetric capacity of R134a and R152a were 35.2-41.4% lower compared to R22. Hence both R134a and R152a requires larger size of compressor compared to R22. Due to above factors R32, R134a and R152a cannot be replaced directly into R22.
- The volumetric capacity of R1270 (3196 kJ/m^3) was very close to that of volumetric capacity of R22 (3251 kJ/m3) and hence same size of compressor can be used for R1270 as that of R22.
- Overall R1270 would be a suitable ecofriendly alternative refrigerant to replace R22 from the stand point of ODP, GWP, volumetric refrigeration capacity, discharge temperature and miscibility with mineral oil although its COP was fairly lower.

Nomenclature

ARI	Air conditioning and refrigeration institute
GWP	Global warming potential
ODP	Ozone depletion potential
RE	Refrigeration effect (kJ/kg)
TR	Ton of refrigeration (kW)
h_1	Enthalpy at the compressor entry (kJ/kg)
h_2	Enthalpy at the compressor exit (kJ/kg)
h_{1d}	Enthalpy at the evaporator inlet (kJ/kg)
$\dot{m}$	Mass flow rate of refrigerant (kg/min)
M	Molecular weight (kg/kmol)
P_c	Critical pressure (MPa)
Q	Refrigeration capacity (kW)
R	Universal gas constant (J/molK)
T_b	Boiling point (°C)
T_c	Critical temperature (K)
T_d	Discharge temperature (°C)
T_e	Evaporating temperature (°C)
T_g	Temperature glide (°C)
T_k	Condensing temperature (°C)
v	Specific volume (m^3/kg)
W_c	Compressor work (kJ/kg)

References

1. M. Mohanraj, S. Jayaraj, C. Muraleedharan, Int. J. Greenh Gas Control 3, 108-119 (2009)
2. United Nations Environmental Programme, Montreal *Protocol on substances that deplete the ozone layer*, Final act. New York: United Nations; 1987
3. RL. Powell, J. Fluorine Chem, 114, 237-250 (2002)
4. D. Sanchez, R. Cabello, R. Llopis, I. Arauzo, J. Catalan-Gil, E. Torrella, Int. J. Refrigeration, 74, 269-282, (2017)
5. Adrian Mota-Babiloni, Pavel Makhnatch, Rahmatollah Khodabandeh, Joaquin Navarro-Esbri, Int. J. Refrigeration, 74, 682-688, (2017)
6. Ki-Jung Park, Dongsoo Jung, J Mech Sci Technol, 22, 532-537, (2008)
7. A. Baskaran, P. Koshy Mathews, IJSRP 2, 1-8, (2012)
8. Wu. Yingwen, Xiangfei Liang, Tu. Xiaoping, Rong Zhuang, Proceedings of the *International Refrigeration and Air Conditioning Conference*, 1189, 1-7, (2012)
9. B.O. Bolaji, Z. Huan, Renew. Sustainable Energy Rev, 18, 49-54, (2013)
10. B.O. Bolaji, Z. Huan, J Eng Thermophys Rus, 23, 148-157, (2014).
11. Martin Joseph J, Yu-Chun Hou, AlChE J, 1, 142-151, (1955)
12. ASHRAE Hand book-Fundamentals (SI), 2009.
13. C P. Arora, Refrigeration and Air conditioning (Tata McGraw-Hill, New Delhi, 2009)

14. T.P. Ashok Babu, V. Samaje Vikas, R. Rajeev, Proceedings of the International Refrigeration and Air Conditioning Conference, 854, 1-8, (2006)
15. R.C. Arora, Refrigeration and Air conditioning (PHI learning Private Limited, New Delhi, 2010)
16. Manohar Prasad, Refrigeration and Air conditioning (New Age International (P) Limited Publishers, New Delhi, 2015)
17. Dalkilic AS Wongwises S. Int. Commun. Heat Mass Transfer, 37, 1340-1349, (2010)

Experimental and numerical investigation of heat dissipation from an electronic component in a closed enclosure

Bobin Saji George[1*], *Ajmal M.*[1], *Deepu S. R.*[1], *Aswin M.*[1], *Ribin D.*[1] and *Akhil J.*[1]

[1] *Department of Mechanical Engineering, Mar Baselios College of Engineering& Technology, Thiruvananthapuram, Kerala*

Abstract. Intensifying electronic component power dissipation levels, shortening product design cycle times, and greater than before requirement for more compact and reliable electronic systems with greater functionality, has heightened the need for thermal design tools that enable accurate solutions to be generated and quickly assessed. The present numerical study aims at developing a computational tool in OpenFOAM that can predict the heat dissipation rate and temperature profile of any electronic component in operation. A suitable computational domain with defined aspect ratio is chosen. For analyzing, "buoyant Boussinesq Simple Foam" solver available with OpenFOAM is used. It was modified for adapting to the investigation with specified initial and boundary conditions. The experimental setup was made with the dimensions taken up for numerical study. Thermocouples were calibrated and placed in specified locations. For different heat input, the temperatures are noted down at steady state and compared with results from the numerical study.

Keywords: Heat transfer, OpenFOAM, Boussinesq, Aspect ratio,

1 Introduction

The computational power of each electronic component increases drastically leading to shrinking the physical size of the component. Whenever current is passed through a conductor heat is produced in it because of the resistance of the conductor. This will happen as such in the case of electronic component also. In the case of circuit boards, the main point of heat generation is the junction where the components are fixed in the circuit board; at that point the resistance is maximum. Due to the continuous miniaturization of the electronic equipment the area of the component also decreases so the rate of heat dissipation also decreases.

Before the manufacturing of the equipment proper study in the heat generation and what type of cooling system is required should be selected according to the study conducted. In many of the electronic component either natural convection or forced convection technique is chosen for the cooling. In avionic component, the cooling system will be either liquid cooling or cooling by heat pipe. Extended surfaces, which are

* Corresponding author: bobin.george@mbcet.ac.in

popularly known as fins, are extensively used in air cooled heat exchangers and electronic component. An insufficient removal of heat can result in temperatures above the equipment operational limits. In these circumstances, the performance, the life and the reliability of the equipment are extremely reduced. During the selection of a cooling technique the environment where the equipment is working should be taken into consideration. The failure rate of the electronic component increases exponentially with temperatures. Without proper design and control, high rate of heat generation result in high operating temperature for electronic equipment, which reduces the safety and reliability

This study deals with the analysis of electronic component in a closed enclosure experimentally and numerically. Here a heater is designed such that it will produce the average heat that electronic component generates. A forced convection technique is used to cool the heater. Thus, the amount of heat dissipated can be studied easily. By varying the flow parameters, different temperature contours are obtained for studying. Numerically the experimental set up is modelled in Open FOAM and analysed.

2 Literature review

Peter Rodgers [1] presents that during the analysis of a normal flow (forced convection) through a cavity having low Re value the normal flow modelling technique with CFD is not applicable. By using normal CFD coding the error in result is nearly 35%. For the accurate result shear stress transport k–ω model is used. RANS is the most suitable approach for analyzing the equipment heat transfer. Domenico Borello [2] states that for obtaining high and low pressure eddies in a flow Large Eddie Simulation (LES) is used. LES allow predicting the pressure rise capability of the fan and can reproduce the flow features. The vortex pattern occur in the blades are due to the thickness distribution of the section at the roof. GiorgiaNardini [3] states that in a square cavity the natural convection becomes more effective when there is 4 cold squares is located in the vertical wall and 1 hot source at the center. Position of heat source plays a significant role in the case of natural convection. As the Ra increases the number of recirculation also increases. Also, the velocity will increase with Ra. D Senthilkumar [4] states that whenever the buoyancy force is smaller than the fluid inertial force the increase in fluid inertial force will not make any change in convective heat and mass transfer. So, inertial force of fluid helps to improve the heat and mass transfer if and only if thermal buoyancy force is of same magnitude or greater than fluid inertial force. ArlindoTribess [5] presents the flow over an electronic equipment is on interest, then the superficial generation is the best approach, because it is more conservative. In the other hand, if the concern is with the temperatures on the equipment surfaces, the volumetric generation is a better approach. Wataru Nakayama [6] presents that thermal design of actual component requires a CFD simulation code to be designed. Many factors that are present in the actual component are not considered while making the model. NesreenGhaddar [7] states the quantity of air infiltrating through a porous member do not affect the external flow pattern over the member. Natural convection enhances ventilation at lower wind. So, if a porous member is placed in between a low wind flow and a heat source, the heat source will experience only natural convection and no forced convection until, the wind velocity reaches a significantly high value. So, when modelling, it is important include natural convection along with forced convection for flow velocities < 3 m/s. Study showed an increase of 12% in heat transfer. Peter J Rodgers [8] validated the predictive accuracy of CFD codes for Heat transfer in electronics components. An experimental set up having a standard PCB with 3 general electronic components mounted in various configurations is made and various parameters were measures under specific working conditions. The results obtained were used to validate the codes. Peter J Rodgers

[9] validated the numerical accuracy of the codes reduced with the increase in complexity of configuration of components on PCB. In case of populated boards, flow models capable of modeling transition were found to be required to capture the entire flow field. Zaghdoudi [10] presents the heat pipe as a medium for heat transfer, it was found that hot spots could be eliminated and the heat was spread better. Using heat pipes the component temperature was reduced significantly. The study was done on AEM (Avionics Electronics Modules) and the limitation for its practical application is the physical design of heat pipe to be accommodated on the small sections between modules. M.A. Hassan [11] presents that, the paper deals with the study of steady state and natural convection of Bingham fluid in a square enclosure discretely heated from bottom wall and symmetrically cooled from side walls. The effects of different values of Bingham number, Ra, Pr and heat transfer characteristics inside the enclosure are studied. A correlation of average Nu as a function of other non–dimensional numbers has been established using the present numerical results. Tunclcoz [12] states the effect of separation distance and pressure drop is significant only in the turbulent case. It is found that as the height of source raises, the heat transfer rate raises due to increase in surface area. Properties of air and water (de–ionized) possess similar trends but among them water has high heat transfer coefficient. Z. Xu [13] validates that to get maximum efficiency, the CPVs should be under a suitable temperature. The rate of increase of efficiency as Re increases is higher at low range of Re values and then get reduced. This is because at low Re values, heat transfer is by forced convection and at high values conduction become dominant. Hakan F Oztop [14] states that, among different geometry shapes, curvilinear shape exerts more control on heat and fluid flow due to natural convection. Heat transfer increases with additional of nanoparticles into the base fluid. Mehmet Arik [15] presents the heat transfer capabilities of both small–scale synthetic and steady jets are studied and found out the value of heat transfer coefficient increases with increasing Re. The distance between jet to heater has more effect on Nu for steady jets than for synthetic jets. At high Re synthetic jets show 30% better heat transfer performance and for the same spacing synthetic jet showing 40% superior heat performance. Yoji Kitamura [16] presents that, as the inclination increases it also enhances the air flow rate. The cooling effect has been improved by inclination of electronic casing for the circulation of air flow inside the casing increases. Mousa M. Mohamed [17] states that, comparatively the effect of air mass flow rates on heat transfer is a positive effect but the effect of channel base temperatures has highly positive effect. This is because as Re increases the Nu increases and thus increasing heat transfer and also at the same air flow rate the heat transfer coefficient is increased with increasing the channel base temperature. Xiaohong Hao [18] validates the heat transfer and pressure drop in serpentine channel heat sinks with 1800 bends. The study is carried out using a thermal resistance model and also pressure drop within the serpentine channel is derived after considering the effects of bends on pressure variation. Then the same model is experimentally validated by measuring temperature and pressure characteristics of heat sinks with different Re and different geometric parameters. Finally, the experimental result and the numerically obtained result (using CFD) are compared and validated. Hamid Reza Seyf [19] analyzes the effect of Nanoencapsulated phase change material (NEPCM) slurry on forced convection heat transfer of steady laminar flow past an isothermal square cylinder using a finite volume method (FVM) on collocated grid arrangement. It was found that for both NEPCM slurry and pure water, local and average heat transfer coefficients increases with increasing Re.

3 Numerical analysis

The computational 2D domain was created in ICEM CFD. After the geometry creation, surface creation followed by part naming was done such as inlet, outlet etc. the entire computational domain was divided into block followed by meshing. Since our region of interest is around the surface of the heating element much finer mesh was created around it. Different bunging laws were applied in order to get the quality mesh. The created mesh was converted into mesh suitable for OpenFOAM. Figure 1 gives the computational domain designed and Figure 2 shows the mesh generated.

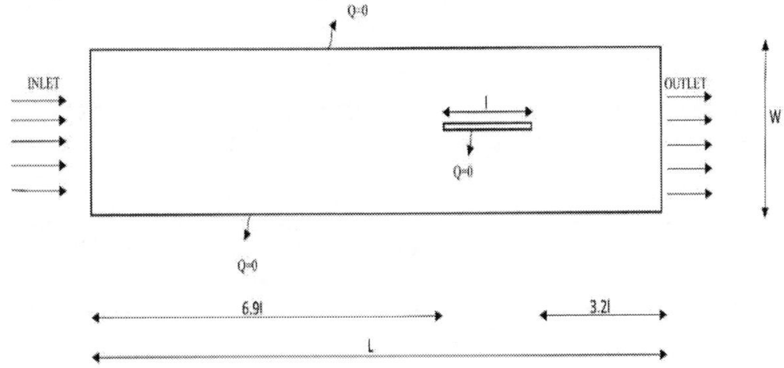

Fig.1. Computational domain

Fig.2. Mesh generated

The domain is designed with an aspect ratio (L/W) of 6. The upstream distance was 6.9l and downstream length was 3.2l. The area of study was around the heater, in order to reduce the disturbances in the flow the heater was placed at the distance of 6.9l. To study the grid independence of the domain, different mesh such as coarse mesh, medium mesh and fine mesh was created for all the power rating and all the velocity. Variation of heat transfer coefficient above the heater is observed for all the cases. The variations of all the cases are shown in the different contours. By analyzing the grid independency mesh with 27460 nodes was selected. For analysis buoyantBoussinesqSimpleFoam solver available with OpenFOAM is used. The mass conservation is given by the equation

$$\frac{\partial \rho}{\partial t} + \nabla.(\rho u) = 0 \tag{1}$$

The momentum conservation is given by the equation

$$\frac{\partial(\rho u)}{\partial t} + \nabla.(\rho u u) = -\nabla \mathrm{p} + \rho g + \nabla.\left(2\mu_{\mathrm{eff}}\mathrm{D(u)}\right) \tag{2}$$

The density ρ in the gravitational term is expressed by the linear function of the temperature,T

$$\rho \approx \rho_o[1 - \beta(T - T_o)] \tag{3}$$

The effective viscosity μ_{eff} is the sum of the molecular and turbulent viscosity and the rate of strain (deformation) tensor D(u) is defined as

$$D(u) = \frac{1}{2}(\nabla \mathrm{u} + (\nabla \mathrm{u})^{\mathrm{T}}) \tag{4}$$

The energy conservation is given by the equation

$$\frac{\partial(\rho e)}{\partial t} + \nabla.(\rho u e) + \frac{\partial(\rho K)}{\partial t} + \nabla.(\rho u K) + \nabla.(p u) = \nabla.\left(\alpha_{eff} \nabla e\right) + \rho u.g \quad (5)$$

Cases with varying power dissipation from heater are studied as the power dissipation of electronic component varies from 10W to 200W for a velocity of range 0.5 m/s to 5 m/s [9].

4 Experimental setup

The test chamber is mainly to enclose the heater into a closed channel of air flow. The entire section is placed on a structure made of ¾ inch 'L' section mild steel bar as shown in Figure 3.

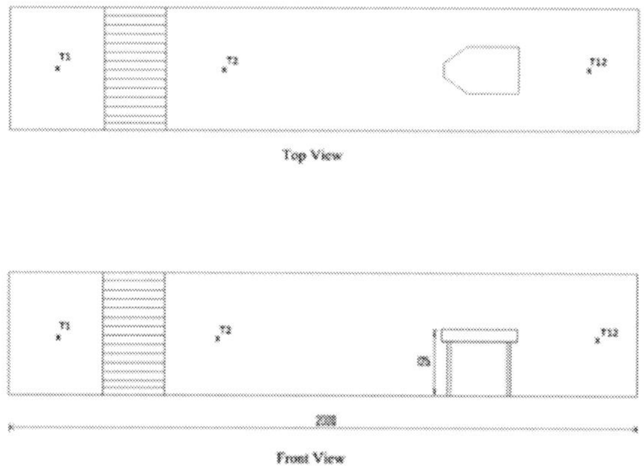

Fig.3. Experimental setup

The function of heater is to simulate an electronic component generating heat 10w to 200w and the heat generation can be controlled. Nickel-Chromium (Nichrome) wire of area 0.035mm2 with thin mica sheet as supporting material is selected as the heater material. Figure 4 shows the heater cross section with thermocouple positions. The heater is placed horizontally on the middle of the test section by supporting it on 3mm bolts. A blower isused for air supply, an autotransformer was used to control the supply voltage to the blower thereby controlling its speed and flow velocity. The honeycomb structure is used to straighten airflow. The honeycomb eliminates whatever unevenness with which flow occurs. Uneven turbulent flows can cause unpredictable forces to be experienced and measured in the test section. The less turbulence there is, the better the experimental setup will simulate actual flow conditions. Honeycomb is very efficient at reducing the lateral turbulence, as the flow pass through long and narrow pipes. But the problem with Honeycombs is, it introduces axial turbulence of the size equal to its diameter which restrains the thickness of the honeycomb. To tackle this problem, Screens are introduced as they reduce longitudinal turbulence very efficiently. The honeycomb can be made of hexagonal cells, like normal honeycomb, but it can also be circular or square cells. The honeycomb is normally placed at the very beginning of the wind tunnel. This is because the honeycomb is more effective when the air is at a lower velocity. In this experiment, a honeycomb structure of circular cells, is employed, which is made of straws. Figure 5 shows photograph of actual experimental setup. Smoke is used as the tracer for air flow.The use of smoke as the tracer particle helps to view the path of the flow. Even then, if the field is illuminated in a plane by appropriate masking of the light source it is possible to examine discrete sections or slices of the flow. So, a laser-lens combo illumination setup is used for visualization.

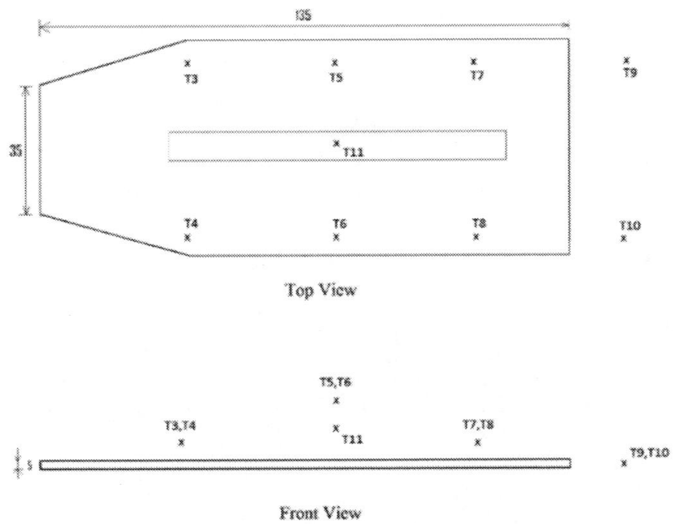

Fig.4. Cross section of heater

Fig.5. Experimental setup

4.1 Experimental procedure

First turn ON the main switch, now turn on the autotransformer switch. Turn the blower autotransformer and using an anemometer the flow velocity is checked. Simultaneously adjust the autotransformer and check flow velocity on anemometer so that the required velocity is obtained. Now turn on the heater switch and adjust the heater autotransformer so that the required power is obtained in the watt meter. Check the ammeter frequently to make sure there is no current over shoot. Check the value of temperature from T1 to T12 at regular intervals of say 10 minutes until steady state is obtained. Once steady state is obtained, the values of temperature from T1 to T12 are noted from the digital temperature indicator. Introduce a smoke source at the inlet of the section. Hold the laser and microscope eye piece on the burette stand and focus the light over the heater. Now observing from the front the flow pattern over the heater can be seen.

For finding the value of Convective heat transfer coefficient from the experimental data, Nusselt number needs to be calculated. Nusselt number is calculated using the following correlations:

Dittus-Boelter correlation, $Nu = 0.023\ Re^{0.8}Pr^{0.4}$ (6)

Gnielinski's correlation

$$Nu_D = \frac{\left(\frac{f}{8}\right)(Re_D - 1000)Pr}{1 + 12.7\left(\frac{f}{8}\right)^{\frac{1}{2}}\left(Pr^{\frac{2}{3}} - 1\right)} \quad (7)$$

$$f = (0.79 \ln(Re_D) - 1.64)^{-2} \quad (8)$$

Boundary layer theory

$$Nu_L = 0.453 \times Re_L^{.5} \times Pr^{0.333} \quad (9)$$

$$\overline{Nu_L} = 2 \times Nu_L \quad (10)$$

With the value of Nusselt number, heat transfer coefficient is calculated using the equation

$$h = \frac{Nu * K}{l} \quad (11)$$

The value of h agreeing closely with numerical results is chosen.

5 Results and discussion

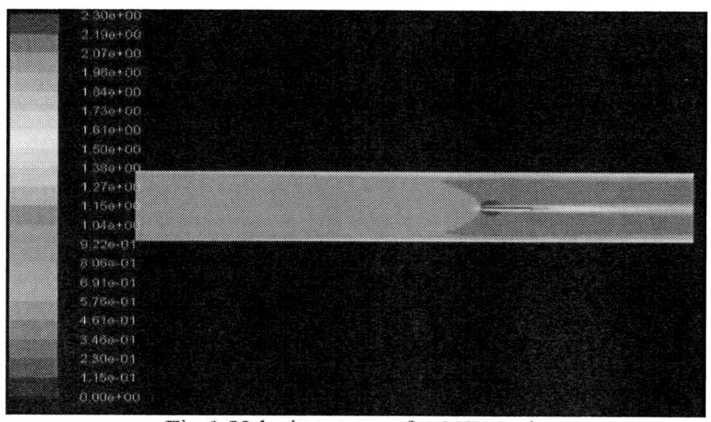

Fig.6. Velocity contour for 80W 2m/s

Figure 6 shows the velocity contour for one of the cases analyzed. The velocity is provided at the inlet. Till the left end of the heater the flow is continuous since there is no obstruction to the flow. But when the flow meets the heater the obstruction will happen and a region of low pressure will form which result in a high velocity at the initial portion. Since no slip condition is provided at the top, bottom and walls of the heater the velocity of the fluid layer adjacent to the wall is zero. As the flow progresses through the surface of the heating element a boundary layer formation will occur. The velocity will increase as the flow proceeds over the heater since the density of the air decreases due to the heating of the surface air. Ambient air temperature and steady state surface temperature of the heater is given at the inlet and the surface of the heating element respectively. The fluid layer adjacent to the surface of the heater will have more temperature.. Under low inlet velocity condition there is a chance for the air surrounding the heater to form convection current due to the variation in density. But at the higher velocities the effect of variation in density is negligible. From the contour the air takes away the majority of the heat generated in the heating element. As the heated air moves to the upstream of the duct the temperature of the air will decrease at a small rate. Figure 7 shows the temperature contour after steady state.

Fig.7. Temperature contour for 80W, 2m/s

Velocity (m/s)	Power (W)	Convective heat transfer coefficient (W/m^{2}K)			
		Analysis	Dittus Boelter correlation	Boundry layer theory	Gnielinski's Correlation
0.5	10	3.348	5.12	10.358	4.816
	40	3.375	5.10	10.357	4.799
	80	3.36	5.08	10.356	4.772
	120	3.613	5.05	10.356	4.741
	160	3.379	5.12	10.339	4.694
	200	3.621	5.21	10.337	4.691
1	10	6.160	8.93	14.649	8.387
	40	6.214	8.91	14.648	8.373
	80	6.177	8.88	14.647	8.347
	120	6.161	8.86	14.646	8.327
	160	6.504	8.83	14.646	8.300
	200	6.482	8.79	14.646	8.269
2	10	11.652	15.53	20.716	14.328
	40	11.717	15.51	20.715	14.313
	80	11.831	15.48	20.714	14.286
	120	11.817	15.45	20.714	14.262
	160	11.290	15.40	20.713	14.218
	200	11.274	15.36	20.713	14.185

5	10	26.75	32.28	32.754	29.06
	40	26.95	32.27	32.753	29.05
	80	26.90	32.24	32.753	29.03
	120	27.44	32.25	32.753	29.04
	160	27.67	32.22	32.752	29.01
	200	27.814	32.19	32.751	28.99

Tab.1. Heat transfer coefficients

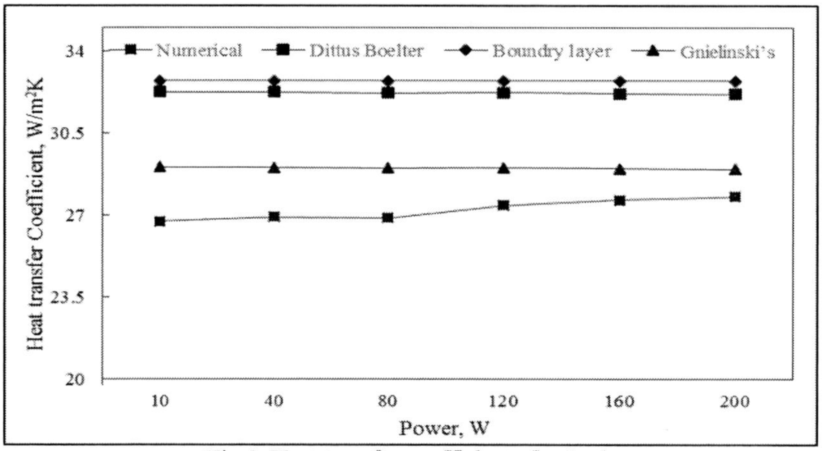

Fig.8. Heat transfer coefficients for 5m/s

From Figure 8, it can inferred that, maximum agreement is between numerically obtained value (hanalysis) and value obtained using Gnielinski's correlation. So, value of heat transfer coefficient obtained using Gnielinski's correlation is taken as hexperimental.

From Figure 9 and 10 of convective coefficient for experimental and numerical results, it is seen that the results show a constant decrease in convective coefficient with increase in heater power for all velocities. This can be justified as the heat generated increases, the heater surface temperature also increases and this heats up the surrounding air much faster. So, the temperature difference between the heater surface and ambient air becomes lesser causing the heat transfer rate to decrease with increase in heat generation. Hence it is very curious to note that for a constant flow, the amount of heat dissipated to air decrease with increase in heat generation. Numerical results also show similar trend with experimental results. But for higher velocities numerical results show some fluctuations. This is mainly because of the limitations in the assumptions made during analysis. Friction was not considered for doing the analysis and frictional effects dominates only at higher velocities. At lower velocities, frictional losses are not very dominant. This justifies the fluctuations at higher velocities. Figure 11 shows the comparison between temperature values from thermocouple and the values obtained from simulation for the thermocouple locations. The agreement between the values validates the numerical model.

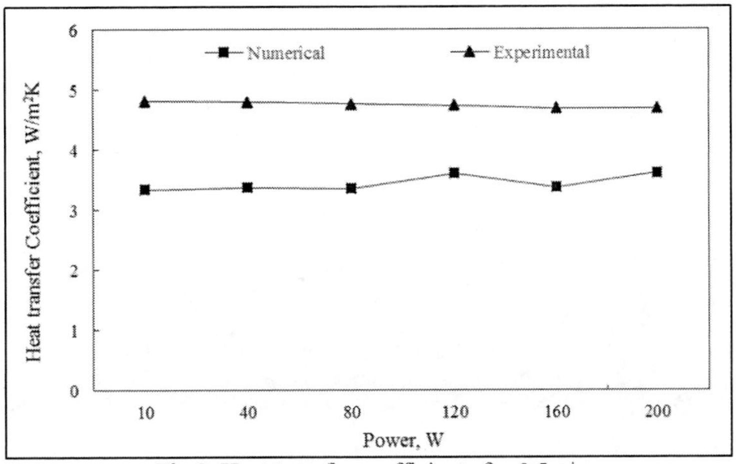

Fig.9. Heat transfer coefficients for 0.5m/s

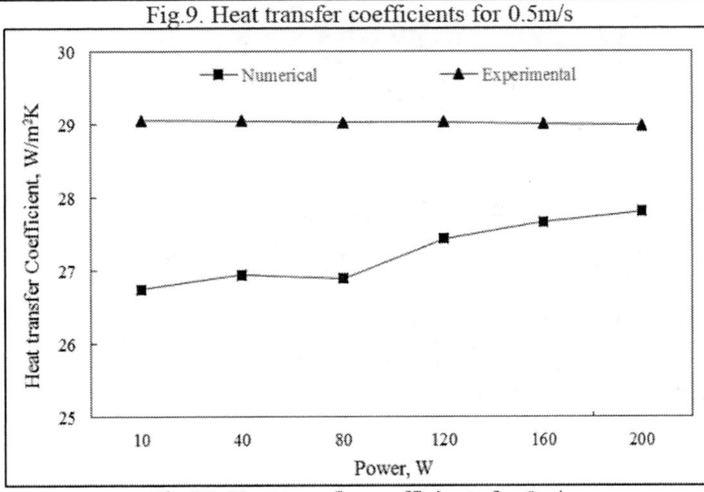

Fig.10. Heat transfer coefficients for 5m/s

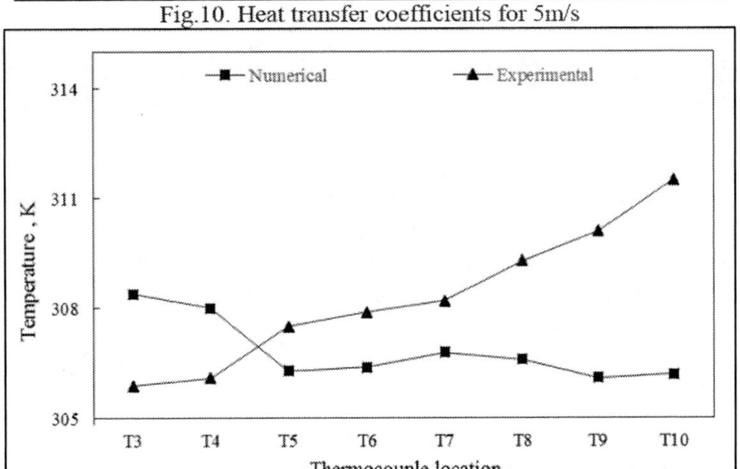

Fig.11. Temperature comparison

Figure 12 shows the comparison between numerical result and the flow visualization observed during experiment.The main feature of the velocity contour is the splitting of the flow stream on hitting the heater and the rising of stream passing above the heater due to density variation caused by the increase in temperature. In experiment also a similar splitting can be observed but some amount of stream is getting diverted to the top and bottom walls. This may be due to absorption of heat by the walls (glass and plywood) where as in analysis the walls are perfectly adiabatic and hence no such stream diversion.

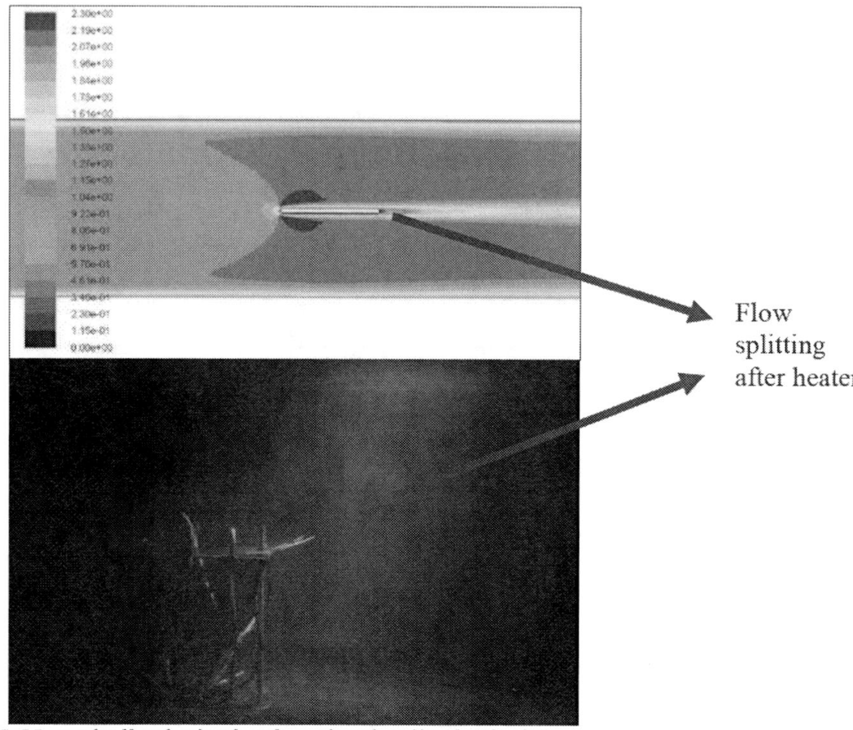

Fig.12. Numerically obtained and smoke visualized velocity contours

6. Conclusions

Numerical and experimental study of heat dissipation from electronic components in closed enclosure has been carried out. The numerical model proposed is validated and can be used to predict the value of average convective heat transfer coefficient for a component with a predictive accuracy of ±3W/m2K, which gives an average error of 13%. The experimental results have been computed using a number of correlations of which Gnielinski's Correlation for flow through ducts gave the most agreeing results. The variation in numerical results from the experimental results is mainly due to heat loss from the source to environment during experiment. Heat loss is not considered in numerical modelling.

References

1. Peter Rodgers, Valerie Eveloy, M.S.J Hashmi, 2005," An investigation into the potential of low Reynolds number Eddy viscosity turbulent flow models to predict electronic component operational temperature" ASME journal of electronic packaging vol.127 pp. 67–75.
2. Domenico Borello, Alessandro Corsini, Giovanni Delibra, Mario Fiorito, Anthony G Sheard, 2013" Large Eddy Simulation of a tunnel ventilation fan" ASME journal of fluid engineering vol.135 pp. 1– 9.

3. Giorgia Nardini, Massimo Paroncini, Raffaella Vitali, 2016," An experimental and numerical analysis of natural convective heat transfer in a square cavity with five discrete heat sources" ASME journal of heat transfer vol.138 pp.1–8.
4. D Senthil Kumar, K Murugesan, Akhilesh Gupta, 2010," Numerical analysis of interaction between inertial and thermosolutal buoyancy forces on convective heat transfer in a lid–driven cavity" ASME journal of heat transfer vol 132 pp.1–11.
5. Arlindo Tribess, Tales Adriano Ferreira, Lorenzzo Victor Schrepel Delmutti, Fernando Stancato, Marcelo Pustelnik, Luis Carlos de Castro Santos, 2005." Analysis of electronic package cooling in an aircraft using CFD" 18th International Congress of Mechanical Engineering
6. Wataru Nakayama, 2013, "Heat conduction in mobile electronic equipment: Study on the effects of some key parameters on heat source temperature based on a 3–layer model" ASME journal of electronic packaging vol. 135 pp.1–5.
7. Nesreen Ghaddar, Kamel Ghali, Mohamad Al–Othmani, Ingvar Holmer, Kalev Kuklane, 2010,"Experimental and theoretical study of ventilation and heat loss from isothermally heated clothed vertical cylinder in uniform flow field" ASME journal of applied mechanics vol.77 pp.1–8.
8. Peter J Rodgers, Valerie C Eveloy, Mark R D Davies,2003, "An experimental assessment of numerical predictive accuracy for electronic component heat transfer in forced Convection–Part 1: Experimental methods and numerical modelling" ASME journal of electronic packaging vol.125 pp.67–75.
9. Peter J Rodgers, Valerie C Eveloy, Mark R D Davies,2003, "An experimental assessment of numerical predictive accuracy for electronic component heat transfer in forced Convection–Part 2: Results and discussion" ASME journals of electronic packaging vol. 125 pp.76–83.
10. M.C Zaghdoudi, A. Teytu., 2000, "Use of heat pipes for Avionic cooling" IEEE, Electronics Packaging Technology Conference pp.425–430
11. M A Hassan, Manabendra Pathak, Mohd. Kaleem Khan,2013, "Natural convection of viscoplastic fluids in a square enclosure" ASME journal of heat transfer vol.135 pp.1–12.
12. Tunc Icoz, Nitin Verma, Yogesh Jaluria, 2006, "Design of air and liquid cooling system for electronic component using concurrent simulation and experiment" Transaction of the ASME vol128 pp.466–478.
13. Z.Xu, C.Kleinstreuer, 2014, "Computational analysis of nano fluid cooling of high concentrstion photovoltaic cells" ASME journal of thermal science and engineering applications vol.6 pp.1–9.
14. Hakan F.Oztop, Fatih Selimefendigil, Eiyad Abu–Nada, Khaled Al–salem , 2016, "Recent developments of computational methods on natural convection in curve linear spaced enclosures" Journal of thermal engineering pp.693–698.
15. Mehmet Arik, Rajdeep Sharma, Jason Lustbaber, Xin He,2013, "Steady and unsteady air impingement heat transfer for electronics cooling applications" ASME journals of heat transfer vol.135 pp.1–8.
16. Yoji Kitamura, Masaru Ishizuka, 2004, "Chimney effect on natural air cooling of electronic equipment under inclination" ASME journal of electronic packaging vol.126 pp.423–428.
17. Mousa M. Mohamed, Mostafa A.Abd El–Baky, 2013, "Air cooling of mini –channel heat sink in electronic devices" journals of electronic cooling and thermal control pp.49–57.

18. Xiaohong Hao, Bei Peng, Gongnan Xie, Yi Chen, 2014, "Thermal Analysis and Experimental Validation of Laminar Heat Transfer and Pressure Drop in Serpentine Channel Heat Sinks for Electronic Cooling" ASME Journal of Electronic Packaging, vol.136 pp.1–9.
19. HR Seyf, M Feizbhakshi, 2012, "Computational analysis of nanofluid effects on convective heat transfer enhancement of micro-pin-fin heat sinks", International Journal of Thermal Sciences, vol 58, pp 168-179.

Solar assisted liquid desiccant cooling using clay based membranes

S. Shanmuga Priya[1*], *Sneha Reddy*[1], *Priyadarshini Balachandar*[1] and *Sanober Wadhwania*[1]

[1]*Department of Chemical Engineering, Manipal Institute of Technology, Manipal Academy of Higher Education, Manipal, Karnataka. India.*

Abstract. The environmental concerns have led to the urge of the usage of non-conventional energy resources like solar, wind, thermal, geothermal etc. which provide enormous source of energy without causing any further diminution of the environment. Instead of the conventional HVAC systems that cause colossal environmental perils, usage of liquid desiccants in coming in vogue whereby reducing ecological threats. Moreover, solar assisted systems provide further impulse to such systems. This paper discusses about the various comparisons between liquid desiccants: Lithium chloride, Potassium formate and Calcium chloride and concludes that potassium formate is the best desiccant to be used among the three. Potassium formate (HCOOK) is used which is cheaper and less corrosive as compared to the other aqueous salts, and has a negative crystallization temperature. Potassium formate is a new liquid desiccant and thus, not much research is available currently. The weather conditions of Manipal provide an appropriate condition for the experimentations of solar aided liquid desiccant evaporative cooling systems due to its humid climate and intense solar radiation obtained. The small scale experimentation also encounters the problem of liquid desiccant carryover by the air flow, with the help of clay based membranes which are again cheap, environmentally benign and obtained in a facile way. The projected system takes complete advantage of pure solar energy aimed at the regeneration of liquid desiccant.

Keywords: Desiccant evaporative cooling – Solar – Liquid desiccant – Clay based membrane

1 Introduction

The rapid increase in world population and economic development has led to the immense deterioration of conventional type of energies with increasing demand for human comfort conditions. The pollution caused by the conventional form of energies and their depletion has headed to the immediate call for alternatives. Hence immerging researches focus on the use of renewable form of energies as the alternative. The usage of air conditioning and heating devices shows a notable role in the emanation of chlorofluorocarbons which causes exhaustion of ozone layer and other environmental concerns. Research is being carried out

* Corresponding author: shan.priya@manipal.edu

in order to decrease the dependency of dehumidification technology on non-renewable resources.

Lowenstein et al. [1] studied on the solar assisted liquid desiccant cooling systems and thereby developed and commercialized a novel design of a liquid desiccant systems to dry and cool supply air of building air conditioners. The new system has a desiccant flooding rate with the factor of 10 to 20 lower than the rates of the presently used conventional packed bed systems which provides a compact system that deeply dries the process air having much lower pressure drop, higher COP and greater cooling effect. Shahab Alizadeh [2] performed the feasibility analysis of a solar assisted liquid desiccant air conditioner developed in Queensland. The arrangement customs an absorber unit of cross flow polymer plate heat exchanger that dries the process air with greater efficiency. The study reveals higher air temperature and relative humidity control on a commercial site of 120 sq. m area in Brisbane. K. Gommed and G. Grossman [3] studied on solar cooling and dehumidification using liquid desiccant and low cost flat plate collectors. Considering several cycle variations, they concluded that the heat and mass transfer in the dehumidification process are considerably affected by the conditions of the inflowing ambient air. They also concluded that the temperature and flow rates of cooling and heating water and the flow rates of the solution through the absorber and regenerator affect the humidity level of the conditioned space. Nia et al. [4] adopted a new approach to study and compare between solid and liquid desiccant cooling systems and concluded that liquid desiccant systems are relatively large without technology improvements, but have lower COP with low capacities on comparison to solid desiccant systems. Mei and Dai [5] reviewed on liquid desiccant dehumidification for air-conditioning applications and concluded that liquid desiccants and easily grasp the humidity ratio, independent control and temperature although they have drawbacks of unstable operation and large dimension. They also concluded that the climatic changes of ambient air condition and solar insolation, hamper in the continuous operation of such systems. Katejanekarn et al. [6] did an experiment on a solar-regenerated liquid desiccant ventilation pre-conditioning system during a period of nine months covering all seasons in the hot and humid climatic conditions of Thailand. Their study reveals that the system could decrease the temperature of the supplied air by about 1.2 C while the humidity ratio was reduced by 0.0042 kgw/kgda equivalent to 11.1% relative humidity reduction. P. Gandhidasan [7] implemented the examination of a solar spaced cooling system using liquid desiccant for ventilation mode and derived an expression to calculate the amount of heat removal from the room to be conditioned through a simplified vapour compression correlation using some known initial parameters. He also conferred the effect of cooling water temperature, ambient air conditions, solution concentration, the efficacy of the dehumidifier and the heat exchanger on the performance of the cooling system. Josh and Jane [8] studied a liquid desiccant storage tank's buoyancy driven mass transfer based on the absorption properties of aqueous calcium chloride for the long term solar thermal storage. The experimentation for certain buoyancy ratios and Rayleigh's numbers produce the corresponding Sherwood numbers needed. Yin and Zang [9] illustrated a relative learning on internally heated and adiabatic regenerators in liquid desiccant air conditioning system and established that internally heated regenerator demonstrated high energy utilization efficiency and could escalate the rate of regeneration. Also, that they can provide comparable regeneration efficiencies and rate of regeneration at lower desiccant flow rates hence proving to be a good alternative to avoid liquid desiccant carryover as compared to adiabatic regenerators. K. Daou et al. [10] reviewed on desiccant air conditioning systems and revealed the feasibility of them in various climatic conditions to prove the benefits offered by them in terms of cost and energy savings. Armando et al. [11] conducted research on novel approach to liquid desiccant assisted cooling for humid environments showing that similar systems have potential to compete with conventional air conditioning units while also being more cost effective and efficient. Li Zhang et al. [12] did general research on a hybrid air cooling system using a liquid desiccant. They found that they received an overall 10% increase in efficiency when the liquid desiccant was used to assist in the cooling process. Zhen Li et al. [13] applied a similar method in a hospital facility in Beijing. They found that along with increased energy efficiency of a desiccant assisted system, they also established a higher quality of air by using more fresh air and using the desiccant to sterilize and clean the air. Huang and Zhang [14] researched on membranes aided in liquid desiccant systems to avoid the carry-over of the liquid desiccant in the air cooling systems. The review collected the data on the recent transport data of various membrane modules and discussed in details about the different types of membranes and several materials used for the membranes. They concluded that the use of membranes was a good option but future work may include multiple stage membrane configuration and a cooling coil in the configuration to offer locally cooling membrane based dehumidifier. Chen et al. [15] executed the experimental analysis on the performance assessment of liquid desiccant dehumidifier and cooling system aided with membrane using calcium chloride solution as the desiccant. They concluded that the performance of the dehumidifier is solely influenced by ambient weather conditions and an increase in cooling output and decrease in effectiveness of the dehumidifier is observed with the increase in the flow rate of air.

After understanding the basics and carrying out a brief literature review, it was observed that calcium chloride and lithium chloride are the most suitable liquid desiccants. They have large saturation absorption capacity, low regeneration temperature, low viscosity, and high heat transfer rate thus they were chosen for experimentation. The performance efficiency of the chosen liquid desiccant was calculated by using the wetted wall column apparatus. Here, the dry bulb and wet bulb temperatures were recorded to find the relative humidity from the psychometric chart.

2 Materials and methods

A clay based membrane was prepared by size reduction followed by drying for a stipulated period of time. Sodium meta silicate, boric acid and sodium carbonate were added to enhance the properties of the membrane. A parabolic trough solar collector is used for the regeneration of the weak desiccant.

Scanning electron microscopic analysis was carried out for the clay based membrane to determine the nature of the membrane. Different desiccants were used as the liquid medium in the wetted wall column set-up. Relative and absolute humidity of the chosen desiccants were determined and compared.

An experimental setup was fabricated that provides dehumidification and air cooling. Efficiencies of the different liquid desiccants were calculated and compared with the values obtained from wetted wall column setup. Performance effectiveness of membrane aided liquid desiccant system was done.

2.1 Preparation of Clay based membrane

Soil brought was dried at 100°C for 24hrs for complete removal of moisture. Then it was crushed with pestle and mortar to remove agglomerations and chucks of sand. This soil was then fed into a ball mill, crushed to fine powder and was sieved with BSS 36 mesh screen sieve to segregate coarse particles. The obtained powder was kept in an air tight container to avoid moisture and stored in a cool, dry place.

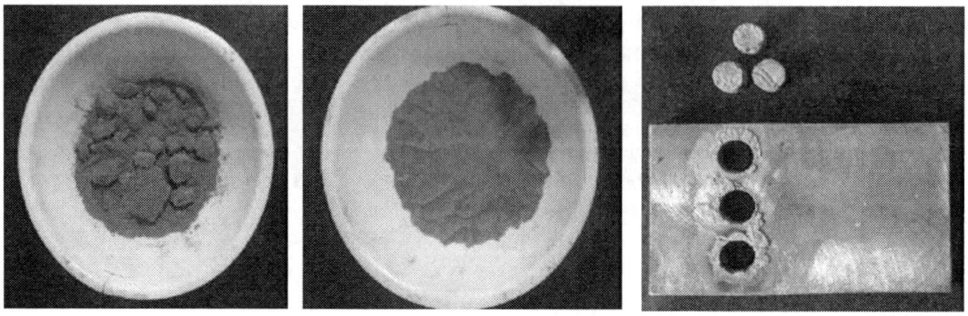

(a) **(b)** **(c)**

Fig. 1. Fabrication of clay based membrane

Fig 1 (a) depicts a sample of the coarse sand that was further processed for the membrane fabrication. Fig 1(b) shows the sand after size reduction techniques. Fig 1 (c) shows the membrane samples in the pellet form.

Low-cost ceramic-based membrane support was arranged. The membrane was prepared by paste casting from the composition of the given materials: clay (70%), sodium carbonate (3%), Sodium meta silicate (1.5%), and Boric acid (1.5%). The ingredients were made into a paste with distilled water(24%) and was casted over an aluminum foil in the shape of a circular molds. After restricted drying in room temperature for 24hrs, the disc was removed carefully and heated at 373K for 12 hours. The cast disc was heated at 1123K for 3 hours. in a furnace. Temperature of the furnace was gradually ventilated to room temperature. The support was finally polished with silicon carbide abrasive paper(C-180), to give it a final form. Sodium meta silicate increases mechanical strength by creating silicate bonds. Boric acid also increases mechanical strength by creating metaborates during sintering. Sodium carbonate improves dispersion properties, thereby creating homogeneity.

2.2 Experimental Unit

A liquid desiccant dehumidification system comprises of an absorber and a regenerator. Inside the dehumidifier, the liquid desiccant absorbs moisture to dehumidify the hot and humid process. A cuboidal shaped container is used for storing the liquid desiccant, to which a fan is attached on one end to draw the hot and humid air. The other end has a combination of fan, heat sink and peltier cooler to cool and dehumidify the air. Lithium chloride and potassium formate are the two desiccants used in this experimental study. Potassium formate is chosen because it is less corrosive and more economical. Lithium chloride is one of the most commonly used desiccants. A comparative study is made between the two desiccants. The weak desiccant flows from the absorber to a flat plate collector. The solar energy is used to evacuate the moisture trapped in the desiccant. The weak desiccant is heated to an elevated temperature so as to remove the moisture absorbed by it.

The ambient room temperature was measured. The peristaltic pumps were switched on to circulate the liquid desiccant through the solar collector. This is done in order to regenerate the liquid desiccant. Power supply is provided to the set up and humid air is drawn in. After a certain time interval, temperature of the hot air at the heat sink and the temperature of the cool air at the peltier cooler are measured. These values are used to measure the percentage humidity and the relative humidity of the air stream.

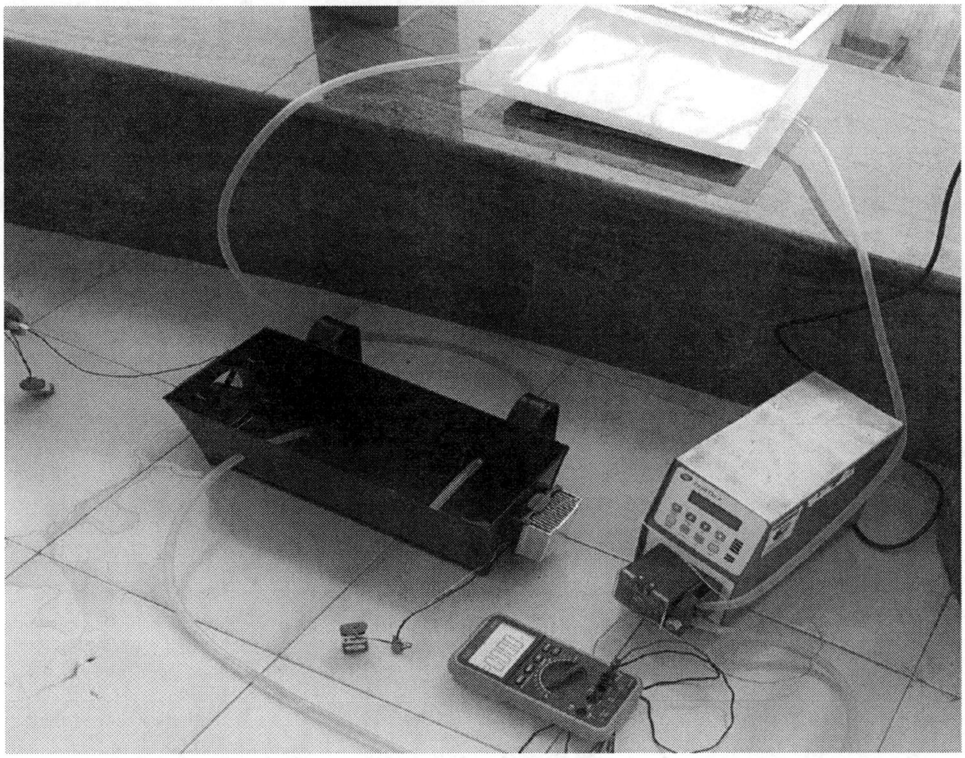

Fig. 2. Experimental set up

The main experimental setup shown in Fig 2 consists of a fan which draws in the hot humid air from outside, a rectangular box which has liquid desiccant to absorb the moisture from the air. On the other side of the box there is a system of fan, heat sink and peltier cooler which draws the heat from the air thus making it cooler. The used desiccant flows to the regenerator which is the flat plate solar collector where the moisture is removed and the desiccant is renewed. Peristaltic pumps are used to force the desiccant to flow through the solar collector.

The multimeter gives us the voltage across the junctions which can be used to determine the temperature of air inside the experimental unit. The fans are attached for inlet and outlet. The fans are powered with normal 12V batteries. The container holds the desiccant and is connected with pump and the solar collector with pipes for desiccant regeneration. The peristaltic pump was used to circulate the desiccant to solar collector and back to the desiccant storage unit.

Fig. 3. Experimental Setup with the membrane

The Fig 3 shows the setup with the membrane which helps preventing the carryover.

3 Results and discussion

3.1 Clay based membrane

The micro-porous clay membrane was fabricated as given in the methodology above. The sample, in a pellet form was given for scanning electron microscope (SEM) analysis.

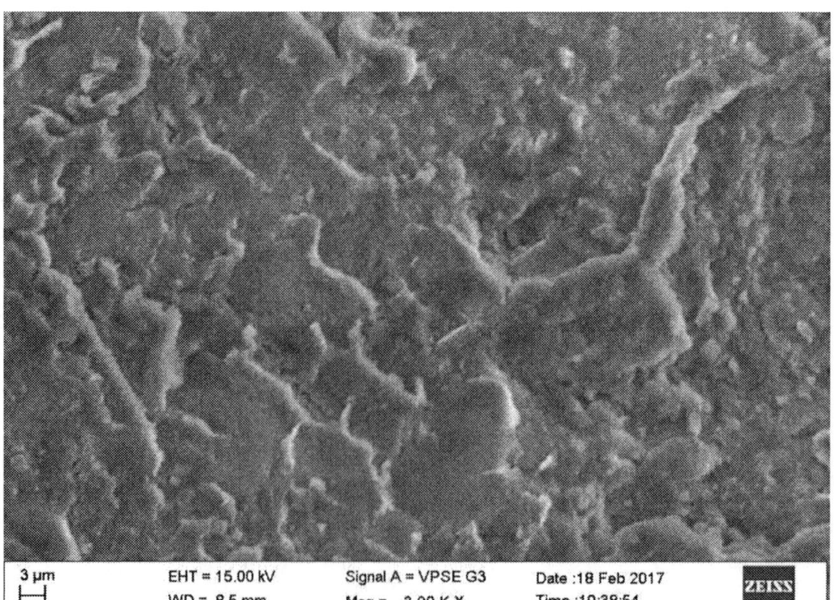

Fig. 4(a). 3 μm SEM analysis

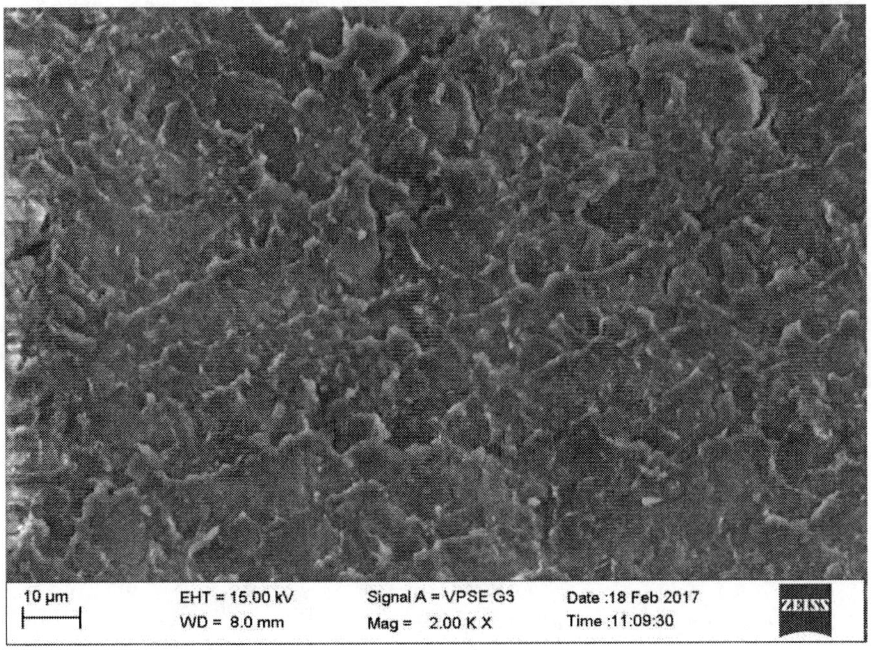

Fig. 4(b). 10 µm SEM analysis

Fig 4a and 4b shows the porous structure of the clay based membrane obtained by the scanning electron microscopic analysis of 3µm and 10µm respectively.

Table 1. ImageJ results of Scanning Electron Microscope images

Trial	Label	Area	Mean	%Area
1	CBM-2-07	703488	106.756	7.541
2	CBM-2-04	703488	134.579	12.66
3	CBM-2-09	671775	97.737	12.804
4	CBM-1-02	697330	127.675	12.973
5	CBM-2-10	705180	97.247	13.59
6	CBM-1-06	702801	97.646	17.316
7	CBM-1-08	697343	169.21	19.792
8	CBM-2-06	697680	102.268	21.492
9	CBM-2-11	697004	90.936	22.309
10	CBM-1-13	695301	115.33	35.065

In Table 1, area refers to the area of the sample that has been taken into consideration for the measurement of porosity. % Area refers to the percentage of the area that is porous/empty in nature.

3.2 Efficiency of solar flat plate collector

The set up consists of an aluminium solar collector with a silicon tube having high heat tolerance that allows the water to flow through and get heated.

The following procedure was adapted to calculate the efficiency of the collector. First the ambient air temperature was measured, then the pipes were filled with water and at different intervals of time the rise in temperature was measured. Efficiency of the flat plate solar collector was measured using the formula:

$$\eta = [m \times Cp \times (T - Ta)] \;/\; [A \times Hb \times t]$$

where:

η = Efficiency of the collector

m = Mass of the water in the pipe in kg

C_p = Specific heat in kJ/kgK

T = Temperature of water in the pipe in K

T_a = Temperature of ambient air in K

A = Collector area in m^2

H_b = Intensity of solar radiation in W/m^2

t = Time in s

Table 2. Efficiency of flat plate collector

TRIAL	VOLUME (l)	TIME (s)	TEMPERATURE (°C)	EFFICIENCY (%)
1	1.50E-05	6	42	5.064
2	1.50E-05	12	44	6.076
3	1.50E-05	18	45	6.583
4	1.50E-05	24	46	7.089
5	1.50E-05	30	46	7.089

Table 2 shows that the efficiencies of the solar flat plate collector generally used is about 50-60% but since we are using a scaled down version of the model the efficiencies lie between 5-7%. Given parameters are T_a as 32°C, C_p as 4200 J/kgK, H_b as 650 W/m^2, A_c as 0.0638 m^2 and ρ as 1000 kg/m^3. The lower efficiencies would not make a difference on the model's efficiency as it will be a continuous operation.

3.3 Experiment with Calcium chloride:

Dry bulb temperature (°C) = 41.00

Wet bulb temperature (°C) = 36.00

Relative humidity (%) = 64.00

P_{ws} (kPa) = 4.33

P_w (kPa) = 4.55

Absolute humidity (g/m³) = 29.90

Table 3. Experimental observations- Calcium chloride

Trial	Voltage (mV)	Dry bulb Temperature (°C)	Wet bulb Temperature (°C)	Relative Humidity (%)	Absolute humidity (g/m³)	P_{ws} (kPa)	P_w (kPa)
1	1.96	37.20	33.20	42.00	22.00	3.15	3.98
2	1.98	37.50	33.50	43.00	23.38	3.35	4.16
3	1.95	37.00	33.00	42.00	22.02	3.15	3.99
4	1.94	36.73	32.73	41.00	20.99	3.00	3.86
5	1.93	36.52	32.52	41.00	21.00	3.00	3.87

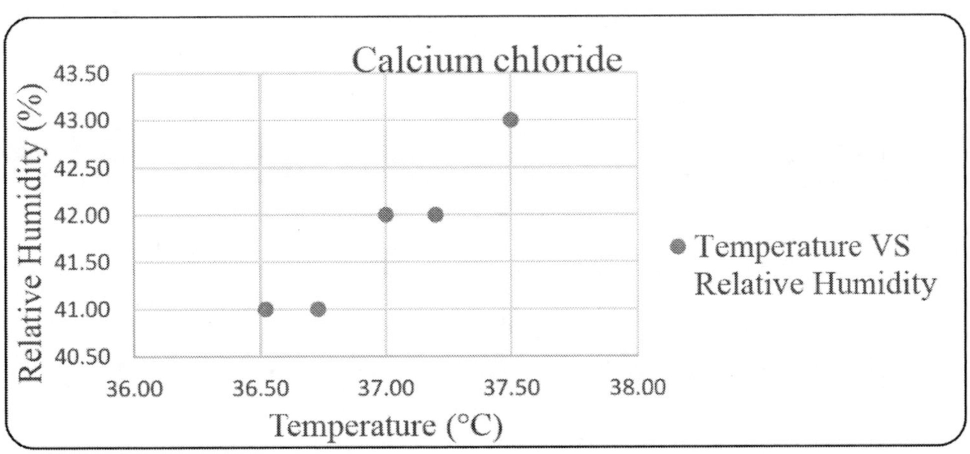

Fig. 5. Relative humidity VS Temperature – Calcium chloride

From the above table 3 and graph of fig.5 it can be inferred that calcium chloride, as a desiccant allows a drop in humidity by 20%. Greater the temperature drop, lower is the relative humidity. Errors in the above results can be due to improper control of flowrates and measurement.

3.4 Experiment with Lithium chloride

Dry bulb temperature (°C)	=	41.00
Wet bulb temperature (°C)	=	36.00
Relative humidity (%)	=	64.00
Pws (kPa)	=	4.33
Pw (kPa)	=	4.55
Absolute humidity (g/m³)	=	29.90

Table 4. Experimental observations- Lithium chloride

Trial	Voltage (mV)	Dry bulb Temperature (°C)	Wet bulb Temperature (°C)	Relative Humidity (%)	Absolute humidity (g/m³)	P_{ws} (kPa)	P_w (kPa)
1	2.13	40.20	36.20	60.00	35.28	5.10	5.09
2	2.14	40.50	36.50	59.00	35.59	5.15	5.18
3	2.09	39.50	35.50	58.50	32.94	4.75	4.85
4	2.08	39.20	35.20	58.00	29.50	4.25	4.37
5	1.95	37.00	33.00	41.00	22.02	3.15	4.04

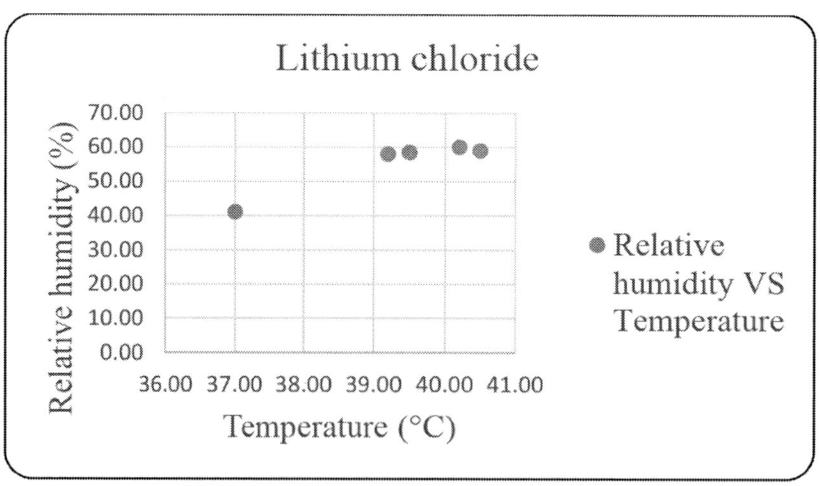

Fig. 6. Relative humidity VS Temperature – Lithium chloride

From tables 4 and graph of fig.6, it can be inferred that lithium chloride, as a desiccant allows a drop in humidity by about 7%. Greater the temperature drop, lower is the relative humidity. Errors in the above results can be due to improper control of flowrates and measurement.

3.5 Experiment with Potassium formate

Dry bulb temerature (°C)	=	41.00
Wet bulb temperature (°C)	=	36.00
Relative humidity (%)	=	64.00
Pws (kPa)	=	4.33
Pw (kPa)	=	4.55
Absolute humidity (g/m³)	=	29.90

Table 5. Experimental observations- Potassium formate

Trial	Voltage (mV)	Dry bulb Temperature (°C)	Wet bulb Temperature (°C)	Relative Humidity (%)	Absolute humidity (g/m³)	P_{ws} (kPa)	P_w (kPa)
1	2.02	38.30	34.30	58.00	30.97	4.45	4.62
2	2.00	37.90	33.90	55.00	29.62	4.25	4.57
3	2.01	38.20	34.20	57.00	30.29	4.35	4.57
4	1.99	37.85	33.85	55.00	27.19	3.90	4.20
5	1.98	37.75	33.75	54.00	26.15	3.75	4.09

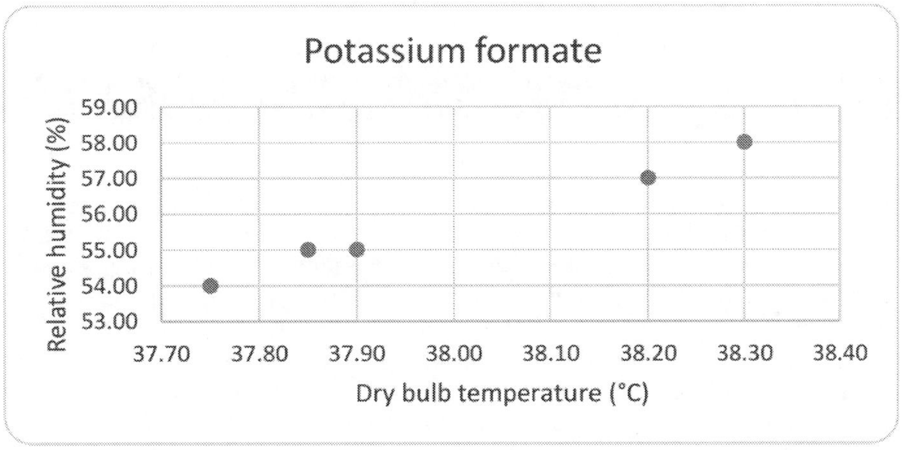

Fig. 7. Relative humidity VS Temperature – Potassium formate

From tables 5 and graph of fig.7, it can be inferred that potassium formate, as a desiccant allows a drop in humidity by 9%. Greater the temperature drop, lower is the relative humidity. Errors in the above results can be due to improper control of flowrates and measurement.

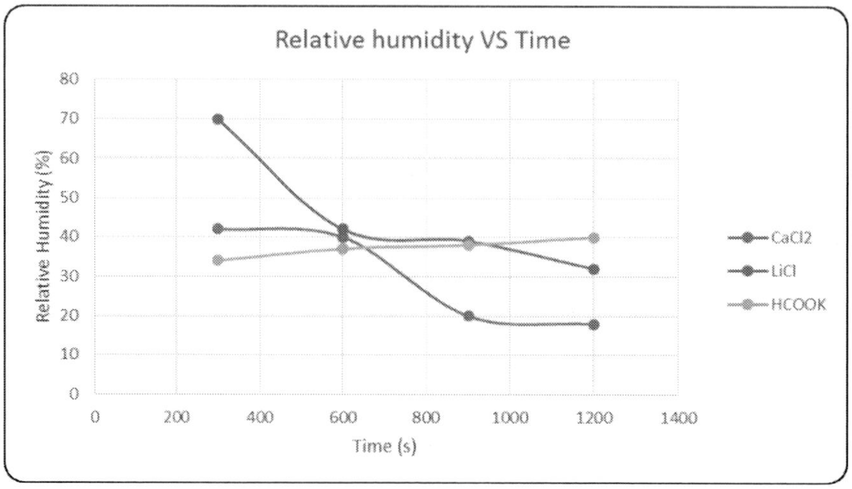

Fig. 8. Comparison of different desiccants with respect to relative humidity

The above graph shows the dehumidification property of the three desiccants with respect to time. Though calcium chloride and lithium chloride show a decrease in relative humidity, potassium formate has been chosen because it is less corossive than the rest. It has been studied that potassium formate's regenration capabilities are better than lithium chloride and calcium chloride.

4 Conclusions

The objective of this project was to reduce the dependency of the cooling systems on conventional and pollution causing energy sources. The utilization of liquid desiccant for air cooling is still emerging. Manipal has hot and humid summers and the temperature reaches values as high as 38°C and this provides an operative ground for the application of solar energy. The assistance of membranes decreases the carry-over of liquid desiccant into the air stream.

After an extensive literature review the liquid desiccants for the project were selected. Calcium chloride was selected because it is cheap and readily available whereas lithium chloride is the most commonly used desiccant. Potassium formate, a new desiccant, was chosen because of its regeneration capacity. The membrane was prepared to reduce the carry over and its topography was seen using the scanning electron microscope. The porosity was found using the IMAGEJ software. The effectiveness of the solar collector was calculated. Experimental Setup was fabricated using pump, solar collector, fans for inlet and outlet and membrane to reduce the carryover. The absolute and relative humidity of the three desiccants were found. The porous nature of the clay based membrane was accounted for with the help of IMAGEJ software. The solar collector has an efficiency value that is appropriate for the scale up of the set up. The fabricated experimental setup was proven to be effective upon running the experiment with the three chosen desiccants.

Acknowledgements

The authors would like to prompt their deep appreciation and indebtness to the Department of Chemical Engineering, Manipal Institute of Technology, Manipal Academy of Higher Education, Manipal by entertaining our requirements and facilitating all the provisions utilized for the project.

References

1. Lowenstein, A., Slayzak, S. and Kozubal, E., 2006, July. "A zero carryover liquid-desiccant air conditioner for solar applications". ISEC.
2. Alizadeh, S., 2008. "A feasibility study of using solar liquid-desiccant air conditioner in Queensland, Australia". Journal of Solar Energy Engineering, 130(2), p.021005.
3. Gommed, K. and Grossman, G., 2004. "A liquid desiccant system for solar cooling and dehumidification". Transactions-american society of mechanical engineers journal of solar energy engineering, 126(3), pp.879-885.
4. Nia, F.E., Phelan, P.E. and van Paassen, D., 2011. "A new approach to study and compare the annual performance of liquid and solid desiccant cooling systems". Journal of Thermal Science and Engineering Applications, 3(2), p.021002.
5. Mei, L. and Dai, Y.J., 2008. "A technical review on use of liquid-desiccant dehumidification for air-conditioning application". Renewable and Sustainable Energy Reviews, 12(3), pp.662-689.
6. Katejanekarn, T., Chirarattananon, S. and Kumar, S., 2009. "An experimental study of a solar-regenerated liquid desiccant ventilation pre-conditioning system". Solar Energy, 83(6), pp.920-933.
7. Gandhidasan, P., 1990, August. "Analysis of a solar space cooling system using liquid desiccants". In Energy Conversion Engineering Conference, 1990. IECEC-90. Proceedings of the 25th Intersociety (Vol. 5, pp. 162-166).
8. Quinnell, J.A. and Davidson, J.H., 2013. "Buoyancy driven mass transfer in a liquid desiccant storage tank". Journal of Solar Energy Engineering, 135(4), p.041009.
9. Yin, Y. and Zhang, X., 2010. "Comparative study on internally heated and adiabatic regenerators in liquid desiccant air conditioning system". Building and Environment, 45(8), pp.1799-1807.
10. Daou, K., Wang, R.Z. and Xia, Z.Z., 2006. "Desiccant cooling air conditioning: a review". Renewable and Sustainable Energy Reviews, 10(2), pp.55-77.
11. Oliveira, Armando C., et al. "Thermal performance of a novel air conditioning system using a liquid desiccant". Applied Thermal Engineering 20 (13) (2000): 1213-1223.
12. Zhang, Li, Chaobin Dang, and Eiji Hihara. "Performance analysis of a no-frost hybrid air conditioning system with integrated liquid desiccant dehumidification". International journal of refrigeration 33(1) (2010): 116-124.
13. Li, Zhen, et al. "New type of fresh air processor with liquid desiccant total heat recovery".Energy and Buildings 37 (6) (2005): 587-593.

Numerical Analysis of Heat transfer Enhancement in a double pipe heat exchanger with a holed twisted tape

Akarsh Kumar[1], *Ujjawal Sureka*[1] and *Shiva Kumar*[1*]

[1]*Dept of Mechanical Engineering, Manipal Institute of Technology, Manipal Acedemy of Higher Education, Manipal- 576104*

Abstract: In the present study numerical analysis of enhancement in heat transfer characteristics in a double pipe heat exchanger is studied using a holed twisted tape.The twisted tape with a constant twist ratio is inserted in a double pipe heat exchanger. Holes of diameter 1mm, 3 mm and 5 mm were drilled at regular pitch throughout the length of the tape. Numerical modeling of a double pipe heat exchanger with the holed twisted tape was constructed considering hot fluid flowing in the inner pipe and cold fluid through the annulus.Simulation was done for varied mass flow rates of hot fluid in the turbulent condition keeping the mass flow rate of cold fluid being constant. Thermal properties like Outlet temperatures, Nusselt number, overall heat transfer coefficient, heat transfer rate and pressure drop were determined for all the cases. Results indicated that normaltwisted tape without holes performed better than the bare tube. In the tested range of mass flow rates the average Nusselt number and heat transfer rate were increased by 85% and 34% respectively. Performance of Twisted tape with holes was slightly reduced than the normal twisted tape and it deteriorated further for higher values hole diameter. Pressure drop was found to be higher for the holed twisted tape than the normal tape.

Key words: Twisted tape, Nusselt number, heat transfer rate, pressure drop

1 Introduction

Double pipe heat exchangers are considered to be the simple heat exchangers with the fact that they are simple in construction and easy to manufacture. The disadvantage is the length of the heat Exchanger for increased heat transfer. The process of improving the performance of heat transfer device is referred to as heat transfer enhancement technique. In the recent years high cost of the energy and material has necessitated the use of efficient heat transfer equipment. The major challenge is to make the heat transfer equipment a compact one and to achieve a high heat transfer with minimum pressure drop. Augmentation techniques in heat transfer devices reduces the thermal resistance and increase the convective heat transfer. Augmentation techniques can be classified into 3 types. Active techniques which involves some external power input for the enhancement in heat transfer. Examples include

[*] Corresponding author: shiva.kumar@manipal.edu

mechanical aids, fluid vibration electrostatic fields, surface vibration and jet impingement [1-2]. Passive methods include modifications in geometry and surface to the flow channel by incorporating inserts or additional devices. They promote higher heat transfer by disturbing or altering the existing fluid flow behavior. Examples of this type includes rough surfaces, extended surfaces inserts, addition of additives etc. In the compound method combination of the above two techniques have been adapted. [3-4]. Smith Eiamsa et al. [5] experimentally studied the heat transfer characteristics in a circular pipe using normal twisted tape and also using alternate clock wise and anti-clock wise twisted tapes. Experiments were conducted in turbulent condition. They analyzed heat transfer rate and friction factor for different conditions of tapes in the turbulent regime. They observed that alternate clock wise and counter clockwise tapes were better than the normal tapes. Thianpong et al. [6] studied heat transfer characteristics using a helically coiled twisted tape. The experiments were conducted using HTTs (Helical twisted tapes) with three twist ratios (y/W) of 2, 2.5 and 3, three helical pitch ratios (p/D) of 1, 1.5 and 2. They concluded that heat transfer rate and friction factor increase as the tape twist ratio and helical pitch ratio decrease, while the thermal performance shows opposite trend. Murugesan et al. [7] carried experiments using V cut twisted tapes and compared the results with the conventional tape. At the same Reynolds number thermo-hydraulic performance of the V cut tapes were better than the normal tapes. Kapatkar et al. [8] carried out experimental study of using a full length twisted tape in a circular pipe under laminar flow conditions. Three different materials were chosen for the twisted tape include aluminum, stainless steel and insulated. Twist ratio of the tapes varied from 5.2 to 3.4. Twisted tapes showed higher nusselt number than smooth ones. For Aluminum tapes, the maximum improvement in Nusselt number range from 50% to 100%, 40 to 94% for stainless steel and 40 -94% for insulated tape. The friction factor for the flow with the twisted tape inserts are 340% to 750 % higher as compared with those of smooth tube flow. According to the author's knowledge literature regarding the holed twisted tape is scare. Hence in the present work twisted tape of uniform twist ratio with constant diameter holes throughout the length were considered for the analysis. Numerical modeling of the above geometry was done using Ansys workbench 14.5 and simulation was done with Reynolds number ranging from 2400- 6000. Heat transfer characteristics like Nusselt number, heat transfer coefficient and pressure drop were calculated. The results so obtained were compared with the results obtained by the simulation of bare tube and the tube with normal twisted tape.

2 CFD modeling

The geometric model of the bare double pipe heat exchanger was constructed using work bench in ANSYS 14.5 environment. Specifications selected for the bare double pipe heat exchanger are given in Table 1. The complete domain consists of 45596 Elements. Quality of the mesh was verified by performing grid independence. Initially at the inlet, temperature and mass flow rate of the cold and hot fluid were specified. At the outlet, a pressure outlet boundary is enforced. SIMPLE algorithm was used to solve pressure velocity. K-€ turbulent model was used in the analysis. For momentum turbulent kinetic energy and turbulent dissipation rate with second order upwind scheme was used. For the energy equation also second order upwind was used. Conservation equations were solved for the control volume to yield the velocity and temperature fields for the water flow in the heat exchanger. Convergence was affected when all the residuals fell below 10^{-6} in the computational domain. A sample residual plot for the above condition is shown in the Fig. 1. Simulations were done by changing the mass flow rates of hot fluid (0.005, 0.01.0.02 and 0.03 kg/s) while mass flow rate of cold fluid being fixed at 0.01kg/s. Heat transfer characteristics like Nusselt number, outlet temperatures were determined. Numerically simulated results were compared with analytical results calculated by using Dittus Boelter equation $Nu = 0.023Re^{0.8}Pr^{0.33}$ for the above mass flow rates. The average error for Nusselt number is within 2.5% and are shown in the Fig. 2. Hence the use of CFD modeling for studying the heat transfer characteristics in a double pipe heat exchanger can be employed with confidence.

Table 1. Specifications of the numerically simulated bare double pipe heat exchanger and twisted tape

S. No.	Specifications	Dimension (mm)
1	Inner diameter	9.5
2	Thickness of the inner tube	1.5
3	Inner diameter of the outer tube	28.5
4	Thickness of the outer tube	2
5	Length of the test section of heat exchanger	500
6	Cross section of the twisted tape	7 mm×0.7 mm
7	Pitch of the tape	60
8	Hole diameters in the twisted tape	1, 3 and 5

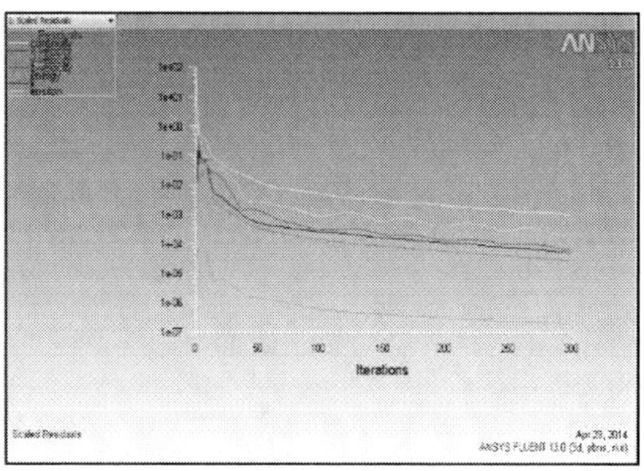

Fig. 1. Residual plot

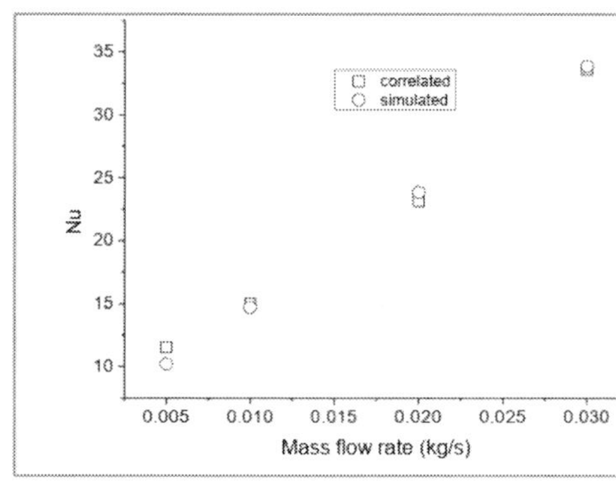

Fig. 2. Comparison of simulated and correlated values of Nusselt number for different mass flow rates of hot liquid

Geometrical modeling of the twisted tape with and without holes in a double pipe heat exchanger was constructed as shown in the Fig. 3 and 4. Specifications of the twisted tape have been mentioned in Table 1. For a holed twisted tape the holes of diameter 1mm, 3 mm and 5 mm drilled centrally along the width of the tape. Pitch of the holes were maintained constant as that of the tape. Meshing is done by using tetrahedral elements. Numerical simulation was done for the double pipe heat exchanger using normal twisted tape, tape with holes of 1mm, 2 mm and 3 mm respectively. Hot water inlet temperature and cold water inlet temperatures together with their mass flow rates were mentioned as the inlet boundary conditions. At the outlet Pressure boundary condition is imposed. Simulation was done for different mass flow rates of hotwater keeping mass flow rate of cold water fixed at 0.01 kg/s as mentioned in the previous section. Outlet cold water temperature, Nusselt number and overall heat transfer coefficient and pressure drop were determined.

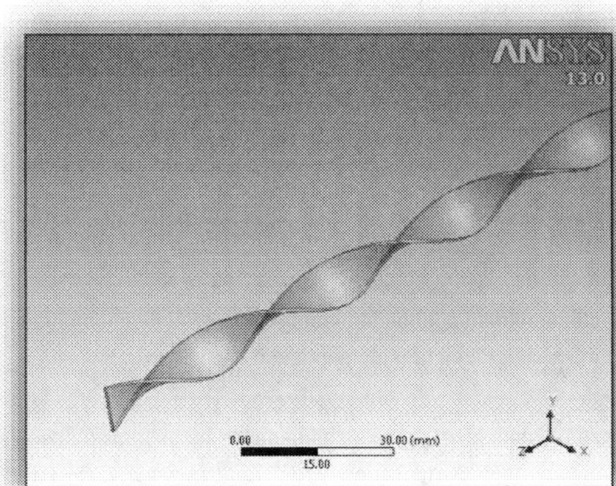

Fig. 3. A twisted tape insert

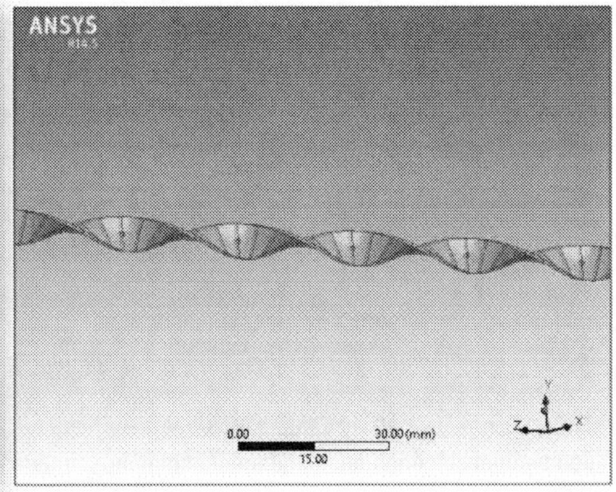

Fig. 4. A twisted tape with holes drilled centrally along the width

3 Results and discussions

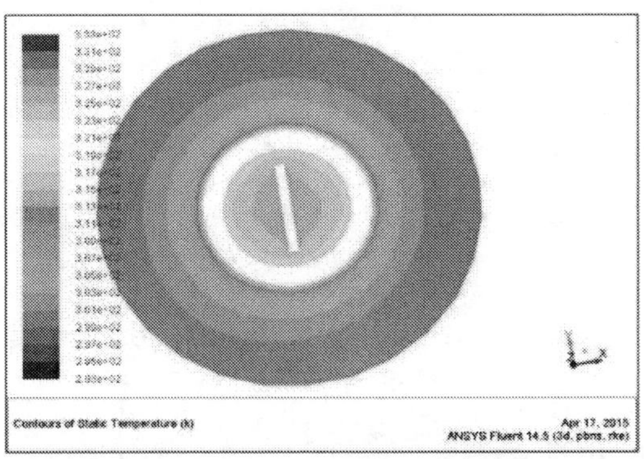

Fig. 5. Temperature profile at the outlet face of the cold water

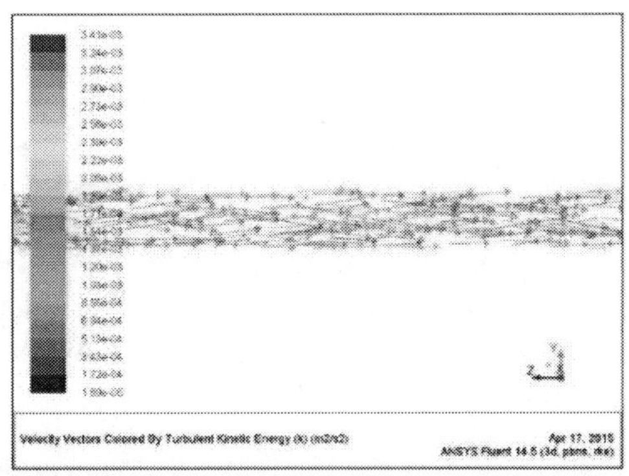

Fig. 6. Vector plot of the velocity on an iso surface showing temperature distribution near

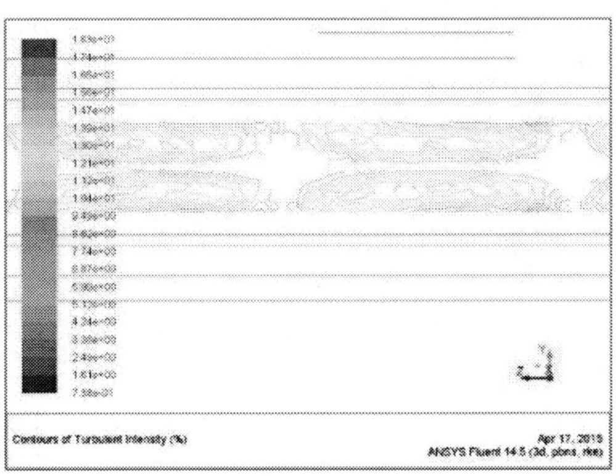

Fig. 7. Turbulent intensity plot on an iso surface in the inner tube with normal twisted tape

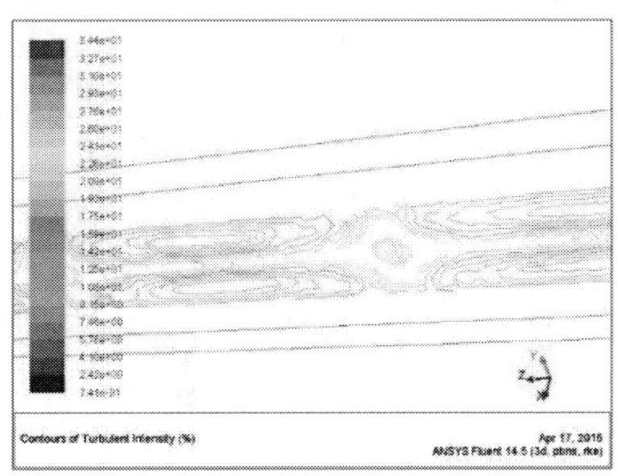

Fig. 8. Turbulent intensity plot on an iso surface in the inner tube with holed twisted tape

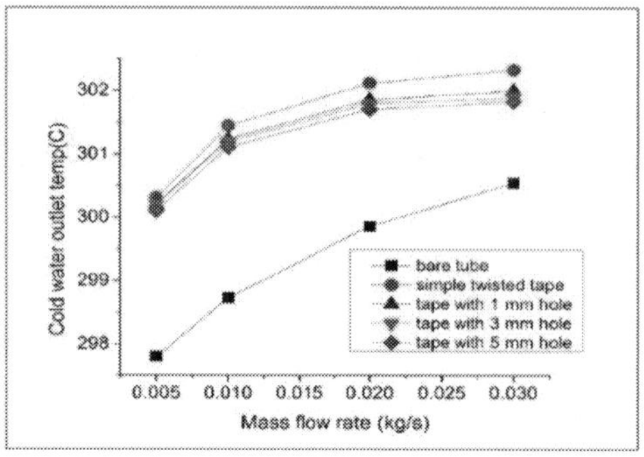

Fig. 9. Variation of cold water outlet temperature with mass flow rates of hot water for bare, twisted tape with and without holes

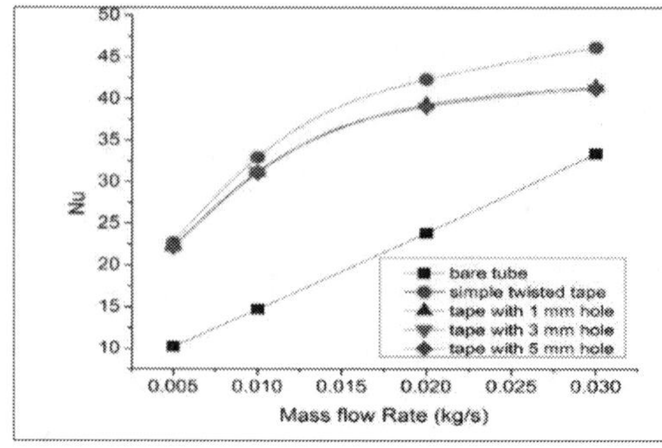

Fig. 10. Variation of Nusselt number with mass flow rates of hot water for bare, twisted tape with and without holes

Numerical simulation reveals that average Cold water outlet temperature is found to be increasing with mass flow rates and is higher for twisted tape than the bare tube. Fig. 5 represents the temperature distribution plot on the cold water outlet face for tube with a twisted tape for a mass flow rate of 0.01 kg/sec for both cold and hot water. Since the geometry of the twisted tape continuously changes this introduces swirl into the bulk flow there by disrupting the boundary layer at the surface. It induces turbulence and superimposed vortex motion causes the higher heat transfer rate. The velocity plot and turbulence plot for the normal twisted tape are shown in Fig. 6 and 7 respectively. Fig. 8 represents the turbulence plot for the twisted tape with holes. For the holed twisted tapeslight increase in turbulence than the normal tape is insufficient to overcome heat conduction that would have been taken place from the region of holes in case of normal twisted tape. This may led to little reduced performance and hence reduced outlet temperature than the normal tape.

Fig. 9 shows the comparison of variation of outlet temperature of the cold water with mass flow rates of bare pipe, pipe with twisted tape and twisted tape with holes under turbulent conditions. It is observed that the average outlet temperature of the cold water is least for bare tube on comparing it with different cases. In case of tube with twisted tape the turbulence induced in the flow path reduces the boundary layer thickness and hence increases the heat transfer with increasing the cold water outlet temperature. The average outlet temperature for a twisted tape is 90% higher than the bare tape under tested conditions. For increased mass flow rates this difference between putting a limit on the Reynolds number for better performance using a twisted tape as is clear from the Fig. 9. For holed twisted tape the outlet temperature was found to be less than the tube with twisted tape for all mass flow rates. On comparing the outlet temperature for the different hole diameters, it was observed outlet temperature drops marginally on increasing the hole diameters. For higher diameters less swirling action which results in reduced turbulence and hence reduced heat transfer rate which resulted in reduced outlet temperatures.

Fig. 10 shows the variation of Nusselt number with mass flow rate for the five cases. Nusselt number was found to increasing with mass flow rates for all the cases. For the bare tube it is found to be lesser than the tube with the twisted tape as well as with holes. For a twisted tape insert the effective hydraulic diameter reduces which alternatively increases the fluid velocity. This in together with the swirling motion increases the Reynold's number and hence Nusselt number. For a holed tube case it is found to be lesser than the normal twisted tube showing its deteriorated performance when compared with the former. As the hole diameter was increased the Nusselt number values decreased marginally showing the poor performance with large diameter holes. It was found that the average Nusselt number increased by 85% for twisted tape than the bare tube.

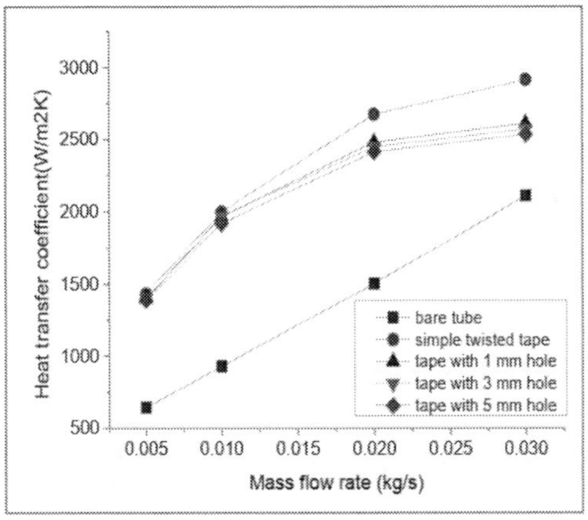

Fig. 11. Variation of heat transfer coefficient with mass flow rates of hot water for bare, twisted tape with and without holes

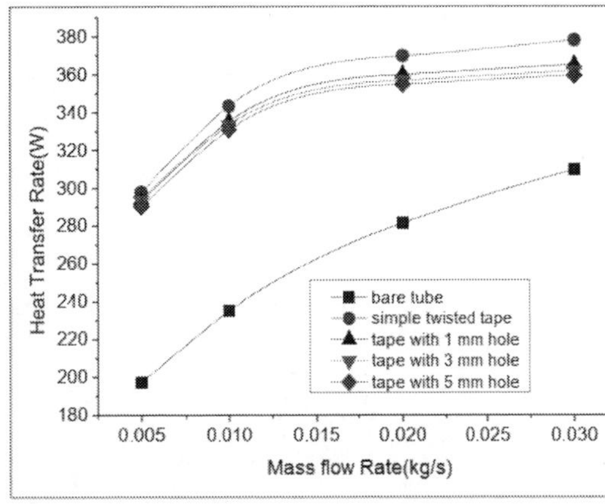

Fig. 12. Variation of heat transfer rate with mass flow rates of hot water for bare, twisted tape with and without holes

Fig. 11 shows the plot of heat transfer coefficient for the different cases. It was again observed that it increases with the mass flow rates of hot fluid. Heat transfer coefficient for the bare tube will be found to be lower than the twisted tape with and without holes. Disturbance in the flow field by introducing a twisted tape will disturb the boundary layer at the tube surface and hence increases the heat transfer coefficient. For a holed twisted tape heat transfer coefficients were found to be marginally lesser than the normal as discussed in the earlier section. This reduces the net heat transfer coefficient slightly for the holed twisted tape. As the hole diameter increases this trend continues further reducing the heat transfer coefficient due to reduced turbulence for higher diameter holes as seen clearly in Fig. 11. For the twisted tape the average heat transfer coefficient has increased by 82% compared with the bare tape.

Fig. 12 shows the heat transfer rates for the different tested cases. It increases with the introduction of the twisted tape into the double pipe heat exchanger for all tested mass flow conditions. The average heat transfer rate is increased by 34% compared to the bare tube. Gain in heat transfer rate decreases for the increased Reynolds number as seen in Fig. 12. For higher mass flow rates primary velocity will be increased and hence swirl flow will be reduced which hence flow boundary will not be affected much. The heat transfer rate for a holed tape is lower due to the reason that increase in the turbulence is not able to overcome the conduction losses. Hence the average heat transfer rate decreases. With respect to different hole diameters there was not much appreciable change in the heat transfer rate showing drilling a single hole in a twisted tape will not influence much heat transfer characteristics in a double pipe heat exchanger.

Fig. 13 shows the variation of the pressure drop for bare, twisted tape with and without holes. Pressure drop for the twisted tape is higher than the bare tube for all tested mass flow rate conditions. It is still higher for holed tape. For a definite configuration it is observed that it increases with mass flow rates. For a normal twisted tape the average pressure drop is 2.85 times than the bare tube. Similarly for the holed twisted tape they are 8.7%, 9.1% and 10.02% for 1 mm, 3 mm and 5 mm holes respectively higher than the normal twisted tape. Increase in pressure drop increases the pumping power. For a holed tape thermal performance decreases in addition to increase in the Pumping power. Hence drilling a single hole along the width of a twisted tape will be least effective in increasing the thermal performance of a twisted tape.

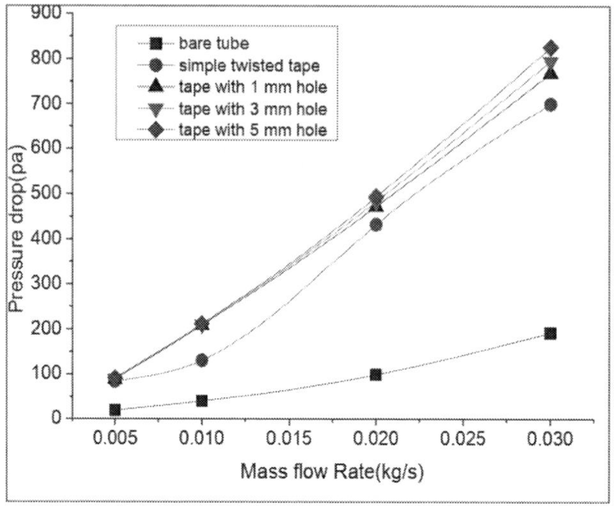

Fig. 13. Variation of pressure drop with mass flow rates of hot water for bare, twisted tape with and without holes

4 Conclusions

In the present study thermal performance of a twisted tape with and without holes in a double pipe heat exchanger was studied numerically. Along the length of the tape diameters of 1 mm, 2 mm and 3 mm were drilled. Performance parameters like nusselt number, heat transfer coefficient and pressure drop were measured and were compared with the bare tube. For a twisted tape Nusselt number, heat transfer coefficient were found to be higher and pressure drop also increased. In case of holed tape Nusselt number and heat transfer coefficient were decreased slightly with respect to normal twisted tape with the increase in pressure drop. Hence it can be concluded that drilling a single hole at regular pitch along the length will not be that effective in increasing turbulence and hence the thermal performance of the tape. Instead of that a large number of small diameter holes can be drilled at a regular pitch to induce more swirl in the flow path which may influence thermal characteristics effectively.

References

1. Chintan Prajapati, Pragna Patel, Jatin Patel3and Umang Patel, Review of heat transfer enhancement using twisted tape, International Journal of Advanced Engineering Research and Studies,Vol ii/Issue I, 2012, pp-162-164.
2. S. Tabatabaeikia, H. A. Mohammed, N. Nik-Ghazali, and B. Shahizare1, Heat Transfer Enhancement by Using Different Types of Inserts, Advances in Mechanical Engineering, Volume 2014, Article ID 250354,
3. S. Pal, S.K. Saha, Laminar flow and heat transfer through a circular tube having integral transverse corrugations and fitted with centre-cleared twisted-tape, Experimental Thermal and Fluid Science, vol. 57, 2014, pp. 388–395.
4. S.W. Chang, M.H. Guo, Thermal performances of enhanced smooth and spiky twisted tapes for laminar and turbulent tubular flows, International Journal of Heat and Mass Transfer, vol. 55, Issue 25–26, 2012, pp. 7651–7667,.
5. Smith Eiamsa-ard a, Pongjetn Promvonge, Performance assessment in a heat exchanger tube with alternate clockwise and counter-clockwise twisted-tape inserts, International Journal of Heat and Mass Transfer, 53, 2010, pp.1364–1372.

6. Eiamsa-ard, K. Yongsiri, K. Nanan, C. Thianpong, Heat transfer augmentation by helically twisted tapes as swirl and turbulence promoters, Chemical Engineering and Processing: Process Intensification, Elsevier, vol. 60, 2012, pp. 42–48,
7. P. Murugesan,a , K. Mayilsamyb, S. Suresh, Heat Transfer and Friction Factor in a Tube Equipped with U-cut Twisted Tape Insert, Jordan Journal of Mechanical and Industrial EngineeringVolume 5, Number 6, 2011.
8. A. V. N. Kapatkar , B. Dr. A. S. Padalkar and C. Sanjay Kasbe, Experimental Investigation on Heat Transfer Enhancement in Laminar Flow in Circular Tube Equipped with Different Inserts, International Journal on Manufacturing and Material Science, Vol. 01, No. 01, 2011.

Principles of energy saving in water supply and sewage systems

Nikolay Makisha[1*] and *Tatiana Kazimirova*[1]

[1]*Moscow state university of civil engineering, Yaroslavskoye shosse, 26, Moscow, Russia, 129337*

Abstract. In this paper, we consider energy efficiency in the system of water supply and sanitation. Particular attention is paid to energy conservation techniques, stages of implementation. The review of energy conservation methods, a table of classification, taking into account the cost. The work is aimed at further more detailed study of energy efficiency techniques in the water supply and sanitation and their application in the regions of Russia.

1 Introduction

The development of economy requires among other things the efficient use of energy and resources of all kind. That approach means significant scientific, technical, organizational and human background. Moreover, reduction of energy and material consumption should not decrease the quality and volumes of production [1, 2].

2 Energy saving and energy efficiency

2.1 Efficiency raise for thermal energy use

If housing and municipal services meant the main directions how to increase efficiency of thermal energy use and energy saving potential are as follows [3]:

- Shutdown of inefficient boilers with appropriate load transfer to other boilers;
- Substitution of inefficient boilers by more efficient;
- Use of gas or local fuel boilers instead of oil boilers;
- Use of hot water instead of steam boilers operation mode;
- Scum and snuff removal from heating surface;
- Application of non-reagent water treatment plants;
- Installation of power generation equipments in boiler plants;
- Introduction of automated control system for boiler plants;
- Thermal rehabilitation of living and public buildings;
- Introduction of metering devices, control and regulation of resource consumption.

* Corresponding author: nmakisha@gmail.com

2.2 Energy saving measures

Energy saving measures can be divided into several groups according to their implementation costs (table 1).

Table 1. Energy saving measures

Cost-free and low cost	Medium costs	High costs
1.Proper operation of water supply and sanitation systems in general and equipments for this systems in particular; 2.Change of graphite to Teflon pump sealing that may increase lifetime up to 6 times; 3. Fittings replacement	1.Pump operation in economy mode 2. Change of pipelines' diameter, use of polymer pipes 3.Saving energy and water during the transition to the circulating water supply systems 4.Elimination scurf in the water supply and sanitation systems 5.Prevention of water leaks 6.Control of water consumption 7. Automated control of operation 8. Social promotion of water and energy saving 9. Analysis of water discharge	1. Automated system of energy consumption control 2. Application of disposal equipment at waste water treatment plant 3. Use of power-saving reserves in hot water systems.

2.2.1 Pump operation in economy mode

There are some recommendations for realizing this measure [1,4-6]:

- Substitution of pumps with low performance or construction that provides additional losses of energy
- Substitution of pump with inappropriate hydraulic characteristics
- Improvement of pump performance by means of new seals or impeller balancing
- Implementation of automated control of pumping equipment operation to provide best performance
- Adjustment of pump performance by means of frequency-regulated electric drive
- If frequency adjustment is not available pumps can be regulated by means of step switching of parallel installed pumps on and off
- Application of accumulation tanks with level sensors for the systems that are designed for maximum capacity at maximum pressure.

2.2.2 Pipelines

If the pipe diameter is increased at 50%, the fluid friction loss can be reduced by 75%. A similar result for energy saving of water supply may be achieved if pipes made from polymers with significantly lower roughness substituted pipes made from traditional materials. As a result, lifetime of pipeline network may increases from 3-10 to 30 years or more. Besides, the hydraulic resistance and the energy costs for pumps operation constant is lower by about 25% for the same pipe diameter and the water flow rate [7,8].

A transfer from direct flow to circulating water supply for cooling systems of energy and technological equipment reduces water consumption from external sources, and the load on the pumps intake system and wastewater treatment plants.

Another big technological problem for pipelines is scurf removal and leakages. There are special automated systems created and widely introduced for treatment with additives of hot or circulation water systems. These systems do not almost change hardness of water and prevent scurf formation

Localization of leakages requires certain labor and technological costs that means special equipment for acoustic search of leaks. Application of water meter devices is an alternative way to control and locate leaks. Going further, consumption control helps to make the water balance, to analyze water schemes and water costs, optimize the cost of water system. All these measures if applied allow the water supply single units performing as a system.

2.2.3 Major energy saving reserves in domestic hot water and recirculation systems

Major energy-saving reserves in domestic hot water systems mean:

- Replacement of water heater with low performance for those, which have a smaller overall size and lower heat loss, as well as simplifying their trim line tube. This leads to a decrease in pump power costs for the heating and circulation of heated water in the heat point.
- Installation of variable frequency drives for circulation and feeding pumps in heat points that allow changing the water flow in the system, without opening or closing of existing valves. These devices help to save 10-30% of electricity.
- Application of balancing valves and water meters that are interlinked and automatically controlled by a dispatching system.
- Significant energy savings reserves are available in the circulation systems, which is discharged into the environment a significant amount of heat energy in many industrial enterprises. The problem of energy saving in the water supply can be solved with the help of heat pumps, which provide the opportunity to return the heat to the production cycle. Such pumps are widely used in Western Europe, USA and Japan [9-11].

3 Auditing of engineering systems

The audit of engineering systems has an aim to create a new program (or rework existing) of data collection system and management of engineering networks and objects, as well as the optimization of energy- water saving measures [12, 13].

This program should include:

- Reasonable proposals for technical and technological changes in the system of engineering networks and objects, and a comparison with consumption standards (utilities);
- A list of equipment for water supply, sewerage and heating;
- The terms of reference for the design of all proposed innovations;
- Justification and calculation of the payback period of the project;
- Introduction of changes in the basis of the developed technical proposals:
- Changes to the напорных characteristics of the pump and network geometry, based on the results of hydraulic calculations;
- Introduction of variable frequency drives with automatic control and regulation;
- Introduction of modern heat generating and heat exchange devices, and thermal resistance materials;
- Introduction of modern metering and control energy- water recourses.

4 Systems of automated control

Measurement system means the number instruments and other technical devices (components of the measuring system) interconnected in a certain way of measuring to form the measuring channels. Measurement system implements the process of measurement and provides automatic (automated) obtaining of measurement results (expressed by means of numbers or codes corresponding to them), time varying and distributed in the space of physical quantities characterizing certain properties (state) measurement object. The measurement system has the main features of measuring devices and their specific features.

As follows from the definition, the components of the measuring system are technical devices, included into its structure to perform one of the functions of the measurement process: measuring, computing or communications [14, 15].

In that way, the measuring components of measurement system are measuring instruments: indicators, measuring converter, measuring switch.

Automated control system has a structure, which consist of three subsystems:

- Subsystem of primary data collection - the lower level;
- Subsystem of primary processing and storage of information - the middle level;
- Subsystem of processing, display, storage and exchange of information - the upper level.

Structurally subsystem of primary data collection includes conversion devices, which measure the parameters of the environment: flow, pressure, temperature and other. The subsystem of middle level normally means controllers. The subsystem of upper level is a specialized computer system with appropriate software. The exchange of information between the subsystems of the lower and middle levels is carried out by measuring channels. Wired links, isolated or switched telephone lines, radio communications may be implemented as communication channels [16].

Economic efficiency of application of automated control system:

- Ensuring payment for energy resources in strict accordance with the actual amount of their consumption;
- Automated registration of energy resources and the control of their parameters;
- Control of the consumption of all energy sources for a certain time interval: from 5 minutes to several years depending to the limits, technological capacity constraints, pressure, flow and temperature;
- Light and sound alarm to notify that measuring parameters go out of allowed limits for making operational decisions.

As soon as the water supply and sewage system equipped with automated control system the payback period can be estimated to register reduction of water consumption; repair cost for pumps and drives; reduction of malfunctions; increase of intervals between repairs. The return of investments begins after the operating order adjusted and launched.

The implementation of such equipment and projects will start obtaining expected results and solving the tasks set. It will also prove the possibility of brand new organization of administrative services and technical personnel of existing enterprises, aimed at addressing the problems of housing and communal services.

The unified information space helps to solve the problem of instant obtaining of information needed for making responsible technological and economic decisions. It also allows carrying out system analysis of the entire data array. Availability of reliable information will provide accurate records of all consuming resources and costs of consumed energy and water resources. The other important peculiarity of the implementation of equipment and automated systems is ability for to systematic and operational management of equipment capacities that has a direct effect on the energy savings (up to 25-40%) [17].

Creating local automated control systems at remote objects, the implementation of the security alarm system and video system will free up staff by 20-40% (a decrease of payment of costs). It becomes possible to plan and develop the housing and communal services, considering the budget revenues and consumers' income; application of economically justified tariffs; provision of actual payment for heat, water and sewage based on metering devices rather than on the "standards of consumption. These results will lead to a significant reduction in the cost of utilities and improve their quality [18].

5 Conclusions

As seen above, energy saving in water supply and sanitation should be based on the following measures: registration of heat and water supply, energy costs and reduction of their losses; reduction in energy consumption by the power optimization; reduction of the level of manual operation; equipment performance increase by means of energy saving technologies; evaluation of automated systems for data collection and the engineering networks management; efficient and optimal control of technological objects; public information on energy saving measures implementation.

References

1. E. Gogina, I. Gulshin, Procedia engineering, 117, 107–113 (2015)
2. A.G. Pervov, A.P. Andrianov, T.P. Gorbunova, A.S. Bagdasaryan, Petroleum chemistry, 55 (10), 879-886 (2015)
3. I. Gulshin, A. Kuzina, International Journal of Applied Engineering Research, 10, 21, 42618–42623 (2015)
4. A. Volkov, V. Chulkov, R. Kazaryan, M. Fachratov, O. Kyzina, R. Gazaryan, Applied mechnics and materials, 580-583, 2281-2284, (2014)
5. N. Makisha, A. Kulakov, MATEC web of conferences, 112, 10018 (2017)
6. A.G. Pervov, A.P. Andrianov, E.B. Yurchevskiy, Petroleum chemistry, 55 (10), 871-878 (2015)
7. E. Gogina, A. Pelipenko, MATEC Web of Conferences, 73, 03007 (2016)
8. V.N. Varapaev, S.A. Doroshenko, A.Y. Trotsko, A.V. Doroshenko, International Journal of Applied Engineering Research, 10, 21, 42588-42592 (2015)
9. O. Kuzina, E. Pankratov, V. Tkachev, MATEC Web of Conferences, 86, 05023 (2016)
10. V. Orlov, A. Andrianov, Applied mechanics and materials, 580–583, 2398-2402 (2014)
11. V.N. Varapaev, A.V. Doroshenko, I.Y. Lantsova, Procedia engineering, 153, 816-823 (2016)
12. N. Makisha, M.Yunchina, MATEC Web of Conferences, 106, 70016 (2017)
13. A.G. Pervov, A.P. Andrianov, Desalination and water treatment, 35 (1-3), 2-9 (2011)
14. E. Gogina, I. Gulshin, Applied mechanics and materials, 580-583, 2367-2369 (2014)
15. Volkov, O. Kuzina, Procedia engineering, 153, 838-843 (2016)
16. A.A. Volkov, A.V. Sedov, P.D. Chelyshkov, D.A. Lysenko, A.V. Doroshenko, International Journal of Applied Engineering Research, 10, 22, 43269-43272 (2015)
17. E. Gogina, I. Gulshin, Procedia engineering, 153, 189-194 (2016)

18. A.G. Pervov, A.P. Andrianov, V.A. Chukhin, R.V. Efremov, International Journal of Applied Engineering Research, 10, 22, 43517-43525 (2015)

Information modeling of windows with the account of thermophysical characteristics

Elena Ignatova[1*]

[1]*Moscow state university of civil engineering, Yaroslavskoye shosse, 26, Moscow, Russia, 129337*

Abstract. The task of improving the buildings energy efficiency is one of the most important tasks in the design of buildings. The buildings energy efficiency can be improved by rational choice of building structures and reduction of the heat transfer. Significant heat transfer of a building occurs through the windows. At the design stage of a building, it is necessary to estimate the magnitude of a heat transfer of different windows. Currently the designers are increasingly using the technology of building information modeling (BIM). 3D-model of the building consists of models of structural elements, which contain information about the different geometric, physical, technological and other characteristics of the structure. The aim of this work is to develop a parametric information model of the window taking into account its heat transfer resistance. In this paper the design of window units with one, two and three sashes are discussed. The value of window-reduced resistance to a heat transfer is calculated inside the window information model and is presented in the table. The method can be applied to different window design. Thermophysical parameter of the window information model can be used to assess the energy costs of operating the building.
Keywords: energy efficiency, modeling, window, heat transfer

1 Introduction

The inflow, utilization and loss of energy are accounted when assessing the energy efficiency of the building. The part of the heat energy goes into the environment as a result of the energy transfer through the outer structure of the building. Significant heat transfer of the building occurs through the windows. To reduce a heat transfer we need to choose a rational design and materials of window construction. A window consists of a frame and several sashes. In turn, a sash consists of sash frames and glazing. Window manufacturers give background information about resistance to a heat transfer for various kind of their glazing and sash frames.

Window heat loss depends on the size and number of sashes, on fastening system of the sashes (Fig.1).

The estimation of the thermal characteristics of window at the design stage can reduce the heat loss of the building and increase its energy efficiency.

* Corresponding author: ignatova@mgsu.ru

Fig 1. The examples of the window structures.

It is common in Russia to use resistance to a heat transfer R ($m^2\cdot°C/W$) for evaluating of the thermal protective performance of structures. A window is a heterogeneous construction and consists of different materials.

The reduced resistance to a heat transfer of the whole window Rw ($m^2\cdot°C/W$) can be calculated using the following formula:

$$Rw = \frac{\sum_i F_i}{\sum_i \frac{F_i}{R_i}},$$

where F_i – *area of the i-th uniform zone,* m^2
R_i – *resistance to a heat transfer of the i-th uniform zone,* $m^2\cdot°C/W$.

To calculate the window reduced resistance to a heat transfer Rw, we need to know the area and the resistance to a heat transfer of each uniform area of the window. Tables of experimental data of the resistance to a heat transfer for different window structures are presented in GOST 23166-99 [1]. Minimum regulatory requirements for Rw are defined in GOST 26602.1-99 [2]. The actual window reduced a heat transfer resistance is determined as the result of certification tests. If the tests do not exist, we can analytically calculate window reduced resistance to a heat transfer.

The problem of evaluation of the building element's heat transfer was considered in the papers [3-5]. In paper [3] various measures to increase the given thermal resistance of the window unit are considered. The analysis value of the reduced resistance to a heat transfer for the window systems of different companies was performed. In paper [4] the dependence of resistance to a heat transfer of the window from the tilt angle, the size of the glazing, selective glass coating and the physical characteristics of the gas filling the space between glass panes were investigated. In paper [5] the effect of wind speed on the change of resistance to a heat transfer of translucent facade structures of tall buildings was investigated.

Analysis of the heat loss for various options of windows can be made by automatically. In paper [6] a methodology and program for calculating the heat loss through the wall with regard to windows was developed. In paper [7] the numerical study of the resistance to a heat transfer of walls with windows was held on the base of software Ansys.

Currently the designers of buildings are increasingly using the technology of building information modeling (BIM). 3D-model of the building consists of models of structural elements, which contain information about the different geometric, physical, technological and other characteristics of the structure. A building information model can serve as a basis for the analysis of the geometry, structure, strength, cost and other characteristics of the building object [8-11]. Technologies of the building information model analysis are described in [12]. Building information model can be used during all stages of the life cycle of construction object.

The most common application of building information modeling is Autodesk Revit [13]. This program has the function of calculating the energy efficiency of a building based on its information model [14]. However, the calculation process is using the cloud service, and interim data calculation is not available. The main parameters of the Autodesk Revit basic window model are length, width, height and material. The basic window model may be useful only for the solution of simple problems of building design. However, Autodesk Revit allows to create new information models of building structures (called families) and allows to add the necessary parameters to them [15,16].

2 Materials and methods

This work is based on the use of the theory of calculation of building heat loss and the theory of building information modeling with the help of Autodesk Revit.

The aim of this work is the creation of Revit information model of a window with automated calculation of the reduced resistance to a heat transfer that depends on the design of the window.

As a result of the research a parametric family of window models was developed. This information model may be used for evaluating the energy efficiency of the building.

Operations with the building information model in Autodesk Revit are based on the creation and application of parametric families of building structures elements [15, 16]. For each element it is needed to create the parameters, to set their values and to define necessary relationships between the parameters. The relationship can be set by analytical expressions. The change of one parameter of the structural element will automatically cause the change of other related parameters. Thus, on the basis of a single parametric family the various options of design elements can be obtained.

Autodesk Revit allows to combine multiple simple families into more complex composite family. The construction of multi-sash window family is based on this principle.

In this study, it is assumed that the parent window family consists of the family of a frame and the families of sashes. In turn, the parametric family of the sash is composed of interconnected families of the glazing and the sash frame (Fig. 2). If the sash is fixed, then there is no sash frame.

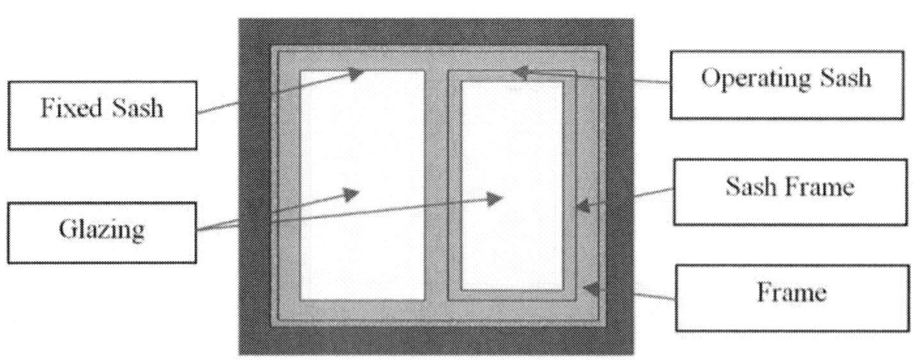

Fig. 2. The elements of the window.

Each family of the glazing or the sash frame has a variable parameter of the material. The parameter of the material includes graphic, strength, thermophysical characteristics, which can be used in 3D and 2D visualization of the window model. They also can be used to define other necessary characteristics of the family. Collections of glazing and frames were developed for window modelling.

The glazing consists of a set of items, such as glass, spacer, desiccant, two sealants, filling gas (Fig. 3). Glazing can have a single glass, two or three glasses. Type and glass thickness conform to industry standards. The resistance to a heat transfer is obtained experimentally for all real types of glass. Therefore, the parameter of thermal resistance is specified numerically for each type of window families. The dimensions of the glazing are parametric and are associated with the size of the sash frame.

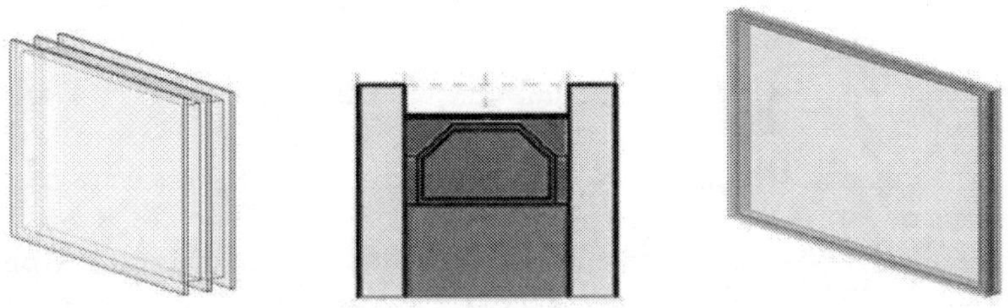

Fig. 3. The structure of glazing.

Plastic, metal and wood can be used as a material of the frames. There are many options of profiles of frames (Fig. 4).

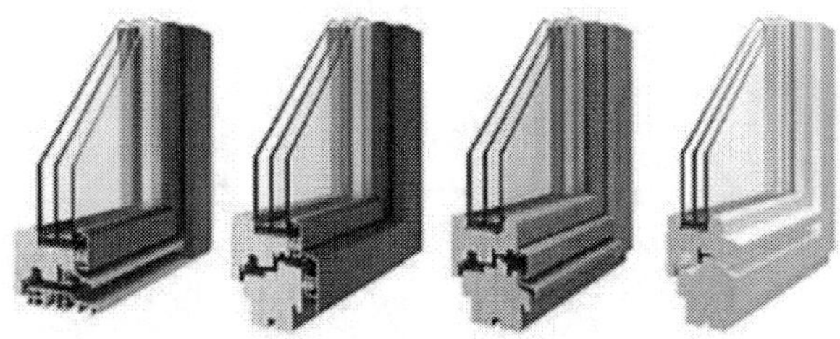

Fig 4. The variants of frames.

Reduced resistance to a heat transfer of the sash depends on the size of the sash. To be more accurate it depends on the area ratio between the glazing and the sash frame. For any size of the sash the calculation of the resistance to a heat transfer is carried out automatically. It is assumed that all sashes in one window have the same profiles of frames and the same glazing.

The reduced resistance to a heat transfer of an operating sash *Rs* ($m^2\cdot{}^\circ C/W$) can be calculated by the formula:

$$Rs = \frac{Sg + Ssf}{\frac{Sg}{Rg} + \frac{Ssf}{Rsf}},$$

where *Sg – area of the glazing,* m^2.
Rg – resistance to a heat transfer of the glazing, $m^2\cdot{}^\circ C/W$.
Ssf – area of the sash frame, m^2;
Rsf – resistance to a heat transfer of the sash frame, $m^2\cdot{}^\circ C/W$.

If the window family consists of a frame family and several sash families, the reduced resistance to a heat transfer *Rw* ($m^2\cdot{}^\circ C/W$) can be calculated by the formula:

$$Rw = \frac{\sum_i Ss_i + Sf}{\sum_i \frac{Ss_i}{Rs_i} + \frac{Sf}{Rf}},$$

where Ss_i – *area of the i-th sash,* m^2;
Rs_i – *reduced resistance to a heat transfer of the i-th sash,* $m^2\cdot{}^\circ C/W$;
Sf – area of frame, m^2;

Rf – resistance to a heat transfer of the frame, $m^2 \cdot °C/W$.

The calculations take into account the structure, the proportions, the materials and thermophysical characteristics of nested families.

3 Results

The proposed method of window modeling and the calculation of its resistance to a heat transfer is based on the technology of building information modeling. Information model of the window was implemented for structures with one, two and three sashes (Fig.5).

Information model of the window allows to set the direction for opening the sashes. Accordingly, the location of the handle and the notation of sash opening are changed. If the fixed sash is selected, the glazing is inserted into the window frame. The size of the glass is parametrically associated with the dimensions of the sash frame. The dimensions of sash frame are parametrically associated with the size of the window frame.

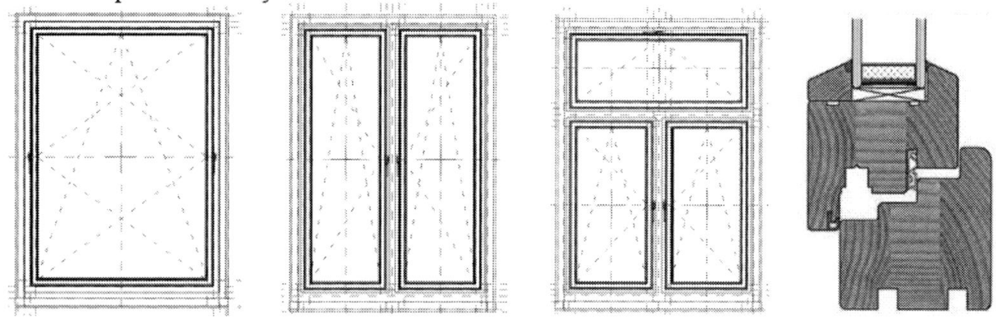

Fig. 5. Models of wood window with one, two and three sashes.

Table 1 shows the results of modelling of different wooden windows. The reduced resistance to a heat transfer *Rw* was compared with data *Ro* from [17] (if it is presented). The size of the window is 1400x2000 mm. The window has a fixed sash with double glazing. Frame design is FloralV-78 Fir. The value of *Rf* equal to 0.71 ($m^2 \cdot °C/W$).

Table 1. Rw of different windows

Number of Sashes	Glazing	*Rg*	*Rw*	*Ro* [17]
1	4M1-8-4M1	0.28	0.334	0.334
1	4M1-Ar8-4M1	0.30	0.355	0.355
2	4M1-8-4M1	0.28	0.325	0.326
2	4M1-Ar8-4M1	0.30	0.347	0.347
3	4M1-8-4M1	0.28	0.352	-
3	4M1-Ar8-4M1	0.30	0.372	-

During creation of the window models with one, two and three sashes, the issue of creating a single universal window family was discussed. The universal window family is the family that can be implemented in all these cases. However, a large number of conditions imposed on the model, and a large number of control parameters starts to hinder the work of the designer.

4 Conclusions

The study deals with the problem of computing the heat loss through the windows to solve the problem of energy efficiency of the building.

The calculation of the reduced resistance to a heat transfer through the window is only the first step in determining the heat loss of a building when calculating its efficiency. For further calculations it is necessary to know the construction area and the climatic characteristics. Also for the design of the building it is necessary to know the minimum allowable values of thermal resistance of building structures.

Thermal insulation properties of the window are enhanced by reducing the area of glazing, by adding inert gases and by increasing the number of glasses. However, improving the thermal insulation properties of the window leads to a reduction in lighting. It is necessary to discuss two problems at the same time. The parameters for evaluating light transmission should be included in the information model of the window in the future.

The method of information modeling implemented on the base of the program of building information modeling Autodesk Revit. Hierarchy of Autodesk Revit families was established for organizing the composite family of the window. The relationships between geometric and thermophysical parameters of the families were determined. The automatic calculation of reduced resistance to a heat transfer of the entire window was organized in the composite family.

The methodology of creation of window information models with the evaluation of the thermophysical characteristics is versatile and can be used for different windows.

References

1. GOST 23166-99. Window structures (1999).
2. GOST 26602.1-99. Blocks of windows and doors. Methods of determining of the resistance to heat transfer (1999).
3. E. S. Konisheva, E. V. Korepanov. The methods of increasing the reduced resistance to a heat transfer of the window. The materials of regional scientifically practical seminar. Energy and resource saving in industry, housing and communal services and agriculture, 119-125 (2016).
4. E. V. Korepanov. *Mathematical modeling of heat transfer through the exterior walls of buildings with windows.(* Izhevsk, Kalashnikov Izhevsk State Technical University, 2011)
5. Y. A. Eldashov, S. G. Sesunin, V. N. Kovrov. Experimental study of the window frame model taking into account the geometric stability and reduced resistance to a heat transfer. Vestnik MGSU, **3**, 146-149 (2009).
6. V. M. Tusnina, Tusnin A.R.Thermal resistance of walls with window openings. Vestnik MGSU, **1-2**, 123-128 (2011).
7. S. S. Golubev. Determination of reduced resistance to a heat transfer of fencing wall based on numerical calculation of distribution of temperature fields, Scientific and Technical Volga region Bulletin, **5**, 93-97 (2011).
8. V. V. Talapov. *Basis of BIM: introduction to building information modeling*. Moscow, (DMK Press, 2011).
9. E. V. Ignatova. Solution of problems on the base of building information models. Vestnik MGSU, Proceedings of Moscow State University of Civil Engineering, **9**, 241—246 (2012).
10. E. V. Ignatova. The effective use of building information model. Abstracts of the 14th International Conference on Computing in Civil and Building Engineering Moscow. June 27-29, 132-133. (2012).

11. E. Ignatova, H. Kirschke, E. Tauscher, K. Smarsly. Parametric geometric modeling in construction planning using industry foundation classes. Proceedings of the 20th International Conference on the Applications of Computer Science and Mathematics in Architecture and Civil Engineering. Weimar, 68-75. (2015).
12. E. V. Ignatova. Technologies of the building information model analysis. Advances in Energy Science and Equipment Engineering II, Taylor & Francis Group, London, 653–657 (2017)
13. NBS National BIM Report 2016. URL: https://www.thenbs.com/knowledge/national-bim-report-2016.
14. Energy Analysis for Autodesk Revit. URL: https://knowledge.autodesk.com/support/revit-products/learn-explore/caas/CloudHelp/cloudhelp/2016/ENU/Revit-Analyze/files/GUID-A4C490A7-A86C-4027-B829-A77AB3211B60-htm.html
15. A. L. Lanzov. *Autodesk Revit 2015. Computer aided design of buildings.* (RIOR Press, 2014).
16. J. Vandezande, E. Krygiel *Mastering Autodesk. Revit architecture 2016.* (SYBEX, 2016).
17. Calculator of windows. URL: http://vbokna.ru/kalkulyatory/r-opr/

Augmentation of Effective Thermal Gain of Solar Air Heater using a Novel Turbulator Design- A CFD Study

Animesh Dhagat[1] and *M. S. Manjunath*[2*]

[1]*Department of Mechanical and Manufacturing Engineering, Manipal Institute of Technology, Manipal Academy of Higher Education, Manipal-576104, India.*
[2]*Department of Mechanical Engineering, School of Engineering and IT, Manipal University, Dubai, UAE.*

Abstract. Augmentation of thermal performance of solar air heater has been the focus of many researchers over the last decades and the use of turbulator or artificial roughness to provide increased fluid mixing in order to achieve augmented heat transfer has been a widely accepted technique. This work aims to evaluate the effect of a novel turbulator design on the effective thermal performance of solar air heater using the methodology of computational fluid dynamics (CFD). A two dimensional CFD analysis is carried out to evaluate the thermal characteristics of solar air heater at various flow Reynolds number conditions for different geometric parameters of the proposed turbulator design. The pitch of the turbulator is varied as 10mm, 20mm, 30mm, 40mm and 50mm for a fixed turbulator height of 2 mm. The Reynolds number is varied from 6,000 to 27,000. The analysis shows that the lower values of pitch produces higher improvement in heat transfer. The maximum increase in Nusselt number is found to be about 2.98 times as compared to the base model for the flow Reynolds number of about 6000. The highest increase in the friction factor is found to be about 3.05 times relative to the base model. The maximum thermal enhancement factor is found to be about 1.99 for the pitch value of 10 mm at a flow Reynolds number of about 6000.

Keywords: Novel turbulator, Nusselt number, CFD, friction factor.

1 Introduction

One of the more abundantly available form of energy, solar energy is also difficult to harness. However, due to the current strain on petroleum and the almost depleted source of coal, there is now increasing research in order to harness this endless pool of energy. Solar Air Heater (SAH) is one such device that has been developed to achieve this very purpose. Though conceptually it's not very new, there is immense scope for improving the conventional device so as to derive the most at the least expense. A lot of research has been made in this particular field and many designs have been tested both experimentally and computationally. Bhagoria and Saini [1] conducted experiments on rectangular duct solar air heaters where they added

* Corresponding author: manjunathnitc@gmail.com

roughness through wedge shaped turbulators having fixed values of relative roughness height and wedge angle. They found that varying the pitch of the roughness affected the heat transfer. They concluded that as the Reynolds numbe0mnr increases, the Nusselt number correspondingly increases while the friction factor subsequently decreases. Another study by Gawande et al. [2] investigating the heat transfer in solar air heater with reverse L-shaped ribs concluded that the average Nusselt number increases with increasing Reynolds number and also that the average Nusselt number increases with decreasing roughness pitch (P/e) for constant height of the roughness (e/D). Certain other studies have pointed towards using corrugated absorber plates. Handoyo et al. [3] studied the effect of delta shaped obstacles in a V-corrugated channel of a SAH. They concluded that the backflow being caused in between the obstacles and the high velocity in the gap between the obstacles and the absorber plate make the flow more turbulent and in turn increase the convective heat transfer. Though this will be at the expense of pressure drop. Yadav and Bhagoria [4] used square sectioned rib roughened duct to induce turbulence in the SAH and found the best thermo-hydraulic performance at roughness pitch (P/e) of 10.71 and roughness height (e/D) of 0.042. Sharma and Kalamkar [5], through their study on the effect of rib turbulators have concluded that the flow reattachment along with unsteady reverse flow taking place in front of the ribs increases the time averaged heat transfer in turbulent flow. They have also found that SST k-w and RNG k-epsilon turbulence models are in good agreement with experimental results and that RNG k-epsilon turbulence model is the more widely used model in analysis involving artificial roughness in the solar air heaters. Further, Yadav and Bhagoria [6] also established that the RNG k-epsilon model yields the best results for two dimensional flows in a conventional air heater. Nowzari et al. [7] tested single pass and double pass air heaters and found that the efficiency of the double pass solar air heater was always greater than the single pass air heater by 5–22.7% for the same air mass flow rate. Alam and Kim [8] investigated the effect of semi elliptic turbulators placed in two arrangements: inline and staggered. For each arrangement, they varied the angle of attack (α) between 30° to 90°. From the results, they found that for the angle of attack of 75°, the staggered arrangement yielded a Nusselt number enhancement of 2.05 while the corresponding enhancement in the case of inlet arrangement was 1.73. They were thus able to conclude that turbulence and the lateral movement of air at an angle of attack of 75° were responsible for the increase in the Nusselt number for both arrangements of turbulators.

Thus, the various works reported in the literature on the use of turbulators reveal the fact that the performance of solar air heater varies with varying turbulator configurations which has led to the development of different designs of turbulator in the quest for achieving increased levels of thermal performance. This work aims to evaluate the effect of a novel turbulator design on the effective thermal performance of solar air heater using the methodology of computational fluid dynamics (CFD). A two dimensional CFD analysis is carried out to evaluate the thermal characteristics of solar air heater at various flow Reynolds number conditions for different geometric parameters of the proposed turbulator design.

2 Computational Domain, Boundary Conditions and Material Properties

The air heater has been divided into three sections namely: inlet, test and outlet section as shown in Fig. 1(a) [9]. The absorber plate of thickness 1.0 mm has been placed at the top of the test section (Fig. 1(a)). The test section has a length of 500 mm and the inlet and outlet have equal length of 245 mm. The height of the inlet, test section and outlet is fixed at 20 mm as shown in Fig. 1(a). The turbulators have been fixed below the absorber plate in the test section as shown in Fig. 1(b). The various configurations of turbulator considered in the analysis are listed in Table 1 while the thermo-physical properties of air are listed in Table 2. Meshing of a portion of the computational domain is shown in Fig. 2.

Table 1. Various turbulator configurations

Rib height (mm)	Pitch (mm)	Hydraulic diameter (mm)
2.0	10	33.33
	20	
	30	
	40	
	50	

Table 2. Thermo-physical properties of air

Properties	Value
Viscosity (Nsm^{-2})	1.85e-5
Density (kgm^{-3})	1.18
Thermal conductivity (W/mK)	0.02624
Specific heat (kJ/kgK)	1006.43

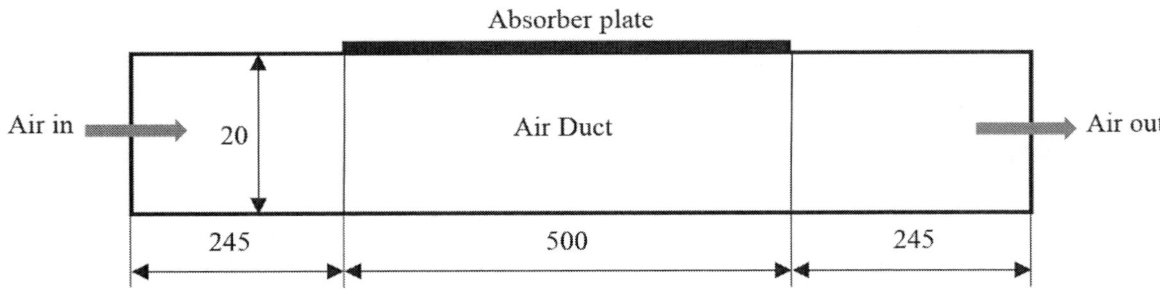

Fig. 1(a). Geometric details of solar air heater with plain absorber plate [9]. (All dimensions are in mm)

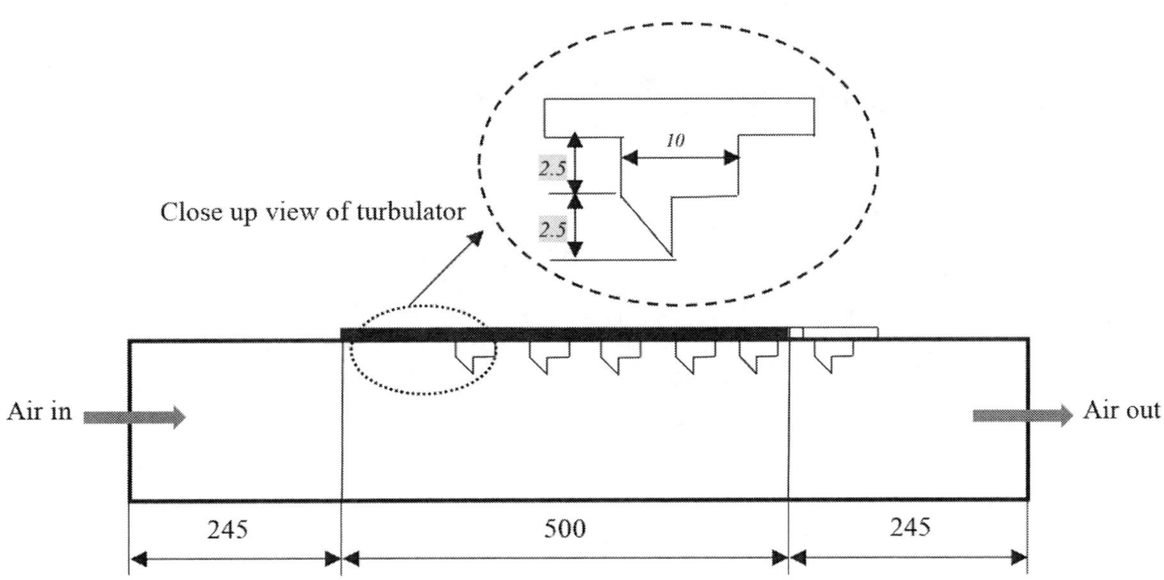

Fig. 1(b). Geometric details of solar air heater with turbulators on the absorber plate (All dimensions are in mm)

A constant heat flux of 1000W/m^2 has been applied at the top surface of the absorber plate. The inlet of the duct has been set to 'mass-flow inlet' condition, and the variation of mass flow rate is between 0.07kg/s to 0.23kg/s which corresponds to the flow Reynolds number range of 6000-24000. The exit of the duct has been set to 'pressure outlet' condition with atmospheric pressure of 101323 Pa. No-slip and impermeable condition is applied at the solid-fluid interface. All other boundaries are set as adiabatic wall. The flow is assumed to be steady and fully developed.

The discretization of the governing equation is carried out using the second order upwind scheme and the convergence of solution is considered when the residuals at the control volumes of the computational domain fell below 10-6 for all the flow parameters. The RNG k-ε turbulence model is used in the analysis [9] to capture the turbulent characteristics of flow.

Based on the grid independence test carried out by the authors on the base model in their previous study [9], a minimum of 0.221 million control volumes are maintained in the computational domain. The authors also carried out a validation study for the numerical solution of the base model in their previous study [9] and found that the numerical solution has good agreement with the Dittus-Boelter correlation for Nusselt number and Blassius solution for friction factor. The deviation of Nusselt number was found to be about 4.5% and for friction factor about 2.1%

3 Results and Discussions

Fig.3 shows the variation of Nusselt number for different flow rate conditions. It is observed that the Nusselt number for heat transfer from the absorber plate to air is generally higher relative to the plain duct. This is due to increased turbulence in the air flow in the presence of turbulators which is clear from Fig. 4 which shows the flow disturbances generated by the turbulator.

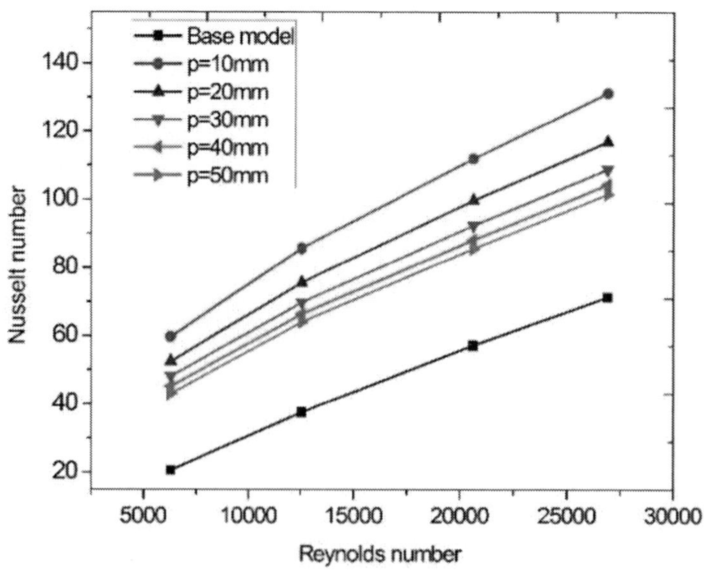

Fig. 3. Velocity vector plot of air flow around the turbulator.

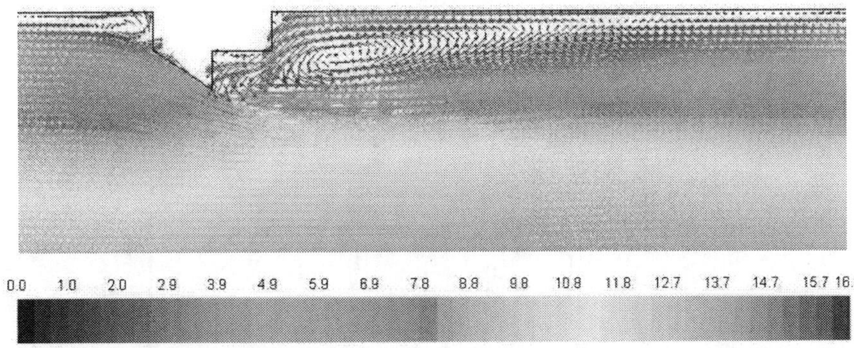

Fig. 4. Velocity vector plot of air flow around the turbulator.

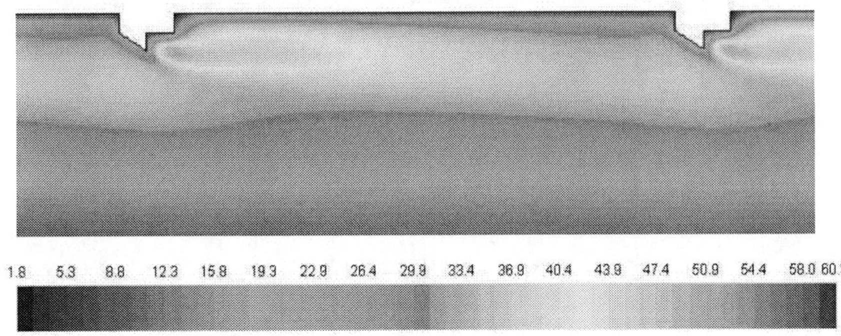

Fig. 5. Distribution of turbulent intensity (%) around the turbulator.

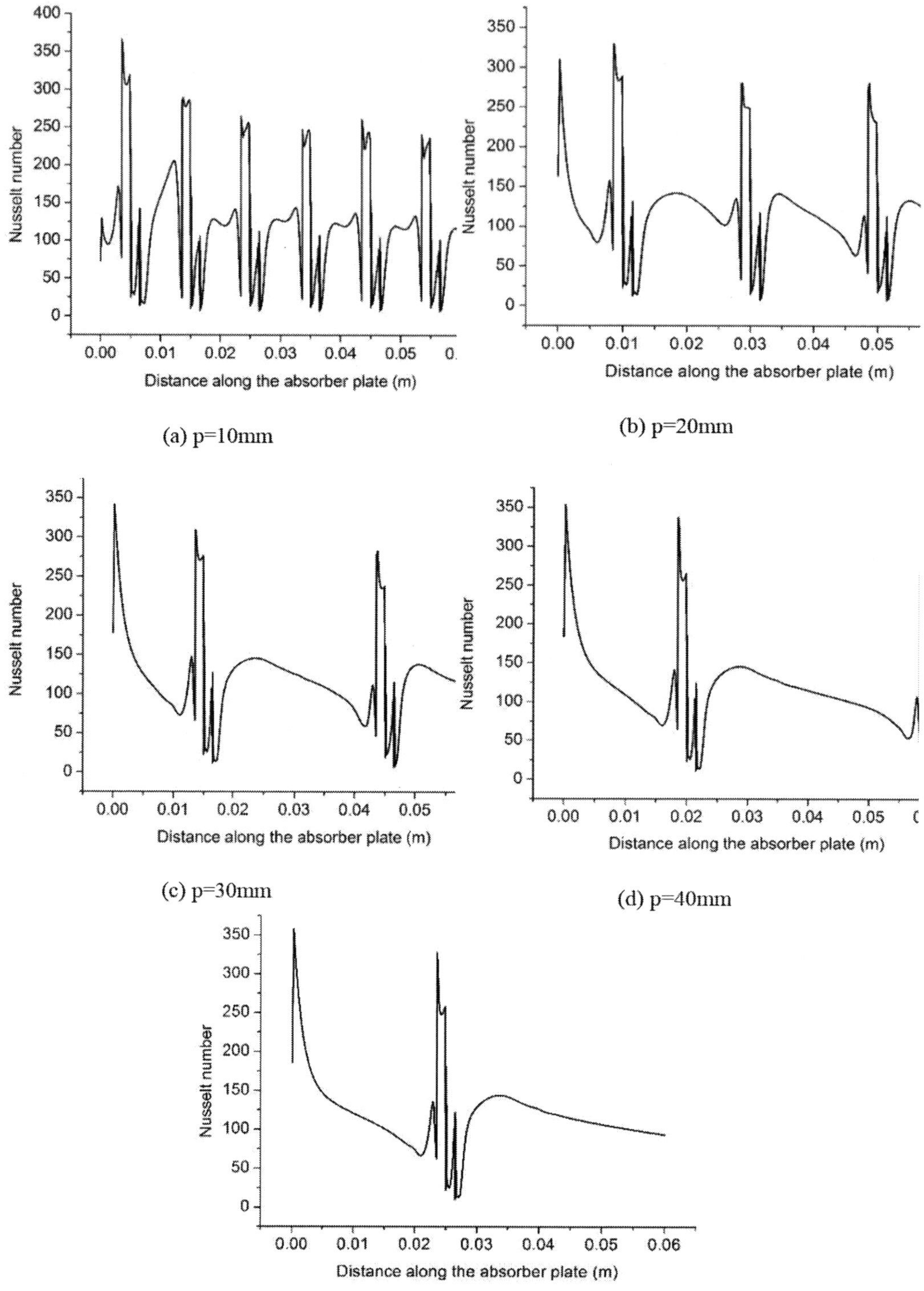

Fig. 6. Variation of Nusselt number along the length of absorber plate

As the incoming air enters the test section, it impinges on the turbulator near the absorber plate. The shape of the turbulator is such that some portion of the air impinges on its vertical edge while some other portion hits the inclined edge and slides downward, as shown in Fig.4. This causes flow separation at the tip of the inclined leading edge of the turbulator surface and destroys the laminar sub-layer. The separated flow tends to reattach with the absorber surface on the downstream side of the turbulator and results in the formation of recirculation zones on the immediate downstream side of the turbulator. The flow recirculation zones interact with the separated flow and provide increased fluid mixing thereby providing higher energy transfer. The recirculation zones also ensure that the relatively colder air of the main flow stream is transported closer to the absorber surface thereby increasing the heat transfer. Fig.5 shows the distribution of turbulent intensity in the air flow around the turbulator. It can be seen that the turbulent intensity is higher around the turbulator region thereby validating the fact that the presence of turbulator enhances fluid mixing and hence heat transfer.

Fig.3 also shows that the Nusselt number increases with reducing pitch values at all flow rates of air. This is due to an increase in the total number of turbulators present in the test section at lower pitch values which produces turbulence in the air flow at a relatively larger number of locations leading to enhanced heat energy transfer. As the pitch increases, the number of turbulators gradually reduce and the number of regions at which the peak turbulence levels exist due to the presence of turbulators decreases thereby reducing the overall fluid mixing and hence energy transfer. Fig.6 shows the variation of Nusselt number along the length of absorber plate for different pitch values of the turbulator. It is clearly seen that the flow disturbances are at its peak around the turbulator region as indicated by a relatively higher values of Nusselt number at locations where the turbulators are attached to the absorber plate. For the pitch value of 10mm, the number of turbulators are large which results in a relatively larger instances of peak values of Nusselt number as shown in Fig. 6(a). As the pitch value increases, the number of instances of peak values of Nusselt number decreases owing to the reduction in the number of turbulators on the absorber plate as shown in Fig. 6(b) to Fig.6(e). Fig.6 also reveals that the highest value of Nusselt number occurs at the leading inclined edge of the turbulator. The reattachment points that exist on the absorber plate in the regions between the turbulators also exhibit peak values of Nusselt number thereby indicating enhanced heat transfer. The maximum increase in Nusselt number is found to be about 2.98 times as compared to the base model for the flow Reynolds number of about 6000. Fig.7 shows the relative variation of friction factor in the presence of turbulator for different pitch values. It is seen that the friction factor is relatively higher in the presence of turbulator owing to an increased obstruction to the air flow which causes greater pressure loss across the duct. The highest increase in the friction factor is found to be about 3.05 times relatives to the base model.

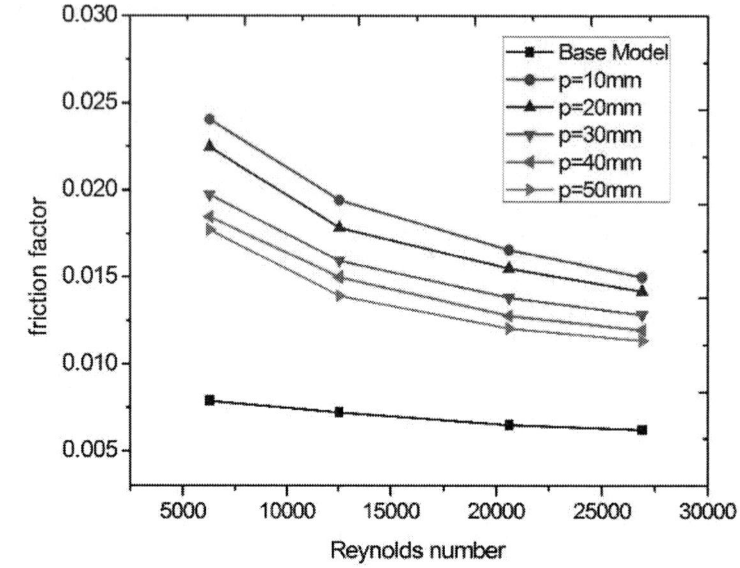

Fig. 7. Velocity vector plot of air flow around the turbulator.

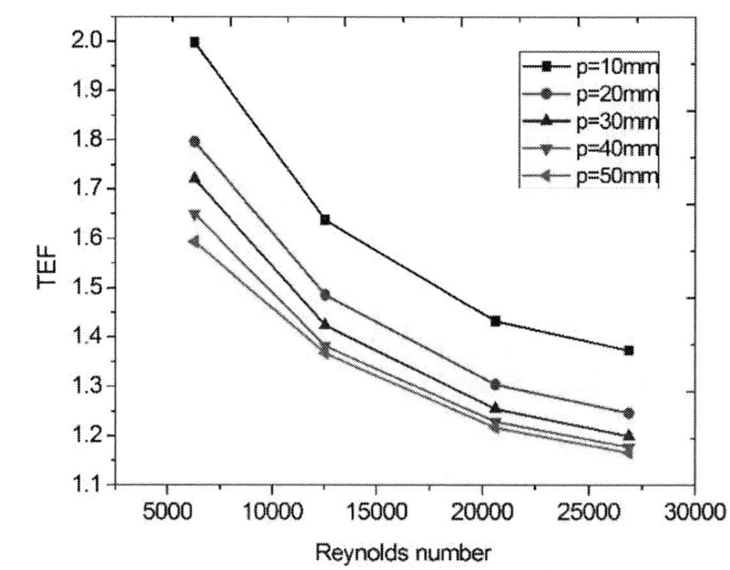

Fig. 8. Comparison of thermal enhancement factor for different pitch values of turbulator.

Fig. 8 shows the variation of thermal enhancement factor (TEF) which is calculated by eq.(1). The thermal enhancement factor indicates the effective thermal gain of a solar air heater by considering the relative increase of both the heat transfer and friction factor.

$$TEF = \frac{\left(\frac{Nu_r}{Nu_s}\right)}{\left(\frac{f_r}{f_s}\right)^{1/3}} \qquad (1)$$

It can be seen that the TEF is higher than 1.0 in the presence of turbulator for all the flow rates and pitch values considered in the analysis. The effective thermal gain is found to be higher for lower pitch values of the turbulator for all the flow rate conditions. This is due to the fact that, at lower flow rates the heat transfer enhancement is found to be significantly higher as compared to higher pitch values. The maximum enhancement in thermal enhancement factor is found to be about 1.99 for the pitch value of 10mm at a flow Reynolds number of about 6000.

4 Conclusions

The following conclusions can be drawn from the analysis:
Lower values of pitch produces higher improvement in heat transfer. The maximum increase in Nusselt number is found to be about 2.98 times as compared to the base model for the flow Reynolds number of about 6000.

- The highest increase in the friction factor is found to be about 3.05 times relatives to the base model.
- The maximum enhancement in thermal enhancement factor is found to be about 1.99 for the pitch value of 10 mm at a flow Reynolds number of about 6000.

Acknowledgement

The computational facilities were provided by the Department of Mechanical and Manufacturing Engineering, Manipal Institute of Technology, Manipal, India which is thankfully acknowledged. This work is not part of any research grant.

References

1. J.L Bhagoria, J.S Saini, S.C Solanki, "Heat transfer coefficient and friction factor correlations for rectangular solar air heater duct having transverse wedge shaped rib roughness on the absorber plate", Renewable Energy, Volume 25, Issue 3, March 2002, Pages 341-369, ISSN 0960-1481, http://dx.doi.org/10.1016/S0960-1481(01)00057-X.
2. Vipin B. Gawande, A.S. Dhoble, D.B. Zodpe, Sunil Chamoli, "Experimental and CFD investigation of convection heat transfer in solar air heater with reverse L-shaped ribs", Solar Energy, Volume 131, June 2016, Pages 275-295, ISSN 0038-092X, http://dx.doi.org/10.1016/j.solener.2016.02.040.
3. Ekadewi A. Handoyo, Djatmiko Ichsani, Prabowo, Sutardi, "Numerical studies on the effect of delta-shaped obstacles' spacing on the heat transfer and pressure drop in v-corrugated channel of solar air heater", Solar Energy, Volume 131, June 2016, Pages 47-60, ISSN 0038-092X, http://dx.doi.org/10.1016/j.solener.2016.02.031.
4. Anil Singh Yadav, J.L. Bhagoria, "A numerical investigation of square sectioned transverse rib roughened solar air heater", International Journal of Thermal Sciences, Volume 79, May 2014, Pages 111-131, ISSN 1290-0729, http://dx.doi.org/10.1016/j.ijthermalsci.2014.01.008.
5. Sanjay K. Sharma, Vilas R. Kalamkar, "Computational Fluid Dynamics approach in thermo-hydraulic analysis of flow in ducts with rib roughened walls – A review", Renewable and Sustainable Energy Reviews, Volume 55, March 2016, Pages 756-788, ISSN 1364-0321, http://dx.doi.org/10.1016/j.rser.2015.10.160.

6. Anil Singh Yadav, J.L. Bhagoria, "Heat transfer and fluid flow analysis of solar air heater: A review of CFD approach", Renewable and Sustainable Energy Reviews, Volume 23, July 2013, Pages 60-79, ISSN 1364-0321, http://dx.doi.org/10.1016/j.rser.2013.02.035.
7. Raheleh Nowzari, L.B.Y. Aldabbagh, F. Egelioglu, "Single and double pass solar air heaters with partially perforated cover and packed mesh", Energy, Volume 73, 14 August 2014, Pages 694-702, ISSN 0360-5442, http://dx.doi.org/10.1016/j.energy.2014.06.069.
8. Tabish Alam, Man-Hoe Kim, "Numerical study on thermal hydraulic performance improvement in solar air heater duct with semi ellipse shaped obstacles", Energy, Volume 112, 1 October 2016, Pages 588-598, ISSN 0360-5442, http://dx.doi.org/10.1016/j.energy.2016.06.105.
9. M.S. Manjunath, K. Vasudeva Karanth, N.Yagnesh Sharma, "Numerical Analysis on the Influence of a Novel Semi-D shaped Turbulator on Heat Transfer Enhancement of Solar Air Heater", Proceedings of 2017 International Conference on Electronics, Computer & Manufacturing Engineering, Singapore, March 29-30, 2017
10. ANSYS FLUENT 17.1, Documentation, ANSYS Inc, 2016.

Production of biogas at wastewater treatment plants and its further application

Nikolay Makisha[1]* and *Daria Semenova*[1]

[1]*Moscow state university of civil engineering, Yaroslavskoye shosse, 26, Moscow, Russia, 129337*

Abstract. This article covered a wide range of questions on the topic of production and use of biogas as alternative energy source. Biogas is produced by anaerobic digestion of biomass due to the breakdown of fats, proteins and carbohydrates, which constitute the bulk of organic matter. The article describes the most common methods of biogas production, their stages and characteristics. In addition, the article describes some of the possible areas of application of biogas on example of different countries. The article also provides information about key environmental and economic benefits in the case of use of biogas: prevention of methane emissions and lower emissions of carbon monoxide and nitrogen oxides into the atmosphere; the possibility of using secondary raw materials to generate electricity.

1 Introduction

The most common method of municipal wastewater treatment around the world is mechanical/biological treatment. Modern sewage treatment facilities show highly efficient wastewater treatment, with virtually no impurities. Sewage sludge is formed as resulting by-product purification, to allow its further use as a source of energy and biomass. However, sludge sewage sludge may only be used after proper treatment, which reduces or removes the high water content (>95%) load, extraneous contamination, pathogenic bacteria and malodorous substances [1, 2].

As an important component to ensure the environmental safety of waste water, modern sludge treatment processes have to meet a number of requirements [3]:

- Effectiveness of capital and operating costs.
- Efficiency of energy use with low consumption and the use of its own reserves.
- Minimal amount of sludge.
- Improvement of the sludge quality. Discharged wastewater that is safe for the environment.

For proper sludge treatment there are two conventional methods of treatment:

- Aerobic sludge stabilization takes place in open sludge tanks with the injection of air. A necessary condition is sufficient volume. Aeration and mixing require a greater energy demand than in the case of anaerobic stabilization [4].

* Corresponding author: nmakisha@gmail.com

• Anaerobic stabilization of sludge is the best solution. In respect of costs, efficiency of energy use and environmental protection, anaerobic stabilization of sludge represents the best solution for sewage treatment plants of medium and high capacity. In addition, the combustible gases resulting from controlled fermentation can be used to produce energy.

Depending on the individual requirements of the sewage treatment plant, the optimum result can be achieved also in the dual stabilization, which involves a combination of both processes [5].

2 Sludge treatment processes

2.1 Anaerobic stabilization of sludge. Main advantages

The growth of energy demand and limitation of CO_2 emissions are global problems that require a prompt solution. And if sludge would be stored in landfills, it would have exacerbated the greenhouse effect, as the contaminating gas from the decomposition of waste that was thrown would be in the atmosphere for decades. And above all, emissions of methane, a particularly harmful impact on the environment, as methane is 25 times more active as greenhouse gas than carbon dioxide.

Anaerobic sludge treatment offers an alternative solution with a number of environmental and economic benefits [6]:

- Reduction of organic content at approx. 50%, and its processing into combustible biogas.
- Production of renewable energy.
- The possibility of independent supply of energy to the sewage treatment plant.
- Lowered production costs.
- Stable and reliable process.
- Reduction of land area taken under construction.
- Reduction in the amount of malodorous substances.
- Sanitary-hygienic measures in relation to sludge (including the reduction of bacteria).
- Climate protection by improving the balance of CO_2 emitted and waste treatment facilities.

Anaerobic treatment of precipitation makes a significant contribution to climate protection.

2.2 Thickening of sludge

Thickening of sludge before stabilization leads to a significant reduction of its volume, reduction of hydraulic load in the reactor, increase of the gas flow rate and temperature increase of emitted energy [2,7].

Biogas plant is a facility that works on any kind of organic material, the target product of which is biogas. It is currently used or being developed about 60 varieties of biogas technology.

The basis of any biogas is bioreactor. Its construction should meet strict requirements. Thus, the case must be solid at absolute air-tightness of walls. Mandatory good insulation of walls and their ability to reliably resist corrosion. Bioreactor construction should provide possibility of loading and discharging of the reactor and access to its internal space.

2.3 Anaerobic digestion of wastewater sludge

The correct temperature range and optimum retention time of sludge in the reactor are the most important initial factors used in the design of anaerobic stabilization.

Classic stabilization occurs in septic sediment trap, which are a type of bioreactor. After about 20 days, the sludge is subjected to controlled fermentation under vacuum, and the organic matter is recycled into water and biogas. Released thus anaerobic digestion gas (biogas) is a mixture of 50-70% methane, 30-40% CO_2, 0-1% of steam and trace elements and is combustible and can produce energy [8,9].

Stabilization can occur in different temperature ranges and is carried out in one or two stages. Mainly used mesophilic stabilization at 30-37 C, which takes about 20 days. However, the future for thermophilic stabilization at a temperature of > 53, which will reduce the period before and at least 15 days.

For optimal process conditions needed to take into account three important factors:

- Sludge mixing by means of a stirrer or pump or injection (bio) gas.
- Avoiding of sludge floating layers.
- Heating of reactors, preferably by the heat of exhaust gases.

2.4 Dewatering, drying and incineration of wastewater sludge

During pre-treatment, the residue should be drained to the maximum extent possible. This leads to a further reduction of the sludge volume, thus reducing the load on the subsequent phases of the process. Depending on the characteristics of the precipitation process, stabilized sewage sludge shows the dry matter content of 2-4%. In General, the stabilization leads to an approximately 30% reduction in volume of a sludge. Subsequent condensation leads to the increase in the concentration of dry solids up to 4-7%. Subsequent dewatering in centrifuges or filter presses, in sludge waste water to achieve a solids content of 20-35%. With additional drying, for example, in the so-called fluidized bed, the content of dry substances reaches 95%. This gives the possibility of further thermal use of sludge as a supplementary fuel, or in agriculture [3,10,11].

2.5 Disinfection of wastewater sludge

Stabilized sewage sludge has a high concentration of nitrogen and phosphorus, which can be used as agricultural fertilizer. But stability, hygiene and transportability of sludge depend entirely on the sanitary safety. In order to reduce the potential impact on the environment, to eliminate health risks and optimize the decomposition there are several additional processes to be used such as heat drying at temperatures > 80 C, the water content reduction to lower than 10%, or treatment and a decrease in the content of lime to the pH-value until at least 12. As a result, accordingly, pathogenic bacteria and viruses are destroyed and inactivated [12].

3 Biogas application

Thermal efficiency of gas produced in reactors under anaerobic stabilization is about 6.5 kWh/cubic m. This corresponds to more than half of the calorific value of natural gas (~ 10 kWh/cubic meter). If the biogas is used for energy production, the electrical and thermal energy can be used for domestic consumption water treatment facilities and (or) the power electric network:

- Production of electricity and heat at thermal powerplants: ~ 35-40% of electricity, about 60% of thermal energy.
- Direct use in gas engines.
- The use of heat when generating steam or hot water.

Efficient use of biogas can provide the treatment plant with electricity for such purposes as aeration tanks of activated sludge, reactor building heating and hot water, i.e. it is guaranteed independent significant energy savings. Large wastewater treatment plants can cover up to 100% of their energy requirements through internal generation of electricity [7, 14].

Thus, biogas is a gas produced from organic waste as a result of "methane fermentation". It occurs in the decomposition of organic substances in the result of the activity of two major groups of microorganisms. One group of microorganisms, commonly called acid-forming bacteria or fermentation of microorganisms, breaks down complex organic compounds (proteins, fats and carbohydrates) into more simple, while in the fermentation environment appear primary products of fermentation - volatile fatty acids, lower alcohols, hydrogen, carbon monoxide, acetic and formic acids, etc [15]. These less complex organic substances are a source of food for the second group of bacteria, methane-producing, which convert organic acids into the desired methane as well as carbon dioxide etc.

Biogas, when treating municipal waste is released during the fermentation of precipitation in the primary clarifiers and excess activated sludge in the digesters at thermophilic mode of its implementation. Biogas at wastewater treatment facilities of Moscow is directed to produce steam in boilers equipped with burners for natural gas and biogas. About 75% of steam is consumed for heating of digesters, the rest for rooms heating and hot water boilers. Biogas digesters are operated in other cities, but the gas is not used at all or partially recycled. Thus, the domestic scale of efficient operation of these facilities is clearly insufficient [16]. However when the fermentation is only half of the produced sludge allows obtaining 350 million m^3 /year of biogas, or 0.3 million tons of conditional fuel by 40-50% and reducing heat and power costs to the treatment plant as a whole. To reduce costs and to increase the degree of stabilization of sludge and biogas there are two basic methods of intensification of digesters work: increasing the concentration of the load and the biomass of microorganisms [17].

The increase the concentration of load (sediment) can be achieved if to use a gravity thickening, flotation, and centrifugal separation. The increase of biomass concentration of microorganisms in digester can be achieved due to its cultivation with the attachment to fill part of the volume structures of the inert porous media (polyurethane, various types of plastic nets, pipes, expanded clay, etc.). The amount of biomass and its activity in these digesters compared to traditional increases in 2-3 times. Respectively, reduced required duration of fermentation (1-2 days) and the volume of digesters increase the biogas yield. One of the varieties of the intensive process of digestion is the separation of the phases of hydrolysis and methane production (using the principle of attachment of the methane bacteria) [9, 18].

The thermal hydrolysis (TG) of wastewater treatment plants sludge developed and operated since 1993 firm "Cambi AS" (Oslo, Norway). The thermal hydrolysis of sludge after dewatering and homogenizing step is achieved by steaming high pressure with subsequent cooling and aging for stabilization. Hydrolysis increases the amount of material available for decomposition of substances for biogas production. After TG, the sediment anaerobic fermented and dried. At the same time its volume is reduced by 50% compared to the original, the yield of methane increases two times compared to traditional fermentation. The plant capacity is 4 thousand tons/year on dry matter. In the UK, sludge is processed by steaming with a temperature of 160-180°C and pressure of 6.3 to 10.2 kg/cm^2, is then sent to anaerobic digestion. The performance of the latter increases in two, and the output of biogas three times. The solid content in the sludge increases from 6 to 12%. Modern technology aeration plants provide heat of anaerobic digestion process of sludge [11, 12, 19].

Modern technology aeration plants provide heat of anaerobic digestion process of sludge. So, on Kuryanovo WWTP (Moscow) spiral heat exchanger (rolled fire hose with a diameter of 10.9 m) set in a separate building. It consists of two blocks of dual-channel spiral with a total length of 630 m, which is moving countercurrent hot (fermented) and cold (heated) precipitation. The temperature of the latter on the inlet of the heat exchanger equal to 20-26°C, at the 38-40°C at a temperature of digested slurry after exit from the apparatus 32 to 34°C. Preheated residue is sent to fermentation. Recent studies showed that biogas from digesters along with the target combustible components (methane and other hydrocarbons) into a number of strong pollutants of the environment. So, at mesophilic/thermophilic fermentation of sludge at the stations of biological treatment of municipal (60%) and industrial wastewater there were revealed 1-100 $\mu g/m^3$ of arsenic and antimony, at 10-1000 mg/m^3 of mercury, tellurium, lead, tin in the analyzed gas-phase samples. These metals are mainly represented by di-, tri - and tetra - methylated compounds, typical of the process of decay of organic matter. Detected concentrations repeatedly exceeded maximum permissible concentrations (MPC) of compounds of interest. In particular, in Russia, the MPC of mercury, tellurium, lead and tin are in the order of 0.0003-0,05 mg/m^3. According to national studies, the nature of contamination is determined by the presence of ammonia and hydrogen sulfide. Their concentration during adverse weather conditions are recorded at a distance of 300-500 m from the aeration station and 500 m of sludge drying beds. The obtained results call for a rethinking of the conditions of biogas utilization as a renewable fuel. Treatment of precipitation in the digesters determines the periodic removal of the fermented mass and adherent sludge with the purpose of depositing or disposal. In this regard, there is a need to stop further gas formation, or the location of the precipitation in the city dump will cause serious problems [20].

4 Conclusions

Summing up, consider the pros and cons of the use and production of biogas. The undoubted advantages of biofuels obtained by means of recycling is its accessibility, especially for rural residents who can organize a closed production cycle on the farm.

Another advantage of biogas is rich in practically exhaustible, self-replenishing raw materials. Thanks to the "omnivorous" biogas plant can be effectively used in large megacity areas, as additional source of energy in combined environmentally friendly systems for the production of renewable energy, as well as in small rural farms, covering almost all of its needs for mineral fertilizers. Another definite plus is that the biogas plant offers one of the most viable answers to the question, who are increasingly concerned about the ecologists: what to do with the trash? The processing of organic waste into highly effective fertilizer, obtaining as by-products of bio-fuel, valuable in the economy of carbon dioxide, as well as at the equipment installation of special filters, clean water, make biogas plants, in fact, out of competition, compared to other units of waste disposal. It is also worth noting that the construction of a biogas plant processing waste on an industrial scale is not too expensive, and combining it with other sources of renewable energy such as wind and solar, increases the efficiency of the equipment for fermentation of biomass almost several times.

Environmental benefits from the production of biogas is enormous:

- Reduction of the emission of methane (a greenhouse gas) resulting from the storage of manure under the open sky;
- Reduction of emissions of carbon dioxide and combustion products: charcoal, firewood and other fuels;
- Reduction of air pollution from nitrogen compounds, have an unpleasant odor;
- Reduction of water pollution manure drains;
- Save the forest from logging;
- Reducing the use of chemical fertilizers;

There are advantages and disadvantages that science has yet to overcome. As not eco-friendly biogas, but this fuel does not completely rule out the greenhouse effect. Burning of biogas at least and minimizes harmful emissions but not eliminate them completely. The second problem of biogas, it is available only in rural areas rich in raw materials for production. This problem is rather of organizational and with proper development of infrastructure will be solved by itself, but while the biogas plant remains more of an exception and not an everyday norm, the problem of poor development of the industrial complex is quite acute. Though installation is relatively not expensive, small and individual biogas plants, still expensive for an ordinary farmer, so to buy them you have to take loans in banks. With the aim of faster to repay the loan and to recoup the investment, farmers are starting to grow energy crops, which in turn, again has a negative impact on the earth's environment. The technology for producing biogas is still not perfect, but still have significant pluses for future. Eliminating defects, reducing costs and increasing production efficiency, biogas, and a biogas plant will play an important role in the normalization of the worsening of the ecological situation on the planet.

References

1. A.G. Pervov, A.P. Andrianov, T.P. Gorbunova, A.S. Bagdasaryan, Petr. Chem., 55(10), 879-886 (2015)
2. Gulshin, A. Kuzina, IJAER, 10, 21, 42618–42623 (2015)
3. E.S. Gogina, O.V. Yantsen, O.A. Ruzhitskaya, AMM, 580–583, 2354–2357 (2014)
4. Volkov, V. Chulkov, R. Kazaryan, M. Fachratov, O. Kyzina, R. Gazaryan, AMM, 580-583, 2281-2284, (2014)
5. A.G. Pervov, A.P. Andrianov, E.B. Yurchevskiy, Petr. Chem., 55 (10), 871-878 (2015)
6. E. Gogina, A. Pelipenko, MATECCONF, 73, 03007 (2016)
7. V.N. Varapaev, S.A. Doroshenko, A.Y. Trotsko, A.V. Doroshenko, IJAER, 10, 21, 42588-42592 (2015)
8. O. Kuzina, E. Pankratov, V. Tkachev, MATECCONF, 86, 05023 (2016)
9. V. Orlov, A. Andrianov, AMM, 580–583, 2398-2402 (2014)
10. V.N. Varapaev, A.V. Doroshenko, I.Y. Lantsova, Pr. Eng., 153, 816-823 (2016)
11. E.S. Gogina, O.A. Ruzhitskaya, O.V. Yantsen. AMR, 919–921, 2145–2148 (2014)
12. N. Makisha, E3sconf, 6, 01002 (2016)
13. A.G. Pervov, A.P. Andrianov, DWT, 35 (1-3), 2-9 (2011)
14. E. Gogina, I. Gulshin, AMM, 580-583, 2367-2369 (2014)
15. Volkov, O. Kuzina, Pr. Eng., 153, 838-843 (2016)
16. N. Makisha, Pr. Eng., 165, 1087-1091 (2016)

17. A.A. Volkov, A.V. Sedov, P.D. Chelyshkov, D.A. Lysenko, A.V. Doroshenko, IJAER, 10, 22, 43269-43272 (2015)
18. N. Makisha, Pr. Eng., 165, 1092-1097 (2016)
19. E. Gogina, I. Gulshin, Pr. Eng., 153, 189-194 (2016)
20. A.G. Pervov, A.P. Andrianov, V.A. Chukhin, R.V. Efremov, IJAER, 10, 22, 43517-43525 (2015)

Effect of bio-ethanol on the performance and emission of a biodiesel fueled compression ignition engine

Shrivathsa Nelly Bhat[1], *Shreyas Shenoy*[2] and *Dinesha P.*[3*]

[1]*Department of Mechatronics Engineering, Manipal Institute of Technology, Manipal Academy of Higher Education, Manipal, India.*
[2]*Department of Computer and Communication Engineering, Manipal Institute of Technology, Manipal Academy of Higher Education, Manipal, India.*
[3]*Department of Mechanical and Manufacturing Engineering, Manipal Institute of Technology, Manipal Academy of Higher Education, Manipal, India.*

Abstract. In the present study investigates the effect of bio-ethanol on the performance and emissions of a biodiesel blend fueled compression ignition engine. The experiments are conducted using pongamia biodiesel blend B20 (20% pongamia biodiesel +80% diesel) with 5, 7.5 and 10% (v/v) of bio-ethanol on a four stroke single cylinder diesel engine. The tests are conducted at different load conditions. Performance and emissions characteristics are investigated for different bio-ethanol compositions. The results show that the brake thermal efficiency is maximum for B20E7.5 blend with a minimum brake specific fuel consumption. Carbon monoxide emission is minimum for B20E7.5 blend and NOx emission decreases as the bio-ethanol percentage is increased from 5 to 7.5%. The study reveals that 7.5% bio-ethanol with B20 pongamia biodiesel blend results better performance and emission characteristics.

Keywords: Biodiesel; bioethanol; diesel engine; emissions; performance.

1 Introduction

Diesel engines produce higher torque at low speeds, which is the key requirement of internal combustion engines. Hence diesel engines are widely used in transportation and industrial sector. Increased use of fossil based fuel in engines cause twin problems viz. energy crisis and environmental degradation. Several attempts being tried by various researchers to overcome these issues [1-3]. Engine modification and fuel modification are the two extensive methods by which these issues can be minimized. Engine modifications include varying the fuel injection parameters such as varying fuel injection pressure and timing [4, 5], combustion chamber designs etc. [6]. Fuel modification method is the tuning of the fuel properties to achieve complete combustion [7]. Use of alternative fuels for compression ignition engines are most promising way to reduce environmental degradation

*Corresponding author: dinesha.p@manipal.edu

without negotiating the engine performance. Literature report reveals that there are several alternative fuels being tried in diesel engines [8-9]. These alternative fuels are derived from vegetable seed oils and animal fat. Due to their higher viscosity, straight oils and fats are converted to methyl esters by transesterification process [10]. The methyl ester produced by transesterification is known as biodiesel. Biodiesel has high potential and is a renewable source of energy. Higher viscosity and lower calorific value of neat biodiesel restricts it direct use in compression ignition engines. The physical properties of the biodiesel can be improved by blending with petro-diesel. A blend of 20% biodiesel with 80% diesel (B20) results better results in terms of engine performance and exhaust emissions. Researchers reported that the B20 blend of biodiesel of various vegetable oils contribute acceptable performance and emission levels results without any engine modifications [11, 12]. Inbuilt oxygen content of biodiesel helps to achieve more complete combustion, this intern results lower carbon monoxide (CO), unburnt hydrocarbon (HC) and smoke emissions. But biodiesel combustion cause higher emission of oxides of nitrogen (NOx) [13].

NOx emission of internal combustion engine is mainly due to availability of oxygen, temperature of residual gas, resident time etc. NOx emission can be minimized by pre-combustion and post-combustion treatments [14, 15]. There are other methods significantly used in reduction of NOx such as use of alcohol as a fuel additive. Due to its higher latent heat of evaporation, lower residual gas temperature is achieved, resulting in lower NOx emission. He et al. [16] reported lower NOx when the ethanol was blended with diesel. Tutak and Bereczky [17] investigated the combustion studies in CI engine when the engine was operated with ethanol-diesel blends. The found lower NOx emission for E85 blend. Use of alcohol as an additive to the fuel blend drastically reduces CO, HC, smoke and NOx emission. This method is also used in biodiesel fuel operated CI engines. Zhu et al. [18] studied the effect of ethanol-biodiesel fuel on the CI engine emissions. They observed lower NOx and smoke emissions for alcohol blended biodiesel blends. Similar results reported by [19]. But the literature available on bio-ethanol blended biodiesel fuels operations are scanty.

In the present investigation, an attempt has been made to investigate the effect of bio-ethanol on the performance and emissions of a CI engine when operated using biodiesel-blend. Pongamia biodiesel B20 blend is selected and three test fuel blends are prepared by adding bio-ethanol in a proportion of 5, 7.5 and 10% by volume to B20 blend. Performance and emission are analyzed at varying load conditions.

2 Materials and Methodology

Pongamia biodiesel is prepared by transesterification process [20]. Bio-ethanol is produced from the cashew apples by fermentation process. The ethanol produced from fermentation process contains small amount of water, it known as raw ethanol. The test fuel blends are made by adding bioethanol to the B20 pongamia biodiesel in a proportion of 5, 7.5 and 10% by volume. The physical properties are determined as per ASTM standard and tabulated in Table 1.

Table 1: Physical properties of Test fuel blends

Property	B20E5	B20E7.5	B20E10
Flash point (°C)	50	41	43
Fire point (°C)	55	49	48
Lower calorific value (MJ/kg)	40.89	40.49	40.09
Kinematic viscosity (cSt)	3.69	3.63	4.21
Density (kg/m^3)	840	848	860

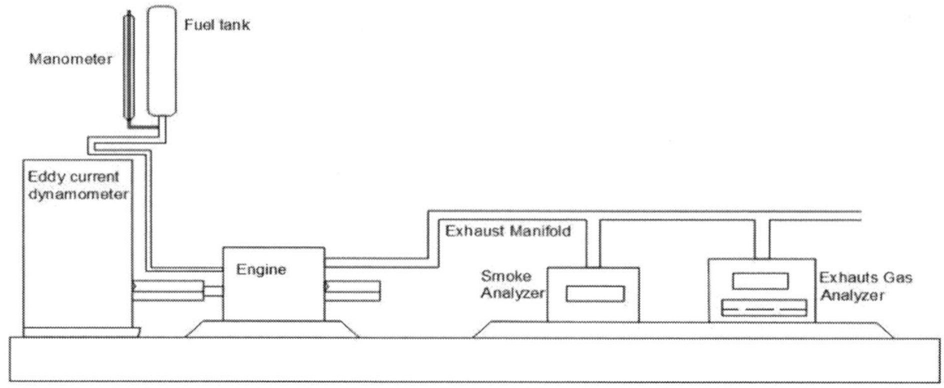

Fig. 1 Schematic representation of the experimental setup

The experiments are conducted on a four stroke single cylinder water cooled diesel engine. Eddy current dynamometer is used to load the engine. Schematic representation of the experimental set up is as shown in Fig. 1. The tests are conducted using three test fuel blends namely B20E5, B20E7.5 and B20E10 at 25, 50, 75 and 100% of full load of the engine. Emission parameters like carbon monoxide (CO) and NOx are measured using AVL Exhaust gas analyzer. The data pertaining to performance and emissions are recorded for these loads. Performance parameters like brake specific energy consumption (BSEC), brake thermal efficiency (BTE) and emissions like NOx and CO are analyzed and discussed.

3 Results and discussion

The engine is run on three different test fuel blends as mentioned above. The performance and engine exhaust emission characteristics are analyzed and discussed.

3.1 Brake thermal efficiency (BTE)

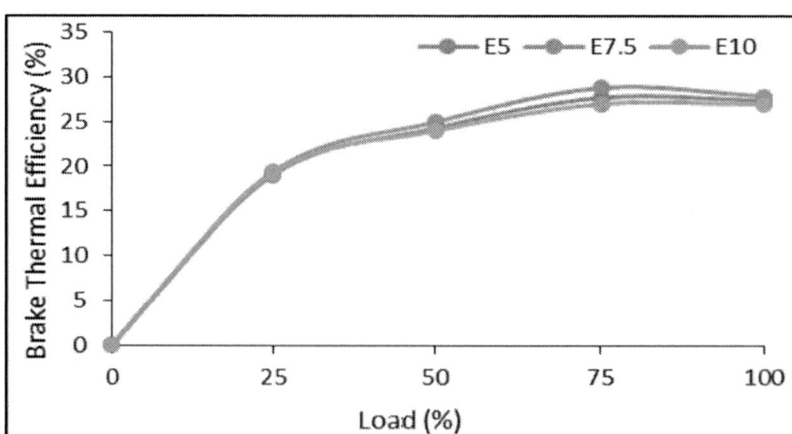

Fig. 2 Brake thermal efficiency verses load

Fig. 2 shows the relation between BTE with engine load for the test fuels. For all fuel blends, BTE increases as the load increased and reach maximum at 75% load. BTE is the ratio of brake power to the heat energy supplied. Hence it depends on the calorific value of the fuel and the combustion quality. The fuel blend B20E7.5 has a BTE of 28.8% at 75% load, which is maximum among the fuels tested. This may be due to the reason that the inbuilt oxygen content of the fuel favors the combustion and more heat will be generated which intern results higher efficiency. B20E5 and B20E10 result lower BTE due to incomplete combustion and lower caloric value respectively. This is also due to increased BSEC of the fuel blends.

3.2 Brake Specific Energy Consumption (BSEC)

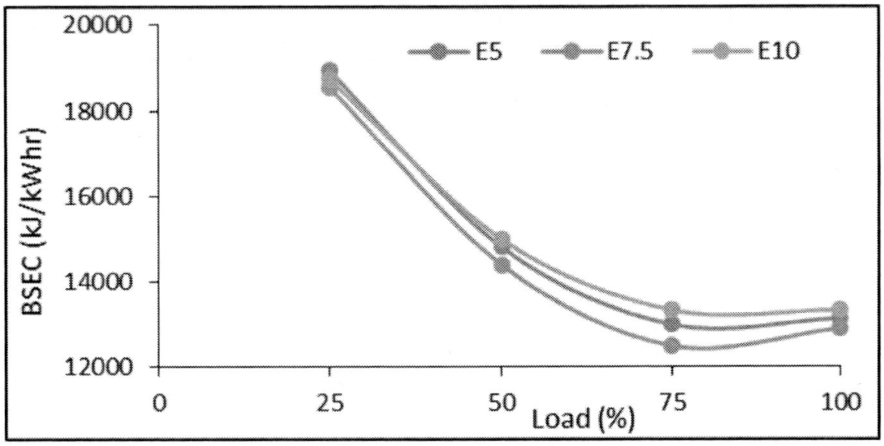

Fig. 3 Brake specific energy consumption verses load

Fig. 3 shows the BSEC for different fuel blends at varying load conditions. BSEC decreases as the load increased and minimum at 75% load. For the fuel blend B20E7.5, BSEC is minimum compared to other fuels tested. This may be due to the complete utilization of the injected fuel and reduced CO and unburnt hydrocarbons by more complete combustion. Even the fuel blend B20E5 has extra oxygen content, more complete combustion is not achieved resulting in increased BSEC. In B20E10, more inbuilt oxygen available, but higher BSEC is obtained. Increase in bio-ethanol percentage in the fuel blend increases the inbuilt oxygen content, but the decreases the calorific value. So higher BSEC is obtained for higher bio-ethanol blend.

3.3 Carbon monoxide emission

In the Fig. 4, Carbon Monoxide (CO) increases as the load increased for all fuel blends due to rich mixture combustion during higher loads. The reasons for higher CO emission are too rich fuel-air ratio, lack of oxygen content, combustion duration etc. Use of bioethanol increases the fuel borne oxygen, which intern promotes combustion. The percentage of bioethanol increased from as the fuel-air ratio increased from 5 to 7.5%, CO emission decreases. Further increment of bioethanol has no significant benefit results higher CO emission. This may be due to higher latent heat of vaporization, which reduces the rate of combustion. More CO formation takes place instead of CO_2 formation.

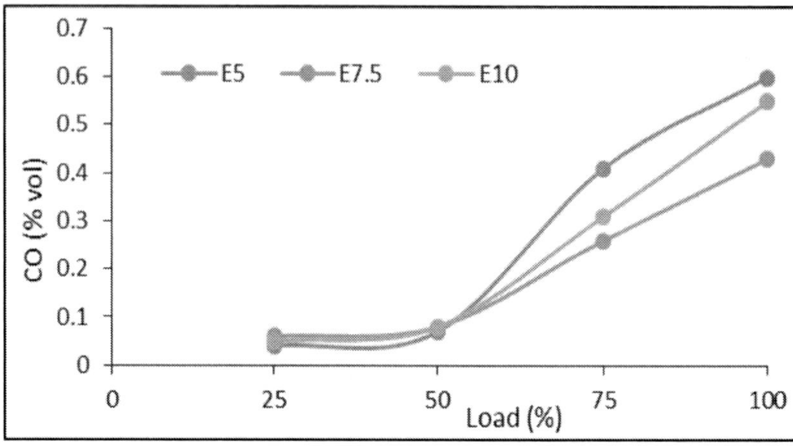

Fig. 4 Carbon monoxide verses load

3.4 NOx emission

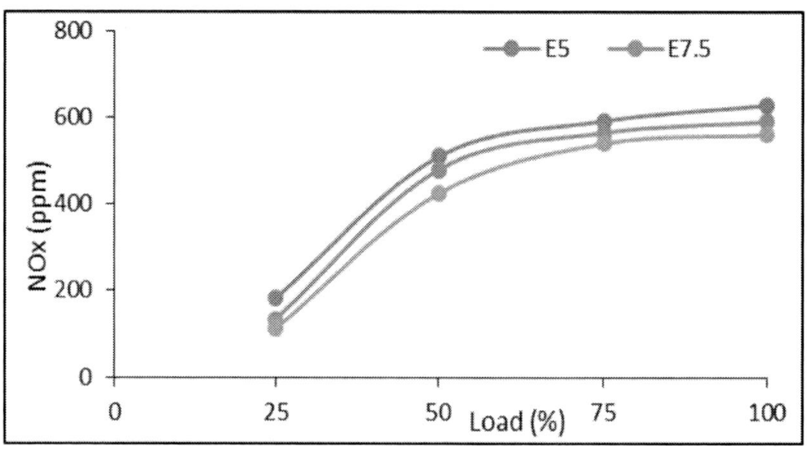

Fig. 5 NOx emission verses load

Fig. 5 shows the variation of NOx emission for various fuel blends at different loads. NOx emission increases as the load increased irrespective of the fuel. There are several reasons for the formation of NOx in CI engine. Higher combustion temperature, availability of oxygen and resident time are the key reasons for the formation of NOx. In this study, NOx decreases as the composition of bio-ethanol is increased. Out of three fuel blends tested, B20E5 gives higher NOx. Increase of bio-ethanol increases the oxygen content, but higher latent heat of vaporization of the bio-ethanol decreases the combustion temperature. This helps in reduction of NOx for higher bio-ethanol blends.

5 Conclusion

In the present study, the effect of bio-ethanol on the performance and emission of B20 pongamia biodiesel fuel blend. An analysis has been made to optimize the bio-ethanol composition in the blend. From the study, the following conclusions have been made.

- Bio-ethanol produced from cashew apples by fermentation is a potential fuel additive to the CI engine operation.
- B20E7.5 has higher BTE compared to other fuel variants tested. It gives a maximum BTE of 28.8% at 75% load.
- Lower BSEC of 12500 kJ/kWhr is obtained for B20E7.5 compared to B20E5 and B20E10 blends.
- B20E7.5 results lower CO emission of 0.43% where as B20E5 and B20E10 respectively give 0.55 and 0.6% at full load condition.
- NOx emission decreases as the percentage of bio-ethanol increased in the blend. A maximum NOx of 628 ppm is obtained for B20E5 blend whereas 590 and 550 ppm are respectively obtained for B20E7.5 and B20E10 blends.

From the study, it can be concluded that a blend having 7.5% of bio-ethanol has significant benefits in terms of performance and emissions.

References

1. Rakopoulos DC, Rakopoulos CD, Giakoumis EG, Dimaratos AM, Kyritsis DC. Effects of butanol–diesel fuel blends on the performance and emissions of a high-speed DI diesel engine. Energy Conversion and Management. 2010, 51, pp.1989-97.
2. Yanfeng G, Shenghua L, Hejun G, Tiegang H, Longbao Z. A new diesel oxygenate additive and its effects on engine combustion and emissions. Applied thermal engineering. 2007, 27, pp.202-7.
3. Sorenson SC, Mikkelsen SE. Performance and emissions of a 0.273 liter direct injection diesel engine fuelled with neat dimethyl ether. SAE Technical Paper. 1995.
4. Dinesha P, Mohanan P. A study of the effect of injection pressure on the combustion, performance, and emission characteristics of cardanol biofuel blend fuelled compression ignition engine. Asia-Pacific Journal of Chemical Engineering. 2015, 10, pp.56-64.
5. Alla GA, Soliman HA, Badr OA, Rabbo MA. Effect of injection timing on the performance of a dual fuel engine. Energy conversion and Management. 2002, 43, pp.269-77.
6. Christensen M, Johansson B, Hultqvist A. The effect of combustion chamber geometry on HCCI operation. SAE Technical Paper, 2002.
7. Durbin TD, Collins JR, Norbeck JM, Smith MR. Effects of biodiesel, biodiesel blends, and a synthetic diesel on emissions from light heavy-duty diesel vehicles. Environmental science and Technology. 2000, 34, pp.349-55.
8. Shivakumar, Dinesha P. Use of alternative fuels in compression ignition engines: a review. Biofuels. 2017, DOI: 10.1080/17597269.2017.1329494.
9. Fukuda H, Kondo A, Noda H. Biodiesel fuel production by transesterification of oils. Journal of bioscience and bioengineering. 2001, 92, pp.405-16.
10. Sarin R, Sharma M, Sinharay S, Malhotra RK. Jatropha–palm biodiesel blends: an optimum mix for Asia. Fuel. 2007, 86, pp.1365-71.
11. Nabi MN, Akhter MS, Shahadat MM. Improvement of engine emissions with conventional diesel fuel and diesel–biodiesel blends. Bioresource Technology. 2006, 97, pp.372-8.
12. Lapuerta M, Armas O, Rodriguez-Fernandez J. Effect of biodiesel fuels on diesel engine emissions. Progress in energy and combustion science. 2008, 34, pp.198-223.

13. Hountalas DT, Mavropoulos GC, Zannis TC, Mamalis SD. Use of water emulsion and intake water injection as NOx reduction techniques for heavy duty diesel engines. SAE Technical Paper. 2006.
14. Koebel M, Elsener M, Kleemann M. Urea-SCR: a promising technique to reduce NO x emissions from automotive diesel engines. Catalysis today. 2000, 59, pp.335-45.
15. He BQ, Shuai SJ, Wang JX, He H. The effect of ethanol blended diesel fuels on emissions from a diesel engine. Atmospheric Environment. 2003, 37, pp.4965-71.
16. Tutak W, Lukács K, Szwaja S, Bereczky Á. Alcohol–diesel fuel combustion in the compression ignition engine. Fuel. 2015, 154, pp.196-206.
17. Zhu L, Cheung CS, Zhang WG, Huang Z. Combustion, performance and emission characteristics of a DI diesel engine fueled with ethanol–biodiesel blends. Fuel. 2011, 90, pp.1743-50.
18. Xing-cai L, Jian-Guang Y, Wu-Gao Z, Zhen H. Effect of cetane number improver on heat release rate and emissions of high speed diesel engine fueled with ethanol–diesel blend fuel. Fuel. 2004, 83, pp.2013-20.
19. Karmee SK, Chandna D, Ravi R, Chadha A. Kinetics of base-catalyzed transesterification of triglycerides from Pongamia oil. Journal of the American Oil Chemists' Society. 2006, 83, pp.873-7.

Vortex dynamics and elliptical structure wake interaction in the proximity of wall using 2-D RANS simulation

Arunkumar H. S.[1*], *Chidanand Mangrulkar*[2] and *Trushar Gohil*[2]
[1]*Deptarment of Mechanical Engineering, Manipal Institute of Technology, Manipal Academy of Higher Education, Manipal, Udupi, Karnataka, India*
[2]*Dept. of Mechanical Engg., Visvesvaraya National Institute of Technology, Nagpur, Maharashtra, India*

Abstract: The 2-D numerical study is performed to analyses the flow characteristic behind the elliptical structure placed near the wall for three different gap ratios as 0.25, 0.5, and 1.0. Computational domain and model is initially validated with the unbounded flow over a cylinder without considering wall effect for Reynolds number of 3900. For flow over the cylinder with near wall, computational domain is modelled as Blasius profile is the input to the area of interest. At different gap ratios the effect of boundary layer on vortex shedding is studied with Reynolds number of 1440. By applying different turbulent model for analysis, study the variation in the results and suggest the suitable model for the present type of study. It has been observed that the wall effect is predominant in case of the gap ratio of 0.25 as compared to other gap ratios.

Keywords: CFD, Elliptical structure, Cross flow, Wake

1 Introduction

Flow across the different type of cylindrical structure is a very common in most of the engineering application which comprises of off shore structures, pipelines, economizer, condenser coils, shell and tube heat exchangers, and Heating, Ventilation, Air Conditioning & Refrigeration (HVAC&R) applications. So the laminar and turbulent viscous flow across these structure is a great interest with immense potential in research. As the flow is viscous, the hydrodynamics associated with the plate viscous layer and turbulence created at the cylinder downstream is studied experimentally and numerically by number of researchers. The flow in most of the application is comparatively low, and Reynolds number is usually less than 2000. Md. Mahbubar Rahman et al. [1], numerically investigated flow over a cylinder using 2-D finite volume basis, and estimated the pressure and drag coefficient. However research also continues in case of the rotating cylinder, as Atul Sharmal et al. [2] investigated numerically heat flow and fluid flow around a rotating cylinder for Reynolds number from 20-160 maintained at constant temperature, and concluded that rotation acts as drag reduction and heat transfer suppression. Gurunath Gandikota et al. [6] investigated

* Corresponding author: arunkumar.hs@manipal.edu

vortex shedding phenomena around a cylinder subjected to heating/cooling and fixed in a vertical channel. Atul Sharma et al. [7] studied effect of channel-confinement of various degree for the case of upwind flow around a heated/cooled square cylinder for Reynolds number = 100 and Prandle number= 0.7.

In majority of the application, the cylinder is often supported of an additional domain wall in terms of adjacent cylinder or an external plate. Such a system is of lower importance and has not been explored much. S. Sarkar et al. [3], studied vortex dynamics of a wake formed by circular cylindrical object near to a plate boundary at Reynolds number equal to 1440, for the different gap ratios (G/D) (G is the gap between the wall and the cylindrical object, and D is the hydraulic diameter) by means of LES simulation. The use of the LES simulation is very costly and time consuming. Hence there is a need to perform the simpler and relatively cheaper analysis by means of the RANS simulation.

Literature study shows the different approach and method to analyse the flow. The aim of the present work is to numerically simulate the flow across the cylinder along with the domain boundary wall and results are analysed with the Large Eddy Simulation results and experimental available data. In the same fluid domain check the effect of different shape (Ellipse) objects effect on the boundary layer formation by means of 2-D RANS simulation. The domain wall is at a distance of 0.25D, 0.5D, and 1.0D from the wall of the object.

2 Methodology

The flow field across a cylinder in two dimensional, with axis of the elliptical body perpendicular to the flow path. The fluid which passing over the elliptical body is air. It is treated to be incompressible and the flow is from left side of the computational domain. Computational domain is developed for unbounded flow over a cylinder by maintaining distance 8D in the inflow and 25D at the outflow and 8D between top and bottom of the wall. The flow over a circular cylinder is performed at a Reynolds number of 3900 with different turbulent models and compare the results with experimental results of Ma et al. (2000) [5].

Similarly computational domain is developed for elliptical structure with the wall at different relative gap distance (G/D =0.25, G/D =0.5 and G/D =1.0).The figure (1) indicates the physical domain of the cylinder along with the wall boundary. Elliptical object dimensions are taken by maintaining same hydraulic diameter as the cylinder which is used for validation of the model with experiment results. Computational domain modelled as 10D in the upstream and 25D at the downstream of the cylinder in the direction of the flowing fluid. The height of the domain is 8.5D for gap ratio (G/D) of 0.25, and 0.5. Computational domain is extended to 9.0D for relative gap (G/D) of 1.0. The above stated computational space is discretized using finite volume method.

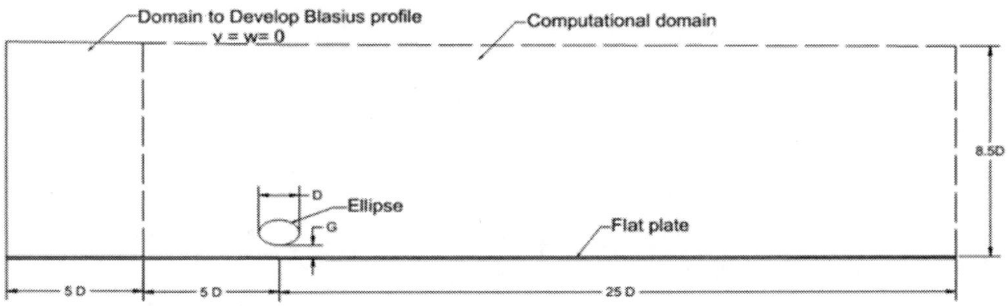

Fig. 1. Schematic diagram of the computational domain

The body fitted Cartesian uniform mesh is generated using partition block method in ICEM CFD ANSYS 16.0. The mesh is much finer nearer to the elliptical structure wall and plate boundary. The grid is formed by regulating the mesh size and structure and quality of mesh is obtained 0.9.Grid independent study is also carried out and typical mesh structure near the elliptical structure and plate as shown in figure 2.

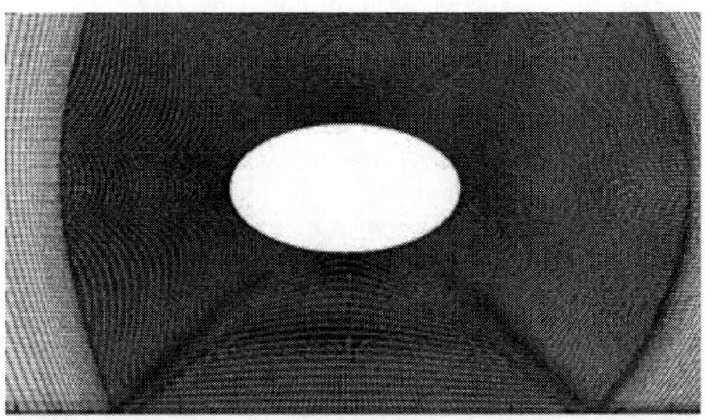

Fig. 2. Schematic diagram of the mesh structure around the Elliptical structure

2.1 Governing equation and boundary condition

The 2-D Navier-Stoke governing equation was discretized and further processed in ANSYS FLUENT 16.0. Very fine mesh around the elliptical object and near to the plat boundaries give better and accurate results. For present study enhanced wall treatment condition is applied by considering viscous boundary layer y+ as 1. The solution accuracy of the CFD simulation is improved significantly by maintaining the y+ =1. The governing equations for momentum is discretized with the second order upwind scheme. The Pressure-velocity coupled algorithm is applied to get the combined effect of pressure and velocity components. In this numerical study, solution residuals are converged of the order of 10-6.

The Reynolds number is maintained at 1440, which depends on the free stream velocity of the working fluid i.e., air. The Blasius profile (v = w = 0) is applied by extending existing computation domain to 5D in upstream direction. The bottom surface of the computation space is treated with no-slip boundary condition, while upper is specified shear.

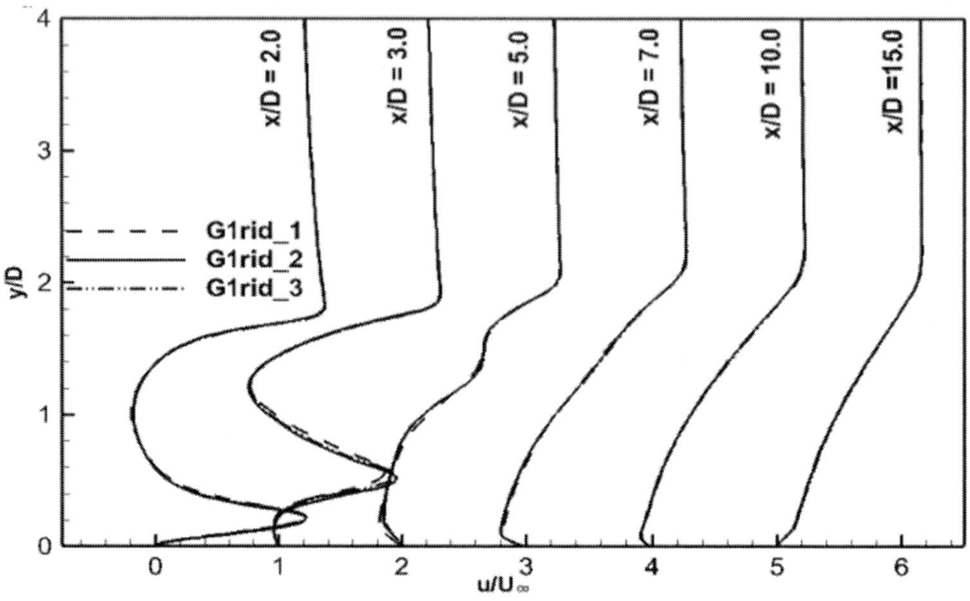

Fig. 3. Grid-independent test for basic model for relative gap ratio = 0.25 and Reynolds number = 1440. (Grid_1=40000 elements; Grid_2=75000 elements; Grid_3 = 95000 elements.)

Figure 3 indicates the grid independence performed by considering three different mesh size, as 40000, 75000, and 95000 number of elements. It has been observed from the graph that the deviation in velocity ratio is nominal as the number of elements increased to 95000 from 75000. Hence the size of the grid comprising of 75000 number of elements is selected for the numerical simulation.

3 Results and Discussion

The numerical analysis is initially performed for the unbounded cylinder by considering the various two equation turbulence model such as RNG k-epsilon, Std. k-epsilon, Realizable k-epsilon, and k-omega models at Reynolds number of 3900.

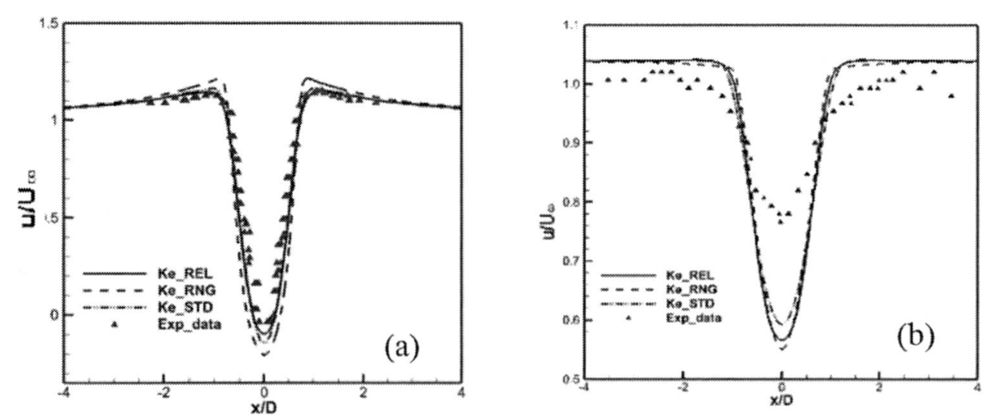

Fig. 4. Free stream velocity profile at (a) y/D=1.06 (b) y/D=7.0

From figure 4, it is clear that the plot for free stream velocity k-ε Realizable model is in close proximity compared to the experimental result obtained to Ma et al. (2000) [5]. Realizable k-ε model gives the good approximation near to the wall i.e. y/D=1.06 to 4, but small deviation is observed at a distance of y/D=7 as indicated in the figure.

To estimate the combined effect of the vortex and boundary layer, in case of near wall for different gap ratio (G/D) similar numerical analysis was performed for Reynolds number 1440.

It has been observed that, unlike unbounded flow other RANS models are unable to capture the all the secondary vortex formed in the cylinder downstream direction. Hence in the case of near wall flow, k-ω SST model is selected with pressure velocity coupling.

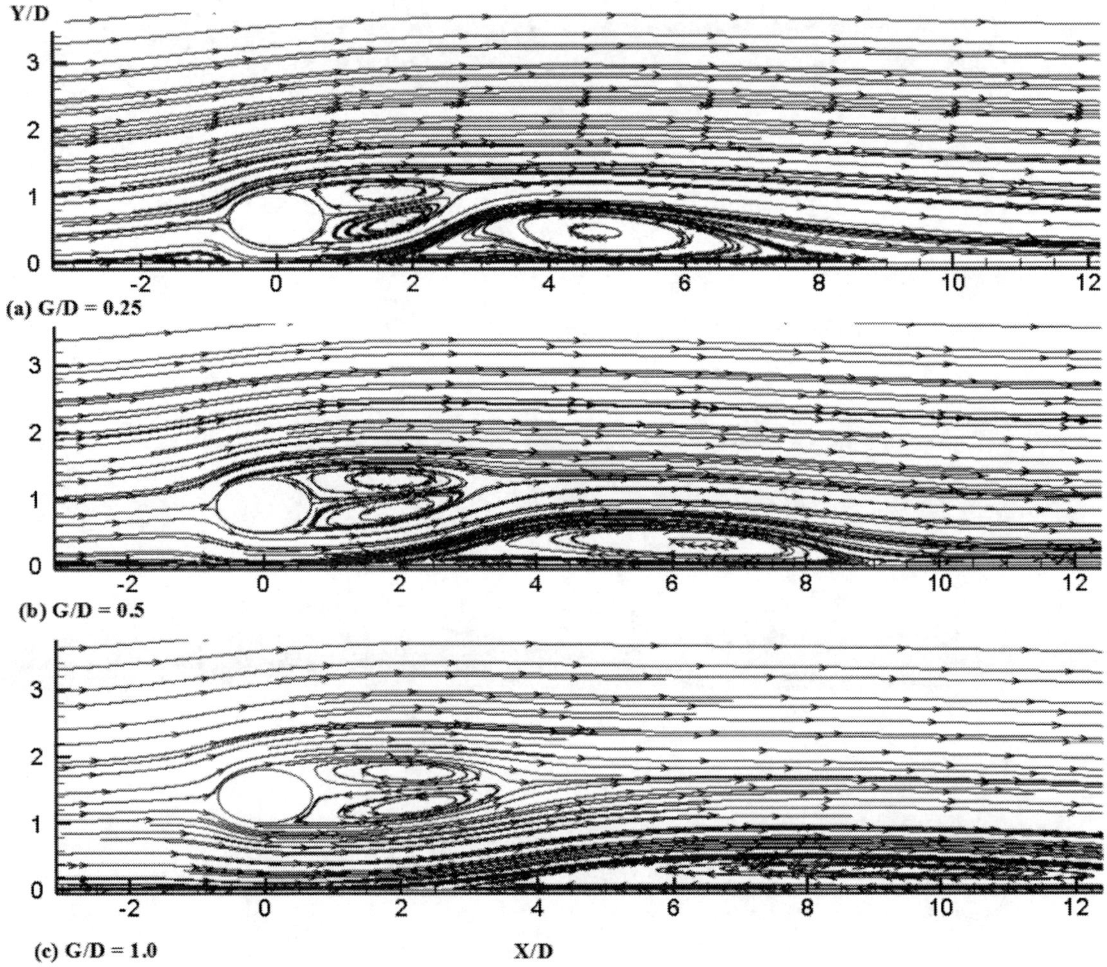

Fig. 5. Mean stream wise velocity contours along with few selected streamlines for different gap ratios

As indicated in the figure 5, k-ω SST turbulence model capture all the primary as well as secondary bubbles in both upstream and downstream direction. As in case of the LES simulation performed by S. Sarkar et al. [3], the upstream bubbles gets reduces in size as the gap ratio increases in case of the present RANS simulation also. Furthermore, it has been observed that the rapid increase in the size of the primary bubble in the downstream direction, as the gap ratio increases. The secondary bubble in the upstream direction gets diminished for highest gap ratio. Primary bubble size increases lengthwise significantly in X direction by reducing its height in Y direction. Figure 6, Figure 7 Figure 8, indicates the dynamic pressure profile at different relative gap ratios.

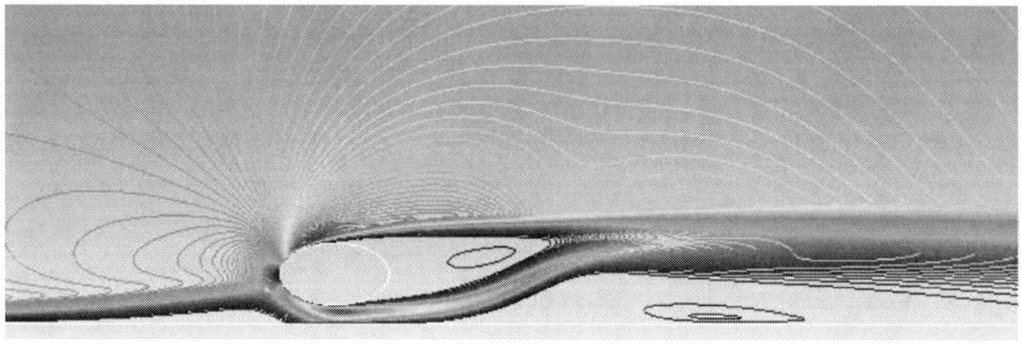

Fig. 6. Dynamic pressure profile at G/D=0.25

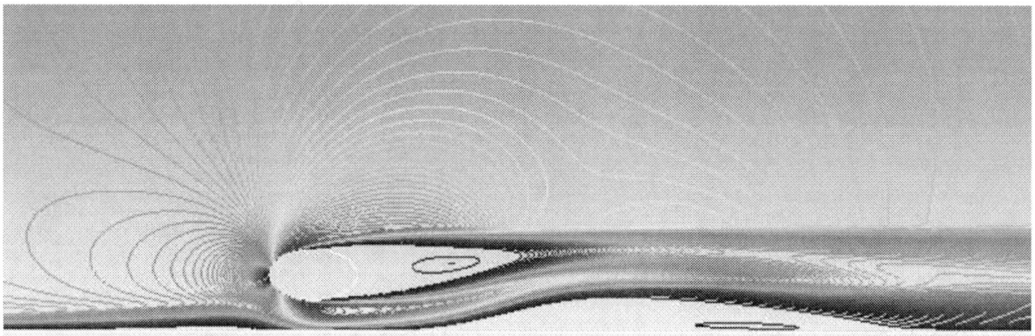

Fig. 7. Dynamic pressure profile at G/D=0.5

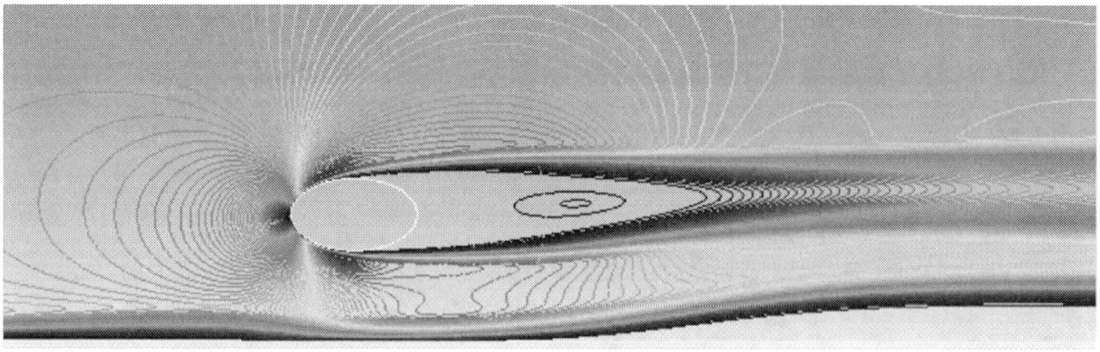

Fig. 8. Dynamic pressure profile at G/D=1.0

A boundary layer may be laminar or turbulent. A laminar boundary layer is one where the flow takes place in layers, i.e., each layer slides past the adjacent layers. Boundary layers appear on the surface of bodies in viscous flow because the fluid seems to attach to the surface. The turbulent kinetic energy is measure of turbulence generated on the air side as the fluid pass over the object.

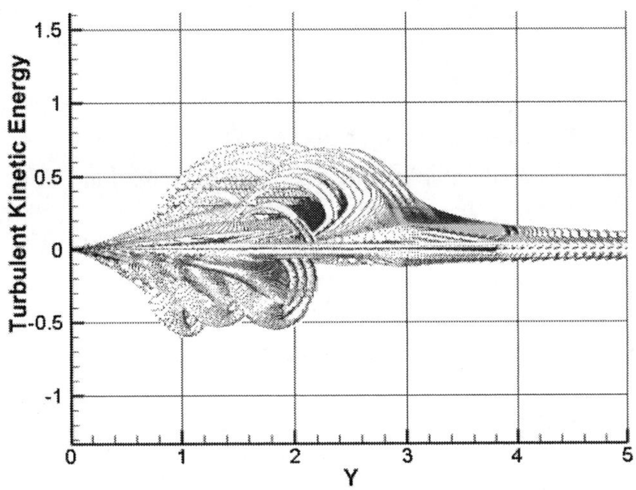

Fig. 9. Turbulance kinetic energy dissipation along the Y axis (G=0.25 ; G=0.5 ; G=1.0)

Figure 9 indicates the variation of turbulence kinetic energy along Y axis. The high variation in the magnitude of the turbulence is often due to the high pressure drop in the air side by increasing the drag at the cylinder downstream. The intensity of the turbulent kinetic energy will further increase if the gap between the domain wall and the side of the square cylinder is reduced. This increased turbulence will accelerate the bubble formation on either side of the object.

4 Conclusions

The RANS numerical analysis of flow past cylinder in the unbounded case and in the vicinity of a wall have been carried. In case of the unbounded case k-ε Realizable model gives good agreement with that of the experimental findings. With reference to the circular cylinder same analysis is carried out for elliptical object with different relative gap distances by maintaining same hydraulic diameter. In case of the object with near wall for different gap ratios the flow is simulated for Reynolds number of 1440. The domain wall has the least effect in case of gap ratio of 1.0, while the gap "ratio" of 0.25 has the maximum impact on the wake and bubble formation. The primary bubble is observed in the upstream direction in case of gap ratio of 0.25 only.

5 Nomenclature :

RANS équations: Reynolds-Averaged Navier–Stokes équations

Acknowledgements

We express our sincere thanks to the Department of Mechanical Engineering, Visvesvaraya National Institute of Technology, Nagpur, for providing the platform through Summer school in CFD 2016.

References

1. M. Rahman, Md. Mashud Karim, Md. Abdul Alim, "Numerical investigation of unsteady flow past a circular cylinder 2-D finite volume method", Journal of Naval Architecture and Marine Engineering, (June 2007), 27-42.
2. B. Sachin Paramane, Atul Sharma, "Numerical investigation of heat and fluid flow across a rotating circular cylinder maintained at constant temperature in 2-D laminar flow regime", International Journal of Heat and Mass Transfer 52 (2009) 3205-3216.
3. S. Sarkar, Sudipto Sarkar, "Vortex dynamics of a cylinder wake in proximity to a wall", Journal of Fluids and Structures 26 (2010) 19–40.
4. A. Ahmed, "On the wake of a circular cylinder with nodal and saddle attachment", Journal of Fluids and Structures 26 (2010) 41–49.
5. Lourenco, L.M., Shih, C., "Characteristics of the plane turbulent near wake of a circular cylinder, a particle image velocimetry study", (2000).
6. G. Gandikota, Sakir Amiroudine, Dipankar Chatterjee, Gautam Biswas, "The Effect of Aiding/Opposing Buoyancy on Two-Dimensional Laminar Flow Across a Circular Cylinder", Numerical Heat Transfer, Part A: Applications: An International Journal of Computation and Methodology, 58: 385–402, (2010).
7. A.Sharma, V. Eswaran, "Effect of channel-confinement and aiding/opposing buoyancy on the two-dimensional laminar flow and heat transfer across a square cylinder", International Journal of Heat and Mass Transfer 48 (2005) 5310–5322

Combined Natural Convection and Surface Radiation Inside Vented Triangular Enclosure - An Experimental Study

Rahul Raj[1*], *Pradyumna K. C.*[2], *Rakshith B. R.*[2], *Nithin R. B.*[2] and *Karthik S. R.*[2]

[1]*Department of Mechanical and Manufacturing Engineering, Manipal Institute of Technology, Manipal Academy of Higher Education, Manipal – 576104, Karnataka, India*
[2]*Department of Mechanical Engineering, B.N.M. Institute of Technology, Banashankari, Bengaluru - 560 070, Karnataka, India*

Abstract. In the present experimental study, the influence of Opening ratio on the chamber temperature in case of vented triangular enclosure with hot base plate has been undertaken. Due to more flow passage, the average air temperature in the enclosure decreases by 12% for OR of 0.25 and 13% for OR of 0.5 compared to OR of 0.125. Similar trend is observed for Rayleigh number with a rise of 73% in OR 0.5 compared to OR of 0.125. Further, the influence of radiation heat transfer in the natural convection based enclosure is also studied. This kind of investigation is useful in the design and analysis of electronic cooling systems.

Keywords: Natural Convection, Triangular Enclosure, Opening Ratio, Radiation heat transfer

1 Introduction

When a fluid or any matter is surrounded on all sides by solid walls, it forms an enclosure. It might be desired to improve or slow down heat transfer by using enclosures depending on the type of application. In electronic enclosures, it is required to increase heat transfer in the enclosure to make sure the electronic components are operating at temperatures lower than the critical temperature of the component. But when we consider a double paned glass window, which is an enclosure trapping air between two glass panes, the heat transfer decreases from the interior. Triangular cavities have a wide range of applications in solar collectors and electronic devices as well as heating and ventilation in buildings with inclined roofs [1].

The geometry of the enclosure used also has an effect on the heat transfer. Use of different shaped enclosures / variation in aspect ratio and changing the inclination of the enclosure influences the heat transfer rate. It has been found that increasing the inclination angle of a rectangular enclosure reduces the heat transfer [2]. Changes in geometry varies the amount of surface area available for convective heat transfer from the enclosure to the surroundings. The convection with respect to the inside walls greatly depends on the geometry of the enclosure especially when only natural convection heat transfer is available.

* Corresponding author: rahulraj2894@gmail.com

The thermal analysis of an enclosure irrespective of its geometry involves finding the temperature and velocity variation of the fluid within the enclosed space. It is important to analyse the flow of the fluid inside the enclosure and its interaction with the enclosure walls. The temperature variation describes the temperature gradient pattern set up inside the enclosure which leads to buoyancy driven flow of the fluid.

Many researchers have studied the natural convection phenomena in enclosures. Mirabedin[3] formulated a correlation for Nusselt number in terms of its aspect ratio and Rayleigh number in case of a triangular enclosure and found that the Nusselt number increased with increase in aspect ratio. Kamiyo et al.[4] studied the heat transfer in an asymmetric triangular enclosure with a hot isothermal base wall. They varied the pitch angle (angle between the hot isothermal wall and cold inclined wall) and found that the heat transfer from hot base wall to inclined cold wall is high at low pitch angles and reduces with increasing pitch angle. This was due to the formation of a multi-cellular flow structure at low pitch angles. With increase in the pitch angle, the number of cells reduced, thereby reducing the heat transfer. Anderson et al.[5] determined the heat transfer coefficient for various aspect ratio in an attic shaped enclosure. They suggested a correlation for Nusselt number in terms of the Grashof number as well as the aspect ratio of the attic shaped enclosure. This correlation covers a wide range of Grashof number than the earlier correlation. Sieres et al.[6] carried out numerical computations for laminar flow in a triangular enclosure with a vertical hot wall by varying the aperture angle. They obtained the maximum temperature at the upper edge of the heated vertical wall and concluded that Nusselt number increases with Rayleigh number till critical state and then reduces with increase in Rayleigh number. Kuznetsov et al.[7] studied the radiative-convective heat transfer in an enclosure. They found that when radiation heat transfer was taken into account, the temperature of the gas cavity increased on average by 11% while measuring temperatures below 200°C. Dubovsky et al.[8] analysed the temperature and flow inside a rectangular enclosure with two openings. Flow visualisation was done by using smoke of incense sticks. He also carried out numerical simulations with the bottom wall as hot wall and found good agreement with the experimental results.

From the above literature review, the concluding remarks are: There has been substantial work done on triangular enclosures with respect to aspect ratio/apex angle/Rayleigh number. The effect of having openings in a triangular enclosure and its size, as well as the influence of variation of the internal hot wall emissivity of the enclosure in the heat transfer has not been studied. Hence the present study is focused on the above mentioned objectives.

2 Methodology

A right angled triangular enclosure was taken with an aspect ratio of 1 as shown in Figure 1. The opening ratio (OR) is the ratio of the sum of the length of the two vents to the length of the vertical edge. The inclined wall and the vertical wall containing the vents were insulated while the bottom wall was maintained at 60°C. The effect of change in the opening ratio was studied experimentally. Further, the hot base plate was blackened and compared to the results when the base was not blackened.

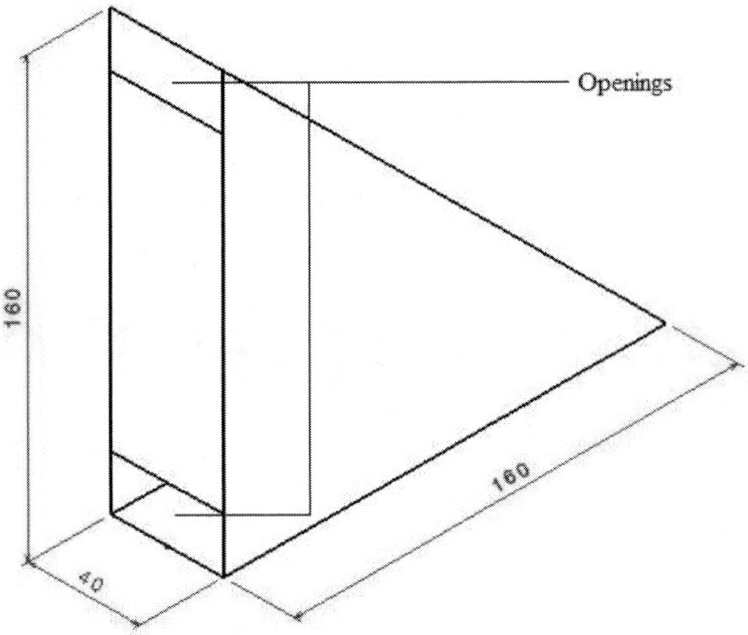

Fig. 1. Schematic of triangular enclosure with openings (dimensions in mm)

3 Experimental setup

As a heat source, a copper plate (160mm×40mm×3mm) was attached to two small 30W heaters through regulators at its ends. Thermal paste was applied between heater and the plate to minimize the thermal contact resistance. This arrangement ensures better heating and a variable uniform temperature system as shown in Figure 2.

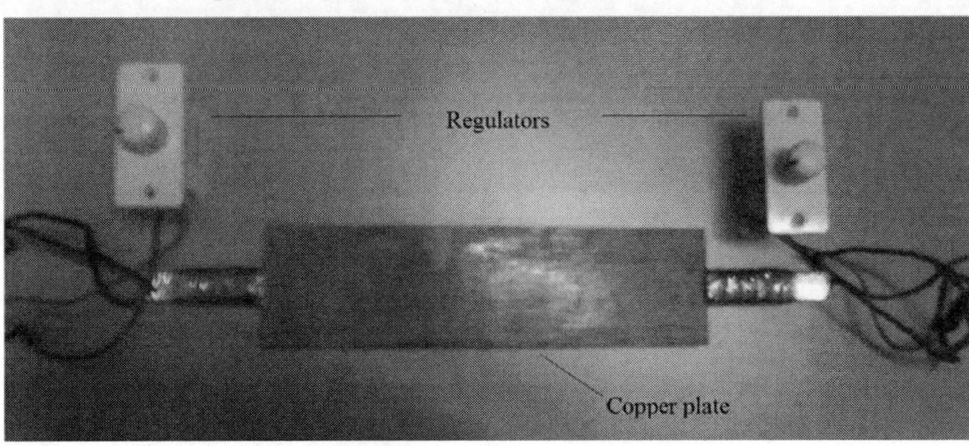

Fig. 2. Heater used in experimental setup

The triangular enclosure was fabricated from galvanised steel sheet. The length of the horizontal and the vertical edge were taken as 160 mm. Two cut-outs were provided, one at the top and the other at the bottom of the vertical wall to act as the openings of the enclosure with a provision for changing opening ratio. 3 different models with opening ratios of 0.125,0.25 and 0.5 were studied. Both inner and outer walls (other than hot wall) were insulated (using glass wool) so that the heat transfer by conduction from the base plate to the triangular vertical sides do not affect the convective heat transfer inside the enclosure. One side of the triangular enclosure is made transparent by using an acrylic sheet to examine the location of thermocouples as shown in Figure 3.

In total, eight K-type thermocouples were used. Two thermocouples were placed on the base plate at equal distances from either edges so that a uniform temperature throughout the base plate can be ensured. The remaining six thermocouples were suspended from the inclined edge of the enclosure and are positioned within the enclosure as shown in Figure 4.

Fig. 3. Experimental Setup

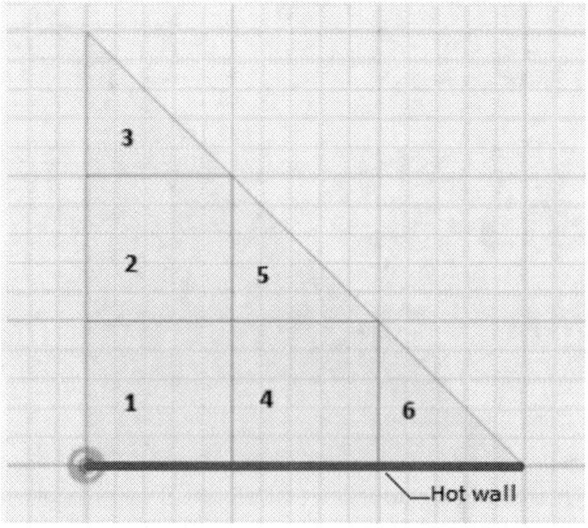

Fig. 4. Triangular enclosure divided into 6 parts

A data acquisition system was used to simultaneously record the temperature readings of all 8 thermocouples. The schematic of the entire arrangement is shown in Figure 5.

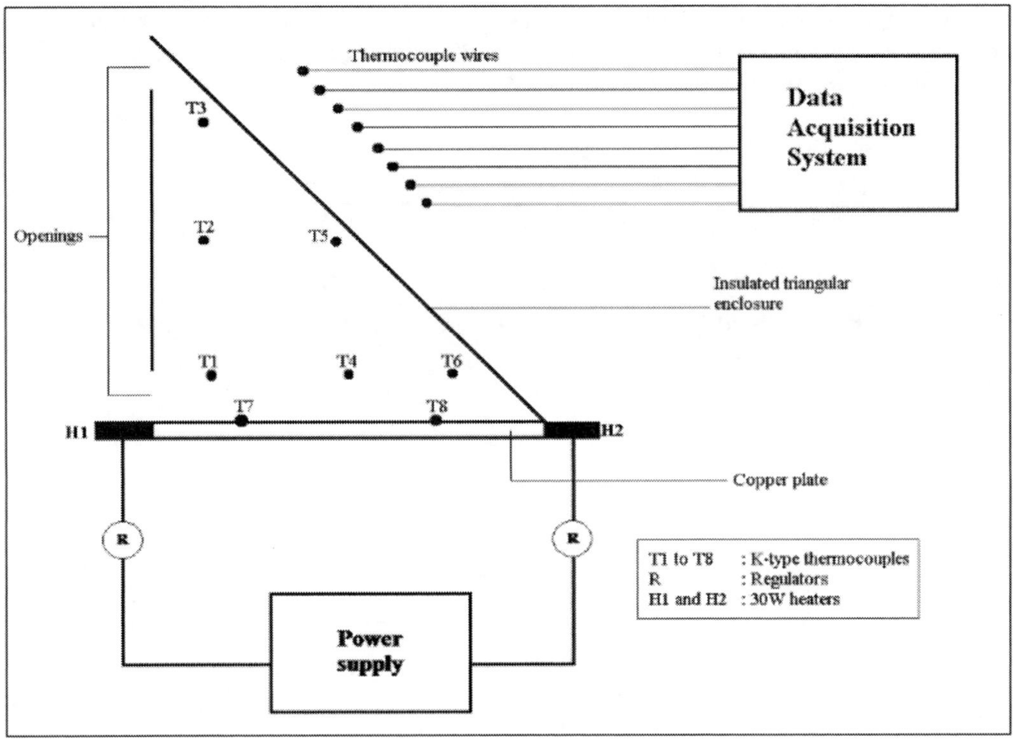

Fig. 5. Schematic of experimental setup

4 Experimental analysis

Rayleigh number is a very important parameter in natural convection heat transfer and affects the Nusselt number. To experimentally determine the Rayleigh number, the bulk mean temperature was taken by the average of the 6 thermocouple readings.

$$Ra = \frac{g\beta(T_s - T_m)L^3}{\nu\alpha} \quad (1)$$

The base plate is initially heated till it reaches a constant temperature, say 60°C. This process is then repeated for different opening ratios. The emissivity of the hot base plate was also altered by blackening the base plate. This increases the radiative heat transfer within the enclosure. It was found that the heat transfer rate is quicker compared to an untreated base. It was seen that the thermocouple readings in the case of the blackened base were much lower due to increased radiative heat flux.

5 Results and discussions

After steady state was reached the temperature readings were obtained. The effect of varying opening ratio on the heat transfer rate is studied. It was found that the Rayleigh number increases with increasing opening ratio and hence leading to better cooling. The temperature at six different locations inside the enclosure is plotted for different opening ratios as shown in Figure 6. As can be observed from the plot, the opening ratio 0.125 has the highest temperature in each zone due to lower natural convection as the opening is very small. The trend of the line plots of the other two opening ratios are similar in the regions away from the openings. Although there is some discrepancy in the thermocouples closer to the vents, this can be due to flow reversal or other errors.

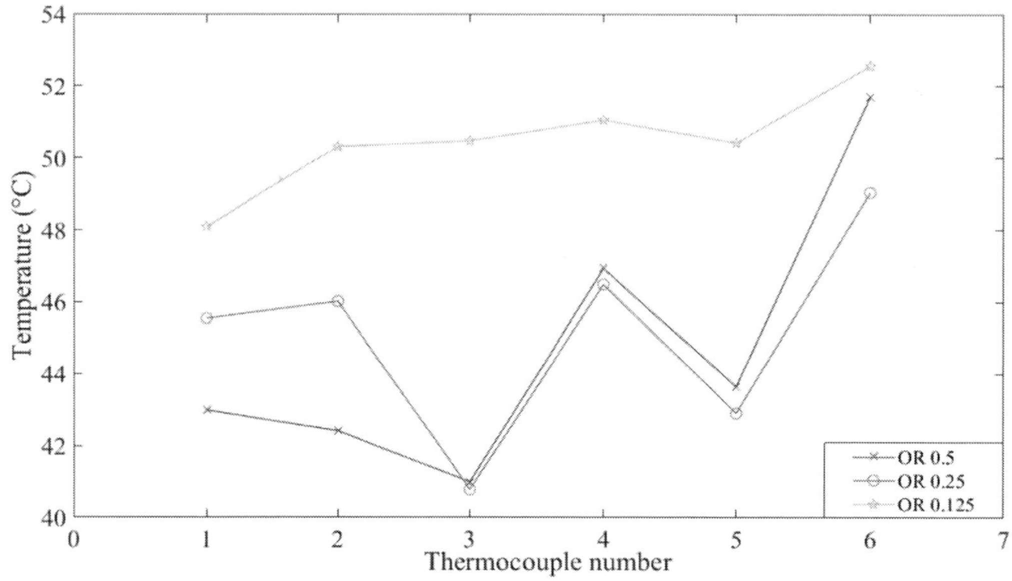

Fig. 6. Temperature variation with opening ratios

The Rayleigh number for the three opening ratios were determined and are shown in Figure 7. From OR 0.125 to 0.25, the Rayleigh number increases steeply. But the variation of Rayleigh number from an OR of 0.25 to 0.5 is very small. This shows that the heat transfer rate increases rapidly when the OR is changed from 0.125 to 0.25 and that it increases insignificantly when the OR is changed from 0.25 to 0.5. Hence, having an OR of 0.5 will give the highest heat transfer rate but similar heat transfer rate can be obtained with a much lower OR of 0.25.

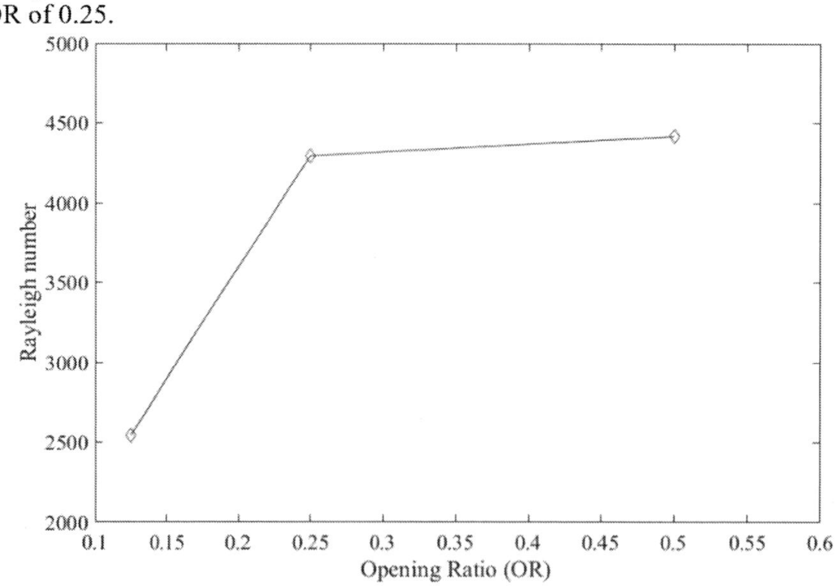

Fig. 7. Rayleigh number variation with Opening Ratio

The effect of change in radiation heat transfer was also considered during experimentation. Hence to increase the radiative heat transfer, the emissivity of the hot base plate was increased. This was done by blackening the top surface of the hot base plate. Figure 8 shows the variation of temperature at different locations of the enclosure with a normal hot base surface as well as with a blackened hot base plate. It was observed that the variation trend was similar in both the cases and the temperature readings inside the enclosure with the blackened base was found to be lower than the one with normal base. The lower temperatures inside the blackened base enclosure is due to the increase in radiative heat transfer between the surfaces.

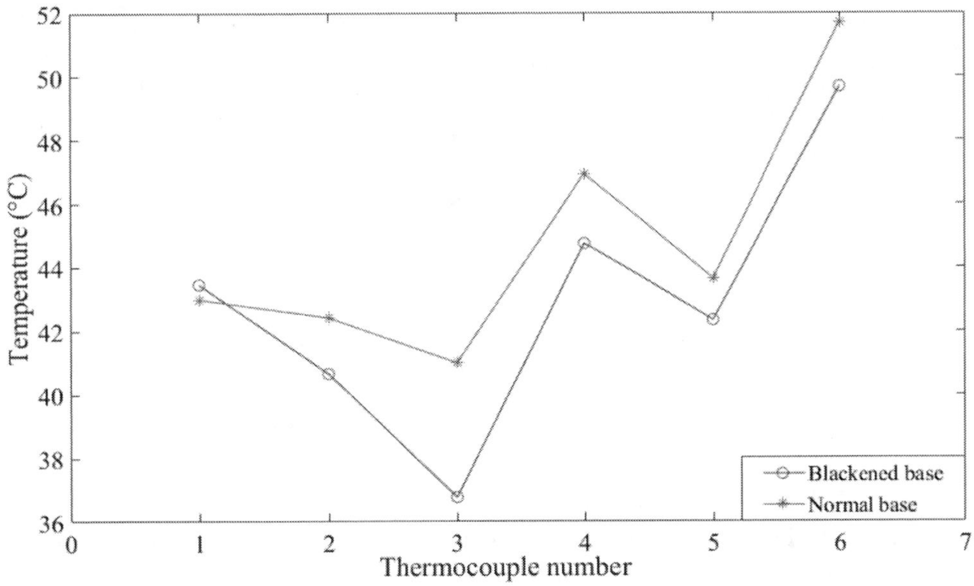

Fig. 8. Enclosure temperature comparison of blackened and normal base plate

6 Conclusions

The outcome of the experimental investigation and analysis is as follows:

1. As natural convection current is directly proportional to the opening ratio, the overall temperature of air in the enclosure is high for opening ratio 0.125, similarly, the optimum opening ratio is 0.25 as no appreciable variation in temperature is observed as in the case of 0.5
2. Since the driving potential is temperature in Ra, OR is proportional to Ra, as a result of this OR of 0.25 maybe optimized
3. The conjugate heat transfer phenomena has been studied which indicates a maximum drop of 10% in temperature locally. This is due to the dominance of radiation heat transfer as the hot surface has been blackened.

7 Nomenclature

g	Gravitational acceleration [m/s^2]	T_m	Mean Temperature [°C]
L	Characteristic length [m]	T_s	Surface temperature [°C]
Nu	Nusselt number	OR	Opening Ratio
Pr	Prandtl number	α	Thermal diffusivity [m^2/s]
Ra	Rayleigh number	β	Thermal expansion coefficient [1/K]
T_a	Ambient temperature [°C]	ν	Kinematic viscosity [m^2/s]

References

1. R. Anderson and F. Kreith, NCAPSTS, Adv. Heat Transf., 18, pp. 1–86, (1987).
2. M. Hasnaoui, P. Vasseur, and E. Bilgen, NCREAFHW, Heat mass Transf., 32, 6, pp. 365–373, (1997).
3. S. Mirabedin, CFD MNCRTE, Int. J. Heat Technol., 34, 3, pp. 503–506, (2016).
4. O. M. Kamiyo, D. Angeli, G. S. Barozzi, and M. W. Collins, NCATEHB, J. Phys. Conf. Ser., 547, pp. 1–11, (2014).
5. T. N. Anderson, M. Duke, and J. K. Carson, SNCHTCAE, Int. Commun. Heat Mass Transf., 37, 8, pp. 984–986, (2010).
6. J. Sieres, A. Campo, and J. Martínez-Súarez, NCFVUTC, Therm. Sci., 20, 5, pp. 1407–1420, (2016).
7. G. V. Kuznetsov and M. A. Sheremet, CNCSRE, Int. J. Heat Mass Transf., 52, 9–10, pp. 2215–2223, (2009).
8. V. Dubovsky, G. Ziskind, S. Druckman, E. Moshka, Y. Weiss, and R. Letan, NCVEHDP, Int. J. Heat Mass Transf., 44, 16, pp. 3155–3168, (2001).
9. Y. A. Çengel, A. J. Ghajar, HMTFA, 3, (2004).

Lithium-Ion Battery Management System: A Lifecycle Evaluation Model for the Use in the Development of Electric Vehicles

Ayush Sisodia[1] and *Jonathan Monteiro*[1*]

[1]*Manipal Institute of Technology, Manipal Academy of Higher Education, Manipal, Karnataka, India*

Abstract. The use of Lithium-ion batteries in the automobile sector has expanded drastically in the recent years. The foreseen increment of lithium to power electric and hybrid electric vehicles has provoked specialists to analyze the long term credibility of lithium as a transportation asset. To give a better picture of future accessibility, this paper exhibits a life cycle model for the key procedures and materials associated with the electric vehicle lithium-ion battery life cycle, on a worldwide scale. This model tracks the flow of lithium and energy sources from extraction, to generation, to on road utilization, and the role of reusing and scrapping. This life cycle evaluation model is the initial phase in building up an examination model for the lithium ion battery production that would enable the policymakers to survey the future importance of lithium battery recycling, and when in time setting up a reusing foundation be made necessary.

Keywords: Lithium-Ion, Batteries, GREET, Electric Vehicles, Hybrid

1 Introduction

In the recent age, in comparison to nickel cadmium and lead acid batteries, Li-ion batteries are more suited for electric vehicles since Li-ion batteries have higher energy density, weighs less, brings down the maintenance, and have an extended battery life. [1] Other options, for example, nickel-metal hydride and sodium nickel chloride batteries, confront similar issues as lead acid and nickel cadmium batteries which have lower energy density, power, and execution. Besides, the alternative nickel batteries may likewise have a more critical effect on nature, giving a disincentive for future improvement. Then again, Lithium particle batteries give a superior option as far as energy density, costs, and ecological effect are concerned and are probably going to be a cutting edge of new innovation. [1] Therefore lead-acid and nickel-cadmium batteries are being eliminated and Li-particle batteries are catching an expanding market of the overall industry for electric vehicles.

Li-ion batteries have taken over, particularly in the latest generation of plug-in electric vehicles (PEVs). Worldwide revenue from Li-ion batteries for EVs was less than $6 billion in 2014 but is expected to grow to $26.1 billion in 2023. [2] This expansion in demand is exceptionally reliant on the stores and resources of lithium. Conservative resource estimates

* Corresponding author: jonathan.m@manipal.edu

of ~11 million tons and there is sufficient ability to fulfill demand until 2050 without actualizing a recycling foundation. [3] In this manner, it is important to not just evaluate the sufficiency of future demand and supply of lithium, but also to consider whether Li-ion batteries can economically control the future of electric vehicles.

For an effective new innovation to endure into the future, it is essential to assess the reserve quantity, lifecycle financial aspects, and potential security issues related with the resource. The initial phase in evaluating the innovation is to build up an entire framework in which the innovation and assets dwell. In a resource constrained world, particularly when resources are not uniformly geologically distributed, it is vital to have the capacity to survey how the potential accessibility of scarce input resources will affect the term suitability of the technology. For any constrained resource, recycling applications may release pressure on natural environment and enhance the financial aspect of the technology that uses the resource. In this manner, an exhaustive comprehensive understanding of the system in which the resource and technology dwell is important to set up resource security, evaluate the advantages of resource reusing, and survey future practicality of the innovation.

2 Currently Available Concept

The battery framework relies upon the different necessities of the vehicle, unique to its size, make, and model. Vehicles ought to introduce a suitable battery size and structure to guarantee their security, lifetime, and performance. Li-ion batteries ideally make up 25% (by weight) of the vehicle and are equipped with a variety of safety features. The lifetime of these batteries relies upon their performance. Better performance guarantees longer battery life, which is important to both consumers and producers.

Battery execution is administered by two critical components: *energy*, which for the most part manages the driving extent or range, and *power*, which is revealed at top speed and acceleration. There is generally a tradeoff between range and performance. Batteries can either have higher energy or higher power, however not both. [5] For instance, batteries in an all-Electric Vehicle (EV) are mostly energy based to guarantee a longer driving range; though, batteries in a Hybrid Electric vehicle (HEV) are power-based for performance, given their capacity to completely charge while driving. Plug in Hybrid Electric vehicle (PHEV) batteries utilize a blend that is both vitality and power based. For shorter driving excursions, they are energy based and when battery is completely drained, they are power based.

Battery condition is another critical basis that guarantees battery's optimum usefulness and is for the most part measured as a State of Charge (SOC). The SOC is expressed as a percent of "maximum battery limit." [5] There are two working modes related with SOC: Charge Depleting (CD), in which the vehicle activity is proceeding to decrease in the battery charge, and Charge Sustaining (CS), which holds a moderately steady charge in the battery for every mode of vehicle. [6] The condition of charge of batteries fluctuates crosswise over various utilizations of EV, HEV, and PHEV. EVs generally keep running in CD mode, HEVs run in CD mode, and PHEVs keep running in both CS and CD mode.
Batteries are the governing part of the vehicle where their selection, size, outline, transfer, and reusing affect the quality, lifetime, and security of the vehicle. [1]

Table 1. Performance Characteristics of Li-ion Batteries in EV, HEV, and PHEV [7]

Parameter	**EV**	**HEV**	**PHEV**
Technology	Energy-based due to longer driving range	Power-based because batteries do not fully charge while driving	Energy-based for shorter driving trips. Power based upon battery depletion
State of Charge	CD	CD	CD & CS at 25% SOC
Energy/Power	2	15-20	3-15
Battery Dimensions	>HEV & >PHEV	1-2 kWh	5-15 kWh

Table 1 summarizes the performances of different types of Electric vehicles. Because of expanding greenhouse emanations and developing risk to resource security at present powering the transportation sector, there is an extreme weight on automakers to devise a new technology that can fulfill the changing needs of the economy. [8]

The improvement of Li-ion batteries utilized in electric and hybrid vehicles are the consequence of that new progression in the economy. The battery is a basic and a significant part of the electric vehicle. The better the battery plays out, the more noteworthy the utility inferred by the two buyers and makers. Different battery chemistries serve one of a kind needs to make and model the vehicle. However, a common factor over all battery technologies is the need to guarantee the long term security of the materials utilized as a part of a battery. That is, there should be sufficient material to meet the present and future requests of the market. Li-ion batteries represent some vulnerability as for the accessibility of Li as a resource. At last, the lifecycle evaluation model will turn out to be a valuable device to decide the measure of Li we need and how Li will be utilized as a resource.

A Lithium work flow model was created after compiling the inventory network of Lithium. The process flow is divided into five sections as shown in Figure 1: Mining/Extraction, Battery production and Assembly, Vehicle Manufacturing, Consumers, and End of Life. Each section details the lithium inflows and outflows throughout the individual stage.

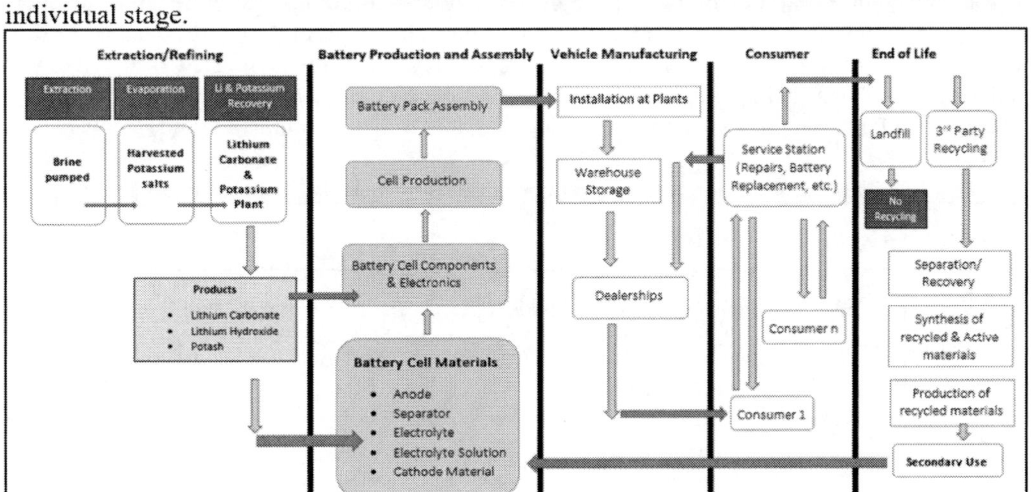

Fig. 1. Lithium Work Flow Model

3 Proposed concept and outcome expected

Once the inventory network for lithium has been investigated and a work flow model has been characterized, the next step is the production of a lifecycle evaluation model. This section plots the methods for making a spreadsheet display, wherein every component of the lithium work flow model is characterized inside a progression of worksheet 'modules.' Every worksheet will take client inputs, for example, amount of metal mined, kind of equipment utilized, and distance traveled from mines to handling plant, and approximate the energy utilization and emissions anticipated. The components in the Lithium work flow chart will be represented by a progression of calculations in the worksheets, and individual cells in the worksheets will be utilized to give assumptions and data values required by the estimations. At each step in the flow chart inside the worksheets, client input factors decide output parameters that proceeds into the following system component. A user interface page will enable the client to determine the procedures that will be utilized and different assumptions made in the model.

For instance, the client will have the option to use a drop-down menu to choose the areas at which Lithium will be mined and prepared and the sorts of mining and transportation that will be utilized in each progression. Whenever finished, the client will likewise have the capacity to basically work backwards through the procedure stream and spreadsheet estimations, by indicating the quantity of Li-ion batteries required. The assumptions inside every module will then give an estimate of the measure of metal to be quarried (in view of mineral richness, handling plant area, process type, process productivity, equipment required at each stage, and so on.) and the energy consumed at each step. Every module is connected to the next module by means of the transportation grid. This transportation grid is required to track fuel utilization, time, and cost of stream of lithium battery product parts as they travel step by step from quarry to definite disposal site. This model will be an open access display for anybody to utilize the information and presumptions can be enhanced with the time.

3.1 Resource Withdrawal Module

The main module, known as the 'Resource Withdrawal Module', comprises of procedures associated with extracting lithium through extraction or mining. Distinctive lithium work flows sub-components are accessible inside this module, depending on whether the lithium is acquired through physical or chemical procedures. Lithium is mostly found in brine solutions and hard rocks and in low sums in clay deposits. Brine sources are additionally named mainland, geothermal, and oilfields and enable clients to make a choice.

Table 2. Input and Output Quantities from Brine and Hard-Rock Process

Process	Input	Output
Brine	Location, type of brine, type of equipment, labor operating hours and wages, equipment operating hours, fuel type, electricity use and mix	Water consumption, energy consumption, labor and capital cost, lithium chloride, lithium carbonate, potash
Hard-Rock	Location, Type of mineral, type of machinery, labor operating hours and wages, machinery operating hours, fuel type, electricity use and mix	Water consumption, energy consumption, labor and capital cost, lithium concentrate, lithium hydroxide, lithium carbonate, potash

The client will have the capacity to indicate the ore of the metal and the equipment engaged with mining and handling as model assumptions. [9] They will also have the alternative to choose from a drop down menu or specifically input quantitative information into relevant cells. For instance, area of the brine solution, sort of the brines, kind of equipment, fuel type, will be accessible as drop down menu parameters. Labor working hours, work cost, and equipment working hours will be estimated after the ore input. The yields will be in tons or kg of material created which can be utilized as a part of Training & Development to model the transport of crude materials.

3.2 Transportation Module

Material distribution and handling will be represented under 'Transportation Mode System' segment which will be displayed with utilizing a GIS (Geographic Information System) and the GREET (Greenhouse gases, Regulated Emissions, and Energy use in Transportation) model. The model, created by Argonne National Lab will be valuable in evaluating air emanations and energy impacts for the duration of the life cycle of Lithium-ion batteries. GREET is isolated in two sections: GREET 1 arrangement assesses the fuel cycle from well to wheels and GREET 2 arrangement assesses the vehicle cycle including pump to wheels, vehicle disposition, and material retrieval.

GIS helps to map household rail and road layers to build up the transportation logistics to limit time and cost. A variety of installed transport chain choices will be accessible for clients to determine in the model inside every module. For instance, prepared metal can be moved by truck, truck-rail-truck, compartment ships, and so on. This segment of the model will be elaborate yet will give clients a lot of adaptability. GREET has a built in transportation and circulation module called Transmission-Distribution (T&D) to represent transporting crude materials to and from the site. Trip distance, measure of material transported, and fuel type are indicated with respect to the method of transportation: truck, tanker, boat, and rail utilized in the procedure to decide the energy utilization (Btu/ton of material transported), energy intensity (Btu/ton-mile), total emissions (grams/ton of material transported) and urban emissions (grams/ton of material transported) by mode. Fuel cost and material transport costs are also represented in the process. The GREET T&D displaying components will be utilized each time material is transported from one module to the next.

3.3 Battery Production and Assembly Module

In this module, all the material and segments required to shape a battery pack will be demonstrated. Lithium and its co-items are sent as inputs to the battery organizations which create cell materials and yields from those organizations will be contributions to cell manufacture and generation organizations. A battery pack is gathered from the outputs of creation and production organizations. This module will comprise a list of all the potential key organizations engaged with creating different parts of the battery. Organizations will be arranged on the basis of their skill in building up a battery segment and will be arranged into classifications. Clients will have the capacity to choose the key battery players and their area to process for transportation vitality utilize and cost.

The GREET 2 model can be utilized to evaluate energy utilization and emanations for battery material preparing and creation, where clients can include percent share of key material structure in particular liquids, and share of electricity utilize and mix in the generation of battery materials. [10] Assembled battery packs are sent as inputs to the vehicle manufacturing process in the following module, delivering vehicles as outputs for buyers. Energy consumption, cost, and emanations from transporting items to installing plant is displayed utilizing T&D and can be found in Table 3.

Table 3. Input and Output Quantities from Battery Manufacturing Module

Process	Input	Output
Battery manufacturing	•Lithium compounds •Cell material •Key players, electricity use and mix, battery material composition	•Cell material •Battery pack •Energy consumption at plant and transport of materials, cost, water use

Lithium Compounds (in tons or kg) got from the 'Resource Withdrawal Module' will go as input to this module alongside other battery material for creation of cell material and number of cells required to make a battery pack will be the input to estimate the battery pack production.

3.4 Vehicle Manufacturing Module

In this module, GREET 2 is utilized to approximate energy use from vehicle generation and vehicle assembly. Material production and composition should be distinguished and are given in GREET. Clients can alter and input composition of material utilized as a part of vehicle production, weight of the materials, weight of the vehicle, battery specific power, and power utilized at installation plant and working productivity of the plant. Produced vehicles are sent to distribution center or yard stockpiling until the point that they are prepared to be sold at dealership. This component will again be displayed again utilizing T&D. Extra vitality utilization and cost related with capacity of parts at the stockroom versus at the dealership will be represented by means of power utilize and power blend.

Energy consumption related with the delivery of vehicles to dealerships by means of cargo or auto transporter trailers will likewise be demonstrated utilizing T&D in GREET. At the point when the vehicle reach the dealership, energy consumption and cost will likewise modelled for time spent at dealership, until bought by customers (refer Table 4).

Table 4. Input and Output Quantities from Vehicle Manufacturing Module

Process	Input	Output
Installation	Vehicle material composition, material weight, battery weight, battery specific power and energy, electricity use and mix, plant operating efficiency	Energy use
Storage	Electricity use and mix, maintenance	Energy consumption, cost,
Delivery	Actual vehicle cost, additional subsidies/benefits, time spent at dealership	Energy consumption, cost of not being sold

3.5 Consumer Module

The battery being used will be demonstrated in fourth module, known as 'Consumer Module'. Vehicles transported from dealership to buyer will create discharges and consume energy; subsequently, T&D parameter will be utilized to estimate those figures. Clients will have the choice to include the number of buyers that vehicle was passed along to in its lifetime, beginning from the first owner of the auto as 'purchaser 1', to the second customer, and so on to consumer n; where n refers to the aggregate number of buyers in the lifetime of the vehicle. The model will be prepared to figure the depreciated value of the original vehicle as it go down and repaired all through vehicle's useful life, in view of the final cost of the vehicle estimated after modelling mishaps, harms caused, battery repair and replacement costs, and other servicing costs, all of which originates from client input. The estimated final value of the battery, and in addition the vehicle, will be compared with the original cost of both to assess and break down the general monetary profit or loss the vehicle had subjected to in its lifetime. Emissions can likewise be computed based on the miles traveled, proficiency of the battery, lifetime of the battery, and fuel intake. Information and output from this module are appeared as in Table 5.

Table 5. Input and Output Quantities from Consumer Module

Process	Input	Output
In-Use	Number of consumers, vehicles per mile travelled, battery efficiency, lifetime of the battery, fuel intake, number of vehicle accidents, vehicle and battery repairs maintenance	Cost of the car after use, energy consumption, in-use emissions

3.6 End of Life Module

The end of life of the batteries when they are never again being used is displayed in fifth module as 'End of Life Module'. Vehicles after their valuable lives are sent to junkyard where they are either sent to landfill, third party reusing or to waste site and have related transportation cost, discharges, and energy utilization from T&D in GREET.

Table 6. Input and Output Quantities from End of Life Module

Process	Input	Output
Landfill	Number of batteries in landfill, pollutant type and quantity	Energy use, Emissions, Cost
Recycling	Reused and recovered material composition, material's second life, electricity use and mix at plant	Price of recycled/recovered material, recycling efficiency, type and quantity of recycled material, energy intensity and use, water use
Hazardous waste site	Pollutant type, concentration of pollutants	Emissions, Environmental and Human health impact

When all the inputs are entered into the model, outputs gives add up of total energy and material utilization from extraction to disposition of Lithium-ion batteries. Clients will have the capability to inspect the practicality of different process components and investigate approaches to limit the effect of the outputs. The input and output parameters from end-of-life module are as given in Table 6. The proposed lifecycle will help policymakers in understanding the material supply and demand and to survey the practicality of reusing and improvement of regulations related with Lithium-particle battery innovation.

4 Conclusions

Lithium-ion batteries are turning into a prevailing battery science to control the transportation sector due to their technical characteristics and their application in electric and hybrid vehicles. As these batteries turn out to be more encouraging with passing time and elaborate research, the long term accessibility of lithium utilized as a part of assembling of Li-ion batteries may turn into a source of concern. Different government and industry specialists have guaranteed huge stores of lithium all around all with varied reporting. The disparity in the information and absence of precision in revealed information is probably going to cause supply demand imperatives later on and hamper the advancement of this technology.

The Lithium work flow chart tracks the flow of lithium from extraction, to battery production, to end of life. When combined with estimates of Li-ion battery demands, the model will be valuable in surveying whether there is sufficient lithium to control the future worldwide demand and when it is important to execute Li-particle battery reusing.

Despite the fact that Li-ion batteries are an exceptionally promising innovation, the specialists should concentrate on battery production, as well as on the fate of these batteries toward the end of life. Reusing battery materials and segments may lessen weight on natural resources by requiring the extraction of less virgin material, diminishing assembling costs by incorporating reused materials, and decreasing general energy consumption related with creation. Thus, this model could demonstrate helpful as an instrument for policymakers to implement and deal with the recycling foundation.

References

1. Budde-Meiwes, H., Drillkens, J., Lunz, B., Muennix, J., Rothgang, S., Kowal, J., Uwe Sauer, D. (2013), "A review of Current automotive battery technology and future prospects.", Journal of Automobile Engineering. 227: 761. http://pid.sagepub.com/content/227/5/761.full.pdf+html
2. Navigant Research. (2014). "Electric Vehicle Batteries: Lithium Ion Batteries for Hybrid, Plug-In Hybrid, and Battery Electric Light Duty Vehicles, Market Analysis and Forecasts". http://www.navigantresearch.com/research/electric-vehicle-batteries.
3. USGS (2014) "Lithium." US Geological Survey, Mineral Commodities Summaries. http://minerals.usgs.gov/minerals/pubs/commodity/lithium/mcs-2014-lithi.pdf.
4. Gaines, Linda and Nelson, Paul. (2009). "Lithium-Ion Batteries: Possible Materials Issue.", Argonne National Lab. http://www.transportation.anl.gov/pdfs/B/583.pdf.
5. MIT Electric Vehicle Team. (2008). "A Guide to Understanding Battery Specifications.", Massachusetts Institute of Technology, http://web.mit.edu/evt/summary_battery_specifications.pdf
6. Pesaran, Ahmad & Markel, Tony. (2007), "Battery Requirements and Analysis for Plug-In Hybrid Vehicles.", National Renewable Energy Laboratory. http://www.nrel.gov/vehiclesandfuels/energystorage/pdfs/42082.pdf
7. Lowe, M., Tokuoka, S., Trigg, T., Gereffi, G. (2010). "Lithium-Ion Batteries for Electric Vehicles: The US Value Chain.",

https://unstats.un.org/unsd/trade/s_geneva2011/refdocs/RDs/LithiumIon%20Batteries %20(Gereffi%20-%20May%202010).pdf

8. Ford Sustainability Report. (2010). "Ford's Electrification Strategy". Ford. http://corporate.ford.com/doc/sr10.pdf
9. Mohr, M., Mudd, G., Guirco, D. (2012). "Lithium Resources and Productions: Critical Assessment and Global Projections." Minerals. http://www.mdpi.com/2075-163x/2/1/65/pdf
10. Burnham, A., Wang, M., Wu, Y. (2006). "Development and Application of GREET 2.7 – The Transportation Vehicle-Cycle Model"

Energy-saving measures in buildings considered as monuments of history and architecture

Alexey Kochev[1], *Mikhail Sokolov*[1], *Elena Kocheva*[1], *Anna Malysheva*[2*] and *Anna Moskaeva*[1]

[1]*Nizhniy Novgorod State University of Architecture and Civil Engineering, 65 Ilyinskaya St., Nizhniy Novgorod, Russia, 603950*
[2]*Moscow State University of Civil Engineering, Yaroslavskoye shosse, 26, Moscow, Russia, 129337*

Abstract. The number of measures is shown in article, that allow to downsize the consumption of thermal energy in the construction, rehabilitation, reconstruction and maintenance of buildings, which are which are recognized today as monuments of history and architecture.

Key words: Unique buildings, Historical and architectural monuments, Energy-saving, Microclimate, Thermal stability

1 Introduction

The engineering systems that existed in the ancient unique buildings built in the XVII-XX centuries almost completely destroyed in connection with the purposeful destruction or lack of qualified operation. Along with the tasks of maintaining the required parameters of the microclimate, should be considered and thermophysical aspects used in the rehabilitation, reconstruction, new construction and the transfer of ancient unique buildings that are monuments of the historical and architectural heritage, to year-round operation [1, 2].

2 Materials and methods

Creation and maintenance of the specified microclimate parameters in historical buildings is possible only if the seasonal heat balances and the characteristics of the heat supply, heating and ventilation systems in the presence of the required heat resistance of enclosing structures and their elements for rational energy use and energy-saving [2].

As a result of the influence of unregulated external natural factors (annual fluctuations in temperature and humidity, the impact of precipitation, soil moisture, groundwater) aging of

* Corresponding author: lushinki@mgsu.ru

the materials takes place, leading to a change in the aesthetic, strength characteristics and temperature stresses of the supporting structures (Fig. 1) [3, 4].

Fig. 1. Type of load-bearing structures of the Cathedral of St. Alexander Nevsky, exposed to moisture, before reconstruction.

Many unique buildings constructed in the central part of Russia in the XVII - XX centuries, according to the thermal characteristics of the exterior walls, are analogues of buildings, which, regulatory requirements should be built with 2003.

This makes it possible to create and maintain the required microclimate parameters in existing, reconstructed and newly constructed buildings, which are historical and architectural monuments, with minimal costs (Fig. 2) [4].

For the protection of buildings and monuments of ancient architecture from the effects of moisture, as well as for more rapid drying walling, exposed to intensive exposure to moisture and water vapor, used electroosmotic device [5, 6].

As a result of electroosmotic dehumidification, the moisture from the layer is filtered to the surface of enclosing structures and evaporated into the volume of the premises, from which it is removed by ventilation systems at the expense of the organized regulated air exchange [3, 8].

3 Researches

In the process of research, was obtained the economic effect of reducing the drying time of thick-walled structures from 1 to 2.5 m thick using the method of electroosmosis, compared to natural drying for 5-8 years to the equilibrium air humidity value.

An important role in the formation of heat and air regimes in unique buildings is provided by window openings. The solution of such problems is based on the study of the temperature regime of the outer walls near the window openings and the temperature regime of the glazing of the window opening, taking into account the calculated parameters of the outside air, as well as taking into account the location of the window openings. We have developed an engineering method for calculating the thermal conditions of the outer walls near the window opening, depending on the position of the window cover relative to the longitudinal axis of the wall and the thickness of the air space between cover of the window block [7] (Fig. 3).

Fig. 2. The walls in the ancient unique buildings have a thickness of more than 1 m.

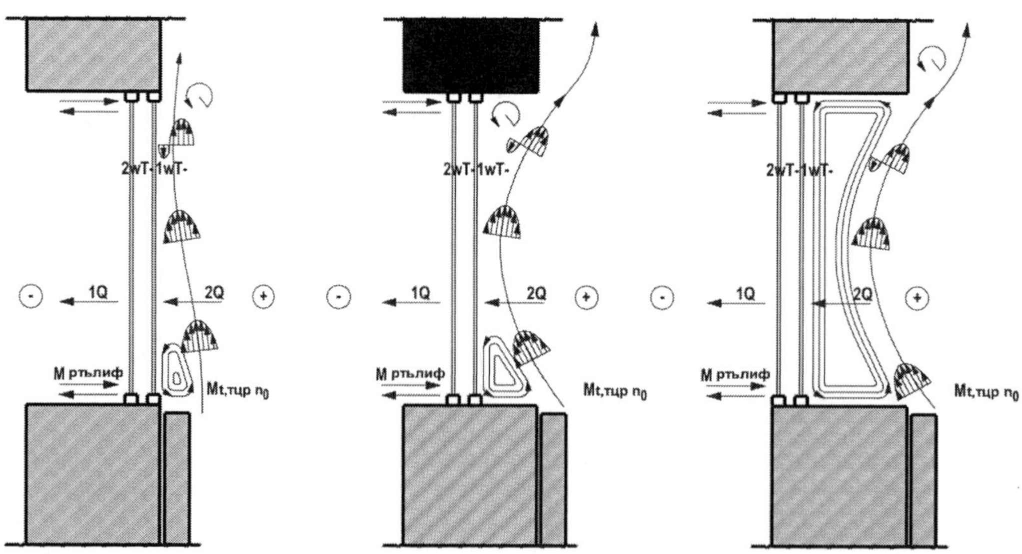

Fig.3. Diagram of convective airflow in glazing, depending on the location of the window sash.

With the glazing of the outer fences of ancient buildings, which are monuments of history and architecture, a reduction of heat loss is achieved within the range of 10÷30% in the process of optimizing the location of window sashes relative to the longitudinal axis of the wall of the order of 3.5÷7%, counting from the total heat loss of the building of the temple, and about 25÷30%, of the heat loss in the conventional arrangement structure of window bindings to ¼ of the wall thickness of the outer surface.

Before designing engineering systems in historic buildings, should ensure the integrity of a building with thermotechnical point of view.

4 Results

We obtained theoretical and experimental results on automatic and manual regulation of natural ventilation, depending on the season of the year, the number of people and the power of the heating system in unique buildings. Gravitational pressure is a self-regulating value in unique buildings with a dominant vertical layout [8, 9, 10].

For each unique building we studied, which is a historical and architectural monument, developed individual card regulation of air exchange on the season of year. Conducting energy-saving measures to regulate the air exchange of gravitational ventilation (aeration) during working hours and non-working hours, can significantly reduce heatconsumption by unique buildings [8].

The use of mechanical ventilation systems in buildings of historical and artistic value, spoils the interior, and leads to a significant increase in operating costs.

5 Conclusions

The unique ancient buildings, unlike modern civil buildings, have a number of design features that affect the heat and mass transfer processes in their premises.

In the design of natural ventilation systems in the unique historic buildings should take into account the particular climatic characteristics of the region.

Calculation of natural ventilation systems is recommended for three accounting periods of the year: warm, transitional and cold, taking into account the different occupancy rates of the ancient unique building by people and the characteristics of the heating system [9, 10, 11, 12].

We have developed measures to ensure the required controlled gravity air exchange and the normalized air permeability of window bindings with seasonal regulation. This saves up to 30% of the heat compared to the absence of automatic regulation and more than 50% compared to the use of mechanical ventilation.

References

1. N.F. Filatov, Dome, looking to the sky: Nizhny Novgorod temple architecture of the XVII-XX (1996)
2. A.G. Kochev, Climate of the Orthodox churches: monograph (2004).
3. A.G. Kochev, O.V. Fedorova, M.M. Sokolov, Proceedings of universities, 2-3, 60-67 (2013)
4. V.N. Kupriyanov, Designing heat insulation of enclosing structures (2011)
5. B.V. Matveyev, Drying walls by electroosmosis (1963)

6. O.M. Fridman, Electroosmotic method for eliminating the dampness of the walls of buildings (1971)
7. A.G. Kochev, A.S. Sergienko, Russian scientific journal, 3 (46), 285-289 (2015)
8. AG Kochev, A.S. Sergienko, Russian scientific journal, 1 (44), 282-286 (2015)
9. M.M. Sokolov, E.A. Kocheva, M.A. Kocheva, A.S. Moskaeva, E.V. Smirnova, Scientific Review. Engineering sciences, 6., 36-38 (2016)
10. M.M. Sokolov, E.A. Kocheva, M.A. Kocheva, A.S. Moskaeva, E.V. Smirnova, Scientific Review. Engineering sciences, 6, 39-42 (2016)
11. Parfentieva N., Samarin O., Lushin K., Paulauskaite S., ICEE 2008 Conference Proceedings, 7, 854-858 (2008).
12. Gagarin V.G., Lushin K.I., Kozlov V.V., Neklyudov A.Yu, Procedia Engineering, 146, 103-111 (2016)

Ways of heat losses reduction in the structural elements of unique buildings

Alexey Kochev[1], *Kirill Lushin*[2*], *Mikhail Sokolov*[1], *Elena Kocheva*[1] and *Marina Kocheva*[1]

[1]*Nizhniy Novgorod State University of Architecture and Civil Engineering, Ilyinskaya St., 65 Nizhniy Novgorod, Russia, 603950*
[2]*Moscow State University of Civil Engineering, Yaroslavskoye shosse, 26, Moscow, Russia, 129337*

Abstract. In this article the results of theoretical and experimental studies of the reduction of heat losses in the elements of external enclosing structures of under construction, reconstructed, reconstructed and currently operated unique buildings (monuments of history and architecture) presents.

Key words: Constructive elements, Unique buildings, Reduction of heat losses, Microclimatic conditions, Historical and architectural monuments, Heat resistance

1 Introduction

The rational energy use and energy saving in buildings with historical and artistic value is possible only if creation and maintenance of the specified microclimate parameters are exactly matched seasonal heat balances, characteristics of the heat supply, heating and ventilation systems, required heat resistance of the enclosing structures and their elements [1].

Many of the ancient unique buildings that are monuments of history and architecture in regions with a design outdoor air temperature ($t_н$) ≤ -25 ^{0}C have basically a single-layered construction of clay brick walls with a wall thickness ($\delta_{ст}$) more than 1 meter [2].

2 Researches

Significant impact on heat loss and the formation of a comfortable environment in the premises have structural elements, in which regions of two-dimensional and three-dimensional temperature fields are formed: joints of walls and ceilings, external corners, window esconsons [1, 3]. The temperature field of the outer wall near the window openings changes markedly. The thicker the wall and the smaller the distance between the window sashes, the more significant this change is (Fig. 1 and Fig. 2).

* Corresponding author: LushinKI@mgsu.ru

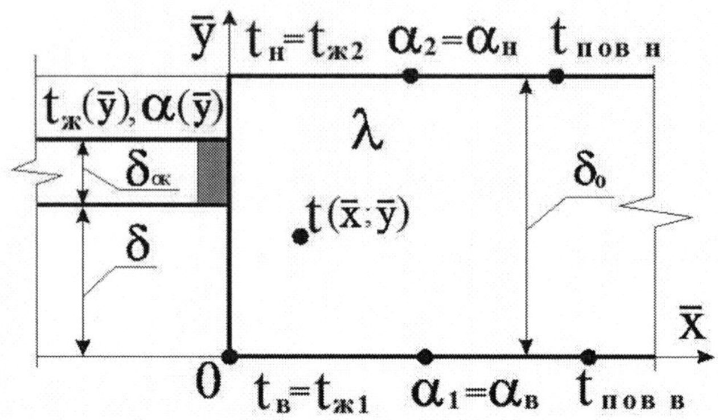

Fig. 1. The scheme of the calculation field of research.

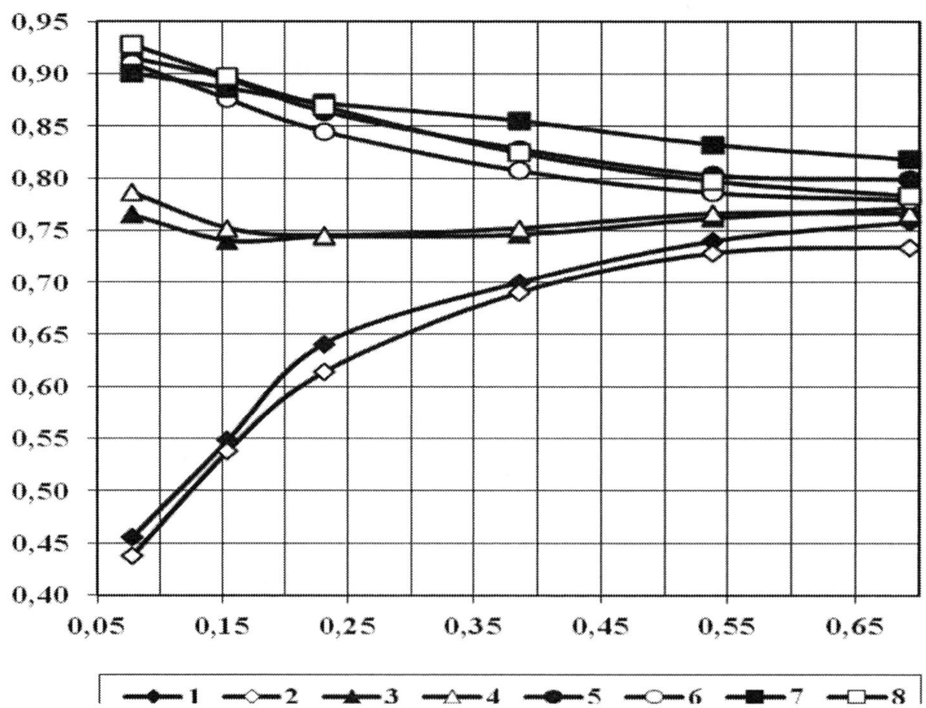

Fig. 2. Dependence of the relative temperature θ(x,y) on the x-coordinate at $T_{ж1}$=14 °C, $T_{ж2}$= -15 °C: 1 - y_1=0,038 (calculation); 2 - y_1=0,038 (experiment); 3 - y_1=0,231 (calculation); 4 - y_1=0,231 (experiment); 5 - y_1=0,5 (calculation); 6 - y_1=0,5 (experiment); 7 - y_1=0,769 (calculation); 8 - y_1=0,769 (experiment).

The study of the temperature regime of window esconsons of ancient unique buildings depending on the temperatures of the external and internal air, the thickness of the wall, the design of the window sash, the position of the window sash relative to the longitudinal axis of the wall is one of the directions for solving the problem of efficient use of thermal energy in the under construction, Restored and reconstructed ancient unique buildings. The received results can be considered at building and operation of buildings of any purpose.

The problem of finding the stationary temperature distribution in the two-dimensional computational domain, including the part of the window and the wall esconson, was solved by the method of finite differences using the establishment method with the realization of the implicit scheme [1, 3].

According to the results of the study of the thermal regime of glazing and window esconsons, the tasks of reducing heat losses and reducing the width of the plane of possible condensation on the surface of window esconsons are solved.

Based on the theoretical and experimental studies carried out, as well as a comparative analysis of the results, we developed the software product «Otkos» («Esconson»), which allows us to find the rational position of the window sash relative to the longitudinal axis of the wall for different wall thicknesses and window blocks, from the condition of minimum width of the possible condensation plane on the surface of the esconson [1, 3].

The loss of heat through the window esconsons of churches is suggested to be calculated taking into account the coefficient of additional heat transfer ΔK. The coefficient of additional heat transfer of the window ΔK is determined by the formula:

$$\Delta K = q_w \cdot \frac{P}{F}, \quad (1)$$

where q_w – Specific (additional) heat flow through the window esconson, Watt /(meter ·^{0}C);

P – perimeter of the window opening, meter;

F – area of the window opening, meter ^{2}.

Theoretical and experimental studies of q_w were carried out depending on the temperature of the internal ($t_в$) and the external air ($t_н$), the thickness of the wall (δ_o) and the position of the double window wooden binding with a thickness of $\delta_{ок}$=0,15 meter relative to the longitudinal axis of the wall $\bar{\delta} = \frac{\delta}{\delta_o}$.

This dependence is most significant for comparable sizes of the width of the standard window opening with the thickness of the fence (δ_o).

The value of the specific additional heat loss through the window esconson (q_W) is determined from the results of the studies in a graphic form (Fig. 3 and Fig. 4).

With periodic exposure to natural factors for a long time, materials are aging, leading to a change in the strength characteristics and temperature stresses of the supporting structures [2, 3].

An important role in the durability of ancient buildings is played by a minimum change in the temperature of the internal air during periodic exposure to radiant and convective heat fluxes during the year.

Electroosmotic devices are used to protect buildings and monuments of ancient architecture from moisture, as well as for faster drying of enclosing structures [4, 5]

Under the influence of water, carbon dioxide, temperature fluctuations in the wall material, various physicochemical processes take place that lead to waterlogging of structures and deterioration of their heat-shielding properties. This changes the ion composition, which increases the water-retaining capacity of the walls, which favors the formation of fungi and the development of microorganisms.

In this regard, the moisted enclosing structures of historical monuments and ancient architecture are subject to mandatory dehumidification, as this saves them from premature destruction and loss of bearing capacity, and also helps maintain the required microclimatic conditions for the effective preservation of structural elements and finishes [5, 6].

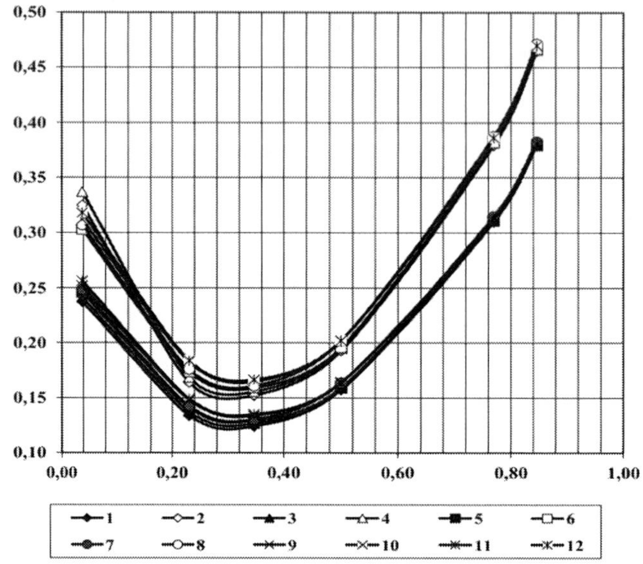

Fig. 3. The results of theoretical (odd figures) and experimental (even numbers) studies of the value of the specific additional heat loss through the window esconson (q_w) as a function of the internal air temperature ($t_в$) = 14 °C and the external air temperature ($t_н$) from -8 °C till -30 °C, wall thickness (δ_o) = 1,04 meter and the position of a double window wooden binding of thickness (δ_o) = 0.15 meter in relation to the longitudinal axis of the wall $\bar{\delta} = \dfrac{\delta}{\delta_o}$.

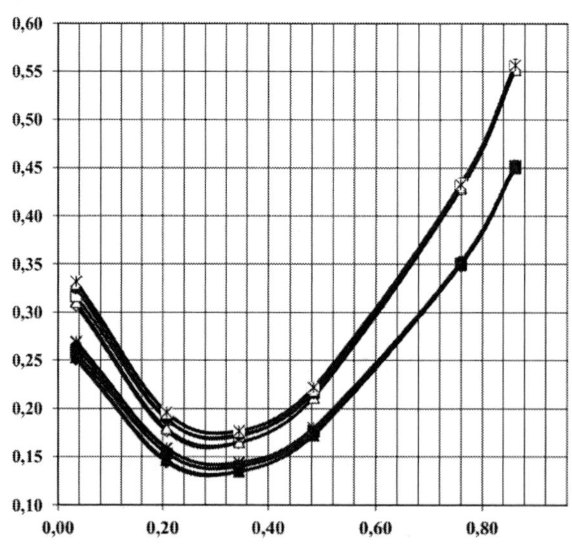

Fig. 4. The results of theoretical (odd figures) and experimental (even numbers) studies of the value of the specific additional heat loss through the window esconson (q_w) as a function of the internal air temperature ($t_в$) = 14 °C and the external air temperature ($t_н$) from -8 °C till -30 °C, wall thickness

(δ_o) = 1,16 meter and the position of a double window wooden binding of thickness (δ_o) = 0.15 meter in relation to the longitudinal axis of the wall $\bar{\delta} = \dfrac{\delta}{\delta_o}$.

The electric current flowing through the wires of the closed circuit when the electroosmosis structures are dried creates around itself a magnetic field, which in turn acts on moving charged particles. "Molecular" current in the wall can be regarded as a directed motion of liquid ions. Consequently, the Lorentz force acts on each particle. The direction of the action of this force is determined by the sign of the charge carrier. The transfer of moisture in the capillaries of building materials is effected by the movement of solutions of positive ions of alkalis, acid medium and diffuse layer, hence, forces are regarded as acting on a positively charged particle moving in a magnetic field.

In a homogeneous electromagnetic field, the transfer of moisture is due to the action of not only the diffusion forces and thermal diffusion, but also the action of thermodynamic forces caused by an inhomogeneous electromagnetic field [6].

3 Results

As a result of electroosmotic dehumidification, moisture falls on the inner surface of the structure, with which it is removed by controlled air exchange. As a result of electroosmotic dehumidification, moisture from the fence surface evaporates into the volume of the room, from which it is removed by ventilation systems due to controlled organized air exchange [6, 7] (Fig. 5).

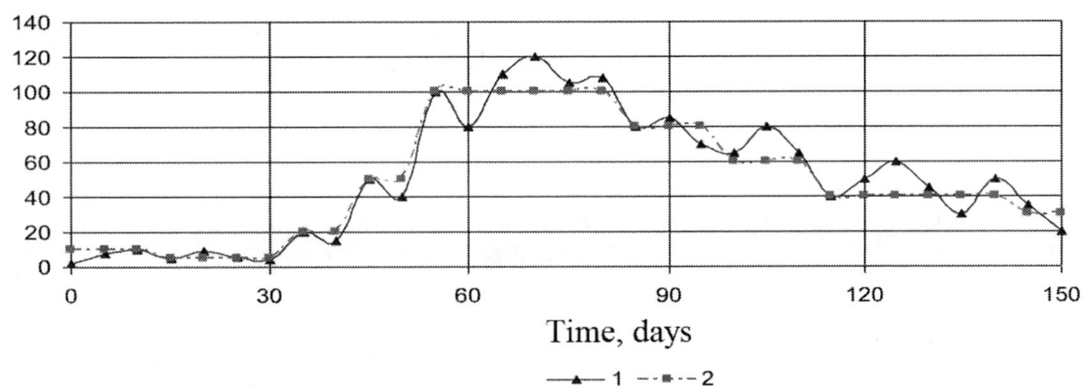

Fig. 5. Fluctuations in the air flow over the drying period, equal to 5 months: 1) theoretical calculation; 2) experimental data.

The task of ventilation is not only to remove moisture from the premises, but also to create the required microclimatic conditions in the premises and cellars of a unique building.

To ensure the required microclimate parameters in buildings that are historical and architectural monuments, it is necessary to restore or redesign heating and ventilation systems while ensuring the required thermal stability of the enclosing structures [8, 9].

We obtained theoretical and experimental results on automatic and manual regulation of natural ventilation, depending on the season of the year, the number of people and the power of the heating system in unique buildings. Gravitational pressure for the operation of ventilation is a self-regulating value in unique buildings with a dominant vertical layout [7, 8, 9].

For each unique building we studied, which is a historical and architectural monument, individual air regulation cards were developed depending on the season of the year. Conducting energy-saving measures to regulate the air exchange of gravity ventilation (aeration) during working hours and non-working hours, can significantly reduce the heat consumption of unique buildings [7, 8, 9].

The use of mechanical ventilation systems in buildings of historical and artistic value spoils the interior and leads to a significant increase in operating costs.

4 Conclusions

Unique ancient buildings, in contrast to modern buildings of general-civil purpose, have a number of design features that affect the heat and mass transfer processes in their premises

When designing natural ventilation systems in unique historical buildings, the characteristics of the climatic characteristics of the region should be taken into account.

It is recommended to calculate the natural ventilation systems for 3 calculation periods of the year: warm, transitional and cold with different occupancy of the ancient unique building by people and the characteristics of the heating system [7, 8, 9].

The construction of natural ventilation systems should be carried out taking into account the exclusion of the formation of compact supply air streams entering the human respiration zone and the most effective removal of polluted air from the room volume by means of structural elements of the exhaust systems that do not spoil the interior and facades of the historic building.

The self-regulation of natural ventilation costs allows saving 20-50% of the heat relative to the mechanical ventilation systems analogous to the maximum air flow rates.

References

1. A.G. Kochev, Climate of the Orthodox churches: monograph (2004).
2. V.N. Kupriyanov, Designing heat insulation of enclosing structures (2011)
3. A.G. Kochev, A.S. Sergienko, Russian scientific journal, 3 (46), 285-289 (2015)
4. B.V. Matveyev, Drying walls by electroosmosis (1963)
5. O.M. Fridman, Electroosmotic method for eliminating the dampness of the walls of buildings (1971)
6. A.G. Kochev, O.V. Fedorova, M.M. Sokolov, Proceedings of universities, 2-3, 60-67 (2013)
7. AG Kochev, A.S. Sergienko, Russian scientific journal, 1 (44), 282-286 (2015)
8. M.M. Sokolov, E.A. Kocheva, M.A. Kocheva, A.S. Moskaeva, E.V. Smirnova, Scientific Review. Engineering sciences, 6., 36-38 (2016)
9. M.M. Sokolov, E.A. Kocheva, M.A. Kocheva, A.S. Moskaeva, E.V. Smirnova, Scientific Review. Engineering sciences, 6, 39-42 (2016)

Fabrication and performance analysis of concentrated hybrid photovoltaic system

Krishna Murthy[1*], *Ajay Daniel*[1], *Lanvin Concessao*[1], *Habbie Alex Roy*[1] and *Ganesha A.*[1]
[1]*Department of Mechanical and Manufacturing Engg, MIT, Manipal Academy of Higher Education, Manipal- 576104, India*

Abstract: Sun is the most important source of renewable source of energy. During the past few decades there has been an ever-increasing interest in Photovoltaic (PV) cells as it directly converts solar radiation into electricity. This paper involves the performance study of photovoltaic system under concentrated solar radiation. The main problem with the concentration solar energy is the drastic increase in temperature of the photovoltaic module resulting in a decrease in performance efficiency of the system. This problem of overheating of the system can be overcome by providing cooling which would ensure operation of the module in the optimal temperature range. Hence, the setup would function as a hybrid model serving the dual purpose of power generation while also utilizing the waste heat for water heating applications. The experimental set up consist of a novel arrangement of concentrator and reflector and the cooling system. The Hybrid Photovoltaic System was repeatedly tested under real time conditions on several days. A comparison was drawn between the results obtained from direct exposure of a standard photovoltaic module to that obtained from the hybrid system in order to better understand the improvement in performance parameters. The study shown a significant improvement of output of standard photovoltaic module under the concentrated solar radiation.

Keywords: Renewable energy, concentrated photovoltaics, Photovoltaic/Thermal hybrid systems, P V module

1 Introduction

Energy is the basic requirement of any civilised society. Renewable energy is the theme of the 21st century. At a time when the world is debating over the use of fossil fuels to generate electricity, it stands united on the idea of implementing sustainable technologies to reduce the carbon footprint. Sun without any doubt is one of the most important sources of energy that we can make use of in our endeavour to achieve our

* Corresponding author: krishna.murthy@manipal.edu

targets of sustainability. Natural gas will give out first, then petroleum, and finally coal. As long as coal is available, it can be converted into gas and oil from the available technology. Some European countries are already feeling the pinch of dwindling fuel supplies. Photovoltaic Technology has come a long way ever since its discovery in 1839 by A. E. Becquerel [1]. They have become a well-established source of energy with them being used in all the imaginable ways. Monocrystalline and polycrystalline Si Photovoltaic modules are the most common in the market with high efficiency multi-junction cells promising a better future.

Photo voltaic (PV) modules are characterised by their fill factor, efficiency and rated power. Fill Factor (FF) is a measure of junction quality and series resistance and its value typically lies between 0.6 and 0.7. For a cell that is reasonably efficient, the fill factor can be expected to be between 0.7 and 0.85. The closer the FF is to unity the higher the quality of the PV module. With poorer quality cells, the difference in voltage between open circuit voltage and voltage at maximum power is greater, and hence the FF is reduced [2, 3]. The light generated current depends linearly on the solar irradiation and is also influenced by the temperature according to the following equation

$$Isc = (Isc,n + KI*\Delta T)*G/Gn \quad (1)$$

$$Voc = Voc,n + (Kv*\Delta T) \quad (2)$$

Where Isc is the short circuit current, G is the irradiation on PV module, Gn is the nominal irradiation, Voc is open circuit voltage, ΔT is the difference between actual and nominal temperature and KI is the short circuit current/temperature coefficient [4]. As can be seen from the equation, current generated increases with irradiation and temperature as KI is positive while Voltage decreases with increase in temperature as Kv is negative. Hence a balance needs to be struck between the level of solar irradiation and temperature of the panel as infrared radiation is an integral part of solar radiation.

A Concentrating Photovoltaic Technology (CPT) makes use of this theory of improvement in output with increased irradiation under controlled temperatures. The practicality of this idea has been tested but all of them made use of solar arrays designed from cells rather than an out of the market standard PV module [5-8]. At present there are many varieties of PV modules are in operation in the market and aim of the present work is to improve their output characteristics. In this context a solar radiation concentrating setup and a heat sink was designed, and its impact on the performance of an off the shelf PV Module was investigated.

2 Experimental Setup

The setup consists of a standard 18W poly crystalline Si PV module in its aluminium casing and its specifications are as given in the Table.1

Table 1. The parameters of the PV module at 1000W/m2, 25 OC and AM 1.5 (STC)

P_m (W)	V_{OC} (V)	I_{SC} (A)	FF (%)
18.06	21.6	1.06	78

The edges of the casing were made leak proof by using a sealant. The junction box was isolated to ensure there was no short circuit once the casing was filled with water. Copper tubes of diameter 8mm were then given a serpentine shape and placed inside the aluminium casing (Fig 1). A back cover was fabricated to prevent any vaporization of water from the enclosure due to direct solar radiation.

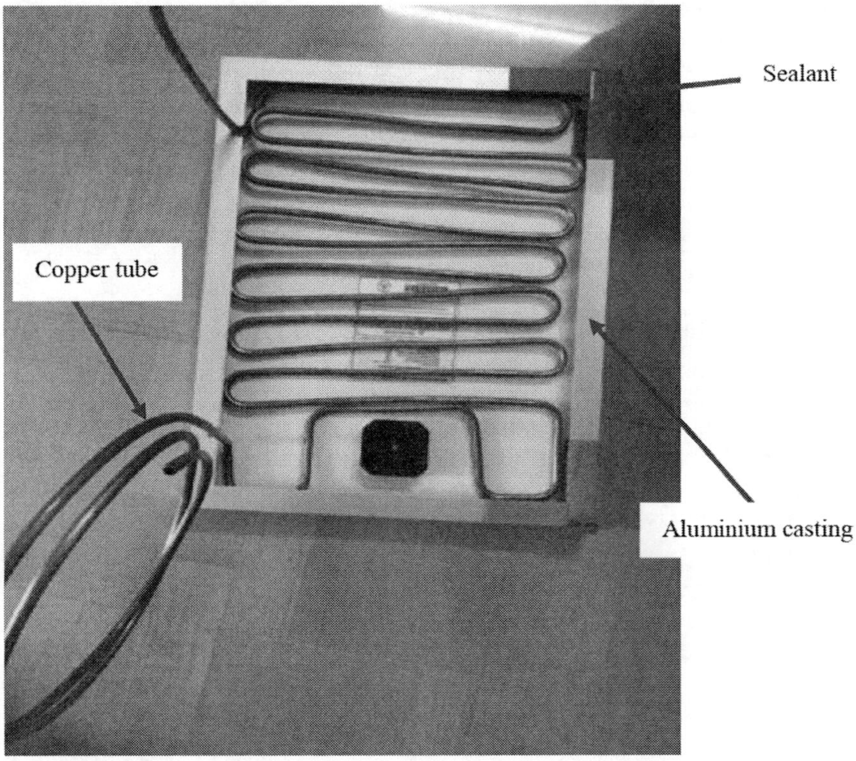

Fig. 1. Copper tubes behind the panel

The concentrator was fabricated using plywood from which three parabolic ribs measuring 1000mm along the length were cut. The ribs were then spaced out at 250mm from each other and supported by rectangular strips placed strategically at three locations. Rectangular strips of mirrors measuring 500mm in width and 25.4 mm in length were then stuck along the parabolic profile to form the parabolic concentrator the focal point of which was 635 mm from the origin of the parabola. Two flat mirror reflectors of dimensions 500x320 mm were placed at the flanks of the parabolic trough with provisions to adjust its inclination in two orthogonal planes as shown in figure 2. The parabolic trough and the two flat mirror reflectors were pivoted on a stainless steel pipe allowing them to be rotated along the width of the trough.

The PV module was supported at the corners and positioned at a height of 800mm from the origin of the parabola to ensure that the entire PV module was lit by the sun and not just a segment. On the day of the experiment the aluminium casing was filled with water which was to serve the purpose of conducting the heat from the PV module to the copper tubes carrying water. One end of the copper tube was fed with water and the warmer water coming out of the other end was collected in a bucket. This formed the cooling mechanism in our concentrating setup.

3 Performance Analysis of the Concentrated PV module

The electrical performance of the concentrated PV module with water cooling is compared to a fixed PV module. At a particular time, readings are taken and current-voltage characteristics are graphically represented. In order to measure the global solar radiation, a solar cell has been used. It is a known fact that the current output of solar cells is a linear function of solar irradiation. Also the current output of solar cell does not depend strongly on the temperature of the solar cell. Therefore solar cell current can be used as a measure of solar irradiation at a given time. Cell current or short circuit current of the module is proportional to Solar Irradiation

$$\text{Solar Irradiation } (W/m2) = K * \text{Cell Current} \quad (3)$$

K is proportionality constant, obtained from the relationship between Solar Irradiation and Cell Current. The cell current is the short circuit current of the solar cell. Irradiation was measured using a standard cell, for which K = 2175.

Under concentrated conditions, the PV module receives more solar radiation as compared to the fixed PV module. Due to this the temperature of the CPV module rises. However, this rise in temperature is compensated by the cooling water passing through the copper tubes that carries away a considerable amount of heat from the module.

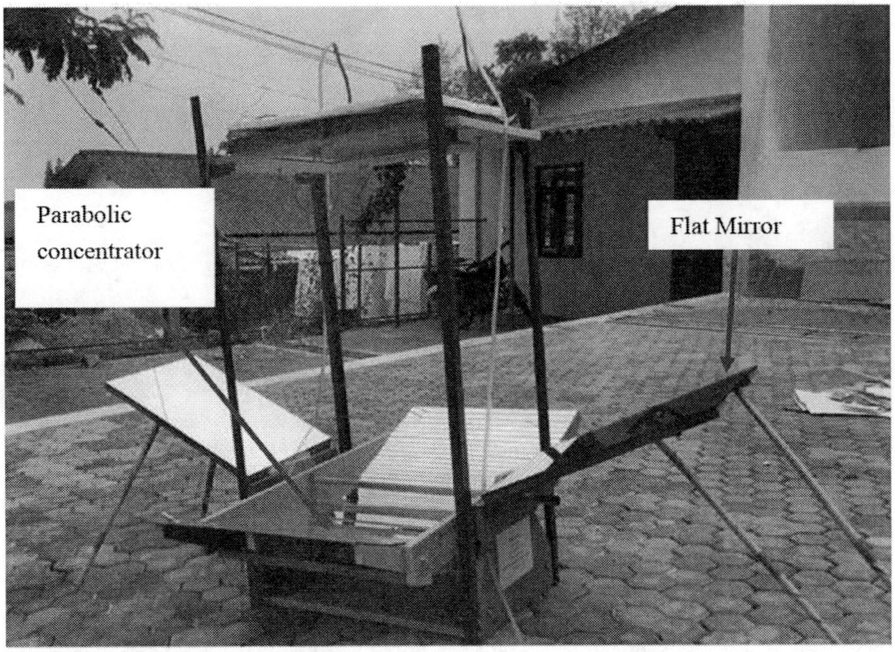

Fig. 2. Concentrated Hybrid Photovoltaic System

The performance of a solar cell is best indicated by its current-voltage characteristics. It is therefore important to know the output voltage and output current and how they vary with respect to each other and the surrounding environment. The current and voltage values for different cases shown are shown in figure 3.

 https://doi.org/10.1051/matecconf/201814404023

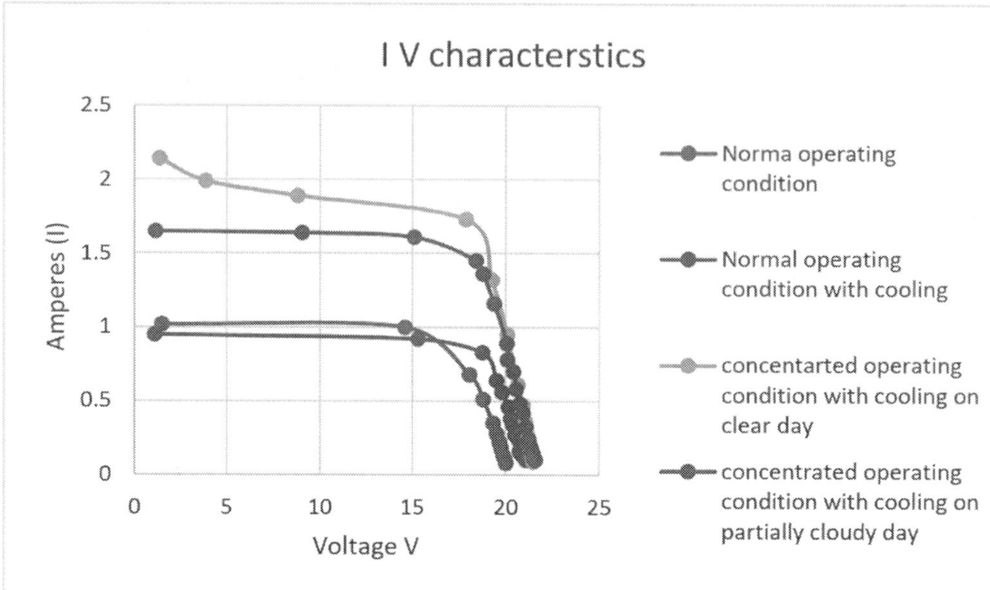

Fig. 3. Current Voltage characteristic of the PV module under different operating conditions

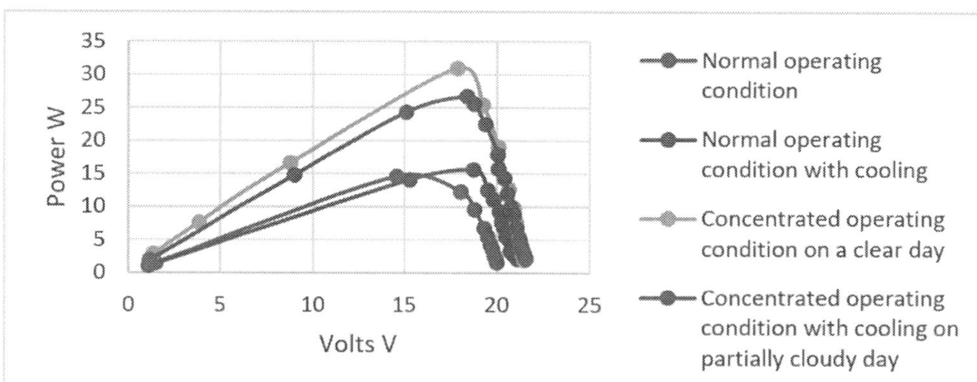

Fig. 4. Power Characteristic of the PV module under different operating conditions

Experimentation under concentration without cooling and with cooling shown an improvement in output current generated by the Concentrated PV module. However the electrical efficiency of this system is reduced. The maximum power produced from the CPV module under the influence of cooling was 30.94 W, while the maximum the maximum power for a fixed PV module without cooling was 15.57 W as shown in figure 4. Comparing the CPV module and fixed PV module, it has been found that the output current of the CPV module was 2.1 times the fixed PV module. The electrical efficiency of the CPV module was 9.2% and of the fixed PV module was 12.63%. The efficiency of both modules are decreased due to high cell temperatures. The solar radiation gain and output from the CPV system can be increased even further, but it could not happen in this experiment due to optical losses of the reflective mirrors, shadow losses due to gaps between the strip of reflective mirrors, the efficiency drop due to the temperature rise of the solar cell and uneven intensity of solar radiation while experiment was being performed.

4 Conclusions

In this work the performance of CPV module subjected to water cooling has been analysed and it is compared with a fixed PV module that is directly exposed to sunlight. It has been found that the output current increases by about 1.5 to 2.1 times, but its electrical efficiency is drops by 2-3%. The CPV module was tested on a clear sunny day and also on a cloudy day. On a cloudy day, the maximum output current was lower, but it indicated a better VOC. This is due to temperature being a function of both voltage and current. On a cloudy day, as the module is subjected to a lower temperature, ISC decreases and VOC increases, in comparison to the CPV module on a sunny day.

Acknowledgement

We would like to express our deepest gratitude to Dr Jagannath K, Head department of Mechanical and Manufacturing Engineering for support and encouragement while carrying out this project. We are also deeply obliged to Mr Jayaraj, Dept. of Electrical and Electronics Engineering, MIT Manipal who has been instrumental in conducting the testing phases of the setup.

References

1. Farrington Daniels, “Solar Energy”, 17th Annual Meeting, American Power Conference, Chicago. Ill., March 20-April 1, 1955.
2. L. El Chaar, L.A. lamont, N. El Zein, “Review of photovoltaic technologies”, Renewable and Sustainable Energy Reviews 15 (2011) page 2165–2175
3. Global Sustainable Energy Solutions (GCES), “Grid-Connected PV Systems Design and Installation”, First Edition India.
4. Marcelo Gradella Villalva, Jonas Rafael Gazoli, and Ernesto Ruppert Filho, “Comprehensive Approach to Modeling and Simulation of Photovoltaic Arrays”, IEEE transactions on power electronics, vol. 24, no. 5, may 2009, page 1198-1208
5. M. Li, G.L. Li, X. Ji, F. Yin, L. Xu, “The performance analysis of the Trough Concentrating Solar Photovoltaic/Thermal system”, Energy Conversion and Management, vol. 52 (2011) page 2378–2383
6. Mohan Kolhea, Du Binb, and Eric Huc, “Water Cooled Concentrated Photovoltaic System”, International Journal of Smart Grid and Clean Energy, vol. 2, no. 2, May 2013, page 159-163
7. Chengdong Kong, Zilin Xu, Qiang Yao, “Outdoor performance of a low-concentrated photovoltaic–thermal hybrid system with crystalline silicon solar cells”, Applied Energy vol 112 (2013), page 618–625, Elsevier.

8. Z. Farhana, Y.M. Irwan, R.M.N. Azimmi, A.R.N. Razliana, N. Gomesh, "Experimental Investigation of Photovoltaic Modules Cooling System", 2012 IEEE Symposium on Computers and Informatics, page 165-169.

Reliability Assessment and Study of Quality Control Tools on Intermixer and Dump Mill of a Tyre Manufacturing Industry

M. Prashanth Pai [1*], *C. G. Ramachandra*[2], *T. R. Srinivas*[3] and *M. J. Raghavendra*[2]

[1]*Department of Mechanical Engineering, P.A College of Engineering, Mangaluru, India.*
[2]*Department of Mechanical Engineering, Srinivas Institute of Technology, Mangaluru, India.*
[3]*Department of Industrial & Production Engineering, SJCE, Mysuru, India.*

Abstract. Quality plays a very important role in today's highly competitive industrial environment. Quality leads to an improvement in the productivity. Quality control tools are used to solve more than 90% of the problems faced by the organizations in improving its effectiveness and meeting the customer expectation on a continuous basis. The success of the quality control procedure depends greatly on the method of data collection and its accuracy. Quality control tools are used worldwide in manufacturing industries for continual improvement and are implemented in different steps of manufacturing process to define the problem, measure its impact, finding out its root cause and its removal to ensure the production of non-defective items. The present work includes collection of failure data of intermix and dump mill of a tyre manufacturing unit located in Karnataka. The failure data collected for one year has been analysed to find out the major causes of breakdown. Quality control tools such as Pareto chart, Bar chart, Histogram, Fish bone diagram have been used to find out the root causes and identifying the action plans to eradicate them. Reliability assessment (theoretical) of intermix and dump mill has been conducted using the failure data. The reliability and maintainability curves speak about the present state of working of maintenance department in regard to addressing of breakdown problems and also help in setting future working standards.

1 Introduction

Machinery and other service facilities of any industry are continuously subjected to deterioration upon their use and exposure to environmental conditions. With the passage of time, machinery will be subjected to wear and tear and their performance as well as reliability decreases. Maintenance helps the plant and machinery to be maintained in good operating condition so as to maximize the output, reliability and profitability [1]. A good

* Corresponding author: shanth.pai@gmail.com

maintenance system also contributes to efficiency, timely delivery of goods and services, quality, safety and customer satisfaction [2]. Preventive, predictive, reliability-centred and total productive maintenance are some of the widely used maintenance techniques in industries. Failure analysis is the process of collecting and analyzing data to identify the root causes of a failure and to determine corrective actions for them. Failure data are the backbone of reliability studies as they provide very useful information to reliability engineers, design engineers and managers [3]. It is a vital tool used in the development of maintenance strategy for any equipment, development of new products and in the improvement of existing products. Various sources of collecting failure data are available, but inspection records, log books generated by maintenance, quality control or manufacturing groups are widely used.

Quality control helps to maintain the desirable product quality in order to increase customer satisfaction and profitability. The quality control tools help to identify the problems, prevent errors due to omissions or carelessness, confirm the effect of improvement and determine whether the assumed cause is right or wrong. Some of the important quality control tools include brainstorming, bar charts and histograms, Pareto charts, Cause and effect diagram, etc. [2]. Brainstorming is a technique of generating a large number of ideas for identifying or solving the root causes of a problem from the members of a team in a structural manner. Cause and effect diagram results from the brainstorming session. The pictorial nature of histogram helps to see patterns that are difficult to see in a simple table of numbers. Pareto charts are very useful in the easy identification of the problems.

Thuleswar Nath [4] has carried out his work on a pouch machine which was used in vertical filling and sealing in a cosmetic plant. The failures of the machine resulted in non-availability of the system and thus affected production. Pareto analysis was used to figure out the major contributors towards downtime losses. Why-why technique was used to better explain the reasons behind breakdowns. Based on the root cause analysis, a new preventive maintenance checklist was prepared for the pouch machine. Prashanth Pai M et al. [5] have carried out their work on a three roll calendar machine in a tyre manufacturing industry. The breakdown details of three roll calendar were analysed to determine the optimal inspection frequency of its critical parts by using a mathematical and hypothetic model. The model was aimed at maximising the profit and minimising the downtime of the selected critical machine.

Praveen Kumar R [6] has carried out his work in view of increasing the availability of 1000 Ton Hydraulic Press machine, maximizing production capacity and improving its preventive maintenance schedule. The breakdown details of the critical parts of the machine were collected, root causes of failure were determined by Fishbone diagram and why-why analysis to reduce idle time and optimize plant performance. Deepak [7] has carried out his work in a bicycle industry to reduce the rejection of bicycle rims by the application of Pareto chart and Fishbone diagram. Process parameters such as heating voltage, temperature and quality of the heated water, time of chemical coating process and chemical composition of material etc., have the influence on the quality of final product which needs to be controlled in order to reduce wastage. Jayant Chandrakar [8] has carried out his work in a food processing plant. Why-why analysis, fish-bone diagram were used to develop and improve preventive maintenance checklist for the machine.

In this work, a critical machine namely 'Intermixer and Dump mill' from a tyre industry (located in the southern part of Karnataka) has been selected for the analysis. The intermixer and dump mill unit (figure 1) had frequent breakdowns in the selected year and hampered the productivity of the plant to a greater extent compared to all other critical machines in the plant and hence it was selected for the analysis. The Intermixer is used to prepare the rubber compound. Measured ingredients like natural or synthetic rubber, sulphur, carbon, oil, zinc oxide and other chemicals are placed onto a conveyor system and the intermixer is charged to initiate the mixing process. A ram pushes the materials into the mixing chamber and two rotors mix rubber with the chemical agents. Mixing will be quick and uniform compound will be obtained. At the end of the mixing cycle, rubber is placed onto a dump mill, which shapes the rubber into flat, long strips by forcing it through two set rolls rotating in different directions at different speeds.

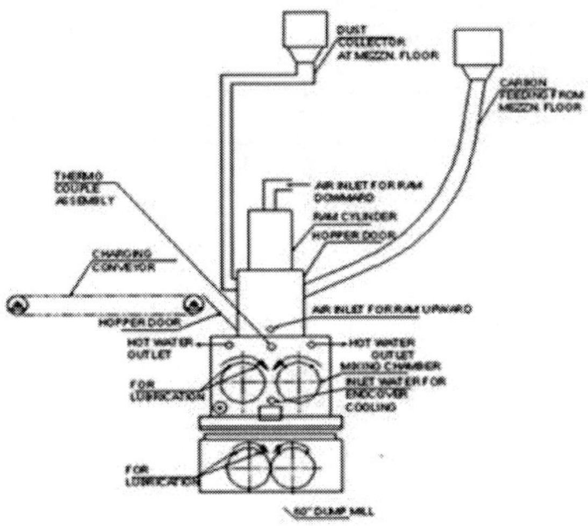

Fig. 1. Block diagram of intermix line with dump mill

2 Methodology

In this work, the breakdown data of intermixer and dump mill for one particular year were collected from the maintenance log books. Monthwise data showing date and time of breakdown, breakdown details, repair time and maintenance action taken were noted down for 12 months of the year. The failure data was analysed with the help of quality control tools such as Bar charts and Histograms, Pareto charts and Fish bone diagram to find out the root causes of failure and the possible ways to eliminate them. Production losses due to important causes of breakdown were also calculated. Brainstorming sessions for the five major subsystems of the machine resulted in a fishbone diagram showing the causes of failure and possible corrective actions. A troubleshooting chart was also developed. Finally reliability assessment (theoretical) of the intermixer and dump mill was carried out by using the failure data.

3 Results and Discussions

3.1 Failure data analysis

For each month, failure data was analysed and Pareto chart was drawn showing each cause of failure, breakdown time in hours and percentage breakdown for each cause. Monthwise total available hours, total breakdown time in hours, total repair time in hours, total inspection time in hours and percentage breakdown were tabulated (table 1) and shown in the bar chart (figure 2). After completing the monthwise analysis, 9 important causes of breakdown were identified and their repetitiveness for the year was tabulated (table 2). The breakdown time and percentage breakdown were tabulated (table 2) for nine causes of breakdown and shown in the Pareto chart (figure 3).

3.2 Production Loss Calculations

Production loss is calculated on the basis of average monthly production in tons to the average available hours per month.
The average monthly production = 1000 tons
The average available hours per month = 7460/12 = 621.66 hrs.

Table 1. Monthwise Breakdown details for the Intermix and Dump mill

Month	Total available Hours	No. of breakdowns in the month	Total breakdown time in Hrs	Total repair time in Hrs	Total Inspection time in Hrs	Percentage breakdown [%]
Jan	632	12	33.5	28.91	4.58	5.3
Feb	560	4	4.75	4	0.75	0.84
Mar	626	10	16.25	13.5	2.75	2.59
Apr	624	6	16.25	14.58	2.16	2.68
May	632	9	17.5	14.83	2.66	2.76
Jun	648	4	10.5	9	1.5	1.62
Jul	648	5	11.5	9.66	1.83	1.77
Aug	632	6	18.5	16.66	1.83	2.92
Sep	592	7	18.5	16	2.5	3.12
Oct	618	6	19	16.5	2.5	3.07
Nov	576	7	25.5	22.91	2.58	4.42
Dec	672	7	36.5	33.25	3.25	5.43
Total	7460	83	228.75	199.8	28.89	

Table 2. Total Outage due to Breakdown Hours for Intermix and Dump mill

Sl. No.	Cause of breakdown	No. of times occurred in the year	Breakdown in hrs	Percentage Breakdown
1	Drop Door Assembly	32	74.25	32.45
2	Ram Assembly	11	24.25	10.60
3	Lubrication System/Oil Circulation	11	19.5	8.52
4	Hopper Door Assembly	05	9.5	4.15
5	Instrumentation/Temperature System	03	11	4.80
6	Water Circulation System	03	6	2.62
7	Pneumatic Circuits	01	3	1.31
8	Rotor/Bearing/Coupling	08	58.25	25.46
9	Dump Mill	09	23	10.05

3.2.1 Lost production per hour

Total breakdown hours in the year equal to 228.75 and the intermix and dump mill was available for 7460 hours in the year. The annual production was 12000 tons.
Average hourly production = (12000x1000)/7460 = 1608.6 kg/hour

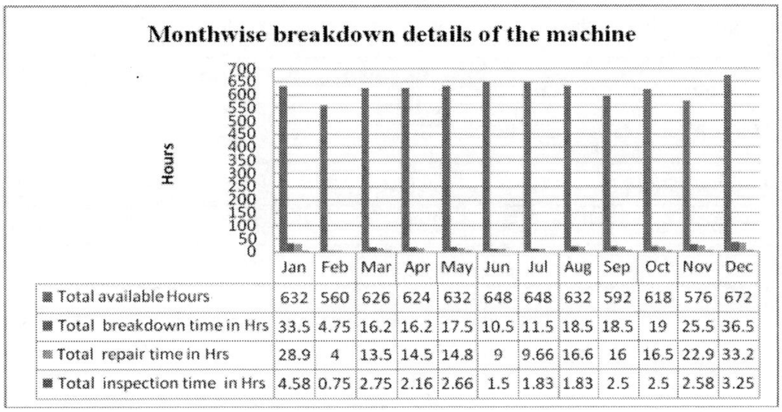

	Jan	Feb	Mar	Apr	May	Jun	Jul	Aug	Sep	Oct	Nov	Dec
Total available Hours	632	560	626	624	632	648	648	632	592	618	576	672
Total breakdown time in Hrs	33.5	4.75	16.2	16.2	17.5	10.5	11.5	18.5	18.5	19	25.5	36.5
Total repair time in Hrs	28.9	4	13.5	14.5	14.8	9	9.66	16.6	16	16.5	22.9	33.2
Total inspection time in Hrs	4.58	0.75	2.75	2.16	2.66	1.5	1.83	1.83	2.5	2.5	2.58	3.25

Fig. 2. Bar chart showing monthwise breakdown details for Intermix and Dump mill

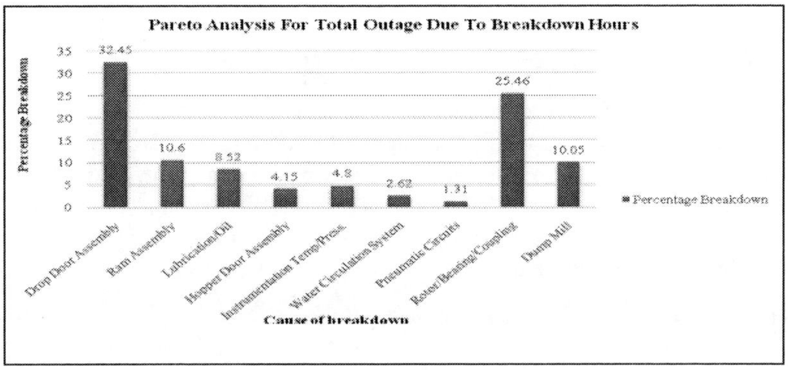

Fig. 3. Pareto chart for Intermix and Dump mill

3.2.2 Total loss in Rupees due to breakdown of 228.75 hrs

This will be equal to the product of Breakdown hours, Production loss/hour and Average profit/kg of compound = 228.75×1608.6×5 = Rs. 18,39,836.

*3.2.3 Total rubber compound lost in the year due to Butyl change***

This will be equal to the product of time lost in butyl change and production loss/hour = 96.25 × 1608.6 = 154.82 tons.

3.2.4 Total loss in rupees due to butyl change

This will be equal to the product of time lost in butyl change, production loss/hour and Average profit/kg of compound = 96.25 × 1608.6 × 5 = Rs. 7,74,139.

** Butyl changeover time is the time taken to clean the intermix so that butyl compound is processed. This is not the machine failure and but is a part of the system.

Similarly production loss due to all nine major repetitive causes of breakdown were found and tabulated (table 3). Details of production loss (in tons) due to spares and inventory cost, man hour losses, utility losses and overhead expenses were not available.

Table 3. Production loss (in tons) for the year for Intermix and Dump mill

Sl. No.	Cause of breakdown	Breakdown in hours	Percentage breakdown	Compound Lost due to breakdown in tons
1	Drop Door Assembly	74.25	32.45	119
2	Ram Assembly	24.25	10.60	39
3	Lubrication oil flow	19.5	8.52	31.36
4	Hopper Door	9.5	4.15	15.28
5	Instrumentation Température/Pressure.	11	4.86	17.69
6	Water circulation	6	2.62	9.65
7	Pneumatic flow	3	1.31	4.82
8	Rotor/bearing Coupling	58.25	25.5	93.7
9	Dump Mill	23	10.05	37

3.3 Cause and Effect Diagram or Fish Bone Diagram

The Pareto analysis and the histograms have identified five subsystems of the intermixer and dump mill as the areas of repetitive failures. They include - Drop door assembly, Ram assembly, Lubrication and process oil circulation, Dump mill and Hopper door. A systematic brainstorming session would yield the different causes of failure and the corrective actions to be taken to eradicate those causes. Brainstorming session for drop door assembly problem resulted in the following causes and corrective actions.

3.3.1 Causes of Failure and Corrective Actions for Drop Door Assembly problem:

As the repetitive breakdowns of the drop door assembly are highest, an effective maintenance strategy has to be planned in order to bring down the frequency of failure of this part. The causes for the failure and their corrective actions are as follows:

- Wear and tear, shearing of drop door of latching pins: It is due to vibrations of ram injector and overloading of the intermixer. Hence stop overfeeding and maintain feeding as specified by the manufacturer. When pins are worn out, arc welding is done and finished to size by grinding. Timings of latch operations have to be inspected periodically to avoid pin shearing. Failed pins must be replaced.
- Improper lubrication: It leads to drop door struck. It is due to malfunctioning of hydraulic system, leakage in clamping cylinder and assembly and malfunctioning of thermocouple. Corrective actions include checking oil circulation system and pressure gauges periodically. Check oil level/water level in thermocouple.
- Scorching of the rubber compound: This leads to heavy loss till the defect is solved. Overheating causes scorch. Corrective action include checking cooling water circulation system and rectifying the defect, checking water pump and valves for proper functioning.

- Hinge block problem: Strucking of hinge blocks during opening or closing of drop door can be corrected by periodically inspecting hydraulic pressure, operating cylinders, pump and motor.

Brainstorming session for the remaining four problems were conducted and resulted causes and corrective actions were identified. A Fishbone Diagram showing five major subsystems and their causes of failure is shown in the figure 4. A troubleshooting chart showing description of the problem and proposed action plan is shown in the table 4.

The bar charts and histograms show the monthwise available hours and total breakdown time for each month which include all the causes of breakdown. Pareto charts help to identify the contributing or major factors for the failure of the equipment, thus focussing our attention on the elimination of these causes. A detailed brainstorming session on each of these major factors help us to identify all possible causes for the failure of the system. These primary and secondary causes can then be used to draw the fishbone diagram and troubleshooting chart, which help to reduce the breakdowns, thus increasing the productivity of the machine.

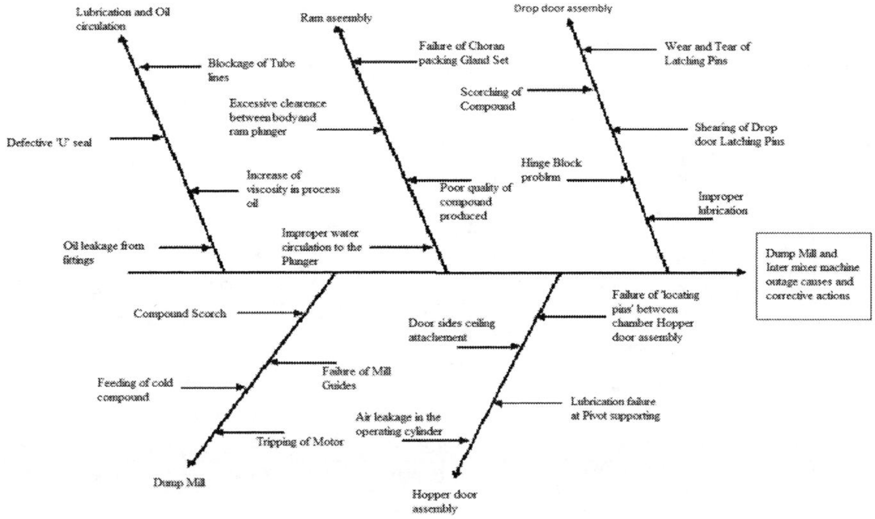

Fig. 4. Cause and Effect diagram or Fishbone diagram for the intermixer and dump mill

3.4 Reliability assessment of the Intermixer and Dump mill (Theoretical)

Total number of available hours, repair hours and breakdown hours in the year are found to be 7460, 199.8 and 228.75 respectively. Total number of breakdowns or number of maintenance actions taken in the year equal to 83. Some of the important factors such as failure rate, Mean time between failures (MTBF), Mean time to repair (MTTR), maintenance action rate, maintainability rate and availability are calculated for assessing the reliability theoretically.

3.4.1 Failure rate

Failure rate, λ = (Total No. of failures)/(Total available hours) = 0.011 breakdowns/hour.

3.4.2 Mean Time Between Failures (MTBF)

MTBF = (Total available time)/(Number of break downs or failures) = 89.87 hours.

3.4.3 Maintenance action rate

Maintenance action rate, μ = (No. of maintenance action)/(Total repair time) = 0.415 per hr.

3.4.4 Mean Time To Repair (MTTR)

MTTR = (Total repair hours)/(No. of maintenance action taken) = 2.40 hours per action.

Table 4. Troubleshooting Chart for Intermixer and Dump Mill

Sl. No.	Problem	Corrective action proposed	Remarks
1	Motor does not start/trips	Check for power, fuse, short circuit	Refer to electrical maintenance.
2	Compound scorch	Uneven water circulation. Improper circulation of water in ram. Scale formation inside the mill roller.	Check water quality if it is a repetitive problem.Remove scales. Check water circulation.
3	Compound coming out of the chamber	U seal may be defective.	Replace if necessary.
4	Compound dust coming out through hopper door	Door sides scaling attachment defective. Failure of flat springs in sealing strips. The dust bag might be blocked. Check for correct insertion of locating pins.	Check and replace if necessary. Clean the dust bag or filter bag. Locating pins may be relocated during overhauling
5	Water leak into the chamber of internal mixer and dump mill	Improper water circulation to the plunger. Check all hoses, nipples.	Recirculate and check for water pressure.
6	Varying oil content in the compound	Viscosity of the process oil might be altered. Check the heating pads.	Make sure that the process oil is maintained at specified temperature.
7	Drop door limit switch	Drop door not operated properly due to malfunctioning of limit switches.	Replace limit switches.
8	Compound smoke coming from U seals and glands	Lubrication failure. Lubrication lines blocked. Wrong selection of lubrication oil.	Check for leaks and select right lubricant. Replace U seals and glands if necessary.
9	Compound lump being thrown out from the mill	Wrong direction of rotation of mill rollers.	Counsel or train the employees properly.
10	Slow Ram movement	Check air pressure and the valves.	Inspect respectively.
11	Excessive vibration of the machine	Check for bearing bushes, ram clearance and rotor pinions.	Apply proper lubricant. Stop the machine to find out the cause.

3.4.5 Maintainability Rate

$$Maintainability\ Rate,\ M = [1\text{-}e^{-\mu t}] \quad (1)$$

where 't' is time (hours) and 'μ'= Maintenance action rate. Maintainability Rate for 1, 2, 3, 5, 10, 20 and 25 hours are found and plotted as shown in maintainability curve (figure 5).

3.4.6 Availability

$$Availability,\ A_e = \{1\text{-}[e^{-(\mu/MTTR)}][1 - e^{-(t/MTBF)}]\} \quad (2)$$

where μ =0.415 repairs/hour, MTTR=2.40 hours/maintenance action, MTBF=89.87 hours and t in Hours. Availability for 1, 5, 10, 25, 50, 100 and 200 hours are found and plotted as shown in the availability curve (figure 6).

3.4.7 Reliability

$$Reliability,\ R = e^{-(\lambda t)} \quad (3)$$

where failure rate, λ = 0.011 breakdowns/hour and 't' time in hours. Reliability for 1, 5, 10, 25, 50, 100 and 200 hours are found and plotted as shown in the reliability curve (figure 7).

From the curves shown in figures 6 and 7, it can be concluded that availability and reliability of the machine decreases with the passage of time, as the machine parts are subjected to wear and tear and deterioration. Availability and reliability may be brought back to the initial level by conducting maintenance activities. From the figure 5 it can be concluded that the maintainability of the machine increases with the passage of time, which in turn increases availability and reliability. Further, it can be seen that reliability and maintainability are the two factors which influence the availability of the plant.

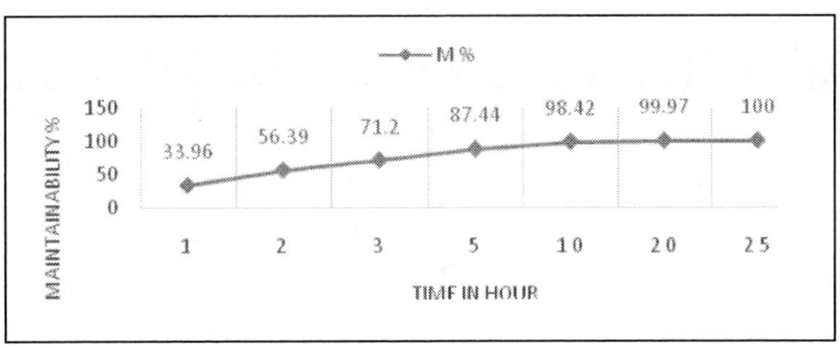

Fig. 5. Maintainability curve for intermixer and dump mill

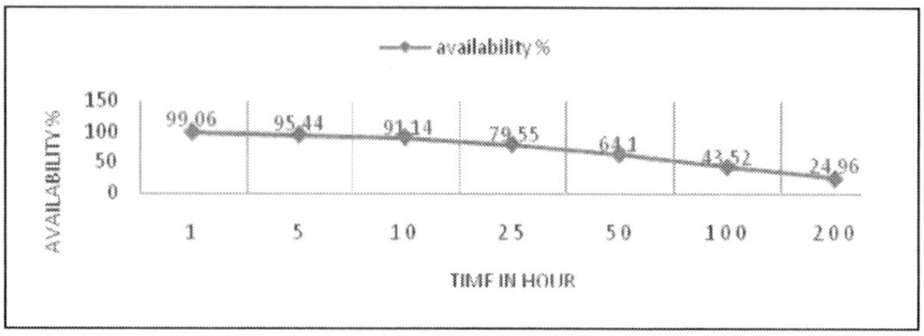

Fig. 6. Availability curve for intermixer and dump mill

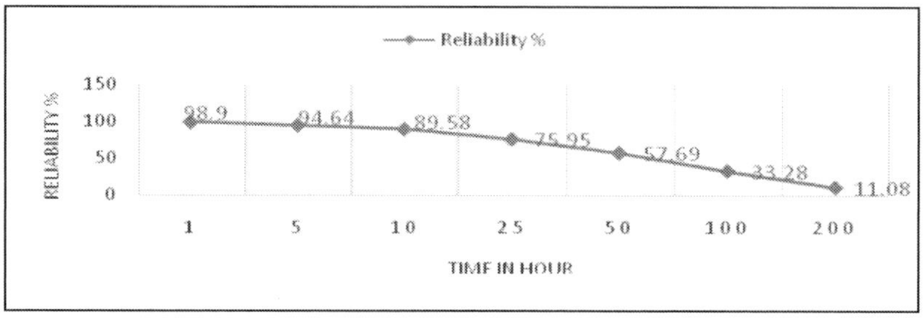

Fig. 7. Reliability curve for intermixer and dump mill

5 Conclusions

This work has been carried out on intermixer and dump mill based on the failure data collected for one year. The major causes for failures have been analyzed with the help of quality control tools such Pareto chart and Fishbone diagram. A bar chart and histogram helps to find out all possible machine breakdown hours for the selected period. A Pareto analysis breaks a big problem down into smaller pieces and identifies the most significant factors. Fishbone diagram helps to identify primary and secondary causes for the failures. Actions plans for the root causes of the problems were suggested and a troubleshooting chart was prepared. Thus, bar chart and histograms, Pareto chart and fish bone diagrams help us in the identification of possible causes of breakdown and to eliminate these causes so that machine availability and productivity can be enhanced. Analysis of the failure data could be further used in preparing the new preventive maintenance inspection schedule for the critical parts of the machine. Existing schedule can be modified based on the failure data analysis.

Reliability and maintainability are the two factors which influence the availability of the plant. Maintainability curves will speak about the present state of working of maintenance department in regard to addressing of breakdown problems. The curves of reliability and availability of the intermixer and the dump mill for the year can be compared to the already set performance standards of the machine in future years. It will be useful to carry out this kind of work on each and every machine in the plant so that any deviation from the set performance standards could be identified and resolved to improve the productivity of the machines.

References

1. L. R. Higgins, Maint. Engg. Handbook, 5 Ed, (1995)
2. A. Kelly, M.J. Harris, Mgmt. of Ind. Maint, (1978)
3. L.S. Srinath, Reliability Engg, 3 Ed, (1991)
4. T. Nath, D. Deka, Int. J.of Engg. Trends Tech, 28, 7 (2015).
5. M. Prashanth Pai, M. S. Prabhuswamy, C. V. Pujar, K. S. Shankar, ICETE-11, 5 (2011)
6. Praveen Kumar, R. Rudramurthy, Int. J. of Em. Tech. Adv. Engg, 3, 10 (2013).
7. Deepak, D. Dhingra, Int. J. of Res. in Engg. Tech, 5, 9 (2016)
8. J. Chandrakar, Rajesh Kumar, Int. J. of Adv. Engg. Res. Studies,4, 3 (2015)

Success of manufacturing industries – Role of Six Sigma

N. Venkatesh[1*] *and C. Sumangala*[2]

[1]*Department of Mechanical Engineering, Canara College of Engineering, Benjanapadavu, Bantwal 574 219, Karnataka, India*
[2]*Department of MBA, University of Mysore, Mysore 570005, Karnataka, India*

Abstract. Six Sigma is a phenomenal quality management concepts which has helped many organizations to overcome quality crisis in the recent past. Six Sigma is observed as a very promising quality management tool for any organization to make its presence felt in the corporate world as it emphasizes on obtaining a fruitful solution to improve accuracy, reduce defect thereby reduce the cost and improve profits. The main objective of this investigation is to unearth the extent to which the companies have been benefitted due to Six Sigma implementation. This article presents the results based on the analysis of collective opinion of employees of various Indian manufacturing industries that have implemented Six Sigma. This research also examines interrelationship among various parameters defined in the research. The research revealed that industries are benefitted irrespective of their nature in terms of their growth, financial benefits, productivity and satisfaction of the customer. However, peoples' equity that deals with the benefits that employees obtain after Six Sigma implementation is not certain. The research also revealed the existence of strong interrelationship among various parameters used to measure the success of Six Sigma.

Keywords: Six Sigma; Qualitative analysis; manufacturing industry.

1 Introduction

The Six Sigma philosophy aims to maintain a process within control limits in order to record no defects (Arendt, 2008). Motorola developed Six Sigma in the 1980s with focus on application in manufacturing entities. In the last two decades, it is observed that global companies have maximised business benefits and gained competitive advantage by implementing Six Sigma (Yang and Hsieh, 2009, Su and Chou, 2008,). Six Sigma has been defined as a systematic and organised method of improving process and development of new products and services that rely on statistical and scientific methods to reduce customer- defined defect rates to a very high level (Linderman et al., 2003). A Six Sigma process approaches 'zero defects', which means only 3.4 Defects Per Million Opportunities (DPMO) might occur. The difference between Six Sigma and other quality programmes is its 'top-down' drive and its rigorous methodology combined with

* Corresponding author: venksati@gmail.com

detailed analysis, factual decisions, and a concrete control plan to ensure that ongoing quality control of a process is achieved (Markarian J, 2004). Besides, Six Sigma methodology enables practitioners to accurately remove hindering issues and demonstrate the improvements using statistical tools such as Pareto Chart and control charts (Kapur and Feng, 2005, Feng and Manuel, 2008).

Researchers have done lot of work regarding the theory and implementation of Six Sigma at various industries in many developed countries. It is therefore required similar kind of studies in other parts of the world as well. Developing nations in particular need to understand factors affecting Six Sigma implementation. Moreover, the report on Six Sigma implementation and its potential benefits in a developing country like India look scanty (Padhy et al., 2011). In the days to come India will face huge competition from china in the manufacturing sector including garments (Chethan Kumar et al., 2010). Many medium scale garment industries in India are not aware of the lean Six Sigma concepts and this implementation will trigger a positive wave across the garment industries and become more competitive (Chethan Kumar et al., 2010). Six Sigma is not being explored by the developing countries to its full potential and India is not an exception in this (Desai and Patel, 2009).

After globalization and liberalization policies of the Indian government during 1990's quality along with productivity has emerged as one of the major areas of concern. It was inevitable for the Indian companies to improve quality or perish in face of stiff competition from global companies. Many Indian industries have reaped success through this breakthrough strategy of business improvement to achieve overall benefits by joining hands with Japanese and western companies via joint ventures. Still it is felt that Six Sigma has not penetrated Indian industries to the level it should have been. Hence, there is an enormous scope for Indian industries, to utilize Six Sigma strategy strengthen their business to have a winning edge in the global market.

2 Literature review

Changes of global industrial scenario compelled the Indian Government to introduce Liberalisation, Privatisation, and Globalisation (LPG), during early 1990's. Due to this several reforms in the form of encouragement for Foreign Direct Investment (FDI), removal of 'quota raj', and supporting IT sector for improved exports were introduced. These changes inspired several multinational automobile and electronics companies to open their units in India. As a result of which Indian automobile industries also changed their stance towards quality of product and processes. Six Sigma was introduced by many Indian manufacturing industries thenceforth (Desai and Patel, 2009).

Kaushik and Khanduja (2009) applied and analysed the application of Six Sigma DMAIC methodology in one of the thermal power plants. Their effort has paid off very well as claimed by them. They targeted the reduction of usage of de-mineralised water in a thermal plant as 0.1% of the extra use, costs the plant an extra 85 lacs. After the implementation of Six Sigma they are able to control the usage of de-mineralised water due to which the company has saved nearly 300 lacs per annum in the form of reduced energy consumption.

Kaushik (2011) in his paper "Relevance of Six Sigma line of attack in SME's: a case study of die casting manufacturing unit", has conducted an analysis of the effect of Six Sigma implementation. The DMAIC methodology applied in the organization to reduce the defect in die casting industry manufacturing Engine Mounting Brackets, resulted in an improvement of sigma level from 1.64 to 5.14. This improvement of sigma level transformed into an annual savings of Rs. 2.6 lacs.

Min Zhang et al. (2014) in their case study at one of the steel plants in China have observed that the implementation of Six Sigma in 470 projects resulted in savings of $100millions.

Srinivasan et al. (2016) have proved in their study how powerful the DMAIC approach is in improving Six sigma level from 3.31 to 3.67.

Thomas, et al. (2016) in their case study of application of Lean Six Sigma (LSS) at an aerospace industry have observed that the company is able to reduce the non value adding times, improved on-time delivery to customer and have estimated a savings of £2millions due to LSS.

Venkatesh and Sumangala et al. (2014) in their qualitative study of application of Six Sigma in an automobile industry highlighted the importance of opinion of six sigma experts in the success of Six Sigma. In their paper they have concluded that employees of different automobile industries have come with the opinion that contribution of Six Sigma is significant in improving the productivity, customer satisfaction, and finances. However, the contribution of Six Sigma is not so significant in improving employees' conditions in automobile industries.

The present study is continuation of qualitative study of manufacturing industries wherein electronics and pharmaceutical industries are also included in the study.

The literatures referred above clearly highlighted Six Sigma as one of the most prominent tools of quality management to achieve required quality. Indian manufacturing industries are putting their efforts to implement Six Sigma and compete at the global level. However, very few studies have tried to explore opinion of employees who have practiced and implemented Six Sigma in manufacturing industries. Under these circumstances there is a strong need of survey of Indian industries, which have implemented Six Sigma. Hence, this research intends to investigate the role of Six Sigma in the success of industries. This in turn helps the non Six Sigma companies to implement the practices that are most suitable for them.

3 Objectives of the study

This paper makes an attempt to examine few companies that have implemented Six Sigma. Industries selected included ten Mechanical based industries, eight Electronics industries and two Pharmaceutical industries.

The objectives of the study are:

1. Investigate the extent to which the companies have been benefitted due to Six Sigma implementation under the following five parameters:
   a) Growth of the company
   b) Financial benefits
   c) Peoples' equity
   d) Productivity
   e) Customer satisfaction
2. Examine the interrelationship among various benefits defined.

4 Research methodology

To start the survey the following variables were identified as benefits that companies can enjoy after implementing Six Sigma.

1. Growth of the company – The growth of the company is measured using nine parameters viz. improved revenue, improved organization structure, expansions, increased recruitments, higher exports, higher share value, better brand image, new product development and competitive edge after introducing Six Sigma.

2. Financial benefits – The financial benefits are measured using eight parameters viz. improved sales, better revenue, reduced Price Of Non Conformance, lesser cost of poor quality, increased profit, rate of return, safe position in the market and whether the benefits are from Six Sigma alone.
3. Productivity – This parameter is measured by considering following eighteen factors: reduced inventory, cost of set up, reduced cost of operations, Standard production methods, appropriate selection of the project, continuous monitoring after every project, removal of non-productive steps, continuous Identification and elimination of defects in the process, reduction of time to complete the project, continuous reduction of variation of the process, responding quickly to changes in demand of the market, lesser defect rate, lesser scrap rate, lesser rejection rate, reduced production lead time, better process capability, considering both short as well as long term goals and utilization of the resources optimally after implementation of Six Sigma.
4. Peoples' equity – This parameter defined by such factors, which are expected to contribute to welfare of employees. Twenty seven factors were identified. They are commitment from top management, resources provided by senior management, proper leadership style, fact based decision making instead of hierarchy, increased employee involvement in making decisions, learning process adopted by employees', knowledge of Six Sigma techniques and tools, training based on needs, Incentive programme, management values high for human beings, improved interpersonal relationships, improved employee participation, recognition and rewards for Employees, Improved morale, better communication skills, improved work culture, increased pride in work, change in work culture, involvement of employees in getting customers' feedback, helping knowledge sharing between employees, Overall commitment from employees, better employee retention, Employees enjoy change, suppliers show improved commitment, enjoy the Six Sigma implementation, and Employees' willingness to work with Six Sigma projects.
5. Customer satisfaction – Customer satisfaction is measured by sixteen different factors viz. Defining needs of customers, better satisfaction of customer, making products available of as per the needs of customers, better customer relationships, better commitment for customers' needs, Completely shifting focus towards customers, customers feedback, developing products based on needs of customers, sharing production plans with customers, making customers a part of the company, setting up of customer redressal mechanism, satisfy customers through continued service, risks and rewards are shared with customers, Reduction in customers' complaints, better customer relationship and achieve Competitive pricing policy through Win - Win situation with customers.

4.1 Hypotheses of the study

Following hypotheses were stated based on the earlier mentioned variables selected for the study.

During discussion with statisticians it was suggested by them to use only null hypotheses as against the normal procedure of stating both null and alternate hypotheses together. The positive form of a null hypothesis is deemed as alternate hypothesis. To make decisions, while testing hypotheses, if the obtained significance value is less than 0.05 from the SPSS software, the null hypotheses are rejected which means the alternate hypotheses are accepted. The hypotheses of the study are:

HA0: Existing practices of Six Sigma do not improve growth of the company.

HB0: Financial status of the company is independent of existing practices of Six Sigma. HC0: Improvement of Productivity is independent of existing

practices of Six Sigma.
HD0: Existing practices of Six Sigma do not improve peoples' equity.
HE0: Existing practices of Six Sigma do not improve customer satisfaction.
HF0: There is no relationship among the benefits of Six Sigma.

4.2 Sample population

Sample Population included 522 employees with direct involvement in implementation of Six Sigma from 20 different manufacturing industries which have implemented Six Sigma in Karnataka, India. While selecting the industry for the survey, it was observed that many industries were in very primitive state with respect to six sigma implementation. Hence, employees were reluctant to share the opinion. Thus, it was decided to select such industries which have implemented six sigma at least for five years and few projects have been completed. Based on this fact, industries selected included ten Mechanical based industries, eight Electronics industries and two Pharmaceutical industries. Again, while selecting the respondents, only such of those employees who are involved in six sigma projects are considered. Because, the data could not be very relevant if it is collected from those whose knowledge on six sigma is limited. The respondents selected were in the cadre of both workers and managers with sufficient training in Six Sigma implementation. Six Sigma training was given to obtain certificates as master black belts, black belts and green belts as well. Out of 522 respondents, 380 were considered as worker and 142 as supervisors or managers.

5 Research design

To begin the study, the collection of data was done using a questionnaire. The questionnaire was prepared by considering five parameters which are believed to be reasons for the success of industries. They are Growth of the company, Financial benefits, Productivity, Peoples' equity and Customer satisfaction.Questionnaire was thoroughly revised and pilot-tested with the help of experts and practitioners of Six Sigma both as academia and industry. The process helped to evaluate and validate the questionnaire. Finally the questionnaire included 09 parameters that define growth of the company, 08 parameters that define the financial benefits, 18 parameters that define productivity, 27 parameters that define peoples' equity and 16 parameters that define customer satisfaction (total 78). Proper care was taken to frame questions around above defined parameters in such a language which even a layman can understand. Even vernacular language of employees was used to explain the purpose of the survey and the content of the questionnaire. A Five point Likert scale with a lowest degree of agreement (strongly disagree=1) to the highest degree of agreement (strongly agree=5) was selected to collect respondents' opinion in an objective manner.

A thorough analysis of the results of the study was done to observe the effect of Six Sigma towards suggested variables. SPSS V16, one of the most commonly used Statistical tool was used to analyze the outcome of the study.

6 Results and interpretations

As a first step towards the analysis, after collecting the data, Cronbach's alpha test was used to test the reliability of the questionnaire. The following table 1 gives details of the reliability analysis.

Table 1. Reliability Statistics

Variable	Number of items	Cronbach's alpha
Growth of the company	9	0.891
Financial benefits	8	0.867
Productivity	18	0.782
Peoples' equity	27	0.892
Customer satisfaction	16	0.894
Overall	**78**	**0.961**

Since all the individual parameters and also the overall alpha value scored the Cronbach's alpha value more than 0.7 (Nunnaly, 1978), the testing instrument is proved to be reliable. Next, to select the type of the test, we conducted the normality test. Table 2 gives the result of Kolmogorov-Smirnov test for normality.

Table 2. Result of Kolmogorov-Smirnov Test of normality

Normal Parameters	Benefits of Six Sigma				
	Customer satisfaction	Financial benefits	Growth of the company	Peoples' equity	Productivity
Mean	61.0820	30.7787	33.4262	96.6557	68.1148
Std. Deviation	7.59468	2.05105	3.20619	11.89539	6.32481
P value	0.142	0.005	0.215	0.047	0.158

It was revealed from K – S test that the distribution follows a pattern of normal curve with significant value (P value) of more than 0.05 in all the cases except Financial benefits. This case has been discussed with the experts who have suggested assuming normal distribution. Hence, we will select t- test for testing the hypotheses.
The following table 3 gives the frequency of responses in percentage for the scale parameters

Table 3. Frequency of responses

Scale parameters	Percentage of responses				
	Growth of the company	Financial benefits	Productivity	Peoples' equity	Customer satisfaction
Strongly disagree	0.18	0.10	1.09	12.58	1.38
Disagree	5.46	7.58	6.33	18.81	6.15
Can't say	24.77	13.63	17.76	29.61	17.16
Agree	61.93	69.67	62.70	33.65	59.94
Strongly agree	7.65	9.02	12.11	5.35	15.37

The frequency distribution Exhibit displays that as per the opinion of respondents all the five variables have been accepted as to be benefitted after the implementation of Six Sigma as more percentage of opinion is towards agree and strongly agree. This can be substantiated with the help of one sample t-test at a significance value of 0.05 for 521 df (N-1).

6.1. Testing of hypotheses

HA0. Existing practices of Six Sigma do not improve growth of the company.

To test the hypothesis it is required to calculate the overall mean and standard deviation for the variable. The following table 4 gives the observed value of the mean for the above variable. The observed mean value for each variable is the total sum of all the individual mean values in each variable. For example, in growth of company, as explained earlier, there are nine factors. Hence, the mean value of growth of company is the sum of all these individual factors. The standard deviation is calculated for N = 522. Same procedure is followed to calculate mean and standard deviation for remaining variables. Since manual calculation of mean and standard deviation for N = 522 is tedious, we have used SPSS V16 software, which will calculate as per the data input. Now this observed mean is compared against the test value as used by the company. A common practice among companies, where the research is done was observed to follow a of practice of accepting the results of a survey when the mean value of each question is more than or equal to 3.6. on a Likert scale of 1 to 5. In this research each of the defined variables has more than one factor, used to measure it. Hence, the overall test value is calculated as (individual test value × number of factors used to measure the variable). For example, in case of growth of company, since there are nine factors, which measure it, the overall test value is $3.6 \times 9 = 32.4$. The same procedure is followed for all the variables.

With this we can observe that the calculated mean value is greater than test value. It gives the impression that implementation of Six Sigma has helped the company to grow. This can be substantiated with the help of one sample t-test conducted at a significance level of 0.05 at 521degrees of freedom (N-1). The following table 4 gives the result of one sample t-test.

Table 4. Mean observed and expected values on growth of the company and results of one sample t-test

	N	Observed value of Mean	Std. Deviation	Expected value of Mean	t-value	p-value
Growth of the company	522	33.42	3.206	32.4	3.535	0.001 (highly significant)

The one sample t-test result shows that the calculated mean value is 33.42 as compared with the test value of 32.4. Since, the significance value p (0.001) < 0.05, the difference between observed and expected mean values on growth of the company is highly significant, t (521) = 3.535, $p < 0.05$. Since, the test statistics revealed significant difference between observed and expected mean values, one can say that there is a significant contribution to the growth of the company due to existing practices of Six Sigma **hence, null hypothesis HA0 is rejected.**

HB0: Financial status of the company is independent of existing practices of Six Sigma.

To test this hypothesis as a first step, it is required to calculate the overall mean and standard deviation for the variable. The following table 5 gives the value of the mean for the above variable. As explained earlier, since there are eight factors that measure the variable financial benefits, the overall test value is 3.6 * 8 = 28.8. With this we can observe that the calculated mean value is more than test value. It seems that implementation of Six Sigma has helped the company to gain in terms of finance. This can be substantiated with the help of one sample t-test conducted at a significance level of 0.05 at 521degrees of freedom (N-1). The following table 5 gives the result of one sample t-test

Table 5. Mean observed and expected values on financial benefits and results of one sample t-test

	N	Observed value of Mean	Std. Deviation	Expected value of Mean	t-value	p-value
Financial benefits	522	30.77	2.04	28.8	10.656	0.000 (highly significant)

The positive value of t signifies that obtained mean is greater than the test value. Since the significance value p (0.000) < p (0.05), this indicates that the contribution of Six Sigma towards the financial benefits is highly significant, t (521) = 10.656, $p < 0.05$. Hence, we can say that the implementation of Six Sigma has contributed significantly to the financial benefits of the organizations as per the opinions of the respondents. **Thus null hypothesis HB0 is rejected.**

HC0: Improvement of Productivity is independent of existing practices of Six Sigma.

To test this hypothesis as a first step, it is required to calculate the overall mean and standard deviation for the variable. The following table 6 gives the value of the mean for the above variable. As a matter of policy by the companies, since there are eighteen factors that define the variable productivity, the overall test value is 3.6 * 18 = 64.8.
With this we can observe that the calculated mean value is more than test value. It looks as if the Six Sigma implementation has helped the company to improve productivity. This can be substantiated with the help of one sample t-test conducted at a significance level of 0.05 at 521degrees of freedom (N-1). The following table 6 gives the result of one sample t-test.

Table 6. Mean observed and expected values on productivity and results of one sample t-test

	N	Observed value of Mean	Std. Deviation	Expected value of Mean	t-value	p-value
Productivity	522	68.1148	6.32481	64.8	5.789	0.000 (highly significant

The positive value of t signifies that obtained mean is greater than the test value. Since the significance value p (0.000) < p (0.05), this indicates that the contribution of Six Sigma towards improvement of productivity is significant, t (521) = 5.789, $p < 0.05$. Hence, we can say that the Six Sigma implementation has contributed significantly to the improvement of the productivity as per the opinions of the respondents. This indicates a significant relationship to improvement of productivity and the existing practices of Six Sigma. **Thus hypothesis HC0 is rejected.**

HD0: Existing practices of Six Sigma do not improve peoples' equity

To test this hypothesis as a first step, it is required to calculate the overall mean and standard deviation for the variable. The following table 7 gives the value of the mean for the above variable. As a matter of policy by the companies, since there are twenty seven factors that define the variable peoples' equity, the overall test value is 3.6 * 27 = 97.2.
With this we can observe that the calculated mean value is slightly less than test value. It appears that implementation of Six Sigma has not helped the company to improve peoples' equity significantly. This can be substantiated with the help of one sample t-test conducted at a significance level of 0.05 at 521 degrees of freedom (N-1).

Table 7. Mean observed and expected values on peoples' equity and results of one sample t-test

	N	Observed value of Mean	Std. Deviation	Expected value of Mean	t-value	p-value
Peoples' equity	522	96.65	11.89539	97.2	1.352	0.179 (non significant)

The one sample t-test results show that the calculated mean value is 96.65 as against the test value of 97.2. Since the significance value p (0.179) > 0.05, the difference between observed and expected mean values on people' equity is not significant, t (521) = 1.352, p >0.05. Hence, the hypothesis fails to get rejected, since the test statistics revealed lower mean value than expected mean value as well as the non significant p value one can say that existing practices of Six Sigma do not contribute to the improving peoples' equity. **Hence, HD0 fails to get rejected.**

HE0: Existing practices of Six Sigma do not improve customer satisfaction.

To test this hypothesis as a first step, it is required to calculate the overall mean and standard deviation for the variable. The following table 8 gives the value of the mean for the above variable. As a matter of policy by the companies, since there are sixteen factors that define the variable customer satisfaction, the overall test value is 3.6 * 16 = 57.6.

With this calculated mean value is observed to be more than the test value. It gives the impression that Six Sigma implementation has helped the companies to improve their customers' satisfaction. This can be substantiated with the help of one sample t-test conducted at a significance level of 0.05 at 521 degrees of freedom (N-1).

Table 8. Mean observed and expected values on customer satisfaction and results of one sample t-test

	N	Observed value of Mean	Std. Deviation	Expected value of Mean	t-value	p-value
Customer satisfaction	522	61.084	7.594	57.6	5.064	0.000 (highly significant)

The positive value of t signifies that obtained mean is greater than the test value. Further the significance value p (0.000) < p (0.05), this indicates that as the obtained mean value is greater than the test value, the implementation of Six Sigma towards the improvement of their customers' satisfaction is significant, t (521) = 5.064, $p < 0.05$. Hence, we can say that the implementation of Six Sigma has significantly contributed to the improvement of customers' satisfaction as per the opinions of the respondents. **Thus hypothesis HE0 is rejected.**

HF0: There is no relationship among the benefits of Six Sigma

Till now we have observed that practices of Six Sigma is not uniform with every company. This could be due to various factors like culture, type of the product, management perception, interrelationship among benefits, etc. The interrelationship among the benefits will reveal the effect of one variable on the other. Thus if the set practice of Six Sigma in one company affects one variable, it would affect all the variables, if those variables are interrelated with each other. We will obtain a correlation matrix for the above test.

The following table 9 gives the details of the relationship among various benefits contained in practice of Six Sigma at a significance level of 0.01 (two tailed)

Table 9. Correlations among the benefits of Six Sigma

		Customer satisfaction	Financial benefits	Growth of the company	Peoples' equity	Productivity
Customer satisfaction	r value	1	.722(**)	.645(**)	.612(**)	.607(**)
	p value	.	.000	.000	.000	.000
Financial benefits	r value		1	.851(**)	.742(**)	.682(**)
	p value		.	.000	.000	.000
Growth of the company	r value			1	.727(**)	.794(**)
	p value			.	.000	.000
Peoples' equity	r value				1	.698(**)
	p value					.000
Productivity	r value					1
	p value					.

The Pearson correlation coefficient (r) values indicate the size of the correlation. Squaring the r value indicates the percentage of affect of one variable over the other.
The following observations can be made with the help of the table 9

a. The customer satisfaction has a significant positive relationship with financial benefits, growth of the company, peoples' equity, and productivity. Its highest significance is seen with financial benefits (52.2%), which means that increase in customer satisfaction will help the company to improve its financial position by 52%.
b. Financial benefits are having significant positive relationship with customer satisfaction, growth of the company, peoples' equity, and productivity. Its highest significance is seen with growth of the company (72.4%), which means the increase in the financial position of the company will affect the growth of the company to an extent of 72%.
c. Growth of the company is having significant positive relationship with customer satisfaction, Financial benefits, peoples' equity, and productivity. Its highest significance is seen with financial benefits (72.4%), which means the growth of the company will be affected to an extent of 72% with the improvement in the financial position of the company.
d. Peoples' equity in a company is having significant positive relationship with

customer satisfaction, financial benefits, growth of the company, and productivity. Its highest significance is seen with financial benefits (55%), which means that the peoples' equity of the company will increase to an extent of 55% with the improvement in the financial position of the company.

e. Productivity of the company is having significant positive relationship with customer satisfaction, financial benefits, growth of the company, and peoples' equity. Its highest significance is seen with financial benefits (63%), which mean that the improved productivity of the company will help the company to grow to an extent of 63%.

Since the correlation matrix revealed a significant interrelationship among all the benefits of Six Sigma, **hypothesis HF0 is rejected**

7 Discussions and conclusions

The Six Sigma is being recognized as critical for the success of any industry. Six Sigma has complemented TQM in achieving continuous improvement by providing new structure and methods.

It is very important for managers to be aware of the cultural values before adopting TQM/Six Sigma initiatives, in their organization that can help to effectively implement multiple TQM/Six Sigma practices in the organization.

A detailed study of the manufacturing industries that have implemented Six Sigma in India, followed by analysis of the results reiterated vital role of Six Sigma in making manufacturing industries achieve success. The study was aimed at understanding the employees' impression about the impact of implementation of Six Sigma on few success variables such as growth of the company, financial benefits, productivity, peoples' equity and customer satisfaction; examine the interrelationship among the benefits of Six Sigma.

The following conclusions are highlighted after the study.

I. Collective responses indicate that the implementation of Six Sigma has done huge impact on the growth of the company. This means that respondents have consented positively towards the fact that due to Six Sigma the brand image of the company has improved and company has obtained a competitive edge due to Six Sigma.

II. Respondents have the opinion that Six Sigma implementation has a positive impact on the improvement of the financial position of the company, as they feel that parameters like reduced cost of poor quality, reduced price of non conformance, improved profit and return on investment have contributed to the improved financial position of the organization.

III. Analysis of the impact of Six Sigma on productivity also indicate a positive relationship as respondents have given thumbs up for improved productivity. They have reacted that factors like setup cost, operating cost, scrap rate, defect rate, rejection rate, production lead time have come down after Six Sigma is implemented, which has resulted in improved process capability. This impact has resulted in improved productivity after Six Sigma implementation.

IV. The variable, peoples' equity that highlighted parameters towards the welfare of the work force do not appears to be in a better shape even after the implementation of Six Sigma. Respondents have observed that factors like human values, morale of the employees, working culture have improved due to which the commitment from the workers side towards the implementation of Six Sigma is positive. But many other factors like top management commitment, leadership style, employee involvement in decision making, continuous need based training, interpersonal relationships, communication skills, incentive schemes, knowledge sharing between employees, employee retention, have not improved substantially. These factors have contributed to the displeasure among employees with Six Sigma. However it is observed and appreciated that they have enjoyed the implementation of Six Sigma and would like to continue in companies that have implemented Six Sigma. This indicates that by giving more attention to the monetary benefits of the workforce the impact of Six Sigma could be enhanced.

V. Finally, the most important variable in the present scenario, the customer satisfaction, is observed to have increased after implementing Six Sigma as per the opinion of the respondents. They have observed that after customers are made part of the product development, they are able to define their requirements properly. Due to reduced defects and rejection rates customers are very happy. Since, the changes suggested by the customers are made quickly, industries are able to achieve win-win situation with the customers. As a result the customer satisfaction level has improved after the implementation of Six Sigma.

VI. There is a significant interrelationship among the benefits of Six Sigma. This is a very important fact that makes the companies to understand that all benefits should be given equal importance as decrease of one will start a cascading effect on all other benefits.

8 Implications for the management and non Six Sigma companies

The survey results bring up the measured opinion of employees who are involved in the implementation of Six Sigma which is very unique. They believe that Six Sigma is a fruitful exercise that helps the manufacturing organizations to achieve success in various facets that includes the growth, financial benefits, satisfaction of customers, and productivity. However, the relationship between peoples' equity and Six Sigma implementation, it is not highly significant. It indicates the employees are not highly benefitted by Six Sigma implementation. This opens up the huge responsibility for the managers to consider this factor seriously. Human resource is the back bone of any organization. Hence, more thoughtfulness is expected to be extended towards the improvement of work force. This will improve the morale of the work force and they will involve to a greater degree in the change process. This starts a chain reaction and results in improving the results of other variables also to a greater extent resulting in greater success of manufacturing industries. Thus the management should look into each of the factor of peoples' equity to obtain a better result.

Finally, the results of the survey acts as a route map for non Six Sigma industries. As the employees of these manufacturing industries observe that Six Sigma implementation yields lot of benefits, not just on paper but also in the minds of employees who have worked on six sigma projects. Hence, the management of non Six Sigma industries should think proactively before the crisis occurs in their respective organization. This survey acts as a motivator for management of non Six Sigma industries to promote Six Sigma in their organization.

9 Future scope of the study

The study which included the benefits of Six Sigma alone cannot be compared with the success due to other manufacturing philosophies such as lean or theory of constraints unless we make a comparative study between such organizations. The organizations where the survey is done follow only Six Sigma methodology. This opens up future scope of the study to compare between Six Sigma and lean or other manufacturing philosophies.

References

1. M. Arendt, "Six Sigma and knowledge management", Economics and Organization of Enterprise, Vol 2, No. 2, pp14–20, 2008.
2. T Yang, C Hsieh, "Six-Sigma project selection using national quality award criteria and Delphi fuzzy multiple criteria decision-making method", Expert Systems with Applications, Vol 36, pp 7594–7603, 2009.
3. CT Su, C.J. Chou, "A systematic methodology for the creation of Six Sigma projects: a case study of semiconductor foundry", Expert Systems with Applications, Vol 34, pp 2693–2703, 2008
4. K Linderman, R.G Schroeder, S Zaheer, A.S Choo, "Six Sigma: a goal-theoretic perspective", Journal of Operations Management, Vol 21, pp 193–203, 2003.
5. J. Markarian, "Six Sigma: quality processing through statistical analysis", Plastics. Additives and Compounds, Vol 6, No.4, pp 28–31, 2004.
6. K.C Kapur, Q Feng, "Integrated optimization models and strategies for the improvement of the Six Sigma process", International Journal of Six Sigma and Competitive Advantage, Vol 1, No.2, pp 210–228, 2005.
7. Q Feng, C.M Manuel, "Under the knife: a national survey of Six Sigma programme in US healthcare organizations", International Journal of Health Care Quality Assurance, Vol 21, No.6, PP535–547, 2008.
8. R.K Padhy, S.Sahu, and R.K. Das, "Implementation of Six Sigma in Indian industries – a Delphi study", Int. J. Computer Aided Engineering and Technology, Vol 3, No.1, pp 19–33, 2011.
9. C. S Chethan Kumar, N V R Naidu, K Ravindranath, "Performance improvement of manufacturing industry by reducing the Defectives using Six Sigma Methodologies" IOSR Journal of Engineering, Vol. 1, No.1, pp 001-009, 2010.
10. D. A Desai, M. B Patel, "Impact of Six Sigma in a developing economy: analysis on benefits drawn by Indian industries", Journal of Industrial Engineering and Management, Vol. 2, No. 3, pp 517-538, 2009.
11. P. Kaushik and D. Khanduja. (2009). Application of Six Sigma DMAIC methodology in thermal power plants: A case study. Total Quality Management, 20(2), 197–207.
12. P Kaushik, "Relevance of Six Sigma line of attack in SME's – a case study of die casting manufacturing unit", Journal of Engineering and Technology, Vol. , No.2, pp 107–112, 2011.
13. Min Zhang, Wei Wang, Thong Ngee Goh, and Zhen He, "Comprehensive Six Sigma application: a case study", Production Planning & Control, Vol.26, No. 3, 2015
14. K Srinivasan, S Muthu, S R Devadasan, and C Sugumaran, "Enhancement of sigma level in the manufacturing of furnace nozzle through DMAIC approach of Six Sigma: a case study", Production Planning & Control, Vol.27, No.10. pp 810 – 822, 2016.
15. A. J. Thomas, M Francis, R Fisher, and P Byard, "Implementing Lean Six Sigma to overcome the production challenges in an aerospace company", Production Planning & Control, Vol 27, No.7-8, 2016.

16. N Venkatesh, and C Sumangala, "Evaluation of Six Sigma in Automobile Manufacturing Industries", Journal of Contemporary Research in Management, Vol.9, No.1, pp 45-54, 2014.
17. J. Nunnaly, "Psychometric theory", New York: McGraw-Hill publishers, 1978

Study of a supply chain system with the presence of bullwhip effect

Ayush Shrivastava[1], *Raghavendra Kamath*[1*], *Himanshu Sharma*[1] *and Ajitesh Gogoi*[1]

[1]*Department of Mechanical and Manufacturing Engineering, Manipal Institute of Technology, Manipal Academy of Higher Education, Manipal – Karnataka-576104, INDIA*

Abstract. A study has been conducted to understand the effect of bullwhip phenomenon in a supply chain system. The hypothetical data collected from the study is used to make a model which can be used to simulate the operation of the system with moderate complexity. System dynamics approach is used to create an open loop system. The phenomenon used is slightly modified and empirically analysed to bring down its adverse effect on the system. The results obtained graphically are emphasising the effect of bullwhip on the various levels of the supply chain system.

1 Introduction

Supply chain management is about managing how goods and services move and it is dependent on flow and storage of raw materials, unfinished goods and finished goods from point of origin to point of consumption. A typical supply chain would consist of three stages- purchase, production and distribution [1]. The systems which are sophisticated, used products may re-enter at any point where residual value is recyclable. The purpose of developing a Generic Supply Chain Model (GSCM) is to provide a deeper understanding of supply chains in terms of both their development and operation management. Models of supply chain seem to concentrate on only one particular aspect like organization structure/strategy or information technology, or human factors. While each of these dimensions is considered crucial in its own right, each factor by itself does not provide a complete and comprehensive view of supply chains and their management. There are many kinds of supply chain systems even though the types of supply chain members are countable [2]. It is believed that feature model gives definitions of such typical members included in a chain and business process activities of them.

An important factor that affects the efficiency of a supply chain system is bullwhip effect. Bullwhip effect is an observed phenomenon addresses the shift of a seemingly steady inventory demand into enhancing demand fluctuation in upstream supply chain. At the order side of supply chain it results in an amplified variation of information regarding demand [3]. The bullwhip effect is a distribution channel phenomenon in which forecasts yield supply chain inefficiencies. This means there will be increasing variations in

* Corresponding author: crkamath77@gmail.com

inventory in response to changes in market demand as one goes higher up the in supply chain. It has been observed that a small fluctuation in point-of-sale demand will be lead to a big change in demand on the supply side. Building upon this conceptual definition of bullwhip, it fundamentally consists of not only a level of demand distortion, but also the extent of variance amplification [4]. They imply while it is important to know if firm is experiencing demand distortion, it is equally important to know if such demand distortion is accelerating along the supply chain. A particular type of a simulation model was used by Chatfield et al to analyse the effect of stochastic lead times, sharing of quality information at regular intervals in a supply chain pertaining to inventory system [5]. It has been found that above parameters are very significant in mitigating the effect of bullwhip in a supply chain system.

Agrawal et al studied a two echelon supply chain system with key parameters as information sharing and lead time in understanding the bullwhip phenomenon [6]. They observed that minimising lead time is more beneficial compared to information sharing in reducing the effect of bullwhip. Some alternative forecasting methods are also proposed in order to reduce bullwhip effect using different simulated demand patterns [7]. The expected performance of a vendor managed inventory supply chain with a traditional "serially linked" supply chain was compared and the emphasis of this investigation is the impact these two alternative structures have on the bullwhip effect generated in the supply chain [8]. Quantifying and mitigating the bullwhip effect in a supply chain system was also carried out in various systems [9, 10].A number of quantitative models for developing a good supply chain management system [11]. Casual loop diagrams are very much essential in having a strong feedback in systems [12,13]. The effect of bullwhip has been quantified in a supply chain system based on the impact of forecasting, lead time and information [14]. Researchers are also working on the effect of information sharing in a supply chain system and its effect on the performance [15, 16, 17].

2 Methodology

This study is carried out based on the model developed by Disney et al using system dynamics approach. This model was used for analysing various ways of minimising bullwhip effect in a manufacturing loop in supply chain system. This loop consists of retailer at one end and supplier at the other end having a manufacturer in between them. The process starts with retailer placing orders with manufacturer based on the needs of the customer. Then manufacturer places order for components with supplier based on retail order. Finally, supplier supplies the components based on the manufacturers order. Once manufacturer receives the components, he produces products as per the retail order and supplies it to retailer. This entire process is explained through a casual loop diagram as shown in figure 1.

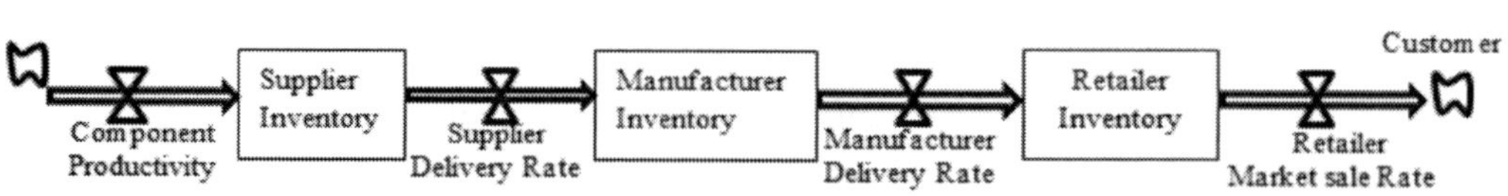

Fig. 1. Casual loop supply diagram of supply chain

Table 1. State Variables of Supply chain model

SI	Supplier Inventory
MI	Manufacturer Inventory
RI	Retailer Inventory

There are some assumptions in the created model which include only one type of product in the system, classic inventory situation is considered, products once sold are not taken back. All the capacities like transport, product life cycle and production are unlimited. The primary and secondary variables of the supply chain model are listed in table 1 and 2.

Table 2. Rate variables of Supply chain model

CP	Components of Productivity
SDR	Supplier Delivery Rate
MDR	Manufacturer Delivery Rate
RMSR	Retailer Market Sales Rate

In a product manufacturing scenario, inventory held by the retailer is determined by the consumption and rate of product delivery by the manufacturer. The manufacturer inventory depends on the rate of component delivery by the supplier and rate of remanufacturing products and rate of supply of products to retailer. The supplier inventory depends on production rate of new components and rate of remanufacturing the components and rate of supplying the components to manufacturer. The needs of retailer, manufacturer and supplier all derived from individual sales forecasting and inventory management. Various variables used at three levels of supply chain system are given in table 3 and 4.

Table 3. Auxiliary variables of Supply chain

POS	Products Order of Supplier
SPDR	Supplier Product Demand Rate
IES	Inventory Expected of Supplier
FSS	Forecast of Sales of supplier
POM	Products Order of Manufacturer
MPDR	Manufacturer Product Demand Rate
IEM	Inventory Expected of Manufacturer
FSM	Forecast of Sales of Manufacturer
POR	Purchase Order of Retailer
REI	Retailer Expected Inventory
RSF	Retailer Sales Forecast

Table 4. Constants in Supply chain

DDS	Delay of Delivery by Supplier
DPS	Delay of Production by Supplier
IAT	Inventory Adjustment Time
STEI	Sustainable Time of expected inventory
MAP	Moving Average Period
DDM	Delivery Delay by Manufacturer
MPD	Manufacturer Production Delay
TD	Transportation Delay
SPC	Supplier Production Capacity
MPC	Manufacturer Production Capacity

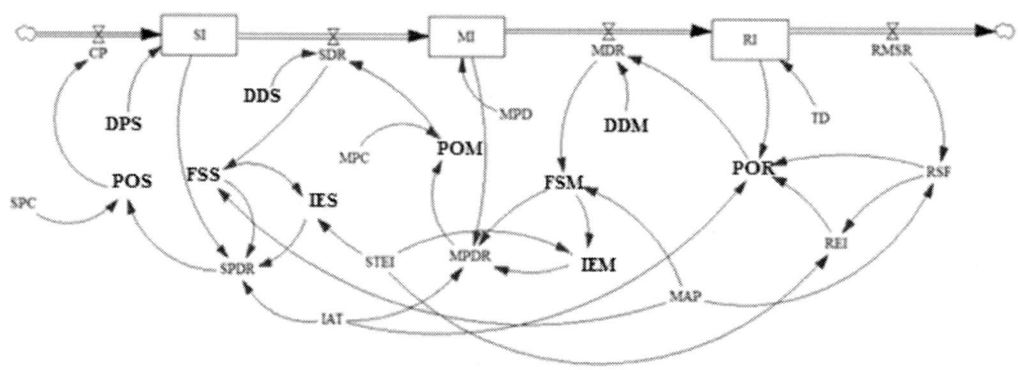

Fig. 2. Supply chain system with an Open loop model

A Supply chain system with an open loop model is shown in figure 2.The purchase order from the retailer is composed of volume of inventory to be adjusted, forecast of sales by retailer. Retailer inventory adjustment is dependent upon the likelihood of having inventory, inventory in real time and inventory adjustment time. Expected inventory by retailer and forecast of sales both are related to the sustainable time of expected inventory. Retailer inventory is a primary variable, the input is the rate of delivery by manufacturer and the output is the consumption in the market. Forecast of sales of retailer is an information delay function of the rate at which sales takes place. Some of the equations used in development of the model for computing retailer, manufacturer and supplier side information are listed below.

Retailer end

$$POR = Max\ (0,\ RSF+(REI-RI)/IAT) \quad (1)$$

$$RI = Integ\ (Delay1\ (MDR,\ TD) - RMSR,\ 3000) \quad (2)$$

$$REI = RSF * STEI \quad (3)$$

$$RSF = Smooth\ (RMSR,\ MAP) \quad (4)$$

Manufacturer end

$$MDR = Delay\ 1\ (RPO,\ DDM) \quad (5)$$

$$MI = Integ(Delay\ 1\ (SDR,\ MPD) - MDR,\ 3000) \quad (6)$$

$$IEM = FSM*STEI \quad (7)$$

$$POM = Min(MPC,\ Max\ (0,\ MPDR) \quad (8)$$

$$MDR = Max(0,\ FSM+(IEM-MI)/IAT) \quad (9)$$

$$FSM = Smooth\ (MDR,\ MAP) \quad (10)$$

Supplier end

$$SDR = Delay\ 1\ (MPO,\ DDS) \quad (11)$$

$$SI = Integ(Delay\ 1\ (PP, DPS) - SDR, 3000) \quad (12)$$
$$IES = SSF*STEI \quad (13)$$
$$POS = Min(SPC, Max\ (0, SPDR) \quad (14)$$
$$SDR = Max(0, SSF+(IES-SI)/IAT) \quad (15)$$
$$SSF = Smooth\ (SDR, MAP) \quad (16)$$

In order to measure the bullwhip effect of the various orders under the different conditions, the bullwhip can be quantified and used as given below [17].

$$BWE = \frac{\sigma_o^{\ 2}}{\sigma_D^{\ 2}}$$

where σ_o^2 represents the variation in the order rate from retail end to the production end and σ_D^2 represents the variation in the demand encountered at retail end.

3 Results and discussions

Lot of information distortion takes places when it moves from one end to the other in a supply chain system. This is one of main reason for presence bullwhip phenomenon. Its detrimental effect can be reduced considerably by sharing of information in an enterprise. When quality information is not shared at any levels be at upstream or at downstream in a supply chain system, enterprise will not have it at the right time and production details can only be predicted based on past data or orders. This has been clearly shown in the simulation of supply chain system obtained using Vensim software. The values in the analysis are obtained by simulation using the software.

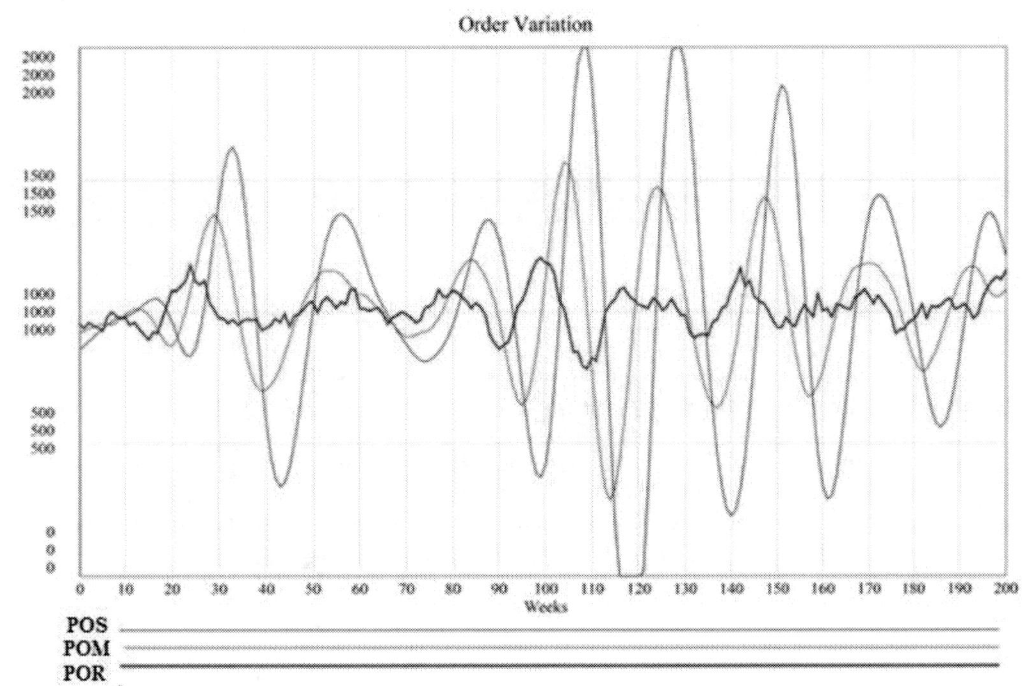

Fig. 4. Baseline model

The results from the simulation are on the lines of the expected results as shown in figure 4. The graph clearly shows that as the movement takes place from customer end to supplier end, the demand distortion increases which represents higher bullwhip effect. The difference in the peak along the time-axis is caused due to the lead times of the respective levels. Greater the lead time, further will the peaks be but it will also have a smoothening effect as there is more time for the market deviations to settle down.

3.1 Effect of variation of different parameters on bullwhip effect

3.1.1 Varying the Delivery Delay (DDM and DDS)

It is observed from the simulation that reducing the magnitude of DDM and DDS effectively reduces the Bullwhip Effect as shown in figure 5. It can be inferred that this is due to the possibility of immediate feedback to latter levels which is absent when large delays are prevailing in the system.

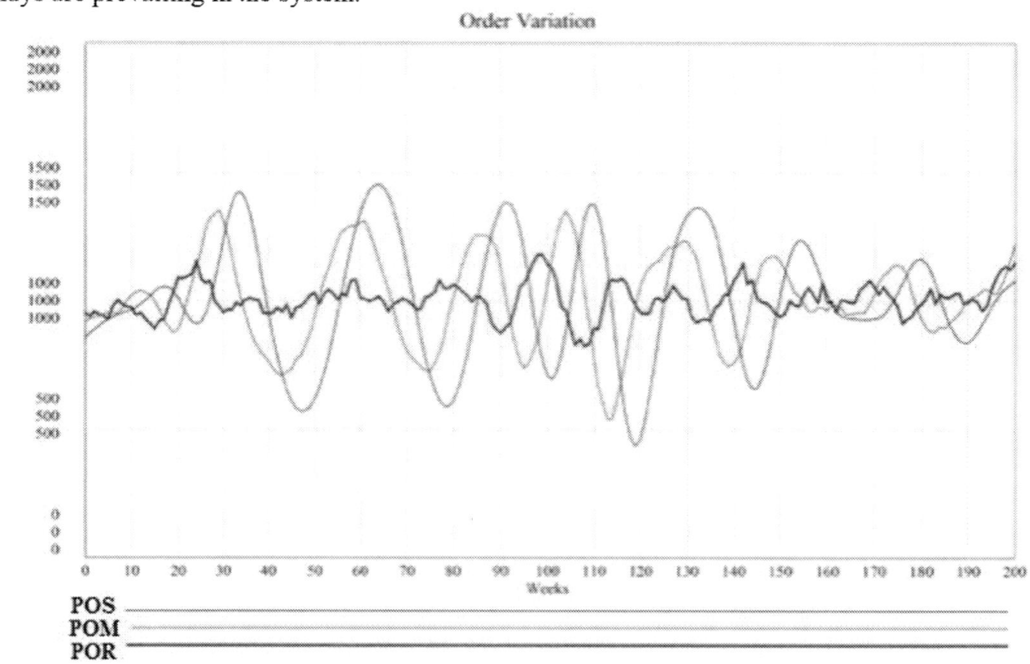

Fig. 5. Effect of varying the delivery delay

3.2 Varying Moving Average Period (MAP)

On analysing the effect of varying the forecast period as shown in figure 6, it is observed that increasing the magnitude of MAP lessens the effective amount of Bullwhip present in the system. The reason for this observation can be explained by the fact that a larger forecast period allows for more accurate predictions of future demands by negating the side-effects from seasonal peaks.

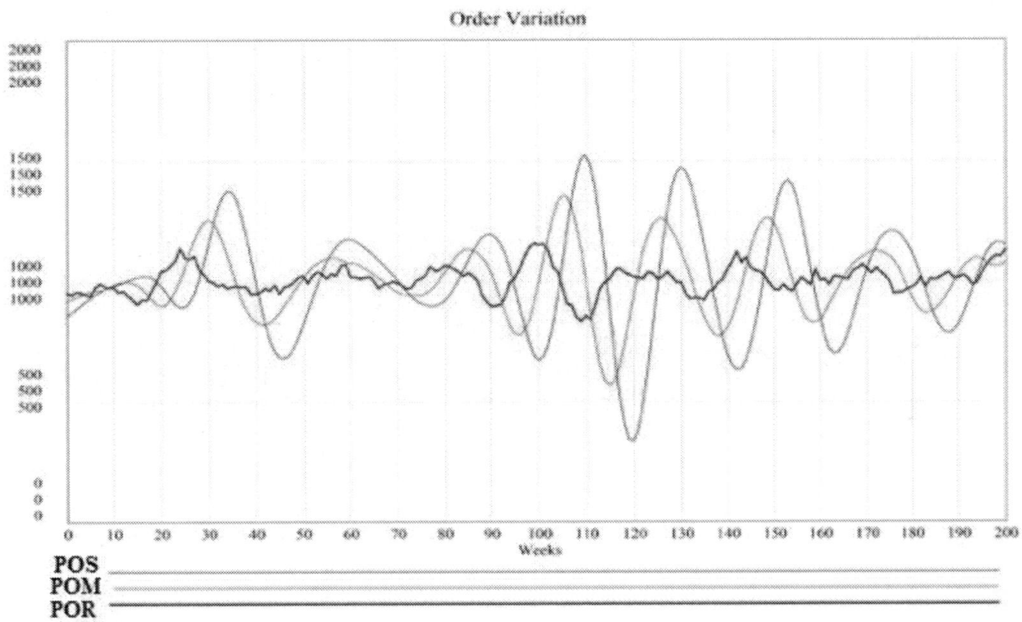

Fig. 6. Effect of varying the forecast period

The impact of different forecast periods on the Bullwhip Effect of a supply chain were observed by conducting a study for a 5-week and a 10-week time frame on the supply chain modelled for 200 weeks. The results for above two time periods are shown in table 5 and 6.

Table 5. Five week time frame

	POR	RMSR
VARIANCE	5095.962	2605.4

BWE = 5095.962/2605.4
= 1.956

Table 6. Ten week time frame

	POR	RMSR
VARIANCE	5105.662	2717.562

BWE = 1.878

In case of purchase order of the retailer (POR), the distortion in demand is minimum but as we move along supply chain demand distortion increases. This is because the distance between customer demand and retailer demand is one step whereas supplier demand and customer demand is three steps. Lead time and distance between customer demand and each level of supply chain affect the purchase order. So POR has minimum demand distortion, thus shows how bullwhip effect exists in three level supply chain as demand distortion in purchase order of supplier (POS) is highest.

Retailer market sales rate (RMSR) is the rate at which retailer sells the products, which in turn affects the retailer forecast and there by influencing the retailer demand as shown in above tables 5 and 6 respectively.

4 Conclusions

A supply chain has been modelled using system dynamics keeping in mind Bullwhip Effect. This has enabled to model a fairly complex supply chain and shed some light on the interactions between its key parameters. An understanding of the bullwhip effect is necessary to effectively reduce cost involved and optimize the whole structure. There are various parameters that can directly and indirectly affect the supply chain to produce a greater or lesser bullwhip effect.

The results obtained in the form of graphs are interesting from a practical point of view and they helped us to visualise the bullwhip effect and understand the various nuances of the parameters.

5 References

1. M. M. Helou, I. N. Caddy, Problems and Perspectives in Management, 4, 4 (2006)
2. S. Umeda, F. Zhang, Proc. Winter Simulation Conference (2010)
3. A. Tanweer, Y. Li, G. Duan, J. Song, 9th Int. Con. Traffic and Transportation Studies, Procedia social and behavioural sciences, 138, 289-297 (2014)
4. A. W. Mackleprang, M. K. Malhotra, J. operations management, 36, 15-32 (2015)
5. D. C. Chatfield, J. G. Kim, T. P. Harrison, J. C. Ha, Production and operations management, 13, 4, 340-353 (2004)
6. S. Agrawal, R.N. Sengupta, K. Shanker, European. J. Operational Research, 192, 576-593 (2009)
7. F. Campuzano-Bolarín, A. G. Frutos, M.C.R. Abellón, A. Lisec, Traffic and transportation, 25, 2, 177-188 (2013)
8. S.M. Disney, D.R. Towill, In. J. Operations & Production Management, 23, 6, 625-651 (2003)
9. D. Fu, C. Ionescu, E. Aghezzaf, R. D. Keyser, Computers and Industrial Engineering, 81, 46-57 (2015)
10. R. Metters, J. Operations Management, 15, 2, 89-100 (1997)
11. G.Srinivasan, Quantitative Models in operations and supply chain management, PHI Learning, 269-273 (2010)
12. A. Radwan, M. Aarabi, Proc. International Conference on Industrial Engineering and Operations Management (2011)
13. J. D. Sterman, Business Dynamics, McGraw Hill, 137-156 (2000)
14. F. Chen, Z. Drezner, J. K. Ryan, D. Simchi-Levi, 46, 3, 436 – 443 (2000)
15. S. E. Fawcett, P. Osterhaus, G. M. Magnan, M. W. McCarter, Int. J. supply chain management, 12, 5, 358-368 (2007)
16. Zahra Lotfi, Muriati Mukhtar, Shahnorbanun Sahran, Ali Taei Zadeh, 4th Int. Conf. Electrical Engineering and Informatics, Procedia Technology, 11, 298-304 (2013)
17. Baris Gunduz, Yaman Barlas,, Ms Thesis (2003)

Implementation of total productive maintenance (TPM) to increase overall equipment efficiency of an hotel industry

Manjunatha B.[1*], *Srinivas T. R.*[1] *and Ramachandra C. G.*[2]

[1]*Department of Industrial and Production Engineering, Sri Jayachamarajendra College of Engineering Mysuru, India*
[2]*Department of Mechanical Engineering, Srinivas Institute of Technology, Mangaluru, India*

Abstract. Hotel industries are one of the fastest growing areas in India, which is attracting more number of jobs & tourists. This also results in direct impact on Indian economy. All Hotels comprises a higher element of total working cost and they are a good source of income. Hospitality is one of the most important components of hotel industry. To give a good hospitality in hotels to customers, we need to use different machines in different departments. For example if we take section of housekeeping it consists of different sub departments like Rooms & corridors, Toilets, Linen, Furniture and furnishings, Gardens, Public areas etc. In this connection a small implementation of TPM tool called focused improvement & planned maintenance, implemented in alternative usage of electrical systems (energy conservation). Result showed a very significant improvement in energy saving. By adopting Small changes show a significant improvement in the overall system. It also indicates the dire need of proper industrial tools in hotels. Total Productive Maintenance (TPM) is one of the pioneering approaches which can be achieved in above said things [1][2].

Keywords: Overall Equipment Efficiency (OEE), Total Productive Maintenance (TPM), Hotel Industries, Maintenance, Focused improvement, Planned Maintenance (PM).

1 Introduction

It is well known fact that in industrial situation huge losses/wastage occur in the manufacturing shop floor. It happens due to operators, maintenance personal, process, tooling problems and non-availability of components in time and many more. Indirect or hidden waste includes idle machines; idle manpower, machine break down, rejected parts etc. are all examples of waste [3]. Quality is very significant factor as they matter the company in terms of time, material and the hard earned reputation of the company. In recent trends companies are leaning towards Zero tolerance for waste, defects, break down & zero accidents in all departments. In order to overcome these problems, a innovative concept such as TPM has been successfully implemented in many industries across the globe to address the above said problems.

*Corresponding author: manjurmech@gmail.com

Total Productive Maintenance (TPM) is a maintenance tool which enables to maintain the plants and equipments [4][5]. The goal of the TPM program is a well defined technique which increases production, efficiency, at the same time, increasing employee morale & job satisfaction.

General classifications of Machines and Equipments in Hotels:

- Air conditioner machines
  - VRF Centralized AC
  - Split AC
  - Cassette AC
  - Fresh air supplier
- Kitchen
  - Exhaust
  - Fresh air supplier
  - Gasoline supplier
  - Cold room
- Water heater and laundry
  - Electrical boilers and solar water heaters
- House keeping
  - Floors scrubbing machines
  - Vacuum cleaner
- Maintenance
  - 125 D5 kVA Transformer
  - 325 kVA Transformer
  - 800 kVA Transformer

2 OEE (Overall Equipment Efficiency)

The main objective of Total Productive Maintenance is to enhance the Overall Equipment Effectiveness of plant equipment.

OEE consists of three basic indicators of equipment performance and reliability. The three main factors of OEE is explained below.

1. Availability or uptime (downtime: planned and unplanned, tool change, tool service, job change etc.)
2. Performance efficiency (actual vs. design capacity)
3. Rate of quality output (Defects and rework)

A - Availability of the machine:

Thus OEE=A*PE*Q

$$\text{Availability} = \frac{(\text{Planned production time} - \text{unscheduled downtime})}{\text{Planned production time}}$$

Production time = Planned production time – Downtime

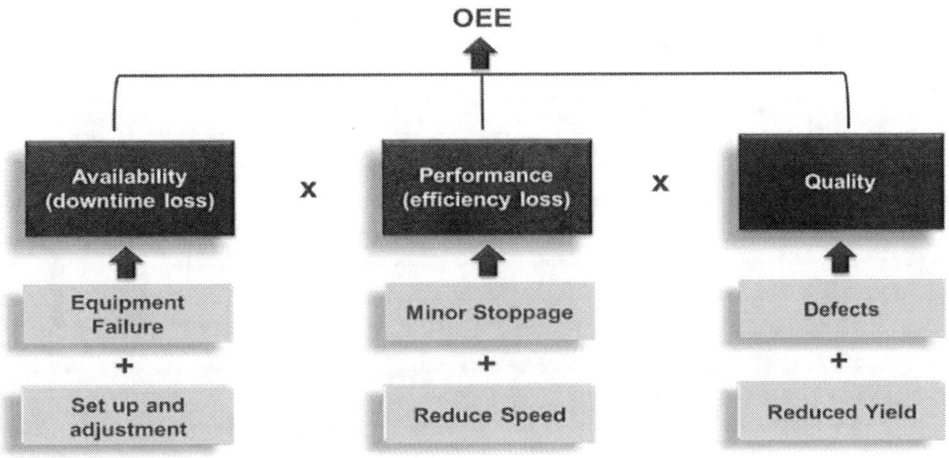

Fig. 1. Overall Equipment Efficiency methodology

3 Observation on electric energy consumption in a hotel

The electricity consumption details of 1 month collected from hotel management. In order to check the day to day variation in energy consumption, $\ddot{X}$ and R chart were plotted using Minitab 17 software. The below tables represents Energy consumptions recorded (table 2) for the month of OCTOBER- Before optimization of electricity units.

Table 2. Electrical Units Consumed/day

Date	Units consumed/day
01/10/2016	650
02/10/2016	732
03/10/2016	450
04/10/2016	385
05/10/2016	700
06/10/2016	714
07/10/2016	692
08/10/2016	701
09/10/2016	809
10/10/2016	980
11/10/2016	997
12/10/2016	1176
13/10/2016	1070
14/10/2016	962
15/10/2016	876
16/10/2016	456
17/10/2016	750
18/10/2016	369
19/10/2016	358
20/10/2016	655
21/10/2016	700
22/10/2016	458
23/10/2016	378
24/10/2016	678

25/10/2016	458
26/10/2016	687
27/10/2016	824
28/10/2016	925
29/10/2016	425
30/10/2016	714
31/10/2016	757
	Total units=21486

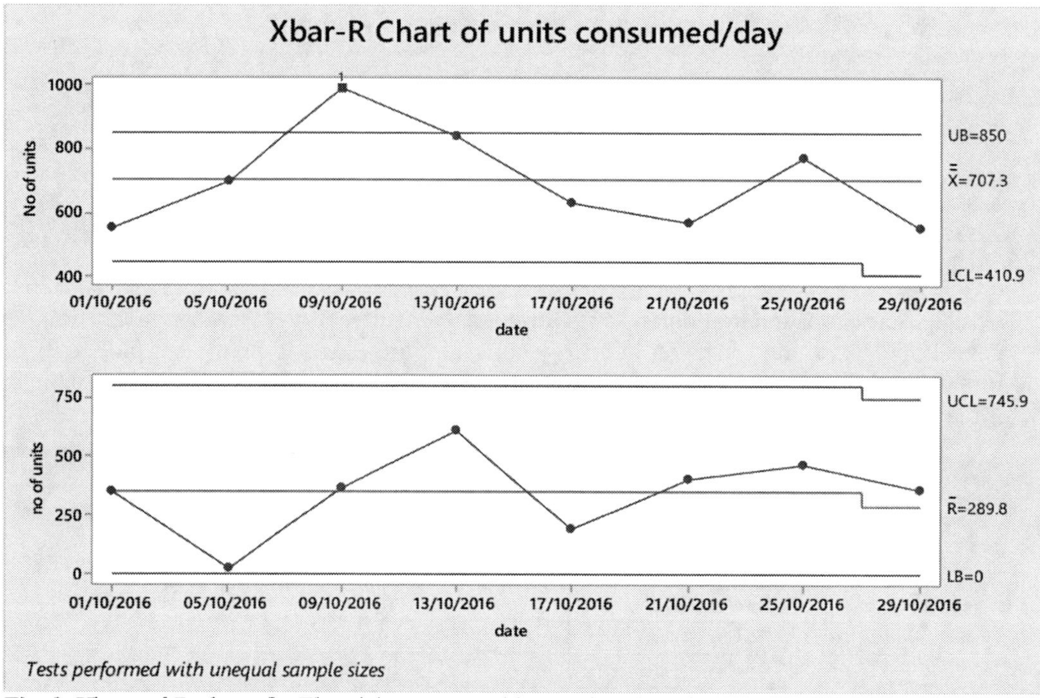

Fig. 2. Xbar and R-charts for Electricity consumed in October (in units)

The above graph represents Xbar-R Chart of electricity consumed during the month of October and it shows that consumption of electricity is not under control

4 Implementation of tpm tool called focused improvement & planned maintenance

In simple words Focused improvement can be defined as a method that identifies the systems problems and then modifies the whole system in order to find the most cost effective, time saving and least disruptive solutions in order to optimize the system[6, 7]. Planned maintenance is another tool where an object, item, or equipment are maintained in a planned manner to increase the efficiency by reducing the downtime.

By adopting above methodologies the following suggestions were given.

1. Replaced incandescent bulbs with LED lights.
2. Introduced Fluorescent Ceiling Lights
3. Replaced the fixtures
4. Installed lower wattage or more efficient lamps

Table 3 shows the differences between different bulbs to different parameters.

Table 3. LED Lights vs. Incandescent Bulbs vs. CFL bulbs

	LED BULBS	CFL BULBS	INCANDESCENT BULBS
Life Span (average)	50,000 hours Approximately, 12years at 10hr/day 20years at 6hr/day	1,200 hours Approximately, 4month at 10hr/day 6month at 6hr/day	8,000 hours Approximately, 2.2years at 10hr/day 3.5years at 6hr/day
Watts of electricity used (equivalent to 60 watt bulb).	6 – 8 watts	60 watts	13 – 15 watts
Kilo-watts of Electricity used	32 KWh/yr. Or 32 Units per year	320 KWh/yr. Or 320 Units per year	55 KWh/yr. Or 55 Units per year.
Annual Operating Cost	Rs. 249.6/yr	Rs.2496/yr	Rs.429/yr

LEDs use less power (watts) per unit of light generated (lumens). LED has the Cheapest Operating Cost and Highest Life Span.

Calculations: at Avg. Electricity cost Rs7.80 per unit, at usage of 10hr per day per year

Frequent maintenance and check-ups of the devices in Hotel were carried out as part of planned maintenance. PM avoids the break-down and maintains the efficiency, productivity, energy consumption of the process. After implementation of the above said suggestions to optimize the energy savings, the data were recorded as showed in Table 4.

November-After implementation of the above suggestions

Table 4. Units consumed/day after implementation

Date	Units consumed/day
01/11/2016	638
02/11/2016	720
03/11/2016	440
04/11/2016	360
05/11/2016	687
06/11/2016	704
07/11/2016	685
08/11/2016	689
09/11/2016	796
10/11/2016	969
11/11/2016	981
12/11/2016	964
13/11/2016	990
14/11/2016	941
15/11/2016	861
16/11/2016	439

17/11/2016	728
18/11/2016	355
19/11/2016	347
20/11/2016	640
21/11/2016	687
22/11/2016	441
23/11/2016	365
24/11/2016	664
25/11/2016	431
26/11/2016	675
27/11/2016	815
28/11/2016	913
29/11/2016	421
30/11/2016	704
01/12/2016	700
	Total units= 20750

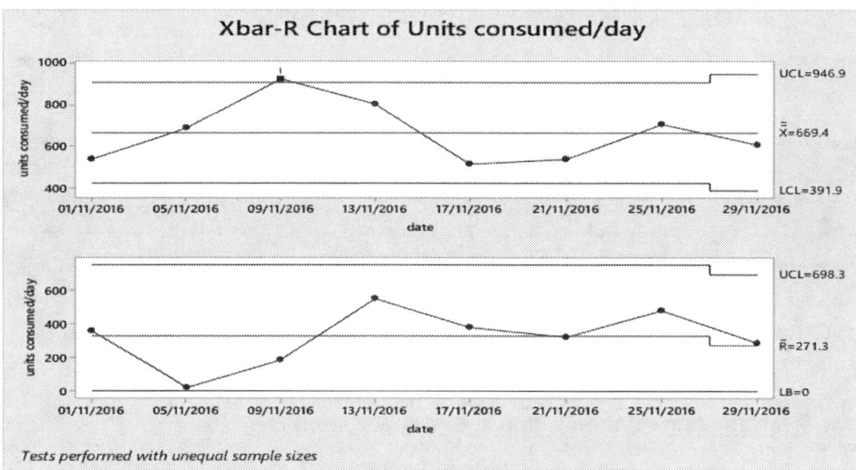

Fig. 3. Xbar and R-charts for Electricity consumed in November (in units)

5 Results and discussions

Figure 3 shows that consumption of electrical units in November are under control. After implementing the suggestions that are stated above, there was significant reduction in power consumption.

Cost reduction in terms of units of electricity saved is calculated:

Total No of units consumed in the month of October 2106 before TPM implementation=21486(a)

Total No of units consumed in the month of November 2106 before TPM implementation= 20750(b)

Reduction in power consumption in units=736 units

Formula: Reduction of power consumption in % $=(\frac{a-b}{a})*100$

$=(\frac{21486-20750}{21486})$

=3.425%

Price/unit of electricity= Rs 7.80

No. of units saved = 736
Total cost reduction/month = Rs 5740
Total cost reduction/year = Rs 68889

The data shows only the one month energy or power consumption record, just by implementing small changes in the system. This small changes implementation has showed a significant improvement and increased profit in the system. The initial investment looks like a burden to the hotel management, but the outcome showed the profit to the Hotel industries. Many areas or departments are needed to be checked at the same time optimization can be achieved by TPM tools.

6 Conclusions

It has been observed that cost reduction and increasing energy efficiency were the two major necessities in the hotel operations. Just by replacing all incandescent bulbs with LED bulbs and using energy star qualified appliances can save up to 3.4% of total cost electricity per month. Result also showed the significant saving in terms of cost. Nowadays, for cost reduction and eliminating wastes it is very essential to go for industrial maintenance tools, techniques and alternative source of energy. It is also got to know that many areas are untouched for optimization in hotel departments. There is lot of need and scope to use the engineering tools to eliminate the issues facing in hotel industries.

References

1. Chan, K.T., Lee, R.H.K., & Burnett, J. (2001). "Maintenance performance: a case study of hospitality engineering systems", Facilities, 19 (13/14) 494 – 504
2. Chan, K. T., Lee, R.H.K., Burnett, J. (2003). “Maintenance Practices and Energy Performance of Hotel Buildings”, Strategic Planning for Energy and the Environment, 23(1) 6-28.
3. Espino-Rodr´ıguez, T., Padr´on-Robaina, V. (2005). “A resource-based view of outsourcing and its implications for organizational performance in the hotel sector” Tourism Management 26 (5), 707–721
4. Forsgren, S., Franchetti, C. (2004). “The marketing role of unique concepts for hotels in Sweden” Tourism and Hospitality Management Master Thesis No. 2004: 53, Goteborg University
5. Guilding, C. (2003). “Hotel owner/operator structures: implications for capital budgeting process”, Management Accounting Research, 14(3) 179-199.
6. Hassanien, A., &Losekoot, E. (2002). "The application of facilities management expertise to the hotel renovation process", Facilities, 20 (7/8), 230 – 238
7. Jordhus-Lier, D., Bergene, A. C., Knutsen, H. M., &Underthun, A. (April 2012). “Hotel workplaces in Oslo and Akershus” Norwegian Institute for Urban and Regional Research, working paper 2012: 106

A Study on Lean Manufacturing Practice Carried Out for Axle Housing Plant

Rishi J. P.[1*], *Abhishek*[1], *Srinivas T. R.*[2] *and Ramachandra C. G.*[3]

[1]*Department of Mechanical Engineering, Vidya Vardhaka College of Engineering, Mysuru, India.*
[2]*Department of Industrial and Production Engineering, Sri Jayachamarajendra College of Engineering, Mysuru, India.*
[3]*Department of Mechanical Engineering, Srinivasa Institute of Technology, Mangaluru, India.*

Abstract. Lean Manufacturing is a way to deal with the elimination of waste through persistent change within the system. The assembling units in India should likewise hope to take its preferences. The Industries are energetically following the Lean assembling strategies to reduce waste and cost. The present work is the study of a manufacturing process for axle housing and to decide how Lean manufacturing process carried out for axle housing plant to improve productivity. The new shot peening machine was installed in a heat treatment plant of a housing line in an automobile rear axle assembly plant and material handling time from storage to this new machine was more and this increased lead time subsequently. To reduce this Kaizens technique was employed and a direct gravity conveyor was chosen and installed between two work stations and the part was directly fed to the new machine installed and this reduces the material handling time considerably. A cost analysis was done to realize the benefits of implementing this lean technique. Therefore reducing lead time and increasing productivity is the Prime benefits of lean manufacturing practices like lean assembly.

Keywords: Lean Assembly, Axle, conveyor, Kaizen.

1 Introduction

The axle housing contains different segments that take into consideration guiding, driving or load bearing. The real motivations behind the utilization of lean assembling are to expand efficiency, enhance product quality and reduce assembling process duration, diminish stock, lessen the lead time and dispense with assembling waste. To accomplish these, the lean assembling logic utilizes a few ideas, for example, one-piece stream, kaizen, cellular fabricating, synchronous assembling, stock administration, poka-burden, institutionalized work, work put association and scrap decrease to diminish producing waste[1]

* Corresponding author: rishijp@vvce.ac.in

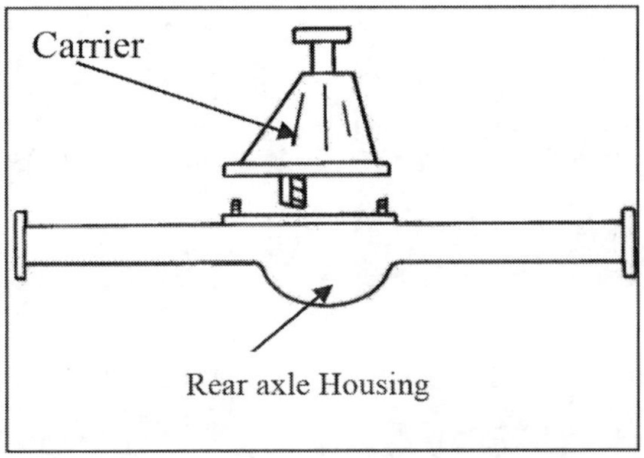

Fig. 1. Rear axle assembly

Sundareshan et al. [1] They carried out a literature review on Lean implementation in industries. The usually utilized lean apparatuses in the different association, most basic obstructions and advantages have been distinguished and recorded in this paper. Harsha Lingareddy et al. [2] They provided finding regarding implementations of 5S techniques in any industry. This system helps in limiting the season of producing and furthermore builds the territory of work put. Vijayakumar et al. [3] Have carried out a case study on lean manufacturing techniques in garment industries. The fundamental topic of his paper has recognizable pieces of proof of huge issues identified with lean waste happens in 5s of clothing have been measured utilizing particular lean devices and strategies. In this association, lean assembling idea is executed as another idea of assembling in the articles of clothing of the examined pieces of the clothing industry. Norani Nordin et al. [4] The reason for this paper is to investigate the degree of LM usage in Malaysian car producing firms. The outcomes demonstrate that a large portion of the respondent firms has actualized lean assembling framework up to a certain degree.

2 Rear Axle Housing Assembly Process

A detailed study of rear axle housing assembly layout was made to analyze various shortcomings of this layout. The 20 manufacturing processes carried out in housing plant are as follows.

Table 1. Overview of variety of Assembly process in a rear axle housing plant

• Shot Blasting	• Friction Welding	• End Facing
• Tack Welding	• Flange Tack Weld	• Key Way Milling
• Seam Welding	• Cooling	• Flange Drilling
• Repair welding and surfacing	• Straightening	• Drain and breather hole drilling
• Straddle Facing	• Drain and breather hole drilling	• Inside Painting
• Ring and Cover Welding	• Carrier and Cover Boring	• Drying
• Swaging	• Spindle Threading	---------

Fig. 2. Shot peening

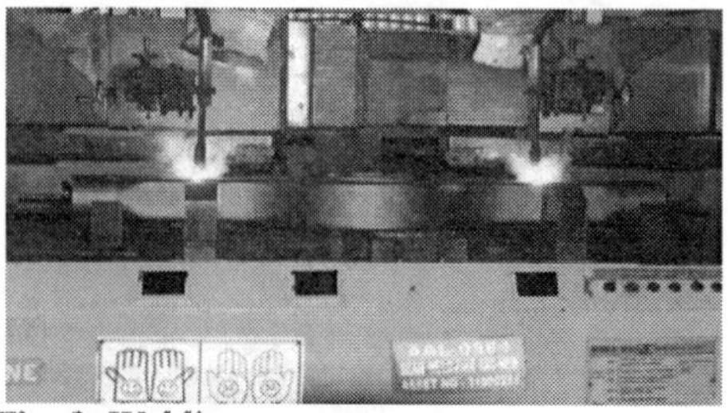

Fig. 3. Welding process

Fig. 4. Friction welding

Fig. 5. Straddle facing

Fig. 6. Flange Drilling

Shot peening here is done to enhance surface finish of carriage assembly Fig. 2. Arc welding is done to join carrier to rear assembly Fig. 3. Friction welding is done to join the shafts Fig. 4.Straddle facing is done to remove any obligate present on the top plane of the surface Fig. 5. Drilling is done on the flange mounting purpose Fig. 6. This set of operations complete the entire assembly process.

3 Analysis of Existing Process in Axle Housing

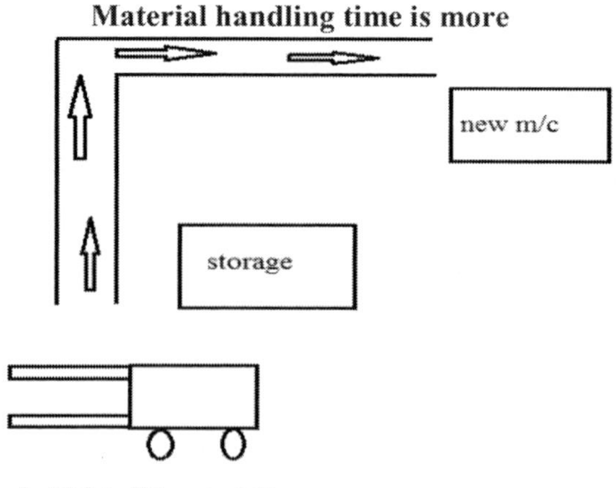

Fig. 7. Layout before (using forklift to lift material).

A New shot peening machine was installed in a housing line of a Heat treatment plant in an Automobile Rear assembly plant at an offset distance. The material transfer for the shot peening machine is done through the forklift and required two workers one to load the material one to operate the fork lift. This Process was time-consuming and labor intensive. Thereby increasing labor cost because two labors are required and due to bottlenecking Lead time of production

4 Kaizen -Solution

Kaizen's is one of the most important techniques used to manage continuous improvement through observation and small modification within the system. Kaizen technique can be used to reduce lead time. In this situation, to eliminate the fork lift and replace it by a simple gravity conveyor which is a conduit kept at an inclination between storage bin and the new machine kept inline so that parts are directly fed to the new machine without waiting for transportation thereby reducing the material handling time and subsequently Lead time. By choosing a simple gravity conveyor both the performances can be achieved.

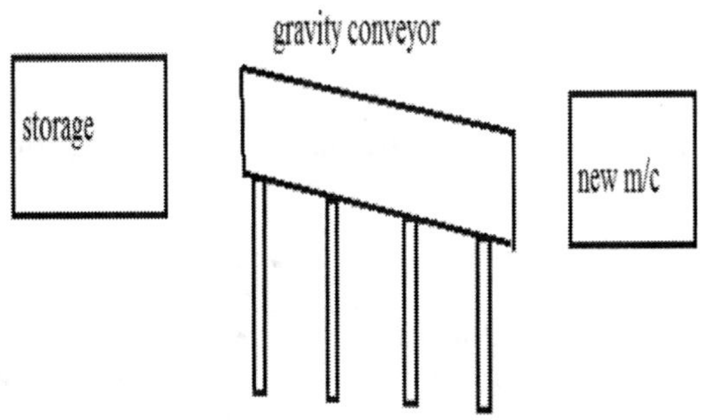

Fig. 8. Layout after (using conveyor)

5 Methodology

In order to realize this improvement following steps are considered while choosing a conveyor.

5.1 Identifying the Problem

The New heat treatment machine was installed in heat treatment section and the material supply for this new machine is done through Forklift. From storage to a new machine it has to cover a lot of certain distance. Hence the material handling time is more [7].

5.2 Workspace Observation

The Workspace between storage bin to the new machine is having a very small Distance. Hence this place can utilize to transport material to new heat treatment machine. (6-7 mts).

5.3 Possible mechanism

By using conveyor, the material can be transported from storage to new heat treatment machine. It is the suits the system for the given small area.

5.4 Given Task

The given task is to make a conveyor without using any external sources.

5.5 Work Volume

The dimensions are measured and the area is calculated and volume is determined.

5.6 Drawing

After taking dimensions, mka detail plan is drawn using AutoCAD drafting software. The plan is a 2D drawing of simple gravity conveyor and its location.

5.7 Implementation

A simple gravity type conveyor is chosen and placed in between the storage bin and new machine as per the plan.

5.8 Cost analysis

A simple cost analysis to determine the benefits of implementing the lean technique.

6 Results and Discussions

The material handling time is more while supplying a material from storage to new machine with the help of the battery operated forklift. There are two workers involved in transferring material i.e. one is for the drive the forklift and another one in storage. By providing gravity conveyor from storage to new machine, manpower is eliminated and material handling time was reduced [8]. Before it took 5 minutes to complete the task.After implementation of Kaizen, the entire task was reduced to 1 minute. So by implementing Kaizen a total of 4 minutes can be saved to accomplish the task. Thereby optimizing labor and enhancing productivity and reducing lead time.

Table 2. Savings in time

Items	Before	After	Difference	percent Improvement
Set up time	5 min	1min	4 min	80%

6.1 Cost Analysis

A simple cost analysis was carried out to understand the benefits of lean assembly as follows. Cost per minute of a line is calculated only for direct labors. Direct labors like operators, helpers, line supervisor and quality Inspectors.The two types of Cost per Minute of an assembly line are Estimated cost per minute of a line and Actual cost per minute of a line.

6.1.1 Estimated cost per minute of a Assembly line: (E) – Existing cost

This cost can be determined by considering total salary, Availablity and line efficiency.

$$E = \frac{\textbf{Total salary of a line for one day}}{\textbf{(Total minutes available x Line efficiency in \%)}}$$

To calculate total salary of a line includes all operators, helpers working in a line. We also include Quality Inspectors and line supervisors and feeder if they are not considered under overhead. So, we have to find following three parameters to calculate estimated cost per minute of a line [5]. Where,

Total minutes available = Nos. of total labors x Daily working hours *60
= 15 × 08 × 60
=7200 mins (for 10 machines)

Total estimated salary of a typical line is INR 8000.00 and total manpower 15 of a 10 machines line and line efficiency is calculated as follows :
(A)Nos of operators = 15 ; (B)Working hrs =08;
(C)Line output production = 25; (D)Standard Allowable Min= 20;
(E) Total minutes produced by the line= C X D=25X20=500 mins
(F) Total minutes attended by all operators =A X B X 60= 15X08X60=7200

$$line\ efficency\ (\%) = \frac{\textbf{Total minutes produced by the line x 100}}{\textbf{total minutes attended}}$$

= 500 X 100 /7200 =6.9 %

Cost per minute of the line = (8000 / (15×480×0.69) = Rs 1.61

6.1.2 Actual cost per minute of a line : (A) –Post Kaisen Implementation

= (Actual salary for the day/ (Total garment produced × Standard Allowed Minutes)
Total actual salary of a typical line is INR 8000.00, the line made 25 pieces and Standard Allowed Minutes is 20 minutes through time study.
Therefore,
Cost per minute of the line = (8000 / (400 × 20)) = Rs. 1

Table 3: Savings in cost

Items	Before	After	Difference	percent Improvement
cost per minute of a line	Rs 1.61	Rs	Rs 0.61	37.8%

But from the implementation, point of view one battery operated forklift and its operating driver was eliminated which can be used in other section for further work [6].

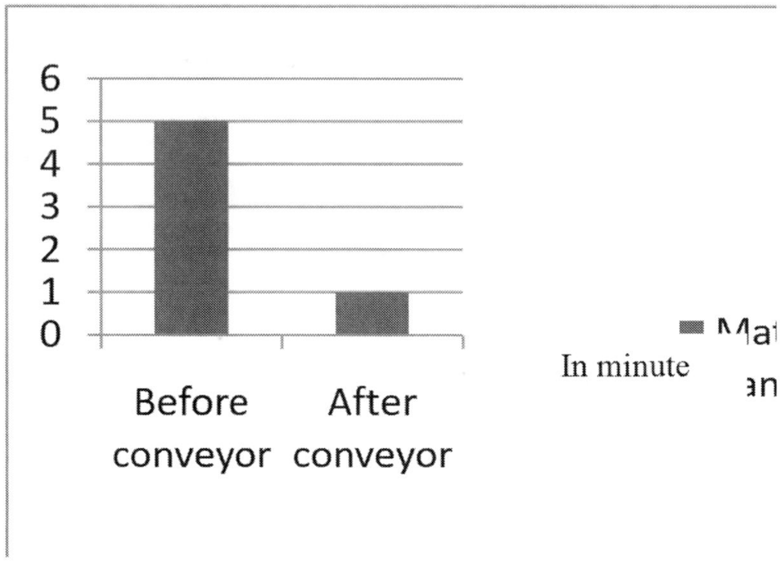

Fig. 9. Effect of using conveyor on assembly time

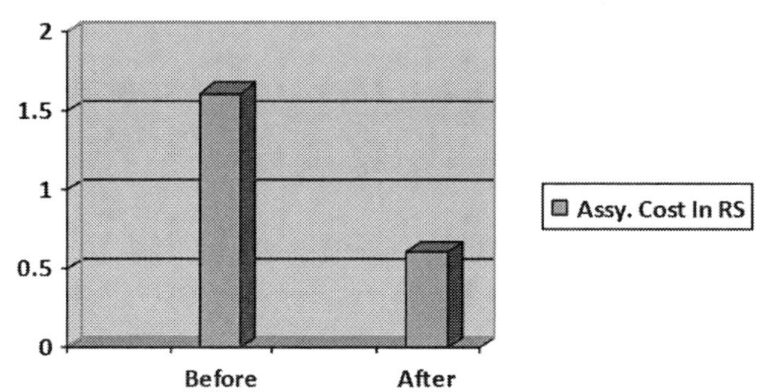

Fig. 10. Effect of using conveyor on assembly cost

7 Conclusions

This paper shows the major difficulties organizations confront in actualizing lean devices and procedures, maybe absence of far reaching and reasonable lean learning related to problems inside the organizations by the chiefs, absence of acknowledgment of lean culture in entire of the association and planning cause the failure inside the usage. Also, a few Manufacturers endeavor to upgrade the implementation by a portion of the lean devices and for the most part attempt to just execute the "ceaseless change" [9]._For the present study by making gravity conveyor from storage to new machine, manpower is eliminated and material handling time was reduced thereby the percentage savings in cost was 37% and saving in time was 80% both being substantial savings in resources.

References

1. Sundareshan S D, "A Literature Review on Lean Implementations – A comprehensive summary", International Journal of Engineering Research and Applications, Vol. 5, Issue 11, (Part - 4) November 2015, pp.73-81.
2. Harsha Lingareddy, "5S as a tool and strategy for improvising the workplace", International Journal of Advanced Engineering Technology, Vol. IV/ Issue II/April-June, 2013/28-30.
3. G. Vijayakumar, "Impacts of lean tools and techniques for improving manufacturing performance in garment industry", International Journal of Advanced Engineering Technology, Vol. VII/Issue II/April-June, 2016/251-260.
4. Norani Nordin, "A survey on lean manufacturing implementation in Malysian automotive industry", International Journal of Innovation, Management and Technology, Vol. 1, No. 4, October 2010.
5. Bhasin, S. and Burcher, P. (2006). Lean viewed as a philosophy. Journal of Manufacturing Technology Management.
6. Dankbaar, B., 1997, "Lean production: denial, confirmation or extension of socio-technical systems design?," Human Relations, 50(5).
7. Hayes, R.H. and Pisano, G.P., 1994. "Beyond world class: the new manufacturing strategy", Harvard Business Review, January-February.
8. Shah, R. and Ward, P.T., 2003. "Lean manufacturing: context, practice bundles, and performance",Journal of Operations Management.
9. Wilson, L. (2010) How To Implement Lean Manufacturing. New York: McGraw-HillWomack, J., Jones, D.T. and Roos, D., 1990,"The machine that changed the world," Rawson Associates, NY.

Evaluation of Manufacturing Process Performance by CONWIP Hybridization of Pull Controlled Production Systems

O. Srikanth[1*], *A. V. Sita Rama Raju*[2] *and B. V. Ramana Murty*[3]

[1]*Dhanekula Institute of Engineering and Technology, Ganguru, Vijayawada, A.P, INDIA*
[2]*Jawaharlal Nehru Technological University Hyderabad, Kukatpally, Hyderabad, T.S, INDIA.*
[3]*Gayatri Vidya Parishad Engineering College, Madhurawada, Visakhapatnam, A.P, INDIA.*

Abstract. The main objective of this paper is pioneering an innovative tactic for the synchronization of multi-stage, multi-line, production system. This tactic is mainly depends on the optimization policy, by means of distinct event simulation process for modeling, analysis and distinction of the execution of two alternatives of Kanban control mechanism namely SEKCS (Simultaneous Extended Kanban Control System) and IEKCS (Independent Extended Kanban Control System). At this juncture the authors putting forward the two variants of Extended Kanban control system with the hybridization of CONWIP control policy to incite HSEKCS (Hybrid Simultaneous Extended Kanban Control System) and HIEKCS (Hybrid Independent Extended Kanban Control System) to make use of pooled benefits of a representative production situation in addition to improve the outcome. Therefore in this study the comparison in between different systems of proposed HEKCS specifically are HSEKCS and HIEKCS compared with the Extended Kanban Control Systems variants SEKCS and IEKCS. Simulation studies were conducted for all the five control policies considered and modeled on a multi-line, multi-stage assembly production control system. The relative performance parameters like Throughput or Production rate, Average Waiting Time and Average Work-in-Process, were assessed by means of exponentially varying demands.

Keywords: CONWIP, Kanban, EKCS, SEKCS, IEKCS, HEKCS HSEKCS, HIEKCS.

1 Introduction

For the last few decades, global market configuration had altered largely due to hasty advancement of technology. Therefore, manufacturing industries are subjected to global antagonism, Which causes them to maintain up with new perceptions and even to practically fit into their every day manufacturing schedule persistently struggle to their insistent benefit meticulously in the hardware and automobile manufacturing sector. These

*Corresponding author: muralisrikanth11@ gmail.com

production industries reacting to meet customer ordering through internet and e-commerce there by changing the make to order situation and re organizable manufacturing resources.

Just in time (JIT) manufacturing depends on real demand prompting the release of the job into the arrangement, and pulling of the job throughout the production system to fulfill the demand order. In such incidents, Kanban control practice has been a kind of revolution. It endeavours at sinking Work in Process (WIP) levels and manufacturing lead times in the industry. Though, the restricted applicability of Kanban control system has aggravated researchers to discover options to this strategy. Consequently new pull control mechanisms have been building up. In pull control systems optimization of manufacturing control is accomplished by functionally comprehending quite a lot of manufacturing activities into various stages after that synchronizing the discharge of parts into every stage, by the entrance of customer demands for the final manufactured goods.

W. Zhang et al. [1] (2001) suggested a mathematical model in a CONWIP production flow system to determine the best possible production quantity and suitable progression, through a precise number of Kanbans to estimate the effectiveness of a multi class generalized Kanban control system. O.S.Wormgoor [2] (2001) build up an analytical practice MVA (Mean value approximation analysis) method to find out the unidentified external arriving rate and performance parameters like WIP and lead times. DRC (Dual Resource Constrained) system conferred by Xu et al. [3] (2010) in that the during scheduling and planning additional technical tackles must be evaluated like worker flexibility, the transfer costs, workers assignment, the transport costs works dispatching. G. Liberopoulos et al. [4] (2000) evolved a unified framework in multi stage manufacturing for a pull production control production system and various features of different Kanban control systems.

Afterwards Y. Dallery et al. [5] (2002) investigated base stock versus WIP limit and make to stock production–inventory systems for the multi stage production system. Little [6] (1961) proposed Little's Law which is fundamentally queuing theory theorem having theoretical and practical importance in inventory control systems. M. L. Spearman [7] (1990) intended the concept of process CONWIP system and discussed the realistic benefits of CONWIP over pull and push systems and conversed numerous theoretical cases of CONWIP system. S H Cheraghi et al. [8] (2001) offered a reasonable effect over the ascendancy of different manufacturing control policies. Diverse input parameters and the type of factory setup influences the superiority of a manufacturing system. By implementing modeling and simulation analysis of different existing arrangements in a manufacturing system.

Yves Dallery et al. [9] (2001) discussed the formation of EKCS, which is a combination of the Base Stock and Kanban control systems and the benefits of EKCS above GKCS. The dynamics of the EKCS are explained and compared with the Kanban and Base Stock control systems. Chu. C et al. [10] (1992) presented a paper to examine how broadly and satisfactorily simulation has been used in studying Just in Time manufacturing. C. Chaouiya et al. [11] (2000) compared the two variants of EKCS namely, SEKCS) Simultaneous Extended Kanban Control System) and IEKCS (Independent Extended Kanban Control System) at what time a part is released into the assembly manufacturing section, releases its Kanban signal, consequently endorsing the production of the next part.

B. J. Berkeley [12] (1992) did an excellent survey concerning Kanban Control System and collected vast amount of literature in the relevant field. The performance of KCS deliberated by J. A. Baynatt et al. [13, 14] (2001 and 2002) in the viewpoint of the multi product manufacturing system. In addition to KCS, he has studied EKCS and GKCS. While enlarging these pull manufacturing control techniques to multiproduct system, differentiated the two diverse subjects, whether Kanbans are dedicated to a single part type or shared among various part types. O. Srikanth et al. [15] (2013) studied the performance evolution on various process parameters in hybrid Kanban control system with the conventional CONWIP system by mathematical modeling and simulation analysis.

The Hybrid mechanisms projected by the authors in which CONWIP is combined with SEKCS and IEKCS mechanisms. That is HSEKCS (Hybrid Simultaneous Extended Kanban Control System) and HIEKCS (Hybrid Independent Extended Kanban Control System) to exploit the collective benefits in addition to study their consequences in a typical production environment. By using the discrete event simulation software Process Model, the simulation studies were executed to assess the considered process performance measures like through put, average Work-in-Process and average waiting times for all the pull production control mechanisms.

2 Problem Definition

A manufacturing assembly system considered as shown in Fig. 1, having three manufacturing flow lines in which three different sub assemblies are produced for the final assembled part. Every flow line consists of three manufacturing facilities, where each line i = 1, 2, 3 and each stage j = 1, 2, 3. One cell is formed by three manufacturing facilities. At the final assembly station these three manufacturing flow lines converges.

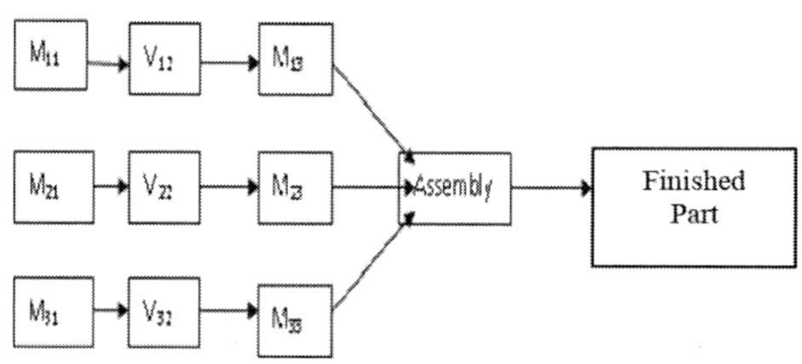

Fig. 1. A multi-line, multi-stage assembly production control system.

There is one production Kanban card for authorizing the production in every production flow line. For the both of network diagrams SEKCS and IEKCS separately the manufacturing assembly system is configured and combined also with CONWIP network. Simulation process conducted by discrete event software Process Model. Considering the processing mean times of 25, 35, 45, 55, 65, 75, 85 and 95 minutes with exponential distribution and also with the degree of imbalance of values 0.1, 0.15, 0.2, 0.25 and 0.3. With 30 duplications the complete assembly manufacturing line is simulated for 1,72,800 minutes (i.e., for 4 months with 3 shifts/day at 8hrs/shift).

For all these five pull production control policies considered explicitly CONWIP, SEKCS, IEKCS, HSEKCS and HIEKCS simulation analysis done for the multi-stage, multi-line assembly manufacturing system by using the Process Model software. The performance parameters related to Production rate or Throughput, Average Waiting Time and Average WIP were computed and comparatively assessed with each other.

3 Assumptions

- The queue lengths capacity is assumed to be of infinite length.
- Raw parts of suitably in great number are accessible that the system at any time should not descend with its deficiency.
- Between the production stages, negligible transpiration time is considered.

- Stochastic technique is presupposed for the inter arrival time of a product demand.
- For the arrival of demands and service times at the manufacturing facilities pursue exponential distribution.
- There are two inventory points considered at every production stage, at the beginning and at the end of the stage.

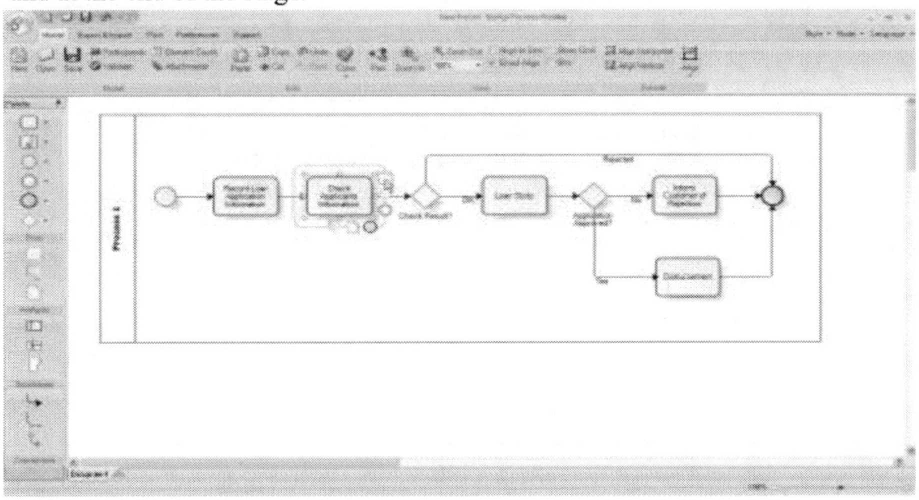

Fig. 2. Process Model Software modeling window

In the Fig. 2 discrete event simulation software Process Model software window is shown for configuring the assembly manufacturing network and for the simulation of assembly system considered.

4 Control Policies

4.1 CONWIP (Constant Work-in-Process) Control System

Constant Work-in-Process (CONWIP) is a generalized structure of Kanban control system shown in the Fig. 3. In this, cards or signals are released into the system like Kanban control system. Here in the system, the cards pass through a path in the complete production line. At the commencement of each production line a card is fastened to a standard bin of parts. Whilst the bin is employed at the ending of the production line, the card is detached and it sends back to the commencement of the line there it will wait in a cards queue and finally to be fastened to another parts container. Where ever the WIP levels can't be acknowledged earlier in a manufacturing line it is very easy to synchronize the production with the constant Work-in-Process than one.

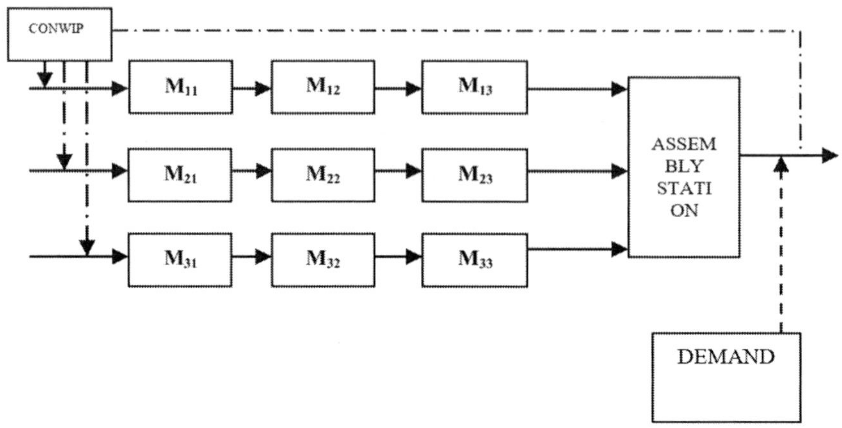

Fig. 3. CONWIP mechanism Schematic Diagram

4.2 EKCS (Extended Kanban Control System)

Whenever a demand arrives from the customer into the system it will be emphasized to all of the stages of the network in the EKCS philosophy. In this way from up stage to down stage the part is released if the production Kanban related with the corresponding stage is accessible. SEKCS and IEKCS are the two variants of Extendended Kanban Control System. The main differences in between SEKCS and IEKCS mechanism are the way the Kanban cards are transported to the production cells. In the IEKCS mechanism Kanbans are independently transported of each other while in the SEKCS mechanism all the Kanbans are transported simultaneously.

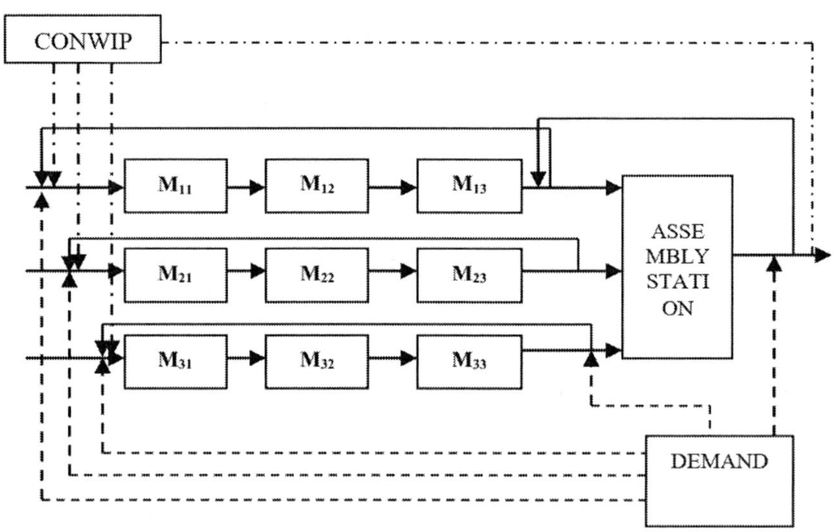

Fig. 4. HIEKCS mechanism schematic diagram.

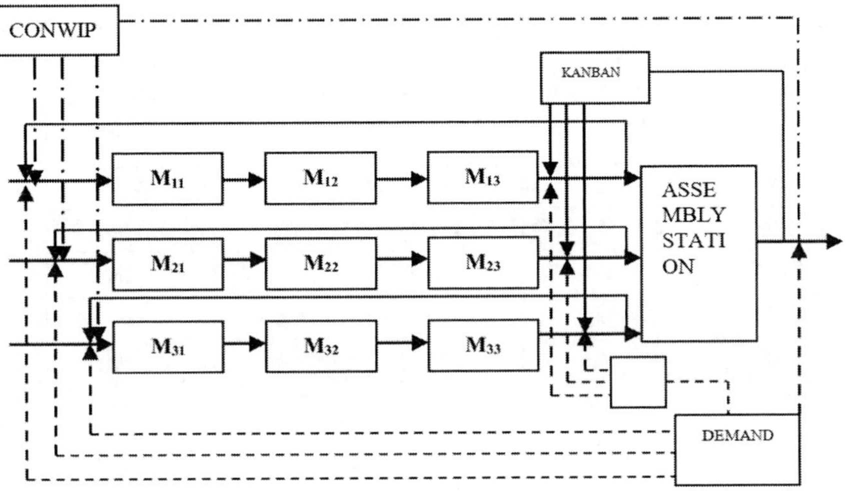

Fig. 5. HSEKCS mechanism schematic diagram

4.3 HEKCS (Hybrid Extended Kanban Control System)

Hybrid of EKCS with CONWIP gives HEKCS (Hybrid Extended Kanban Control System) pull control policy. Thus forms two variants namely HSEKCS and HIEKCS. These mechanisms in the Fig. 4 and Fig. 5 shows may have the collective benefits of EKCS and CONWIP policies.

5 Simulation Results of Various Control Policies

The Simulation results of the different policies considered are presented in the Tables 1, 2 and 3 and the respective graphs 6 to 11 plotted and presented for a mean time of 35minutes. Process Model is a discrete event simulation software which uses basic flow charting techniques when an activity time, action logic, or routing decision are considered.

Simulation studies were accomplished to evaluate the significance of EKCS, HEKCS and CONWIP in multi-line, multi-stage manufacturing system. The significance of production rate, average WIP and average waiting times for the processes CONWIP, EKCS, SEKCS, IEKCS, HSEKCS and HIEKCS are revealed in the individual figures. All the performance parameters are effected by change in the demand rate.

Out of all of the models considered, once the demand rate is rises the throughput or production rate and also increases progressively whereas average waiting times and average Work-in-Process decline when the demand rate is rises.

Production rate increases slowly and when the demand rate increases and average WIP and average waiting times decline for all the five pull control manufacturing mechanisms. It is observed that for both the variants of Hybrid Extended Kanban Control Systems even the average WIP and the average waiting time of HSKCS is nearly 50% of HIEKCS for almost the same the production rates.

Table 1. Relative Production Rate performance assessment of the Pull production control systems.

Exponential Demand	CONWIP (in units)	SEKCS (in units)	HSEKCS (in units)	IEKCS (in units)	HIEKCS (in units)
E(95)	480	388	330	360	322
E(85)	521	431	367	409	349
E(75)	566	481	395	438	377
E(65)	594	508	450	488	412
E(55)	590	550	486	540	490
E(45)	603	590	540	599	519
E(35)	594	596	583	636	572
E(25)	608	552	601	674	600

Table 2. Relative Average Waiting Time performance assessment of the Pull production control systems.

Exponential Demand	CONWIP (in min)	SEKCS (in min)	HSEKCS (in min)	IEKCS (in min)	HIEKCS (in min)
E(95)	334.1	675.6	175.8	526.98	308
E(85)	331.1	616.6	157.8	537.9	344
E(75)	328.4	552.1	142.1	496.2	344
E(65)	297.8	512.8	115.4	425.9	282
E(55)	283.9	511.6	111.3	376.18	254
E(45)	278	493.1	96.6	324.6	240
E(35)	279.2	517.2	93.1	316.19	240
E(25)	286.2	528.2	92.5	326.18	245

Table 3. Relative Average WIP performance assessment of the Pull production control systems.

Exponential Demand	CONWIP (in min)	SEKCS (in min)	HSEKCS (in min)	IEKCS (in min)	HIEKCS (in min)
E(95)	12.3	14.28	5.62	16.02	11.55
E(85)	12.45	14.57	5.88	16.02	11.74
E(75)	12.27	14.48	5.94	16.4	11.91
E(65)	12.72	14.29	5.96	16.54	12
E(55)	12.52	14.59	6	16.61	12.18
E(45)	12.63	15.3	6.35	16.83	12.49
E(35)	12.79	17.7	6.31	17.1	12.61
E(25)	12.9	18.16	6.42	16.8	12.79

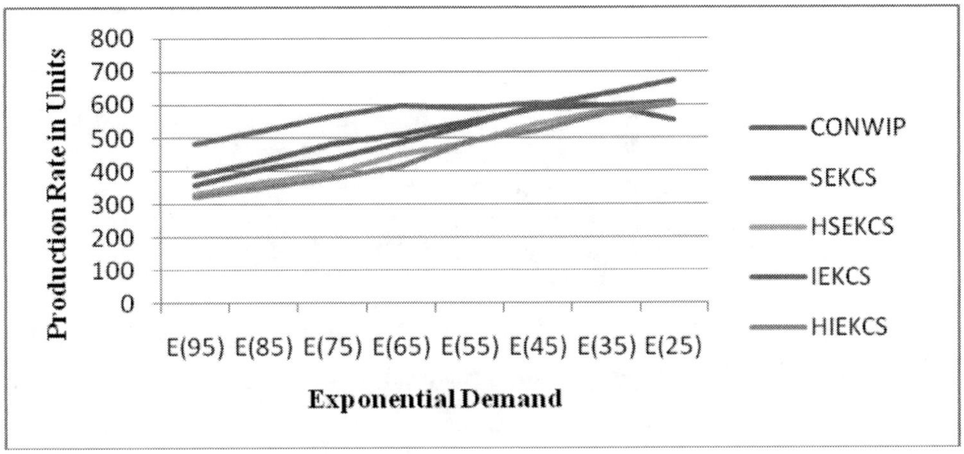

Fig. 6. Relative Production Rate performance assessment of the Pull production control systems.

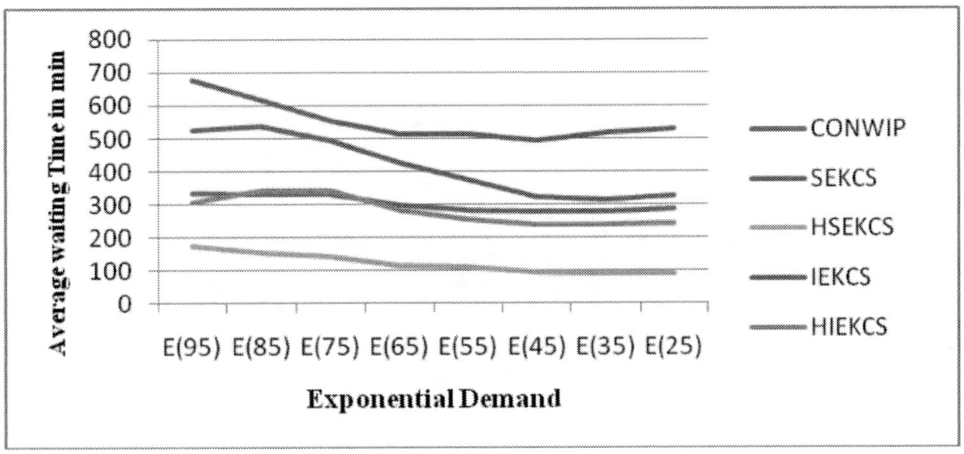

Fig. 7. Relative Average Waiting Time performance assessment of the Pull production control systems.

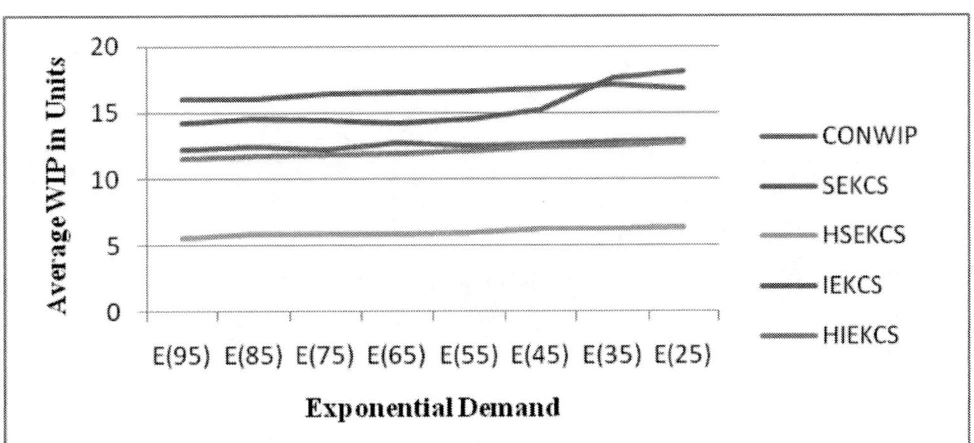

Fig. 8. Relative Average WIP performance assessment of the Pull production control systems.

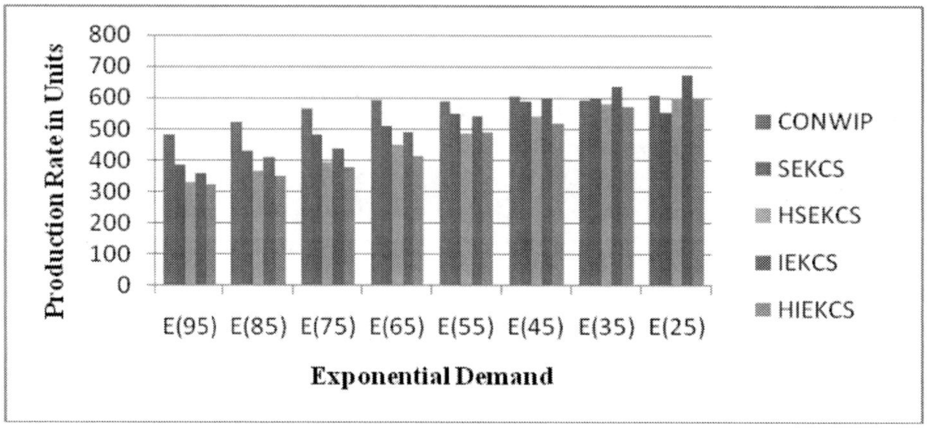

Fig. 9. Relative Production Rate performance assessment of the Pull production control systems.

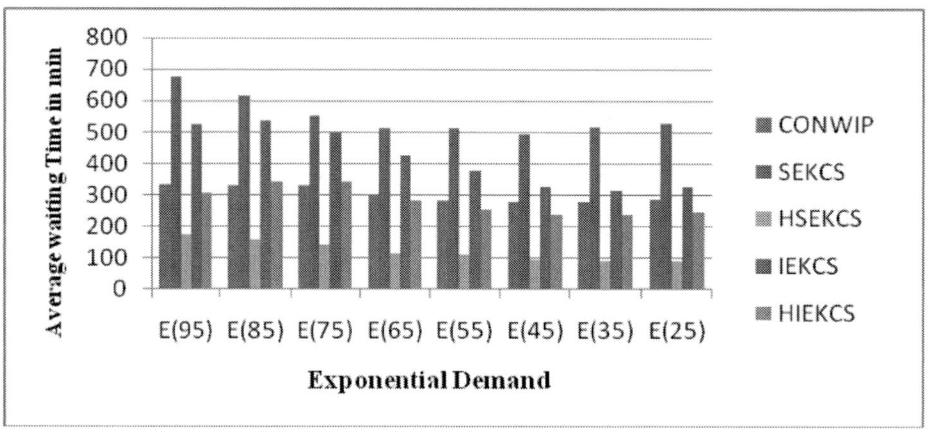

Fig. 10. Relative Average Waiting Time performance assessment of the Pull production control systems.

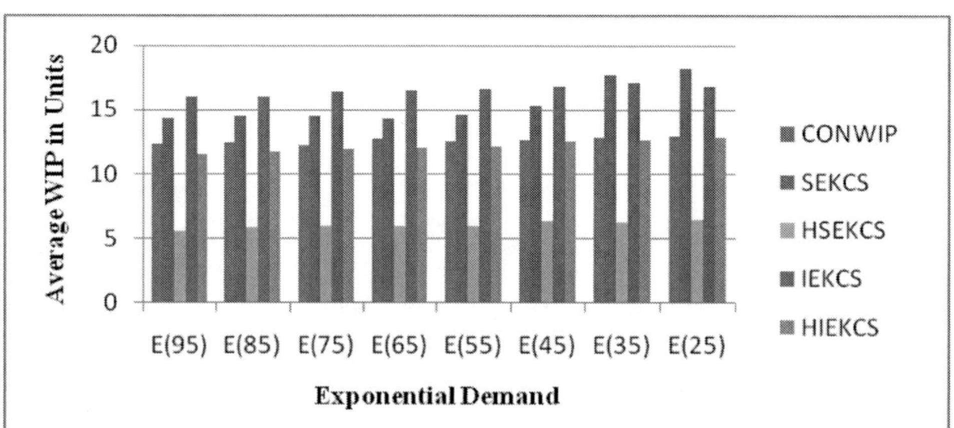

Fig. 11. Relative Average WIP performance assessment of the Pull production control systems.

6 Conclusions

The work intended here is paying awareness on the assessment of substitute policies in the planning and organizing of a pull manufacturing system. Here an effort is made to speak about some pull manufacturing control strategies for diverse arrangements of a Just-in-Time manufacturing system. The proposed hybrid production control policy may be regarded as one of the preference mechanism for a distinctive pull controlled production system.

1. It is found Extended Kanban Control Systems performs better in one instance compared to CONWIP. Proposed Hybrid Kanban Control system appears better performance with the other control policies. At the demand of E(20) it is observed for Production Rate in units for mean time of 25 min is more for IEKCS where as the average waiting time and average WIP are relatively more. For these values HIEKCS gives 40-65% reduction compared to EKCS. In case of Hybrid systems, even though the production rate is less while comparing with CONWIP, the average waiting time and average WIP are observed to be very less.
2. The effect of degree of imbalance was studied and assessed. The assembly system in each flow line harmonizes with demand at all stages of the arrangement therefore the flow lines are not independent. Consequently there is no effect of the degree of imbalance taking place the system.
3. HEKCS, the Hybrid of CONWIP and EKCS control policy has shown improved results than CONWIP and EKCS individually relating to average WIP, average waiting time and production rate. HEKCS shows good results compared to the three proposed control mechanisms (IEKCS, CONWIP and HIEKCS) evaluated. Through the observation, it is found that for all the considered five pull production mechanisms the average waiting time and average WIP are declined when demand rate increases. While related to CONWIP and HEKCS strategies, HSEKCS shows a lot of augmentation in the process parameters average WIP and average waiting time.
4. By increase in the demand rates of all of these five pull control policies, average waiting time the Work in Process increases. Relatively improved performance parameters established in the Hybrid Extended Kanban Control policies while assessing merely with CONWIP and EKCS policies. In the proposed two hybrid policies, HSEKCS shows much improved outcomes contrast to HIEKCS. Even though the average WIP and average waiting times in HIEKCS are nearly twice compared to HSEKCS, however the throughput rates are equal in both the hybrid policies.

References

1. Zhang. W and Chen. M. Y., "A mathematical programming model for production planning using CONWIP", International Journal of Production Research, Vol. 39, No.12, 2001, pp. 2723-2734.
2. Wormgoor O. S., "Performance evaluation of generalized kanban systems". M. S. Thesis, University of Twente, Netherlands, 2000.
3. Xu, X. Xu and S. Q. Xie, "Recent developments in Dual Resource Constrained (DRC) system research", European Journal of Operational Research, Vol. 215, No. 2, 2011, pp. 309–318.
4. Liberopoulos, G., and Dallery, Y., "A unified framework for pull control mechanisms in multi-stage manufacturing systems". Annals of Operations Research, Vol.93, 2000, pp. 325–355.

5. Liberopoulos, G., and Dallery, "Base stock versus WIP cap in single-stage make to stock production –inventory systems. IIE Transactions, Vol.34, No.7, 2002, pp. 627-636.
6. Little. J. D. C, "A proof for the queuing formula: L = λW," International journal of operations Research, Vol. 9, Issue 3, 1961, pp. 383-387.
7. W. J. Hopp and Mark L. Spearman, "To Pull or Not to Pull: What Is the Question?" International Journal of Manufacturing & Service Operations ManagementVol.6, No.2, 2004, pp.133-148.
8. S. Hossein Cheraghi, Mohammad Dadashzadeh, Mahesh Soppin, "Comparative Analysis of Production Control Systems Through Simulation," International Journal of Business and Economics Research, Vol. 6, No. 5, 2008, pp.87-103.
9. Yves Dallery and George Liberopoulos "Extended Kanban Control Systems: Combining Kanban and Base stock", IIE Transactions, Vol.32, 2000, pp.369-386.
10. Chu. C. and Shih. W. "Simulation studies in JIT production." International Journal of production Research, Vol.30, No.11, 1992, pp.2573-2586.
11. Claudine Chaouiya, George Liberopoulos and Yves Dallery, "The extended kanban control system for production coordination of assembly manufacturing systems" IIE Transactions Vol. 32, 2000, pp.999-1012.
12. Berkley, "A review of the kanban production control research literature," International Journal of Production and Operations Management, Vol.1, No.4, 1992, pp.393-411.
13. B. Baynat, Y. Dallery, M. Di Mascolo and Y. Frein, "A multi class approximation technique for the analysis of kanban like controlled systems", International Journal of production research, Vol.39, No.2, 2001, pp.307-328.
14. Baynat, J. A. Buzacott and Y. Dallery, "Multi-product Kanban-like control systems," International Journal of Production Research, Vol.40, 2002, pp.4225-4255.
15. O. Srikanth, B. V. Ramana Murty, N. Selvaraj "Evaluation of Performance Parameters by Hybrid Modelling of Pull Production Control System and by Simulation Approach," International Journal of Emerging Technology and Advanced Engineering, 2013, Volume 3, Special Issue 2, January 2013, pp.69-73.

MATEC Web of Conferences **144**, 05007 (2018) https://doi.org/10.1051/matecconf/201814405007
RiMES 2017

Solutions for the disposal of industrial and rain wastewater of transport enterprises in Moscow

Galina Varyushina[1*]
[1]*Moscow State University of Civil Engineering, Yaroslavskoye shosse, 26, Moscow, Russia, 129337*

Abstract. The article gives results of the surveys water management systems, auto repair companies, bus companies, taxis, fleets of trucks, production and technical enterprises. Studied technological processes of primary production: cleaning engines, flushing of radiators and parts of batteries, testing of fuel tanks, cleaning of parts and units, exterior washing and painting of motor vehicles. There were carried out field examinations of local industrial wastewater treatment facilities industrial wastewater and onsite construction runoff, the laboratory-production work to identify quantitative and qualitative indicators of the incoming and treated effluents, including petroleum products, suspended matter, salinity, synthetic surfactants, heavy metals etc. Proposed sustainable solutions for the improvement of the work of treatment facilities of rain sewage by mechanical and physical and chemical methods, including horizontal sand catchers, septic tanks – oil separator, a cascade of pressure filters with granular mineral over-loading and high-performance sorption materials, the effectiveness of treatment meets the requirements of water for its re-use for technical needs, and discharge to water bodies.

1 Introduction

The problem of disposal of industrial and rain wastewater of truck companies due to the continuous increase in the number of cars that consider as high toxic sources of pollution both air and water basins of Moscow city.

The main factors arising from the operation of various types of vehicles and driving to the negative impact on the atmosphere and hydrosphere of the city include:

- Aging transport that fail to meet ecological requirements, which leads to the fact that road transport is the main pollutant of the atmosphere, contributing more than 80 % of all emissions;
- Low-quality motor fuels and lubricants;

* Corresponding author: nocviv@mgsu.ru

- decrease of the capacity of the road network during peak times, as vehicles moving with low speed, a lot of maneuvering, the engines are stressed and, consequently, increases the amount of products of combustion of motor fuel and increased wear of the tread rubber tyres of cars, tire dust, which contains carcinogenic substances (N-nitrosamines) and benzopyrene, flushed by storm water;
- contaminants (more than 200 types, in particular, carbon monoxide (CO), oxides of nitrogen (NO_x) and sulfur (SO_2), polycyclic aromatic hydrocarbons (benzopyrene ($C_{20}H_{12}$), hydrocarbons, halogen-containing, phenols, aldehydes, and ketones of aliphatic, alkylaromatic esters, heavy metals, sulfur-containing deposited substances, soot, etc.) received from vehicles in the atmospheric air, soil-and water Wu in the movement, storage, launching, cleaning, maintenance and repair, as well as disposing of the car and its elements.

2 Features of transport enterprises sewage

Modern dynamics of the growth rate of the car amount leads to the need to develop organizational and technical solutions aimed at minimizing the negative impact of the functioning transport enterprises (TE) on the environment of the city. Depending on the functional purpose of TE installed a water consumption rate of drinking and industrial water for production needs and requirements for its sanitary and chemical composition, as well as the conditions of discharge of industrial wastewater in municipal drainage network.

Under the new Rules of water supply and sanitation, approved by Decree No. 644 dated 29.11.2013 of the Russian Government changed the standard indicators of General properties of wastewater and concentrations of pollutants in wastewater are allowed to discharge to a centralized sewerage system. In particular, the maximum permissible concentration of dissolved petroleum products is 10 mg/l, salinity – 3,000 mg/l, fats – 50 mg/l, volatile organic compounds (toluene, benzene, acetone, methanol) 20 mg/l, suspended solids – 300 mg/l, synthetic surfactants – 10 mg/l and phenols 0.25 mg/l [1-3]. In addition, the new Rules provided production processes, the implementation of the enterprise central systems are required to have local treatment facilities and discharge of treated wastewater. These businesses include motor, in production processes, which formed oil, acid, alkali, heavy metal ions, organic solvents, volatile organic compounds, synthetic surfactants, organic chlorine compounds deposited mineral inclusions hydraulic size greater than 2 mm/s, floating substances with particle size more than 20 mm/s. To achieve the modern requirements to the wastewater before discharge into the centralized Sewerage system, it seems advisable to collect and diversion of industrial runoff of TE to separate the Sewerage scheme and subsequent local cleaning each thread. As you know, it processes all the water drinking and technical water on the TE is used for household needs and showers, sanitary washing of floors in buildings and wash works to clean the territory, in garage, technological processes for the repair and maintenance of motor vehicles and feeding of circulating technical water supply systems and boilers [4-7].

Peculiarity of formation of the industrial wastewater is a complex of TE with a wide technological base is a multi-component composition of wastewater generated in the basic processes:

- Paint shop for painting body parts with the use of the painting chambers, strainers and mechanical treatment of water for removal of paints, varnishes and solvents and water reuse;

- Workshop for electroplating component elements of the vehicles with local treatment from the armed removal of heavy metals (copper, chromium, zinc, iron, etc.);
- Workshop for cleaning of engines, chassis and components of vehicles of cleaning solutions and their preparation for repair and testing on the stands with a working system of multiple use of cleaning agents and disposal of petroleum products and sediments;
- Workshop for repair and service of batteries with internal structures to neutralize acidic wastewater;
- cooling of technological equipment;
- Production areas for oil change and mechanical filters with the disposal of effluent in onsite wastewater treatment plant for industrial and rain waste water;
- Workshop external washing of vehicles and their salons with turnaround system of technical water supply;
- surface waste water (rainwater, snowmelt, snow) from the roofs of buildings and site of TE, parsed samples introduced for on-site wastewater treatment plants prior to their diversion into the municipal rainwater drainage networks.

Studies of water management systems on the TE of the city of Moscow are performed by the laboratory processing the city, industrial and rain sewage and sludge of "Institute MosvodokanalNIIproekt" from 1990 to the present time that allows you to organize and summarize quantitative and qualitative indicators of efficiency of use of water and operation of local treatment facilities. The results of surveys of TE indicate that the effect of sewage water purification from dissolved organics, paint strainers cameras the methods of sedimentation and filtration does not satisfy the requirements of the water or its reuse in the paint shops, and any discharge to a centralized sewerage system. For the removal of contaminants needs to upgrade the existing technological scheme of water purification by applying highly efficient coagulants and flocculants with the purification of the mineral sorbents. Evaluating the work of treatment facilities to remove chromium and other heavy metals Metalworking shops, it was established that widespread chemical and electrocoagulation methods show the effect of treatment of 72 – 85% recovery of chromium, forming a hard residue neutralizes and spending a significant amount of power [8-12].

As practice shows, the best way to clean chrome wastewater is ion-exchange method, the essence of which is to exchange part of ions in the result of which purified water meets the requirements of water for reuse in the shops for metal processing. The most contaminated streams of industrial wastewater on TE contain oil products, suspended substances and settled, scale, caustic soda, soda ash are the wastewater plant washing parts and assemblies of cars is a necessary operation before the implementation of office repair and maintenance [13,14]. For the purpose of rational use of valuable raw materials for TE provides multiple use of cleaning solutions consisting of high concentrated solutions of soda ash, trisodium phosphate and sodium tripolyphosphate, in the recirculating system with the preliminary removal of oil in emulsified and dissolved condition. The most widely methods of neutralization of stable emulsions by demulsifiers resulting in delamination of the suspension and, as a consequence, sediments and floating oil film. The removal efficiency of oil is 65 – 90 % [15-18].

The most water-intensive manufacturing process TE should include mechanized and manual cleaning of vehicles carried out daily according to the sanitary-epidemiological requirements and regulations of the contents of the vehicle. Depending on the car category, its dimensions, the technical condition of the roads, season of year and type of applied anti-icing agents in a wide range of change of concentration of pollution of wastewater from the car wash: settling and suspended solids, mineral oils, gasoline and diesel fractions, full BOD, COD, detergents, sulfates, chlorides, pH. It should be noted that wastewater from the car wash are characterized by a high degree of bacterial material contamination, in this connection, according to the sanitary requirements of the technological scheme should be included bactericidal installation for disinfection of treated wastewater used in the system of technical water supply local company [1, 4, 7].

Based on the results of chemical analyses of initial and treated waters obtained in the course system of optical monitoring of the efficiency treatment of the complexes made, we can conclude that in the original water act of technogenic contamination of the basic technological processes of TE, as evidenced by the excess of concentration of oil products, detergents and other substances. Table 1 shows the data for the control of wastewater main production processes of TE provided in accordance with industry technical standards for all enterprises that perform work on maintenance, repair and service of vehicles.

Table 1. Transport enterprises sewage features

Type of process	Concentration [mg/l]							
	TSS	Petrochem	pH	Sulphates	Chlorides	synthetic surfactants	Cr	COD
Car parts painting	1400 - 2300	50 - 380	7.5 – 9.0	970 - 1600	2400 - 3900	25 - 40	0.05 – 2.4	1430 - 2600
Metal processing	630 – 820	4.6 – 10.5	6.2 – 7.1	120 - 230	220 - 460	1.5 – 3.0	310 - 400	600 - 830
Engine washing	3000 - 5200	2600 – 5600	9 – 12	2100 - 3400	4100 - 4700	260 – 390	1-10	1300 – 2800
Batteries maintenance	110 – 230	12- 35	3 – 11	250 – 400	650 – 920	0.5 – 2.0	0.1 – 0.5	55 – 83
Car washing	870 – 3200	100 – 200	7 – 9	190 – 320	590 – 740	58 – 170	1 – 5	240 – 580
Surface sewage	380 – 1600	35 – 180	7 – 8	50 – 120	260 – 410	3 – 18	0.5 – 1.0	350 – 560

3 Solutions for wastewater treatment

Collection and transportation of industrial wastewaters is carried out by a separate scheme with local treatment plants and reuse of treated water for technical needs. To clean storm water and wastewater from washing vehicles there were process solutions and project documentation, relevant for nowadays environmental and sanitary requirements, according to which in Moscow and Moscow region, there were built dozens of treatment systems in bus and tram depots, fleets and taxi companies, repair bases, and road construction equipment. Technological scheme of purification of rainwater sewage includes the following facilities mechanical, physical, and chemical treatment: horizontal sand traps, oil traps, clarifiers, three-stage pressure filters with granular mineral feed, and tertiary treatment facilities [5-8]. Typical pollution runoff include suspended solids, oil products, sulphates, chlorides and organic material of natural origin washing into catchment areas of rain and melt waters. To achieve the fisheries management requirements of the discharge rain sewage into water objects of Moscow there were developed projects that built facilities for deep cleaning of the cumulative type.

A method of forming and cleaning runoff is as follows: collection, averaging and cleaning exposed to the entire volume of rainwater falling on the territory of the catchment. Calculation of the quantity of rainwater runoff from 1 ha of the catchment is determined by the results of observations of meteorological service in the area for producing the monitoring of quantitative parameters of precipitation. Systematization and analysis of these data allows determining the quantity of fallen rain, their intensity, frequency and duration. Rain wastewater is collected in sewers inside drains through grids located in the territory in accordance with the vertical layout and go to the underground regulating reservoir.

Regulatory capacity is composed of rectangular sections into which the polluted wastewater coming consistently through the overflow window, located in the upper part of the section. For their abduction in gravity mode wastewater treatment plant located in the lower areas of the catchment, with the low grid determining the filling of the regulating tanks. After passing through the grids located in the main and after filling of the regulating tanks by the estimated volume of one rain, non-nominal volume of 5 %, overflows to urban collector or pond. Facility design eliminates the possibility of discharge into reservoirs of oil slicks, dropped out sediment and other coarse impurities.

Industrial water mechanical cleaning method of settling goes within 2 – 3 days, after which the clarified effluent is directed to the physical-chemical treatment including a cascade of three successive series pressure filters with mineral filter and sorption materials.

The main criteria for the filters include the following: developed specific surface of the grains, large porosity and dirt-holding capacity, mechanical strength, low friability, high sorption capacity, regeneration ability without the use of reagents in the process of operation, the possibility of environmentally friendly disposal after the complete exhaustion of the initial properties, long life cycle and low cost [12, 14, 15].

During the research, it was found that the most effective filter materials meeting the above requirements should be classified as quartz sand, basalt, zeolite, crushed anthracite and haydite chips (fractional composition of 1 – 2 mm). As sorption material there were recommended fine-grained activated carbon of various grades. High sorption capacity of activated carbons leads to a profound purification of industrial wastewater of trace amounts of suspended solids, dissolved oil products, residues, heavy metal salts, synthetic toxicants. Rinsing in counter-current mode of filter and sorption materials purified water accumulated in the collecting tank leads to the effective removal of dirt from the surface of the grains and of the interporous space. Contaminated rinsing water is returned to the regulating tank for further settling and post-treatment on pressure filters.

Based on the results of laboratory and production control treatment plants during the different seasons of the year, the efficiency of treatment of the complexes corresponds to the design performance, subject to the rules of operation are developed and agreed in the project documentation. The program of laboratory and production control and monitoring of the effectiveness of local treatment facilities is developed by the environmental service of the company based on the environmental regulations and demands to the used technological equipment installed by the manufacturer. The program includes the following sections: list of studied indicators of the composition of water and sediments; sampling frequency; sampling; quantitative and qualitative characteristics of the resulting sludge. Analysis of the main indicators of the structures with regular observations of the processes of industrial wastewater purification allows making fast technical and organizational solutions aimed at improving the efficiency of contaminants removal and minimizing energy consumption for the operation of the facilities.

Table 2 shows the performance of rain sewage treatment facilities of tram depot, located in the Northwestern administrative district of Moscow.

Table 2. The efficiency of cleaning of storm water at the treatment facilities of the tram depot

Indicators	**Initial concentration [mg/l]**	**Treated water concentration [mg/l]**	**Discharge limits [mg/l]**
Aluminum	3.557	0.023	0.04
BOD	8.63	2.19	3.0
Suspended solids	256.0	3.0	3.0
Iron	9.714	0.047	0.1
Cadmium	0.972	0,001	0.005
Manganese	2.153	0.006	0.01
Copper	1.067	0.001	0.001
Petrochemicals	64.32	0.03	0.05
Nickel	1.295	0.007	0.01
Total salts	659.37	640.22	1000.0
pH	7.5	7.3	6.5-8.5
Lead	2.689	0.002	0.006
synthetic surfactants	5.732	0.005	0.01
Sulphates	90.41	83.62	100.0
Chlorides	207.83	190.54	300.0
COD	42.65	10.78	30.0
Chromium +3	0.874	0.019	0.07
Chromium +6	0.163	0.015	0.02
Cyanides	0,927	0.036	0.05
Zinc	1.538	0.008	0.01

Fig. 1. Wastewater facilities for treatment of bus washing wastewater

For disposal of industrial wastewater formed in water-intensive process of mechanized and manual cleaning of vehicles, the reverse process water supply system with reuse of treated water for car wash was developed. Technological solutions based on physical and chemical methods of water purification and processing of the resulting precipitate. On the first stage (oil trap), the process of gravity removal of coarse impurities and floating oil goes; on the second stage - filters with granular mineral downloads and germicidal lamps - purification from emulsified oil and suspended solids to the requirements of water for reuse. For dewatering and collection of the resulting precipitate, there is a filter press and storage bin to export of filter cake for disposal twice a year. Fig. 1 shows the room of pressure filters of facilities for treatment of wastewater from bus washing, located in the Northern administrative district of the Moscow [19-20].

4 Conclusions

1. As the results of surveys of the water systems of the motor transport enterprise, it is necessary to continue improving the work of local treatment facilities of industrial sewage and rain wastewater on-site objects, as trucking companies continue to remain highly toxic man-made objects in both air and water basins of Moscow city.
2. There were treatment facilities developed for disposal of wastewater from washing vehicles of different categories: cars, trucks, buses, road and construction equipment.
3. There were high efficiency technology developed of treatment of industrial and rain wastewater of transport enterprises, to meet current regulations for discharge to water bodies.

References

1. E. Gogina, I. Gulshin, Pr. Eng., **117**, 107–113 (2015)
2. A.G. Pervov, A.P. Andrianov, T.P. Gorbunova, A.S. Bagdasaryan, Petr. Chem., **55** (10), 879-886 (2015)
3. I. Gulshin, A. Kuzina, IJAER, **10**, 21, 42618–42623 (2015)
4. A. Volkov, V. Chulkov, R. Kazaryan, M. Fachratov, O. Kyzina, R. Gazaryan, AMM, **580-583,** 2281-2284, (2014)
5. N. Makisha, A. Kulakov, MATEC web of conferences, **112**, 10018 (2017)
6. A.G. Pervov, A.P. Andrianov, E.B. Yurchevskiy, Petr. Chem., **55** (10), 871-878 (2015)
7. E. Gogina, A. Pelipenko, MATECCONF, **73,** 03007 (2016)
8. V.N. Varapaev, S.A. Doroshenko, A.Y. Trotsko, A.V. Doroshenko, IJAER, **10**, 21, 42588-42592 (2015)
9. N. Makisha, Pr. Eng., 165, 1092-1097 (2016)
10. O. Kuzina, E. Pankratov, V. Tkachev, MATECCONF, **86,** 05023 (2016)
11. V. Orlov, A. Andrianov, AMM, **580–583**, 2398-2402 (2014)
12. V.N. Varapaev, A.V. Doroshenko, I.Y. Lantsova, Pr. Eng., **153**, 816-823 (2016)
13. N. Makisha, M.Yunchina, MATEC Web of Conferences, **106**, 70016 (2017)
14. A.G. Pervov, A.P. Andrianov, DWT, **35** (1-3), 2-9 (2011)
15. E. Gogina, I. Gulshin, AMM, **580-583**, 2367-2369 (2014)
16. A. Volkov, O. Kuzina, Pr. Eng., **153**, 838-843 (2016)
17. A.A. Volkov, A.V. Sedov, P.D. Chelyshkov, D.A. Lysenko, A.V. Doroshenko, IJAER, **10**, 22, 43269-43272 (2015)

18. E. Gogina, I. Gulshin, Pr. Eng., **153**, 189-194 (2016)
19. A.G. Pervov, A.P. Andrianov, V.A. Chukhin, R.V. Efremov, IJAER, **10**, 22, 43517-43525 (2015)
20. A. Kulakov, N. Makisha, MATEC Web of Conferences, 112, 10019 (2017)

Environmental situation of Baikal Lake

Varvara Shmalko[1*] *and Elena* Gogina[1]

[1]*Moscow State University of Civil Engineering (MSUCE), Yaroslavskoye shosse, 26, Moscow, Russia, 129337*

Abstract. This article considers the ecological situation, which has developed around a unique natural object - Lake Baikal. There are studied physical and geographical characteristics of this lake, because it is one of the greatest lakes in the world and the largest freshwater lake in Russia. Lake Baikal is 1.7 times larger the Lake Ladoga, the largest lake in Europe. Baikal is the largest fresh water lake in the world. Its volume is 23 600 cubic kilometers, which is about 20% of the planets' fresh lake water - much more than in all the freshwater lakes of the world. There are described the most large-scale tributaries, which fill the lake, as well as the only river, flowing from it. There are studied basin areas of this valuable lake, which are under protection. There are four main periods identified by analyzing the dynamics of wastewater performances.

Keywords: Baikal, ecology, basin, water supply

1 Introduction

Lake Baikal (fig.1) and its catchment basin belong to the unique geological systems of the world. Baikal is located in the central part of Eastern Siberia, near the conventional geographic center of Asia. The mountain hollow of the lake represents the most important natural boundary of Siberia. The borders of various floristic and faunal areas converge in this region, creating biogeocenoses that have no analogues.

Baikal is one of the greatest lakes of our planet: the deepest (1637 meters) and the oldest (about 25 million years old), containing the largest number of endemics (more than 1,000 species) and representatives of flora and fauna (more than 2600 species), inhabiting the fresh water bodies of the Earth. The lake has a unique reserve of fresh water (23 615 cubic kilometers) and quality (20% of world reserves).

The antiquity of Baikal is the most remarkable feature. Taking into consideration deep relict endemism of the lake fauna, the huge amount of researchers determine its age in 20-30 million years. The vast majority of lakes, especially the lakes with glacial and bayou origin, live 10-15 thousand years, then they are filled with sediments, dragged out by fusions, and sooner or later they turn into swamps and then dry up. Investigations of recent years have allowed geophysicists to put forward a hypothesis that Baikal, on the contrary, is a nascent ocean. This hypothesis is confirmed by the fact that its shores diverge at a speed of up to 2 cm per year, just as the continents of Africa and South America, the shores of the Mediterranean and Red Seas, etc. In the Baikal region, significant magnetic anomalies along its axis were noted, alongside with the active movements of the earths crust. On a scale, these

* Corresponding author: goginaes@mgsu.ru

anomalies are comparable with similar anomalies in the area of the mid-Atlantic fault. The lake has many features inherent to the ocean - abyssal depths, huge mass of water, internal waves and seiches, tides, strong storms, high waves, expansion of the basin due to the expansion of the shores, large values of magnetic anomalies, etc.

Fig. 1. Baikal Lake

The lake lies in the Baikal basin - a bottomless stone bowl, surrounded on all sides by mountains. The mid-altitude mountain ridges on the western side - Primorsky and Baikal, from the east and southeast - Barguzinsky (with a maximum height of 2840 m.) and Khamar-Daban, frame the cavity. The height of the mountains above it, the depth of the lake and the thickness of the loose sediments lining its bottom determine the depth of the depression. The layer of these lake sediments in some places reaches 6000 meters, and their volume is twice the volume of the lake and reaches 46,000 cubic kilometers. It is not difficult to calculate that the depth of the crystalline bed of Lake Baikal reaches 8 to 9 kilometers.

The basin of Baikal is the central link of the Baikal rift zone, which arose and develops simultaneously with the world rift system [1-3].

The deepest point of Baikal's basal bath lies about 7000 meters below the ocean level. The Baikal basin is the deepest basin on earth. Its "roots" dissect the entire earth's crust and go into the upper mantle to a depth of 50-60 kilometers.

Annually about 60 cubic kilometers of beautiful and unique water are reproduced in Baikal, which in some cases can be used instead of distilled. The rare purity of water is provided by the vital activity of its unique flora and fauna. The main properties of Baikal water are characterized as follows: there are very few dissolved and suspended mineral substances, negligible organic impurities, and a lot of oxygen. The total mineralization of water in Lake Baikal is 120 milligrams per liter, while in many other lakes it reaches 400 or more milligrams per liter. The total content of ions in the lake water is 96.7 milligrams per liter.

Transparency depends from the purity of water. Baikal not only extremely clean, but also is the most transparent lake in the world. Baikal water transparency reaches 40 meters (tens of times more than in many other lakes), it happens in spring, after breaking of the ice. The water of the Sargasso Sea is a gold standard of the highest transparency, which is approaching the transparency of distilled water. Here the disc of Secchi disappears from sight at a record depth of 65 meters. Recent investigations have shown that at depths from 250 to 1200 meters the transparency of Baikal water is not less than in the Sargasso Sea.

The catchment basin of Lake Baikal covers an area of 541 thousand square kilometers (without the area of the Baikal water area - 31.5 thousand square kilometers). There are 240500 square kilometers of the basin of surface and underground runoff into Baikal is located on the territory of Russia. The rest of the catchment basin (300.5 thousand square kilometers) is within Mongolia [13].

The entire lake basin is a peculiar and very fragile natural geosystem, the basis of which is the lake system with its natural formation process of the purest waters of drinking quality [1, 4].

2 The water basin of Lake Baikal

Baikal collects water from a vast territory - rivers flow into it, starting far in the mountains of northern latitudes, and on the plains of the south. More than 300 rivers flow into the lake. The exact number is unknown, since in dry years the number of rivers decreases, and in rainy - accordingly increases. Among them, there are large navigable and floating rivers, which spread far, mainly to taiga areas, but there are also small ones that can be crossed.

The largest of the rivers flowing into Baikal is the deep-water Selenga, which is the main artery of the southeastern part of Transbaikalia. The origins of the Selenga River are in the steppe expanses of the Mongolian People's Republic. Passing the state border, the river flows at a distance of 415 km along the territory of the Buryat-Mongolian ASSR and flows into Baikal. It is navigable on a large extent; large tributaries - Hilok and Chikoy - are navigable too.

There is one astonishing feature peculiar to Selenge. The water in Selenga all year round and especially in summer is cloudy, unlike the other rivers flowing into the lake, where the water is usually clean and transparent. This is explained by the fact that in its origins – in the steppes of Mongolia – the climate is sharply continental, almost no vegetation and all year round strong winds are blowing. Winds pick up small grains of sand, grind them into the finest powder and spread them to large spaces. Rainwater captures fine dust and demolishes into the rivers - the tributaries of Selenga.

Despite the relatively slow Selenga current, fine dust particles do not settle to the bottom and the flow of water carries them to Baikal. Selenga's turbidity is also explained by the fact that the river flows through the territory where loesslike loams are widespread. A number of small mountain rivers flow into Selenga in addition to the steppe tributaries; their violent flows produce a great destructive work and carry huge masses of sand and larger particles. One part of this material is deposited in the channel of the Selenga, and the other is carried to the lake, where the particles are first dispersed by waves, and then gradually fall to the bottom, forming a wide shallow water near the river delta. So the muddy waters of the Selenga River carry a huge amount of suspended, small particles to Lake Baikal, which are deposited in the lake as in a gigantic settler. Because of this process, which lasts for millennia, Baikal had to retreat in the region of the confluence of the river far to the west. On the site of the vast bay of the lake, due to the material brought in, formed an even, vast inland area of the delta; At present, wide tracts of mowing, fields, several dozen settlements are spread over it. Not every inhabitant of the villages will believe that once in the distant past the waves raged at the place where his house stands.

The next largest tributary of the Baikal - the Barguzin River - has a length of about 400 km. In the upper reaches, it is as calm as all the rivers of the plains. Twisting, breaking into many channels, the river slowly flows along the wide flat-bottomed Barguzin depression to the mountain lintel, which separates the depression from the lake. Having reached it, the waters of the river, quickly gaining speed, rush towards Baikal. In the lower reaches of the river it is already like a mountain river - fast, dotted with rushes.

From the north, another large river - the Upper Angara - falls into Baikal. It is less abundant than Selenga and Barguzin, its length is 439 km. The river flows through two large basins, framed by mountain ranges from the west to the east.

Other rivers flowing into the Baikal are less abundant and less noticeable. The largest of them - Kichera, Turka, Snezhnaya, Goloustnaya – are less than 170 km lengthwise. The remaining rivers are small, shallow, but fast and turbulent. They start from the slopes of ridges, a dense ring surrounding the lake [2,7,8].

In 2014, the annual flow to Baikal was below the average long-term values of 41.9 cubic kilometers (1.33 thousand m^3/s cubic metres per second), which is 21% less than in 2013, when the runoff was 52.98 cubic kilometers (1.67 thousand m^3/s) [9,10,13].

Many rivers carry their waters into Baikal; the only river that flows out of it is the Angara [2]. The river results in the source the formation of river flow in the Baikal watershed and the processes of its purification by the ecosystem of Lake Baikal. The average annual volume of annual runoff from the lake is 60 cubic kilometers, which corresponds to a water discharge of 1.9 thousand m^3/s [13].

The length of this "daughter of Baikal" is about 1860 km. Every year, Angara takes out 60.9 cubic kilometers of water from Lake Baikal, and all its tributaries bring 58.75 cubic kilometers per year. It is noteworthy that the Angara at the mouth, at the confluence of the Yenisei, brings 120 cubic kilometers of water in a year, and the Yenisei in front of the confluence of the Angara has a water discharge of only about 100 cubic kilometers. The source of the river is located at the level of Baikal, i.e. at an altitude of 456 m above the level of the world ocean, and the mouth - at a height of 76 m. The difference is 380 m, which is used by a cascade of hydroelectric power stations built on the Angara. The width of the Angara in the source is about 1 km, the depth varies from 0.5 to 6 m, the flow velocity along the channel is 1-2 m/sec [2].

3 Protected natural territories in the basin of Lake Baikal

In the Baikal region, there is a network of specially protected areas. Within the boundaries of the Baikal Natural Territory (BNT), 5 preserves, 3 national parks, 23 reserves, 128 nature monuments, 1 botanical garden, medical and health areas and resorts represent the network of specially protected natural areas (SPA). The Botanical Garden with an area of 27.1 hectares is located in Irkutsk and is managed by the Irkutsk State University. Within the boundaries of the BPT, there are five recreational areas – «Baikal Priboy-Kultushnaya» and «Lemasovo» in the Kabansky District, «North Baikal» in the Severobaikalsk District, «Barguzinsky Coast of Baikal» in Barguzinsky District and «Lake Shchuchye» in Selenginskoye District. These recreational areas are located in the Republic of Buryatia and are managed by the administration of municipalities [11,12].

The SPA area within the BPT is 39.7 thousand square kilometers, which is 10% of the BPT area. Within the boundaries of the CES of the BTT and the site of the World Natural Heritage «Lake Baikal», the SPA occupy 25.6 thousand square kilometers (29% of the CES area).

4 Sewage waters and polluting surface substances

Analysis of wastewater performance dynamics in the Irkutsk region for 1981-2003 years identified four main periods. In the first period (1981-1987), there was a relative stability of wastewater volumes, including polluted runoff.

In the second period (1988-1989), there was a rapid increase in the volume of polluted runoff, with a slight increase in the volume of sewage, which is due to a certain objectivization of statistical reporting, primarily the registration of enterprises of the military-industrial complex and the well-founded trans-over of a significant part of normatively clean and normatively cleaned wastewater into the category of contaminated.

A hallmark of the third period (1990-1998) is the reduction of wastewater, including polluted runoff, caused by a general decline of industrial production in the region.

The fourth period (1999-2003) is characterized by the stabilization of the wastewater indicators at the lowest level in recent decades. Indicators of the volumes of wastewater discharged into water bodies (2.21 cubic kilometers), including polluted runoff (1.92 cubic kilometers) fixed in 1990, decreased 1.8 and 2.3 times respectively by the year 2003 [5, 6].

5 Conclusions

The Lake Baikal has a unique hydrological and chemical regime. The uniqueness of the hydrological regime lies in the fact that water exchange occurs throughout the water column. The lake is surrounded by mountain-taiga landscapes and specially protected natural areas, mostly preserved in the natural state and representing additional value.

In an effort to preserve original form of this natural object and not to damage the exceptional endemic fauna and flora, it is necessary to investigate deeper both the Baikal waters and the surrounding territories.

It is necessary to increase the number of observation posts, coordinate work on exploring, reproducing, using and protecting the natural resources of the Baikal natural territory, preserving biological diversity, ensuring environmental safety, and solving the region's social and economic objectives on the principles of sustainable development.

The Lake Baikal is a cosmoplanetary phenomenon, the real natural monument, which has outstanding value from science and aesthetics point of view. The beauty of Baikal is an irreplaceable and unrenewable wealth. Sooner or later progress of science and technology will help to find a substitute for raw materials, but nothing will help us to reproduce the aesthetic appeal of Lake Baikal.

References

1. http://www.obaykale.ru/baykal-obshie-dannye_04.htm
2. S.G. Sargsyan, Gos. publishing of geographical literature (1955)
3. Present and future of the Baikal region (opportunities for sustainable development) (1996)
4. Ryashchenko S.V., Savenkova T.P., Snytko V.A., Geogr. and natur. Resources 3, 44-49 (1998)
5. Znamensky V.A., Wat. Res., 3. 76, (1975)
6. Natural resources and environmental protection / Statistical compilation (2004)
7. Bezrukov L.A., Misyurkev Yu.A. - Geographical and cartographic assessment of the use, pollution and protection of the Irkutsk region waters // Geogr. and nat. res., 2 (1995)
8. N. Makisha, Pr. Eng., 165, 1087-1092 (2016)
9. N. Makisha, I. Panteleeva, MATEC Web of Conferences, 106, 70015 (2017)
10. http://baikalarea.ru/pribaikal/villages/baikalsk/a8793.htm
11. Certificate of the results of inspection of facilities and equipment of JSC «BPPM» from May 6, 2014.
12. https://03grb.ru/news/43403

13. http://ekogradmoscow.ru/novosti/pochemu-net-ochistnykh-sooruzhenij-na-bajkale
14. Garmaeva E.P., Dambayev T.S., Sheluheev A.A., General plan of the «Tanghoi» Kabansky District of the Republic of Buryatia - Explanatory note - 3406 - pz.
15. N. Makisha, E3S Web of Conferences, 6, 01002 (2016)

Root cause analysis of rough conical seat grinding problem in fuel pump cylinder head by Shainin methodology

Sanket Shankar Kadam[1*], *Virupakshappa N. M.* [2] and *U. Achutha Kini* [1]

[1]*Department of Mechanical and Manufacturing Engineering, Manipal Institute of Technology, Manipal Academy of Higher Education, Manipal, Karnataka, India*
[2]*Process planning- CPx components, Bosch Ltd., Bidadi, 562109, Karnataka, India*

Abstract. Problem solving plays a major role in quality engineering and it is popularly adopted by manufacturing industry. This study is aimed to determine root cause for rejection of cylinder head due to rough conical grinding problem by Shainin method. This has two major parts, first is diagnostic part which involves finding out the root cause and second is a remedial action part which consist of actual application of findings to rule out the problem. Wear in tool holder assembly is identified as root cause for rough grinding problem. Converging to root cause and validating the same achieved by successful implementation of Shainin tools.

1 Introduction

Fuel injection pump is the important part of the common rail system (CRS). To generate the pressure demanded by CRS it is necessary to produce components of pump with highest quality .Problem solving comes under lean efforts taken by organization [1]. It plays a major role in both ways to improve quality and reduce the rejections and losses [2]. Shainin is convergent method of problem solving, involves continuously keep on eliminating the variables to converge to the most influencing variable [3]. Shainin methodology based on identifying contrast between good and bad pats, that's why it is simple and fast [4]. It makes use of specific tools to converge to the root cause quickly [5].

2 Experimental approach: identification of root cause

2.1 Problem identification and understanding

Cylinder head is one of the component for the CPx fuel injection pump, plunger reciprocates in the barrel which sucks fuel from low pressure (LP) port and delivers high pressure fuel from high pressure (HP) port. Conical grinding of HP port is done as illustrated in Fig. 1 for the depth of 40 μm. Tool can be lifted up by 0.20 to 0.40 mm with

* Corresponding author: sanket.kadam0200@gmail.com

each one micron downward movement during grinding, this value can be manually controlled and known as tool lift off value. Scrap cost per annum and rework cost per annum is high for high pressure (HP) grinding operation than rejection from any other defect in cylinder head. Due to this problem statement selected was to find and eliminate root cause (Red X) for HP seat grinding rejections in cylinder head seat grinding process

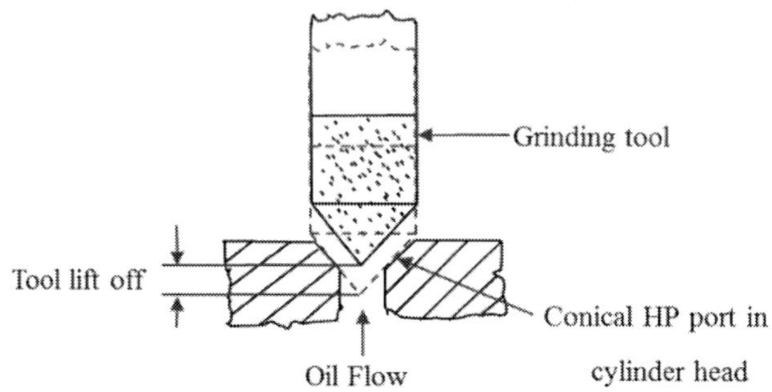

Fig. 1. Schematic representation of conical grinding operation

In HP seat grinding rejections, three kind of rejections are observed and those are rough lines on HP port, patches on HP port and cone offset. Amongst these, rejections due to rough lines as indicated in Fig. 2 are more than patches and cone offset. Rough lines on high pressure seat leads to leakage of diesel from high pressure area to low pressure area. Rough lines on HP port are indicated in Fig. 3.

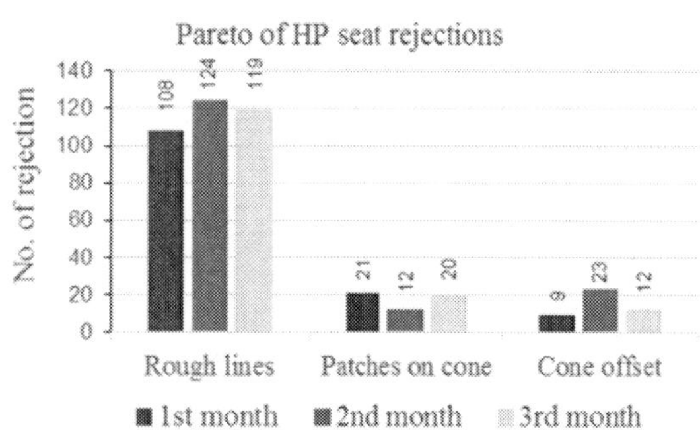

Fig. 2. Rejections Pareto

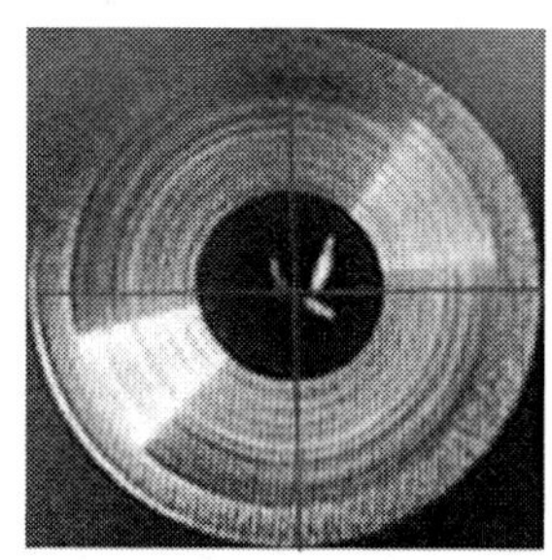

Fig. 3. Rough lines on HP port

2.2 Clue generation

Clue generation is done in two stages first one involves finding Green Y in Project definition tree and in second stage finding contrast between good and bad parts which further converges analysis. Project definition tree as shown in Fig.4, starts with project definition. Next step will be categorization of problem into one of the four categories Event, Feature, Property and Defect. In present study problem comes under defect category in which rough lines problem is dominant. Green Y (effect of problem on part) for problem is rough lines on HP seat. Clue generation in Shainin methodology is done by micro to macro approach. In second stage of clue generation, contrast is observed in families of variations such as, point to point, region to region, part to part etc. between best of best and worst of worst parts [6].

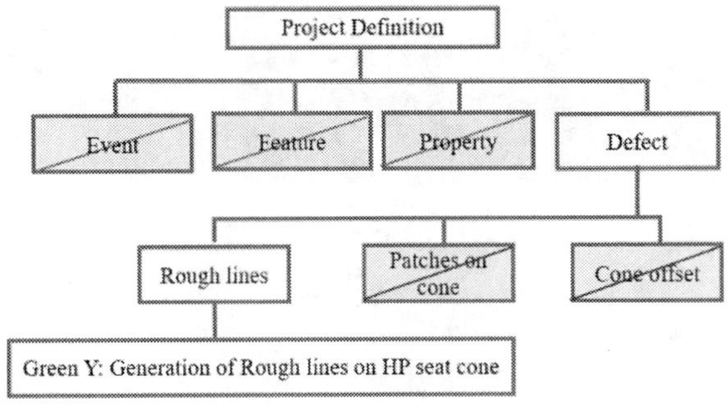

Fig. 4. Project definition tree

In this study contrast checked for region to region by dividing HP grinding face of cylinder head in four equal parts as shown in Fig.3 and found that rough marks are there in all four sectors in rejected parts which indicates low contrast. There are three machines which are used for HP seat grinding operation, Machine to Machine rejection was checked and high variation was observed (Machine 1: 3%, Machine 2: 2%, Machine 3: 6%).

2.3 Root cause identification

In Shainin methodology Green Y (effect of problem on part) is output and Red X (root cause) is input. For root cause identification tool used in Shainin methodology is solution tree. In solution tree shown in Fig.5, variables are eliminated one by one to identify most influencing variable amongst all. In first stage either measurement system variation (ΔM) or process variation (ΔP) has to be eliminated. Isoplot tool is utilized to check measurement variation, descrimination ratio more than 6.0 in Isoplot ensures elimination of measurement variation [7]. Here measurement variation is eliminated as descrimination ratio is 7.6 (more than 6.0) from Isoplot as shown in Fig.6.

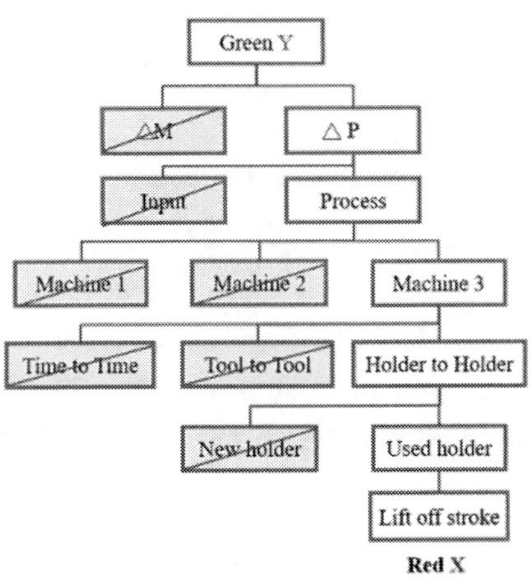

Fig. 5. Solution Tree

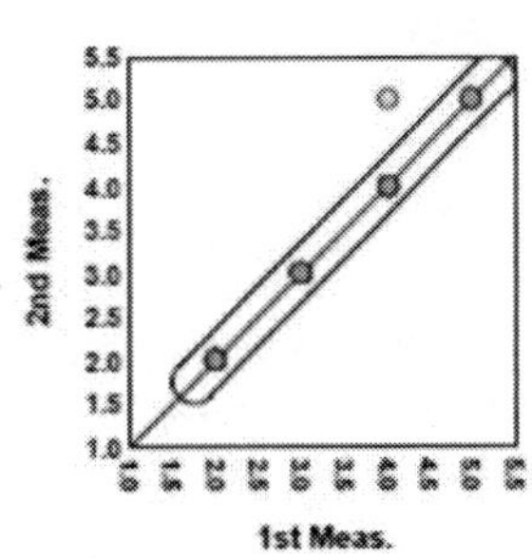

Fig. 6. Isoplot

Fig. 7. Sensory scoring system

As defect is visual, for quantifying defect severity sensory scoring system is used as shown in Fig. 7. In second stage of solution tree both supplier and machine variation are checked by active concentration chart, for this batch of 100 parts fed to three machines. Input is eliminated because only variable available for input is supplier to supplier variation and it doesn't show large contrast from active concentration chart shown in Table 1. In process only variation showing large contrast from clue generation is machine to machine variation. Machine 3 was giving more rejections than other two machines, hence further focus is given to machine 3.

Table 1. Active concentration chart

Machine No.	Machine 1										Machine 2										Machine 3									
Supplier No.	1					2					1					2					1					2				
Batch No.	B1	B2	B3	B4	B5	B1	B2	B3	B4	B5	B1	B2	B3	B4	B5	B1	B2	B3	B4	B5	B1	B2	B3	B4	B5	B1	B2	B3	B4	B5
Rejection Quantity	0	0	0	0	0	1	0	0	0	0	0	0	0	0	2	0	1	0	0	1	1	5	0	0	1	0	8	0	0	0

In machine 3, possible variations are time to time, tool to tool and holder (tool holder) to holder. Time to time variation is checked by collecting rejection data for one month as shown in Fig. 8, which doesn't give any clear contrast hence it is eliminated.

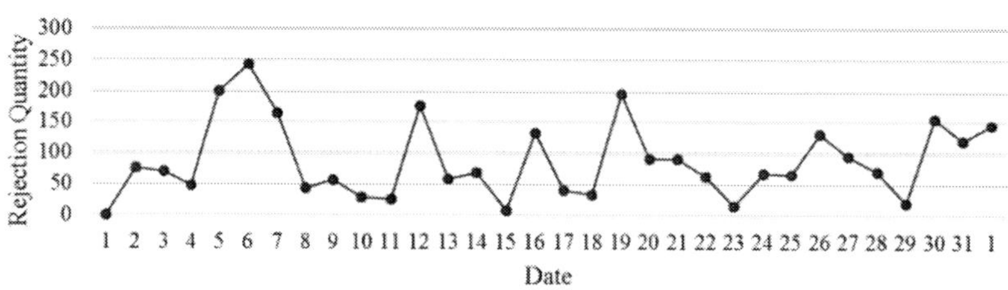

Fig. 8. Time to time rejection

For every 100 parts tool life is getting over and tool is changed automatically. Tool to tool variation is checked by collecting rejection data for multiple tools for its whole life shown in Table 2.

Table 2. Tool to Tool rejection

Shift 1			Shift 2		
Tool 1	Tool 2	Tool 3	Tool 1	Tool 2	Tool 3
2	1	2	1	1	2

To check the tool holder variation, holders of best machine (Machine 1) and worst machine (Machine 3) are interchanged to check the effect (best and worst machines are selected on the basis of active concentration chart). From this exercise it is observed that rejection behaviour of machine 1 and machine 3 is reversed. This clearly indicates large contrast hence for further study both holders are disassembled. It is observed that holder associated with more rejections has an additional axial wear of 0.15 mm in tool holder assembly than new tool holder assembly. It is found that due to wear in tool holder assembly tool lift off value is ineffective, and hence tool holder wear is the root cause candidate.

3 Results and Discussion

B vs. W is technique in Shainin method to verify root cause. For confirmation of root cause candidate B vs. W test was conducted. For this test, three parts are fed to machine for conical grinding operation with worn out tool holder as batch 1 and three parts are fed to same machine with new tool holder as batch 2. Result of this exercise is shown in Fig.9. Where Green Y is surface roughness (profile peak), for which maximum acceptable value is 2µm. It is observed that parts fed to the worn out tool holder gives all worst of worst parts and parts fed to new tool holder gives all best of best parts. B vs. W test indicates that Red X candidate drives Green Y hence axial wear in tool holder is confirmed as root cause for rough conical grinding problem.

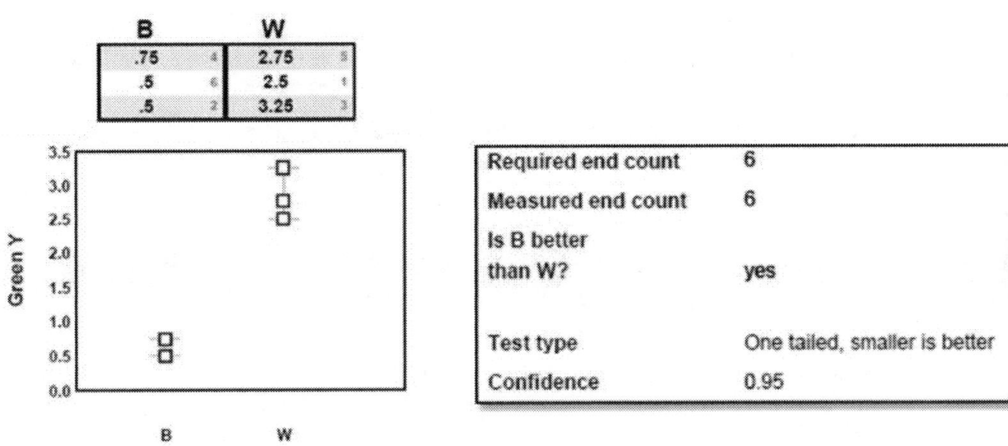

Fig. 9. B vs. W test

Contrast between good and bad parts speeds the Red X hunt which is important part of clue generation in Shainin method of problem solving. It is observed that Shainin® methodology can be successfully used in manufacturing related problems which helps to converge from large number of variables to most influential variable (root cause) by systematic use of Shainin tools.

4 Conclusions

Experimental investigation by Shainin method revealed additional axial wear of 0.15 mm in tool holder assembly is root cause (Red X) for HP port rough grinding issue of cylinder head. Wear in tool holder assembly makes tool lift off value ineffective, which restricts flow of oil between grinding tool and HP port surface during tool lift off. Identifying contrast between good and bad parts which is basic philosophy of Shainin method helps not just in clue generation but to identify the variables. Tools like active concentration chart and Isoplot in Shainin method helped eliminating variables and converging to most influential variable. B vs. W test helped to validate root cause candidate. Further in this study root cause identified can be eliminated by applying design of experiments by taking multiple levels of tool lift off as variable.

References

1. J. D. Mast, J. Lokkerbol, Int. J.Prod. Econ. **139**, 604, (2012)
2. R. D. Shainin, Qty. Engg. **5**, 433-448, (1993)
3. P Bhaskara, R Prasad, K M Subbaiah, G L Shekar, IJMET, **5**, 200, (2014)
4. N. Logothetis, Qty & Relia. Engg. Int.**6**, 195, (1990)
5. S. H. Steiner, R. J. Mackay, J. S. Ramberg, Qty. Engg**.** **20**, 19, (2008)
6. A. J. Jegadeesan, L. Karunamoorthy, N. Arunkumar, JMST **29**, 1731, (2015)
7. A Prashar, Prod. Plan. Ctrl. **27**, 83, (2016)

Influence of Farmer demographics and Supply Chain Issues on Organized Retailer Performance

Aruna Prabhu[1]*, *B. Giridhar Kamath*[2], *A. Karthik*[1] and *Kushagra Sinha*[1]

[1]*Department of Mechanical and Manufacturing Engineering, Manipal Institute of Technology, Manipal Academy of Higher Education, Manipal, India*
[2]*Department of Humanities and Management, Manipal Institute of Technology, Manipal Academy of Higher Education, Manipal, India*

Abstract. Advances in information technology and increased social networking have increased health awareness among people in this era. The increased health awareness has prompted people to eat healthy and stay fit. In this regard, the food sector has seen tremendous demand in terms of good quality vegetables/fruits, cereals, with hygiene, nutrition, proteins etc. With the consumer willing to pay an extra premium for the quality expected, it has been observed that the producers are still offered low prices for a good quality produce whereas the intermediaries grab a huge share of the price paid by the consumer. This research tried to understand the supply chain with the farmers in focus and explored the influence of farmer demographics and the supply chain issues on the performance of the organized retail. Linear regression analysis was used to test the linear relationship of supply chain issues with organized retailer performance. The outcome of the study implies that independent variables like demographic characteristics and supply chain issues had no control over the retailer's performance.

Keywords: Supply chain issues, Demographic factors, Statistical analysis, Organized retailer performance

1 Introduction

In terms of production of fruits and vegetables, India is in the second place after china [1]. In India, fruits and vegetables (F&V) is a fast growing sector and comprises around 90 per cent of the total horticulture produce. It is believed that, in future the F&V segment has high potential to drive agricultural growth with the rising demand for these commodities. The F&V plays a vital role in improving the income of farming community, generating employment opportunities and thereby contributing to India's economy.

India is considered to be most suitable for farming various types of fruits and vegetables due to the condition of soil, climate change and geography [1]. Numerous aspects such as growing population, changing life styles, nuclear families, working women, increasing urbanization are favourably influencing the food supply chains hinting at a better future.

* Corresponding author: aruna.prabhu@manipal.edu

However, increasing health awareness has influenced the food consumption pattern with a transition from the consumption of food grain to fruit and vegetables. This has increased the demand for protein rich, high value nutritional food. Perishability and very short shelf life of fruit and vegetables need a proper supply chain to make sure that the produce is still fresh when it reaches the customer. Wastage is a major issue threatening the Fruit and Vegetable sector. Fruit and Vegetable sector accounts for about 2 lakh crore Rupees worth wastage every year due to the lack of proper cold chain infrastructure [2] [3].

Demand for fresh fruits and vegetables is gradually increasing with rising population. However, the shelf life of these is a concern. Increase in demand for the processed food has created various opportunities in production, supply chain and marketing of agricultural commodities. This has attracted attention of several players in the industry and the last few years particularly has seen several private food retail players in India entering the food processing and supply chain sector with huge investments as they expect attractive growth opportunities in the future [4]. To facilitate the development of this sector, the government of India has initiated different schemes and incorporated policy changes at different levels. Various researchers have highlighted the pathetic conditions of current agri-food supply chains, and also pointed out the various issues, challenges and opportunities in the sector. In an attempt to address this issue, this study will focus on the following objectives:

1. To analyse the influence of farmer demographics on organized retailer performance.
2. To identify the supply chain issues and its impact on organized retailer performance.

2 Literature review

2.1 Supply chain in fruits and vegetables sector in India

The supply chain management is to ensure that the product is reached to the proper destination at exact time by optimizing the existing inventory [5]. Supply chain management (SCM) is also defined as "a well-established of practices employed to efficiently integrate all stake holders like vendors, retailers, producers, and stores, thus the goods are created and circulated to the right places, at the right cost and at the right time, so as to reduce total costs while fulfilling service level necessities". The supply chain management concepts are widely used in manufacturing and service fields, but not applied in the agriculture sector even though the agriculture commodities are the raw materials for most of the industries and also major part of world economy. In some papers, the supply chain management of agri-fresh produce is referred as agri-fresh supply chain management (FSCM), comprises all the process involved from the production to the delivery of agri-fresh produce. FSCM is complicated when compared with other SCMs because it carries items which are perishable in nature, short shelf life, demand and price fluctuations, dependence on climate conditions, and consumer concerns for food safety [6]. The existing supply chains in F&V sector consists of lot of intermediaries who grab about 75 per cent of the total net margin accumulated throughout the supply chain [7]. The agri-fresh produce has to cross about six to seven different distribution channels from farmer to consumer [8]. The F&V supply chain carries number of intermediaries who perform various tasks such as transportation, maintenance and preservation of quantity and quality, transfer of ownership of commodities, payment to the seller and delivering commodity to the buyer [4].

2.2 Value chain model in fruits and vegetables supply chain

Famers, organized retailers and customers are the players in this supply chain model. Reliance Retail Ltd is following a value chain model and purpose of this model is to buy agri-produce directly from the farmers and sell it to customers thereby avoiding intermediaries. The agri-produce movement takes place from farmer or contract farmer or lease famers to wholesaler hubs through agents or consolidation centres, wholesaler's hubs to small stores, and then small stores to customers. Organized retailers receive the agri-produce directly from small farmers, contract farmers and lease farmers. Contract farmers and lease farmers supply agri-produce on sale trade agreement with the retailer. The Value chain model is as shown in Fig. 1.

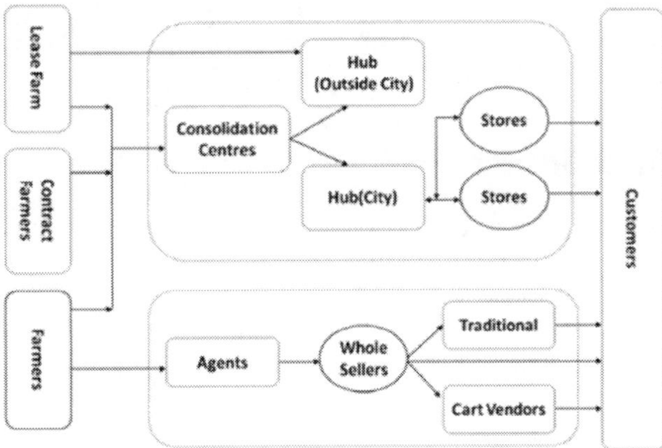

Fig. 1. Value chain model (Source: Halder and Pati, 2012)

Initially, farmers send their vegetables to the consolidation centres based on the trade agreements. Then these vegetables are transported to hubs and quality and are well maintained with a proper cold chain. Depending upon the requirements, vegetables are transported from hub to stores twice a day in a fresh state. Customers visit to the nearest retail stores to purchase the fresh vegetables [3] [4].

2.3 Issues in fresh fruits and vegetables supply chain

Fruits and vegetable supply chain need a proper infrastructure from farm level to consumer. Various researchers have identified several issues in this supply chain. Some of these are: lack of a proper infrastructure for maintaining agri-produce in fresh state i.e. cold chain, poor transportation facilities; small farmers are fragmented type and can only produce less quantity of agri-produce and cannot afford latest technologies; difficulty in integrating the stakeholders; and, uneducated farmers are unaware of demand or market information. These farmers cannot incorporate technologies for quality processed and packaged food which is in demand now days. Farmers also have lack of knowledge about post-harvest losses [1]. Wastages in Indian fruits and vegetables supply chain takes place mainly due to lack of cold storage, poor logistics, and unscientific harvesting/processing methods, lack of communication and coordination, inappropriate demand information and lot of intermediaries in supply chain[9]. Age, gender, marital status, education level, field experience, earnings are important variables that influence the work place performance of an employee [10]. Factors influencing the fruits and vegetable supply chain in India are population and demographic changes, integration in Agri-sector, emergence of organized retail, emergence of technologies, globalization, and government role [11].

2.4 Organized Retailer Performance

Organized retailers performance is measured in four important dimensions such as financial, operational, environmental and social performance. Sales growth, increase in market share and profit are relating to economic outcomes. Efficiency, productivity, inventory level, & optimum utilization of capacity reflect operational performance. Reduced emissions and pollution, savings in power consumptions are linked to environmental performance. Similarly organizations people friendly initiatives and practices like employee morale, company image, and community relations are treated as social performance [12] [13].

3 Research Methodology

The main purpose of this research was to study the impact of supply chain factors on organized retailer performance. A detailed qualitative investigation was conducted in the area of agri-supply chain management. Journal papers related to the field of agri-supply chain were surveyed to identify the factors influencing vegetable supply chain and organized retailer performance. Questionnaire was designed based on the prior studies in the field of agricultural supply chain. Factors influencing the supply chain were identified and questions were framed to obtain answers to research objectives. A questionnaire is used as measuring instrument which was developed after the completion of literature survey [14]. The data collection was done with closed-ended, structured questionnaire from various farmers of Udupi district, India. Questionnaires were served to the farmers who supply their vegetables to organized retailers like Bigbazar, More, Neelgiries and Super Bazar etc. Educated farmers were asked to fill the questionnaire and uneducated farmers were interviewed to obtain the accurate responses. In this survey, 87 completely filled questionnaires were collected.

3.1 Population and Sampling

Our study was focused on vegetable supplying farmers to organized retailers so the size of population was estimated to be 120 farmers. This population size of farmers was decided approximately based on the feedback obtained from various vegetable vendors of district. Following formula can be used to calculate the sample size (n) is given by: [15].

$$n = N / (1+Ne^2) \quad (1)$$
$$= 93 \text{samples}$$

where, $N = 120$ persons and
$e = 0.05$ (is desired level of precision).

3.2 Hypotheses

Following hypotheses are framed based on the literature studies,
H_1: Demographic factors are influencing organized retailer performance.
H_2: Supply chain issues are related to organized retailer performance.

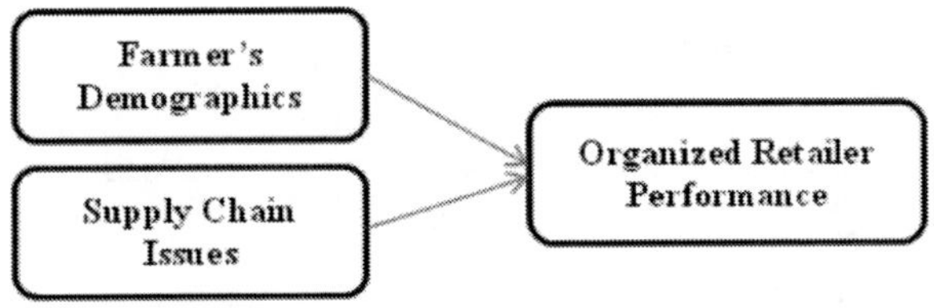

Fig. 2. Conceptual framework (Source: Authors)

Conceptual framework is developed based on the literature study. Organized retailer performance is dependent variable which is influenced by two independent variables demographic characteristics and supply chain issues.

4 Data analysis

Table 1. Frequency table

Variable	Groups	Frequency	Percent
Gender	Male	68	78.2
	Female	19	21.8
Age	20-30 years	02	2.3
	31-40 years	08	9.2
	41-50 years	26	29.9
	Above 50 years	51	58.6
Marital Status	Unmarried	08	9.2
	Married	79	90.8
Education Level	Illiterate	27	31
	Primary School	46	52.9
	High school	12	13.8
	Above PU	02	2.3
Field Experience	lessthan 5 years	02	2.3
	5-10 years	03	3.4
	10-15 years	09	10.3
	Above 15 years	73	83.9
Income Level (in Rupees)	lessthan 1 Lakh	29	33.3
	1 to 3 Lakhs	49	56.3
	3 to 5 Lakhs	09	10.3
	Above 5 Lakhs	00	00

Statistical Package for the Social Science (SPSS) software was used to analyse the data collected from the survey. The information from each questionnaire was entered and coded carefully in SPSS software prior to the analysis. Demographics of data collection are highlighted in the table 1. From the above table, the maximum number of samples in gender variable are male respondents i.e.68%; the maximum number of samples of age variable is above 50 years old is 58.6%; Married people respondent is highest in marital status variable i.e.91%; respondents with primary school education level is highest i.e. 52.9% followed by illiterate i.e.31%; 83.4% of respondents are with field experience more than 15 years; 56.3% of respondents are with an income level in rupees between 1 lakh to 3 lakhs is highest followed by 33.3% of respondents with less than 1lakh income.

4.1 Hypotheses testing results

4.1.1 The analyses of influence of demographic factors on organized retailer performance

A. The results of organized retailer performance is influenced by demographic variable farmer's gender
Independent sample t-test is used for testing the hypothesis 1.1
Hypothesis 1.1
$H_0$1.1.1: Farmer's gender doesn't influence organized retailer performance.
H_a1.1.2: Farmer's gender influence organized retailer performance.
The study result indicates that test t-value is 1.018 and significance value i.e. p-value is 0.316. As the significance value is greater than 0.05, it is unable to reject null hypothesis at 5% significance level.

Table 2. The analyses of influence of demographic factors on organized retailer performance

A. The results of organized retailer performance is influenced by farmer's gender – Independent Sample t-Test					
Variable	*Gender*	*Mean*	*Standard Deviation (S.D)*	*Statistics*	*Significance Value*
Organized Retailer Performance	Male	23.83	2.86	1.018	0.316
	Female	22.10	3.12		
B. The results of organized retailer performance is influenced by farmer's age- One Way Annova Test					
Variable	*Age*	*Mean*	*Standard Deviation (S.D)*	*Statistics*	*Significance Value*
Organized Retailer Performance	20-30	27	1.41	1.891	0.137
	31-40	24.75	2.65		
	41-50	23.65	3.02		
	above 50	23.45	2.96		
C. The results of organized retailer performance is influenced by farmer's marital status - Independent Sample t-Test					
Variable	*Marital Status*	*Mean*	*Standard Deviation (S.D)*	*Statistics*	*Significance Value*
Organized Retailer Performance	Unmarried	25.5	2.61	1.276	0.262
	Married	23.25	2.96		
D. The results of organized retailer performance is influenced by farmer's level of education -One Way Annova Test					
Variable	*Education Level*	*Mean*	*Standard Deviation (S.D)*	*Statistics*	*Significance Value*
Organized Retailer Performance	Illiterate	23.62	2.73	1.194	0.317
	Primary School	23.13	3.09		
	High School	23.75	3.16		
	PU and above	27	1.41		
E. The results of organized retailer performance is influenced by farmer's field experience - One Way Annova Test					
Variable	*Field Experience*	*Mean*	*Standard Deviation (S.D)*	*Statistics*	*Significance Value*

Organized Retailer Performance	Less than 5 Years	27	1.41	1.082	0.361
	5 to 10 Years	24.33	2.51		
	10 to 15 Years	23.55	3.46		
	Above 15 Years	23.31	2.95		
F. The results of organized retailer performance is influenced by farmer's income - One Way Annova Test					
Variable	*Income Level (Rupees)*	*Mean*	*Standard Deviation (S.D)*	*Statistics*	*Significance Value*
Organized Retailer Performance	Up to 1 Lakh	22.72	2.75	2.369	0.1
	1 to 3 Lakhs	23.59	3.08		
	3 to 5 Lakhs	25.11	2.71		

Hence, gender variable does not influence organized retailer performance at 5% significance level as described in table 2.

B. The results of organized retailer performance is influenced by demographic variable farmer's age
One way Annova test is used for testing the hypothesis 1.2
Hypothesis 1.2
$H_0$1.2.1: Farmer's age doesn't influence organized retailer performance.
H_a1.2.2: Farmer's age influences organized retailer performance.
The study result indicates that test F-value is 1.891 and significance value i.e. p-value is 0.137. As the significance value is greater than 0.05, it is unable to reject null hypothesis at 5% significance level. Hence, age variable does not influence organized retailer performance at 5% significance level as described in table 2.

C. The results of organized retailer performance is influenced by demographic variable farmer's marital status
Independent sample t-test is used for testing the hypothesis 1.3
Hypothesis 1.3
$H_0$1.3.1: Farmer's marital status doesn't influence organized retailer performance.
H_a1.3.2: Farmer's marital status influences organized retailer performance.
The study result indicates that test t-value is 1.276 and significance value i.e. p-value is 0.262. As the significance value is greater than 0.05, it is unable to reject null hypothesis at 5% significance level. Hence, marital status variable does not influence organized retailer performance at 5% significance level as described in table 2.

D. The results of organized retailer performance is influenced by demographic variable farmer's education level
One way Annova test is used for testing the hypothesis 1.4
Hypothesis 1.4
$H_0$1.4.1: Farmer's education level doesn't influence organized retailer performance.
H_a1.4.2: Farmer's education level influences organized retailer performance.
The study results indicate that test F-value is 1.194 and significance value i.e. p-value is 0.317. As the significance value is greater than 0.05, it is unable to reject null hypothesis at 5% significance level. Hence, education level variable does not influence organized retailer performance at 5% significance level as described in table 2.

E. The results of organized retailer performance is influenced by demographic farmer's field experience
One way Annova test is used for testing the hypothesis 1.5
Hypothesis 1.5
$H_0$1.5.1: Farmer's field experience doesn't influence organized retailer performance.
H_a1.5.2: Farmer's field experience influences organized retailer performance.
The study result indicates that test F-value is 1.082 and significance value i.e. p-value is 0.361. As the significance value is greater than 0.05, it is unable to reject null hypothesis at 5% significance level. Hence, farmer's field experience variable does not influence organized retailer performance at 5% significance level as described in table 2.

F. The results of organized retailer performance is influenced by demographic variable Farmer's income level
One way Annova test is used for testing the hypothesis 1.6
Hypothesis 1.6
$H_0$1.6.1: Farmer's income level doesn't influence organized retailer performance.
H_a1.6.2: Farmer's income level influences organized retailer performance.
The study result indicates that test F-value is 2.369 and significance value i.e. p-value is 0.1. As the significance value is greater than 0.05, it is unable to reject null hypothesis at 5% significance level. Hence, farmer's income level variable does not influence organized retailer performance at 5% significance level as described in table 2.

G. Summary of influence of demographic factors on organized retailer performance
The statistical analysis is conducted with a significance level 5% and the results reveal that none of demographic factor stated in the study influence organized retailer performance. So this concludes that farmer's demographic factors not to be considered for measuring the supply chain performance of the organised retailer.

4.1.2 Results of supply chain issues are related to organized retailer performance.

Linear regression analysis is used for testing the hypothesis 2.1
Hypothesis 2.1
$H_0$2.1.1: Supply chain issues are not related to organized retailer performance.
H_a2.1.2: Supply chain issues are related to organized retailer performance.
Regression analysis is performed to find the linear relationship between variables supply chain issues and organized retailer performance.

Table 3. Regression analysis-model summary

Dependent variable	Organized Retailer Performance (Y)
Independent Variable	Supply Chain issues (X)
Multiple R-value	0.179
R-Square Value	0.032
Adjusted R-Square Value	0.021
F-value	2.807
p-value	<0.098

Table 3 shows the results of regression analysis, R^2=0.032 and Adjusted R^2=0.021, which reveals that the independent variable, supply chain issues describes 3.2% of the variability of the dependent variable, organized retailer performance in a given population. Adjusted R^2 value also an estimate of the effect size which is at 0.021 i.e.2.1% is indicative of a small effect size to Cohen's (1988) classification [16]. The regression model is statistically insignificant with p-value=0.098 at 5% significance level, this indicates that overall model is not significant to predict the dependent variable.

Table 4. Regression analysis-coefficients

Model		**(Constant)**	**Total SCI**
Unstandardized Coefficients	B	29.914	-0.240
	Standard Error	3.865	0.143
Standardized Coefficients	Beta	-----	-0.179
Significance value		0.000	0.098

Regression analysis coefficients are shown in table 4 and regression model can be fit using the above results. The linear regression model is Y=29.91+3.865X or Organized retailer performance=29.914+3.865Supply chain issues). F-value is 2.807 and significance value is 0.098 which is greater than 0.05, at a significance level of 5%. Hence, it indicates that there is lack of linear relationship between supply chain issues and organized retailer performance.

5 Conclusions

The statistical analysis was performed in the study to determine the effect of various factors such as farmer's demographics and supply chain issues influencing performance of organized retailer. First hypothesis test results indicate that farmer's demographics have no influence on organized retailer performance. Second hypothesis analyse the impact of supply chain problems faced by famers and its influence on retailer performance. Linear regression analysis was used to test the linear relation between the variables. Results of the study show the lack of linear relationship and hence supply chain issues have no influence on organized retailer performance. The outcome of the study implies that independent variables stated have no control over the retailer's performance. Management has to look in to other aspects which imbalance the supply chain leading to not meeting the customer requirements at right time. This study may help the retailers to understand the supplier problems and to monitor the supplier performance to control variables like delivery time, wastage, quality, capacity utilization.

References

1. S. Negi and N. Anand, "Issues and challenges in the supply chain of fruits and vegetable sector in India-A Review", International Journal of Managing Value and Supply Chains (IJMVSC) Vol. 6, No. 2, June 2015.
2. ASSOCHAM "Horticulture Sector in India – State Level Experience", The Associated Chambers of Commerce and Industry of India, New Delhi, 2013.
3. S Negi and N Anand, "Cold Chain: A Weak Link in the Fruits and Vegetables Supply Chain in India", The IUP Journal of Supply Chain Management, Vol. 52 XII, No. 1, 2015.
4. Piali Halder and Simayan Pati, "A need for paradigm shift to improve supply Chain management of fruits & vegetables in India", Asian Journal of Agriculture and Rural Development, 1(1), 1-20, 2012.
5. Adi Djoko Guritno, Rika Fujianti, Dinovita Kusumasari, "Assessment of the Supply Chain Factors and Classification of Inventory Management in Suppliers' Level of Fresh Vegetables", Agriculture and Agricultural Science Procedia 3 51 – 55, 2015.

6. Manish Shukla, Sanjay Jharkharia, "Agri-fresh produce supply chain management: a state-of-the art literature review", International Journal of Operations & Production Management, Vol. 33 Issue: 2, pp.114-158, 2013.
7. Modi P, Mishra D, Gulati H and Murugesan K "Uttarakhand State Cooperative Federation: Can it Help The Horticulture Farmers?", Vision—The Journal of Business Perspective, Vol. 13, No. 2, pp. 53-61, 2009.
8. Viswanadham N "Can India be the Food Basket for the World?", Working Paper Series, Indian School of Business, Hyderabad, India, 2007.
9. M. Balaji, K. Arshinder, "Modeling the causes of food wastage in Indian perishable food supply chain" Resources, Conservation and Recycling 114, 153–167, 2016.
10. L. Wanxian, L. Xinmei, and W. Weiwu, "Demographic Effects of Work Values and Their Management Implications", Journal of Business Ethics, 2008, 81:875–885, 2007.
11. Sunil Bhardwaj and Indrani Palaparthy, "Factors Influencing Indian Supply Chains of Fruits and Vegetables: A Literature Review", The Icfai University Journal of Supply Chain Management, Vol. V, No. 3, 2008.
12. Raghuram Naga Venkata J & Ravilochanan P, "The Role of Supply Chain Management on Organized and Unorganized Grocery Retailers-A Review on Indian Perspective", Asian Social Science; Vol. 10, No. 22; 2014.
13. Ruoqi Geng, S. Afshin Mansouri, Emel Aktas, "The relationship between green supply chain management and performance: A meta-analysis of empirical evidences in Asian emerging economies" Int. J. Production Economics 183, 245–258, 2017.
14. N. R. Kenneth, "Quantitative research methods in educational planning" UNESCO International Institute for Educational Planning, 2005.
15. Israel, Glenn D., "Sampling The Evidence Of Extension Program Impact. Program Evaluation and Organizational Development", IFAS,University of Florida. PEOD-5. 1992.
16. Kotrlik, Williams, & Jabor, "Reporting and Interpreting Effect Size in Quantitative Agricultural Education Research", Journal of Agricultural Education Volume 52, Number 1, pp. 132–142, 2011.
17. C.R. Kothari, "A Text Book On Research Methodology-Methods and Techniques", 2004.

The Effect of Lean Techniques on Elimination of Waste in Composite Panel Production Using Paired t-test

Pavan Hiremath[1], *S. Narayanan*[1*] and *Manjunath Shettar*[1]

[1]*Department of Mechanical and Manufacturing Engineering, Manipal Institute of Technology, Manipal Academy of Higher Education, Manipal-576104, Karnataka, India.*

Abstract. 'A penny saved is a penny earned' as this phrase says it all, the key for increased profit lies in the elimination of non-value adding actions in any production process. This study concentrates mainly on the procedures to identify and to reduce the different types of wastes in a production process. Initially the data related to types of waste and their classification according lean techniques were analyzed and implemented on a production process. The data were recorded for a whole month before implementation and for a whole month after implementation. Later the influence of the lean techniques was analyzed on two different sections of waste, for 5% significance value using paired t-test. It was found that there is a positive impact of lean techniques on some areas of production flow.

1 Introduction

There is a phrase, "you can't make something out of nothing." Which means there is s need of resource to produce anything which is small or huge. But, there are only very few instances where the transformation of resource/raw material to finished entity take place effectively, and the rest leads to the creation of Muda (waste). Wastes in a production system may be generated due to using wrong resources, failing to get necessary resources, using a resource in a fruitless way etc. The money invested, time consumed, man power used etc. on these resources also lead to wastes which some time goes unnoticed. In production structure a waste which is created at any stage, will have its effect twice or thrice at the end and leaves the customer and manufacturer dissatisfied.

The main purpose of lean manufacturing technique is to eliminate different wastes which are present in a company through various improvement activities. So, what is waste? It can be in many forms, but the basic idea is to eliminate anything and everything that does not add value in a manufacturing forum. Some forms of wastes are: waiting, is the time taken by a component to proceed for the next process. Transport, is the time taken for unnecessary movements of the raw material. Motion, is the unproductive movements of the people. Over-processing, is more processing of the raw material than required. Excess

* Corresponding author: s.narayanan@manipal.edu

inventory, is the quantity of raw material stored than required and the defective part production as well as over production adds up to this list.

2 Lean techniques for Waste Treatment

In most of the fabrication or R&D sectors, the type and quantity of the resource utilized my not be same in each interval of time. In this context this research focuses on a particular family of materials namely, composite panels. To introduce lean concepts anywhere one should know all the corners of it. Identification of type and source of waste holds the top requisite for planning the steps to eliminate it. According to Justyna Rybicka et al. [1] there is a need of classification of waste and arranging them in a specific order. Figure 1 shows the hierarchy of wastes (by looking at this arrangement, one should be able to identify what should be done next) from prevention which is the most desirable (non-waste) scenario, through to waste disposal, which is considered the most wasteful currently known process. Implementation of these ideas to the fabrication area plays a crucial part. There are several challenges faced during the implementation, few among them are stressed in table 1.

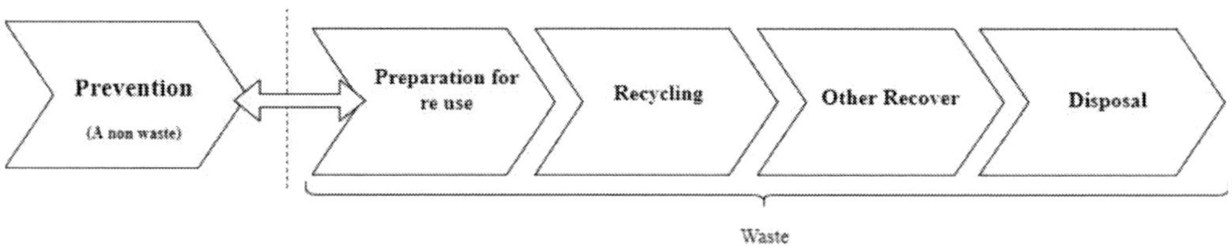

Fig. 1. Waste hierarchy [1].

Table 1. Challenges faced during implementation of lean manufacturing [2].

Management	funding, time available, training quality, bonding with labors, size of the plant, cost of implementation, difficulties in collecting data, job satisfaction of employees, quality of the labor etc.
Workers	Lack of interest, negative attitude, friction for changing the habits, lack of awareness, negative feedback from colleagues, less skilled labors, lack of interest in learning, unable to judge the outcomes etc.
Machinery	Insufficient machines, poor advanced machinery, insufficient technical resources etc.

To overcome the obstacles in implementing lean philosophy to reduce waste, needs recording the flow of the resources (flow mapping) from a supplier till final product's quality assurance. Waste management in the field of production is a quick growing field and it contributes largely by increasing efficiency, economy as well as a systematic approach [3]. Tools such as mapping is widely used for optimization in manufacturing. Mapping representing the present scenario in a well-arranged way. It acts as a bench mark for continuous actives [4]. Process mapping is one of the efficient and convenient way to capture and draft the data related to waste mapping. It allows to identify the material flow path and corners of waste generation in a manufacturing process. The objectives of mapping such as product quality improvement, waste elimination, maintaining inventory under control and operation and financial control [5]. The benefits of IDEF0 (process modelling tool) and MEW (material energy, waste mapping) on manufacturing process improvement and improvement opportunities for economic development through resource minimization was described by Bolin and Smith [6]. The tools that have been used for process mapping with intent to outline waste are: value stream mapping, Sankey diagrams, material flow analysis, IDEF0, MEW and flow chats. [7-8].

3 Identification of wastes in process flow

With all the available and analyzed data, lean techniques were initially implemented on one segment of a unit to check its significant influence on the reduction of waste. Some of the techniques were; Kanban, just in time production, multi-functional worker and one-piece production. Figure 2 shows the process flow and identified wastes (1 to 12) during each process. Figure 3 shows the variation in time recorded before and after implementation of lean techniques for a month. For other identified wastes (4, 6 and 11), they were subdivided into different further categories such as waste (consumables) in fiber waste, breather, polyethene bags etc.

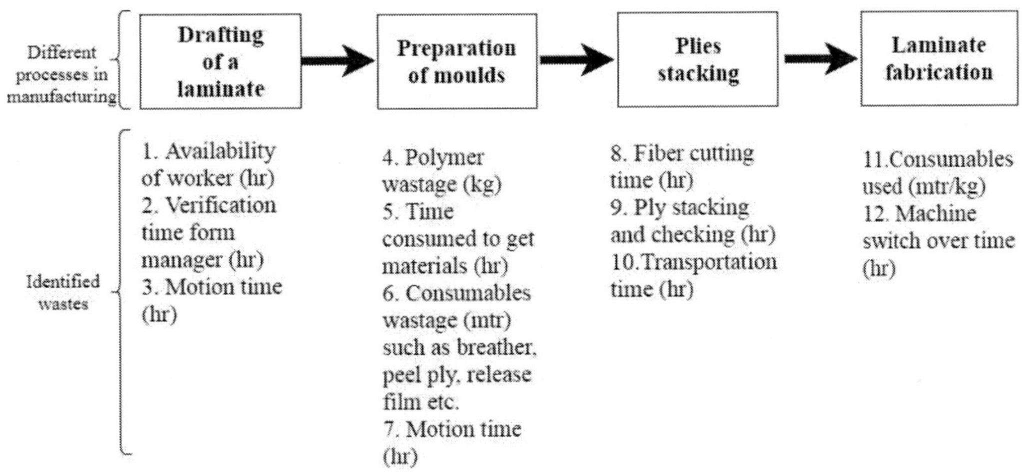

Fig. 2. Process flow chart and identified wastes.

4 Hypothesis Testing

The implement of lean techniques had an impact on the composite panel fabrication. By considering the availability of the worker (1) as shown in the figure 2 for production process for a time span of one-month hypotheses were drawn. After implementation, the worker has to undergo several training processes which include: training, meetings, audit and improvement etc. Table 2 shows the time, a worker available near the machine during the implementation. Later the influence on the availability of worker as well as consumables used for 5% significance was checked using paired t-test.

Hypothesis formed were:
Ha1: There is a significant influence of lean implementation on the availability of worker near the machine.
Ho1: There is no significant influence of lean implementation on the availability of worker near the machine.

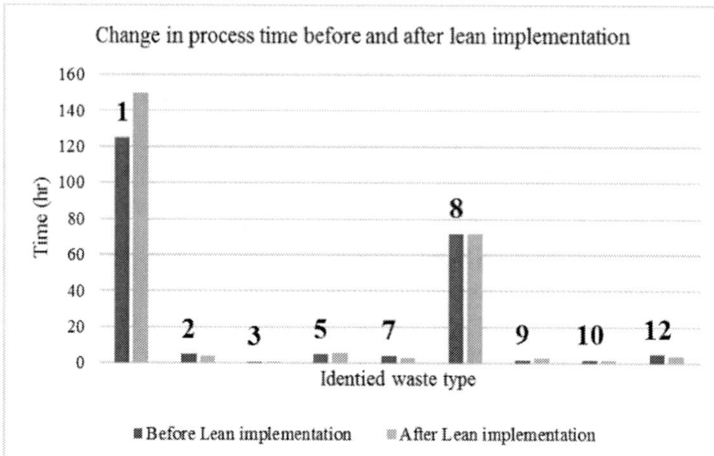

Fig. 3. Time consumed for different processes before and after implementation of lean.

Table 2. Availability of worker (hr) for a month.

Day	1	2	3	4	5	6	7	8	9	10	11	12	13	14	15	16	17	18	19	20	21	22
Before	5	6	4	4	8	6	5	6	8	6	4	5	7	4	3	8	7	5	6	8	6	4
After	7	7	7	8	6	7	6	7	8	4	8	8	4	7	6	7	7	8	5	8	7	8

Later the hypothesis was further extended towards the consumables used (4, 6 and 11 of figure 2). Table 3 shows some of the consumables used before and after lean implementation.

Hypothesis formed :
Ha2: There is a significant influence of lean implementation on the consumables used.
Ho2: There is no significant influence of lean implementation on the consumables used.

Table 3. Consumables used (mtr) before and after lean implementation.

Consumable	**Flow mesh**	**Breather**	**Coremate**	**Fiber**	**Poly Bags**	**Peel ply**	**Sealant**
Before	120	200	90	1200	600	550	2400
After	115	180	90	1100	600	560	2500

5 Results and discussion

Table 4 shows the results obtained from paired t-test for two different entities, where table 4(a) is the output for availability of worker and 4(b) is the output for consumables used. For a time span of one month (22 working days) the data were collected and the influence of the lean techniques was analyzed. It was found that, the availability of worker had a P-value of 0.019 by rejecting null hypothesis Ho1. As there was an increase in total availability of the worker by 20% after a month, the time consumed for meetings, trainings, audit etc. have not affected the aspects related to time and the lean techniques like Kanban, standardization has influenced considerably. In the case of consumables used the P-value was0.926 which accepts null hypo thesis Ho2. Which implies the employees may need more training on lean techniques or the Kanban must be implemented in a very effective way.

Table 4. Results of paired t-test

(a). Output for availability of worker

Paired T-Test and CI: before, after Descriptive Statistics

Sample	N	Mean	StDev	SE Mean
before	22	5.682	1.524	0.325
after	22	6.818	1.220	0.260

Estimation for Paired Difference

Mean	StDev	SE Mean	95% CI for μ_difference
-1.136	2.100	0.448	(-2.067, -0.205)

μ_difference: mean of (before - after)

Test

Null hypothesis H_0: μ_difference = 0

Alternative hypothesis H_1: μ_difference ≠ 0

T-Value	P-Value
-2.54	0.019

(b). Output for consumables

Paired T-Test and CI: before, after Descriptive Statistics

Sample	N	Mean	StDev	SE Mean
before	7	737	829	313
after	7	735	857	324

Estimation for Paired Difference

Mean	StDev	SE Mean	95% CI for μ_difference
2.1	58.4	22.1	(-51.9, 56.2)

μ_difference: mean of (before - after)

Test

Null hypothesis H_0: μ_difference = 0

Alternative hypothesis H_1: μ_difference ≠ 0

T-Value	P-Value
0.10	0.926

6 Conclusions

In a production unit, lean ideas were implemented on a small unit for a time span of one month. It is evident that mapping the process flow helps in identifying the different non-value adding activities in a process flow easily. Lean techniques such as Kanban and standardized operations reduces the confusions of a worker. It takes longer time for a worker to adjust to new environment, the trainings, meetings, audit etc. by pulling down the worker's availability near the machine. Finally, there is a significant influence of lean implementation on the availability of worker near the machine and there is no significant influence of lean implementation on the consumables used when considered for a month of 22 working days.

References

1. J. Rybicka, A. Tiwari, P. A. Del Campo, J. Howarth, J. of Clea. Prod., **91**, 251-261 (2015).
2. M. K. Wyrwicka, B. Mrugalska, Pro. Engg., **182**, 780-785 (2017).
3. G. Mouzon, M. B. Yildirim, J. Twomey, Int. J. of Prod. Res., **45(18-19)**, 4247-4271 (2007).
4. L. Smith, P. Ball, Int.l J. of Pro. Econ., **140(1)**, 227-238 (2012).
5. F. A. Abdulmalek, J. Rajgopal, Int. J. of prod. Econ., **107(1)**, 223-236 (2007).

6. C. A. Bolin, S. Smith, J. of Clea. Prod., **19(6)**, 620-629 (2011).
7. R. Kahhat, E. Williams, Cons. and Rec., **67**, 67-74 (2012).
8. A. R. Rahani, M. Al-Ashraf, Proce. Engg., **41**, 1727-1734 (2012).
9. E. Antane, D. Blumberga, Energy Procedia, **128**, 358-362. (2017).
10. M. Shettar, P. Hiremath, A. Kini, S. Sharama, In Mat. Sci. For., **904**, 137-141. Trans Tech Publications (2017).
11. K. W. Stier, J. of Ind. Tech. **19(4)**, 1-6 (2003).
12. T. L. Doolen, M. E. Hacker, J. of Mfg. syst., **24(1)**, 55-67 (2005).
13. S. J. Pavnaskar, J. K. Gershenson, A. B. Jambekar, Inte. Jo. of Prod. Res., **41(13)**, 3075-3090 (2003).

AUTHOR INDEX

AUTHOR INDEX

AUTHOR INDEX

EDP Sciences – Web of Conferences
17, Avenue du Hoggar
Parc d'Activité de Courtabouf
BP 112
F-91944 Les Ulis Cedex A
France

ISBN 978-1-5108-5581-6

9 781510 855816

CURRAN ASSOCIATES INC.
proceedings.com